The Entire Electromagnetic Spectrum

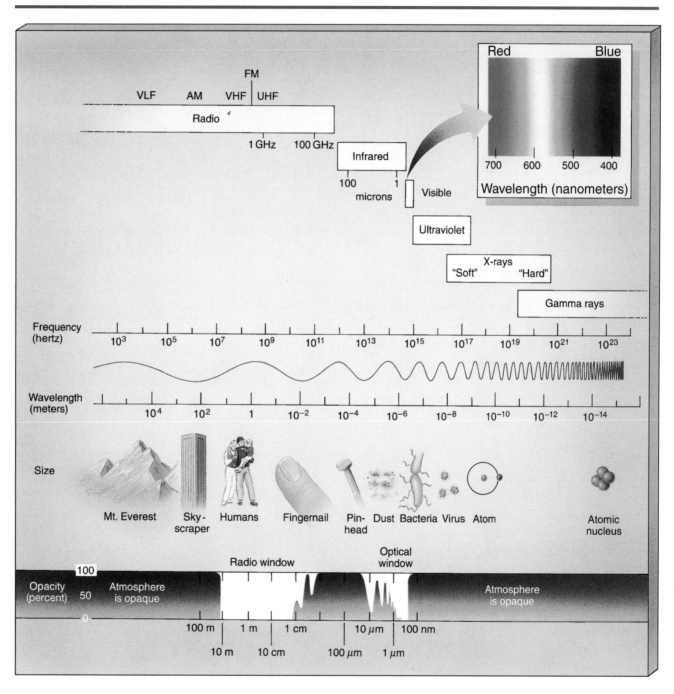

ASTRONOMY TODAY

ABOUT THE AUTHORS

ERIC CHAISSON

Eric holds a doctorate in Astrophysics from Harvard University, where he spent ten years on the faculty of Arts and Sciences. For five years, Eric was a Senior Scientist and Director of Educational Programs at the Space Telescope Science Institute and Adjunct Professor of Physics at Johns Hopkins University. He joined Tufts University, where he is now Professor of Physics, Professor of Education, and Director of the Wright Center for Innovative Science Education. He has written nine books on astronomy, which have received such literary awards as the Phi Beta Kappa Prize, two American Institute of Physics Awards, and Harvard's Smith Prize for Literary Merit. He has published more than 100 scientific papers in professional journals, and has also received Harvard's Bok Prize for original contributions to astrophysics.

STEVE McMILLAN

Steve holds a bachelor's and master's degree in Mathematics from Cambridge University and a doctorate in Astronomy from Harvard University. He held post-doctoral positions at the University of Illinois and Northwestern University, where he continued his research in theoretical astrophysics, star clusters, and numerical modeling. Steve is currently Distinguished Professor of Physics at Drexel University and a frequent visiting researcher at Princeton's Institute for Advanced Study and the University of Tokyo. He has published over 40 scientific papers in professional journals.

ASTRONOMY TODAY

Third Edition

Eric Chaisson
TUFTS UNIVERSITY

Steve McMillan
DREXEL UNIVERSITY

Prentice Hall
Upper Saddle River, New Jersey 07458

Chaisson, Eric.
   Astronomy today / Eric Chaisson, Steve McMillan. — 3rd ed.
     p.   cm.
   Includes bibliographical references and index.
   ISBN 0-13-085863-3
   1. Astronomy.   I. McMillan, S. (Stephen).   II. Title.
QB43.2.C44   1999
520—dc21   98-28516
                                        CIP

Executive Editor: Alison Reeves
Editor in Chief: Paul F. Corey
Editorial Director: Tim Bozik
Assistant Vice President of Production and Manufacturing: David W. Riccardi
Executive Managing Editor: Kathleen Schiaparelli
Production Editor: Alison Aquino
Manufacturing Manager: Trudy Pisciotti
Director, Creative Services: Paula Maylahn
Assistant Manager, Formatting: Richard Foster
Assistant Manager, Formatting (2nd Shift): Amy Peltier
Assistant Managing Editor, Media: Erik Unhjem
Media Production Editor: Kim Gundling
Page Layout: Richard Foster, Amy Peltier, Karen Noferi, Shari Toron, Jim Sullivan, Alison Aquino
Production Assistance: Karen Stephens
CD Production: Jeff Henn, Rolando Corujo, Steven Gagliostro, Amy Peltier, Lawrence La Raia, Peter Moraites, Ray Caramanna
Testing Manager: David Moles

Art Manager: Gus Vibal
Art Editor: Karen Branson
Artists: Rolando Corujo, Patricia Gutierrez, Charles Pelletreau
Art Studio: Page Two, Inc.
Associate Creative Director: Amy Rosen
Interior Design: Carole Anson
Cover Artist: Bill McClosky
CD Silkscreen Design: Paul Gourhan
Development Editor: Dan Schiller
Assistant Editor: Wendy Rivers
Editorial Assistant: Gillian Kieff
Image Resource Center Director: Lori Morris-Nantz
Image Permission Supervisor: Kay Dellosa
Image Coordinator: Zina Arabia
Copy Editor: Barbara Liguori
Director of Market Development: Kelly McDonald
Director of Marketing: John Tweeddale
Executive Marketing Manager: Leslie Cavaliere
Marketing Assistant: Connie Wu

 © 1999, 1996, 1993 by Prentice-Hall, Inc.
Simon & Schuster/A Viacom Company
Upper Saddle River, NJ 07458

Printed in the United States of America

10 9 8 7 6

ISBN 0-13-085863-3

Prentice Hall International (UK) Limited, London
Prentice-Hall of Australia Pty. Limited, Sydney
Prentice-Hall Canada Inc., Toronto
Prentice-Hall Hispanoamericana, S.A., Mexico
Prentice-Hall of India Private Limited, New Delhi
Prentice-Hall of Japan, Inc., Tokyo
Simon & Schuster Asia Pte. Ltd., Singapore
Editora Prentice-Hall do Brasil, Ltda., Rio de Janeiro

BRIEF CONTENTS

CONTENTS

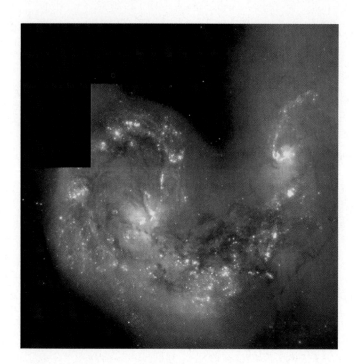

PREFACE

Astronomy continues to enjoy a golden age of exploration and discovery. Fueled by new technologies and novel theoretical insights, the study of the cosmos has never been more exciting. We are pleased to have the opportunity to present in this book a representative sample of the known facts, evolving ideas, and frontier discoveries in astronomy today.

This book is written for students who have taken no previous college science courses and who will likely not major in physics or astronomy. The text is suitable for both one-semester and two-semester courses. We present a broad view of astronomy, straightforwardly descriptive and without complex mathematics. The absence of sophisticated mathematics, however, in no way prevents discussion of important concepts. Rather, we rely on qualitative reasoning as well as analogies with objects and phenomena familiar to the student to explain the complexities of the subject without oversimplification. We have tried to impart the enthusiasm that we feel about astronomy, and to awaken students to the marvelous universe around us.

In teaching astronomy to non-scientists, as in writing this book, we are not seeking to convert students to careers in astronomy or even science. Instead, we strive to reach the wider audience of students who are majoring in many other worthwhile fields. We want to encourage these students to become scientifically literate members of modern society—to appreciate new developments in the world of science, to understand what scientists do for a living and why it is important, to make informed judgements regarding national initiatives in science and the public funding of scientific projects, and to vote intelligently in our democratic, increasingly technological world.

We are very gratified that the first two editions of this text have been so well received by many in the astronomy education community. In using those earlier texts, many of you—teachers and students alike—have sent us your helpful feedback and constructive criticisms. From these, we have learned to communicate better both the fundamentals and the excitement of astronomy. Many improvements inspired by your comments have been incorporated into this new edition.

Organization and Approach

Our overall organization follows the popular and effective "Earth-out" progression. We have found that most students, especially those with little scientific background, are much more comfortable studying the (relatively familiar) solar system before tackling stars and galaxies. Thus, Earth is the first object we discuss in detail. With Earth and the Moon as our initial planetary models, we move through the solar system, drawing on comparative planetology to provide an understanding of the many varied worlds we encounter. We conclude our coverage of the solar system with a discussion of its formation, a line of investigation that leads directly into a study of our Sun.

With the Sun as our model star, we then broaden the scope of our discussion to include stars in general—their properties, their evolutionary histories, and their varied fates. This journey naturally leads us to coverage of the Milky Way Galaxy, which in turn serves as an introduction to our treatment of other galaxies, both normal and active. Finally, we reach the subject of cosmology and the large-scale structure and dynamics of the universe as a whole. Throughout, we strive to emphasize the dynamic nature of the cosmos—virtually every major topic, from planets to quasars, includes a discussion of how those objects formed and how they evolve.

We continue to place much of the needed physics in the early chapters—an approach derived from years of experience teaching thousands of students. Additional physical principles are developed as needed later, both in the text narrative and in the boxed "More Precisely" features (described below). We feel strongly that this is the most economical and efficient means of presentation. However, we acknowledge that not all instructors feel the same way. Accordingly, we have made the treatment of physics, as well as the more quantitative discussions, as modular as possible, so that these topics can be deferred to later stages of an astronomy course if desired. In addition, we have included as much modern astronomy as possible in the introductory chapters. These chapters are likely to engage students only if they are made to realize how simple physical principles provide the keys to our understanding of a vast and otherwise incomprehensible universe.

New and Revised Material

The third edition has been substantially updated in content since the second edition, and several chapters have seen significant internal reorganization. Among the many changes are:

• Updated material in Chapter 5 on adaptive optics at Keck and Haleakala, VLT, the Arecibo upgrade, optical interferometry (COAST) and the EUVE and planned SIRTF mission.

• New information in Chapter 6 on Galileo, Cassini, and Mars *Pathfinder/Sojourner* in the section on "Spacecraft Exploration"

• New material on the rotation of Earth's inner core in Chapter 7

• New material on *Lunar Prospector* in *Interlude 8-1* on Lunar Exploration, and a new Interlude on *Clementine*

• A new *More Precisely* box on synodic and sidereal periods in Chapter 9

• Expanded coverage of *Pathfinder/Sojourner*, and new information on Mars *Global Surveyor*, the Martian meteorite, and the 1997 discovery of the Martian magnetic field in Chapter 10

• A significant amount of new *Galileo* material in Chapter 11

• Coverage of the two new moons of Uranus in Chapter 13

• Extensively revised coverage of Pluto's origin and discussion of a surface map of the planet in Chapter 13

• Reorganized sections in Chapter 14 on asteroids and comets as well as new information on the NEAR mission and an image of asteroid Mathilde

• New Interludes on Lunar and Martian meteorites and Comets Hyakutake and Hale–Bopp in Chapter 14

• A new Interlude on extrasolar planets in Chapter 15, including late-breaking news of what may be the first image ever obtained of a planet outside our own solar system

• A new Interlude on the Kuiper belt and new *Hubble Space Telescope* material on Beta Pictoris in Chapter 15

• A new Interlude on the SOHO mission in Chapter 16

• A new Interlude on *Hipparcos* in Chapter 17

• The discussion of the magnitude scale in Chapter 17 has been moved to the main body of the text and a new More Precisely box on magnitudes with more detailed information has been added.

• An extensive revision in the subsection on stellar mass determination in Chapter 17

• Updated coverage on the Eagle nebula and related information in Chapter 18

• A new Interlude on EUVE and the Local Bubble in Chapter 18

• New coverage of the *HST* M4 white-dwarf sequence in Chapter 20

• Extensive new subsection on gamma-ray bursts in Chapter 22

• New coverage of Sgr A* in Chapter 23

• New material on "Building Blocks of Galaxies" in the section on galaxy formation in Chapter 24

• A new *HST* image of Antennae in the Interlude on galaxy collisions in Chapter 24

• Updated discussion of Hubble's constant in Chapter 24, with figures and distances altered throughout the text to reflect this change

• New Interludes on HDF and the "age controversy" in Chapter 27

• New coverage in Chapters 26 and 27 of the recent observations based on Type I supernovae that the expansion of the universe may not be slowing as expected

• Updated coverage on such topics as Europa, Mars, and extrasolar planets in Chapter 28

In addition, the layout of the book has been redesigned to give the text a somewhat more open look and feel.

The Illustration Program

Visualization plays an important role in both the teaching and the practice of astronomy, and we continue to place strong emphasis on this aspect of our book. We have tried to combine aesthetic beauty with scientific accuracy in the artist's conceptions that adorn the text, and we have sought to present the best and latest imagery of a wide range of cosmic objects. Each illustration has been carefully crafted to enhance student learning; each is pedagogically sound and tied tightly to nearby discussion of important scientific facts and ideas.

FULL-SPECTRUM COVERAGE AND SPECTRUM ICONS

Increasingly, astronomers are exploiting the full range of the electromagnetic spectrum to gather information about the cosmos. Throughout this book, images taken at radio, infrared, ultraviolet, X ray, or gamma ray wavelengths are used to supplement visible-light images. As it is sometimes difficult (even for a professional) to tell at a glance which images are visible-light photographs and which are false-color images created with other wavelengths, each photo in the text is provided with an icon that identifies the wavelength of electromagnetic radiation used to capture the image and reinforces the connection between wavelength and radiation properties.

It is rare that a single image, be it a photograph or an artist's conception, can capture all aspects of a complex subject. Wherever possible, multiple-part figures are used in an attempt to convey the greatest amount of information in the most vivid way:

- Visible images are often presented along with their counterparts captured at other wavelengths.
- Interpretive line drawings are often superimposed on or juxtaposed with real astronomical photographs, helping students to really "see" what the photographs reveal.
- Breakouts—often multiple ones—are used to zoom in from wide-field shots to closeups, so that detailed images can be understood in their larger context.

Figure 19.9 (a) The constellation Orion, with the region around its famous emission nebula marked by a rectangle. The Orion Nebula is the middle "star" of Orion's sword. The framed region is enlarged in part (b), suggesting how the nebula is partly surrounded by a vast molecular cloud. Various parts of this cloud are probably fragmenting and contracting, with even smaller sites forming protostars. The three frames at right show some of the evidence for those protostars. (c) A false-color radio image of some intensely emitting molecular sites. (d) A nearly real color visible image of embedded nebular "knots" thought to harbor protostars. (e) A high-resolution image of several young stars surrounded by disks of gas and dust where planets might ultimately form.

▲ EXPLANATORY CAPTIONS

Students often review a chapter by "looking at the pictures." For this reason, the captions in this book are often a bit longer and more detailed than those in other texts.

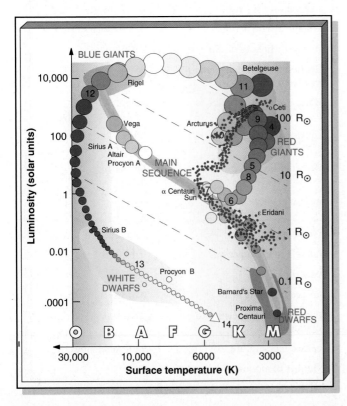

◀ H–R DIAGRAMS AND ACETATE OVERLAYS

All of the book's H-R diagrams, assembled and drawn by Lola Judith Chaisson, are presented in a uniform format, using real data wherever possible. In addition, a unique set of transparent acetate overlays dramatically demonstrate to students how the H-R diagram helps us to organize our information about the stars and track their evolutionary histories.

Other Pedagogical Features

As with many other parts of our text, adopting instructors have helped guide us toward what is most helpful for effective student learning. With their assistance, we have revised both our in-chapter and end-of-chapter pedagogical apparatus to increase its utility to students.

LEARNING GOALS   Studies indicate that beginning students often have trouble prioritizing textual material. For this reason, a few (typically 5 or 6) well-defined Learning Goals are provided at the start of each chapter. These help students to structure their reading of the chapter and then test their mastery of key facts and concepts. The Goals are numbered and cross-referenced to key sections in the body of each chapter. This in-text highlighting of the most important aspects of the chapter also helps students to review. The Goals are organized and phrased in such a way as to make them objectively testable, affording students a means of gauging their own progress.

KEY TERMS   Like all subjects, astronomy has its own special vocabulary. To aid student learning, the most important astronomical terms are boldfaced at their first appearance in the text. Each boldfaced Key Term is also incorporated in the appropriate chapter summary, together with the page number where it was defined. In addition, a full alphabetical glossary, defining each Key Term and locating its first use in the text, appears at the end of the book.

PLANETARY DATA boxes, new in this edition, provide ▶ concise summaries of planetary properties within the body of each chapter. As with previous editions, the data are repeated in the Appendix for easy reference. Similar data boxes are provided for both the Moon and the Sun. For this edition, all planetary data are drawn self-consistently from the database maintained by the Solar System Dynamics Group at Jet Propulsion Laboratory.

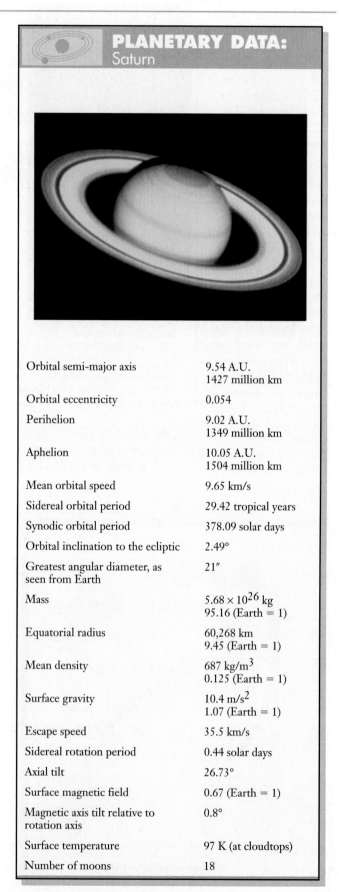

PLANETARY DATA:
Saturn

Orbital semi-major axis	9.54 A.U. 1427 million km
Orbital eccentricity	0.054
Perihelion	9.02 A.U. 1349 million km
Aphelion	10.05 A.U. 1504 million km
Mean orbital speed	9.65 km/s
Sidereal orbital period	29.42 tropical years
Synodic orbital period	378.09 solar days
Orbital inclination to the ecliptic	2.49°
Greatest angular diameter, as seen from Earth	21″
Mass	5.68×10^{26} kg 95.16 (Earth = 1)
Equatorial radius	60,268 km 9.45 (Earth = 1)
Mean density	687 kg/m^3 0.125 (Earth = 1)
Surface gravity	10.4 m/s^2 1.07 (Earth = 1)
Escape speed	35.5 km/s
Sidereal rotation period	0.44 solar days
Axial tilt	26.73°
Surface magnetic field	0.67 (Earth = 1)
Magnetic axis tilt relative to rotation axis	0.8°
Surface temperature	97 K (at cloudtops)
Number of moons	18

INTERLUDE boxes explore ▶
a wide variety of interesting
supplementary topics, such as
*An Armada of Spacecraft Bound
for Mars, Lunar and Martian
Meteorites on Earth, SOHO:
Eavesdropping on the Sun,
Observations of Brown Dwarfs,*
and *The Age Controversy
Revisited.*

 INTERLUDE 15-1   The Discovery of Planets Beyond the Solar System

Many claims of evidence of extrasolar planets have been reported (and even published) over the past few decades, but virtually none have been confirmed and most have been discredited. Only within the past few years have we seen genuine advances in this fascinating astronomical specialty. These advances have come not through any dramatic scientific or technical breakthrough but through steady improvements in both telescope and detector technology and computerized data analysis.

It is not yet possible to image any of these newly discovered planets. The techniques used to find them are indirect, based on analysis of light from the parent star, not light from the unseen planet. As a planet orbits a star, gravitationally pulling first one way and then the other, the star wobbles slightly. The more massive the planet or the less massive the star, the greater the star's movement. If the wobble happens to occur along our line of sight to the star, then we see small fluctuations in the star's velocity, which can be measured using the Doppler effect. ∞ (Sec. 3.5) Alternatively, if the wobble is predominantly perpendicular to our line of sight, then the star's position in the sky changes slightly from night to night. The star's motion is very small, which is why unambiguous measurements have been so hard to obtain. However, *both* types of wobble have now been seen and confirmed.

The first figure shows two sets of data that betray the presence of planets. The top graph shows the line-of-sight velocity of the star 51 Pegasi, a near-twin to our Sun lying some 40 light years away. These data were acquired in 1994 by Swiss astronomers using the 1.9-m telescope at Haute-Provence Observatory in France. The regular 50-m/s fluctuations in the star's velocity have been confirmed by several groups of astronomers and imply that a planet at least half the mass of Jupiter orbits 51 Pegasi with a period of just 4.2 days. The bottom graph in the figure plots the declination (see *More Precisely 1-2*) of the star Lalande 21185, corrected for the motion of Lalande through the Milky Way Galaxy and the

motion of Earth around the Sun over the last half century. A 30-year-period wobble at the level of about 0.01″ s can be seen—about the amount of wobble that would be produced in our own Sun by the planets of the solar system. These observations have triggered a minirevolution in astronomy and have doubtless paved the way for a flood of new planet searches (and reinterpretation of old data) in the next few years.

Currently, these techniques can be used only to detect planets that are at least as large as Jupiter. Thus far, none of the reports of nearby extrasolar planets has claimed an object of Earth dimensions. In fact, many of the findings to date indicate "hot Jupiters"—gigantic planets surprisingly close (within 1 A.U.) to their parent star. Furthermore, only one such unseen planet has been found (and confirmed) in each

---

 MORE PRECISELY 22-2   Tests of General Relativity

Special relativity is the most thoroughly tested and most accurately verified theory in the history of science. General relativity, however, is on somewhat less firm experimental ground.

The problem with verifying general relativity is that its effects on Earth and in the solar system—the places where we can most easily perform tests—are very small. Just as special relativity produces major departures from Newtonian mechanics only when velocities approach the speed of light, general relativity predicts large departures from Newtonian gravity only when extremely strong gravitational fields are involved—in effect, when orbit speeds and escape velocities become relativistic.

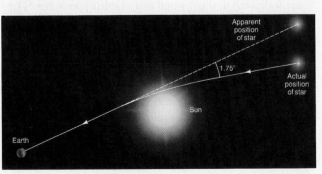

We will encounter other experimental and observational tests of general relativity elsewhere in this chapter. Here we consider just two "classical" tests of the theory. These tests are solar system observations that helped ensure

in excellent agreement with the prediction of general relativity. Virtually overnight Einstein became world famous. His previous major accomplishments notwithstanding, this single prediction assured him a permanent position as the best-known scientist on Earth! Recently, the high-precision *Hipparcos* satellite (*Interlude 17-1*) has observed shifts in the apparent positions of many stars, even those whose line of sight is far from the Sun. The shifts are exactly as predicted by Einstein's theory.

Another prediction of general relativity is that planetary orbits should deviate slightly from the perfect ellipses of Kepler's laws. Again, the effect is greatest where gravity is strongest—that is, closest to the Sun. Thus, the largest relativistic effects are found in the orbit of Mercury. Relativity predicts that Mercury's orbit is not a closed ellipse. Instead, its orbit should rotate slowly, as shown in the (highly exaggerated) diagram at right. The amount of rotation is very small—only 43″ per century—but Mercury's orbit is so well charted that even this tiny effect is measurable.

In fact, the observed rotation rate is 574″ per century, much greater than that predicted by relativity. However, when other (nonrelativistic) gravitational influences, primarily the perturbations due to the other planets, are taken into account, the rotation is in complete agreement with the foregoing prediction.

◀ MORE PRECISELY boxes
provide quantitative treatments
of subjects discussed qualitatively
in the text, or explore the
physics of astronomical
processes in greater detail.
Removing these more
challenging topics from the
main flow of the narrative and
placing them within a separate
modular element of the chapter
design (so that they can be
covered in class, assigned as
supplementary material, or
simply left as optional reading
for those students who find
them of interest) affords
instructors greater flexibility
in setting the level of their
coverage.

CROSS-LINKS   In astronomy, as in many scientific disciplines, almost every topic seems to have some bearing on almost every other. In particular, the connection between the specifically astronomical material and the physical principles set forth early in the text is crucial. Practically everything in Chapters 6–28 of this text rests on the foundation laid in the first five chapters. It is important that students, when they encounter, say, the cosmological redshift in Chapter 24, recall what they learned about the Doppler shift in Chapter 3. Similarly, the discussions of the mass of binary star components (Chapter 17) and of galactic rotation (Chapter 23) both depend on the discussion of Kepler's and Newton's Laws in Chapter 2. Throughout, the discussion of new astronomical objects relies heavily on comparison with topics introduced earlier.

It is essential to remind students of these links so that they can recall the principles on which later discussions rest, and if necessary review them. To reinforce these connections, "cross-links" have been inserted throughout the text—symbols that mark key intellectual bridges between material in different chapters. The links, denoted by the symbol ∞, together with a section reference (a hyperlink on the accompanying CD-ROM), signal to students that the topic under discussion is related in some significant way to ideas developed earlier, and direct them to material that they might wish to review before proceeding.

CHAPTER SUMMARIES   The Chapter Summaries, a primary review tool for the student, have been revised and improved for the third edition. All Key Terms introduced in each chapter are listed again, in context and in boldface, along with page references to the text discussion.

QUESTIONS, PROBLEMS, AND PROJECTS   Other elements of the end-of-chapter material have also seen substantial reorganization:

• Each chapter incorporates 30 revised Self-Test Questions, equally divided between "true/false" and "fill-in-the-blank" formats, designed to allow students to assess their understanding of the chapter material. Answers to all questions appear at the end of the book.
• Each chapter also has 15 Review and Discussion Questions, which may be used for in-class review or for assignment. As with the Self-Test Questions, the material needed to answer Review Questions may be found within the chapter. The Discussion Questions explore particular topics more deeply, often asking for opinions, not just facts. As with all discussions, these questions usually have no single "correct" answer. An answer key for the self-test questions is provided at the back of the book.
• New: The addition of 3–5 more challenging Problems per chapter now provide a wider range of difficulty in the end-of-chapter problem sets. Answers to the Problems, particularly for the more sophisticated ones, are not necessarily contained verbatim within the chapter but all the information required to solve the problems is in the text. Solutions for all Problems are in the Instructor's Manual.

8.7 Interiors

THE MOON

5 The Moon's average density, about 3300 kg/m^3, is similar to the measured density of lunar surface rock, virtually eliminating any chance that the Moon has a large, massive, and very dense nickel–iron core like that of Earth. In fact, the low density implies that the entire Moon is actually deficient in iron and other heavy metals relative to our planet. In addition, there is no evidence for any lunar magnetic field. As we saw in Chapter 7, researchers believe that planetary magnetism requires a rapidly rotating liquid metal core, like Earth's. ∞ (Sec. 7.4) Thus, the absence of a lunar

• Each chapter ends with a few (2–4) Projects meant to get the student out of the classroom and looking at the sky, although some entail research in libraries or other extracurricular activities.

CD-ROM

Following the enthusiastic response to the *Astronomy Today 2/e 1997 Media Edition*, the third edition comes complete with a free CD-ROM in the back of the text. This CD-ROM contains a fully hyperlinked version of the text to help the reader quickly find related information and assist in review. It also contains nearly 70 videos and animations integrated with the text to help make text figures and concepts more meaningful, and links to our companion web site, which is organized by text chapter and updated on a monthly basis. The availability of a related video/animation or web material is indicated by a starburst icon in both the electronic and print-based versions of the text, and all videos and animations included in the electronic version of the text are indicated in the printed text's table of contents. The CD for the third edition employs a new easier-to-use navigation system to facilitate moving within and across chapters and the videos and animations are now viewable on the same screen as corresponding text figures/discussions.

As with the second edition CD, the third edition CD also contains a separately executable Multimedia Study Guide program which provides review questions for each chapter. The computer automatically grades and scores all questions. For the third edition, the Multimedia Study Guide now includes labeling, sequencing, numerous multiple-choice and matrix questions.

The CD-ROM material can be used on Macintosh and PC computers using any standard browser (such as Netscape Navigator or Microsoft Explorer). For those students who do not already have a browser, Microsoft's Internet Explorer is included on the CD. A script for using the CD on a UNIX system is available at

ftp://ftp.prenhall.com/pub/esm/physics.s-085/chaissonbg/

WORLD WIDE WEB SITE   For both teachers and students we have a companion web site specifically for *Astronomy Today 3/e* at

http://www.prenhall.com/chaisson

This powerful resource organizes material from a variety of sources on the web on a chapter-by-chapter basis, is updated monthly, and provides interactive on-line exercises for each chapter.

Each chapter of the web site for *Astronomy Today 3/e* has the following three kinds of material:

- *Online Archives*—annotated images, videos, animations and free downloadable software
- *Online Destinations*—annotated links to relevant web sites
- *Online Exercises*—interactive questions for students to answer on-line; scoring and feedback are provided immediately.

In addition, the *Astronomy Today 3/e* web site includes our new Syllabus Builder tools, which allow instructors to easily set up an online syllabus with links to assignments on the companion web site.

Supplementary Material

This edition of *Astronomy Today* is accompanied by an outstanding set of instructional aids.

COMETS   Published annually at the beginning of each academic year, Comets is a unique kit available free to adopters that includes a collection of slides, videos, and *New York Times* articles on events and discoveries that have occurred since the publication of the prior year's Comets kit. The specific contents of the Comets kit for Fall 1998 are:

- A VHS tape with 10 new animations created by Alan Sill of Texas Tech University to go with specific text figures, as well as 15 new simulations, animations and videos from the Space Telescope Science Institute, JPL, the Goddard Space Flight Center, and NCSA. In addition, the tape continues to include the twenty-seven custom animations created at the Wright Center exclusively for Prentice Hall by renowned artists Dana Berry and Jim Palmer. These are not currently available elsewhere on VHS tape.
- A slide kit with 28 new slides from STScI/NASA, the COBE/DIRBE Science Team/NASA, the Smithsonian Astrophysical Observatory, the High-z Supernova Search Team/NASA, and JPL/NASA.

- The Astronomy Themes of the Times for Fall 1998, which is the fall issue of a biannual publication of articles on astronomy from *The New York Times*. The Fall 1998 issue contains 28 articles and is available free in quantity for students using the *Astronomy Today* text.
- A newsletter which provides a cross-reference between all the materials in the Comets kit and corresponding text chapters, as well as annotations describing the subject and source of each slide and video in the kit.
ISBN: 0-13-081206-4

 PRENTICE HALL/THE NEW YORK TIMES THEMES OF THE TIMES   Timely and relevant articles from recent editions of *The New York Times* compiled into a customized newspaper format. It enhances text coverage by helping students relate what they learn in class to exciting new theories and discoveries. Published twice yearly and available free up to the quantity of adoption.

INSTRUCTOR'S MANUAL   by Leo Connolly (California State University at San Bernardino). This manual provides an overview of each chapter; pedagogical tips, useful analogies, and suggestions for classroom demonstrations; answers to the end-of-chapter review and discussion questions and problems; and a list of selected readings.
ISBN: 0-13-081212-9

PRESENTATION MANAGER CD-ROM   This flexible, easy-to-use tool contains a wealth of photographs, line art, animations, and videos to use in class lectures. With the *Presentation Manager* system, instructors can easily search, access, and organize the materials according to their lecture outlines and add their own visuals and lecture notes. The CD contains all the line art and photographs from *Astronomy Today 3/e*, as well as all the animations and videos that are on the CD in the back of the student text. In addition, the *Presentation Manager* incorporates over 80 slides from the current and past editions of Comets.
ISBN: 0-13-081205-6 (Mac and Windows [hybrid])

ACETATES AND SLIDES   An extensive set of color acetates and a comprehensive 35mm slide set are available free to qualified adopters.
ISBN: 0-13-081204-8 (Slide set)
ISBN: 0-13-081203-X (Transparency pack)

TEST ITEM FILE   An extensive file of test questions, newly compiled for the third edition by Rex Joyner (Indiana Institute of Technology), is offered free upon adoption. Available in both printed and electronic form (Macintosh or IBM-compatible formats). The third edition Test Item File has been substantially revised, including many new True-False and Short Answer questions, a much sharper distinction between Short Answer and Essay questions, and suggested answers for Essay questions. Overall approximately 400 new questions have been added.
ISBN: 0-13-081211-0

PRENTICE HALL CUSTOM TEST Prentice Hall Custom Test is based on the powerful testing technology developed by Engineering Software Associates, Inc. (ESA). Available for Windows, Macintosh, and DOS, Prentice Hall Custom Test allows educators to create and tailor the exam to their own needs. With the Online Testing option, exams can also be administered online and data can then be automatically transferred for evaluation. A comprehensive desk reference guide is included, along with on-line assistance.

ISBN: 0-13-081202-1 (Mac)
ISBN: 0-13-081200-5 (Windows)

STUDENT OBSERVATION GUIDE WITH LABORATORY EXERCISES by Michael Seeds and Joseph Holzinger (Franklin and Marshall College). The 2nd edition of this useful supplement contains 42 classic labs and observational activities, along with cardboard cutout instruments that students can build and use for observations. Available for sale to students.

ISBN: 0-13-644196-3

BASIC ASTRONOMY LABS by Jay Huebner and Terry Smith (University of North Florida, Jacksonville), and Michael Reynolds (Chabot Observatory and Science Center). A collection of 40 laboratory exercises, including a wide range of both traditional and innovative topics (such as the nature of human vision, radioactivity and time, and astronomy on the Internet). Detailed introductions provide a fully-developed context for each exercise. Available for sale to students.

ISBN: 0-13-376336-6

ASTRONOMY ON THE INTERNET by Andrew Stull and Alan Sill (Texas Tech University). A guide to general astronomy resources on the Internet. Everything you need to know to get yourself online and browsing the World Wide Web!

ISBN: 0-13-890112-0

Acknowledgments

Throughout the many drafts that have led to this edition, we have relied on the critical analysis of many colleagues. Their suggestions ranged from the macroscopic issue of the book's overall organization to the minutiae of the technical accuracy of each and every sentence. We have also benefited from much good advice and feedback from many users of the second edition of the text and from our shorter book, *Astronomy: A Beginner's Guide to the Universe*. To these many helpful colleagues, we offer our sincerest thanks.

Stephen G. Alexander
Miami University

Robert H. Allen
University of Wisconsin, La Crosse

Timothy C. Beers
University of Evansville

William J. Boardman
Birmingham Southern College

Donald J. Bord
University of Michigan, Dearborn

Elizabeth P. Bozyan
University of Rhode Island

Anne Cowley
Arizona State Unversity

Bruce Cragin
Richland College

Norman Derby
Bennington College

Kimberly Engle
Drexel University

Martin Goodson
Delta College

David J. Griffiths
Oregon State University

Donald Gudehus
Georgia State University

Marilynn Harper
Delaware County Community College

Clint D. Harper
Moorpark College

Joseph Heafner
Catawaba Valley Community College

Fred Hickok
Catonsville Community College

Darren L. Hitt
Loyola College (Maryland)

F. Duane Ingram
Rock Valley College

Steven D. Kawaler
Iowa State University

William Keel
University of Alabama

Mario Klairc
Midlands Technical College

Kristine Larsen
Central Connecticut State University

Andrew R. Lazarewicz
Boston College

Robert J. Leacock
University of Florida

Larry A. Lebofsky
University of Arizona

M.A. Lohdi
Texas Tech University

Michael C. LoPresto
Henry Ford Community College

Phillip Lu
Western Connecticut State University

Milan Mijic
California State University, Los Angeles

Richard Nolthenius
Cabrillo College

Gregory W. Ojakangas
University of Minnesota, Duluth

Ronald Olowin
Saint Mary's College of California

Robert S. Patterson
Southwest Missouri State University

Cynthia W. Peterson
University of Connecticut

James A. Roberts
University of North Texas

Malcolm P. Savedoff
University of Rochester

John C. Schneider
Catonsville Community College

Harry L. Shipman
University of Delaware

C.G. Pete Shugart
Memphis State University

Stephen J. Shulik
Clarion University

Don Sparks
Los Angeles Pierce College

Maurice Stewart
Williamette University

Jack W. Sulentic
University of Alabama

Andrew Sustich
Arkansas State University

Stephen R. Walton
California State University, Northridge

Peter Wehinger
Arizona State University

Louis Winkler
Pennsylvania State University

We would also like to acknowledge our gratitude to Leo Connolly for preparing the hyperlinks for the end-of-chapter questions as well as all the content for the Multimedia Study Guide; to Ray Villard of the Space Telescope Science Institute for compiling the Comets supplement; and to Alan Sill of Texas Tech University for creating and maintaining our world wide web site, as well as for the new animations accompanying the text.

The publishing team at Prentice Hall has assisted us at every step along the way in creating this text. Much of the credit for getting the project completed on time goes to our Executive Editor, Alison Reeves, who has successfully navigated us through the many twists, turns, and "absolute final deadlines" of the publishing world. We have been fortunate to work once again with Production Editor Alison Aquino, who has done a truly excellent job of tying together the threads of this very complex project, made all the more complex by the necessity of combining text, art, and electronic media into a coherent whole. Special thanks go to Elizabeth Bozyan at the University of Rhode Island for her painstaking work in checking the accuracy of the text; to Development Editor Dan Schiller for his assistance throughout the production process; to Rolando Corujo and the Prentice Hall art staff, for their skillful execution of the innumerable last-minute corrections we made to the art program; and to Erik Unhjem, Assistant Managing Editor, Media. The intricate and crucial task of page makeup was led with aplomb by Richard Foster; he and the layout group deserve the greatest credit for their extraordinary patience, diligence, and craftsmanship. We are also grateful to Wendy Rivers, Assistant Editor, for managing the development and timely publication of the Instructor's Manual, Test Item File, and Slide and Transparencies set as well as to Gillian Kieff, Editorial Assistant, for facilitating the flow of various materials and information to everyone. Finally, we would like to express our gratitude to renowned space artist Dana Berry for allowing us to use many of his strikingly beautiful renditions of astronomical scenes and to Lola Judith Chaisson for creating all the H–R diagrams (including the acetate overlays) used in the text.

ASTRONOMY TODAY

A

B

C

D

1 CHARTING THE HEAVENS

The Foundations of Astronomy

LEARNING GOALS

Studying this chapter will enable you to:

1 Explain the concept of the celestial sphere and the conventions of angular measurement that enable us to locate objects in the sky.

2 Describe how the Sun, the Moon, and the stars appear to change their positions from night to night and from month to month.

3 Account for these apparent motions in terms of the actual motions of Earth and the Moon.

4 Show how the relative motions of Earth, the Sun, and the Moon lead to eclipses.

5 Explain the simple geometric reasoning that allows astronomers to measure the distances and sizes of faraway objects.

(Opposite page, background) One of the most easily recognizable star fields in the winter nighttime sky, the familiar constellation Orion. This field of view spans roughly 100 light-years, or 10^{15} kilometers. (See also Figure 1.6.)

(Inset A) If we magnify the wide view of the Orion constellation, shown at left, by a million times, we enter into the realm of the largest stars, with sizes of about a billion kilometers. Such a star is seen in this false-color image of the red-giant star Betelgeuse (which is actually the bright star at the upper left of the Orion constellation).

(Inset B) Another magnification of a million brings us to the scale of typical moons—roughly 1,000 kilometers—represented here by Ariel, one of the many moons of Uranus.

(Inset C) With yet another million-times magnification, we reach scales of meters, represented here by an astronomer at the controls of her telescope.

(Inset D) At a final magnification of an additional million, we reach the scale of molecules (about 10^{-6} meter), represented by this coiled DNA molecule of a rat's liver.

Nature offers no greater splendor than the starry sky on a clear, dark night. Silent and jeweled with the constellations of ancient myth and legend, the night sky has inspired wonder throughout the ages—a wonder that leads our imaginations far from the confines of Earth and the pace of the present day, and out into the distant reaches of space and cosmic time itself. Astronomy, born in response to that wonder, is built on two of the most basic traits of human nature: the need to explore and the need to understand. Through the interplay of curiosity, discovery, and analysis—the keys to exploration and understanding—people have sought answers to questions about the universe since the earliest times. Astronomy is the oldest of all the sciences, yet never has it been more exciting than it is today.

1.1 Our Place in Space

Of all the scientific insights attained to date, one stands out boldly: Earth is neither central nor special. We inhabit no unique place in the universe. Astronomical research, especially within the past few decades, strongly suggests that we live on what seems to be an ordinary rocky *planet* called Earth, which is one of nine known planets orbiting an average *star* called the Sun, which is one star near the edge of a huge collection of stars called the Milky Way *Galaxy*, which is one galaxy among countless billions of others spread throughout the observable universe. To begin to get a feel for the relationships among these very different objects, consult Figures 1.1 through 1.4; put them in perspective by studying Figure 1.5.

We are connected to the most distant realms of space and time not only by our imaginations, but also through a common cosmic heritage: most of the chemical elements in our bodies were created billions of years ago in the hot centers of long-vanished stars. Their fuel supply spent, these giant stars died in huge explosions, scattering afar the elements created deep within their cores. Eventually, this matter collected into clouds of gas that slowly collapsed to give birth to a new generation of stars. In this way, the Sun and its family of planets formed nearly 5 billion years ago. Everything on Earth embodies atoms from other parts of the universe and from a past far more remote than the beginning of human evolution. Elsewhere, other beings, perhaps with intelligence much greater than our own, may at this very moment be gazing in wonder at their own night sky. Our own Sun may be nothing more than an insignificant point of light to them, if it is visible at all. Yet if such beings exist, then they must share our cosmic origin.

Simply put, the **universe** is the totality of all space, time, matter, and energy. **Astronomy** is the study of the universe. It is a subject unlike any other, for it requires us to change profoundly our view of the cosmos and to consider matter on scales totally unfamiliar from everyday experience. Look again at the galaxy in Figure 1.3. This galaxy, whose catalog name is M83, is a swarm of about a hundred billion stars—more stars than the number of people who have ever lived on Earth. The entire assemblage is spread across a vast expanse of space some 100,000 **light years** in diameter. A light year is the *distance* traveled by light, at a speed of about 300,000 kilometers per second, in a year. It equals about 10 trillion kilometers (or around 6 trillion miles). Typical galactic systems are truly "astronomical" in size! The light year is a convenient unit introduced by astronomers to help them describe immense distances. We will encounter many such "custom" units in our studies (see Appendix 2).

A thousand (1000), a million (1,000,000), a billion (1,000,000,000), and even a trillion (1,000,000,000,000)—these words occur regularly in everyday speech. But let's take a moment to understand the magnitude of these numbers and to appreciate the differences among them. One thousand is easy enough to understand; at the rate of one number per second, you could count to a thousand in about 16 minutes. However, if you wanted to count to a million, you would need more than 2 weeks of counting at the rate of one number per second, 16 hours per day (allowing 8 hours per day for sleep). To count from 1 to a billion at the same rate of one number per second and 16 hours per day would take nearly 50 years—the better part of an entire human lifetime.

In this text we consider spatial domains spanning not just billions of kilometers but billions of light years, objects containing not just trillions of atoms but trillions of stars, time intervals of not just billions of seconds or hours but billions of years. You will need to become familiar with—and comfortable with—such enormous numbers. A good way to begin is to try and recognize just how much larger than a thousand is a million, and how much larger still is a billion. (Appendix 1 explains the convenient method used by scientists for writing and manipulating very large and very small numbers.)

1.2 The Obvious View

How have we come to know the universe around us? How do we know the proper perspective sketched in Figure 1.5? Our study of the universe, the science of astronomy, begins by examining the sky.

CONSTELLATIONS IN THE SKY

Between sunset and sunrise on a clear night we can see some 3000 points of light. Include the view from the opposite side of Earth, and nearly 6000 stars are visible to the unaided eye. A natural human tendency is to see patterns and relationships between objects even when no true connection

15,000 kilometers

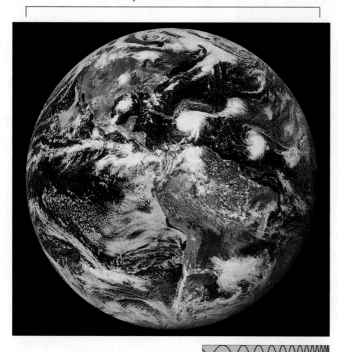

R I V U X G

Figure 1.1 Earth is a planet, a mostly solid object, though it has some liquid in its oceans and its core, and gas in its atmosphere. (In this view, you can clearly see the North and South American continents.)

1,500,000 kilometers

R I V U X G

Figure 1.2 The Sun is a star, a very hot ball of gas. Much bigger than Earth, the Sun is held together by its own gravity.

About 100,000 light years

R I V U X G

Figure 1.3 A typical galaxy is a collection of a hundred billion stars, each separated by vast regions of nearly empty space. This galaxy is the 83d entry in the catalog compiled by the eighteenth-century French astronomer Charles Messier—M83 for short. Our Sun is a rather undistinguished star near the edge of another such galaxy, called the Milky Way.

About 100,000,000 light years

R I V U X G

Figure 1.4 This photograph shows a typical cluster of galaxies, roughly a billion light years from Earth. Each galaxy contains hundreds of billions of stars, probably planets, and possibly living creatures.

Local Group
of galaxies

Local stellar
neighborhood

Earth

20

15,000

10,000

1,000,000

500,000

Milky Way Galaxy

Solar System

Figure 1.5 This artist's conception puts each of the previous figures in perspective. The bottom of this figure shows spacecraft (and astronauts) in Earth orbit, a view that widens progressively in each of the next five cubes drawn from bottom to top—Earth, the planetary system, the local neighborhood of stars, the Milky Way Galaxy, and the closest cluster of galaxies. The numbers indicate approximately the increase in scale between successive images.

exists, and people long ago connected the brightest stars into configurations called **constellations**, which ancient astronomers named after mythological beings, heroes, and animals—whatever was important to them. Figure 1.6 shows a constellation especially prominent in the nighttime sky from October through March: the "hunter" named Orion. Orion was a mythical Greek hero famed, among other things, for his amorous pursuit of the Pleiades, the seven daughters of the giant Atlas. According to Greek mythology, in order to protect the Pleiades from Orion, the gods placed them among the stars, where Orion nightly stalks them across the sky. Many constellations have similarly fabulous connections with ancient lore.

Perhaps not surprisingly, the patterns seen have a strong cultural bias—the astronomers of ancient China saw mythical

figures different from those seen by the ancient Greeks, the Babylonians, and the people of other cultures, even though they were all looking at the same stars in the night sky. Interestingly, though, different cultures often made the same basic *groupings* of stars, despite widely varying interpretations of what they saw. For example, the group of seven stars usually known in North America as "the Dipper" is known as "the Wagon" or "the Plough" in Western Europe. The ancient Greeks regarded these same stars as the tail of "the Great Bear," the Egyptians saw them as the leg of an ox, the Siberians as a stag. Some Native Americans saw two mythical brothers, others an ermine, others still a funeral procession.

Early astronomers had very practical reasons for studying the sky. Some constellations served as navigational guides. For example, the star Polaris, part of the Little Dipper, indicates north, and the near-constancy of its location in the sky, from hour to hour and night to night, has aided travelers for centuries. Other constellations served as primitive calendars to predict planting and harvesting seasons. For example, many cultures knew well that the appearance of certain stars on the horizon just before daybreak signaled the beginning of spring and the end of winter.

In many societies, people came to believe that there were other benefits in being able to trace the regularly changing positions of heavenly bodies. The relative positions of stars and planets at a person's birth were carefully studied by *astrologers*, who used the data to make predictions about that person's destiny. Thus, in a sense, astronomy and astrology arose from the same basic impulse—the desire to "see" into the future—and, indeed, for a long time they were indistinguishable from one another. Today most people recognize that astrology is nothing more than an amusing diversion (although millions still study their horoscope in the newspaper every morning!). Nevertheless, the ancient astrological terminology—the names of the constellations and

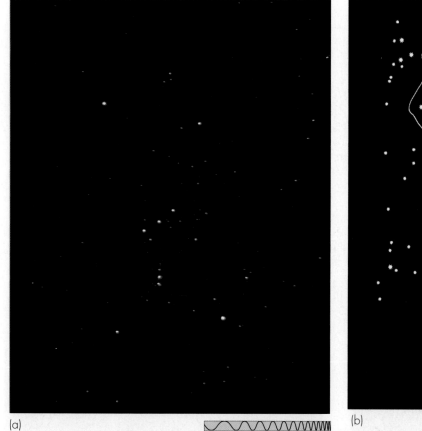

(a)

R I V U X G

(b)

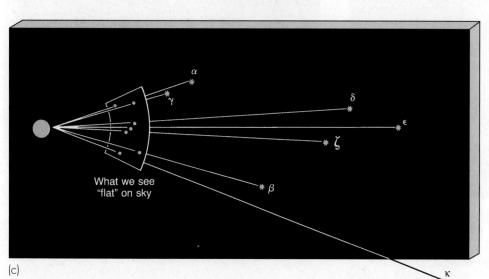

(c)

Figure 1.6
(a) A photograph of the group of bright stars that make up the constellation Orion. (b) The stars are connected to show the pattern visualized by the Greeks: the outline of a hunter. You can easily find this constellation in the winter sky by identifying the line of three bright stars in the hunter's "belt." (c) The true relationships among the stars, in three dimensions.

many terms used to describe the locations and motions of the planets, for example—is still used throughout the astronomical world.

Generally speaking, as illustrated in Figure 1.6(c) for the case of Orion, the stars that make up any particular constellation are not actually close to one another in space, even by astronomical standards. They merely are bright enough to observe with the naked eye and happen to lie in roughly the same direction in the sky as seen from Earth. Nevertheless, the constellations provide a convenient means for astronomers to specify large areas of the sky, much as geologists use continents, or politicians use voting precincts to identify certain localities on planet Earth. In all, there are 88 constellations, most of them visible from North America at some time during the year.

THE CELESTIAL SPHERE

1 Over the course of a night, the constellations seem to move smoothly across the sky from east to west, but ancient sky-watchers were well aware that the *relative* locations of stars remain unchanged as this nightly march takes place. It was natural for those observers to conclude that the stars must be firmly attached to a **celestial sphere** surrounding Earth—a canopy of stars resembling an astronomical painting on a heavenly ceiling. Figure 1.7 shows how early astronomers pictured the stars as moving with this celestial sphere as it turned around a fixed, unmoving Earth. Figure 1.8 shows how all stars appear to move in circles around a point very close to the star Polaris (better known as the Pole Star, or the North Star). To the ancients, this point represented the axis around which the entire celestial sphere turned.

From our modern standpoint, the apparent motion of the stars is the result of the spin, or **rotation**, not of the celestial sphere but of Earth. Polaris indicates the direction—due north—in which Earth's rotation axis points. Even though we now know that the celestial sphere is an incorrect description of the heavens, we still use the idea as a convenient fiction that helps us visualize the positions of stars

(a)

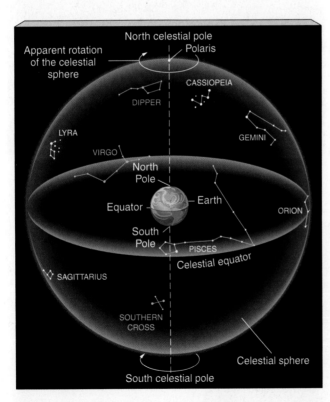

Figure 1.7 Planet Earth sits fixed at the hub of the celestial sphere, which contains all the stars. This is one of the simplest possible models of the universe, but it doesn't agree with all the facts that astronomers know about the universe.

(b)

R   I   V   U   X   G

Figure 1.8 Time-lapse photographs of the northern sky. Each trail is the path of a single star across the nighttime sky. The duration of the exposure in (a) is about 1 hour, that in (b) is about 5 hours. (How can you tell that this is so?) The center of the concentric circles is near the north star, Polaris, whose short, bright arc is prominently visible in (b).

in the sky. The point where Earth's axis intersects the celestial sphere in the Northern Hemisphere is known as the **north celestial pole**, and it is directly above Earth's North Pole. In the Southern Hemisphere, the extension of Earth's axis in the opposite direction defines the **south celestial pole**, directly above Earth's South Pole. Midway between the north and south celestial poles lies the **celestial equator**, representing the intersection of Earth's equatorial plane with the celestial sphere. These parts of the celestial sphere are marked on Figure 1.7.

When discussing the locations of stars "on the sky," astronomers naturally talk in terms of *angular* positions and separations. *More Precisely 1-1* presents some basic information on angular measure. *More Precisely 1-2* discusses in more detail the system of coordinates used to specify stellar positions.

1.3 The Motion of the Sun and the Stars

DAY-TO-DAY CHANGES

2 We measure time by the Sun. Because the rhythm of day and night is central to our lives, it is not surprising that the period of time from one sunrise (or noon, or sunset) to the next, the 24-hour **solar day**, is our basic social time unit. The daily progress of the Sun and the other stars across the sky is known as *diurnal motion*. As we have just seen, it is a consequence of Earth's rotation. But the stars' positions in the sky do *not* repeat themselves exactly from one night to the next. Each night, the whole celestial sphere appears to be shifted a little relative to the

MORE PRECISELY 1-1 Angular Measure

The size and scale of things are often specified by measuring lengths and angles. The concept of *length measurement* is fairly intuitive. We've already noted the dimensions of some objects found in the universe, ranging from subatomic particles with sizes of 10^{-15} m to whole galaxies with sizes of 10^{21} m or more. The concept of *angular measurement* may be less intuitive, but it too can become second nature if you remember a few simple facts.

A full circle contains 360 *arc degrees* (or just 360°). Therefore, the half-circle that stretches from horizon to horizon, passing directly overhead and spanning the portion of the sky visible to one person at any one time, contains 180°. Each 1° increment can be further subdivided into fractions of an arc degree, called *arc minutes*; there are 60 arc minutes in 1 arc degree. Both the Sun and the Moon project an angular size of 30 arc minutes on the sky. Your little finger, held at arm's length, does just about the same, covering a small (actually, about 40 arc minutes) slice of the entire 180 arc degrees from horizon to horizon. Finally, an arc minute can be further sliced into 60 equal *arc seconds*. Put another way, an arc minute is 1/60 of an arc degree, and an arc second is $1/60 \times 1/60 = 1/3600$ of an arc degree. An arc second is an extremely small unit of angular measure. It is, in fact, the angle subtended (projected) by a centimeter-sized object at a distance of about 2 kilometers or approximately the angle subtended by a dime when seen from a mile away. The accompanying figure illustrates this subdivision of the circle into progressively smaller units.

Don't be confused by the units used to measure angles. Arc minutes and arc seconds have nothing to do with the measurement of time, and arc degrees have nothing to do with temperature. Arc degrees, arc minutes, and arc seconds are simply ways to measure the size and position of objects in the universe. An arc degree is usually written as 1°, 1 arc minute is denoted by 1′, and 1 arc second is 1″.

The angular size of an object depends both on its actual size and on its distance from us. For example, the Moon, at its present distance from Earth, has an angular diameter of 0.5°, or 30′. If the Moon were twice as far away, it would appear half as big—15′ across—even though its actual size would be the same. Similarly, if the Moon were one-third as far away, its angular diameter would be three times as great—1.5° and so on. Thus, *angular size by itself is not enough to determine the actual diameter of an object—the distance must also be known.*

MORE PRECISELY 1-2 Celestial Coordinates

The simplest method of locating stars in the sky is to specify their constellation and then rank the stars in it in order of brightness. The brightest star is denoted by the Greek letter α (alpha), the second brightest by β (beta), and so on. Thus, the two brightest stars in the constellation Orion—Betelgeuse and Rigel—are also known as α Orionis and β Orionis, respectively. (Precise recent observations show that Rigel is actually brighter than Betelgeuse, but the names are now permanent.) Because there are many more stars in any given constellation than there are letters in the Greek alphabet, this method is of limited utility. However, for naked-eye astronomy, where only bright stars are involved, it is quite satisfactory.

For more precise measurements, astronomers find it helpful to lay down a system of *celestial coordinates* on the sky. If we think of the stars as being attached to the celestial sphere centered on Earth, then the familiar system of latitude and longitude on Earth's surface extends naturally to cover the sky. The celestial analogs of latitude and longitude on Earth's surface are called *declination* and *right ascension*, respectively. The accompanying figure illustrates the meanings of right ascension and declination on the celestial sphere and compares them with longitude and latitude on Earth.

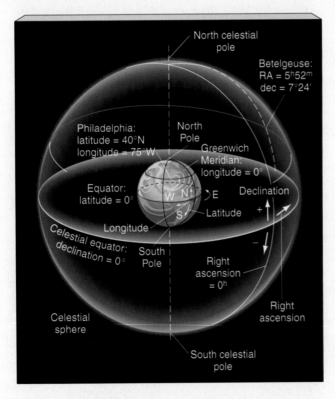

Declination (dec) is measured in degrees (°) north or south of the celestial equator, just as latitude is measured in degrees north or south of Earth's equator. (See *More Precisely 1-1* for a discussion of angular measure.) Thus, the celestial equator is at a declination of 0°, the north celestial

pole is at +90°, and the south celestial pole is at −90° (the minus sign here just means "south of the celestial equator"). Right ascension (RA) is measured in units called *hours*, *minutes*, and *seconds*, and it increases in the eastward direction. The angular units used to measure right ascension are constructed to parallel the units of time; the two sets of units are connected by the rotation of Earth (or of the celestial sphere). In 24 hours, Earth rotates once on its axis, or through 360°. Thus, in a time period of 1 hour, Earth rotates through $360°/24 = 15°$, or 1^h. In 1 minute of time, Earth rotates through 1^m; in 1 second, Earth rotates through 1^s. The choice of zero right ascension is quite arbitrary—it is conventionally taken to be the position of the Sun in the sky *at the instant of the vernal equinox*.

Just as longitude and latitude allow us to locate a point on Earth's surface, right ascension and declination specify locations on the sky. For example, to find Philadelphia on Earth, look 75° west of the Greenwich Meridian and 40° north of the equator (as marked on the figure). Similarly, to locate the star Betelgeuse on the celestial sphere, look 5^h54^m east of the vernal equinox (the line on the sky with a right ascension of zero) and 7°24′ north of the celestial equator.

The units of right ascension were originally defined to assist astronomical observation, but the names are rather unfortunate, as these are angular measures, *not* units of time. Moreover, the angular units used for right ascension are not the same units as defined in *More Precisely 1-1*. In fact, right ascension $1^m = 15°/60 = 0.25°$, or 15 arc minutes (15′). Similarly, right ascension $1^s = 15$ arc seconds (15″). If you remember that the units of right ascension are used *only* for that one purpose, and that all angular measurements except right ascension use *arc minutes* (′) and *arc seconds* (″), you should avoid undue confusion.

Just as latitude and longitude are tied to Earth, right ascension and declination are fixed on the celestial sphere. Although the stars appear to move across the sky because of Earth's rotation, their celestial coordinates remain *constant* over the course of a night. Thus, we have a quantitative alternative to the use of constellations in specifying the positions of stars in the sky. For example, the stars Rigel and Betelgeuse mentioned earlier can be precisely located by looking in the directions $5^h13^m36^s$ (RA), −8°13′ (dec) and $5^h52^m0^s$ (RA), 07°24′ (dec), respectively. The coordinates of Betelgeuse on the celestial sphere are marked on the figure.

Actually, because right ascension is tied to the position of the vernal equinox, the celestial coordinates of any given star drift slowly due to Earth's precession (see Section 1.4). Since one cycle of precession takes some 26,000 years to complete, this represents a shift of a little over 0.1″ on a night-to-night basis—a small angle, but one that nevertheless must be taken into account in high-precision astronomical measurements. Rather than deal with slowly changing coordinates for every object in the sky, astronomers conventionally correct their observations to the location of the vernal equinox at some standard epoch (for example, January 1, 1950, or January 1, 2000).

horizon, compared with the night before. The easiest way to confirm this difference is by noticing the stars that are visible just after sunset or just before dawn. You will find that they are in slightly different locations from those of the previous night. Because of this shift, a day measured by the stars—called a **sidereal day** after the Latin word *sidus*, meaning "star"—differs in length from a solar day. Evidently, there is more to the apparent motion of the heavens than simple rotation.

The reason for the difference between a solar day and a sidereal day is sketched in Figure 1.9. It is a result of the fact that Earth moves in two ways simultaneously: it rotates on its central axis while at the same time **revolving** around the Sun. Each time Earth rotates once on its axis, it also moves a small distance along its orbit about the Sun. Earth therefore has to rotate through slightly more than 360° (360 degrees—see *More Precisely 1-1*) for the Sun to return to the same apparent location in the sky. Thus, the interval of time between noon one day and noon the next (a solar day) is slightly greater than one true rotation period (one sidereal day). Our planet takes 365 days to orbit the Sun, so the ad-

ditional angle is 360°/365 = 0.986°. Because Earth takes about 3.9 minutes to rotate through this angle, the solar day is 3.9 minutes longer than the sidereal day (that is, one sidereal day is roughly 23ʰ56ᵐ long.)

SEASONAL CHANGES

Because Earth orbits the Sun, the Sun appears to move relative to the background stars. The apparent motion of the Sun on the sky over the course of a year, relative to the stars, defines a path on the celestial sphere known as the **ecliptic**. As illustrated in Figure 1.10(a), the ecliptic forms a great circle on the celestial sphere, inclined at an angle of about 23.5° to the celestial equator. In reality, as illustrated in Figure 1.10(b), the plane of the ecliptic is *the plane of Earth's orbit around the Sun*. Its tilt is a consequence of the *inclination* of our planet's rotation axis to its orbital plane.

The point on the ecliptic where the Sun is at its northernmost point above the celestial equator is known as the **summer solstice** (from the Latin words *sol*, meaning "sun," and *stare*, "to stand"). As shown in Figure 1.10, it represents the point in Earth's orbit where our planet's North Pole points closest to the Sun. This occurs on or near June 21—the exact date varies slightly from year to year because the actual length of a year is not a whole number of days. As Earth rotates, points north of the equator spend the greatest fraction of their time in sunlight on that date, so the summer solstice corresponds to the longest day of the year in the Northern Hemisphere and the shortest day in the Southern Hemisphere. Six months later, the Sun is at its southernmost point, or the **winter solstice** (December 21)—the shortest day in the Northern Hemisphere and the longest in the Southern Hemisphere. These two effects—the height of the Sun above the horizon and the length of the day— combine to account for the **seasons** we experience. In summer in the Northern Hemisphere, the Sun is high in the sky and the days are long, so temperatures are generally much higher than in winter, when the Sun is low and the days are short.

The two points where the ecliptic intersects the celestial equator are known as **equinoxes**. On those dates, day and night are of equal duration. (The word *equinox* derives from the Latin for "equal night.") In the fall (in the Northern Hemisphere), as the Sun crosses from the Northern into the Southern Hemisphere, we have the **autumnal equinox** (on September 21). The **vernal equinox** occurs in northern spring, on or near March 21, as the Sun crosses the celestial equator moving north. Because of its association with the end of winter and the start of a new growing season, the vernal equinox was particularly important to early astronomers and astrologers. It also plays an important role in human timekeeping. The interval of time from one vernal equinox to the next—365.242 solar days—is known as one **tropical year**.

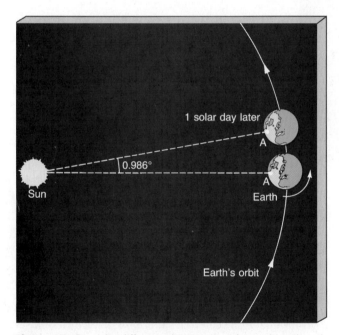

Figure 1.9 The difference between a solar and a sidereal day can be easily explained once we understand that Earth revolves around the Sun at the same time as it rotates on its axis. A solar day is the time from one noon to the next. In that time, Earth also moves a little in its solar orbit. Because Earth completes one circuit (360°) around the Sun in 1 year (365 days), it moves through nearly 1° in 1 day. Thus, between noon at point A on one day and noon at the same point the next day, Earth actually rotates through about 361°. Consequently, the solar day exceeds the sidereal day (360° rotation) by about 4 minutes. Note that the diagram is not drawn to scale, so the true 1° angle is in reality much smaller than shown here.

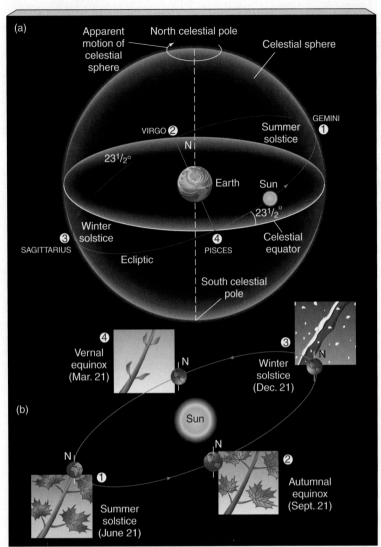

Figure 1.10   (a) The apparent path of the Sun on the celestial sphere and (b) its actual relation to Earth's rotation and revolution. The seasons result from the changing height of the Sun above the horizon. At the summer solstice (the points marked 1), the Sun is highest in the sky, as seen from the Northern Hemisphere, and the days are longest. In the "celestial sphere" picture (a), the Sun is at its northernmost point on its path around the ecliptic; in reality (b), the summer solstice corresponds to the point on Earth's orbit where our planet's North Pole points most nearly toward the Sun. The reverse is true at the winter solstice (point 3). At the vernal and autumnal equinoxes, day and night are of equal length. These are the times when, as seen from Earth (a), the Sun crosses the celestial equator. They correspond to the points in Earth's orbit when our planet's axis is perpendicular to the line joining Earth and Sun (b).

Animation

SUMMER AND WINTER CONSTELLATIONS

Figure 1.11 (a) illustrates the major stars visible from most locations in the United States on clear summer evenings. The brightest stars—Vega, Deneb, and Altair—form a conspicuous triangle high above the constellations Sagittarius and Capricornus, which are low on the southern horizon. In the winter sky, however, these stars are replaced, as shown in Figure 1.11(b), by several well-known constellations that include Orion, Leo, and Gemini. In the constellation Canis Major lies Sirius (the Dog Star), the brightest star in the sky. Year after year, the same stars and constellations return, each in its proper season. Every winter evening, Orion is high overhead; every summer, it is gone. (For more detailed maps of the sky at different seasons, consult the star charts at the end of the book.)

The reason for these regular seasonal changes is Earth's revolution around the Sun: Earth's darkened hemisphere faces in a slightly different direction in space each evening. The change in direction is only about 1° per night

(see Figure 1.9)—too small to be easily noticed with the naked eye from one evening to the next but clearly noticeable over the course of weeks and months, as illustrated in Figure 1.12. After 6 months, Earth has reached the opposite side of its orbit, and at night we face an entirely different group of stars and constellations. Ancient astronomers would have said that the Sun had moved to the opposite side of the celestial sphere, so that a different set of stars was visible at night.

The 12 constellations through which the Sun passes as it moves along the ecliptic—that is, the constellations we would see looking in the direction of the Sun, if they weren't overwhelmed by the Sun's light—had special significance for astrologers of old. These constellations are collectively known as the **zodiac**. The time required for the constellations to complete one cycle around the sky and to return to their starting points as seen from a given point on Earth is one **sidereal year**. Earth completes exactly one orbit around the Sun in this time. One sidereal year is 365.256 solar days long, about 20 minutes longer than a tropical year.

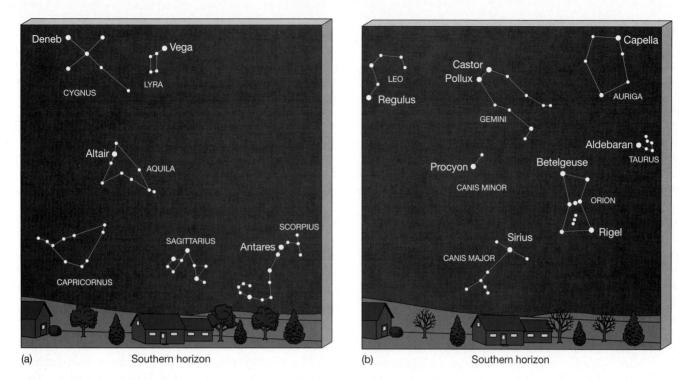

(a)   Southern horizon

(b)   Southern horizon

Figure 1.11 (a) A typical summer sky above the United States. Some prominent stars (labeled in larger print) and constellations (labeled in small capital letters) are shown. (b) A typical winter sky above the United States.

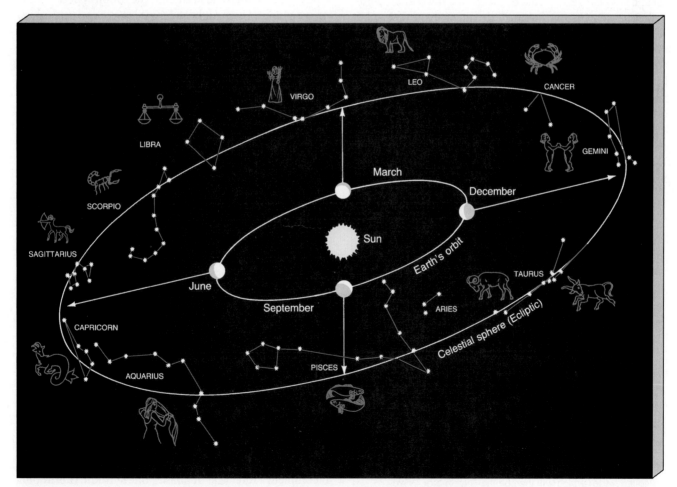

Figure 1.12 The view of the night sky changes as Earth moves in its orbit about the Sun. As drawn here, the night side of Earth faces a different set of constellations at different times of the year.

LONG-TERM CHANGES

3 Earth has many motions—it spins on its axis, it travels around the Sun, and it moves with the Sun through the galaxy. We have just seen how some of these motions can account for the changing nighttime sky and the changing seasons. In fact, the situation is even more complicated. Like a spinning top that rotates rapidly on its own axis while that axis slowly revolves about the vertical, Earth's axis changes its *direction* over the course of time (although the angle between the axis and a line perpendicular to the plane of the ecliptic remains close to 23.5°). This change is called **precession.** Figure 1.13 illustrates Earth's precession, which is caused mostly by the gravitational pulls of the Moon and the Sun. During a complete cycle of precession, taking about 26,000 years, Earth's axis traces out a cone.

Because of Earth's precession, the length of time from one vernal equinox to the next—one tropical year—is not quite the same as the time required for Earth to complete one orbit—one sidereal year. Recall that the vernal equinox occurs when Earth's rotation axis is perpendicular to the line joining Earth and Sun, and the Sun is crossing the celestial equator moving from south to north. In the absence of precession, this would occur exactly once per orbit, and the tropical and sidereal years would be identical. However, because of the slow precessional shift in the orientation of Earth's rotation axis, the instant when the axis is next perpendicular to the line from Earth to the Sun occurs slightly *sooner* than we would otherwise expect. Consequently, the vernal equinox drifts slowly around the zodiac over the course of the precession cycle. This is the cause of the 20-minute discrepancy between the two "years" mentioned earlier.

The tropical year is the year that our calendars measure. If our timekeeping were tied to the sidereal year, the seasons would slowly march around the calendar as Earth precessed—13,000 years from now, summer in the North-

ern Hemisphere would be at its height in late February! By using the tropical year instead, we ensure that July and August will always be (northern) summer months. However, in 13,000 years' time, Orion will be a summer constellation.

1.4 The Motion of the Moon

2 3 The Moon is our nearest neighbor in space. Apart from the Sun, it is by far the brightest object in the sky. Like the Sun, the Moon appears to move relative to the background stars. Unlike the Sun, however, the Moon really does revolve around the Earth. It crosses the sky at a rate of about 12° per day, which means it moves an angular distance equal to its own diameter—30 arc minutes—in about an hour.

LUNAR PHASES

The Moon's appearance undergoes a regular cycle of changes, or **phases**, taking a little more than 29 days to complete. (The word *month* is derived from the word *Moon.*) Figure 1.14 illustrates the appearance of the Moon at different times in this monthly cycle. Starting from the so-called **new Moon**, which is all but invisible in the sky, the Moon appears to *wax* (or grow) a little each night and is visible as a growing *crescent* (panel 1 of Figure 1.14). One week after new Moon, half of the lunar disk can be seen (panel 2). This phase is known as a **quarter Moon.** During the next week, the Moon continues to wax, passing through the *gibbous* phase (panel 3) until, 2 weeks after new Moon, the **full Moon** (panel 4) is visible. During the next 2 weeks, the Moon *wanes* (or shrinks), passing in turn through the gibbous, quarter, and crescent phases (panels 5–7), eventually becoming new again.

The Moon doesn't actually change its size and shape on a monthly basis, of course; its full circular disk is present at all times. Why then don't we always see a full Moon? The answer to this question lies in the fact that, unlike the Sun and the other stars, the Moon emits no light of its own. Instead, it shines by reflected sunlight. As illustrated in Figure 1.14, half of the Moon's surface is illuminated by the Sun at any instant. However, not all of the Moon's sunlit face can be seen because of the Moon's position with respect to Earth and the Sun. When the Moon is full, we see the entire "daylit" face because the Sun and the Moon are in opposite directions from Earth in the sky. In the case of a new Moon, the Moon and the Sun are in almost the same part of the sky, and the sunlit side of the Moon is oriented away from us. At new Moon, the Sun must be almost behind the Moon, from our perspective.

As it revolves around Earth, the Moon's position in the sky changes with respect to the stars. In one **sidereal month** (27.3 days), the Moon completes one revolution and returns to its starting point on the celestial sphere, having traced out a great circle in the sky. The time required for the Moon to complete a full cycle of phases, one **synodic month**, is a little longer—about 29.5 days. The synodic month is a little longer than the sidereal month for the same reason that

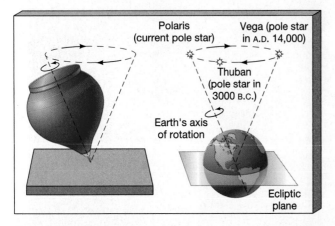

Figure 1.13   Earth's axis currently points nearly toward the star Polaris. Some 12,000 years from now—nearly halfway through one cycle of precession—Earth's axis will point toward a star called Vega, which will then be the "North Star." Five thousand years ago, the North Star was a star named Thuban in the constellation Draco.

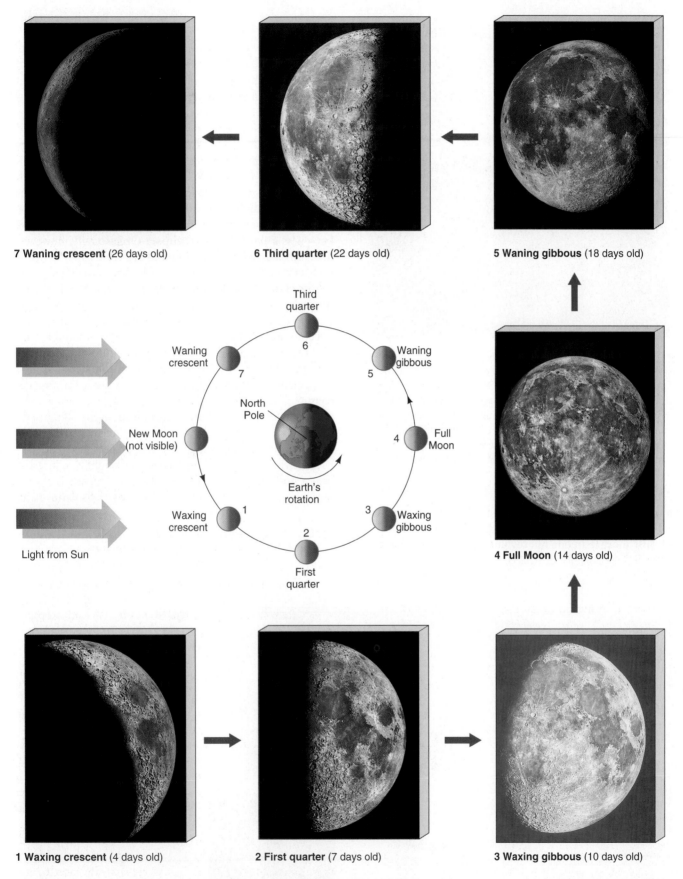

7 Waning crescent (26 days old)

6 Third quarter (22 days old)

5 Waning gibbous (18 days old)

4 Full Moon (14 days old)

1 Waxing crescent (4 days old)

2 First quarter (7 days old)

3 Waxing gibbous (10 days old)

Figure 1.14   Because the Moon orbits Earth, the visible fraction of the sunlit face differs from night to night. The complete cycle of lunar phases takes 29 days to complete.

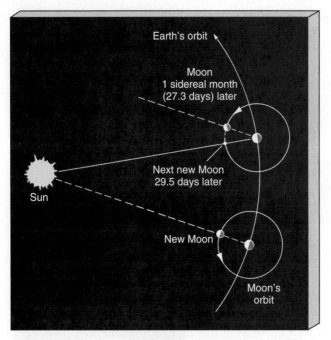

Figure 1.15 The difference between a synodic and a sidereal month stems from the motion of Earth relative to the Sun. Because Earth orbits the Sun in 365 days, in the 29.5 days from one new Moon to the next (one synodic month), Earth moves through an angle of approximately 29°. Thus the Moon must revolve more than 360° between new Moons. The sidereal month, which is the time taken for the Moon to revolve through exactly 360°, relative to the stars, is about 2 days shorter.

a solar day is slightly longer than a sidereal day: because of Earth's motion around the Sun, the Moon must complete slightly more than one full revolution to return to the same phase in its orbit (Figure 1.15).

ECLIPSES

4 From time to time—but only at new or full Moon—the Sun and the Moon line up precisely as seen from Earth, and we observe the spectacular phenomenon known as an **eclipse.** When the Sun and the Moon are in exactly *opposite* directions, as seen from Earth, Earth's shadow sweeps across the Moon, temporarily blocking the Sun's light and darkening the Moon in a **lunar eclipse,** as illustrated in Figure 1.16. From Earth, we see the curved edge of Earth's shadow begin to cut across the face of the full Moon and slowly eat its way into the lunar disk. Usually, the alignment of the Sun, Earth, and Moon is imperfect, so the shadow never completely covers the Moon. Such an occurrence is known as a **partial lunar eclipse.** Occasionally, however, the entire lunar surface is obscured in a **total lunar eclipse,** such as that shown in the inset of Figure 1.16. Total lunar eclipses last only as long as is needed for the Moon to pass through Earth's shadow—no more than about 100 minutes. During that time, the Moon often acquires an eerie, deep red coloration—the result of a small amount of sunlight that is refracted (bent) by Earth's atmosphere onto the lunar surface, preventing the shadow from being completely black.

When the Moon and the Sun are in exactly the *same* direction, as seen from Earth, an even more awe-inspiring event occurs. The Moon passes directly in front of the Sun, briefly turning day into night in a **solar eclipse.** In a **total solar eclipse,** when the alignment is perfect, planets and some stars become visible in the daytime as the Sun's light is reduced to nearly nothing. We can also see the Sun's ghostly outer atmosphere, or *corona* (Figure 1.17a).* In a **partial solar eclipse,**

Actually, although a total solar eclipse is undeniably a spectacular occurrence, the visibility of the corona is probably the most important astronomical aspect of such an event today. It enables us to study this otherwise hard-to-see part of our Sun.

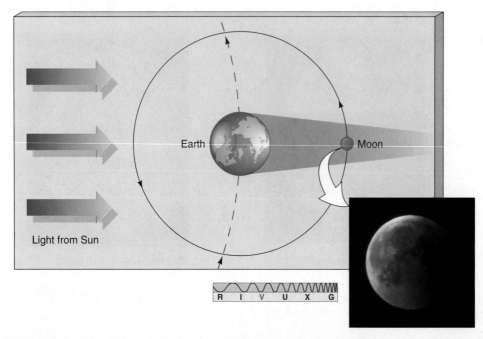

Figure 1.16 A lunar eclipse occurs when the Moon passes through Earth's shadow. At these times we see a darkened, copper-colored Moon, as shown in the inset photograph. The coloration is caused by sunlight deflected by Earth's atmosphere onto the Moon's surface. (Note that this figure is not drawn to scale.)

R I V U X G

(a)   R I V U X G   (b)   R I V U X G

Figure 1.17 (a) During a total solar eclipse the Sun's corona becomes visible as an irregularly shaped halo surrounding the blotted-out disk of the Sun. This was the July 1991 eclipse, as seen from the Baja Peninsula. (b) During an annular eclipse, the Moon fails to completely hide the Sun, so a thin ring of light remains. No corona is seen in this case because even the small amount of the Sun still visible completely overwhelms the corona's faint glow. This was the December 1973 eclipse, as seen from Algiers. (The gray fuzzy areas at top left and right are clouds in Earth's atmosphere.)

the Moon's path is slightly "off center," and only a portion of the Sun's face is covered. In either case, the sight of the Sun apparently being swallowed up by the black disk of the Moon is disconcerting even today. It must surely have inspired fear in early observers. Small wonder, then, that the ability to predict such events was a highly prized skill.

Unlike a lunar eclipse, which is simultaneously visible from all locations on Earth's night side, a total solar eclipse can be seen from only a small portion of Earth's daytime side. The Moon's shadow on Earth's surface is about 7000 kilometers wide—roughly twice the diameter of the Moon (Figure 1.18). Outside of that shadow, no eclipse is seen. However, only within the central region of the shadow, called the **umbra**, is the eclipse total. Within the shadow but outside the umbra, in the **penumbra**, the eclipse is partial, with less and less of the Sun obscured the farther one travels from the shadow's center. The connections between the umbra, the penumbra, and the relative locations of

Figure 1.18 Photograph taken by an Earth-orbiting weather satellite of the Moon's shadow projected onto Earth's surface (near the Baja Peninsula) during the total solar eclipse of July 11, 1991.

Earth, Sun, and Moon are illustrated in Figure 1.19. One of the reasons that total solar eclipses are rare is that although the penumbra is some 7000 kilometers across, the umbra is always very small—even under the most favorable circumstances, its diameter never exceeds 270 kilometers. Because the shadow sweeps across Earth's surface at over 1700 kilometers per hour, the duration of a total eclipse at any given point can never exceed 7.5 minutes.

The Moon's orbit around Earth is not exactly circular. Thus, the Moon may be far enough from Earth at the mo-

Animation

Figure 1.19   The Moon's shadow on Earth during a solar eclipse consists of the umbra, where the eclipse is total, and the penumbra, where the Sun is only partially obscured. If the Moon is too far from Earth at the moment of the eclipse, there is no region of totality; instead, an annular eclipse is seen.

ment of an eclipse that its disk fails to cover the disk of the Sun completely, even though their centers coincide. In that case, there is no region of totality—the umbra never reaches Earth at all, and a thin ring of sunlight can still be seen surrounding the Moon. Such an occurrence, called an **annular eclipse**, is depicted in Figures 1.17(b) and 1.19. Roughly half of all solar eclipses are annular.

Why isn't there a solar eclipse at every new Moon and a lunar eclipse at every full Moon? The answer is that the Moon's orbit is slightly inclined to the ecliptic (at an angle of 5.2°), so the chance that a new (or full) Moon will occur just as the Moon happens to cross the ecliptic plane (so Earth, Moon, and Sun are perfectly aligned) is quite low. Figure 1.20 illustrates some possible configurations of the

(a)

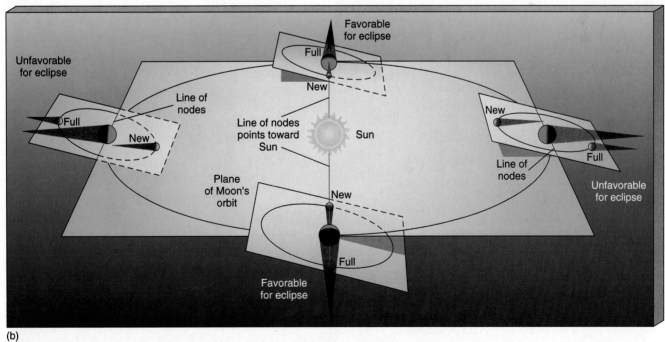

(b)

Figure 1.20 (a) An eclipse occurs when Earth, Moon, and Sun are precisely aligned. If the Moon's orbital plane lay in exactly the plane of the ecliptic, this alignment would occur once a month. However, the Moon's orbit is inclined at about 5° to the ecliptic, so not all configurations are actually favorable for producing an eclipse. (b) For an eclipse to occur, the line of intersection of the two planes must lie along the Earth–Sun line. Thus, eclipses can occur only at specific times of the year.

three bodies. If the Moon happens to lie above or below the plane of the ecliptic when new (or full), a solar (or lunar) eclipse cannot occur. Such a configuration is termed *unfavorable* for producing an eclipse. In a *favorable* configuration, on the other hand, the Moon is new or full just as it crosses the ecliptic plane, and eclipses are seen. Unfavorable configurations are much more common than favorable ones, so eclipses are relatively rare events.

As indicated on Figure 1.20(b), the two points on the Moon's orbit where it crosses the ecliptic plane are known as the *nodes* of the orbit. The line joining them, which is also the line of intersection of Earth's and the Moon's orbital planes, is known as the *line of nodes*. Times when the line of nodes is not directed toward the Sun are unfavorable for eclipses. However, when the line of nodes briefly lies along Earth–Sun line, eclipses are possible. These two periods, known as *eclipse seasons*, are the only times at which an eclipse can occur. Notice that there is no guarantee that an eclipse *will* occur. For a solar eclipse, we must have a new Moon during an eclipse season. Similarly, a lunar eclipse can occur only at full Moon during an eclipse season. Because we know the orbits of Earth and the Moon to great accuracy, we can predict eclipses far into the future. Figure 1.21 shows the location and duration of all total and annular eclipses of the Sun from 1995 to 2005.

The solar eclipses that we do see highlight a remarkable cosmic coincidence. Although the Sun is many times farther away from Earth than is the Moon, it is also much larger. In fact, the ratio of distances is almost exactly the same as the ratio of sizes, so the Sun and the Moon both have roughly the *same* angular diameter—about half a degree seen from Earth. Thus, the Moon covers the face of the Sun almost exactly. If the Moon were larger, we would never see annular eclipses, and total eclipses would be much more common. If the Moon were a little smaller, we would see only annular eclipses.

The gravitational tug of the Sun causes the Moon's orbital orientation, and hence the line of nodes, to change slowly with time. The result is that the eclipse seasons gradually progress backward through the calendar, occurring about 20 days earlier each year and taking 18.6 years to make one complete circuit. This phenomenon is known as the *regression of the line of nodes*. In 1991 the eclipse seasons were in January and July; on July 11 a total eclipse actually occurred, visible in Hawaii, Mexico, and parts of Central and South America. Three years later, in 1994, the eclipse seasons were in May and October; on May 10 an annular eclipse was visible across much of the continental United States. The regression of the line of nodes of the Moon's orbit causes Earth's rotation axis to wobble slightly, chang-

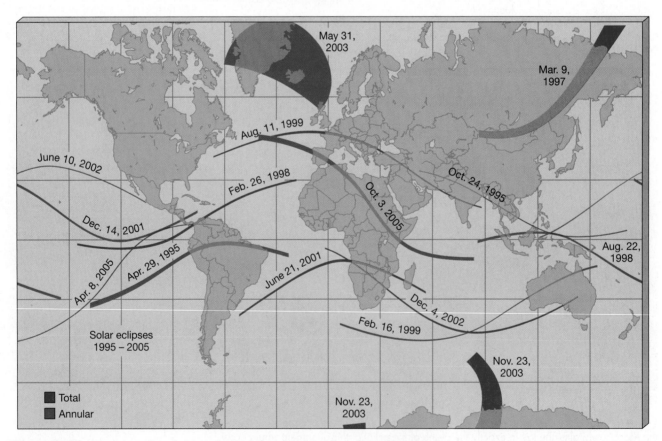

Figure 1.21 Regions of Earth that will see total or annular solar eclipses between the years 1995 and 2005. Each track represents the path of the Moon's umbra across Earth's surface during an eclipse.

ing the angle between Earth's axis and the ecliptic by plus or minus 9 arcseconds every 18.6 years. This additional motion, which is superimposed on Earth's precession, is known as *nutation*.

1.5 The Measurement of Distance

5 We have seen a little of how ancient astronomers tracked and recorded the positions of the stars on the sky and how modern astronomers make similar observations. But knowing the directions in which objects lie is only part of the information needed to locate them in space. Before we can make a systematic study of the heavens, we must find a way of measuring *distances*, too. One distance-measurement method, called **triangulation**, is based on the principles of Euclidean geometry and finds widespread application today in both terrestrial and astronomical settings. Today's engineers, especially surveyors, use these age-old geometric ideas to measure indirectly the distance to faraway objects. Triangulation forms the foundation of the family of distance-measurement techniques that together make up the **cosmic distance scale**.

Imagine trying to measure the distance to a tree on the other side of a river. The most direct method is to lay a tape across the river, but that's not the simplest way. A smart surveyor would make the measurement by visualizing an *imaginary* triangle (hence *triangulation*), sighting the tree

on the far side of the river from two positions on the near side, as illustrated in Figure 1.22. The simplest possible triangle is a right triangle, in which one of the angles is exactly 90°, so it is usually convenient to set up one observation position directly opposite the object, as at point A. The surveyor then moves to another observation position at point B, noting the distance covered between points A and B. This distance is called the **baseline** of the imaginary triangle. Finally, the surveyor, standing at point B, sights toward the tree and notes the angle at point B between this sightline and the baseline. No further observations are required. Knowing the value of one side (AB) and two angles (the right angle itself, at point A, and the angle at point B) of the right triangle, the surveyor can geometrically construct the remaining sides and angles and so establish the distance from A to the tree.

To use triangulation to measure distances, a surveyor must be familiar with *trigonometry*, the mathematics of geometrical angles. However, even if we know no trigonometry at all, we can still solve the problem by graphical means, as shown in Figure 1.23. Suppose that we pace off the baseline AB, measuring it to be 450 meters, and measure the angle between the baseline and the line from B to the tree to be 52°, as illustrated in the figure. We can transfer the problem to paper by letting one box on our graph represent 25 meters on the ground. Drawing the line AB on paper, completing the other two sides of the triangle, at angles of 90° (at A) and 52° (at B), we measure the distance on paper from A to the tree to be 23 boxes—that is, 575 meters. We have solved the real problem by *modeling* it on paper. The point to remember here is this: nothing more complex than basic geometry is needed to infer the distance, the size, and

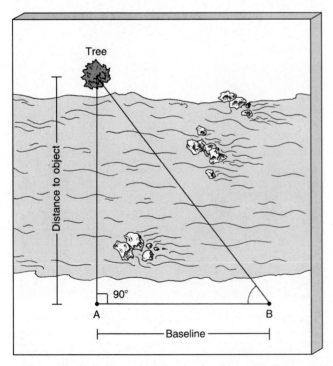

Figure 1.22 Surveyors often use simple geometry and trigonometry to estimate the distance to a faraway object.

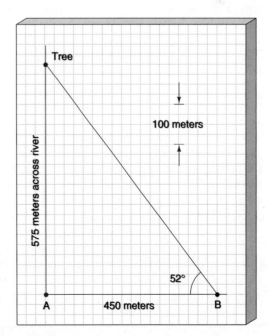

Figure 1.23 We don't even need trigonometry to estimate distances indirectly. Scaled estimates, like this one on a piece of paper, often suffice.

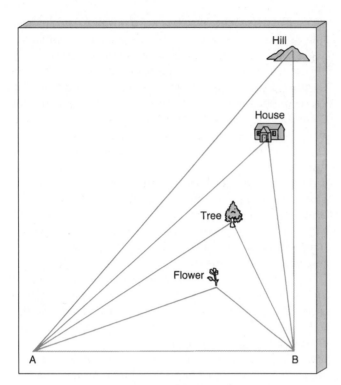

Figure 1.24 A triangle of fixed baseline (distance between points A and B) is narrower the farther away the object. As shown here, the imaginary triangle is much thinner when estimating the distance to a remote hill than it is when estimating the distance to a nearby flower.

even the shape of an object too far away or too inaccessible for direct measurement.

Triangles with larger baselines are needed if we are to measure greater distances. Figure 1.24 shows a triangle having a fixed baseline between two observation positions at points A and B. Note how the triangle becomes narrower as an object's distance becomes progressively greater. Narrow triangles cause problems because the angles at points A and B are hard to measure accurately. The measurements can be made easier by "fattening" the triangle—in other words, by lengthening the baseline.

Now consider an imaginary triangle extending from Earth to a nearby object in space, perhaps a neighboring planet. The imaginary triangle is extremely long and narrow, even for the nearest cosmic objects. Figure 1.25(a) illustrates the case in which the longest baseline on Earth—Earth's diameter, measured from point A to point B—is used. In principle, two observers could sight the planet from opposite sides of Earth, measuring the triangle's angles at points A and B. In practice, though, these angles cannot be measured with sufficient precision. It is actually easier to measure the third angle of the imaginary triangle, namely, the very small one near the planet. Here's how.

The observers on either side of Earth sight toward the planet, taking note of its position *relative to some distant stars* seen on the plane of the sky. The observer at point A sees the planet at apparent location A′ relative to those

stars, as indicated in Figure 1.25(a). The observer at B sees the planet at point B′. If each observer takes a photograph of the appropriate region of the sky, the planet will appear at slightly different places in the two images. In other words, the planet's photographic image is slightly displaced, or shifted, relative to the field of distant background stars, as shown in Figure 1.25(b). The background stars themselves appear undisplaced because of their much greater distance from the observer. This apparent displacement of a foreground object relative to the background as the observer's location changes is known as **parallax**. The size of the shift in Figure 1.25(b),

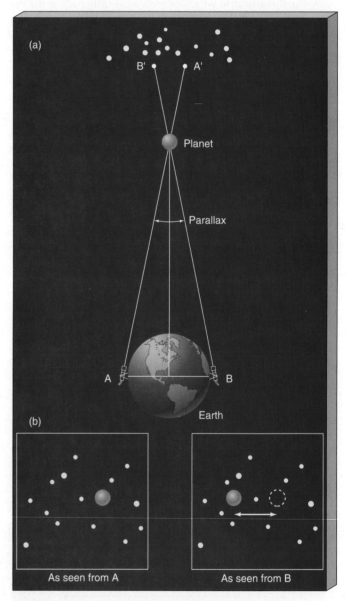

Figure 1.25 (a) This imaginary triangle extends from Earth to a nearby object in space (such as a planet). The group of stars at the top represents a background field of very distant stars. (b) Hypothetical photographs of the same star field showing the nearby object's apparent displacement, or shift, relative to the distant, undisplaced stars.

measured as an angle on the celestial sphere, is equal to the third, small angle shown in Figure 1.25(a).

The closer an object is to the observer, the larger the parallax. To see this for yourself, hold a pencil vertically in front of your nose, as sketched in Figure 1.26. Concentrate on some far-off object—a distant wall, say. Close one eye, then open it while closing the other. By blinking in this way, you should be able to see a large shift of the apparent position of the pencil projected onto the distant wall—a large parallax. In this example, one eye corresponds to point A, the other eye to point B, the distance between your eyeballs to the baseline, the pencil to the planet, and the distant wall to a remote field of stars. Now hold the pencil at arm's length, corresponding to a more distant object (but still not as far away as the distant stars). The apparent shift of the pencil will be less. By moving the pencil farther away, we are nar-

rowing the triangle and decreasing the parallax (and, in the process, making its accurate measurement more difficult). If you were to paste the pencil to the wall, corresponding to the case where the object of interest is as far away as the background star field, blinking would produce no apparent shift of the pencil at all.

The amount of parallax is thus inversely proportional to an object's distance. Small parallax implies large distance, and large parallax implies small distance. Knowing the amount of parallax (as an angle) and the length of the baseline, we can easily derive the distance through triangulation.

Once we know the distance to an object, we can determine many other properties. In particular, by measuring the object's *angular diameter* we can calculate its size. Figure 1.27 illustrates the geometry involved. Notice that this is basically the same picture as Figure 1.25(a), except that the angle (the angular diameter) and the distance are known, instead of the angle (the parallax) and the baseline. We compute the object's diameter by noting that the ratio of the diameter to the circumference of the circle centered on the observer and passing through the object (2π times the distance) must be equal to the ratio of its angular diameter to one full revolution, 360°:

$$\frac{\text{diameter}}{2\pi \times \text{distance}} = \frac{\text{angular diameter}}{360°}.$$

Surveyors of the land routinely use such simple geometric techniques to map out planet Earth (see *More Precisely 1-3* for an early example). As surveyors of the sky, astronomers use the same basic principles to chart the universe.

Figure 1.26 Parallax is inversely proportional to an object's distance. An object near your nose has a much larger parallax than an object held at arm's length.

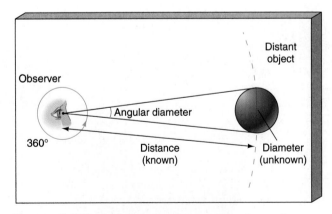

Figure 1.27 If the angular diameter of a distant object can be measured and its distance is known, its true diameter may be calculated by simple geometry.

MORE PRECISELY 1-3 Earth Dimensions

In about 200 B.C. a Greek philosopher named Eratosthenes (276–194 B.C.) used simple geometric reasoning to calculate the size of Earth. The logic he employed still provides the basis for all measurements of distance outside of our own solar system. Eratosthenes knew that at noon on the first day of summer, observers in the city of Syene (now called Aswan), in Egypt, saw the Sun pass directly overhead. This was evident from the fact that vertical objects cast no shadows and sunlight reached to the very bottoms of deep wells, as shown in the insets in the accompanying figure. However, at noon of the same day in Alexandria, a city 5000 *stadia* to the north, the Sun was seen to be displaced slightly from the vertical. (The *stadium* was a Greek unit of length, believed to have been about 0.16 km, although the exact value is uncertain—not all Greek stadia were the same size!) Using the simple technique of measuring the length of the shadow of a vertical stick and applying elementary trigonometry, Eratosthenes determined the angular displacement of the Sun from the vertical at Alexandria to be 7.2°.

What could have caused this discrepancy between the two measurements? It was not the result of measurement error—the same results were obtained every time the observations were repeated. Instead, as illustrated in the figure, the explanation is simply that Earth's surface is not flat but is actually *curved*. Our planet is a sphere. Eratosthenes was not the first person to realize that Earth is spherical—the philosopher Aristotle had done that over 100 years earlier (see *Interlude 2-2*)—but he was apparently the first to build on this knowledge, combining geometry with direct measurement to infer the size of our planet. Here's how he did it.

Rays of light reaching Earth from a very distant object, such as the Sun, travel almost parallel to one another. Consequently, as shown in the figure, the angle measured at Alexandria between the Sun's rays and the vertical (that is, the line joining Alexandria to the center of Earth) is equal to the angle between Syene and Alexandria, as seen from Earth's center. (For the sake of clarity, this angle has been exaggerated in the drawing.) The size of this angle in turn is proportional to the fraction of Earth's circumference that lies between Syene and Alexandria. Since 7.2° is 1/50 of a full circle (360°), so Earth's entire circumference can be estimated by multiplying the distance between the two cities by a factor of 50. We can express this reasoning as follows:

$$\frac{7.2°}{360°} = \frac{5000 \text{ stadia}}{\text{Earth's circumference}}.$$

(Notice that this is precisely the same geometric reasoning as that presented in Section 1.5.) Earth's circumference is therefore 50 × 5000, or 250,000 stadia. If we take the stadium unit to be 0.16 km, we find that Eratosthenes estimated Earth's circumference to be about 40,000 km. Since the circumference C of a circle is related to its radius r by the relation $C = 2\pi r$, it follows that Earth's radius is $250,000/2\pi$ stadia, or 6366 km. The correct values for Earth's circumference and radius, now measured accurately by orbiting spacecraft, are 40,070 km and 6378 km, respectively.

Eratosthenes' reasoning was a remarkable accomplishment. More than 20 centuries ago, he estimated the circumference of Earth to within 1 percent accuracy, using only simple geometry. Even if our modern value for the size of one stadium turns out to be incorrect, the real achievement—that a person making measurements on only a small portion of Earth's surface was able to compute the size of the entire planet on the basis of observation and logic—is undiminished.

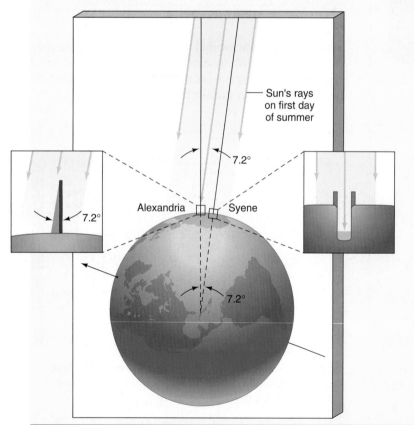

Sun's rays on first day of summer

7.2°

Alexandria   Syene

7.2°

7.2°

Chapter Review ✹WWW✹

SUMMARY

The **universe** (p. 4) is the totality of all space, time, matter, and energy. **Astronomy** (p. 4) is the study of the universe. A widely used unit of distance in astronomy is the **light year** (p. 4), the distance traveled by a beam of light in one year. Early observers grouped the thousands of stars visible to the naked eye into patterns called **constellations** (p. 6). These patterns have no physical significance, although they are a very useful means of labeling regions of the sky. The nightly motion of the stars across the sky is the result of Earth's **rotation** (p. 8) on its axis. Early astronomers, however, imagined that the stars were attached to a vast **celestial sphere** (p. 8) centered on Earth and that the motions of the heavens were caused by the rotation of the celestial sphere about a fixed Earth. The points where Earth's rotation axis intersects the celestial sphere are called the **north** and **south celestial poles** (p. 9). The line where Earth's equatorial plane cuts the celestial sphere is the **celestial equator** (p. 9).

The time from one sunrise to the next is called a **solar day** (p. 9). The time between successive risings of any given star is one **sidereal day** (p. 11). Because of Earth's **revolution** (p. 11) around the Sun, the solar day is a few minutes longer than the sidereal day.

The Sun's yearly path around the celestial sphere, or, equivalently, the plane of Earth's orbit around the Sun, is called the **ecliptic** (p. 11). Because Earth's axis is inclined to the ecliptic plane, we experience **seasons** (p. 11), depending on which hemisphere (Northern or Southern) happens to be "tipped" toward the Sun. At the **summer solstice** (p. 11), the Sun is highest in the sky, and the length of the day is greatest. At the **winter solstice** (p. 11), the Sun is lowest, and the day is shortest. At the **vernal** (p. 11) and **autumnal equinoxes** (p. 11), Earth's rotation axis is perpendicular to the line joining Earth to the Sun, and so day and night are of equal length. The interval of time from one vernal equinox to the next is one **tropical year** (p. 11).

Because Earth orbits the Sun, we see different constellations at different times of the year. The constellations lying along the ecliptic are collectively called the **zodiac** (p. 12). The time required for the same constellations to reappear at the same location in the sky, as viewed from a given point on Earth, is one **sidereal year** (p. 12). In addition to its rotation about its axis and its revolution around the Sun, Earth has may other motions. One of the most important of these is **precession** (p. 14), the slow "wobble" of Earth's axis due to the influence of the Moon. As a result, the particular constellations that happen to be visible on any given night change over the course of many years.

The Moon emits no light of its own. It shines by reflected sunlight. As the Moon orbits Earth, we see **lunar phases** (p. 14) as the amount of the Moon's sunlit face visible to us varies. At **full Moon** (p. 14), we can see the entire illuminated side. At **quarter Moon** (p. 14), only half the sunlit side can be seen. At **new Moon** (p. 14), the sunlit face points away from us, and the Moon is all but invisible from Earth. The time between successive full Moons is one **synodic month** (p. 14). The time taken for the Moon to return to the same position in the sky, relative to the stars, is one **sidereal month** (p. 14). Because of Earth's motion around the Sun, the synodic month is about 2 days longer than the sidereal month.

A **lunar eclipse** (p. 16) occurs when the Moon enters Earth's shadow. The eclipse may be **total** (p. 16), if the entire Moon is (temporarily) darkened, or **partial** (p. 16), if only a portion of the Moon's surface is affected. A **solar eclipse** (p. 16) occurs when the Moon passes between Earth and the Sun, so that a small part of Earth's surface is plunged into shadow. For observers in the **umbra** (p. 17), the entire Sun is obscured, and the solar eclipse is total. In the **penumbra** (p. 17), a **partial solar eclipse** (p. 16) is seen. If the Moon happens to be too far from Earth for its disk to completely hide the Sun, an **annular eclipse** (p. 19) occurs. Because the Moon's orbit around Earth is slightly inclined with respect to the ecliptic, solar and lunar eclipses do not occur every month, but only a few times per year.

Surveyors on Earth use **triangulation** (p. 21) to determine the distances to distant objects. Astronomers use the same technique to measure the distances to planets and stars. The **cosmic distance scale** (p. 21) is the family of distance-measurement techniques by which astronomers chart the universe. **Parallax** (p. 22) is the apparent motion of a foreground object relative to a distant background as the observer's position changes. The larger the **baseline** (p. 21), the distance between the two observation points, the greater the parallax. Astronomers use parallax when measuring the distances to the planets by triangulation. The same geometric reasoning is used to determine the sizes of objects whose distances are known.

SELF-TEST: TRUE OR FALSE?

_____ **1.** The light year is a measure of distance.

_____ **2.** The number 2×10^6 is equal to 2 billion.

_____ **3.** The stars in a constellation are physically close to each other.

_____ **4.** Some constellations, at one time, were used as simple calendars.

_____ **5.** Constellations are no longer of any use to astronomers.

_____ **6.** The solar day is longer than the sidereal day.

_____ **7.** The constellations lying immediately adjacent to the north celestial pole are collectively referred to as the zodiac.

_____ **8.** The seasons are caused by the precession of Earth's axis.

_____ **9.** The vernal equinox marks the beginning of fall.

_____ **10.** The new phase of the Moon cannot be seen because it always occurs during the daytime.

_____ **11.** A lunar eclipse can occur only during the full phase.

____ **12.** Solar eclipses are possible during any phase of the Moon.

____ **13.** An annular eclipse is a type of eclipse that occurs every year.

____ **14.** Eclipses can occur only during winter and summer months.

____ **15.** The parallax of an object is inversely proportional to its distance.

SELF-TEST: FILL IN THE BLANK

1. A _____ is a collection of hundreds of billions of stars.

2. The light year is a unit of _____.

3. Rotation is the term used to describe the motion of a body around some _____.

4. To explain the daily and yearly motions of the heavens, ancient astronomers imagined that the Sun, Moon, stars, and planets were attached to a rotating _____.

5. The solar day is measured relative to the Sun; the sidereal day is measured relative to the _____.

6. The apparent path of the Sun across the sky is known as the _____.

7. On December 21, known as the _____, the Sun is at its _____ point on the celestial sphere.

8. Declination measures the position of an object north or south of the _____.

9. An arc second is _____ (give the fraction) of an arc minute.

10. When the Sun, Earth, and Moon are positioned to form a right angle at Earth, the Moon is seen in the _____ phase.

11. A _____ eclipse can be seen by about half of Earth at once.

12. As seen from Earth, the Sun and the Moon have roughly the same _____.

13. To measure distances to nearby stars by parallax, a baseline equal to Earth's _____ is used.

14. The size of an object can be determined, if we know its distance, by measuring its _____.

15. The radius of _____ was first measured by Eratosthenes in 200 B.C.

REVIEW AND DISCUSSION

1. Compare the size of Earth with that of the Sun, the Milky Way, and the entire universe.

2. What does an astronomer mean by "the universe"?

3. What is a constellation?

4. Why does the Sun rise in the east and set in the west each day? Does the Moon also rise in the east and set in the west? Why? Do stars do the same? Why?

5. How many times in your life have you orbited the Sun?

6. Why are there seasons on Earth?

7. Why do we see different stars in summer and in winter?

8. How and why does a day measured by the Sun differ from a day measured by the stars?

9. If one complete hemisphere of the Moon is always lit by the sun, why do we see different phases of the Moon?

10. What causes a lunar eclipse? A solar eclipse?

11. Why aren't there lunar and solar eclipses every month?

12. What is precession, and what is its cause?

13. What is parallax? Give an everyday example.

14. Why is it necessary to have a long baseline when using triangulation to measure the distances to objects in space?

15. If you traveled to the outermost planet in our solar system, do you think the constellations would appear to change their shapes? What would happen if you traveled to the next-nearest star? If you traveled to the center of our galaxy, could you still see the familiar constellations found in Earth's night sky?

PROBLEMS

1. (a) Write the following numbers in scientific notation (see Appendix 1 if you are unfamiliar with this notation): 1000; 0.000001; 1001; 1,000,000,000,000,000; 123,000; 0.000456. (b) Write the following numbers in "normal" numerical form: 3.16×10^7; 2.998×10^5; 6.67×10^{-11}; 2×10^0. (c) Calculate: $(2 \times 10^3) + 10^{-2}$; $(1.99 \times 10^{30}) \div (5.97 \times 10^{24})$; $(3.16 \times 10^7) \times (2.998 \times 10^5)$.

2. In 1 second, light leaving Los Angeles will reach approximately as far as (a) San Francisco (about 500 km), (b) London (roughly 10,000 km), (c) the Moon (400,000 km), (d) Venus (0.3 AU from Earth at closest approach), or (e) the nearest star (about 1 pc from Earth). Which is correct?

3. How would the length of the solar day change if Earth's rotation were suddenly to reverse direction?

4. What would be the length of the month if the Moon's sidereal orbital period were (a) 1 week (7 solar days)? (b) 1 (sidereal) year?

5. The vernal equinox is now just entering the constellation Aquarius. In what constellation will it lie in the year A.D. 10,000?

6. Through how many degrees, arc minutes, or arc seconds does the Moon move in (a) 1 hour of time, (b) 1 minute, (c) 1 second? How long does it take for the Moon to move a distance equal to its own diameter?

7. A surveyor wishes to measure the distance between two points on either side of a river, as illustrated in Figure 1.22. She measures the distance AB to be 250 m and the angle at B to be 30°. What is the distance between the two points?

8. At what distance is an object if its parallax, as measured from either end of a 1000-km baseline, is (a) 1°, (b) 1′, (c) 1″?

9. Given that the distance to the Moon is 384,000 km and its angular size is 0.5°, calculate the Moon's diameter.

10. What angle would Eratosthenes have measured (see *More Precisely 1-3*) had Earth been flat?

PROJECTS

1. Go to a country location on a clear dark night. Imagine patterns among the stars, and name the patterns yourself. Note (or better yet, draw) location of these stars with respect to trees or buildings in the foreground. Do this every week or so for a couple of months. Be sure to look at the same time every night. What happens?

2. Find the star Polaris, also known as the North Star, in the evening sky. Identify any separate pattern of stars in the same general vicinity of the sky. Wait several hours, at least until after midnight, and then locate Polaris again. Has Polaris moved? What has happened to the nearby pattern of stars? Why?

3. Hold your little finger out at arm's length. Can you cover the disk of the Moon? The Moon projects an angular size of 30′ (half a degree); your finger should more than cover it. How can you apply this fact in making sky measurements?

2 THE COPERNICAN REVOLUTION

The Birth of Modern Science

LEARNING GOALS

Studying this chapter will enable you to:

1 Relate how some ancient civilizations attempted to explain the heavens in terms of Earth-centered models of the universe.

2 Summarize the role of Renaissance science in the history of astronomy.

3 Explain how the observed motions of the planets led to our modern view of a Sun-centered solar system.

4 Sketch the major contributions of Galileo and Kepler to the development of our understanding of the solar system.

5 State Kepler's laws of planetary motion.

6 Explain how Kepler's laws enable us to construct a scale model of the solar system, and explain the technique used to determine the actual size of the planetary orbits.

7 State Newton's laws of motion and universal gravitation and explain how they account for Kepler's laws.

8 Explain how the law of gravitation enables us to measure the masses of astronomical bodies.

(**Opposite page, background**) In an early twentieth-century classroom at Radcliffe College, as at other schools, astronomy came into its own as a viable and important subject.

(**Inset A**) Harlow Shapley (1885–1972), discovered our place in the "suburbs" of the Milky Way, dispelling the notion that the Sun resides at the center of the universe.

(**Inset B**) Annie Cannon (1863–1941), one of the greatest astronomical cataloguers of all time, carefully analyzed photographic plates to classify nearly a million stars over the course of fifty years of work at Harvard.

(**Inset C**) Maria Mitchell (1818–1889) in her observatory on Nantucket Island, made important contributions to several areas of astronomy.

(**Inset D**) Edwin Hubble (1889–1953), here posing in front of one of the telescopes on Mount Wilson, in California, is often credited with having discovered the expansion of the universe.

Living in the Space Age, we have become accustomed to the modern view of our place in the universe. Images of our planet taken from space leave little doubt that Earth is round, and no one seriously questions the idea that we orbit the Sun. Yet there was a time, not so long ago, when our ancestors maintained that Earth was flat and lay at the center of all things. Our view of the universe—and of ourselves—has undergone a radical transformation since those early days. Earth has become a planet like many others, and humankind has been torn from its throne at the center of the cosmos and relegated to a rather unremarkable position on the periphery of the Milky Way Galaxy. But we have been amply compensated for our loss of prominence—we have gained a wealth of scientific knowledge in the process. The story of how all this came about is the story of the rise of the scientific method and the genesis of modern astronomy.

2.1 Ancient Astronomy

1 Many ancient cultures took a keen interest in the changing nighttime sky. The records and artifacts that have survived until the present make that abundantly clear. But unlike today, the major driving force behind the development of astronomy in those early societies was probably neither scientific nor religious in nature. Instead, it was decidedly practical and very down to earth. Seafarers needed to navigate their vessels, and farmers had to know when to plant their crops. In a very real sense, then, human survival depended on knowledge of the heavens. As a result, the ability to predict the arrival of the seasons, as well as other astronomical events, was undoubtedly a highly prized, and perhaps also jealously guarded, skill.

In Chapter 1 we saw that the human brain's ability to perceive patterns in the stars led to the "invention" of constellations as a convenient means of labeling regions of the celestial sphere. ∞ (Sec. 1.2) The realization that these patterns returned to the night sky at the same time each year met the need for a practical means of tracking the seasons. Many separate cultures, all over the world, built large and elaborate structures to serve, at least in part, as primitive calendars. In some cases, the keepers of the secrets of the sky eventually enshrined their knowledge in myth and ritual, and these astronomical sites were often also used for religious rites.

Perhaps the best known such site is *Stonehenge*, located on Salisbury Plain, in England, and shown in Figure 2.1. This ancient stone circle, which today is one of the most popular tourist attractions in Britain, dates from the Stone Age. Researchers believe it was an early astronomical observatory of sorts—not in the modern sense of the term (a place for making new observations and discoveries) but rather a kind of three-dimensional calendar or almanac,

Figure 2.1 Stonehenge was probably constructed as a primitive calendar and almanac. The fact that the largest stones were carried to the site from many miles away attests to the importance of this structure to its Stone Age builders. The inset shows sunrise at Stonehenge on the summer solstice. As seen from the center of the stone circle, the Sun rose directly over the "heel stone" on the longest day of the year.

(a)

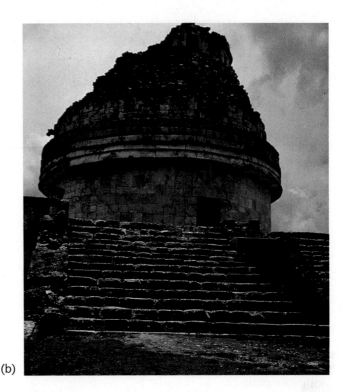

(b)

Figure 2.2 (a) The Big Horn Medicine Wheel, in Wyoming, was built by the Plains Indians. Its spokes and other features are aligned with risings and settings of the Sun and other stars. (b) Caracol temple in Mexico. The many windows of this Mayan construct are aligned with astronomical events, indicating that at least part of Caracol's function was to keep track of the seasons and the heavens.

enabling its builders and their descendants to identify important dates by means of specific celestial events. Its construction apparently spanned a period of some 17 centuries, beginning around 2800 B.C. Additions and modifications continued up to about 1100 B.C., indicating its ongoing importance to the Stone Age and later Bronze Age people who built, maintained, and used Stonehenge. The largest stones shown in Figure 2.1 weigh up to 50 tons and were transported from quarries many miles away.

Many of the stones are aligned so that they point toward important astronomical events. For example, the line joining the center of the inner circle to the so-called heel stone, set off at some distance from the rest of the structure, points in the direction of the rising Sun on the summer solstice. Other alignments are related to the rising and setting of the Sun and the Moon at various other times of the year. The accurate alignments (within a degree or so) of the stones of Stonehenge were first noted in the eighteenth century, but it is only relatively recently—in the second half of the twentieth century, in fact—that the scientific community has credited Stone Age technology with the ability to carry out such a precise feat of engineering. While some of Stonehenge's purposes remain uncertain and controversial, the site's function as an astronomical almanac seems well established. Although Stonehenge is the most impressive and the best preserved, other stone circles, found all over Europe, are believed to have performed similar functions.

Many other cultures are now known to have been capable of similarly precise accomplishments. The Big Horn Medicine Wheel in Wyoming (Figure 2.2a) is similar to Stonehenge in design—and, presumably, intent—although

it is somewhat simpler in execution. The Medicine Wheel's alignments with the rising and setting Sun and with some bright stars indicate that its builders—the Plains Indians— had much more than a passing familiarity with the changing nighttime sky. Figure 2.2(b) shows the Caracol temple, built by the Mayans around A.D. 1000 on Mexico's Yucatan peninsula. This temple is much more sophisticated than Stonehenge, but it probably played a similar role as an astronomical observatory. Its many windows are accurately aligned with astronomical events, such as sunrise and sunset at the solstices and equinoxes and the risings and settings of the planet Venus. Astronomy was of more than mere academic interest to the Mayans, however. Caracol was also the site of countless human sacrifices, carried out when Venus appeared in the morning or evening sky.

The ancient Chinese too observed the heavens. Their astrology attached particular importance to "omens" such as comets and "guest stars"—stars that appeared suddenly in the sky and then slowly faded away—and they kept careful and extensive records of such events. Twentieth-century astronomers still turn to the Chinese records to obtain observational data recorded during the Dark Ages (roughly from the fifth to the tenth century A.D.), when turmoil in Europe largely halted the progress of Western science. Perhaps the best-known guest star was one that appeared in A.D. 1054 and was visible in the daytime sky for many months. We now know that the event was actually a *supernova*: the explosion of a giant star, which scattered most of its mass into space. It left behind a remnant that is still detectable today, nine centuries later. The Chinese data are a prime source of historical information for supernova research.

Figure 2.3 Arab astronomers at work, as depicted in a medieval manuscript.

A vital link between the astronomy of ancient Greece and that of medieval Europe was provided by Arab astronomers (Figure 2.3). For six centuries, from the depths of the Dark Ages to the beginning of the Renaissance, Islamic astronomy flourished and grew, preserving and augmenting the knowledge of the Greeks. The Arab influence on modern astronomy is subtle but quite pervasive. Many of the mathematical techniques involved in trigonometry were developed by Muslim astronomers in response to very practical problems, such as determining the precise dates of holy days or the direction of Mecca from any given location on Earth. Astronomical terms like "zenith" and "azimuth" and the names of many stars, such as Rigel, Betelgeuse, and Vega, all bear witness to this extended period of Muslim scholarship.

Astronomy is not the property of any one culture, civilization, or era. The same ideas, the same tools, and even the same misconceptions have been invented and reinvented by human societies all over the world, in response to the same basic driving forces. Astronomy came into being because people believed that there was a practical benefit in being able to predict the positions of the stars, but its roots go much deeper than that. The need to understand where we came from, and how we fit into the cosmos, is an integral part of human nature.

2.2 The Geocentric Universe

The Greeks of antiquity, and undoubtedly civilizations before them, built models of the universe. The study of the workings of the universe on the very largest scales is called **cosmology**. Today, cosmology entails looking at the universe on scales so large that even entire galaxies can be regarded as mere points of light scattered throughout space. To the Greeks, however, the universe was basically the *solar system*—namely, the Sun, Earth, Moon, and the planets known at that time. The stars beyond were surely part of the universe, but they were considered to be fixed, unchanging beacons on a mammoth celestial dome. The Greeks did not consider the Sun, the Moon, and the planets to be part of the celestial sphere, however. Those objects had patterns of behavior that set them apart.

Greek astronomers observed that over the course of a night, the stars slid smoothly across the sky. Over the course of a month, the Moon moved smoothly and steadily along its path on the sky relative to the stars, passing through its familiar cycle of phases. Over the course of a year, the Sun progressed along the ecliptic at an almost constant rate, varying little in brightness from day to day. In short, the behavior of both Sun and Moon seemed fairly simple and orderly. But ancient astronomers were also aware of five other bodies in the sky—the planets Mercury, Venus, Mars, Jupiter, and Saturn—whose behavior was not so easy to grasp. Their motions ultimately led to the downfall of an entire theory of the solar system and to a fundamental change in humankind's view of the universe.

Planets do not behave in as regular and predictable a fashion as the Sun, Moon, and stars. They vary in brightness, and they don't maintain a fixed position in the sky. Unlike the Sun and the Moon, the planets seem to wander around the celestial sphere—indeed, the word planet derives from the Greek word *planetes*, meaning "wanderer." Planets never stray far from the ecliptic and generally traverse the celestial sphere from west to east, as the Sun does. However, they seem to speed up and slow down during their journeys, and at times they even appear to loop back and forth relative to the stars, as shown in Figure 2.4. In other words, there are periods when a planet's eastward motion (relative to the stars) stops, and the planet appears to move westward in the sky for a month or two before reversing direction again and continuing on its eastward journey. Motion in the eastward sense is usually referred to as *direct*, or *prograde*, motion; the backward (westward) loops are known as **retrograde motion**.

Like the Moon, the planets produce no light of their own; instead, they shine by reflected sunlight. Ancient astronomers correctly reasoned that the apparent brightness of a planet in the night sky is related to its distance from Earth—planets appear brightest when closest to us. However, the planets Mars, Jupiter, and Saturn are always brightest during the retrograde portions of their orbits. The challenge facing astronomers was to explain the observed motions of the planets and to relate those motions to the variations in planetary brightness.

The earliest models of the solar system followed the teachings of the Greek philosopher Aristotle (384–322 B.C.) and were **geocentric** in nature, meaning that Earth lay at the center of the universe and that all other bodies moved around it. (Figures 1.7 and 1.10a illustrate the basic geocentric view.) ∞ (Sec. 1.2) These models employed what

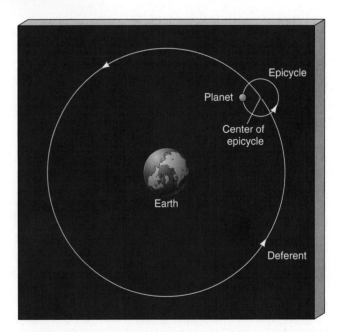

Figure 2.4 Most of the time, planets move from west to east relative to the background stars. Occasionally, however, they change direction and temporarily undergo retrograde motion before looping back. The art at left shows an actual retrograde loop in the motion of the planet Mars. The inset above depicts the movements of several planets over the course of several years, as reproduced on the inside dome of a planetarium. The motion of the planets relative to the stars (represented as unmoving points) produces continuous streaks on the planetarium "sky."

Aristotle, and Plato before him, had taught was the perfect form: the circle. The simplest possible description—uniform motion around a circle having Earth at its center—provided a fairly good approximation to the orbits of the Sun and the Moon, but it could not account for the observed variations in planetary brightness or their retrograde motion. A more complex model was needed to describe the planets.

In the first step toward this new model, each planet was taken to move uniformly around a small circle, called an **epicycle**, whose *center* moved uniformly around Earth on a second and larger circle, known as the **deferent** (Figure 2.5). The motion was now composed of two separate circular orbits, creating the possibility that, at some times, the planet's apparent motion could be retrograde. Also, the distance from the planet to Earth would vary, accounting for changes in brightness. By tinkering with the relative sizes of epicycle and deferent, with the planet's speed on the epicycle, and with the epicycle's speed along the deferent, early astronomers were able to bring this "epicyclic" motion into fairly good agreement with the observed paths of the planets in the sky. Moreover, this model had good predictive power, at least to the accuracy of observations at the time.

However, as the number and the quality of observations increased, it became clear that the simple epicyclic model was not perfect. Small corrections had to be introduced to bring it into line with new observations. The center of the deferents had to be shifted slightly from Earth's center, and the motion of the epicycles had to be imagined uniform with respect not to Earth but to yet another point in space. Around A.D. 140, a Greek astronomer named Ptolemy constructed perhaps the best geocentric model of all time. Illustrated in simplified

Figure 2.5 In the geocentric model of the solar system, the observed motions of the planets made it impossible to assume that they moved on simple circular paths around Earth. Instead, each planet was thought to follow a small circular orbit (the epicycle) about an imaginary point that itself traveled in a large, circular orbit (the deferent) about Earth.

form in Figure 2.6, it explained remarkably well the observed paths of the five planets then known, as well as the paths of the Sun and the Moon. However, to achieve its explanatory and predictive power, the full **Ptolemaic model** required a

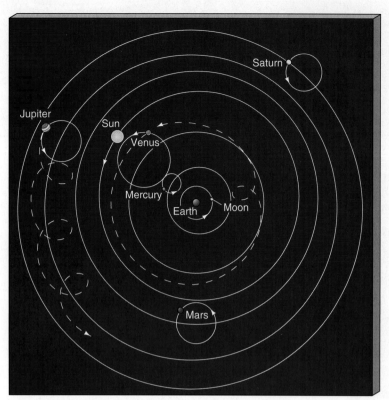

Figure 2.6 The basic features, drawn roughly to scale, of the geocentric model of the inner solar system that enjoyed widespread popularity prior to the Renaissance. To avoid confusion, we have drawn partial paths (dashed) of only two planets, Venus and Jupiter.

series of no fewer than 80 distinct circles. To account for the paths of the Sun, the Moon, and all the nine planets (and their moons) that we know today would require a vastly more complex set. Nevertheless, Ptolemy's text on the topic, *Syntaxis* (better known today by its Arabic name *Almagest*—"the greatest"), provided the intellectual framework for all discussion of the universe for well over a thousand years.

Today, our scientific training leads us to seek simplicity, because simplicity in the physical sciences has so often proved to be an indicator of truth. We would regard the intricacy of a model as complicated as the Ptolemaic system as a clear sign of a fundamentally flawed theory. With the benefit of hindsight, we now recognize that the major error lay in the assumption of a geocentric universe. This was compounded by the insistence on uniform circular motion, whose basis was largely philosophical, rather than scientific, in nature.

Actually, history records that some ancient Greek astronomers reasoned differently about the motions of heavenly bodies. Foremost among them was Aristarchus of Samos (310–230 B.C.), who proposed that all the planets, including Earth, revolve around the Sun and, furthermore, that Earth rotates on its axis once each day. This, he argued, would create an *apparent* motion of the sky—a simple idea that is familiar to anyone who has ridden on a merry-go-round and watched the landscape appear to move past in the opposite direction. However, Aristarchus's description of the heavens, though essentially correct, did not gain widespread acceptance during his lifetime. Aristotle's influence was too strong, his followers too numerous, his writings too comprehensive. The geocentric model went largely unchallenged until the sixteenth century A.D.

The Aristotelian school did present some simple and (at the time) compelling arguments in favor of their views. First, of course, Earth doesn't *feel* as if it's moving. And if it were, wouldn't there be a strong wind as we moved at high speed around the Sun? Then again, considering that the vantage point from which we view the stars changes over the course of a year, why don't we see stellar parallax? Nowadays we might be inclined to dismiss the first two points as merely naive, but the third is a valid argument and the reasoning is essentially sound. We now know that there *is* stellar parallax as Earth orbits the Sun. However, because the stars are so distant, it amounts to less than 1″, even for the closest stars. Early astronomers simply would not have noticed it. We will encounter many other instances in astronomy where correct reasoning has led to the wrong conclusions because it was based on inadequate data.

2.3 The Heliocentric Model of the Solar System

2 The Ptolemaic picture of the universe survived, more or less intact, for almost 13 centuries, until a sixteenth-century Polish cleric, Nicholas Copernicus (Figure 2.7), rediscovered Aristarchus's **heliocentric** (Sun-centered) model and showed how, in its harmony and organization, it provided a more natural explanation of the observed facts than did the tangled geocentric cosmology. Copernicus asserted that Earth spins on its axis and, like the other planets, orbits

Figure 2.7   Nicholas Copernicus (1473–1543).

the Sun. Only the Moon, he said, orbits Earth. Not only does this model explain the observed daily and seasonal changes in the heavens, as we saw in Chapter 1, but it also naturally accounts for planetary retrograde motion and brightness variations. The critical realization that Earth is not at the center of the universe is now known as the **Copernican revolution**. The seven crucial statements that form its foundation are summarized in *Interlude 2-1*.

Figure 2.8 shows how the Copernican view explains both the varying brightness of a planet (in this case, Mars) and its observed looping motions. If we suppose that Earth moves faster than Mars, then every so often Earth "overtakes" that planet. Mars will then appear to move backward in the sky, in much the same way as a car we overtake on the highway seems to slip backward relative to us. Notice that in the Copernican picture the planet's looping motions are only apparent; in the Ptolemaic view, they were real.

Copernicus's major motivation for introducing the heliocentric model was simplicity. Even so, he was still influenced by Greek thinking and clung to the idea of circles to model the planets' motions. As a result, in order to bring his theory into agreement with observations, he was forced to retain the idea of epicyclic motion, although with the deferent

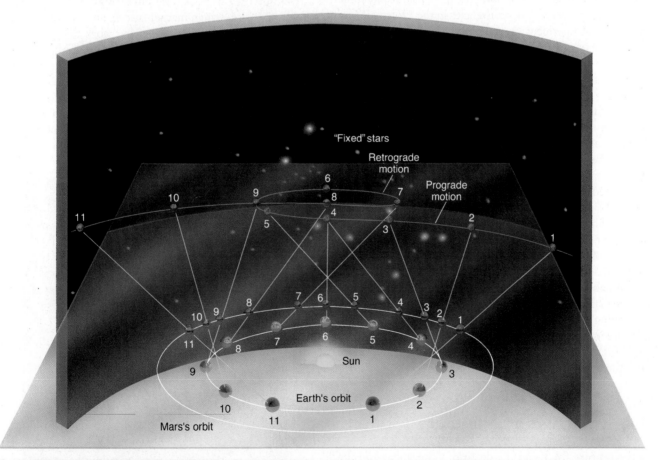

Figure 2.8   The Copernican model of the solar system explains the varying brightnesses of the planets, something the Ptolemaic system largely ignored. Here, for example, when Earth and Mars are relatively close to each other in their respective orbits (as at position 6), Mars seems brighter; when farther away (as at position 1), Mars seems dimmer. Also, because the line of sight from Earth to Mars changes as the two planets smoothly orbit the Sun, Mars appears to loop back and forth, undergoing retrograde motion. The line of sight changes because Earth, on the inside track, moves faster in its orbit than Mars moves along its path.

INTERLUDE 2-1 The Foundations of the Copernican Revolution

The following seven points are essentially Copernicus's own words. The italicized material is additional explanation.

1. The celestial spheres do not have just one common center. *Specifically, Earth is not at the center of everything.*

2. The center of Earth is not the center of the universe but is instead only the center of gravity and of the lunar orbit.

3. All the spheres revolve around the Sun. *By spheres, Copernicus meant the planets.*

4. The ratio of Earth's distance from the Sun to the height of the firmament is so much smaller than the ratio of Earth's radius to the distance to the Sun that the distance to the Sun is imperceptible when compared with the height of the firmament. *By firmament, Copernicus meant the distant stars. The point he was making is that the stars are very much farther away than the Sun.*

5. The motions appearing in the firmament are not its motions but those of Earth. Earth performs a daily rotation around its fixed poles while the firmament remains immobile as the highest heaven. *Because the stars are so far away, any apparent motion we see in them is the result of Earth's rotation.*

6. The motions of the Sun are not its motions but the motion of Earth. *Similarly, the Sun's apparent daily and yearly motion are actually due to the various motions of Earth.*

7. What appears to us as retrograde and forward motion of the planets is not their own but that of Earth. *The heliocentric picture provides a natural explanation for retrograde planetary motion, again as a consequence of Earth's motion.*

---

centered on the Sun rather than on Earth, and with smaller epicycles than in the Ptolemaic picture. Thus, he retained unnecessary complexity and actually gained little in accuracy over the geocentric model. The heliocentric model did rectify some small discrepancies and inconsistencies in the Ptolemaic system, but for Copernicus, the primary attraction of heliocentricity was its simplicity, its being "more pleasing to the mind." His theory was more something he *felt* than he could *prove*. To the present day, scientists still are guided by simplicity, symmetry, and beauty in modeling all aspects of the universe.

Despite the support of some observational data, neither his fellow scholars nor the general public easily accepted Copernicus's model. For the learned, heliocentricity went against the grain of much previous thinking and violated many of the religious teachings of the time, largely because it relegated Earth to a noncentral and undistinguished place within the solar system and the universe. And Copernicus's work had little impact on the general populace of his time, at least in part because it was published in Latin (the standard language of academic discourse at the time), which most people could not read. Only after Copernicus's death, when others—notably Galileo Galilei—popularized his ideas, did the Roman Catholic church take them seriously enough to bother banning them. Copernicus's writings on the heliocentric universe were placed on the *Index of Prohibited Books* in 1616, 73 years after they were first published. They remained there until the end of the eighteenth century.

2.4 The Laws of Planetary Motion

In the century following the death of Copernicus and the publication of his theory of the solar system, two scientists—Johannes Kepler and Galileo Galilei—made indelible imprints on the study of astronomy. Contemporaries, they were aware of each other's work and corresponded from time to

time about their theories. Each achieved fame for his discoveries and, in his own way, made great strides in popularizing the Copernican viewpoint, yet in their approaches to astronomy they were as different as night and day. Kepler (Figure 2.9), a German mathematician and astronomer, was a pure theorist. His groundbreaking work that so clarified our knowledge of planetary motion was based almost entirely on the observations of others (partly because of Kepler's own poor eyesight). In contrast, Galileo was in many ways the first "modern" astronomer. He used emerging technology, in the form of the telescope, to achieve new insights into the universe. We will study some of Galileo's accomplishments, and their consequences, in Section 2.5.

Figure 2.9   Johannes Kepler (1571–1630).

BRAHE'S COMPLEX DATA

Kepler's work was based on an extensive collection of data compiled by Tycho Brahe (1546–1601), Kepler's employer and arguably one of the greatest observational astronomers that has ever lived. Tycho, as he is often called, was both an eccentric aristocrat and a skillful observer. Born in Denmark, he was educated at some of the best universities in Europe, where he studied astrology, alchemy, and medicine. Most of his observations, which predated the invention of the telescope by several decades, were made at his own observatory, named *Uraniborg*, in Denmark (Figure 2.10). There, using instruments of his own design, Tycho maintained meticulous and accurate records of the stars, planets, and other noteworthy celestial events.

In 1597, having fallen out of favor with the Danish court, Tycho moved to Prague, which happens to be fairly close to Graz, in Austria, where Kepler lived and worked. Kepler joined Tycho in Prague in 1600. There Kepler was put to work trying to find a theory that could explain Brahe's planetary data. When Tycho died a year later, Kepler inherited not only Brahe's position as Imperial Mathematician of the Holy Roman Empire (then located in Eastern Europe) but also his priceless possession: the accumulated observations of the planets, spanning several decades. Tycho's observations, though made with the naked eye, were nevertheless of very high quality. In most cases, his measured positions of stars and planets were accurate to within about 1′. Kepler set to work seeking a unifying principle to explain in detail the motions of the planets, without the need for epicycles. The effort was to occupy much of the remaining 29 years of his life.

Kepler had already accepted the heliocentric picture of the solar system. His goal was to find a simple and elegant description of the solar system, within the Copernican framework, that fit Tycho's complex mass of detailed observations. In the end, he found it necessary to abandon Copernicus's original simple idea of circular planetary orbits. However, an even greater simplicity emerged as a result. After long years studying Brahe's planetary data, and after many false starts and blind alleys, Kepler developed the laws of planetary motion that now bear his name.

Kepler determined the shape of each planet's orbit by triangulation—not from different points on Earth, but from different points on Earth's orbit, using observations made at many different times of the year. ∞ (Sec. 1.9) By using a portion of Earth's orbit as a baseline, Kepler was able to measure the relative sizes of the other planetary orbits. Noting where the planets were on successive nights, he found the speeds at which the planets move. We do not know how many geometric shapes Kepler tried for the orbits before he hit upon the correct one. His difficult task was made even more complex because he had to determine Earth's own orbit, too. Nevertheless, he eventually succeeded in summarizing the motions of all the known planets, including Earth, in just three laws, the **laws of planetary motion**.

KEPLER'S SIMPLE LAWS

5 *Kepler's first law* has to do with the *shapes* of the planetary orbits:

> The orbital paths of the planets are elliptical (*not* circular), with the Sun at one focus.

An **ellipse** is simply a flattened circle. Figure 2.11 illustrates a means of constructing an ellipse using a piece of string and two thumbtacks. Each point at which the string is pinned is called a **focus** (plural: *foci*) of the ellipse. The long axis of the ellipse, containing the two foci, is known as the *major axis*. Half the length of this long axis is referred to as the **semi-major axis**; it is a measure of the ellipse's size. The **eccentricity** of the ellipse is the ratio of the distance between the foci to the length of the major axis. The length of the semi-major axis and the eccentricity are all we need to describe the size and shape of a planet's orbital path (see *More Precisely 2-1*). A circle is a special kind of ellipse in which the two foci happen to coincide, so the eccentricity is zero. The semi-major axis of a circle is simply its radius.

Figure 2.10 Tycho Brahe in his observatory *Uraniborg*, on the island of Hveen, in Denmark.

In fact, no planet's elliptical orbit is nearly as elongated as the one shown in Figure 2.11. With two exceptions (the paths of Mercury and Pluto), planetary orbits in our solar system have such small eccentricities that our eyes

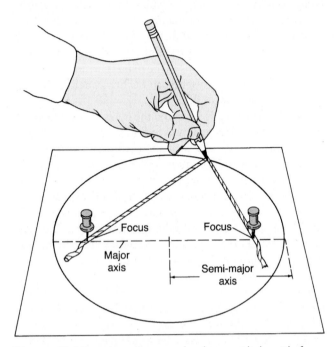

Figure 2.11 Any ellipse can be drawn with the aid of a string, a pencil, and two thumbtacks. The wider the separation of the foci, the more elongated, or eccentric, is the ellipse. In the special case where the two foci are at the same place, the drawn curve is a circle.

would have trouble distinguishing them from true circles. Only because the orbits are so nearly circular were the Ptolemaic and Copernican models able to come as close as they did to describing reality.

Kepler's substitution of elliptical for circular orbits was no small advance. It amounted to abandoning an aesthetic bias—the Aristotelian belief in the perfection of the circle—that had governed astronomy since Greek antiquity. Even Galileo Galilei, not known for his conservatism in scholarly matters, clung to the idea of circular motion and never accepted that the planets move on elliptical paths.

Kepler's second law, illustrated in Figure 2.12, addresses the speed at which a planet traverses different parts of its orbit:

An imaginary line connecting the Sun to any planet sweeps out equal areas of the ellipse in equal intervals of time.

While orbiting the Sun, a planet traces the arcs labeled *A*, *B*, and *C* in Figure 2.12 in equal times. Notice, however, that the distance traveled by the planet along arc *C* is greater than the distance traveled along arc *A* or arc *B*. Because the time is the same and the distance is different, the speed must vary. When a planet is close to the Sun, as in sector *C*, it moves much faster than when farther away, as in sector *A*.

By taking into account the relative speeds and positions of the planets in their elliptical orbits about the Sun, Kepler's first two laws explained the variations in planetary brightness and some observed peculiar nonuniform motions that could not be accommodated within the assumption of

---

 MORE PRECISELY 2-1 Some Properties of Planetary Orbits

Two numbers—semi-major axis and eccentricity—are all that are needed to describe the size and shape of a planet's orbital path. From them we can derive many other useful quantities. Two of the most important are the planet's *perihelion* (its point of closest approach to the Sun) and its *aphelion* (greatest distance from the Sun). From the definitions presented in the text, it follows that if the planet's orbit has semi-major axis *a* and eccentricity *e*, its perihelion is at a distance $a(1-e)$ from the Sun, while its aphelion is at $a(1+e)$. These points and distances are illustrated in the accompanying figure.

Note that, while the Sun resides at one focus, the other focus is empty and has no particular significance. Thus, for example, a (hypothetical) planet with a semi-major axis of 400 million km and an eccentricity of 0.5 (the eccentricity of the ellipse shown in the diagram) would range between $400 \times (1 - 0.5) = 200$ million km and $400 \times (1 + 0.5) = 600$ million km from the Sun over the course of one complete orbit. With $e = 0.9$, the range would be 40–760 million km, and so on.

No planet has an orbital eccentricity as large as 0.5— the planet with the most eccentric orbit is Pluto, with $e = 0.248$ (see Table 2.1). However, many meteoroids, and

all comets (see Chapter 14) have eccentricities considerably greater than this. In fact, most comets visible from Earth have eccentricities very close to 1. Their highly elongated orbits approach within a few A.U. of the Sun at perihelion, yet these tiny frozen worlds spend most of their time far beyond the orbit of Pluto.

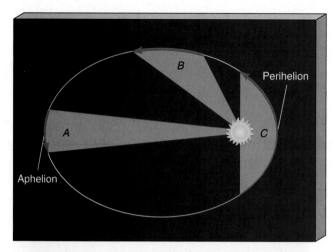

Figure 2.12 A diagram illustrating Kepler's second law: Equal areas are swept out in equal intervals of time. The three shaded areas (A, B, and C) are equal. Note that an object would travel the length of each of the three arrows in the same amount of time. Therefore, planets move faster when closer to the Sun.

circular motion, even with the inclusion of epicycles. Gone at last were the circles within circles that rolled across the sky. Kepler's modification of the Copernican theory to allow the possibility of elliptical orbits both greatly simplified the model of the solar system and at the same time provided much greater predictive accuracy than had previously been possible. Note, by the way, that these laws are not restricted to planets. They apply to *any* orbiting object. Spy satellites, for example, move very rapidly as they swoop close to Earth's surface not because they are propelled with powerful on-board rockets but because their highly eccentric orbits are governed by Kepler's laws.

Kepler published his first two laws in 1609, stating that he had proved them only for the orbit of Mars. Ten years later, he extended them to all the then-known planets (Mercury, Venus, Earth, Mars, Jupiter, and Saturn) and added a

third law relating the size of a planet's orbit to its sidereal orbital **period**—the time needed for the planet to complete one circuit around the Sun. *Kepler's third law* states that

> The square of a planet's orbital period is proportional to the cube of its semi-major axis.

This law becomes particularly simple when we choose the (Earth) year as our unit of time and the *astronomical unit* as our unit of length. One **astronomical unit** (A.U.) is the semi-major axis of Earth's orbit around the Sun—essentially the average distance between Earth and the Sun. Like the light year, the astronomical unit is custom-made for the vast distances encountered in astronomy. Using these units for time and distance, we can write Kepler's third law for any planet as

$$P^2 \text{ (in Earth years)} = a^3 \text{ (in astronomical units)},$$

where P is the planet's sidereal orbital period, and a is the length of its semi-major axis. The law implies that a planet's "year" P increases more rapidly than does the size of its orbit a. For example, Earth, with an orbital semi-major axis of 1 A.U., has an orbital period of 1 Earth year. The planet Venus, orbiting at a distance of roughly 0.7 A.U., takes only 0.6 Earth years—about 225 days—to complete one circuit. By contrast, Saturn, almost 10 A.U. out, takes considerably more than 10 Earth years—in fact, nearly 30 years—to orbit the Sun just once.

Table 2.1 presents basic data describing the orbits of the nine planets now known. Renaissance astronomers knew these properties for the innermost six planets and used them to construct the currently accepted heliocentric model of the solar system. The second column presents each planet's orbital semi-major axis, measured in astronomical units; the third column gives the orbital period, in Earth years. The fourth column lists the planets' orbital eccentricities. For purposes of verifying Kepler's third law, the rightmost column lists the ratio P^2/a^3. As we have just seen, in the units

TABLE 2.1 Some Solar System Dimensions

PLANET	ORBITAL SEMI-MAJOR AXIS, a (astronomical units)	ORBITAL PERIOD, P (Earth years)	ORBITAL ECCENTRICITY	P^2/a^3
Mercury	0.387	0.241	0.206	1.002
Venus	0.723	0.615	0.007	1.001
Earth	1.000	1.000	0.017	1.000
Mars	1.524	1.881	0.093	1.000
Jupiter	5.203	11.86	0.048	0.999
Saturn	9.537	29.42	0.054	0.998
Uranus	19.19	83.75	0.047	0.993
Neptune	30.07	163.7	0.009	0.986
Pluto	39.48	248.0	0.249	0.999

used in the table, the third law implies that this number should equal 1 in all cases.

The main points to be grasped from Table 2.1 are these: (1) with the exception of Mercury and Pluto, the planets' orbits are very nearly circular (that is, their eccentricities are close to zero), and (2) the farther a planet is from the Sun, the greater is its orbital period, in agreement with Kepler's third law to within the accuracy of the numbers in the table. (The small but significant deviations of P^2/a^3 from 1 in the cases of Uranus and Neptune are caused by the gravitational attraction between those two planets; see Chapter 13.) Most important, note that Kepler's laws are obeyed by *all* the known planets, *not just by the six on which he based his conclusions.*

2.5 The Birth of Modern Astronomy

At about the same time as Kepler was developing his laws of planetary motion, Galileo Galilei (Figure 2.13) was finding fame—and notoriety—as an outspoken proponent of the Copernican system. Galileo was an Italian mathematician and philosopher. By being willing to perform experiments to test his ideas—a rather radical approach in those days (see *Interlude 2-2*)—and by embracing the brand-new technology of the telescope, he revolutionized the way in which science was done, so much so that he is now widely regarded as the father of experimental science.

GALILEO'S HISTORIC OBSERVATIONS

4 The telescope was invented in Holland in the early seventeenth century. Hearing of the invention (but without having seen one), Galileo built a telescope for himself in 1609 and aimed it at the sky. What he saw conflicted greatly with the philosophy of Aristotle and provided much new data to support the ideas of Copernicus.*

Using his telescope, Galileo discovered that the Moon had mountains, valleys, and craters—terrain in many ways reminiscent of that on Earth. Looking at the Sun (something that should *never* be done directly, and which eventually blinded Galileo), he found imperfections—dark blemishes now known as *sunspots*. Furthermore, by noting the changing appearance of these sunspots from day to day, he inferred that the Sun *rotates*, approximately once per month, around an axis roughly perpendicular to the ecliptic plane. These observations ran directly counter to the orthodox wisdom of the day.

In studying the planet Jupiter, Galileo saw four small points of light, invisible to the naked eye, orbiting it, and realized that they were moons. To Galileo, the fact that

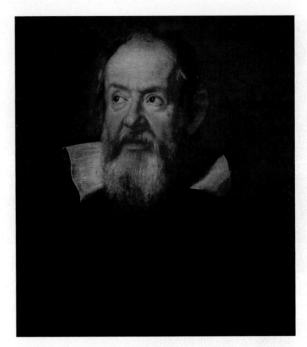

Figure 2.13 Galileo Galilei (1564–1642).

another planet had moons provided the strongest support for the Copernican model; clearly, Earth was not the center of all things. He also found that Venus shows a complete cycle of phases, like those of our Moon, a finding that could be explained only by the planet's motion around the Sun (see Figure 2.14). These observations were further strong evidence that Earth is not the center of the solar system, and that at least one planet orbited the Sun.

Galileo published his findings, and his controversial conclusions supporting the Copernican theory, in 1610, in a book called *Sidereus Nuncius* (*The Starry Messenger*). In reporting these wondrous observations made with his new telescope, Galileo was directly challenging the scientific establishment and religious dogma of the time. He was (literally) playing with fire—he must certainly have been aware that only a few years earlier, in 1600, the astronomer Giordano Bruno had been burned at the stake in Rome for his heretical teaching that Earth orbited the Sun. However, by all accounts, Galileo delighted in publicly ridiculing and irritating his Aristotelian colleagues. In 1616 his ideas were judged heretical, Copernicus's works were banned by the Roman Church, and Galileo was instructed to abandon his cosmological pursuits.

But Galileo would not desist. In 1632 he raised the stakes by publishing *Dialogue Concerning the Two Chief World Systems*, which compared the Ptolemaic and Copernican models. The book presented a discussion among three people: one of them a dull-witted Aristotelian, whose views time and again were roundly defeated by the arguments of one of his two companions, an articulate proponent of the heliocentric system. To make the book accessible to a wide popular audience, Galileo wrote it in Italian rather than Latin.

In fact, Galileo had already abandoned Aristotle in favor of Copernicus, although he had not published these beliefs at the time he began his telescopic observations.

(a)

(b)

Figure 2.14 (a) The phases of Venus, rendered at different points in the planet's orbit. If Venus orbits the Sun and is closer to the Sun than is Earth, as Copernicus maintained, then Venus should display phases, much as our Moon does. As shown here, when directly between Earth and the Sun, Venus's unlit side faces us, and the planet is invisible to us. As Venus moves in its orbit (at a faster speed than Earth moves in its orbit), progressively more of its illuminated face is visible from Earth. Note also the connection between orbital phase and the apparent size of the planet. Venus seems much larger in its crescent phase than when it is full because it is much closer to us during its crescent phase. (The insets at bottom left and right are actual photographs of Venus at two of its crescent phases.) (b) The Ptolemaic model (see also Figure 2.6) is unable to account for these observations. In particular, the full phase of the planet cannot be explained. Seen from Earth, Venus reaches only a "fat crescent" phase, then begins to wane as it nears the Sun.

These actions brought Galileo into direct conflict with the Church. Eventually, the Inquisition forced him, under threat of torture, to retract his claim that Earth orbits the Sun, and he was placed under house arrest in 1633; he remained imprisoned for the rest of his life. Not until 1992 were Galileo's "crimes" publicly forgiven by the Church. But the damage to the orthodox view of the universe was done, and the Copernican genie was out of the bottle once and for all.

THE ASCENDANCY OF THE COPERNICAN SYSTEM

Although Renaissance scholars were correct, none of them could *prove* that our planetary system is centered on the Sun, or even that Earth moves through space. Direct evidence for this was obtained only in the early eighteenth century, when astronomers discovered the *aberration of starlight*—a slight (20″) shift in the observed direction to a star, caused by Earth's motion perpendicular to the line of

sight. Additional proof came in the mid-nineteenth century, with the first unambiguous measurement of stellar parallax. Further verification of the heliocentricity of the solar system came gradually, with innumerable observational tests that culminated with the expeditions of our unmanned space probes of the 1960s, 1970s, and 1980s. The development and eventual acceptance of the heliocentric model were milestones in human thinking. This removal of Earth from any position of great cosmic significance is generally known, even today, by the term *Copernican principle*.

The Copernican episode is a good example of how the scientific method, though affected at any given time by the subjective whims, human biases, and even sheer luck of researchers, does ultimately lead to a definite degree of objectivity. Over time, many groups of scientists checking, confirming, and refining experimental tests can neutralize the subjective attitudes of individuals. Usually one generation of scientists can bring sufficient objectivity to

INTERLUDE 2-2 The Scientific Method

Most ancient philosophers held firmly to the belief that, whatever the reasons for the motions of the heavens, Earth in general and humankind in particular were absolutely central to the workings of the universe. Modern science, by contrast, has arrived at a diametrically opposite view. Our present-day outlook is that Earth, the solar system, and (some would argue) humanity are ordinary in every way. This idea is often (and only half-jokingly) called the "principle of mediocrity," and it is deeply embedded in modern scientific thought. It is a natural extension of the Copernican principle discussed in Sections 2.3 and 2.4 (see also *Interlude 2-1*). Nowadays, any theory or observation that even appears to single out Earth, the solar system, or the Milky Way Galaxy as in some way special is immediately regarded with great suspicion in scientific circles.

The principle of mediocrity extends far beyond mere philosophical preference, however. Simply put, if we do not make this assumption, then we cannot make much headway in science, and we cannot do astronomy at all. Virtually every statement made in this text rests squarely on the premise that the laws of physics, as we know them here on Earth, apply everywhere else too, without modification and without exception.

This transformation in the perception of humanity's place in the universe went hand in hand with a gradual—but radical—shift in the way philosophers and scientists conducted their investigation of the cosmos. The earliest known models of the universe were based largely on imagination and pure reasoning, with little attempt to explain the workings of the heavens in terms of known earthly experience. However, history shows that some philosophers did come to realize the importance of careful observation and testing to the formulation of their theories. The success of their ap-

proach changed, slowly but surely, the way science was done and opened the door to a fuller understanding of nature.

As knowledge from all sources was sought and embraced for its own sake, the influence of logic and reasoned argument grew, and the power of myth diminished. People began to inquire more critically about themselves and the universe. They realized that thinking about nature was no longer sufficient; looking at it was also necessary. Experiments and observations became a central part of the process of inquiry. To be effective, a theory—the framework of ideas and assumptions used to explain some set of observations and make predictions about the real world—must be continually tested. If experiments and observations favor it, a theory can be further developed and refined, but if they do not, it must be rejected, no matter how appealing it originally seemed. The process is illustrated schematically in the accompanying figure. This new approach to investigation, combining thinking and doing—that is, theory and experiment—is often known as the *scientific method*. It lies at the heart of modern science.

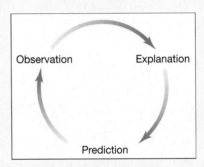

Notice, incidentally, that there is no "end point" to the process depicted in the figure. A theory can be invalidated

bear on a problem, though some especially revolutionary concepts are so swamped by tradition, religion, and politics that more time is necessary. In the case of heliocentricity, objective confirmation was not obtained until about three centuries after Copernicus published his work and more than 2000 years after Aristarchus had proposed the concept. Nonetheless, that objectivity *did in fact* eventually prevail.

2.6 The Dimensions of the Solar System

6 Kepler's laws allow us to construct a scale model of the solar system, with the correct shapes and *relative* sizes of all the planetary orbits, but they do not tell us the *actual* size of any orbit. We can express the distance to each planet only in terms of the distance from Earth to the Sun. Why is this? Because Kepler's triangulation measurements all used a portion of Earth's orbit as a baseline, distances could be expressed only relative to the size of that orbit, which was not

itself determined. Thus our model of the solar system would be analogous to a road map of the United States showing the *relative* positions of cities and towns but lacking the all-important scale marker indicating distances in kilometers or miles. For example, we would know that Kansas City is about three times more distant from New York than it is from Chicago, but we would not know the actual mileage between any two points on the map.

If we could somehow determine the value of the astronomical unit—in kilometers, say—we would be able to add the vital scale marker to our map of the solar system and compute the exact distances between the Sun and each of the planets. We might propose using triangulation to measure the distance from Earth to the Sun directly. However, we would find it impossible to measure the Sun's parallax using Earth's diameter as a baseline. The Sun is too bright, too big, and too fuzzy for us to distinguish any apparent displacement relative to a field of distant stars. To measure the Sun's distance from Earth, we must resort to some other method.

Before the middle of the twentieth century, the most accurate measurements of the astronomical unit were made using triangulation on the planets Mercury and by a single wrong prediction, but no amount of observation or experimentation can ever prove it correct. Theories simply become more and more widely accepted as their predictions are repeatedly confirmed.

In astronomy we are rarely afforded the luxury of performing experiments to test our theories, so observation becomes vitally important. One of the first documented uses of the scientific method in an astronomical context was performed by Aristotle nearly 25 centuries ago. He noticed that during a lunar eclipse, when Earth is positioned between the Sun and the Moon, it casts a curved shadow onto the surface of the Moon. The following figure shows a series of photographs taken during a recent lunar eclipse. The Earth's shadow, projected onto the Moon's surface, is indeed slightly curved. This is what Aristotle must have seen and recorded so long ago.

Because the observed shadow seemed always to be an arc of the same circle, Aristotle concluded that Earth, the cause of the shadow, must be round. On the basis of this hypothesis—this possible explanation of the observed facts—he then went on to predict that any and all future lunar eclipses would show that Earth's shadow was curved, regardless of the orientation of our planet. That prediction has been tested every time a lunar eclipse has occurred. It has

yet to be proved wrong. Aristotle was not the first person to argue that Earth is round, but he was apparently the first to offer a proof of it using the lunar-eclipse method.

The reasoning procedure Aristotle used forms the basis of all scientific inquiry today. He first made an observation. He then formulated a hypothesis to explain that observation. Finally, he tested the validity of his hypothesis by making predictions that could be confirmed or refuted by further observations. Observation, theory, and testing—these are the cornerstones of the scientific method, a technique whose power will be demonstrated again and again throughout our text.

Scientists throughout the world today use an approach that relies heavily on testing ideas. They gather data, form a working hypothesis that explains the data, and then proceed to test its predictions using experiment and observation. Experiment and observation are integral parts of the process of scientific inquiry. Theories unsupported by such evidence rarely gain any measure of acceptance in scientific circles. Used properly over a period of time, this rational, methodical approach enables us to arrive at conclusions that are mostly free of the personal bias and human values of any one scientist. The scientific method is designed to yield an objective view of the universe we inhabit.

R I V U X G

Venus during their rare *transits* of the Sun—that is, during the brief periods when those planets passed directly between the Sun and Earth (as shown for the case of Mercury in Figure 2.15). Because the time at which a transit occurs can be determined with great precision, astronomers can use this information to make very accurate measurements of a planet's position in the sky. They can then use simple geometry to compute the distance to the planet by combining observations made from different locations on Earth, as discussed earlier in Chapter 1. ∞ (Sec. 1.5) For example, the parallax of Venus at closest approach to Earth, as seen from two diametrically opposite points on Earth (separated by about 13,000 km), is about 1 arc minute—at the limit of naked-eye capabilities but easily measurable telescopically. This parallax represents a distance of 45 million km.

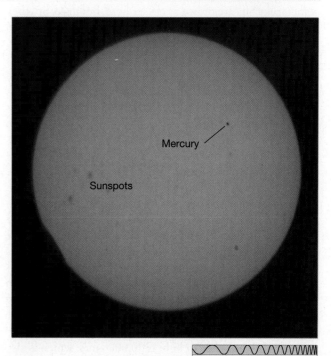

Mercury

Sunspots

R I V U X G

Figure 2.15 A solar transit of Mercury. Such transits happen only about once per decade, because Mercury's orbit does not quite coincide with the plane of the ecliptic. Transits of Venus are even rarer, occurring only about twice per century.

Figure 2.16 Simplified geometry of the orbits of Earth and Venus as they move around the Sun. The wavy lines represent the paths along which radar signals might be transmitted toward Venus and received back at Earth at the moment when Venus is at its minimum distance from Earth. Because the radius of Earth's orbit is 1 A.U. and that of Venus is about 0.7 A.U., we know that this distance is 0.3 A.U. Thus, radar measurements allow us to determine the astronomical unit in kilometers.

Knowing the distance to Venus, we can compute the magnitude of the astronomical unit. Figure 2.16 is an idealized diagram of the Sun–Earth–Venus orbital geometry. The planetary orbits are drawn as circles here, but in reality they are slight ellipses. This is a subtle difference, and we can correct for it using detailed knowledge of orbital motions. Assuming for the sake of simplicity that the orbits are perfect circles, we see from the figure that the distance from Earth to Venus at closest approach is approximately 0.3 A.U. Knowing that 0.3 A.U. is 45,000,000 km makes determining 1 A.U. straightforward—the answer is 45,000,000/0.3, or 150,000,000 km.

The modern method for deriving the absolute scale of the solar system uses radar rather than triangulation. The word **radar** is an acronym for **ra**dio **d**etection **a**nd **r**anging. In this technique, radio waves are transmitted toward an astronomical body, such as a planet. (We cannot use radar ranging to measure the distance to the Sun directly because radio signals are absorbed at the solar surface and are not reflected to Earth.) The returning echo indicates the body's direction and range, or distance, in absolute terms—that is, in kilometers rather than in astronomical units. Multiplying the round-trip travel time of the radar signal (the time elapsed between transmission of the signal and reception of the echo) by the speed of light (300,000 km/s, which is also

the speed of radio waves), we obtain twice the distance to the target planet.

Venus, whose orbit periodically brings it closest to Earth, is the most common target for radar ranging. The round-trip travel time (for example, at closest approach, as indicated by the wavy lines on Figure 2.16) can be measured with high precision—in fact, well enough to determine the planet's distance to an accuracy of about 1 km. In this way, the astronomical unit is now known to be 149,597,870 km. We will use the rounded-off value of 1.5×10^8 km in this text. (For more on the use of scientific notation to represent very large or very small numbers, see Appendix 1.)

Having determined the value of the astronomical unit, we can reexpress the sizes of the other planetary orbits in terms of more familiar units, such as miles or kilometers. The entire scale of the solar system can then be calibrated to high precision.

2.7 Newton's Laws

7 Kepler's three laws, which so simplified the solar system, were discovered *empirically*. In other words, they resulted solely from the analysis of observational data and were not derived from any theory or mathematical model. Indeed, Kepler did not have any appreciation for the physics underlying his laws. Nor did Copernicus understand the basic reasons *why* his heliocentric model of the solar system worked. Even Galileo, often called the father of modern physics, failed to understand why the planets orbit the Sun.

What prevents the planets from flying off into space or from falling into the Sun? What causes them to revolve about the Sun, apparently endlessly? To be sure, the motions of the planets obey Kepler's three laws, but only by considering something more fundamental than those laws can we really understand these motions. The heliocentric system was secured when, in the seventeenth century, the British mathematician Isaac Newton (Figure 2.17) developed a deeper understanding of the way *all* objects move and interact with one another as they do.

THE LAWS OF MOTION

Isaac Newton was born in Lincolnshire, England, on Christmas Day in 1642, the year that Galileo died. Newton studied at Trinity College of Cambridge University, but when the bubonic plague reached Cambridge in 1665, he returned to the relative safety of his home for 2 years. During that time he made probably the most famous of his discoveries, the law of gravity (although it is but one of the many major scientific advances Newton is responsible for). However, either because he regarded the theory as incomplete or possibly because he was afraid that he would be attacked or plagiarized by his colleagues, he did not tell anyone of his monumental achievement for almost 20 years. It was not until 1684, when Newton was discussing with complex dynamic behavior we see on Earth and throughout the universe. Newton's laws are listed in *More Precisely 2-2*.

Figure 2.18 illustrates *Newton's first law of motion*. The first law simply states that a moving object will move forever in a straight line unless some external **force** changes its direction of motion. For example, the object might glance off a brick wall or be hit with a baseball bat; in either case, a force changes the original motion of the object. The tendency of an object to keep moving at the same speed and in the same direction unless acted upon by a force is known as **inertia**. A familiar measure of an object's inertia is its **mass**—loosely speaking, the total amount of matter it contains. The greater an object's mass, the more inertia it has, and the greater is the force needed to change its state of motion.

Newton's first law contrasts sharply with the view of Aristotle, who maintained (incorrectly) that the natural state of an object was to be *at rest*—most probably an opinion based on Aristotle's observations of the effect of friction. In our discussion we will neglect friction—the force that slows balls rolling along the ground, blocks sliding across tabletops, and baseballs moving through the air. In any case, it is not an issue for the planets because there is no appreciable friction in outer space. The fallacy in Aristotle's argument was first realized and exposed by Galileo, who conceived of the notion of inertia long before Newton formalized it into a law.

The rate of change of the velocity of an object—speeding up, slowing down, or simply changing direction—is called its **acceleration**. *Newton's second law* states that the acceleration of an object is directly proportional to the applied force and inversely proportional to its mass—that is, the greater the force acting on the object, or the smaller the mass of the object, the greater its acceleration. Thus, if two objects are pulled with the same force, the more massive one will accelerate less; if two identical objects are pulled with different forces, the one experiencing the greater force will accelerate more.

Finally, *Newton's third law* simply tells us that forces cannot occur in isolation—if body A exerts a force on body B,

Figure 2.17   Sir Isaac Newton (1642–1727).

Edmund Halley (of Halley's comet fame) the leading astronomical problem of the day—*Why* do the planets move according to Kepler's laws?—that he astounded his companion by revealing that he had solved the problem in its entirety nearly two decades before!

Prompted by Halley, Newton published his theories in perhaps the most influential physics book ever written: *Philosophiae Naturalis Principia Mathematica* (*The Mathematical Principles of Natural Philosophy*—what we would today call "science"), usually known simply as Newton's *Principia*. The ideas expressed in that work form the basis for what is now known as **Newtonian mechanics**. Three basic laws of motion, the law of gravity, and a little calculus (which Newton also invented) are sufficient to explain and quantify virtually all the

Figure 2.18   An object at rest will remain at rest (a) until some force acts on it (b). It will then remain in that state of uniform motion until another force acts on it. The arrow in (c) shows a second force acting at a direction different from the first, which causes the object to change its direction of motion.

MORE PRECISELY 2-2 Newton's Laws of Motion and Gravitation

1. Every body continues in a state of rest or in a state of uniform motion in a straight line unless it is compelled to change that state by a force acting on it.

It requires no force to maintain motion in a straight line with constant velocity—that is, motion with constant speed and constant direction in space. The tendency of a body to remain in a state of uniform motion is usually called inertia. When velocity does vary (the speed increases or decreases, or the direction of motion changes), its rate of change is called acceleration. The relation of acceleration to any forces acting on a body is the subject of the second law of motion:

2. When a force F acts on a body of mass m, it produces in it an acceleration a equal to the force divided by the mass. Thus, $a = F/m$, or $F = ma$.

In honor of Newton, the SI unit of force is named after him. By definition, 1 newton (N) is the force required to cause a mass of 1 kilogram to accelerate at a rate of 1 meter per second every second.

Newton's third law relates the forces acting between separate bodies:

3. To every action there is an equal and opposite reaction.

This law means, for example, that you attract Earth with exactly the same force as it attracts you (a force known as your weight). This attraction is governed by one final law:

THE LAW OF UNIVERSAL GRAVITATION ("NEWTON'S LAW OF GRAVITY")

Every particle of matter in the universe attracts every other particle with a force that is directly proportional to the product of the masses of the particles and inversely proportional to the square of the distance between them.

In other words, two bodies of masses m_1 and m_2, separated by a distance r, attract each other with a force F that is proportional to $(m_1 \times m_2)/r^2$. The constant of proportionality is known as the gravitational constant, or often simply as Newton's constant, and is always denoted by the letter G. We can then express the law of gravity as

$$F = \frac{Gm_1m_2}{r^2}.$$

The value of G has been measured in extremely delicate laboratory experiments. In SI units, its value is 6.67×10^{-11} newton meter^2/kilogram^2 (N m^2/kg^2).

then body B necessarily exerts a force on body A that is equal in magnitude, but oppositely directed.

Only in extreme circumstances do Newton's laws break down, and this fact was not realized until the twentieth century, when Albert Einstein's theories of relativity once again revolutionized our view of the universe (see Chapter 22). Most of the time, however, Newtonian mechanics provides an excellent description of the motion of planets, stars, and galaxies through the cosmos.

GRAVITY

Forces can act *instantaneously* or *continuously*. To a good approximation, the force from a baseball bat that hits a home run can be thought of as being instantaneous in nature. A good example of a continuous force is the one that prevents the baseball from zooming off into space—**gravity**, the phenomenon that started Newton on the path to the discovery of his laws. Newton hypothesized that any object having mass always exerts an attractive *gravitational force* on all other massive objects. The more massive an object, the stronger its gravitational pull.

Consider a baseball thrown upward from Earth's surface, as illustrated in Figure 2.19. In accordance with Newton's first law, the downward force of Earth's gravity continuously modifies the baseball's velocity, slowing the

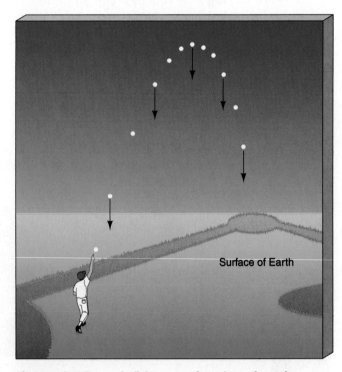

Figure 2.19 A ball thrown up from the surface of a massive object such as a planet is pulled continuously by the gravity of that planet (and, conversely, the gravity of the ball continuously pulls the planet).

initial upward motion and eventually causing the ball to fall back to the ground. Of course, the baseball, having some mass of its own, also exerts a gravitational pull on Earth. By Newton's third law, this force is equal and opposite to the weight of the ball (the force with which Earth attracts it). But, by Newton's second law, Earth has a much greater effect on the light baseball than the baseball has on the much more massive Earth. The ball and Earth feel the same gravitational force, but Earth's *acceleration* is much smaller.

Now consider the trajectory of the same baseball batted from the surface of the Moon. The pull of gravity is about one-sixth as great on the Moon as on Earth, so the baseball's velocity changes more slowly—a typical home run in a ballpark on Earth would travel nearly half a mile on the Moon. The Moon, less massive than Earth, has less gravitational influence on the baseball. The magnitude of the gravitational force, then, depends on the *masses* of the attracting bodies. Theoretical insight, as well as delicate laboratory experiments, tells us that the force is in fact directly proportional to the product of the two masses.

Studying the motions of the planets reveals a second aspect of the gravitational force. At locations equidistant from the Sun's center, the gravitational force has the same strength, and it is always directed toward the Sun. Furthermore, detailed calculation of the planets' accelerations as they orbit the Sun reveals that the strength of the Sun's gravitational pull decreases in proportion to the *square* of the distance from the Sun. (Newton himself is said to have first realized this fact by comparing the accelerations not of the planets but of the Moon and an apple falling to the ground—the basic reasoning is the same in either case.) The force of gravity is said to obey an **inverse-square law**. As shown in Figure 2.20, inverse-square forces decrease rapidly with distance from their source. For example, tripling the distance makes the force $3^2 = 9$ times weaker, while multiplying the distance by 5 results in a force that is $5^2 = 25$ times weaker. Despite this rapid decrease, the force never quite reaches zero. The gravitational pull of an object having some mass can never be completely extinguished.

We can combine the preceding statements about mass and distance to form a law of gravity that dictates the way in which *all* material objects attract one another. As a proportionality, Newton's law of gravity is

$$\text{gravitational force} \propto \frac{\text{mass of object \#1} \times \text{mass of object \#2}}{\text{distance}^2}.$$

The symbol $\propto$ here means "is proportional to." (See *More Precisely 2-2* for a fuller statement of this law.) This relationship is a compact way of stating that the gravitational pull between two objects is directly proportional to the product of their masses and inversely proportional to the square of the distance separating them.

To Newton, gravity was a force that acted at a distance, with no obvious way in which it was actually transmitted from place to place. Newton was not satisfied with this explana-

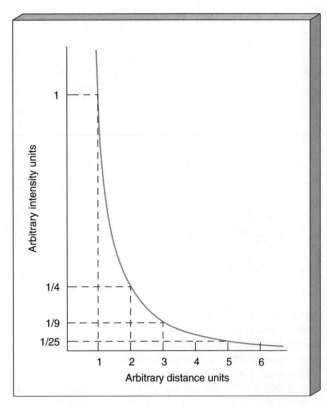

Figure 2.20 Inverse-square forces rapidly weaken with distance from their source. The strength of the gravitational force decreases with the square of the distance from the Sun. The force never quite diminishes to zero, however, no matter how far away from the Sun we go.

tion, but he had none better. To appreciate the modern view of gravity, consider any piece of matter having some mass—it could be smaller than an atom or larger than a galaxy. Extending outward from this object in all directions is a **gravitational field** produced by the matter. We now regard such a field as a property of space itself—a property that determines the influence of one massive object on another. All other matter "feels" the field as a gravitational force.

PLANETARY MOTION

8 The mutual gravitational attraction of the Sun and the planets, as expressed by Newton's law of gravity, is responsible for the observed planetary orbits. As depicted in Figure 2.21, this gravitational force continuously pulls each planet toward the Sun, deflecting its forward motion into a curved orbital path. Because the Sun is much more massive than any of the planets, it dominates the interaction. We might say that the Sun "controls" the planets, not the other way around.

The Sun–planet interaction sketched here is analogous to what occurs when you whirl a rock at the end of a string above your head. The Sun's gravitational field is your hand and the string, and the planet is the rock at the end of that string. The tension in the string provides the force necessary for the rock to move in a circular path. If

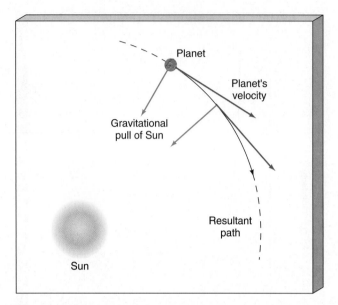

Figure 2.21 The Sun's inward pull of gravity on a planet competes with the planet's tendency to continue moving in a straight line. These two effects combine, causing the planet to move smoothly along an intermediate path, which continually "falls around" the Sun. This unending tug-of-war between the Sun's gravity and the planet's inertia results in a stable orbit.

you were suddenly to release the string—which would be like eliminating the Sun's gravity—the rock would fly away along a tangent to the circle, in accordance with Newton's first law.

In the solar system, at this very moment, Earth is moving under the combined influence of these two effects: the competition between gravity and inertia. The net result is a stable orbit, despite our continuous rapid motion through space. In fact, Earth orbits the Sun at a speed of about 30 km/s, or some 70,000 mph. (Verify this for yourself by calculating how fast Earth must move to complete a circle of radius 1 A.U.—and hence of circumference 2π A.U., or 940 million km—in 1 year, or 3.2×10^7 seconds. The answer is 9.4×10^8 km / 3.1×10^7s, or 30.3 km/s.) *More Precisely 2-3* describes how astronomers can use Newtonian mechanics and the law of gravity to measure the masses of Earth, the Sun, and many other astronomical objects by studying the orbits of objects near them.

KEPLER'S LAWS RECONSIDERED

Newton's laws of motion and the law of universal gravitation provided a theoretical explanation for Kepler's empirical laws of planetary motion. Just as Kepler modified Copernicus's model by introducing ellipses rather than circles, so too did Newton make corrections to Kepler's first and third laws. It turns out that a planet does not orbit the exact center of the Sun. Instead, both the planet and the Sun orbit their common **center of mass**. Because the Sun and the planet feel equal and opposite gravitational forces (by Newton's third law), the Sun must also move (by Newton's first law), driven by the gravitational influence of the planet. The Sun is so much more massive than any planet that the center of mass of the planet–Sun system is very close to the center of the Sun, which is why Kepler's laws are so accurate. Thus, Kepler's first law becomes

The orbit of a planet around the Sun is an ellipse, with the center of mass of the planet–Sun system at one focus.

 MORE PRECISELY 2-3 Weighing the Sun

We can use Newtonian mechanics to calculate the relationship between the distance (r) and the speed (v) of a planet moving in a circular orbit around the Sun (of mass m). By calculating the force required to keep the planet moving in a circle and comparing it with the gravitational force due to the Sun, it can be shown that the circular speed is

$$v = \sqrt{\frac{Gm}{r}},$$

where the gravitational constant G is defined in *More Precisely 2-2*. Dividing this speed into the circumference of the orbit ($2\pi r$), we obtain the modified form of Kepler's third law (equivalent to the formula presented in the text):

$$P = 2\pi \sqrt{\frac{r^3}{Gm}},$$

where $P = 2\pi r/v$ is the orbital period.

Because we have measured G in the laboratory on Earth and because we know the length of a year and the size of the astronomical unit, we can use Newtonian mechanics to *weigh* the Sun. Inserting the known values of $v = 30$ km/s, $r = 1$ A.U. $= 1.5 \times 10^{11}$ m, and $G = 6.7 \times 10^{-11}$ N m²/kg² in the equation, we can calculate the mass of the Sun to be 2.0×10^{30} kg—an enormous mass by terrestrial standards. Similarly, knowing the distance to the Moon and the length of the (sidereal) month, we can measure the mass of Earth to be 6.0×10^{24} kg.

In fact, this is how basically *all* masses are measured in astronomy. Because we can't just go out and weigh an astronomical object when we need to know its mass, we must look for its gravitational influence on something else. This principle applies to planets, stars, galaxies, and even clusters of galaxies—very different objects but all subject to the same physical laws.

As shown in Figure 2.22, however, the center of mass for two objects of comparable mass does not lie within either object. For identical masses (Figure 2.22a), the orbits are identical ellipses, with a common focus located midway between the two objects. For unequal masses (as in Figure 2.22b), the elliptical orbits still share a focus and both have the same eccentricity, but the more massive object moves more slowly and on a tighter orbit. (Note that Kepler's second law, as stated earlier, continues to apply without modification to each orbit separately, but the *rates* at which the

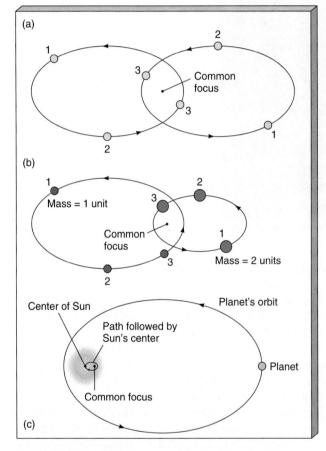

Figure 2.22 (a) The orbits of two bodies (stars, for example) with equal masses, under the influence of their mutual gravity, are identical ellipses with a common focus. That focus is not at the center of either star but instead is located at the center of mass of the pair, midway between them. The positions of the two bodies at three different times are indicated by the pairs of numbers. (Notice that a line joining the bodies always passes through the common focus.) (b) The orbits of two bodies, one of which is twice as massive as the other. Again, the elliptical orbits have a common focus, and the two ellipses have the same eccentricity. However, in accordance with Newton's laws of motion, the more massive body moves more slowly, and in a smaller orbit, staying closer to the center of mass (at the common focus). In this particular case, the larger ellipse is twice the size of the smaller one. (c) In this extreme case of a hypothetical planet orbiting the Sun, the common focus of the two orbits lies inside the Sun.

two orbits sweep out area are different.) In the extreme case of a planet orbiting the much more massive Sun (see Figure 2.22c), the path traced out by the Sun's center lies entirely within the Sun itself.

The change to Kepler's third law is also small in the case of a planet orbiting the Sun but very important in other circumstances, such as the orbital motion of two stars that are gravitationally bound to each other. Following through the mathematics of Newton's theory, we find that the true relationship between the semi-major axis a (measured in astronomical units) of the planet's orbit relative to the Sun and its orbital period P (in Earth years) is

$$P^2 \text{ (in Earth years)} = \frac{a^3 \text{ (in astronomical units)}}{M_{\text{total}} \text{ (in solar masses)}}$$

where M_{total} is the *combined* mass of the two objects. Notice that Newton's restatement of Kepler's third law preserves the proportionality between P^2 and a^3, but now the proportionality includes M_{total}, so it is *not* quite the same for all the planets. The Sun's mass is so great, however, that the differences in M_{total} among the various combinations of the Sun and the other planets are almost unnoticeable, and so Kepler's third law, as originally stated, is a very good approximation. This modified form of Kepler's third law is true in all circumstances, inside or outside the solar system.

ESCAPING FOREVER

The law of gravity that describes the orbits of planets around the Sun applies equally well to natural moons and artificial satellites orbiting any planet. All our Earth-orbiting, human-made satellites move along paths governed by a combination of the inward pull of Earth's gravity and the forward motion gained during the rocket launch. If the rocket initially imparts enough speed to the satellite, it can go into orbit. Satellites not given enough speed at launch (such as intercontinental ballistic missiles, ICBMs) fail to achieve orbit and fall back to Earth (see Figure 2.23). (Technically, ICBMs actually do orbit Earth's attracting center, but their orbits intersect Earth's surface.)

Some space vehicles, such as the robot probes that visit the other planets, attain enough speed to escape our planet's gravitational field and move away from Earth forever. This speed, known as the **escape speed**, is about 41 percent greater (actually, $\sqrt{2} = 1.414...$times greater) than the speed of a circular orbit at any given radius.* At less than the escape speed, the adage "what goes up must come down" (or at least stay in orbit) still applies. At more than the escape speed, however, a spacecraft will leave Earth for good (neglecting the effect of air resistance on its way through Earth's

---

*In terms of the formula presented earlier (see More Precisely 2-3), the escape speed is given by $v_{\text{escape}} = \sqrt{\dfrac{2Gm}{r}}$.

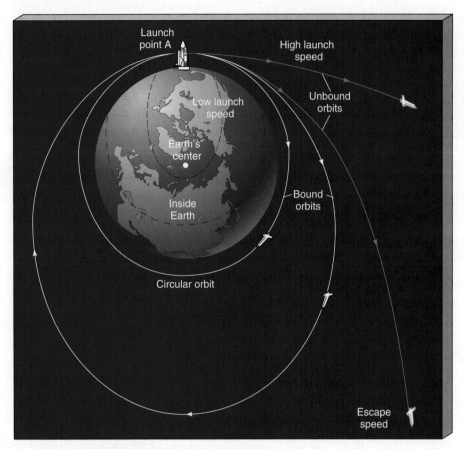

Figure 2.23 The effect of launch speed on the trajectory of a satellite. With too low a speed at point A the satellite will simply fall back to Earth. Given enough speed, however, the satellite will go into orbit—it "falls around Earth." As the initial speed at point A is increased, the orbit will become more and more elongated. When the initial speed exceeds the escape speed, the satellite will become unbound from Earth and will escape along a hyperbolic trajectory.

atmosphere and assuming that we don't turn the craft around using an on-board rocket motor). Planets, stars, galaxies—all gravitating bodies—have escape speeds. No matter how massive the body, gravity decreases with distance. As a result, the escape speed diminishes with increasing separation. The farther we go from Earth (or any gravitating body), the easier it becomes to escape.

The speed of a satellite in a circular orbit just above Earth's atmosphere is 7.9 km/s (roughly 18,000 mph). The satellite would have to travel at 11.2 km/s (about 25,000 mph) to escape from Earth altogether. If an object exceeds the escape speed, its motion is said to be **unbound**, and the orbit is no longer an ellipse. In fact, the path of the spacecraft relative to Earth is a related geometric figure called a *hyperbola*. If we simply change the word *ellipse* to *hyperbola*, the modified version of Kepler's first law still applies, as does Kepler's second law. (Kepler's third law does not extend to unbound orbits because it doesn't make sense to talk about a period in those cases.)

Newton's laws explain the paths of objects moving at any point in space near any gravitating body. These laws provide a firm physical and mathematical foundation for Copernicus's heliocentric model of the solar system and for Kepler's laws of planetary motions. But they do much more than that. Newtonian gravitation governs not only the planets, moons, and satellites in their elliptical orbits but also the stars and galaxies in their motion throughout our universe—as well as apples falling to the ground.

Chapter Review ✺WWW✺

SUMMARY

Many ancient cultures constructed elaborate structures that served as calendars and astronomical observatories. The study of the universe on the very largest scales is called **cosmology** (p. 32).

Unlike the Sun and the Moon, planets sometimes appear to temporarily reverse their direction of motion (from night to night) relative to the stars and then resume their normal "forward" course. This phenomenon is called **retrograde motion** (p. 32).

Geocentric (p. 32) models of the universe were based on the assumption that the Sun, the Moon, and the planets all orbit Earth. The most successful and long-lived of these was the **Ptolemaic model** (p. 33). To account for retrograde motion within the geocentric picture, it was necessary to suppose that planets moved on small circles called **epicycles** (p. 33), whose centers orbited Earth on larger circles called **deferents** (p. 33). The **heliocentric** (p. 34) view of the solar system holds that Earth, like all the planets, orbits the Sun. This model accounts for retrograde motion and the observed size and brightness variations of the planets in a much more natural way than the geocentric model. The widespread realization during the Renaissance that the solar system is Sun centered, and not Earth centered, is known as the **Copernican revolution** (p. 35), in honor of Nicholas Copernicus, who laid the foundations of the modern heliocentric model.

Johannes Kepler improved on Copernicus's model with his three **laws of planetary motion** (p. 37): (1) Planetary orbits are **ellipses** (p. 37), with the Sun at one **focus** (p. 37). (2) A planet moves faster as its orbit takes it closer to the Sun. (3) The **semi-major axis** (p. 37) of the orbit is related in a simple way to the planet's orbit **period** (p. 39). Most planets move on orbits whose **eccentricities** (p. 37) are quite small, so their paths differ only slightly from perfect circles.

Galileo Galilei is often regarded as the father of experimental science. His telescopic observations of the Moon, the Sun, Venus, and Jupiter played a crucial role in supporting and strengthening the Copernican picture of the solar system.

The distance from Earth to the Sun is called the **astronomical unit** (p. 39). Nowadays, the astronomical unit is determined by bouncing **radar** (p. 44) signals off the planet Venus and measuring the time taken for the signal to return.

Isaac Newton succeeded in explaining Kepler's laws in terms of a few general physical principles, now known as **Newtonian mechanics** (p. 45). The tendency of a body to keep moving at constant velocity is called **inertia** (p. 45). The greater the body's **mass** (p. 45), the greater its inertia. To change the velocity, a **force** (p. 45) must be applied. The rate of change of velocity, called **acceleration** (p. 45), is equal to the applied force divided by the body's mass. To explain planetary orbits, Newton postulated that **gravity** (p. 46) attracts the planets to the Sun. Every object with any mass is surrounded by a **gravitational field** (p. 47), whose strength decreases with distance according to an **inverse-square law** (p. 47). This field determines the gravitational force exerted by the object on any other body in the universe.

Newton's laws imply that a planet does not orbit the precise center of the Sun but instead that both the planet and the Sun orbit the common **center of mass** (p. 49) of the two bodies.

For an object to escape from the gravitational pull of another, its speed must exceed the **escape speed** (p. 49) of the second body. In this case, the motion is said to be **unbound** (p. 50), and the orbital path is no longer an ellipse, although it is still described by Newton's laws.

SELF-TEST: TRUE OR FALSE?

_____ **1.** Historical records show that it was Aristotle who first proposed that all planets revolve around the Sun.

_____ **2.** The teachings of Aristotle remained unchallenged until the eighteenth century A.D.

_____ **3.** Ptolemy was responsible for a geocentric model that was successful at predicting the positions of the planets, Moon, and the Sun.

_____ **4.** The heliocentric model of the universe held that Earth was at the center, and everything else moved around it.

_____ **5.** Kepler's discoveries regarding the orbital motion of the planets were based on his own observations.

_____ **6.** The Sun's location in a planet's orbit is at the center.

_____ **7.** The semi-major axis of an orbit is half the major axis.

_____ **8.** A circle has an eccentricity of zero.

_____ **9.** The astronomical unit is a distance equal to the semi-major axis of Earth's orbit around the Sun.

_____ **10.** The speed of a planet orbiting the Sun is independent of the planet's position in its orbit.

_____ **11.** Kepler's laws work for only the six planets known in his time.

_____ **12.** Kepler never knew the true distances between the planets and the Sun, only their relative distances.

_____ **13.** Galileo's observations of the sky were made with the naked eye.

_____ **14.** Using his laws of motion and gravity, Newton was able to prove Kepler's laws.

_____ **15.** You throw a baseball to someone. Before the ball is caught, it is temporarily in orbit around Earth's center.

SELF-TEST: FILL IN THE BLANK

1. Stonehenge was used as a _____ by people in the Stone Age.
2. Accurate records of comets and "guest" stars were kept over many centuries by _____ astrologers.
3. The astronomical knowledge of ancient Greece was kept alive and augmented by _____ astronomers.
4. The apparent "backward" (westward motion) of the planets Mars, Jupiter, or Saturn in the sky relative to the stars is known as _____ motion.
5. Observation, theory, and testing are the cornerstones of the _____.
6. The heliocentric model was reinvented by _____.
7. Central to the heliocentric model is the assertion that the observed motions of the planets and the Sun are the result of _____ motion around the Sun.
8. Kepler's laws were based on observational data obtained by _____.
9. Kepler discovered that the shape of an orbit is an _____, not a _____, as had previously been believed.
10. Kepler's third law relates the _____ of the orbital period to the _____ of the semi-major axis.
11. Galileo discovered _____ orbiting Jupiter, the _____ of Venus, and the Sun's rotation from observations of _____.
12. The modern method of measuring the astronomical unit uses _____ measurements of a planet or asteroid.
13. Newton's first law states that a moving object will continue to move in a straight line with constant speed unless acted upon by a _____.
14. Newton's law of gravity states that the gravitational force between two objects depends on the _____ of their masses and inversely on the _____ of their separation.
15. Newton discovered that, in Kepler's third law, the orbital period depends on the semi-major axis and on the sum of the _____ of the two objects involved.

REVIEW AND DISCUSSION

1. What contributions to modern astronomy were made by Chinese and Islamic astronomers during the Dark Ages of medieval Europe?
2. Briefly describe the geocentric model of the universe.
3. The benefit of our current knowledge lets us see flaws in the Ptolemaic model of the universe. What is its basic flaw?
4. What was the great contribution of Copernicus to our knowledge of the solar system? What was still a flaw in the Copernican model?
5. When were Copernicus's ideas finally accepted?
6. What is the Copernican principle?
7. What discoveries of Galileo helped confirm the views of Copernicus?
8. Briefly describe Kepler's three laws of orbital motion.
9. If radio waves cannot be reflected from the Sun, how can radar be used to find the distance from Earth to the Sun?
10. What is meant by the statement that Kepler's laws are empirical in nature?
11. List the two modifications made by Newton to Kepler's laws.
12. Why does a baseball fall toward Earth and not Earth toward the baseball?
13. Why would a baseball go higher if it were thrown up from the surface of the Moon than if it were thrown with the same velocity from the surface of Earth?
14. What is the meaning of the term *escape speed*?
15. What would happen to Earth if the Sun's gravity were suddenly "turned off"?

PROBLEMS

1. Tycho Brahe's observations of the stars and planets were accurate to about 1 arc minute ($1'$). To what distance does this angle correspond (a) at the distance of the Moon, (b) of the Sun, (c) of Saturn (at closest approach)?
2. Assume for simplicity that Earth and Mars move on circular orbits of radii 1.0 A.U. and 1.5 A.U., respectively, in exactly the same plane. To an observer on Earth, through what angle will Mars appear to move relative to the stars over the course of 24 hours, when the two planets are at closest approach? Will the apparent motion be prograde or retrograde?
3. Using the data in Table 2.1 show that Pluto is closer to the Sun at perihelion (the point of closest approach to the Sun in its orbit) than Neptune is at any point in its orbit. For simplicity, you may assume that Neptune has a circular orbit, which is not far from the truth.
4. An asteroid has a perihelion distance of 2.0 A.U. and an aphelion distance of 4.0 A.U. Calculate its orbital semi-major axis, eccentricity, and period.
5. Halley's comet has a perihelion distance of 0.6 A.U. and an orbital period of 76 years. What is its aphelion distance from the Sun?

6. How long will a radar signal take to complete a round trip between Earth and Mars when the two planets are 0.7 A.U. apart?

7. Jupiter's moon Callisto orbits the planet at a distance of 1.88 million km. Callisto's orbital period about Jupiter is 16.7 days. What is the mass of Jupiter? [Assume that Callisto's mass is negligible compared with that of Jupiter, and use the modified version of Kepler's third law (Section 2.7).]

8. The acceleration due to gravity at Earth's surface is 9.80 m/s². What is the acceleration at altitudes of (a) 100 km? (b) 1000 km? (c) 10,000 km?

9. Use Newton's law of gravity to calculate the force of gravity between you and Earth. Convert your answer, which will be in newtons, to pounds using the conversion 4.45 N equals 1 pound. What do you normally call this force?

10. The Moon's mass is 7.35×10^{22} kg and its radius is 1700 km. What is the speed of a spacecraft moving in a circular orbit just above the lunar surface? What is the escape speed from the Moon?

PROJECTS

1. Look in an almanac for the date of opposition of one or all of these bright planets: Mars, Jupiter, and Saturn. At opposition, these planets are at their closest points to Earth and are at their largest and brightest in the night sky. Observe these planets. How long before opposition does each planet's retrograde motion begin? How long afterward does it end?

2. Draw an ellipse. (See Figure 2.11, p. 38.) You'll need two pins, a piece of string, and a pencil. Tie the string in a loop and place it around the pins. Place the pencil inside the loop and run it around the inside of the string, holding the loop taut. The two pins will be at the foci of the ellipse. What is the eccentricity of the ellipse you have drawn?

3. Use a small telescope to replicate Galileo's observations of Jupiter's four largest moons. Note the moons' brightnesses and their locations with respect to Jupiter. If you watch over a period of several nights, draw what you see; you'll notice that these moons change their positions as they orbit the giant planet. Check the charts given monthly in *Astronomy* or *Sky & Telescope* magazines to identify each moon you see.

3 RADIATION

Information from the Cosmos

LEARNING GOALS

Studying this chapter will enable you to:

1 Discuss the nature of electromagnetic radiation, and tell how that radiation transfers energy and information through interstellar space.

2 List the major regions of the electromagnetic spectrum and explain how the properties of Earth's atmosphere affect our ability to make astronomical observations at different wavelengths.

3 Explain what is meant by the term "blackbody radiation" and describe the basic properties of such radiation.

4 Tell how we can determine the temperature of an object by observing the radiation that it emits.

5 Show how the relative motion of a source of radiation and its observer can change the perceived wavelength of the radiation, and explain the importance of this phenomenon to astronomy.

(Opposite page, background) The Ring Nebula in the constellation Lyra is one of the most magnificent sights in the nighttime sky. Seen here glowing in the light of its own emitted radiation, the nebula is actually the expanding outer atmosphere of a nearly dead star.

(Inset A) This nearly true color view clearly shows the dying dwarf star at the center of the expanding gas cloud, which is really a three-dimensional shell and not a ring.

(Inset B) False-color images can sometimes enhance certain features. Here some of the fine structure in the shell of gas can be seen more clearly.

(Inset C) Contour images, like this one derived from red light emitted by hydrogen gas in the nebula's shell and star's lingering atmosphere, can be used to map regions of relative brightness.

Astronomical objects are more than just things of beauty in the night sky. Planets, stars, and galaxies are of vital significance if we are to understand fully our place in the big picture—the "grand design" of the universe. Each object is a source of information about the material aspects of our universe—its state of motion, its temperature, its chemical composition, even its past history. When we look at the stars, the light we see actually began its journey to Earth decades, centuries—even millennia—ago. The faint rays from the most distant galaxies have taken billions of years to reach us. The stars and galaxies in the night sky show us the far away and the long ago. In this chapter we begin our study of how astronomers extract information from the light emitted by astronomical objects. These basic concepts of radiation are central to modern astronomy.

3.1 Information from the Skies

Figure 3.1 shows a galaxy in the constellation Andromeda. On a dark, clear night, far from cities or other sources of light, the Andromeda Galaxy, as it is generally called, can be seen with the naked eye as a faint, fuzzy patch on the sky, comparable in diameter to the full Moon. Yet the fact that it is visible from Earth belies this galaxy's enormous distance from us. It lies roughly 3 million *light years* away. An object at such a distance is truly inaccessible in any realistic human sense. Even if a space probe could miraculously travel at the speed of light, it would need 3 million years to reach this galaxy and 3 million more to return with its findings. Considering that civilization has existed on Earth for fewer than 10,000 years (and its prospects for the next 10,000 are far from certain), even this unattainable technological feat would not provide us with a practical means of exploring other galaxies—or even the farthest reaches of our own, "only" a few tens of thousands of light years distant.

LIGHT AND RADIATION

Given the impossibility of traveling to such remote parts of the universe, how do astronomers know *anything* about objects far from Earth? How do we obtain detailed information about any planet, star, or galaxy too distant for a personal visit or any kind of controlled experiment? The answer is that we use the laws of physics, as we know them here on Earth, to interpret the **electromagnetic radiation** emitted by these objects. *Radiation* is any way in which energy is transmitted through space from one point to another without the need for any physical connection between those two locations. The term *electromagnetic* just means that the energy is carried in the form of rapidly fluctuating *electric* and *magnetic* fields (to be discussed in more detail later in Section 3.2). Virtually all we know about the universe beyond Earth's atmosphere has been gleaned from painstaking analysis of electromagnetic radiation received from afar. **Visible light** is the particular type of electromagnetic radiation to which our human eyes happen to be sensitive. As light enters our eye, the cornea and lens focus it onto the retina, whereupon small chemical reactions triggered by the incoming energy send electrical impulses to the brain, producing the sensation of sight. But there is also *invisible* electromagnetic radiation, which goes completely undetected by our eyes. **Radio**, **infrared**, and **ultraviolet** waves, as well as **X rays** and **gamma rays**, all fall into this category.

Recognize that, despite the different names, the words *light*, *rays*, *radiation*, and *waves* really all refer to the same thing. The names are just historical accidents, reflecting the fact that it took many years for scientists to realize that these apparently very different types of radiation are in reality one and the same physical phenomenon. Throughout this text, we will use the general terms "light" and "electromagnetic radiation" more or less interchangeably.

WAVE MOTION

Despite the early confusion still reflected in the modern terminology, scientists now know that all types of electromagnetic radiation travel through space in the form of *waves*.

Figure 3.1 The pancake-shaped Andromeda Galaxy is almost 3 million light years away, according to the most recent distance measurements. It contains a few hundred billion stars.

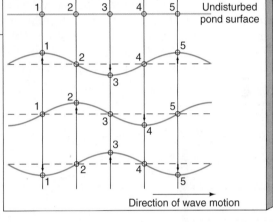

Figure 3.2 The passage of a wave across a pond causes the surface of the water to bob up and down, but there is no movement of water from one part of the pond to another. Here waves ripple out from the point where a pebble has hit the water to the point where a twig is floating. The inset shows schematically a cross section through part of the pond surface as the wave passes. The points numbered 1 through 5 represent nearby particles on the surface.

To understand the behavior of light, then, we must know a little about wave motion.

Simply stated, a **wave** is a way in which energy is transferred from place to place without physical movement of material from one location to another. In wave motion, the energy is carried by a disturbance of some sort. This *disturbance*, whatever its nature, occurs in a distinctive repeating pattern. Ripples on the surface of a pond, sound waves in air, and electromagnetic waves in space, despite their many obvious differences, all share this basic defining property.

As a familiar example, imagine a twig floating in a pond. A pebble thrown into the pond at some distance from the twig disturbs the surface of the water, setting it into up-and-down motion. This disturbance will propagate outward from the point of impact in the form of waves. When the waves reach the twig, some of the pebble's energy will be imparted to it, causing the twig to bob up and down. In this way, both energy and *information*—the fact that the pebble entered the water—are transferred from the place where the pebble landed to the location of the twig. We could tell that a pebble (or, at least, some object) had entered the water just by observing the twig. With a little additional physics, we could even estimate the pebble's energy.

A wave is not a physical object. No water traveled from the point of impact of the pebble to the twig—at any location

on the surface, the water surface simply moved up and down as the wave passed. What, then, *did* move across the surface of the pond? As illustrated in Figure 3.2, the answer is that the wave was the *pattern* of up-and-down motion, and it is this pattern that was transmitted from one point to the next as the disturbance moved across the water.

Figure 3.3 shows how wave properties are quantified and establishes some standard terminology. We characterize waves not only by the speed with which they move but also by the length of their cycle. How many seconds does it take for a wave to repeat itself at some point in space? This is its **wave period**. How many meters does it take for the wave to repeat itself at a given moment in time? This is its **wavelength**, defined as the length of an individual wave

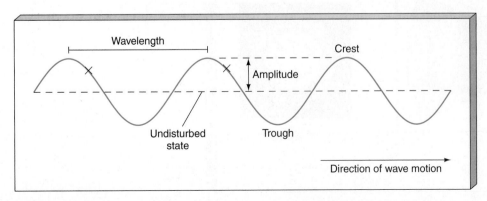

Figure 3.3 Representation of a typical wave, showing its direction of motion, wavelength, and amplitude.

cycle. The wavelength can be measured as the distance be-tween two adjacent wave *crests*, two adjacent wave *troughs*, or any other two similar points on adjacent wave cycles (such as the two points marked by an "×" in the figure). The maximum departure of the wave from the undisturbed state—still air or the flat pond surface—is called its **amplitude**.

If a wave moves at high speed, then the number of crests or cycles passing any given point per unit time—the wave's **frequency**—is high. Conversely, if a wave moves slowly, with only a few crests passing per unit time, we say that it has a low frequency. The frequency of a wave is just 1 divided by the wave's period:

$$\text{wave frequency} = \frac{1}{\text{wave period}}.$$

Frequency is expressed in units of inverse time, or cycles per second, termed hertz (Hz) in honor of the nineteenth-century German scientist Heinrich Hertz, who studied the properties of radio waves. Thus, a wave having a period of 5 s, meaning that one wave crest passes any given point every 5 seconds, has a frequency of (1/5) cycles/s = (1/5) Hz = 0.2 Hz.

Wavelength and wave frequency are *inversely* related. Doubling the frequency halves the wavelength, halving the frequency doubles the wavelength, and so on. This inverse relationship is easily understood. For a given wave speed, if the wave crests are close together, then more of them pass by a given point each second; when the crests are far apart, few of them pass by per unit time. The *wave speed* is simply the product of the wavelength and the frequency:

$$\text{wavelength} \times \text{frequency} = \text{wave speed}.$$

Put another way, a wave crest moves a distance equal to one wavelength in one wave period. For example, if the wave in our earlier example has a wavelength of 0.5 m, we can calculate its speed to be (0.5 m) × (0.2 Hz) = 0.1 m/s.

DIFFRACTION AND INTERFERENCE

Light exhibits two key properties that are characteristic of all forms of wave motion: *diffraction* and *interference*. **Diffraction** is the deflection, or "bending," of a wave as it passes a corner or moves through a narrow gap. As depicted in Figure 3.4, a sharp-edged hole in a barrier seems at first glance to

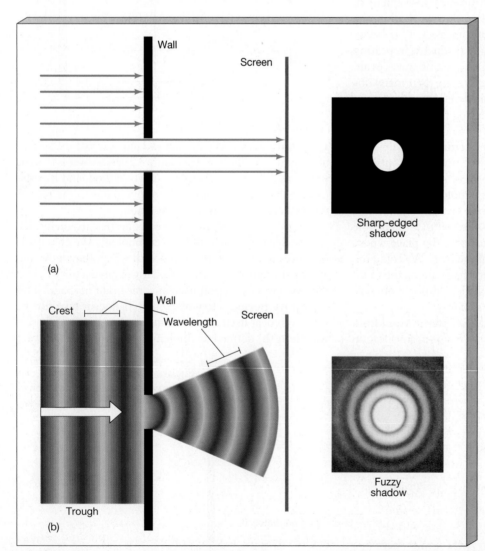

(a)

Wall

Screen

Sharp-edged shadow

(b)

Crest

Wall

Wavelength

Screen

Trough

Fuzzy shadow

Figure 3.4 Diffraction of a light wave. (a) If radiation were composed of rays or particles moving in perfectly straight lines, no bending would occur as a beam of light passed through a circular hole in a barrier, and the outline of the hole, projected onto a screen, would have perfectly sharp edges. (b) In fact, light is diffracted through an angle that depends on the ratio of the wavelength of the wave to the size of the gap. The result is that the outline of the hole becomes "fuzzy," as shown in this actual photograph of the diffraction pattern produced by a small circular opening.

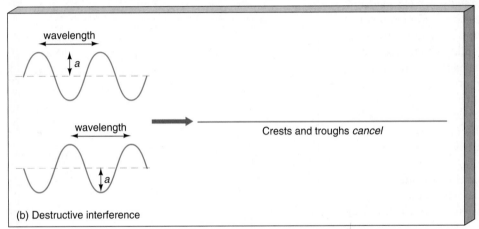

Figure 3.5 Interference of two identical waves: (a) constructive and (b) destructive. In constructive interference, the two waves (of amplitude *a*) reinforce each other to produce a larger-amplitude wave of the same wavelength. However, in destructive interference, the two waves exactly cancel out.

produce a sharp shadow. Closer inspection, however, reveals that the shadow actually has a "fuzzy" edge. We are not normally aware of such effects in everyday life because diffraction is generally very small for visible light. For any wave, the amount of diffraction is proportional to the ratio of the wavelength to the width of the gap. The longer the wavelength and/or the smaller the gap, the greater the angle through which the wave is diffracted. Thus, visible light, with its extremely short wavelengths, shows perceptible diffraction only when passing through very narrow openings. (The effect is much more noticeable for sound waves, however—no one thinks twice about our ability to hear people even when they are around a corner and out of our line of sight.)

Interference is the ability of two or more waves to reinforce or cancel each other. Figure 3.5 shows two identical waves moving through the same region of space. In Figure 3.5(a) the waves are positioned so that their crests and troughs exactly coincide. The net effect is that the two wave motions reinforce each other, resulting in a wave of greater amplitude. This is known as *constructive interference*. In Figure 3.5(b) the two waves exactly cancel, so that no net motion remains. This is *destructive interference*. As with diffraction, interference between waves of visible light is not noticeable in everyday experience; however, it is easily measured in the laboratory.

In the chapters that follow we will find that both diffraction and interference play important roles in many areas of observational astronomy.

3.2 Waves in What?

1 Waves of radiation differ fundamentally from water waves, sound waves, or any other waves that travel through a material medium. Radiation needs *no* such medium. When light radiation travels from a distant galaxy, or from any other cosmic object, it moves through the virtual vacuum of space. Sound waves, by contrast, cannot do this; if we were to remove all the air from a room, oral conversation would be impossible. Communication by flashlight or radio, however, would be entirely feasible.

The ability of light to travel through empty space was once a great mystery. The idea that light, or any other kind of radiation, could move as a wave through nothing at all seemed to violate common sense, yet it is now a cornerstone of modern physics.

INTERACTIONS BETWEEN CHARGED PARTICLES

To understand more about the nature of light, consider for a moment an *electrically charged* particle, such as an **electron** or a **proton**. Like mass, electrical charge is a fundamental property of matter. Electrons and protons are elementary particles—"building blocks" of atoms and all matter—that carry the basic unit of charge. Electrons are said to carry a *negative* charge, whereas protons carry an equal and opposite *positive* charge. Just as a massive object exerts a gravitational force on any other massive body (as we saw in Chapter 2), an electrically charged particle exerts an *electrical* force on every other charged particle in the universe. ∞ (Sec. 2.7) Buildup of electrical charge (a net imbalance of positive over negative, or vice versa) is what causes "static cling" on your clothes when you take them out of a hot clothes dryer, or the shock you sometimes feel when you touch a metal door frame on a particularly dry day.

Unlike the gravitational force, which is always attractive, electrical forces can be either attractive or repulsive. As illustrated in Figure 3.6(a), particles with *like* charges (that is, both negative or both positive—for example, two electrons or two protons) repel one another. Particles with *unlike* charges (that is, having opposite signs—an electron and a proton, say) attract. How is the electrical force transmitted through space? Extending outward in all directions from any charged particle is an **electric field**, which determines the electrical force exerted by the particle on all other charged particles in the universe (Figure 3.6b). The strength of the electric field, like the strength of the gravitational field, decreases with increasing distance from the charge according to an inverse-square law. By means of the electric field, the particle's presence is "felt" by all other charged particles, near and far.

Now suppose our particle begins to vibrate, perhaps because it becomes heated or collides with some other particle. Its changing position causes its associated electric field to change, and this changing field in turn causes the electrical force exerted on other charges to vary (Figure 3.6c). If we measure the change in the force on these other charges, we learn about our original particle. Thus, *information about the particle's state of motion is transmitted through space via a changing electric field*. This *disturbance* in the particle's electric field travels through space as a wave.

ELECTROMAGNETIC WAVES

The laws of physics tell us that a **magnetic field** must accompany every changing electric field. Magnetic fields govern the influence of *magnetized* objects on one another, much as electric fields govern interactions between charged particles. The fact that a compass needle always points to magnetic north is the result of the interaction between the magnetized needle and Earth's magnetic field (Figure 3.7). Magnetic fields also exert forces on *moving* electric charges (that is, electric currents)—electric meters and motors rely on this basic fact. Conversely, moving charges *create* magnetic fields (electromagnets are a familiar example). In short, electric and magnetic fields are inextricably linked to one another: a change in either one necessarily creates the other.

Thus, as illustrated in Figure 3.8, the disturbance produced by our moving charge actually consists of vibrating electric *and* magnetic fields, always oriented perpendicular to one another and moving together through space. These fields do not exist as independent entities; rather, they are

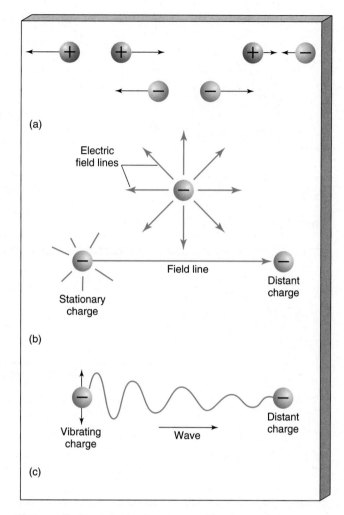

Figure 3.6 (a) Particles carrying like electrical charges repel one another, whereas particles carrying unlike charges attract. (b) A charged particle is surrounded by an electric field, which determines the particle's influence on other charged particles. We represent the field as a series of field lines. (c) If a charged particle begins to vibrate back and forth, its electric field changes. The resulting disturbance travels through space as a wave.

different aspects of a single physical phenomenon: **electromagnetism**. Together, they constitute an *electromagnetic wave* that carries energy and information from one part of the universe to another.

Now consider a real cosmic object—a star, say. When some of its charged contents move around, their electric fields change, and we can detect that change. The resulting electromagnetic ripples travel outward in waves, requiring no material medium in which to travel. Small charged particles, either in our eyes or in our experimental equipment, eventually respond to the electromagnetic field changes by vibrating in tune with the received radiation. This response is how we detect the radiation—how we see.

Figure 3.9 shows a more familiar example of information being transferred by electromagnetic radiation. A television transmitter causes electric charges to oscillate up and

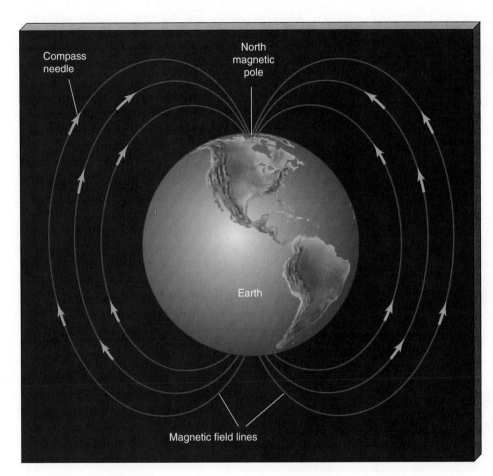

Figure 3.7 Earth's magnetic field interacts with a magnetic compass needle, causing the needle to become aligned with the field.

down a metal rod near the tower's top, thereby generating electromagnetic radiation. This radiation can be detected by rooftop antennas. In the metal rods of the receiving antenna, electric charges respond by vibrating in time with the transmitted wave frequency. The information carried by the pattern of vibrations is then converted into sound and pictures by your TV set.

How *quickly* does one charge feel the change in the electromagnetic field when another begins to move? This is an important question, because it is equivalent to asking how

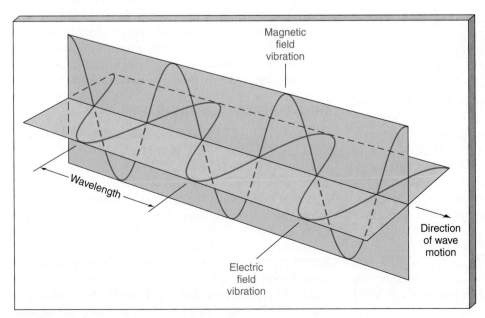

Figure 3.8 Electric and magnetic fields vibrate perpendicular to each other. Together they form an electromagnetic wave that moves through space at the speed of light.

Figure 3.9   Charged particles in an ordinary household television antenna vibrate in response to electromagnetic radiation broadcast by a distant transmitter. The radiation is produced when electric charges are made to oscillate in the transmitter's emitting antenna. The vibrations in the receiving antenna "echo" the oscillations in the transmitter, allowing the original information to be retrieved.

(labels in figure: Incoming wave; Motion of charged particles; Antenna)

fast an electromagnetic wave travels. Does it propagate at some measurable speed, or is it instantaneous? Both theory and experiment tell us that all electromagnetic waves move at a very specific speed—the **speed of light** (always denoted by the letter c). Its exact value is 299,792.458 km/s in a vacuum (and somewhat less in material substances such as air or water). We will round this value off to $c = 3.00 \times 10^5$ km/s. This is an extremely high speed. In the time needed to snap your fingers (about a tenth of a second) light can travel three quarters of the way around our planet! If the currently known laws of physics are correct, then the speed of light is the fastest speed possible.

The speed of light is very large, but it is still *finite*. That is, light does not travel instantaneously from place to place. This fact has some interesting consequences for our study of distant objects. It takes time—often lots of time—for light to travel through space. The light we see from the nearest large galaxy—the Andromeda Galaxy, shown in Figure 3.1—left that object about 3 million years ago—around the time our first human ancestors appeared on planet Earth. We can know nothing about this galaxy as it exists today. For all we know, it may no longer even exist! Only our descendants, 3 million years into the future, will know if it exists now. So as we study objects in the cosmos, remember that the light now seen left those objects long ago. We can never observe the universe as it is—only as it was.

3.3 The Electromagnetic Spectrum

2 White light is a mixture of colors, which we conventionally divide into six major hues—red, orange, yellow, green, blue, and violet. As shown in Figure 3.10, we can identify each of these basic colors by passing light through a prism. In principle, the original beam of white light could be restored by passing the entire red-to-violet range of colors—called a *spectrum* (plural: *spectra*)—through a second, oppositely oriented prism to recombine the colored beams. This experiment was first reported by Isaac Newton over 300 years ago.

THE COMPONENTS OF VISIBLE LIGHT

What determines the color of a beam of light? The answer is its wavelength (or, equivalently, its frequency). We see different colors because our eyes react differently to electromagnetic waves of different wavelengths. A prism splits a beam of light up into the familiar "rainbow" of colors because light rays of different wavelengths are bent, or refracted, slightly differently as they pass through the prism—red light the least, violet light the most. Red light has a frequency of roughly 4.3×10^{14} Hz, corresponding to a wavelength of about 7.0×10^{-7} m. Violet light, at the other end of the visible range, has nearly double the frequency—7.5×10^{14} Hz—and (since the speed of light is the same in

Figure 3.10   While passing through a prism, white light splits into its component colors, spanning red to violet in the visible part of the electromagnetic spectrum. The slit narrows the beam of radiation. The image on the screen is just a series of different-colored images of the slit.

either case) just over half the wavelength—4.0×10^{-7} m. The other colors we see have frequencies and wavelengths intermediate between these two extremes, spanning the entire **visible spectrum** shown in Figure 3.10; radiation outside this range is invisible to human eyes.

Astronomers often use a unit called the *nanometer* (nm) when describing the wavelength of light (see Appendix 2). There are 10^9 nanometers in 1 meter. An older unit called the *angstrom* ($1\text{Å} = 10^{-10}$ m $= 0.1$ nm) is also widely used. (The unit is named after the nineteenth-century Swedish physicist Anders Ångstrom—pronounced "ongstrem.") However, in SI units, the nanometer is preferred. Thus, the visible spectrum covers the wavelength range from 400 to 700 nm (4000 to 7000 Å). The radiation to which our eyes are most sensitive has a wavelength near the middle of this range, at about 550 nm (5500 Å), in the yellow-green region of the spectrum. It is no coincidence that this wavelength falls within the range of wavelengths at which the Sun emits most of its electromagnetic energy—our eyes have evolved to take greatest advantage of the available light.

THE FULL RANGE OF RADIATION

Figure 3.11 plots the entire range of electromagnetic radiation, illustrating the relationships among the different "types" of electromagnetic radiation listed earlier. We see that the only characteristic that distinguishes one from another is its wavelength, or frequency. To the low-frequency, long-wavelength side of visible light lie *radio* and *infrared*

radiation. Radio frequencies include radar, microwave radiation, and the familiar AM, FM, and TV bands. We perceive infrared radiation as heat. At higher frequencies (shorter wavelengths) are the domains of *ultraviolet*, *X-ray*, and *gamma-ray* radiation. Ultraviolet radiation, lying just beyond the violet end of the visible spectrum, is responsible for suntans and sunburns. X rays are perhaps best known for their ability to penetrate human tissue and reveal the state of our insides without resorting to surgery. Gamma rays are the shortest-wavelength radiation. They are often associated with radioactivity and are invariably damaging to living cells they encounter.

All these spectral regions, including the visible spectrum, collectively make up the **electromagnetic spectrum**. Remember that, despite their greatly differing wavelengths and the very different roles they play in everyday life on Earth, all are basically the same phenomenon, and all move at the same speed—the speed of light, *c*.

Figure 3.11 is worth studying carefully, as it contains a great deal of information. Note that wave frequency (in hertz) increases from left to right, and wavelength (in meters) increases from right to left. These wave properties behave in opposite ways because, as noted earlier, they are inversely related. When picturing wavelengths and frequencies, this book will adhere to the convention that frequency increases toward the right. Notice that the wavelength and frequency scales in Figure 3.11 do not increase by equal increments of 10. Instead, successive values marked on the horizontal axis differ by *factors of 10*—each is 10 times greater

Figure 3.11 The entire electromagnetic spectrum.

than its neighbor. This type of scale, called a *logarithmic* scale, is often used in science in order to condense a very large range of some quantity into a manageable size. Had we used a linear scale for the wavelength range shown in Figure 3.11, the figure would have been many light years long! Throughout the text we will often find it convenient to use a logarithmic scale in order to compress a wide range of some quantity onto a single, easy-to-view plot.

Figure 3.11 shows that wavelengths extend from the size of mountains for radio radiation to the size of an atomic nucleus for gamma-ray radiation. The box at the upper right emphasizes how small the visible portion of the electromagnetic spectrum is. Most objects in the universe emit large amounts of invisible radiation. Indeed, many of them emit only a tiny fraction of their total energy in the visible range. A wealth of extra knowledge can be gained by studying the invisible regions of the electromagnetic spectrum. To remind you of this important fact and to identify the region of the electromagnetic spectrum in which a particular observation was made, we have attached a spectrum icon—an idealized version of the wavelength scale in Figure 3.11—to every astronomical image presented in this text.

ATMOSPHERIC BLOCKAGE

Our eyes are sensitive to only a minute portion of the many different kinds of radiation known. In addition, only a small fraction of the radiation produced by astronomical objects actually reaches our eyes, in part because of the *opacity* of Earth's atmosphere. **Opacity** is the extent to which radiation is blocked by the material through which it is passing—in this case, air. The more opaque an object is, the less radiation gets through it. Opacity is the opposite of transparency. At the bottom of Figure 3.11, the atmospheric opacity is plotted along the wavelength and frequency scales. The extent of shading is proportional to the opacity. Where the shading is greatest, no radiation can get in or out. Where there is no shading at all, the atmosphere is almost completely transparent, so that extraterrestrial radiation can reach Earth's surface, and terrestrial radiation from human transmissions can pass virtually unhindered into space.

What causes opacity to vary along the spectrum? Certain atmospheric gases are known to absorb radiation very efficiently at some wavelengths. For example, water vapor (H_2O) and oxygen (O_2) absorb radio waves having wavelengths less than about a centimeter, whereas water vapor and carbon dioxide (CO_2) are strong absorbers of infrared radiation. Ultraviolet, X-ray, and gamma-ray radiation are completely blocked by the *ozone layer* high in Earth's atmosphere (see Sec. 7.3). A passing but unpredictable source of atmospheric opacity in the visible part of the spectrum is the blockage of light by atmospheric clouds. In addition, the interaction between the Sun's ultraviolet radiation and the upper atmosphere produces a thin, electrically conducting layer at an altitude of about 100 km. The *ionosphere*, as this layer is known, reflects long-wavelength radio waves (wavelengths greater than about 10 m) as well as a mirror reflects visible light. In this way, extraterrestrial waves are kept out, and terrestrial waves—such as those produced by AM radio stations—are kept in. (That is why it is possible to transmit some radio frequencies beyond the horizon—the broadcast waves bounce off the ionosphere.)

The effect of all this blockage is that there are only a few *windows*, at well-defined locations in the electromagnetic spectrum, where Earth's atmosphere is transparent. In much of the radio and in the visible portions of the spectrum, the opacity is low, and we can study the universe at those wavelengths from ground level. In parts of the infrared range, the atmosphere is partially transparent, so we can make certain infrared observations from the ground. Moving to the tops of mountains, above as much of the atmosphere as possible, improves observations. In the rest of the spectrum, however, the atmosphere is opaque. Ultraviolet, X-ray, and gamma-ray observations can be made only from above the atmosphere, from orbiting satellites.

3.4 The Distribution of Radiation

3 *All* macroscopic objects—fires, ice cubes, people, stars—emit radiation at all times, regardless of their size, shape, or chemical composition. They radiate mainly because the microscopic charged particles they are made up of are in constantly varying random motion, and whenever charges change their state of motion, electromagnetic radiation is emitted. The **temperature** of an object is a direct measure of the amount of microscopic motion within it (see *More Precisely 3-1*). The hotter the object—that is, the higher its temperature—the faster its constituent particles move and the more energy they radiate.

THE BLACKBODY SPECTRUM

Intensity is a term often used to specify the amount or strength of radiation at any point in space. Like frequency and wavelength, intensity is a basic property of radiation. No natural object emits all its radiation at just one frequency. Instead, the energy is generally spread out over a range of frequencies. By studying the way in which the intensity of this radiation is distributed across the electromagnetic spectrum, we can learn much about the object's properties.

Figure 3.12 illustrates schematically the distribution of radiation emitted by any object. The curve peaks at a single, well-defined frequency and falls off to lesser values above and below that frequency. Note that the curve is not shaped like a symmetrical bell that declines evenly on either side of the peak. The intensity falls off more slowly from the peak to lower frequencies than it does from the peak to high frequencies. This overall shape is characteristic of the radiation emitted by *any* object, regardless of its size, shape, composition, or temperature.

The curve drawn in Figure 3.12 is the radiation-distribution curve for a mathematical idealization known as a

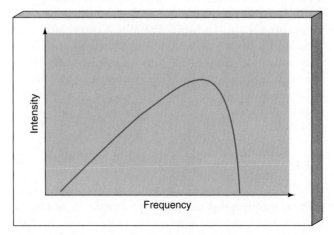

Figure 3.12 The blackbody, or Planck, curve represents the distribution of the intensity of radiation emitted by any heated object.

MORE PRECISELY 3-1 The Kelvin Temperature Scale

The atoms and molecules that make up any piece of matter are in constant random motion. This motion represents a form of energy known as *thermal energy*—or, more commonly, *heat*. The quantity we call *temperature* is a direct measure of this internal motion: the higher an object's temperature, the faster, on average, the random motion of its constituent particles. More precisely, the temperature of a piece of matter specifies the average thermal energy of the particles it contains.

Our familiar Fahrenheit temperature scale, like the archaic English system in which length is measured in feet and weight in pounds, is of somewhat dubious value. In fact, the "degree Fahrenheit" is now a peculiarity of American society. Most of the world uses the Celsius scale of temperature measurement (also called the centigrade scale). In the Celsius system, water freezes at 0 degrees (0°C) and boils at 100 degrees (100°C), as illustrated in the accompanying figure.

There are, of course, temperatures below the freezing point of water. Although we know of no matter anywhere in the universe that is actually this cold, temperatures can in theory reach as low as −273.15°C. This is the temperature at which atomic and molecular motion all but ceases. It is convenient to construct a temperature scale based on this lowest possible temperature, or *absolute zero*. Scientists commonly use such a scale, called the *Kelvin scale* in honor of the nineteenth-century British physicist Lord Kelvin. Since it takes absolute zero as its starting point, the Kelvin scale differs from the Celsius scale by 273.15°. In this book, we round off the decimal places and simply use

kelvins = degrees Celsius + 273.

Thus,

■ Translational motion ceases at 0 kelvins (0 K).
■ Water freezes at 273 kelvins (273 K).
■ Water boils at 373 kelvins (373 K).

Note that the unit is "kelvins," or "K," *not* "degrees kelvin" or "°K." (Occasionally, the term "degrees absolute" is used instead.)

blackbody—an object that absorbs all radiation falling on it. In a steady state, a blackbody must reemit the same amount of energy as it absorbs. The **blackbody curve** shown in the figure describes the distribution of that reemitted radiation. (The curve is also known as the *Planck curve*, after Max Planck, whose mathematical analysis of such thermal emission in 1900 played a key role in modern physics.) No real object absorbs and radiates as a perfect blackbody. However, in many cases, the blackbody curve is a very good approximation to reality, and the properties of blackbodies provide important insights into the behavior of real objects.

THE RADIATION LAWS

4 The blackbody curve shifts toward higher frequencies (shorter wavelengths) and greater intensities as an object's temperature increases. Even so, the *shape* of the curve remains the same. This shifting of radiation's peak frequency with temperature is familiar to us all: very hot glowing objects, such as toaster filaments or stars, emit visible light. Cooler objects, such as warm rocks or household radiators, produce invisible radiation—warm to the touch but not

glowing hot to the eye. These latter objects emit most of their radiation in the lower-frequency infrared part of the electromagnetic spectrum.

Imagine a piece of metal placed in a hot furnace. At first, the metal becomes warm, although its visual appearance doesn't change. As it heats up, it begins to glow dull red, then orange, brilliant yellow, and finally white. How do we explain this? As illustrated in Figure 3.13, when the metal is at room temperature (300 K—see *More Precisely 3-1* for a discussion of the Kelvin temperature scale), it emits only invisible infrared radiation. As the metal becomes hotter, the peak of its blackbody curve shifts toward higher frequencies. At 1000 K, for instance, most of the emitted radiation is still infrared, but now there is also a small amount of visible (dull red) radiation being emitted (note in Figure 3.13 that the high-frequency portion of the 1000 K curve just overlaps the visible region of the graph).

As the temperature continues to rise, the peak of the metal's blackbody curve moves through the visible spectrum, from red (the 4000 K curve) through yellow. The metal eventually becomes white hot because, when its blackbody curve peaks in the blue or violet part of the spectrum

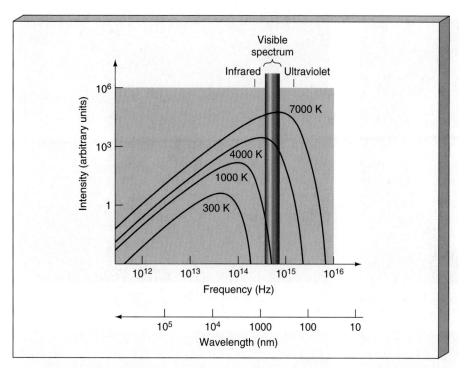

Visible spectrum
Infrared — Ultraviolet

Figure 3.13 As an object is heated the radiation it emits peaks at higher and higher frequencies. Shown here are curves corresponding to temperatures of 300 K (room temperature), 1000 K (beginning to glow deep red), 4000 K (red hot), and 7000 K (white hot).

(the 7000 K curve), the low-frequency tail of the curve extends through the entire visible spectrum (to the left in Figure 3.13), meaning that substantial amounts of green, yellow, orange, and red light are also emitted. Together, all these colors combine to produce white.

From studies of the precise form of the blackbody curve we obtain a very simple connection between the wavelength at which most radiation is emitted and the absolute temperature (that is, the temperature measured in kelvins) of the emitting object:

$$\text{wavelength of peak emission} \propto \frac{1}{\text{temperature}}.$$

This relationship, called **Wien's law**, is discussed in more detail in *More Precisely 3-2*.

Simply put, Wien's law tells us that the hotter the object, the bluer its radiation. For example (see Figure 3.15), an object with a temperature of 6000 K emits most of its energy in the visible part of the spectrum, with a peak wavelength of 480 nm. At 600 K, the object's emission would peak at a wavelength of 4800 nm, well into the infrared portion of the spectrum. At a temperature of 60,000 K, the peak would move all the way through the visible spectrum to a wavelength of 48 nm, in the ultraviolet range.

It is also a matter of everyday experience that, as the temperature of an object increases, the *total* amount of energy it radiates (summed over all frequencies) increases rapidly. For example, the heat given off by an electric heater increases very sharply as it warms up and begins to emit visible light. Careful experimentation leads to the conclusion that the total amount of energy radiated per unit time is actually proportional to the fourth power of the object's temperature:

$$\text{total energy radiated} \propto \text{temperature}^4.$$

This relation is called **Stefan's law**. Again, it is discussed further in *More Precisely 3-2*. From this form of Stefan's law we can see that the energy emitted by a body rises dramatically as its temperature increases. Doubling the temperature, for example, causes the total energy radiated to increase by a factor of 16.

ASTRONOMICAL APPLICATIONS

No known natural terrestrial objects reach temperatures high enough to emit very-high-frequency radiation. Only human-made thermonuclear explosions are hot enough for their spectra to peak in the X- or gamma-ray range. (Most human inventions that produce short-wavelength, high-frequency radiation, such as X-ray machines, are designed to emit only a specific range of wavelengths and do not operate at high temperatures.*) Many extraterrestrial objects, however, do emit copious quantities of ultraviolet, X-ray, and even gamma-ray radiation. Figure 3.14 shows a familiar object—our Sun—as it appears when viewed using radiation from different regions of the electromagnetic spectrum. Although most sunlight is visible, a great deal of information about our parent star can be obtained by studying it in other parts of the electromagnetic spectrum.

Astronomers often use blackbody curves as thermometers to determine the temperatures of distant objects. For example, study of the solar spectrum makes it possible to measure the temperature of the Sun's surface. Observations of the radiation from the Sun at many frequencies yield a curve shaped somewhat like that shown in Figure 3.12. The Sun's curve peaks in the visible part of the electromagnetic

They are said to produce a nonthermal spectrum of radiation.

(a) R I V U X G

(b) R I V U X G

(c) R I V U X G

(d) R I V U X G

Figure 3.14 Four images of the Sun, made using (a) visible light, (b) ultraviolet light, (c) X rays, and (d) radio waves. By studying the similarities and differences among these views of the same object, astronomers can find important clues to its structure and composition.

spectrum; the Sun also emits a lot of infrared and a little ultraviolet radiation. Using Wien's law, we find that the temperature of the Sun's surface is approximately 6000 K. (A more precise measurement, applying Wien's law to the blackbody curve that best fits the solar spectrum, yields a temperature of 5800 K.)

Other cosmic objects have surfaces very much cooler or hotter than the Sun's, emitting most of their radiation in invisible parts of the spectrum (Figure 3.15). For example, the relatively cool surface of a very young star may measure 600 K and emit mostly infrared radiation. Cooler still is

Figure 3.15 Comparison of blackbody curves for four cosmic objects having different temperatures. Peak frequencies and wavelengths are marked. (a) A cool, invisible galactic gas cloud called Rho Ophiuchi. At a temperature of 60 K, it emits mostly low-frequency radio radiation. (b) A dim, young star (shown here in red) near the center of the Orion Nebula. The star's atmosphere, at 600 K, radiates primarily in the infrared. (c) The Sun's surface, at 6000 K, is brightest in the visible region of the electromagnetic spectrum. (d) A cluster of very bright stars, called Omega Centauri, as observed by a telescope aboard a space shuttle. At a temperature of 60,000 K, these stars radiate strongly in the ultraviolet.

MORE PRECISELY 3-2 More About the Radiation Laws

As mentioned in Section 3.4, Wien's law relates the temperature T of an object to the wavelength λ_{max} at which it emits the most radiation. (The Greek letter λ—lambda—is conventionally used to denote wavelength.) Mathematically, if we measure T in kelvins and λ_{max} in centimeters, we find that

$$\lambda_{max} = \frac{0.29 \text{ cm}}{T}.$$

We could also convert Wien's law into a statement about frequency (f), using the relation $f = c/\lambda$ to obtain $f_{max} = 1.0 \times 10^{11} \, T$ Hz. However, the law is most commonly stated in terms of wavelength and is probably easier to remember that way. Thus at 6000 K, the wavelength of maximum intensity is 0.29/6000 cm, or 480 nm, corresponding to the yellow-green part of the visible spectrum. (Recall that 1 nanometer—1 nm—is equal to 10^{-9} m; see Appendix 2.)

We can also give Stefan's law a more precise mathematical formulation. With T measured in kelvins, the total amount of energy emitted per square meter of its surface per second (a quantity known as the *energy flux F*) is given by

$$F = \sigma T^4.$$

The constant σ (the Greek letter sigma) is known as the *Stefan-Boltzmann constant*, or often just Stefan's constant, after Josef Stefan, the Austrian scientist who formulated the equation.

The SI unit of energy is the *joule* (J). Probably more familiar is the closely related unit called the *watt* (W), which measures power—the *rate* at which energy is emitted or expended by an object. One watt is the emission of one joule per second; for example, a 100-W light bulb emits energy (mostly in the form of infrared and visible light) at a rate of 100 J/s. In these units, the Stefan-Boltzmann constant has the value $\sigma = 5.67 \times 10^{-8}$ W/m²·K⁴.

Notice just how rapidly the energy flux increases with increasing temperature. A piece of metal in a furnace, when at a temperature of 3500 K, radiates energy at a rate of about 850 W for every square centimeter of its surface area. Doubling its temperature to 7000 K (so that it becomes yellow to white hot, by Wien's law) increases the energy emitted by a factor of 16 (four "doublings"), to 13.6 *kilo*watts (kW) (13,600 W) per square centimeter.

---

the interstellar gas cloud from which the star formed; at a temperature of 60 K, such a cloud emits mainly long-wavelength radiation in the radio and infrared parts of the spectrum. The brightest stars, in contrast, have surface temperatures as high as 60,000 K and hence emit mostly ultraviolet radiation.

3.5 The Doppler Effect

5 Imagine a rocket ship launched from Earth with enough fuel to allow it to accelerate to speeds comparable to the speed of light. As the ship's speed increased, a remarkable thing would happen (Figure 3.16). Passengers would notice that the light from the star system toward which they were traveling seemed to be getting *bluer*. In fact, *all* stars in front of the ship would appear bluer than normal, and the greater the ship's speed, the greater the color change would be. Furthermore, stars behind the vessel would seem *redder* than normal, while stars to either side would be unchanged in appearance. As the spacecraft slowed down and came to rest relative to Earth, all stars would resume their usual appearance. The travelers would have to conclude that the stars had changed their colors not because of any real change in their physical properties but because of the spacecraft's own *motion*.

This phenomenon is not restricted to electromagnetic radiation and fast-moving spacecraft. Waiting at a railroad crossing for an express train to pass, most of us have had

the experience of hearing the pitch of the engine's horn shift from high shrill to low blare as the train approaches and then recedes. High-pitched sound (treble) has shorter wavelengths than low-pitched sound (bass); this change in wavelength is analogous to the color change observed by our space travelers. Indeed, if the train had a light atop its en-

Figure 3.16 Observers in a fast-moving spacecraft will see the stars ahead of them seem bluer than normal, while those behind are reddened. The stars have not changed their properties—the color changes are the result of the motion of the spacecraft relative to the stars.

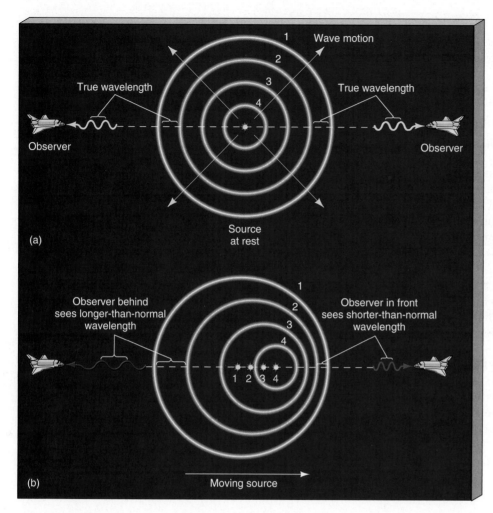

Figure 3.17 (a) Wave motion from a source toward an observer at rest with respect to the source. The four numbered circles represent successive wave crests emitted by the source; at the instant shown, the fifth wave crest is just about to be emitted. As seen by the observer, the source is not moving, so the wave crests are just concentric spheres (shown here as circles). (b) Waves from a moving source tend to "pile up" in the direction of motion and be "stretched out" on the other side. (The numbered points indicate the location of the source at the instant each wave crest was emitted.) As a result, an observer situated in front of the source measures a shorter-than-normal wavelength—a blueshift—while an observer behind the source sees a redshift. In this diagram the source is shown in motion. However, the same general statements hold whenever there is any relative motion between source and observer.

gine, we could, in principle, witness the light change color, from bluer than normal to redder than normal, as the train passed by. (In practice, however, the color variation would be impossible to discern because trains travel far too slowly for us to perceive the effect.)

This motion-induced change in the observed wavelength of any wave—be it electromagnetic (light) or acoustical (sound)—is known as the **Doppler effect**, in honor of Christian Doppler, the nineteenth-century Austrian physicist who first explained it. Note that in our example of the spaceship, the *observer* was in motion, whereas, in the train example, it was the *source* of the wave that was moving. For electromagnetic radiation, the result is the same in either case—only the *relative* motion of source and observer matters.

To understand how the Doppler effect occurs, imagine a wave moving from the place where it is generated toward an observer who is not moving with respect to the wave source, as shown in Figure 3.17(a). By counting the number of wave crests passing per unit time, the observer could determine the wavelength of the emitted wave. Suppose now that the wave source begins to move. As illustrated in Figure 3.17(b), because the source moves between the times of

emission of one wave crest and the next, successive wave crests in the direction of motion of the source will be seen to be *closer together* than normal, whereas crests behind the source will be more widely spaced. An observer in front of the source will therefore measure a *shorter* wavelength than normal, while one behind will see a *longer* wavelength. (The numbers indicate successive wave crests emitted by the source and the location of the source at the instant each wave crest was emitted.)

The greater the relative speed of the source and the observer, the greater the observed change in wavelength. Furthermore, only motion *along* the line joining the source to the observer (called *radial* motion) contributes to the Doppler effect. Motion *across* the line of sight (known as *transverse* motion) has no effect on the perceived wavelength.* In terms of the net velocity of *recession* between source and observer (that is, the radial component of their relative velocity, with a positive value meaning that the two

---

*In fact, Einstein's Theory of Relativity (see Chapter 22) implies that when the transverse velocity is comparable to the speed of light, a wavelength change, called the transverse Doppler shift, does occur. For most terrestrial and astronomical applications, however, this shift is negligibly small, and we will ignore it here.

are moving apart, a negative value that they are approaching), the apparent wavelength (measured by the observer) is related to the true wavelength as follows:

$$\frac{\text{apparent wavelength}}{\text{true wavelength}} = 1 + \frac{\text{recession velocity}}{\text{wave speed}}.$$

In the case of electromagnetic radiation, the wave speed is the speed of light c. Because c is so large—300,000 km/s—the Doppler effect is *extremely* small for everyday terrestrial velocities. Even with its source receding at Earth's orbital speed of 30 km/s, a beam of blue light will be shifted by only 0.01 percent, from 400 nm to 400.04 nm—a very small change indeed, and one that the human eye cannot distinguish (but one that is easily detectable with modern instruments.)

The radiation measured by an observer situated in front of a moving source is said to be **blueshifted**, because blue light has a shorter wavelength than red light. Similarly, an observer situated behind the source will measure a longer-than-normal wavelength—the radiation is said to be **redshifted**. This terminology is used even for invisible radiation, for which "red" and "blue" have no meaning: any shift toward shorter wavelengths is called a blueshift, and any shift toward longer wavelengths is called a redshift. For example, ultraviolet radiation might be blueshifted into the X-ray part of the spectrum or redshifted into the visible, infrared radiation could be redshifted into the microwave range, and so on.

Astronomers can use the Doppler effect to measure the speed of any cosmic object along the line of sight simply by determining the extent to which its light is red- or blueshifted. The motions of nearby stars and distant galaxies—even the expansion of the universe itself—have all been measured in this way. Motorists stopped for speeding on the highway have experienced another, much more down-to-earth, application: police radar measures speed by means of the Doppler effect, as do the radar guns used to clock the velocity of a pitcher's fastball or a tennis player's serve. Notice, incidentally, that the Doppler effect depends only on the relative motion of source and observer; it does not depend on *distance* in any way.

In practice, it is hard to measure the Doppler shift of an entire blackbody curve, simply because it is widely spread over many wavelengths, making small shifts hard to determine with any accuracy. However, if the radiation were more narrowly defined and took up just a narrow "sliver" of the spectrum, then precise measurements of Doppler effect *could* be made. We will see in the next chapter that in many circumstances this is precisely what does happen, making the Doppler effect one of the observational astronomer's most powerful tools.

Chapter Review

SUMMARY

Visible light (p. 56) is a particular type of **electromagnetic radiation** (p. 56) and travels through space in the form of a **wave** (p. 57). A wave is characterized by the **wave period** (p. 57), the length of time taken for one complete cycle; the **wavelength** (p. 57), the distance between successive wave crests; and the wave **amplitude** (p. 58), which measures the size of the disturbance associated with the wave. The wave **frequency** (p. 58) is simply 1 divided by the wave period—it counts the number of wave crests that pass a given point in one second. **Diffraction** (p. 58) is the tendency of a wave to spread out after passing through an opening or to bend around a corner. **Interference** (p. 59) is the ability of two waves to reinforce or partially cancel each other.

Electrons (p. 59) and **protons** (p. 59) are elementary particles that carry equal and opposite electrical charges. Any electrically charged object is surrounded by an **electric field** (p. 60) that determines the force it exerts on other charged objects. Like gravitational fields, electric fields decrease as the square of the distance from their source.

When a charged particle moves, information about that motion is transmitted throughout the universe by the particle's changing electric field. The information travels in the form of a wave at the **speed of light** (p. 62). Both electric and **magnetic fields** (p. 60) are involved, so the phenomenon is known as **electromagnetism** (p. 60).

A beam of white light is bent, or refracted, as it passes through a prism. Different frequencies of light within the beam are refracted by different amounts, so the beam is split up into its component colors—the **visible spectrum** (p. 63). The color of visible light is simply a measure of its wavelength—red light has a longer wavelength than blue light. The entire **electromagnetic spectrum** (p. 63) consists of (in order of increasing frequency) **radio waves, infrared radiation, visible light, ultraviolet radiation, X rays**, and **gamma rays** (p. 56).

The **opacity** (p. 65) of Earth's atmosphere—the extent to which it absorbs radiation—varies greatly with wavelength. Only radio waves, some infrared wavelengths, and visible light can penetrate the atmosphere and reach the ground from space.

The **temperature** (p. 65) of an object is a measure of the speed with which its constituent particles move. The **intensity** (p. 65) of radiation of different frequencies emitted by a hot object has a characteristic distribution, called a **blackbody curve** (p. 66), that depends only on the temperature of the object. **Wien's law** (p. 67) tells us that the wavelength at which the object radiates most energy is inversely proportional to its temperature. **Stefan's law** (p. 67) states that the total amount of energy radiated is proportional to the fourth power of the temperature.

Our perception of the wavelength of a beam of light can be altered by our velocity relative to the source. This motion-induced change in the observed frequency of a wave is called the **Doppler effect** (p. 71). Any net motion away from the source causes a **redshift** (p. 72)—a shift to lower frequencies—in the received beam. Motion toward the source causes a **blueshift** (p. 72). The extent of the shift is directly proportional to the observer's radial velocity relative to the source.

SELF-TEST: TRUE OR FALSE?

_____ **1.** Light, radio, ultraviolet, and gamma rays are all forms of electromagnetic radiation.

_____ **2.** Sound is a familiar type of electromagnetic wave.

_____ **3.** The amount of diffraction increases with increasing wavelength.

_____ **4.** Interference occurs when one wave is brighter than another; the fainter wave cannot be observed.

_____ **5.** Electromagnetic waves cannot travel through a perfect vacuum.

_____ **6.** Electromagnetic waves all travel at the same speed, the speed of light.

_____ **7.** Visible light makes up the greatest part of the entire electromagnetic spectrum.

_____ **8.** Ultraviolet light has the shortest wavelength of any electromagnetic wave.

_____ **9.** A blackbody emits all its radiation at one wavelength or frequency.

_____ **10.** A perfect blackbody emits exactly as much radiation as it absorbs from outside.

_____ **11.** The shape of a blackbody curve depends on the temperature of the body.

_____ **12.** The frequency at which the blackbody curve peaks increases with temperature.

_____ **13.** Objects moving away from an observer are redshifted because they actually turn red.

_____ **14.** The Doppler effect occurs for all types of wave motion.

_____ **15.** An object emitting radiation, moving transverse to the line of sight, produces no Doppler effect.

SELF-TEST: FILL IN THE BLANK

1. The speed of light is _____ km/s.
2. The _____ of a wave is the distance between any two adjacent wave crests.
3. The _____ of a wave is measured in units of hertz (Hz).
4. _____ is the ability of a wave to "bend around corners."
5. When a charged particle moves, information about this motion is transmitted through space by means of its changing _____ and _____ fields.
6. The visible spectrum ranges from _____ to _____ in wavelength.
7. Light with a wavelength of 700 nm is perceived to be _____ in color.
8. Earth's atmosphere has low opacity for three forms of electromagnetic radiation. They are _____, _____, and _____.
9. The peak of an object's emitted radiation occurs at a frequency or wavelength determined by the object's _____.
10. The lowest possible temperature is _____ K.
11. Water freezes at _____ K.
12. Because the Sun emits its peak amount of radiation at about 480 nm, its temperature must be about _____ K.
13. Two identical objects have temperatures of 1000 K and 1200 K, respectively. It is observed that one of the objects emits twice as much radiation as the other. Which one is it? _____.
14. If an astronomical object is observed to emit X rays, it is reasonable to assume its temperature is very _____.
15. When an observer and/or an object emitting radiation move toward each other, the observer sees the radiation shifted to _____ wavelengths.

REVIEW AND DISCUSSION

1. Define the following wave properties: period, wavelength, amplitude, frequency.
2. What is the relationship between wavelength, wave frequency, and wave velocity?
3. What is diffraction, and how does it relate to the behavior of light as a wave?
4. What's so special about c?
5. Compare and contrast the gravitational force with the electric force.
6. Describe the way in which light radiation leaves a star, travels through the vacuum of space, and finally is seen by someone on Earth.
7. Name the colors that combine to make white light. What is it about the various colors that causes us to perceive them differently?
8. What do radio waves, infrared radiation, visible light, ultraviolet radiation, X rays, and gamma rays have in common? How do they differ?
9. In what regions of the electromagnetic spectrum is the atmosphere transparent enough to allow observations from the ground?
10. What is a blackbody? What are the characteristics of the radiation emitted by a blackbody?
11. What does Wien's law reveal about stars in the sky?
12. What does Stefan's law tell us about the radiation emitted by a blackbody?
13. In terms of its blackbody curve, describe what happens as a red-hot glowing coal cools off.
14. What is the Doppler effect, and how does it alter the way in which we perceive radiation?
15. If Earth were completely blanketed with clouds and we couldn't see the sky, could we learn about the realm beyond the clouds? What other forms of radiation might be received?

PROBLEMS

1. A sound wave moving through water has a frequency of 256 Hz and a wavelength of 5.77 m. What is the speed of sound in water?
2. What is the wavelength of a 100-MHz ("FM 100") radio signal?
3. What would be the frequency of an electromagnetic wave having a wavelength equal to Earth's diameter? In what part of the electromagnetic spectrum would such a wave lie?
4. According to Wien's law, how many times hotter is an object whose blackbody emission spectrum peaks in the ultraviolet, at a wavelength of 200 nm, than an object whose spectrum peaks in the red, at 650 nm? According to Stefan's law, how much more energy does it radiate per unit area per second?
5. Normal human body temperature is about 37°C. What is this temperature in kelvins? What is the peak wavelength emitted by a person with this temperature? In what part of the spectrum does this lie?
6. The Sun has a temperature of 5800 K, and its blackbody emission peaks at a wavelength of approximately 550 nm. At what wavelength does a protostar with a temperature of 1000 K radiate most strongly?
7. Two otherwise identical bodies have temperatures of 300 K and 1500 K, respectively. Which one radiates more energy, and by what factor does its emission exceed the emission of the other body?

8. According to Stefan's law, how much energy is radiated into space per unit time by each square meter of the Sun's surface (see *More Precisely 3-2*)? If the Sun's radius is 696,000 km, what is the total power output of the Sun?

9. At what radial velocity, and in what direction, would a spacecraft have to be moving for a radio station transmitting at 100 MHz to be picked up by a radio tuned to 99.9 MHz?

10. Imagine you are observing a spacecraft moving in a circular orbit of radius 100,000 km around a distant planet. You happen to be located in the plane of the spacecraft's orbit. You find that the spacecraft's radio signal varies periodically in wavelength between 2.99964 m and 3.00036 m. Assuming that the radio is broadcasting normally, at a constant wavelength, what is the mass of the planet?

PROJECTS

1. Locate the constellation Orion. Its two brightest stars are Betelgeuse and Rigel. Which of these is the hotter star? Which is cooler? How can you tell? Which of the other stars scattered across the night sky are hot, and which are cool?

2. Stand near (but not too near!) a train track or busy highway and wait for a train or traffic to pass by. Can you notice the Doppler effect in the pitch of the engine noise or whistle blowing? How does the sound frequency depend on (a) speed? (b) the motion toward or away from you?

4 SPECTROSCOPY

The Inner Workings of Atoms

LEARNING GOALS

Studying this chapter will enable you to:

1 Describe the characteristics of continuous, emission, and absorption spectra and the conditions under which each is produced.

2 Explain the relation between emission and absorption lines and what we can learn from these lines.

3 Discuss the observations that led scientists to conclude that light has particle as well as wave properties.

4 Specify the basic components of the atom and describe our modern conception of its structure.

5 Explain how electron transitions within atoms produce unique emission and absorption features in the spectra of those atoms.

6 Describe the general features of the spectra produced by molecules.

7 List and explain the kinds of information that can be obtained by analyzing the spectra of astronomical objects.

(Opposite page, background) When a prism is placed in the path of light captured by a telescope, the resulting photograph can look almost psychedelic. Each of the stars and other sources of radiation in the picture has had its light split into its component colors, from red to violet. Here, in this strange image, we see a star-forming region (called the Eta Carinae nebula) containing many stars and much loose gas—all of whose light has been colorfully dispersed.

(Inset A) This is the same celestial object, now photographed without the prism. The result is Eta Carinae in its true colors—mostly red, due to vast quantities of hydrogen gas spread across the 900-square-light-year area of this photo.

(Inset B) Looking more carefully at spectra from individual stars, we often see two things: a colorful continuous spectrum extending from red to violet, and a series of thin, dark lines across the spectrum. This is the spectrum of the brightest star in the sky, Sirius A.

(Inset C) This is the slightly different spectrum of Vega, the bright star in the constellation Lyra.

(Inset D) Examining spectra of stars even more closely, we find ever more dark lines. Here, in this spectrum of the star Arcturus, we can see many such lines in the yellow region alone. The tracing below records precise details about these lines that would be difficult or impossible to obtain just by looking at the spectrum. Spectra like these, which tell astronomers virtually all that we know about stars, are the subject of this chapter.

In the previous chapter we saw how light behaves as a continuous wave and how this description of electromagnetic radiation allows us to begin to decipher the information reaching us from the cosmos in the form of visible and invisible light. However, early in the twentieth century, it became clear that the wave theory of electromagnetic phenomena was incomplete—some aspects of light simply could not be explained purely in wave terms. When radiation interacts with matter on atomic scales, it does so not as a continuous wave but in a jerky, discontinuous way—in fact, as a particle. With this discovery, scientists quickly realized that atoms, too, must behave in a discontinuous way, and the stage was set for a scientific revolution that has affected virtually every area of modern life. In astronomy, the observational and theoretical techniques that enable researchers to determine the nature of distant atoms by the way they emit and absorb radiation are now the indispensable foundation of modern astrophysics.

4.1 Spectral Lines

1 In Chapter 3 we saw something of how astronomers can analyze electromagnetic radiation received from space to obtain information about distant objects. A vital step in this process is the formation of a *spectrum*—splitting the incoming radiation into its component wavelengths. But in reality, *no* cosmic object emits a perfect blackbody spectrum like those discussed earlier. ∞ (Sec. 3.4) All spectra deviate from this idealized form—some by only a little, others by a lot. Far from invalidating our earlier studies, however, these deviations contain a wealth of detailed information about physical conditions in the source of the radiation. Because spectra are so important, let's examine in more detail how astronomers obtain and interpret them.

Radiation can be analyzed with an instrument known as a **spectroscope.** In its most basic form, this device consists of an opaque barrier with a slit in it (to define a beam of light), a prism (to split the beam into its component colors), and an eyepiece or screen (to allow the user to view the resulting spectrum). Figure 4.1 shows such an arrangement.

The research instruments (called *spectrographs*, or *spectrometers*) used by professional astronomers are rather more complex, consisting of a telescope (to capture the radiation), a dispersing device (to spread it out into a spectrum), and a detector (to record the result). Despite their greater sophistication, however, their basic operation is conceptually similar to the simple spectroscope shown in the figure.

In many large instruments the prism is replaced by a device called a *diffraction grating*, consisting of a sheet of transparent material with many closely spaced parallel lines ruled on it. The spaces between the lines act as many tiny openings, and light is diffracted as it passes through them. Because different wavelengths of electromagnetic radiation are diffracted by different amounts as they pass through a narrow gap, the effect of the grating is to split a beam of light into its component colors. ∞ (Sec. 3.1)

EMISSION LINES

The spectra we encountered in Chapter 3 are examples of **continuous spectra.** A light bulb, for example, emits radiation of all wavelengths (mostly in the visible range), with an in-

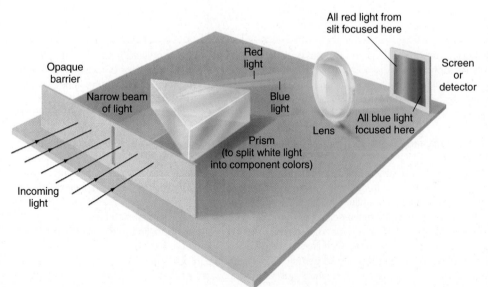

Labels: Opaque barrier · Narrow beam of light · Incoming light · Prism (to split white light into component colors) · Red light · Blue light · Lens · All red light from slit focused here · All blue light focused here · Screen or detector

Figure 4.1   Diagram of a simple spectroscope. A small slit in the mask on the left allows a narrow beam of light to pass. The light passes through a prism and is split up into its component colors. The resulting spectrum can be viewed through an eyepiece or simply projected onto a screen.

tensity distribution that is well described by the blackbody curve corresponding to the bulb's temperature. ∞ (Sec. 3.4) Viewed through a spectroscope, the spectrum of the light from the bulb would show the familiar rainbow of colors, from red to violet, without interruption, as presented in Figure 4.2(a).

Not all spectra are continuous, however. For instance, if we took a glass jar containing pure hydrogen gas and passed an electrical discharge through it (a little like a lightning bolt arcing through Earth's atmosphere), the gas would begin to glow—that is, it would emit radiation. If we were to examine that radiation with our spectroscope, we would find that its spectrum consists of only a few bright lines on an otherwise dark background, quite unlike the continuous spectrum described for the incandescent light bulb. Figure 4.2(b) shows this schematically. A more detailed rendering of the spectrum of hydrogen appears in the top panel of Figure 4.3. The light produced by the hydrogen in this experiment does *not* consist of all possible colors but instead includes only a few narrow, well-defined **emission lines**—narrow "slices" of the continuous spectrum. The black background represents all the wavelengths *not* emitted by hydrogen.

After further experimentation we would also find that although we could alter the *intensity* of the lines (for example, by changing the amount of hydrogen in the jar or the strength of the electrical discharge), we could not alter their *color* (in other words, their frequency or wavelength). The pattern of spectral emission lines is a property of the element hydrogen. Whenever we perform this experiment, the same characteristic colors result.

By the early nineteenth century, scientists had carried out similar experiments on many different gases. By vaporizing solids and liquids in a flame, they extended their inquiries to include materials that are not normally found in the gaseous state. Sometimes the pattern of lines was fairly simple, sometimes it was very complex. Always, though, it was *unique* to that element. Even though the origin of the lines was a mystery, scientists quickly realized that the lines provided a one-of-a-kind "fingerprint" of the substance under investigation. Scientists could deduce the presence of a particular atom or molecule (a group of atoms held together by chemical bonds—see Sec. 4.4) solely through the study of the light it emitted.

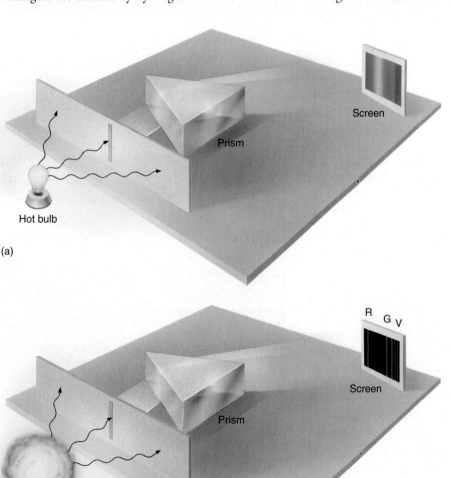

Figure 4.2 When passed through a slit and split up by a prism, light from a source of continuous radiation (a) gives rise to the familiar rainbow of colors. By contrast, the light from excited hydrogen gas (b) consists of a series of distinct spectral lines.

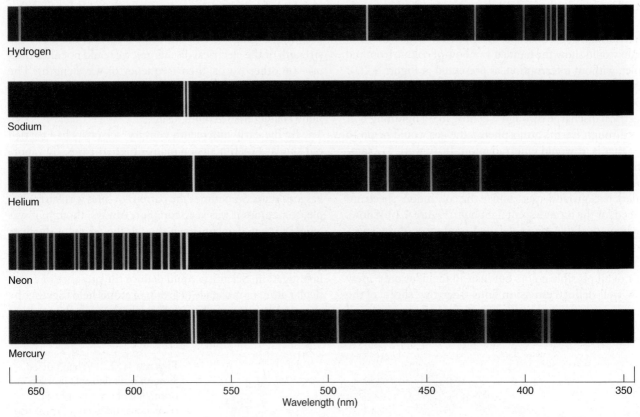

Hydrogen

Sodium

Helium

Neon

Mercury

650　600　550　500　450　400　350
Wavelength (nm)

Figure 4.3　The emission spectra of some well-known elements.

Scientists have accumulated extensive catalogs of the specific wavelengths at which many different hot gases emit radiation. For gas of a given chemical composition, the particular pattern of the light it emits is known as its **emission spectrum.** Examples of the emission spectra of some common substances are shown in Figure 4.3.

ABSORPTION LINES

When sunlight is split by a prism, at first glance it appears to produce a continuous spectrum. However, closer scrutiny with a spectroscope shows that the solar spectrum is interrupted by a large number of narrow dark lines, as shown in Figure 4.4. We now know that many of these lines represent wavelengths of light that have been removed (absorbed) by gases present either in the outer layers of the Sun or in Earth's atmosphere. These gaps in the spectrum are called **absorption lines.**

The English astronomer William Wollaston first noticed the solar absorption lines in 1802. They were studied

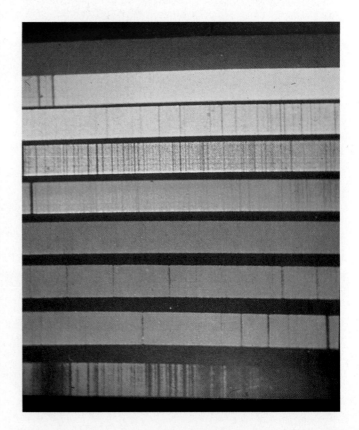

Figure 4.4　This visible spectrum of the Sun shows hundreds of dark absorption lines superimposed on a bright continuous spectrum. Here, the scale extends from long wavelengths (red) at the upper left to short wavelengths (blue) at the lower right.

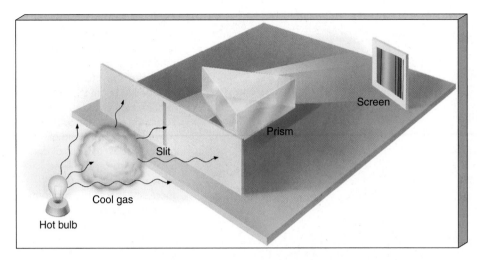

Figure 4.5 When cool gas is placed between a source of continuous radiation, such as a hot light bulb, and a detector, the resulting spectrum consists of a continuous spectrum crossed by a series of dark absorption lines. These lines are formed when the intervening gas absorbs certain wavelengths (colors) from the original beam. The absorption lines appear at precisely the same wavelengths as the emission lines that would be produced if the gas was heated to high temperatures.

in greater detail about 10 years later by the German physicist Joseph Fraunhofer, who measured and cataloged over 600 of them. They are now referred to collectively as **Fraunhofer lines.** Although the Sun is by far the easiest star to study, and so has the most extensive set of observed absorption lines, similar lines are known to exist in the spectra of all stars.

At around the same time as the solar absorption lines were discovered, scientists found that absorption lines could also be produced in the laboratory by passing a beam of light from a source that produces a continuous spectrum through a cool gas, as shown in Figure 4.5. They quickly observed an intriguing connection between emission and absorption lines: the absorption lines associated with a given gas occur at precisely the *same* wavelengths as the emission lines produced when the gas is heated.

As an example, consider the element sodium, whose emission spectrum appears in Figure 4.3. When heated to high temperatures, a sample of sodium vapor emits visible light strongly at just two wavelengths—589.9 nm and 589.6 nm—lying in the yellow part of the spectrum. When a continuous spectrum is passed through some relatively cool sodium vapor, two sharp, dark absorption lines appear at precisely the same wavelengths. The emission and absorption spectra of sodium are compared in Figure 4.6, clearly showing the relation between emission and absorption features.

KIRCHHOFF'S LAWS

2 The analysis of the ways in which matter emits and absorbs radiation is called **spectroscopy.** One early spectroscopist, the German physicist Gustav Kirchhoff, summarized the observed relationships among the three types of spectra—continuous, emission line, and absorption line—in 1859. He formulated three spectroscopic rules, now known as **Kirchhoff's laws,** governing the formation of spectra:

1. A luminous solid or liquid, or a sufficiently dense gas, emits light of all wavelengths and so produces a *continuous spectrum* of radiation.

2. A low-density hot gas emits light whose spectrum consists of a series of bright *emission lines*. These lines are characteristic of the chemical composition of the gas.

3. A cool thin gas absorbs certain wavelengths from a continuous spectrum, leaving *dark absorption lines* in their place superimposed on the continuous spectrum. Once again, these lines are characteristic of the composition of the intervening gas—they occur at precisely the same wavelengths as the emission lines produced by that gas at higher temperatures.

Figure 4.7 illustrates Kirchhoff's laws and the relationship between absorption and emission lines. When viewed di-

Figure 4.6 (a) The characteristic emission lines of sodium. The two bright lines in the center appear in the yellow part of the spectrum. (b) The absorption spectrum of sodium. The two dark lines appear at exactly the same wavelengths as the bright lines in the sodium emission spectrum.

Figure 4.7 A source of continuous radiation, here represented by a light bulb, is used to illustrate Kirchhoff's laws of spectroscopy. (a) The unimpeded beam shows the familiar continuous spectrum of colors. (b) When the source is viewed through a cloud of hydrogen gas, a series of dark hydrogen absorption lines appears in the continuous spectrum. These lines are formed when the gas absorbs some of the bulb's radiation and reemits it in random directions. Because most of the reemitted radiation does not go through the slit, the effect is to remove the absorbed radiation from the light that reaches the screen at left. (c) When the gas is viewed from the side, a fainter hydrogen emission spectrum is seen, consisting of reemitted radiation. The absorption lines in (b) and the emission lines in (c) have the same wavelengths.

rectly, the light source, a hot solid (the filament of the bulb), has a continuous (blackbody) spectrum. When the light source is viewed through a cloud of cool hydrogen gas, a series of dark absorption lines appear, superimposed on the spectrum at wavelengths characteristic of hydrogen. The lines appear because the light at those wavelengths is absorbed by the hydrogen. As we will see later in this chapter, the absorbed energy is subsequently reradiated into space, but in all directions, not just the original direction of the beam. Consequently, when the cloud is viewed from the side against an otherwise dark background, a series of faint emission lines is seen. These lines contain the energy lost by the forward beam. If the gas was heated to incandescence, it would produce stronger emission lines at precisely the same wavelengths.

ASTRONOMICAL APPLICATIONS

By the late nineteenth century, spectroscopists had developed a formidable arsenal of techniques for interpreting the radiation received from space. Once astronomers knew that spectral lines were indicators of chemical composition, they set about identifying the observed lines in the solar spectrum. Almost all the lines in light from extraterrestrial sources could be attributed to known elements (for example, many of the Fraunhofer lines in sunlight are associated with the element iron). However, some new lines also appeared in the solar spectrum. In 1868, astronomers

realized that those lines must correspond to a previously unknown element. It was given the name helium, after the Greek word *helios*, meaning "Sun." Only in 1895, almost three decades after its detection in sunlight, was helium discovered on Earth. (A laboratory spectrum of helium is part of Figure 4.3.)

Yet for all the information that nineteenth-century astronomers could extract from observations of stellar spectra, they still lacked a theory explaining how the spectra themselves arose. Despite their sophisticated spectroscopic equipment, they knew scarcely any more about the physics of stars than did Galileo or Newton. To understand how spectroscopy can be used to extract detailed information about astronomical objects from the light they emit, we must delve more deeply into the processes that produce line spectra.

4.2 The Formation of Spectral Lines

By the start of the twentieth century, physicists had accumulated evidence that light sometimes behaves in a manner that cannot be explained by the wave theory. As we have just seen, the production of absorption and emission lines involves only certain very specific frequencies or wavelengths of light. This would not be expected if light behaved like a continuous wave and matter always obeyed the laws of Newtonian mechanics. Other experiments conducted around the same time strengthened the conclusion that the notion of radiation as a wave was *incomplete*. It became clear that when light interacts with matter on very small scales, it does so not in a continuous way but in a discontinuous, stepwise manner. The challenge was to find an explanation for this unexpected behavior. The eventual solution revolutionized our view of nature and now forms the foundation not just for physics and astronomy but for virtually all modern science.

ATOMIC STRUCTURE

4 To explain the formation of emission and absorption lines, we must understand not just the nature of light but also something of the structure of **atoms**—the microscopic building blocks from which all matter is constructed. Let us start with the simplest atom of all—hydrogen. A hydrogen atom consists of an *electron*, with a negative electrical charge, orbiting a *proton*, which carries a positive charge. The proton forms the central **nucleus** (plural: *nuclei*) of the atom. The hydrogen atom as a whole is electrically neutral. The equal and opposite charges of the proton and the orbiting electron produce an electrical attraction that binds them together within the atom.

How does this picture of the hydrogen atom relate to the characteristic emission and absorption lines associated with hydrogen gas? If an atom emits some energy in the form of radiation, that energy has to come from somewhere within the atom. Similarly, if energy is absorbed, it must cause some internal change. It is reasonable (and correct) to suppose that the energy emitted or absorbed by the atom is associated with changes in the motion of the orbiting electron.

The first theory of the atom to provide an explanation of hydrogen's observed spectral lines was propounded by the Danish physicist Niels Bohr. This theory is now known simply as the **Bohr model** of the atom. Its essential features are as follows. First, there is a state of lowest energy—the **ground state**—which represents the "normal" condition of the electron as it orbits the nucleus. Second, there is a maximum energy that the electron can have and still be part of the atom. Once the electron acquires more than that maximum energy, it is no longer bound to the nucleus, and the atom is said to be *ionized*; an atom missing one or more of its electrons is called an **ion.** Third, and most important (and also least intuitive), between those two energy levels, the electron can exist only in certain sharply defined energy states, often referred to as *orbitals.*

This description of the atom contrasts sharply with the predictions of Newtonian mechanics, which would permit orbits with *any* energy, not just at certain specific values. In the atomic realm such discontinuous behavior is the norm. In the jargon of the field the orbital energies are said to be **quantized.** The rules of *quantum mechanics*, the branch of physics governing the behavior of atoms and subatomic particles, are far removed from everyday experience.

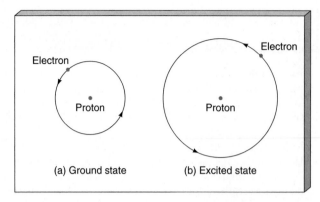

Figure 4.8 An early conception of the hydrogen atom pictured its electron orbiting the central proton in a well-defined orbit, rather like a planet orbiting the Sun. Two electron orbits of different energies are shown. The left-hand figure represents the ground state, the right-hand figure an excited state.

In Bohr's model each electron orbital was pictured as having a specific radius, much like a planetary orbit in the solar system, as shown in Figure 4.8. However, the modern view is not so simple. Although each orbital *does* have a precise energy, the electron is now envisioned as being smeared out in an "electron cloud" surrounding the nucleus, as illustrated in Figure 4.9. It is common to speak of the *mean* (average) distance from the cloud to the nucleus as the "radius" of the electron's orbit. When a hydrogen atom is in its ground state, the radius of the orbit is about 0.05 nm (0.5 Å). As the orbital energy increases, the radius increases, too. For the sake of clarity in the diagrams that follow, we will represent electron orbitals as solid lines, but bear in mind always that Figure 4.9 is a more accurate depiction of reality.

Atoms do not always remain in their ground state. An atom is said to be in an **excited state** when an electron occupies an orbital at a greater than normal distance from its parent

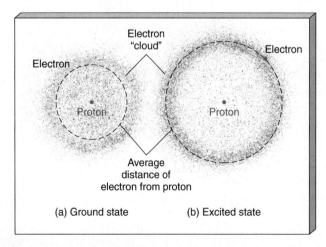

Figure 4.9 The modern view of the hydrogen atom sees the electron as a "cloud" surrounding the nucleus. The same two energy states are shown as in Figure 4.8.

MORE PRECISELY 4-1 The Photoelectric Effect

Einstein developed his insight into the nature of radiation partly as a means of explaining a puzzling experimental result known as the *photoelectric effect*. This effect can be demonstrated by shining a beam of light on a metal surface (as shown in the accompanying figure). When high-frequency ultraviolet light is used, bursts of electrons are dislodged from the surface by the beam, much like when one billiard ball hits another, knocking it off the table. Curiously, the speed with which the particles are ejected from the metal is found to depend only on the *color* of the light, not on its intensity. For lower-frequency light—blue, say—an electron detector still records bursts of electrons, but now their speeds, and hence their *energies*, are less. For even lower frequencies—red or infrared light—*no* electrons are kicked out of the metal surface at all.

These results were very difficult to reconcile with a wave model of light. Einstein realized that the only way to explain the presence of an abrupt cutoff at the detector—that is, that the detector registered nothing when the frequency of the incoming radiation dropped below a certain level—and the increase in electron speed with light frequency was to envision radiation as traveling as "bullets," or particles, or *photons*. To account for the experimental findings, the energy of any photon had to be proportional to the *frequency* of the radiation:

photon energy ∝ radiation frequency.

If we suppose that some minimum amount of energy is needed just to "unglue" the electrons from the metal, then we can see why no electrons are emitted below some critical frequency—red photons just don't carry enough energy. Above the critical frequency, photons have enough energy to dislodge the electrons; moreover, any energy they possess above the necessary minimum is imparted to the electrons as kinetic energy, the energy of motion. Thus, as the fre-

quency of the radiation increases, so too does the photon energy, and hence the speed of the electrons that they liberate from the metal.

The constant of proportionality in the foregoing relation is now known as *Planck's constant*, in honor of the German physicist Max Planck, who first determined its numerical value. It is always denoted by the symbol h. The equation relating photon energy E to radiation frequency f is thus

$$E = hf.$$

As noted in Chapter 3, the SI unit of energy is the *joule* (J). ∞ (*More Precisely 3-2*) The value of Planck's constant is a very small number—6.63×10^{-34} joule seconds (J·s). Consequently, the energy of a single photon is tiny—even a very-high-frequency gamma ray (the most energetic type of electromagnetic radiation) with a frequency of 10^{22} Hz has an energy of only 7×10^{-12} joule—about the same energy as is carried by a flying gnat. Nevertheless, this energy is more than enough to damage a living cell. The basic reason that gamma rays are so much more dangerous to life than, say, visible light, is that each gamma-ray photon carries billions of times more energy than a photon of visible radiation.

The realization and acceptance of the fact that light can behave both as a wave and as a particle is another example of the scientific method at work. Despite the enormous success of the wave theory of radiation in the nineteenth century, the experimental evidence led scientists to the inevitable conclusion that the theory was incomplete—it had to be modified to allow for the fact that light sometimes exhibits particle characteristics. In addition to bringing about the birth of a whole new branch of physics—the field of quantum mechanics—this new theory radically changed the way physicists view light and all other forms of radiation.

nucleus. An atom in such an excited state has a greater than normal amount of energy. The excited state with the lowest energy (that is, the one closest in energy to the ground state) is called the *first excited state*, that with the second-lowest energy the *second excited state*, and so on. An atom can become excited in one of two ways: by absorbing some energy from a source of electromagnetic radiation or by colliding with some other particle—another atom, for example. However, the electron cannot stay in a higher orbital forever; the ground state is the only level where it can remain indefinitely. After about 10^{-8} s, an excited atom returns to its ground state.

THE PARTICLE NATURE OF RADIATION

3 Here now is the crucial point that links atoms to radiation and allows us to interpret emission and absorption spectra. Because electrons may exist only in orbitals having specific energies, atoms can absorb only specific amounts of energy as their electrons are boosted into excited states. Likewise, they can emit only specific amounts of energy as their electrons fall back to lower energy states. Thus, the amount of light energy absorbed or emitted in these processes *must correspond precisely to the energy difference between two orbitals.* The quantized nature of the atom's energy levels requires that light must be absorbed and emitted in the form of little "packets" of electromagnetic radiation, each carrying a very specific amount of energy. We call these packets **photons.** A photon is, in effect, a "particle" of electromagnetic radiation.

The idea that light sometimes behaves not as a continuous wave but as a stream of particles was first proposed by Albert Einstein in 1905 in order to explain a number of puzzling experimental results (especially the *photoelectric effect*—see *More Precisely 4-1*). Further, Einstein was able to quantify the relationship between the two aspects of light's double nature. He found that the energy carried by a photon is proportional to the *frequency* of the radiation. Thus, for example, a "red" photon having a frequency of 4×10^{14} Hz (corresponding to a wavelength of approximately 750 nm, or 7500 Å) has 4/7 the energy of a "blue" photon of frequency of 7×10^{14} Hz (and a wavelength of 400 nm, or 4000 Å).

Many people are confused by the idea that light can behave in two such different ways. To be truthful, modern physicists don't yet fully understand *why* nature displays this wave-particle duality. Nevertheless, there is irrefutable experimental evidence for both of these aspects of radiation. Environmental conditions ultimately determine which description—wave or stream of particles—better fits the behavior of electromagnetic radiation. As a general rule of thumb, in the macroscopic realm of everyday experience, radiation is more usefully described as a wave, whereas in the microscopic domain of atoms it is best characterized as a series of particles.

The equivalence between photon energy and photon frequency, or wavelength, completes the connection between atomic structure and atomic spectra. Atoms absorb and emit radiation at characteristic wavelengths determined by their own particular internal structure. Because this structure is *unique* to each element, the colors of the absorbed and emitted photons—that is, the spectral lines we observe—are characteristic of that element *and only that element.*

THE SPECTRUM OF HYDROGEN

Absorption and emission of photons by a hydrogen atom are illustrated schematically in Figure 4.10. Figure 4.10(a) shows the atom absorbing a photon and making a transition from the ground state to the first excited state, then emitting a photon of precisely the same energy and dropping back to the ground state. The energy difference between the two states corresponds to an *ultraviolet* photon, of wavelength 121.6 nm (1216 Å).

Absorption may also boost an electron into an excited state higher than the first excited state. Figure 4.10(b) depicts the absorption of a more energetic (higher-frequency, shorter-wavelength) ultraviolet photon, this one having a wavelength of 102.6 nm (1026 Å). Absorption of this photon causes the atom to jump to the *second* excited state. As before, the atom returns rapidly to the ground state, but this time it can do so in one of two possible ways:

1. It can proceed directly back to the ground state, in the process emitting an ultraviolet photon identical to the one that excited the atom in the first place.

2. Alternatively, the electron can *cascade* down one orbital at a time. If this occurs, the atom will emit *two* photons: one with an energy equal to the difference between the second and first excited states, and the other with an energy equal to the difference between the first excited state and the ground state.

The second step of this cascade process produces a 121.6-nm ultraviolet photon, just as in Figure 4.10(a). However, the first transition—the one from the second to the first excited state—produces a photon with a wavelength of 656.3 nm (6563 Å), which is in the visible part of the spectrum. This photon is seen as red light. An individual atom—if one could be isolated—would emit a momentary red flash. This is the origin of the red line (often called the Hα line—see *More Precisely 4-2*) in the hydrogen spectrum shown in Figure 4.3. The inset in Figure 4.10 shows an astronomical object whose red coloration is the result of precisely this process. As ultraviolet photons from a young, hot star pass through the surrounding cool hydrogen gas out of which the star recently formed, some photons are absorbed by the gas, boosting its atoms into excited states. The 656.3-nm red glow characteristic of excited hydrogen gas results as the atoms cascade back to their ground states.

Absorption of additional energy can boost the electron to even higher orbitals within the atom. As the excited electron cascades back down to the ground state, the atom may emit many photons, each with a different energy and hence a different wavelength, and the resulting spectrum shows

Figure 4.10 (a) Diagram of a photon being absorbed by a hydrogen atom (left), causing the momentary excitation of that atom (center) into its first excited state. Eventually, the atom returns to its ground state, accompanied by the emission of a photon of the same energy as the original photon (right). (b) Absorption of a higher-energy photon may also boost the atom into a higher excited state, from which there may be several possible paths back to the ground state. (The sharp lines used for the orbitals here and in similar figures that follow are intended merely as a schematic representation of the electron energy levels and are not meant to be taken literally. In actuality, electron orbitals are "clouds," as shown in Figure 4.9.) As ultraviolet photons from a hot star pass through surrounding hydrogen gas, many are absorbed by the gas, boosting its atoms into excited states. Electrons in the second excited state can fall to the first excited state on their way back to the ground state (the upper path in part b). This transition produces radiation in the visible region of the spectrum—the 656.3-nm red glow that is characteristic of excited hydrogen gas. The object shown in the inset, designated NGC 2440, is an emission nebula: an interstellar cloud consisting largely of hydrogen gas excited by an extremely hot star (the white dot in the center).

many spectral lines. In a sample of heated hydrogen gas, at any instant, atomic collisions ensure that atoms are found in many different excited states; the complete emission spectrum therefore consists of wavelengths corresponding to all possible transitions between those states and states of lower energy. In the case of hydrogen, it so happens that transitions from higher states back to the first excited state produce all the spectral lines in the visible range (Figure 4.3). As we have just seen, transitions ending at the ground state produce ultraviolet photons. The energy levels and spectrum of hydrogen are discussed in more detail in *More Precisely 4-2*.

KIRCHHOFF'S LAWS EXPLAINED

5 Let's reconsider our earlier discussion of emission and absorption lines in terms of the model just presented. In Figure 4.7 a beam of continuous radiation shines through a cloud of hydrogen gas. The beam contains photons of all energies, but most of them cannot interact with the gas— the gas can absorb only those photons having just the right energy to cause a change in an electron's orbit from one state to another. All other photons in the beam—with energies that cannot produce a transition—do not interact with the gas at all but pass through it unhindered. Photons having the right energies are absorbed, excite the gas, and are removed from the beam. This is the cause of the dark absorption lines in the spectrum of Figure 4.7(b). These lines are direct indicators of the energy differences between orbitals in the atoms making up the gas.

The excited gas atoms return rapidly to their original states, each emitting one or more photons in the process. We might think, then, that although some photons from the beam are absorbed by the gas, they are quickly replaced by reemitted photons, with the result that we could never observe the effects of absorption. This is not the case, however, for two important reasons. First, while the photons not absorbed by the gas continue on directly to the detector, the reemitted photons can leave in *any* direction. Most of the reemitted photons leave at angles that do not take them through the slit and on to the detector, and so they are effectively lost from the original beam. Second, as we have just seen, electrons can cascade back to the ground state, emitting several lower-energy photons instead of a single photon equal in energy to the one originally absorbed.

The net result of these processes is that some of the original energy is channeled into photons of many different colors, moving in many different directions. A second detector looking at the cloud from the side would record the reemitted energy as an emission spectrum, as in Figure 4.7(c). (A spectrum of the object shown in the inset of Figure 4.10, called an *emission nebula*, would show the same thing.) Like the absorption spectrum, the emission spectrum is characteristic of the gas, not of the original beam.

Absorption and emission spectra are created by the same atomic processes. They correspond to the same atomic transitions. They contain the same information about the composition of the gas cloud. In the laboratory we can move our detector and can measure both. In astronomy we are not able to change our vantage point, so the type of spectrum we see depends on our chance location with respect to both the source and the intervening gas cloud.

MORE COMPLEX SPECTRA

All hydrogen atoms have basically the same structure—a single electron orbiting a single proton—but, of course, there are many other kinds of atoms, each kind having a unique internal structure. The number of protons in the nucleus of an atom determines the **element** that it represents. That is, just as all hydrogen atoms have a single proton, all oxygen atoms have 8 protons, all iron atoms have 26 protons, and so on.

The next simplest element after hydrogen is helium. The central nucleus of the most common form of helium is made up of two protons and two **neutrons** (another kind of elementary particle having a mass slightly larger than that of a proton but having no electrical charge at all). About this nucleus orbit two electrons. As with hydrogen and all other atoms, the "normal" condition for helium is to be electrically neutral, with the negative charge of the orbiting electrons exactly canceling the positive charge of the nucleus (Figure 4.11).

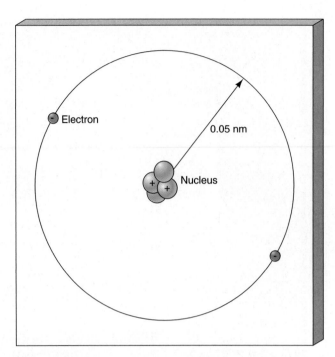

Figure 4.11   A helium atom in its normal, ground state. Two electrons occupy the lowest-energy orbital around a nucleus containing two protons and two neutrons.

More complex atoms contain more protons (and neutrons) in the nucleus and have correspondingly more orbiting electrons. For example, an atom of carbon, shown in Figure 4.12, consists of six electrons orbiting a nucleus containing six protons and six neutrons. As we progress to heavier and heavier elements, the number of orbiting electrons

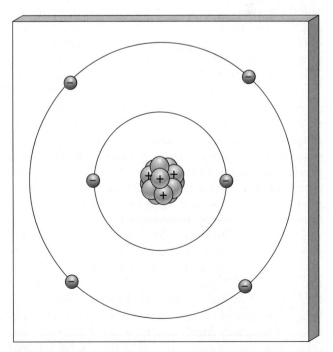

Figure 4.12   A carbon atom in its normal, ground state. Six electrons orbit a six-proton, six-neutron nucleus, two in an inner orbital, the other four at a greater distance from the center.

Figure 4.13 The visible spectrum of the hot gases in a nearby star-forming region known as the Omega nebula (M17). Shining by the light of several very hot stars, the nebula produces a complex spectrum of bright and dark lines (bottom), also shown here as an intensity trace from red to blue (center).

increases, and the number of possible electronic transitions rises rapidly. The result is that very complicated spectra can be produced. The complexity of atomic spectra generally reflects the complexity of the atoms themselves. A good example is the element iron, which contributes several hundred of the Fraunhofer absorption lines seen in the solar spectrum (see Figure 4.4). The many possible transitions of its 26 orbiting electrons yield an extremely rich line spectrum.

Even more complex spectra can occur when many different gases are mixed together. Here the power of spectroscopy is most apparent, as it enables us to study one kind of atom to the exclusion of all others, simply by focusing on specific wavelengths of radiation. For example, a cool intervening gas cloud containing many elements will produce a very complicated absorption spectrum in the light received from a background continuous source. Nevertheless, by identifying the (superimposed) absorption spectra of many different atoms, we can determine the cloud's composition. Figure 4.13 shows an actual spectrum observed coming from a cosmic object.

Spectral lines occur throughout the entire electromagnetic spectrum. Usually, electron transitions among the lowest orbitals of the lightest elements (such as hydrogen and helium) produce visible and ultraviolet spectral lines. Transitions among very highly excited states of hydrogen and other elements can produce spectral lines in the infrared and radio parts of the electromagnetic spectrum. Conditions on Earth make it all but impossible to detect these radio and infrared features in the laboratory, but they can be routinely observed coming from space. Electron transitions among lower energy levels in heavier, more complex elements produce X-ray spectral lines. These lines have also been observed in the laboratory; some have also been observed in stars and other cosmic objects.

4.3 Molecules

6 A **molecule** is a tightly bound group of atoms held together by interactions among their orbiting electrons—interactions that we call *chemical bonds*. Much like atoms, molecules can exist only in certain well-defined energy states, and again like atoms, molecules produce emission or absorption spectral lines when they make a transition from one state to another. Because molecules are more complex than individual atoms, the rules of molecular physics are also much more complex. Nevertheless, as with atomic spectral lines, painstaking experimental work over many decades has determined the precise frequencies (or wavelengths) at which millions of molecules emit and absorb radiation.

In addition to the lines resulting from electron transitions, molecular lines result from two other kinds of change not possible in atoms: molecules can *rotate*, and they can

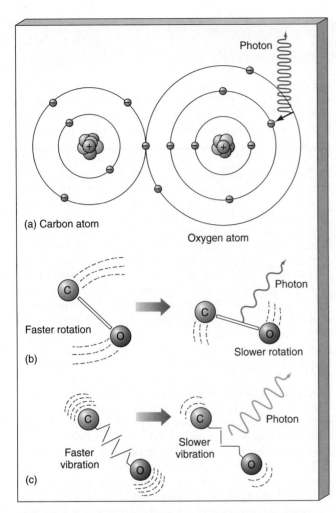

(a) Carbon atom

Oxygen atom

Photon

Faster rotation

Slower rotation

Photon

(b)

Faster vibration

Slower vibration

Photon

(c)

Figure 4.14 Molecules can change in three ways while emitting or absorbing electromagnetic radiation. Sketched here is the molecule carbon monoxide (CO) experiencing (a) a change in electron arrangement in which an electron in the outermost orbital of the oxygen atom drops to a lower-energy state, (b) a change in rotational state, and (c) a change in vibrational state.

vibrate. Figure 4.14 illustrates these basic molecular motions. Molecules rotate and vibrate in specific ways. Just as with atomic states, only certain spins and vibrations are allowed by the rules of molecular physics. When a molecule *changes* its rotational state or its vibrational state, a photon is emitted or absorbed. Spectral lines characteristic of the specific kind of molecule result. These lines are molecular fingerprints, just like their atomic counterparts, enabling researchers to identify and study one kind of molecule to the exclusion of all others.

As a rule of thumb, we can say that

- *Electron transitions* within molecules produce visible and ultraviolet spectral-line features.
- Changes in molecular *vibration* produce infrared spectral features.
- Changes in molecular *rotation* produce spectral lines in the radio part of the electromagnetic spectrum.

Molecular lines usually bear little resemblance to the spectral lines associated with their component atoms. For example, Figure 4.15(a) shows the emission spectrum of the simplest molecule known—molecular hydrogen. Notice how different it is from the spectrum of atomic hydrogen shown in part (b) of the figure.

4.4 Spectral-Line Analysis

Astronomers apply the laws of spectroscopy in analyzing radiation from beyond the Earth. A nearby star or a distant galaxy takes the place of the light bulb in our previous examples. A galactic cloud or a stellar (or even planetary) atmosphere plays the role of the intervening cool gas. And a spectrograph attached to a telescope replaces our simple prism and detector. We began our study of electromagnetic radiation by stating that virtually all we know about planets, stars, and galaxies is gleaned from studies of the light we receive from them, and we have presented some of the ways in which that knowledge is obtained. Here we describe a few of the ways in which the properties of emitters and absorbers can be determined by careful analysis of radiation received on (or near) Earth. We will encounter other important examples as our study of the cosmos unfolds.

(a)

(b)

Figure 4.15 (a) The spectrum of molecular hydrogen. Notice how it differs from the spectrum of the simpler atomic hydrogen (b).

MORE PRECISELY 4-2 The Energy Levels of the Hydrogen Atom

By observing the emission spectrum of hydrogen, Niels Bohr determined early in the twentieth century what the energy differences between the various energy levels must be. Using that information, he was then able to infer the actual energies of the excited states of hydrogen.

A unit of energy often used in atomic physics is the *electron volt* (eV). Its name derives from the amount of energy imparted to an electron by accelerating it through an electric potential of 1 volt. For our purposes, however, it is just a convenient quantity of energy, numerically equal to 1.6×10^{-19} joule—roughly half the energy carried by a single photon of red light. The minimum amount of energy needed to ionize hydrogen from its ground state is 13.6 eV. Bohr numbered the energy levels of hydrogen, with level 1 the ground state, level 2 the first excited state, and so on. He found that by assigning zero energy to the ground state, the energy of any state (the *n*th, say) could be written as follows:

$$E_n = 13.6 \left(1 - \frac{1}{n^2} \right) \text{eV}.$$

Thus, the ground state has energy $E_1 = 0$ (by our definition), the first excited state has energy $E_2 = 10.2$ eV, the second excited state has energy $E_3 = 12.1$ eV, and so on. Notice that there are infinitely many excited states between the ground state and the energy at which the atom is ionized, crowding closer and closer together as n becomes large and E_n approaches 13.6 eV.

Knowing the energy of each electron orbital, we can calculate the energy associated with a transition between any two given states. For example, to boost an electron from

the second state to the third, an atom must be supplied with $E_3 - E_2 = 1.9$ eV of energy, or 3.0×10^{-19} joule. Using the formula $E = hf$ presented in *More Precisely 4-1*, we find that this corresponds to a photon with a frequency of 4.6×10^{14} Hz, having a wavelength of 656.3 nm (6563 Å), and lying in the red portion of the spectrum. Similarly, the jump from level 3 to level 4 requires $E_4 - E_3 = 0.66$ eV of energy, corresponding to an infrared photon with a wavelength of 1890 nm (18,900 Å), and so on. The accompanying diagram summarizes the structure of the hydrogen atom. The various energy levels are shown as a series of circles of increasing radius, representing increasing energy (but remember that the electron orbitals are actually rather fuzzy and do not really have well-defined radii). The electronic transitions between these levels (indicated by arrows) are conventionally grouped into families, named after their discoverers, that define the terminology used to identify specific spectral lines. (Note that the spacings of the energy levels are not drawn to scale here, to provide room for all labels on the diagram. In reality, the circles should become more and more closely spaced as we move outward.)

Transitions down to (or up from) the ground state (level 1) from (or to) higher, excited levels form the *Lyman series*. The first is *Lyman alpha* (Lyα), corresponding to the transition between the first excited state (level 2) and the ground state. As we have seen, the energy difference is 10.2 eV, and the Lyα photon has a wavelength of 121.6 nm (1216 Å). The Lyβ (beta) transition, between level 3 (the second excited state) and the ground state, corresponds to an energy change of 12.1 eV and a photon of wavelength 102.6 nm (1026 Å). Lyγ (gamma) corresponds to a jump from level 4 to level 1, and so on. You can calculate the energies, frequencies, and wavelengths of all the photons in the Lyman series using

LINE WAVELENGTHS

Stars are very hot, especially deep down in their cores, where the temperature is measured in millions of kelvins. Because of the heat, the atoms are ionized, and the spectrum of radiation is continuous. However, at the relatively cool surface of a star, some atoms retain a few, or even most, of their orbital electrons. By matching the spectral lines we see with the laboratory spectra of known atoms, ions, and molecules, we can determine the chemical composition of the star.

As we have already seen, literally thousands of dark absorption lines cover the Sun's visible spectrum; nearly 800 of them are produced by variously excited atoms and ions of just one element: iron. Atoms of a single element, such as iron, can yield many lines for two reasons. First, the 26 electrons of a normal iron atom can make an enormous number of different transitions among energy levels. Second, many iron atoms are ionized, with some of their 26 electrons stripped away. Because the removal of electrons alters an atom's electromagnetic structure, the energy levels of ionized iron are quite *different* from those of neutral iron. Each new level of ionization introduces a whole new set of spectral lines. Besides

iron, many other elements, also in different stages of excitation and ionization, absorb photons at visible wavelengths. When we observe the entire Sun, all these atoms and ions absorb simultaneously to yield the rich spectrum we see.

The spectra of many atoms and ions are well known from laboratory measurements. Often, however, a familiar pattern of lines appears, but the lines are displaced from their expected locations. In other words, a set of spectral lines may be recognized as belonging to a particular element, except that they are all offset—blueshifted or redshifted—by the same amount from their normal wavelengths. Such shifts are produced by the Doppler effect; they thus allow astronomers to find out how fast the source of the radiation is moving along the line of sight from the observer (its *radial velocity*). ∞ (Sec. 3.5)

LINE INTENSITY

Still more information can be obtained from detailed study of the lines themselves. Because the intensity of a line is proportional to the number of photons emitted or absorbed by the atoms, the intensity of a particular line depends in part the formulas given in the text. All Lyman series energies lie in the ultraviolet region of the spectrum.

The next series of lines, the *Balmer series*, involves transitions down to (or up from) level 2, the first excited state. All the Balmer series lines lie in or close to the visible portion of the electromagnetic spectrum. Because they form the most easily observable part of the hydrogen spectrum and were the first to be discovered, the Balmer series lines are often referred to simply as the "Hydrogen" series and denoted by the letter H. As with the Lyman series, the individual transitions are labeled with Greek letters. An Hα photon (level 3 to level 2) has a wavelength of 656.3 nm (6563 Å) and is red, Hβ (level 4 to level 2) has a wavelength of 486.1 nm (4861 Å) (green), Hγ (level 5 to level 2) has a wavelength of 434.1 nm (4341 Å) (blue), and so on. The most energetic Balmer series photons have energies that place them just beyond the blue end of the visible spectrum, in the near ultraviolet.

The classification continues with the *Paschen series* (transitions down to or up from the second excited state), the *Brackett series* (third excited state), and the *Pfund series* (fourth excited state). Beyond that point, infinitely many other families exist, moving farther and farther into the infrared and radio regions of the spectrum, but they are not commonly referred to by any special names. A few of the transitions making

up the Lyman, Balmer, and Paschen series are marked on the figure. Astronomically, the most important series are the Lyman and Balmer (Hydrogen) sequences.

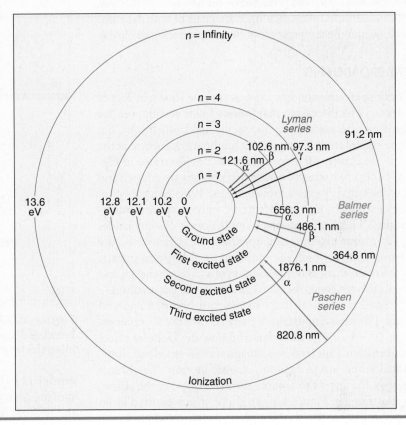

on the *number* of atoms giving rise to the line. The more atoms present to emit or absorb the photons corresponding to a given line, the stronger (brighter or darker, depending on whether it is seen in emission or absorption) that line is.

But intensity also depends on the *temperature* of the atoms—that is, the temperature of the entire gas of which the atoms are members—because temperature determines what fraction of the atoms at any instant are in the right orbital to undergo any particular transition. Consider the absorption of radiation by hydrogen atoms in an interstellar gas cloud or in the outer atmosphere of a star. If all the hydrogen were in its ground state—as it would be if the temperature were relatively low—the only transitions that could occur would be the Lyman series (see *More Precisely 4-2*), resulting in absorption lines in the ultraviolet portion of the spectrum. Thus, astronomers would observe *no* visible hydrogen absorption lines (for example, the Balmer series) in the spectrum of this object, not because there was no hydrogen, but because there would be no hydrogen atoms in the first excited state (as is required to produce visible absorption features).

The spectrum of our own Sun is a case in point. Because the temperature of the Sun's atmosphere is a relatively

cool 6000 K (as we saw in Chapter 3), few hydrogen atoms have electrons in any excited state. ∞ (Sec. 3.4.1) Hence, in the Sun, visible hydrogen lines are relatively weak—that is, of low intensity compared with the same lines in many other stars—even though hydrogen is by far the most abundant element there.

As the temperature rises, atoms move faster and faster. More and more energy becomes available in the form of collisions, and more and more electrons are boosted into an excited state. At any instant, then, some atoms are temporarily in an excited state and so are capable of absorbing at visible or longer wavelengths. As the number of atoms in the first excited state increases, lines in the Balmer series become more and more evident in the spectrum. Eventually, a temperature is reached at which *most* of the atoms are in the first excited state, simply because of their frequent, energetic collisions with other atoms in the gas. At this point the Balmer lines are at their strongest (and the Lyman lines are much weaker).

At even higher temperatures, most atoms are kicked beyond the first excited state into higher-energy orbitals, and new series of absorption lines are seen, while the strength of the Balmer series declines again. Eventually, the temperature becomes so high that most hydrogen is *ionized*, and no spectral lines are seen at all.

Over the years, astronomers have developed mathematical formulas that relate the number of emitted or absorbed photons to the temperature of the atoms as well as to their number. Once an object's spectrum is measured, astronomers can interpret it by matching the observed intensities of the spectral lines with those predicted by the formulas. In this way, astronomers can refine their measurements of both the composition and the temperature of the gas producing the lines.

LINE BROADENING

Consider an emission line, such as the one shown in Figure 4.16(a). (The discussion that follows holds equally well for absorption features.) The line seems uniformly bright, but more careful study shows that its brightness is greatest at the center and tapers off toward either side, as illustrated in Figure 4.16(b). We stressed earlier that photons are emitted and absorbed at very precise wavelengths. Why aren't spectral lines extremely narrow, occurring only at specific wavelengths? This *line broadening* is not the result of some inadequacy of our experimental apparatus. It is caused by the *environment* in which the emission or absorption occurs, which often changes our perception of a photon's energy, and it tells us a lot about the physical state of the gas involved.

Several physical mechanisms can broaden spectral lines. The most important of these involve the Doppler effect. ∞ (Sec. 3.5) To understand how the Doppler effect can broaden a spectral line, imagine a hot gas cloud. Individual atoms are in random, chaotic motion. The hotter the gas, the faster the random *thermal motions* of the atoms, as illustrated in Figure 4.17(a). If a photon is emitted by an atom *in motion*, the wavelength of the detected photon is

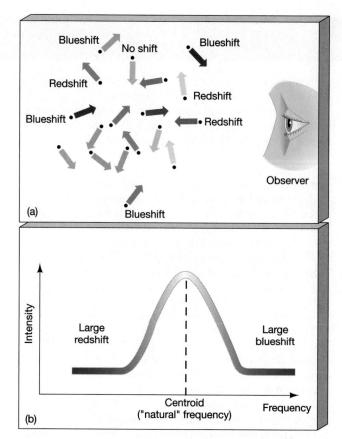

Figure 4.17   Atoms moving randomly (a) produce broadened spectral lines (b) as their individual redshifted and blueshifted emission lines merge in our detector.

changed by the Doppler effect. For example, if an atom is moving away from our eye or from our detector while in the process of emitting a photon, that photon is redshifted. The photon is not recorded at the precise wavelength predicted by atomic physics but rather at a slightly longer wavelength. The extent of this redshift is proportional to the velocity away from the detector. Similarly, if the atom is moving toward us, its light is blueshifted.

In a cloud of gas, atoms are in constant thermal motion. Some atoms move toward us, some away from us. Still others are moving transverse to our line of sight and are unaffected by the Doppler effect (at least, from our perspective). Throughout the whole cloud, atoms move in every possible direction. The result is that many atoms emit or absorb photons at slightly different wavelengths than would normally be the case if all the atoms were motionless. Most atoms in a typical cloud have very small thermal velocities. As a result, most atoms emit or absorb radiation that is Doppler-shifted only a little, and very few atoms have large shifts. So, the center of any spectral line is much more pronounced than either of its "wings." The result is a bell-shaped spectral feature like that in Figure 4.17(b). Thus, even if all atoms emitted and absorbed photons at only one specific wavelength, the effect of their thermal motion would be to smear the line out over a small range of wavelengths. The hotter the gas, the larger the spread of Doppler motions and the greater the width of

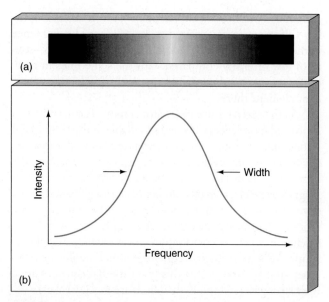

Figure 4.16   By tracing the changing brightness across a typical emission line (a) and expanding the scale, we obtain a graph of its intensity plotted versus wavelength (b).

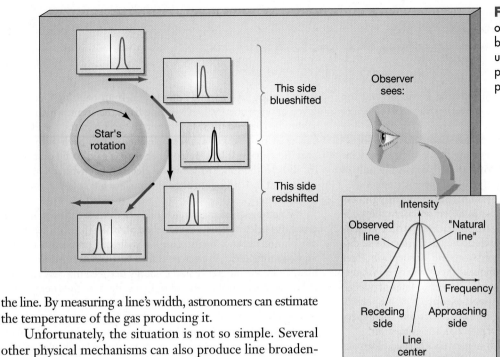

Figure 4.18 The rotation of a star can cause spectral line broadening. Since most stars are unresolved, light rays from all parts of the star merge to produce wide lines.

the line. By measuring a line's width, astronomers can estimate the temperature of the gas producing it.

Unfortunately, the situation is not so simple. Several other physical mechanisms can also produce line broadening. One such mechanism is gas *turbulence*, which exists when the gas in a cloud is not at rest or flowing smoothly but instead is seething and churning in eddies and vortices of many sizes. Motion of this type causes Doppler-shifting of spectral lines, but lines from different parts of the cloud are shifted more or less randomly. Very often, the cloud is so small or far away that our equipment cannot distinguish, or *resolve*, different parts from one another—the light from the entire cloud is blended together in our detector. When averaged over the whole cloud, the net effect appears rather similar to the thermal broadening just discussed. However, it has nothing to do with the temperature of the gas.

Rotation produces a similar effect. Consider a star or a gas cloud oriented so that we see it spinning. Photons emitted from the side spinning toward us are blueshifted by the Doppler effect. Photons emitted from the side spinning away from us are redshifted. As with turbulence, if our equipment is unable to resolve the object, a net broadening of its observed spectral lines results, as illustrated in Figure 4.18. Like the effect of turbulence, line broadening due to rotation has nothing to do with the temperature of the gas producing the lines.

Other broadening mechanisms do not depend on the Doppler effect at all. For example, if electrons are moving between orbitals while their parent atom is colliding with another atom, the energy of the emitted or absorbed photons changes slightly, thus "blurring" the spectral lines. This mechanism occurs most often in dense gases, where collisions are most frequent. It is usually referred to as *collisional broadening*. The amount of broadening increases as the density of the emitting or absorbing gas rises. Yet another cause of spectral-line broadening is *magnetism*. The electrons and nuclei within atoms behave as tiny, spinning magnets. As a result, the basic emission and absorption rules of atomic physics change slightly whenever atoms are immersed in a magnetic field, as is found in many stars to a

greater or lesser degree. Generally, the greater the magnetic field, the more pronounced the spectral-line broadening.

Given sufficiently sensitive equipment, there is almost no end to the wealth of data that can be obtained from starlight. Table 4.1 lists some basic measurable properties of an incoming beam of radiation, and indicates what sort of information can be obtained from them. It is important to realize, however, that deciphering the extent to which each of the factors just described influences a spectrum can be a very difficult task. Typically, the spectra of many elements are superimposed on one another, and several competing physical effects are occurring simultaneously, each modifying the spectrum in its own way. The challenge facing astronomers is to unravel the extent to which each mechanism contributes to spectral-line profiles and so obtain meaningful information about the source of the lines. In the next chapter we will discuss some of the means by which astronomers obtain the data they need in their quest to understand the cosmos.

TABLE 4.1 Spectral Information from Starlight

OBSERVED SPECTRAL CHARACTERISTIC	INFORMATION PROVIDED
Peak frequency or wavelength (continuous spectra only)	Temperature (Wien's law)
Lines present	Composition, temperature
Line intensities	Composition, temperature
Line width	Temperature, turbulence, rotation speed, density, magnetic field
Doppler shift	Line-of-sight velocity

Chapter Review *www*

SUMMARY

A **spectroscope** (p. 78) is a device for splitting a beam of radiation into its component frequencies and delivering them to a screen or detector for detailed study.

Many hot objects emit a **continuous spectrum** (p. 78) of radiation, containing light of all wavelengths. A hot gas may instead produce an **emission spectrum** (p. 80), consisting only of a few well-defined **emission lines** (p. 79) of specific frequencies, or colors. Passing a continuous beam of radiation through cool gas will produce **absorption lines** (p. 80) at precisely the same frequencies as are present in the gas's emission spectrum. **Kirchhoff's laws** (p. 81) describe the relationships among these different types of spectra. The emission and absorption lines produced by each element are unique—they provide a "fingerprint" of that element. The study of the spectral lines produced by different substances is called **spectroscopy** (p. 81). Spectroscopic studies of the **Fraunhofer lines** (p. 81) in the solar spectrum yield detailed information on the Sun's composition.

Atoms (p. 83) are made up of negatively charged electrons orbiting a positively charged heavy **nucleus** (p. 83) consisting of positively charged protons and electrically neutral **neutrons** (p. 87). In normal circumstances the number of orbiting electrons equals the number of protons in the nucleus, and the atom as a whole is electrically neutral. The number of protons in the nucleus determines the type of **element** (p. 87) that the atom represents. The **Bohr model** (p. 83) of the atom was an early attempt to explain how atoms can produce emission and absorption line spectra. An atom has a minimum-energy **ground state** (p. 83),

representing its "normal" condition. If an orbiting electron is given enough energy, it can escape from the atom, which is then said to be **ionized** (p. 83). Between these two states, the electron can exist only in certain well-defined states each with a very specific energy. The electron's energy is said to be **quantized** (p. 83). In the modern view the electron is envisaged as being spread out in a "cloud" around the nucleus but still with a sharply defined energy.

Electromagnetic radiation exhibits both wave and particle properties. Particles of radiation are called **photons** (p. 85). The energy of a photon depends on its color—it is directly proportional to the photon's frequency. As electrons move between energy levels within an atom, the difference in the energy between the states is emitted or absorbed in the form of photons. Because the energy levels have definite energies, the photons also have definite energies, and hence colors, which are characteristic of the type of atom involved.

Molecules (p. 88) are groups of two or more atoms bound together by electromagnetic forces. Like atoms, molecules exist in energy states that obey rules similar to those governing the internal structure of atoms. Again like atoms, when molecules make transitions between energy states, they emit or absorb a characteristic spectrum of radiation that identifies them uniquely.

Astronomers apply the laws of spectroscopy in analyzing radiation from beyond Earth. Several physical mechanisms can broaden spectral lines. The most important is the Doppler effect, which occurs because stars are hot and their atoms are in motion, or because the object being studied is rotating or in turbulent motion.

SELF-TEST: TRUE OR FALSE?

_____ **1.** Emission spectra are characterized by narrow, bright lines of different colors.

_____ **2.** Imagine an emission spectrum produced by a container of hydrogen gas. Changing the amount of hydrogen in the container will change the colors of the lines in the spectrum.

_____ **3.** In the previous question, changing the gas in the container from hydrogen to helium will change the colors of the lines occurring in the spectrum.

_____ **4.** An absorption spectrum appears as a continuous spectrum that is interrupted by a series of dark lines.

_____ **5.** The wavelengths of the emission lines produced by an element are different from the wavelengths of the absorption lines produced by the same element.

_____ **6.** Gustav Kirchhoff is credited with the discovery of blackbody radiation.

_____ **7.** The density of the hot gas producing an emission spectrum must be very high.

_____ **8.** The energy of a photon is inversely proportional to the wavelength of the radiation.

_____ **9.** The ground state of an atom is that in which the electron is in its lowest energy level (orbital).

_____ **10.** An electron can have any energy within an atom so long as it is above the ground state energy.

_____ **11.** An atom can remain in an excited state indefinitely.

_____ **12.** Emission and absorption lines correspond to the specific energy differences between orbitals in an atom.

_____ **13.** The number of electrons in an atom or ion determines the identity of the element it represents.

_____ **14.** More than one element or molecule can have the same emission or absorption spectrum.

_____ **15.** Spectral lines of hydrogen are relatively weak in the Sun because the Sun contains relatively little hydrogen.

SELF-TEST: FILL IN THE BLANK

1. A _____ is a glass wedge that disperses light into a spectrum.

2. blackbody radiation is an example of a _____ spectrum.

3. Fraunhofer discovered absorption lines in the _____.

4. A continuous spectrum can be produced by a luminous solid, liquid, or _____ gas.

5. An absorption spectrum is produced when a _____ gas lies in front of a source of continuous radiation.

6. Light behaves both as a wave and as a _____.

7. The experiment that demonstrated as the _____ caused Einstein to realize that light does not always behave like a wave.

8. Protons carry a _____ charge; electrons carry a _____ charge.

9. When one or more electrons are stripped from the atom, the atom is said to be _____.

10. When an electron moves to a higher energy level in an atom it _____ a photon of a specific energy.

11. When an electron moves to a lower energy level in an atom it _____ a photon of a specific energy.

12. The "specific energy" of the photon referred to in the last two questions is exactly equal to the energy _____ between the energy levels involved.

13. Changes in molecular vibration states produce spectral features in the _____ part of the electromagnetic spectrum.

14. Changes in molecular rotational states produce spectral features in the _____ part of the electromagnetic spectrum.

15. High temperatures, rotation, and magnetic fields all tend to _____ spectral lines.

REVIEW AND DISCUSSION

1. What is spectroscopy? Why is it so important to astronomers?

2. Describe the basic components of a simple spectroscope.

3. What is a continuous spectrum? An absorption spectrum?

4. Why are gamma rays generally harmful to life forms, but radio waves generally harmless?

5. In the particle description of light, what is color?

6. Give a brief description of a hydrogen atom.

7. What is the normal condition for atoms? What is an excited atom? What are orbitals?

8. Why do excited atoms absorb and reemit radiation at characteristic frequencies?

9. How are absorption and emission lines produced in a stellar spectrum? What information might absorption lines in the spectrum of a star reveal about a cloud of cool gas lying between us and the star?

10. According to Kirchhoff's laws, what are the necessary conditions for a continuous spectrum to be produced?

11. Why is the Hα absorption line of hydrogen in the Sun relatively weak, even though the Sun has abundant hydrogen?

12. How do molecules produce spectral lines unrelated to the movement of electrons between energy levels?

13. How does the intensity of a spectral line yield information about the source of the line?

14. How does the Doppler effect cause broadening of a spectral line?

15. List three properties of a star that can be determined from observations of its spectrum.

PROBLEMS

1. What is the energy (in joules and electron volts) of a 450-nm blue photon? A 200-nm ultraviolet photon?

2. What are the frequency and wavelength of a 2-eV red photon? Repeat your calculation for an 0.1-eV infrared photon and a 5000-eV (5 keV) X ray.

3. How many times more energy has an 0.0001-nm gamma ray than a 10-MHz radio photon?

4. Calculate the wavelength and frequency of the radiation emitted by the electronic transition from the 10th to the 9th excited state of hydrogen. In what part of the electromagnetic spectrum does this radiation lie? Repeat the question for transitions from the 100th to the 99th excited state, and from the 1000th to the 999th excited state.

5. How many different photons (that is, photons of different frequencies) can be emitted as a hydrogen atom in the third excited state falls back, directly or indirectly, to the ground state? What are their wavelengths?

6. The Hα line of a certain star is received on Earth at a wavelength of 655 nm. What is the star's radial velocity with respect to Earth?

7. In a demonstration of the photoelectric effect suppose that a minimum energy of 5×10^{-19} J (3.1 eV) is required to dislodge an electron from the metal surface. What is the minimum frequency (or longest wavelength) of radiation for which the detector registers a response?

8. At a temperature of 5800 K, hydrogen atoms in the solar atmosphere have typical random speeds of about 12 km/s. Use this information to estimate the thermal width (in nanometers) of the 656.3-nm solar Hα line.

9. Turbulent motion in a radio-emitting gas cloud broadens a 1.2-GHz (1 gigahertz = 10^9 Hz) radio line to a width of 0.5 MHz. Estimate the speed at which the gas within the cloud is moving.

10. Estimate how fast, in revolutions per day, the Sun would have to rotate in order for rotational broadening to be comparable to thermal broadening in determining line widths (see problem #8).

PROJECTS

1. Find a spectrum of the Sun that also has a wavelength scale alongside. Figure 16.8 is a good example; however, you may want to enlarge it on a copying machine. Select various absorption lines and determine their wavelengths by interpolation. Now, try to identify the element that produced these lines. Use a reference of lines such as found in Moore's *A Multiplet Table of Astrophysical Interest*. Other references may be found in the astronomy, chemistry, or physics sections of your library. Work with the darkest lines first before trying the fainter lines.

2. Use a handheld spectroscope, available through Learning Technologies Inc. While in the shade, point the spectroscope at a white cloud or white piece of paper that is in direct sunlight. Look for the absorption lines in the Sun's spectrum. Note their wavelength from the scale inside the spectroscope. Compare your list with the Fraunhofer lines given in many physics, astronomy, or chemistry reference books.

5 TELESCOPES

The Tools of Astronomy ✦ *www*

LEARNING GOALS

Studying this chapter will enable you to:

1 Sketch and describe the basic designs of the major types of optical telescopes used by astronomers.

2 Explain why very large telescopes are needed for most astronomical study, and specify the particular advantages of reflecting telescopes for astronomical use.

3 Describe how Earth's atmosphere affects astronomical observations, and discuss some of the current efforts to improve ground-based astronomy.

4 Discuss the advantages and disadvantages of radio astronomy compared to optical observations.

5 Explain how interferometry can enhance the usefulness of radio observations.

6 List the other types of nonvisible radiation currently being exploited for astronomical observation, and summarize the advantages, limitations, and chief uses of each.

7 Tell why it is important to make astronomical observations in different regions of the electromagnetic spectrum.

(Opposite page, background) This semi-transparent illustration shows some of the main features of the *Hubble Space Telescope*. The large blue disk at center of the spacecraft is the primary mirror, and the red gadgets to its rear are the sensors that guide the pointing of the telescope. The open aperture door is at upper left. The huge solar panels are shown in yellow-checkered blue at left and partly obscured at right. Looking inside the aft bay of the vehicle, we can see key components of each of the science instruments—the spectrometers are shown in copper and blue (in the foreground), the cameras in pink, green and lavender (mostly in the background).

(Insets A, B, C, and D) The four small insets are computer-rendered views of *Hubble* in orbit. Despite having the size of a city bus, *Hubble* is designed to move in space with the grace of a prima ballerina. All these illustrations are taken from video animations made by the astronomy artist and animator Dana Berry.

At its heart, astronomy is an observational science. More often than not, observations of cosmic phenomena precede any clear theoretical understanding of their nature. As a result, our detecting instruments—our telescopes—have evolved to observe as broad a range of wavelengths as possible. Until the middle of the twentieth century, telescopes were limited to visible light. Since then, technological advances have expanded our view of the universe to all regions of the electromagnetic spectrum. Some telescopes are sited on Earth, whereas others must be placed in space, and design considerations vary widely from one part of the spectrum to another. Whatever the details of their construction, however, telescopes are devices whose basic purpose is to collect electromagnetic radiation and deliver it to a detector for detailed study.

5.1 Optical Telescopes

1 In essence, a **telescope** is a "light bucket" whose primary function is to capture as many photons as possible from a given region of the sky and concentrate them into a focused beam for analysis. An *optical* telescope is one designed specifically to collect the wavelengths that are visible to the human eye. Optical telescopes have a long history, reaching back to the days of Galileo in the early seventeenth century. They are probably also the best-known type of telescope, so it is fitting that we begin our study of astronomical hardware with these devices. Later we will turn our attention to telescopes designed to capture and analyze radiation in other, *invisible*, regions of the electromagnetic spectrum.

Modern astronomical telescopes have evolved a long way from Galileo's simple apparatus. Their development over the years has seen a steady increase in *size* for one simple, but very important, reason: Large telescopes can gather and focus more radiation than can their smaller counterparts, allowing astronomers to study fainter objects and to obtain more detailed information about bright ones. This fact has played a central role in determining the design of contemporary instruments.

REFLECTING AND REFRACTING TELESCOPES

Optical telescopes fall into two basic categories—*reflectors* and *refractors*. Figure 5.1 shows how a **reflecting telescope** uses a curved mirror to gather and concentrate a beam of light. The mirror, usually called the *primary mirror* because telescopes often contain more than one mirror, is constructed so that all light rays arriving parallel to its axis (the imaginary line through the center of and perpendicular to the mirror), regardless of their distance from that axis, are reflected to pass through a single point, called the *focus*. The distance between the primary mirror and the focus is the *focal length*. In astronomical contexts, the focus of the primary mirror is referred to as the **prime focus.**

A **refracting telescope** uses a lens to focus the incoming light. **Refraction** is the bending of a beam of light as it passes from one transparent medium (for example, air) into another (such as glass). For example, consider how a pencil half immersed in a glass of water looks bent. The pencil is straight, of course, but the light by which we see it is bent—refracted—as that light leaves the water and enters the air. When that light then enters our eyes, we perceive the pencil as being bent. Figure 5.2(a) illustrates the process and shows how a prism can be used to change the direction of a beam of light. As illustrated in Figure 5.2(b), we can think of a lens as a series of prisms combined in such a way that all light rays striking the lens parallel to the axis are refracted to pass through the focus.

Astronomical telescopes are often used to make **images** of their field of view. Figure 5.3 illustrates how this is accomplished, in this case by the mirror in a reflecting telescope. Light from a distant object (in this case, a comet) reaches us

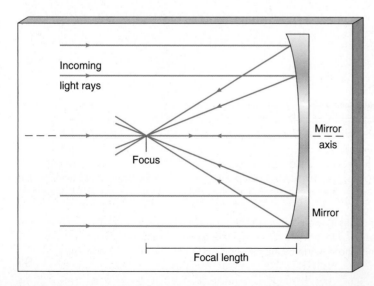

Figure 5.1   A curved mirror can be used to focus to a single point all rays of light arriving parallel to the mirror axis. Light rays traveling along the axis are reflected back along the axis, as indicated by the arrowheads pointing in both directions. Off-axis rays are reflected through greater and greater angles the farther they are from the axis, so that they all pass through the same point—the focus.

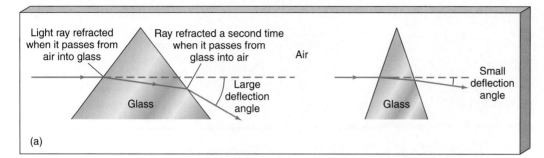

(a)

Figure 5.2   (a) Refraction by a prism changes the direction of a light ray by an amount that depends on the angle between the faces of the prism. (b) A lens can be thought of as a series of prisms. A light ray traveling along the axis of a lens is unrefracted as it passes through the lens. Parallel rays arriving at progressively greater distances from the axis are refracted by increasing amounts, in such a way that all are focused to a single point.

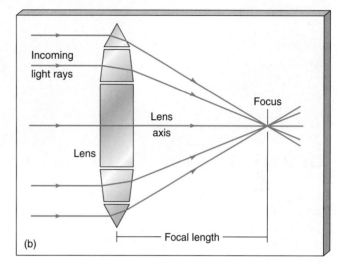

as parallel, or very nearly parallel, rays. Any ray of light entering the instrument parallel to the telescope's axis strikes the mirror and is reflected through the prime focus. Light coming from a slightly different direction—inclined slightly to the axis—is focused to a slightly different point. In this way, an image is formed near the prime focus. Each point on the image corresponds to a different point in the field of view.

The prime-focus images produced by large telescopes are actually quite small—the image of the entire field of view may be as little as 1 cm across. Often, the image is magnified with a lens known as an *eyepiece* before being observed by eye or, more likely, recorded as a photograph or digital image. Figure 5.4(a) shows the basic design of a simple reflecting telescope, illustrating how a small secondary

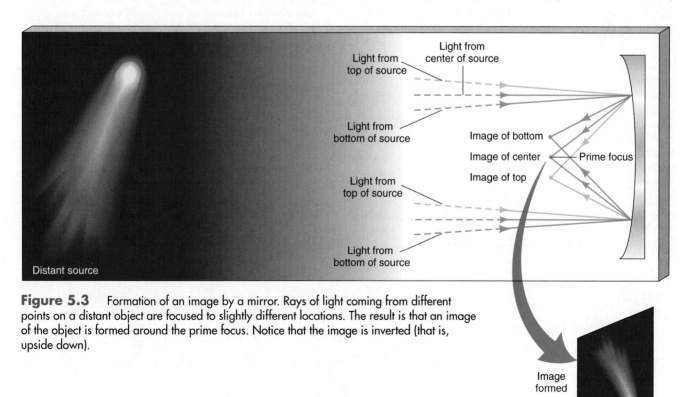

Figure 5.3   Formation of an image by a mirror. Rays of light coming from different points on a distant object are focused to slightly different locations. The result is that an image of the object is formed around the prime focus. Notice that the image is inverted (that is, upside down).

Figure 5.4 Comparison of (a) reflecting and (b) refracting telescope systems. Both types are used to gather and focus cosmic radiation—to be observed by human eyes or recorded on photographs or in computers. In both cases the image formed at the focus is viewed with a small magnifying lens called an eyepiece.

mirror and eyepiece are used to view the image. Figure 5.4(b) shows how a refracting telescope accomplishes the same function.

The two telescope designs shown in Figure 5.4 achieve the same result—light from a distant object is captured and focused to form an image. On the face of it, then, it might appear that there is little to choose between the two in deciding which type to buy or build. However, as telescope size has steadily increased over the years (for reasons to be discussed in Section 5.2), a number of important factors have tended to favor reflecting instruments over refractors:

■ The fact that light must pass through the lens is a major disadvantage of refracting telescopes. Large lenses cannot be constructed in such a way that light passes through them uniformly. Just as a prism disperses white light into its component colors, the lens in a refracting telescope focuses red and blue light differently. This deficiency is known as **chromatic aberration**. Figure 5.5 shows how chromatic aberration occurs and indicates how it affects the image of a star. Careful design and choice of materials can largely correct chromatic aberration, but it is very difficult to eliminate entirely.

Figure 5.5 Chromatic aberration. A prism bends blue light more than it bends red light, so the blue component of light passing through a lens is focused slightly closer to the lens than is the red component. As a result, the image of an object acquires a colored "halo," no matter where we place our detector.

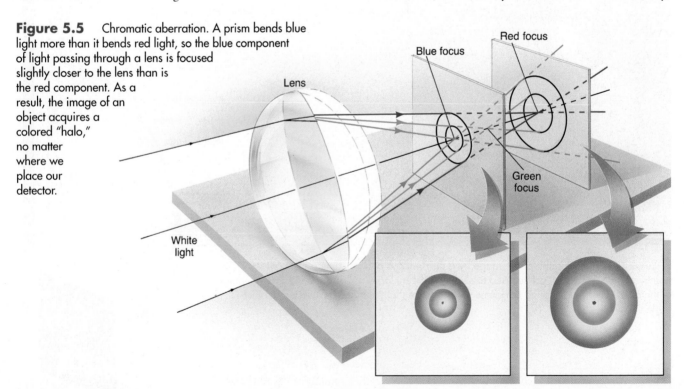

■ As light passes through the lens, some of it is absorbed by the glass. This absorption is a relatively minor problem for visible radiation, but it can be severe for infrared and ultraviolet observations because glass blocks most of the radiation coming from those regions of the electromagnetic spectrum. This problem obviously does not affect mirrors.

■ A large lens can be quite heavy. Because it can be supported only around its edge (so as not to block the incoming radiation), the lens tends to deform under its own weight. A mirror does not have this drawback because it can be supported over its entire back surface.

■ A lens has two surfaces that must be accurately machined and polished—which can be very difficult—but a mirror has only one.

For these reasons, *all* large modern telescopes use mirrors as their primary light gatherers. Figure 5.6 shows the world's largest refractor, installed in 1897 at the Yerkes Observatory in Wisconsin and still in use today. It has a lens diameter of 1 m (about 40 inches). By contrast, some new reflecting telescopes have mirror diameters in the 10-m range, and larger instruments are on the way.

TELESCOPE DESIGN

Figure 5.7 shows some basic reflecting telescope designs. Radiation from a star enters the instrument, passes down the main tube, strikes the primary mirror, and is reflected back toward the prime focus, near the top of the tube. Sometimes astronomers place their recording instruments at the

Figure 5.6 Photograph of the Yerkes Observatory's 1-m-diameter refracting telescope.

(a) Prime focus

(b) Newtonian focus

(c) Cassegrain focus

(d) Coudé focus

Figure 5.7 Four reflecting telescope designs: (a) prime focus, (b) Newtonian focus, (c) Cassegrain focus, and (d) coudé focus. Each uses a primary mirror at the bottom of the telescope to capture radiation, which is then directed along different paths for analysis. Notice that the secondary mirrors shown in (c) and (d) are actually slightly diverging, so that they move the focus outside the telescope.

Figure 5.8 (a) Artist's illustration of the 5-m-diameter Hale optical telescope on Mount Palomar in California. (b) A photograph of the telescope. (c) Astronomer Edwin Hubble in the observer's cage at the Hale prime focus.

(b)

(c)

prime focus. However, it can be very inconvenient, or even impossible, to suspend bulky pieces of equipment there. More often, the light is intercepted on its path to the focus by a *secondary mirror* and redirected to a more convenient location, as in Figure 5.7(b) through (d).

In a **Newtonian telescope** (named after Sir Isaac Newton, who invented this particular design), the light is intercepted before it reaches the prime focus and is deflected by 90°, usually to an eyepiece at the side of the instrument. This is a particularly popular design for smaller reflecting telescopes, such as those used by amateur astronomers.

Alternatively, astronomers may choose to work on a rear platform where they can use equipment, such as a spectroscope, that is too heavy to hoist to the prime focus. In this case, light reflected by the primary mirror toward the prime focus is intercepted by a smaller secondary mirror, which

reflects it back down through a small hole at the center of the primary mirror. This arrangement is known as a **Cassegrain telescope** (after Guillaume Cassegrain, a French lensmaker). The point behind the primary mirror where the light from the star finally converges is called the *Cassegrain focus*.

A more complex observational configuration requires starlight to be reflected by several mirrors. As in the Cassegrain design, light is first reflected by the primary mirror toward the prime focus and reflected back down the tube by a secondary mirror. A third, much smaller, mirror then reflects the light into an environmentally controlled laboratory. Known as the *coudé* room (from the French word for "bent"), this laboratory is separate from the telescope itself, enabling astronomers to use very heavy and finely tuned equipment that could not possibly be lifted to either the prime focus or the Cassegrain focus. The light path to the coudé room lies along the axis of the telescope's mount—that is, the axis around which the telescope rotates as it tracks objects across the sky—so that the light path does not change as the telescope moves.

To illustrate some of these points, Figure 5.8 depicts the Hale 5-m-diameter optical telescope on California's Mount Palomar. As the size of the person drawn in the observer's cage at the prime focus indicates, this is indeed a very large telescope. In fact, for almost three decades after its dedication in 1948, the Hale telescope was the largest in the world. It has been at or near the forefront of astronomical research for much of the last half century. Observations can be made at the prime, the Cassegrain, or the coudé focus, depending on the needs of the user.

IMAGES AND DETECTORS

Large reflectors are good at forming images of narrow fields of view, where all the light that strikes the mirror surface moves almost parallel to the axis of the instrument. However, if the light enters at an appreciable angle, it cannot be accurately focused, degrading the overall quality of the image. The effect (called *coma*) worsens as we move farther from the center of the field of view. Eventually, the image quality is reduced to the point where it is no longer usable. The distance from the center to where the image becomes unacceptable defines the useful field of view of the telescope—typically, only a few arc minutes for large instruments.

A design that overcomes this problem is the *Schmidt telescope*, named after its inventor, Bernhard Schmidt, who built the first such instrument in the 1930s. The telescope uses a correcting lens, which sharpens the final image of the entire field of view. Consequently, a Schmidt telescope is well suited to producing wide-angle photographs, covering several degrees of the sky. Because the design of the Schmidt telescope results in a curved image that is not suitable for viewing with an eyepiece, the image is recorded on a specially shaped piece of photographic film. For this reason, the instrument is often called a *Schmidt camera* (Figure 5.9). The Palomar Observatory Schmidt camera, one of the largest in the world (with a 1.8-m mirror and a 1.2-m lens), performed a survey of the entire northern sky in the 1950s. The Palomar Observatory Sky Survey, as it is known, has for decades been an invaluable research tool for professional observers. (A new, and much more detailed, digital sky survey is now being carried out by a consortium of U.S. research institutions. It should be completed by 2004.)

When a photographic plate is placed at the focus to record an image of the field of view, the telescope is acting in effect as a high-powered camera. However, this is by no means the only light-sensitive device that can be placed at the focus to analyze the radiation received from space. When very accurate and rapid measurements of light intensity are required, a device known as a **photometer** is used. A photometer measures the total amount of light received in all or part of the image. When only part of the image is under study, the region of interest is selected simply by masking out the rest of the field of view. Using a photometer often means "throwing away"

Figure 5.9 The Schmidt camera of the European Southern Observatory in Chile.

spatial detail, but in return more information is obtained about the intensity and time variability of a particular source, such as a pulsating star or a supernova explosion.

Often, astronomers want to study the *spectrum* of the incoming light. Large **spectrometers** frequently work in tandem with optical telescopes. Light collected by the primary mirror may be redirected to the underground coudé room, defined by a narrow slit, passed through a prism or a diffraction grating, and projected onto a screen—a process not so different from the operation of the simple spectroscope described in Chapter 4. ∞ (Sec. 4.1) The spectrum can be studied in real time (that is, as it happens) or stored on a photographic plate (or, more commonly nowadays, on a computer disk) for later analysis.

5.2 Telescope Size

2 Astronomers generally prefer large telescopes over small ones, for two main reasons. The first has to do with the amount of light a telescope can gather—its *light-gathering power*. The second is related to the amount of detail that can be seen—the telescope's *resolving power*.

LIGHT-GATHERING POWER

One reason for using a larger telescope is simply that it has a greater **collecting area**—the total area of a telescope capable of capturing radiation. The larger the telescope's reflecting mirror (or refracting lens), the more light it collects, and the easier it is to measure and study an object's radiative properties. Astronomers spend much of their time observing very distant—and hence very *faint*—cosmic sources. If we wish to make detailed observations of objects far from our own cosmic neighborhood, very large telescopes are essential tools. Figure 5.10 illustrates the effect of increasing telescope size by comparing images of the Andromeda Galaxy taken with two different instruments. Large collecting area is particularly important for spectroscopic work, as the received radiation in that case must be split into its component wavelengths for further analysis.

The observed brightness of an astronomical object is directly proportional to the area of our telescope's mirror and therefore to the *square* of the mirror diameter. Thus, a 5-m telescope will produce an image 25 times as bright as a 1-m instrument because a 5-m mirror has $5^2 = 25$ times the collecting area of a 1-m mirror. We can also think of this relationship in terms of the length of *time* required for a telescope

to collect enough energy to create a recognizable image on a photographic plate. Our 5-m telescope will produce an image 25 times faster than the 1 m device because it gathers energy at a rate 25 times greater. Expressed in another way, a 1-hour time exposure with a 1-m telescope is roughly equivalent to a 2.4-minute time exposure with a 5-m instrument.

RESOLVING POWER

A second advantage of large telescopes is their finer **angular resolution.** In general, *resolution* refers to the ability of any device, such as a camera or a telescope, to form distinct, separate images of objects lying close together in the field of view. The finer the resolution, the better we can distinguish the objects and the more detail we can see. In astronomy, where we are always concerned with angular measurement, "close together" means "separated by a small angle on the sky," so angular resolution is the factor that determines our ability to see fine structure. Figure 5.11 illustrates how the appearance of two objects—stars, say—might change as the angular resolution of our telescope varies. Figure 5.12 illustrates the result of increasing resolving power with views of the Andromeda Galaxy at several different resolutions.

What limits a telescope's resolution? One important factor is *diffraction*, the tendency of light, and all other waves for

R I V U X G

Figure 5.10 Effect of increasing telescope size on an image of the Andromeda Galaxy. Both photographs had the same exposure time; the bottom image was taken with a telescope twice the size of that used to make the top image. Fainter detail can be seen as the diameter of the telescope mirror increases because larger telescopes are able to collect more photons per unit time.

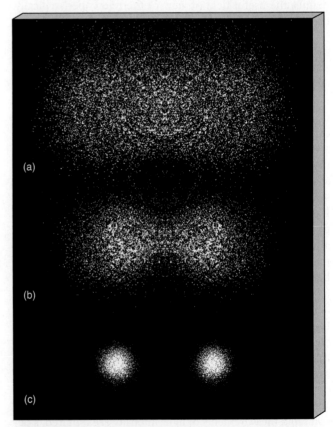

Figure 5.11 Two comparably bright light sources become progressively clearer when viewed at finer and finer angular resolution. When the angular resolution is much poorer than the separation of the objects, as at the top, the objects appear as a single fuzzy "blob." As the resolution improves, the two sources become discernible as separate objects.

(a)

(b)

(c)

(d)

R I V U X G

Figure 5.12 Detail becomes clearer in the Andromeda Galaxy as the angular resolution is improved some 600 times, from (a) 10′, to (b) 1′, (c) 5″, and (d) 1″.

that matter, to bend around corners. ∞ (Sec. 3.1) Because of diffraction, when a parallel beam of light enters a telescope, the rays spread out slightly, making it impossible to focus the beam to a sharp point, even with a perfectly constructed mirror. Diffraction introduces a certain "fuzziness," or loss of resolution, into the system. The degree of fuzziness—the minimum angular separation that can be distinguished—determines the angular resolution of the telescope. The amount of diffraction is proportional to the wavelength of the radiation divided by the diameter of the telescope mirror. As a result we can write, in convenient units,

$$\text{angular resolution (arc sec)} = 0.25 \frac{\text{wavelength (μm)}}{\text{diameter (m)}},$$

where 1 μm (1 micron) = 10^{-6} m (see Appendix 2).

For a given telescope size, the amount of diffraction increases in proportion to the wavelength used, and observations in the infrared or radio range are often limited by its effects. For example, in an otherwise perfect observing environment, the best possible angular resolution of blue light (with a wavelength of 400 nm) that can be obtained using a 1-m telescope observing is about 0.1″. This quantity is known as the *diffraction-limited* resolution of the telescope. But if we were to use our 1-m telescope to make observations in the near infrared range, at a wavelength of 10 μm, the best resolution we could obtain would be only 2.5″. A 1-m radio telescope operating at a wavelength of 1 cm would have an angular resolution of slightly under 1°.

For light of any given wavelength, large telescopes produce less diffraction than small ones. A 5-m telescope observing in blue light would have a diffraction-limited resolution five times finer than the 1-m telescope just discussed—about 0.02″. A 0.1-m (10-cm) telescope would have a diffraction limit of 1″ and so on. For comparison, the angular resolution of the human eye is about 0.5′.

BUILDING VERY LARGE TELESCOPES

In the push toward larger and larger telescopes, refractors were abandoned long ago, in large part because of the difficulty in constructing them. However, large reflectors also present problems.

Conventional telescope mirrors are made from large blocks of quartz, glass, or some other type of polishable material capable of withstanding large temperature changes

(a)

(b)

Figure 5.13 (a) The world's highest ground-based observatory, at Mauna Kea, Hawaii, is perched atop an extinct volcano more than 4 km above sea level. Among the domes visible in the picture are those that house the Canada–France–Hawaii 3.6-m telescope, the 2.2-m telescope of the University of Hawaii, Britain's 3.8-m infrared facility, and the twin 10-m Keck telescopes. To the right of the twin Kecks is the Japanese 8.3-m Subaru telescope, still under construction. The thin air at this high-altitude site guarantees less atmospheric absorption of incoming radiation and hence a clearer view than at sea level, but the air is so thin that astronomers must occasionally wear oxygen masks while working. (b) The 10-m mirror in the first Keck telescope. Note the technician in orange coveralls at center.

with little expansion or contraction. The construction of such a mirror begins by pouring molten material into a large cast, then cooling it slowly over the course of several years to keep it from cracking or developing internal stresses as it solidifies. It takes years more to grind and polish the surface to the required curvature. Finally, the surface is coated with a thin film of aluminum to provide a reflecting surface. All these stages are slow, painstaking, and, unfortunately, not always successful. In fact, since the construction of the 5-m Palomar instrument in 1948, only one larger conventional telescope (a 6-m instrument in Russia) has been completed, and this telescope, regrettably, suffers from several optical defects, resulting in poor-quality images.

Until the 1980s the accepted wisdom was that telescopes with mirrors larger than 5 or 6 m in diameter were simply too expensive and impractical to build. However, new high-tech manufacturing techniques, coupled with radically new mirror designs, now make the construction of telescopes in the 8- to 12-m range almost a routine matter. Experts can now make large mirrors much lighter for their size than had previously been believed feasible and can combine many smaller mirrors into the equivalent of a much larger single-mirror telescope. Several large-diameter instruments are now under construction, and many more are planned.

Currently, the largest operating optical telescopes are the twin Keck instruments atop Mauna Kea in Hawaii (see Figure 5.13), administered jointly by the California Institute of Technology and the University of California. Each telescope combines 36 hexagonal 1.8-m mirrors into the equivalent collecting area of a single 10-m reflector. The first Keck telescope became fully operational in 1992; the second was completed in 1996. The high altitude and large size of these telescopes makes them particularly well suited for detailed spectroscopic studies of very faint objects. The Very Large Telescope (VLT), now under construction at the European Southern Observatory in Chile, will consist of four separate 8.2-m mirrors that will function as a single instrument.

5.3 High-Resolution Astronomy

3 Even large telescopes have their limitations. For example, according to the discussion in the preceding section, the 5-m Hale telescope should have an angular resolution of around 0.02″. In practice, however, it cannot do better than about 1″. In fact, apart from instruments using special techniques developed to examine some particularly bright stars,

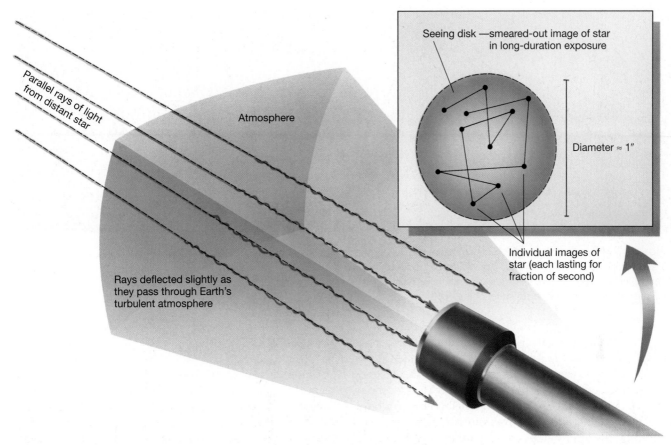

Figure 5.14 Individual photons from a distant star strike the detector in a telescope at slightly different locations because of turbulence in Earth's atmosphere. Over time, the individual photons cover a roughly circular region on the detector, and even the pointlike image of a star is recorded as a small disk, called the seeing disk.

no ground-based optical telescope built before 1990 can resolve astronomical objects to much better than 1″. The reason is Earth's turbulent atmosphere, which blurs the image even before the light reaches our instruments. In recent years, great strides have been made in overcoming this obstacle. Telescopes have been placed above the atmosphere, and computers are playing an increasingly important role in both telescope operation and image processing.

ATMOSPHERIC BLURRING

As we observe a star, atmospheric turbulence produces continual small changes in the optical properties of the air between the star and our telescope (or eye). The light from the star is refracted slightly, and the stellar image dances around on the detector (or on our retina). This continual deflection is the cause of the well-known "twinkling" of stars. It occurs for the same reason that objects appear to shimmer when viewed across a hot roadway on a summer day.

On a good night at the best observing sites, the maximum amount of deflection produced by the atmosphere is slightly less than 1″. Consider taking a photograph of a star. After a few minutes' exposure time (long enough for the intervening atmosphere to have undergone many small, ran-

dom changes), the image of the star has been smeared out over a roughly circular region an arc second or so in diameter. Astronomers use the term **seeing** to describe the effects of atmospheric turbulence. The circle over which a star's light (or the light from any other astronomical source) is spread is called the **seeing disk.** Figure 5.14 illustrates the formation of the seeing disk for a small telescope.*

To achieve the best possible seeing, telescopes are sited on mountaintops (to get above as much of the atmosphere as possible) in regions of the world where the atmosphere is known to be fairly stable and relatively free of dust, moisture, and light pollution from cities. In the continental United States, these sites tend to be in the desert Southwest. The U.S. National Observatory for optical astronomy in the Northern Hemisphere, completed in 1973, is located high on Kitt Peak near Tucson, Arizona. The site was chosen because of its many dry, clear nights. Seeing of 1″ from such a location is regarded as good, and seeing of a few arc seconds is tolerable for many purposes. Even better condi-

---

In fact, for a large instrument—more than about 1 m in diameter—the situation is more complicated, because rays striking different parts of the mirror have actually passed through different turbulent atmospheric regions. The end result is still a seeing disk, however.

Figure 5.15 Located in the Andes Mountains of Chile, the European Southern Observatory at La Silla is run by a consortium of European nations. Numerous domes house optical telescopes of different sizes, each with varied support equipment, making this one of the most versatile observatories south of the equator.

tions are found on Mauna Kea, Hawaii, and at Cerro Tololo and La Silla in the Andes Mountains of Chile (Figure 5.15)—which is why many large telescopes have recently been constructed at those two exceptionally clear locations.

An optical telescope placed in orbit about Earth or on the Moon could obviously overcome the limitations imposed by the atmosphere on ground-based instruments. Without atmospheric blurring, extremely fine resolution—close to the diffraction limit—can be achieved, subject only to the engineering restrictions of building or placing large structures in space. The *Hubble Space Telescope* (*HST;* named for one of America's most notable astronomers, Edwin Hubble) was launched into Earth orbit by NASA's space shuttle *Discovery* in 1990 (see *Interlude 5-1.*). This telescope has a 2.4-m mirror, with a diffraction limit of only 0.05″, giving astronomers a view of the universe as much as 20 times sharper than that normally available from even much larger ground-based instruments.

IMAGE PROCESSING

Computers play an important role in observational astronomy. Most large telescopes today are controlled either by computers or by operators who rely heavily on computer assistance, and images and data are recorded in a form that can be easily read and manipulated by computer programs.

It is becoming rare for photographic equipment to be used as the primary means of data acquisition at large observatories. Instead, electronic detectors known as **charge-coupled devices,** or **CCDs,** are in widespread use. Their output goes directly to a computer. A CCD (Figure 5.16) consists of a wafer of silicon divided into a two-dimensional array of many tiny picture elements, known as **pixels**. When light strikes a pixel, an electric charge builds up on the device. The amount of charge is directly proportional to the number of photons striking each pixel—in other words, to the intensity of the light at that point. The charge buildup

Figure 5.16 A charge-coupled device consists of hundreds of thousands, or even millions, of tiny light-sensitive cells, or pixels, usually arranged in a square array. Light striking a pixel causes an electrical charge to build up on it. By electronically reading out the charge on each pixel, a computer can reconstruct the pattern of light—the image—falling on the chip. (a) Detail of a CCD array. (b) A CCD chip mounted for use at the focus of a telescope.

(a)   (b)

(a) (b) (c) (d)

Figure 5.17 (a) Ground-based view of the star cluster R136, a group of stars in the Large Magellanic Cloud (a nearby galaxy). (b) The "raw" image of this same region as seen by the *Hubble Space Telescope* in 1990, before the repair mission. (c) The same image after computer processing that partly compensated for imperfections in the mirror. (d) The same region as seen by the repaired *HST* in 1994.

is monitored electronically, and a two-dimensional image is obtained. A CCD is typically a few square centimeters in area and may contain several million pixels, generally arranged on a square grid. As the technology improves, both the areas of CCDs and the number of pixels they contain continue to increase. Incidentally, the technology is not limited to astronomy—many home video cameras contain CCD chips similar in basic design to those in use at the great astronomical observatories of the world.

CCDs have two important advantages over photographic plates, which were the staple of astronomers for over a century. First, CCDs are much more *efficient* than photographic plates, recording as many as 75 percent of the photons striking them, compared with less than 5 percent for photographic methods. This means that a CCD image can show objects 10 to 20 times fainter than can a photograph made using the same telescope and the same exposure time. Alternatively, a CCD can record the same level of detail in less than a tenth of the time required by photographic techniques, or record that detail with a much smaller telescope. Second, CCDs produce a faithful representation of an image in a digital format that can be placed directly on magnetic tape or disk, or even sent across a computer network to an observer's home institution.

Computers are also widely used to reduce *background noise* in astronomical images. Noise is anything that corrupts the integrity of a message, such as static on an AM radio or "snow" on a television screen. The noise corrupting telescopic images has many causes. In part, it results from faint, unresolved sources in the telescope's field of view and from light scattered into the line of sight by Earth's atmosphere. It can also be caused by electronic "hiss" within the detector. Whatever the origin of noise, its characteristics can be determined (for example, by observing a part of the sky where there are no known sources of radiation) and its effects partially removed with the aid of high-speed computers, allowing astronomers to see features that would otherwise remain hidden.

Using computer processing, astronomers can also compensate for known instrumental defects and even correct some effects of bad seeing. In addition, the computer can often carry out many of the relatively simple but tedious and time-consuming chores that must be performed before an image (or spectrum) reaches its final "clean" form. Figure 5.17 illustrates how computerized image-processing techniques were used to correct for known instrumental problems in the *Hubble Space Telescope*, allowing much of the planned resolution of the telescope to be recovered even before its repair in 1993.

NEW TELESCOPE DESIGN

An exciting development that is bringing about striking improvements in the resolution of ground-based optical telescopes takes these ideas of computer control and image processing one stage further. If an image can be analyzed while the light is still being collected (a process that can take many minutes, or even hours, in some cases), it is possible to adjust the telescope from moment to moment to correct for the effects of mirror distortion, temperature changes, and bad seeing. In principle, the telescope might even come close to the theoretical (diffraction-limited) resolution.

Some of these techniques, collectively known as **active optics,** are already in use in the New Technology Telescope (NTT), located at the European Southern Observatory in Chile (NTT is the most prominent instrument visible in Figure 5.15). This 3.5-m instrument, employing the latest in real-time telescope controls, achieves resolution of about 0.5″ by making minute modifications to the tilt of the mirror as its temperature and orientation change, thus maintaining the best possible focus at all times (see Figure 5.18a). The Keck 10-m instruments, one of whose hexagonal mirrors is shown in Figure 5.18b, also employ these methods and may ultimately achieve resolution as fine as 0.25″ by these means.

(a)

R I V U X G

(b)

Figure 5.18 (a) These false-color infrared photographs of part of the star cluster R136—the same object shown in Figure 5.17—contrast the resolution obtained without the active optics system (left image) with that achievable when the active optics system is in use (right image). (b) A hexagonal mirror segment destined for one of the Keck telescopes undergoes shaping and polishing. The unusually thin glass will be backed by push-pull pistons that can adjust the precise configuration of the segment during observations so as to attain improved resolution.

An even more ambitious undertaking is known as **adaptive optics.** This technique actually deforms the shape of the mirror's surface, under computer control, while the image is being exposed. The intent is to undo the effects of atmospheric turbulence. Adaptive optics presents formidable theoretical and technological problems, but the rewards are so great that they are presently the subject of intense research. Declassified SDI ("Star Wars") technology has provided an enormous boost to this effort. In the experimental system shown in Figure 5.19(a), lasers probe the atmosphere above the telescope, returning information about the air's swirling motion to a computer that modifies the mirror thousands of times per second to compensate for poor seeing.

Figure 5.19(b) compares the results of a pair of observations of a nearby double star called Castor. Spectroscopic observations long ago revealed the double nature of this object, which appears to be a binary-star system. The image on the left shows the star system as seen through an ordinary, moderate-sized telescope—an oblong blur combining the light from the two stars, spread over several arc seconds. On the right, with the adaptive optics system turned on, one star is clearly distinguishable from the other. The two stars are separated by less than an arc second; the resolution, with adaptive optics turned on, is about 0.1".

Adaptive optics systems are planned for many existing large telescopes. For example (if development goes as planned), the Keck telescopes will, by the year 2000, incorporate adaptive-optics instrumentation capable of producing diffraction-limited (0.02" resolution) images at near-infrared wavelengths. In the next decade, it may well be possible to have the "best of both worlds," achieving with a large ground-based telescope the kind of resolution presently attainable only from space.

5.4 Radio Astronomy

4 In addition to the visible radiation that penetrates Earth's atmosphere on a clear day, radio radiation also reaches the ground. Indeed, as indicated in Figure 3.11, the radio window in the electromagnetic spectrum is much wider than the optical window. ∞ (Sec. 3.3) Because the atmosphere is no hindrance to long-wavelength radiation, radio astronomers have built many ground-based **radio telescopes** capable of detecting cosmic radio waves. These devices have all been constructed since the 1950s—radio astronomy is a much younger subject than optical astronomy.

The field originated with the work of Karl Jansky at Bell Labs in 1931, but only after the technological push of World War II did it grow into a distinct branch of astronomy. Jansky was engaged in a study of shortwave radio interference when he discovered a faint static "hiss" that had no apparent terrestrial source. He noticed that the strength of the hiss varied in time and that its peak occurred about 4 minutes earlier each day. He soon realized that the peaks were coming exactly one *sidereal day* apart and correctly inferred that the hiss was not of terrestrial origin but came from a definite direction in space. That direction is now known to correspond to the center of our Galaxy. It took over a decade, and the realization by astronomers that interstellar gas could actually be observed at radio wavelengths, for the full importance of his work to be appreciated, but today Jansky is widely regarded as the father of radio astronomy.

(a)

(b)

Figure 5.19   (a) Until mid-1991 the Starfire Optical Range at Kirtland Air Force Base in New Mexico was one of the U.S. Air Force's most closely guarded secrets. Here, beams of laser light probe the atmosphere above the system's 1.5-m telescope, allowing minute computer-controlled changes to be made to the mirror surface thousands of times each second. (b) The improvement in seeing produced by such systems can be dramatic, as can be seen in these images acquired at another military observatory atop Mount Haleakala in Maui, Hawaii, employing similar technology. The uncorrected image (left) of the double star Castor is a blur spread over several arc seconds, with little hint of its binary nature. With adaptive compensation applied (right), the resolution is improved to a mere 0.1″ and the two components are clearly resolved.

ESSENTIALS OF RADIO TELESCOPES

Figure 5.20 shows a fairly typical radio telescope, the large 43-m (140-foot)-diameter telescope located at the National Radio Astronomy Observatory in West Virginia. Although much larger than reflecting optical telescopes, most radio telescopes are built in basically the same way. They have a large, horseshoe-shaped mount supporting a huge, curved metal dish that serves as the collecting area. The dish captures cosmic radio waves and reflects them to the focus, where a receiver detects the signals and channels them to a computer. Conceptually, the operation of a radio telescope is similar to the operation of an optical reflector with the detecting instruments placed at the prime focus (Figure 5.7a). However, unlike optical instruments, which can detect all visible wavelengths simultaneously, radio detectors normally register only a narrow band of wavelengths at any one time. To observe radiation at another radio frequency, we must retune the equipment, much as we tune a television set to a different channel.

Large radio telescopes are very sensitive instruments and can detect even very faint radio sources. However, their angular resolution is generally poor compared with that of optical counterparts, despite the enormous size of many radio

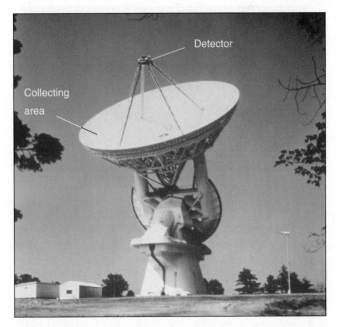

Figure 5.20   The 43-m-diameter radio telescope at the National Radio Astronomy Observatory in Green Bank, West Virginia.

INTERLUDE 5-1   The Hubble Space Telescope

The *Hubble Space Telescope* (*HST*) is the largest, most complex, most sensitive observatory ever deployed in space. At over $3 billion (including the cost of two missions to service and refurbish the system), it is also the most expensive scientific instrument ever constructed. Built jointly by NASA and the European Space Agency, *HST* is designed to allow astronomers to probe the universe with at least 10 times finer resolution and with some 30 times greater sensitivity to light than existing Earth-based devices. *HST* is operated remotely from the ground; there are no astronauts aboard the telescope, which orbits Earth about once every 95 minutes, at an altitude of about 600 km (380 miles).

The telescope's overall dimensions approximate those of a city bus or railroad tank car—13 m (43 feet) long, 12 m (39 feet) across with solar arrays extended, and 11,000 kg (12.5 tons when weighed on the ground). The heart of *HST* is a 2.4-m (94.5-inch) diameter mirror designed to capture optical, ultraviolet, and infrared radiation before it reaches Earth's murky atmosphere. The accompanying figure shows the telescope being lifted out of the cargo bay of the space shuttle *Discovery* in the spring of 1990.

The optical system and scientific instruments aboard *HST* are compact and pioneering. The telescope reflects light from its large mirror back to a smaller, 0.3-m (12-inch), secondary mirror, which in turn sends the light through a hole in the doughnut-shaped main mirror and into the aft bay of the spacecraft. There, any of five major scientific instruments wait to analyze the incoming radiation. Most of these instruments are about the size of a telephone booth. They include two cameras to image (or electronically photograph)

various regions of the sky, two spectrographs to split the radiation into its component wavelengths, and a group of fine guidance sensors to measure the positions of stars in the sky. Most of these instruments have been upgraded or replaced by NASA astronauts since the telescope's launch.

Soon after launch, astronomers discovered that the telescope's primary mirror had been polished to the wrong shape. The mirror is too flat by 2 μm, about 1/50 the width of a human hair. Even though it is the smoothest mirror ever made, this imperfection makes it impossible to focus light as well as expected. This optical flaw (known as *spherical aberration*) meant that *HST* was not as sensitive as designed, although it could still see many objects in the universe with unprecedented resolution. In late 1993, astronauts aboard the space shuttle *Endeavour* visited *HST* and succeeded in repairing some of its ailing equipment. They replaced *Hubble*'s gyroscopes to help the telescope point more accurately and installed sturdier versions of the solar arrays that power the telescope's electronics. They also partly corrected *Hubble*'s flawed vision by inserting an intricate set of small mirrors (each about the size of a coin) to compensate for the faulty primary mirror—in much the

dishes. It is not Earth's atmosphere that is to blame—the radio wavelengths normally studied pass through air without any significant distortion. The problem is that the typical wavelengths of radio waves are about a million times longer than those of visible light, and these longer wavelengths impose a corresponding crudeness in angular resolution because of the effects of diffraction. Recall from Section 5.2 that the longer the wavelength, the greater the diffraction.

The best angular resolution obtainable with a single radio telescope is about 10″ (for the largest instruments operating at millimeter wavelengths)—at least 10 times coarser than the capabilities of the largest optical mirrors. The resolution

varies widely, depending on the wavelength being observed. The 43-m radio telescope shown in Figure 5.20 can achieve resolution of about 1′ when receiving radio waves having wavelengths of around 1 cm. However, it was designed to operate most efficiently (that is, it is most sensitive to radio signals) at wavelengths closer to 5 cm, where the resolution is only about 6′, or 0.1°.

Radio telescopes are large in part because that is the only way they can achieve good resolution. But, as with optical telescopes, there is another reason: light-gathering power. The amount of energy arriving at Earth in the form of radio radiation is extremely small. In fact, the total amount of radio same way we use eyeglasses or contact lenses to help humans see better.

Since the 1993 repair mission *Hubble*'s resolution is close to the original design specifications, and the telescope has regained much of its lost sensitivity, enabling it to see very faint objects. In early 1997 a second servicing mission replaced several instruments with more sensitive models, performed maintenance on the satellite's fine-guidance system, and upgraded some of the telescope's data systems. The next service mission is scheduled for 1999.

A good example of *Hubble*'s scientific capabilities today can be seen by comparing the two images of the spiral galaxy M100, shown here. On the left is perhaps the best ground-based photograph of this beautiful galaxy, showing rich detail and color in its spiral arms. On the right, to the same scale and orientation, is an *HST* image showing improvement in both resolution and sensitivity. (The chevron-shaped field of view is caused by the corrective optics inserted into the telescope; an additional trade-off is that *Hubble*'s field of view is smaller than those of ground-based telescopes.) The inset shows *Hubble*'s exquisite resolution of small fields of view.

Given that *Hubble* was so expensive to build, was meant to be the flagship of a whole new generation of NASA spacecraft, and was greeted with such public fanfare, it is perhaps understandable that the news media sensationalized so many aspects of the telescope's problems and of the repair mission. The bottom line, however, is that *HST* was never really "broken," nor is it now completely "fixed." Despite the fact that it still does not operate as originally designed, *Hubble* has been a superb telescope since its deployment in space in 1990, and it remains so today. In fact, especially due to the heroic efforts of the astronaut repair crew, *Hubble* is arguably the best telescope built by humans to date. Many spectacular examples of its remarkable data appear throughout this book.

energy received by Earth's entire surface is less than a trillionth of a watt. Compare this with the roughly 10 *million* watts our planet's surface receives in the form of infrared and visible light from any of the bright stars visible in the night sky.

Radio telescopes can be built so much larger than their optical counterparts because their reflecting surface need not be as smooth as is needed for shorter-wavelength light waves. Provided that surface irregularities (dents, bumps, and the like) are much smaller than the wavelength of the waves to be detected, the surface will reflect them without distortion. Because the wavelength of visible radiation is short (approximately 10^{-6}

m), very smooth mirrors are needed to reflect the waves properly, and it is difficult to construct very large mirrors to such exacting tolerances. However, even rough metal surfaces can accurately focus 1-cm waves, and radio waves of wavelength a meter or more can be reflected and focused perfectly well by surfaces having irregularities even as large as your fist.

Figure 5.21 shows the world's largest and most sensitive radio telescope, located in Arecibo, Puerto Rico. Approximately 300 m (1000 feet) in diameter, the surface of the Arecibo telescope spans nearly 20 acres. Constructed in 1963 in a natural depression in the hillside, the dish was originally surfaced with chicken wire, which was lightweight

Figure 5.21 An aerial photograph of the 300-m-diameter dish at the National Astronomy and Ionospheric Center near Arecibo, Puerto Rico. The receivers that detect the focused radiation are suspended nearly 300 m (about 80 stories) above the center of the dish. One insert shows a close-up of the radio receivers hanging high above the dish. The other insert shows technicians adjusting the dish surface.

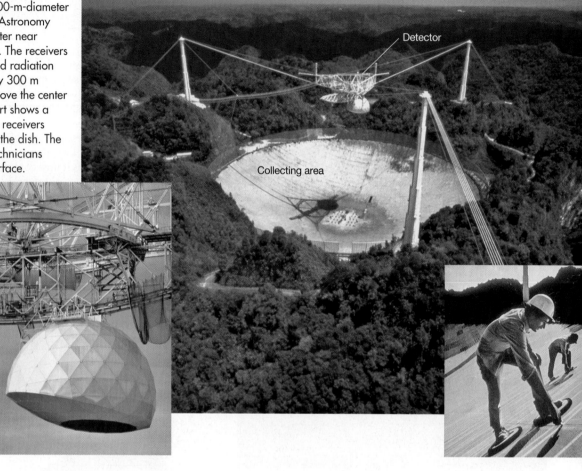

and cheap. Although fairly rough, the chicken wire was adequate for proper reflection because the openings between adjacent strands of wire were much smaller than the long-wavelength radio waves to be detected. The entire Arecibo dish was resurfaced in 1974 with thin metal panels, and upgraded in 1997, so that it can now be used to study shorter-wavelength radio radiation. Since the 1997 upgrade the panels can be adjusted to maintain a precise spherical shape to an accuracy of about 3 mm over the entire surface. At a frequency of 5 GHz (corresponding to a wavelength of 6 cm—the shortest wavelength that can be studied given the properties of the dish surface), the telescope's angular resolution is about 1′. The huge size of the dish creates one distinct disadvantage, however: the Arecibo telescope cannot be pointed very well to follow cosmic objects across the sky; its observations are restricted to those objects that happen to pass within about 20° of overhead as Earth rotates.

Arecibo is an example of a rough-surfaced telescope capable of detecting long-wavelength radio radiation. At the other extreme, Figure 5.22 shows the 36-m-diameter Haystack dish in northeastern Massachusetts. It is constructed of polished aluminum and maintains a parabolic curve to an accuracy of about a millimeter all the way across its solid surface. It can reflect and accurately focus radio radiation with wavelengths as short as a few millimeters. The telescope is contained within a protective shell, or radome, that protects the surface from the harsh wind and weather of New England. It acts much like the protective dome of an optical telescope, except that there is no slit through which the telescope "sees." Incoming cosmic radio signals pass virtually unimpeded through the radome's fiberglass construction.

THE VALUE OF RADIO ASTRONOMY

Despite the inherent disadvantage of relatively poor angular resolution, radio astronomy enjoys many advantages. Radio telescopes can observe 24 hours a day. Darkness is not needed for receiving radio signals because the Sun is a relatively weak source of radio energy, so its emission does not swamp radio signals arriving at Earth from elsewhere in the sky. In addition, radio observations can often be made through cloudy skies, and radio telescopes can detect the longest-wavelength radio waves even during rain or snowstorms. Poor weather causes few problems because the wavelength of most radio waves is much larger than the typical size of atmospheric raindrops or snowflakes. Optical astronomy cannot be done under these conditions because the wavelength of visible light is smaller than a raindrop, a snowflake, or even a minute water droplet in a cloud.

However, perhaps the greatest value of radio astronomy (and, in fact, of all invisible astronomies) is that it opens up a whole new window on the universe. There are two lengths but they allow us to see whole new classes of objects that would otherwise be completely unknown.

Figure 5.23 shows an optical photograph of the Orion Nebula (a huge cloud of interstellar gas) taken with the 4-m telescope on Kitt Peak. Superimposed on the optical image is a radio map of the same region, obtained by scanning a radio telescope back and forth across the nebula and taking many measurements of radio intensity. The map is drawn as a series of contour lines connecting locations of equal radio brightness, similar to pressure contours drawn by meteorologists on weather maps or height contours drawn by cartographers on topographic maps. The inner contours represent stronger radio signals, the outside contours weaker signals.

The radio map in Figure 5.23 has many similarities to the visible-light image of the nebula. For instance, the radio emission is strongest near the center of the optical image and declines toward the nebular edge. But there are also subtle differences between the radio and optical images. The two differ mainly toward the upper left of the main cloud, where visible light seems to be absent, despite the existence of radio waves. How can radio waves be detected from locations not showing any light emission? The answer is that this particular nebular region is known to be especially dusty in its top left quadrant. The dust obscures the short-wavelength visible light but not the long-wavelength radio radiation. Thus, our radio map allows us to see the true extent of this cosmic source.

Figure 5.22 Photograph of the Haystack dish, inside its protective radome. For scale, note the engineer standing at the bottom. Also note the dull shine on the telescope surface, indicating its smooth construction. Haystack is a poor optical mirror but a superb radio telescope. Accordingly, it can be used to reflect and accurately focus radiation having short radio wavelengths, even as small as a fraction of a centimeter.

R I V U X G

Figure 5.23 The Orion Nebula is a star-forming region about 1500 light years from Earth. (The nebula is located in the constellation Orion and can be seen in Figure 1.6.) The bright regions in this photograph are stars and clouds of glowing gas. The dark regions are not empty, but their visible emission is obscured by interstellar matter. Superimposed on the optical image is a radio contour map of the same region. Each curve of the contour map represents a different intensity of radio emission. The resolution of the optical image is about 1″; that of the radio map 1′.

main reasons for this. First, just as objects that are bright in the visible part of the spectrum (the Sun, for example) are not necessarily strong radio emitters, many of the strongest radio sources in the universe emit little or no visible light. Second, visible light may be strongly absorbed by interstellar dust along the line of sight to a source. Radio waves, on the other hand, are generally unaffected by intervening matter. Many parts of the universe cannot be seen at all by optical means but are easily detectable at radio wavelengths. The center of the Milky Way Galaxy is a prime example of such a totally invisible region—our knowledge of the Galactic Center is based almost entirely on radio and infrared observations. Thus, radio observations not only afford us the opportunity to study the same objects at different wave-

5.5 Interferometry

5 The main disadvantage of radio astronomy compared with optical work is its relatively poor angular resolution. However, in some circumstances, radio astronomers can overcome this problem by using a technique known as **interferometry.** This technique makes it possible to produce radio images of much higher angular resolution than can be achieved with even the best optical telescopes, on Earth or in space.

In interferometry, two or more radio telescopes are used in tandem to observe the *same* object at the *same* wavelength and at the *same* time. The combined instruments together make up an **interferometer.** Figure 5.24 shows a large interferometer—many separate radio telescopes working together as a team. By means of electronic cables or radio links, the signals received by each antenna in the array making up the interferometer are sent to a central computer that combines and stores the data. The technique works by analyzing how the waves interfere with each other when added together. ∞ (Sec. 3.1) If the detected radio waves are in step, they combine constructively to produce a strong signal. If the signals are not in step, they destructively interfere and cancel each other. As the antennas track their target, a pattern of peaks and troughs emerges. After extensive computer processing, this pattern translates into a high-resolution image of the target object.

An interferometer is essentially a substitute for a single huge antenna. As far as resolving power is concerned, the effective diameter of an interferometer is the distance between its outermost dishes. In other words, two small dishes can act as opposite ends of an imaginary but huge single radio telescope, dramatically improving the angular resolution. For example, resolution of a few arc seconds can be achieved at typical radio wavelengths (such as 10 cm), either by using a single radio telescope 5 km in diameter (which is impossible to build) or by using two or more much smaller dishes separated by 5 km and connected electronically. The larger the distance separating the telescopes—the longer the *baseline* of the interferometer—the better the resolution attainable. Large interferometers like the instrument shown in Figure 5.24 now routinely attain radio resolution comparable to that of optical images. Figure 5.25 compares an interferometric radio map of a nearby galaxy with a photograph of that same galaxy made using a large optical telescope. The radio clarity is superb—much better than the radio contour map of Figure 5.23.

Astronomers have created radio interferometers spanning very great distances, first across North America and later between continents. A typical very-long-baseline interferometry experiment (usually known by the acronym VLBI) might use radio telescopes in North America, Europe, Australia, and Russia to achieve angular resolution on the order of 0.001″, about 1000 times better than images produced by most current optical telescopes. It seems that even Earth's diameter is no limit. Radio astronomers have successfully used an antenna in orbit, together with several antennas on the ground, to construct an even longer baseline and achieve still better resolution. Proposals exist to place interferometers entirely in Earth orbit, and even on the Moon.

(a)

(b)

Figure 5.24 This large interferometer is made up of 27 separate dishes spread along a Y-shaped pattern about 30 km across on the Plain of San Augustin in New Mexico. The most sensitive radio device in the world, it is called the Very Large Array or VLA, for short. (b) A close-up view from ground level of some of the VLA antennas. Notice that the dishes are mounted on railroad tracks so that they can be repositioned easily.

Figure 5.25 (a) VLA radio "photograph" (or radiograph) of the spiral galaxy M51, observed at radio frequencies with an angular resolution of a few arc seconds (a); shows nearly as much detail as (b) an actual (light) photograph of that same galaxy made with the 4-m Kitt Peak optical telescope.

Although the technique was originally developed by radio astronomers, interferometry is no longer restricted to the radio domain. Radio interferometry became feasible when electronic equipment and computers achieved speeds great enough to combine and analyze radio signals from separate radio detectors without loss of data. As the technology has improved, it has become possible to apply the same methods to higher-frequency radiation. Millimeter-wavelength interferometry has already become an established and important observational technique, and infrared interferometry will become commonplace in the next few years.

Optical interferometry is the subject of intensive research. In 1997 a group of scientists in Cambridge, England, succeeded in combining the light from three small optical telescopes to produce a single, remarkably clear, image. Each telescope of the Cambridge Optical Aperture Synthesis Telescope (COAST) is only 0.4 m in diameter, but when the equipment is positioned 6 m apart, the resulting resolution is a stunning 0.01″. This enables a "splitting" of the binary star Capella, whose two member stars are separated by only 0.05″ and which therefore is normally seen from the ground only as a slightly oblong blur. With the new Cambridge telescope array, the two stars are cleanly and individually separated. Even *HST*, operating above the atmosphere in orbit, cannot match these test observations with this brand-new device. The Keck telescopes on Mauna Kea are also designed to be used for infrared—and perhaps someday for optical—interferometric work.

5.6 Other Astronomies

6 The electromagnetic spectrum consists of far more than just visible light and radio waves. Optical and radio astronomy are the oldest and best-established branches of astronomy, but since the 1970s there has been a virtual explosion of observational techniques covering the rest of the electromagnetic spectrum. Today, all portions of the spectrum are studied, from radio waves to gamma rays, to maximize the amount of information available about astronomical objects. As noted earlier, the types of astronomical objects that can be observed differ quite markedly from one wavelength range to another. Full-spectrum coverage is essential not only to see things more clearly, but even to see some things at all.

Because of the transmission characteristics of Earth's atmosphere, astronomers must study most wavelengths other than optical and radio from space. The rise of these "other astronomies" has therefore been closely tied to the development of the space program.

INFRARED ASTRONOMY

Infrared studies are a very important component of modern observational astronomy. Generally, **infrared telescopes** resemble optical telescopes (indeed, many optical telescopes are also used for infrared work), but their detectors are sensitive to longer-wavelength radiation. Although most infrared radiation is absorbed by the atmosphere (primarily by

Figure 5.26   An optical photograph (a) taken near San Jose, California, and an infrared photo (b) of the same area taken at the same time. Longer-wavelength infrared radiation can penetrate smog much better than short-wavelength visible light.

water vapor), there are a few windows in the high-frequency part of the infrared spectrum where the opacity is low enough to allow ground-based observations (see Figure 3.11). ∞ (Sec. 3.3) Some of the most useful infrared observing is done from the ground, even though the radiation is somewhat diminished in intensity by our atmosphere. Because of its 4-km altitude, Mauna Kea (Figure 5.13a) is one of the finest locations on Earth for both optical and infrared ground-based astronomy. The thin air at this high altitude guarantees less atmospheric absorption of incoming radiation and hence a clearer view than is possible from sea level.

As with radio observations, the longer wavelength of infrared radiation often enables us to perceive objects partially hidden from optical view. As a terrestrial example of the penetrating properties of infrared radiation, Figure 5.26 shows a dusty and hazy region in California, hardly viewable optically, but easily seen using infrared radiation.

Astronomers can make still better infrared observations if they can place their instruments above most or all of Earth's atmosphere. Improvements in balloon-, aircraft-, rocket-, and satellite-based telescope technologies have made infrared research a very powerful tool with which to study the universe (see Figure 5.27). However, as might be expected, the infrared telescopes that can be carried above the atmosphere are considerably smaller than the massive instruments found in ground-based observatories. The most advanced facility to function in this part of the spectrum is the *Infrared Astronomy Satellite* (*IRAS*), shown in Figure 5.27(b). Launched into Earth orbit in 1983 but now inoperative, this British–Dutch–U.S. satellite housed a 0.6-m mirror with an angular resolution as fine as 30″. (As usual, the resolution depended on the precise wavelength observed.) Its sensitivity was greatest for radiation in the 10- to 100-μm range. During its 10-month lifetime (and long

Figure 5.27   (a) A gondola containing a 1-m infrared telescope (lower left) is readied for its balloon-borne ascent to an altitude of about 30 km (100,000 feet), where it will capture infrared radiation that cannot penetrate the atmosphere. (b) An artist's conception of the *Infrared Astronomy Satellite*, placed in orbit in 1983. This 0.6-m telescope surveyed the infrared sky at wavelengths ranging from 10 to 100 μm. During its 10 months of operation it greatly increased astronomers' understanding of many different aspects of the universe, from the formation of stars and planets to the evolution of galaxies.

(a)   (b)

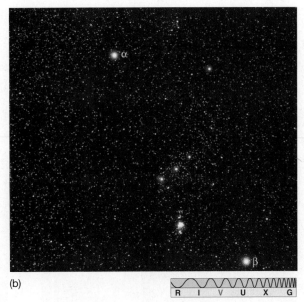

Figure 5.28 (a) This infrared image of the Orion Nebula and its surrounding environment was made by the *Infrared Astronomy Satellite*. The whiter regions denote greater strength of infrared radiation; the false colors denote different temperatures, descending from white to red to black. (b) The same region photographed in visible light. The labels α and β refer, respectively, to Betelgeuse and Rigel, the two brightest stars in the constellation. Note how the red star Betelgeuse can be seen in the infrared (part a), but the blue star Rigel cannot.

afterward—the data archives are still heavily used even today), *IRAS* contributed greatly to our knowledge of clouds of Galactic matter destined to become stars, and possibly planets. These regions of interstellar gas are composed of warm gas that cannot be seen with optical telescopes or adequately studied with radio telescopes. Because much of the material between the stars has a temperature between a few tens and a few hundreds of kelvins, Wien's law tells us that the infrared domain is the natural portion of the electromagnetic spectrum in which to study it. ∞ (Sec. 3.4) Throughout the text we will encounter many findings made by this satellite about comets, stars, galaxies, and the scattered dust and rocky debris found among the stars.

Figure 5.28(a) shows an *IRAS* image of the Orion region. At about 1′ angular resolution, the fine details of the Orion nebula evident in the visible portion of the earlier image (Figure 5.23) cannot be perceived. Nonetheless, astronomers can extract much useful information about this object and others like it from infrared observations. Star-forming clouds of warm dust and gas and extensive groups of bright young stars, completely obscured at visible wavelengths, are seen.

Unfortunately, by Wien's law, telescopes themselves also radiate strongly in the infrared unless they are cooled to nearly absolute zero. The end of *IRAS*'s mission came not because of any equipment malfunction or unexpected mishap but simply because its supply of liquid helium coolant ran out. *IRAS*'s own thermal emission then overwhelmed the radiation it was built to detect. In November 1995 the European Space Agency launched the 0.6-m *Infrared Space Observatory* (*ISO*). It is now in orbit, refining and extending the groundbreaking work begun by *IRAS*. NASA plans to deploy the 0.85-m *Space Infrared Telescope*

Facility (*SIRTF*) in 2002. Both instruments are designed to operate at wavelengths in the 3- to 200-μm range.

ULTRAVIOLET ASTRONOMY

To the short-wavelength side of the visible spectrum lies the ultraviolet domain. This region of the spectrum, extending in wavelength from 400 nm (4000 Å, blue light) down to a few nanometers ("soft" X rays), has only recently begun to be explored. Because Earth's atmosphere is partially opaque to radiation below 400 nm and is totally opaque below about 300 nm (in part because of the ozone layer), astronomers cannot conduct any useful ultraviolet observations from the ground, not even from the highest mountaintop. Rockets, balloons, or satellites are therefore essential to any **ultraviolet telescope**—a device designed to capture and analyze this high-frequency radiation.

One of the most successful ultraviolet space missions was the *International Ultraviolet Explorer* (*IUE*), placed in Earth orbit in 1978 and shut down for budgetary reasons in late 1996 (see *Interlude 18-1*). Like all ultraviolet telescopes, its basic appearance and construction were quite similar to optical and infrared devices. Several hundred astronomers from all over the world have used *IUE* to explore a variety of phenomena in planets, stars, and galaxies. In subsequent chapters we will learn what this relatively new window on the universe has shown us about the activity and even the violence that seems to pervade the cosmos. The *Hubble Space Telescope* (*Interlude 5-1*), best known as an optical telescope, is also a superb ultraviolet instrument.

Figure 5.29(a) shows an image of a supernova remnant— the remains of a violent stellar explosion that occurred some

(a) R I V U X G

(b) R I V U X G

Figure 5.29   (a) A camera on board the *Extreme Ultraviolet Explorer* satellite captured this image of the Vela supernova remnant, the result of a massive star blowing itself virtually to smithereens. The release of energy is prodigious, and the afterglow lingers for millennia. The glowing field of debris shown here lies some 1500 light years from Earth. Based on the velocity of the outflowing debris, astronomers estimate that the explosion itself must have occurred about 12,000 years ago. For scale, the inset shows the angular size of the Moon. (b) This false-color image of the spiral galaxy M74 was made by an ultraviolet telescope aboard the *Astro* payload carried by a space shuttle in 1990 and 1995.

12,000 years ago—obtained by the *Extreme Ultraviolet Explorer* (*EUVE*) satellite, which was launched in 1992. *EUVE* operates at the short-wavelength (1–50 nm) end of the ultraviolet range, making it sensitive to phenomena involving high temperatures (hundreds of thousands, even millions, of kelvins) or other energetic events. Since its launch *EUVE* has mapped out our local cosmic neighborhood as it presents itself in the far ultraviolet and has radically changed astronomers' conception of interstellar space in the vicinity of the Sun.

An alternative means of placing astronomical payloads into (temporary) Earth orbit is provided by NASA's space shuttles. In December 1990 and again in March 1995, a shuttle carried aloft the *Astro* package of three ultraviolet telescopes. An *Astro* image of a nearby galaxy is shown in Figure 5.29(b). Astronomical shuttle missions offer a potentially very flexible way for astronomers to get instruments into space, without the long lead times and great expense of permanent satellite missions like *HST*.

HIGH-ENERGY ASTRONOMY

High-energy astronomy studies the universe as it presents itself to us in X rays and gamma rays—the types of radiation whose photons have the highest frequencies and hence the greatest energies. How do we detect radiation of such short wavelengths? First, it must be captured high above Earth's atmosphere because none of it reaches the ground. Second, its detection requires the use of equipment basically differ-

ent in design from that used to capture the relatively low energy radiation discussed up to this point. The basic difference in the design of **high-energy telescopes** comes about because X and gamma rays cannot be reflected easily by any kind of surface. Rather, these rays tend to pass straight through, or be absorbed by, any material they strike. When X rays barely graze a surface, however, they can be reflected from it in a way that yields an image, although the mirror design is fairly complex (see Figure 5.30). For gamma rays (with wavelengths of less than about 0.01 nm), no such method of producing an image has yet been devised. Present-day gamma-ray telescopes simply point in a specified direction and count photons received.

In addition, X-ray and gamma-ray detection methods using photographic plates or CCD devices do not work well. Instead, individual X-ray and gamma-ray photons are counted by electronic detectors on board an orbiting device, and the results are then transmitted to the ground for further processing and analysis. Furthermore, the number of photons in the universe seems to be inversely related to frequency. Trillions of visible (starlight) photons reach the detector of an optical telescope on Earth each second, but hours or even days are often needed for a single gamma-ray photon to be recorded. Not only are these photons hard to focus and measure, they are also very scarce.

Figure 5.31(a) is a photograph of the second *High-Energy Astronomy Observatory* (*HEAO-2*, also known as the *Einstein Observatory*). Launched in 1978, this was the first X-ray

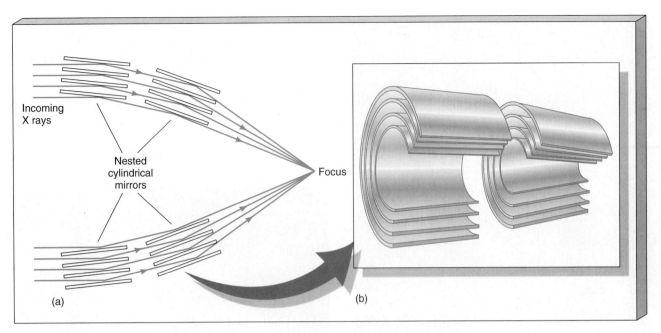

Figure 5.30 The arrangement of mirrors in an X-ray telescope allows X rays to be reflected at grazing angles and focused to form an image.

Figure 5.31 (a) *HEAO-2*, also known as the *Einstein Observatory*, the first imaging X-ray telescope. (b) *AXAF*, the *Advanced X-ray Astrophysics Facility*, is a much larger device, shown here being assembled in 1998.

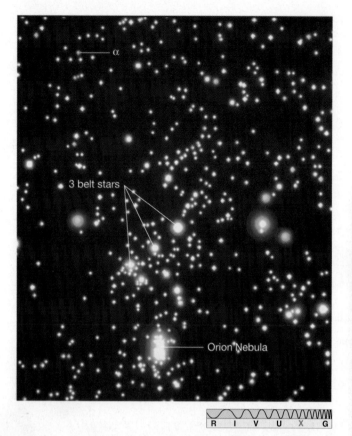

Figure 5.32 An X-ray image of the Orion region, taken by the *ROSAT* X-ray satellite. (Compare with Figures 1.6, 5.23, and 5.28.) Objects that are most intensely emitting X-rays are colored blue, including the three stars of Orion's belt and the glowing nebula below them at bottom center in the photograph.

telescope capable of forming an image of its field of view. During its 2-year lifetime, this spacecraft made major advances in our understanding of high-energy phenomena throughout the universe; its observational database is still heavily used. The most recent major X-ray satellite is the German *ROSAT* (short for *Röntgen Satellite*, after Wilhelm Röntgen, the discoverer of X rays), launched in 1990 by a US Air Force rocket. With more sensitivity, a wider field of view, and better resolution than Einstein, *ROSAT* is providing high-energy astronomers with new levels of observational detail (Figure 5.32). Even more powerful will be NASA's *Advanced X-ray Astrophysics Facility* (*AXAF*), another long-duration orbiting observatory (in the spirit of *IUE* and *HST*) that is scheduled to become operational in 1999 (see Figure 5.31b).

Gamma-ray astronomy is the youngest entrant into the observational arena. As just mentioned, imaging gamma-ray telescopes do not exist, so only fairly coarse (1° resolution) observations can be made. Nevertheless, even at this resolution, there is much to be learned. Cosmic gamma rays were originally detected in the 1960s by the U.S. *Vela* series of satellites, whose primary mission was to monitor illegal nuclear detonations on Earth. Since then, several X-ray tele-

scopes have also been equipped with gamma-ray detectors. By far the most advanced instrument is the *Gamma Ray Observatory* (*GRO*), launched by space shuttle in 1991. This satellite can scan the sky and study individual objects in much greater detail than previously attempted. Figure 5.33 shows *GRO* in low Earth orbit, along with a false-color gamma-ray image of a highly energetic outburst in the nucleus of a distant galaxy.

(a)

(b)

Figure 5.33 (a) This photograph of the 17-ton *Gamma-Ray Observatory* (also called the *Compton Observatory*, after an American gamma-ray pioneer) was taken by an astronaut during the satellite's deployment from the space shuttle *Atlantis* over the Pacific Coast of the United States. (b) A typical false-color gamma-ray image—this one showing a violent event in the distant galaxy 3C279, also known as a "gamma-ray blazar."

5.7 Full-Spectrum Coverage

7 Table 5.1 lists the basic regions of the electromagnetic spectrum and describes objects typically studied in each frequency range. Bear in mind that the list is far from exhaustive and that many astronomical objects are now routinely observed at many different electromagnetic wavelengths. As we proceed through the text, we will discuss more fully the wealth of information that high-precision astronomical instruments can provide us.

It is reasonable to suppose that the future holds many further improvements in both the quality and the availability of astronomical data and that many new discoveries will be made. The current and proposed pace of technological progress presents us with the following very exciting prospect: early in the 21st century, if all goes according to plan, it will be possible, for the first time ever, to make *simultaneous* high-quality measurements of any astronomical object at *all* wavelengths, from radio to gamma ray. The consequences of this development for our understanding of the workings of the universe may be little short of revolutionary.

TABLE 5.1 Astronomy at Many Wavelengths

WAVELENGTH/FREQUENCY RANGE	GENERAL CONSIDERATIONS	COMMON APPLICATIONS (CHAPTER REFERENCE)
Radio	Can penetrate dusty regions of interstellar space Earth's atmosphere largely transparent to these wavelengths Can be detected in the daytime as well as at night High resolution at long wavelengths requires very large telescopes or interferometers	Radar studies of planets (2,9) Planetary magnetic fields (11) Interstellar gas clouds (18) Center of the Milky Way Galaxy (23) Galactic structure (23, 24) Active galaxies (25) Cosmic background radiation (27)
Infrared	Can penetrate dusty regions of interstellar space Earth's atmosphere only partially transparent to IR radiation, so some observations must be made from space	Star formation (19,20) Cool stars Center of the Milky Way Galaxy (23) Active galaxies (25) Large-scale structure in the universe (24, 27)
Visible	Earth's atmosphere transparent to visible light	Planets (7–14) Stars and stellar evolution (17, 20, 21) Galactic structure (23, 24) Galaxies (24) Active galaxies (25) Large-scale structure in the universe (24, 27)
Ultraviolet	Earth's atmosphere opaque to UV radiation, so observations must be made from space	Interstellar medium (19) Hot stars
X ray	Earth's atmosphere opaque to X rays, so observations must be made from space Special mirror configurations needed to form images	Stellar atmospheres (16) Neutron stars and black holes (22) Hot gas in galaxy clusters (24) Active galactic nuclei (25)
Gamma ray	Earth's atmosphere opaque to gamma rays, so observations must be made from space Cannot form images	Neutron stars (22) Active galactic nuclei (25)

As a preview of the sort of comparison that full-spectrum coverage allows, Figure 5.34 shows a series of images of our own Milky Way Galaxy. They were made by several different instruments, at wavelengths ranging from radio to gamma ray, over a period of about 5 years. By comparing the features visible in each, we immediately see how multi-wavelength observations can complement one another, greatly extending our perception of the universe around us.

Figure 5.34   The Milky Way Galaxy, as it appears (from top to bottom) at (a) radio, (b) infrared, (c) visible, (d) X-ray, and (e) gamma-ray wavelengths. Each frame is a panoramic view covering the entire sky. The center of our Galaxy, which lies in the direction of the constellation Sagittarius, is at the center of each map.

Chapter Review *WWW*

SUMMARY

A **telescope** (p. 98) is a device designed to collect as much light as possible from some distant source and deliver it to a detector for detailed study. **Reflecting telescopes** (p. 98) use a mirror to concentrate and focus the light. **Refracting telescopes** (p. 98) use a lens; **refraction** (p. 98) is the bending of light as it passes from one medium to another. The **prime focus** (p. 98) of the telescope is the point where the incoming beam is focused and where analysis instruments may be placed. The **Newtonian** (p. 102) and **Cassegrain telescope** (p. 102) designs employ secondary mirrors to avoid placing heavy equipment at the prime focus.

The lenses used in refracting telescopes suffer from a number of problems, among them **chromatic aberration** (p. 101), the tendency of lenses to focus different colors to different prime foci. These problems become more difficult to correct the larger the lens is, with the result that all astronomical telescopes larger than about 1 m in diameter use mirrors in their design.

The light collected by a telescope may be processed in a number of ways. It can be made to form an **image** (p. 98), a **photometer** (p. 103) may be used to make detailed measurements of the energy received, or a **spectrometer** (p. 103) may study its spectrum.

The light-gathering power of a telescope depends on its **collecting area** (p. 103), which is proportional to the square of the mirror diameter. To study the faintest sources of radiation, astronomers must use large telescopes.

An important aspect of a telescope is its **angular resolution** (p. 104), the ability to distinguish between light sources lying close together on the sky. One limitation on the resolution of a telescope is diffraction, which makes it impossible to focus a beam perfectly. The amount of diffraction is proportional to the wavelength of the radiation under study and inversely proportional to the size of the mirror. Thus, at any given wavelength, larger telescopes suffer least from the effects of diffraction.

The resolution of most ground-based optical telescopes is actually limited by **seeing** (p. 107)—the blurring effect of Earth's turbulent atmosphere, which smears the pointlike images of stars out into **seeing disks** (p. 107) a few arc seconds in diameter. Radio and space-based telescopes do not suffer from atmospheric effects, so their resolution is determined by the effects of diffraction.

Most modern telescopes now use **charge-coupled devices, or CCDs** (p. 108), instead of photographic plates to collect their data. The field of view is divided into an array of millions of **pixels** (p. 108) that accumulate an electric charge when light strikes them. CCDs are many times more sensitive than photographic plates, and the resultant data are easily saved directly on disk or tape for later image processing.

Using **active optics** (p. 109), in which a telescope's environment and focus are carefully monitored and controlled, and **adaptive optics** (p. 110), in which the blurring effects of atmospheric turbulence are corrected for in real time, it may soon be possible to achieve diffraction-limited resolution in ground-based optical instruments.

Radio telescopes (p. 110) are conceptually similar in construction to optical reflectors. However, radio telescopes are generally much larger than optical instruments, for two reasons. First, the amount of radio radiation reaching Earth from space is tiny compared with optical wavelengths, so a large collecting area is essential. Second, the long wavelengths of radio waves mean that diffraction severely limits the resolution unless large instruments are used.

In order to increase the effective area of a telescope, and hence improve its resolution, several separate instruments may be combined into a device called an **interferometer** (p. 116). Using **interferometry** (p. 116), radio telescopes can produce images much sharper than those from the best optical equipment. Infrared interferometers are under construction, and optical interferometric systems are under active development.

Infrared telescopes (p. 117) and **ultraviolet telescopes** (p. 119) are similar in their basic design to optical systems. Infrared studies in some parts of the infrared range can be carried out using large ground-based systems. Ultraviolet astronomy must be carried out from space.

High-energy telescopes (p. 120) study the X- and gamma-ray regions of the electromagnetic spectrum. X-ray telescopes can form images of their field of view, although the mirror design is more complex than for lower-energy instruments. Gamma-ray telescopes simply point in a certain direction and count photons received. Because the atmosphere is opaque at these short wavelengths, both types of telescopes must be placed in space.

Radio and other nonoptical telescopes are essential to studies of the universe because they allow astronomers to probe regions of space that are completely opaque to visible light and to study the many objects that emit little or no optical radiation at all.

SELF-TEST: TRUE OR FALSE?

_____ **1.** The primary purpose of any telescope is to produce an enormously magnified image of the field of view.

_____ **2.** A refracting telescope cannot form an image of its field of view.

_____ **3.** A Newtonian telescope has no secondary mirror.

_____ **4.** A Cassegrain telescope has a hole in the middle of the primary mirror to allow light reflected from its secondary mirror to reach a focus behind the primary mirror.

_____ **5.** The term "seeing" is used to describe how faint an object can be detected by a telescope.

_____ **6.** The primary advantage to using the *Hubble Space Telescope* is the increased amount of "night" time available to it.

_____ **7.** One of the primary advantages of CCDs over photograph plates is their high efficiency in detecting light.

_____ **8.** The *Hubble Space Telescope* can observe objects in the optical, infrared, and ultraviolet parts of the spectrum.

_____ **9.** The Keck telescopes contain the largest single mirrors ever produced.

_____ **10.** Radio telescopes are large, in part to improve their angular resolution, which is poor because of the long wavelengths at which they observe.

_____ **11.** Radio telescopes are large, in part because the sources of radio radiation they observe are very faint.

_____ **12.** Radio telescopes have to have surfaces as smooth as those in optical telescope mirrors.

_____ **13.** Infrared astronomy must be done from space.

_____ **14.** Because the ozone layer absorbs ultraviolet light, astronomers must make observations in the ultraviolet from the highest mountain tops.

_____ **15.** Gamma-ray telescopes employ the same basic design as optical instruments.

SELF-TEST: FILL IN THE BLANK

1. A telescope that uses a lens to focus light is called a _____ telescope.

2. A telescope that uses a mirror to focus light is called a _____ telescope.

3. All large modern telescopes are of the _____ type.

4. The light-gathering power of a telescope is determined by the _____ of its mirror or lens.

5. The angular resolution of a telescope is limited by the _____ of the telescope and the _____ of the radiation being observed.

6. The angular resolution of ground-based optical telescopes is more seriously limited by Earth's _____ than by diffraction.

7. Optical telescopes on Earth can see angular detail down to about _____ arc second.

8. CCDs produce images in _____ form that can be easily transmitted, stored, and later processed by computers.

9. Active optics and adaptive optics are both being used to improve the _____ of ground-based optical telescopes.

10. All radio telescopes are of the _____ design.

11. An _____ is two or more telescopes used in tandem to observe the same object, in order to improve angular resolution.

12. _____, _____, and _____ astronomy can be done only from above Earth's atmosphere.

13. An object with a temperature of 300 K would be best observed with an _____ telescope.

14. The mirrors in X-ray telescopes are different in design from those in optical instruments because X rays tend to be _____, rather than reflected, by solid surfaces.

15. Gamma-ray telescopes are unable to form _____ of their fields of view.

REVIEW AND DISCUSSION

1. Cite two reasons why astronomers are continually building larger and larger telescopes.

2. What are three advantages of reflecting telescopes over refracting telescopes?

3. How does Earth's atmosphere affect what is seen by an optical telescope?

4. What advantages does the _Hubble Space Telescope_ have over ground-based telescopes? List some disadvantages.

5. What are the advantages of a CCD over a photographic plate?

6. What is image processing?

7. Describe some ways in which optical astronomers can compensate for the blurring effects of Earth's atmosphere.

8. Why do radio telescopes have to be very large?

9. What kind of astronomical objects can we best study with radio techniques?

10. What is interferometry, and what problem in radio astronomy does it address?

11. Compare the highest resolution attainable with optical telescopes with the highest resolution attainable with radio telescopes (including interferometers).

12. What special conditions are required to conduct observations in the infrared?

13. Compared with optical astronomy, what new problems arise when we wish to make observations in the high-energy domain?

14. What is the main advantage of studying objects at different wavelengths of radiation?

15. Our eyes can see light with an angular resolution of 1′. Suppose our eyes detected only infrared radiation, with 1° angular resolution. Would we be able to make our way around on Earth's surface? To read? To sculpt? To create technology?

PROBLEMS

1. A certain telescope has a 10′ × 10′ field of view that is recorded using a CCD chip having 1024 × 1024 pixels. What angle on the sky corresponds to 1 pixel? What would be the diameter of a typical seeing disk (2″ radius), in pixels?

2. Estimate the fraction of incoming starlight obscured by the observer and observing cage at the prime focus of the Hale 5-m telescope (see Figure 5.8).

3. A 2-m telescope can collect a given amount of light in 1 hour. Under the same observing conditions, how much time would be required for a 6-m telescope to perform the same task? A 12-m telescope?

4. A certain space-based telescope can achieve (diffraction-limited) angular resolution of 0.05′ for red light (of wavelength 700 nm). What would its resolution be (a) in the infrared, at 3.5 μm, and (b) in the ultraviolet, at 140 nm?

5. The photographic equipment on a telescope is replaced by a CCD. If the photographic plate records 5 percent of the light reaching it, while the CCD records 75 percent, how much time will the new system take to collect as much information as the old detector recorded in a 1-hour exposure?

6. The Andromeda Galaxy lies about 2.9 million light years or 900 kpc away. To what distances do the angular resolutions of the Hale telescope (1″), *HST* (0.05″), and a radio interferometer (0.001″) correspond to at that distance?

7. Based on collecting areas, how much more sensitive would you expect the Arecibo telescope (Figure 5.21) to be, compared with the 43-m Green Bank instrument (Figure 5.20)?

8. What would be the equivalent single-mirror diameter of a telescope constructed from two separate 10-m mirrors? Four separate 8-m mirrors?

9. Estimate the angular resolutions of (a) a radio interferometer with a 5000-km baseline, operating at a frequency of 5 GHz, and (b) an infrared interferometer with a baseline of 50 m, operating at a wavelength of 1 μm.

10. An instrument aboard the *Gamma Ray Observatory* detects a burst of energy coming from some distant source. Simultaneously, an X-ray telescope with an effective diameter of 0.5 m observes the same event in the 2 keV energy range (see *More Precisely 4-2* for a definition of the electron volt, or eV). Both detectors record the location of the event on the sky, to within the accuracy of their respective measurements. You are about to make a follow-up observation using a large optical telescope with a 5′ field of view. Which position—the X-ray or the gamma-ray measurement—should you use, and why?

PROJECTS

1. Here's how to take some easy pictures of the night sky. You will need a location with a clear, dark sky, a 35mm camera with a standard 50-mm lens, tripod, and cable release, a watch with a seconds display visible in the dark, and a roll of high-speed color film.

Set your camera to the "bulb" setting for the exposure and attach the cable release so you can take a long exposure. Set the focus on infinity. Point the camera to a favored constellation, seen through your viewfinder, and take a 20- to 30-second exposure. Don't touch any part of the camera or hold on to the cable release during the exposure; minimize all vibration. Keep a log of your shots. When finished, have the film developed in the standard way.

2. For some variations, vary your exposure times, use different films, take hours-long exposures for star trails, use different lenses such as wide-angle or telephoto, place the camera piggyback on a telescope that is tracking, and take exposures that are a few minutes long. Experiment and have fun!

Venus

Earth

Jupiter

Jupiter
Sun   Mercury
Earth
Venus
Mars

Saturn

Uranus

Neptune

Saturn

Uranus

Neptune

6 THE SOLAR SYSTEM

An Introduction to Comparative Planetology www

LEARNING GOALS

Studying this chapter will enable you to:

1 Discuss the importance of comparative planetology to solar system studies.

2 Describe the overall scale and structure of the solar system.

3 Summarize the basic differences between the terrestrial and the jovian planets.

4 Identify and describe the major nonplanetary components of the solar system.

5 Describe some of the spacecraft missions that have contributed significantly to our knowledge of the solar system.

(Opposite page, background) *Voyager 2*, one of the most remarkable robotic space probes to travel through the solar system, sent back a wealth of scientific data about several of our neighboring planets. As it sped toward interstellar space after its close encounter with Neptune, it took this "family portrait" of our planetary system. The diagram depicts the positions at which numerous photographs were taken, superposed on a sketch of the solar system. The inserts show some of the actual images radioed back to Earth.

In less than a single generation, we have learned more about the solar system—the Sun and everything that orbits it—than in all the centuries that went before. By studying the planets—the major bodies that orbit the Sun and reflect its light—the moons that orbit them, and the countless fragments of material that move in interplanetary space, astronomers have gained a richer outlook on our own home in space. Instruments aboard unmanned robots have taken close-up photographs of the planets and their moons and in some cases have made on-site measurements. The discoveries of the past few decades have revolutionized our understanding not only of our cosmic neighborhood but also its history, for our solar system is filled with clues to its own origin and evolution.

6.1 An Inventory of the Solar System

The Greeks and other astronomers of old were aware of the Moon, the stars, and five planets—Mercury, Venus, Mars, Jupiter, and Saturn—in the night sky. ∞ (Sec. 2.2) They also knew of two other types of heavenly objects that were clearly neither stars nor planets. *Comets* appear as long, wispy strands of light in the night sky that remain visible for periods of up to several weeks, then slowly fade from view. *Meteors,* or "shooting stars" are sudden bright streaks of light that flash across the sky, usually vanishing less than a second after they first appear. These transient phenomena must have been very familiar to ancient astronomers, but their role in the "big picture" of the solar system was not understood until much later.

Human knowledge of the basic content of the solar system remained largely unchanged from ancient times until the early seventeenth century, when the invention of the

telescope made more detailed observations possible. Galileo Galilei was the first to capitalize on this new technology (his simple telescope is shown in Figure 6.1). Galileo's discovery of the phases of Venus and of four moons orbiting Jupiter early in the seventeenth century helped change forever humankind's vision of the universe. ∞ (Sec. 2.5) As technological advances continued, knowledge of the solar system improved rapidly. Astronomers began discovering objects invisible to the unaided human eye. By the end of the nineteenth century, astronomers had found Saturn's rings (1659), the planets Uranus (1758) and Neptune (1846), many planetary moons, and the first *asteroids*—"minor planets" orbiting the Sun, mostly in a broad band (called the *asteroid belt*) lying between Mars and Jupiter. Ceres, the largest asteroid and the first to be sighted, was discovered in 1801. A large telescope of mid-nineteenth-century vintage is shown in Figure 6.2.

The twentieth century brought continued improvements in optical telescopes. One more planet (Pluto) was discovered, along with three more planetary ring systems,

Figure 6.1 The telescope with which Galileo made his first observations was simple, but its influence on astronomy was immeasurable.

Figure 6.2 By the mid-nineteenth century, telescopes had improved enormously in both size and quality. Shown here is the telescope built and used by Irish nobleman and amateur astronomer the Earl of Rosse.

Figure 6.3 An *Apollo* astronaut doing some lunar geology— prospecting near a huge boulder near the Mare Serenitatus.

Figure 6.4 The Mars rover *Sojourner*, seen here using a device called an Alpha Proton X-Ray Spectrometer to determine the chemical composition of a large Martian rock, nicknamed Yogi by mission scientists. For scale, the rock is about 1 m tall, and lies some 10 m from *Sojourner's* mother ship, *Pathfinder*.

dozens of moons, and thousands of asteroids. The century also saw the rise of both nonoptical astronomy—especially radio and infrared—and spacecraft exploration, each of which has made vitally important contributions to the field of planetary science. Astronauts have carried out experiments on the Moon (see Figure 6.3), and numerous unmanned probes have left Earth and traveled to all but one of the other planets. Figure 6.4 shows the *Sojourner* robot roaming the Martian surface in 1997.

As currently explored, our **solar system** is known to contain 1 star (the Sun), 9 planets, 63 moons (at last count) orbiting those planets, 6 asteroids larger than 300 km in diameter, more than 7000 smaller (but well-studied) asteroids, myriad comets a few kilometers in diameter, and countless *meteoroids* less than 100 m across. The list will undoubtedly grow as we continue to explore our cosmic neighborhood.

1 As we proceed through the solar system in the next few chapters, we will seek to understand how each planet compares with our own and what each contributes to our knowledge of the solar system as a whole. We will use the powerful and still emerging perspective of **comparative planetology**—comparing and contrasting the properties of the diverse worlds we encounter—to understand better the conditions under which planets form and evolve. Our goal is to develop a comprehensive theory of the origin and evolution of our planetary system that explains all, or at least most, of its observed properties. We will try to answer basic questions such as: Why did planet X evolve in one way, while planet Y turned out completely differently? and Why are the planets' orbits so orderly, when their individual properties are not? In addressing these issues we will find many similarities and common features among planets, but each

will also present us with new questions and unique insights. Whatever the answers, the comparison enriches our knowledge of the ways planets work. As we unravel the origin of the solar system, we also learn about the nature of planetary systems beyond our own.

6.2 Planetary Properties

Table 6.1 lists some basic orbital and physical properties of the nine planets, with a few other solar-system objects (the Sun, the Moon, an asteroid, and a comet) included for comparison. Most of these quantities can be determined using methods described in Chapters 1 and 2.

The distance of each planet from the Sun is known from Kepler's laws once the scale of the solar system is set by radar-ranging on Venus. ∞ (Sec. 2.6) A planet's (sidereal) orbital period is easily measurable from repeated observations of its location on the sky, so long as Earth's own motion around the Sun is properly taken into account. The masses of planets with moons may be calculated by application of Newton's laws of motion and gravity, just by observing the moons' orbits around the planets. ∞ (*More Precisely 2-3*) The sizes of those orbits, like the sizes of the planets themselves, are found by measuring their angular sizes and then applying elementary geometry. ∞ (Sec. 1.5)

The masses of Mercury and Venus (as well as those of our own Moon and the asteroid Ceres) are a little harder to determine accurately because these bodies have no natural satellites of their own. Nevertheless, it is possible to measure

TABLE 6.1 Properties of Some Solar System Objects

OBJECT	ORBITAL SEMI-MAJOR AXIS (A.U.)	ORBIT PERIOD (Earth years)	MASS (Earth masses)	RADIUS (Earth radii)	NUMBER OF KNOWN MOONS	ROTATION PERIOD* (days)	AVERAGE DENSITY (kg/m³)	(g/cm³)
Mercury	0.39	0.24	0.055	0.38	0	59	5400	5.4
Venus	0.72	0.62	0.82	0.95	0	−243	5200	5.2
Earth	1.0	1.0	1.0	1.0	1	1.0	5500	5.5
Moon	—	—	0.012	0.27	—	27.3	3300	3.3
Mars	1.5	1.9	0.11	0.53	2	1.0	3900	3.9
Ceres (asteroid)	2.8	4.7	0.00015	0.073	0	0.38	2700	2.7
Jupiter	5.2	11.9	318	11.2	16	0.41	1300	1.3
Saturn	9.5	29.4	95	9.5	18	0.44	700	0.7
Uranus	19.2	84	15	4.0	17	−0.72	1300	1.3
Neptune	30.1	164	17	3.9	8	−0.67	1600	1.6
Pluto	39.5	248	0.002	0.2	1	−6.4	2100	2.1
Comet Hale-Bopp	180	2400	1.0×10^{-9}	0.004	—	0.47	100	0.1
Sun	—	—	332,000	109	—	25.8	1400	1.4

*A negative rotation period indicates retrograde (backward) rotation relative to the sense in which all planets orbit the Sun.

their masses by careful observations of their influence on other planets or nearby bodies. Mercury and Venus produce small but measurable effects on each other's orbits, as well as that of Earth. The Moon causes small "wobbles" in Earth's motion as the two bodies orbit their common center of mass. Only in the case of Ceres is the mass still poorly known, mainly because that asteroid's gravity is so weak. All the techniques necessary for determining mass were available to astronomers well over a century ago. Today, the masses of most of the objects in Table 6.1 have been very accurately measured through their gravitational interaction with artificial satellites and space probes launched from Earth.

As indicated in Table 6.1, the Sun, having more than 1000 times the mass of the next most massive object (the planet Jupiter), is clearly the dominant member of the solar system. In fact, the Sun contains about 99.9 percent of all solar system material. The planets—including our own—are insignificant in comparison.

A planet's rotation period is determined simply by watching surface features appear and disappear again as the planet rotates. For some planets this is difficult to do, as their surfaces are hard to see or may even be nonexistent! The surface of Venus is completely obscured by clouds, whereas Jupiter, Saturn, Uranus, and Neptune have no solid surfaces at all—their atmospheres simply thicken, and eventually become liquid, as we descend deeper and deeper below the visible clouds. We will describe the methods used to measure the rotation periods of these planets in later chapters.

The final columns in Table 6.1 lists a property called **density.** This is a measure of the "compactness" of the matter within an object. It is computed by dividing the object's mass (in kilograms, say) by its volume (in cubic meters, for example). Dividing Earth's mass (determined from observations of the Moon's orbit) by its volume (which we know because we know Earth's radius—see *More Precisely 1-3*), we obtain an average density of approximately 5500 kg/m³. *On average*, then, there are about 5500 kilograms of Earth matter in every cubic meter of Earth volume. For comparison, the density of ordinary water is 1000 kg/m³, rocks on Earth's surface have densities in the range 2000–3000 kg/m³, and iron has a density of some 8000 kg/m³. Earth's atmosphere has a density of only a few kilograms per cubic meter. Because most working astronomers are more familiar with the cgs (centimeter, gram, second) units of density (grams per cubic centimeter—g/cm³—where 1 kg/m³ = 1000 g/cm³), Table 6.1 lists density in both SI and cgs units.

6.3 The Overall Layout of the Solar System

2 By earthly standards, the solar system is immense. The distance from the Sun to Pluto is about 40 A.U., almost a million times Earth's radius and roughly 15,000 times the distance from Earth to the Moon. Yet, despite the solar system's vast extent, the planets all lie very close to the Sun, astronomically speaking. Even the diameter of the largest orbit, that of Pluto, is less than 1/1000 of a light year, whereas the next nearest star is several light years distant.

The planet closest to the Sun is Mercury. Moving outward, we encounter in turn Venus, Earth, Mars, Jupiter, Saturn, Uranus, Neptune, and Pluto. In Chapter 2 we saw the basic properties of the planets' orbits. Their paths are all ellipses, with the Sun at (or very near) one focus. ∞ (Sec. 2.4) Most planetary orbits have low eccentricities. The exceptions are the innermost and the outermost worlds, Mercury and Pluto. Accordingly, we can reasonably think of most planets' orbits as circles centered on the Sun. The orbits of the major bodies in the solar system are illustrated in Figure 6.5.

All the planets orbit the Sun in the same sense—counterclockwise as seen from above Earth's North Pole—and in nearly the same plane as Earth (that is, the ecliptic plane). Mercury and Pluto deviate somewhat from this rule. Their orbital planes lie at 7° and 17° to the ecliptic, respectively. Still, we can think of the solar system as being quite flat. Its "thickness" perpendicular to the plane of the ecliptic is less than 1/50 the diameter of Pluto's orbit. If we were to view

the planets' orbits from a vantage point in the ecliptic plane about 50 A.U. from the Sun, only Pluto's orbit would be noticeably tilted. Figure 6.6 is a photograph of planets Mercury, Venus, Mars, and Jupiter taken during the July 1991 solar eclipse. These four planets happen to be visible in this one photograph in large part because their orbits lie nearly in the same plane.

The planetary orbits are not evenly spaced: the orbits get farther and farther apart as we move farther out from the Sun. Nevertheless, there is a certain regularity in their spacing. In the eighteenth century a fairly simple rule, now known as the Titius-Bode law (*Interlude 6-1*), seemed to "predict" the radii of the planetary orbits remarkably well. Even the asteroid belt between Mars and Jupiter appeared to have a place in the scheme, which excited great interest among astronomers and numerologists alike. There is apparently no simple explanation for this empirical "law." Today, it is regarded more as a curiosity than as a fundamental property of the solar system, although many theorists still feel that it may hold some deep, but as yet undiscovered, insight into the way the solar system formed.

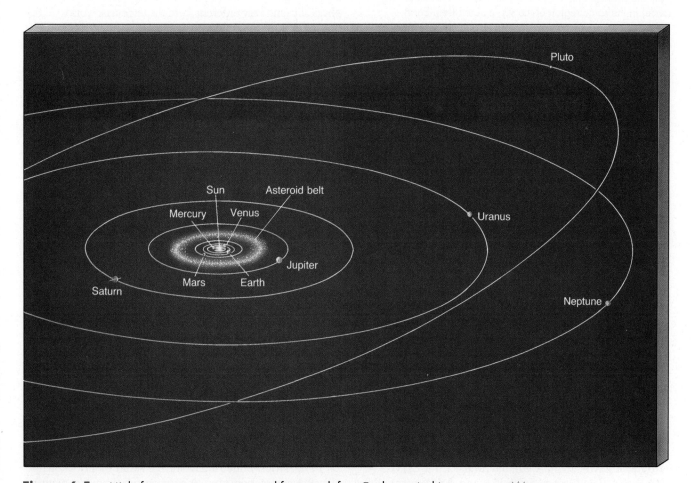

Figure 6.5   Might future space voyagers travel far enough from Earth to gain this perspective on our solar system? Except for Mercury and Pluto, the orbits of the planets lie nearly in the same plane. As we move out from the Sun, the distance between the orbits of the planets increases. The entire solar system spans nearly 80 A.U.

Figure 6.6 Taken from Hawaii during the July 1991 eclipse of the Sun, this single photograph shows four planets—Mercury, Venus, Mars, and Jupiter. Because they all orbit in nearly the same plane, it is possible for them all to appear (by chance) in the same region of the sky, as seen from Earth.

6.4 Terrestrial and Jovian Planets

3 On large scales, the solar system presents us with a sense of orderly motion. The planets move nearly in a plane, on almost concentric (and nearly circular) elliptical paths, in the same direction around the Sun, at steadily increasing orbital intervals. However, the individual properties of the planets themselves are much less regular.

Figure 6.7 compares the major planets with one another and with the Sun. A clear distinction can be drawn between the inner and the outer members of our planetary system based on densities and other physical properties. The inner planets—Mercury, Venus, Earth, and Mars—are small, dense, and *rocky* in composition. The outer worlds—Jupiter, Saturn, Uranus, and Neptune (but not Pluto)—are large, of low density, and *gaseous*.

Because the physical and chemical properties of Mercury, Venus, and Mars are somewhat similar to Earth's, the four innermost planets are called the **terrestrial planets**. (The word *terrestrial* derives from the Latin word *terra*, meaning "land" or "earth.") The larger outer planets—Jupiter, Saturn, Uranus, and Neptune—are all similar to one another chemically and physically (and very different from the terrestrial worlds). They

 INTERLUDE 6-1 The Titius–Bode "Law"

A close look at the list of semi-major axes of the known planets (see Table 6.1) reveals considerable regularity in their orbits. The *spacing* of the orbits increases more or less geometrically as we move out from the Sun: at any point in the list, the distance to the next planet out is about twice that to the next planet in.

Is there some underlying structure to the solar system? In 1766, in search of an answer to that question, the German astronomer Johann Titius came up with a simple formula that "predicted" quite well the orbits of the then-known planets, Mercury through Saturn. Johann Bode, a better-known astronomer of the day, later popularized this relationship among the planets' orbits, so that it now is usually known as the Titius–Bode law, or even just Bode's law. We must emphasize that it is not a law at all, in the scientific sense, but rather just a rule for determining the approximate orbital semi-major axes of the planets. Nevertheless, the "law" part of the name has stuck.

The rule for determining the planets' orbits is as follows. Start with 0.4, the distance (in A.U.) from the Sun to Mercury. Then add to it, successively, 0.3 (to arrive at Venus, with an orbital semi-major axis of 0.7 A.U.), 0.3 (to reach Earth, at 1.0 A.U.), 0.6 (Mars: 1.6 A.U.), 1.2 (?: 2.8 A.U.), 2.4 (Jupiter: 5.2 A.U.), and 4.8 (Saturn: 10.0 A.U.) to arrive at the (approximate) orbital distances of the known planets, plus one extra "planet" between Mars and Jupiter. The relation between successive orbits is easily

seen—after Venus, the added term (that is, the interplanetary spacing) simply doubles at each step. Even the fictitious extra "planet" is acceptable—it corresponds to the middle of the asteroid belt! Thus it was that this rule appeared to contain some deep insight into our planetary system.

The test of any theory is its predictive power. According to the Titius–Bode rule, the next three planets beyond Saturn should lie at 19.6, 38.8, and 77.2 A.U. When Uranus was discovered in 1781, it fell close to the prediction, at 19.2 A.U. Speculation was rife that the next planet would also lie where the "law" decreed it should. Sadly, the rule fails for Neptune, which lies only 30 A.U. from the Sun. However, if we ignore Neptune, Pluto is in just about the right place!

Few people today take the rule sufficiently seriously to bother to look for a scenario in which Neptune started out in the "right" orbit and somehow got moved farther in toward the Sun, perhaps leaving Pluto behind. Still, the regularity of the "law" was appealing to astronomers of the eighteenth and nineteenth centuries, as it suggested some fundamental harmony in the structure of the solar system, and, to some extent, that appeal remains today.

The Titius–Bode rule is not the result of any known planetary interaction, nor is it even very accurate. Yet many astronomers suspect that it may be telling us *something* about the formation of the solar system—they just don't know quite *what*.

Figure 6.7 Diagram, drawn to scale, of the relative sizes of the planets and our Sun. Notice how much larger the jovian planets are than Earth and the other terrestrials and how much larger still is the Sun.

are labeled the **jovian planets,** after Jupiter, the largest member of the group. (The word *jovian* comes from *Jove,* another name for the Roman god Jupiter.) The jovian worlds are all much larger than the terrestrial planets, and quite different from them in both composition and structure.

The four terrestrial planets all lie within about 1.5 A.U. of the Sun. All are small and of relatively low mass, and all have generally rocky composition and solid surfaces. Beyond that, however, the similarities end. When we take into account how the weight of overlying layers compresses the interiors of the planets to different extents (greatest for Earth, least for Mercury), we find that the average *uncompressed densities* of the terrestrial worlds—that is, the densities they would have in the absence of any compression—decrease steadily as we move farther from the Sun. This decrease in density indicates that the overall compositions of these planets differ significantly one from the other.

There are many more differences among the terrestrial worlds. All have atmospheres, but the atmospheres are about as dissimilar as we could imagine, ranging from a near-vacuum on Mercury to a hot, dense inferno on Venus. Earth alone has oxygen in its atmosphere (as well as liquid water on its surface). The present-day conditions on the surfaces of the four planets are also quite distinct from one another. Earth

and Mars spin at roughly the same rate—one rotation every 24 (Earth) hours—but Mercury and Venus both take months to rotate just once, and Venus rotates in the opposite sense from the others. Earth and Mars have moons, but Mercury and Venus do not. Earth and Mercury have measurable magnetic fields, of very different strengths, whereas Venus and Mars have none. Finding the common threads in the evolution of four such diverse worlds is no simple task! Comparative planetology will be our indispensable guide as we proceed through the coming chapters.

Yet for all their differences, the terrestrial worlds still seem very similar when compared with the jovian planets. Perhaps the simplest way to express the major differences between the terrestrial and jovian worlds is to say that the jovian planets are everything the terrestrial planets are not. Table 6.2 compares and contrasts some key properties of these two planetary classes.

The terrestrial worlds lie close together, near the Sun; the jovian worlds are widely spaced through the outer solar system. The terrestrial worlds are small, dense, and rocky; the jovian worlds are large and gaseous, being made up predominantly of hydrogen and helium (the lightest elements), which are rare on the inner planets. The terrestrial worlds have solid surfaces; the jovian worlds have none (their dense

TABLE 6.2 Comparison of the Terrestrial and Jovian Planets

TERRESTRIAL PLANETS	JOVIAN PLANETS
close to the Sun	far from the Sun
closely spaced orbits	widely spaced orbits
small masses	large masses
small radii	large radii
predominantly rocky	predominantly gaseous
solid surface	no solid surface
high density	low density
slower rotation	faster rotation
weak magnetic fields	strong magnetic fields
few moons	many moons
no rings	many rings

atmospheres thicken with depth, eventually merging with their liquid interiors). The terrestrial worlds have weak magnetic fields, if any; the jovian worlds all have strong magnetic fields. The terrestrial worlds have only three moons among them; the jovian worlds have many moons each, no two of them alike and none of them like our own. Furthermore, all the jovian planets have *rings*, a feature unknown on the terrestrial planets. Despite their greater size, the jovian worlds all rotate much faster than any terrestrial planet.

Finally, beyond the outermost jovian planet, Neptune, lies one more small world, frozen and mysterious. Pluto doesn't fit well into either planetary category. Indeed, there is debate among planetary scientists as to whether it should be classified as a planet at all. In both mass and composition, it has much more in common with the icy jovian moons than with any terrestrial or jovian planet. Astronomers speculate that it may in fact be the largest member of a newly recognized class of solar system objects that reside beyond the jovian worlds.

6.5 Interplanetary Debris

4 In the vast space among the nine known major planets move countless chunks of rock and ice, all orbiting the Sun, many on highly eccentric paths. This final component of the solar system is the collection of *interplanetary matter*— cosmic "debris" ranging in size from the relatively large asteroids, through the smaller comets and even smaller meteoroids, down to the smallest grains of interplanetary dust that litter our cosmic environment.

The dust arises when larger bodies collide and break apart into smaller pieces that, in turn, collide again and are slowly ground into microscopic fragments, which eventually settle into the Sun or are swept away by the *solar wind*, a stream of energetic charged particles that continually flows outward from the Sun and pervades the entire solar system. The dust is quite difficult to detect in visible light, but infrared studies reveal that interplanetary space contains surprisingly large amounts of it. Our solar system is an extremely good vacuum by terrestrial standards, but positively dirty by the standards of interstellar or intergalactic space.

Asteroids (Figure 6.8a) and meteoroids are generally rocky in composition, somewhat like the outer layers of the terrestrial planets. The distinction between the two is simply a matter of size—anything larger than 100 m in diameter (corresponding to a mass of about 10,000 tons) is conven-

(a)

(b)

Figure 6.8 (a) Asteroids, like meteoroids, are generally composed of rocky material. This asteroid, Gaspra, is about 20 km long; it was photographed by the *Galileo* spacecraft on its way to Jupiter. (b) Comet Hale-Bopp, seen as it approached the Sun in 1997. Most comets are composed largely of ice, and so tend to be relatively fragile. Its vaporized gas and dust form the tail, here extending away from the Sun for nearly a quarter of the way across the sky. (See also the chapter-opening photo for Chapter 14.)

tionally termed an asteroid, anything smaller is a meteoroid. Their total mass is much less than that of Earth's Moon, so these objects play no important role in the present-day workings of the planets or their moons. Yet they are of crucial importance to our studies, for they are the keys to answering some very fundamental questions about our planetary environment and what the solar system was like soon after its birth. Many of these bodies are made of material that has evolved hardly at all since the early days of the solar system. (In addition, they often conveniently deliver themselves right to our doorstep, in the form of meteorites, allowing us to study them in detail without having to fetch them from space.)

Comets are quite distinct from the other small bodies in the solar system. They are generally icy rather than rocky in composition (although they do contain some rocky material) and typically have diameters in the 1–10 km range. They are quite similar in chemical makeup to some of the icy moons of the outer planets. Even more so than the asteroids and meteoroids, comets represent truly ancient material—the vast majority have probably not changed in any significant way since their formation long ago along with the rest of the solar system. Comets striking Earth's atmosphere do not reach the surface intact, so we do not have actual samples of cometary material. However, they do vaporize and emit radiation as their highly elongated orbits take them near the Sun (see Figure 6.8b). Astronomers can determine a comet's makeup by spectroscopic study of the radiation it gives off as it is destroyed. ∞ (Sec. 4.2)

6.6 Spacecraft Exploration of the Solar System

5 Since the 1960s dozens of unmanned space missions have visited the other planets of the solar system. All the planets but Pluto have been visited and probed at close range. The impact of these missions on our understanding of our planetary system has been nothing short of revolutionary. In the next few chapters we will see many examples of the marvelous images radioed back to Earth. Here, we focus on just a few of these remarkable technological achievements.

THE *MARINER 10* FLYBYS OF MERCURY

In 1974, the U.S. spacecraft *Mariner 10* came within 10,000 km of the surface of Mercury, sending back high-resolution images of the planet. These photographs, which showed surface features as small as 150 m across, dramatically increased our knowledge of the planet. For the first time, we saw Mercury as a heavily cratered world, in many ways reminiscent of our own Moon.

Mariner 10 was launched from Earth in November 1973 and was placed in an eccentric 176-day orbit about the Sun, aided by a gravitational assist (see *Interlude 6-2*) from the planet Venus (Figure 6.9). In that orbit, *Mariner 10*'s nearest point to the sun (perihelion) is close to Mercury's path, and its farthest point away (aphelion) lies between the

Figure 6.9 The path of the *Mariner 10* probe to Mercury included a gravitational boost from Venus. The spacecraft (inset) returned data from March 1974 until March 1975, providing astronomers with a wealth of information on the planet Mercury.

INTERLUDE 6-2 Interplanetary Navigation

Celestial mechanics is the study of the motions of gravitationally interacting objects, such as planets and stars. These days, people studying the subject bring powerful computers to bear on Newton's laws of motion (see Chapter 2) to understand the intricate movements of astronomical bodies. Computerized celestial mechanics enables astronomers to calculate the orbits of the planets with high precision, taking their small gravitational influences on one another into account. Even before the computer age, the discovery of one of the outermost planets, Neptune, came about almost entirely through studies of the distortions of Uranus's orbit that were caused by Neptune's gravity.

Celestial mechanics has become an essential tool as scientists and engineers navigate manned and unmanned spacecraft throughout the solar system. Robot probes can now be sent on stunningly accurate trajectories, expressed in the trade with such slang phrases as "sinking a corner shot on a billion-kilometer pool table." In fact, near-flawless rocket launches, aided by occasional midcourse changes in flight paths, now enable interplanetary navigators to steer remotely controlled spacecraft through an imaginary "window" of space just a few kilometers wide and a billion kilometers away.

Sophisticated knowledge of celestial mechanics also aids navigation of a single space probe toward several planets. For example, in 1974 the *Mariner 10* spacecraft was guided into the part of Venus's gravitational field that would swing the ship around to precisely the right path for an additional trek toward Mercury. In other words, Venus itself propelled the probe in a new direction, a course alteration that required no fuel.

The accompanying figure illustrates how such a *gravitational slingshot* works. From the point of view of the planet, the spacecraft arrives along a hyperbolic (unbound) trajectory, passes close by, and then escapes along the same trajectory, in a new direction and with the same speed relative to the planet. However, if the planet itself is moving, some of the planet's momentum is transferred to the spacecraft as it passes by. If the orbit of the spacecraft is chosen correctly, the craft can gain energy, and speed up, as a result of the encounter. Of course, there is no "free lunch"—the extra energy acquired by the spacecraft comes from the planet's motion, causing its orbit to change ever so slightly. However, since planets are so much more massive than spacecraft, the effect is tiny—*Mariner 10* had no measurable effect on Venus's orbit.

Such a slingshot maneuver has been used several times. For example, the *Voyager 2* spacecraft (launched in late 1977)

closely passed by Jupiter (in 1979), Saturn (in 1981), Uranus (in 1986), and Neptune (in 1989). It is now moving through the outermost reaches of the solar system. The gravitational fields of the giant planets whipped the craft around at each visitation, enabling flight controllers to get considerable extra "mileage" out of the probe.

The Galileo mission to Jupiter which was launched in 1989 and arrived at its target in 1995, received *three* gravitational assists en route—one from Venus and two from Earth. Once in the Jupiter system, one of its probes used the gravity of Jupiter and its moons to propel it through a complex series of maneuvers designed to bring it close to all the major moons as well as to the planet itself. Every encounter with a moon had a slingshot effect—sometimes accelerating and sometimes slowing the probe, but each time moving it onto a different orbit—and every one of these effects was carefully calculated long before Galileo ever left Earth.

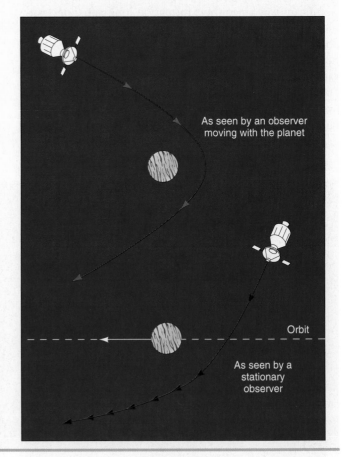

As seen by an observer moving with the planet

Orbit

As seen by a stationary observer

orbits of Venus and Earth. The 176-day period is exactly two Mercury years, so the spacecraft revisits Mercury roughly every 6 months. However, only on the first three encounters—in March 1974, September 1974, and March 1975—did the spacecraft return data. After that, the craft's

supply of maneuvering fuel was exhausted. In total, over 4000 photographs, covering about 45 percent of the planet's surface, were radioed back to Earth during the mission's active lifetime. The remaining 55 percent of Mercury is still unexplored.

EXPLORATION OF VENUS

In all, some 20 spacecraft have visited Venus since the 1970s, far more than have spied on any other planet. The Soviet space program took the lead role in exploring Venus's atmosphere and surface, while American spacecraft have performed extensive radar mapping of the planet from orbit. The American *Mariner 2* and *Mariner 5* missions passed within 35,000 km of the planet in 1962 and 1967, and *Mariner 10* grazed Venus at a distance of 6000 km en route to Mercury.

During roughly the same period, the Soviet *Venera* (derived from the Russian word for Venus) program got under way, and the Soviet *Venera 4* through *Venera 12* probes parachuted into the planet's atmosphere between 1967 and 1978. The early *Venera probes* were destroyed by enormous atmospheric pressures before reaching the surface. Then, in 1970, *Venera 7* (Figure 6.10) became the first spacecraft to soft-land on the planet. During the 23 minutes it survived on the surface, it radioed back information on atmospheric pressure and temperature. Since then, a number of *Venera* landers have transmitted photographs of the surface back to Earth and have analyzed the atmosphere and the soil. None of them survived for more than an hour in the planet's hot, dense atmosphere. The data they sent back make up the entirety of our direct knowledge of Venus's surface. In 1983 the *Venera 15* and *Venera 16* orbiters sent back detailed radar maps (at about 2-km resolution) of large portions of Venus's northern hemisphere.

The U.S. *Pioneer Venus* mission in 1978 placed an orbiter at an altitude of some 150 km above Venus's surface and dispatched a "multiprobe" consisting of five separate instrument packages into the planet's atmosphere. During its hour-long descent to the surface, the probe returned information on the variation of density, temperature, and chemical composition with altitude in the atmosphere. The orbiter's radar produced images of most of the planet's surface.

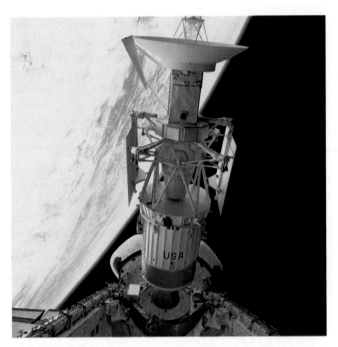

Figure 6.11 The U.S. *Magellan* spacecraft is launched from the space shuttle *Atlantis* in May 1989.

The most recent U.S. mission was the *Magellan* probe (shown in Figure 6.11), which entered orbit around Venus in August 1990. The spacecraft began sending back spectacular data (see Chapter 9) in September 1990. It completed its first 243-day mapping cycle (the time required for Venus to rotate once beneath the probe's orbit) in May 1991. *Magellan's* spatial resolution was at least 10 times better than the best images previously obtained. It could distinguish objects as small as 120 m across and measure vertical distances to within less than 50 m. The probe covered the entire surface of Venus with unprecedented clarity, rendering all previous data virtually obsolete. Many theories of the processes shaping the planet's surface have had to be radically altered or abandoned completely because of *Magellan's* data.

EXPLORATION OF MARS

Both NASA and the Soviet (now Russian) space agency have Mars exploration programs that began in the 1960s. However, the Soviet effort was plagued by a string of technical problems, along with a liberal measure of plain bad luck. As a result, almost all the detailed planetary data we have on Mars has come from unmanned U.S. probes.

The first spacecraft to reach the Red Planet was *Mariner 4*, which flew by Mars in July 1965. The images sent back by the craft showed large numbers of craters caused by impacts of meteoroids with the planet's surface, but nothing of the Earthlike terrain some scientists had expected to find. Flybys in 1969 by *Mariner 6* and *Mariner 7* confirmed these findings, leading to the conclusion that Mars was a geologically dead planet having a heavily cratered, old surface. Studies of Mars received an enormous boost with the ar-

Figure 6.10 One of the Soviet *Venera* landers that reached the surface of Venus. The design was essentially similar for all the surface missions. Note the heavily armored construction, necessary to withstand the harsh conditions on the planet's surface.

rival in November 1971 of the *Mariner 9* orbiter. The craft mapped the entire Martian surface at a resolution of about 1 km, and it rapidly became clear that here was a world far more complex than the dead planet imagined only a year or two previously. *Mariner 9*'s maps revealed vast plains, volcanoes, drainage channels, and canyons. All these features were completely unexpected, given the data provided by the earlier missions. These new findings paved the way for the next step—actual landings on the planet's surface.

The two U.S. *Viking* spacecraft arrived at Mars in mid-1976. *Viking 1* and *Viking 2* each consisted of two parts. An orbiter mapped the surface at a resolution of about 100 m (about the same as the resolution achieved by *Magellan* on Venus), and a lander (see Figure 6.12) descended to the surface and performed a wide array of geological and biological experiments. *Viking 1* touched down on Mars on July 20, 1976. *Viking 2* arrived in September of the same year. By any standards, the *Viking* mission was a complete success; the orbiters and landers returned a wealth of long-term data on the Martian surface and atmosphere. *Viking 2* stopped transmitting data in April 1980. *Viking 1* continued to operate until November 1982.

In August 1993 the first U.S. probe since *Viking*—*Mars Observer*, which was designed to radio back detailed images of the planet's surface and provide data on the Martian atmosphere—exploded just before entering Mars' orbit. However, since that time, the U.S. Mars program has moved into high gear (see *Interlude 6-3*). A replacement for *Mars Observer*, called *Mars Global Surveyor*, was launched from Earth in 1996, arriving at Mars in late 1997. This spacecraft is currently orbiting the planet, scanning it with cameras and other sensors, and charting the Martian landscape.

Mars Global Surveyor was followed (and in fact overtaken) by *Mars Pathfinder*, whose launch date was moved up from 1998 to December 4, 1996. *Pathfinder* arrived at Mars in late June, 1997. On July 4 it parachuted an instrument package to the Martian surface. Near the ground, the parachute fell away and huge airbags deployed, enabling the robot to bounce softly to a safe landing. Side panels opened

and out came a small six-wheeled minirover, called *Sojourner*, (shown in Figure 6.4). For almost three months *Sojourner* roamed the Martian countryside controlled from Earth, at a rate of a few meters per day, taking geological and atmospheric measurements far and wide near the mother ship.

NASA also has ambitious plans for manned missions to Mars. However, the enormous expense (and danger) of such an undertaking, coupled with the belief of many astronomers that unmanned missions are economically and scientifically preferable to manned missions, make the future of these projects uncertain at best.

MISSIONS TO THE OUTER PLANETS

Two pairs of U.S. spacecraft launched in the 1970s—the *Pioneer* and *Voyager* missions—revolutionized our knowledge of Jupiter and the jovian planets. *Pioneer 10* and *Pioneer 11* were launched in March 1972 and April 1973, respectively, arriving at Jupiter in December 1973 and December 1974. The *Pioneer* spacecraft took many photographs and made numerous scientific discoveries. Their orbital trajectories also allowed them to observe the polar regions of Jupiter in much greater detail than later missions would achieve. In addition to their many scientific accomplishments, the *Pioneer* craft also played an important role as "scouts" for the later *Voyager* missions. The *Pioneer* series demonstrated that spacecraft could travel the long route from Earth to Jupiter without colliding with debris in the solar system. They also discovered—and survived—the perils of Jupiter's extensive radiation belts (somewhat like Earth's Van Allen belts, but on a much larger scale). In addition, *Pioneer 11* used Jupiter's gravity to propel it along the same trajectory to Saturn that the *Voyager* controllers planned for *Voyager 2*'s visit to Saturn's rings.

The two *Voyager* spacecraft (see Figure 6.13) left Earth in 1977, reaching Jupiter in March (*Voyager 1*) and July (*Voyager 2*) of 1979 to study the planet and its major

Figure 6.12 A *Viking* lander, here being tested in the Mojave Desert prior to launch.

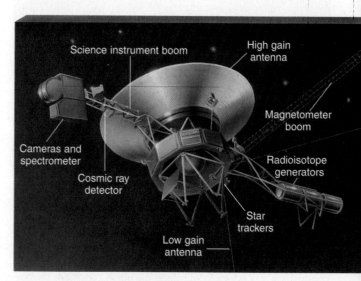

Figure 6.13 The *Voyager* spacecraft. *Voyager 1* and *Voyager 2* (shown here) were identical.

INTERLUDE 6-3 An Armada of Spacecraft Bound for Mars

Even before the 1996 announcement that fossilized life may have been found on Mars (see *Interlude 10-2*), several robotic spacecraft destined to visit the Red Planet were already in the design and construction phase. Now the desire to confirm or deny the claims for life on Mars have given even greater urgency to those missions. Indeed, some of the spacecraft now being readied for launch after the year 2000 may be modified to increase their capability to search for signs of Martian life.

Typically, the relative positions of Mars and Earth in their respective orbits means that it is most favorable to launch a spacecraft from Earth every 2 years; it then takes 6 to 9 months for the craft to arrive at Mars. *Mars Observer* exploded mysteriously while entering Martian orbit in 1993. The replacement mission, *Mars Global Surveyor*, is currently surveying Mars from orbit. No part of this craft will land on the planet's surface. A Russian craft was launched in November 1996, to arrive at Mars in mid-1997. The previous time the Russians tried to rendezvous with the Red Planet, the twin spacecraft of the Soviet mission *Phobos* lost contact with Earth just as both craft were entering Mars's orbit. This time, unfortunately, fortune was no kinder. The rocket's final-stage booster failed to fire, and the probe ended up in the South Pacific Ocean.

As described in the text, the most recent craft to land on Mars was *Sojourner*, part of the *Mars Pathfinder* mission. The accompanying image shows the *Sojourner* robot shortly after landing on July 4, 1997. The robot is still inside part of its parent ship; airbags and the parachute surround the vehicle on the surface. New data about Mars's rock and soil chemistry were among *Sojourner*'s main findings. Surprisingly, many of the rocks sampled resembled those on Earth; they have more silicon than expected, suggesting that they came from volcanic activity on Mars, yet Mars shows little evidence of plate tectonics or other forms of geological activity. Weather conditions were also sampled over the course of several weeks, and a host of new images were obtained of this alien world.

A significant controversy has developed regarding the *Pathfinder* mission, and especially its lander robot, *Sojourner*. The latter was not sterilized before leaving Earth. The only other spacecraft from Earth to land on Mars—two *Viking* robots in 1976—were sterilized in order not to in-troduce any kind of terrestrial bacteria into the Martian environment. Now that there are claims of life on Mars—even fossilized life—it might not be a good idea to run the risk of contaminating Mars with Earth life. This is one of those areas of debate that has become socio-scientific in nature.

The next spacecraft scheduled for touchdown on Mars, in late 1999, is *Mars Surveyor Lander*. The target area is Mars's south pole, a landing site thought to contain alternating layers of ice and dust. This robot lander is designed to drill into the layered terrain, permitting scientists to reconstruct some of the climatic history of Mars, in an attempt to understand conditions when the planet was warmer and wetter. It will also set up a completely automated meteorological station to study the changing Martian weather and will have a 2-m-long robotic arm to probe surface soil, ice, and rocks. Spacecraft builders are now investigating whether they may be able to add a device to this lander that will search for organic compounds typical of terrestrial life; even at the cold polar regions, such Martian life forms could conceivably have found a safe haven just below the surface. The need to sterilize *Mars Surveyor Lander* is still being debated.

These and other missions are seen as a prelude to one that will return samples of Martian matter to Earth robotically. Planned for the time frame 2005–2008, a landing craft on Mars will scoop up some dirt and rocks, leave that planet, then return the surface material at least to Earth orbit. Whether those pieces of Mars will then be brought to Earth's surface for detailed study is another part of the sterilization issue that needs to be thoroughly discussed. Will we contaminate the Mars rocks with terrestrial life by intentionally inviting them onto our home planet? Worse, will we be contaminating Earth with some sort of odd Martian bugs harmful to our own species? Our decisions could have important long-term consequences.

All these above missions to Mars utilize automated robots only. No humans are involved in situ. Nor are there any missions planned at this time that will require astronauts to go along for the ride. Manned missions to Mars are currently considered too expensive and too risky, and it is unclear what humans could do in that harsh environment that robots could not.

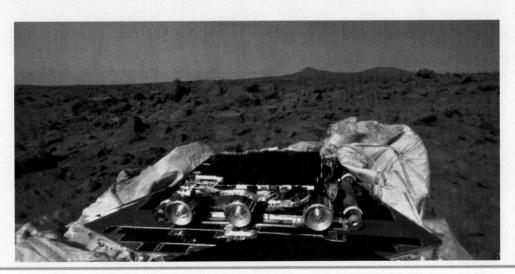

satellites in detail. Each craft carried sophisticated equipment to study the planet's magnetic field, as well as radio, visible-light, and infrared sensors to analyze its reflected and emitted radiation. Both *Voyager 1* and *Voyager 2* used Jupiter's gravity to send them on to Saturn. *Voyager 1* was programmed to visit Titan, Saturn's largest moon, and so did not come close enough to the planet to receive a gravity-assisted boost to Uranus. However, *Voyager 2* went on to visit both Uranus and Neptune in a spectacularly successful "Grand Tour" of the outer planets. The data returned by the two craft are still being analyzed today. Like *Pioneer 11*, the two *Voyager* craft are now headed out of the solar system, still sending data as they race toward interstellar space. Figure 6.14 shows the past and present trajectories of the *Voyager* spacecraft.

The most recent mission to Jupiter is the U.S. *Galileo* probe, launched by NASA in 1989 (see Figure 6.4). It arrived at its target in 1995 after a rather roundabout route (shown in Figure 6.15) involving a gravity assist from Venus and two from Earth itself. In the process, *Galileo* provided astronomers with spectacular views of Earth and Moon, as well as the only close-up photographs of asteroids ever obtained.

The *Galileo* mission consisted of an orbiter and an atmospheric probe. The probe descended into Jupiter's atmosphere, slowed by a heat shield and a parachute, making

measurements and chemical analyses as it went. The orbiter executed a complex series of gravity-assisted maneuvers through Jupiter's moon system, returning to some moons already studied by Voyager and visiting others for the first time. Some of *Galileo*'s main findings are described in Chapter 11.

Shortly after launch, *Galileo* mission controllers discovered that the craft's main antenna, needed to radio its findings back to scientists on Earth, had failed to fully deploy. Despite repeated efforts to open it, the instrument remains jammed in an almost closed position. As a result, all data had to be transmitted via a small secondary antenna, greatly reducing the amount of information that could be returned. Nevertheless, with improved data processing techniques, most of the scientific objectives of the mission were still met. So successful was the original mission, in fact, that it has now been extended for two more years, mainly to study Jupiter's innermost moons Io and Europa.

In October 1997 NASA launched the *Cassini* mission to Saturn. The launch (from Cape Canaveral) sparked controversy because of fears that the craft's plutonium power source might contaminate parts of our planet following an accident either during launch or during a subsequent gravity assist from Earth in 1999—one of four needed for the craft to reach Saturn's distant (9.5 A.U.) orbit. (Two of the

Figure 6.14 The paths taken by the two *Voyager* spacecraft to reach the outer planets. *Voyager 1* is now high above the plane of the solar system, having been deflected up and out of the ecliptic plane following its encounter with Saturn. *Voyager 2* continued on for a "Grand Tour" of the four jovian planets. It is now outside the orbit of Pluto.

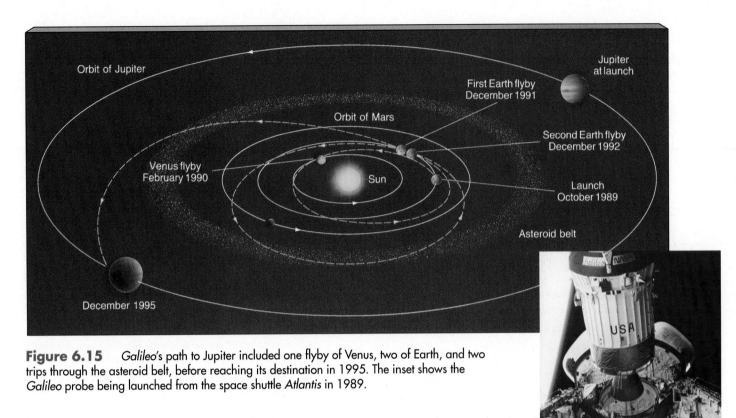

Figure 6.15 *Galileo's* path to Jupiter included one flyby of Venus, two of Earth, and two trips through the asteroid belt, before reaching its destination in 1995. The inset shows the *Galileo* probe being launched from the space shuttle *Atlantis* in 1989.

other assists will come from Venus, the final one from Jupiter.) Once *Cassini* arrives at Saturn in 2004, it will dispatch a probe into the atmosphere of Titan, Saturn's largest moon, and orbit among the planet's moons for 4 years, much like *Galileo* at Jupiter. If experience with *Galileo* is any guide, *Cassini* will likely resolve many outstanding questions about the Saturn system, but it is sure to pose many new ones too.

Chapter Review

SUMMARY

The **solar system** (p. 131) consists of the Sun and everything that orbits it, including the nine major planets, the moons that orbit them, and the many small bodies found in interplanetary space. The science of **comparative planetology** (p. 131) compares and contrasts the properties of the diverse bodies found in the solar system, to understand better the conditions under which planets form and develop.

The asteroids, or "minor planets," are small bodies, none of them larger than Earth's Moon, most of which orbit in a broad band called the asteroid belt between the orbits of Mars and Jupiter. Comets are chunks of ice found mostly in the outer solar system. Their importance to planetary astronomy lies in the fact that they

are thought to be "leftover" material from the formation of the solar system and therefore contain clues to the very earliest stages of its development.

The major planets orbit the Sun in the same sense—counterclockwise as viewed from above Earth's North Pole—on roughly circular orbits that lie close to the ecliptic plane. The orbits of the innermost planet, Mercury, and the outermost, Pluto, are the most eccentric and have the greatest orbital inclination. The spacing between planetary orbits increases as we move outward from the Sun.

Density (p. 132) is a convenient measure of the compactness of any object. The average density of a planet is obtained by dividing the planet's total mass by its volume. The innermost four planets in the solar system have average densities comparable to Earth's and are generally rocky in composition. The outermost planets have much lower densities and, with the exception of Pluto, are made up mostly of gaseous or liquid hydrogen and helium.

Planetary scientists divide the eight large planets (excluding Pluto) in the solar system on the basis of their densities and composition, into the rocky **terrestrial planets** (p. 134)—Mercury, Venus, Earth, and Mars—which lie closest to the Sun, and the gaseous **jovian planets** (p. 135)—Jupiter, Saturn, Uranus, and Neptune— which lie at greater distances. Compared with the terrestrial worlds, the jovian planets are larger and more massive, rotate more rapidly, and have stronger magnetic fields. In addition, the jovian planets all have ring systems and many moons orbiting them.

All the major planets, with the exception of Pluto, have been visited by unmanned space probes. Spacecraft have landed on Venus and Mars. In many cases, the spacecrafts' trajectories have included "gravitational assists" from one or more planets to reach their destinations.

SELF-TEST: TRUE OR FALSE?

____ **1.** Moons are small bodies that orbit the Sun.

____ **2.** Most planets orbit the Sun in nearly the same plane as Earth.

____ **3.** The Titius–Bode law, giving the spacing of planetary orbits, is not a true law at all and is not very accurate.

____ **4.** The largest planets also have the largest densities.

____ **5.** The total mass of all the planets is about half the mass of the Sun.

____ **6.** A planet with a density of 5000 kg/m³ most likely has a gaseous composition.

____ **7.** Most terrestrial planets rotate more slowly than Earth.

____ **8.** Most jovian planets rotate more rapidly than Earth.

____ **9.** The solar wind occurs on Earth during days of exceptional solar heating.

____ **10.** *Mars Observer* was successful in mapping the surface of Mars to a resolution of 100 m.

____ **11.** All planets have moons.

____ **12.** The landing of the Soviet *Venera 7* on the planet Venus was notable because it was the first time any spacecraft had ever landed on a planet.

____ **13.** *Mars Pathfinder* is the last U.S. spacecraft scheduled to visit Mars.

____ **14.** Both *Voyager* missions visited all four jovian planets, but *Voyager 2* then went on to visit Pluto.

____ **15.** The *Galileo* mission was designed to explore the atmosphere and moon system of Jupiter.

SELF-TEST: FILL IN THE BLANK

1. The major bodies orbiting the Sun are known as _____.

2. The _____ are bodies that orbit the Sun between the orbits of Mars and Jupiter

3. The two planets with the highest eccentricities and orbital tilts are _____ and _____.

4. The inner planets Mercury, Venus, Earth, and Mars are known as the _____ planets.

5. The outer planets Jupiter, Saturn, Uranus, and Neptune are known as the _____ planets.

6. The largest and most massive planet in the solar system is _____.

7. Asteroids and meteoroids have a _____ composition, in contrast with comets, which have an _____ composition.

8. Asteroids are similar in composition to the _____ planets.

9. Comets have compositions similar to the _____ moons of the _____ planets.

10. The *Mariner 10* spacecraft was sent by _____ (country) to photograph the planet _____ .

11. The U.S *Magellan* probe mapped the entire surface of Venus using _____.

12. The *Viking 1* and *Viking 2* missions sent orbiters and landers to the planet _____.

13. Jupiter was first visited by the U.S. spacecraft _____.

14. The U.S. spacecraft _____ is the only probe to have visited each of the giant outer planets.

15. Spacecraft visiting one planet often use a _____ assist to visit another planet.

REVIEW AND DISCUSSION

1. Name and describe all the different types of objects found in the solar system. Give one distinguishing characteristic of each. Include a mention of interplanetary space!

2. What is the order of the planets, from the closest to the farthest from the Sun?

3. What is comparative planetology? Why is it useful? What is its ultimate goal?

4. Why has our knowledge of the solar system increased greatly in recent years?

5. Compare and contrast Kepler's laws with the Titius–Bode "law" (*Interlude 6-1*). Why are Kepler's laws considered to be true natural laws, whereas the Titius–Bode "law" is not?

6. In what sense is the solar system "flat"?

7. Which are the terrestrial planets? Why are they given this name? Repeat the question for the jovian planets. What property actually defines these two different types of planets?

8. Name three important differences between the terrestrial planets and the jovian planets.

9. Compare the properties of Pluto given in Table 6.1 with the properties of the terrestrial and jovian planets in Table 6.2. What do you conclude regarding the classification of Pluto as either a terrestrial or jovian planet?

10. Why are asteroids and meteoroids important to planetary scientists?

11. Comets generally vaporize upon striking Earth's atmosphere. How then do we know their composition?

12. How and why do scientists use gravity assists to propel spacecraft through the solar system?

13. Which planets have been visited by spacecraft from Earth? On which ones have spacecraft actually landed?

14. Why do you think *Galileo* and *Cassini* took such circuitous routes to Jupiter and Saturn, while *Pioneer* and *Voyager* did not?

15. How do you think NASA's new policy of building less complex, smaller and cheaper spacecraft, but with shorter times between design and launch, will affect future exploration of the outer planets? Will missions like *Galileo* and *Cassini* be possible?

PROBLEMS

1. Choose two of the objects listed in Table 6.1 and calculate their densities. Compare your results with those given in the table. Assume that all objects are spherical.

2. Use Newton's law of gravity (see *More Precisely 2-2*) to compute your weight (a) on Earth, (b) on Mars, (c) on the asteroid Ceres, and (d) on Jupiter (neglecting temporarily the absence of a solid surface on this planet!).

3. Only Mercury, Mars, and Pluto have orbits that deviate significantly from circles. Calculate the perihelion and aphelion distances from the Sun (see *More Precisely 2-1*) of these planets.

4. According to the Titius–Bode "law," where should the ninth, tenth, and eleventh planets in the solar system lie?

5. Suppose the average mass of each of the 7000 asteroids in the solar system is about 10^{17} kg. Compare the total mass of all asteroids with the mass of Earth. Assuming a roughly spherical shape and a density of 3000 kg/m³, estimate the diameter of an asteroid having this average mass.

6. A *short-period comet* is conventionally defined as a comet having an orbital period of less than 200 years. What is the maximum possible aphelion distance for a short-period comet with a perihelion of 0.5 A.U.? Where does this place the comet relative to the outer planets?

7. How many times has *Mariner 10* now orbited the Sun?

8. A spacecraft has an orbit that just grazes Earth's orbit at perihelion and that of Mars at aphelion. What is its orbital eccentricity and semi-major axis? What is its orbital period? (This is the so-called minimum energy orbit for a craft leaving Earth and reaching Mars. Assume circular planetary orbits for simplicity.)

9. Earth and Mars were at closest approach in March 1997. *Mars Pathfinder* was launched on December 4, 1996 and arrived at Mars in late June, 1997. Sketch the orbits of the two planets and the trajectory of the spacecraft. Be sure to indicate on your diagram the location of Mars at launch and of Earth when *Pathfinder* reached Mars. (For simplicity, neglect the eccentricity of Mars's orbit in your sketch.)

10. How long would it take for a radio signal to complete the round trip between Earth and Saturn? Assume that Saturn is at its closest point to Earth. How far would a spacecraft orbiting the planet in a circular orbit of radius 100,000 km travel in that time? Do you think mission control could maneuver the spacecraft in real time—that is, control all its functions directly from Earth?

PROJECTS

1. You can begin to visualize the ecliptic—the plane of the planets' orbits—just by noticing the path of the Sun throughout the day and of the full Moon in the course of a single night. It helps if you watch from one spot, such as your backyard or a rooftop. It's also good to have a general notion of direction (west is where the Sun sets). You will see that the movements of the Sun and Moon are confined to a narrow pathway across our sky. The planets also travel along this path. The motion of the Sun, Moon, and planets is a two-dimensional reflection of the three-dimensional plane of our solar system.

2. Once you get a feeling for the whereabouts of the ecliptic, try locating the North Star. Knowing the direction to celestial north makes it easier to imagine the motion of the planets in the plane of the solar system. Don't worry about being too precise. Just get a sense of the ecliptic as a kind of merry-go-round of planets—that we on Earth also ride!

3. Go to your library and find what planetary missions are in progress or planned, other than those described in the text. Apart from the United States, what other countries have space agencies actively engaged in planetary exploration? What non-planetary missions are in progress or planned?

EARTH

Our Home in Space www

LEARNING GOALS

Studying this chapter will enable you to:

1 Summarize the overall physical properties of planet Earth.

2 Explain how Earth's atmosphere helps to heat us as well as protect us.

3 Outline our current model of Earth's interior structure and describe some of the experimental techniques used to establish this model.

4 Discuss the nature and origin of Earth's magnetosphere.

5 Summarize the evidence for the phenomenon of "continental drift" and discuss the physical processes that drive it.

6 Describe how both the Moon and the Sun influence Earth's oceans.

(Opposite page, background) This beautiful image of the whole Earth was taken by a weather satellite hovering over the equator. One can clearly see Africa and the Middle East (with desert areas prominent in both), as well as Europe (at top).

(Inset A) A band of cirrus clouds cuts across Egypt, the Nile River, and the Red Sea (upper left) in this (southward) view taken by a crewed spaceship in Earth orbit.

(Inset B) The Red Sea, separating Egypt (right) from Saudi Arabia (left) and the Sinai Peninsula is a new ocean being formed as the African continent begins to drift away from Asia Minor.

(Inset C) Here, the southern tip of the Sinai Peninsula is centered, with the Red Sea at the top, the Gulf of Aqaba at lower left, and the Gulf of Suez at right.

(Inset D) An even closer view of desolate sand dunes in Arabia. The long dunes parallel the prevailing wind direction and can be up to 150 meters high and 600 kilometers long without a break.

Earth is the best-studied terrestrial planet. From the matter of our world sprang life, intelligence, culture, and all the technology we now use to explore the cosmos. We ourselves are made of "earthstuff" as much as are rocks, trees, and air. Now, as humanity begins to explore the solar system, we can draw on our knowledge of Earth to aid our understanding of the other planets. By cataloging Earth's properties and attempting to explain them, we set the stage for our comparative study of the solar system. Every piece of information we can glean about the structure and history of our own world plays a potentially vital role in helping us understand the planetary system in which we live. If we are to appreciate the universe, we must first come to know our own planet. Our study of astronomy begins at home.

7.1 Earth in Bulk

1 The radius of our planet has been known since the time of the ancient Greeks to be about 6400 km (see *More Precisely 1-3*). More precise measurements are now routinely made by spacecraft, yielding a value of 6378 km. We saw in Chapter 2 how we can measure the mass of Earth, or any other astronomical body, by observing its gravitational influence on some other nearby object and applying Newton's law of gravity. ∞ (Sec. 2.7) By studying the dynamic behavior of objects near our planet—be they baseballs, rockets, or the Moon—we can compute Earth's mass to be approximately 6.0×10^{24} kg—6000 billion billion metric tons. A more accurately measured value is 5.974×10^{24} kg. The Earth Data box on p.152 lists in detail some of Earth's physical properties; however, we will use rounded-off numbers throughout the body of the text.

Dividing Earth's mass by its volume yields an average density of around 5500 kg/m³. ∞ (Sec. 6.2) This simple measurement allows us to make a very important deduction about the interior of our planet. The water that makes up much of Earth's surface has a density of 1000 kg/m³, and the rock beneath us on the continents, as well as on the seafloor, has a density in the range 2000–4000 kg/m³. We can immediately conclude that, because the surface layers have densities much less than the average, much denser material must lie deeper, under the surface. Hence we should expect that much of Earth's interior is made up of very dense matter, far more compact than the densest continental rocks on the surface.

Based on measurements made in many different ways—using aircraft in the atmosphere, satellites in orbit, gauges on the land, submarines in the ocean, and drilling gear below the rocky crust—scientists have built up the following overall picture of our planet. As indicated in Figure 7.1, our planet may be divided into six main regions. In Earth's interior, a thick **mantle** surrounds a smaller, two-part **core**. At the surface we have a relatively thin **crust**, comprising the solid continents and the seafloor, and the **hydrosphere**, which contains the liquid oceans and accounts for some 70 percent of our planet's total surface area. An **atmosphere** of air lies just above the surface. At much greater altitudes, a zone of charged particles trapped by our planet's magnetic field forms Earth's **magnetosphere**. Virtually all our planet's mass is contained within the surface and interior. The gaseous atmosphere and the magnetosphere contribute hardly anything—less than 0.1 percent—to the total.

7.2 Earth's Atmosphere

From a human perspective, probably the most important aspect of Earth's atmosphere is that we can breathe it. Air is a mixture of gases, the most common of which are nitrogen (78 percent by volume), oxygen (21 percent), argon (0.9 percent), and carbon dioxide (0.03 percent). Water vapor is a variable constituent, making up anywhere from 0.1 percent to 3 percent, depending on location and climatic conditions. The presence of a large amount of oxygen makes our atmosphere unique in the solar system, and the presence of even trace amounts of water and carbon dioxide plays a vital role in the workings of our planet.

ATMOSPHERIC STRUCTURE

Figure 7.2 shows a cross section of our planet's atmosphere. Compared with Earth's overall dimensions, the extent of the atmosphere is not great. Half of it lies within 5 km of the surface, and all but 1 percent of it is found below 30 km. The portion of the atmosphere below about 12 km is called the **troposphere**. Above it, extending up to an altitude of 40 to 50 km, lies the **stratosphere**. Between 50 and 80 km from the surface lies the **mesosphere**. Above about 80 km, in the **ionosphere**, the atmosphere is kept partly ionized by solar ultraviolet radiation. These various atmospheric regions are distinguished from one another by the temperature gradient (decreasing or increasing with altitude) in each.

Atmospheric density decreases steadily with increasing altitude, and as the right-hand vertical axis in Figure 7.2 shows, so does pressure. Climbing even a modest mountain—4 or 5 km high, say—clearly demonstrates the thin-

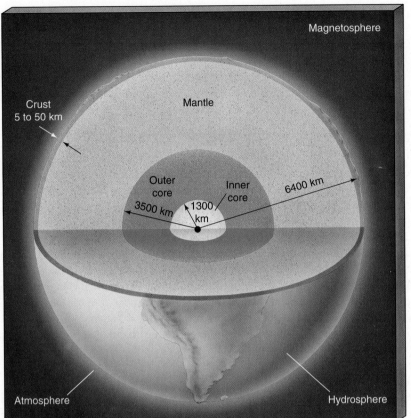

Figure 7.1 The main regions of planet Earth. At Earth's center lies our planet's inner core, about 2600 km in diameter. Surrounding the inner core is an outer core, some 7000 km across. Most of the rest of Earth's 13,000-km diameter is taken up by the mantle, which is topped by a thin crust only a few tens of kilometers thick. The liquid portions of Earth's surface make up the hydrosphere. Above the hydrosphere and solid crust lies the atmosphere, most of it within 50 km of the surface. Earth's outermost region is the magnetosphere, extending thousands of kilometers out into space.

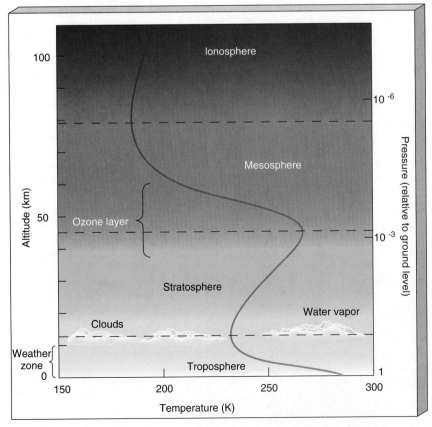

Figure 7.2 Diagram of Earth's atmosphere, showing the changes of temperature and pressure from the surface to the bottom of the ionosphere.

ning of the air in the troposphere. Climbers must wear oxygen masks when scaling the tallest peaks on Earth.

The troposphere is the region of Earth's (or any other planet's) atmosphere where *convection* occurs, driven by the heat of Earth's warm surface. **Convection** is the constant upwelling of warm air and the concurrent downward flow of cooler air to take its place, a process that physically transfers heat from a lower (hotter) to a higher (cooler) level. In Figure 7.3, part of Earth's surface is heated by the Sun. The air immediately above the warmed surface is heated, expands a little, and becomes less dense. As a result, the hot air becomes buoyant and starts to rise. At higher altitudes, the opposite effect occurs: the air gradually cools, grows denser, and sinks back to the ground. Cool air at the surface rushes in to replace the hot buoyant air. In this way, a circulation pattern is established. These *convection cells* of rising and falling air not only contribute to atmospheric heating but are also responsible for surface winds. This constant churning motion is responsible for all the weather we experience.

Atmospheric convection can also create clear-air turbulence—the bumpiness we sometimes experience on aircraft flights. Ascending and descending parcels of air, especially below fluffy clouds (themselves the result of convective processes), can cause a choppy ride. For this reason, passenger aircraft tend to fly above most of the turbulence, at the top of the troposphere or in the lower stratosphere, where the atmosphere is stable and the air is calm.

Straddling the boundary between the stratosphere and the mesosphere is the **ozone layer** where, at an altitude of around 50 km, incoming solar ultraviolet radiation is absorbed by atmospheric oxygen, ozone, and nitrogen. (Ozone is a form of oxygen, consisting of three oxygen atoms combined into a single molecule.) The ozone layer is one of the insulating spheres that serve to shield life on Earth from the harsh realities of outer space. Not so long ago, scientists judged space to be hostile to advanced life forms because of

what is missing out there—breathable air and a warm environment. Now, most scientists regard outer space harsh because of what is *present* out there—fierce radiation and energetic particles, both of which are injurious to human health. Without the protection of the ozone layer, advanced life (at least on Earth's surface) would be at best unlikely and at worst impossible.

Above about 100 km, in the ionosphere, the atmosphere is significantly ionized by the high-energy portion of the Sun's radiation spectrum, which breaks down molecules into atoms and atoms into ions. The degree of ionization increases with altitude. The presence of many free electrons makes this region of the upper atmosphere a good conductor of electricity, and this conductivity renders the ionosphere highly reflective to certain radio wavelengths. ∞ (Sec. 3.3) The reason that AM radio stations can be heard well beyond the horizon is that their signals bounce off the ionosphere before reaching the receiver. FM signals cannot be received from stations over the horizon because the ionosphere is transparent to the somewhat shorter wavelengths of radio waves in the FM band.

SURFACE HEATING

2 Much of the Sun's radiation manages to penetrate Earth's atmosphere, eventually reaching the ground. Most of this energy takes the form of visible radiation—ordinary sunlight. Essentially all of the solar radiation not absorbed by or reflected from clouds in the upper atmosphere is absorbed by Earth's surface. The result is that our planet's surface and most objects on it heat up considerably during the day. Earth cannot absorb this solar energy indefinitely, however. If it did, the surface would soon become hot enough to melt, and life on our planet would not exist.

As it heats up, Earth's surface reradiates much of the absorbed energy. This reemitted radiation follows the

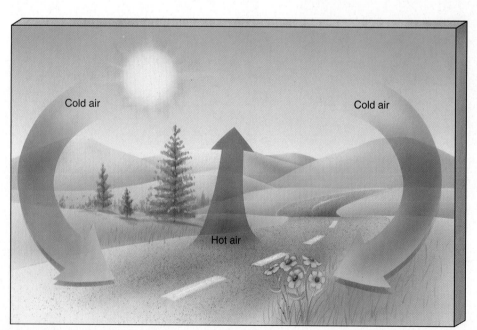

Figure 7.3 Convection occurs whenever cool fluid overlies warm fluid. The resulting circulation currents are familiar to us as the winds in Earth's atmosphere caused by the solar-heated ground. Hot air rises, cools, and falls repeatedly. Eventually, steady circulation patterns with rising and falling currents are established and maintained, provided that the source of heat (the Sun in the case of the atmosphere) remains intact.

Cold air

Cold air

Hot air

MORE PRECISELY 7-1   Why Is the Sky Blue?

Is the sky blue because it reflects the color of the ocean, or is the ocean blue because it reflects the color of the surrounding sky? The answer is the latter, and the reason has to do with the way that light is *scattered* by air molecules and minute dust particles. By scattering we mean the process by which radiation is absorbed and then reradiated by the material through which it passes.

As sunlight passes through our atmosphere, it is scattered by gas molecules in the air. The British physicist Lord Rayleigh first investigated this phenomenon about a century ago, and today it bears his name—it is known as Rayleigh scattering. The process turns out to be very sensitive to the wavelength of the light involved.

Rayleigh found that blue light is much more easily scattered than red light, in essence because the wavelength of blue light (400 nm) is closer to the size of air molecules than is the wavelength of red light (700 nm). He went on to prove that the amount of scattering is actually inversely proportional to the *fourth* power of the wavelength, so that blue light is scattered about nine times (that is, $[700/400]^4$) more efficiently than red light. Dust particles also preferentially scatter blue light, but the amount of scattering by dust is just inversely proportional to the wavelength, so the contrast between red and blue scattering is only a factor of (700/400), or 1.75.

Consequently, with the Sun at a reasonably high elevation, the blue component of incoming sunlight will scatter much more than any other color component. Thus, some blue light is removed from the line of sight between us and the Sun and may scatter many times in the atmosphere before eventually entering our eyes, as shown in the accompanying figure. Red or yellow light is scattered relatively little and arrives at our eyes predominantly along the line of sight to the Sun. The net effect is that the Sun is "reddened" slightly, because of the removal of blue light, while the sky away from the Sun appears blue. In outer space, where there is no atmosphere, there is no Rayleigh scattering of sunlight, and the sky is black (although, as we will see in Chapter 18, light from distant stars is reddened in precisely the same way as it passes through clouds of interstellar gas and dust.)

At dawn or dusk, with the Sun near the horizon, sunlight must pass through much more atmosphere before reaching our eyes—so much so, in fact, that the blue component of the Sun's light is almost entirely scattered out of the line of sight, and even the red component is diminished in intensity. Accordingly, the Sun itself appears orange—a combination of its normal yellow color and a reddishness caused by the subtraction of virtually all of the blue end of the spectrum—and dimmer than at noon. At the end of a particularly dusty day, when weather conditions or human activities during the daytime hours have raised excess particles into the air, short-wavelength Rayleigh scattering can be so heavy that the Sun appears brilliantly red. Reddening is often especially evident when we look at the westerly "sinking" summer Sun over the ocean, where seawater molecules have evaporated into the air, or during the weeks and months after an active volcano has released huge quantities of gas and dust particles into the air—as was the case in North America when the Philippine volcano Mount Pinatubo erupted in 1991.

usual blackbody curve discussed in Chapter 3. ∞ (Sec. 3.4) As the surface temperature rises, the amount of energy radiated increases rapidly, according to Stefan's law. Eventually, Earth radiates as much energy back into space as it receives from the Sun, and a stable balance is struck.

In the absence of any complicating effects, this balance would be achieved at an average surface temperature of about 250 K (−23°C). At that temperature, Wien's law tells us that most of the reemitted energy is in the form of infrared (heat) radiation.

PLANETARY DATA:
EARTH

Orbital semi-major axis	1.00 A.U.
	149.6 million km
Orbital eccentricity	0.017
Perihelion	0.98 A.U.
	147.1 million km
Aphelion	1.02 A.U.
	152.1 million km
Mean orbital speed	29.79 km/s
Sidereal orbital period	1.000038 tropical years
Orbital inclination to the ecliptic	0.01°
Mass	5.97×10^{24} kg
Equatorial radius	6378 km
Mean density	5520 kg/m^3
Surface gravity*	9.80 m/s^2
Escape speed	11.2 km/s
Sidereal rotation period	0.9973 solar days
Axial tilt†	23.45°
Magnetic axis tilt relative to rotation axis	11.5°
Mean surface temperature	290 K
Number of moons	1

*Acceleration due to gravity at Earth's surface

†Angle between Earth's rotation axis and the perpendicular to Earth's orbital plane (the plane of the ecliptic).

But there are complications. Infrared radiation is partially blocked by Earth's atmosphere. The primary reason for this is the presence of molecules of water vapor and carbon dioxide, which absorb very efficiently in the infrared portion of the spectrum. Even though these two gases are only trace constituents of our atmosphere, they manage to absorb a large fraction of all the infrared radiation emitted from the surface. Consequently, only some of that radiation escapes back into space. The remainder is trapped within our atmosphere, causing the temperature to increase.

This partial trapping of solar radiation is known as the **greenhouse effect**. The name comes from the fact that a very similar process operates in a greenhouse. Sunlight passes relatively unhindered through glass panes, but much of the infrared radiation reemitted by the plants is blocked by the glass and cannot get out. Consequently, the interior of the greenhouse heats up, and flowers, fruits, and vegetables can grow even on cold wintry days.* The radiative processes that determine the temperature of Earth's atmosphere are illustrated in Figure 7.4. Earth's greenhouse effect makes our planet almost 40 K hotter than would otherwise be the case.

The magnitude of the greenhouse effect is very sensitive to the concentration of so-called *greenhouse gases* (that is, gases that absorb infrared radiation efficiently) in the atmosphere. Carbon dioxide and water vapor are the most important of these, although other atmospheric gases also contribute. The amount of carbon dioxide in Earth's atmosphere is increasing, largely as a result of the burning of fossil fuels (principally oil and coal) in the industrialized world. Carbon dioxide levels have increased by over 20 percent in the last century, and they are continuing to rise at a present rate of 4 percent per decade. In Chapter 9 we will see how a runaway increase in carbon dioxide levels in the atmosphere of the planet Venus has radically altered conditions on its surface, causing its temperature to rise to over 700 K. Although no one is predicting that Earth's temperature will ever reach that of Venus, many scientists now believe that this increase, if left unchecked, may result in global temperature increases of several kelvins over the next half-century—enough to cause dramatic, and possibly catastrophic, changes in our planet's climate.

ORIGIN OF EARTH'S ATMOSPHERE

Why is our atmosphere made up of its present constituents? Why is it not composed entirely of nitrogen, say, or of carbon dioxide, like the atmospheres of Venus and Mars? The origin and development of Earth's atmosphere was a fairly complex and lengthy process.

---

*Although this process does contribute to warming the interior of a greenhouse, it is not the most important effect. A greenhouse works mainly because its glass panes prevent convection from carrying heat up and away from the interior. Nevertheless, the name "greenhouse effect" to describe the heating effect due to Earth's atmosphere has stuck.

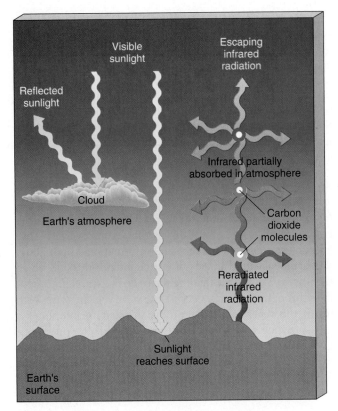

Figure 7.4 The greenhouse effect. Sunlight that is not reflected by clouds reaches Earth's surface, warming it up. Infrared radiation reradiated from the surface is partially absorbed by water vapor and carbon dioxide in the atmosphere, causing the overall surface temperature to rise.

When Earth first formed, any *primary atmosphere* it might have had would have consisted of the gases most common in the early solar system. These were light gases, such as hydrogen, helium, methane, ammonia, and water vapor—a far cry from the atmosphere we enjoy today. Almost all this light material, and especially any hydrogen or helium, escaped into space during the first half-billion or so years after Earth was formed. (For more information on how planets retain or lose their atmospheres, consult *More Precisely 8-1*.)

Subsequently, Earth developed a *secondary atmosphere*, which was *outgassed* from the planet's interior as a result of volcanic activity. Volcanic gases are rich in water vapor, methane, carbon dioxide, sulfur dioxide, and compounds containing nitrogen (such as nitrogen gas, ammonia, and nitric oxide). Solar ultraviolet radiation decomposed the lighter, hydrogen-rich gases, allowing the hydrogen to escape, and liberated much of the nitrogen from its bonds with other elements. As Earth's surface temperature fell, the water vapor condensed and oceans formed. Much of the carbon dioxide and sulfur dioxide became dissolved in the oceans or combined with surface rocks. Oxygen is such a reactive gas that any free oxygen that appeared at early times was removed as quickly as it formed. An atmosphere consisting largely of nitrogen slowly appeared.

The final major development in the story of our planet's atmosphere is known so far to have occurred only on Earth. *Life* appeared in the oceans more than 3.5 billion years ago, and organisms eventually began to produce atmospheric oxygen. The ozone layer formed, shielding the surface from the Sun's harmful radiation. Eventually, life spread to the land and flourished. The fact that oxygen is a major constituent of the present-day atmosphere is a direct consequence of the evolution of life on Earth.

7.3 Earth's Interior

3 Although we reside on Earth, we cannot easily probe our planet's interior. Drilling gear can penetrate rock only so far before breaking. No substance used for drilling—even diamond, the hardest known material—can withstand the pressure below a depth of about 10 km. That's rather shallow compared with Earth's 6400-km radius. Fortunately, geologists have developed other techniques that indirectly probe the deep recesses of our planet.

SEISMIC WAVES

A sudden dislocation of rocky material near Earth's surface—an **earthquake**—causes the entire planet to vibrate a little. Earth literally rings like a bell. These vibrations are not random, however. They are systematic waves, called **seismic waves** (after the Greek word for "earthquake"), that move outward from the site of the quake. Like all waves, they carry information. This information can be detected and recorded using sensitive equipment—a *seismograph*—designed to monitor Earth tremors.

Decades of earthquake research have demonstrated the existence of many kinds of seismic waves. Two are of particular importance to the study of Earth's internal structure. First to arrive at a monitoring site after a distant earthquake are the primary waves, or *P-waves*. These are *pressure* waves, a little like ordinary sound waves in air, that alternately expand and compress the material medium through which they move. Seismic P-waves usually travel at speeds ranging from 5 to 6 km/s and can travel through both liquids and solids. Some time later (the actual delay depends on the distance from the earthquake site), secondary waves, or *S-waves*, arrive. These are *shear* waves. Unlike P-waves, which vibrate the material through which they pass back and forth along the direction of travel of the wave, S-waves cause side-to-side motion, more like waves in a guitar string. The two types of waves are illustrated in Figure 7.5. S-waves normally travel through Earth's interior at 3 to 4 km/s; they cannot travel through liquid, which absorbs them.

The speeds of both P- and S-waves depend on the density of the matter through which the waves are traveling. Consequently, if we can measure the time taken for the waves to move from the site of an earthquake to one or more monitoring stations on Earth's surface, we can determine the

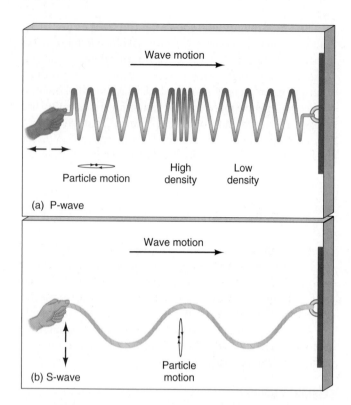

Figure 7.5 (a) A pressure (P-) wave traveling through Earth's interior causes material to vibrate in a direction parallel to the direction of motion of the wave. Material is alternately compressed and expanded. (b) A shear (S-) wave produces motion perpendicular to the direction in which the wave travels, pushing material from side to side. Also shown is the motion of one typical particle. In case (a) the particle oscillates forward and backward about its initial position. In (b) the particle moves from side to side.

density of matter in the interior. Figure 7.6 illustrates some P- and S-wave paths away from the site of an earthquake. Seismographs located around the world measure the times of arrival as well as the strengths of the seismic waves. Both observations contain much useful information—both about the earthquake itself and about Earth's interior through which the waves pass. Notice that the waves do not travel in straight lines through the planet. Because the wave velocity varies with depth, the waves bend as they move through the interior.

A particularly important result emerged after numerous quakes were monitored several decades ago: seismic stations on the side of Earth opposite a quake never detect S-waves—these waves are blocked by material within Earth's

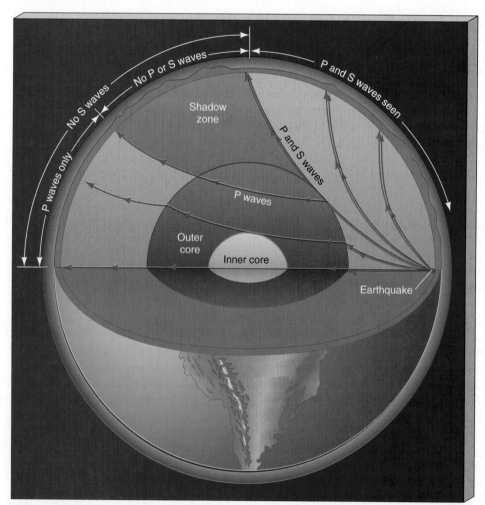

Figure 7.6 Earthquakes generate pressure (*P*, or primary) and shear (*S*, or secondary) waves that can be detected at seismographic stations around the world. S-waves are not detected by stations "shadowed" by the liquid core of Earth. P-waves do reach the side of Earth opposite the earthquake, but their interaction with Earth's core produces another shadow zone, where no P-waves are seen.

Seismic waves racing out from the site of earthquakes allow geologists to map the interior makeup of planet Earth. ∞ (Sec. 7.3) Recent analyses of old earthquake data suggest that Earth's inner core is spinning slightly *faster* than the rest of our planet. The whole planet rotates in the same direction (west to east), but the inner core takes about 2/3 of a second less to complete its daily rotation than the bulk of the matter surrounding it—which means that it gains a quarter turn each century.

Geologists arrived at this rather surprising result by carefully timing seismic P-waves as they moved through Earth's interior. Actually, the scientists used historical data acquired over the past few decades, mainly from 38 earthquakes that occurred in the South Atlantic Ocean and whose seismic vibrations had to pierce the inner core on their way to monitoring stations in Alaska. As discussed in the text, only P-waves can make it through the liquid outer core, to be detected on the opposite side of Earth.

For many years, geologists have known that P-waves travel through the inner core about 3 or 4 percent faster along north–south pathways than along those close to the plane of the equator. This difference probably results from the orientation of the iron atoms in the solid inner core; the atoms there must be aligned, as in a crystal lattice. Where the atomic alignment parallels the motion of the P-waves, those P-waves travel just a little faster—a little like going "with the grain" rather than against it.

The new finding is that the route of the fastest seismic waves through Earth's deep interior is gradually shifting eastward, implying that the inner core has some motion slightly different from the rest of the planet. In the accompanying figure, the red region is the solid inner core, the orange region the liquid outer core. The dashed lines passing through the core depict the paths of the fastest seismic waves from 1900 to 1996. The data indicate that the axis of the "fast track" for the P-waves is shifting by about 1° per year relative to the crust above. That axis now intersects Earth's surface in the Arctic Ocean northeast of Siberia; some 30 years ago it was 33° farther west. Gradual as it may be, this shift amounts to about 20 km at the edge of the

inner core—around a million times faster than the typical rates (centimeters per year) at which the continents creep across Earth's surface.

These discoveries may help us understand how the inner core is continuing to grow by condensing from the molten outer core as Earth cools, and how these two regions interact to produce Earth's magnetic field. As noted elsewhere in this chapter, the dynamo theory of magnetism requires rotating, electrically conducting matter within Earth's interior. ∞ (Sec. 7.4) Detailed modeling had also predicted that a rapidly moving jet stream of partly molten matter—a "mushy zone" —is probably established at the base of the outer core, causing the inner core to "super-rotate." Now it seems that this differential rotation pattern has been found.

interior. Further, while P-waves always arrive at stations diametrically opposite the quake, there are parts of Earth's surface that they cannot reach (see Figure 7.6). Most geologists believe that S-waves are absorbed by a liquid core at Earth's center and that P-waves are "refracted" at the core boundary, much as light is refracted by a lens. The result is the S- and P-wave "shadow zones" we observe. The fact that every earthquake exhibits these shadow zones is the best evidence that the core of our planet is hot enough to be in the liquid state. *Interlude 7-1* presents some more seismic data on the structure of Earth's core.

The radius of the core, as determined from seismic data, is about 3500 km. In fact, very faint P-waves *are* observed in the P-wave shadow zone indicated in Figure 7.6. These are be-

lieved to be reflected off the surface of a solid **inner core**, of radius 1300 km, lying at the center of the liquid **outer core**.

MODELING EARTH'S INTERIOR

Because earthquakes occur often and at widespread places across the globe, geologists have accumulated a large amount of data about shadow zones and seismic-wave properties. They have used these data, along with direct knowledge of surface rocks, to build mathematical models of Earth's interior.

Figure 7.7 presents a model that most scientists accept. Earth's outer core is surrounded by a thick mantle and topped with a thin crust. The mantle is about 3000 km thick

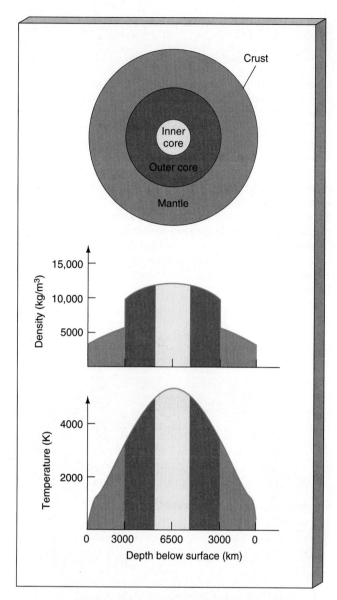

Figure 7.7 Computer models of Earth's interior imply that the density and temperature vary considerably through the mantle and the core. Note the sharp density discontinuity between Earth's core and mantle.

(whose densities under surface conditions are around 8000 kg/m³) can be compressed to the high densities predicted by the model. The sharp density increase at the mantle–core boundary results from the difference in composition between the two regions. The mantle is composed of dense but *rocky* material, compounds of silicon and oxygen. The core consists primarily of even denser *metallic* elements. There is no similar jump in density or temperature at the inner core boundary—the material there simply changes from the liquid to the solid state.

The model suggests that the core must be a mixture of nickel, iron, and some other lighter element, possibly sulfur. Without direct observations, it is difficult to be absolutely certain of the light component's identity. All geologists agree that much of the core must be liquid. The existence of the shadow zone demands that (and, as we will see, our current explanation of Earth's magnetic field relies on it). However, despite the high temperature, the pressure near the center—about 4 million times atmospheric pressure at Earth's surface—is high enough to force the material there into the solid state.

Despite the fact that no experiment has yet succeeded in piercing Earth's crust to recover a sample of the mantle, we are not entirely ignorant of the mantle's properties. In a *volcano*, hot lava upwells from below the crust, bringing a little of the mantle to us and providing some inkling of Earth's interior. The chemical makeup and physical state of newly emerged lava are generally consistent with predictions based on the model sketched in Figure 7.7.

The composition of the upper mantle is probably quite similar to the iron–magnesium–silicate mixtures known as *basalt*. You may have seen some dark gray basaltic rocks scattered across Earth's surface, especially near volcanoes. Basalt is formed as mantle material upwells from Earth's interior as lava, then cools and solidifies. With a density between 3000 and 3300 kg/m³, basalt contrasts with the lighter *granite* (density 2700–3000 kg/m³) that constitutes much of the rest of Earth's crust. Granite is richer than basalt in the light elements silicon and aluminum, which explains why the surface continents do not sink into the interior. Their low-density composition lets the crust "float" atop the denser matter of the mantle and core below.

DIFFERENTIATION

Earth, then, is not a homogeneous ball of rock. Instead, it has a layered structure, with a low-density crust at the surface, intermediate-density material in the mantle, and a high-density core. Such variation in density and composition is known as **differentiation**.

Why isn't our planet just one big, rocky ball of uniform density? The answer appears to be that much of Earth was *molten* at some time in the past. As a result, the higher-density matter sank to the core, and the lower-density material was displaced toward the surface. A remnant of this ancient heating exists today: Earth's central temperature is nearly equal to the surface temperature of the Sun. What processes were

and accounts for the bulk (80 percent) of our planet's volume. The crust has an average thickness of only 15 km—a little less (around 8 km) under the oceans and somewhat more (20 to 50 km) under the continents. The average density of crust material is around 3000 kg/m³. Density and temperature both increase with depth. Specifically, from Earth's surface to its very center, the density increases from roughly 3000 kg/m³ to a little more than 12,000 kg/m³, while the temperature rises from just under 300 K to well over 5000 K. Much of the mantle has a density midway between the densities of the core and crust: about 5000 kg/m³.

The high central density suggests to geologists that the inner parts of Earth must be rich in nickel and iron. Under the heavy pressure of the overlying layers, these metals responsible for heating the entire planet to this extent? To answer this question, we must try to visualize the past.

We will see in Chapter 15 that when Earth formed, it did so by capturing material from its surroundings, growing in mass as it swept up "preplanetary" chunks of matter in its vicinity. As the young planet grew, its gravitational field strengthened and the speed with which newly captured matter struck its surface increased. This process generated a lot of heat—so much, in fact, that Earth may already have been partially or wholly molten by the time it reached its present size. As Earth began to differentiate and heavy material sank to the center, even more gravitational energy was released, and the interior temperature must have increased still further.

Later, Earth continued to be bombarded with debris left over from the formation process. At its peak some 4 billion years ago, this secondary bombardment was probably intense enough to keep the surface molten, but only down to a depth of a few tens of kilometers. Erosion by wind and water has long since removed all trace of this early period from the surface of Earth, but the Moon still bears visible scars of the onslaught.

A second important process for heating Earth soon after its formation was **radioactivity**—the release of energy by certain rare, heavy elements, such as uranium, thorium, and plutonium (see *More Precisely 7-2*). These elements emit energy as their complex, heavy nuclei decay into simpler, lighter ones. While the energy produced by the decay of a single radioactive atom is tiny, Earth contained a lot of radioactive atoms, and a lot of time was available. Rock is such a poor conductor of heat that the energy would have taken a very long time to reach the surface and leak away into space, so the heat built up in the interior, adding to the energy left there by Earth's formation.

Provided that enough radioactive elements were originally spread throughout the primitive Earth, rather like raisins in a cake, the entire planet—from crust to core—could have melted and remained molten for about a billion years. That's a long time by human standards, but not so long in the cosmic scheme of things. Measurements of the ages of some surface rocks indicate that Earth finally began to solidify roughly a billion years after it originally formed. Radioactive heating did not stop after the first billion years, however. It continued even after Earth's surface cooled and solidified. But radioactive decay works in only one direction, always producing lighter elements from heavier ones. Once gone, the heavy and rare radioactive elements cannot be replenished.

The early source of heat diminished with time, allowing the planet to cool over the past 3.5 billion years. In so doing, it cooled from the outside in, much like a hot potato, since regions closest to the surface could most easily unload their excess heat into space. In this way, the surface developed a solid crust, and the differentiated interior attained the layered structure now implied by seismic studies.

7.4 Earth's Magnetosphere

Discovered by artificial satellites launched in the late 1950s, Earth's magnetosphere extends far above the atmosphere. Simply put, the magnetosphere is the region around a planet that is influenced by that planet's magnetic field. Sketched in Figure 7.8, Earth's magnetic field is similar in overall structure to the field of a gigantic bar magnet and completely surrounds our planet. The magnetic field lines, which indicate the strength and direction of the field at any

Figure 7.8 Earth's magnetic field resembles that of an enormous bar magnet situated inside our planet. The arrows on the field lines indicate the direction in which a compass needle would point.

MORE PRECISELY 7-2 Radioactive Dating

In Chapter 4 we saw that atoms are made up of electrons and nuclei, and that nuclei are composed of protons and neutrons. ∞ (Sec. 4.2) The number of protons in a nucleus determines which element it represents. However, the number of neutrons can vary. In fact, most elements can exist in several *isotopic* forms, all containing the same number of protons but different numbers of neutrons in their nuclei. The particular nuclei we have encountered so far—the most common forms of hydrogen, helium, carbon, iron—are all *stable*. For example, left alone, a carbon-12 nucleus, consisting of six protons and six neutrons, will remain unchanged forever. It will not break up into smaller pieces, nor will it turn into anything else.

Not all nuclei are stable, however. Many nuclei—for example, carbon-14 (containing 6 protons and 8 neutrons), thorium-232 (90 protons, 142 neutrons), uranium-235 (92 protons, 143 neutrons), uranium-238 (92 protons, 146 neutrons), and plutonium-241 (94 protons, 147 neutrons)—are inherently *unstable*. Left alone, they will eventually break up into lighter "daughter" nuclei, in the process emitting some elementary particles and releasing some energy. The change happens spontaneously, without any external influence. This instability is known as *radioactivity*. The energy released by the disintegration of the radioactive elements just listed is the basis for nuclear fission reactors (and atomic bombs).

Unstable heavy nuclei achieve greater stability by disintegrating into lighter nuclei, but they do not do so immediately. Each type of "parent" nucleus takes a characteristic amount of time to decay. The *half-life* is the name given to the time required for half of a sample of parent nuclei to disintegrate. Notice that this is really a statement of probability. We cannot say *which* nuclei will decay in any given half-life interval, only that half of them are expected to do so. Thus, if we start with a billion radioactive nuclei embedded in a sample of rock, a half-billion nuclei will remain after one half-life, a quarter-billion after two half-lives, and so on.

Every radioactive isotope has its own half-life, and most of them are now well known from studies conducted since the 1950s. For example, the half-life of uranium-235 is 713 million years, and that of uranium-238 is 4.5 billion years. Some radioactive elements decay much more rapidly, others much more slowly, but these two types of uranium are particularly important to geologists because their half-lives are comparable to the age of the solar system. The accompanying figure illustrates the half-lives and decay reactions for four unstable heavy nuclei.

The decay of unstable radioactive nuclei into more stable *daughter* nuclei provides us with a useful tool for measuring the ages of any rocks we can get our hands on. The first step is to measure the amount of stable nuclei of a given kind (for example, lead-206, which results from the decay of uranium-238). This amount is then compared with the amount of remaining unstable parent nuclei (in this case, uranium-238) from which the daughter nuclei descended. Knowing the rate (or half-life) at which the disintegration occurs, the age of the rock then follows directly. For example, if half of the parent nuclei of some element have decayed, so that the number of daughter nuclei equals the number of parents, the age of the rock must be equal to the half-life of the radioactive nucleus studied. Similarly, if only a quarter of the parent nuclei remain (three times as many daughters as parents), the rock's age is twice the half-life of that element, and so on. In practice, ages can be determined by these means to within an accuracy of a few percentage points. The most ancient rocks on Earth are dated at

---

point in space, run from south to north, as indicated by the white arrowheads in the figure.

The north and south *magnetic poles*, where the magnetic field lines intersect Earth's surface vertically, are roughly aligned with Earth's spin axis. Neither pole is fixed relative to our planet, however—both drift at a rate of some 10 km per year—nor are the poles symmetrically placed. At present, Earth's magnetic north pole lies in northern Canada, at a latitude of about 80°N, almost due north of the center of North America; the magnetic south pole lies at a latitude of about 60°S, just off the coast of Antarctica south of Adelaide, Australia.

Earth's magnetosphere contains two doughnut-shaped zones of high-energy charged particles, one located about 3000 km and the other 20,000 km above Earth's surface. These zones are named the **Van Allen belts**, after the American physicist whose instruments on board one of the first satellites first detected them. We call them "belts" because they are most pronounced near Earth's equator and because they completely surround the planet. Figure 7.9 shows how these invisible regions envelop Earth except near the North and South Poles.

The particles that make up the Van Allen belts originate in the solar wind. Traveling through space, neutral particles and electromagnetic radiation are unaffected by Earth's magnetism, but electrically charged particles are strongly influenced. As illustrated in Figure 7.10, a magnetic field exerts a force on a moving charged particle, causing the particle to spiral around the magnetic field lines. In this way, charged particles—mainly electrons and protons—from the solar wind can become trapped by Earth's magnetism. Earth's magnetic field exerts electromagnetic control over these particles, herding them into the Van Allen belts. The outer belt contains mostly electrons; the much heavier protons accumulate in the inner belt.

We could never survive unprotected in the Van Allen belts. Unlike the lower atmosphere, on which humans and other life forms rely for warmth and protection, much of the magnetosphere is subject to intense bombardment by large numbers of high-velocity, and potentially very harmful, charged particles. Colliding violently with an unprotected human body, these particles would deposit large amounts of energy wherever they made contact, causing severe damage to living organisms. Without sufficient shielding on the *Apollo* spacecraft, for

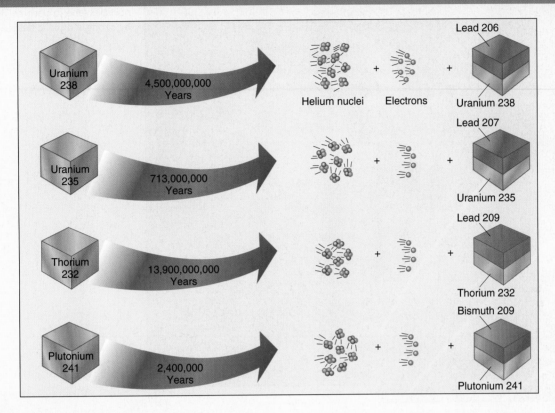

3.9 billion years old. These rare specimens have been found in Greenland and Labrador.

The radioactive-dating technique rests on the assumption that the rock has remained *solid* while the radioactive decays have been going on. If the rock melts, there is no reason to expect the daughter nuclei to remain in the same locations their parents had occupied, and the whole method fails. Thus, radioactive dating indicates the time that has elapsed since the last time the rock in question solidified. Hence this 3.9 billion year value represents only a portion—a lower limit—of the true age of our planet. It does not measure the duration of Earth's molten existence.

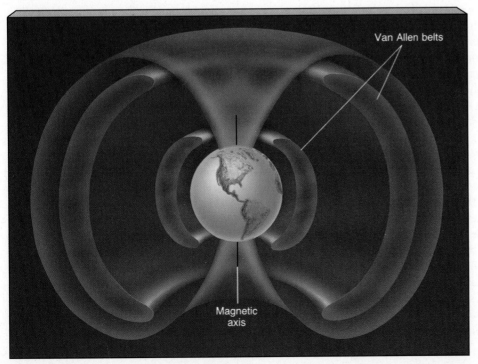

Figure 7.9 High above Earth's atmosphere, the magnetosphere (lightly shaded blue area) contains at least two doughnut-shaped regions (heavily shaded violet areas) of magnetically trapped charged particles. These are the Van Allen belts.

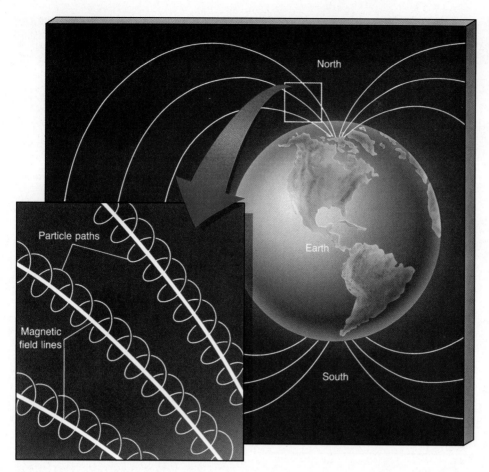

Figure 7.10 A charged particle in a magnetic field spirals around the field lines. Thus, charged particles tend to become "trapped" by strong magnetic fields.

North

Earth

South

Particle paths

Magnetic field lines

example, the astronauts might not have survived the passage through the magnetosphere on their journey to the Moon.

Particles from the Van Allen belts often escape from the magnetosphere near Earth's north and south magnetic poles, where the field lines intersect the atmosphere. Their collisions with air molecules create a spectacular light show called an **aurora** (Figure 7.11). This colorful display results when atmospheric molecules, excited upon collision with the charged particles, fall back to their ground states and emit visible light. Many different colors are produced because each type of atom or molecule can take one of several possible paths as it returns to its ground state. ∞ (Sec. 4.2) Aurorae are most brilliant at high latitudes, especially inside the Arctic and Antarctic circles. In the north, the spectacle is called the *aurora borealis*, or *Northern Lights*. In the south, it is called the *aurora australis*, or *Southern Lights*.

Occasionally, particularly after a storm on the Sun (see Chapter 16), the Van Allen belts can become distorted by the solar wind and overloaded with many more particles than normal, allowing some particles to escape prematurely and at lower latitudes. For example, in North America, the aurora borealis is normally seen with any regularity only in northern Canada and Alaska. However, at times of greatest solar activity, the display has occasionally been seen as far south as the southern United States.

In reality, Earth's magnetosphere is not nearly as symmetrical as depicted in Figure 7.8. Satellites have mapped its

true shape. As shown in Figure 7.12, the entire region of trapped particles is quite distorted, forming a teardrop-shaped cavity. On the sunlit (daytime) side of Earth, the magnetosphere is compressed by the flow of high-energy particles in the solar wind. The boundary between the magnetosphere and this flow is known as the *magnetopause*. It is found at about 10 Earth radii from our planet. On the side opposite the Sun, the field lines are extended away from Earth, with a long tail often reaching beyond the orbit of the Moon.

What is the origin of the magnetosphere and the Van Allen belts within it? Earth's magnetism is not really the result of a huge bar magnet lying within our planet. In fact, geophysicists believe that Earth's magnetic field is not a "permanent" part of our planet at all. Instead, it is thought to be continuously generated within the outer core and exists only because Earth is rotating. As in the dynamos that run industrial machines, Earth's magnetism is produced by the spinning, electrically conducting, liquid metal core deep within our planet. The theory that explains planetary (and other) magnetic fields in terms of rotating, conducting material flowing in the planet's interior is known as **dynamo theory**. Both rapid rotation *and* a conducting liquid core are needed for this mechanism to work. This connection between internal structure and magnetism is very important for studies of the other planets in the solar system: we can tell a lot about a planet's interior simply by measuring its magnetic field.

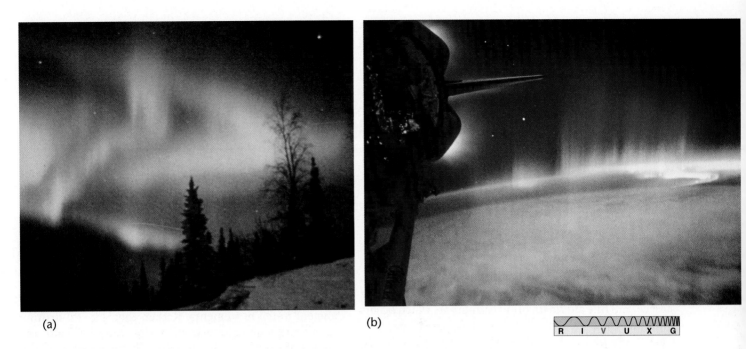

(a)

(b)

Figure 7.11 (a) A colorful aurora rapidly flashes across the sky like huge wind-blown curtains glowing in the dark. The aurora is created by the emission of light radiation after magnetospheric particles collide with atmospheric molecules. The colors are produced as excited atoms and molecules return to their ground states. (b) The aurora high above Earth, as photographed from a space shuttle (visible at left).

Figure 7.12 Earth's real magnetosphere is actually greatly distorted by the solar wind, with a long tail extending from the nighttime side of Earth well into space.

Earth's magnetic field plays an important role in controlling many of the potentially destructive charged particles that venture near our planet. Without the magnetosphere, Earth's atmosphere—and perhaps the surface, too—would be bombarded by harmful particles, possibly damaging many forms of life on our planet. Some researchers have even suggested that had the magnetosphere not existed in the first place, life might never have arisen at all on our planet.

7.5 Surface Activity

Earth is geologically alive today. Its interior seethes and its surface constantly changes. Figure 7.13 shows two indicators of surface geological activity: a volcano, where molten rock and hot ash upwell through fissures or cracks in the surface, and (the aftermath of) an earthquake, which occurs when the crust suddenly dislodges under great pressure. Catastrophic volcanoes and earthquakes are relatively rare events these days, but geological studies of rocks, lava, and other surface features imply that surface activity must have been more frequent, and probably more violent, long ago.

CONTINENTAL DRIFT

5 Many traces of past geological events are scattered across our globe. Erosion by wind and water has wiped away much of the evidence for ancient activity, but modern exploration has documented the sites of most of the recent activity, such as earthquakes and volcanic eruptions. Figure 7.14 is a map of the currently active areas of our planet. The red dots represent sites of volcanism or earthquakes. Nearly all these sites

have experienced surface activity within this century, some of them suffering much damage and the loss of many lives.

The intriguing aspect of Figure 7.14 is that the active sites are not spread evenly across our planet. Instead, they trace well-defined lines of activity, where crustal rocks dislodge (as in earthquakes) or mantle material upwells (as in volcanoes). In the mid-1960s, it became clear that these lines are really the outlines of gigantic "plates," or slabs of Earth's surface. Most startling of all, the plates are slowly moving—literally drifting around the surface of our planet. These plate motions have created the surface mountains, the oceanic trenches, and the other large-scale features across the face of planet Earth. In fact, plate motions have shaped the continents themselves. The process is popularly known as "continental drift." The technical term for the study of plate movement and its causes is **plate tectonics**. The major plates of the world are marked on Figure 7.14.

The plates are not simply slowing to a stop after some ancient initial movements. Rather, they are still drifting today, although at an extremely slow rate. Typically, the speeds of the plates amount to only a few centimeters per year—about the same rate as your fingernails grow. This is well within the measuring capabilities of modern equipment. Curiously, one of the best ways of monitoring plate motion on a global scale is by making accurate observations of very distant astronomical objects. Quasars (see Chapter 25), lying many hundreds of millions of light years from Earth, will never show any measurable apparent motion on the sky stemming from their own motion in space. Thus any apparent change in their position (after correction for Earth's motion, of course) can be interpreted as arising from the motion of the telescope—or of the plate on which it is located. On smaller scales, laser-ranging and other techniques now

(a)

(b)

Figure 7.13 (a) An active volcano on Mount Kilauea in Hawaii. Kilauea seems to be a virtually ongoing eruption. Other eruptions, such as that of Mount St. Helens in Washington State on May 18, 1980, are catastrophic events that can release more energy than the detonation of a thousand nuclear bombs. (b) The aftermath of an earthquake that claimed more than 5000 lives and caused billions of dollars' worth of damage in Kobe, Japan, in January 1995.

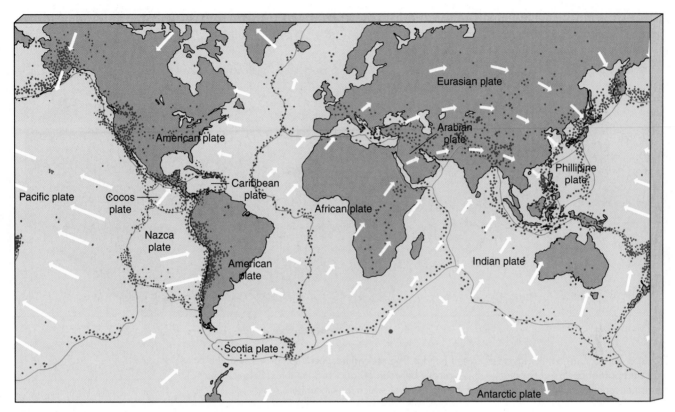

Figure 7.14   Red dots represent active sites where major volcanoes or earthquakes have occurred in the twentieth century. Taken together, the sites outline vast "plates" that drift around on the surface of our planet. The arrows show the general directions of the plate motions.

routinely track the relative motion of plates in many populated areas, such as California.

During the course of Earth history, each plate has had plenty of time to move large distances, even at its sluggish pace. For example, a drift rate of only 2 cm per year can cause two continents (for example, Europe and North America) to separate by some 4000 km over the course of 200 million years. That may be a long time by human standards, but it represents only about 5 percent of the age of Earth.

A common misconception is that the plates are the continents themselves. Some plates are indeed made mostly of continental landmasses, but other plates are made of a continent plus a large part of an ocean. For example, the Indian plate includes all of India, much of the Indian Ocean, and all of Australia and its surrounding south seas (see Figure 7.14). Still other plates are mostly ocean. The seafloor itself is the slowly drifting plate, and the oceanic water merely fills in the depressions between continents. The southeastern portion of the Pacific Ocean, called the Nazca plate, contains no landmass at all. For the most part, the continents are just passengers riding on much larger plates. Taken together, the plates make up Earth's **lithosphere**, which contains both the crust and a small part of the upper mantle. The lithosphere is the portion of Earth that undergoes

tectonic activity. The semisolid part of the mantle over which the lithosphere slides is known as the **asthenosphere**. The relationships between these regions of Earth are shown in Figure 7.15.

EFFECTS OF PLATE MOTION

As the plates drift around, we might expect collisions to be routine. Indeed, plates do collide. But unlike two automobiles that collide and then stop, the surface plates are driven by enormous forces. They do not stop easily. Instead, they just keep crunching into one another. Figure 7.16(a) shows a collision currently occurring between two continental landmasses: the subcontinent of India, on the prow of the northward-moving Indian plate, is crashing into the landmass of Asia, located on the Eurasian plate (see Figure 7.14). The resulting folds of rocky crust create mountains—in this case the snow-covered Himalayan mountain range at upper right. A peak like Mount Everest (Figure 7.16b) represents a portion of Earth's crust that has been lifted over 8800 m by the slow but inexorable force produced when one plate plows into another.

Not all colliding plates produce mountain ranges. At other collision locations, called *subduction zones*, one plate slides under the other, ultimately to be destroyed as it sinks

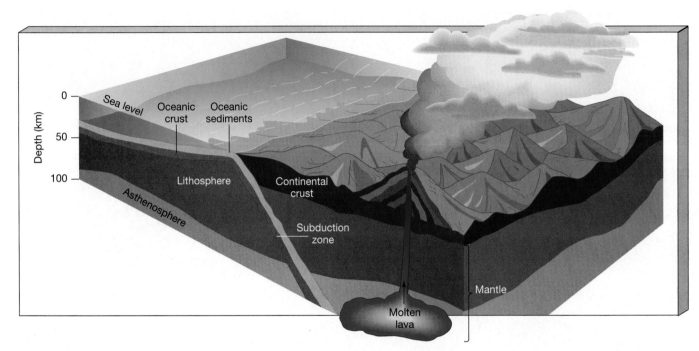

Figure 7.15 The outer layers of Earth's interior. The rocky lithosphere comprises both the crust and part of Earth's upper mantle. It is typically between 50 and 100 km thick. Below it lies the asthenosphere, a relatively soft part of the mantle over which the lithosphere slips.

(a)

(b)

Figure 7.16 Mountain building results largely from plate collisions. (a) The subcontinent of India, imaged here in infrared light from orbit, lies at the northernmost tip of the Indian plate. As this plate drifts northward, the Indian landmass collides with Asia, on the Eurasian plate. The impact causes Earth's crust to buckle and fold, thrusting up the Himalayan mountain range (covered with snow at the upper right). (b) The results of this process can be seen in this view of Mount Everest.

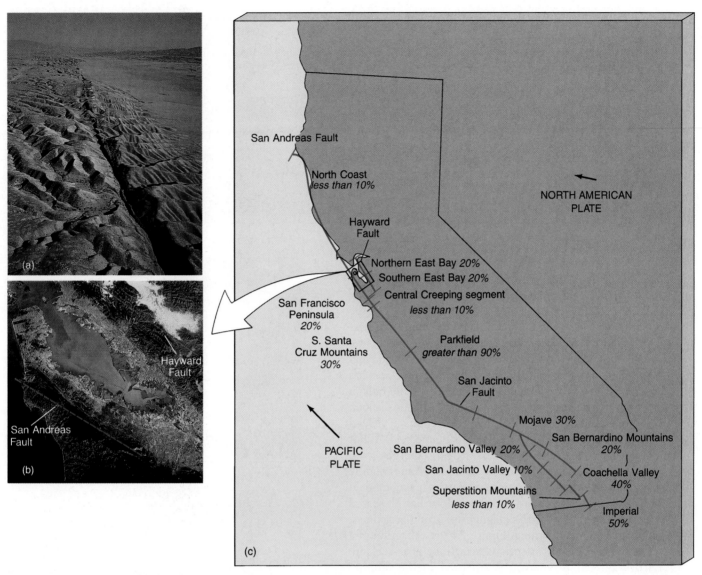

Figure 7.17 The San Andreas and associated faults in California result as the North American and Pacific plates slide past each other. The Pacific plate, which includes a large slice of the California coast, is drifting to the northwest with respect to the North American plate. (a) A small part of the fault line separating the two plates. (b) A satellite photo of the San Francisco Bay area, showing the location of two fault lines. (c) A larger-scale view of the fault system. The numbers represent estimates of the probability of a major earthquake occurring during a 30-year period at various locations along the faults.

into the mantle. Subduction zones are responsible for most of the deep trenches in the world's oceans.

Nor do all plates experience head-on collisions. As noted by the arrows of Figure 7.14, many plates slide or shear past one another. A good example is the most famous active region in North America—the San Andreas Fault in California (Figure 7.17). The site of much earthquake activity, this fault marks the boundary where the Pacific and North American plates are rubbing past each other. The motion of these two plates, like that of moving parts in a poorly oiled machine, is neither steady nor smooth. The sudden jerks that occur when they do move against each other are often strong enough to cause major earthquakes.

At still other locations, the plates are moving apart. As they recede, new mantle material wells up between them, forming *midocean ridges*. Notice in Figure 7.14 the major boundary separating the North and South American plates from the Eurasian and African plates, marked by the thin strip down the middle of the Atlantic Ocean. Discovered after World War II by oceanographic ships studying the geography of the seafloor, this giant fault is called the Mid-Atlantic Ridge. It extends, like a seam on a giant baseball, all the way from Scandinavia in the North Atlantic to the latitude of Cape Horn at the southern tip of South America. The entire ridge is a region of seismic and volcanic activity, but the only major part of it that rises above sea level is the island of Iceland.

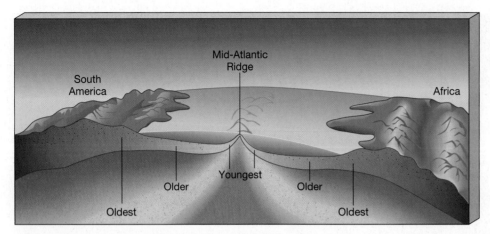

Figure 7.18 Samples of ocean floor retrieved by oceanographic vessels are youngest close to the Mid-Atlantic Ridge and progressively older farther away.

Robot submarines have retrieved samples of the ocean floor at a variety of locations on either side of the Mid-Atlantic Ridge, and the ages of the samples have been measured using radioactive dating techniques. As depicted in Figure 7.18, the ocean floor closest to the ridge is relatively young, whereas material farther away, on either side, is older—exactly as we would expect if hot molten matter is upwelling and solidifying as the plates on either side drift apart. The Atlantic Ocean has apparently been growing in this way for the past 200 million years, the oldest age found on any part of the Atlantic seafloor.

Other studies of the Mid-Atlantic Ridge have yielded important information about Earth's magnetic field. As hot mantle material (carrying traces of iron) emerges from cracks in the oceanic ridges and solidifies, it becomes slightly magnetized, retaining an imprint of Earth's magnetic field *at the time of cooling*. Thus, the ocean floor has preserved within it a record of Earth's magnetism during past times, rather like a tape recording. Figure 7.19 is a diagram of a small portion of the Atlantic Ocean floor near Iceland. Earth's current magnetism is oriented in the familiar north–south fashion, and when samples of ocean floor close to the ridge are examined, the iron deposits are oriented just as expected—north–south. This is the "young" basalt that upwelled and cooled fairly recently. However, samples retrieved farther from the ridge, corresponding to older material that upwelled long ago, are often magnetized with the *opposite* orientation. As we move away from the ridge, the imprinted magnetic field flips back and forth, more or less regularly and symmetrically on either side of the ridge.

Scientists believe that these different magnetic orientations were caused by reversals in Earth's magnetic field that occurred as the plates drifted away from the central ridge. Working backward, we can use the fossil magnetic field to infer the past positions of the plates, as well as the orientation of Earth's magnetism. In addition to providing strong support for the idea of seafloor spreading, these measurements, when taken in conjunction with the data on seafloor age, allow us to time our planet's magnetic reversals. On average, Earth's magnetic field reverses itself roughly every half-million years. Dynamo theory suggests that such reversals are part of the way in which all planetary

Figure 7.19 Samples of basalt retrieved from the ocean floor often show Earth's magnetism to have been oriented oppositely from the current north–south magnetic field. This simplified diagram shows the ages of some of the regions in the vicinity of the Mid-Atlantic Ridge (see Figure 7.14), together with the direction of the fossil magnetic field. The colored areas have the current orientation; they are separated by regions of reversed magnetic polarity.

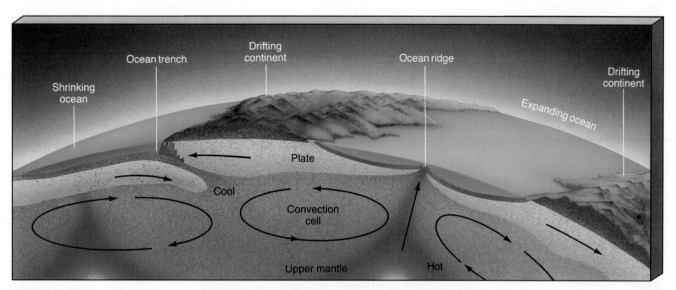

Figure 7.20 Plate drift is probably caused by convection—in this case, giant circulation patterns in the upper mantle that drag the plates across the surface.

magnetic fields are generated. As we will see in Chapter 16, a similar phenomenon (with a reversal time of approximately 11 years) is also observed on the Sun.

WHAT DRIVES THE PLATES?

What process is responsible for the enormous forces that drag plates apart in some locations and ram them together in others? The answer is probably convection—the same physical process we encountered earlier in our study of the atmosphere.

Figure 7.20 is a cross-sectional diagram of the top few hundred kilometers of our planet's interior. It depicts roughly the region in and around a midocean ridge. There the ocean floor is covered with a layer of sediment—dirt, sand, and dead sea organisms that have fallen through the seawater for millions of years. Below the sediment lies about 10 km of granite, the low-density rock that makes up the crust. Deeper still lies the upper mantle, whose temperature increases with depth. Below the base of the lithosphere, at a depth of perhaps 50 km, the temperature is sufficiently high that the mantle is soft enough to flow, very slowly, although it is not molten. This region is the asthenosphere.

This is a perfect setting for convection—warm matter underlying cool matter. The warm mantle rock rises, just as hot air rises in our atmosphere. Sometimes, the rock squeezes up through cracks in the granite crust. Every so often, such a fissure may open in the midst of a continental landmass, producing a volcano such as Mount St. Helens or possibly a geyser like those at Yellowstone National Park. However, most such cracks are underwater. The Mid-Atlantic Ridge is a prime example.

Not all the rising warm rock in the upper mantle can squeeze through cracks and fissures. Some warm rock cools and falls back down to lower levels. In this way, large circu-

lation patterns become established within the upper mantle, as depicted in Figure 7.20. Riding atop these convection patterns are the plates. The circulation is extraordinarily sluggish. Semisolid rock takes millions of years to complete one convection cycle. Although the details are far from certain and remain controversial, many researchers believe that it is the large-scale circulation patterns near plate boundaries that ultimately drive the motion of the plates.

PAST CONTINENTAL DRIFT

Figure 7.21 illustrates how all the continents nearly fit together like pieces of a puzzle. Note especially how the Brazilian coast meshes with the Ivory Coast of Africa. In fact, most of the continental landmasses in the Southern Hemisphere fit together remarkably well. Following the arrows on Figure 7.14 backward, we can see that the fits are roughly consistent with the present motions of the plates involved. The fits would be even better if we were to consider the entire continental shelves (the continental borders, which are underwater) instead of just the portions that happen to stick up above sea level.

Geologists now believe that sometime in the past a single gargantuan landmass dominated our planet. This ancestral supercontinent, known as Pangaea (meaning "all lands"), is shown in Figure 7.21(a). The rest of the planet was presumably covered with water. The present locations of the continents, along with measurements of their current drift rates, suggest that Pangaea was the major land feature on Earth approximately 200 million years ago. Dinosaurs, which were then the dominant form of life, could have sauntered from Russia to Texas via Boston without getting their feet wet. Indeed, it was the existence of similar-looking fossils on opposite sides of the Atlantic Ocean, together with the striking geographical fit of the continents, that first led scientists to suggest the idea of

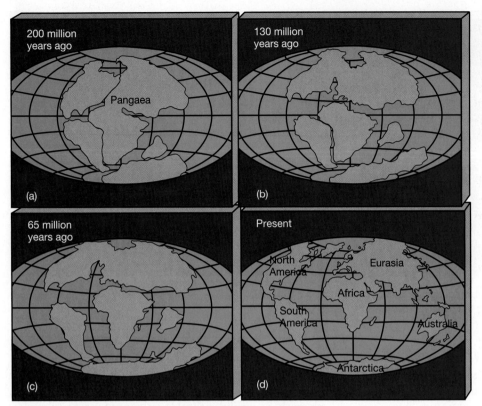

Figure 7.21 Given the currently estimated drift rates and directions of the plates, we can trace their movements back into the past. About 200 million years ago, they would have been at the approximate positions shown in (a). The continents' current positions are shown in (d).

continental drift early in the twentieth century. The other frames in Figure 7.21 show how Pangaea split apart, eventually leading to the familiar continents of today.

Of course, there is nothing particularly special about a time 200 million years in the past. We do not suppose that Pangaea remained intact for some 4 billion years after the crust first formed, only to break up so suddenly and so (relatively) recently. Much more likely, Pangaea itself came into existence after an earlier period during which other plates, carrying widely separated continental masses, were driven together by tectonic forces, merging their landmasses to form a single supercontinent. It is quite possible that there has been a long series of "Pangaeas" stretching back in time over much of Earth's history. There will probably be many more.

7.6 The Tides

Earth is unique among the planets in that it has large quantities of liquid water on its surface. Approximately three quarters of Earth's surface is covered by water, to an average depth of about 3.6 km. Only 2 percent of the water is contained within lakes, rivers, clouds, and glaciers. The remaining 98 percent is in the oceans.

A familiar hydrospheric phenomenon is the daily fluctuation in ocean level known as the **tides**. At most coastal locations on Earth, there are two low tides and two high tides each day. The "height" of the tides—the magnitude of the variation

in sea level—can range from a few centimeters to many meters, depending on the location on Earth and time of year. The height of a typical tide on the open ocean is about a meter, but if this tide is funneled into a narrow opening such as the mouth of a river, it can become much higher. For example, at the Bay of Fundy, on the Maine–Canada border, the high tide can reach nearly 20 m (approximately 60 feet, or the height of a six-story building) above the low-tide level. An enormous amount of energy is contained in the daily motion of the oceans. This energy is constantly eroding and reshaping our planet's coastlines. In some locations, it has been harnessed as a source of electrical power for human activities.

GRAVITATIONAL DEFORMATION

6 What causes the tides? A clue comes from the observation that they exhibit daily, monthly, and yearly cycles. In fact, the tides are a direct result of the gravitational influence of the Moon and the Sun on Earth. We have already seen how gravity keeps Earth and Moon in orbit about each other, and both in orbit around the Sun. ∞ (Sec. 2.7) For simplicity, let us first consider just the interaction between Earth and the Moon.

Recall that the strength of the gravitational force depends on the distance separating any two objects. Thus, the Moon's gravitational attraction is greater on the side of Earth that faces the Moon than on the opposite side, some 12,800 km (Earth's diameter) farther away. This difference in the gravitational force is small—only about 3 percent—but it produces a noticeable effect—a **tidal bulge**. As illustrated in Figure 7.22, Earth becomes slightly

Figure 7.22 This exaggerated illustration shows how the Moon induces tides on both the near and the far sides of Earth. The lengths of the straight arrows indicate the relative strengths of the Moon's gravitational pull on various parts of Earth. (a) The lunar gravitational forces acting on several different locations on and in Earth. The force is greatest on the side nearest the Moon and smallest on the opposite side. (b) The difference between the lunar forces experienced at those same locations and the force acting on Earth's center. The arrows represent the force with which the Moon tends to pull matter away from, or squeeze it toward, the center of our planet. Material on the side of Earth nearest the Moon tends to be pulled away from the center, and material on the far side is "left behind," and a bulge is formed. High and low tides result, twice per day, as Earth rotates beneath the bulges in the oceans.

elongated, with the long axis of the distortion pointing toward the Moon.

Earth's oceans undergo the greatest deformation because liquid can most easily move around on our planet's surface. (A bulge *is* actually raised in the solid material of Earth, but it is about a hundred times smaller than the oceanic bulge.) Thus, the ocean becomes a little deeper in some places (along the line joining Earth to the Moon) and shallower in others (perpendicular to this line). The daily tides we experience result as Earth rotates beneath this deformation.

Notice in Figure 7.22 that the side of Earth opposite the Moon also experiences a tidal bulge. The different gravitational pulls—greatest on that part of Earth closest to the Moon, weaker at Earth's center, and weakest of all on Earth's opposite side—cause average tides on opposite sides of our planet to be approximately equal in height. On the side nearer the Moon, the ocean water is pulled slightly toward the Moon. On the opposite side, the ocean water is left behind as Earth is pulled closer to the Moon. Thus, high tide occurs *twice*, not once, each day at any given location.

Both the Moon and the Sun exert tidal forces on our planet. Even though the Sun is roughly 375 times farther away from Earth than is the Moon, its mass is so much greater (by a factor of about 27 million) that its tidal influence is still significant—about half that of the Moon. Thus, instead of one tidal bulge, there are actually two—one pointing toward the Moon, the other toward the Sun—and the interaction between them accounts for the changes in the height of the tides over the course of a month or a year. When Earth, the Moon, and the Sun are roughly lined up (Figure 7.23a), the gravitational effects reinforce one another, and so the highest tides are generally found at times of new and full moons. These tides are known as *spring tides*. When the Earth–Moon line is perpendicular to the Earth–Sun line (at the first and third quarters; Figure 7.23b), the daily tides are smallest. These are termed *neap tides*.

The variation of the Moon's gravity across Earth is an example of a *differential force*, or **tidal force**. The *average* gravitational interaction between two bodies determines their orbit around each other. However, the *tidal* force, superimposed on that average, tends to deform the bodies themselves. The tidal influence of one body on another diminishes very rapidly with increasing distance—in fact, as the inverse *cube* of the separation. For example, if the distance from Earth to the Moon were to double, the tides resulting from the Moon's gravity would decrease by a factor of 8. We will see many situations in this book where tidal forces are critically important in understanding astronomical phenomena. Notice that we still use the word *tidal* in these other contexts, even though we are not discussing oceanic tides and possibly not even planets at all.

EFFECT OF TIDES ON EARTH'S ROTATION

According to the Earth Data box on p. 152, Earth rotates once on its axis (relative to the stars) in $23^h 56^m$—1 sidereal day. However, we know from fossil measurements that Earth's rotation is gradually slowing down, causing the length of the day to increase by about 1.5 milliseconds (ms) every century—not much on the scale of a human lifetime, but over millions of years, this steady slowing of Earth's spin adds up. At this rate, half a billion years ago, the day was just over 22 hours long and the year contained 397 days.

A number of natural biological clocks lead us to the conclusion that Earth's spin rate is decreasing. For example, each day a growth mark is deposited on a certain type of coral in the reefs off the Bahamas. These growth marks are similar to the annual rings found in tree trunks, except that in the case of coral, the marks are made daily, in response to the day–night cycle of solar illumination. However, they also show yearly variations as the coral's growth responds to Earth's seasonal changes, allowing us to perceive annual cycles. Coral growing today shows 365 marks per year, but ancient coral shows many more growth deposits per year. Fossilized reefs that are five hundred million years old contain coral with nearly 400 deposits per year of growth.

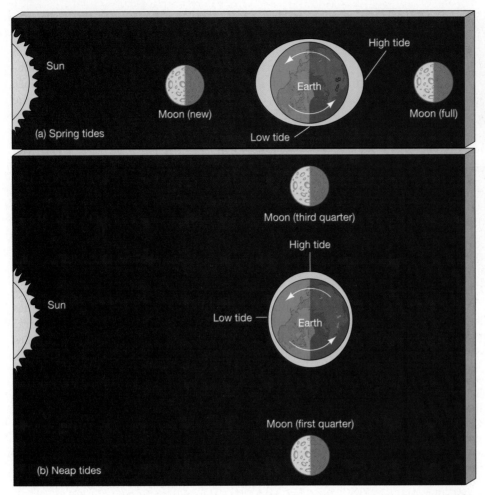

High tide

Sun

Moon (new)

Earth

Moon (full)

Low tide

(a) Spring tides

Moon (third quarter)

High tide

Sun

Low tide

Earth

Moon (first quarter)

(b) Neap tides

Figure 7.23 The combined effects of the Sun and the Moon produce variations in the high and low tides. (a) When the Moon is either full or new, Earth, Moon, and Sun are approximately aligned, and the tidal bulges raised in Earth's oceans by the Moon and the Sun reinforce one another. (b) At first or third-quarter Moon, the tidal effects of the Moon and the Sun partially cancel, and the tides are smallest. The Moon's tidal effect is greater than that of the Sun, and so the net bulge points toward the Moon.

Why is Earth's spin slowing? The main reason is the tidal effect of the Moon. In reality, the tidal bulge raised in Earth by the Moon does *not* point directly at the Moon, as was shown in Figure 7.22. Instead, because of the effects of friction, both between the crust and the oceans and within Earth itself, Earth's rotation tends to drag the tidal bulge around with it, causing the bulge to be displaced by a small angle from the Earth–Moon line, in the same direction as Earth's spin (Figure 7.24). The net effect of the Moon's gravitational pull on this slightly offset bulge is to *reduce* our planet's rotation rate. At the

same time, the Moon is spiraling slowly away from Earth, increasing its average distance from our planet by about 4 cm per year.

This process will continue until Earth rotates on its axis at exactly the same rate as the Moon orbits Earth. At that time the Moon will always be above the same point on Earth and will no longer lag behind the bulge it raises. Earth's rotation period will be 47 of our present days, and the distance to the Moon will be 550,000 km (about 43 percent greater than at present). However, this will take a very long time—many billions of years—to occur.

Ocean

Earth's rotation

Tidal bulge offset

Moon

Figure 7.24 The tidal bulge raised in Earth by the Moon does not point directly at the Moon. Instead, because of the effects of friction, the bulge points slightly "ahead" of the Moon, in the direction of Earth's rotation. (The magnitude of the effect is greatly exaggerated in this diagram.) Because the Moon's gravitational pull on the near-side part of the bulge is greater than the pull on the far side, the overall effect is to decrease Earth's rotation rate.

Chapter Review ✳ *WWW*

SUMMARY

The six main regions of Earth are (from inside to outside) a central metallic **core** (p. 148), which is surrounded by a thick rocky **mantle** (p. 148) and topped with a thin **crust** (p. 148). The liquid oceans on our planet's surface make up the **hydrosphere** (p. 148). Above the surface is the **atmosphere** (p. 148), which is composed primarily of nitrogen and oxygen. Higher still lies the **magnetosphere** (p. 148), where charged particles from the Sun are trapped by Earth's magnetic field.

Earth's atmosphere, composed primarily of nitrogen (78 percent), oxygen (21 percent), argon (0.9 percent), and carbon dioxide (0.03 percent), thins rapidly with altitude. **Convection** (p. 150) is the process by which heat is moved from one place to another by the upwelling or downwelling of a fluid, such as air or water. Convection occurs in the **troposphere** (p. 148), the lowest region of Earth's atmosphere. It is the cause of surface winds and weather. Above the troposphere, in the **mesosphere** (p. 148) and **stratosphere** (p. 148), the air is calm. Straddling the stratosphere and mesosphere is the **ozone layer** (p. 150), where incoming solar ultraviolet radiation is absorbed. At even higher altitudes is the **ionosphere** (p. 148), where the atmosphere is kept ionized by high-energy radiation and particles from the Sun.

The **greenhouse effect** (p. 152) is the absorption and trapping by atmospheric gases (primarily carbon dioxide and water vapor) of infrared radiation emitted by Earth's surface. Incoming visible light from the Sun is not significantly absorbed by these gases. By making it more difficult for Earth to radiate its energy back into space, the greenhouse effect makes our planet's surface some 40 K warmer than would otherwise be the case. The air we breathe is not Earth's original atmosphere. It originated in material that was outgassed from our planet's interior by volcanoes and was then altered by solar radiation and, finally, by the emergence of life.

We study Earth's interior by observing how **seismic waves** (p. 153), produced by **earthquakes** (p. 153) just below Earth's surface, travel through the mantle. We can also study the upper mantle by analyzing the material brought to the surface when a volcano erupts. Seismic studies and mathematical modeling indicate that Earth's iron core consists of a solid **inner core** (p. 155) surrounded by a liquid **outer core** (p. 155). Earth's center is extremely hot—about the same temperature as the surface of the Sun. The density

at Earth's center is much greater than the density of surface rocks. The process by which heavy material sinks to the center of a planet while lighter material rises to the surface is called **differentiation** (p. 156). The differentiation of Earth implies that our planet must have been at least partially molten in the past. One way in which this could have occurred is by the heat released during Earth's formation and subsequent bombardment by material from interplanetary space. Another possibility is the energy released by the decay of **radioactive** (p. 157) elements present in the material from which Earth formed.

Earth's magnetic field extends far beyond the surface of our planet. Charged particles from the solar wind are trapped by Earth's magnetic field lines to form the **Van Allen belts** (p. 158) that surround our planet. When particles from the Van Allen belts hit Earth's atmosphere, they heat and ionize the atoms there, causing them to glow in an **aurora** (p. 160). According to **dynamo theory** (p. 160), planetary magnetic fields are produced by the motion of rapidly rotating, electrically conducting fluid (such as molten iron) in the planet's core.

Earth's surface is made up of about a dozen enormous slabs, or plates. The slow movement of these plates across the surface is called continental drift, or **plate tectonics** (p. 162). Earthquakes, volcanism, and mountain building are associated with plate boundaries, where plates may collide, move apart, or rub against one another. The motion of the plates is thought to be driven by convection in Earth's mantle. The rocky upper layer of Earth that makes up the plates is the **lithosphere** (p. 163). The semisolid region in the upper mantle over which the plates slide is called the **asthenosphere** (p. 163). Evidence for past plate motion can be found in the geographical fit of continents, in the fossil record, and in the ages and magnetism of surface rocks.

The daily **tides** (p. 168) in Earth's oceans are caused by the gravitational effect of the Moon and the Sun, which raise **tidal bulges** (p. 168) in the hydrosphere. The tidal effect of the Moon is almost twice that of the Sun. The size of the tides depends on the orientations of the Sun and the Moon relative to Earth. A differential gravitational force is always called a **tidal force** (p. 168), even when no oceans or even planets are involved. The tidal interaction between Earth and the Moon is causing Earth's spin to slow.

SELF-TEST: TRUE OR FALSE?

____ **1.** Earth's average density is less than the density of water.

____ **2.** The ozone layer is the warmest part of Earth's atmosphere.

____ **3.** Earth's atmosphere is composed primarily of oxygen.

____ **4.** Most of Earth's atmosphere lies within 30 km of the surface.

____ **5.** A continued rise in the level of greenhouse gases in Earth's atmosphere will cause our planet's temperature to decrease.

____ **6.** P-waves can travel through both liquid and solid material; S-waves travel only through solid material.

____ **7.** Geologists obtain most of their information about Earth's mantle by drilling deep into our planet's interior.

____ **8.** Earth's core is hotter than the surface of the Sun.

____ **9.** Samples of Earth's core are available from volcanoes.

____ **10.** Earth's magnetic field is the result of our planet's large, permanently magnetized iron core.

____ **11.** Earth's magnetosphere extends about 5000 km above our planet's surface.

____ **12.** Motion of the crustal plates is driven by convection in Earth's upper mantle.

____ **13.** When plates collide, they simply come to rest and fuse together.

____ **14.** There is one high tide and one low tide per day at any given coastal location on Earth.

____ **15.** Because of the tides, Earth's rotation rate is slowing down.

SELF-TEST: FILL IN THE BLANK

1. Earth's radius is roughly _____ km.
2. Of Earth's crust, mantle, outer core, and inner core, which layer is the thinnest? _____
3. Seventy-eight percent of Earth's atmosphere is _____; 21 percent is _____.
4. The troposphere is where the process of _____ occurs.
5. Sunlight is absorbed by Earth's surface and is reemitted in the form of _____ radiation.
6. The primary greenhouse gases in Earth's atmosphere are _____ and _____.
7. Earth's secondary atmosphere was outgassed by _____.
8. Oxygen in Earth's atmosphere is the result of the appearance of _____.
9. Observations of S- and P-waves have confirmed that Earth's inner core is _____, and the outer core is _____.
10. Crustal rocks are made up primarily of low-density _____; the upper mantle is composed of slightly higher density _____.
11. For differentiation to have occurred, Earth's interior must, at some time in the past, have been largely _____.
12. When trapped electrons and protons in the magnetosphere eventually collide with the upper atmosphere, they produce an _____.
13. Continental drift, volcanism, earthquakes, faults, and mountain building can all be explained by the process known as _____.
14. Earth is unique among the planets in that it has _____ on its surface.
15. Tides are due to the _____ in the gravitational forces due to the Moon and the Sun from one side of Earth to the other.

REVIEW AND DISCUSSION

1. By comparison with Earth's average density, what do the densities of the water and rocks in Earth's crust tell us about Earth's interior?
2. What is Rayleigh scattering? What is its most noticeable effect for us on Earth?
3. Give a brief description of Earth's magnetosphere, and tell how it was discovered.
4. Compare and contrast P-waves and S-waves, and say how they are useful.
5. How would our knowledge of Earth's interior change if our planet were geologically dead, like the Moon?
6. Give two reasons why geologists believe that part of Earth's core is in a liquid state.
7. What clue does Earth's differentiation provide to our planet's history?
8. What is convection? What effect does it have on (a) Earth's atmosphere? (b) Earth's interior?
9. How did radioactive decay heat Earth early in its history? When did this heating end?
10. What conditions are needed to create a dynamo in Earth's interior? What effect does this dynamo have?
11. What process has created the surface mountains, oceanic trenches, and other large-scale features on Earth's surface?
12. Discuss how distant quasars, lying hundreds of millions of light years from Earth, are used to monitor the motion of Earth's tectonic plates.
13. How do we know that Earth's magnetic field has undergone reversals in the past? How do you think Earth's magnetic field reversals might have affected the evolution of life on our planet?
14. If the Moon had oceans like Earth's, what would the tidal effect be like there? How many high and low tides would there be during a "day"? How would the variations in height compare with those on Earth?
15. Is the greenhouse effect operating in Earth's atmosphere helpful or harmful? Give examples. What are the consequences of an enhanced greenhouse effect?

PROBLEMS

1. Verify that Earth's orbital perihelion and aphelion, mean orbital speed, surface gravity, and escape velocity are correct as listed in the Earth Data box on p. 152.
2. What would Earth's surface gravity and escape velocity be if the entire planet had a density equal to that of the crust—3000 kg/m^3, say?

3. Approximating Earth's atmosphere as a layer of gas 7.5 km thick, with uniform density 1.3 kg/m³, calculate its total mass. Compare this with Earth's mass.

4. As discussed in the text, without the greenhouse effect, Earth's average surface temperature would be about 250 K. With the greenhouse effect, it is some 40 K higher. Use this information and Stefan's Law to calculate the fraction of infrared radiation leaving Earth's surface that is absorbed by greenhouse gases in the atmosphere.

5. Most of Earth's ice is found in Antarctica, whose permanent ice caps cover approximately 0.5 percent of Earth's surface and are 2–4 km thick, on average. Earth's oceans cover roughly 71 percent of our planet, to an average depth of 3.6 km. Assuming that water and ice have roughly the same density, estimate by how much sea level would rise if global warming were to cause the Antarctic ice caps to melt.

6. Following an earthquake, how long would it take a P-wave, moving in a straight line with a speed of 5 km/s, to reach the opposite side of Earth?

7. Based on the data presented in the text, estimate the fractions of Earth's volume represented by (a) the inner core, (b) the outer core, (c) the mantle, and (d) the crust. Repeat your calculations for mass instead of volume.

8. At 3 cm/yr, how long would it take a typical plate to traverse the present width of the Atlantic Ocean, about 6000 km?

9. You are standing on Earth's surface, and the Moon is directly overhead. By what fraction is your weight decreased due to the Moon's tidal gravitational force?

10. Compare the magnitude of the tidal gravitational force on Earth due to Jupiter with that due to the Moon. Assume an Earth-Jupiter distance of 4.2 A.U. Based on your answer, do you think that the tidal stresses caused by a "cosmic convergence"—a chance alignment of the four jovian planets, so that they all appear from Earth to be in exactly the same direction in the sky—would have any noticeable affect on our planet?

PROJECTS

1. Go to a sporting goods store and get a tide table; many stores near the ocean provide them free. Choose a month and plot the height of one high and one low tide versus the day of the month. Now mark the dates when the primary phases of the Moon occur. How well does the phase of the Moon predict the tides?

2. Measure Earth's radius. You will need a friend or colleague (or another astronomy student with a project assignment!) who lives a few hundred kilometers due north or south of you. On the day of the first quarter moon, right at sunset, you should both estimate, to within a tenth of a degree, the angular distance of the moon above your southern horizon. Compare the angles you obtain; they should be different. Call this difference Θ (the Greek letter theta). Determine the exact distance

between your two locations using a map; call this d. The radius of Earth can then be computed from the equation $r = 57.3\ d/\Theta$. Many details of how to do this experiment have been left for you to figure out. While you are at it, show where the formula comes from!

3. Go to a library and read about global warming. How much carbon dioxide is produced each year by human activities? How does this compare with the total amount of carbon dioxide in Earth's atmosphere? What natural processes tend to reduce the level of atmospheric carbon dioxide? Do all scientists agree that global warming is an inevitable consequence of carbon dioxide production? What political initiatives are presently underway to address the problem?

8 THE MOON AND MERCURY

Scorched and Battered Worlds

LEARNING GOALS

Studying this chapter will enable you to:

1 Specify the general characteristics of the Moon and Mercury, and compare them with those of Earth.

2 Explain how the Moon's rotation is influenced by its orbit around Earth, and Mercury's by its orbit around the Sun.

3 Describe the surface features of the Moon and Mercury, and recount how they were formed by dynamic events early in their history.

4 Explain how observations of cratering can be used to estimate the age of a body's surface.

5 Compare the Moon's interior structure with that of Mercury.

6 Summarize the various theories for the formation of the Moon, and indicate which is presently considered most likely.

7 Discuss how astronomers have pieced together the story of the Moon's evolution, and compare its evolutionary history with that of Mercury.

(Opposite page, background) The *Apollo 15* mission to the Moon explored a geological fault, called Hadley Rille, where molten lava once flowed. In the large photograph printed here, the rille—a system of valleys—runs along the base of the Apennine Mountains (lower right) at the edge of Mare Imbrium (to the left). For scale, the lower of the two large craters, Autolycus, spans 40 km. The shadow-sided, most prominent peak at lower right, Mount Hadley, rises almost 5 km high.

(Inset A) An astronaut embarks from his lunar rover, preparing to explore Hadley Rille in the distance. The width of the rille is about 1.5 km and its depth averages 300 m.

(Inset B) Photograph of an astronaut and a lunar rover in front of Mount Hadley.

(Inset C) An astronaut collects samples at the edge of Hadley Rille.

(Inset D) This photographic mosaic of a small part of an interior wall of Hadley Rille shows evidence for subsurface horizontal layering. Each distinct layer (denoted A, B, C) is a few meters deep and presumably represents successive lava flows that helped form the extensive lava plain called Mare Imbrium.

The Moon is Earth's only natural satellite. Mercury, the smallest terrestrial world, is the planet closest to the Sun. Despite their different environments, however, these two bodies have many similarities—indeed, at first glance, you might even mistake one for the other. Both have heavily cratered, ancient surfaces, littered with boulders and pulverized dust. With no atmosphere to moderate variations in solar heating, each experiences wild temperature swings from day to night. Both are geologically dead. In short, the Moon and Mercury differ greatly from Earth, but it is precisely those differences that make these desolate worlds so interesting to planetary scientists. Why is the Moon so unlike our own planet, despite its nearness to us, and why does planet Mercury apparently have so much more in common with Earth's Moon than with Earth itself? In this chapter we explore the properties of these two worlds as we begin our comparative study of the planets and moons that make up our solar system.

8.1 Orbital Properties

We begin our study of the Moon and Mercury by examining their orbits. Knowledge of these will, in turn, aid us in determining and explaining many of the other properties of these worlds. Detailed orbital and physical data are presented in the Moon Data box (p. 178) and the Mercury Data box (p. 193).

THE MOON

Parallax methods, described in Chapter 1, can provide us with quite accurate measurements of the distance to the Moon, using Earth's diameter as a baseline. ∞ (Sec. 1.5) Radar yields a more accurate distance. The Moon is much closer than any of the planets, and the radar echo bounced off the Moon's surface is strong. A radio telescope receives the echo after about a 2.56-second wait. Dividing this time by 2 (to account for the round trip taken by the signal) and multiplying it by the speed of light (300,000 km/s) gives us a mean distance of 384,000 km. (The actual distance at any specific time depends on the Moon's location in its slightly elliptical orbit around Earth.)

Current laser-ranging technology, using reflectors placed on the lunar surface by *Apollo* astronauts to reflect laser beams fired from Earth, allows astronomers to measure the round-trip time to submicrosecond accuracy. Repeated measurements have allowed astronomers to determine the Moon's orbit to within a few centimeters. This precision is needed for programming unmanned spacecraft to land successfully on the lunar surface.

MERCURY

Viewed from Earth, Mercury never strays far from the Sun: the angular distance between the Sun and Mercury is never greater than 28°. Consequently, the planet is visible to the naked eye only when the Sun's light is blotted out—just before dawn or just after sunset (or, much less frequently, during a total solar eclipse), and it is not possible to follow Mercury through a full cycle of phases. In fact, although Mercury was well known to ancient astronomers, they originally believed that this companion to the Sun was two different objects, and the connection between the planet's morning and evening appearances took some time to establish. However, later Greek astronomers were certainly aware that the "two planets" were really different alignments of a single body. Figure 8.1, a photograph taken just after sunset, shows Mercury above the western horizon, along with three other planets and the Moon.

Because Earth rotates at a rate of 15° per hour, Mercury is visible for at most 2 hours on any given night, even under the most favorable circumstances. For most observers at most times of the year, Mercury is considerably less than 28° above the horizon, so it is generally visible for a much shorter period (see Figure 8.2). Nowadays, large telescopes can filter out the Sun's glare and observe Mercury even during the daytime, when the planet is higher in the sky and the atmosphere's effects are reduced. (The amount of air that the

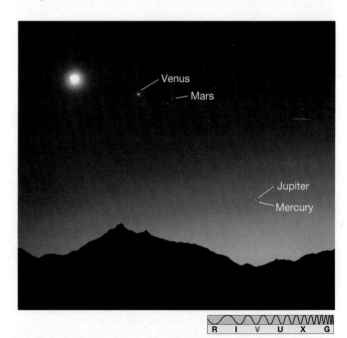

R I V U X G

Figure 8.1 Four planets, together with the Moon, are visible in this photograph taken shortly after sunset. To the right of the Moon (top left) is the brightest planet, Venus. A little farther to the right, at top center, is Mars, with the star Regulus just below and to its left. At the lower right, at the edge of the Sun's glare, is Jupiter, with Mercury just below it. (The moon appears round rather than crescent-shaped because the "dark" portion of its disk is indirectly illuminated by sunlight reflected from Earth. This "earthshine," relatively faint to the naked eye, is exaggerated in the overexposed photographic image.)

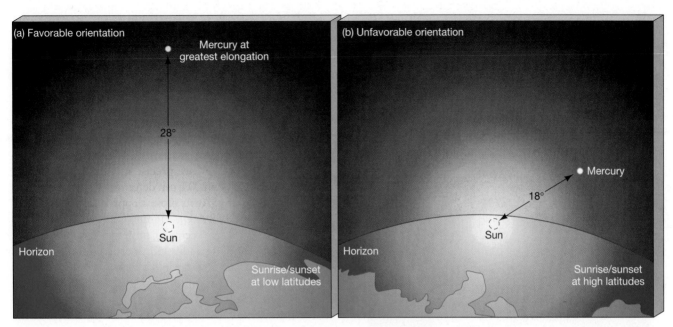

Figure 8.2 Favorable and unfavorable orientations of Mercury's orbit result from different Earth orientations and observer locations. At the most unfavorable orientations, Mercury is close to both the Sun and the horizon.

light from the planet has to traverse before reaching our telescope decreases as the height above the horizon increases.) In fact, some of the best views of Mercury have been obtained in this way. The naked-eye or amateur astronomer is generally limited to nighttime observations, however.

In all cases, it becomes progressively more difficult to view Mercury the closer (in the sky) its orbit takes it to the Sun. The best images of the planet therefore show a "half Mercury," close to its maximum angular separation from the Sun, or *maximum elongation*, as illustrated in Figure 8.3.

Figure 8.3 Some views of Mercury at different points along its orbit. The best images of the planet are taken when it is at its maximum elongation (greatest apparent distance from the Sun) and show a "half Mercury." (Compare with Figure 2.14a.)

MOON DATA

Orbital semi-major axis*	384,000 km
Orbital eccentricity	0.055
Perigee†	363,000 km
Apogee‡	406,000 km
Mean orbital speed	1.02 km/s
Sidereal orbital period	27.3 solar days
Synodic orbital period	29.5 solar days
Orbital inclination to the ecliptic	5.2°
Greatest angular diameter, as seen from Earth	32.9'
Mass	7.35×10^{22} kg 0.012 (Earth = 1)
Equatorial radius	1738 km 0.27 (Earth = 1)
Mean density	3340 kg/m³ 0.61 (Earth = 1)
Surface gravity	1.62 m/s² 0.17 (Earth = 1)
Escape speed	2.38 km/s
Sidereal rotation period	27.3 solar days
Axial tilt	6.7°
Surface magnetic field	(no detectable field)
Surface temperature	100–400 K

*Orbit around Earth

†Closest approach to Earth

‡Greatest distance from Earth

8.2 The Moon and Mercury in Bulk

PHYSICAL PROPERTIES

1 From Earth, the Moon's angular diameter is about 0.5°. Knowing that and the distance to the Moon, we can easily calculate its true size, as discussed in Chapter 1. ∞ (Sec. 1.5) The Moon's radius is about 1700 km, roughly one-fourth that of Earth. More precise measurements yield a lunar radius of 1738 km. We can determine Mercury's radius by similar reasoning. At its closest approach to Earth, at a distance of about 0.52 A.U., Mercury's angular diameter is measured to be 13″ (arc seconds), implying a radius of about 2450 km, or 0.38 Earth radii. More accurate measurements by unmanned space probes yield a result of 2440 km.

As with all astronomical objects, we determine the masses of the Moon and Mercury by studying their gravitational pull on other objects in their vicinity, such as Earth and human-made spacecraft. ∞ (More Precisely 1-2) As mentioned in Chapter 6, even before the Space Age, the masses of both the Moon and Mercury were already quite well known from studies of their effects on Earth's orbit. ∞ (Sec. 6.2) The mass of the Moon is 7.3×10^{22} kg, approximately 1/80 the mass of Earth. The mass of Mercury is 3.3×10^{23} kg—about 0.055 Earth masses.

The Moon's average density of 3300 kg/m³ contrasts with the average Earth value of about 5500 kg/m³, suggesting that the Moon contains fewer heavy elements (such as iron) than Earth. However, despite its many other similarities to the Moon, Mercury's mean density is 5400 kg/m³, only slightly less than that of Earth. Assuming that surface rocks on Mercury are of similar density to surface rocks on Earth and the Moon, we are led to the conclusion that the interior of Mercury must contain a lot of high-density material, most probably iron. In fact, since Mercury is considerably less massive than Earth, its interior is squeezed less by the weight of overlying material, so Mercury's iron core must actually contain a much larger fraction of the planet's mass than is the case for our own planet.

Because the Moon and Mercury are so much less massive than Earth, their gravitational fields are also weaker. The force of gravity on the lunar surface is only about one-sixth that on Earth; Mercury's surface gravity is a little stronger—about 0.4 Earth gravity. Thus, an astronaut weighing 180 lb on Earth would weigh a mere 30 lb on the Moon and 72 lb on Mercury. Those bulky spacesuits used by the *Apollo* astronauts on the Moon were not nearly as heavy as they appeared!

ATMOSPHERES

Astronomers have never observed any appreciable atmosphere on either the Moon or Mercury, either spectroscopically from Earth or during close approaches by spacecraft. As discussed in *More Precisely 8-1*, this is a direct consequence of these bodies' weak gravitational fields. Massive objects have a better chance of retaining their atmosphere, because the more massive an object, the larger are the speeds needed for atoms and molecules to escape. The Moon's escape speed is only 2.4 km/s, compared with 11.2 km/s for Earth; Mercury's escape speed is 4.2 km/s. Any primary atmospheres these worlds had initially, or secondary atmospheres that appeared later, are gone forever.

During its flybys of Mercury in 1974 and 1975, the U.S. space probe *Mariner 10* found traces of what was at first thought to be an atmosphere on Mercury. ∞ (Sec. 6.6) However, this gas is now known to be temporarily trapped hydrogen and helium stolen from the solar wind. Mercury captures this gas and holds it for just a few weeks. Earth-based observations have also found an extremely tenuous envelope of sodium and potassium around the planet. Scientists believe that these atoms are torn out of the surface rocks following impacts with high-energy particles in the solar wind; they do not constitute a true atmosphere in any sense. Thus, neither the Moon nor Mercury has any protection against the harsh environment of interplanetary space. This fact is crucial in understanding their surface evolution and present-day appearance.

Lacking the moderating influence of an atmosphere, both the Moon and Mercury experience wide variations in surface temperature. Noontime temperatures on the Moon can reach 400 K, well above the boiling point of water. Because of its proximity to the Sun, Mercury's daytime temperature is even higher—radio observations of the planet's thermal emission indicate that it can reach 700 K. ∞ (Sec. 3.4) But at night or in the shade, temperatures on both worlds fall to about 100 K, well below water's freezing point. Mercury's 600 K temperature range is the largest of any planet or moon in the solar system.

SURFACE FEATURES

The first observers to point their telescopes at the Moon, most notable among them Galileo Galilei, noted large dark areas, resembling (they thought) Earth's oceans. They also saw light-colored areas resembling the continents. Both types of regions are clear in Figure 8.4(a), a mosaic of the full Moon. The light and dark surface features are also evident to the naked eye, creating the face of the familiar "man in the moon."

Figure 8.4 (a) A photographic mosaic of the full Moon, north pole at the top. Because the Moon emits no visible radiation of its own, we can see it only by the reflected light of the Sun. (b) The Moon near third quarter. Notice that surface features are much more visible near the terminator (the line separating light from dark), where sunlight strikes at a sharp angle, and shadows highlight the topography. The light-colored crater just below the lower left corner of the breakout box is Copernicus. (c) Magnified view of a region near the terminator, as seen from Earth through a large telescope. The central dark area is Mare Imbrium, ringed at bottom right by the Apennine mountains. (d) An enlargement of a portion of (c). The crater at the bottom left is Eratosthenes; Archimedes is at top center.

MORE PRECISELY 8-1   Why Air Sticks Around

Some planets and moons in the solar system—for example, Venus and Titan (the largest moon of Saturn)—have atmospheres thicker than Earth's. Other objects—such as Mars—have thinner atmospheres. Still others—the Moon and Mercury, for example—have virtually no atmospheres at all. The jovian planets have atmospheres rich in hydrogen, while gaseous hydrogen is rare on the terrestrial worlds.

Why do some planets and moons have atmospheres, while others do not, and what determines the composition of the atmosphere if one exists? Why does a layer of air, made up mostly of nitrogen and oxygen, lie just above Earth's surface? After all, experience shows that most gas naturally expands to fill all the volume available. Perfume in a room, fumes from a poorly running engine, and steam from a tea kettle all rapidly disperse until we can hardly sense them. Why doesn't our planet's atmosphere similarly disperse by floating away into space?

The answer is that *gravity* holds it down. Earth's gravitational field exerts a pull on all the atoms and molecules in our atmosphere, preventing them from escaping. However, gravity is not the only influence acting, for if it were, all of Earth's air would have fallen to the surface long ago. *Heat* competes with gravity to keep the atmosphere buoyant. All gas molecules are in constant random motion. The temperature of the gas is a direct measure of this motion—the hotter the gas, the faster the molecules are moving (see *More Precisely 3-1*). The Sun continuously supplies heat to our planet's atmosphere, and the resulting rapid movement of heated molecules produces pressure. This pressure tends to oppose the force of gravity, exerting a net upward force and preventing our atmosphere from collapsing under its own weight. Let's explore this competition between gravity and heat in a little more detail.

An important measure of the strength of a body's gravity is its *escape speed*—the speed needed for any object to escape forever from its surface. ∞ (Sec. 2.7) This speed increases with increased mass or decreased radius of the parent body (often a moon or a planet). Mathematically, it can be expressed as follows:

$$\text{escape speed} \propto \sqrt{\frac{\text{mass of parent object}}{\text{radius of parent object}}}.$$

Thus, if the *mass* of the parent were to quadruple, the escape speed would double. If the parent's *radius* were to quadruple, then the escape speed would be halved, and so on.

To determine whether a planet will retain an atmosphere, we must compare the planet's escape speed with the *molecular speed*, which is the average speed of the gas particles making up the atmosphere. This speed actually depends not only on the temperature of the gas but also on the mass of the individual molecules—the hotter the gas or the smaller the molecular mass, the higher the average speed of the molecules:

$$\text{average molecular speed} \propto \sqrt{\frac{\text{temperature of gas}}{\text{molecular mass}}}.$$

Thus, increasing the temperature of a sample of gas by a factor of 4—for example, from 100 K to 400 K—doubles the average speed of its constituent molecules; at a given temperature, molecules of hydrogen (molecular mass 2) in air move, on average, 4 times faster than molecules of oxygen (molecular mass 32), which are 16 times heavier.

For nitrogen (molecular mass 28) and oxygen in Earth's atmosphere, where the temperature near the surface is nearly 300 K, the typical molecular speed is about 0.5 km/s (actually, 0.52 km/s), far smaller than the 11.2 km/s needed for a molecule to escape into space (see the Earth Data box on p. 152). As a result, Earth is able to re-

---

Today we know that the dark areas are not oceans but extensive flat areas that resulted from lava flows during a much earlier period of the Moon's evolution. Nevertheless, they are still called **maria**, a Latin word meaning "seas" (singular: *mare*). There are 14 maria, all roughly circular. The largest of them (Mare Imbrium) is about 1100 km in diameter. The lighter areas, originally dubbed *terrae*, from the Latin word for "land," are now known to be elevated several kilometers above the maria. Accordingly, they are usually called the lunar **highlands**.

The smallest lunar features we can distinguish with the naked eye are about 200 km across. Telescopic observations further resolve the surface into numerous bowl-shaped depressions, or **craters**. Most craters apparently formed eons

ago primarily as the result of meteoritic impact. Craters are particularly clear in Figures 8.4(b) and (c) near the *terminator* (the line that separates day from night on the surface), where the Sun is low in the sky and casts long shadows that enable us to distinguish quite small surface details.

Due to the blurring effects of our atmosphere, the smallest lunar objects that telescopes on Earth's surface can resolve are about 1 km across (see Figure 8.4d). Much more detailed photographs have been taken by orbiting spacecraft and, of course, by visiting astronauts. Figure 8.5 is a view of some lunar craters taken from an orbiting spacecraft, showing features as small as 500 m across. Craters are found everywhere on the Moon's surface, although they are much more prevalent in the highlands. They come in all tain its nitrogen–oxygen atmosphere. On the whole, our planet's gravity simply has more influence than the heat of our atmosphere.

In reality the situation is a little more complicated than this simple comparison of speeds. Atmospheric molecules can gain or lose speed by bumping into one another or by colliding with objects near the ground. Thus, although we can characterize a gas by its average molecular speed, the molecules do not *all* move at the same speed, as illustrated in the accompanying figure. A tiny fraction of the molecules in any gas have speeds much greater than average—one molecule in 2 million has a speed more than three times the average, and one in 10^{16} exceeds the average by more than a factor of 5. This means that at any instant, *some* molecules are moving fast enough to escape, even when the average molecular speed is much less than the escape speed. The result is that all planetary atmospheres slowly leak away into space.

Don't be alarmed—the leakage is usually very gradual! As a rule of thumb, if the escape speed from a planet exceeds the average speed of a given type of molecule by a factor of 6 or more, then molecules of that type will not have escaped from the planet's atmosphere in significant quantities in the 4.6 billion years since the solar system formed. Conversely, if the escape speed is less than six times the average speed of molecules of a given type, then most of them will have escaped by now, and we should not expect to find them in the atmosphere.

For air on Earth, the mean molecular speeds of oxygen and nitrogen are comfortably below one-sixth of the escape speed. However, if the Moon originally had an Earth-like atmosphere, that lunar atmosphere would have been heated by the Sun to much the same temperature as Earth's air today, so the average molecular speed would have been about 0.5 km/s. Because the Moon's escape speed is only 2.4 km/s, any original lunar atmosphere long ago dispersed into interplanetary space. Similarly, Mercury's escape speed is 4.2 km/s. Its peak surface temperature is around 700 K, corresponding to an average molecular speed for nitrogen or oxygen of about 0.8 km/s, more than one-sixth of the escape speed, so there has been ample time for those gases to escape.

We can also use these arguments to understand some aspects of atmospheric *composition*. For example, hydrogen molecules move, on average, at about 2 km/s in Earth's atmosphere at sea level, so they have had plenty of time to escape since our planet formed (6×2 km/s $= 12$ km/s, which is greater than Earth's 11.2 km/s escape speed). Consequently, we find very little hydrogen in Earth's atmosphere today. However, on the planet Jupiter, with a lower temperature (about 100 K), the speed of hydrogen molecules is correspondingly lower—about 1.1 km/s. At the same time, Jupiter's escape speed is 60 km/s, over five times higher than on Earth. For those reasons, Jupiter has retained its hydrogen—in fact, hydrogen is the dominant ingredient of Jupiter's atmosphere.

sizes—the largest are hundreds of kilometers in diameter; the smallest are microscopic.

All the Moon's significant surface features have names. The 14 maria bear fanciful Latin names—Mare Imbrium ("Sea of Showers"), Mare Nubium ("Sea of Clouds"), Mare Nectaris ("Sea of Nectar"), and so on. Most mountain ranges in the highlands bear the names of terrestrial mountain ranges—the Alps, the Carpathians, the Apennines, the Pyrenees, and so on. Most of the craters are named after great scientists or philosophers, such as Plato, Aristotle, Eratosthenes, and Copernicus.

Because the Moon rotates once on its axis in exactly the same time as it takes to complete one orbit around Earth, the Moon has a "near" side, which is always visible from

Earth, and a "far" side, which never is (see Section 8.3). To the surprise of most astronomers, when the far side of the Moon was mapped, first by Soviet and later by U.S. spacecraft (see *Interlude 8-1*), no major maria were found there. The lunar far side (Figure 8.6) is composed almost entirely of highlands. This fact has great bearing on our theory of how the Moon's surface terrain came into being, for it implies that the processes involved could *not* have been entirely internal in nature. Earth's presence must somehow have played a role.

Mercury is difficult to observe from Earth because of its closeness to the Sun. Even with a fairly large telescope, we see it only as a slightly pinkish disk. Figure 8.7 is one of the few photographs of Mercury taken from

Figure 8.5 The Moon, as seen from the *Apollo 8* orbiter during the first human circumnavigation of the Moon in 1968. Craters ranging in size from 50 km to 500 m (also the width of the long fault lines) can be seen.

Earth that shows any evidence of surface markings. Astronomers could only speculate about the faint, dark markings in the days before *Mariner 10*'s arrival. We now know that these markings are much like those seen when gazing casually at Earth's Moon. The largest ground-based telescopes can resolve surface features on Mercury about as well as we can perceive features on the Moon with our unaided eyes.

In 1974, *Mariner 10* approached within 10,000 km of the surface of Mercury, sending back high-resolution images of the planet. ∞ (Sec. 6.6) These photographs, which showed surface features as small as 150 m across, revolutionized our knowledge of the planet. Figures 8.8 (a) and (b) show views of Mercury taken by *Mariner 10* from a distance of about 200,000 km. Together, these two mosaics cover the known surface of Mercury. No similar

Figure 8.6 The western hemisphere of the Moon, as seen by the *Galileo* probe en route to Jupiter. The large, dark region is part of the face visible from Earth. This image shows only one or two small maria on the far side.

Figure 8.7 Photograph of Mercury taken from Earth with one of the largest ground-based optical telescopes. Only a few surface features are discernible.

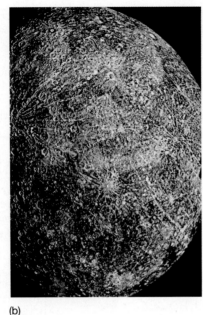

(a)   (b)

Figure 8.8   (a) Mercury is imaged here as a mosaic of photographs taken by the *Mariner 10* spacecraft in the mid-1970s during its approach to the planet. At the time, the spacecraft was some 200,000 km away. (b) *Mariner 10*'s view of Mercury as it sped away from the planet after each encounter. Again, the spacecraft was about 200,000 km away when the photographs making up this mosaic were taken.

photographs exist of the hemisphere that happened to be in shadow during the encounters. Figure 8.9 shows a higher-resolution photograph of the planet from a distance of 20,000 km. The similarities to the Moon are very striking. We see no sign of clouds, rivers, dust storms, or other aspects of weather. Much of the cratered surface bears a strong resemblance to the Moon's highlands. Mercury, however, shows few extensive lava flow regions akin to the lunar maria.

Figure 8.9   Another photograph of Mercury by *Mariner 10*, this time from about 20,000 km above the planet's surface. The double-ringed crater at the upper left, named C. Bach, is about 100 km across; it exemplifies many of the large craters on Mercury, which tend to have double, rather than single, rings. The reason is not yet understood.

8.3 Rotation Rates

THE ROTATION OF THE MOON

2 As just mentioned, the Moon's rotation period is precisely equal to its period of revolution about Earth—27.3 days—so the Moon keeps the same side facing Earth at all times (see Figure 8.10). To an astronaut standing on the Moon's near-side surface, Earth would appear almost stationary in the sky (although its daily rotation would be clearly evident). This condition, in which the spin of one body is precisely equal to (or *synchronized* with) its revolution around another body, is known as a **synchronous orbit**. The fact that the Moon is in a synchronous orbit around Earth is no accident. It is an inevitable consequence of the gravitational interaction between those two bodies.

Just as the Moon raises tides on Earth, Earth also produces a tidal bulge in the Moon. Because Earth is so much more massive, the tidal force on the Moon is about 20 times greater than that on Earth, and the Moon's tidal bulge is correspondingly larger. In Chapter 7 we saw how lunar tidal forces are causing Earth's spin to slow and how, as a result, Earth will eventually rotate on its axis at the same rate as the Moon revolves around Earth. ∞ (Sec. 7.6) Earth's rotation will not become synchronous with the Earth–Moon orbital period for hundreds of billions of years. In the case of the Moon, however, the process has already gone to completion. The Moon's much larger tidal deformation caused it to evolve into a synchronous orbit long ago, and the Moon is said to have become *tidally locked* to Earth. Most of the moons in the solar system are similarly locked by the tidal fields of their parent planets.

In fact, the size of the lunar bulge is too great to be produced by Earth's present-day tidal influence. The explanation

INTERLUDE 8-1 Lunar Exploration

The Space Age began in earnest on October 4, 1957, with the launch of the Soviet satellite *Sputnik 1*. Thirteen months later, on January 4, 1959, the Soviet *Luna 1*, the first human-made craft to escape Earth's gravity, passed the Moon. *Luna 2* crash-landed on the surface in September of that year, and *Luna 3* returned the first pictures of the far side a month later. The long-running *Luna* series established a clear Soviet lead in the early "space race" and returned volumes of detailed information about the Moon's surface. Several of the *Luna* missions landed and returned surface material to Earth.

The U.S. lunar exploration program got off to a rocky start. The first six attempts in the *Ranger* series, between 1961 and 1964, failed to accomplish their objective of just hitting the Moon. The last three were successful, however. *Ranger 7* collided with the lunar surface (as intended) on June 28, 1964. Five U.S. *Lunar Orbiter* spacecraft, launched in 1966 and 1967, were successfully placed in orbit around the Moon, and they relayed back to Earth high-resolution images of much of the lunar surface. Between 1966 and 1968, seven *Surveyor* missions soft-landed on the Moon and performed detailed analyses of the surface.

Many of these unmanned U.S. missions were performed in support of the manned *Apollo* program. On May 25, 1961, at a time when the U.S. space program was in great disarray, President John F. Kennedy declared that the United States would "send a man to the Moon and return him safely to Earth" before the end of the decade, and the *Apollo* program was born. On July 20, 1969, less than 12 years after *Sputnik* and only 8 years after the statement of the program's goal, *Apollo 11* commander Neil Armstrong became the first human to set foot on the Moon, in Mare Tran-

quilitatis (Sea of Tranquility). Three and a half years later, on December 14, 1972, scientist-astronaut Harrison Schmitt, of *Apollo 17*, was the last.

The astronauts who traveled in pairs to the lunar surface in each lunar lander (see the accompanying photograph) performed numerous geological and other scientific studies on the surface. The later landers brought with them a "lunar rover"—a small golf cart–sized vehicle that greatly expanded the area the astronauts could cover. Probably the most important single aspect of the *Apollo* program was the collection of samples of surface rock from various locations on the Moon. In all, some 382 kg of material was returned to

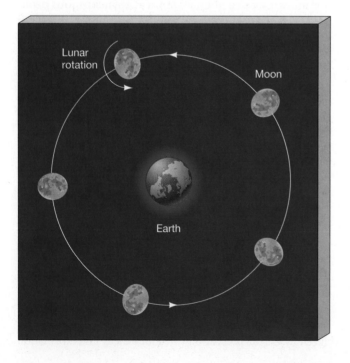

Figure 8.10 The Moon is slightly elongated in shape, with its long axis perpetually pointing toward Earth. (The elongation is highly exaggerated in this diagram.)

seems to be that, long ago, the distance from Earth to the Moon may have been as little as two-thirds of its current value, or about 250,000 km. Earth's tidal force on the Moon would then have been more than three times greater than it is today and could have accounted for the Moon's elongated shape. The resulting distortion could have "set" when the Moon solidified, thus surviving to the present day, while at the same time accelerating the synchronization of the Moon's orbit.

MEASUREMENT OF MERCURY'S SPIN

In principle, the ability to discern surface features on Mercury should allow us to measure its rotation rate simply by watching the motion of a particular region around the planet. In the mid-nineteenth century, an Italian astronomer

Earth. Chemical analysis and radioactive dating of these samples revolutionized our understanding of the Moon's surface history—no amount of Earth-based observations could have achieved the same results.

Each *Apollo* lander also left behind a nuclear-powered package of scientific instruments called *ALSEP* (*Apollo* Lunar Surface Experiments Package) to monitor the solar wind, measure heat flow in the Moon's interior, and, perhaps most important, record lunar seismic activity. With several *ALSEP*s on the surface, scientists can determine the location of "moonquakes" by triangulation and can map the Moon's inner structure, obtaining information critical to our understanding of the Moon's evolution.

By any standards, the *Apollo* program was a spectacular success. It represents a towering achievement of the human race. The project goals were met on schedule and within budget, and our knowledge of the Moon, Earth, and the solar system increased enormously. But the "Age of Apollo" was short-lived. Public interest quickly waned. Over half a billion people breathlessly watched on television as Neil Armstrong set foot on the Moon, yet barely 3 years later, when the program was abruptly canceled for largely political (rather than scientific, technological, or economic) reasons, the landings had become so routine that they no longer excited the interest of the American public. Unmanned space science moved away from the Moon and toward the other planets, and the manned space program foundered. Perhaps one of the most amazing—and saddest—aspects of *Apollo* is that today, nearly three decades later, *no* nation on Earth (including the United States) has the desire, the capability, or the money to repeat the feat.

In 1994, the small U.S. military satellite *Clementine* (see *Interlude 8-2*) was placed in lunar orbit, and performed a detailed survey of the lunar surface. In January 1998, NASA returned to the Moon for the first time in a quarter century with the launch of *Lunar Prospector*, another small satellite on a 1-year mission to study the Moon's structure and origins. *Prospector*'s first task was to survey craters near the lunar poles, to search for water ice hinted at by *Clementine*'s detectors a few years earlier. Since the Sun never rises more than a few degrees above the horizon, as seen from the Moon's polar regions, temperatures on the permanently shaded floors of craters near the poles never exceed about 50 K. Consequently, scientists had theorized, ice there could have remained permanently frozen since the very early days of the solar system, never melting or vaporizing and hence never escaping into space. In early March, 1998, *Prospector* mission scientists announced that sensitive equipment on board the spacecraft had indeed detected large amounts of water ice—perhaps hundreds of millions of tons, in the form of tiny crystals mixed with the lunar regolith.

Plans do exist to establish permanent human colonies on the Moon, either for commercial ventures, such as mining, or for scientific research. Proposals have also been made to site large optical, radio, and other telescopes on the lunar surface. Such instruments, which could be constructed larger than Earth-based devices, would enjoy perfect seeing and no light pollution. None of these projects is scheduled to become reality in the near future, although the discovery of water on the lunar surface alleviates at least one major logistical problem associated with such an undertaking. After a brief encounter with humankind, the Moon is once again a lifeless, unchanging world.

named Giovanni Schiaparelli did just that. He concluded that Mercury always keeps one side facing the Sun, much as our Moon perpetually presents only one face to Earth. The explanation suggested for this synchronous rotation was the same as for the Moon—the tidal bulge raised in Mercury by the Sun had modified the planet's rotation rate until the bulge always pointed directly at the Sun. Although the surface features could not be seen clearly, the combination of Schiaparelli's observations and a plausible physical explanation was enough to convince most astronomers, and the belief that Mercury rotates synchronously with its revolution about the Sun (that is, once every 88 Earth days) persisted for almost half a century.

In 1965, astronomers making radar observations of Mercury from the Arecibo radio telescope in Puerto Rico (see Figure 5.21) discovered that this long-held view was in error. The technique they used is illustrated in Figure 8.11, which shows a radar signal reflecting from the surface of a hypothetical planet. Let's imagine, for the purpose of this discussion, that the pulse of outgoing radiation is of a single frequency.

The returning pulse bounced off the planet is very much weaker than the outgoing signal. Beyond this change, the reflected signal can be modified in two important ways. First, the signal as a whole may be redshifted or blueshifted as a consequence of the Doppler effect, depending on the overall radial velocity of the planet with respect to Earth. ∞ (Sec. 3.5) Let's assume for simplicity that this velocity is zero, so that, on average, the frequency of the reflected signal is the same as the outgoing beam. Second, if the planet is rotating, the radiation reflected from the side of the planet moving toward us returns at a slightly higher frequency than the radiation reflected from the receding side. (Think of the two hemispheres as being separate sources of radiation and moving at slightly different velocities, one toward us and one away.) The effect is very similar to the rotational line broadening discussed in Chapter 4 (see Figure 4.18), except that in this case the radiation we are measuring was not emitted by the planet but only reflected from its surface. ∞ (Sec. 4.4) What we see in the reflected signal is a spread of frequencies on either side of the original frequency. By measuring the

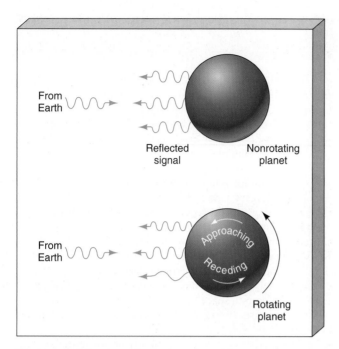

Figure 8.11 A radar beam reflected from a rotating planet yields information about both the planet's overall motion and its rotation rate.

extent of that spread, we can determine the planet's rotational speed.

In this way, the Arecibo researchers found that the rotation period of Mercury is not 88 days, as had previously been believed, but 59 days, exactly two-thirds of the planet's orbital period. Because there are exactly three rotations for every two revolutions, we say that there is a 3:2 **spin–orbit resonance** in Mercury's motion. In this con-

text, the term *resonance* just means that two characteristic times—here Mercury's day and year—are related to each other in a simple way. An even simpler example of a spin–orbit resonance is the Moon's orbit around Earth. In that case, the rotation is synchronous with the revolution, so the resonance is said to be 1:1.

Figure 8.12 illustrates some implications of Mercury's curious rotation for a hypothetical inhabitant of the planet. Mercury's solar day—the time from noon to noon, say—is actually 2 Mercury years long! The Sun stays "up" in the black Mercury sky for almost 3 Earth months at a time, after which follows nearly 3 months of darkness. At any given point in its orbit, Mercury presents the same face to the Sun not every time it revolves, but *every other* time.

EXPLANATION OF MERCURY'S ROTATION

Mercury's 3:2 spin–orbit resonance did not occur by chance. What mechanism establishes and maintains it? In the case of the Moon orbiting Earth, the 1:1 resonance is explained as the result of tidal forces. In essence, the lunar rotation period, which probably started off much shorter than its present value, has lengthened so that the tidal bulge created by Earth is fixed relative to the body of the Moon. Tidal forces (this time due to the Sun) are also responsible for Mercury's 3:2 resonance, but in a much more subtle way.

Mercury cannot settle into a 1:1 resonance because its orbit around the Sun is quite eccentric. By Kepler's second law, Mercury's orbital speed is greatest at perihelion (closest approach to the Sun) and least at aphelion (greatest distance from the Sun). ∞ (Sec. 2.3) A moment's thought shows that because of these variations in the planet's orbital speed, there is no way that the planet (rotating at a constant rate) can remain in a synchronous orbit. If its rotation

Figure 8.12 Mercury's orbital and rotational motions combine to produce a day that is 2 years long. The arrow represents an observer standing on the surface of the planet. At day 0, it is noon for our observer, and the Sun is directly overhead. By the time Mercury has completed one full orbit around the Sun and moved from day 0 to day 88, it has rotated on its axis exactly 1.5 times, so that it is now midnight at the observer's location. After another complete orbit, it is noon once again. The eccentricity of Mercury's orbit is not shown in this simplified diagram.

were synchronous near perihelion, it would be too rapid at aphelion, while synchronism at aphelion would be too slow at perihelion.

Tidal forces always act to try to synchronize the rotation rate with the instantaneous orbital speed, but such synchronization cannot be maintained over Mercury's entire orbit. What happens? The answer is found when we realize that tidal effects diminish very rapidly with increasing distance. The tidal forces acting on Mercury at perihelion are much greater than those at aphelion, and perihelion "won" the struggle to determine the rotation rate. In the 3:2 resonance, Mercury's orbital and rotational motion are almost exactly synchronous *at perihelion*, so that particular rotation rate was naturally "picked out" by the Sun's tidal influence on the planet. Notice that even though Mercury rotates through only 180° between one perihelion and the next (see Figure 8.12), the appearance of the tidal bulge is the *same* each time around.

The motion of Mercury is one of the simplest nonsynchronous resonances known in the solar system. Astronomers now believe that these intricate dynamical interactions are responsible for much of the fine detail observed in the motion of the solar system. Examples of resonances can be found in the orbits of many of the planets, their moons, their rings, and in the asteroid belt.

The Sun's tidal influence also causes Mercury's rotation axis to be exactly perpendicular to its orbit plane. As a result, and because of Mercury's eccentric orbit and the spin–orbit resonance, some points on the surface get much hotter than others. In particular, the two (diametrically opposite) points on the surface where the Sun is directly overhead at perihelion get hottest of all. They are called the *hot longitudes*. The peak temperature of 700 K mentioned earlier occurs at noon at those two locations. At the *warm longitudes*, where the Sun is directly overhead at aphelion, the peak temperature is about 150 K cooler—a mere 550 K! By contrast, the Sun is always on the horizon as seen from the planet's poles. Recent Earth-based radar studies suggest that Mercury's polar temperatures may be as low as 125 K and that, despite the planet's scorched equator, the poles may be covered with extensive sheets of water ice. (See *Interlude 8-1* for similar findings regarding the Moon.)

8.4 Lunar Cratering

3 On Earth, the combined actions of wind and water erode our planet's surface and reshape its appearance almost on a daily basis. Coupled with the never-ending motion of Earth's surface plates, this means that most of the ancient history of our planet's surface is lost to us. The Moon, on the other hand, has no air, no water, no plate tectonics, and no ongoing volcanic or seismic activity. Consequently, features dating back almost to the formation of the Moon itself are still

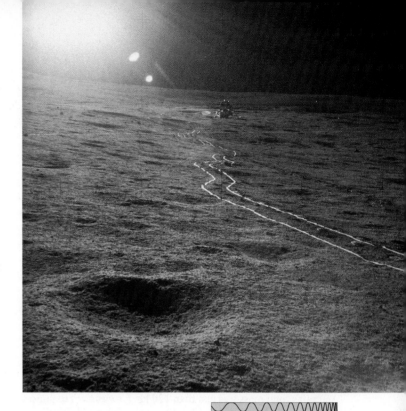

Figure 8.13 Despite the complete lack of wind and water on the airless Moon, the surface has still eroded a little under the constant "rain" of impacting meteoroids, especially micrometeoroids. The twin tracks were made by the *Apollo* lunar rover.

visible today. The lunar surface is not entirely changeless, however. There is ample evidence for erosion—for example, the soft edges of the craters visible in the foreground of Figure 8.13. In the absence of erosion, those features would still be as jagged and angular today as they were just after they formed. Something must have worn them down to their present condition.

METEORITIC IMPACTS

The primary source of erosion on the Moon is interplanetary debris, in the form of small *meteoroids* that collide with the lunar surface. This material, much of it rocky or metallic in composition, is strewn throughout the solar system. ∞ (Sec. 6.5) It orbits the Sun in interplanetary space, perhaps for billions of years, until it happens to collide with some planet or moon. On Earth, most meteoroids burn up in the atmosphere, producing the streaks of light known as *meteors*, or "shooting stars." But the Moon, without an atmosphere, has no protection against this onslaught. Large and small meteoroids just zoom in and collide with the surface, sometimes producing huge craters. Over billions of years, these collisions have scarred, cratered, and sculpted the landscape. Craters are still being formed today—even as you read this—all across the lunar surface.

Meteoroids generally strike the Moon at speeds of several kilometers per second. At these speeds, even a small piece of matter carries an enormous amount of energy—for

INTERLUDE 8-2 A Remarkable Robot Named Clementine

The *Clementine* satellite was sent to the Moon in 1994 by the U.S. Defense Department, largely to test some new sensing devices developed for the ballistic missile defense program. This was the first lunar mission by any nation since the crew of *Apollo 17* left the Moon in 1972. Now, some of the data from this mission have been made public, and the results are remarkable—not only for the high quality of the science returned but also because *Clementine* was a miniature satellite that cost just $70 million, a small fraction of the cost of most other planetary spacecraft, which normally carry price tags of hundreds of millions, sometimes even billions, of dollars.

Clementine originally was a code word for a military-classified space project known as the Deep Space Program Science Experiment. The vehicle and its onboard suite of instruments were designed to test the feasibility of miniaturizing a complex spacecraft, its engineering subsystems, and its sophisticated sensors for use in deep space. The spacecraft's total mass was less than 150 kg. *Clementine*'s technical design was a product of the Strategic Defense Initiative—the Star Wars program—now known as the Ballistic Missile Defense Organization. Its target—the Moon—was of no interest to the military, other than being a convenient, known subject in the cosmic neighborhood.

The lunar mission was a spectacular success, although a follow-on mission to map an asteroid was canceled. *Clementine* was able to make the first digital global map of the Moon, and it did so with very high resolution. In 2 months of operation, its sensors took over 2.5 million images with a clarity at least 10 times better than NASA's most sophisticated planetary camera in the 1990s—the one onboard the *Galileo* mission to Jupiter—and *Clementine*'s highest-resolution camera imaged the Moon with up to 100 times better resolution than *Galileo*

can achieve. *Clementine* was able to obtain global coverage of the Moon at visible, ultraviolet, and infrared wavelengths, and to sense the Moon in 11 different spectral bands. In addition, the small craft also carried lidar devices (the visible equivalent of radar), able to pulse the lunar surface and listen for echoes. In all, more than 50 advanced lightweight technologies were demonstrated on this powerful dwarf spacecraft. Maybe "smaller, faster, cheaper" really is better.

The first figure shows a mosaic of about 1500 images centered on the Moon's south pole. The bottom half is part of the near side of the Moon, as seen from Earth; the top half is the far side that we never see from home. For scale, the double-ringed crater at upper left, called Schroedinger, has an outer diameter of 320 km. The dark region at the pole is an old, permanently shadowed depression. Radar echoes captured by *Clementine* from within it suggested deposits of water ice at a depth of a few meters. As reported in *Interlude 8-1*, these findings have since been confirmed by NASA's *Lunar Prospector* mission.

The second illustration is not an image; it is an altimetry map—that is, an elevation map of the mountains and valleys—for nearly the same south polar region. The colors indicate the ups and downs in the terrain—a good example of how the *Clementine* orbiter found that the Moon is a good deal rougher than was previously thought. The big, dark feature at center is named Aitken Basin. Its rim (yellow, outlined with black dashed circle) spans some 2000 km; the basin itself (mostly purple) averages 10 km deep. This huge depression is the largest impact basin known in the solar system. The violent collision that created this gargantuan scar must have thrown quadrillions of tons of lunar debris far into space, some of which, perhaps, eventually landed on Earth.

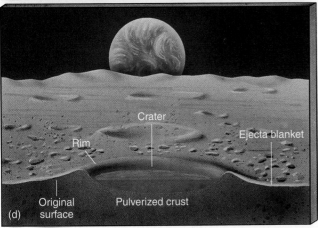

Figure 8.14   Several stages in the formation of a crater by meteoritic impact. (a) A meteoroid strikes the surface, releasing a large amount of energy. (b, c) The resulting explosion ejects material from the impact site and sends shock waves through the underlying surface. (d) Eventually, a characteristic crater surrounded by a blanket of ejected material results.

example, a 1-kg object hitting the Moon's surface at 10 km/s releases as much energy as the detonation of 10 kg of TNT. As illustrated in Figure 8.14, impact of a meteoroid with the surface causes sudden and tremendous pressures to build up, heating the normally brittle rock and deforming the ground like heated plastic. The ensuing explosion pushes previously flat layers of rock up and out, forming a crater.

The diameter of the eventual crater is typically 10 times that of the incoming meteoroid; the crater depth is about twice the meteoroid's diameter. Thus, our 1-kg meteoroid, measuring perhaps 10 cm across, would produce a crater about 1 m in diameter and 20 cm deep. Shock waves from the impact pulverize the lunar surface to a depth many times that of the crater itself. The material thrown out by the explosion surrounds the crater in a layer called an *ejecta blanket*. The ejected debris ranges in size from fine dust to large boulders. Figure 8.15(a) shows the result of one particularly large meteoritic impact on the Moon. As shown in Figure 8.15(b), the larger pieces of ejecta may themselves form secondary craters.

CRATERING RATES

4 In addition to the bombardment by meteoroids with masses of a gram or more, a constant "rain" of *micrometeoroids* (debris with masses ranging from a few micrograms up to about 1 gram) also eats away at the structure of the lunar surface, contributing to the overall erosion process. Some examples can be seen in Figure 8.16, a photomicrograph of glassy "beads" brought back to Earth by the *Apollo* astronauts. The beads themselves were formed during the explosion following a meteoroid impact, when surface rock was melted, ejected, and rapidly cooled. However, several of them also display fresh miniature craters caused by micrometeoroids that struck the beads after they had cooled and solidified.

The *rate* of cratering decreases rapidly with crater size—fresh large craters are scarce, but small craters are very common. The reason for this is simple: there just aren't very many large chunks among the interplanetary debris, so their collisions with the Moon are rare. At the present average rates, one new 10-km (diameter) lunar crater is formed roughly every 10 million years, a new meter-sized crater is created about once a month, and centimeter-sized craters are formed every few minutes.

Despite this constant barrage from space, the Moon's present-day erosion rate is still very low—about 1/10,000

(a) R I V U X G   (b) R I V U X G

Figure 8.15 (a) A large lunar crater, called the Orientale Basin. The impact that produced this crater thrust up much surrounding matter, which can be seen as concentric rings of cliffs called the Cordillera Mountains. The outermost ring is nearly 1000 km in diameter. (b) Two smaller craters called Reinhold and Eddington sit amid the secondary cratering resulting from the impact that created the 90-km-wide Copernicus crater (near the horizon) about a billion years ago. The ejecta blanket from crater Reinhold, 40 km across, and in the foreground, can be seen clearly. View is looking northeast from the lunar module during the *Apollo 12* mission.

that on Earth. Wind and water on Earth are far more effective erosive agents than is meteoritic bombardment on the Moon. For example, the Barringer Meteor Crater (shown in Figure 8.17) in the Arizona desert, one of the largest meteoroid craters on Earth, is only 25,000 years old but has already undergone noticeable erosion. It will probably disappear completely in a mere million years, quite a short time geologically. If a crater that size had formed on the Moon even 4 billion years ago, it would still be plainly visible today.

When the *Apollo* astronauts visited several lunar sites and brought back rock samples, it became possible to measure the ages of the highlands and the maria using radioactive dating techniques, and astronomers can now use the known ages of Moon rocks to estimate the rate of cratering in the past. The highlands are typically more than 4 billion years old, whereas the maria have ages ranging from 3.2 to 3.9 billion years. Thus the much more heavily cratered highlands are indeed older than the less cratered maria, but the difference in cratering is not simply a matter of exposure

Figure 8.16 Craters of all sizes litter the lunar landscape. Some shown here, embedded in glassy beads retrieved by *Apollo* astronauts, measure only 0.01 mm across. (The scale at the top is in millimeters.)

dence of the violent shock waves and high temperatures produced in meteoritic impacts.

Geologists believe that the type of rock found in the maria arose on the Moon much as basalt did on Earth, through the upwelling of molten material through the lunar crust. The great basins that comprise the maria are thought to have been created during the final stages of the heavy meteoritic bombardment just described, between about 4.1 and 3.9 billion years ago. Subsequent volcanic activity filled the craters with lava, creating the formations we see today. In a sense, then, the maria *are* oceans—ancient seas of molten lava, now solidified.

Not all these great craters became flooded with lava, however. One of the youngest craters is the Orientale Basin (Figure 8.15a), which formed about 3.9 billion years ago. It did not undergo much subsequent volcanism, and so we can recognize its structure as an impact crater rather than as another mare. Similar "unflooded" basins are seen on the lunar far side.

LUNAR DUST

Meteoroid collisions with the Moon are the main cause of the layer of pulverized ejecta—also called lunar dust, or *regolith* (meaning "fine rocky layer")—that covers the lunar landscape to an average depth of about 20 m. This microscopic dust has a typical particle size of about 0.01 mm. In consistency, it is rather like talcum powder or ready-mix dry mortar. Figure 8.18 shows an astronaut's bootprint in the regolith. Owing to the very low rate of lunar erosion, even these shallow bootprints will remain intact for millions of years. The regolith is thinnest on the maria (10 m) and thickest on the highlands (over 100 m deep in places).

In contrast to Earth's soil, the lunar regolith contains no organic matter like that produced by biological organisms. No life whatsoever exists on the Moon. Nor were any

Figure 8.17 The Barringer Meteor Crater, near Winslow, Arizona, is 1.2 km in diameter and 0.2 km deep. Geologists think a large meteoroid made it about 25,000 years ago. The meteoroid probably was about 50 m across and weighed around 200,000 tons.

time. Astronomers now believe that the Moon, and presumably the entire inner solar system, experienced a sudden sharp drop in meteoritic bombardment about 3.9 billion years ago. The highlands solidified and received most of their craters before that time, while the maria solidified afterward. The rate of cratering has been roughly constant ever since.

8.5 Lunar Surface Composition

SURFACE ROCKS

3 The *Apollo* program demonstrated clear differences in composition between the lunar highlands and the maria. The highlands are made largely of rocks rich in aluminum, making them lighter in color and lower in density (2900 kg/m^3). The maria's basaltic matter contains more iron, giving it a darker color and greater density (3300 kg/m^3). Loosely speaking, the highlands represent the Moon's crust, while the maria are made of mantle material. Many of the rock samples brought back by the *Apollo* astronauts show patterns of repeated shattering and melting—direct evi-

Figure 8.18 Photograph of an *Apollo* astronaut's bootprint in the lunar dust. The astronaut's weight has compacted the regolith to a depth of a few centimeters.

R I V U X G

Figure 8.19 This "chain" of well-ordered craters was photographed by an *Apollo 14* astronaut. The largest crater, called Davy, is located on the western edge of Mare Nubium. The entire field of view measures about 100 km across.

fossils found in *Apollo* samples. Lunar rocks are barren of life and apparently always have been. NASA was so confident of this that the astronauts were not even quarantined on their return from the last few *Apollo* landings. Furthermore, all the lunar samples returned by the U.S. and Soviet Moon programs were bone dry. Apart from the regolith near the poles (see *Interlude 8-1*), lunar rock doesn't even contain minerals having water molecules locked within their crystal structure. Terrestrial rocks, by contrast, are almost always 1 or 2 percent water.

VOLCANISM

Only a few decades ago, debate raged in scientific circles as to the origin of lunar craters, with most scientists holding the opinion that the craters were the result of volcanic activity. We now know that almost all lunar craters are in fact meteoritic in origin. However, a few of them apparently are not. For example, Figure 8.19 shows an intriguing alignment of several craters in a *crater-chain* pattern so straight that it is very unlikely to have been produced by the random collision of meteoroids with the surface. Instead, the crater chain probably marks the location

of a subsurface fault—a place where cracking or shearing of the surface once allowed molten matter to well up from below. As the lava cooled, it formed a solid "dome" above each fissure. Subsequently, the underlying lava receded and the centers of the domes collapsed, forming the craters we see today. Similar features have been observed on Venus by the orbiting *Magellan* probe (see Chapter 9).

Many other examples of lunar volcanism are known, both in telescopic observations from Earth and in the close-up photographs taken during the *Apollo* missions. Figure 8.20 shows a volcanic **rille**, a ditch where molten lava once flowed. There is good evidence for surface volcanism early in the Moon's history, and the volcanism explains the presence of the lava that formed the maria. However, whatever volcanic activity once existed on the Moon ended long ago. The measured ages for rock samples returned from the Moon are all greater than 3 billion years. (Recall from Chapter 7 that the radioactivity clock doesn't start "ticking" until the rock solidifies.) ∞ *(More Precisely 7-2)* Apparently, the maria solidified over 3 billion years ago and the Moon has been dormant ever since.

R I V U X G

Figure 8.20 A volcanic rille, photographed from the *Apollo 15* spacecraft orbiting the Moon, can be seen clearly here (bottom and center) winding its way through one of the maria. Called Hadley Rille, this system of valleys runs along the base of the Apennine Mountains (lower right) at the edge of the Mare Imbrium (to the left). Autolycus, the large crater closest to the center, spans 40 km. The shadow-sided, most prominent peak at the lower right, Mount Hadley, rises almost 5 km high. (See also the chapter–opening image on p. 174.)

8.6 The Surface of Mercury

3 Like the Moon's craters, almost all craters on Mercury are the result of meteoritic bombardment. However, Mercury's craters are less densely packed than their lunar counterparts, and there are extensive **intercrater plains**. The crater walls are generally not as high as those on the Moon, and the ejected material appears to have landed closer to the impact site, exactly as we would expect on the basis of Mercury's greater surface gravity.

One likely explanation for Mercury's relative lack of craters is that the older craters were filled in by volcanic activity, in much the same way as the Moon's maria filled in older craters as they formed. However, the intercrater plains do not look much like maria—they are much lighter in color and not as flat. Still, most geologists believe that volcanism did occur in Mercury's past, obscuring the old craters. The details of how Mercury's landscape came to look the way it does remain unexplained. The apparent absence of rilles or other obvious features associated with very-large-scale lava flows, along with the light color of the lava-flooded regions, suggest that Mercury's volcanic past was different from the Moon's.

Mercury has at least one type of surface feature not found on the Moon. Figure 8.21 shows a **scarp**, or cliff, on the surface that does not appear to be the result of volcanic or other familiar geological activity. The scarp cuts across several craters, which indicates that whatever produced it occurred *after* most of the meteoritic bombardment was over. Mercury shows no evidence for crustal motions like plate tectonics on Earth. The scarps, of which several are known from the *Mariner* images, probably formed when the planet's interior cooled and shrank long ago, much as wrinkles form on the skin of an old shrunken apple. If we can apply to Mercury the cratering age estimates we use for the Moon, the scarps probably formed about 4 billion years ago.

Figure 8.22 shows what may have been the last great geological event in the history of Mercury—an immense bull's-eye crater called the Caloris Basin, formed eons ago by the impact of a large asteroid. (The basin is so called because it lies in Mercury's "hot longitudes"—see Section 8.3—close to the planet's equator; *calor* is the Latin word for "heat.") Because of the orientation of the planet during *Mariner 10*'s flybys, only half of the basin was visible. The center of the crater is off the left-hand side of the photograph. Compare this basin with the Orientale Basin on the Moon (Figure 8.15a). The impact-crater structures are quite similar, but even here there is a mystery: the patterns visible on the Caloris floor are unlike any seen on the Moon. Their origin, like the composition of the floor itself, is unknown.

So large was the impact that created the Caloris basin that it apparently sent strong seismic waves reverberating throughout the entire planet. On the opposite side of Mercury from Caloris there is a region of oddly rippled and wavy surface features, often referred to as *weird* (or *jumbled*) terrain. Scientists believe that this terrain was produced

PLANETARY DATA: MERCURY

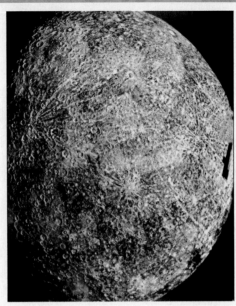

Orbital semi-major axis	0.39 A.U. 57.9 million km
Orbital eccentricity	0.206
Perihelion	0.31 A.U. 46.0 million km
Aphelion	0.47 A.U. 69.8 million km
Mean orbital speed	47.9 km/s
Sidereal orbital period	88.0 solar days 0.241 tropical years
Synodic orbital period	115.9 solar days
Orbital inclination to the ecliptic	7.00°
Greatest angular diameter, as seen from Earth	13″
Mass	3.30×10^{23} kg 0.055 (Earth = 1)
Equatorial radius	2440 km 0.38 (Earth = 1)
Mean density	5430 kg/m^3 0.98 (Earth = 1)
Surface gravity	3.70 m/s^2 0.38 (Earth = 1)
Escape speed	4.2 km/s
Sidereal rotation period	58.6 solar days
Axial tilt	0.0°
Surface magnetic field	0.011 (Earth = 1)
Magnetic axis tilt relative to rotation axis	<10°
Mean surface temperature	100–700 K
Number of moons	0

R I V U X G

Figure 8.21 Discovery Scarp on Mercury's surface. This appears to be a compressional feature that formed when the planet's crust cooled and contracted early in its history, causing a crease in the surface. This scarp, running diagonally across the center of the frame, is several hundred kilometers long and up to 3 km high in places. (Another scarp can be seen in the top left of the figure, close to the horizon.)

when seismic waves from the Caloris impact traveled around the planet and converged on the diametrically opposite point, causing large-scale disruption of the surface there, as illustrated in Figure 8.23.

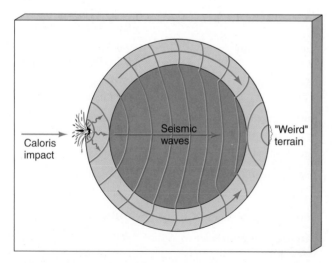

Figure 8.23 The refocusing of seismic waves after the Caloris Basin impact may have created the weird terrain on the opposite side of the planet.

R I V U X G

Figure 8.22 Mercury's most prominent geological feature—the Caloris Basin—measures about 1400 km across and is ringed by concentric mountain ranges that reach more than 3 km high in places. This huge circular basin, only half of which can be seen (at the left) in this *Mariner 10* photo, is similar in size to the Moon's Mare Imbrium and spans more than half of Mercury's radius.

8.7 Interiors

THE MOON

5 The Moon's average density, about 3300 kg/m³, is similar to the measured density of lunar surface rock, virtually eliminating any chance that the Moon has a large, massive, and very dense nickel–iron core like that of Earth. In fact, the low density implies that the entire Moon is actually deficient in iron and other heavy metals relative to our planet. In addition, there is no evidence for any lunar magnetic field. As we saw in Chapter 7, researchers believe that planetary magnetism requires a rapidly rotating liquid metal core, like Earth's. ∞ (Sec. 7.4) Thus, the absence of a lunar magnetic field could be a consequence of the Moon's slow rotation, the absence of a liquid core, or both.

Our knowledge of the details of the Moon's interior is very limited. Models based on the available data indicate that the Moon's interior is of rather uniform density. As depicted schematically in Figure 8.24, the models suggest a core of perhaps 200-km radius surrounded by a roughly 500-km-thick inner mantle of semisolid rock having properties similar to Earth's asthenosphere. The core is probably somewhat more iron-rich than the rest of the Moon (although it is still iron-poor compared with Earth's core).

Near the center, the temperature may be as low as 1500 K, too cool to melt rock. However, seismic data collected by sensitive equipment left on the surface by *Apollo* astronauts (see *Interlude 8-1*) suggest that the inner parts of the core may be at least partially molten, implying a somewhat higher temperature.

Above these regions lies an outer mantle of solid rock, some 900–950 km thick, topped by a 60–150-km crust (considerably thicker than that of Earth). Together these layers constitute the Moon's lithosphere. The crust material, which forms the lunar highlands, is lighter than the mantle, which is similar in chemical composition to the lunar maria.

The crust on the lunar far side is *thicker* than that on the side facing Earth. If we assume that lava takes the line of least resistance in getting to the surface, then we can readily understand why the far side of the Moon has no large maria—volcanic activity did not occur on the far side simply because the crust was too thick to allow it to occur there. But *why* is the far-side crust thicker? The answer is probably related to Earth's gravitational pull. Just as heavier material tends to sink to the center of Earth, the denser lunar mantle tended to sink below the lighter crust in Earth's gravitational field. The effect of this was that the crust and the mantle became slightly off center with respect to each other. The mantle was pulled a little closer to Earth, while the crust moved slightly away. Thus, the crust became thinner on the near side and thicker on the far side.

MERCURY

Mercury's magnetic field, discovered by *Mariner 10*, is about 1/100 that of Earth. Actually, the discovery that Mercury has any magnetic field at all came as a surprise to planetary scientists. Having detected no magnetic field in the Moon (and, in fact, none in Venus or Mars, either), they had expected Mercury to have no measurable magnetism. Mercury certainly does not rotate rapidly, and it may lack a liquid metal core, yet a magnetic field undeniably surrounds it. Although weak, the field is strong enough to deflect the solar wind and create a small magnetosphere around the planet.

Scientists have no clear understanding of the origin of Mercury's magnetic field. If it is produced by ongoing dynamo action, as in Earth, Mercury's core must be at least partially molten. Yet the absence of any recent surface geological activity suggests that the outer layers are solid to a considerable depth, as on the Moon. It is difficult to reconcile these two considerations in a single theoretical model of Mercury's interior. If the field is being generated dynamically, Mercury's slow rotation may at least account for the field's weakness. Alternatively, Mercury's current weak magnetism may simply be the remnant of an extinct dynamo—the planet's iron core may have solidified long ago but still bears a permanent magnetic imprint of the past. The models are inconclusive on this issue, and no spacecraft is scheduled to revisit Mercury in the foreseeable future.

Mercury's magnetic field and large average density together imply that the planet is differentiated. Even without

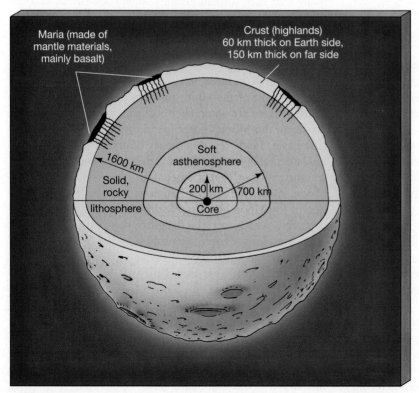

Figure 8.24 Cross-sectional diagram of the Moon. Unlike Earth's rocky lithosphere, the Moon's is very thick—about 1000 km. Below the lithosphere is the inner mantle, or lunar asthenosphere, similar in properties to that of Earth. At the center lies the core, which may be partly molten.

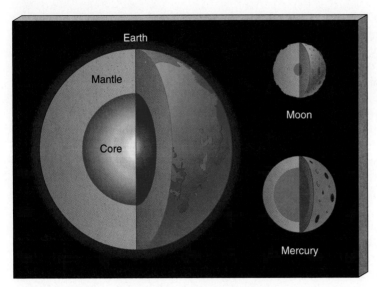

Figure 8.25 The internal structures of Earth, the Moon, and Mercury, drawn to the same scale. Note how large a fraction of Mercury's interior is core.

Animation

the luxury of seismographs on the surface, we can infer that most of its interior must be dominated by a large, heavy, iron-rich core with a radius of perhaps 1800 km. Whether that core is solid or liquid remains to be determined. Probably a less dense, lunarlike mantle lies above this core, to a depth of about 500 to 600 km. Thus, about 40 percent of the volume of Mercury, or 60 percent of its mass, is contained in its iron core. The ratio of core volume to total planet volume is greater for Mercury than for any other object in the solar system. Figure 8.25 illustrates the relative sizes and internal structures of Earth, the Moon, and Mercury.

8.8 The Origin of the Moon

6 The origin of the Moon is uncertain, although several theories have been advanced to account for it. However, both the similarities *and* the differences between the Moon and Earth conspire to confound many promising attempts to explain the Moon's existence.

One theory (the *sister*, or *coformation*, theory) suggests that the Moon formed as a separate object near Earth and in much the same way as our own planet formed—the "blob" of material that eventually coalesced into Earth gave rise to the Moon at about the same time. The two objects thus formed as a double-planet system, each revolving about the common center of mass. Although once favored by many astronomers, this idea suffers from a major flaw: the Moon differs in both density and composition from Earth, making it hard to understand how both could have originated from the same preplanetary material.

A second theory (the *capture* theory) maintains that the Moon formed far from Earth and was later captured by it. In this way, the density and composition of the two objects need not be similar, for the Moon presumably materialized in a quite different region of the early solar system. The ob-

jection to this theory is that the Moon's capture would be an extraordinarily difficult event; it might even be an impossible one. Why? Because the mass of our Moon is so large relative to that of Earth. It's not that our Moon is the largest natural satellite in the solar system, but it is unusually large compared with its parent planet. Mathematical modeling suggests that it is quite implausible that Earth and the Moon could have interacted in just the right way for the Moon to have been captured during a close encounter sometime in the past. Furthermore, although there are indeed significant composition differences between our world and its companion, there are also many similarities—particularly between the mantles of the two bodies—that make it unlikely that they formed entirely independently of one another.

A third, older, theory (the *daughter*, or fission, theory) speculates that the Moon originated out of Earth itself. The Pacific Ocean basin has often been mentioned as the place from which protolunar matter may have been torn—the result, perhaps, of the rapid spin of a young, molten Earth. Indeed, there are some chemical similarities between the matter in the Moon's outer mantle and that in Earth's Pacific basin. However, this theory offers no solution to the fundamental mystery of how Earth could possibly have been spinning so fast that it ejected an object as large as our Moon. Also, computer simulations indicate that the ejection of the Moon into a stable orbit simply would not have occurred. As a result, the daughter theory, in this form at least, is no longer taken seriously.

Today, many astronomers favor a hybrid of the capture and daughter themes. This idea—often called the *impact* theory—postulates a collision by a large, Mars-sized object with a youthful and molten Earth. Such collisions may have been quite frequent in the early solar system (see Chapter 15). The collision presumed by the impact theory would have been more a glancing blow than a direct impact. The matter dislodged from our planet then reassembled to form the Moon.

Computer simulations of such a catastrophic event show that most of the bits and pieces of splattered Earth could have coalesced into a stable orbit. Figure 8.26 shows some of the stages of one such calculation. If Earth had already formed an iron core by the time the collision occurred, then the Moon would indeed have ended up with a composition similar to that of Earth's mantle. During the collision, any

iron core in the impacting object itself would have been left behind in Earth, eventually to become part of Earth's core. Thus both the Moon's overall similarity to that of Earth's mantle and its lack of a dense central core are naturally explained. Over the past decade, planetary scientists have come to realize that collisions such as this probably played very important roles in the formation of all the terrestrial planets.

8.9 Evolutionary History of the Moon and Mercury

THE MOON

7 Given all the data, can we construct a reasonably consistent history of the Moon? The answer seems to be yes. Many specifics are still debated, but a consensus exists. Refer to Figure 8.27 while studying the following details.

The Moon apparently formed about 4.6 billion years ago. The approximate age of the oldest rocks discovered in the lunar highlands is 4.4 billion years, so we know that at least part of the crust must already have solidified by that time and survived to the present. At formation, the Moon was already depleted in heavy metals compared with Earth.

During the earliest phases of the Moon's existence—roughly the first half billion years or so—meteoritic bombardment must have been frequent enough to heat and remelt most of the *surface* layers of the Moon, perhaps to a depth of 400 km. The early solar system was surely populated with lots of interplanetary matter, much of it in the form of boulder-sized fragments that were capable of generating large amounts of energy on collision with planets and their moons. But the intense heat derived from such collisions could not have penetrated very far into the lunar interior. Rock simply does not conduct heat well.

This situation resembles the surface melting we suspect occurred on Earth from meteoritic impacts during the first billion years or so. But the Moon is much less massive than Earth, and it does not contain enough radioactive elements to heat it much further. Radioactivity probably heated the Moon a little, but not sufficiently to transform it from a warm, plastic object to a completely liquid one. The chemical differentiation now inferred in the Moon's interior must have occurred during this period. If the Moon has a small iron core, it also formed at this time.

About 3.9 billion years ago, around the time that Earth's crust solidified, the heaviest phase of the meteoritic bombardment ceased. The Moon was left with a solid crust, which would ultimately become the highlands, dented with numerous large basins, soon to flood with lava and become the maria. Between 3.9 and 3.2 billion years ago, lunar volcanism filled the maria with the basaltic material we see today. The age of the youngest maria—3.2 billion years—indicates the time when the volcanic activity subsided. The maria are the sites of the last extensive lava flows on the

Figure 8.26 This sequence shows a simulated collision between Earth and an object the size of Mars. The sequence proceeds top to bottom and zooms out dramatically. The arrow in the final frame shows the newly formed Moon.

(a) 4 billion years ago

(b) 3 billion years ago

(c) Today

Figure 8.27   Paintings of the Moon (a) about 4 billion years ago, after much of the meteoritic bombardment had subsided and the surface had somewhat solidified; (b) about 3 billion years ago, after molten lava had made its way up through surface fissures to fill the low-lying impact basins and create the smooth maria; and (c) today, with much of the originally smooth maria now heavily pitted with craters formed at various times within the past 3 billion years.

Moon, over 3 billion years ago. Their smoothness, compared with the older, more rugged highlands, disguises their great age.

Small objects cool more rapidly than large ones because their interior is closer to the surface, on average. Being so small, the Moon rapidly lost its internal heat to space. As a consequence, it cooled much faster than did Earth. As the Moon cooled, the volcanic activity ended as the thickness of the solid surface layer increased. With the exception of a few meters of surface erosion from eons of meteoritic bombardment, the lunar landscape has remained more or less structurally frozen for the past 3 billion years. The Moon is dead now, and it has been dead for a long time.

MERCURY

7 Like the Moon, Mercury seems to have been a geologically dead world for much of the past 4 billion years. On both the Moon and Mercury, the lack of ongoing geological activity is a consequence of a thick solid mantle that prevents volcanism or tectonic motion. Because of the *Apollo* program, the Moon's early history is much better understood than Mercury's, which remains somewhat speculative.

What we do know about Mercury's history is learned mostly through comparison with the Moon.

When Mercury formed some 4.5 billion years ago, it was already depleted of lighter, rocky material. We will see later that this was largely a consequence of its location in the hot inner regions of the early solar system, although possibly a collision stripped away some of its light mantle. During the next half-billion years, it melted and differentiated, like the other terrestrial worlds. It suffered the same intense meteoritic bombardment as the Moon. Being more massive than the Moon, Mercury cooled more slowly, so its crust was thinner and volcanic activity more common at early times. More craters were erased, resulting in the intercrater plains found by *Mariner 10*.

As the planet's large iron core formed and then cooled, the planet began to shrink, compressing the crust. This compression produced the scarps seen on Mercury's surface and may have prematurely terminated volcanic activity by squeezing shut the cracks and fissures on the surface. Thus Mercury did not experience the extensive volcanic outflows that formed the lunar maria. Despite its larger mass and greater internal temperature, Mercury has probably been geologically inactive for even longer than the Moon.

Chapter Review *WWW*

SUMMARY

Both the Moon and Mercury are airless, virtually unchanging worlds that experience extremes in temperature. Mercury has no permanent atmosphere, although it does have a thin envelope of gas temporarily trapped from the solar wind. The main surface features on the Moon are the dark **maria** (p. 180) and the lighter-colored **highlands** (p. 180). The surfaces of both the Moon and Mercury are covered with **craters** (p. 180) of all sizes, caused by impacting meteoroids. Meteoritic impacts are the main source of erosion on the surfaces of both worlds. The lunar highlands are older than the maria and are much more heavily cratered. The rate at which craters are formed decreases rapidly with increasing crater size.

The high day-side temperatures and cold night-side temperatures on the Moon and Mercury result from the absence of significant heat conduction or atmospheric blanketing on the planet. Sunlight strikes the polar regions of both the Moon and Mercury at such an oblique angle that temperatures there are very low, with the result that both bodies may have amounts of water ice near the poles.

The tidal interaction between Earth and the Moon is responsible for the Moon's **synchronous orbit** (p. 183), in which the same side of the Moon always faces our planet. The large lunar equatorial bulge probably indicates that the Moon once rotated more rapidly and orbited closer to Earth. Mercury's rotation rate is strongly influenced by the tidal effect of the Sun. Because of Mercury's eccentric orbit, the planet rotates not synchronously but exactly three times for every two orbits around the Sun. The condition in which a body's rotation rate is simply related to its orbit period around some other body is known as a **spin–orbit resonance** (p. 186).

The Moon's surface consists of both rocky and dusty material. Highland rocks, which are less dense than rocks from the maria, are believed to represent the Moon's crust. Maria rocks are believed to have originated in the mantle of the Moon. Lunar dust, called regolith, is made mostly of pulverized lunar rock, mixed with a small amount of material from impacting meteorites. Evidence for past volcanic activity on the Moon is found in the form of solidified lava channels called **rilles** (p. 192). Mercury's surface features bear a striking similarity to those of the Moon. The planet is heavily cratered, much like the lunar highlands. Among the differences between Mercury and the Moon are Mercury's lack of lunarlike maria, its extensive **intercrater plains** (p. 193), and the great cracks, or **scarps** (p. 193), in its crust. The plains were caused by extensive lava flows early in Mercury's history. The scarps were apparently formed when the planet's core cooled and shrank, causing the surface to crack. Mercury's evolutionary path was similar to that of the Moon for half a billion years after they both formed. Mercury's volcanic period probably ended before that of the Moon.

The absence of a lunar atmosphere and any present-day lunar volcanic activity are both consequences of the Moon's small size. Lunar gravity is too weak to retain any gases, and lunar volcanism was stifled by the Moon's cooling mantle shortly after extensive lava flows formed the maria more than 3 billion years ago. The crust on the near side of the Moon is substantially thicker than the crust on the far side. As a result, there are almost no maria on the lunar far side. Mercury has a large impact crater called the Caloris basin, whose diameter is comparable to the radius of the planet. The impact that formed it apparently sent violent shock waves around the entire planet, buckling the crust on the opposite side.

The Moon's average density is not much greater than that of its surface rocks, probably because the Moon cooled more rapidly than the larger Earth and solidified sooner, so there was insufficient time for significant differentiation to occur. The lunar crust is too thick and the mantle too cool for plate tectonics to occur. Mercury's average density is considerably greater—similar to that of Earth—implying that Mercury contains a large high-density core, probably composed primarily of iron. The Moon has no measurable magnetic field, a consequence of its slow rotation and lack of a molten metallic core. Mercury's weak magnetic field seems to have been "frozen in" long ago when the planet's iron core solidified.

The most likely explanation for the formation of the Moon is that the newly formed Earth was struck by a large (Mars-sized) object. Part of the impacting body remained behind as part of our planet. The rest ended up in orbit as the Moon.

SELF-TEST: TRUE OR FALSE?

_____ **1.** Laser-ranging can determine the distance to the Moon to an accuracy of a few centimeters.

_____ **2.** Mercury and Earth have about the same average density.

_____ **3.** The Moon rotates on its axis about once per week.

_____ **4.** Mercury has a very small orbital eccentricity.

_____ **5.** Neither Mercury nor the Moon has any detectable atmosphere.

_____ **6.** Both the Moon and Mercury have nighttime low temperatures of 250 K, well below the freezing point of water.

_____ **7.** Telescopes on Earth could see the astronauts and the lunar landers on the Moon during the *Apollo* missions.

_____ **8.** Mercury has few lava-flow regions like the lunar maria.

_____ **9.** Mercury's solar day is actually longer than its solar year.

_____ **10.** Large craters are formed frequently today on the surface of the Moon.

_____ **11.** Some volcanic activity continues today on the surface of the Moon.

_____ **12.** Mercury's craters are more densely packed than are craters on the Moon.

_____ **13.** Scarps are found only on Mercury, not on the Moon.

_____ **14.** The Moon has an extensive iron core.

_____ **15.** Unlike the Moon, Mercury is differentiated.

SELF-TEST: FILL IN THE BLANK

1. The most accurate method for determining the distance to the Moon is by _____.

2. Mercury can be seen only just before _____ or just after _____.

3. The radius of the Moon is about _____ Earth's radius; the radius of Mercury is about _____ that of Earth. (Give your answers as simple fractions, not decimals.)

4. Because the Moon's average density is so much lower than Earth's average density, we can infer that the Moon must contain less _____.

5. Mercury's iron core contains a _____ fraction of the planet's mass than does Earth's core.

6. Mercury's daytime temperature is higher than the Moon's because it is _____.

7. The _____ on the Moon are dark, flat, roughly circular regions hundreds of kilometers in diameter.

8. Craters on the Moon and Mercury are primarily due to _____.

9. Mercury's rotation rate was first measured using _____.

10. Although Mercury's daytime temperatures are always very hot, it may still be possible for it to have sheets of water ice at its _____.

11. The crater produced by the impact of a meteoroid on the Moon has a typical diameter about _____ times the diameter of the meteoroid.

12. The lunar crust is significantly _____ than Earth's crust.

13. The lunar maria's dark, dense rock originally was part of the lunar _____.

14. Mercury's _____ is about 1/100 that of Earth and was originally thought not to exist at all.

15. The most likely scenario for the formation of the Moon is a collision between Earth and a _____.

REVIEW AND DISCUSSION

1. Why is Mercury seldom seen with the naked eye?

2. Employ the concept of escape speed to explain why the Moon and Mercury have no significant atmospheres.

3. What does it mean to say that Mercury has a 3:2 spin–orbit resonance? Why didn't Mercury settle into a 1:1 spin–orbit resonance with the Sun as the Moon did with Earth?

4. Why is the surface of Mercury often compared with that of the Moon? List two similarities and two differences between the surfaces of Mercury and the Moon.

5. What is a scarp? How are scarps thought to have formed? Why do scientists believe that the scarps formed after most meteoritic bombardment ended?

6. What is the primary source of erosion on the Moon? Why is the average rate of lunar erosion so much less than on Earth?

7. Name two pieces of evidence indicating that the lunar highlands are older than the maria.

8. In contrast with Earth, the Moon and Mercury undergo extremes in temperature. Why?

9. How is Mercury's evolutionary history like that of the Moon? How is it different?

10. Describe the theory of the Moon's origin favored by many astronomers.

11. Because the Moon always keeps one face toward Earth, an observer on the moon's near side would see Earth appear almost stationary in the lunar sky. Still, Earth would change its appearance as the Moon orbited Earth. How would Earth's appearance change?

12. The best place to aim a telescope or binoculars on the Moon is along the terminator line, the line between the Moon's light and dark hemispheres. Why? If you were standing on the lunar terminator, where would the Sun be in your sky? What time of day would it be if you were standing on Earth's terminator line?

13. Where on the Moon would be the best place from which to make astronomical observations? What would be this location's advantage over locations on Earth?

14. Explain why Mercury is never seen overhead at midnight in Earth's sky.

15. How is the varying thickness of the lunar crust related to the presence or absence of maria on the Moon?

PROBLEMS

1. How long does a radar signal take to travel from Earth to Mercury and back when Mercury is at its closest point to Earth?

2. The Moon's mass is 1/80 that of Earth, and the lunar radius is 1/4 Earth's radius. Based on these figures, calculate the total weight on the Moon of a 100-kg astronaut with a 50-kg spacesuit and backpack, relative to his weight on Earth.

3. Based on the data presented in the Moon Data box (p. 178), verify the values given for the Moon's perigee (minimum distance from Earth) and apogee (maximum distance from Earth), and estimate the Moon's minimum and maximum angular diameter, as seen from Earth. Compare these values with the angular diameter of the Sun (of actual diameter 1.4 million km), as seen from a distance of 1 A.U.

4. The *Hubble Space Telescope* has a resolution of about 0.05″. What is the size of the smallest feature it can distinguish on the surface of the Moon? On Mercury? Give your answers in meters.

5. What was the orbit period of the *Apollo 11* command module, orbiting 10 km above the lunar surface?

6. What is the angular diameter of the Sun, as seen from Mercury, at perihelion?

7. Compare the gravitational tidal effect of the Sun on Mercury with the tidal effect of Earth on the Moon (see Section 7.6).

8. Mercury's average orbital speed around the Sun is 48 km/s. Use Kepler's second law (see Section 2.4) to calculate Mercury's speed (a) at perihelion and (b) at aphelion. Convert these speeds to angular speeds (in degrees per day), and compare them with Mercury's 6.1°/day rotation rate.

9. Using the rate given in the text for the formation of 10-km craters on the Moon, estimate how long would be needed for the entire Moon to be covered with new craters of this size. How much higher must the cratering rate have been in the past to cover the entire lunar surface with such craters in the 4.6 billion years since the Moon formed?

10. Assume that a planet will have lost its initial atmosphere by the present time if the average molecular speed exceeds one-sixth of the escape speed (see *More Precisely 8-1*). What would Mercury's mass have to be in order for it still to have a nitrogen (molecular weight 28) atmosphere?

PROJECTS

1. Observe the Moon during an entire cycle of phases. When does the Moon rise, set, and appear highest in the sky at each major phase? What is the interval of time between each phase?

2. If you have binoculars, turn them on the Moon when it appears at twilight and when it appears high in the sky. Draw pictures of what you see. What differences do you notice in your two drawings? What color is the Moon seen near the horizon? What color is the Moon seen high in the sky? Why is there a difference?

3. Watch the Moon over a period of hours on a night when you can see one or more bright stars near it. Estimate how many moon diameters it moves per hour. Knowing the Moon is about 0.5° in diameter, how many degrees per hour does it move? What is your estimate of its orbital period?

4. Try to spot Mercury in morning or evening twilight. (Hint: As seen from the Northern Hemisphere, the best evening apparitions of the planet take place in the spring, and the best morning apparitions take place in the fall.)

A

B

C

D

9 VENUS

Earth's Sister Planet

LEARNING GOALS

Studying this chapter will enable you to:

1 Summarize Venus's general orbital and physical properties.

2 Explain why Venus is hard to observe from Earth and how we have obtained more detailed knowledge of the planet.

3 Compare the surface features and geology of Venus with those of Earth and the Moon.

4 Describe the characteristics of Venus's atmosphere and contrast it with that of Earth.

5 Explain why the greenhouse effect has produced conditions on Venus very different from those on Earth.

6 Describe Venus's magnetic field and internal structure.

(Opposite page, background) The *Magellan* robot spacecraft is seen here in artist's conception orbiting Venus in the early 1990s. When close to the veiled planet, its radar system scans the surface; when farther away, it sends data home to Earth.

(Inset A) Each of the insets shows a three-dimensional, computer-enhanced view of surface features on Venus, artificially colored and vertically exaggerated to enhance small-scale structure. This is the volcanic Gula Mons area, with extensive lava flows across otherwise fractured plains.

(Inset B) A distant view of Maat Mons (above rear), an 8-kilometer high volcano, now apparently dormant.

(Inset C) A closer view of Maat Mons, showing clearly more lava flows across the fractured terrain in the foreground.

(Inset D) The Danu Montes comprise a mountain range rising 1 to 3 kilometers above the plains—probably formed ages ago by surface compression, much like the Andes and Appalachian mountain ranges on Earth.

In its bulk properties at least, Venus seems almost a carbon copy of our own world. The two planets are similar in size, density, and chemical composition. They orbit at comparable distances from the Sun. At formation, they must have been almost indistinguishable from each other. Yet they are now about as different as two terrestrial planets can be. Whereas Earth is a vibrant world, teeming with life, Venus is an uninhabitable inferno, with a dense, hot atmosphere of carbon dioxide, lacking any trace of oxygen or water. Somewhere along their respective evolutionary paths, Venus and Earth diverged, and diverged radically. How did this occur? What were the factors leading to Venus's present condition? Why are Venus's surface, atmosphere, and interior so different from Earth's? In answering these questions we will discover that a planet's environment, as well as its composition, can play a critical role in determining its future.

9.1 Orbital Properties

1 Venus is the second planet from the Sun. Its orbit lies within Earth's, so Venus, like Mercury, is always found fairly close to the Sun in the sky—Venus is never seen more than 47° from the Sun. Given Earth's rotation rate of 15° per hour, this means that Venus is visible above the horizon for at most 3 hours before the Sun rises or after it sets. Because we can see Venus from Earth only just before sunrise or just after sunset, the planet is often called the "morning star" or the "evening star," depending on where it happens to be in its orbit. Figure 9.1 shows Venus in the western sky just after sunset. The Venus Data box on p. 207 lists some of the planet's orbital and physical properties.

Venus is the third brightest object in the entire sky (after the Sun and the Moon). It appears more than 10 times brighter than the brightest star, Sirius. You can see Venus even in the daytime, if you know just where to look. On a moonless night away from city lights, Venus casts a faint shadow. The planet's brightness stems from the fact that Venus is highly reflective. Nearly 70 percent of the sunlight reaching Venus is reflected back into space (compare this with roughly 10 percent in the case of Mercury and the Moon). Most of the sunlight is reflected from clouds high in the planet's atmosphere.

We might expect Venus to appear brightest when it is "full"—that is, when we can see the entire sunlit side. However, because Venus orbits between Earth and the Sun, Venus is full when it is at its greatest distance from us, 1.7 A.U. away on the other side of the Sun (an alignment known as *superior conjunction*, where the term "conjunction" simply indicates that two objects are close together on the sky), as illustrated in Figure 9.2. When Venus is closest to us, the planet is at the new phase, lying between Earth and the Sun (*inferior conjunction*), and we again can't see it, because now the sunlit side faces away from us—only a thin ring of sunlight, caused by refraction in Venus's atmosphere, surrounds the planet. As Venus moves away from inferior conjunction, more and more of it becomes visible, but its distance from us also continues to increase. Venus's maximum brightness, as seen from Earth, actually occurs about 36 days before or after closest approach to our planet. At that time, Venus is about 39° from the Sun, and we see it as a rather fat crescent.

9.2 Venus in Bulk

RADIUS, MASS, AND DENSITY

We can determine Venus's radius from simple geometry, just as we did for Mercury and the Moon. ∞ (Sec. 8.2) At closest approach, when Venus is only 0.28 A.U. from us, its angular diameter is 64″. From this observation, we can determine its radius to be about 6000 km. More accurate measurements from spacecraft give a value of 6052 km, or 0.95 Earth radii.

Like Mercury, Venus has no moon. Before the Space Age, astronomers calculated its mass by indirect means—through studies of its small gravitational effect on the orbits of the other planets, especially Earth. Now that spacecraft have orbited the planet, we know its mass very accurately: 4.9×10^{24} kg, or 0.82 the mass of Earth.

From its mass and radius we find that Venus's average density is 5200 kg/m³. As far as these bulk properties are concerned, then, Venus seems very similar to Earth. If the

Figure 9.1 The Moon and Venus in the western sky just after sunset. Venus clearly outshines even the brightest stars in the sky.

Figure 9.2   Venus appears full when it is at its greatest distance from Earth, on the opposite side of the Sun from us (superior conjunction). As its distance decreases, less and less of its sunlit side becomes visible. When it is closest to Earth, it lies between us and the Sun (inferior conjunction), so we cannot see the sunlit side of the planet at all. Venus appears brightest when it is about 39° from the Sun. (Compare Figure 2.14.)

Animation

planet's overall composition were similar to Earth's, we could then reasonably conclude that the planet's internal structure and evolution were basically Earthlike. We will review what little evidence there is on this subject later in this chapter.

ROTATION RATE

The same clouds whose reflectivity makes Venus so easy to see in the night sky also make it impossible for us to discern any surface features on the planet, at least in visible light. As a result, until the advent of suitable radar techniques in the 1960s, astronomers did not know the rotation period of Venus. Even when viewed through a large optical telescope, the planet's cloud cover shows few features (see Figure 9.5), and attempts to determine Venus's rotation by observing the cloud deck were frustrated by the rapidly changing nature of the clouds themselves. Some astronomers argued for a 25-day period, while others favored a 24-hour cycle. Controversy raged until, to the surprise of all, radar observers announced that the Doppler broadening of their returned echoes implied a sluggish 243-day rotation period. ∞ (Sec. 8.3) Furthermore, Venus's spin was found to be *retrograde*—that is, in a sense opposite that of Earth and most other solar system objects, and opposite that of Venus's orbital motion.

Planetary astronomers conventionally define "north" and "south" for individual members of the solar system by the condition that planets *always* rotate from west to east.

With this definition, Venus's retrograde spin means that the planet's north pole lies *below* the ecliptic plane, unlike any of the other terrestrial worlds. Venus's axial tilt—the angle between its equatorial and orbital planes—is 177.4° (compared with 23.5° in the case of Earth). Note, however, that astronomical images of solar system objects conventionally place objects lying above the ecliptic plane at the top of the frame. Thus, with the preceding definition of north and south, all the images of Venus shown in this chapter have the *south* pole at the top. Figure 9.3 illustrates Venus's retrograde rotation and compares it with the rotation of its neighbors, Mercury, Earth, and Mars. Because of Venus's slow, retrograde rotation, the planet's solar day (noon to noon) is quite different from its sidereal rotation period of 243 Earth days (the time for one "true" rotation relative to the stars). ∞ (Sec. 1.3) In fact, as illustrated in Figure 9.4, one Venus day is a little more than half a Venus year (225 Earth days). *More Precisely 9-1* discusses this interplay between orbital and rotational motion in a little more detail.

Why is Venus rotating "backward," and why so slowly? At present, the best explanation planetary scientists can offer is that early in Venus's evolution, the planet was struck by a large body, much like the one that may have hit Earth and formed the Moon, and that impact was sufficient to reduce the planet's spin almost to zero. ∞ (Sec. 8.8) Whatever its cause, the planet's rotation poses practical problems for Earthbound observers. It turns out that Venus rotates almost exactly five times between one closest approach to Earth

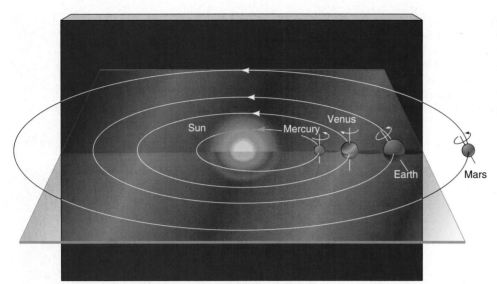

Figure 9.3 The inner four planets of the solar system—Mercury, Venus, Earth, and Mars—display widely differing rotational properties. Although all orbit the Sun in the same sense and in nearly the same plane, Mercury's rotation is slow and prograde (in the same sense as the orbital motion), the rotations of Earth and Mars are fast and prograde, while Venus's is slow and retrograde. Venus rotates clockwise as seen from above the plane of the ecliptic in the direction of the north celestial pole.

and the next. As a result, *Venus always presents nearly the same face to Earth at closest approach.* This means that observations of the planet's surface cover one side—the one facing us at closest approach—much more thoroughly than the other side, which we can see only when the planet is close to its maximum distance from Earth.

Astronomers hate to appeal to coincidence to explain their observations, but the case of Venus's rotation appears to be just that. The near-perfect 5:1 "resonance" between Venus's rotation and orbital motion is reminiscent of the Moon's synchronous orbit around Earth, and Mercury's 3:2 spin–orbit resonance with the Sun. ∞ (Sec. 8.3) However, no known

interaction between Earth and Venus can account for this odd state of affairs. Earth's tidal effect on Venus is tiny, and is much less than the Sun's tidal effect in any case. Furthermore, the key word in the above sentence is *near*. A resonance, if it existed, would require that the number of rotations per orbit be *exactly* five. The discrepancy amounts to less than 3 hours in 584 days (Venus's synodic period relative to Earth—see *More Precisely 9-1*), but it appears to be real, and if that is so, then no resonance exists. For now we are simply compelled to accept this strange coincidence without explanation.

9.3 Long-Distance Observations of Venus

2 Because Venus, of all the other planets, most nearly matches Earth in size, mass, and density, and because its orbit is closest to us, it is often called Earth's sister planet. But unlike Earth, Venus has a dense atmosphere and thick clouds that are opaque to visible radiation, making its surface completely invisible from the outside at optical wavelengths. Figure 9.5 shows one of the best photographs of Venus taken with a large telescope on Earth. The planet resembles a white-yellow disk and shows occasional hints of cloud circulation.

Atmospheric patterns are much more evident when examined with equipment capable of detecting ultraviolet radiation. Some of Venus's atmospheric constituents absorb this high-frequency radiation, greatly increasing the cloud contrast. Figure 9.6 is an ultraviolet image taken in 1979 by the U.S. *Pioneer Venus* spacecraft at a distance of 200,000 km from the planet. The large, fast-moving cloud patterns resemble Earth's high-altitude jet stream more than the great cyclonic whirls characteristic of Earth's low-altitude clouds. The upper deck of clouds on Venus rotates around the planet in just 4 days—much faster than the planet itself.

Figure 9.4 Venus's orbit and retrograde rotation combine to produce a Venus solar day equal to 117 Earth days, or slightly more than half a Venus year. The numbers in the figure mark time in Earth days.

Animation

PLANETARY DATA:
VENUS

Orbital semi-major axis	0.72 A.U.
	108.2 million km
Orbital eccentricity	0.007
Perihelion	0.72 A.U.
	107.5 million km
Aphelion	0.73 A.U.
	108.9 million km
Mean orbital speed	35.0 km/s
Sidereal orbital period	224.7 solar days
	0.615 tropical years
Synodic orbital period	583.9 solar days
Orbital inclination to the ecliptic	3.39°
Greatest angular diameter, as seen from Earth	64″
Mass	4.87×10^{24} kg
	0.82 (Earth = 1)
Equatorial radius	6052 km
	0.95 (Earth = 1)
Mean density	5240 kg/m^3
	0.95 (Earth = 1)
Surface gravity	8.87 m/s^2
	0.91 (Earth = 1)
Escape speed	10.4 km/s
Sidereal rotation period	−243.0 solar days*
Axial tilt	177.4°
Surface magnetic field	<0.001 (Earth = 1)
Magnetic axis tilt relative to rotation axis	—
Mean surface temperature	730 K
Number of moons	0

A negative sign denotes retrograde rotation.

Figure 9.5 This photograph, taken from Earth, shows Venus with its creamy yellow mask of clouds.

Figure 9.6 Venus as it was photographed by the *Pioneer* spacecraft's cameras some 200,000 km from the planet. This image was made by capturing solar ultraviolet radiation reflected from the planet's clouds, which are probably composed mostly of droplets of sulfuric acid (the highly corrosive acid used in car batteries).

Early spectroscopic studies of sunlight reflected from Venus's clouds revealed the presence of large amounts of carbon dioxide but provided little evidence for any other atmospheric gases. Until the 1950s, astronomers generally believed that observational difficulties alone prevented them from seeing other atmospheric components. The hope lingered that Venus's clouds were actually predominantly water vapor, like those on Earth, and that below the cloud cover Venus might be a habitable planet similar to our own. Indeed, in the 1930s scientists had measured the temperature of the atmosphere spectroscopically at about 240 K, not much different from our own upper atmosphere. ∞ (Sec. 4.4) Calculations of the surface temperature—taking into account the cloud cover and Venus's proximity to the Sun, and assuming an atmosphere much like our own—suggested that Venus should have a surface temperature only 10 or 20 degrees higher than Earth's.

These hopes for an Earth-like Venus were dashed in 1956, when radio observations of the planet were used to measure its thermal energy emission. Unlike visible light, radio waves easily penetrate the cloud layer, and they gave the first indication of conditions on or near the surface. The radiation emitted by the planet has a blackbody spectrum characteristic of a temperature near 730 K! ∞ (Sec. 3.4) Almost overnight, the popular conception of Venus changed from that of a lush tropical jungle to an arid, uninhabitable desert.

Radar observations of the surface of Venus are now routinely carried out from Earth using the Arecibo radio telescope. ∞ (Sec. 5.4) This instrument can achieve a resolution of a few kilometers, but it can adequately cover only a small fraction (roughly 25 percent) of the planet. The telescope's view of Venus is limited by the planet's peculiar near-resonance, as just described, and also because radar reflections from regions near the "edge" of the planet are hard to obtain. However, the Arecibo data have been usefully combined with information from probes orbiting Venus to build up a detailed picture of the planet's surface. Only quite recently, with the arrival of the *Magellan* probe, were more accurate data obtained.

9.4 The Surface of Venus

3 Although its clouds are thick and the terrain below them totally shrouded, we are by no means ignorant of Venus's surface. Radar astronomers have bombarded the planet with radio signals, both from Earth and from the *Venera, Pioneer Venus*, and, most recently, the *Magellan* spacecraft. ∞ (Sec. 6.6) Analysis of the radar echoes yields a map of the planet's surface; except for the last two figures, all the views of Venus in this section are "radargraphs" (not photographs) created in this way. As Figure 9.7(a) illustrates, the early maps of Venus suffered from poor resolution; however, more recent probes, especially *Magellan*, have provided much sharper views. As in all the *Magellan* images, the light areas in Figure 9.7(b) represent regions where the surface is rough

(a)

(b)

R   I   V   U   X   G

Figure 9.7   (a) This image of the surface of Venus was made by a radar transmitter and receiver onboard the *Pioneer* spacecraft, which is still in orbit about the planet but is now inoperative. The two continent-sized landmasses are named Ishtar Terra (upper left) and Aphrodite (lower right). Colors represent altitude: blue is lowest, red highest. The spatial resolution is about 25 km. (b) A planetwide mosaic of *Magellan* images. The largest "continent" on Venus, Aphrodite Terra, is at the center of the image.

and efficiently scatters *Magellan*'s sideways-looking radar beam back to the detector. Smooth areas tend to reflect the beam off into space instead and so appear dark.

LARGE-SCALE TOPOGRAPHY

Figure 9.8(a) shows basically the same *Pioneer Venus* data of Venus as Figure 9.7(a), except that it has been flattened out into a more conventional map. The altitude of the surface relative to the average radius of the planet is indicated by the use of color, with white representing the highest elevations, blue the lowest. (Note that the blue has nothing to do with oceans, and white does not indicate snow-capped mountains!) A few of the planet's main features are labeled in Figure 9.8(c).

Figure 9.8 (a) Radar map of the surface of Venus, based on the *Pioneer Venus* data. Color represents elevation according to the scale at the right. (b) A similar map of Earth, at the same spatial resolution. (c) Another version of (a), with major surface features labeled.

MORE PRECISELY 9-1   Synodic Periods and Solar Days

Recall from Chapter 1 that a body's *sidereal* orbital (or rotational) period is the time taken for it to complete one orbit (or rotation) relative to the "fixed" stars. ∞ (Sec. 1.3) The *synodic* orbital period is defined to be the time taken for the body to return to the same configuration relative to the Sun, *taking Earth's own motion into account*. ∞ (Sec. 1.4) Examples are the time from one full Moon to the next, or between successive inferior conjunctions (times of closest approach) of Venus. A closely related concept is the *solar day*, the time from noon to noon, which in general differs from the sidereal day because of a planet's motion around the Sun. ∞ (Sec. 1.3) In the case of Earth, the difference between a sidereal and a solar day is relatively small, as is the difference between a sidereal and a synodic month. However, as we have seen, Mercury's sidereal and solar days differ significantly, as do the sidereal and synodic periods of all the planets. Let's take a moment to look at this topic in a little more detail.

Calculating the synodic period is easy. Denoting Earth's sidereal orbital period (365.26 solar days) by P_E and that of the body in question by P, we see then that the *angular speeds* of Earth and the other body (in degrees per day, say) are, respectively, $360°/P_E$ and $360°/P$, since there are 360 degrees in one complete revolution. Thus, the rate at which the body outruns Earth is $360°/P - 360°/P_E$. The body's synodic period S is by definition the time taken for the body to "lap" Earth (that is, outstrip it by 360°), so it follows that

$$\frac{1}{S} = \frac{1}{P} - \frac{1}{P_E},$$

where S is positive for the interior planets Mercury and Venus (having P less than P_E) and negative for the exterior

planets Mars, Jupiter, and so on (meaning that Earth overtakes them). For example, the Moon has $P = 27.3$ days, so its synodic period is $S = 29.5$ days; for Venus, $P = 224.7$ days, yielding a synodic period $S = 583.9$ days.

Similar reasoning applied to the competition between the planet's sidereal rotation period R and its sidereal orbital period P leads to the following expression for the planetary solar day D:

$$\frac{1}{D} = \frac{1}{R} - \frac{1}{P}.$$

Earth has $R = 0.9973$ solar days and $P = P_E = 365.26$ solar days, which gives $D = 1$ solar day, as expected. The difference between Earth's solar and sidereal days is small because Earth rotates on its axis much more rapidly than it revolves around the Sun. For Mercury and Venus, however, the rotation and revolution periods are comparable and the difference between solar and sidereal days is much greater. Using the data from the Mercury Data box, p. 193 we find, for Mercury, $R = 58.6$ (Earth) solar days, $P = 88.0$ solar days, so $D = 176.0$ solar days, or two sidereal years. ∞ (Sec. 8.2) Venus has $R = -243.0$ solar days and $P = 224.7$ solar days, so $D = -116.7$ solar days (the negative signs, as usual, indicate retrograde rotation). The extreme example would be a planet in a synchronous orbit around the Sun (or the Moon orbiting Earth), with $R = P$. In that case, D would be infinite—the Sun would never set!

Sidereal periods are "physical," in the sense that they are the quantities to which the laws of Kepler and Newton refer. However, from the point of view of timekeeping, skywatching, or scheduling a mission to another planet, synodic periods are in many ways just as important.

Figure 9.8(b) shows a map of Earth to the same scale and at the same spatial resolution.

The surface of Venus appears to be relatively smooth, resembling rolling plains with modest highlands and lowlands. Two continent-sized features, called Ishtar Terra and Aphrodite Terra (named after the Babylonian and Greek counterparts, respectively, of Venus, the Roman goddess of love), adorn the landscape, and these contain mountains comparable in height to those on Earth. The elevated "continents" occupy only 8 percent of Venus's total surface area. For comparison, continents on Earth make up about 25 percent of the surface. The remainder of Venus's surface is classified as lowlands (27 percent) or rolling plains (65 percent), although there is probably little geological difference between the two terrains. Note that, while Earth's tectonic plate boundaries are plainly evident in Figure 9.8(b), no similar features can be seen in Figure 9.8(a). ∞ (Sec. 7.5) There appears to be no large-scale plate tectonics on Venus.

Ishtar Terra ("Land of Ishtar") lies in the southern high latitudes (at the *top* of Figure 9.8a—recall our earlier discussion of Venus's retrograde rotation). The projection of the map makes Ishtar Terra appear much larger than it really

is—it is actually about the same size as Australia. This landmass is dominated by a great plateau known as Lakshmi Planum (Figure 9.9), which is some 1500 km across at its widest point. This plain is ringed by mountain ranges that include the highest peak on the planet, Maxwell Mons, which rises some 14 km above the level of Venus's deepest surface depressions. Again for comparison, the highest point on Earth (the summit of Mount Everest) lies about 20 km above the deepest section of Earth's ocean floor (Challenger Deep, at the bottom of the Marianas Trench on the eastern edge of the Philippines plate).

Figure 9.9(a) shows a large-scale *Venera* image of Lakshmi Planum, at a resolution of about 2 km. The "wrinkles" are actually chains of mountains, hundreds of kilometers long and tens of kilometers apart. The red area immediately to the right of the plain is Maxwell Mons. On the western (right-hand) slope of the Maxwell range lies a great crater, called Cleopatra, about 100 km across. Figure 9.9(b) shows a *Magellan* image of Cleopatra. Close-up views of the crater's structure have led planetary scientists to conclude that it is meteoritic in origin, although some volcanic activity was associated with its formation. Notice the dark

(a)

(b)

Figure 9.9 (a) A *Venera* orbiter image of a plateau in Ishtar Terra known as Lakshmi Planum. Maxwell Mons (red) lies on the western margin of the plain, near the right-hand edge of the image. A meteor crater named Cleopatra is visible on the western slope of the Maxwell mountain range. Note the two larger craters in the center of the plain itself. (b) A *Magellan* image of Cleopatra showing a double-ringed structure that identifies it to geologists as an impact crater.

(smooth) lava flow emerging from within the inner ring and cutting across the outer rim at the upper right.

It is now conventional to name features on Venus after famous women—Aphrodite, Ishtar, Cleopatra, and so on. However, the early nonfemale names (for example, Maxwell Mons, named after the Scottish physicist James Clerk Maxwell) predating this convention have stuck, and they are unlikely to change.

Venus's other continent-sized formation, Aphrodite Terra, is located on the planet's equator and is comparable in size to Africa. Before *Magellan*'s arrival, some researchers had speculated that Aphrodite Terra might have been the

site of something akin to seafloor spreading on Earth—a region where two lithospheric plates moved apart and molten rock rose to the surface in the gap between them, forming an extended ridge. With the low-resolution data then available, the issue could not be settled at the time.

The *Magellan* images now seem to rule out even this small-scale tectonic activity, and the Aphrodite region shows no signs of spreading. Figure 9.10(a) shows a portion of Aphrodite Terra called Ovda Regio. The crust appears buckled and fractured, with ridges running in two distinct directions across the image, suggesting that large compressive forces are distorting the crust. There seem to have been repeated peri-

(a)

(b)

Figure 9.10 (a) A *Magellan* image of Ovda Regio, part of Aphrodite Terra. The intersecting ridges indicate repeated compression and buckling of the surface. The dark areas represent regions that have been flooded by lava upwelling from cracks like those shown in (b), detected by *Magellan* in another part of Aphrodite Terra. This network of fissures is about 50 km long.

ods when extensive lava flows occurred. The dark regions are probably solidified lava flows. Some narrow lava channels, akin to rilles on the Moon, also appear. Figure 9.10(b) shows a series of angular cracks in the crust, thought to have formed when lava welled up from a deep fissure, flooded the surrounding area, and then retreated back below the planet's surface. As the molten lava withdrew, the thin new crust of solidified material collapsed under its own weight, forming the cracks we now see. Even taking into account the temperature and composition differences between Venus's crust and Earth's, this terrain is not at all what we would expect at a spreading site similar to the Mid-Atlantic Ridge. ∞ (Sec. 7.5)

Although there is no evidence for plate tectonics on Venus, it is likely that the stresses in the crust that led to the large mountain ranges were caused by convective motion within Venus's mantle, the same basic process that drives Earth's plates. ∞ (Sec. 7.5) Lakshmi Planum, for example, is probably the result of a "plume" of upwelling mantle material that raised and buckled the planet's surface.

VOLCANISM AND CRATERING

Many areas of Venus have extensive volcanic features. Figure 9.11 shows a series of seven pancake-shaped **lava domes**, each about 25 km across. They probably formed when lava oozed out of the surface, formed the dome, and then withdrew, leaving the crust to crack and subside. Lava domes such as these are found in numerous locations on Venus. The most common volcanoes on the planet are of the type known as **shield volcanoes**. Two large shield volcanoes, called Sif Mons and Gula Mons, are shown (in false color) in Figure 9.12. Shield volcanoes, such as the Hawaiian Islands on Earth, are not associated with plate boundaries. Instead, they form when

lava wells up through a "hot spot" in the crust, and are built up over long periods of time by successive eruptions and lava flows. A characteristic of shield volcanoes is the formation of a *caldera*, or crater, at the summit when the underlying lava withdraws and the surface collapses.

The largest volcanic structures on Venus are huge, roughly circular regions known as **coronae**. A large corona, called Aine, can be seen in Figure 9.13, another large-scale mosaic of *Magellan* images. Coronae are unique to Venus. They appear to have been caused by upwelling mantle material, similar to the uplift that resulted in Lakshmi Planum, but on a somewhat smaller scale. They generally have volcanoes both in and around them, and closer inspection of the rims usually shows evidence for extensive lava flows into the plains below.

There is overwhelming evidence for past surface activity on Venus. Has this activity now stopped, or is it still going on? Two pieces of indirect evidence suggest that volcanism continues today. First, the level of sulfur dioxide above Venus's clouds shows large and fairly frequent fluctuations. It is quite possible that these variations result from volcanic eruptions on the surface. If so, volcanism may be the primary cause of Venus's thick cloud cover. Second, both the *Pioneer Venus* and the *Venera* orbiters observed bursts of radio energy from Aphrodite and other regions of the planet's surface. These bursts are similar to those produced by lightning discharges that often occur in the plumes of erupting volcanoes on Earth, again suggesting ongoing activity. However, while these pieces of evidence are quite persuasive, they are still only circumstantial. No "smoking gun" (or erupting volcano) has yet been seen, so the case for active volcanism is not yet complete.

Not all the craters on Venus are volcanic in origin. Some, like Cleopatra (Figure 9.9b), were formed by meteoritic im-

(a)

R I V U X G

(b)

Figure 9.11   (a) These dome-shaped structures resulted when viscous molten rock bulged out of the ground and then retreated, leaving behind a thin solid crust that subsequently cracked and subsided. *Magellan* has found features like this in several locations on Venus. (b) A three-dimensional representation of four of the domes. The computer view is looking toward the right from near the center of the image in part (a). Color is based on data returned by Soviet *Venera* landers.

(a)

Radius
(km)

(b)

R I V U X G

(c)

R I V U X G

Figure 9.12 (a) Two larger volcanoes, known as Sif Mons (left) and Gula Mons, appear in this *Magellan* image. Color indicates height, ranging from purple (the level of the surrounding plain) to orange (corresponding to an altitude of about 4 km). The two volcanic calderas at the summits are about 100 km across. (b) A computer-generated view of Sif Mons, as seen from ground level. (c) Gula Mons, as seen from ground level. In (b) and (c), the colors are based on data returned from Soviet landers, and the vertical scales have been greatly exaggerated; Venus is actually a remarkably flat place.

Figure 9.13 This corona, called Aine, lies in the plains south of Aphrodite Terra; it is about 300 km across. It is probably the result of mantle material that upwelled, causing the surface to bulge outward. Note the fractures in the crust, and the many large impact craters with their surrounding white (rough) ejecta blankets that stud the region.

pact. Large impact craters on Venus are generally circular, but those less than about 15 km in diameter can be quite asymmetric in appearance. Figure 9.14(a) shows a *Magellan* image of a relatively small meteoritic impact crater, about 10 km across, in Venus's southern hemisphere. Geologists believe the light-colored region to be the ejecta blanket—material ejected from the crater following the impact. The odd shape may be the result of a large meteoroid's breaking up just before impact into pieces that hit the surface near one another. This seems to be a fairly common fate for medium-sized bodies (1 km or so in diameter) that plow through Venus's dense atmosphere. Numerous impact craters, again identifiable by their ejecta blankets, can also be discerned in Figure 9.13.

Venus's atmosphere is sufficiently thick that small meteoroids do not reach the ground, so there are no impact craters smaller than about 3 km across. Atmospheric effects probably also account for the observed scarcity of impact craters less than 25 km in diameter. Overall, the rate of formation of large-diameter craters on Venus's surface seems to be only about 1/10 that in the lunar maria. Applying the same crater-age estimates to Venus as we do to Earth and the Moon suggests that much of the surface of Venus is quite young—around a billion years old. Some planetary scientists have suggested that some areas, such as the region shown in Figure 9.12 are even younger—perhaps as little as 200 or 300 million years old. Although erosion by the planet's atmosphere may play some part in obliterating surface features, the main agent is volcanism, which appears to "resurface" the planet every few hundred million years.

(a)

Figure 9.14 (a) A *Magellan* image of an impact crater in Venus's southern hemisphere. The peculiar kidney shape seems to be the result of a meteoroid that fragmented just prior to impact. The dark regions in the crater may be pools of solidified lava. (b) Some more regular looking craters. (c) Three-dimensional computer-generated representations of the craters shown in part (b).

(b)

(c)

Figure 9.15   The first direct view of the surface of Venus, radioed back to Earth from the Soviet *Venera 9* spacecraft, which made a soft landing on the planet in 1975. The amount of sunlight penetrating Venus's cloud cover apparently resembles that on a heavily overcast day on Earth.

Figure 9.16   Another view of Venus, in true color, from *Venera 14.* Flat rocks like those visible in Figure 9.15 are present, but there are also many smaller rocks and even fine soil on the surface. This landing site is not far from the *Venera 9* site shown in the previous figure. The peculiar filtering effects of whatever light does penetrate the clouds make Venus's air and ground peach-colored.

DATA FROM THE SOVIET LANDERS

The 1975 soft landings of the Soviet *Venera 9* and *Venera 10* spacecraft established directly that Venus's surface is very dry and dusty. Figure 9.15 shows one of the first photographs of the surface of Venus radioed back to Earth. Each craft lasted only about an hour before overheating, their electronic circuitry literally melting in this planetary oven. Typical rocks in the photo measure about 50 cm by 20 cm across—a little like flagstones on Earth. Having sharp edges and a slablike character, they show little evidence of erosion. Apparently they are quite young rocks, again supporting the idea of ongoing surface activity of some kind.

Later *Venera* missions took more detailed photographs, as shown in Figure 9.16. The presence of small rocks and finer material indicates the effects of erosive processes. These later missions also performed simple chemical analyses of the surface of Venus. The samples studied by *Venera 13* and *Venera 14* were predominantly basaltic in nature, again implying a volcanic past. However, not all the rocks were found to be volcanic. The *Venera 17* and *Venera 18* landers also found surface material resembling terrestrial granite, probably (as on Earth) part of the planet's ancient crust.

9.5 The Atmosphere of Venus

ATMOSPHERIC STRUCTURE

4 Measurements made by the *Venera* and *Pioneer Venus* spacecraft have allowed us to paint a fairly detailed picture of Venus's atmosphere. ∞ (Sec. 6.6) Figure 9.17 shows the run of temperature and pressure with height. Compare this figure with Figure 7.4, which gives similar information for Earth. The atmosphere of Venus is about 90 times more massive than Earth's, and it extends to a much greater height above the surface. On Earth, 90 percent of the atmosphere lies within about 10 km of sea level. On Venus the corresponding (90 percent) level is found at an altitude of 50 km instead. The surface temperature and pressure of Venus's atmosphere are much greater than Earth's. However, the temperature drops more rapidly with altitude, and the upper atmosphere of Venus is actually colder than our own.

Venus's troposphere extends up to an altitude of nearly 100 km. The reflective clouds that block our view of the surface lie between 50 and 70 km above the surface. Data from the *Pioneer Venus* multiprobe indicate that the clouds

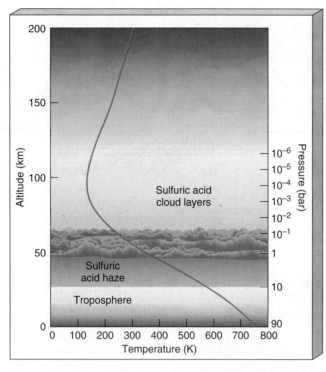

Figure 9.17   The structure of the atmosphere of Venus, as determined by U.S. and Soviet probes. (One bar is the atmospheric pressure at sea level on Earth.)

responsible for the rapidly moving cloud patterns seen in ultraviolet light. Figure 9.18 shows a sequence of three ultraviolet images of Venus in which the variations in the cloud patterns can be seen. Note the characteristic V-shaped appearance of the clouds—a consequence of the fact that, despite their slightly lower speeds, the winds near the poles have a shorter distance to travel in circling the planet and so are always forging ahead of winds at the equator. Near the surface, the dense atmosphere moves more sluggishly—indeed, the fluid flow bears more resemblance to Earth's oceans than to its air. Surface wind speeds are typically less than 2 m/s (roughly 4 mph).

ATMOSPHERIC COMPOSITION

As we saw earlier, observations from Earth revealed the presence of carbon dioxide in Venus's atmosphere but were inconclusive about other possible constituents. We now know that carbon dioxide is in fact the dominant component of the atmosphere, accounting for 96.5 percent of it by volume. Almost all of the remaining 3.5 percent is nitrogen. Trace amounts of other gases, such as water vapor, carbon monoxide, sulfur dioxide, and argon, are also present. This composition is clearly radically different from Earth's atmosphere. The absence of oxygen is perhaps not surprising, given the absence of life (recall our discussion of Earth's atmosphere in Chapter 7). ∞ (Sec. 7.2) However, there is no sign of the water vapor that we might expect to find if a volume of water equivalent to Earth's oceans had evaporated. If Venus started off with Earthlike composition, something has happened to its water—it is now a very dry planet.

For a long time, the chemical makeup of the reflective cloud layer surrounding Venus was unknown. At first scientists assumed the clouds were water vapor or ice, as on

may actually be separated into three distinct layers within that altitude range. Below the clouds, extending down to an altitude of some 30 km, is a layer of haze. Below 30 km, the air is clear. Above the clouds, a high-speed "jet stream" blows from west to east at about 300–400 km/h, fastest at the equator and slowest at the poles. This high-altitude flow is

Figure 9.18   Three ultraviolet views of Venus, taken by the *Pioneer Venus* orbiter, showing the changing cloud patterns in the planet's upper atmosphere. The wind flow is in the direction opposite the "V" in the clouds. Notice the motion of the dark region marked by the arrow. Venus's retrograde rotation means that north is at the bottom of these images, and west is to the right. The time difference between the left and right photographs is about 20 hours.

Earth, but the reflectivity of the clouds at different wavelengths didn't match that of water ice. Later infrared observations carried out in the 1970s showed that the clouds (or at least the top layer of clouds) are actually composed of sulfuric acid, created by reactions between water and sulfur dioxide. Sulfur dioxide is an excellent absorber of ultraviolet radiation and could be responsible for many of the cloud patterns seen in ultraviolet light. Spacecraft observations confirmed the presence of these compounds in the atmosphere. They also indicated that there may be particles of sulfur suspended in and near the cloud layers, which may account for Venus's characteristic yellowish hue.

THE GREENHOUSE EFFECT ON VENUS

5 Given the distance of Venus from the Sun, the planet was not expected to be such a pressure cooker. Calculations based on Venus's orbit and reflectivity indicated a temperature not much different from Earth's, and early measurements of the cloud temperatures seemed to concur. Certainly, scientists reasoned, Venus could be no hotter than the sunward side of Mercury, and it should probably be much cooler. This reasoning was obviously seriously in error.

Why is Venus's atmosphere so hot? And if, as we believe, Venus started off like Earth, why is it now so different? The answer to the first question is fairly easy: given the present composition of its atmosphere, Venus is hot because of the greenhouse effect. Recall from our discussion in Chapter 7 that "greenhouse gases" in Earth's atmosphere, particularly water vapor and carbon dioxide, serve to trap heat from the Sun. ∞ (Sec. 7.2) By inhibiting the escape of infrared radiation reradiated from Earth's surface, these gases serve to increase the planet's equilibrium temperature, in much the same way as an extra blanket keeps you warm on a cold night. Continuing the analogy a little further, the more blankets you place on the bed, the warmer you will become. Similarly, the more greenhouse gases there are in the atmosphere, the hotter the surface will be.

The same effect naturally occurs on Venus, whose dense atmosphere is made up almost entirely of a primary greenhouse gas, carbon dioxide. As illustrated schematically in Figure 9.19, the thick carbon dioxide blanket absorbs nearly 99 percent of all the infrared radiation released from the surface of Venus, and it is the immediate cause of the planet's sweltering 730 K surface temperature. Furthermore, the temperature is nearly as high at the poles as at the equator,

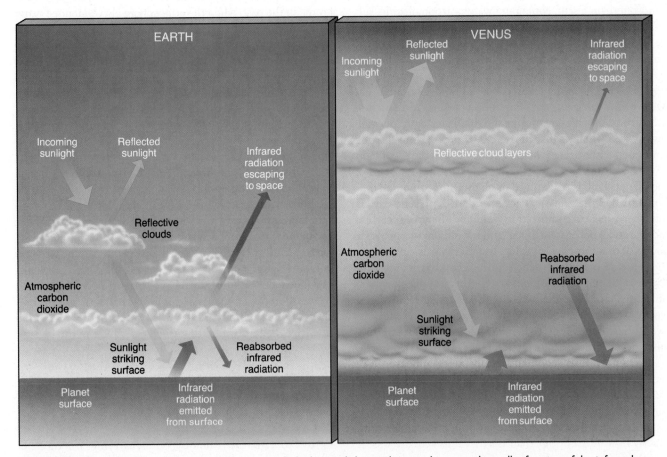

Figure 9.19   Because Venus's atmosphere is much thicker and denser than Earth's, a much smaller fraction of the infrared radiation leaving the planet's surface actually escapes into space. The result is a much stronger greenhouse effect than on Earth and a correspondingly hotter planet. The outgoing infrared radiation is not absorbed at a single point in the atmosphere; instead, absorption occurs at all atmospheric levels. (The arrows are meant to indicate only that absorption occurs, not that it occurs at one specific level.)

and there is not much difference between the temperatures on the day and night sides. The circulation of the atmosphere spreads energy very efficiently around the planet, making it impossible to escape the blazing heat, even during the planet's 2-month-long night.

THE RUNAWAY GREENHOUSE EFFECT

But *why* is Venus's atmosphere so different from Earth's? Why is there so much carbon dioxide in the atmosphere of Venus, and why is the atmosphere so dense? To address these questions, we must consider the processes that created the atmospheres of the terrestrial planets and then determined their evolution. In fact, we can turn the question around and ask instead, "Why is there so *little* carbon dioxide in Earth's atmosphere compared with that of Venus?"

We believe that Earth's atmosphere has evolved greatly since it first appeared. Our planet's secondary atmosphere was outgassed from the interior by volcanic activity 4 billion years ago. ∞ (Sec. 7.2) Since then it has been reprocessed, in part by living organisms, into its present form. On Venus, the initial stages probably took place in more or less the same way, so that at some time in the past, Venus might well have had an atmosphere similar to the primitive secondary atmosphere on Earth, containing water, carbon dioxide, sulfur dioxide, and nitrogen-rich compounds. What happened on Venus to cause such a major divergence from subsequent events on our own planet?

On Earth, nitrogen was released into the air by the action of sunlight on the chemical compounds containing it. Meanwhile, the water condensed into oceans, and much of the carbon dioxide and sulfur dioxide eventually became dissolved in them. Most of the remaining carbon dioxide combined with surface rocks. Thus, much of the secondary outgassed atmosphere quickly became part of the surface of the planet. If all the dissolved or chemically combined carbon dioxide were released back into Earth's present-day atmosphere, its new composition would be 98 percent carbon dioxide and 2 percent nitrogen, and it would have a pressure about 70 times its current value. In other words, apart from the presence of oxygen (which appeared on Earth only after the development of life) and water (the absence of which on Venus will be explained shortly), Earth's atmosphere would be a lot like that of Venus! The real difference between Earth and Venus, then, is that Venus's greenhouse gases never left the atmosphere the way they did on Earth.

When Venus's secondary atmosphere appeared, the temperature was higher than on Earth, simply because Venus is closer to the Sun. However, the Sun was probably somewhat dimmer then (see Chapter 22)—perhaps only half its present brightness—so there is some uncertainty as to exactly how much hotter than Earth Venus actually was. If the temperature was already so high that no oceans condensed, the outgassed water vapor and carbon dioxide would have remained in the atmosphere, and the full greenhouse effect would have gone into operation immediately. If oceans did form and most of the greenhouse gases left the atmosphere, the temperature must still have been sufficiently high that a process known as the **runaway greenhouse effect** came into play.

To understand the runaway greenhouse effect, imagine that we took Earth from its present orbit and placed it in Venus's orbit, some 30 percent closer to the Sun. At that distance from the Sun, the amount of sunlight striking Earth's surface would be about twice its present level, so the planet would warm up. More water would evaporate from the oceans, leading to an increase in atmospheric water vapor. At the same time, the ability of both the oceans and surface rocks to hold carbon dioxide would diminish, allowing more carbon dioxide to enter the atmosphere. As a result, the greenhouse heating would increase, and the planet would warm still further, leading to a further increase in atmospheric greenhouse gases, and so on. Once started, the process would "run away," eventually leading to the complete evaporation of the oceans, restoring all the original greenhouse gases to the atmosphere. Although the details are quite complex, basically the same thing would have happened on Venus long ago, ultimately leading to the planetary inferno we see today.

The greenhouse effect on Venus was even more extreme in the past, when the atmosphere also contained water vapor. By intensifying the blanketing effect of the carbon dioxide, the water vapor helped the surface of Venus reach temperatures perhaps twice as hot as at present. At those high temperatures, the water vapor was able to rise high into the planet's upper atmosphere—so high that it was broken up by solar ultraviolet radiation into its components, hydrogen and oxygen. The light hydrogen rapidly escaped, the reactive oxygen quickly combined with other atmospheric gases, and all water on Venus was lost forever.

Although it is highly unlikely that global warming will ever send Earth down the path taken by Venus, this episode highlights the relative fragility of the planetary environment. No one knows how close to the Sun Earth could have formed before a runaway greenhouse effect would have occurred. But in comparing our planet with Venus, we have come to understand that there is an orbital limit, presumably between 0.7 and 1.0 A.U., inside of which Earth would have suffered a similar catastrophic runaway. We must consider this "greenhouse limit" when assessing the likelihood that planets harboring life formed elsewhere in the Galaxy.

9.6 Venus's Magnetic Field and Internal Structure

In 1962, *Mariner 2* flew by Venus, carrying, among other instruments, magnetometers to measure the strength of the planet's magnetic field. None was detected, and subsequent Soviet and U.S. missions, carrying more sensitive detectors, have confirmed this finding. Venus, with an average density similar to Earth's, probably has a similar overall composition and a partially molten iron-rich core. The lack of any detectable magnetic field on Venus, then, is almost surely the result of the planet's extremely slow rotation and consequent lack of dynamo action. ∞ (Sec. 7.4)

Having no magnetosphere, Venus has no protection from the solar wind. Its upper atmosphere is continually bombarded by high-energy particles from the Sun, keeping the topmost layers permanently ionized. However, the great thickness of the atmosphere prevents any of these particles from reaching the surface.

None of the *Venera* landers carried seismic equipment, so no direct measurements of the planet's interior have been made, and theoretical models of the interior have very little hard data to constrain them. However, to many geologists the surface of Venus resembles the young Earth, at an age of perhaps a billion years. At that time, volcanic activity had already begun, but the crust was still relatively thin and the convective processes in the mantle that drive plate tectonic motion were not yet fully established.

Why has Venus remained in that immature state and not developed plate tectonics like Earth? That question remains to be answered. Some planetary geologists have speculated that the high surface temperature on Venus has inhibited the planet's evolution by slowing the planet's cooling. Possibly the high surface temperature has made the crust too soft for Earth-style plates to develop. Or perhaps the high temperature and soft crust have led to more volcanism, tapping the energy that might otherwise go into convective motion. It may also be that the presence of water plays an important role in lubricating mantle convection and plate motion, so that arid Venus could not evolve along the same path as Earth.

Chapter Review WWW

SUMMARY

The interior orbit of Venus means that it never strays far from the Sun in the sky. Because of its highly reflective cloud cover, Venus is brighter than any star in the sky, as seen from Earth. It is so bright that it can be seen even in the daytime.

Venus's rotation is slow and retrograde, most likely because of a collision between Venus and some other solar system body during the late stages of the planet's formation.

The extremely thick atmosphere of Venus is nearly opaque to visible radiation, making the planet's surface invisible from the outside. Spectroscopic examination of sunlight reflected from the planet's cloud tops shows the presence of large amounts of carbon dioxide. Venus's atmosphere is nearly 100 times denser than Earth's. The temperature of the upper atmosphere is much like that of Earth's upper atmosphere, but the surface temperature is 730 K.

Venus's surface cannot be seen in visible light from Earth, but it has been thoroughly mapped by radar from Earth-based radio telescopes and orbiting satellites. The most recent and most thorough survey has been carried out by the U.S. *Magellan* satellite.

Venus's surface is mostly smooth, resembling rolling plains with modest highlands and lowlands. Two elevated continent-sized regions are called Ishtar Terra and Aphrodite Terra. There is no evidence for plate tectonic activity as on Earth.

Many **lava domes** (p. 212) and **shield volcanoes** (p. 212) have been found by *Magellan* on Venus's surface, but none of the volcanoes has yet been proven to be currently active. The planet's surface shows no sign of plate tectonics. Features called **coronae** (p. 212) are thought to have been caused by an upwelling of mantle material that, for unknown reasons, never developed into full convective motion. The surface of the planet appears to be relatively young, resurfaced by volcanism every few hundred million years. Some craters on Venus are due to meteoritic impact, but the majority are volcanic in origin. The evidence for currently active volcanoes on Venus includes surface features resembling those produced in earthly volcanism, fluctuating levels of sulfur dioxide in Venus's atmosphere, and bursts of radio energy similar to those produced by lightning discharges that often occur in the plumes of erupting volcanoes on Earth. However, no actual eruptions have been seen.

Soviet spacecraft that landed on Venus photographed surface rocks with sharp edges and a slablike character. Some rocks on Venus appear predominantly basaltic in nature, implying a volcanic past. Other rocks resemble terrestrial granite and are probably part of the planet's ancient crust.

Venus is comparable in both mass and radius to Earth, suggesting that the two planets started off with fairly similar surface conditions. However, the atmospheres of Earth and Venus are now very different. The total mass of Venus's atmosphere is about 90 times greater than Earth's. The greenhouse effect stemming from the large amount of carbon dioxide in Venus's atmosphere is the basic cause of the planet's current high temperatures. Almost all the water vapor and carbon dioxide initially present in Earth's early atmosphere quickly became part of the oceans or surface rocks. Because Venus orbits closer to the Sun than does Earth, surface temperatures were initially higher, and the planet's greenhouse gases never left the atmosphere. On Venus, the **runaway greenhouse effect** (p. 218) caused all the planet's greenhouse gases—carbon dioxide and water vapor—to end up in the atmosphere, leading to the extreme conditions we observe today.

Venus has no detectable magnetic field, almost certainly because the planet's rotation is too slow for any appreciable dynamo effect to occur. To some planetary geologists, Venus's interior structure suggests that of the young Earth, before convection became established in the mantle.

SELF-TEST: TRUE OR FALSE?

_____ **1.** Venus appears brighter in the sky than any star.

_____ **2.** Venus has the least eccentric orbit of any of the planets in the solar system.

_____ **3.** Venus is brightest when it is in its full phase.

_____ **4.** Although Venus is slightly smaller than Earth, its atmosphere is significantly more extensive.

_____ **5.** The average surface temperature of Venus is about 260 K.

_____ **6.** Numerous large surface features on Venus can be seen from Earth-based observations made in the ultraviolet part of the spectrum.

_____ **7.** The top cloud layers of Venus are composed of sulfuric acid.

_____ **8.** Other than lacking oxygen, Venus's atmosphere is quite similar to that of Earth.

_____ **9.** Venus has roughly the same temperature at its equator as at its poles.

_____ **10.** The surface of Venus is relatively rough, compared with the surface of Earth, with higher highs and lower lows.

_____ **11.** Images from _Magellan_ show signs of tectonic activity on Venus.

_____ **12.** Lava flows are common on the surface of Venus.

_____ **13.** There is strong circumstantial evidence that active volcanism continues on Venus.

_____ **14.** Venus has a magnetic field similar to that of Earth.

_____ **15.** Soviet spacecraft are still sending back data from the surface of Venus.

SELF-TEST: FILL IN THE BLANK

1. Venus's mass has now been well determined through the use of _____ orbiting it.

2. Because Venus has a mass and average _____ only slightly lower than Earth's, we might expect that its internal structure and evolution should be Earthlike.

3. Venus's rotation is unusual because it is _____.

4. The most abundant gas in the atmosphere of Venus is _____.

5. Water vapor is found in _____ amounts in the atmosphere of Venus.

6. The process that makes Venus so hot is known as the _____.

7. The runaway greenhouse effect on Venus was a result of the planet's being _____ to the Sun than is Earth.

8. The surface of Venus has been mapped using _____.

9. Ishtar Terra and Aphrodite Terra are two _____ on the surface of Venus.

10. Most craters on the surface of Venus are the result of _____.

11. Huge, roughly circular regions on the surface of Venus are known as _____.

12. A meteoroid less than about 1 km in diameter will most likely _____ as it passes through Venus's atmosphere.

13. The scarcity of large impact craters on the surface of Venus indicates that its surface is quite _____.

14. The surface of Venus appears to be resurfaced by _____ every few hundred million years.

15. The main difficulties in using landers to study Venus's surface are the planet's extremely high _____ and _____.

REVIEW AND DISCUSSION

1. Why does Venus appear so bright to the eye? Upon what factors does the brightness of Venus depend?

2. Explain why Venus is always found in the same region of the sky as the Sun.

3. If you were standing on Venus, how would Earth look?

4. How did radio observations of Venus made in the 1950s change our conception of the planet?

5. What did ultraviolet images returned by _Pioneer Venus_ show about the planet's high-level clouds?

6. Name three ways in which the atmosphere of Venus differs from that of Earth.

7. What are the main constituents of Venus's atmosphere? What are clouds in the upper atmosphere made of?

8. What component of Venus's atmosphere causes Venus to be so hot? Explain why there is so much of this gas in the atmosphere of Venus, compared with Earth's. What happened to all the water that Venus must have had when formed?

9. Earth and Venus are nearly alike in size and density. What primary fact caused one planet to evolve as an oasis for life, while the other became a dry and inhospitable inferno?

10. If Venus had formed at Earth's distance from the Sun, what do you imagine its climate would be like today? Why do you think this?

11. How do the "continents" of Venus differ from earthly continents?

12. How are the impact craters of Venus different from those found on other bodies?

13. What evidence exists that volcanism of various types has changed the surface of Venus? What is the evidence for active volcanoes?

14. Given that Venus, like Earth, probably has a partially molten iron-rich core, why doesn't it also have a magnetic field?

15. Do you think there might be life on Venus? Explain your answer.

PROBLEMS

1. Using the data given in the text, calculate Venus's angular diameter, as seen by an observer on Earth, when the planet is (a) at its brightest, (b) at greatest elongation, and (c) at the most distant point in its orbit.

2. How long does a radar signal take to travel from Earth to Venus and back when Venus is brightest (see problem 1)? Compare this with the round-trip time when Venus is at its closest point to Earth.

3. Using the data presented in the Venus Data box on p. 207, calculate the length of a solar day on Venus (that is, the time from one sunrise to the next). How would your answer change if Venus's rotation were prograde instead of retrograde?

4. Compare the magnitude of the tidal gravitational force on Venus due to Earth at closest approach with that due to the Sun. Assume circular orbits for both planets.

5. What is the size of the smallest feature that can be distinguished on the surface of Venus by the Arecibo radio telescope, at an angular resolution of $1'$?

6. *Pioneer Venus* observed high-level clouds moving around Venus's equator in 4 days. What was their speed in km/h? In mph?

7. Venus's atmosphere is 90 times more massive than Earth's atmosphere. Approximating Venus's atmosphere as a layer of gas of uniform density 70 kg/m^3, calculate its effective thickness, in kilometers. (Compare Chapter 7, problem 3.)

8. According to Stefan's law (see Section 3.4), how much more radiation—per square meter, say—is emitted by Venus's surface at 750 K than is emitted by Earth's surface at 300 K?

9. In the absence of any greenhouse effect, Venus's average surface temperature, like Earth's, would be about 250 K. In fact, it is about 750 K. Use this information and Stefan's law to estimate the fraction of infrared radiation leaving Venus's surface that is absorbed by carbon dioxide in the planet's atmosphere.

10. Calculate the orbital period of the *Magellan* spacecraft, moving around Venus on an elliptical orbit with a minimum altitude of 294 km and a maximum altitude of 8543 km above the planet's surface. In 1993 the spacecraft's orbit was changed to have minimum and maximum altitudes of 180 km and 541 km, respectively. What was the new period?

PROJECTS

1. Is Venus in the morning or evening sky right now? Look for it every few days, over the course of several weeks. Draw a picture of the planet with respect to foreground trees or buildings. If you always observe at the same time every day, you may begin to notice that the planet is getting higher or lower in the sky.

2. Consult an almanac to determine the next time Venus will pass between Earth and Sun. How many days before and after this event can you glimpse the planet with the eye alone?

3. Consult the almanac again to find out the next time Venus will pass on the far side of the Sun from Earth. How many days before and after this event can you see the planet with the naked eye?

4. When Venus ornaments the predawn sky, try keeping track of the planet with your eye alone until it appears in a blue sky, after sunrise. As always, be careful not to look at the Sun!

5. Using a powerful pair of binoculars or a small telescope, examine Venus as it goes through its phases. Note the phase and the relative size of it. (You can compare its size to the field of view in a telescope; always use the same eyepiece for this.) Look at it every few days or once a week. Make a table of the shape of the phase, the size, and the relative brightness to the naked eye. After you have observed it through a significant change in phase, can you see the correlations between these three properties first recognized by Galileo?

10 MARS

A Near Miss for Life?

LEARNING GOALS

Studying this chapter will enable you to:

1 Summarize the general orbital and physical properties of Mars.

2 Describe the observational evidence for seasonal changes on Mars.

3 Compare the surface features and geology of Mars with those of the Moon and Earth, and account for these characteristics in terms of Martian history.

4 Discuss the evidence that Mars once had a much denser atmosphere and running water on its surface.

5 Compare the atmosphere of Mars with those of Earth and Venus, and explain why the evolutionary histories of these three worlds diverged so sharply.

6 Describe the characteristics of Mars's moons and explain their probable origin.

(Opposite page, background) The red planet, Mars, shown in true color here, displays an array of fascinating surface features. Most prominent in this *Viking*-mission image is the Mariner Valley, a vast "canyon" named after an earlier spacecraft, *Mariner*, that discovered it. The valley extends for approximately 4000 km—roughly the width of the United States.

(Inset) A close-up view of the rock-strewn Martian surface, as seen from a camera onboard *Mars Pathfinder*, which landed in the Ares Vallis region of Mars some 1000 km north of the equator. The landing site lies in an outflow channel—a vast trench carved in the Martian surface some 3 billion years ago by catastrophic floods which carried millions of tons of material from the planet's southern highlands down into the northern plains. The *Sojourner* rover can be seen surveying the debris field near the center of the image. For reference, the largest boulders near the rover are about 30 cm high; the horizon is about 1 km away.

Named by the ancient Romans for their bloody god of war, Mars is for many people the most intriguing of all celestial objects. Over the years it has inspired much speculation that life—perhaps intelligent, and possibly hostile—may exist there. With the dawn of the Space Age, those notions had to be abandoned. Visits by robot spacecraft have revealed no signs of life of any sort, even at the microbial level, on Mars. Even so, the planet's properties are close enough to those of Earth that Mars is still widely regarded as the next most hospitable environment for the appearance of life in the solar system, after Earth itself. At about the same time as Earth's "twin," Venus, was evolving into a searing inferno, the Mars of long ago may have had running water and blue skies. If life ever arose there, however, it must be long extinct. The Mars of today appears to be a dry, dead world.

10.1 Orbital Properties

1 Mars is the fourth planet from the Sun and the outermost of the four terrestrial worlds in the solar system. It lies outside Earth's orbit, as illustrated in Figure 10.1(a), which shows the orbits of both planets drawn to scale. Because of its exterior orbit, Mars ranges in our sky from a position close to the Sun (for example, when Earth and Mars are at the points marked A in the figure) to one far from the Sun (with the planets at points B). Contrast this orbit with the nighttime appearances of Mercury and Venus, whose interior orbits ensure that we never see them far from the Sun. From our earthly viewpoint, Mars appears to traverse a great circle in the sky, keeping close

to the ecliptic and occasionally executing retrograde loops. ∞ (Sec. 2.2) The Mars Data box on p. 225, lists some detailed orbital and physical data on the planet.

Mars's orbital eccentricity is 0.093, much larger than that of most other planets—only the innermost and the outermost planets, Mercury and Pluto, have more elongated orbits. Because of this, Mars's perihelion distance from the Sun— 1.38 A.U. (207 million km)—is substantially smaller than its aphelion distance—1.67 A.U. (249 million km), resulting in a large variation in the amount of sunlight striking the planet over the course of its year. In fact, the intensity of sunlight on the Martian surface is almost 45 percent greater when the planet is at perihelion than when it is at aphelion. As we will see, this has a substantial effect on the Martian climate.

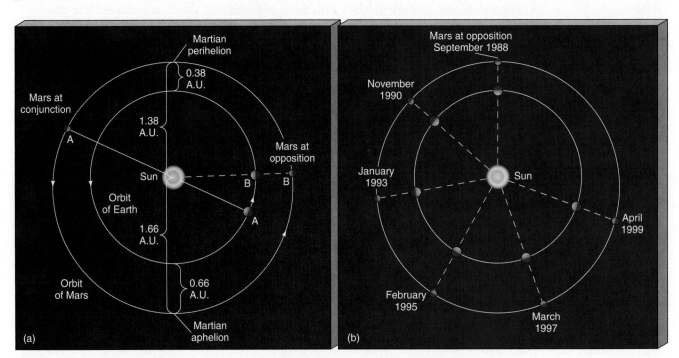

Figure 10.1 (a) The orbit of Mars compared with that of Earth. Notice that Mars's orbit is noticeably elliptical, unlike Earth's, whose eccentricity is barely perceptible. When the planets are on opposite sides of the Sun, as at the points marked A, Mars is said to be at conjunction. The planets are at their closest at opposition, when Earth and Mars are aligned and on the same side of the Sun, as at the points marked B. (Note that to get from point A to point B, Earth must travel for nearly 13 months—all the way around its orbit and then some.) (b) Several oppositions of Mars, including the particularly favorable configuration of September 1988 and the unfavorable opposition of March 1997.

Mars is at its largest and brightest when it is at *opposition*—that is, when Earth lies between Mars and the Sun (location B on Figure 10.1). If this happens to occur near a Martian perihelion, the two planets can come within 0.38 A.U. (56 million km) of each other. The angular size of Mars under those circumstances is about 25″. Ground-based observations of the planet at those times can distinguish surface features as small as 100 km across—about the same resolution as the unaided human eye can achieve when viewing the Moon. Such a coincidence of opposition with Martian perihelion is relatively rare. It last happened in September 1988 and will next occur in August 2003, although Mars will not be quite so close to perihelion then as it was in 1988. The dates and configurations of some oppositions of Mars are illustrated in Figure 10.1(b), which shows the locations of Earth and Mars at six oppositions (at 780-day intervals—Mars's synodic period) between September 1988 and April 1999.

Although Mars is quite bright and easily seen at opposition, the planet is still considerably fainter than Venus. This faintness results from a combination of three factors. First, Mars is more than twice as far from the Sun as is Venus, and each square meter on the Martian surface receives less than one-quarter the amount of sunlight that strikes each square meter on Venus. Second, the surface area of Mars is only about 30 percent that of Venus, so there are fewer square meters to intercept the sunlight. Finally, Mars is much less reflective than Venus—only about 15 percent of the sunlight striking the planet is reflected back into space, compared with nearly 70 percent in the case of Venus. Still, at its brightest, Mars is brighter than any star. Its characteristic red color, visible even to the naked eye, makes it easily identifiable in the night sky.

10.2 Mars in Bulk

As with Mercury and Venus, we can determine the radius of Mars by means of simple geometry. From the data given earlier for the planet's size and distance, we obtain a radius of about 3400 km. More accurate measurements give a result of 3397 km, or 0.53 Earth radii.

Unlike Mercury and Venus, Mars has two small moons in orbit around it; they are visible (through telescopes) from Earth. Named Phobos (Fear) and Deimos (Panic) for the horses that drew the Roman war god's chariot, these moons are little more than large rocks trapped by the planet's gravity. We will return to their individual properties at the end of this chapter. The larger of the two, Phobos, orbits at a distance of just 9380 km from the center of the planet once every 459 minutes. Applying the modified version of Kepler's third law (which states that the square of a moon's orbital period is proportional to the cube of its orbital semi-major axis divided by the mass of the planet it orbits), we find that the mass of Mars is 6.4×10^{23} kg, or 0.11 that of

PLANETARY DATA:
MARS

Orbital semi-major axis	1.52 A.U.
	227.9 million km
Orbital eccentricity	0.093
Perihelion	1.38 A.U.
	206.6 million km
Aphelion	1.67 A.U.
	249.2 million km
Mean orbital speed	24.1 km/s
Sidereal orbital period	686.9 solar days
	1.881 tropical years
Synodic orbital period	779.9 solar days
Orbital inclination to the ecliptic	1.85°
Greatest angular diameter, as seen from Earth	24.5″
Mass	6.42×10^{23} kg
	0.11 (Earth = 1)
Equatorial radius	3394 km
	0.53 (Earth = 1)
Mean density	3930 kg/m³
	0.71 (Earth = 1)
Surface gravity	3.72 m/s²
	0.38 (Earth = 1)
Escape speed	5.0 km/s
Sidereal rotation period	1.026 solar days
	24ʰ 37ᵐ
Axial tilt	23.98°
Surface magnetic field	approximately 1/800 (Earth = 1)
Magnetic axis tilt relative to rotation axis	—
Mean surface temperature	210 K
	(range 150–310 K)
Number of moons	2

Earth. ∞ (Sec. 2.7) Naturally, the orbit of Deimos yields the same result.

From the mass and radius, we find that the average density of Mars is 3900 kg/m³, only slightly greater than that of the Moon. If we assume that the Martian surface rocks are similar to those on the other terrestrial planets, this average density suggests the existence of a substantial core of higher than average density within the planet. Planetary scientists now believe that this core is composed largely of iron sulfide (a compound about twice as dense as surface rock) and has a diameter of about 2500 km.

Surface markings easily seen on Mars allow astronomers to track the planet's rotation. Mars rotates once on its axis every 24.6 hours. One Martian day is thus very similar in length to one Earth day. The planet's equator is inclined to the orbit plane at an angle of 24.0°, again very similar to Earth's inclination of 23.5°. Thus, as Mars orbits the Sun, we find both daily and seasonal cycles, just as on Earth. In the case of Mars, however, the seasons are complicated somewhat by variations in solar heating due to the planet's eccentric orbit.

10.3 Earth-Based Observations of Mars

2 At opposition, when Mars is closest to us and most easily observed, we see it as full, so the Sun's light strikes the surface almost vertically, casting few shadows and preventing us from seeing any topographic detail, such as craters or mountains. Even through a large telescope Mars appears only as a reddish disk, with some light and dark patches and prominent polar caps. These surface features undergo slow seasonal changes over the course of a Martian year. We saw in Chapter 1 how the inclination of Earth's axis produces similar seasonal changes. ∞ (Sec. 1.3) Figure 10.2 shows some of the best images of Mars ever made from Earth (or Earth orbit), along with a photograph taken by one of the U.S. *Viking* spacecraft en route to the planet.

Viewed from Earth, the most obvious Martian surface features are the bright polar caps (see Figure 10.2a). They grow or diminish according to the seasons, almost disappearing at the time of Martian summer. The dark surface features on Mars also change from season to season, although their variability probably has little to do with the melting of the polar ice caps. To the more fanciful observers around the start of the twentieth century, these changes suggested the seasonal growth of vegetation on the planet. It was but a small step from seeing polar ice caps and speculating about teeming vegetation to imagining a planet harboring intelligent life, perhaps not unlike us.

But those speculations and imaginings were not to be confirmed. The caps are mostly frozen carbon dioxide (that is, dry ice), not water ice, as at Earth's North and South Poles. The polar caps do contain water, but it remains permanently frozen, and the dark markings seen in Figures 10.2, once claimed to be part of a network of "canals" dug by Martians for irrigation purposes (see *Interlude 10-1*), are actually highly cratered and eroded areas around which surface dust occasionally blows. Repeated covering and uncovering of these landmarks gives the impression (from a distance) of surface variability, but it's only the thin dust cover that changes.

The powdery Martian surface dust is borne aloft by strong winds that often reach hurricane proportions (hundreds of kilometers per hour). In fact, when the American *Mariner 9* spacecraft went into orbit around Mars in 1971, a planetwide dust storm obscured the entire landscape. Had the craft been on a flyby mission (for a quick look) instead

(a)

(b)

R I V U X G

(c)

Figure 10.2   (a) A deep red (800 nm) image of Mars, taken in 1991 at Pic du Midi, an exceptionally clear site in the French Alps. One of the planet's polar caps appears at the top. (b) A visible-light *Hubble Space Telescope* image of Mars, taken while the planet was near opposition in 1997. (c) A view of Mars taken from a *Viking* spacecraft during its approach in 1976. Some of the planet's surface features can be seen clearly at a level of detail completely invisible from Earth.

Video

The year 1877 was an important one in the human study of the planet Mars. The Red Planet came unusually close to Earth, affording astronomers an especially good view. Of particular note was the discovery, by U.S. Naval Observatory astronomer Asaph Hall, of the two moons circling Mars. But most exciting was the report of the Italian astronomer Giovanni Schiaparelli (the man responsible for erroneously reporting the synchronous orbit of Mercury) on his observation of a network of linear markings that he termed *canali*. In Italian, *canali* means simply "grooves" or "channels," but the word was translated into English as "canals," suggesting that the grooves had been constructed by intelligent beings. Observations of these features became sensationalized in the world's press (especially in the United States), and some astronomers began drawing elaborate maps of Mars, showing oases and lakes where canals met in desert areas.

Percival Lowell (see photo), a successful Boston businessman (and brother of the poet Amy Lowell and Harvard president Abbott Lawrence Lowell), became fascinated by these reports. He abandoned his business and purchased a clear-sky site at Flagstaff, Arizona, where he built a major observatory. He devoted his life to achieving a better understanding of the Martian "canals." In doing so, he championed the idea that Mars was drying out and that an intelligent society had constructed the canals to transport water from the wet poles to the arid equatorial deserts.

Alas, the Martian valleys and channels photographed by robot spacecraft during the 1970s are far too small to be

the "canali" that Schiaparelli, Lowell, and others thought they saw on Mars. The entire episode represents a classic case in the history of science—a case in which well-intentioned observers, perhaps obsessed with the notion of life on other worlds, let their personal opinions and prejudices seriously affect their interpretations of reasonable data. The accompanying figures of Mars show how surface features (which were probably genuinely observed by astronomers at the turn of the century) might have been imagined to be connected. The figure on the left is a photograph of how Mars actually looked in a telescope at the end of the nineteenth century. The sketch at right is an interpretation (done at the height of the canal hoopla) of the pictured view. The human eye, under physiological stress, tends to connect dimly observed yet distinctly separated features. Humans saw patterns and canals where none in fact existed.

The chronicle of the Martian canals illustrates how the scientific method requires scientists to acquire new data to sort out sense from nonsense, fact from fiction. Rather than simply believing the claims about the Martian canals, other scientists demanded further observations to test Lowell's hypothesis. Eventually, improved observations, climaxing in the *Mariner* and *Viking* exploratory missions to the Red Planet nearly a century after all the fuss began, totally disproved the existence of canals. Although it often takes time, the scientific method does in fact lead to progress toward understanding reality.

of an orbiting mission (for a longer view), its visit would have been a failure. Fortunately, the storm subsided, enabling the craft to radio home detailed information about the surface.

10.4 The Surface of Mars

3 Maps of the surface of Mars returned by orbiting spacecraft show a wide range of geological features. Mars has huge volcanoes, deep canyons, vast dune fields, and many other geological wonders. Orbiters have performed large-scale surveys of much of the planet's surface; lander data

have complemented these planetwide studies with detailed information on (so far) three specific sites. ∞ (Sec. 6.6)

LARGE-SCALE TOPOGRAPHY

A striking feature of the terrain of Mars is the marked difference between the northern and southern hemispheres. The northern hemisphere is made up largely of rolling volcanic plains, not unlike the lunar maria. These extensive lava plains—much larger than those found on Earth or the Moon—were formed by eruptions involving enormous volumes of lava. They are strewn with blocks of volcanic rock as well as with boulders blasted out of impact areas by infalling meteoroids (the Martian atmosphere is too thin to offer much resistance to incoming debris). The southern hemisphere

(a)

(b)

R I V U X G

Figure 10.3 (a) The northern hemisphere of Mars consists of rolling, volcanic plains. (b) The southern Martian highlands are heavily cratered (true color).

consists of heavily cratered highlands lying several kilometers above the level of the lowland north. Most of the dark regions visible from Earth are mountainous regions in the south. Figure 10.3 contrasts typical terrains in the two hemispheres.

The northern plains are much less cratered than the southern highlands. On the basis of arguments presented in Chapter 8, this smoother surface suggests that the northern surface is younger. ∞ (Sec. 8.4) Its age is perhaps 3 billion years, compared with 4 billion in the south. In places, the boundary between the southern highlands and the northern plains is quite sharp. The surface level can drop by as much as 4 km in a distance of 100 km or so. Most scientists assume that the southern terrain is the original crust of the planet. How most of the northern hemisphere could have been lowered in elevation and flooded with lava remains a mystery.

The major geological feature on the planet is the Tharsis bulge (visible in Figure 10.4, a *Viking* mosaic of Mars). Roughly the size of North America, Tharsis lies on the Martian equator, and rises some 10 km higher than the rest of the Martian surface. To its east lies Chryse Planitia (the "Plains of Gold"); to the west a region called Isidis Planitia (the "Plains of Isis," an Egyptian goddess). These features are wide depressions, hundreds of kilometers across and up to 3 km deep. If we wished to extend the idea of "continents" from Earth and Venus to Mars, we would conclude that Tharsis is the only continent on the Martian surface.

However, as on Venus, there is no sign of plate tectonics—the continent of Tharsis is not drifting like its Earthly counterparts. ∞ (Sec. 7.5)

Tharsis appears to be even less heavily cratered than the northern plains, making it the youngest region on the planet. It is estimated to be only 2 to 3 billion years old.

VOLCANISM

Mars contains the largest known volcanoes in the solar system. Four very large volcanoes are found on the Tharsis bulge, two of them visible on the left-hand side of Figure 10.4. The largest

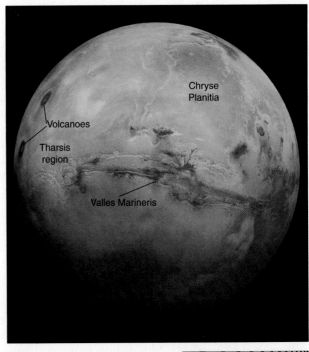

Chryse
Planitia

Volcanoes

Tharsis
region

Valles Marineris

R I V U X G

Figure 10.4 The Tharsis region of Mars, some 5000 km across, bulges out from the planet's equatorial region, rising to a height of about 10 km. The large volcanoes on the left mark the approximate peak of the bulge. One of the plains flanking the Tharsis bulge, Chryse Planitia, is toward the right. Dominating the center of the field of view is a vast "canyon" known as Valles Marineris. (See also the chapter-opening image, p. 222.)

Figure 10.5 Olympus Mons, the largest volcano known on Mars or anywhere else in the solar system. Nearly three times taller than Mount Everest on Earth, this Martian mountain measures about 700 km across the base and extends 25 km at the peak. It seems currently inactive and may have been extinct for at least several hundred million years. By comparison, the largest volcano on Earth, Hawaii's Mauna Loa, measures a mere 120 km across and peaks just 9 km above the Pacific Ocean floor.

volcano of all is Olympus Mons (Figure 10.5), on the north-western slope of Tharsis, lying just over the left (western) horizon of Figure 10.4. It measures some 700 km in diameter at its base—only slightly smaller than the state of Texas—and rises to a height of 25 km above the surrounding plains. The caldera, or crater, at its summit, measures 80 km across. The other three large volcanoes are a little smaller—a mere 18 km high—and lie near the top of the bulge.

Like Maxwell Mons on Venus, these volcanoes are not associated with plate motion but instead are shield volcanoes (see Section 9.4), sitting atop a hot spot in the underlying Martian mantle. All four show distinctive lava channels and other flow features very similar to those found on shield volcanoes on Earth. The *Viking* images of the Martian surface reveal many hundreds of volcanoes. Most of the largest are associated with the Tharsis bulge, but many smaller volcanoes are also found in the northern plains.

The great height of Martian volcanoes is a direct consequence of the planet's low surface gravity. As lava flows and spreads to form a shield volcano, its eventual height depends on the new mountain's ability to support its own weight. The lower the gravity, the less the weight and the higher the mountain. It is no accident that Maxwell Mons on Venus and the Hawaiian shield volcanoes on Earth rise to roughly the same height (about 10 km) above their respective bases—Earth and Venus have similar surface gravity. Mars's surface gravity is only 40 percent that of Earth, so volcanoes rise roughly 2.5 times as high.

Are these volcanoes still active? Scientists have found no direct evidence for recent or ongoing eruptions. However, if these volcanoes were around since the Tharsis uplift (as the formation of the Tharsis bulge is known) and were active as recently as 100 million years ago (an age estimate based on the extent of impact cratering on their slopes), some of them may still be at least intermittently active. Millions of years, though, may pass between eruptions.

IMPACT CRATERING

The *Mariner* spacecraft found that the surfaces of Mars and its two moons are pitted with impact craters formed by meteoroids falling in from space. As on our Moon, the smaller craters are often filled with surface matter—mostly dust—confirming that Mars is a dry, desert world. However, Martian craters are filled in considerably faster than their lunar counterparts. On the Moon, ancient craters less than 20 m deep (corresponding to diameters of about 100 m) have been obliterated, primarily by meteoritic erosion. ∞ (Sec. 8.4) On Mars, there is also a lack of small craters, but extending to craters 5 km in diameter. The Martian atmosphere is an efficient erosive agent, transporting dust from place to place and erasing surface features much faster than meteoritic impacts alone can obliterate them.

As on the Moon, the extent of large impact cratering (that is, craters too big to have been filled in by erosion since they formed) serves as an age indicator for the Martian surface. The ages quoted earlier, ranging from 4 billion years for the southern highlands to a few hundred million years in the youngest volcanic areas, were obtained in this way.

The detailed appearance of Martian impact craters provides an important piece of information about conditions just below the planet's surface. The ejecta blankets surrounding many Martian craters look quite different from their lunar counterparts. Figure 10.6 compares the Copernicus crater on the Moon with the (fairly typical) crater Yuty on Mars. The material surrounding the lunar crater is just what one would expect from an explosion ejecting a large volume of dust, soil, and boulders. However, the ejecta blanket on Mars gives the distinct impression of a liquid that has splashed or flowed out of the crater. Geologists believe that this **fluidized ejecta** crater indicates that a layer of **permafrost**, or water ice, lies just under the surface. The explosive impact heated and liquefied the ice, resulting in the fluid appearance of the ejecta.

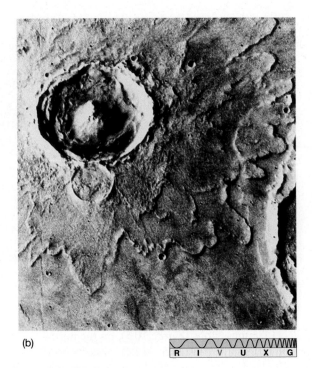

(a)

(b)

Figure 10.6 (a) The large lunar crater Copernicus is quite typical of those found on Earth's Moon. Its ejecta blanket appears to be composed of dry, powdery material. (b) The ejecta from Mars's crater Yuty (18 km in diameter) evidently was liquid in nature. This type of crater is sometimes called a "splosh" crater.

THE MARTIAN "GRAND CANYON"

Yet another feature associated with the Tharsis bulge is a great "canyon" known as Valles Marineris (the Mariner Valley). Shown in its entirety in Figure 10.4 and in more detail in Figure 10.7, it is not really a canyon in the terrestrial sense, because running water played no part in its formation. Planetary astronomers believe that it was formed by the same crustal forces that forced the entire Tharsis region to bulge outward, causing the surface to split and crack. These cracks, called *tectonic fractures*, are found all around the Tharsis bulge. The Valles Marineris is the largest of them. Cratering studies suggest that the cracks are at least 2 billion years old. Similar (but much smaller) cracks, with similar causes, have been found in the Aphrodite Terra region of Venus, as we saw in Chapter 9. ∞ (Sec. 9.4)

Valles Marineris runs for almost 4000 km along the Martian equator, about one-fifth of the way around the planet. At its widest, it is some 120 km across, and it is as deep as 7 km in places. Like many Martian surface features, it simply dwarfs Earthly competition. The Grand Canyon in Arizona would easily fit into one of its side "tributary" cracks. Valles Marineris is so large that it can even be seen from Earth—in fact, it was one of the few "canals" observed by nineteenth-century astronomers (see *Interlude 10-1*) that actually corresponded to a real feature on the planet's surface (it was known as the Coprates canal). We must reemphasize, however, that this Martian feature was not constructed by intelligent beings, nor was it carved by a river, nor is it a result of Martian plate tectonics. For some reason, the crustal forces that formed it never developed into full-fledged plate motion as on Earth.

EVIDENCE FOR RUNNING WATER

Although the great surface cracks in the Tharsis region are not really canyons and were not formed by running water, photographic evidence reveals that liquid water once existed in great quantity on the surface of Mars. Two types of flow feature are seen: the *runoff channels* and the *outflow channels*.

The **runoff channels** (one of which is shown in Figure 10.8) are found in the southern highlands. They are extensive systems—sometimes hundreds of kilometers in total length—of interconnecting, twisting channels that seem to merge into larger, wider channels. They bear a strong resemblance to river systems on Earth, and it is believed by geologists that this is just what they are—the dried-up beds of long-gone rivers that once carried rainfall on Mars from the mountains down into the valleys. These runoff channels speak of a time 4 billion years ago (the age of the Martian highlands), when the atmosphere was thicker, the surface warmer, and liquid water widespread.

The **outflow channels** (Figure 10.9) are probably relics of catastrophic flooding on Mars long ago. They appear only in equatorial regions and generally do not form the extensive interconnected networks that characterize the runoff channels. Instead, they are probably the paths taken by huge volumes of water draining from the southern high-

Figure 10.7 (a) The Mariner Valley is an incredibly huge canyon, 120 km wide and 7 km deep. Its length is about 4000 km, or nearly the full breadth of the continental United States. (b) A close-up view shows the complexity of the valley walls and dry tributaries. (c) A comparison, to scale, with Earth's Grand Canyon, which is a mere 20 km wide and 2 km deep, suggests just how big the Mariner Valley is.

Figure 10.8 (a) This runoff channel on Mars measures about 400 km long and 5 km wide. Here, we compare it with a photograph of the Red River (b) running from Shreveport, Louisiana, to the Mississippi River. The two differ mainly in that there is currently no liquid water in this, or any other, Martian valley.

Figure 10.9 An outflow channel near the Martian equator bears witness to a catastrophic flood that occurred about 3 billion years ago.

lands into the northern plains. The onrushing water arising from these flash floods probably also formed the odd teardrop-shaped "islands" (resembling the miniature versions seen in the wet sand of our beaches at low tide) that have been found on the plains close to the ends of the outflow channels (Figure 10.10). Judging from the width and depth of the channels, the flow rates must have been truly enormous—perhaps as much as a hundred times greater than the 10^5 tons per second carried by the Amazon river, the largest river system on Earth. Flooding shaped the outflow channels about 3 billion years ago, about the same time as the northern volcanic plains formed.

Planetary astronomers find no evidence for liquid water anywhere on Mars today, and the amount of water vapor in the Martian atmosphere is tiny. Yet the extent of the outflow channels indicates that a huge total volume of water existed on Mars in the past. Where did all that water come from? And where did it all go? The answer may be that virtually all the water on Mars is now locked in the permafrost layer under the surface, with perhaps a little more contained in the polar caps. Four billion years ago, as climatic conditions changed, the running water that formed the runoff channels began to freeze, forming the permafrost and drying out the river beds. Mars remained frozen for about a billion years, until volcanic (or some other) activity heated large regions of the surface, melting the subsurface ice and causing the flash floods that created the outflow channels. Subsequently, volcanic activity subsided, the water refroze, and Mars once again became a dry world.

POLAR CAPS

We have already noted that Mars's polar caps are composed predominantly of carbon dioxide frost—dry ice—and show seasonal variations. Each cap in fact consists of two distinct parts—the **seasonal cap**, which grows and shrinks each Martian year, and the **residual cap**, which remains permanently

frozen. At maximum size, in southern midwinter, the southern seasonal cap is some 4000 km across. Half a Martian year later, the northern cap is at its largest, reaching a diameter of roughly 3000 km. The two seasonal polar caps do not have the same maximum size because of the eccentricity of Mars's orbit around the Sun. During southern winter, Mars is considerably farther from the Sun than half a year later, in northern winter. The southern winter season is longer and colder than that of the north, and the polar cap grows correspondingly larger.

The seasonal caps are composed entirely of carbon dioxide. Their temperatures are never greater than about 150 K ($-120°C$), the point at which dry ice can form. During the Martian summer, when sunlight striking a cap is most intense, carbon dioxide evaporates into the atmosphere, and the cap shrinks. In the winter, atmospheric carbon dioxide refreezes, and the cap reforms. As the caps grow and shrink, they cause substantial variations (up to 30 percent) in the Martian atmospheric pressure—a large fraction of the planet's atmosphere freezes out and evaporates again on a yearly basis. From studies of these atmospheric fluctuations, scientists can estimate the amount of carbon dioxide in the seasonal polar caps. The maximum thickness of the seasonal caps is thought to be about 1 m.

Figure 10.10 The onrushing water that carved out the outflow channels was responsible for forming these oddly shaped "islands" as the flow encountered obstacles—impact craters—in its path. Each "island" is about 40 km long.

(a) (b)

R I V U X G

Figure 10.11 The southern (a) and northern (b) polar caps of Mars are shown to scale in these mosaics of *Mariner 9* images. These are the residual caps, seen here during their respective summers half a Martian year apart. The southern cap is only 350 km across and made up mostly of frozen carbon dioxide. The northern cap is about 1000 km across and composed mostly of water ice.

Video

The residual caps, shown in Figure 10.11, are smaller and brighter than the seasonal caps and show an even more marked north–south asymmetry. The southern residual cap is about 350 km across and, like the seasonal caps, is made mostly of carbon dioxide, although it may contain some water ice. Its temperature remains below 150 K at all times. The northern residual cap is much larger—about 1000 km across—and warmer, with a temperature that can exceed 200 K in northern summertime. Planetary scientists believe that the northern cap is made mostly of water ice, an opinion strengthened by the observed increase in the concentration of water vapor above the north pole in northern summer as some small fraction of its water ice evaporates in the Sun's heat. It is quite possible that the northern residual polar cap is a major storehouse for water on Mars.

Why is there such a temperature difference (at least 50 K) between the two residual polar caps? The reason seems to be related to the giant dust storms that envelop the planet during southern summer. These storms, which last for a quarter of a Martian year (about 6 Earth months), tend to blow the dust from the warmer south into the cooler northern hemisphere. The northern ice cap becomes dusty and less reflective. As a result, it absorbs more sunlight and warms up.

THE VIEW FROM THE MARTIAN LANDERS

Viking 1 landed in Chryse Planitia, the broad depression to the east of Tharsis. The view that greeted its cameras (Figure 10.12) was a windswept, gently rolling, rather desolate plain, littered with rocks of all sizes, not unlike a high desert on Earth. The surface rocks visible in Figure 10.12 are probably part of the ejecta blanket of a nearby impact crater. *Viking 2* landed somewhat farther north, in a region of Mars called Utopia, chosen in part because mission planners anticipated greater seasonal climatic variations

Figure 10.12 This is the view from the perspective of the *Viking 1* spacecraft now parked on the surface of Mars. The fine-grained soil and the rock-strewn terrain stretching toward the horizon are reddish. Containing substantial amounts of iron ore, the surface of Mars is literally rusting away. The sky is a pale yellow-pink color, the result of airborne dust.

R I V U X G

Figure 10.13 Another view of the Martian surface, this one rock-strewn and flat, as seen through the camera aboard the *Viking 2* robot that soft-landed on the northern Utopian plains. The discarded canister is about 20 cm long. The 0.5-m scars in the "dirt" were made by the robot's shovel.

R I V U X G

there. The plain on which *Viking 2* landed was flat and featureless. From space, the landing site appeared smooth and dusty. In fact, the surface turned out to be very rocky, even rockier than the Chryse site, and without the dust layer the mission directors had expected (Figure 10.13). The views that the two landers recorded may turn out to be quite typical of the low-latitude northern plains.

The *Viking* landers performed numerous chemical analyses of the Martian regolith. One important finding of these studies was the high iron content of the planet's surface. Chemical reactions between the iron-rich surface soil and free oxygen in the atmosphere is responsible for the iron oxide ("rust") that gives Mars its characteristic color. Although the surface layers are rich in iron relative to Earth's surface, the overall abundance is similar to Earth's average iron content. On Earth, much of the iron has differentiated

to the center. Chemical differentiation does not appear to have been nearly so complete on Mars.

The most recent visitor to the Martian surface was *Mars Pathfinder*, along with its robot rover *Sojourner*. ∞ (Sec. 6.6) During the unexpectedly long lifetime of its mission (which lasted almost 3 months instead of the anticipated 1), the lander performed measurements of the Martian atmosphere and atmospheric dust while *Sojourner* carried out chemical analyses of the soil and rocks within about 50 m of the parent craft. In addition, over 16,000 images of the region were returned to Earth.

The *Pathfinder* landing site (Figure 10.14) had been carefully chosen to lie near the mouth of an outflow channel, and the size distribution and composition of the many rocks and boulders surrounding the lander were consistent with their having been deposited there by floodwaters.

Figure 10.14 This 360° panorama of Mars was taken by the *Pathfinder* robot lander in 1997. Its small roving vehicle, *Sojourner*, can be seen up against a rock at center (see also Fig. 6.4). The "Twin Peaks" on the left-center horizon are 1–2 km away.

R I V U X G

In addition, the presence of numerous rounded pebbles strongly suggested the erosive action of running water at some time in the past. The soil chemistry in the vicinity of the landing site was similar to that found by the two *Viking* landers. Analyses of the nearby rocks revealed a chemical makeup different from that of the Martian meteorites found on Earth (see *Interlude 10-2*).

10.5 The Martian Atmosphere

COMPOSITION

4 Long before the arrival of the *Mariner* and *Viking* spacecraft, astronomers knew from Earth-based spectroscopy that the Martian atmosphere was quite thin and composed primarily of carbon dioxide. In 1964, *Mariner 4* confirmed these results, finding that the atmospheric pressure is only about 1/150 the pressure of Earth's atmosphere at sea level and that carbon dioxide makes up at least 95 percent of the total atmosphere. With the arrival of *Viking*, more detailed measurements of the Martian atmosphere could be made. Its composition is now known to be 95.3 percent carbon dioxide, 2.7 percent nitrogen, 1.6 percent argon, 0.13 percent

oxygen, 0.07 percent carbon monoxide, and about 0.03 percent water vapor. The level of water vapor is quite variable. Weather conditions encountered by *Mars Pathfinder* were quite similar to those found by *Viking 1*.

As the *Viking* landers descended to the surface they made measurements of the temperature and pressure at various heights. The results are shown in Figure 10.15. The Martian atmosphere contains a troposphere (the lowest-lying atmospheric zone, where convection and "weather" occur), which varies both from place to place and from season to season. ∞ (Sec. 7.2) The variability of the troposphere arises from the variability of the Martian surface temperature. At noon in the summertime, surface temperatures may reach 300 K. Atmospheric convection is strong, and the top of the troposphere can reach an altitude of 30 km. At night, the atmosphere retains little heat, and the temperature can drop by as much as 100 K. Convection then ceases and the troposphere vanishes.

On average, surface temperatures on Mars are about 50 K cooler than on Earth. The low early-morning temperatures often produce water-ice "fog" in the Martian canyons (Figure 10.16). Temperatures in the stratosphere are low enough for carbon dioxide to solidify, giving rise to a high-level layer of carbon dioxide clouds and haze.

For most of the year, there is little day-to-day variation in the Martian weather: The Sun rises, the surface warms up, and light winds blow until sunset, when the temperature drops again. Only in the southern summer does the daily routine change. Strong surface winds (without rain or snow)

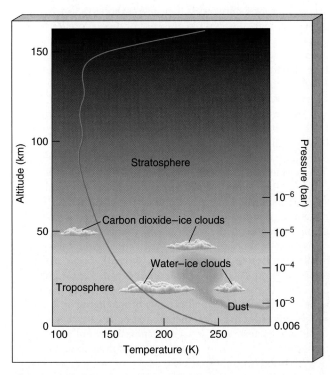

Figure 10.15 Structure of the Martian atmosphere, as determined by *Viking*. The troposphere rises to an altitude of about 30 km in the daytime. It occasionally contains clouds of water ice or, more frequently, dust during the planetwide dust storms that occur each year. Above the troposphere lies the stratosphere. Note the absence of a higher-temperature zone in the stratosphere, indicating the absence of an ozone layer.

R I V U X G

Figure 10.16 Fog in the Martian canyons, photographed by the *Viking* orbiter. As the Sun's light reaches and heats the canyon floor, it drives water vapor from the surface. When this vapor comes in contact with the colder air above the surface, it condenses again, and a temporary water–ice "fog" results.

INTERLUDE 10-2 Life on Mars?

Even before the *Viking* missions reached Mars in 1976, astronomers had abandoned hope of finding life on the planet. Scientists knew there were no large-scale canal systems, no surface water, almost no oxygen in the atmosphere, and no seasonal vegetation changes. The present lack of liquid water on Mars especially dims the chances for life there now. However, running water and possibly a dense atmosphere in the past may have created conditions suitable for the emergence of life long ago. In the hope that some form of microbial life might have survived to the present day, the *Viking* landers carried out experiments designed to detect biological activity. The accompanying pair of photographs show the robot arm of one of the landers before and after digging a shallow trench to scoop up soil samples.

All three *Viking* biological experiments assumed some basic similarity between hypothetical Martian bacteria and those found on Earth. A *gas-exchange experiment* offered a nutrient broth to any residents of a sample of Martian soil and looked for gases that would signal metabolic activity. A *labeled-release experiment* added compounds containing radioactive carbon to the soil, then waited for results signaling that Martian organisms had either eaten or inhaled this carbon. Finally, a *pyrolitic-release experiment* added radioactively tagged carbon dioxide to a sample of Martian soil and atmosphere, waited a while, then removed the gas and tested the soil (by heating it) for signs that something had absorbed the tagged gas.

Initially, all three experiments appeared to be giving positive signals! However, subsequent careful studies showed that all the results could be explained by inorganic (that is, nonliving) chemical reactions. Thus, we have no clear evidence for even microbial life on the Martian surface. The

Viking robots detected peculiar reactions that mimic in some ways the basic chemistry of living organisms, but they did not detect life itself.

A criticism of the *Viking* experiments is that they searched only for life now living. Today, Mars seems locked in an ice age—the kind of numbing cold that would prohibit sustained life as we know it. If bacterial life did arise on an Earthlike early Mars, however, then we might be able to find its fossilized remains preserved on or near the Martian surface. Surprisingly, one place to look for life on Mars is right here on Earth. Scientists think that some meteorites found on Earth's surface come from the Moon and from Mars. These meteorites were apparently blasted off these bodies long ago during an impact of some sort, thrown into space, and eventually trapped by Earth's gravity, ultimately to fall to the ground. The most fascinating of these rocks are surely those from the Red Planet—for one of them may harbor fossil evidence for past life on Mars!

The figure at right top shows ALH84001, a blackened, 2-kg meteorite about 17 cm across, found in 1984 in Antarctica. Based on estimates of the cosmic-ray exposure it received before reaching Earth, the rock is thought to have been blasted off Mars about 16 million years ago. Looking at this specimen through a microscope, inset, scientists can see rounded orange-brown "globules" of carbonate minerals on the rock's shiny crust. Because carbonates form only in the presence of water, the presence of these globules suggests that carbon dioxide gas and liquid water existed near ground level at some point in Mars's history, a conclusion that planetary scientists had earlier drawn from studies of *Viking's* orbital images of valleys apparently carved by water when the Martian climate was wetter and warmer.

In a widely viewed press conference in Washington, D.C., in the summer of 1996, a group of scientists argued, based on all the data accumulated from studies of ALH84001, that they had discovered fossilized evidence for life on Mars. The key pieces of evidence they presented for primitive Martian life were as follows: (1) Bacteria on Earth can produce structures similar to the globules shown in the inset. (2) The meteorite contains traces of *polycyclic aromatic hydrocarbons*—a tongue-twisting name for a class of complex organic molecules (usually abbreviated PAHs) that, although not directly involved in known biological cycles on Earth, occur among the decay products of plants and other organisms. (3) High-powered electron microscopes show that ALH84001 contains tiny, teardrop-shaped crystals of magnetite and iron sulfide embedded in places where the carbonate has dissolved. On Earth, bacteria are known to manufacture similar chemical crystals. (4) On very small scales, elongated and egg-shaped structures are seen within the carbonate globules. The researchers interpret these minute structures as fossils of primitive organisms.

The photomicrograph at right bottom shows this fourth, and most controversial, piece of evidence—curved, rodlike structures that resemble bacteria on Earth. Scale is crucial here, however. The structures are only about 0.5 µm across, 30 times smaller than ancient bacterial cells found fossilized

R I V U X G

(Before)   (After)

on Earth. Furthermore, several key tests have not yet been done, such as cutting through the suspected fossilized tubes to search for evidence of cell walls or semipermeable membranes, or of any internal cavities where body fluids would have resided. Nor has anyone yet found in ALH84001 any amino acids, the basic building blocks of life as we know it (see Chapter 28).

These results are very controversial. Many experts do not agree that life has been found on Mars—not even fossilized life. The skeptics maintain that all the evidence could be the result of chemical reactions not requiring any kind of biology. Carbonate compounds are common in all areas of chemistry; PAHs are found in many lifeless places (glacial ice, asteroid-belt meteorites, interstellar clouds, and even the exhaust fumes of automobiles); bacteria are not needed to produce crystals; and it remains unclear whether the tiny tubular structures shown are animal, vegetable, or merely mineral. In addition, there is the huge problem of contamination—after all, ALH84001 was found on Earth and apparently sat in the Antarctic ice fields for 13,000 years before being picked up by meteorite hunters.

As things now stand, it's a matter of interpretation—at the frontiers of science, issues are not always as clear-cut as we would hope. Only additional analysis and new data will tell for sure if primitive Martian life existed long ago. Should the claim of life on Mars hold up against much skepticism in the scientific community, these findings will go down in history as one of the greatest scientific discoveries of all time. A fantastic result: we are—or at least were—not alone in the universe! Maybe.

sweep up the dry dust, carry it high into the stratosphere, and eventually deposit it elsewhere on the planet. At its greatest fury, a Martian storm floods the atmosphere with dust, making the worst storm we could imagine on Earth's Sahara Desert seem inconsequential by comparison. The dust can remain airborne for months at a time. The blown dust forms systems of sand dunes similar in appearance to those found on Earth.

EVOLUTION

5 Although there is some superficial similarity in composition between the atmospheres of Mars and Venus, the two planets obviously have quite different atmospheric histories—Mars's "air" is over 10,000 times thinner than that on Venus. As with the other planets we have studied, we can ask *why* the Martian atmosphere is as it is.

Presumably, Mars acquired a secondary outgassed atmosphere quite early in its history, just as the other terrestrial worlds did. Around 4 billion years ago, as indicated by the runoff channels in the highlands, Mars may have had a fairly dense atmosphere, complete with blue skies and rain. Despite Mars's distance from the Sun, the greenhouse effect would have kept conditions fairly comfortable, and a surface temperature of over 0°C seems quite possible. Sometime during the next billion years, most of the Martian atmosphere disappeared. Possibly some of it was lost because of impacts with other large (Moon-sized) bodies in the early solar system. More likely, the Martian atmosphere became unstable, in a kind of reverse runaway greenhouse effect.

On Venus, as we have seen, the familiar greenhouse effect ran away to high temperatures and pressures. ∞ (Sec. 9.4) Planetary scientists theorize that the early Martian atmosphere also ran away, but in the opposite direction. The presence of liquid water would have caused much of the atmospheric carbon dioxide to dissolve in Martian rivers and lakes (and oceans, if any), ultimately to combine with Martian surface rocks. Recall that on Earth most carbon dioxide is presently found in surface rocks. Calculations show that much of the Martian atmospheric carbon dioxide could have been depleted in this way in a relatively short period of time, in perhaps as little as a few hundred million years, although some of it might have been replenished by volcanic activity, possibly extending the "comfortable" lifetime of the planet to a half-billion years or so.

As the level of carbon dioxide declined and the greenhouse-heating effect diminished, the planet cooled. The water froze out of the atmosphere, lowering still further the level of atmospheric greenhouse gases, and accelerating the cooling. (Recall from Section 9.4 that water vapor also contributes to the greenhouse effect.)

Since those early times the overall density of the Martian atmosphere has continued to decline through the steady loss of carbon dioxide, nitrogen, and water vapor as solar ultraviolet radiation in the upper atmosphere splits the molecules of these gases into their component atoms,

providing some with enough energy to escape. The present level of water vapor in the Martian atmosphere is the maximum possible, given the atmosphere's present density and temperature. Estimates of the total amount of water stored as permafrost or in the polar caps are quite uncertain, but it is likely that if all the water on Mars were to become liquid, it would cover the surface to a depth of several meters.

10.6 Martian Internal Structure

3 The *Viking* landers carried seismometers to probe the internal structure of the planet. However, one failed to work, and the other was unable to clearly distinguish seismic activity from the buffeting of the Martian wind. As a result, no seismic studies of the Martian interior have yet been carried out. On the basis of studies of the stresses that occurred during the Tharsis uplift, astronomers estimate the thickness of the crust to be about 100 km.

During its visit to Mars in 1965, *Mariner 4* detected no planetary magnetic field, and for many years the most that could be said about the Martian magnetic field was that its strength was no more than a few thousandths the strength of Earth's field (the level of sensitivity of *Mariner*'s instruments). The *Viking* spacecraft were not designed to make magnetic measurements. In September 1997, *Mars Global Surveyor* succeeded in detecting a very weak Martian field, about 800 times weaker than Earth's; this is probably a local anomaly, not a global field. Because Mars rotates rapidly, the weakness of the magnetic field is taken to mean that the planet's core is nonmetallic, or nonliquid, or both. ∞ (Sec. 7.4)

The small size of Mars means that any radioactive (or other internal) heating of its interior would have been less effective at heating and melting the planet than similar heating on Earth. The heat was able to reach the surface and escape more easily than in a larger planet such as Earth or Venus. The evidence we noted earlier for ancient surface activity, especially volcanism, suggests that at least parts of the planet's interior must have melted and possibly differentiated at some time in the past. But the lack of current activity, the absence of any significant magnetic field, the relatively low density (3900 kg/m^3), and an abnormally high abundance of iron at the surface all suggest that Mars never melted as extensively as did Earth.

The history of Mars appears to be that of a planet where large-scale tectonic activity almost started but was stifled by the planet's rapidly cooling outer layers. The large upwelling of material that formed the Tharsis bulge might have developed on a larger, warmer planet into full-fledged plate tectonic motion, but the Martian mantle became too rigid and the crust too thick for that to occur. Instead, the upwelling continued to fire volcanic activity, almost up to the present day, but geologically much of the planet apparently died 2 billion years ago.

10.7 The Moons of Mars

6 Unlike Earth's moon, Mars's moons are tiny compared with their parent planet and orbit very close to it, relative to the planet's radius. Discovered by American astronomer Asaph Hall in 1877, the two Martian moons—Phobos and Deimos—are only a few kilometers across. Their composition is quite unlike that of the planet. Astronomers generally believe that Phobos and Deimos did not form along with Mars but instead are asteroids that were slowed and captured by the outer fringes of the early Martian atmosphere (which, as we have just seen, was probably much denser than the atmosphere today). It is even possible that they are remnants of a single object that broke up during the capture process. They are quite difficult to study from Earth because their proximity to Mars makes it hard to distinguish them from their much brighter parent. The *Mariner* and *Viking* orbiters, however, studied them in great detail.

Both moons, shown in Figure 10.17, are quite irregularly shaped and heavily cratered. The larger of the two is Phobos (Figure 10.17a), which is about 28 km long and 20 km wide and dominated by an enormous 10-km-wide crater named Stickney (after Angelina Stickney, Asaph Hall's wife, who encouraged him to persevere in his observations). The smaller Deimos (Figure 10.17b) is only 16 km long by 10 km wide. Its largest crater is 2.3 km in diameter. The fact that both moons have quite dark surfaces, reflecting no more than 6 percent of the light falling on them, contributes to the difficulty in observing them from Earth.

Phobos and Deimos move in circular, equatorial orbits, and they rotate synchronously (that is, they keep the same face permanently turned toward the planet). All these characteristics are direct consequences of the tidal influence of Mars. Both moons orbit Mars in the prograde sense—that is, in the same sense (counterclockwise, as seen from above the north celestial pole) as the planet orbits the Sun and rotates on its axis.

Phobos lies only 9378 km (less than three planetary radii) from the center of Mars and, as we saw earlier, has an orbital period of 7 hours and 39 minutes. This orbit period is much less than a Martian day, so an observer standing on the Martian surface would see Phobos move "backward" across the Martian sky—that is, in a direction opposite the apparent daily motion of the Sun. Because the moon moves faster than the observer, it overtakes the planet's rotation, rising in the west and setting in the east, crossing the sky from horizon to horizon in about 5.5 hours. Deimos lies somewhat farther out, at 23,459 km, or slightly less than seven planetary radii, and orbits in 30 hours and 18 minutes. Because it completes its orbit in more than a Martian day, it moves "normally," as seen from the ground (that is, from east to west), taking almost 3 days to traverse the sky.

On the basis of measurements of their gravitational effect on the *Viking* orbiters, astronomers have estimated the masses of the two moons. Their densities are around 2000 kg/m³, far less than that of any world we have yet encountered in our outward journey through the solar system. This is an important reason why astronomers do not believe these moons formed along with Mars. If they are indeed captured asteroids, Phobos and Deimos represent material left over from the earliest stages of the solar system. Astronomers study them not to gain insight into Martian evolution but rather because the moons contain information about the very early solar system, before the major planets had formed.

(a)

(b)

R I V U X G

Figure 10.17 Mars is accompanied in its trek around the Sun by two tiny moons. Phobos (Greek for "fear") and Deimos ("panic") measure about 28 and 16 km long, respectively. Both are shaped like potatoes, with lots of cratering. (a) This is a *Mariner 9* photograph of the irregularly shaped Phobos, not much larger than Manhattan Island. (b) Like Phobos, the smaller moon, Deimos, has a composition unlike that of Mars. Both moons are probably captured asteroids. This photograph was taken by the *Viking* orbiter. The field of view is only 2 km across, with most of the boulders shown about the size of a house.

Chapter Review www

SUMMARY

Mars lies outside Earth's orbit, so it traverses the entire ecliptic plane, as seen from Earth. Its orbit is more elliptical than Earth's, so its distance from the Sun varies more. Mars rotates at almost the same rate as Earth, and its rotation axis is inclined to the ecliptic at almost the same angle as Earth's axis. Because of its axial tilt, Mars has daily and seasonal cycles much like those on our own planet, but they are more complex than those on Earth because of Mars's eccentric orbit.

From Earth, the most obvious Martian surface features are the polar caps, which grow and diminish as the seasons change on Mars. The appearance of the planet also changes because of seasonal dust storms that obscure its surface.

Like the atmosphere of Venus, Mars's atmosphere is composed primarily of carbon dioxide. However, unlike Venus's, the density of the cool Martian atmosphere is less than 1 percent that of Earth's. Mars may once have had a dense atmosphere, but it was lost, partly to space and partly to surface rocks and subsurface **permafrost** (p. 229) and polar caps. Even today, the thin atmosphere is slowly leaking away. Surface temperatures on Mars average about 50 K cooler than those on Earth. Otherwise, Martian weather is reminiscent of that on Earth, with dust storms, clouds, and fog.

The two polar caps on Mars consist of a **seasonal cap** (p. 232), composed of carbon dioxide, which grows and shrinks, and a **residual cap** (p. 232), of water ice, which remains permanently frozen.

There is a marked difference between the two Martian hemispheres. The northern hemisphere consists of rolling volcanic plains and lies several kilometers below the level of the heavily cratered southern hemisphere. The lack of craters in the north suggests that this region is younger. The cause of the north–south asymmetry is not known.

In 1971, *Mariner 9* mapped the entire Martian surface, revealing plains, volcanoes, channels, and canyons. *Viking 1* and *Viking 2* reached Mars in 1976 and returned a wealth of data on the planet's surface and atmosphere. Experiments onboard the *Viking* landers detected no evidence for Martian life.

Martian craters differ from those on the Moon by the presence of **fluidized ejecta** (p. 227), which provide direct evidence for the permafrost layer beneath the surface.

Mars's major surface feature is the Tharsis bulge, located on the planet's equator. It may have been caused by a "plume" of upwelling material in the youthful Martian mantle. Associated with the bulge is Olympus Mons, the largest known volcano in the solar system, and a huge crack, called the Valles Marineris, in the planet's surface. The height of the Martian volcanoes is a direct consequence of Mars's low surface gravity. No evidence for recent or ongoing eruptions has been found.

There is clear evidence that water once existed in great quantity on Mars. Mars may have had a brief "Earthlike" phase early on in its evolution. The **runoff channels** (p. 230) are the remains of ancient Martian rivers. The **outflow channels** (p. 230) are the paths taken by flash floods that cascaded from the southern highlands into the northern plains. Today, a large amount of that water may be locked up in the polar caps and in the layer of permafrost lying under the Martian surface.

Convection in the Martian interior seems to have been stifled 2 billion years ago by the planet's rapidly cooling and solidifying mantle.

Mars has an extremely weak magnetic field. Since the planet rotates rapidly, this implies that its core is nonmetallic, nonliquid, or both. The lack of current volcanism, the absence of any significant magnetic field, the planet's relatively low density, and a high abundance of surface iron all suggest that Mars never melted and differentiated as extensively as did Earth.

Mars's moons, Phobos and Deimos, are probably asteroids captured by Mars early in its history. Their densities are far less than that of any planet in the inner solar system. They may be representative of conditions in the early solar system.

SELF-TEST: TRUE OR FALSE?

_____ **1.** It is possible, over time, to see Mars at any angular distance from the Sun, at any time of night.

_____ **2.** Seen from Earth, Mars goes through phases, just like Venus and Mercury.

_____ **3.** Mars at times is engulfed by global dust storms.

_____ **4.** Because Mars has such a thin atmosphere, the planet has no significant surface winds.

_____ **5.** Seasonal changes in the appearance of Mars are caused by vegetation on the surface.

_____ **6.** Mars has the largest volcanoes in the solar system.

_____ **7.** The northern hemisphere of Mars is much older than the southern hemisphere.

_____ **8.** Olympus Mons is the largest impact crater on Mars.

_____ **9.** There are many indications of past plate tectonics on Mars.

_____ **10.** Valles Marineris is similar in size to Earth's Grand Canyon.

_____ **11.** The orange-red color of the surface of Mars is primarily due to rust (iron oxide) in its soil.

_____ **12.** Daytime temperatures on Mars can reach 300 K.

_____ **13.** The *Viking* landers are still sending data back to Earth.

_____ **14.** No spacecraft have landed on Mars since the 1970s.

_____ **15.** One of the two moons of Mars is larger than Earth's Moon.

SELF-TEST: FILL IN THE BLANK

1. When Mars is on the opposite side of the Sun from Earth, it is said to be at _____.

2. The radius of Mars is about _____ that of Earth. (Give an answer as a simple fraction.)

3. The length of the Martian day is about _____ hours.

4. The seasonal polar ice caps of Mars are composed of _____.

5. The southern hemisphere of Mars consists of heavily _____ highlands.

6. The Tharsis region of Mars is a large equatorial _____.

7. Several large _____ lie at approximately the center of Tharsis.

8. The great height of Martian volcanoes is a direct result of the planet's low _____.

9. The fluidized ejecta surrounding Martian impact craters is evidence of a layer of _____ just under the surface.

10. Runoff channels carried _____ from the southern mountains into the valleys.

11. Outflow channels are the result of catastrophic _____.

12. Water flowed on the surface of Mars a few _____ years ago.

13. Mars's northern residual polar ice caps consists mostly of _____.

14. Most of the carbon dioxide on Mars is now found in the planet's _____.

15. The most recent claims of life on Mars have been based on analyses of _____ found on Earth.

REVIEW AND DISCUSSION

1. When is the best time to see Mars from Earth?

2. What is the evidence that water once flowed on Mars?

3. Is there water on Mars today?

4. For a century, there was speculation that intelligent life had constructed irrigation canals on Mars. What did the "canals" turn out to be?

5. Imagine that you will be visiting the southern hemisphere of Mars during its summer. Describe the atmospheric conditions you might face.

6. Describe the two Martian polar caps, their seasonal and permanent composition, and the differences between them.

7. Why is Mars red?

8. Why couldn't you breathe on Mars?

9. Why were Martian volcanoes able to grow so large?

10. How were the masses of Mars's moons measured, and what did these measurements tell us about their origin?

11. What is the evidence that Mars never melted as extensively as did Earth?

12. How would Earth look from Mars?

13. If humans were sent to Mars to live, what environmental factors would have to be considered? What resources might Mars provide, and which would have to come from Earth?

14. Since Mars has an atmosphere, and it is composed mostly of carbon dioxide, why isn't there a significant greenhouse effect to warm its surface?

15. Compare and contrast the evolution of the atmospheres of Mars, Venus, and Earth.

PROBLEMS

1. Verify that the surface gravity on Mars is 40 percent that of Earth.

2. Calculate Earth's synodic period, as seen from Mars.

3. What is the maximum elongation of Earth, as seen from Mars? (For simplicity, assume circular orbits for both planets.)

4. A certain star, observed from Earth, has a parallax of 0.1″. What would be its parallax as seen from Mars? How might observations from Mars-based telescopes be superior to those made from Earth?

5. Using the data from Section 10.7, compute the synodic periods of the moons Phobos and Deimos, as seen from the Martian surface (see *More Precisely 9-1*).

6. How long would it take the wind in a Martian dust storm, moving at a speed of 150 km/h, to encircle the planet?

7. The mass of the Martian atmosphere is about 1/150 the mass of Earth's atmosphere and is composed mainly (95 percent) of carbon dioxide. Using the result of problem 3, Chapter 7, to determine the mass of Earth's atmosphere, estimate the total mass of carbon dioxide in the atmosphere of Mars. Compare this with the mass of a seasonal polar cap, approximated as a circular sheet of frozen carbon dioxide ("dry ice," having a density of 1600 kg/m^3) of diameter 3000 km and thickness 1 m.

8. Assume that a planet will have lost its initial atmosphere by the present time if the molecular speed exceeds one-sixth of the escape speed (see *More Precisely 8-1*). What would the mass of Mars have to have been in order for it to have retained an Earthlike atmosphere?

9. The outflow channel shown in Figure 10.9 is about 10 km across and 100 m deep. If it carried 10^7 metric tons (10^{10} kilograms) of water per second, as stated in the text, estimate the speed at which the water must have flowed.

10. Using the data given in the text, calculate the maximum angular sizes of Phobos and Deimos, as seen by an observer standing on the Martian surface directly under their orbits. Compare these sizes with the angular diameter of Earth's Moon, as seen from Earth. Would a Martian observer ever see a total solar eclipse?

PROJECTS

1. Track the motion of the Red Planet in front of the stars for several months following its return to the predawn sky. (Consult an almanac to determine where Mars will be in the sky this year.) You will see that Mars moves rapidly in front of the stars, crossing many constellation boundaries.

2. Several months before opposition, Mars begins retrograde motion. Chart the planet's motion in front of the stars to determine when it stops moving eastward and begins moving toward the west.

3. Notice the increase in Mars's brightness as it approaches opposition. Why is it getting brighter? What other planets appear in the sky now? How do their brightnesses compare with that of Mars?

4. Look at Mars with as large a telescope as is available to you; binoculars will not be of use. Prepare ahead of time and find out the Martian season that will be occurring at the time of your observation, which hemisphere will be tilted in Earth's direction, and what longitude will be pointing toward Earth at the time of observation. This information can be obtained from many different computer almanacs. Sketch what you see. Look very carefully and take your time. Afterward, try to identify the various features you have seen with known objects on Mars.

11 JUPITER

Giant of the Solar System

LEARNING GOALS

Studying this chapter will enable you to:

1   Specify the ways in which Jupiter differs from the terrestrial planets in its physical and orbital properties.

2   Discuss the processes responsible for the appearance of Jupiter's atmosphere.

3   Describe Jupiter's internal structure and composition and explain how these are inferred from external measurements.

4   Summarize the characteristics of Jupiter's magnetosphere.

5   Discuss the orbital properties of the Galilean moons of Jupiter and describe the appearance and physical properties of each.

6   Explain how tidal forces can produce enormous internal stresses in a jovian moon, and discuss the results of those stresses.

(Opposite page, background) In this close-up image of Jupiter, taken by the *Voyager 2* spacecraft, the moon Io is clearly visible. One of the most peculiar moons in the solar system, and certainly the most active, Io's multihued surface is caused by various chemical compounds deposited by currently active volcanos.

(Inset A) Photo of Io taken by *Voyager 1* with a remarkable resolution of 7 km. The red color results largely from sulfur compounds, the white areas covered probably with sulfur dioxide frost.

(Inset B) At top left, a plume from an erupting volcano, known as Prometheus, is clearly seen at the limb of Io against the blackness of space.

(Inset C) Many lava flows can be seen on Io, like this one stretching for several hundred kilometers across the moon's surface.

Beyond the orbit of Mars, the solar system is very different from our own backyard. The outer solar system presents us with a totally unfamiliar environment—huge gas balls, peculiar moons, complex ring systems, and a wide variety of physical and chemical phenomena, many of which are still only poorly understood. Although the jovian planets—Jupiter, Saturn, Uranus, and Neptune—differ from one another in many ways, we will find that they have much in common, too. As with the terrestrial planets, we will learn from their differences as well as from their similarities. Our study of these alien places begins with the jovian planet closest to Earth—Jupiter, the largest planet in the solar system and a model for the other jovian worlds.

11.1 Jupiter in Bulk

ORBITAL PROPERTIES

1 Named after the most powerful god of the Roman pantheon, Jupiter is by far the largest planet in the solar system. Ancient astronomers could not have known the planet's true size, but their choice of names was very apt. The Jupiter Data box on p. 249 presents some orbital and physical data on the planet.

Jupiter is the third-brightest object in the night sky (after the Moon and Venus), making it very easy to locate and study. As in the case of Mars, Jupiter is brightest when it is near opposition. When this happens to occur near perihelion, the planet can be up to 50″ across, and a lot of detail can be discerned through even a small telescope.

Figure 11.1 shows two of the best views of Jupiter ever obtained from Earth's distance—one from the ground and one from space. Figure 11.1(a) shows a ground-based photograph of the planet, and Figure 11.1(b) shows a *Hubble Space Telescope* image taken during the opposition of December 1990. Notice the alternating light and dark bands that cross the planet parallel to its equator and the prominent reddish orange spot at the lower right of Figure 11.1(b).

MASS AND RADIUS

Unlike any of the terrestrial planets, Jupiter has many moons, with a wide range of sizes and properties. The four largest are visible from Earth with a small telescope (or even with the naked eye). They are known as the **Galilean moons**, after Galileo Galilei, who first observed them early in the seventeenth century. ∞ (Sec. 2.5) Since astronomers have been able to study the motion of the Galilean moons for quite some time, Jupiter's mass has long been known. It is 1.9×10^{27} kg, or 318 Earth masses. Jupiter has more than twice the mass of all the other planets combined.

Jupiter is such a large planet that many celestial mechanicians—those researchers concerned with the motions of interacting cosmic objects—regard our solar system as containing only two important objects—the Sun and Jupiter. To be sure, in this age of sophisticated and precise spacecraft navigation, the gravitational influence of all the planets must be considered, but in the broadest sense, our solar system is a two-object system with a lot of debris. As massive as Jupiter is, though, it is still only a thousandth the mass of the Sun. This makes studies of Jupiter all the more important, for here we have an object intermediate in size between the Sun and the terrestrial planets.

(a)

R I V U X G

(b)

R I V U X G

Figure 11.1 (a) Photograph of Jupiter made with a large ground-based telescope, showing several of its moons. (b) A *Hubble Space Telescope* image of Jupiter, in true color. Features as small as a few hundred kilometers across are resolved.

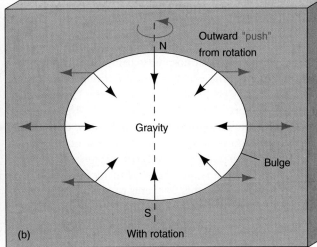

Figure 11.2   All spinning objects tend to develop an equatorial bulge because rotation causes matter to push outward against the inward-pulling gravity. The size of the bulge depends on the mechanical strength of the matter and the rate of rotation. The inward-pointing arrows denote gravity, the outward arrows the "push" due to rotation.

Knowing Jupiter's distance and angular size, we can easily determine its radius. It is 71,500 km, or 11.2 Earth radii. More dramatically stated, more than 1400 Earths would be needed to equal the volume of Jupiter. From the size and mass, we derive an average density of 1300 kg/m^3 for the planet. Here (as if we needed it) is yet another indicator that Jupiter is radically different from the terrestrial worlds. It is clear that, whatever Jupiter's composition it is not made up of the same material as the inner planets (recall from Chapter 7 that Earth's average density is 5500 kg/m^3). Studies of the planet's internal structure indicate that Jupiter must be composed primarily of hydrogen and helium. The enormous pressures in the planet's interior greatly compress these light gases, producing the average density we observe—very high for hydrogen, but still considerably lower than the densities of the terrestrial planets.

ROTATION RATE

As with other planets, we can attempt to determine Jupiter's rotation rate simply by timing a surface feature as it moves around the planet. However, in the case of Jupiter (and, indeed, all the gaseous outer planets), there is a catch—Jupiter has no solid surface. All we see are cloud features in the planet's upper atmosphere. With no solid surface to "tie them down," different parts of Jupiter's atmosphere move independently of one another. Visual observations and Doppler-shifted spectral lines indicate that the equatorial zones rotate a little faster (9^h50^m period) than the higher latitudes (9^h55^m period). Jupiter exhibits **differential rotation**—the rotation rate is not constant from one location to another. Differential rotation is not possible in solid objects like the terrestrial planets, but it is normal for fluid bodies such as Jupiter.

Observations of Jupiter's magnetosphere provide a more meaningful measurement of the rotation period. The plan-

et's magnetic field is strong and emits radiation at radio wavelengths as charged particles accelerate in response to the planet's magnetic field. Careful studies show a periodicity of 9^h55^m in this radio emission. We assume that this measurement matches the rotation of the planet's interior, where the magnetic field arises. ∞ (Sec. 7.4) Thus, Jupiter's interior rotates at the same rate as the clouds at its poles. The equatorial zones rotate more rapidly.

A rotation period of 9^h55^m is fast for such a large object. In fact, Jupiter has the fastest rotation rate of any planet in the solar system, and this rapid spin has altered Jupiter's shape. As illustrated in Figure 11.2, a spinning object tends to develop a bulge around its midsection. The more loosely the object's matter is bound together, or the faster it spins, the larger the bulge becomes. In objects like Jupiter, which are made up of gas or loosely packed matter, high spin rates can produce a quite pronounced bulge. Jupiter's equatorial radius (71,500 km) exceeds its polar radius (66,900 km) by about 6.5 percent.*

But there is more to the story of Jupiter's shape. Jupiter's observed equatorial bulge also tells us something very important about the planet's deep interior. Careful calculations indicate that Jupiter would be *more* flattened than it actually is if its core were composed of hydrogen and helium alone. To account for the planet's observed shape, we must assume that Jupiter has a dense, probably rocky, core that is between 10 and 20 times the mass of Earth. This is one of the few pieces of data we have on Jupiter's internal structure.

Earth also bulges slightly at the equator because of rotation. However, our planet is much more rigid than Jupiter, and the effect is much smaller—the equatorial diameter is only about 40 km larger than the distance from pole to pole, a tiny difference compared with Earth's full diameter of nearly 13,000 km. Relative to its overall dimensions, Earth is smoother and more spherical than a billiard ball.

11.2 The Atmosphere of Jupiter

2 Jupiter is visually dominated by two features: one is a series of ever-changing atmospheric bands arranged parallel to the equator; the other is an oval atmospheric blob called the **Great Red Spot**, or often just the "Red Spot." The cloud bands, clearly visible in Figure 11.1, display many colors—pale yellows, light blues, deep browns, drab tans, and vivid reds among others. Shown in more detail in Figure 11.3, a close-up photograph taken as *Voyager 1* sped past in 1979, the Red Spot is the largest of many features associated with Jupiter's weather. It seems to be an Earth-sized hurricane that has persisted for hundreds of years.

ATMOSPHERIC COMPOSITION

Spectroscopic studies of sunlight reflected from Jupiter gave astronomers their first look at the planet's atmospheric composition. Radio, infrared, and ultraviolet observations later provided more details. The most abundant gas is molecular hydrogen (86.1 percent by number), followed by helium (13.8 percent). Together these two gases make up over 99 percent of Jupiter's atmosphere. Small amounts of atmospheric methane, ammonia, and water vapor are also found.

Unlike the gravitational pull of the terrestrial planets, the gravity of the much more massive jovian planets is strong enough to have retained even hydrogen. ∞ *(More Precisely 8-1)* Little, if any, of Jupiter's original atmosphere has escaped.

Researchers generally accept that hydrogen and helium also make up the bulk of the planet's interior. This belief is based not on direct evidence of the interior—until the

collision of a comet with Jupiter, described in Section 14.2, there was virtually none—but largely on theoretical studies of the internal structure of the planet, such as we have already seen in the discussions of Jupiter's density and shape.

ATMOSPHERIC BANDS

Astronomers generally describe Jupiter's banded appearance—and, to a lesser extent, the appearance of the other jovian worlds as well—as a series of bright **zones** and dark **belts** crossing the planet. The zones and belts vary in both latitude and intensity during the year, but the general pattern remains. These variations are not seasonal in nature—Jupiter has no seasons—but instead appear to be the result of dynamic motion in the planet's atmosphere. The light-colored zones lie above upward-moving convective currents in Jupiter's atmosphere. The dark belts are regions representing the other part of the convection cycle, where material is generally sinking downward, as illustrated schematically in Figure 11.4.

Because of the upwelling material below them, the zones are regions of high pressure; the belts, conversely, are low-pressure regions. Thus, belts and zones are Jupiter's equivalents of the familiar high- and low-pressure systems that cause our weather on Earth. However, a major difference between Jupiter and Earth is that Jupiter's rapid rotation has caused these systems to wrap all the way around the planet, instead of forming localized circulating storms, as on our own world. Because of the pressure difference between the two, the zones lie slightly higher in the atmosphere than the belts. The associated temperature differences (recall that temperature increases with depth) and the resulting differences in chemical reactions are the basic reasons for their different colors.

Underlying the bands is an apparently very stable pattern of eastward and westward wind flow, referred to as Jupiter's **zonal flow**. This zonal flow is evident in Figure 11.5, which shows the wind speed at different planetary latitudes measured relative to the rotation of the planet's interior (determined from studies of Jupiter's magnetic field). As mentioned earlier, the equatorial regions of the atmosphere rotate faster than the planet, with an average flow speed of some 85 m/s, or about 300 km/h, in the easterly direction. The speed of this equatorial flow is quite similar to that of the jet stream on Earth. At higher latitudes, there are alternating regions of westward and eastward flow, roughly symmetric about the equator, with the flow speed generally diminishing toward the poles.

As Figure 11.5 shows, the belts and zones are closely related to Jupiter's zonal flow pattern. However, closer inspection shows that the simplified picture presented in Figure 11.4, with wind direction alternating between adjacent bands as Jupiter's rotation deflects surface winds into eastward or westward streams, is really too crude to describe the actual flow. Scientists now believe that the interaction between convective motion in Jupiter's atmosphere and the planet's rapid rotation channels the largest eddies into the observed zonal

R I V U X G

Figure 11.3 *Voyager 1* took this photograph of Jupiter's Red Spot (upper right) from a distance of about 100,000 km. The resolution is about 100 km. Note the complex turbulence patterns to the left of both the Red Spot and the smaller white oval vortex below it. (For scale, planet Earth is about the size of the white oval.)

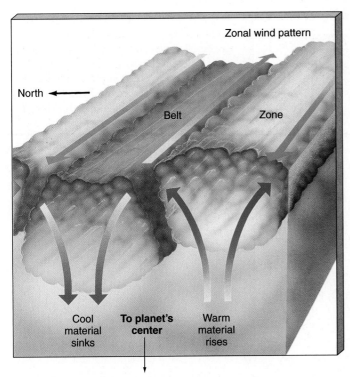

Figure 11.4 The colored bands in Jupiter's atmosphere are associated with vertical convective motion. Upwelling warm gas results in the lighter-colored zones; the darker bands lie atop lower-pressure regions where cooler gas is sinking back down into the atmosphere. As on Earth, surface winds tend to blow from high- to low-pressure regions. Jupiter's rotation channels these winds into an east–west flow pattern, as indicated.

pattern, but that smaller eddies tend to cause irregularities in the flow. Near the poles, where the zonal flow disappears, the band structure vanishes also.

ATMOSPHERIC STRUCTURE AND COLOR

None of the atmospheric gases listed earlier can, by itself, account for Jupiter's observed coloration. For example, frozen ammonia and water vapor would simply produce white clouds, not the many colors actually seen. Scientists believe that the cloud colors are the result of complex chemical processes occurring in the planet's turbulent upper atmosphere, although the details are still not fully understood. When we observe Jupiter's colors, we are actually looking down to many different depths in the planet's atmosphere.

Based on the best available data and mathematical models, Figure 11.6 is a cross-sectional diagram of Jupiter's atmosphere. Since the planet lacks a solid surface to use as a reference level for measuring altitude, the top of the troposphere is conventionally taken to lie at 0 km. As on all planets, weather on Jupiter is the result of convection in the troposphere, so the clouds, which are associated with planetary weather systems, all lie at negative altitudes in the diagram. Just above the troposphere lies a thin, faint layer of haze created by photochemical reactions (that is, reactions involving sunlight) similar to those that cause smog on Earth. The temperature at this level is about 110 K; it increases with altitude as the atmosphere absorbs solar ultraviolet radiation.

Figure 11.5 The wind speed in Jupiter's atmosphere, measured relative to the planet's internal rotation rate. The alternations in wind direction are associated with the atmospheric band structure.

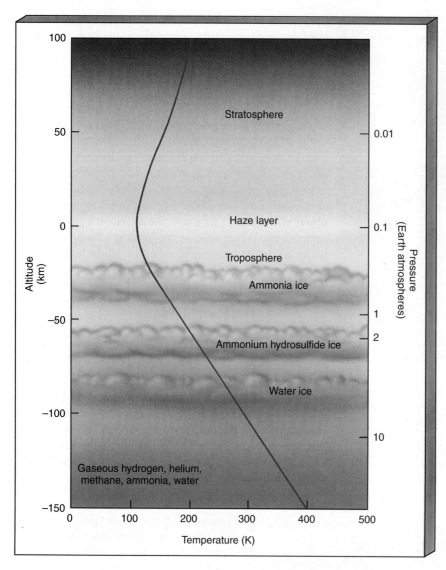

Figure 11.6 The vertical structure of Jupiter's atmosphere. Jupiter's clouds are arranged in three main layers, each with quite different colors and chemistry. The colors we see in photographs of the planet depend on the cloud cover. The white regions are the tops of the upper ammonia clouds. The yellows, reds, and browns are associated with the second cloud layer, which is composed of ammonium hydrosulfide ice. The lowest cloud layer is water ice and bluish in color. However, the overlying layers are sufficiently thick that this level is not seen in visible light.

Jupiter's clouds are arranged in three main layers. Below the haze, at a depth of about 40 km (shown as −40 km in Figure 11.6), lies a layer of white wispy clouds made up of ammonia ice. The temperature here is approximately 125–150 K; it increases quite rapidly with increasing depth. A few tens of kilometers below the ammonia clouds, the temperature is a little warmer—over 200 K—and the clouds are probably made up mostly of droplets or crystals of ammonium hydrosulfide, produced by reactions between ammonia and hydrogen sulfide in the planet's atmosphere. At deeper levels in the atmosphere, the ammonium hydrosulfide clouds give way to clouds of water ice or water vapor. This lowest cloud layer, which is not seen in visible-light images of Jupiter, lies some 80 km below the top of the troposphere.

Instead of being white (the color of ammonium hydrosulfide on Earth), Jupiter's middle cloud layer is tawny in color. This is the level at which atmospheric chemistry begins to play a role in determining Jupiter's appearance. Many planetary scientists believe that molecules containing the element sulfur, and perhaps even sulfur itself, are important in influencing the cloud colors—particularly the reds, browns,

and yellows, all colors associated with sulfur or sulfur compounds. It is also possible that compounds containing the element phosphorus contribute to the coloration.

Deciphering the detailed causes of Jupiter's distinctive colors is a difficult task, however. The cloud chemistry is complex and very sensitive to small changes in atmospheric conditions, such as pressure and temperature, as well as to chemical composition. The atmosphere is in incessant, churning motion, causing these conditions to change from place to place and from hour to hour. In addition, the energy that powers the reactions comes in many different forms: the planet's own interior heat, solar ultraviolet radiation, aurorae in the planet's magnetosphere, and lightning discharges within the clouds themselves. All these factors combine to keep a complete explanation of Jupiter's appearance beyond our present grasp.

This description of Jupiter's atmosphere, which was based largely on *Voyager* data, was put to the test in December 1995 when the *Galileo* atmospheric probe arrived at the planet. The probe survived for about an hour before being crushed by atmospheric pressure at an altitude of −150 km (that is, right

PLANETARY DATA:
Jupiter

Orbital semi-major axis	5.20 A.U. 778.4 million km
Orbital eccentricity	0.048
Perihelion	4.95 A.U. 740.7 million km
Aphelion	5.46 A.U. 816.1 million km
Mean orbital speed	13.1 km/s
Sidereal orbital period	11.86 tropical years
Synodic orbital period	398.88 solar days
Orbital inclination to the ecliptic	1.31°
Greatest angular diameter, as seen from Earth	50″
Mass	1.90×10^{27} kg 317.8 (Earth = 1)
Equatorial radius	71,492 km 11.21 (Earth = 1)
Mean density	1330 kg/m^3 0.241 (Earth = 1)
Surface gravity	24.8 m/s^2 2.53 (Earth = 1)
Escape speed	59.5 km/s
Sidereal rotation period	0.41 solar days
Axial tilt	3.08°
Surface magnetic field	13.89 (Earth = 1)
Magnetic axis tilt relative to rotation axis	9.6°
Surface temperature	124 K (at cloudtops)
Number of moons	16

at the bottom of Figure 11.6). Initially, the data from the probe appeared to contradict many of the facts we have just reported! Preliminary analysis of the data indicated that Jupiter was much windier, hotter, and drier than expected, and severely depleted in helium. Furthermore, no clear evidence of the three-layered cloud structure depicted in Figure 11.6 was found.

Most of these discrepancies are now known to have been the result of improperly calibrated instruments; *Galileo*'s revised findings on wind speed, temperature, and composition are in good agreement with the picture just presented. In addition, the probe's entry location was in Jupiter's equatorial zone and, as luck would have it, coincided with an atypical hole almost devoid of upper-level clouds (see Figure 11.7). The warm temperature readings—425 K at 150 km depth, a little higher than indicated by Figure 11.6—are consistent with the probe's having entered a relative clearing in Jupiter's cloud decks, where convective heat can more readily rise (and thus be detected). The low water content may also be normal for the hot, windy regions near Jupiter's equator.

Most experts were somewhat surprised by the depth to which Jupiter's winds continued; *Galileo*'s probe measured high wind speeds throughout its descent into the clouds and not just at the cloud tops, implying that heat deep within

Animation

R I V U X G

Figure 11.7 The arrow on this *Hubble* image shows where the *Galileo* atmospheric probe plunged into Jupiter's cloud deck on December 7, 1995. The entry location was in Jupiter's equatorial zone, and was apparently almost devoid of upper-level clouds. Until its demise, the probe took numerous meteorological measurements, transmitting those signals to the mother ship overhead, which then relayed them to Earth.

the planet, rather than sunlight, drives Jupiter's weather patterns. Finally, complex organic molecules were sought but not found. Some simple carbon-based molecules, such as ethane (C_2H_6), were detected by one of the onboard spectrometers, but nothing suggesting prebiotic compounds or bacteria floating in the atmosphere. That same instrument also detected traces of phosphine (PH_3), which may be a key coloring agent for Jupiter's clouds.

THE WEATHER ON JUPITER

In addition to the zonal flow pattern, Jupiter has many "small-scale" weather patterns. The Great Red Spot (Figure 11.3) is a prime example. The Great Red Spot was first reported by the British scientist Robert Hooke in the mid-seventeenth century, and we can be reasonably sure that it has existed continuously, in one form or another, for over 300 years. It may well be much older. *Voyager* observations showed the spot to be a region of swirling, circulating winds, rather like a whirlpool or a terrestrial hurricane—a persistent and vast atmospheric storm. The size of the Spot varies, although it averages about twice the diameter of Earth. Its present dimensions are roughly 25,000 km by 15,000 km. It rotates around Jupiter at a rate similar to that of the planet's interior, perhaps suggesting that its roots lie far below the atmosphere.

The origin of the Spot's red color is unknown, as is its source of energy, although it is generally supposed that the Spot is somehow sustained by Jupiter's large-scale atmospheric motion. Repeated observations show that the gas flow around the Spot is counterclockwise, with a period of about 6 days. Turbulent eddies form and drift away from its edge. The Spot's center, however, remains quite tranquil in appearance, like the eye of a hurricane on Earth. The zonal motion north of the Spot is westward, whereas that to the south is eastward (see Figure 11.8), supporting the idea that the Spot is confined and powered by the zonal flow. However, the details of how this occurs are still a matter of conjecture. Computer simulations of the complex fluid dynamics of Jupiter's atmosphere are only now beginning to hint at answers.

Storms, which as a rule are much smaller than the Red Spot, may be quite common on Jupiter. Spacecraft photographs of the dark side of the planet reveal both auroral activity and bright flashes resembling lightning. The *Voyager* mission discovered many smaller light- and dark-colored spots that are also apparently circulating storm systems. Note the several **white ovals** in Figures 11.3 and 11.8, south of the Red Spot. Like the Red Spot, they rotate counterclockwise. Their high cloud tops give them their color. These particular white ovals are known to be at least 40 years old. Figure 11.9 shows a **brown oval**, a "hole" in the clouds that allows us to look down into the lower atmosphere. For unknown reasons, brown ovals appear only in latitudes around 20°N. Although not as long-lived as the Red Spot, these systems can persist for many years or even decades.

We cannot explain their formation, but we can offer at least a partial explanation for the longevity of storm systems on Jupiter. On Earth a large storm, such as a hurricane, forms over the ocean and may survive for many days, but it dies quickly once it encounters land. Earth's continental landmasses disrupt the flow patterns that sustain the storm. Jupiter has no continents, so once a storm becomes established and reaches a size at which other storm systems cannot destroy it, apparently little affects it. The larger the system, the longer its lifetime.

R I V U X G

Figure 11.8 These *Voyager 2* close-up views of the Red Spot, taken 4 hours apart, show clearly the turbulent flow around its edges. The general direction of motion of the gas north of (above) the Spot is westward (to the left), whereas gas south of the Spot flows east. The Spot itself rotates counterclockwise, suggesting that it is being "rolled" between the two oppositely directed flows. The colors have been exaggerated somewhat to enhance the contrast.

R I V U X G

Figure 11.9 A brown oval in Jupiter's northern hemisphere. Its color comes from the fact that it is actually a break in the upper cloud layer, allowing us to see deeper into the atmosphere. The oval's length is approximately equal to the diameter of Earth.

11.3 Internal Structure

3 On the basis of Jupiter's distance from the Sun, astronomers had expected to find the temperature of the cloud tops to be around 105 K. At that temperature, they reasoned, Jupiter would radiate back into space exactly the same amount of energy as it received from the Sun. When radio and infrared observations were first made of the planet, however, astronomers found that its blackbody spectrum corresponded to a temperature of 125 K instead. Subsequent measurements, including those made by *Voyager* and *Galileo*, have verified that finding. Although a difference of 20 K may seem small, recall from Chapter 5 that the energy emitted by a planet grows as the *fourth* power of the surface temperature (in Jupiter's case, the temperature of the cloud tops). ∞ (Sec. 3.4) A planet at 125 K therefore radiates $(125/105)^4$, or about twice as much energy as a 105 K planet. Put another way, Jupiter actually emits about twice as much energy as it receives from the Sun. Thus, unlike any of the terrestrial planets, Jupiter must have its own internal heat source.

What is responsible for Jupiter's extra energy? It is not the decay of radioactive elements within the planet—that must be occurring, but not nearly at the rate necessary to produce the temperature we record. Nor is it the process that generates energy in the Sun, nuclear fusion—the temperature in

Jupiter's interior, high as it is, is far too low for that (see *Interlude 11-1*). Instead, astronomers theorize that the source of Jupiter's excess energy is the slow escape of gravitational energy released during the planet's formation. As the planet took shape, some of its gravitational energy was converted into heat in the interior. That heat is still slowly leaking out through the planet's heavy atmospheric blanket, resulting in the excess emission we observe. Despite the huge amounts of energy involved—Jupiter's emits about 4×10^{17} watts more energy than it receives from the Sun—the loss is quite slight compared with the planet's total energy. A simple calculation indicates that the average temperature of the interior of Jupiter falls by only about a millionth of a kelvin per year.

Jupiter's clouds, with their complex chemistry, are probably less than 200 km thick. Below them, the temperature and pressure steadily increase as the atmosphere becomes the "interior" of the planet. Much of our knowledge of Jupiter's interior comes from theoretical modeling. Planetary scientists use all available bulk data on the planet—mass, radius, composition, rotation, temperature, and so on—to construct a model of the interior that agrees with observations. Our statements about Jupiter's interior are, then, really statements about the model that best fits the facts. However, because the interior consists largely of hydrogen and helium—two simple gases whose physics we think we understand well—we can be fairly confident that Jupiter's internal structure is now understood.

Both the temperature and the density of Jupiter's atmosphere increase with depth below the cloud cover. However, no "surface" of any kind exists anywhere inside. Instead, Jupiter's atmosphere just becomes denser and denser, because of the pressure of the overlying layers. At a depth of a few thousand kilometers, the gas makes a gradual transition into the liquid state (see Figure 11.10). By a depth of about 20,000 km, the pressure is about 3 million times greater than atmospheric pressure on Earth. Under those conditions, the hot liquid hydrogen is compressed so much that it undergoes another transition, this time to a "metallic" state with properties in many ways similar to those of a liquid metal. Of particular importance for Jupiter's magnetic field (see Section 11.4) is that this metallic hydrogen is an excellent conductor of electricity.

As mentioned earlier, Jupiter's observed flattening requires that there be a relatively small, dense core at its center, containing perhaps 15 times the mass of Earth. The core's exact composition is unknown, but planetary scientists think that it contains much denser materials than the rest of the planet. Present best estimates indicate that it consists of "rocky" materials, similar to those found in the terrestrial worlds. (Note that the term *rocky* here refers to the chemical composition of the core, not to its physical state. At the high temperatures and pressures found deep in the jovian interiors, the core material bears little resemblance to rocks found on Earth's surface.) In fact, it now appears that all four jovian planets contain large rocky cores, and that the formation of such a large "terrestrial" planetary core is a necessary stage in the process of building up a gas giant.

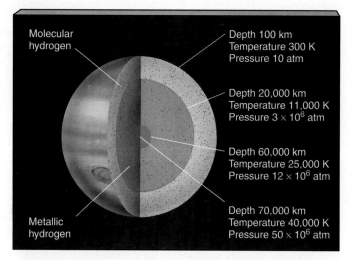

Figure 11.10   Jupiter's internal structure, as deduced from *Voyager* measurements and theoretical modeling. The outer radius represents the top of the cloud layers, some 70,000 km from the planet's center. The density and temperature increase with depth, and the atmosphere gradually liquefies at a depth of a few thousand kilometers. Below a depth of 20,000 km, the hydrogen behaves like a liquid metal. At the center of the planet lies a large rocky core, somewhat terrestrial in composition but much larger than any of the inner planets. Although the values are very uncertain, the temperature and pressure at the center are probably about 40,000 K and 50 million (Earth) atmospheres, respectively.

Because of the enormous pressure at the center of Jupiter—approximately 50 million times that on Earth's surface, or 10 times that at Earth's center—the core must be compressed to a very high density (perhaps twice the core density of Earth). It is probably not much more than 20,000 km in diameter, and the central temperature may be as high as 40,000 K.

11.4 Jupiter's Magnetosphere

4   For decades, ground-based radio telescopes monitored radiation leaking from Jupiter's magnetosphere, but only when the *Pioneer* and *Voyager* spacecraft reconnoitered the planet in the mid-1970s did astronomers realize the full extent of the planet's magnetic field. Jupiter, it turns out, is surrounded by a vast sea of energetic charged particles, mostly electrons and protons, somewhat similar to Earth's Van Allen belts but much, much larger. The radio radiation detected on Earth is emitted when these particles are accelerated to very high speeds—close to the speed of light—by Jupiter's powerful magnetic field. This radiation is several thousand times more intense than that produced by Earth's magnetic field and represents a potentially serious hazard for manned and unmanned space vehicles alike.

Sensitive electronic equipment (not to mention even more sensitive human bodies) would require special protective shielding to operate for long in this hostile environment.

Direct spacecraft measurements show Jupiter's magnetosphere to be almost 30 million km across, roughly a million times more voluminous than Earth's magnetosphere, and far larger than the entire Sun. As with Earth's, the size and shape of Jupiter's magnetosphere is determined by the interaction between the planet's magnetic field and the solar wind. Jupiter's magnetosphere has a long tail extending away from the Sun at least as far as Saturn's orbit (over 4 A.U. farther out from the Sun), as sketched in Figure 11.11. However, on the sunward side, the *magnetopause*—the boundary of Jupiter's magnetic influence on the solar wind—lies only 3 million km from the planet.

The outer magnetosphere appears to be quite unstable, sometimes deflating in response to "gusts" in the solar wind, then reexpanding as the wind subsides. In the inner magnetosphere, Jupiter's rapid rotation has forced most of the charged particles into a flat *current sheet*, lying on the planet's magnetic equator. The portion of the magnetosphere close to Jupiter is sketched in Figure 11.12. Notice that the planet's magnetic axis is not exactly aligned with its rotation axis but is inclined to it at an angle of approximately 10°. The planet's magnetic field happens to be oriented opposite Earth's (see also Figure 11.18).

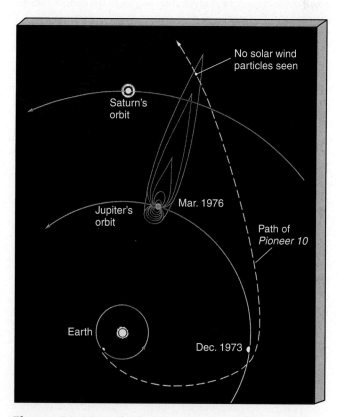

Figure 11.11   The *Pioneer 10* spacecraft did not detect any solar particles while moving behind Jupiter. Accordingly, as sketched here, Jupiter's magnetosphere apparently extends beyond the orbit of Saturn.

INTERLUDE 11-1   Almost a Star?

Jupiter has a starlike composition—predominantly hydrogen and helium, with a trace of heavier elements. Did Jupiter ever come close to becoming a star itself? Might the solar system have formed as a double-star system? Probably not. Unlike a star, Jupiter is cold. Its central temperature is far too low to ignite the nuclear fires that power our Sun. Jupiter's mass would have to increase 80-fold before its central temperature would rise to the point where nuclear reactions could begin, converting Jupiter into a small, dim star. Even so, it is interesting to note that although Jupiter's present-day energy output is very small (by solar standards, at least), it must have been much greater in the distant past while the planet was still contracting rapidly toward its present size. For a brief period of time—a few hundred million years—Jupiter might actually have been as bright as a faint star, although its brightness never came within a factor of 100 of the Sun's. Seen from Earth at that time, Jupiter would have been about 100 times brighter than the Moon!

What might have happened had our solar system formed as a double-star system? Conceivably, had Jupiter been massive enough, its radiation might have produced severe temperature fluctuations on all the planets, perhaps to the point of making life on Earth impossible. Even if Jupiter's brightness were too low to cause us any problems, its gravitational field (which would be 1/12 that of the Sun if its mass were 80 times greater) might have made the establishment of stable, roughly circular planetary orbits a fairly improbable event. The size of the "Jupiter," or second largest body, in a newborn planetary system may be a very important factor in determining the likelihood of the appearance of life.

Both ground- and space-based observations of the radiation emitted from Jupiter's magnetosphere imply that the *intrinsic* strength of the planet's magnetic field is nearly 20,000 times greater than Earth's. The existence of such a strong field further supports our theoretical model of the interior; the conducting liquid interior that is thought to make up most of the planet should combine with Jupiter's rapid rotation to produce a large dynamo effect and a strong magnetic field, just as observed. ∞ (Sec. 7.4)

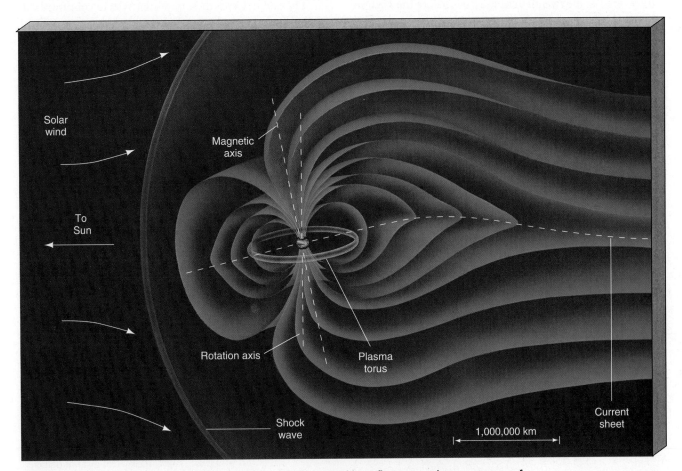

Figure 11.12   Jupiter's inner magnetosphere is characterized by a flat current sheet, consisting of charged particles squeezed into the magnetic equatorial plane by the planet's rapid rotation. The plasma torus is a ring of charged particles associated with the moon Io; it is discussed in Section 11.5.

11.5 The Moons of Jupiter

5 Jupiter has at least 16 moons. The four largest—the *Galilean satellites*—are each comparable in size to Earth's Moon. ∞ (Sec. 2.5) Moving outward from Jupiter, the four are named Io, Europa, Ganymede, and Callisto, after the mythical attendants of the Roman god Jupiter. They move in nearly circular orbits about their parent planet. When the *Voyager 1* spacecraft passed close to the Galilean moons in 1979, it sent some remarkably detailed photographs back to Earth, allowing planetary scientists to discern fine surface detail on each. More recently, in the late 1990s, the *Galileo* mission expanded still further our knowledge of these small but complex worlds. We will consider the Galilean satellites in more detail in a moment.

Within the orbit of Io lie four small satellites, all but one discovered by *Voyager* cameras. The largest of the four, Amalthea, is less than 300 km across and is irregularly shaped. E. E. Barnard discovered it in 1892. It orbits at a distance of 181,000 km from Jupiter's center—only 110,000 km above the cloud tops. Its rotation, like that of most of Jupiter's satellites, is synchronous with its orbit because of Jupiter's strong tidal field. Amalthea rotates once per orbit period, every 11.7 hours.

Beyond the Galilean moons lie eight more small satellites, all discovered in the twentieth century but before the *Voyager* missions. They fall into two groups of four moons each. The moons in the inner group move in eccentric, inclined orbits, about 11 million km from the planet. The outer four moons lie about 22 million km from Jupiter. Their orbits too are fairly eccentric, but *retrograde*, moving in a sense opposite all the other moons' orbit (and Jupiter's rotation). It is very likely that each group represents a single body that was captured by Jupiter's strong gravitational field long after the planet and its larger moons originally formed. Both bodies subsequently broke up, either during or after the capture process, resulting in the two families of similar orbits we see today. The masses, and hence the densities, of these small worlds are unknown. However, their appearance and sizes suggest compositions more like asteroids than their larger Galilean companions. Table 11.1 presents the general properties of Jupiter's moons.

THE GALILEAN MOONS AS A MODEL OF THE INNER SOLAR SYSTEM

If we think of Jupiter's moon system as a scaled-down system, then the Galilean moons correspond to the terrestrial planets. Their orbits are direct (that is, in the same sense as Jupiter's rotation), roughly circular, and lie close to Jupiter's equatorial plane. They range in size from slightly smaller than Earth's Moon (Europa) to slightly larger than Mercury (Ganymede). Figure 11.13 compares the appearances and sizes of the four Galilean satellites. Figure 11.14 shows two of the moons photographed against the background of their parent planet.

The parallel with the inner solar system continues with the realization that the moons' densities decrease with increasing distance from Jupiter. Based largely on detailed measurements made by *Galileo* of the moons' gravitational fields,

TABLE 11.1 The Moons of Jupiter

NAME	DISTANCE FROM JUPITER (km)	(planet radii)	ORBIT PERIOD (days)	SIZE (longest diameter, km)	MASS (Earth Moon masses)	DENSITY (kg/m³)	(g/cm³)
Metis	128,000	1.79	0.29	40			
Adastea	129,000	1.80	0.30	20			
Amalthea	181,000	2.54	0.50	260			
Thebe	222,000	3.10	0.67	100			
Io	422,000	5.90	1.77	3640	1.22	3500	3.5
Europa	671,000	9.38	3.55	3130	0.65	3000	3.0
Ganymede	1,070,000	15.0	7.15	5270	2.02	1900	1.9
Callisto	1,880,000	26.3	16.7	4800	1.46	1900	1.9
Leda	11,100,000	155	239	10			
Himalia	11,500,000	161	251	170			
Lysithea	11,700,000	164	259	24			
Elara	11,700,000	164	260	80			
Ananke	21,200,000	297	631*	20			
Carme	22,600,000	316	692*	30			
Pasiphae	23,500,000	329	735*	36			
Sinope	23,700,000	332	758*	28			

Indicates a retrograde orbit.

Figure 11.13 The *Voyager 1* spacecraft photographed each of the four Galilean moons of Jupiter. Shown here to scale, as they would appear from a distance of about 1 million km, they are, clockwise from upper left, Io, Europa, Callisto, and Ganymede.

R   I   U   X   G

researchers have built up fairly detailed pictures of the moons' compositions and internal structures (Figure 11.15). The innermost two Galilean moons, Io and Europa, have thick rocky mantles, possibly similar to the crusts of the ter-restrial planets, surrounding iron/iron sulfide cores. Io's core accounts for about half that moon's total radius. Europa has a thin water/ice outer shell about 150 km thick. The two outer moons, Ganymede and Callisto, are clearly deficient in rocky materials. Lighter materials, such as water ice, may account for as much as half of their total mass. Ganymede appears to have a relatively small metallic core topped by a rocky mantle and a thick icy outer shell. Callisto seems to be a largely undifferentiated mixture of rock and ice.

Many astronomers think that the formation of Jupiter and the Galilean satellites may in fact have mimicked on a small scale the formation of the Sun and the inner planets. For that reason, studies of the Galilean moon system may provide us with valuable insight into the processes that created our own world. We will return to this parallel in Chapter 15. So interested are mission planners in learning more about the Galilean moon system that the already highly successful *Galileo* mission has been extended for two more years, to concentrate first on Europa, then on Jupiter's atmosphere and magnetosphere, and finally on Io. The two moons will be studied at resolutions as fine as a few meters during extremely close passages by the spacecraft.

Not all the properties of the Galilean moons find analogs in the inner solar system, however. For example, because of Jupiter's tidal effect, all four Galilean satellites are in states of synchronous rotation, so they all keep one face permanently pointing toward their parent planet. By contrast, of the terrestrial planets, only Mercury is strongly influenced by

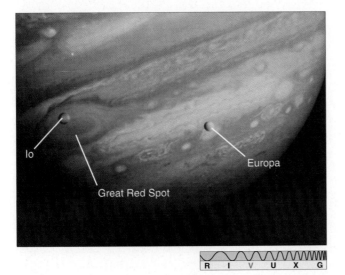

Io

Great Red Spot

Europa

R   I   U   X   G

Figure 11.14 *Voyager 1* took this photo of Jupiter with ruddy Io on the left and pearl-like Europa toward the right. Note the scale of objects here: Both Io and Europa are comparable in size to our Moon, and the Red Spot (seen here to the left bottom) is roughly twice as big as Earth. (See also the chapter-opening image on p. 242.)

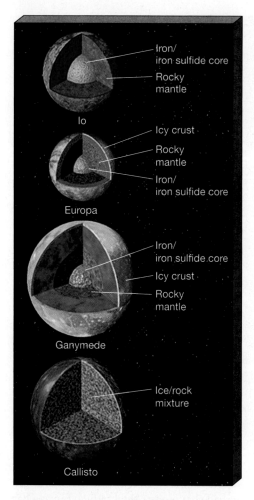

Figure 11.15 Cutaway diagrams showing the interior structure of the four Galilean satellites. Moving outward from Io to Callisto, the moons' densities steadily decrease as the composition shifts from rocky mantles and metallic cores in Io and Europa, to a thick icy crust and smaller core in Ganymede, to an almost uniform rock–ice mix in Callisto.

Animation

Io
- Iron/iron sulfide core
- Rocky mantle

Europa
- Icy crust
- Rocky mantle
- Iron/iron sulfide core

Ganymede
- Iron/iron sulfide core
- Icy crust
- Rocky mantle

Callisto
- Ice/rock mixture

the Sun's tidal force, and even its orbit is not synchronous. And, of course, the Jupiter system has no analogs of the jovian planets. Finally, inspection of Table 11.1 shows a remarkable coincidence in the orbit periods of the three inner Galilean moons. Their periods are almost exactly in the ratio 1:2:4—a kind of "Bode's law" for Jupiter. ∞ *(Interlude 6-1)* This is most probably the result of a complex, but poorly understood, three-body resonance in the Galilean moon system, something not found among the terrestrial worlds.

IO

6 Io, the densest of the Galilean moons, is the most geologically active object in the entire solar system. Its mass and radius are fairly similar to those of Earth's own Moon, but there the resemblance ends. Shown in Figure 11.16, Io's surface is a collage of reds, yellows, and blackish browns—resembling a giant pizza in the minds of some startled *Voyager* scientists. As the spacecraft glided past Io an outstanding discovery was made: Io has active volcanoes! *Voyager 1* photographed eight erupting volcanoes, and six were still erupting when *Voyager 2* passed by 4 months later. In Figure 11.17, one volcano is seen ejecting matter to an altitude of over 200 km. The gases are spewed forth at speeds up to

(a)

(b) *Animation*

R I V U X G

Figure 11.16 Jupiter's innermost moon, Io, is quite different in character from the other three Galilean satellites. Its surface is kept smooth and brightly colored by the moon's constant volcanism. The resolution of the photograph in (a) is about 7 km. In the more detailed image (b), features as small as 2 km across can be seen.

Figure 11.17 One of Io's volcanoes was caught in the act of erupting while the *Voyager 1* spacecraft flew past this fascinating moon. Surface features here are resolved to within a few kilometers. The volcano's umbrella-like profile shows clearly against the darkness of space. The plume measures about 100 km high and 300 km across.

2 km/s, quite unlike the (relatively) sluggish ooze that emanates from Earth's insides.

The orange color immediately surrounding the volcano most likely results from sulfur compounds in the ejected material. In stark contrast to the other Galilean moons, Io's surface is neither cratered nor streaked. (The circular features visible in Figures 11.16 and 11.17 are volcanoes.) Its surface is exceptionally smooth, apparently the result of molten matter that constantly fills in any "dents and cracks." Accordingly, we can conclude that this remarkable moon has the youngest surface of any known object in the solar system. Of further significance, Io also has a thin, temporary atmosphere made up primarily of sulfur dioxide, presumably the result of gases ejected by volcanic activity. By the time *Galileo* arrived, several of the volcanoes observed by *Voyager* had subsided; however, many new ones were seen.

Io's volcanism has a major effect on Jupiter's magnetosphere. All the Galilean moons orbit within the magnetosphere and play some part in modifying its properties, but Io's influence is particularly marked. Although many of the charged particles in Jupiter's magnetosphere come from the solar wind, there is strong evidence that Io's volcanism is the primary source of heavy ions in the inner regions. Jupiter's magnetic field continually sweeps past Io, gathering up the particles its volcanoes spew into space and accelerating them to high speed. The result is the *Io plasma torus* (Figure 11.18; see also Figure 11.12), a doughnut-shaped region of energetic heavy ions that follows Io's orbital track,

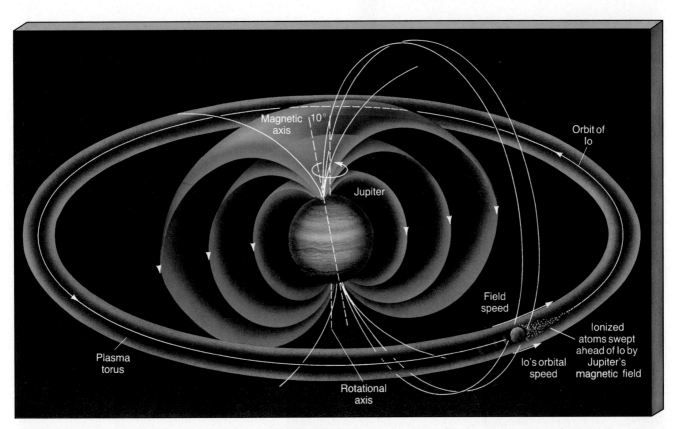

Figure 11.18 The Io plasma torus is the result of material being ejected from Io's volcanoes and swept up by Jupiter's rapidly rotating magnetic field. Spectroscopic analysis indicates that the torus is composed primarily of sodium and sulfur atoms.

completely encircling Jupiter. (A plasma is a gas that has been heated to such high temperatures that all its atoms are ionized.) The plasma torus is quite easily detectable from Earth, but before *Voyager* its origin was unclear. *Galileo* made detailed studies of the plasma's dynamic and rapidly varying magnetic field. Spectroscopic analysis shows that sulfur is indeed one of the torus's major constituents, strongly implicating Io's volcanoes as its source. As a hazard to spacecraft—manned or unmanned—the plasma torus is formidable. The radiation levels there are lethal.

What causes such astounding volcanic activity on Io? Surely that moon is too small to have geological activity like Earth. Io should be long dead, like our own Moon. At one time, some scientists suggested that Jupiter's magnetosphere might be the culprit—perhaps the (then-unknown) processes creating the plasma torus were somehow also stressing the moon. We now know that this is not the case. The real source

of Io's energy is gravity—Jupiter's gravity. Io orbits very close to Jupiter—only 422,000 km, or 5.9 Jupiter radii, from the center of the planet. As a result, Jupiter's huge gravitational field produces strong tidal forces on the moon. If Io were the only satellite in the Jupiter system, it would long ago have come into a state of synchronous rotation with the planet, just like our own Moon, for the reasons discussed in Chapter 8. ∞ (Sec. 8.3) In that case, Io would move in a perfectly circular orbit, with one face permanently turned toward Jupiter. The tidal bulge would be stationary with respect to the moon, and there would be no internal stresses and hence no volcanism.

But Io is not alone. As it orbits it is constantly tugged by the gravity of its nearest large neighbor, Europa. These tugs are small and not enough to cause any great tidal effect, but they are sufficient to make Io's orbit slightly noncircular, preventing the moon from settling into a precisely synchronous state. The reason for this effect is exactly the same as in the case of Mercury, as discussed in Chapter 8. ∞ (Sec. 8.3) In a noncircular orbit, the moon's speed varies from place to place as it revolves around its planet, but its rate of rotation on its axis remains constant. Thus it cannot keep one face always turned toward Jupiter. Instead, as seen from Jupiter, Io rocks or "wobbles" slightly from side to side as it moves. The large tidal bulge, however, always points directly toward Jupiter, so it moves back and forth across Io's surface as the moon wobbles. These conflicting forces result in enormous tidal stresses that continually flex and squeeze Io's interior.

Just as repeated back-and-forth bending of a piece of wire can produce heat through friction, the ever-changing

Figure 11.19 The second Galilean moon is Europa. Its icy surface is only lightly cratered, indicating that some ongoing process must be obliterating impact craters soon after they are formed. The origin of the cracks crisscrossing the surface is uncertain. The resolution of the *Voyager 2* mosaic in (a) is about 5 km. The two images below it (b and c) display even finer detail. (d) At 20-m resolution—the width of a typical house—this image from the *Galileo* spacecraft shows a smooth yet tangled surface resembling the huge ice flows that cover Earth's polar regions.

(a)

(b)

(c)

(d)

Videos

R I V U X G

distortion of Io's interior constantly energizes the moon. This generation of large amounts of heat within Io ultimately causes huge jets of gas and molten rock to squirt out of the surface. *Galileo*'s sensors indicated extremely high temperatures in the outflowing material. It is likely that much of Io's interior is soft or molten, with only a relatively thin solid crust overlying it. In fact, Io's volcanoes are probably more like geysers on Earth, but the term volcano has stuck. Researchers estimate that the total amount of heat generated within Io as a result of tidal flexing is about 100 million megawatts. This phenomenon makes Io one of the most fascinating objects in our solar system.

EUROPA

Europa (Figure 11.19) is a very different world from Io. Lying outside Io's orbit, 671,000 km (9.4 Jupiter radii) from Jupiter, it has relatively few craters on its surface, suggesting geologic youth, perhaps just a few million years. Recent activity must have erased the scars of ancient meteoritic impacts. Europa's surface displays a vast network of lines crisscrossing bright, clear fields of water ice. Some of these linear "bands," or fractures, appear to extend halfway around the satellite and resemble in some ways the pressure ridges that develop in ice floes on Earth's polar oceans.

Before *Galileo*'s arrival, some researchers had theorized that Europa is completely covered by an ocean of liquid water whose top is frozen at the low temperatures that prevail so far from the Sun. In this view, the cracks in the surface are attributed to the tidal influence of Jupiter and the gravitational pulls of the other Galilean satellites, although these forces are considerably weaker than those powering Io's violent volcanic activity. However, other planetary scientists had contended

that Europa's fractured surface was instead related to some form of tectonic activity, one involving ice rather than rock. High-resolution *Galileo* observations now appear to support the former idea. Figure 11.19(d) is a *Galileo* image of this weird moon, showing what look like "icebergs"—flat chunks of ice that have been broken apart and reassembled, perhaps by the action of water currents below. Mission scientists speculate that Europa's ice may be several kilometers thick and that there may be a 100-km-deep ocean below it.

If Europa does have an ocean of liquid water below its surface ice, it opens up many interesting avenues of speculation about the possible development of life there. In the rest of the solar system, only Earth has liquid water on or near its surface, and most scientists agree that water played a key role in the appearance of life here (see Chapter 28). However, bear in mind that the existence of water does not *necessarily* imply the emergence of life. Europa is a very hostile environment compared with Earth. The surface temperature on Europa is just 130 K, and the atmospheric pressure is only a billionth the pressure on our planet. Nevertheless, the possibility, however remote, of life on Europa was an important motivating factor in the decision to extend the *Galileo* mission for two more years.

GANYMEDE AND CALLISTO

The two outermost Galilean moons are Ganymede (at 1.1 million km, or 15 planetary radii, from the center of Jupiter) and Callisto (at 1.9 million km, or 26 Jupiter radii). The density of each is only about 2000 kg/m³, suggesting that they harbor substantial amounts of ice throughout and are not just covered by thin icy or snowy surfaces. Ganymede, shown in Figure 11.20, is the largest moon in the solar system, exceeding not only Earth's Moon but also the planets Mercury

(a)   (b) *Videos*   R I V U X G

Figure 11.20   Jupiter's largest moon, Ganymede, is also the largest satellite in the solar system. The dark regions on the surface are the oldest and probably represent the original icy crust of the moon. The largest dark region visible in the *Voyager 2* image in (a) is called Galileo Regio. It spans some 320 km. The lighter, younger regions are the result of flooding and freezing that occurred within a billion years or so of Ganymede's formation. The light-colored spots are recent impact craters. The resolution of the detailed image in (b) is about 3 km.

and Pluto in size. It has many impact craters on its surface and patterns of dark and light markings that are reminiscent of the highlands and maria on Earth's own Moon. In fact, Ganymede's history has many parallels with that of the Moon (with water ice replacing lunar rock). The large, dark region clearly visible in Figure 11.20 is called Galileo Regio.

As with the inner planets, we can estimate ages on Ganymede by counting craters. We learn that the darker regions, like Galileo Regio, are the oldest parts of Ganymede's surface. These regions are the original icy surface of the moon, just as the ancient highlands on our own Moon are its original crust. The surface darkens with age as micrometeorite dust slowly covers it. The light-colored parts of Ganymede are much less heavily cratered, so they must be younger. They are Ganymede's "maria" and probably formed in a manner similar to the way that maria on the Moon were created. Intense meteoritic bombardment caused liquid water—Ganymede's counterpart to our own Moon's molten lava—to upwell from the interior and flood the impacting regions before solidifying.

Not all of Ganymede's surface features follow the lunar analogy. Ganymede has a system of grooves and ridges (shown in Figure 11.21) that may have resulted from crustal tectonic motion, much as Earth's surface undergoes mountain building and faulting at plate boundaries. Ganymede's large size indicates that its original radioactivity probably helped heat and differentiate its interior, after which the moon cooled and the crust cracked. Ganymede seems to have had some early plate tectonic activity, but the process stopped about 3 billion years ago when the cooling crust became too thick. The *Galileo* data suggest that the surface of Ganymede may be older than was previously thought. With the improved resolution of that spacecraft's images (Figure 11.21c), some regions believed to have been smooth, and hence young, are now seen to be heavily splintered by fractures and thus probably very old. *Galileo* also detected a weak magnetosphere surrounding Ganymede, making it the first moon in the solar system on which a magnetic field has been observed.

Callisto, shown in Figure 11.22, is in many ways similar in appearance to Ganymede, although it has more craters and fewer fault lines. Its most obvious feature is a huge series of concentric ridges surrounding each of two large basins. The larger of the two, on Callisto's Jupiter-facing side, is named Valhalla and measures some 3000 km across. It is clearly visible in Figure 11.22. The ridges resemble the ripples made as a stone hits water, but on Callisto they probably

(a)

(b)

(c)

Figure 11.21 (a) "Grooved terrain" on Ganymede may have been caused by a process similar to plate tectonics on Earth. (b) The resolution of this detailed *Voyager 2* image is about 3 km. (c) The image on the left was taken by *Voyager 2* in 1979, that on the right by *Galileo* in 1996. The 55-by-35-km area shown here reveals a multitude of ever-smaller ridges, valleys, and craters, right down to the resolution limit of *Galileo's* camera (about 100 m, the length of a football field).

Figure 11.22 Callisto, the outermost Galilean moon of Jupiter, is similar to Ganymede in composition but is more heavily cratered. The large series of concentric ridges visible on the left of the image is known as Valhalla. Extending nearly 1500 km from the basin center, they formed when "ripples" from a large meteoritic impact froze before they could disperse completely. The resolution in this *Voyager 2* image is around 10 km.

resulted from a cataclysmic impact with an asteroid or comet. The upthrust ice was partially melted, but it resolidified quickly, before the ripples had a chance to subside. Today, both the ridges and the rest of the crust are frigid ice and show no obvious signs of geological activity (such as the grooved terrain on Ganymede). Apparently, Callisto froze before plate tectonic or other activity could start. The density of impact craters on the Valhalla basin indicates that it formed long ago, perhaps 4 billion years in the past.

11.6 Jupiter's Ring

Yet another remarkable finding of the 1979 *Voyager* missions was the discovery of a faint ring of matter encircling Jupiter in the plane of the planet's equator (see Figure 11.23). This ring lies roughly 50,000 km above the top cloud layer of the planet, inside the orbit of the innermost moon. A thin sheet of material may extend all the way down to Jupiter's cloud tops, but most of the ring is confined within a region only a few thousand kilometers across. The outer edge of the ring is quite sharply defined. In the direction perpendicular to the equatorial plane, the ring is only a few tens of kilometers thick. The small, dark particles that make up the ring may be fragments chipped off by meteorite impacts from two small moons—Metis and Adastrea, discovered by *Voyager*—that lie very close to the ring itself.

Despite differences in appearance and structure, Jupiter's ring can perhaps be best understood by studying the most famous planetary ring system—that of Saturn—so we will postpone further discussion of ring properties until the next chapter.

Figure 11.23 Jupiter's faint ring, as photographed (nearly edge-on) by *Voyager 2*. The ring, made up of dark fragments of rock and dust, possibly chipped off the innermost moons by meteorites, was unknown before the two *Voyager* spacecraft arrived at the planet. It lies in Jupiter's equatorial plane, only 50,000 km above the cloud tops.

Chapter Review www

SUMMARY

Jupiter is the largest planet in the solar system. Its mass is more than twice the mass of all the other planets combined, although it is still about 1/1000 the mass of the Sun. It is composed primarily of hydrogen and helium.

Jupiter rotates very rapidly, producing a pronounced equatorial flattening. The amount of flattening allows astronomers to infer the presence of a large rocky core in its interior. Jupiter displays **differential rotation** (p. 245)—because the planet has no solid surface, the rotation rate varies from place to place in the atmosphere. Measurements of radio emission from Jupiter's magnetosphere provide a measure of the planet's interior rotation rate.

Jupiter's atmosphere consists of three main cloud layers. The colors we see are the result of chemical reactions, fueled by the planet's interior heat, solar ultraviolet radiation, auroral phenomena, and lightning, at varying depths below the cloud tops, seen through "holes" in the overlying clouds. The cloud layers on Jupiter, as on all the jovian worlds, are arranged into bands of bright **zones** (p. 246) and darker **belts** (p. 246) crossing the planets parallel to the equator. The bands are the result of convection in the planet's interior and the planet's rapid rotation. The lighter zones are the tops of up-welling, warm currents, and the darker bands are cooler regions where gas is sinking. Underlying them is a stable pattern of eastward or westward wind flow called the **zonal flow** (p. 246). The wind direction alternates as we move north or south away from the equator.

The main weather pattern on Jupiter is the **Great Red Spot** (p. 246), an Earth-sized hurricane that has been raging for at least three centuries. Other, smaller, weather systems—the **white** (p. 250) and **brown ovals** (p. 250)—are also observed. They can persist for decades.

Jupiter's atmosphere becomes hotter and denser with depth, eventually becoming liquid. Interior pressures are so high that the hydrogen is "metallic" in nature near the center. The planet has a large "terrestrial" core 10–20 times the mass of Earth. Jupiter radiates about twice as much energy into space as it receives from the Sun. The source of this energy is most likely heat left over from the planet's formation 4.6 billion years ago.

The magnetosphere of Jupiter is about a million times more voluminous than Earth's magnetosphere, and the planet has a long magnetic "tail" extending away from the Sun to at least the distance of Saturn's orbit. Energetic particles spiral around magnetic field lines, accelerated by Jupiter's rotating magnetic field, producing intense radio radiation.

Jupiter and its system of moons resemble a small solar system. Sixteen moons have been discovered so far. The outermost eight moons resemble asteroids and have retrograde orbits, suggesting that they may have been captured by Jupiter's gravity long after the planets and largest moons formed. Jupiter's four major moons are called the **Galilean satellites** (p. 244), after their discoverer, Galileo Galilei. Their densities decrease with increasing distance from the planet.

The innermost Galilean moon, Io, has active volcanoes powered by the constant flexing of the moon by Jupiter's tidal forces. As Io orbits Jupiter the moon "wobbles" because of the gravitational pull of Europa. The ever-changing distortion of its interior energizes Io, and geyserlike volcanoes keep its surface smooth with constant eruptions. The material ejected by these volcanoes forms the Io plasma torus in Jupiter's inner magnetosphere.

Europa has a cracked, icy surface that may conceal an ocean of liquid water. Its fields of ice are nearly devoid of craters but have extensive fractures, which are most likely due to the tidal influence of Jupiter and the gravitational effects of the other Galilean satellites. Ganymede and Callisto have ancient, heavily cratered surfaces. Ganymede, the largest moon in the solar system, shows some evidence of past geological activity, but it is now unmoving rock and ice. Callisto apparently froze before tectonic activity could start there.

Jupiter has a faint, dark ring extending down to the planet's cloud tops. It was discovered in 1979 by *Voyager 1.*

SELF TEST: TRUE OR FALSE?

___ **1.** Jupiter has over 300 times the mass of Earth, and twice the mass of all the other planets combined.

___ **2.** The solid surface of Jupiter lies just below the cloud layers visible from Earth.

___ **3.** There is no evidence to suggest that Jupiter has a rocky core.

___ **4.** Jupiter has only one large storm system.

___ **5.** In general, a storm system in Jupiter's atmosphere is much longer lived than storms in Earth's atmosphere.

___ **6.** The element helium plays an important role in producing the colors in Jupiter's atmosphere.

___ **7.** The magnetosphere of Jupiter is similar in intensity to Earth's magnetosphere.

___ **8.** Most of Jupiter's moons rotate synchronously with their orbits.

___ **9.** The densities of the Galilean moons increase with increasing distance from Jupiter.

___ **10.** Io has a noticeable lack of impact craters on its surface.

___ **11.** The surface of Europa is completely covered by water ice.

___ **12.** Ganymede shows evidence of ancient plate tectonics.

___ **13.** Jupiter's ring is made up of icy particles a few meters across.

___ **14.** Most of the small moons of Jupiter have diameters of several hundred kilometers.

___ **15.** The thickness of Jupiter's cloud layer is less than 1 percent of the planet's radius.

SELF TEST: FILL IN THE BLANK

1. The _____ of Jupiter indicates that its overall composition differs greatly from that of the terrestrial planets.

2. The main constituents of Jupiter are _____ and _____.

3. Jupiter's rapid _____ produces a significant equatorial bulge.

4. Jupiter's Great Red Spot has similarities to _____ on Earth.

5. The diameter of the Great Red Spot is about _____ that of Earth .

6. Jupiter's clouds consist of a series of bright _____ and dark _____.

7. Jupiter emits about _____ more radiation than it receives from the Sun.

8. Although often referred to as a gaseous planet, Jupiter is mostly _____ in its interior.

9. Jupiter's magnetic field is generated by its rapid rotation and the element _____, which becomes metallic at the high pressure found in the planet's deep interior.

10. Because of Jupiter's strong tidal field, most of Jupiter's satellites rotate _____ with their orbits around the planet.

11. The Galilean moons make up _____ of the _____ moons of Jupiter. (Give the numbers.)

12. The Galilean moon _____ is larger than the planet Mercury.

13. In contrast with the inner Galilean moons, the outer two Galilean moons have compositions that include significant amounts of _____.

14. Io is the only moon in the solar system with active _____.

15. Scientists speculate that _____ may have liquid water below its frozen surface.

REVIEW AND DISCUSSION

1. In what sense does our solar system consist of only two important objects?

2. What is differential rotation, and how is it observed on Jupiter?

3. Describe some of the ways in which the *Voyager* mission changed our perception of Jupiter.

4. What is the Great Red Spot? What is known about the source of its energy?

5. What is the cause of the colors in Jupiter's atmosphere?

6. Why has Jupiter retained most of its original atmosphere?

7. Explain the theory that accounts for Jupiter's internal heat source.

8. What is Jupiter thought to be like beneath its clouds? Why do we think this?

9. What is responsible for Jupiter's enormous magnetic field?

10. In what sense are Jupiter and its moons like a miniature solar system?

11. How does the density of the Galilean moons vary with increasing distance from Jupiter? Is there a trend to this variation? If so, why?

12. What is the source of Io's volcanic activity?

13. How does the amount of cratering vary among the Galilean moons? Does it depend on their location? If so, why?

14. Why is there speculation that the Galilean moon Europa might be an abode for life?

15. Water is relatively uncommon among the terrestrial planets. Is it common among the moons of Jupiter?

PROBLEMS

1. How does the force of gravity at Jupiter's cloudtops compare with the force of gravity at Earth's surface?

2. How long does it take Jupiter's equatorial winds to circle the planet, relative to the interior?

3. Given Jupiter's current atmospheric temperature, what is the smallest possible mass the planet could have and still retain its hydrogen atmosphere? (See *More Precisely 8-1*.)

4. If Jupiter had been just massive enough to fuse hydrogen (see *Interlude 11-1*), calculate what the planet's gravitational force on Earth would have been at closest approach, relative to the gravitational pull of the Sun. Also estimate what the magnitude of the planet's tidal effect on our planet would have been, again relative to that of the Sun.

5. Calculate the ratio of Jupiter's mass to the total mass of the Galilean moons. Compare this with the ratio of Earth's mass to that of the Moon.

6. Io orbits Jupiter at a distance of six planetary radii in 42 hours. At what distance would a satellite orbit Jupiter in the time taken for Jupiter to rotate exactly once (10 hours, say) so that the satellite would appear "stationary" above the planet? (Use Kepler's third law; see Chapter 2.)

7. Estimate the strength of Jupiter's gravitational tidal force on Io, relative to the moon's own surface gravity.

8. Calculate the synodic period of Europa, as seen from Io, and the strength of Europa's gravitational pull on Io at closest approach, relative to Jupiter's gravitational attraction.

9. What is the speed needed to escape from Io?

10. Compare the apparent sizes of the Galilean moons, as seen from Jupiter's cloudtops, with the angular diameter of the Sun at Jupiter's distance. Would you expect ever to see a total solar eclipse from Jupiter's cloudtops?

PROJECTS

1. Are there any stars in the night sky that look as bright as Jupiter? What other difference do you notice between Jupiter and the stars?

2. Use binoculars to peer at Jupiter. Be sure to hold them steady (try propping your arms up on the hood of a car, or sitting down and bracing them against your knees). Can you see any of Jupiter's four largest moons? If you come back the following evening, the moons' relative positions will have changed. Have some changed more than others?

3. Through a telescope, you should be able to see the red-and-tan cloud bands of Jupiter, and you can clearly see some moons. Do the moons orbit in the equatorial plane? Before observing, look up the positions of the Galilean moons in a current magazine such as *Astronomy* or *Sky & Telescope*. Identify each of the moons. Watch Io over a period of at least an hour or more. Can you see its motion? Do the same for Europa.

12 SATURN

Spectacular Rings and Mysterious Moons *www*

LEARNING GOALS

Studying this chapter will enable you to:

1 Summarize the orbital and physical properties of Saturn, and compare them with those of Jupiter.

2 Describe the composition and structure of Saturn's atmosphere and interior.

3 Explain why Saturn's internal heat source and magnetosphere differ from those of Jupiter.

4 Describe the structure and composition of Saturn's rings.

5 Define the Roche limit and explain its relevance to the origin of Saturn's rings.

6 Summarize the general characteristics of Titan and discuss the chemical processes in its atmosphere.

7 Discuss some of the orbital and geological properties of Saturn's smaller moons.

(Opposite page, background) Through the cameras onboard the *Hubble Space Telescope*, the planet Saturn appears much as our naked eyes would see it if it were only six times as far away as the Moon. Resolution here is 670 km, good enough to see clearly the band structure on the ball of the planet, as well as several gaps in the rings. True-color images like this one are made by combining separate images taken in red, green, and blue light, as shown in the inserts.
(Inset A) Saturn as seen through a red filter, centered at a wavelength of 718 nm.
(Inset B) Saturn as seen through a green filter, centered at a wavelength of 547 nm.
(Inset C) Saturn as seen through a blue filter, centered at a wavelength of 439 nm.

Saturn is one of the most beautiful and enchanting of all astronomical objects. Its rings are a breathtaking sight when viewed through even a small telescope, and they are probably the planet's best-known feature. Aside from its famous rings, however, Saturn presents us with another good example of a giant gaseous planet. Saturn is in many ways similar to its larger neighbor, Jupiter, in terms of composition, size, and structure. Yet when we study the two planets in detail, we find that there are important differences as well. A comparison between Saturn and Jupiter provides us with valuable insight into the structure and evolution of all the jovian worlds.

12.1 Saturn in Bulk

PHYSICAL PROPERTIES

1 Saturn was the outermost planet known to ancient astronomers. Named after the father of Jupiter in Greek and Roman mythology, Saturn orbits the Sun at almost twice the distance of Jupiter. The planet's sidereal orbital period of 29.4 Earth years was the longest natural unit of time known to the ancient world. At opposition, when Saturn is at its brightest, it can lie within 8 A.U. of Earth. However, its great distance from the Sun still makes it considerably fainter than either Jupiter or Mars. Saturn ranks behind Jupiter, the inner planets, and several of the brightest stars in the sky in terms of apparent brightness.

The Saturn Data box on p. 271 presents orbital and physical data for the planet. Less than one-third the mass of Jupiter, Saturn is still an enormous body, at least by terrestrial standards. As with Jupiter, Saturn's many moons allowed an accurate determination of the planet's mass long before the arrival of the *Pioneer* and *Voyager* missions. Saturn's mass is 5.7×10^{26} kg, or 95 times the mass of Earth. From Saturn's distance and angular size, the planet's radius—and hence the average density—quickly follow. Saturn's equatorial radius is 60,000 km, or 9.5 Earth radii. The average density is 700 kg/m³—less than the density of water (which is 1000 kg/m³). Here we have a planet that would float in the ocean—if only Earth had one big enough! Saturn's low average density indicates that, like Jupiter, it is composed primarily of hydrogen and helium. Saturn's lower mass, however, results in lower interior pressure, so these gases are less compressed than in Jupiter's case.

ROTATION RATE

Saturn, like Jupiter, rotates very rapidly and differentially. The rotation period of the interior (as measured from magnetospheric outbursts, which trace the rotation of the planet's core) and at high planetary latitudes (determined by tracking weather features observed in Saturn's atmosphere) is 10^h40^m. The rotation period at the equator is 10^h14^m, or about 26 minutes shorter. Because of Saturn's lower density, this rapid rotation makes Saturn even more flattened than Jupiter. Saturn's polar radius is only 54,000 km, about 10 percent less than the equatorial radius. Careful calculations show that this degree of flattening is less than would be expected for a planet composed of hydrogen and

helium alone. Astronomers believe that Saturn also has a rocky core, perhaps twice the mass of Jupiter's.

RINGS

Saturn's best-known feature is its spectacular *ring system*. Because the rings lie in the equatorial plane, their appearance (as seen from Earth) changes in a seasonal manner, as shown in Figure 12.1. Saturn's rotation axis is significantly tilted with respect to the planet's orbit plane—the axial tilt is 27°, similar to that of both Earth and Mars. Consequently, as Saturn orbits the Sun the angles at which the rings are illuminated and at which we view them vary. When the planet's north or south pole is tipped toward the Sun, during Saturn's summer or winter, the highly reflective rings are at their brightest. During Saturn's spring and fall, the rings are close to being edge-on, both to the Sun and to us, so they seem to disappear altogether. One important deduction that we can make from this simple observation is that the rings are very thin. In fact, we now know that their thickness is less than a few hundred meters, even though they are over 200,000 km in diameter.

12.2 Saturn's Atmosphere

2 Saturn is much less colorful than Jupiter. Figure 12.2 shows yellowish and tan cloud belts that parallel the equator, but these regions display less atmospheric structure than do the belts on Jupiter. No obvious large "spots" or "ovals" adorn Saturn's cloud decks. Bands and storms do exist, but the color changes that distinguish them on Jupiter are largely absent on Saturn.

COMPOSITION AND COLORATION

Astronomers first observed methane in the spectrum of sunlight reflected from Saturn in the 1930s, about the same time that it was discovered on Jupiter. However, it was not until the early 1960s, when more sensitive observations became possible, that ammonia was finally detected. In Saturn's cold upper atmosphere, most ammonia is in the solid or liquid form, with relatively little of it present as a gas to absorb sunlight and create spectral lines. Astronomers finally made the first accurate determinations of the hydrogen and helium content in the late 1960s. These Earth-based measurements were later confirmed with the arrival of the *Pioneer* and *Voyager* spacecraft in the 1970s.

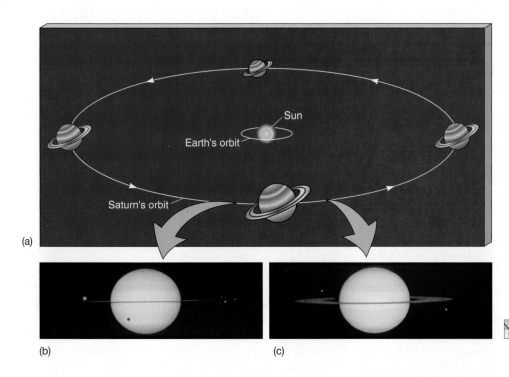

(a)

(b)   (c)

Figure 12.1   (a) Over time, Saturn's rings change their appearance to terrestrial observers as the tilted ring plane orbits the Sun. At some times during Saturn's 29.5-year orbital period the rings seem to disappear altogether as Earth passes through their plane and we view them edge-on. The two images are separated by a few months, having been taken (in true color) by the *Hubble Space Telescope* in the summer and fall of 1995. (b) The ring plane is precisely edge-on as viewed from Earth; several moons are visible, including Titan's shadow (black dot). (c) The rings have begun to tilt.

Saturn's atmosphere consists of molecular hydrogen (92.4 percent), helium (7.4 percent), methane (0.2 percent), and ammonia (0.02 percent). As on Jupiter, hydrogen and helium dominate—these most abundant elements never escaped from Saturn's atmosphere because of the planet's large mass and low temperature (see *More Precisely 8-1*). However,

the fraction of helium on Saturn is far less than is observed on Jupiter (where, as we saw, helium accounts for nearly 14 percent of the atmosphere) or in the Sun. It is extremely unlikely that the processes that created the outer planets preferentially stripped Saturn of nearly half its helium or that the missing helium somehow escaped from the planet while

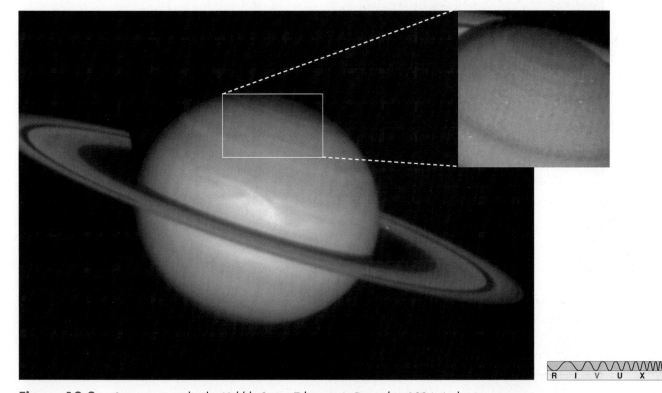

Figure 12.2   Saturn as seen by the *Hubble Space Telescope* in December 1994. At the time, a rare storm was visible near the planet's equator. The bland colors are approximately true—that is, as the human eye sees things. The insert shows the northern polar region at an earlier time.

the lighter hydrogen remained behind. Instead, astronomers believe that at some time in Saturn's past the heavier helium began to sink toward the center of the planet, reducing its abundance in the outer layers and leaving them relatively hydrogen-rich. We will return to the reasons for this differentiation and its consequences in a moment.

Figure 12.3 illustrates Saturn's atmospheric structure (compare with the corresponding diagram for Jupiter, Figure 11.6). In many respects, Saturn's atmosphere is quite similar to Jupiter's, except that the temperature is a little lower because of its greater distance from the Sun and because its clouds are somewhat thicker. Since Saturn, like Jupiter, lacks a solid surface, we take the top of the troposphere as our reference level and set it to 0 km. The top of the visible clouds lies about 50 km below this level. As on Jupiter, the clouds are arranged in three distinct layers, composed (in order of increasing depth) of ammonia, ammonium hydrosulfide, and water ice. Above the clouds lies a layer of haze formed by the action of sunlight on Saturn's upper atmosphere.

The total thickness of the three cloud layers in Saturn's atmosphere is roughly 200 km, compared with about 80 km on Jupiter, and each layer is itself somewhat thicker than its counterpart on Jupiter. The reason for this difference is Saturn's weaker gravity. At the haze level, Jupiter's gravitational field is nearly two and a half times stronger than Saturn's, so Jupiter's atmosphere is pulled much more powerfully toward the center of the planet. Thus Jupiter's atmosphere is compressed more than Saturn's, and the clouds are squeezed more closely together. The colors of Saturn's cloud layers, as well as the planet's overall butterscotch hue, are due to the same basic cloud chemistry as on Jupiter. However, because Saturn's clouds are thicker, there are few holes and gaps in the top layer, so we rarely glimpse the more colorful levels below. Instead, we see only different levels in the topmost layer, which accounts for Saturn's rather uniform appearance.

WEATHER

Saturn has atmospheric wind patterns that are in many ways reminiscent of those on Jupiter. There is an overall east–west zonal flow, which is apparently quite stable. Computer-enhanced images of the planet that bring out more cloud contrast (see Figure 12.4) clearly show the existence of bands, oval storm systems, and turbulent flow patterns looking very much like those seen on Jupiter. Scientists believe that

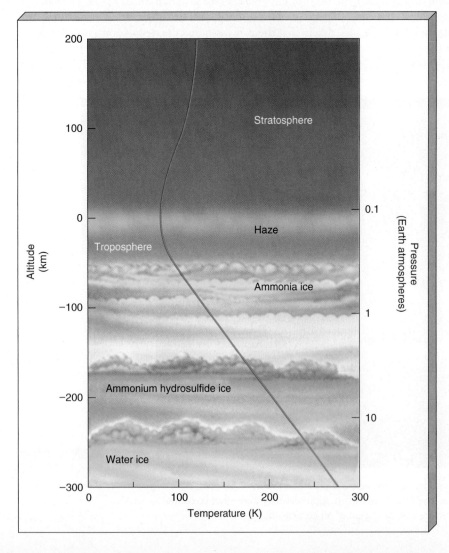

Figure 12.3 The vertical structure of Saturn's atmosphere. As with Jupiter, there are several cloud layers, but Saturn's weaker gravity results in thicker clouds and a more uniform appearance.

Figure 12.4 We see more structure in Saturn's cloud cover when computer processing and artificial color are used to enhance the image contrast, as in these *Voyager 1* images of the entire gas ball and of a smaller, magnified piece of it.

Saturn's bands and storms have essentially the same cause as does Jupiter's weather. Ultimately, the large-scale flows and small-scale storm systems are powered by convective motion in Saturn's interior and the planet's rapid rotation.

The zonal flow on Saturn is considerably faster than on Jupiter and shows fewer east–west alternations, as can be seen from Figure 12.5 (compare with Figure 11.5). The equatorial eastward jet stream, which reaches a speed of about 400 km/h

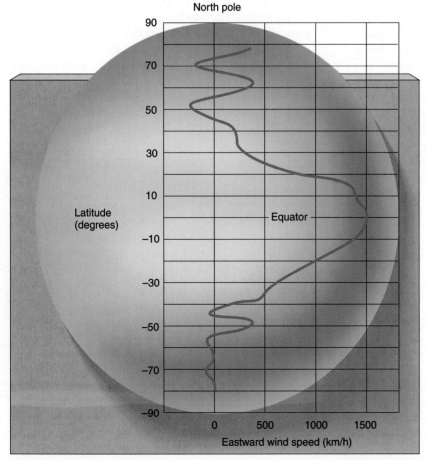

Figure 12.5 Winds on Saturn reach speeds even greater than those on Jupiter. As on Jupiter, the visible bands appear to be associated with variations in wind speed.

Figure 12.6 Circulating and evolving cloud systems on Saturn, imaged by the *Hubble Space Telescope.*

on Jupiter, moves at a brisk 1500 km/h on Saturn, and extends to much higher latitudes. Not until latitudes 40° north and south of the equator are the first westward flows found. Latitude 40° north also marks the strongest bands on Saturn and the most obvious ovals and turbulent eddies. Astronomers still do not fully understand the reasons for the differences between Jupiter's and Saturn's flow patterns.

In September 1990, amateur astronomers detected a large white spot in Saturn's southern hemisphere, just below the equator. In November of that year, when the *Hubble Space Telescope* imaged the phenomenon in more detail, the spot had developed into a band of clouds completely encircling the planet's equator. Some of these images are shown in Figure 12.6. Astronomers believe that the white coloration arose from crystals of ammonia ice formed when an upwelling plume of warm gas penetrated the cool upper cloud layers. Because the crystals were freshly formed, they had not yet been affected by the chemical reactions that color the planet's other clouds.

Such spots are relatively rare on Saturn. The previous one visible from Earth appeared in 1933, but it was much smaller than the 1990 system and much shorter lived, lasting for only a few weeks. Another storm system is visible in Figure 12.2, taken in 1994. The turbulent flow patterns seen around the 1990 white spot have many similarities to the flow around Jupiter's Great Red Spot. Scientists speculate that these white spots represent long-lived weather systems on Saturn and hope that routine observations of such temporary atmospheric phenomena on the outer worlds will enable them to gain greater insight into the dynamics of planetary atmospheres.

12.3 Saturn's Interior and Magnetosphere

INTERIOR STRUCTURE AND INTERNAL HEATING

3 Figure 12.7 depicts Saturn's internal structure (compare with Figure 11.10 for Jupiter). This picture has been pieced together by planetary scientists using the same tools—*Voyager* observations and theoretical modeling—that they used to infer Jupiter's inner workings. Saturn has the same basic internal

parts as Jupiter, but their relative proportions are somewhat different: Saturn's metallic hydrogen layer is thinner, and its core is larger. Because of its lower mass, Saturn has a less extreme core temperature, density, and pressure than Jupiter. The central pressure is around a tenth of Jupiter's—not too different from the pressure at the center of Earth.

Infrared measurements indicate that Saturn's surface (that is, cloud-top) temperature is 97 K, substantially higher than the temperature at which Saturn would reradiate all the energy it receives from the Sun. In fact, Saturn radiates almost three times more energy than it absorbs. Thus Saturn, like Jupiter, has an internal energy source. ∞ (Sec. 11.3) But the explanation behind Jupiter's excess energy—that the planet has a large reservoir of heat left over from its formation—doesn't work for Saturn. Saturn is smaller than Jupiter and so must have cooled more rapidly—rapidly enough that its original supply of energy was long ago used up. What, then, is happening inside Saturn to produce this extra heat?

The explanation for this strange phenomenon also explains the mystery of Saturn's apparent helium deficit. At the temperatures and high pressures found in Jupiter's interior, liquid helium *dissolves* in liquid hydrogen. In Saturn, where the internal temperature is lower, the helium doesn't dissolve so easily and tends to form droplets instead. The phenomenon

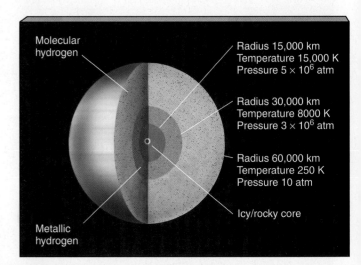

Molecular hydrogen

Radius 15,000 km
Temperature 15,000 K
Pressure 5 × 10⁶ atm

Radius 30,000 km
Temperature 8000 K
Pressure 3 × 10⁶ atm

Radius 60,000 km
Temperature 250 K
Pressure 10 atm

Icy/rocky core

Metallic hydrogen

Figure 12.7 Saturn's internal structure, as deduced from *Voyager 1* and *2* observations and computer modeling.

is familiar to cooks, who know that it is generally much easier to dissolve ingredients in hot liquids than in cold ones. Saturn probably started out with a fairly uniform solution of helium dissolved in hydrogen, but the helium tended to condense out of the surrounding hydrogen, much as water vapor condenses out of Earth's atmosphere to form a mist. The amount of helium condensation was greatest in the planet's cool outer layers, where the mist turned to rain about 2 billion years ago. A light shower of liquid helium has been falling through Saturn's interior ever since. This **helium precipitation** is responsible for depleting the outer layers of their helium content.

So we can account for the unusually low abundance of helium in Saturn's atmosphere—much of it has rained down to lower levels. But what about the excess heating? The answer is simple: as the helium sinks toward the center the planet's gravitational field compresses it and heats it up. The gravitational energy thus released is the source of Saturn's internal heat. In the distant future the helium rain will stop, and Saturn will cool until its outermost layers radiate only as much energy as they receive from the Sun. When that happens, the temperature at Saturn's cloud tops will be 74 K. As Jupiter cools, it too may someday experience helium precipitation in its interior, which will cause its surface temperature to rise once again.

MAGNETOSPHERIC ACTIVITY

Saturn's electrically conducting interior and rapid rotation produce a strong magnetic field and an extensive magnetosphere. Probably because of the considerably smaller mass of Saturn's metallic hydrogen zone, the planet's basic magnetic field strength is only about 1/20 that of Jupiter, or about 1000 times greater than that of Earth. The magnetic field at Saturn's cloud tops (roughly 10 Earth radii from the planet's center) is approximately the same as at Earth's surface. *Voyager* measurements indicate that, unlike Jupiter's and Earth's magnetic axes, which are slightly tilted, Saturn's magnetic field is not inclined with respect to its rotation axis. Saturn's magnetic field, like Jupiter's, is oriented opposite that of Earth—that is, an Earth compass needle would point toward Saturn's south pole rather than its north.

Saturn's magnetosphere extends about 1 million km toward the Sun and is large enough to contain the planet's ring system and the innermost 16 small moons. Saturn's largest moon, Titan, orbits about 1.2 million km from the planet, so it is sometimes found just inside the outer magnetosphere and sometimes just outside, depending on the intensity of the solar wind (which tends to push the sunward side of the magnetosphere closer to the planet). Because no major moons lie deep within Saturn's magnetosphere, the details of its structure are different from those of Jupiter's magnetosphere. For example, there is no equivalent of the Io plasma torus. ∞ (Sec. 11.5) Like Jupiter, Saturn emits radio waves, but as luck would have it, they are reflected from Earth's ionosphere (they lie in the AM band) and were not detected until the *Voyager* craft approached the planet.

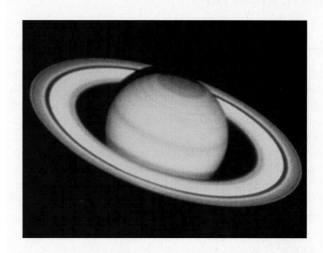

Orbital semi-major axis	9.54 A.U. 1427 million km
Orbital eccentricity	0.054
Perihelion	9.02 A.U. 1349 million km
Aphelion	10.05 A.U. 1504 million km
Mean orbital speed	9.65 km/s
Sidereal orbital period	29.42 tropical years
Synodic orbital period	378.09 solar days
Orbital inclination to the ecliptic	2.49°
Greatest angular diameter, as seen from Earth	21″
Mass	5.68×10^{26} kg 95.16 (Earth = 1)
Equatorial radius	60,268 km 9.45 (Earth = 1)
Mean density	687 kg/m³ 0.125 (Earth = 1)
Surface gravity	10.4 m/s² 1.07 (Earth = 1)
Escape speed	35.5 km/s
Sidereal rotation period	0.44 solar days
Axial tilt	26.73°
Surface magnetic field	0.67 (Earth = 1)
Magnetic axis tilt relative to rotation axis	0.8°
Surface temperature	97 K (at cloudtops)
Number of moons	18

12.4 Saturn's Spectacular Ring System

THE VIEW FROM EARTH

4 The most obvious aspect of Saturn's appearance is, of course, its **planetary ring system**. Astronomers now know that all the jovian planets have rings, but Saturn's are by far the brightest, the most extensive, and the most beautiful. Galileo saw them first in 1610, but he did not recognize what he saw as a planet with a ring. At the resolution of his small telescope, the rings looked like bumps on the planet, or perhaps components of a triple system of some sort. In 1659 the Dutch astronomer Christian Huygens realized what the "bump" was—a thin, flat ring, completely encircling the planet.

In 1675 the French–Italian astronomer Giovanni Domenico Cassini discovered the first ring feature, a dark band about two-thirds of the way out from the inner edge. From Earth, the band looks like a gap in the ring (an observation that is not too far from the truth, although we now know that there is actually some ring material within it). This "gap" is named the **Cassini Division**, in honor of its discoverer. Careful observations from Earth show that the inner "ring" is in reality also composed of two rings. From the outside in, the three rings are known somewhat prosaically as the **A**, **B**, and **C rings**. The Cassini Division lies between A and B. The much narrower **Encke gap**, some 300 km wide, is found in the outer part of the A ring. These ring features are marked on Figure 12.8. No finer ring details are visible from our Earthly vantage point. Of the three main rings, the B ring is brightest, followed by the somewhat fainter A ring, and then

TABLE 12.1 The Rings of Saturn

| RING | INNER RADIUS | | OUTER RADIUS | | WIDTH |
	(km)	(planet radii)	(km)	(planet radii)	(km)
D	67,000	1.11	74,700	1.24	7,700
C	74,700	1.24	92,000	1.53	17,300
B	92,000	1.53	117,500	1.95	25,500
Cassini Division	117,500	1.95	122,300	2.03	4800
A	122,300	2.03	136,800	2.27	14,500
Encke gap*	133,400	2.22	133,700	2.22	300
F	140,300	2.33	140,400	2.33	100
E	180,000	3.00	480,000	8.00	300,000

The Encke gap lies within the A ring.

by the almost translucent C ring. A more complete list of ring properties appears in Table 12.1. (The D, E, and F rings listed in the table are discussed later in this section.)

WHAT ARE SATURN'S RINGS?

A fairly obvious question—and one that perplexed the best scientists and mathematicians on Earth for almost 2 centuries—is, What are Saturn's rings made of? By the middle of the nineteenth century, various dynamical and thermodynamic arguments had conclusively proved that the rings

Encke gap
A ring
Cassini Division
B ring
C ring

R I V U X G

Figure 12.8 Much fine structure, especially in the rings, appears in this image of Saturn taken while the *Voyager 2* spacecraft approached the planet. (One of Saturn's moons appears at bottom, and a second casts a black shadow on the cloud tops.) The banded structure of Saturn's atmosphere is also more evident in this photograph. Note the absence of the vivid colors that characterize Jupiter's atmospheric cloud layers.

could not be solid, liquid, or gas! What is left? In 1857 Scottish physicist James Clerk Maxwell, after showing that a solid ring would become unstable and break up, suggested that the rings are composed of a great number of small particles, all independently orbiting Saturn, like so many tiny moons. That inspired speculation was verified in 1895, when Lick Observatory astronomers measured the Doppler shift of sunlight reflected from the rings and showed that the velocities thus determined were exactly what would be expected from separate particles moving in circular orbits in accordance with Newton's law of gravity.

What sort of particles make up the rings? The fact that they reflect most (over 80 percent) of the sunlight striking them had long suggested to astronomers that they were made of ice, and infrared observations in the 1970s confirmed that water ice is indeed a prime ring constituent. Radar observations and later *Voyager* studies of scattered sunlight showed that the diameters of the particles range from fractions of a millimeter to tens of meters, with most particles being about the size (and composition) of a large snowball on Earth.

We now know that the rings are truly thin—perhaps only a few tens of meters thick in places. Stars can occasionally be seen through them, like automobile headlights penetrating a snowstorm. Why are the rings so thin? The answer seems to be that collisions between ring particles tend to keep them

all moving in circular orbits in a single plane. Any particle that tries to stray away from this orderly motion finds itself in an orbit that soon runs into other ring particles. Over long periods of time the ensuing jostling serves to keep all particles moving in circular, planar orbits. The asymmetric gravitational field of Saturn (the result of its flattened shape) sees to it that the rings lie in the planet's equatorial plane.

THE ROCHE LIMIT

5 But why a ring of particles at all? What process produced the rings in the first place? To answer these questions, consider the fate of a small moon orbiting close to a massive planet such as Saturn. The moon is held together by internal forces—its own gravity, for example. As we bring our hypothetical moon closer to the planet the tidal force on it increases. Recall from Chapter 7 that the effect of such a tidal force is to stretch the moon along the direction to the planet—that is, to create a tidal bulge. Recall also that the tidal force increases rapidly with decreasing distance from the planet. ∞ (Sec. 7.6) As the moon is brought closer to the planet it reaches a point where the tidal force tending to stretch it out becomes *greater* than the internal forces holding it together. At that point the moon is torn apart by the planet's gravity, as shown in Figure 12.9. The pieces of the satellite then pursue their

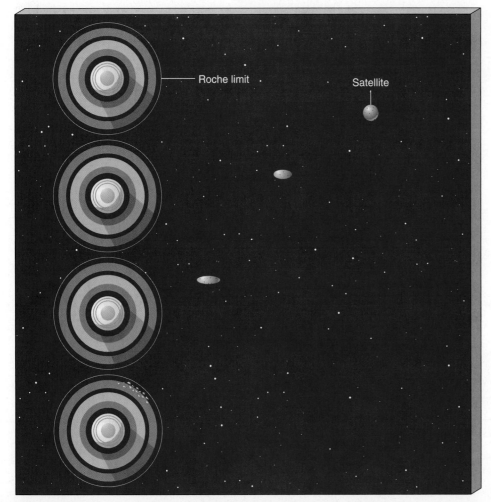

Figure 12.9 The increasing tidal field of a planet first distorts, and then destroys, a moon that strays too close. (The distortion is exaggerated in the second and third panels.)

own individual orbits around that planet, eventually spreading all the way around it in the form of a ring.

For any given planet and any given moon, this critical distance, inside of which the moon is destroyed, is known as the *tidal stability limit*, or the **Roche limit**, after the nineteenth-century French mathematician Edouard Roche, who first calculated it. If our hypothetical moon is held together by its own gravity and its average density is the same as that of the parent planet (both reasonably good approximations for Saturn's larger moons), then the Roche limit is just 2.4 times the radius of the planet. Thus, for Saturn, no moon can survive within a distance of 144,000 km of the planet's center, about 7000 km beyond the outer edge of the A ring. The rings of Saturn occupy the region inside Saturn's Roche limit.

These considerations apply equally well to the other jovian worlds. Figure 12.10 shows the location of the ring system of each jovian planet relative to the planet's Roche limit. Given the approximations in our assumptions, we can conclude that all the rings are found within the Roche limit of their parent planet. Notice that the calculation of this limit applies only to moons massive enough for their own gravity to be the dominant force binding them together.

Sufficiently small moons can survive even within the Roche limit because they are held together mostly by interatomic (electromagnetic) forces, not by gravity.

THE VIEW FROM VOYAGER

Thus it was, as *Voyagers 1* and *2* approached Saturn, that scientists on Earth were fairly confident that they understood the nature of the rings. However, there were many surprises in store. The *Voyager* missions changed forever our view of this spectacular region in our cosmic backyard, revealing the rings to be vastly more complex than astronomers had imagined.

As the *Voyager* probes approached Saturn it became obvious that the main rings are actually composed of tens of thousands of narrow **ringlets** (shown in Figure 12.11). Although *Voyager* cameras did find several new gaps in the rings, the ringlets are generally not separated from one another by empty space. Instead, the rings contain concentric regions of alternating high and low concentrations of ring particles—the ringlets are just the high-density peaks. Although the process is not fully understood, it seems that the mutual gravitational attraction of the ring particles (as well as

Figure 12.10   The rings of Jupiter, Saturn, Uranus, and Neptune. All distances are expressed in planetary radii. The red line represents the Roche limit. In all cases the rings lie within the Roche limit of the parent planet.

R I V U X G

Figure 12.11 The *Voyager 2* cameras took this close-up of the ring structure just before plunging through the tenuous outer rings of Saturn. The ringlets in the B ring, spread over several thousand kilometers, are resolved here to about 10 km. As *Voyager* approached Saturn, more and more of these tiny ringlets became noticeable in the main rings. The (false) color variations probably indicate different sizes and compositions of the particles making up the thousands of rings. Earth is superposed, to proper scale, for a size comparison.

the effects of Saturn's inner moons) enables waves of matter to form and move in the plane of the rings, rather like ripples on the surface of a pond. The wave crests typically wrap around the rings, forming tightly wound spiral patterns called *spiral density waves* that resemble grooves in a huge celestial phonograph record.

Although the ringlets are probably the result of spiral waves in the rings, the true gaps are not. The narrower gaps—about 20 of them—are most likely swept clean by the action of small moonlets embedded in them. These moonlets are larger (perhaps 10 or 20 km in diameter) than the largest true ring particles, and they simply "sweep up" ring material through collisions as they go. Despite many careful searches of *Voyager* images, only one of these moonlets has so far been found—in 1991, after 5 years of exhaustive study, NASA scientists confirmed the discovery of the eighteenth moon of Saturn (now named Pan) in the Encke gap. Astronomers have found indirect evidence for embedded moonlets, in the form of "wakes" that they leave behind them in the rings, but no other direct sightings have occurred. Despite their elusiveness, however, moonlets are still regarded as the best explanation for the small gaps.

Voyager 2 found a series of faint rings, now known collectively as the **D ring**, inside the inner edge of the C ring, stretching down almost to Saturn's cloud tops. The D ring contains relatively few particles and is so dark that it is completely invisible from Earth. Another faint ring, also a *Voyager 2* discovery, lies well outside the main ring structure. Known as the **E ring**, it appears to be associated with volcanism on the moon Enceladus.

The *Voyager 2* cameras revealed one other completely unexpected feature. A series of dark radial "spokes" formed on the B ring, moved around the planet for about one ring orbit period, and then disappeared (Figure 12.12). Careful scrutiny of these peculiar drifters showed that they were

composed of very fine (micron-sized) dust hovering a few tens of meters *above* the plane of the rings. Scientists believe that this dust was held in place by electrostatic forces generated in the ring plane, perhaps resulting from particle collisions there. The electrical fields slowly dispersed, and the spokes faded as the ring revolved. We expect that the creation and dissolution of such spokes is a regular occurrence in the Saturn ring system.

R I V U X G

Figure 12.12 Saturn's B ring showed a series of dark temporary spokes as *Voyager 2* flew by at a distance of about 4 million km. The spokes were caused by small particles suspended just above the ring plane.

Cassini Division

Figure 12.13 Close inspection by *Voyager 2* revealed that the Cassini Division (shown here as the darker color) is not completely empty. It contains a series of faint ringlets and gaps, assumed to be caused by unseen embedded satellites. The density of material in the division is very low, accounting for its dark appearance from Earth.

ORBITAL RESONANCES AND SHEPHERD SATELLITES

Voyager images show that the largest gap in the rings, the Cassini Division, is not completely empty of matter. In fact, as shown in Figure 12.13, the Division contains a series of faint ringlets and gaps (and, presumably, embedded moonlets too). The overall concentration of ring particles in the division as a whole is, however, much lower than in the A and B rings. Although its small internal gaps probably result from embedded satellites, the Division itself does not. It owes its existence to another solar system *resonance*, this time involving particles orbiting in the Division and Saturn's innermost major moon, Mimas. ∞ (Sec. 8.3)

A ring particle moving in an orbit within the Cassini Division has an orbital period exactly half that of Mimas. Particles in the Division thus complete exactly two orbits around Saturn in the time taken for Mimas to orbit once—a configuration known as a 2:1 resonance. The effect of this resonance is that particles in the Division feel a gravitational tug from Mimas at exactly the same location in their orbit every other time around. Successive tugs reinforce one another, and the initially circular trajectories of the ring particles soon get stretched out into ellipses. In their new orbits these particles collide with other particles and eventually find their way into new circular orbits at other radii. The net effect is that the number of ring particles in the Cassini Division is greatly reduced.

Particles in "nonresonant" orbits (that is, at radii where the orbital period is not simply related to the period of Mimas) also experience Mimas's gravitational pull. But the times when the force is greatest are spread uniformly around

the orbit, and the tugs cancel out. It's a little like pushing a child on a swing—pushing at the same point in the swing's motion each time produces much better results than do random shoves. Thus, Mimas (or any other moon) has a large effect on the ring at those radii where a resonance exists and little or no effect elsewhere.

We now know that resonances between ring particles and moons play a very important role in shaping the fine structure of Saturn's rings. For example, the sharp outer edge of the A ring is governed by a 3:2 resonance with Mimas (three ring orbits in two Mimas orbital periods). Most theories of planetary rings predict that the ring system should spread out with time, basically because of collisions among ring particles. Instead, the A ring's outer edge is "patrolled" by a small satellite named Atlas, held in place by the gravity of Mimas, which prevents ring particles from diffusing away.

Outside the A ring lies the strangest ring of all. The faint, narrow **F ring** (shown in Figure 12.14) was discovered by *Pioneer 11* in 1979, but its full complexity became evident only when *Voyager 1* took a closer look. Unlike the inner major rings, the F ring is narrow, less than a hundred kilometers wide. It lies just inside Saturn's Roche limit, separated from the A ring by about 3500 km. Its narrowness by itself is unusual, as is its slightly eccentric shape, but its oddest feature is that it looks as though it is made up of several separate strands braided together! This remarkable discovery sent dynamicists scrambling in search of an explanation. It now seems as though the ring's intricate structure, as well as its thinness, arise from the influence of two small moons, known as **shepherd satellites**, that orbit on either side of it (Figure 12.15).

These two small, dark satellites, each about 50 km in diameter, are called Prometheus and Pandora. They orbit

Figure 12.14 Saturn's narrow F ring appears to contain kinks and braids, making it unlike any of Saturn's other rings.

R I V U X G

Figure 12.15   The F ring's thinness, and possibly its other peculiarities too, can be explained by the effects of two shepherd satellites that orbit a few hundred kilometers inside and outside the ring. This photo shows one of the shepherding satellites, roughly 100 km in length, at right.

about 1000 km on either side of the F ring, and their gravitational influence on the F-ring particles keeps the ring tightly confined in its narrow orbit. As shown in Figure 12.16, any particle straying too far out of the F ring is gently guided back into the fold by one or the other of the moons. (The moon Atlas confines the A ring in a somewhat similar way.) However, the details of how Prometheus and Pandora produce the braids in the F ring and why the two moons are there at all, in such similar orbits, remain unclear. There is some evidence that other eccentric rings found in the gaps

in the A, B, and C rings may also result from the effects of shepherding moonlets.

THE ORIGIN OF THE RINGS

Two possible origins have been suggested for Saturn's rings. Astronomers estimate that the total mass of ring material is no more than 10^{15} tons—enough to make a satellite about 250 km in diameter. If such a satellite strayed inside Saturn's Roche limit or was destroyed (perhaps by a collision) near that radius, a ring could have resulted. An alternative view is that the rings represent material left over from Saturn's formation stage 4.6 billion years ago. In this scenario, Saturn's tidal field prevented any moon from forming inside the Roche limit, and so the material has remained a ring ever since. Which view is correct?

All the dynamic activity observed in Saturn's rings suggests to many researchers that the rings must be quite young—perhaps no more than 50 million years old, or 100 times younger than the solar system. There is just too much going on for them to have remained stable for billions of years, so they probably aren't left over from the planet's formative stages. If this is so, then either the rings must be continuously replenished, perhaps by fragments of Saturn's moons chipped off by meteorites, or they are the result of a relatively recent, possibly catastrophic, event in the planet's system—perhaps a small moon that was hit by a large comet, or even by another moon.

Astronomers normally prefer not to invoke catastrophic events to explain observed phenomena, but the more we learn of the universe, the more we realize that catastrophe probably plays an important role. For now, the details of the formation of Saturn's ring system simply aren't known.

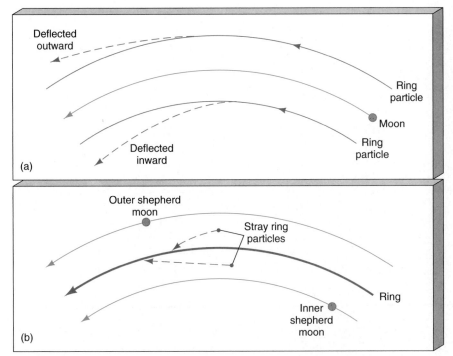

(a) Deflected outward / Deflected inward / Ring particle / Moon / Ring particle

(b) Outer shepherd moon / Stray ring particles / Ring / Inner shepherd moon

Figure 12.16   (a) Strange as it may seem, the net effect of the interactions between a moon and ring particles is that the moon tends to push those particles away from it. (b) The F ring shepherd satellites operate by forcing errant F ring particles back into the main ring. Each moon operates as in part (a) so that the ring is confined between the two moons. As a consequence of Newton's third law of motion, the satellites themselves slowly drift away from the ring.

12.5 The Moons of Saturn

GENERAL FEATURES

Saturn has the most extensive, and in many ways the most complex, system of natural satellites of all the planets. Saturn's 18 named moons are listed in Table 12.2. In 1995, researchers using the *Hubble Space Telescope* announced the discovery of two more moons (see *Interlude 12-1*). However, these observations now seem to have been in error, and the official moon count remains as listed in the table. The reflected light from Saturn's moons suggests that most are covered with snow and ice. Many of them are probably made almost entirely of water ice. Even so, they are a curious and varied lot.

The satellites fall into three fairly natural groups. First, there are the "small" moons—irregularly shaped chunks of ice, all less than 300 km across—that exhibit a bewildering variety of complex and fascinating motion. Second, there are six "medium-sized" moons—spherical bodies with diameters ranging from about 400 to 1500 km—that offer clues to the past and present state of the environment of Saturn while presenting many puzzles in their own appearance and history. Finally, there is Saturn's single "large" moon—Titan—which, at 5150 km in diameter, is the second-largest satellite in the solar system (Jupiter's Ganymede is a little bigger). It has an atmosphere denser than Earth's and (some scientists think) surface conditions possibly conducive to life. Notice, incidentally, that Jupiter has no "medium" moons, as just defined. The Galilean satellites are large, like Titan, and all of Jupiter's other satellites are small—no more than 200 km in diameter.

TITAN

6 Perhaps the most intriguing of all Saturn's moons, Titan was discovered by Christian Huygens in 1655. Even through a large Earth-based telescope, this moon is visible only as a barely resolved reddish disk. However, long before the *Voyager* missions, astronomers already knew (from spectroscopic observations) that the moon's reddish coloration is caused by something quite special—an atmosphere. So anxious were mission planners to obtain a closer look that they programmed *Voyager 1* to pass very close to Titan, even though that meant the spacecraft could not then use Saturn's gravity to continue on to Uranus and Neptune. (Instead, *Voyager 1* left the Saturn system on a path taking the craft out of the solar system well above the ecliptic plane.)

Scientists believe that Titan's internal composition and structure must have similarities to those of Ganymede and Callisto, because these three moons have quite similar masses and radii, and hence average densities (Titan's density is 1900 kg/m³). ∞ (Sec.11.5) Titan probably contains a rocky core surrounded by a thick mantle of water ice. However, in view of *Galileo*'s discovery of significant differences between

TABLE 12.2 The Moons of Saturn

NAME	DISTANCE FROM SATURN (km)	(planet radii)	ORBIT PERIOD (days)	SIZE (longest diameter, km)	MASS (Earth Moon masses)	DENSITY (kg/m³)	(g/cm³)
Pan	133,600	2.22	0.57	20	4×10^{-8}		
Atlas	138,000	2.28	0.60	37			
Prometheus	139,000	2.31	0.61	148	1.9×10^{-6}		
Pandora	142,000	2.35	0.63	110	1.8×10^{-6}		
Epimetheus	151,000	2.51	0.69	138	7.5×10^{-6}		
Janus	151,000	2.51	0.69	199	2.7×10^{-5}		
Mimas	186,000	3.08	0.94	398	0.00051	1100	1.1
Enceladus	238,000	3.95	1.37	498	0.00099	1100	1.1
Calypso	295,000	4.89	1.89	30			
Telesto	295,000	4.89	1.89	30			
Tethys	295,000	4.89	1.89	1060	0.0085	1000	1.0
Dione	377,000	6.26	2.74	1120	0.014	1400	1.4
Helene	377,000	6.26	2.74	32			
Rhea	527,000	8.74	4.52	1530	0.032	1200	1.2
Titan	1,220,000	20.3	16.0	5150	1.83	1900	1.9
Hyperion	1,480,000	24.6	21.3	370			
Iapetus	3,560,000	59.1	79.3	1440	0.022	1000	1.0
Phoebe	13,000,000	215	550*	230			

Indicates a retrograde orbit.

In 1995, researchers using the *Hubble Space Telescope* announced the discovery of several new moons orbiting Saturn. The findings were reported with great fanfare in the press, for these would have been the first such moons spotted anywhere by *Hubble*. Since then, however, it has become unclear if these sightings are new moons, or even real objects at all.

The original images were processed by removing stray light from near the edge of Saturn's ring, which was minimal, since the observations were made during a ring-crossing event in 1995, when the rings virtually disappeared from view (see Figure 12.1). After accounting for the positions of all of Saturn's known moons, the field of view still displayed four blobs of light, in addition to the body of Saturn itself. The accompanying four-picture sequence spans 30 minutes of observing time and shows how astronomers might have been fooled into thinking they had made a discovery. Saturn itself appears as a bright white disk at the far right, heavily overexposed in the search for faint new moons; its edge-on rings extend diagonally to the upper left and are barely discernible. What remains is one known moon—Epimetheus, the brightest small blob at left—and several other fainter blobs. The blobs in these images formed the basis for the claims of new moons in the Saturn system.

Subsequent observations by another team of astronomers failed to confirm some of the faint blobs of light in these images but reported some new blobs. What is going on here? It seems that if the blobs are indeed real, then they are probably just clouds of ice and dust orbiting near the planet's rings. Conceivably, the strong tidal pull of Saturn on some of its small moons broke off a piece large enough to be spotted for a time as a transient piece of debris before ultimately dispersing. Perhaps this is even how Saturn's rings

are maintained. As for the blobs that cannot be confirmed, they may be only instrumental artifacts, caused by small-scale imperfections in *Hubble's* mirror, by "hot pixels" within the onboard cameras, or by the computer methods used by astronomers to massage their data.

In any case, most astronomers are now of the opinion that these faint blobs, even if real, do not represent new moons around Saturn. For now, the number of moons officially orbiting Saturn remains the 18 noted in the text.

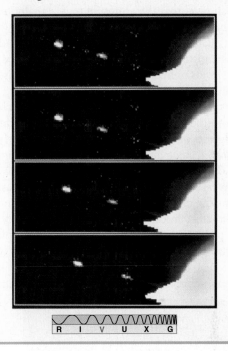

R I V U X G

---

the internal structures of Ganymede and Callisto, the degree of differentiation within Titan is presently unknown. In addition to probing Titan's atmosphere and surface, another important objective of the *Cassini* mission on reaching Saturn

in 2004 will be to make detailed measurements of the moon's interior. ∞ (Sec. 6.6)

A *Voyager 1* image of Titan is shown in Figure 12.17. Unfortunately, despite the spacecraft's close pass, the moon's

(a)

R I V U X G

(b)

R I V U X G

Figure 12.17 (a) Titan, larger than the planet Mercury and roughly half the size of Earth, was photographed in visible light from only 4000 km away as the *Voyager 1* spacecraft passed by in 1980. All we can see here is Titan's upper cloud deck. For unknown reasons, the northern hemisphere appears slightly brighter than the southern. (b) In infrared, however, as captured by the *Hubble Space Telescope*, we can see large-scale surface features. The most prominent bright area shown here is nearly 400 km across, about the size of Australia.

surface remains a mystery. A thick, uniform haze layer, similar to the photochemical smog found over many cities on Earth, that envelops the moon completely obscured the spacecraft's view. *Voyager 1* was able to provide mission specialists with detailed atmospheric data, however.

Titan's atmosphere is thicker and denser even than Earth's, and it is certainly far more substantial than that of any other moon. Prior to *Voyager 1*'s arrival in 1980, only methane and a few other simple hydrocarbons had been conclusively detected on Titan. (Hydrocarbons are molecules consisting solely of hydrogen and carbon atoms; methane, CH_4, is the simplest). Radio and infrared observations from *Voyager 1* showed that the atmosphere is actually made up mostly of nitrogen (roughly 90 percent) and argon (at most 10 percent), with a small percentage of methane. In addition, complex chemistry in Titan's atmosphere maintains steady (but trace) levels of hydrogen gas, the hydrocarbons ethane and propane, and carbon monoxide.

Titan's atmosphere seems to act like a gigantic chemical factory. Powered by the energy of sunlight, it is undergoing a complex series of chemical reactions that ultimately result in the observed smog and trace chemical composition. The upper atmosphere is thick with aerosol haze, and the unseen surface may be covered with organic sediment that has settled down from the clouds. Speculation runs the gamut from oceans of liquid hydrocarbons, especially ethane, to icy valleys laden with petrochemical sludge. Future spacecraft exploration of Titan may present scientists with an opportunity to study the kind of chemistry thought to have occurred billions of years ago on Earth—the prebiotic chemical reactions that eventually led to life on our own planet (see Chapter 28).

Based largely on *Voyager* measurements, Figure 12.18 shows the probable structure of Titan's atmosphere. Despite the moon's low mass (a little less than twice that of Earth's Moon) and hence its low surface gravity (one-seventh of Earth's), the atmospheric pressure at ground level is 60 percent greater than on Earth. Titan's atmosphere contains about 10 times more gas than Earth's atmosphere. Because of Titan's weaker gravitational pull, the atmosphere extends some 10 times farther into space than does our own. The top of the main haze layer lies about 200 km above the surface, although there are additional layers, seen primarily through their absorption of ultraviolet radiation, at 300 km and 400 km (Figure 12.19). Below the haze the atmosphere is reasonably clear, although rather gloomy, because so little sunlight gets through.

Titan's surface temperature is a frigid 94 K, roughly what we would expect on the basis of its distance from the Sun. At the temperatures typical of the lower atmosphere, methane and ethane may behave rather like water on Earth, raising the possibility of methane rain, snow, and fog and even ethane oceans! At higher levels in the atmosphere the temperature rises, the result of photochemical absorption of solar radiation.

Why does Titan have such a thick atmosphere, when similar moons of Jupiter such as Ganymede and Callisto have none? The answer seems to be a direct result of Titan's greater distance from the Sun. The moons of Saturn formed at considerably lower temperature than did those of Jupiter. Those conditions would have enhanced the ability of the water ice that makes up the bulk of Titan's interior to absorb methane and ammonia, both of which were present in abun-

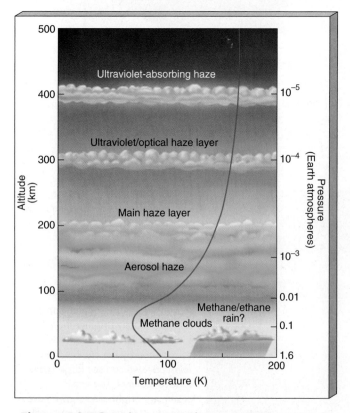

Figure 12.18 The structure of Titan's atmosphere, as deduced from *Voyager 1* observations.

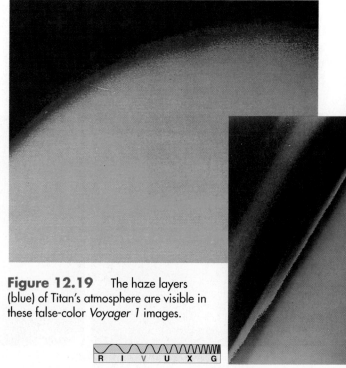

Figure 12.19 The haze layers (blue) of Titan's atmosphere are visible in these false-color *Voyager 1* images.

R I V U X G

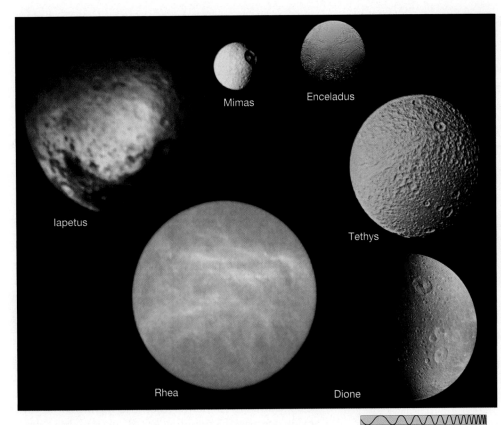

Figure 12.20 Saturn's six medium-sized satellites, to scale. All are heavily cratered. Iapetus shows a clear contrast between its light-colored (icy) trailing surface, at the top and center in this image, and its dark leading hemisphere, at the bottom. The light-colored "wisps" covering essentially the whole central portion of this image of Rhea are thought to be water that was released from the moon's interior during some long-ago period of activity and then froze on the surface. Dione and Tethys show evidence of ancient geological activity of some sort. Enceladus appears to still be volcanically active, but the cause of the volcanism is unexplained. Mimas's main surface feature is the large crater Herschel, plainly visible in this *Voyager 2* image; the impact that caused this crater must have come very close to shattering the moon.

dance at those early times. As a result, Titan was initially laden with much more methane and ammonia gas than either Ganymede or Callisto. As Titan's internal radioactivity warmed the moon the ice released the trapped gases, forming a thick methane–ammonia atmosphere. Sunlight split the ammonia into hydrogen, which escaped into space, and nitrogen, which remained in the atmosphere. The methane, which was less easily broken apart, survived intact. Together with argon outgassed from Titan's interior, these gases form the basis of the atmosphere we see today.

SATURN'S MEDIUM-SIZED MOONS

7 Saturn's complement of midsized moons consists (in order of increasing distance from the planet) of Mimas (at 3.1 planetary radii), Enceladus (at 4.0), Tethys (at 4.9), Dione (at 6.3), Rhea (at 8.7), and Iapetus (at 59.1). They are shown, to proper scale, in Figure 12.20. All six were known from Earth-based observations long before the space age. The inner five move on circular trajectories, and all are tidally locked by Saturn's gravity into synchronous rotation (so that one side always faces the planet). They therefore all have permanently "leading" and "trailing" faces as they move in their orbits, a fact that is important in understanding their often asymmetrical surface markings.

Unlike the densities of the Galilean satellites of Jupiter, the densities of these six moons do not show any correlation with distance from Saturn. Their densities are all between 1000 and 1400 kg/m³, implying that nearness to the central planetary heat source was a less important influence during their formation than it was in the Jupiter system. Scientists believe that the midsized moons are composed largely of rock and water ice, as is Titan. Their densities are lower than Titan's, primarily because their lower masses produce less compression of their interiors.

The largest of the six, Rhea, has a mass only 1/30 that of Earth's Moon, and its icy surface is very reflective and heavily cratered (Figure 12.21). At the low temperatures found on its surface, water ice is very hard and behaves rather like rock on the inner planets. For that reason, Rhea's surface craters look very much like craters on the Moon or Mercury. The crater density is similar to that in the lunar highlands, indicating that the surface is old, and there is no evidence of extensive geological activity. Rhea's only real riddle is the presence of so-called *wispy terrain*—prominent light-colored streaks—on its trailing side. The leading face, by contrast, shows no such markings, only craters. Astronomers believe that the wisps were caused by some event in the distant past during which water was released from the interior and condensed on the surface. Any similar markings on the leading side have presumably been long since obliterated by cratering, which should be much more frequent on the satellite's forward-facing surface.

Inside Rhea's orbit lie the orbits of Tethys and Dione. These two moons are comparable in size and have masses somewhat less than half the mass of Rhea. Like Rhea, they

R I V U X G

Figure 12.21 Rhea's north polar region, seen here at a superb resolution of only 1 km. The heavily cratered surface resembles that of the Moon, except that we see here craters in bright ice, rather than dark lunar rock.

have reflective surfaces that are heavily cratered, but each shows signs of surface activity, too. Dione's trailing face has prominent bright streaks which are probably similar to Rhea's wispy terrain. Dione also has "maria" of sorts, where flooding appears to have obliterated the older craters. The cracks on Tethys may have been caused by cooling and shrinking of the surface layers or, more probably, by meteoritic bombardment.

The innermost, and smallest, medium-sized moon is Mimas. Despite its low mass—only 1 percent the mass of Rhea—its closeness to the rings causes resonant interactions with the ring particles, resulting most notably in the Cassini Division, as we have already seen. Possibly because of its proximity to the rings, Mimas is heavily cratered. The moon's chief surface feature is an enormous crater, known as Herschel, on the leading face. Its diameter is almost one-third that of the moon itself. The impact that formed Herschel must have come very close to destroying Mimas completely. It is quite possible that the debris produced by such impacts is responsible for creating or maintaining the spectacular rings we see.

Enceladus orbits just outside Mimas. Its size, mass, composition, and orbit are so similar to those of Mimas that one might guess that the two moons would also be very similar to each other in appearance and history. This is not so. Enceladus is so bright and shiny—it reflects virtually 100 percent of the sunlight falling on it—that astronomers believe its surface must be completely coated with fine crystals of pure ice, which may be the icy "ash" of water "volcanoes" on Enceladus. The moon bears visible evidence of large-scale volcanic activity. Much of

its surface is devoid of impact craters, which seem to have been erased by what look like lava flows, except that the "lava" is water, temporarily liquefied during recent internal upheavals and now frozen again.

Although no geysers or volcanoes have actually been observed on Enceladus, there seems to be strong circumstantial evidence of volcanism on the satellite. In addition, the nearby thin cloud of small, reflective particles that makes up Saturn's E ring is known to be densest near Enceladus. Calculations indicate that the E ring is unstable because of the disruptive effects of the solar wind, supporting the view that volcanism on Enceladus continually supplies new particles to maintain the ring.

Why is there so much activity on a moon so small? No one knows. Attempts have been made to explain Enceladus's water volcanism in terms of tidal stresses. (Recall the role that Jupiter's tidal stresses play in creating volcanism on Io.) ∞ (Sec. 11.5) However, Saturn's tidal force on Enceladus is only one quarter the force exerted by Jupiter on Io, and there are no nearby large satellites to force Enceladus away from a circular trajectory. Thus, the ingredients that power Io's volcanoes may not be present on Enceladus. For now, the mystery of Enceladus's internal activity remains unresolved.

The outermost midsized moon is Iapetus (Figure 12.20). It orbits Saturn on a somewhat eccentric, inclined orbit with a semi-major axis of 3.6 million km. Its mass is about three quarters that of Rhea. Iapetus is a two-faced moon. The dark, leading face reflects only about 3 percent of the sunlight reaching it, whereas the icy trailing side reflects 50 percent. Spectral studies of the dark regions seem to indicate that the material originates on Iapetus, in which case the moon is not simply sweeping up dark material as it orbits. Similar dark deposits seen elsewhere in the solar system are thought to be organic (carbon-containing) in nature; they can be produced by the action of solar radiation on hydrocarbon (for example, methane) ice. But how the dark markings can adorn only one side of Iapetus in that case is still unknown.

THE SMALL SATELLITES

7 Finally we come to Saturn's dozen or so small moons. Their masses are poorly known (they are inferred mainly from their gravitational effects on the rings), but they are thought to be similar in composition to the small moons of Jupiter. The outermost small moons, Hyperion and Phoebe, were discovered in the nineteenth century, in 1848 and 1898, respectively. The others were first detected in the second half of the twentieth century. Only the moons in or near the rings themselves were actually discovered by the *Voyager* spacecraft.

Just 10,000 km beyond the F ring lie the so-called *co-orbital satellites* Janus and Epimetheus. As the name implies, these two satellites "share" an orbit, but in a very strange way. At any given instant, both moons are in circular orbits about Saturn, but one of them has a slightly smaller orbital radius than the other. Each satellite obeys Kepler's laws, so

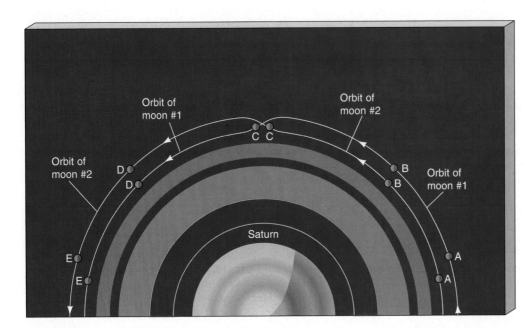

Figure 12.22 The peculiar motion of Saturn's co-orbital satellites, Janus and Epimetheus, which play a never-ending game of tag as they move on their orbits around the planet. The labeled points represent the locations of the two moons at a few successive times. From A to C, satellite 2 gains on satellite 1. However, before it can overtake it, the two moons swap orbits, and satellite 1 starts to pull ahead of satellite 2 again, through points D and E. The whole process is repeated ad infinitum.

the inner satellite orbits slightly faster than the outer one and slowly catches up to it. The inner moon takes about four Earth years to "lap" the outer one. As the inner satellite gains ground on the outer one a strange thing happens. As illustrated in Figure 12.22, when the two get close enough to begin to feel each other's weak gravity, they switch orbits—the new inner moon (which used to be the outer one) begins to pull away from its companion, and the whole process begins again! No one knows how the co-orbital satellites came to be engaged in this curious dance. Possibly they are portions of a single moon that broke up, perhaps after a meteoritic impact, leaving the two pieces in almost the same orbit.

In fact, several of the other small moons also share orbits, this time with larger moons. Telesto and Calypso have orbits that are synchronized with the orbit of Tethys, always remaining fixed relative to the larger moon, lying precisely 60° ahead of and 60° behind it as it travels around Saturn (see Figure 12.23). The moon Helene is similarly tied to Dione. These 60° points are known as **Lagrange points**, after the French mathematician Joseph Louis Lagrange, who first studied them. Later we will see further examples of this special 1:1 orbital resonance in the motion of some asteroids about the Sun, trapped in the Lagrange points of Jupiter's orbit.

The strangest motion of all is that of the moon Hyperion, which orbits between Titan and Iapetus, at a distance of 1.5 million km from the planet. Unlike most of Saturn's

moons, its rotation is not synchronous with its orbital motion. Because of the gravitational effect of Titan, Hyperion's orbit is not circular, so synchronous rotation cannot occur. In response to the competing gravitational influences of Titan and Saturn, this irregularly shaped satellite constantly changes both its rotation speed and its rotation axis, in a condition known as **chaotic rotation**. As Hyperion orbits Saturn it tumbles apparently at random, never stopping and never repeating itself, in a completely unpredictable way. Since the 1970s the study of chaos on Earth has revealed new classes of unexpected behavior in even very simple systems. Hyperion is one of the few other places in the universe where this behavior has been unambiguously observed.

Figure 12.23 The orbits of the moons Telesto and Calypso are tied to the motion of the moon Tethys. The combined gravitational pulls of Saturn and Tethys keep the small moons exactly 60° ahead and behind the larger moon at all times, so that all three moons share an orbit and never change their relative positions.

Chapter Review

SUMMARY

Saturn was the outermost planet known to ancient astronomers. Its rings and moons were not discovered until after the invention of the telescope. Saturn is smaller than Jupiter but is still much larger than any of the terrestrial worlds. Like Jupiter, Saturn rotates rapidly, producing a pronounced flattening, and displays differential rotation. Strong radio emission from the planet's magnetosphere allows the rotation rate of the interior to be determined.

Weather systems are seen on Saturn, as on Jupiter, although they are less distinct. Short-lived storms are occasionally seen. Saturn has weaker gravity and a more extended atmosphere than Jupiter. The planet's overall butterscotch hue is due to cloud chemistry similar to that occurring in Jupiter's atmosphere. Saturn, like Jupiter, has bands, ovals, and turbulent flow patterns powered by convective motion in the interior.

Like Jupiter, Saturn emits far more radiation into space than it receives from the Sun. Unlike Jupiter's, Saturn's excess energy emission is the result of **helium precipitation** (p. 271) in the planet's interior, where helium liquefies and forms droplets, which then fall toward the center of the planet. This process is also responsible for Saturn's observed helium deficit.

Saturn's interior is theoretically similar to that of Jupiter, but with a thinner layer of metallic hydrogen and a larger core. Its lower mass gives Saturn a less extreme core temperature, density, and pressure than Jupiter's. Saturn's conducting interior and rapid rotation produce a strong magnetic field and an extensive magnetosphere that contains the planet's ring system and the innermost 16 moons.

Saturn's rings lie in the planet's equatorial plane, so their appearance from Earth changes as Saturn orbits the Sun. From Earth, the main visible features of the rings are the **A, B,** and **C rings** (p. 272), the **Cassini Division** (p. 272), and the **Encke gap** (p. 272). The Cassini Division is a dark region between the A and B rings. The Encke gap lies near the outer edge of the A ring. The rings are made up of trillions of icy particles ranging in size from dust grains to boulders, all orbiting Saturn like so many tiny moons. Their total mass is comparable to that of a small moon. Both divisions are dark because they are almost empty of ring particles.

When the *Pioneer* and *Voyager* probes reached Saturn, they found that the rings are actually made up of tens of thousands of

narrow **ringlets** (p. 274). Interactions between the ring particles and the planet's inner moons are responsible for much of the fine structure observed in the main rings.

The **Roche limit** (p. 274) of a planet is the distance within which the planet's tidal field would overwhelm the internal gravity of a moon, tearing it apart and forming a ring. All known planetary ring systems lie inside their parent planets' Roche limits.

Saturn's narrow **F ring** (p. 276), discovered by *Pioneer 2*, lies just outside the A ring. It has a kinked, braided structure, apparently caused by two small **shepherd satellites** (p. 276) that orbit close to the ring and prevent it from breaking up. *Voyager 2* discovered the faint **D ring** (p. 275), lying between the C ring and Saturn's cloud layer, and the **E ring** (p. 275), associated with the moon Enceladus. Planetary rings may have lifetimes of only a few tens of millions of years. If so, the fact that we see rings around all four jovian planets means that they must constantly be re-formed or replenished, perhaps by material chipped off moons by meteoritic impact or by the tidal destruction of entire moons.

Saturn's large moon Titan is the second-largest moon in the solar system. Its thick atmosphere obscures the moon's surface and may be the site of complex cloud and surface chemistry. The existence of Titan's atmosphere is a direct consequence of the cold conditions that prevailed at the time of the moon's formation.

The medium-sized moons of Saturn are made up predominantly of rock and water ice. They show a wide variety of surface terrains and are also heavily cratered. They are all tidally locked by the planet's gravity into synchronous orbits. The innermost mid-sized moon Mimas exerts influence over the structure of the rings. The Cassini Division, now known to contain faint ringlets and gaps, is the result of resonance between its particles and Mimas. The moon Iapetus has a marked contrast between its leading and trailing faces, while Enceladus has a highly reflective appearance, possibly the result of water "volcanoes" on its surface.

Saturn's small moons exhibit a wide variety of complex motion. Several moons "share" orbits, in some cases lying at the **Lagrange points** (p. 283) 60° ahead of and behind the orbit of a larger moon. The moon Hyperion undergoes **chaotic rotation** (p. 283)—tumbling in an unpredictable way as it orbits the planet.

SELF-TEST: TRUE OR FALSE?

_____ **1.** Saturn's orbit is almost twice the size of Jupiter's orbit.

_____ **2.** The rotation of Saturn is unlike that of Jupiter; it is slow and shows little differential rotation.

_____ **3.** Saturn's large axial tilt produces strong seasonal variations in the planet's appearance.

_____ **4.** Saturn probably does not have a rocky core.

_____ **5.** The atmosphere of Saturn contains only half the helium of Jupiter's atmosphere, but, overall, Saturn probably has a normal helium abundance.

_____ **6.** No storm systems have ever been seen on Saturn.

_____ **7.** The magnetic field of Saturn is much less than Jupiter's and, overall, is about as strong as Earth's magnetic field.

_____ **8.** Saturn is the only planet with a ring system.

_____ **9.** A typical ring particle is 100 m in diameter.

_____ **10.** The composition of the ring particles of Saturn is predominantly water ice.

_____ **11.** Although Saturn's ring system is tens of thousands of kilometers wide, it is only a few tens of meters thick.

_____ **12.** Saturn has small and mid-sized moons, but no large moons.

_____ **13.** Water ice predominates in Saturn's moons.

_____ **14.** Titan's atmosphere is 10 times denser than Earth's.

_____ **15.** Titan's surface is obscured by thick clouds of water ice.

SELF-TEST: FILL IN THE BLANK

1. Saturn is best known for its spectacular _____ system.
2. Features in the atmosphere of Saturn, as on Jupiter, are mostly hidden by an upper layer of frozen _____ clouds.
3. Saturn's cloud layers are thicker than those of Jupiter because of Saturn's weaker _____.
4. Relative to Jupiter's atmosphere, Saturn's atmosphere is deficient in _____.
5. Saturn's excess energy emission is caused by _____.
6. As viewed from Earth, Saturn's ring system is conventionally divided into _____ broad rings.
7. The Cassini Division lies between the _____ and _____ rings.
8. The _____ ring of Saturn is the brightest.
9. The rings exist because they lie within Saturn's _____.
10. Two small moons, known as _____ satellites, are responsible for the unusually complex form of the F ring.
11. The composition of Titan's atmosphere is 90 percent _____.
12. Titan's surface may be covered with _____ sediment.
13. Except for Titan, most of the moons of Saturn are heavily cratered, even though their surfaces are mostly water ice. The low _____ of the ice makes it hard as rock.
14. Saturn's _____ ring is probably related to volcanism on the icy moon Enceladus.
15. When small moons share an orbit with a large moon, they are found at the _____ points of the large moon's orbit.

REVIEW AND DISCUSSION

1. Seen from Earth, Saturn's rings sometimes appear broad and brilliant but at other times seem to disappear. Why?
2. Why does Saturn have a less varied appearance than Jupiter?
3. What does Saturn's shape tell us about its deep interior?
4. Compare and contrast the atmospheres and weather systems of Saturn and Jupiter, and tell how the differences affect each planet's appearance.
5. Compare the thicknesses of Saturn's various layers (clouds, molecular hydrogen, metallic hydrogen, and core) with the equivalent layers in Jupiter.
6. What is the relevance of the Roche limit to planetary rings?
7. What evidence supports the idea that a relatively recent catastrophic event was responsible for Saturn's rings?
8. What effect does Mimas have on the rings?
9. When *Voyager 1* passed Saturn in 1980, why didn't it see the surface of Saturn's largest moon, Titan?
10. Compare and contrast Titan with Jupiter's Galilean moons.
11. Why does Titan have a dense atmosphere when other large moons in the solar system don't?
12. What is the evidence for geological activity on Enceladus?
13. What mystery is associated with Iapetus?
14. Describe the behavior of Saturn's co-orbital satellites.
15. Imagine what the sky would look like from Saturn's moon Hyperion. Would the Sun rise and set in the same way it does on Earth? How do you imagine Saturn might look?

PROBLEMS

1. What is the angular diameter of Saturn's A ring, as seen from Earth at closest approach?
2. How long does it take for Saturn's equatorial flow, moving at 1500 km/h, to encircle the planet? Compare this with the wind circulation time on Jupiter.
3. If Saturn's surface temperature is 97 K and the planet radiates three times more energy than it receives from the Sun, use Stefan's law to calculate what the surface temperature would be in the absence of any internal heat source.
4. Based on the data given in Section 12.3, estimate the mass of Saturn's core and compare it with Earth's mass.
5. The text states that the total mass of material in Saturn's rings is about 10^{15} tons (10^{18} kg). Suppose the average ring particle is 6 cm in radius (a large snowball) and has a density of 1000 kg/m^3. How many ring particles would there be?
6. What is the orbital speed of ring particles at the inner edge of the B ring, in km/s? Compare this with the speed of a satellite in low Earth orbit (500 km altitude, say). Why are these speeds so different?
7. Show that Titan's surface gravity is about one-seventh of Earth's, as stated in the text. What is Titan's escape speed?
8. Calculate the orbital radii of ring particles having the following properties: (a) in a 2:1 orbital resonance with Mimas, that is, orbiting Saturn twice for every orbit of Mimas; (b) in a 3:2 resonance with Mimas (three orbits for every two of Mimas); (c) in a 3:1 resonance with Enceladus.
9. Compare Saturn's tidal gravitational effect on Mimas with Mimas's own surface gravity.
10. Sunlight reflected back to Earth from a particle in Saturn's rings is Doppler-shifted twice—first because of the relative motion of the source of the radiation (the Sun) and the ring particle, and then again by the relative motion of the particle and the observer on Earth (see Section 3.5). As a result, if Earth, Saturn and the Sun are roughly aligned (that is, Saturn is near opposition), the observed Doppler shift corresponds to *twice* the particle's orbital speed. A certain solar spectral line, of wavelength 656 nm, is reflected from the rings and is observed on Earth. If the rings happen to be seen almost edge-on, what is the line's observed wavelength in light reflected from (a) the approaching inner edge of the B-ring? (b) the receding inner edge of the B-ring? (c) the approaching outer edge of the A-ring? (d) reflected from the receding outer edge of the A-ring?

PROJECTS

1. Saturn moves more slowly among the stars than any other visible planet. How many degrees per year does it move? Look in an almanac to see where the planet is now. What constellation is it in now? Where will it be in one year?
2. Binoculars may not reveal the rings of Saturn, but most small telescopes will. Use a telescope to look at Saturn. Does Saturn appear oblate? Examine the rings. How are they tilted? Can you see a dark line in the rings? This is the Cassini Division.

It once was thought to be a gap in the rings, but the *Voyager* spacecraft discovered that it is filled with tiny ringlets. Can you see the shadow of the rings on Saturn?

3. While looking at Saturn through a telescope can you see any of its moons? They line up with the rings; Titan is often the farthest out, and always the brightest. How many moons can you see? Use an almanac to identify each one you find.

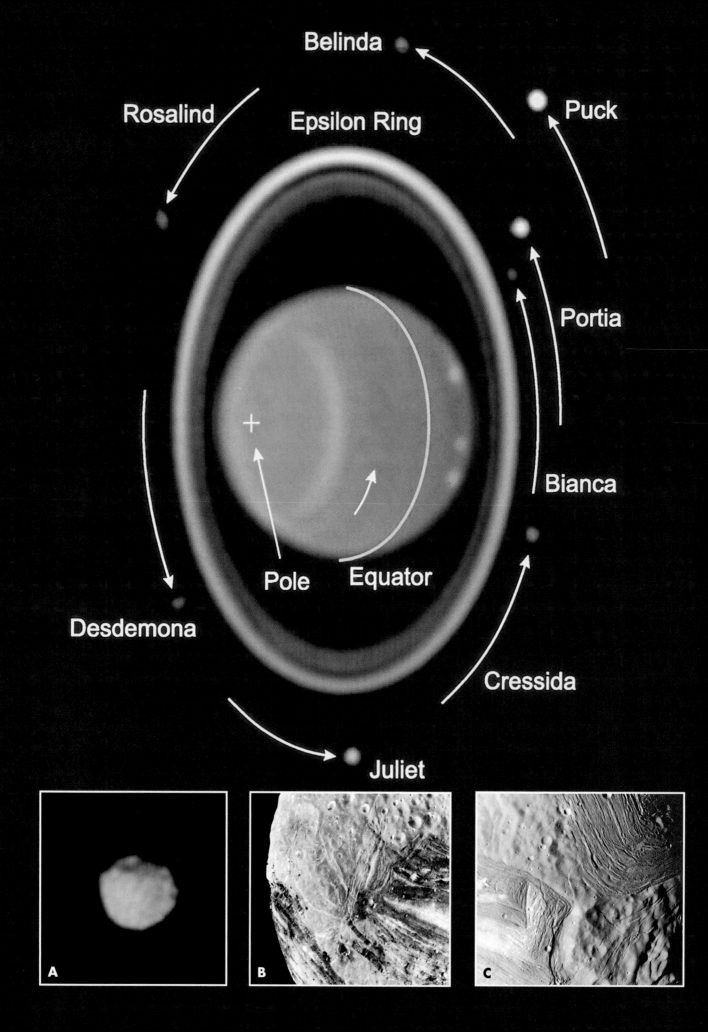

Belinda

Puck

Rosalind

Epsilon Ring

Portia

Bianca

Pole   Equator

Desdemona

Cressida

Juliet

A   B   C

13 URANUS, NEPTUNE, AND PLUTO

The Outer Worlds of the Solar System

WWW

LEARNING GOALS

Studying this chapter will enable you to:

1 Describe how both calculation and chance played a major role in the discoveries of the outer planets.

2 Summarize the similarities and differences between Uranus and Neptune and compare these planets with the other two jovian worlds.

3 Explain what the moons of the outer planets tell us about their past.

4 Contrast the rings of Uranus and Neptune with those of Jupiter and Saturn.

5 Summarize the orbital and physical properties of Pluto and explain how the Pluto–Charon system differs fundamentally from all the other planets.

(Opposite page, background) This photo, taken by the *Hubble* telescope, shows the planet Uranus with its many rings and several of its inner moons.
(Inset A) The largest of Uranus's inner moons (just beyond the field of view in the main image) is Miranda, seen here from the perspective of the oncoming *Voyager* spacecraft. It contains less than one percent of the mass of Earth's Moon.
(Insets B and C) Progressively closer views of Miranda, radioed back as *Voyager* skirted to within 30,000 kilometers of this peculiar moon, show several of its prominent features. So many major terrain types are evident in these photos that some scientists have likened Miranda to a cosmic geology museum. Currently, astronomers have no clear consensus about how Miranda's unusual surface features originated, but it seems clear that some very violent events must have been involved, probably about 4 billion years ago.

*T*he three outermost planets were unknown to the ancients—all were discovered by telescopic observations: Uranus in 1781, Neptune in 1846, and Pluto in 1930. Uranus and Neptune have very similar bulk properties, so it is natural to consider them together; they are part of the jovian family of planets. Pluto, by contrast, is not a jovian world. It is very much smaller than even the terrestrial planets and generally seems much more moonlike than planetlike in character. Indeed, at one time, astronomers even speculated that Pluto was a one-time moon that had somehow escaped from one of the outer planets, most likely Neptune. However, it now seems more probable that Pluto is really the best-known representative of a newly recognized class of objects residing in the outer solar system. Whatever its origin, because of Pluto's location and its similarity to the jovian moons, we study it here along with its larger jovian neighbors.

13.1 The Discovery of Uranus

1 The planet Uranus was discovered by British astronomer William Herschel in 1781. Herschel was engaged in charting the faint stars in the sky when he came across an odd-looking object that he described as "a curious either nebulous star or perhaps a comet." Repeated observations showed that it was neither. The object appeared as a disk in Herschel's 6-inch telescope and moved relative to the stars, but it traveled too slowly to be a comet. Herschel soon realized that he had found the seventh planet in the solar system. Since this was the first new planet discovered in well over 2000 years, the event caused quite a stir at the time. The story goes that Herschel's first instinct was to name the new planet "Sidus Georgium" (Latin for "George's star") after his king, George III of England. The world was saved from a planet named George by the wise advice of another astronomer, Johann Bode. He suggested instead that the tradition of using names from Greco–Roman mythology be continued and that the planet be named after Uranus, the father of Saturn.

Uranus is in fact just barely visible to the naked eye, if you know exactly where to look. At opposition, it has a maximum

angular diameter of 4.1″ and shines just above the unaided eye's threshold of visibility. It looks like a faint, undistinguished star. No wonder it went unnoticed by the ancients. Even today, few astronomers have seen it without a telescope.

Through a large Earth-based optical telescope (see Figure 13.1), Uranus appears hardly more than a tiny pale greenish disk. With the flyby of *Voyager 2* in 1986, our detailed knowledge of Uranus increased dramatically. Figure 13.2 is a combination of close-up visible-light images of the planet and one

Figure 13.2 A montage of Uranus and one of its satellites, composed of photos sent back to Earth by the *Voyager 2* spacecraft as it whizzed past this giant planet at 10 times the speed of a rifle bullet. The image of Uranus, taken from a distance of about 100,000 km, shows the planet's blue yet featureless upper atmosphere. The image in the foreground is Miranda, one of Uranus's moons. (Uranus's rings, not visible in the photograph, have been added by an artist.)

Figure 13.1 Details are barely visible on photographs of Uranus made with large Earth-based telescopes. (Arrows point to three of its moons.)

of its moons. The apparently featureless atmosphere of Uranus contrasts sharply with the bands and spots visible on all the other jovian worlds. Some orbital and physical properties of Uranus are presented in the Uranus Data box on p. 290.

13.2 The Discovery of Neptune

1 Once Uranus was discovered, astronomers set about charting its orbit and quickly discovered a small discrepancy between the planet's predicted position and where they actually observed it. Try as they might, astronomers could not find an elliptical orbit that fit the planet's trajectory to within the accuracy of their measurements. Half a century after Uranus's discovery, the discrepancy had grown to a quarter of an arc minute, far too big to be explained away as observational error.

The logical conclusion was that an unknown body must be exerting a gravitational force on Uranus—much weaker than that of the Sun, but still measurable. But what body could this be? Astronomers realized that there had to be *another* planet in the solar system perturbing Uranus's motion.

In the 1840s two mathematicians independently solved the difficult problem of determining this new planet's mass and orbit. A British astronomer, John Adams, reached the solution in September 1845; in June of the following year, the French mathematician Urbain Leverrier came up with essentially the same answer. British astronomers seeking the new planet found nothing during the summer of 1846. In September a German astronomer named Johann Galle began his own search from the Berlin Observatory, using a newly completed set of more accurate sky charts. He found the new planet within one or two degrees of the predicted position—on his first attempt. After some wrangling over names and credits, the new planet was named Neptune, and Adams and Leverrier (but not Galle!) are now jointly credited with its discovery.

Neptune's orbital and physical properties are listed in the Neptune Data box on p. 300. With an orbital semi-major axis

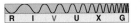

Figure 13.3 Neptune and two of its moons, Triton (large arrow) and Nereid (small arrow), imaged with a large Earth-based telescope.

of 30.1 A.U. and an orbital period of 163.7 years, Neptune has not yet completed one revolution since its discovery. Unlike Uranus, distant Neptune cannot be seen with the naked eye, although it can be seen with a small telescope—in fact, according to his notes, Galileo might actually have seen Neptune, although he had no idea what it really was at the time. Through a large telescope, Neptune appears as a bluish disk, with a maximum angular diameter of 2.4″ at opposition.

Figure 13.3 shows a long Earth-based exposure of Neptune and its largest moon, Triton. Neptune is so distant that surface features are virtually impossible to discern. Even under the best observing conditions, only a few markings can be seen. These are suggestive of multicolored cloud bands—light bluish hues seem to dominate. With *Voyager 2*'s arrival, much more detail emerged, as shown in Figure 13.4. Superficially, at least, Neptune resembles a blue-tinted Jupiter, with atmospheric bands and spots clearly evident.

Figure 13.4 (a) Neptune as seen by *Voyager 2*, from a distance of roughly 1 million km. A closer view (b), resolved to about 10 km, shows cloud streaks ranging in width from 50 to 200 km.

(a)

(b)

PLANETARY DATA:
Uranus

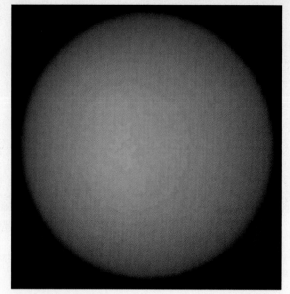

Orbital semi-major axis	19.19 A.U.
	2871 million km
Orbital eccentricity	0.047
Perihelion	18.29 A.U.
	2736 million km
Aphelion	20.10 A.U.
	3006 million km
Mean orbital speed	6.80 km/s
Sidereal orbital period	83.75 tropical years
Synodic orbital period	369.66 solar days
Orbital inclination to the ecliptic	0.77°
Greatest angular diameter, as seen from Earth	4.1″
Mass	8.68×10^{25} kg
	14.54 (Earth = 1)
Equatorial radius	25,559 km
	4.01 (Earth = 1)
Mean density	1271 kg/m^3
	0.230 (Earth = 1)
Surface gravity	8.87 m/s^2
	0.91 (Earth = 1)
Escape speed	21.3 km/s
Sidereal rotation period	−0.72 solar days (retrograde)
Axial tilt	97.92°
Surface magnetic field	0.74 (Earth = 1)
Magnetic axis tilt relative to rotation axis	58.6°
Mean surface temperature	58 K
Number of moons	17

13.3 Uranus and Neptune in Bulk

MASSES AND RADII

2 Figure 13.5 shows Uranus and Neptune to scale, along with Earth for comparison. The two giant planets are quite similar in their bulk properties. The radius of Uranus is 4.0 times that of Earth; that of Neptune, 3.9 Earth radii. Their masses (first determined from terrestrial observations of their larger moons and later refined by *Voyager 2*) are 14.5 Earth masses for Uranus and 17.1 Earth masses for Neptune. Uranus's average density is 1300 kg/m^3, and Neptune's is 1600 kg/m^3. These densities imply that large rocky cores constitute a greater fraction of the planets' masses than do the cores of either Jupiter or Saturn. The cores themselves are probably comparable in size, mass, and composition to those of the two larger giants.

ROTATION RATES

Like the other jovian planets, Uranus has a short rotation period. Earth-based observations of the Doppler shifts in spectral lines first indicated that Uranus's "day" was between 10 and 20 hours long. The precise value of the planet's rotation period—accurately determined when *Voyager 2* timed radio signals associated with its magnetosphere—is now known to be 17.2 hours. Again as with Jupiter and Saturn, the planet's atmosphere rotates differentially, but Uranus's atmosphere actually rotates *faster* at the poles (where the period is 14.2 hours) than near the equator (where the period is 16.5 hours).

Each planet in our solar system seems to have some outstanding peculiarity, and Uranus is no exception. Unlike all the other planets, whose spin axes are roughly perpendicular to the ecliptic plane, Uranus's rotation axis lies almost within that plane—98° from the perpendicular, to be precise. (Note that because the north pole lies below the ecliptic plane, the rotation of Uranus, like that of Venus, is classified as retrograde.) Relative to the other planets, we might say that Uranus lies tipped over on its side. As a result, the "north" (spin) pole of Uranus, at some time in its orbit, points almost directly toward the Sun.* Half a "year" later, its "south" pole faces the Sun, as illustrated in Figure 13.6. When *Voyager 2* encountered the planet in 1986, the north pole happened to be pointing nearly at the Sun, so it was midsummer in the northern hemisphere.

The strange orientation of Uranus's rotation axis produces some extreme seasonal effects. Starting at the height of northern summer, when the north pole points closest to the Sun, an observer near that pole would see the Sun move in gradually increasing circles in the sky, completing one circuit (counterclockwise) every 17 hours and dipping slightly lower in the sky each day. Eventually the Sun would begin to set and rise again in a daily cycle, and the nights would grow progressively longer with each passing day. Twenty-one years after the summer solstice, the autumnal equinox would occur, with day and night each 8.5 hours long.

*As in Chapter 9, we adopt the convention that a planet's rotation is always counterclockwise as seen from above the north pole (that is, planets always rotate from west to east). ∞ (Sec. 9.2.)

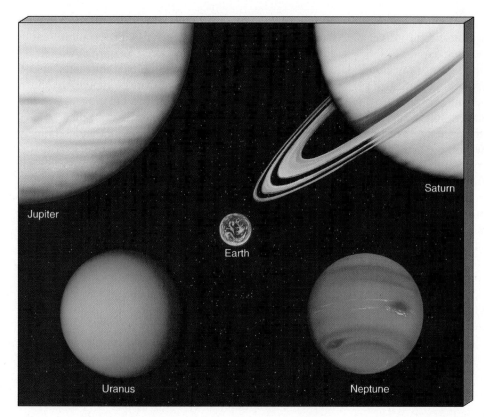

Figure 13.5   Jupiter, Saturn, Uranus, Neptune, and Earth, drawn to scale. Uranus and Neptune are quite similar in their bulk properties. Each probably contains a core about 10 times more massive than Earth. Jupiter and Saturn are each substantially larger, but their rocky cores are probably comparable in mass to those of Uranus and Neptune.

The days would continue to shorten until one day the Sun would fail to rise at all. The ensuing period of total darkness would be equal in length to the earlier period of constant daylight, plunging the northern hemisphere into the depths of winter. Eventually, the Sun would rise again; the days would lengthen through the vernal equinox and beyond, and in time the observer would again experience a long summer of uninterrupted (though dim) sunshine.

From the point of view of an observer on the equator, by contrast, summer and winter would be almost equally

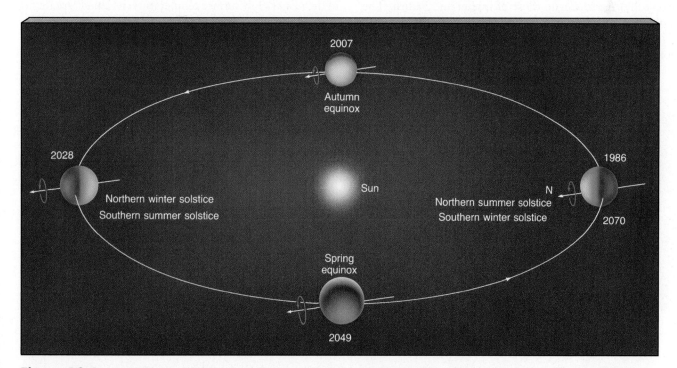

Figure 13.6   Uranus's 98° axial tilt places its equator almost perpendicular to the ecliptic. As a result, the planet experiences the most extreme seasons known in the solar system. The equatorial regions experience two warm seasons (around the two equinoxes) and two cold seasons (at the solstices) each year; the poles are alternately plunged into darkness for 42 years at a time.

cold seasons, with the Sun never rising far above the horizon. Spring and fall would be the warmest times of year, with the Sun passing almost overhead each day.

No one knows why Uranus is tilted in this way. Some scientists have speculated that a catastrophic event, such as a grazing collision between the planet and another planet-sized body, might have altered the planet's spin axis. There is no direct evidence for such an occurrence, however, and no theory to tell us how we should seek to confirm it.

Neptune's clouds show more variety and contrast than do those of Uranus, and Earth-based astronomers studying them determined a rotation rate for Neptune even before *Voyager 2*'s flyby in 1989. The average rotation period of Neptune's atmosphere is 17.3 hours (quite similar to that of Uranus). Measurements of Neptune's radio emission by *Voyager 2* showed that the magnetic field of the planet, and presumably also its interior, rotates once every 16.1 hours. Thus, Neptune is unique among the jovian worlds in that its atmosphere rotates *more slowly* than its interior. Neptune's rotation axis is inclined 29.6° to a line perpendicular to its orbital plane, quite similar to the 27° tilt of Saturn.

13.4 The Atmospheres of Uranus and Neptune

COMPOSITION

Spectroscopic studies of sunlight reflected from Uranus's and Neptune's dense clouds indicate that the two planets' outer atmospheres (the parts we actually measure spectroscopically) are similar to those of Jupiter and Saturn. The most abundant element is molecular hydrogen (84 percent), followed by helium (about 14 percent) and methane, which is more abundant on Neptune (about 3 percent) than on

Uranus (2 percent). Ammonia, which plays such an important role in the Jupiter and Saturn systems, is not present in any significant quantity in the outermost jovian worlds.

The abundances of gaseous ammonia and methane vary systematically among the jovian planets. Jupiter has much more gaseous ammonia than methane, but moving outward from the Sun, we find that the more distant planets have steadily decreasing amounts of ammonia and relatively greater amounts of methane. The reason for this variation is temperature. Ammonia gas freezes into ammonia ice crystals at about 70 K. This is cooler than the cloud-top temperatures of Jupiter and Saturn but warmer than those of Uranus (58 K) and Neptune (59 K). Thus, the outermost jovian planets have little or no *gaseous* ammonia in their atmospheres, so their spectra (which record atmospheric gases only) show just traces of ammonia.

The increasing amounts of methane are largely responsible for the outer jovian planets' blue coloration. Methane absorbs long-wavelength red light quite efficiently, so sunlight reflected from the planets' atmospheres is deficient in red and yellow photons and appears blue-green or blue. As the concentration of methane increases, the reflected light should appear bluer. This is just the trend observed: Uranus, with less methane, looks blue-green, while Neptune, with more, looks distinctly blue.

WEATHER

Voyager 2 detected only a few features in Uranus's atmosphere, and even these became visible only after extensive computer enhancement (see Figure 13.7). Uranus apparently lacks any significant internal heat source, and because of the planet's low surface temperature, its clouds are found only at low-lying, warmer levels in the atmosphere. The absence of high-level clouds means that we must look deep into the planet's atmosphere to see any structure, so the

(a)   (b)   (c)   (d)

R   I   V   U   X   G

Figure 13.7 (a) This *Voyager* view of Uranus approximates the planet's true color but shows little else. Parts (b), (c), and (d) are *Hubble Space Telescope* photographs made at roughly 4-hour intervals, showing the motion of a pair of bright clouds (labeled A and B) in the planet's southern hemisphere. (The numbers at the top give the time of each photo.)

Figure 13.8 (a) Close-up views, taken by *Voyager 2* in 1989, of the Great Dark Spot of Neptune, a large storm system in the planet's atmosphere, possibly similar in structure to Jupiter's Great Red Spot. Resolution in the photo on the right is about 50 km; the entire dark spot is roughly the size of planet Earth. (b) These three *Hubble Space Telescope* views of Neptune were taken about 10 days apart in 1994, when the planet was some 4.5 billion km from Earth. The aqua color is caused by absorption of red light by methane; the cloud features (mostly methane ice crystals) are tinted pink here because they were imaged in the infrared but are really white in visible light. Neptune apparently has a remarkably dynamic atmosphere that changes over just a few days. Notice that the Great Dark Spot has now disappeared.

bands and spots that characterize flow patterns on the other jovian worlds are largely "washed out" on Uranus by intervening stratospheric haze.

From computer-processed images astronomers have learned that Uranus's atmospheric clouds and flow patterns move around the planet in the same sense as the planet's rotation, with wind speeds in the range 200–500 km/h. Tracking these clouds allowed the measurement of the differential rotation mentioned earlier. Despite the odd angle at which sunlight is currently striking the surface (recall that it is just after midsummer in the northern hemisphere), the planet's rapid rotation still channels the wind flow into bands reminiscent of those found on Jupiter and Saturn. Even though the predominant wind flow is in the east–west direction, Uranus's atmosphere seems to be quite efficient at transporting energy from the heated north to the unheated southern hemisphere. Although the south is currently in total darkness, the temperature there is only a few kelvins less than in the north.

Neptune's clouds and band structure are much more easily seen. Although it lies at a greater distance from the Sun,

Neptune's upper atmosphere is actually slightly warmer than that of Uranus. Like Jupiter and Saturn, but unlike Uranus, Neptune has an internal energy source—in fact, it radiates 2.7 times more heat than it receives from the Sun. The cause of this heating is still uncertain. Some scientists have suggested that Neptune's excess methane has helped "insulate" the planet, tending to maintain its initially high internal temperature. If that is so, then the source of Neptune's internal heat is the same as Jupiter's—it is energy left over from the planet's formation. ∞ (Sec. 11.3) The combination of extra heat and less haze may be responsible for the greater visibility of Neptune's atmospheric features (see Figure 13.8), as its cloud layers lie at higher levels in the atmosphere than do those of Uranus.

Neptune sports several storm systems similar in appearance to those seen on Jupiter (and assumed to be produced and sustained by the same basic processes). The largest such storm, known simply as the **Great Dark Spot**, is shown in Figure 13.8(a). Discovered by *Voyager 2* in 1989, the Spot was about the size of Earth, was located near the planet's equator, and exhibited many of the same general characteristics as the

Great Red Spot on Jupiter. The flow around it was counter-clockwise, as with the Red Spot, and there appeared to be turbulence where the winds associated with the Great Dark Spot interacted with the zonal flow to its north and south. The flow around this and other dark spots may drive updrafts to high altitudes, where methane crystallizes out of the atmosphere to form the high-lying cirrus clouds. Astronomers did not have long to study the Dark Spot's properties, however. As shown in Figure 13.8(b), when the *Hubble Space Telescope* viewed Neptune in 1994, the Spot had disappeared.

13.5 Magnetospheres and Internal Structure

Voyager 2 found that both Uranus and Neptune have fairly strong internal magnetic fields—about 100 times stronger than Earth's field and 1/10 as strong as Saturn's. However,

because the radii of Uranus and Neptune are so much larger than the radius of Earth, the magnetic fields at the cloud tops—spread out over far larger volumes than the field on Earth—are actually comparable in strength to Earth's field. Uranus and Neptune each have substantial magnetospheres, populated largely by electrons and protons either captured from the solar wind or created from ionized hydrogen gas escaping from the planets themselves.

When *Voyager 2* arrived at Uranus, it discovered that the planet's magnetic field is tilted at about 60° to the axis of rotation. On Earth, such a tilt would put the north magnetic pole somewhere in the Caribbean. Furthermore, the magnetic field lines are *not* centered on the planet. It is as though Uranus's field were due to a bar magnet that is tilted with respect to the planet's rotation axis and displaced from the center by about one-third the radius of the planet. Figure 13.9 compares the magnetic field structures of Earth and those of the four jovian planets. The locations and orientations of the bar magnets represent the observed

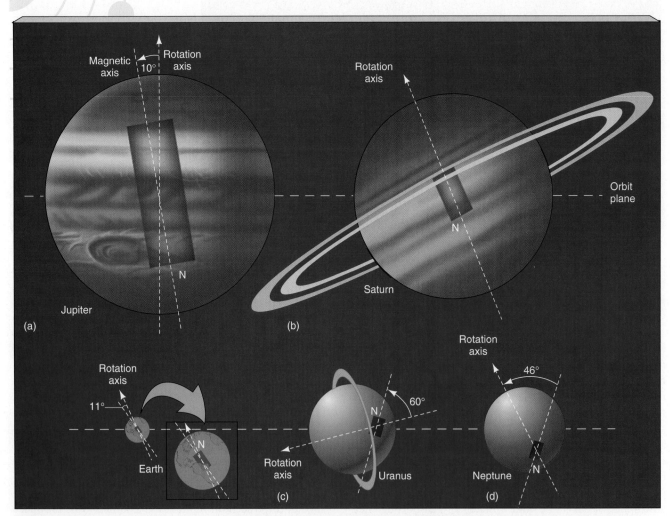

Figure 13.9   A comparison of the magnetic field strengths, orientations, and offsets in the four jovian planets: (a) Jupiter, (b) Saturn, (c) Uranus, (d) Neptune. The planets are drawn to scale, and in each case the magnetic field is represented as though it came from a simple bar magnet. The size and location of each magnet represent the strength and orientation of the planetary field. Notice that the fields of Uranus and Neptune are significantly offset from the center of the planet and are significantly inclined to the planet's rotation axis. Earth's magnetic field is shown for comparison.

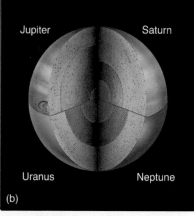

Figure 13.10   A comparison of the interior structures of the four jovian planets. (a) The planets drawn to scale. (b) The relative proportions of the various internal zones.

Animation

planetary fields, and the sizes of the bars indicate magnetic field strength.

Because dynamo theories generally predict that the magnetic axis should be roughly aligned with the rotation axis—as on Earth, Jupiter, Saturn, and the Sun—the misalignment on Uranus suggested to some researchers that perhaps the planet's field had been caught in the act of reversing. ∞ (Sec. 7.5) Another possibility was that the oddly tilted field was in some way related to the planet's axial tilt—perhaps one catastrophic collision skewed both axes at the same time. Those ideas evaporated in 1989 when *Voyager 2* found that Neptune's field is also inclined to the planet's rotation axis, at an angle of 46° (see Figure 13.9d), and also substantially offset from the center. It now appears that the internal structures of Uranus and Neptune are different from those of Jupiter and Saturn, and this difference changes the nature of the field-generation process.

Theoretical models indicate that Uranus and Neptune have rocky cores similar to those found in Jupiter and Saturn—about the size of Earth and perhaps 10 times more massive. However, the pressure outside the cores of Uranus and Neptune (unlike the pressure within Jupiter and Saturn) is too low to force hydrogen into the metallic state, so hydrogen stays in its molecular form all the way in to the planets' cores. Astronomers theorize that deep below the cloud layers, Uranus and Neptune may have high-density, "slushy" interiors containing thick layers of water clouds. It is also possible that much of the planets' ammonia is dissolved in the water, accounting for the absence of ammonia at higher cloud levels. Such an ammonia solution would provide a thick, electrically conducting ionic layer that could conceivably explain the planets' misaligned magnetic fields.

At present, we don't know enough about the interiors of Uranus and Neptune to assess the correctness of this picture. Our current state of knowledge is summarized in Figure 13.10, which compares the internal structures of the four jovian worlds.

13.6 The Moon Systems of Uranus and Neptune

3 Like Jupiter and Saturn, both Uranus and Neptune have extensive moon systems, each consisting of a few large moons, long known from Earth, and many smaller moonlets, most of them discovered by *Voyager 2*.

URANUS'S MOONS

William Herschel discovered and named the two largest of Uranus's five major moons, Titania and Oberon, in 1789. British astronomer William Lassell found the next largest, Ariel and Umbriel, in 1851. Gerard Kuiper found the smallest, Miranda, in 1948. In order of increasing distance from the planet, they are Miranda (at 5.1 planetary radii), Ariel (at 7.5), Umbriel (at 10.4), Titania (at 17.1), and Oberon (at 22.8). Ten smaller moons discovered by *Voyager 2* all lie inside the orbit of Miranda. Many of them are intimately related to Uranus's ring system. All these moons revolve in Uranus's skewed equatorial plane, almost perpendicular to the ecliptic, in circular, tidally locked orbits. Their properties are listed in Table 13.1. Because these satellites share Uranus's odd orientation, they experience the same extreme seasons as their parent planet.

In September 1997 two new moons were discovered orbiting Uranus. Interestingly, they were found not by visiting spacecraft (there have been none since *Voyager 2*, and none are planned), nor by the latest generation of telescopes, such as *Hubble* or Keck. Instead, these newest satellites were discovered by astronomers using one of the oldest instruments still in daily use—the 5-m Hale Telescope on Mount Palomar in California. ∞ (Sec. 5.1) Orbiting much farther (about 6 million km) from Uranus than the other known moons, on retrograde, highly inclined orbits, these moons,

TABLE 13.1 The Moons of Uranus

NAME	DISTANCE FROM URANUS (km)	(planet radii)	ORBIT PERIOD (days)	SIZE (longest diameter, km)	MASS (Earth Moon masses)	DENSITY (kg/m³)	(g/cm³)
Cordelia	49,800	1.95	0.34	26			
Ophelia	53,800	2.10	0.38	32			
Bianca	59,200	2.31	0.43	44			
Cressida	61,800	2.42	0.46	66			
Desdemona	62,700	2.45	0.47	58			
Juliet	64,400	2.52	0.49	84			
Portia	66,100	2.59	0.51	110			
Rosalind	69,900	2.74	0.56	58			
Belinda	75,300	2.94	0.62	68			
Puck	86,000	3.36	0.76	150			
Miranda	130,000	5.08	1.41	480	0.00090	1100	1.1
Ariel	191,000	7.48	2.52	1160	0.018	1600	1.6
Umbriel	266,000	10.4	4.14	1170	0.016	1400	1.4
Titania	436,000	17.1	8.71	1580	0.048	1700	1.7
Oberon	583,000	22.8	13.5	1520	0.041	1600	1.6
S/1997U1*	5,700,000	223	400	80			
S/1997U2	5,750,000	225	−415†	160			

*Orbital parameters not well determined.
†Indicates a retrograde orbit.

like the outer moons of Jupiter, and Phobos and Deimos of Mars, are most likely debris captured from interplanetary space following a grazing encounter with the planet's atmosphere. S/1997U1 and S/1997U2 are just 80 km and 160 km in diameter, respectively, and are the faintest moons ever to have been imaged with a ground-based telescope.

The five largest Uranian moons are similar in many respects to the six midsized moons of Saturn. Their densities lie in the range 1100–1700 kg/m³, suggesting a composition of ice and rock, like Saturn's moons, and their diameters range from 1600 km for Titania and Oberon, to 1200 km for Umbriel and Ariel, to 480 km for Miranda. Uranus has no moons comparable to the Galilean satellites of Jupiter, nor to Saturn's single large moon, Titan. Figure 13.11 shows Uranus's five large moons to scale, along with Earth's Moon for comparison.

The two outermost moons, Titania and Oberon, are heavily cratered and show little indication of geological activity. Their overall appearance (and quite possibly their history) is comparable to that of Saturn's moon Rhea, except that they lack Rhea's wispy streaks. ∞ (Sec. 12.5) Also, like all Uranus's moons, they are considerably less reflective than Saturn's satellites, suggesting that their icy surfaces are quite dirty.

One possible reason for this may simply be that the planetary environment in the vicinity of Uranus and Neptune contains more small "sooty" particles than does the solar system closer to the Sun. An alternative explanation, now considered more likely by many planetary scientists, cites the effects of radiation and high-energy particles that strike the surfaces of these moons. These impacts tend to break up the molecules on the moons' surfaces, eventually leading to chemical reactions that slowly build up a layer of dark, organic material. This **radiation darkening** is thought to contribute to the generally darker coloration of many of the moons and rings in the outer solar system. In either case, the longer a moon has been inactive and untouched by meteoritic impact, the darker its surface should be.

The darkest of the moons of Uranus is Umbriel. It displays little evidence of any past surface activity; its only mark of distinction is a bright spot about 30 km across, of unknown origin, on its sunward side. By contrast, Ariel, similar in size to Umbriel but closer to Uranus, does appear to have experienced some activity in the past. It shows signs of resurfacing in places and exhibits surface cracks a little like those seen on another of Saturn's moons, Tethys. However, unlike Tethys, whose cracks are probably due to meteoritic impact, Ariel's activity probably occurred when internal forces and external tidal stresses (due to the gravitational pull of Uranus) distorted the moon and cracked its surface.

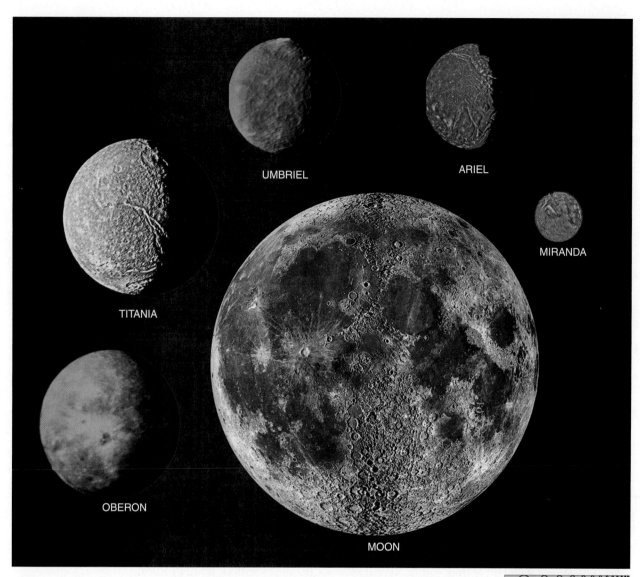

Figure 13.11 The five largest moons of Uranus, to scale. In order of increasing distance from the planet, they are Miranda, Ariel, Umbriel, Titania, and Oberon. Earth's moon is shown for comparison. The appearance, structure, and history of Titania and Oberon may be quite similar to those of Saturn's moon Rhea. The smallest details visible on both moons are about 15 km across. Umbriel is one of the darkest bodies in the solar system, although it has a bright white spot on its sunward side. Ariel is similar in size but has a brighter surface, with signs of past geological activity. Resolution is approximately 10 km.

Strangest of all Uranus's icy moons is Miranda, shown in Figure 13.12. Before the *Voyager 2* encounter, astronomers expected that Miranda would resemble Mimas, the moon of Saturn whose size and location it most closely approximates. However, instead of being a relatively uninteresting cratered, geologically inactive world, Miranda displays a wide range of surface terrains, including ridges, valleys, large oval faults, and many other tortuous geological features.

In order to explain why Miranda seems to combine so many different types of surface features, some researchers have hypothesized that this baffling object has been catastrophically disrupted several times (from within or with-

out), with the pieces falling back together in a chaotic, jumbled way. Certainly, the frequency of large craters on the outer moons suggests that destructive impacts may once have been quite common in the Uranus system. It will be a long time, though, before we can obtain more detailed information to test this theory.

NEPTUNE'S MOONS

From Earth we can see only two moons orbiting Neptune. William Lassell discovered the inner moon, Triton, in 1846. The outer moon, Nereid, was located by Gerard Kuiper in 1949. *Voyager 2* discovered six additional moons, all less than

R I V U X G

Figure 13.12 Miranda, an asteroid-sized moon of Uranus photographed by *Voyager 2*, has a strange, fractured surface suggestive of a violent past, but the cause of the grooves and cracks is presently unknown. The resolution here is about 1 km.

a few hundred kilometers across and all lying within Nereid's orbit. Neptune's known moons are listed in Table 13.2.

In its moons we find Neptune's contribution to our list of solar system peculiarities. Unlike the other jovian worlds, Neptune has no regular moon system—that is, no moons on roughly circular, equatorial, prograde orbits. The larger moon, Triton, is 2700 km in diameter and occupies a circular retrograde orbit 355,000 km (14.3 planetary radii) from the planet, inclined at about 20° to Neptune's equatorial plane. It is the only large moon in our solar system to have a retrograde orbit. The other moon visible from Earth, Nereid, is only 340 km across. It orbits Neptune in the prograde sense, but on an elongated trajectory that brings it as close as 1.4 million km to the planet and as far away as 9.7 million km.

Nereid is probably similar in both size and composition to Neptune's small inner moons.

Voyager 2 approached to within 24,000 km of Triton's surface, providing us with essentially all that we now know about that distant, icy world. Astronomers redetermined the moon's radius (which was corrected downward by about 20 percent) and measured its mass for the first time. Along with Saturn's Titan and the four Galilean moons of Jupiter, Triton is one of the six large moons in the outer solar system. Triton is the smallest of them, with about half the mass of the next smallest, Jupiter's Europa.

Lying 4.5 billion km from the Sun, and with a fairly reflective surface, Triton has a surface temperature of just 37 K. It has a tenuous nitrogen atmosphere, perhaps a hundred

TABLE 13.2 The Moons of Neptune

NAME	DISTANCE FROM NEPTUNE (km)	(planet radii)	ORBIT PERIOD (days)	SIZE (longest diameter, km)	MASS (Earth Moon masses)	DENSITY (kg/m³)	(g/cm³)
Naiad	48,200	1.95	0.29	58			
Thalassic	50,100	2.02	0.31	80			
Designate	52,500	2.12	0.33	150			
Galatea	62,000	2.50	0.43	160			
Larissa	73,500	2.97	0.55	210			
Proteus	118,000	4.75	1.12	440			
Triton	355,000	14.3	−5.88*	2710	0.292	2100	2.1
Nereid	5,510,000	223	360	340	0.0000034	1200	1.2

*Indicates a retrograde orbit

R I V U X G

Figure 13.13   The south polar region of Triton, showing a variety of terrains, ranging from deep ridges and gashes to what appear to be frozen water lakes, all indicative of past surface activity. The pinkish region at the right is nitrogen frost, forming the moon's polar cap. Resolution is about 4 km.

thousand times thinner than Earth's, and a surface that most likely consists primarily of water ice. A *Voyager 2* mosaic of Triton's south polar region is shown in Figure 13.13. The moon's low temperatures produce a layer of nitrogen frost that forms and evaporates over the polar caps, a little like the carbon dioxide frost responsible for the seasonal caps on Mars. The frost is visible as the pinkish region on the right of the figure.

Overall, there is a marked lack of cratering on Triton, presumably indicating that surface activity has obliterated the evidence of most impacts. There are many other signs of an active past. Triton's face is scarred by large fissures similar to those seen on Ganymede, and the moon's odd cantaloupe-like terrain may indicate repeated faulting and deformation over the moon's lifetime. In addition, Triton has numerous frozen "lakes" of water ice (Figure 13.14), which are believed to be volcanic in origin.

Triton's surface activity is not just a thing of the past. As *Voyager 2* passed the moon its cameras detected two great jets of nitrogen gas erupting from below the surface, rising several kilometers into the sky. It is thought that these "geysers" result when liquid nitrogen below Triton's surface is heated and vaporized by some internal energy source, or perhaps even by the Sun's feeble light. Vaporization produces high pressure, which forces the gas through cracks and fissures in the crust, creating the displays *Voyager 2* saw. Scientists conjecture that nitrogen geysers may be very common on Triton and are perhaps responsible for much of the moon's thin atmosphere.

The event or events that placed Triton on a retrograde orbit and Nereid on such an eccentric path are unknown, but they are the subject of considerable speculation. Triton's peculiar orbit and surface features suggest to some astronomers that the moon did not form as part of the Neptune system but instead was captured, perhaps not too long ago. Other astronomers, basing their views on Triton's chemical composition, maintain that it formed as a "normal" moon but was later kicked into its abnormal orbit by some catastrophic event, such as an interaction with another similar-sized body. It has even been suggested that the planet Pluto may have played a role in this process (see Section 13.10), although no really convincing demonstration of such an encounter has ever been presented.

The surface deformations on Triton certainly suggest fairly violent and relatively recent events in the moon's past.

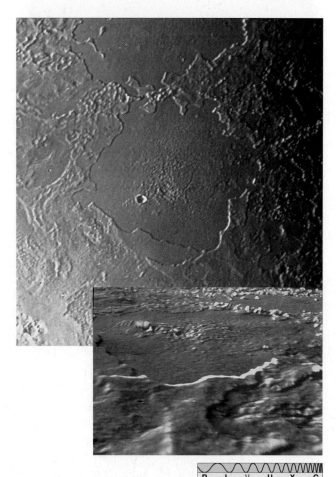

R I V U X G

Figure 13.14   Scientists believe that this lakelike feature on Triton may have been caused by the eruption of an ice volcano. The water "lava" has since solidified, leaving a smooth surface. The absence of craters indicates that this eruption was a relatively recent event in Triton's past. The nearly circular feature at the center of this image spans some 200 km in diameter; its details are resolved to a remarkable 1 km. The insert is a computer-generated view illustrating the topographic relief of the same area.

PLANETARY DATA:
Neptune

Orbital semi-major axis	30.07 A.U. 4498 million km
Orbital eccentricity	0.009
Perihelion	29.81 A.U. 4460 million km
Aphelion	30.33 A.U. 4537 million km
Mean orbital speed	5.43 km/s
Sidereal orbital period	163.7 tropical years
Synodic orbital period	367.49 solar days
Orbital inclination to the ecliptic	1.77°
Greatest angular diameter, as seen from Earth	2.4″
Mass	1.02×10^{26} kg 17.15 (Earth = 1)
Equatorial radius	24,766 km 3.88 (Earth = 1)
Mean density	1638 kg/m³ 0.297 (Earth = 1)
Surface gravity	11.14 m/s² 1.14 (Earth = 1)
Escape speed	23.5 km/s
Sidereal rotation period	0.67 solar days
Axial tilt	29.6°
Surface magnetic field	0.43 (Earth = 1)
Magnetic axis tilt relative to rotation axis	46.0°
Mean surface temperature	59 K
Number of Moons	8

However, they were most likely caused by the tidal stresses produced in Triton as Neptune's gravity circularized its orbit and synchronized its spin, and they give little indication of the processes responsible for the orbit.

Whatever its past, Triton's future is fairly clear. Because of its retrograde orbit, the tidal bulge Triton raises on Neptune tends to make the moon spiral *toward* the planet rather than away from it (as our Moon moves away from Earth). ∞ (Sec. 7.6) Thus, Triton is doomed to be torn apart by Neptune's tidal gravitational field, probably in no more than 100 million years or so, the time required for the moon's inward spiral to bring it inside Neptune's Roche limit. ∞ (Sec. 12.4) By that time, it is conceivable that Saturn's ring system may have disappeared, so Neptune will then be the planet in the solar system with spectacular rings!

13.7 The Rings of the Outermost Jovian Planets

THE RINGS OF URANUS

4 All the jovian planets have rings. The ring system surrounding Uranus was discovered in 1977, when astronomers observed it passing in front of a bright star, momentarily dimming its light. Such a **stellar occultation** (Figure 13.15) happens a few times per decade and allows astronomers to measure planetary structures that are too small and faint to be detected directly. The 1977 observation was actually aimed at studying the planet's atmosphere by watching how it absorbed starlight. However, 40 minutes before and after Uranus itself occulted the star, the flickering starlight revealed the presence of a set of rings. The discovery was particularly exciting because, at the time, only Saturn was known to have rings. Jupiter's rings went unseen until *Voyager 1* arrived there in 1979, and those of Neptune were unambiguously detected only in 1989, by *Voyager 2*.

The ground-based observations revealed the presence of a total of nine thin rings. The main rings, in order of increasing radius, are named Alpha, Beta, Gamma, Delta, and Epsilon, and they range from 44,000 to 51,000 km from the planet's center. All lie within the Roche limit of Uranus, which is about 62,000 km from the planet's center. A fainter ring, known as the Eta ring, lies between the Beta and Gamma rings, and three other faint rings, known as 4, 5, and 6, lie between the Alpha ring and the planet itself. In 1986, *Voyager 2* discovered two more even fainter rings, one between Delta and Epsilon and one between ring 6 and Uranus. The main rings are shown in Figure 13.16. More details on the rings are provided in Table 13.3.

Uranus's rings are quite different from those of Saturn. Whereas Saturn's rings are bright and wide with relatively narrow gaps between them, the rings of Uranus are dark, narrow, and widely spaced. With the exception of the Epsilon ring and the diffuse innermost ring, the rings of Uranus are all less than about 10 km wide, and the spacing between them

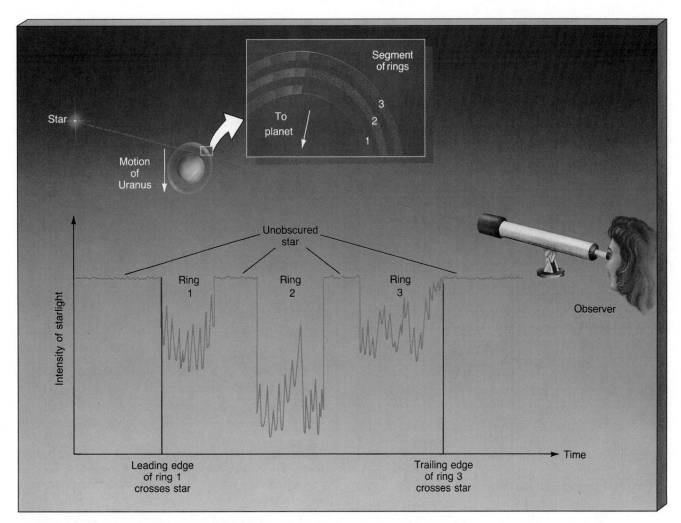

Figure 13.15   Occultation of starlight allows astronomers to detect fine detail on a distant planet. The rings of Uranus were discovered using this technique.

Figure 13.16   The main rings of Uranus, as imaged by *Voyager 2*. All the rings known before *Voyager 2*'s arrival can be seen in this photo. From the inside out, they are 6, 5, 4, Alpha, Beta, Eta, Gamma, Delta, and Epsilon. Resolution is about 10 km, which is just about the width of most of these rings. The two rings discovered by *Voyager 2* are too faint to be seen here.

TABLE 13.3   The Rings of Uranus

RING	INNER RADIUS (km)	(planet radii)	OUTER RADIUS* (km)	(planet radii)	WIDTH (km)
1986U2R	37,000	1.45	39,500	1.55	2500
6	41,800	1.64			2
5	42,200	1.65			2
4	42,600	1.67			3
Alpha	44,700	1.75			4–10
Beta	45,700	1.79			5–11
Eta	47,200	1.83			2
Gamma	47,600	1.86			1–4
Delta	48,300	1.90			3–7
1986U1R	50,000	1.96			2
Epsilon	51,200	2.00			20–100

Most of Uranus's rings are so thin that there is little difference between their inner and outer radii.

Figure 13.17 A close-up of Uranus's Epsilon ring, showing some of its internal structure. The width of the ring averages 30 km; special image processing has magnified the resolution to about 100 m.

ranges from a few hundred to about a thousand kilometers. However, like Saturn's rings, all Uranus's rings are less than a few tens of meters thick (that is, measured in the direction perpendicular to the ring plane).

The density of particles within the rings themselves is comparable to that found in Saturn's A and B rings. The particles that make up Saturn's rings range in size from dust grains to boulders, but in the case of Uranus, the particles show a much smaller spread—few if any are smaller than a centimeter or so in diameter. The ring particles are also considerably less reflective than Saturn's ring particles, possibly because they are covered with the same dark material as Uranus's moons. The Epsilon ring (shown in detail in Figure 13.17), exhibits properties a little like those of Saturn's F ring. It is slightly eccentric (its eccentricity is 0.008) and of variable width, although no braids are found. It also appears to be composed of ringlets.

Like the F ring of Saturn, Uranus's narrow rings require shepherding satellites to keep them from diffusing away. In fact, the theory of shepherd satellites was first worked out to explain the rings of Uranus, which had been detected by stellar occultation even before *Voyager 2*'s Saturn encounter. Thus, the existence of the F ring did not come as quite such a surprise as it might have otherwise! Presumably, many of the small inner satellites of Uranus play some role in governing the appearance of the rings. *Voyager 2* detected the shepherds of the Epsilon ring, Cordelia and Ophelia (see Figure 13.18). Many other, undetected, shepherd satellites must also exist.

THE RINGS OF NEPTUNE

As shown in Figure 13.19 and presented in more detail in Table 13.4, Neptune is surrounded by five dark rings. Three are quite narrow, like the rings of Uranus; the other two are broad and diffuse, more like Jupiter's ring. The dark coloration probably results from radiation darkening, as discussed earlier in the context of the moons of Uranus. All the rings lie within Neptune's Roche limit. The outermost (Adams) ring is noticeably clumped in places. From Earth we see not a com-

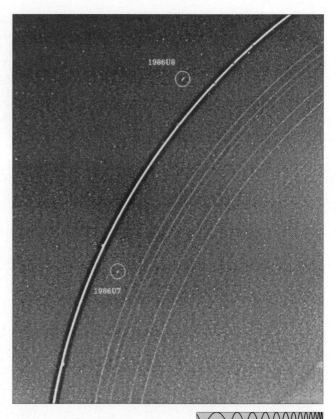

Figure 13.18 These two small moons, discovered by *Voyager 2* in 1986 and now named Cordelia and Ophelia, tend to "shepherd" Uranus's Epsilon ring, keeping it from diffusing away into space.

plete ring but only partial arcs—the unseen parts of the ring are simply too thin (unclumped) to be detected. The connection between the rings and the planet's small inner satellites has not yet been firmly established, but many astronomers believe that the clumping is caused by shepherd satellites.

TABLE 13.4 The Rings of Neptune

RING	INNER RADIUS		OUTER RADIUS*		WIDTH
	(km)	(planet radii)	(km)	(planet radii)	(km)
Galle (1989N3R)	40,900	1.65	42,900	1.73	2000
Leverrier (1989N2R)	53,200	2.15			100
Lassell (1989N4R)†	53,200	2.15	57,200	2.31	4000
Arago (1989N4R)†	57,200	2.31			100
Adams (1989N1R)	62,900	2.54			50

*Three of Neptune's rings are so thin that there is little difference between their inner and outer radii.
†Lassell and Arago were originally identified as a single ring.

Figure 13.19 Neptune's faint rings. In this long-exposure image, the planet (center) is heavily overexposed and has been artificially blotted out to make the rings easier to see. One of the two faint rings lies between the inner bright ring (Leverrier) and the planet. The others lie between the Leverrier ring and the outer bright ring (known as the Adams ring).

Although all the jovian worlds have ring systems, the rings themselves differ widely from planet to planet. Is there some "standard" way in which rings form around a planet? And is there a standard manner in which ring systems evolve? Or do the processes of ring formation and evolution depend entirely on the particular planet in question? If, as now appears to be the case, ring systems are relatively shortlived, their formation must be a fairly common event. Otherwise, we would not expect to find rings around all four jovian planets at once. There are also many indications that the individual planetary environment plays an important role in determining a ring system's appearance and longevity. Although many aspects of ring formation and evolution are now understood, it must be admitted that no comprehensive theory yet exists.

13.8 The Discovery of Pluto

 By the end of the nineteenth century, observations of the orbits of Uranus and Neptune suggested that Neptune's influence was not sufficient to account for all the irregularities in Uranus's motion. Furthermore, it seemed that Neptune itself might be affected by some other unknown body. Following their success in the discovery of Neptune, astronomers hoped to pinpoint the location of this new planet using similar techniques. One of the most ardent searchers was Percival Lowell, a capable, persistent observer and one of the best-known astronomers of his day. (Recall that he was also the leading proponent of the theory that the "canals" on Mars were constructed by an intelligent race of Martians—see *Interlude 10-1*.)

Basing his investigation primarily on the motion of Uranus (Neptune's orbit was still relatively poorly determined at the time), and using techniques similar to those developed earlier for Neptune by Adams and Leverrier, Lowell set about calculating where the supposed ninth planet should be. He searched for it, without success, during the decade preceding his death in 1916. Not until 14 years later did American astronomer Clyde Tombaugh, working with improved equipment and photographic techniques at the Lowell Observatory, finally succeed in finding Lowell's ninth planet, only 6° away from Lowell's predicted position. The new planet was named Pluto for the Roman god of the dead who presided over eternal darkness (and also because its first two letters and its astrological symbol ♇ are Lowell's initials). Its discovery was announced on March 13, 1930, Percival Lowell's birthday.

On the face of it, the discovery of Pluto looked like another spectacular success for celestial mechanics. Unfortunately, it now appears that the supposed irregularities in the motions of Uranus and Neptune did not exist and that the mass of Pluto, not measured accurately until the 1980s, is far too small to have caused them anyway. The discovery of Pluto owed much more to simple luck than to elegant mathematics!

Some orbital and physical data for Pluto are presented in the Pluto Data box on p. 305. Unlike the paths of the other outer planets, Pluto's orbit is quite elongated, with an eccentricity of 0.25. It is also inclined at 17.2° to the plane of the ecliptic. Here already we have some indication that Pluto is unlike its jovian neighbors. Because of its substantial orbital eccentricity, Pluto's distance from the Sun varies consider-

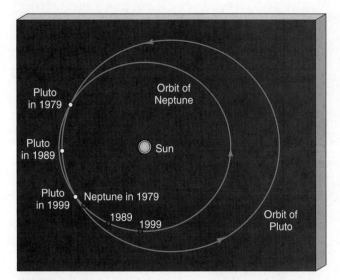

Figure 13.20 The orbits of Neptune and Pluto cross, although Pluto's orbital inclination and a 3:2 resonance prevent the planets from actually coming close to each other. Between 1979 and 1999, Pluto was inside Neptune's orbit, making Neptune the most distant planet from the Sun.

Animation

ably. At perihelion, it lies 29.7 A.U. (4.4 billion km) from the Sun, inside the orbit of Neptune. At aphelion, the distance is 49.3 A.U. (7.4 billion km), well outside Neptune's orbit. Pluto last passed perihelion in 1989, and it will remain inside Neptune's orbit until February, 1999. Its sidereal period is 248.0 years, so the next perihelion passage will not occur until the middle of the 23rd century.

Pluto's orbital period is apparently exactly 1.5 times that of Neptune—the two planets are locked into a 3:2 resonance (two orbits of Pluto for every three of Neptune) as they orbit the Sun. As a result, even though their orbits appear to cross, Pluto and Neptune are in no danger of colliding with each other. Because of the orbital resonance and Pluto's tilted orbit plane, the distance between the two planets at closest approach is actually about 17 A.U. (compare with Pluto's closest approach to Uranus of just 11 A.U.). The orbits of Neptune and Pluto are sketched in Figure 13.20.

At nearly 40 A.U. from the Sun, Pluto is often hard to distinguish from the background stars. As the two photographs of Figure 13.21 indicate, the planet is actually considerably fainter than many stars in the sky. Like Neptune, it is never visible to the naked eye.

13.9 Pluto in Bulk

5 Pluto is so far away that little is known of its physical nature. Until the late 1970s, studies of sunlight reflected from its surface suggested a rotation period of nearly a week, but measurements of its mass and radius were very uncertain. All this changed in July 1978, when astronomers at the U.S. Naval Observatory discovered that Pluto has a satellite. It is now named Charon, after the mythical boatman who ferried the dead across the river Styx into Hades, Pluto's domain. The discovery photograph of Charon is shown in Figure 13.22(a). Charon is the small bump near

Figure 13.21 These two photographs, taken one night apart, show motion of the planet Pluto (arrow) against a field of much more distant stars. Most of Pluto's apparent motion in these two frames is actually due to the orbital motion of Earth rather than that of Pluto.

Charon

(a)

(b)

R I V U X G

Figure 13.22 (a) The discovery photograph of Pluto's moon, Charon. The moon is the small bump on the top right portion of the image. (b) The Pluto–Charon system, to the same scale, as seen by the *Hubble Space Telescope*. The angular separation of the planet and its moon is about 0.9″.

Video

PLANETARY DATA:
Pluto

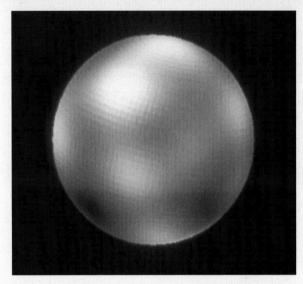

Orbital semi-major axis	39.48 A.U. 5906 million km
Orbital eccentricity	0.249
Perihelion	29.66 A.U. 4437 million km
Aphelion	49.31 A.U. 7376 million km
Mean orbital speed	4.74 km/s
Sidereal orbital period	248.0 tropical years
Synodic orbital period	366.72 solar days
Orbital inclination to the ecliptic	17.15°
Greatest angular diameter, as seen from Earth	0.11″
Mass	1.27×10^{22} kg 0.0021 (Earth = 1)
Equatorial radius	1137 km 0.18 (Earth = 1)
Mean density	2060 kg/m^3 0.374 (Earth = 1)
Surface gravity	0.66 m/s^2 0.067 (Earth = 1)
Escape speed	1.2 km/s
Sidereal rotation period	−6.387 solar days (retrograde)
Axial tilt	118°
Surface magnetic field	unknown
Magnetic axis tilt relative to rotation axis	—
Surface temperature	40 to 60 K
Number of moons	1

the top of the image. Knowing the moon's orbital period of 6.4 days, astronomers could determine the mass of Pluto to much greater accuracy than had previously been possible. It is 0.0021 Earth masses (1.3×10^{22} kg), far smaller than any earlier estimate—more like the mass of a moon than of a planet. In 1990 the *Hubble Space Telescope* imaged the Pluto–Charon system (Figure 13.22b). The improved resolution of that instrument clearly resolved the two bodies and allowed even more accurate measurements of their properties.

The discovery of Charon also allowed astronomers to measure Pluto's radius very precisely. Pluto's angular size is much less than 1″, so its true diameter is blurred by the effects

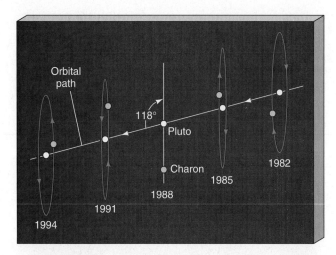

Figure 13.23 The orbital orientation of Charon produced a series of eclipses between 1985 and 1991. Observations of eclipses of Charon by Pluto and of Pluto by Charon have provided detailed information about the sizes and orbits of both bodies.

Animation

of Earth's turbulent atmosphere. ∞ (Sec. 5.3) However, Charon's orbit has given astronomers new insight into the system. By pure chance, Charon's orbit over the 6-year period from 1985 to 1991 (less than 10 years after the moon was discovered) happened to be oriented in such a way that viewers on Earth saw a series of eclipses. Pluto and Charon repeatedly passed in front of each other, as seen from our vantage point. Figure 13.23 sketches this orbital configuration. With more good fortune, these eclipses took place while Pluto was closest to the Sun, making for the best possible Earth-based observations.

Basing their calculations on the variations in reflected light as Pluto and Charon periodically hid each other, astronomers computed their masses and radii and determined their orbit plane. Additional studies of Pluto's surface brightness indicate that the two are tidally locked as they orbit each other. Pluto's diameter is 2270 km, about one-fifth the size of Earth. Charon is about 1300 km across and orbits at a distance of 19,700 km from Pluto. If planet and moon have the same composition (probably a reasonable assumption), Charon's mass must be about one-sixth that of Pluto, giving the Pluto–Charon system by far the largest satellite-to-planet mass ratio in the solar system.

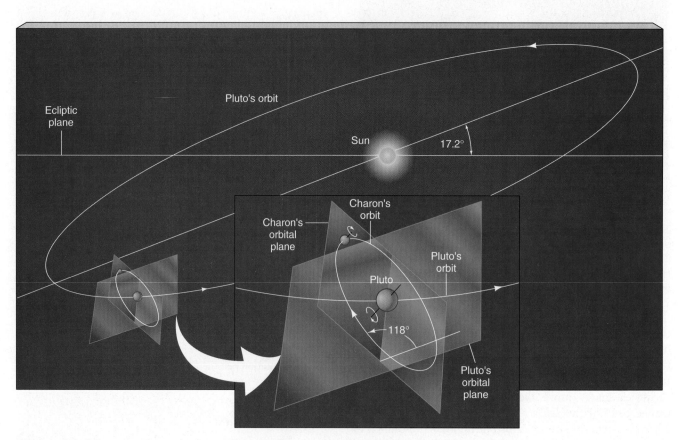

Figure 13.24 Charon's path around Pluto is circular, synchronous, and inclined at 118° to the orbit plane of the Pluto–Charon system about the Sun, which is itself inclined at 17° to the plane of the ecliptic.

As shown in Figure 13.24, Charon's orbit is inclined at an angle of 118° to the plane of Pluto's orbit around the Sun. Since the spins of both planet and moon are perpendicular to the plane of Charon's orbit around Pluto, the geographic "north" poles of both bodies lie below the plane of Pluto's orbit. Thus, Pluto is the third planet in the solar system (along with Venus and Uranus) found to have retrograde rotation.

The known mass and radius of Pluto allow us to determine its average density, which is 2100 kg/m³—too low for a terrestrial planet, but far too high for a mixture of hydrogen and helium of that mass. Instead, the mass, radius, and density of Pluto are just what we would expect for one of the icy moons of a jovian planet. In fact, Pluto is quite similar in both mass and radius to Neptune's large moon, Triton. The planet is almost certainly made up mostly of water ice.

Spectroscopy reveals the presence of *frozen* methane as a major surface constituent. Pluto is the only planet in the solar system on which methane exists in the solid state, implying that the surface temperature on Pluto is no more than 50 K. Pluto may also have a thin methane atmosphere, associated with the methane ice on its surface. Recent computer-generated maps have begun to hint at surface features on Pluto (see *Interlude 13-1*). Similar studies indicate that Charon may have bright polar caps, but their composition and nature are as yet unknown.

13.10 The Origin of Pluto

Because Pluto is neither terrestrial nor jovian in its make-up, and because of its similarity to the ice moons of the outer planets, some researchers suspect that Pluto is not a "true" planet at all. This view is bolstered by Pluto's eccentric, inclined orbit, which is quite unlike the orbits of the other known planets.

Soon after Pluto was discovered, its orbital association with Neptune suggested to some theorists that a catastrophic encounter of some sort might have ejected Pluto from its original orbit around Neptune and perhaps even simultaneously knocked Triton onto its present retrograde path. Although this was an attractive theory at the time, a plausible sequence of events that could account in detail for the present orbits of those bodies proved elusive. Furthermore, since 1978 the picture has been greatly complicated by the presence of Charon. It was much easier to suppose that Pluto was an escaped moon before we learned that it had a moon of its own. As a result, few astronomers still regard this theory as a likely explanation of Pluto's origin.

Pluto may be just what it seems—a planet that formed in its current orbit, possibly even with its own moon right from the outset. Because we know so little about the environment in the outer solar system, we cannot yet rule out the possibility that planets beyond Neptune should simply look like Pluto. However, there is mounting evidence that this is not the case either. Astronomers now know of numerous large chunks of ice circulating in interplanetary space beyond the orbits of the jovian planets (see Chapter 14), and many researchers believe that there may have been thousands of Pluto-sized objects initially present in the outer solar system. Indeed, as we will see in Chapter 15, such objects are the natural building blocks of planets, according to current theories of solar system formation.

Almost all these Pluto-sized objects have since been "kicked" out to larger distances from the Sun following gravitational interactions with Uranus and Neptune. The capture of a few of them by the giant planets could explain some of the strange moons of the outer worlds, especially Triton, and if there were enough moon-sized chunks originally orbiting beyond Neptune, it is quite plausible that Pluto could have captured Charon following a collision (or near-miss) between the two. Pluto remained behind after the other bodies were captured or ejected, perhaps because its orbital resonance with Neptune prevented it from ever experiencing a close encounter with that (or any other) planet.

At present, our scant knowledge of the compositions of the two bodies does not allow us to confirm or disprove either the "coformation" or the "capture" theory of the Pluto–Charon system. Unfortunately, this uncertainty may persist for some time—there is no present or proposed space mission that might suddenly and radically improve our understanding of these distant worlds.

INTERLUDE 13-1 Surface Detail on Planet Pluto

Pluto is the only planet for which we have very little detailed surface information, because Pluto is the only planet to date that hasn't been visited by a robot spacecraft. Now new models of Pluto's surface have been constructed, based mainly on images taken by the *Hubble Space Telescope*. The first figure shows two samples of a dozen snapshots of Pluto, taken at visible and ultraviolet wavelengths throughout the planet's 6.4-day rotation period. The raw images are the small inserts at the upper left of each frame. The clearer ball of Pluto shown in each of these frames is a modeled view, that is, a computer-generated sketch of the planet's surface based on modeling of all available data.

The second figure is a complete surface map of Pluto, created by carefully collating and analyzing all the data. It is good confirmation of rougher maps made over the previous decade using ground-based observations of eclipses of Pluto by its moon, Charon. Clearly, we have now progressed beyond the stage of seeing Pluto merely as a fuzzy, distant dot of light. With maps like these, astronomers can monitor its surface detail about as well as you can study nearby Mars with a small backyard telescope.

These models reveal quite a lot of large-scale contrast—in fact, surprisingly for a body composed almost entirely of ice, more contrast than on any other planet except Earth. About a dozen "provinces" can be identified, including icy-bright polar cap regions, another bright spot seen rotating with the planet itself, a cluster of dark spots, and a few peculiar linear markings. Some of these spots could turn out to be craters or impact basins, such as on Earth's Moon, but much of Pluto's contrast is probably caused by frosts and snows that migrate across the planet as it goes through its seasonal cycles. The reflectivities of the lighter areas resemble that of fresh snow, whereas the darker areas are more akin to the subdued brightness of dirty snow. Let's just hope that people don't start talking about the odd linear markings as being reminiscent of canals!

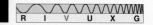

Chapter Review *WWW*

SUMMARY

The outer planets Uranus, Neptune, and Pluto were unknown to ancient astronomers. Uranus was discovered in the eighteenth century, by chance. Neptune was discovered after mathematical calculations of Uranus's slightly non-Keplerian orbit revealed the presence of an eighth planet.

At opposition, Uranus is barely visible to the unaided eye. It appears as a pale green disk through a telescope. Neptune cannot be seen with the naked eye, but a telescope shows it as a tiny bluish disk. Today we know the giant planets Uranus and Neptune mainly through data taken by *Voyager 2*. Small, remote Pluto has not been visited by a spacecraft, and our knowledge of it stems from painstaking observations from Earth and the *Hubble Space Telescope*.

The masses of the outer planets are determined from measurements of their orbiting moons. The radii of Uranus and Neptune were relatively poorly known until the *Voyager 2* flybys in the 1980s. Uranus and Neptune have similar bulk properties; their densities imply large, rocky cores making up a greater fraction of the planets' masses than in either Jupiter or Saturn.

Surface features are barely discernible on Uranus, but computer-enhanced images from *Voyager 2* revealed atmospheric clouds and flow patterns moving beneath Uranus's haze. Neptune, although farther away from us, has atmospheric features that are clearer because of warmer temperatures and less haze. **The Great Dark Spot** (p. 293) on Neptune had many similarities to Jupiter's Red Spot. It disappeared in 1994.

For unknown reasons, Uranus's spin axis lies nearly in the ecliptic plane, leading to extreme seasonal variations in solar heating on the planet as it orbits the Sun.

Unlike the other jovian planets, Uranus has no excess heat emission. The source of Neptune's excess energy, like that of Jupiter's, is most likely heat left over from the planet's formation.

Both Uranus and Neptune have substantial magnetospheres. *Voyager 2* discovered that the magnetic fields of both planets are tilted at large angles to the planets' rotation axes. The reason for this is not known.

All but two of Uranus's moons revolve in the planet's equatorial plane, almost perpendicular to the ecliptic, in circular, synchronous orbits. Like the moons of Saturn, the medium-sized moons of Uranus are made up predominantly of rock and water ice. Many of them are heavily cratered and in some cases must have come close to being destroyed by the meteoritic impacts whose craters we now see. The strange moon Miranda has geological features that suggest repeated violent impacts in the past.

Neptune's moon Triton has a fractured surface of water ice and a thin atmosphere of nitrogen, probably produced by nitrogen "geysers" on its surface. Triton is the only large moon in the solar system to have a retrograde orbit around its parent planet. This orbit is unstable and will eventually cause Triton to be torn apart by Neptune's gravity.

Uranus has a series of dark, narrow rings, first detected from Earth by **stellar occultation** (p. 300)—their obscuration of the light received from background stars. Shepherd satellites are responsible for the rings' thinness. Neptune has three narrow rings like Uranus's and one broad ring, like Jupiter's. They were discovered by *Voyager 2*. The dark coloration of both the rings and the moons of the outer giant planets may be due to **radiation darkening** (p. 296), whereby exposure to solar high-energy radiation slowly causes a dark hydrocarbon layer to build up on a body's icy surface.

Pluto was discovered in the twentieth century after a laborious search for a planet that was supposedly affecting Uranus's orbital motion. We now know that Pluto is far too small to have any detectable influence on Uranus's path.

Pluto has a moon, Charon, whose mass is about one-sixth that of Pluto itself. Studies of Charon's orbit around Pluto have allowed the masses and radii of both bodies to be accurately determined. Pluto is too small for a terrestrial planet. Its properties are far more moonlike than planetlike. Most astronomers believe that it is probably the largest (or nearest) member of a class of icy asteroids found in the outer solar system.

SELF-TEST: TRUE OR FALSE?

_____ **1.** Uranus was discovered by Galileo.

_____ **2.** After the discovery of Uranus, astronomers started looking for other planets and quickly discovered Neptune.

_____ **3.** Since its discovery Uranus has completed just two and a half orbits of the Sun.

_____ **4.** Neptune has a smaller radius than Uranus and therefore a smaller mass and density.

_____ **5.** The rotation rates of Uranus and Neptune are almost identical.

_____ **6.** During the northern summer of Uranus, an observer near the north pole would observe the Sun high and almost stationary in the sky.

_____ **7.** Both Uranus and Neptune have a layer of metallic hydrogen surrounding their central cores.

_____ **8.** Relative to the size of the planet, the rocky cores of Uranus and Neptune are larger than the cores of Jupiter and Saturn.

_____ **9.** The system of moons around Uranus have orbits that share the tilt of the planet.

_____ **10.** Uranus has no large moons.

_____ **11.** The surfaces of the largest Uranian moons are darker than similar moons of Saturn.

_____ **12.** Triton's surface has a marked lack of cratering, indicating significant amounts of surface activity.

_____ **13.** Pluto's moon, Charon, is more than half the size of Pluto.

_____ **14.** Pluto is larger than Earth's Moon.

_____ **15.** Astronomers have firm evidence that Pluto is an escaped moon of Neptune.

SELF-TEST: FILL IN THE BLANK

1. Uranus is about _____ more distant from the Sun than is Saturn.

2. Uranus's rotation axis is almost _____ to the ecliptic plane.

3. Whereas the abundance of hydrogen and helium for the jovian planets remains more or less constant among the four planets, the relative abundance of _____ increases with increasing distance from the Sun.

4. The planet _____ is blue-green in color, and virtually featureless.

5. The planet _____ is dark blue in color, with white cirrus clouds and visible storm systems.

6. The _____ of both Uranus and Neptune are highly tilted relative to their rotation axis and significantly offset from the planets' centers.

7. The Uranian moon that shows the greatest amount of geological activity and disruption over time is _____.

8. *Voyager 2* discovered six of the _____ known moons of Neptune.

9. Triton's orbit is unusual because it is _____.

10. Nereid's orbit is unusual because it has a high orbital _____ .

11. Like only one other moon in the solar system, Triton has an _____.

12. Nitrogen geysers are found on _____.

13. The orbit of Pluto is locked into a 3:2 resonance with _____.

14. Overall, Pluto is most similar to which object in the solar system? _____.

15. The radii of Pluto and Charon were determined accurately by observing a fortuitous series of _____ during the late 1980s.

REVIEW AND DISCUSSION

1. How was Uranus discovered?

2. Why did astronomers suspect an eighth planet beyond Uranus?

3. What is unusual about the rotation of Uranus?

4. What is responsible for the colors of Uranus and Neptune?

5. How are the interiors of Uranus and Neptune thought to differ from those of Jupiter and Saturn?

6. What is odd about the magnetic fields of Uranus and Neptune?

7. What is unique about Miranda? Give a possible explanation for this uniqueness.

8. How does Neptune's moon system differ from those of the other jovian worlds? What do these differences suggest about the origin of the moon system?

9. What is the predicted fate of Triton?

10. The rings of Uranus are dark, narrow, and widely spaced. Which of these properties makes them different from the rings of Saturn?

11. How do the rings of Neptune differ from those of Uranus and Saturn?

12. How was Pluto discovered?

13. How were the mass and radius of Pluto determined?

14. In what respect is Pluto more like a moon than a jovian or terrestrial planet?

15. Why was the discovery of Uranus in 1781 so surprising? Might there be similar surprises in store for today's astronomers?

PROBLEMS

1. Calculate the time between successive closest approaches of (a) Neptune and Uranus (b) Pluto and Neptune. Compare your answers with Neptune's sidereal orbital period.

2. What is the gravitational force exerted on Uranus by Neptune, at closest approach? Compare your answer with the Sun's gravitational force on Uranus.

3. What is the angular diameter of the Sun, as seen from Neptune? Compare it with the angular diameter of Triton. Would you expect solar eclipses to be common on Neptune?

4. From Wien's law, at what wavelength does Pluto's thermal emission peak? In what part of the electromagnetic spectrum does this lie? (See *More Precisely 3-2* and Figure 3.11.)

5. Assuming a spherical shape and a uniform density of 2000 kg/m^3, calculate how small an icy moon of one of the outer planets would have to be before a fastball pitched at 40 m/s (about 90 mph) could escape.

6. Add up the masses of all the moons of Uranus, Neptune, and Pluto. (Neglect the masses of the small moons—they contribute little to the result.) How does this sum compare with the mass of Earth's Moon, and with the mass of Pluto?

7. Astronomers on Earth are observing the occultation of a star by Uranus and its rings (see Figure 13.18). It so happens that the event is occurring when Uranus is at opposition, and the center of the planet appears to pass directly across the star. Assuming circular planetary orbits and, for simplicity, taking the rings to be face-on, calculate (a) how long the 90-km-wide Epsilon ring takes to cross the line of sight, and (b) the time interval between the passage of the Alpha ring and that of the Epsilon ring. (Hint: Most of Uranus's apparent motion is actually due to Earth's motion in its orbit around the Sun.)

8. What would be your weight on Pluto? On Charon?

9. How close is Charon to Pluto's Roche limit?

10. What is the round-trip travel time of light from Earth to Pluto (at a distance of 40 A.U.)? How far would a spacecraft orbiting Pluto at a speed of 0.5 km/s travel during that time?

PROJECTS

1. The major astronomy magazines *Sky and Telescope* and *Astronomy* print charts showing the whereabouts of the planets in their January issues. Consult one of these charts, and locate Neptune and Uranus in the sky. Uranus may be visible to the naked eye, but binoculars make the search much easier. (Hint: Uranus shines more steadily than the background stars.) With the eye alone, can you detect a color to Uranus? Through binoculars?

2. The search for Neptune requires a much more determined effort! A telescope is best, but high-powered binoculars mounted on a steady support will do. If you can see both planets through a telescope—and they will remain close together on the sky for the rest of this century—contrast their colors. Which planet appears bluer? Through a telescope, does Uranus show a disk? Can you see that Neptune shows a disk, or does it look more like a point of light?

14 SOLAR SYSTEM DEBRIS

Keys to Our Origin ✦ *www*

LEARNING GOALS

Studying this chapter will enable you to:

1   Describe the orbital properties of the major groups of asteroids.

2   Summarize the composition and physical properties of a typical asteroid.

3   Explain the effect of orbital resonances on the structure of the asteroid belt.

4   Detail the composition and structure of a typical comet, and explain the formation and appearance of its tail.

5   Discuss the characteristics of cometary orbits and what they tell us about the probable origin of comets.

6   Distinguish among the terms meteor, meteoroid, and meteorite.

7   Summarize the orbital and physical properties of meteoroids, and explain what these suggest about their probable origin.

(Opposite page) A truly great comet—one of the brightest of the twentieth century—illuminated the skies of the northern hemisphere in early 1997. Comet Hale–Bopp is shown here in this wide-angle photograph, taken at the moment of perihelion, or closest approach to the Sun, yet still some 0.9 A.U. away. Note how the gas (glowing bluish light) boiling off the comet extends straight away from the Sun, while the comet's heavier dust particles (reflected whitish light) lag behind, forming a more gently curved tail. The full extent of the tail measured nearly 40 degrees, sweeping a huge arc across the nighttime sky. Almost surely, Hale–Bopp was the most-photographed comet in history.

According to classical definitions, there are only nine planets in the solar system. But several thousand other celestial bodies are also known to revolve around the Sun. These minor objects—the asteroids and comets—are small and of negligible mass compared with the planets and their major moons. Yet each is a separate world, with its own story to tell about the early solar system. On the basis of statistical deductions, astronomers estimate that there are more than a billion such objects still to be discovered. They may seem to be only rocky and icy "debris," but more than the planets themselves, they hold a record of the formative stages of our planetary system. Many are nearly pristine, unevolved bodies with much to teach us about our local origins.

14.1 Asteroids

Asteroids are relatively small, rocky objects that revolve around the Sun. Their name literally means "starlike bodies," but asteroids are definitely not stars. They are too small even to be classified as planets. Astronomers often refer to them as "minor planets," or sometimes "planetoids."

Asteroids differ from planets in both their orbits and their size. They generally move on quite eccentric trajectories between Mars and Jupiter, unlike the almost circular paths of the major planets. Few are larger than 300 km in diameter, and most are far smaller—as small as a tenth of a kilometer across. The largest known asteroid, Ceres, is just 1/10,000 the mass of Earth and measures only 940 km across. Taken together, the known asteroids amount to less than 1/10 the mass of the Moon, so they do not contribute significantly to the total mass of the solar system.

ORBITAL PROPERTIES

1 European astronomers discovered the first asteroids early in the nineteenth century as they searched the sky for an additional planet orbiting between Mars and Jupiter, where the Titius–Bode "law" (see *Interlude 6-1*) suggested one might be found. Italian astronomer Giuseppe Piazzi was the first to discover an asteroid. He detected Ceres in 1801 and measured its orbital semi-major axis to be 2.8 A.U.—exactly where the "law" predicted. Within a few years, three more asteroids—Pallas (at 2.8 A.U.), Juno (at 2.7 A.U.), and Vesta (at 3.4 A.U.)—were discovered.

By the start of the twentieth century, astronomers had cataloged several hundred asteroids with well-determined orbits. Now, at the end of that century, the list numbers over 7000. The total number of potentially visible asteroids (that is, asteroids that would be visible through present-day telescopes if we knew just where and when to look) may exceed 100,000. The vast majority of these bodies are found in a region of the solar system known as the **asteroid belt**, located between 2.1 and 3.3 A.U. from the Sun—roughly midway between the orbits of Mars (at 1.5 A.U.) and Jupiter (at 5.2 A.U.). All but one of the known asteroids revolve about the Sun in prograde orbits, in the same sense as the planets. The overall layout of the asteroid belt is sketched in Figure 14.1.

Such a compact concentration of asteroids in a well-defined belt suggests that they are either the fragments of a planet broken up long ago or primal rocks that never managed to accumulate into a genuine planet. On the basis of the best evidence currently available, researchers favor the latter view. There is far too little mass in the belt to constitute a planet, and the marked chemical differences among individual asteroids strongly suggest that they could not all have originated in a single body. Instead, astronomers believe that the strong gravitational field of Jupiter continuously disturbs the motions of these chunks of primitive matter, nudging and pulling at them, preventing them from aggregating into a planet.

PHYSICAL PROPERTIES

2 With few exceptions, asteroids are too small to be resolved by Earth-based telescopes, so astronomers must rely on indirect methods to find their sizes, shapes, and composition. Consequently, only a few of their physical and chemical properties are accurately known. To the extent that astronomers can determine their compositions, asteroids have been found to differ not only from the nine known planets and their many moons, but also among themselves.

Asteroids are classified in terms of their spectroscopic properties. The darkest, or least reflective, asteroids contain a large fraction of carbon in their makeup. They are known as *C-type* (or *carbonaceous*) asteroids. The more reflective *S-type* asteroids contain silicate, or rocky, material. Generally speaking, S-type asteroids predominate in the inner portions of the asteroid belt, and the fraction of C-type bodies steadily increases as we move outward. Overall, about 15 percent of all asteroids are S-type, 75 percent are C-type, and 10 percent are other types (mainly the *M-type* asteroids, containing large fractions of nickel and iron). Many planetary scientists believe that the carbonaceous asteroids consist of very primitive material representative of the earliest stages of the solar system and have not experienced significant heating or chemical evolution since they first formed 4.6 billion years ago.

In most cases, astronomers estimate the sizes of asteroids from the amount of sunlight they reflect and the amount of heat they radiate. These observations are difficult, but size measurements have been obtained in this way for more than 1000 asteroids. On rare occasions, astronomers witness an asteroid occulting a star, allowing them to determine its size and shape with great accuracy. The largest asteroids are roughly spherical, but the smaller ones can be highly irregular.

The three largest asteroids, Ceres, Pallas, and Vesta, have diameters of 940, 580, and 540 km, respectively. Only two dozen or so asteroids are more than 200 km across, and

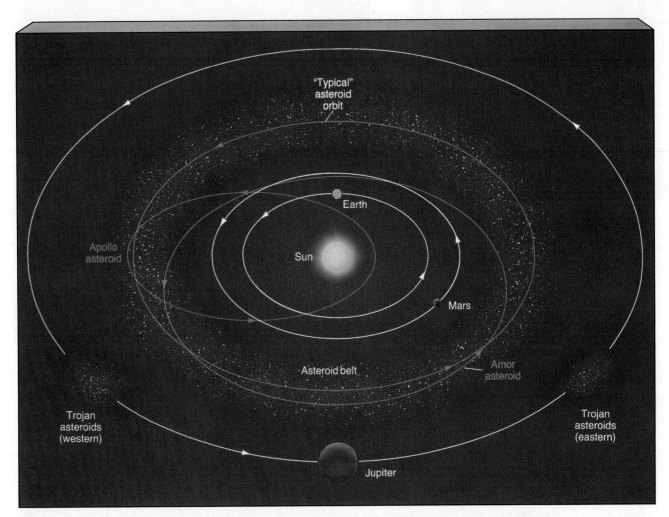

Figure 14.1   The asteroid belt, along with the orbits of Earth, Mars, and Jupiter (not drawn to scale). The main belt, the Trojan asteroids, and some Apollo (Earth-crossing) and Amor (Mars-crossing) orbits are shown.

most are much smaller. Almost assuredly, there exist hundreds of thousands more awaiting discovery. However, observers estimate that they are mostly very small. Probably 99 percent of all asteroids larger than 100 km are known and cataloged, and at least 50 percent of asteroids larger than 10 km are accounted for. Although the vast majority of asteroids are probably less than a few kilometers across, most of the *mass* in the asteroid belt resides in objects greater than a few tens of kilometers in diameter.

The first close-up views of asteroids were provided by the Jupiter probe *Galileo* which, on its rather roundabout path to the giant planet, passed twice through the asteroid belt, making close encounters with asteroid Gaspra in October 1991 and asteroid Ida in August 1993 (Figure 14.2). ∞ (Sec. 6.6) Both Gaspra and Ida are S-type asteroids. Technical problems limited the amount of data that could be sent back from the spacecraft during the flybys. Nevertheless, the images produced by *Galileo* show far more detail than any photographs made from Earth.

Gaspra and Ida are irregularly shaped bodies with maximum diameters of about 20 and 60 km, respectively. They are pitted with craters ranging in size from a few hundred

meters to 2 km across and are covered with a layer of dust of variable thickness. Ida is much more heavily cratered than Gaspra, in part because it resides in a denser part of the asteroid belt. Also, scientists believe that Ida has suffered more from the ravages of time. Ida is about a billion years old, far older than Gaspra, which is estimated to have an age of just 200 million years, based on the extent of cratering. Both asteroids are thought to be fragments of much larger objects that broke up into many smaller pieces following violent collisions long ago.

To the surprise of most mission scientists, closer inspection of the Ida image (Figure 14.2b) revealed the presence of a tiny *moon*, now named Dactyl, just 1.5 km across, orbiting the asteroid at a distance of about 90 km. A few such *binary asteroids* had previously been observed from Earth (for example, the second-largest asteroid, Pallas, is a binary). However, the rare binary systems known before the Ida flyby were all much larger than Ida—a moon the size of Dactyl cannot be detected from the ground. Scientists believe that, given the relative congestion of the asteroid belt, collisions between asteroids may be quite common, providing a source of both interplanetary dust and smaller asteroids and possibly

(a)

(b)

R I V U X G

Figure 14.2 (a) The S-type asteroid Gaspra as seen from a distance of 1600 km by the probe *Galileo* on its way to Jupiter. (b) The S-type asteroid Ida, photographed by *Galileo* from a distance of 3400 km. (Ida's moon, Dactyl, is visible at the right of the photo.) The resolution in these photographs is on the order of 100 m. True-color images showed the surfaces of both bodies to be a fairly uniform shade of gray. Sensors onboard the spacecraft indicated that the amount of infrared radiation absorbed by these surfaces varies from place to place, probably as a result of variations in the thickness of the dust layer blanketing them.

deflecting one or both of the bodies involved onto eccentric, Earth-crossing orbits. The less violent collisions may be responsible for the binary systems we see.

By studying the *Galileo* images, astronomers were able to obtain limited information on Dactyl's orbit around Ida and hence (using Newton's law of gravity—see *More Precisely 2-3*) to estimate Ida's mass at about $5–10 \times 10^{16}$ kg. This information in turn allowed them to measure Ida's density as 2200–2900 kg/m³, a range consistent with its rocky, S-type classification.

As of 1998 only one other asteroid has been studied at such close range. In June 1997 the *Near Earth Asteroid Rendezvous (NEAR)* spacecraft visited the C-type asteroid Mathilde on its way to the mission's main target, the asteroid Eros. Shown in Figure 14.3, Mathilde is some 60 km across. By sensing its gravitational pull, *NEAR* measured Mathilde's mass to be about 10^{17} kg, implying a density of just 1400 kg/m³. To account for this low density, scientists speculate that the asteroid's interior must be quite porous, perhaps like styrofoam in structure. The interior's relatively soft consistency may also help explain the unexpectedly large size of many of the craters observed on Mathilde's surface.

On reaching Eros in February 1999, *NEAR* will go into orbit around it, coming as close as 24 km to the surface. For a period of 1 year, the spacecraft will make detailed measurements, at many different electromagnetic wavelengths, of the S-type asteroid's size, shape, gravitational field, composition, surface and internal structure, and magnetic field. Scientists back on Earth are eagerly awaiting the most detailed view yet of one of Earth's smallest neighbors.

Apart from those of Ida, Mathilde, and (soon) Eros, most asteroid masses are unknown. However, a few of the largest asteroids do have strong enough gravitational fields that their effects on their neighbors can be measured and their masses thereby determined to reasonable accuracy. Their computed densities are generally compatible with the rocky or carbonaceous compositions just described.

EARTH-CROSSING ASTEROIDS

The orbits of most asteroids have eccentricities lying in the range 0.05–0.3, ensuring that they always remain between the orbits of Mars and Jupiter. Very few asteroids have eccentricities greater than 0.4. Those that do are of particular interest to us, however, as their orbits may intersect Earth's orbit, leading to the possibility of a collision with our planet. They are collectively known as **Earth-crossing asteroids**. Those stray asteroids having very elliptical orbits or orbits well inside the main asteroid belt have probably been influenced by the gravitational fields of nearby Mars and especially Jupiter. These planets can disturb normal asteroid orbits, deflecting them into the inner solar system. Asteroids

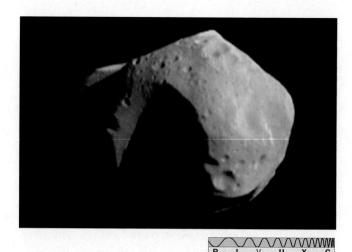

R I V U X G

Figure 14.3 The C-type asteroid Mathilde, imaged by the *NEAR* spacecraft en route to the near-earth asteroid Eros. It measures some 60×50 km, and rotates every 17.5 days. The largest craters visible in this image are about 20 km across—much larger than the craters seen on either Gaspra or Ida. The reason may be the asteroid's low density (approximately 1400 kg/m³) and rather soft composition.

Figure 14.4 The asteroid Icarus has an orbit that passes within 0.2 A.U. of the Sun, well within the orbit of Earth. Icarus occasionally comes close to our own planet, making it one of the better-studied asteroids in the solar system. The asteroid's motion relative to the stars makes it appear as a streak (marked) in this long-exposure photograph, taken in the 1970s.

whose paths cross Earth's orbit are termed *Apollo asteroids* (after the first known Earth-crossing asteroid, Apollo) if their orbital semi-major axes exceed 1 A.U., and *Aten asteroids* otherwise. Those crossing only the orbit of Mars are known as *Amor asteroids*. (See again Figure 14.1.)

Although we are currently aware of only a few dozen Earth-crossing asteroids, they are among the best known because of their occasional close encounters with Earth. For example, the perihelion of the Aten asteroid Icarus (Figure 14.4) is within 0.2 A.U. of the Sun. On its way past Earth in 1968 it missed our planet by "only" 6 million km—a very close call by cosmic standards. More recently, in 1991 an unnamed asteroid (designated 1991 BA) came much closer, passing only 170,000 km from Earth, less than half the distance to the Moon. In December 1994 the Apollo asteroid 1994 XM1 missed us by a mere 105,000 km.

The potential for collision with Earth is real. Calculations imply that most Earth-crossing asteroids will eventually collide with Earth. On average during any given million-year period, our planet is struck by about three asteroids. Because Earth is largely covered with water, on average two of those impacts should occur in the ocean and only one on land. Several dozen large land basins and eroded craters on our planet are suspected to be sites of ancient asteroid collisions (see, for example, Figure 14.17 later in this chapter). The many large impact craters on the Moon, Venus, and Mars are direct evidence of similar events on other worlds.

Most known Earth-crossing asteroids are relatively small—about 1 km in diameter (although one 10 km in diameter has

been identified). Even so, a visit of even a kilometer-sized asteroid to Earth could be catastrophic by human standards. Such an object packs enough energy to devastate an area some 100 km in diameter. The explosive power would be equivalent to about a million 1-megaton nuclear bombs, a hundred times more than all the nuclear weapons currently in existence on Earth. A fatal blast wave would doubtless affect a much larger area still. Should an asteroid hit our planet hard enough, it might even cause the extinction of entire species—indeed, many scientists think that the extinction of the dinosaurs was the result of just such an impact (see *Interlude 14-1*). Some astronomers take the prospect of an asteroid impact seriously enough to advocate an "asteroid watch"—an effort to catalog and monitor all Earth-crossing asteroids in order to maximize our warning time of any impending collision.

ORBITAL RESONANCES

3 Although most asteroids orbit in the main belt, between about 2 and 3 A.U. from the Sun, an additional class of asteroids, called the **Trojan asteroids**, orbit at the distance of Jupiter. Several hundred such asteroids are now known. They are locked into a 1:1 orbital resonance with Jupiter by that planet's strong gravity, just as some of the small moons of Saturn share orbits with the medium-sized moons Tethys and Dione, as described in Chapter 12. ∞ (Sec. 12.5)

Calculations first performed by the French mathematician Joseph Louis Lagrange in 1772 show that there are exactly five places in the solar system where a small body can orbit the Sun in synchrony with Jupiter (or, for that matter, with any other planet) subject to the combined gravitational influence of both large bodies. These places are known as the *Lagrange points* of Jupiter's orbit. As illustrated in Figure 14.5, three of these points (referred to as L_1, L_2, and L_3) lie on the line joining Jupiter and the Sun (or its

Figure 14.5 The Lagrange points of the Jupiter–Sun system, where a third body could orbit in synchrony with Jupiter on a circular trajectory. Only the L_4 and L_5 points are stable. They are the locations of the Trojan asteroids.

extension in either direction). The other two—L$_4$ and L$_5$—are located on Jupiter's orbit, exactly 60° ahead of and behind the planet. All five Lagrange points revolve around the Sun at the same rate as Jupiter.

In principle, an asteroid placed at any of the Lagrange points will circle the Sun in lockstep with Jupiter, always maintaining the same position relative to the planet. However, the three Lagrange points in line with Jupiter and the Sun are known to be *unstable*—a body displaced, however slightly, from any of those points will tend to drift slowly away from it, not back toward it. Since matter in the solar system is constantly subjected to small perturbations—by the planets, the asteroids, and even the solar wind—matter does not accumulate in these regions. No asteroids orbit near the L$_1$, L$_2$, or L$_3$ points of Jupiter's orbit.

This is not the case for the other two Lagrange points, L$_4$ and L$_5$. They are both *stable*—matter placed near them tends to remain in the vicinity. Consequently, asteroids tend to accumulate near these points. For unknown reasons, Trojan asteroids tend to be found near Jupiter's leading (L$_4$) Lagrange point rather than the trailing (L$_5$) point. Recently, a few small asteroids have been found similarly trapped in the Lagrange points of Venus, Earth, and Mars.

The main asteroid belt also has structure—not so obvious as the Trojan orbits or the prominent gaps and ringlets in Saturn's ring system, but nevertheless of great dynamic significance. A graph of the number of asteroids having various orbital semi-major axes (Figure 14.6a) shows that there are several prominently underpopulated regions in the distribution. These "holes" are known as the **Kirkwood gaps**, after their discoverer, the nineteenth-century American astronomer Daniel Kirkwood.

The Trojan asteroids share an orbit with Jupiter—they orbit in 1:1 resonance with that planet. The Kirkwood gaps result from other, more complex, orbital resonances with Jupiter. For example, an asteroid with a semi-major axis of 3.3 A.U. would (by Kepler's third law) orbit the Sun in exactly half the time taken by Jupiter. ∞ (Sec. 2.4) The gap

(a)

(b)

Figure 14.6 (a) The distribution of asteroid semi-major axes shows some prominent gaps caused by resonances with Jupiter's orbital motion. Note, for example, the prominent gap at 3.3 A.U., which corresponds to the 2:1 resonance—the orbital period is 5.9 years, exactly half that of Jupiter. (b) An asteroid in a 2:1 resonance with Jupiter receives a strong gravitational tug from the planet each time they are closest together (as in panels 1 and 3). Because the asteroid's period is precisely half that of Jupiter, the tugs come at exactly the same point in every other orbit, and their effects reinforce each other.

at 3.3 A.U., then, corresponds to a 2:1 resonance. An asteroid at that particular resonance feels a regular, periodic tug from Jupiter at the same point in every other orbit (Figure 14.6b). The cumulative effect of those tugs is to deflect the asteroid into an elongated orbit—one that crosses the orbit of Mars or Earth. Eventually, the asteroid collides with one of those two planets or comes close enough that it is pushed onto an entirely different trajectory. In this way, Jupiter's gravity creates the Kirkwood gaps, and some of the cleared-out asteroids become Apollo or Amor asteroids.

Notice that, although there are many similarities between this mechanism and the resonances that produce the gaps in Saturn's rings (see Chapter 12), there are differences too. ∞ (Sec. 12.4) Unlike Saturn's rings, where eccentric orbits are rapidly circularized by collisions among ring particles, there are no *physical* gaps in the asteroid belt. The in-and-out motion of the belt asteroids as they travel in their eccentric orbits around the Sun means that no part of the belt is actually empty. Only when we look at semi-major axes (or, equivalently, at orbital *energies*) do the gaps become apparent.

14.2 Comets

Comets are usually discovered as faint, fuzzy patches of light on the sky while still several astronomical units away from the Sun. Traveling in a highly elliptical orbit with the Sun at one focus, a comet brightens and develops an extended **tail** as it nears the Sun. (The name "comet" derives from the Greek word *kome*, meaning "hair.") As the comet departs from the Sun's vicinity, its brightness and its tail diminish until it once again becomes a faint point of light receding into the distance. Like the planets, comets emit no visible light of their own—they shine by reflected (or reemitted) sunlight.

COMET APPEARANCE AND STRUCTURE

4 The various parts of a typical comet are shown in Figure 14.7. Even through a large telescope, the **nucleus**, or main solid body, of a comet is no more than a minute point of light. A typical cometary nucleus is extremely small—only a few kilometers in diameter. During most of the comet's orbit, far from the Sun, only this frozen nucleus exists. When a comet comes within a few astronomical units of the Sun, however, its icy surface becomes too warm to remain stable. Part of it becomes gaseous and expands into space, forming a diffuse **coma** ("halo") of dust and evaporated gas around the nucleus. The coma becomes larger and brighter as the comet nears the Sun. At maximum size, the coma can measure 100,000 km in diameter—almost as large as Saturn or Jupiter.

Engulfing the coma, an invisible **hydrogen envelope**, usually distorted by the solar wind, stretches across millions of kilometers of space. The comet's tail, most pronounced when

(a)

(b) *Animation*

R I V U X G

Figure 14.7 (a) Diagram of a typical comet, showing the nucleus, coma, hydrogen envelope, and tail. The tail is not a sudden streak in time across the sky, as in the case of meteors or fireworks. Instead, it travels through space along with the rest of the comet (so long as the comet is sufficiently close to the Sun for the tail to exist). (b) Halley's comet in 1986, about 1 month before perihelion.

the comet is closest to the Sun and the rate of sublimation* of material from the nucleus is greatest, is much larger still, sometimes spanning as much as 1 A.U. From Earth, only the coma and tail of a comet are visible to the naked eye. Despite the size of the tail, however, most of the light comes from the coma; most of the comet's mass resides in the nucleus.

*Sublimation is the process by which a solid changes directly into a gas without passing through the liquid phase. Frozen carbon dioxide—dry ice—is an example of a solid that undergoes sublimation rather than melting and subsequent evaporation. In space, sublimation is the rule, rather than the exception, for the behavior of ice when exposed to heat.

(b)

(a)

Figure 14.8 (a) A comet with a primarily ion tail. Called comet Giacobini–Zinner and seen here in 1959, its coma measured 70,000 km across; its tail was well over 500,000 km long. (b) Photograph of a comet having (mostly) a dust tail, showing both its gentle curvature and inherent fuzziness. This is comet West, in 1976, whose tail stretched 13° across the sky.

Two types of comet tails may be distinguished. The **ion tails** are approximately straight, often made up of glowing, linear streamers like those seen in Figure 14.8(a). Their spectra show emission lines of numerous *ionized* molecules— molecules that have lost some of their normal complement of electrons—including carbon monoxide, nitrogen, and water among many others. ∞ (Sec. 4.2) The **dust tails** are usually broad, diffuse, and gently curved (Figure 14.8b). They are rich in microscopic dust particles that reflect sunlight, making the tail visible from afar.

The tails are in all cases directed *away* from the Sun by the solar wind (the invisible stream of matter and radiation escaping the Sun). Consequently, as depicted in Figure 14.9, the tail always lies outside the comet's orbit and actually *leads* the comet during the portion of the orbit that is outbound from the Sun.

The ion tails and dust tails differ in shape because of the different responses of gas and dust to the forces acting in interplanetary space. Every tiny particle in space in our solar system—including those in comet tails—follows an orbit determined by gravity and the solar wind. If gravity alone were acting, the particle would follow the same curved path as its parent comet, in accordance with Newton's laws of motion. ∞ (Sec. 2.7) If the solar wind were the only influence, the tail would be swept up by it and would trail radially outward from the Sun. The ion tails are much more strongly influenced by the solar wind than by the Sun's gravity, so those tails always point directly away from the Sun. The heavier dust particles have more of a tendency to follow the comet's orbit, giving rise to the slightly curved dust tails.

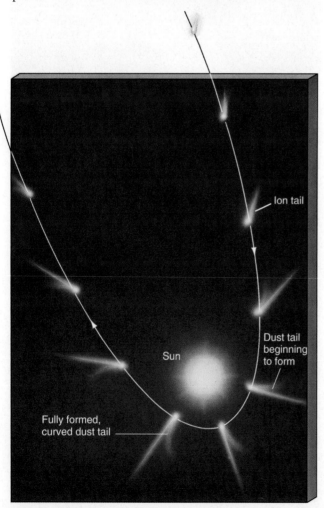

Ion tail

Dust tail beginning to form

Sun

Fully formed, curved dust tail

Figure 14.9 Diagram of part of the orbit of a typical comet. As the comet approaches the Sun it develops an ion tail, which is always directed away from the Sun. Closer in, a curved dust tail, also directed generally away from the Sun, may also appear. Notice that although the ion tail always points directly away from the Sun on both the inbound and the outgoing portions of the orbit, the dust tail has a marked asymmetry, always tending to "lag behind" the ion tail.

Animation

COMET ORBITS

5 Comets that survive a close encounter with the Sun—some break up entirely—continue their outward journey to the edge of the solar system. Their highly elliptical orbits take many comets far beyond Pluto, perhaps even as far as 50,000 A.U., where, in accord with Kepler's second law, they move more slowly and so spend most of their time. ∞ (Sec. 2.4) The majority of comets take hundreds of thousands, some even millions, of years to complete a single orbit around the Sun. However, a few *short-period* comets (conventionally defined as those having orbital periods of less than 200 years) return for another encounter within a relatively short time. According to Kepler's third law, the short-period comets do not venture far beyond the distance of Pluto at aphelion.

Unlike the orbits of the other solar system objects we have studied so far, the orbits of comets are *not* necessarily confined to within a few degrees of the ecliptic plane. Short-period comets do tend to have prograde orbits lying close to the ecliptic, but long-period comets exhibit all inclinations and all orientations, both prograde and retrograde, roughly uniformly distributed in all directions from the Sun.

Astronomers believe that the short-period comets originate beyond the orbit of Neptune, in a region of the solar system called the **Kuiper belt** (after Gerard Kuiper, a pioneer in infrared and planetary astronomy). A little like the asteroids in the inner solar system, most Kuiper belt comets move in roughly circular orbits between about 30 and 100 A.U. from the Sun, never venturing inside the orbits of the jovian planets. Occasionally, however, a close encounter between two comets, or (more likely) the cumulative gravitational influence of one of the outer planets, "kicks" a Kuiper belt comet into an eccentric orbit that brings it into the inner solar system, and into our view. The observed orbits of these comets reflect the flattened structure of the Kuiper belt.

What of the long-period comets? How do we account for their apparently random orbital orientations? Only a tiny portion of a typical long-period cometary orbit lies within the inner solar system, so it follows that for every comet we see, there must be many more similar objects at great distances from the Sun. On these general grounds, many astronomers reason that there must be a huge "cloud" of comets far beyond the orbit of Pluto, completely surrounding the Sun. This region, which may contain trillions of comets, of total mass comparable to the mass of the inner planets, is named the **Oort cloud**, after the Dutch astronomer Jan Oort, who first wrote (in the 1950s) of the possibility of such a vast and distant reservoir of inactive, frozen comets. The Kuiper belt and the orbits of some typical Oort cloud comets are sketched in Figure 14.10.

The observed orbital properties of long-period comets, have led researchers to believe that the Oort cloud may be up to 100,000 A.U. in diameter. Like those of the Kuiper belt, however, most of the comets of the Oort cloud never come anywhere near the Sun. Indeed, Oort cloud comets rarely approach even the orbit of Pluto, let alone that of Earth. Only when the gravitational field of a passing star happens to deflect a comet into an extremely eccentric orbit that passes through the inner solar system do we actually get to see one of these objects. Because the Oort cloud sur-

Figure 14.10 (a) Diagram of the Oort cloud, showing a few cometary orbits. Most Oort cloud comets never come close to the Sun. Of all the orbits shown, only the most elongated ellipse represents a comet that will actually enter the solar system (which is smaller than the dot at the center of the figure on this scale) and possibly become visible from Earth. (b) The Kuiper belt, believed to be the source of the short-period comets.

INTERLUDE 14-1 What Killed the Dinosaurs?

The name *dinosaur* derives from the Greek words, *deinos* (terrible) and *sauros* (lizard). Dinosaurs were no ordinary reptiles. In their prime, roughly 100 million years ago, the dinosaurs were the all-powerful rulers of Earth. Their fossilized remains have been uncovered on all the world's continents. Despite their dominance, according to the fossil record, these creatures vanished from Earth quite suddenly about 65 million years ago. What happened to them?

Until fairly recently, the prevailing view among paleontologists—scientists who study prehistoric life—was that dinosaurs were rather small-brained, cold-blooded creatures. In chilly climates, or even at night, the metabolisms of these huge reptiles would have become sluggish, making it difficult for them to move around and secure food. The suggestion was that they were poorly equipped to adapt to sudden changes in Earth's climate, so that they eventually died out. However, a competing, and still controversial, view of dinosaurs has emerged. Recent fossil evidence suggests that many of these monsters may in fact have been warm-blooded and relatively fast moving creatures—not at all the dull-witted, slow-moving giants of earlier conception. In any case, no species able to dominate Earth for more than 100 million years could have been too poorly equipped for survival. For comparison, humans have thus far dominated for a little over 2 million years.

If the dinosaurs didn't die out simply because of stupidity and inflexibility, then what happened to cause their sudden and complete disappearance? Many explanations have been offered for the extinction of the dinosaurs. Devastating plagues, magnetic field reversals, increased tectonic activity, severe climate changes, and supernova explosions have all been proposed. In the 1980s it was suggested that a huge extraterrestrial object collided with Earth 65 million years ago, and this is now (arguably) the leading explanation for the demise of the dinosaurs, although it is by no means universally accepted. According to this idea, a 10- to 15-km-wide asteroid or comet struck Earth, releasing as much energy as 10 million

or more of the largest hydrogen bombs humans have ever constructed and kicking huge quantities of dust (including the pulverized remnants of the asteroid itself) high into the atmosphere (see the first figure below). The dust may have shrouded our planet for many years, virtually extinguishing the Sun's rays during this time. On the darkened surface, plants could not survive. The entire food chain was disrupted, and the dinosaurs, at the top of that chain, eventually became extinct.

Although we have no direct astronomical evidence to confirm or refute this idea, we can estimate the chances that a large asteroid or comet will strike Earth today, on the basis of observations of the number of objects presently on Earth-crossing orbits. The second figure shows the likelihood of an impact as a function of the size of the impacting body. The horizontal scale indicates the energy released by the collision, measured in *megatons* of TNT. (The megaton—

rounds the Sun in all directions, instead of being confined near the ecliptic plane like the Kuiper belt, the long-period comets we see can come from any direction in the sky. Despite their great distances and long orbital periods, however, the Oort cloud comets are still gravitationally bound to the Sun. Their orbits are governed by precisely the same laws of motion that control the planets.

A VISIT TO HALLEY'S COMET

Probably the most famous comet of all is Halley's comet. (Two more recent and widely publicized contenders for that title are described in *Interlude 14-2*.) In 1705 the British astronomer Edmund Halley realized that the 1682 appearance of this comet was not a one-time event. Basing his work on previous sightings of the comet, Halley calculated its path and found that the comet orbited the Sun with a period of 76 years. He

predicted its reappearance in 1758. Halley's successful determination of the comet's trajectory and his prediction of its return was an early triumph of Newton's laws of motion and gravity. Although Halley did not live to see his calculations proved correct, the comet was named in his honor.

Once astronomers knew the comet's period, they traced its appearances back in time. Historical records from many ancient cultures show that Halley's comet has been observed at every passage since 240 B.C. A spectacular show, the tail of Halley's comet can reach almost a full astronomical unit in length, stretching many tens of degrees across the sky. Figure 14.11(a) shows Halley's comet as seen from Earth in 1910. Its most recent appearance, in 1986 (Figure 14.11b and also Figure 14.7b), was not ideal for terrestrial viewing, but the comet was closely scrutinized by spacecraft. The comet's orbit is shown in Figure 14.12; its next scheduled visit to the inner solar system is in 2061.

4.2×10^{16} joules, the explosive yield of a large nuclear warhead—is the only common terrestrial measure of energy adequate to describe the violence of these occurrences.) We see that 100-million-megaton events, like the planetwide catastrophe that supposedly wiped out the dinosaurs, are very rare, occurring only once every 10 million years or so. However, smaller impacts, equivalent to "only" a few tens of kilotons of TNT (about one MX missile warhead from the U.S. nuclear arsenal), could happen every few years—we may be long overdue for one. The most recent large impact was the Tunguska explosion in Siberia, in 1908, which packed a roughly 1-megaton punch (see Figure 14.18).

The main geological evidence supporting this theory is a layer of clay enriched with the element iridium. This layer is found in 65-million-year-old rocky sediments all around our planet. Iridium on Earth's surface is rare because most of it sank into our planet's interior long ago. The abundance of iridium in this one layer of clay is about 10 times greater than in other terrestrial rocks, but it matches closely the abundance of iridium found in meteorites (and, we assume, in asteroids and comets too).

This theory has its problems, however. The amount of iridium in this clay layer varies greatly from place to place across the globe, and there is no complete explanation of why that should be so. And if this body was so massive as to kick up enough dust to darken the entire planet, then where is its impact crater? One leading candidate has recently been identified in the Yucatan Peninsula in Mexico, where possible evidence for a heavily eroded, but not completely obliterated, crater has been found. However, its identity as the site of the dinosaur-killing collision has not yet been conclusively proven. Perhaps, some scientists argue, the iridium layer was laid down by volcanoes and had nothing to do with an extraterrestrial impact at all.

Another potential difficulty concerns the speed at which the dinosaurs disappeared. The fossil and geologic records are rather imprecise in the ages they yield, and this translates into uncertainties in just how long the extinction process took. It seems to be no more than a million or so years, but that is still a very long time. If the dinosaurs vanished as a direct result of a collision and subsequent explosion, we might expect their disappearance to have been complete in a matter of decades, or perhaps centuries. It is hard to see how the process could have spanned tens or hundreds of millennia. Greatly improved geologic age—measurement techniques will be needed to settle this issue decisively.

Whatever killed the dinosaurs, dramatic environmental change of some sort was almost surely responsible. It is important that we continue the search for the cause of their extinction, for there's no telling if and when that sudden change may strike again. As the dominant species on Earth, we are the ones who now stand to lose the most.

(a)

(b)

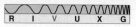

Figure 14.11   (a) Halley's comet as it appeared in 1910. Top, on May 10, with a 30° tail, bottom, on May 12, with a 40° tail. (b) Halley, upon return and photographed with higher resolution, on March 14, 1986.

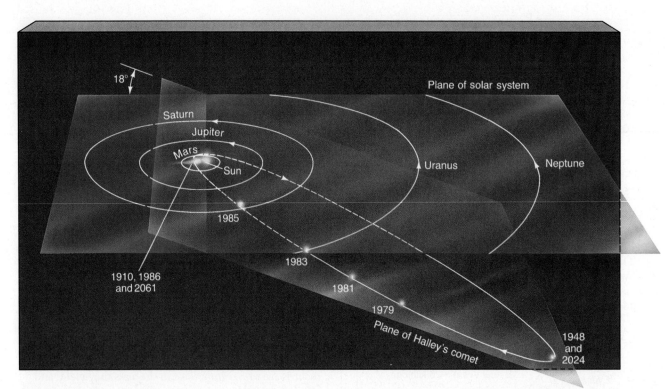

Figure 14.12   Halley's comet has a smaller orbital path and a shorter period than most comets, but its orbital orientation is not typical of a short-period comet. Sometime in the past the comet must have encountered a jovian planet (probably Jupiter itself), which threw it into a tighter orbit that extends not to the Oort cloud but merely a little beyond Neptune. Halley applied Newton's law of gravity to predict this comet's return.

When Halley's comet rounded the Sun in 1986, a small armada of spacecraft launched by the USSR, Japan, and a group of western European countries went to meet it. One of the Soviet craft, *Vega 2*, traveled through the comet's coma, coming to within some 8000 km of the nucleus. Using positional knowledge of the comet gained from the Soviet craft encounter, the European *Giotto* spacecraft (named after the Italian artist who painted an image of Halley's comet not long after its appearance in the year A.D. 1301) was navigated to within 600 km of the nucleus. This was a daring trajectory, since at 70 km/s—the speed of the craft relative to the comet—a colliding dust particle becomes a devastating bullet. Debris did in fact damage *Giotto*'s camera, but not before it sent home a wealth of data. Figure 14.13 shows *Giotto*'s view of the comet's nucleus, along with a sketch of its structure.

The results of the Halley encounters were somewhat surprising. Halley's nucleus is an irregular, potato-shaped object, larger than astronomers had estimated. Spacecraft measurements showed it to be 15 km long by as much as 10 km wide. Also, the nucleus appeared almost jet black—as dark as finely ground charcoal. This solid nucleus was enveloped by a cloud of dust, which scattered light throughout the coma. Partly because of this scattering and partly because of dimming by the dust, no spacecraft was able to discern much surface detail on the nucleus.

The visiting spacecraft found direct evidence for several jets of matter streaming from the nucleus. Instead of evaporating uniformly from the whole surface to form the comet's coma and tail, gas and dust apparently vent from small areas on the sunlit side of Halley's nucleus. The force of these jets may be largely responsible for the comet's 53-hour rotation period. Like maneuvering rockets on a spacecraft, such jets can cause a comet to change its rotation rate and even to veer away from a perfectly elliptical orbit. Astronomers had hypothesized the existence of these nongravitational forces on the basis of slight deviations from Kepler's laws observed in some cometary trajectories. However, only during the Halley encounter did astronomers actually see these jets at work.

PHYSICAL PROPERTIES OF COMETS

4 The mass of a comet can sometimes be estimated by watching how it interacts with other solar system objects or by determining the size of the nucleus and assuming a density characteristic of icy composition. These methods yield typical cometary masses ranging from 10^{12} to 10^{16} kg, comparable to the masses of small asteroids. A comet's mass decreases with time because some material is lost each time the comet rounds the Sun. For comets that travel within an astronomical unit of the Sun, this evaporation rate can reach as high as 10^{30} molecules per second—about 30 tons of cometary material lost for *every second* the comet spends near the Sun (within Earth's orbit, say). Astronomers have estimated that this loss of material will destroy Halley's comet in about 5000 orbits, or 40,000 years.

(a)

R I V U X G

(b)

Figure 14.13 (a) The *Giotto* spacecraft resolved Halley's comet, showing its nucleus to be very dark, although heavy dust in the area obscured any surface features. Resolution here is about 50 m—half the size of a football field. At the time this image was made, in March 1986, the comet was within days of perihelion, and the Sun was toward the top. The brightest parts of the images are jets of evaporated gas and dust spewing from the comet's nucleus. (b) A diagram of Halley's nucleus, showing its size, shape, jets, and other physical and chemical properties.

Animation

In seeking the physical makeup of a cometary body itself, astronomers are guided by the observation that comets have dust that reflects light, as well as gas that emits spectral lines of hydrogen, nitrogen, carbon, and oxygen. Even as the atoms, molecules, and dust particles boil off, creating the coma and tail, the nucleus itself remains a cold mixture of gas and dust,

hardly more than a ball of loosely packed ice with a density of about 100 kg/m^3 and a temperature of only a few tens of kelvins. Experts now consider cometary nuclei to be largely made of dust particles trapped within a mixture of methane, ammonia, and ordinary water ice. (These constituents should be fairly familiar to you as the main components of most of the small moons in the outer solar system, discussed in Chapters 12 and 13.) Because of this composition, comets are often described as "dirty snowballs."

A COMETARY IMPACT

In July 1994, skywatchers were treated to an exceedingly rare event that greatly increased our knowledge of comet composition and structure—the collision of a comet (called Shoemaker–Levy 9, after its discoverers) with the planet Jupiter! When it was discovered in March 1993, Shoemaker–Levy 9 appeared to have a curious, "squashed" appearance. Higher-resolution images (see Figure 14.14a) revealed that the comet's flattened nucleus was really made up of several pieces, the largest no more than 1 km across. All the pieces were following the same orbit, but they were spread out along the comet's path like a string of pearls a million kilometers long.

What could have caused such an unusual object? Tracing the orbit backward in time, researchers calculated that early in July 1992 the comet had approached to within about 100,000 km of Jupiter, well inside the planet's Roche limit. ∞ (Sec. 12.4) They realized that the objects shown in Figure 14.14a were the fragments produced when a previously "normal" comet was captured by Jupiter and torn apart by its strong gravitational field.

On its next approach to Jupiter, in July 1994, Shoemaker–Levy 9 struck the planet's upper atmosphere, plowing into it at a speed of more than 60 km/s and causing a series of enormous explosions (Figure 14.14b). Every major telescope on Earth, the *Hubble Space Telescope*, *Galileo* (which was only 1.5 A.U. from the planet at the time), and even *Voyager 2* were watching. Each impact created, for a period of a few minutes, a brilliant fireball hundreds of kilometers across and having a temperature of many thousands of kelvins. The energy released in each explosion was comparable to a billion terrestrial nuclear detonations, rivaling in violence the prehistoric impact suspected of causing the extinction of the dinosaurs on Earth 65 million years ago (see *Interlude 14-1*).

The effects on the planet's atmosphere and the vibrations produced throughout Jupiter's interior were observable for days after the impact. The fallen material from the impacts spread slowly around Jupiter's bands and after 5 months reached completely around the planet. It probably took years for all the cometary matter to settle into Jupiter's interior.

As best we can determine, none of the cometary fragments breached the jovian clouds. Only *Galileo* had a direct view of the impacts on the back side of Jupiter, and in every case the explosions seemed to occur high in the atmosphere, above the uppermost cloud layer. Most of the dark material seen in the images is probably pieces of the comet rather than parts of Jupiter. Spectral lines from silicon, magnesium, and

Toward
Sun

Toward
Jupiter

(a)   S   G   D

(b)

(c)

Figure 14.14 (a) Comet Shoemaker–Levy 9 is seen here approaching Jupiter a couple of months before its mid-1994 collision. Its many fragments are strung out like the pearls of a necklace 1,000,000 km long. (b) One of the largest parts of SL-9, fragment G, produced this fireball on the southwest limb of Jupiter; it is seen here some 10 minutes after impact, radiating strongly in the infrared (that is, giving off lots of heat). Also visible is the small, warm cloud on the southeast limb left over from the impact of fragment A, which hit the planet on the previous day. (c) The collisions caused several "black eyes" roughly the size of Earth in Jupiter's southern hemisphere. One of the most prominent impact sites, caused by fragment G, is shown in this true-color, visible-light photo. Taken nearly 2 hours after impact, it also shows a large dark arc some 6000 km from the impact site—the result of plume material falling back onto Jupiter.

iron were detected in the aftermath of the collisions, and the presence of these metals might explain the dark material observed near some of the impact sites (Figure 14.14c). Water vapor was also detected spectroscopically, again apparently from melted and vaporized comet—which really did resemble a loosely packed snowball.

14.3 Meteoroids

6 On a clear night it is possible to see a few *meteors*—"shooting stars"—every hour. A **meteor** is a sudden streak of light in the night sky caused by friction between air molecules in Earth's atmosphere and an incoming piece of interplanetary matter—an asteroid, comet, or **meteoroid**. This friction heats and excites the air molecules, which then emit light as they return to their ground states, producing the characteristic bright streak shown in Figure 14.15. Note that the brief flash that is a meteor is in no way similar to the broad, steady swath of light associated with a comet's tail. A meteor is a fleeting event

in Earth's atmosphere, whereas a comet tail exists in deep space and can be visible in the sky for weeks or even months. Recall from Section 6.5 that the distinction between an asteroid and a meteoroid is simply a matter of size. Both are chunks of rocky interplanetary debris; meteoroids are conventionally taken to be less than 100 m in diameter.

Before encountering the atmosphere, the piece of debris causing a meteor was almost certainly a meteoroid, simply because these small interplanetary fragments are far more common than either asteroids or comets. Any piece of interplanetary debris that survives its fiery passage through our atmosphere and finds its way to the ground is called a **meteorite**.

COMETARY FRAGMENTS

Smaller meteoroids are mainly the rocky remains of broken-up comets. Each time a comet passes near the Sun, some cometary fragments dislodge from the main body. The fragments initially travel in a tightly knit group of dust or pebble-sized objects, called a **meteoroid swarm**, moving

(a)

(b)

| R | I | V | U | X | G |

Figure 14.15 A bright streak called a meteor is produced when a fragment of interplanetary debris plunges into the atmosphere, heating the air to incandescence. (a) A small meteor photographed against a backdrop of stars. (b) An auroral display provides the background for a brighter meteor trail.

Table 14.1 lists some prominent meteor showers, the dates they are visible from Earth, and the comet from which they are thought to originate. Meteor showers are usually named for their *radiant*, the constellation from whose direction they appear to come. For example, the Perseid shower is seen to emanate from the constellation Perseus. It can last for several days but reaches maximum every year on the morning of August 12, when upward of 50 meteors per hour can be observed.

Astronomers can use the speed and direction of a meteor's flight to compute its interplanetary trajectory. This is how certain meteoroid swarms have come to be identified with well-known comet orbits. For example, the Perseid shower shares the same orbit as comet 1862III, the third comet discovered in the year 1862 (also known as comet Swift–Tuttle).

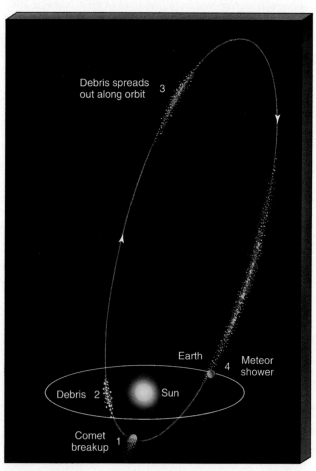

Figure 14.16 A meteoroid swarm associated with a given comet intersects Earth's orbit at specific locations, giving rise to meteor showers at specific times of the year. We imagine that a portion of the comet breaks up near perihelion, at the point marked 1. The fragments continue along the original comet orbit, gradually spreading out as they go (points 2 and 3). The rate at which the debris disperses around the orbit is actually much slower than depicted here—it takes many orbits for the material to spread out as shown. Eventually, the fragments extend all around the orbit, more or less uniformly. If the orbit happens to intersect Earth's orbit, a meteor shower is seen each time Earth passes through the intersection (point 4).

in nearly the same orbit as the parent comet. Over the course of time, the swarm gradually disperses along the orbit, and eventually the **micrometeoroids**, as these small meteoroids are known, become more or less smoothly spread all the way around the parent comet's orbit. If Earth's orbit happens to intersect the orbit of such a young cluster of meteoroids, a spectacular **meteor shower** can result. Earth's motion takes it across a given comet's orbit at most twice a year (depending on the precise orbit of each body). Intersection occurs at the same time each year (see Figure 14.16), so the appearance of certain meteor showers is a regular and (fairly) predictable event.

INTERLUDE 14-2 Comets Hyakutake and Hale–Bopp

One of the most spectacular comets in recent years was comet Hyakutake 1996. Named after a Japanese amateur astronomer who noticed it as "something odd and out of place" while scanning the skies with a pair of binoculars, Hyakutake grew from a small smudge while still far from the Sun into a splendid display comprising a huge coma nearly the apparent size of the Moon and a tail that eventually stretched a third of the way across the sky. The accompanying figure shows a *Hubble Space Telescope* image of Hyakutake taken in March 1996, when the comet passed closest to Earth—only 15 million km (0.1 A.U.) away. The comet's icy nucleus, the brightest point in the image, is unresolved here (the field of view is about 1000 km across), but radar pulses sent toward Hyakutake did return an echo, indicating that the diameter of the nucleus was 1 to 3 km. In the image, the Sun is out of the frame at bottom right, and the innermost part of the comet's tail is at upper left (on the side opposite the Sun, as explained in the text).

Other images showed sporadic jets pointing mostly sunward. These jets were gases gushing from the side of the comet closest to the Sun before wrapping around to become part of the graceful tail. The comet was examined at every conceivable wavelength, but perhaps the most surprising result was the intense X rays emitted from its head. Even the sunward side of the comet was far too cool to emit X rays, which are usually associated with very-high-temperature phenomena. Astronomers speculate that the X rays were produced by shock waves created as the solar wind hit the leading edge of the comet's coma.

Hot on the heels of Hyakutake came another interplanetary vagabond—comet Hale–Bopp 1995. Discovered in 1995 by two American amateur astronomers, it reached its maximum brightness in the spring of 1997. Outshining everything in the night sky except the Moon and the brightest planets and stars, Hale–Bopp was probably the most widely viewed and studied comet in history.

This comet's unusual brightness and long (20°) tail was probably caused by a huge nucleus—about 30–40 km in diameter. That's a very large ball of dirty ice, compared with

TABLE 14.1 Some Prominent Meteor Showers

MORNING OF MAXIMUM ACTIVITY	SHOWER NAME	ROUGH HOURLY COUNT	PARENT COMET
Jan. 3	Quadrantid	40	—
Apr. 21	Lyrid	10	1861I (Thatcher)
May 4	Eta Aquarid	20	Halley
June 30	Beta Taurid	25	Encke
July 30	Delta Aquarid	20	—
Aug. 11	Perseid	50	1862III (Swift–Tuttle)
Oct. 9	Draconid	up to 500	Giacobini–Zinner
Oct. 20	Orionid	30	Halley
Nov. 7	Taurid	10	Encke
Nov. 16	Leonid	12*	1866I (Tuttle)
Dec. 13	Geminid	50	3200 Phaeton†

Every 33 years, as Earth passes through the densest region of this meteoroid swarm, we see intense showers that can reach 1000 meteors per minute for brief periods of time. This is next expected to occur in 1999.

†*Phaeton is actually an asteroid and shows no signs of cometary activity, but its orbit matches the meteoroid paths very well.*

the average comet core, which measures some 3–5 km across. The comet or asteroid that struck the Earth 65 million years ago (see *Interlude 14.1*), perhaps causing the extinction of the dinosaurs, is thought to have been about 10–15 km in size. At perihelion, Hale–Bopp's hydrogen envelope was enormous—1 A.U. across.

The accompanying figure shows a series of *HST* observations of the inner part of comet Hale–Bopp. The colors are artificial, with white representing the brightest parts of the comet, and red less bright; resolution is nearly 500 km, far larger than the estimated size of the icy nucleus itself. These eight pictures were taken over the course of a year, from September 1995 to October 1996 as the comet neared the Sun.

Astronomers used multiple images like these to chronicle the evolution of the comet's nucleus. Especially interesting is the dust outburst in the first photo at top left—even when Hale–Bopp was still far beyond the orbit of Jupiter. At bottom right, as the comet neared the Sun, multiple jets are seen emanating from the surface. The dramatic increase in activity as the comet approached the Sun is clearly evident in this time sequence. *Hubble* was unable to follow the comet even closer to the Sun (which it rounded on April 1997, at a perihelion distance of 0.9 A.U.), for fear of damaging its delicate optics.

Apart from its unusual size, Hale–Bopp seems to be quite representative of many long-period comets. Just about all the molecules previously known from other comets were seen in Hale–Bopp's spectrum. New infrared observing techniques allowed, for the first time, identification of many of the minerals making up the comet's dust, which was found to be very similar to the interplanetary dust found in near-Earth space and collected in Earth's upper atmosphere. What was unique about Hale–Bopp was that, being so large, it could be observed for a very long period of time, even when far from the Sun, allowing observers to keep a continuous and detailed record of its evolution.

Hale–Bopp is now moving out of the inner solar system, its incoming orbit modified by the forces from its jets. Astronomers plan to monitor the comet for several more years as it moves far from the Sun's warmth, and its activity winds down. After a trip well outside the orbit of Pluto, extending some 350 A.U. from the Sun, the comet will return to the inner solar system in about 2400 years.

STRAY ASTEROIDS

Larger meteoroids—more than a few centimeters in diameter—are usually *not* associated with swarms of cometary debris. Generally regarded as small bodies that have strayed from the asteroid belt, possibly as the result of asteroid collisions, these objects have produced most of the cratering on the surfaces of the Moon, Mercury, Venus, Mars, and some of the moons of the jovian planets. When these large meteoroids enter Earth's atmosphere with a typical velocity of nearly 20 km/s, they produce energetic shock waves, or "sonic booms," as well as a bright sky streaks and dusty trails of discarded debris. Such large meteors are sometimes known as *fireballs*. The greater the speed of the incoming object, the hotter its surface becomes and the faster it burns up. A few large meteoroids enter the atmosphere at such high speed (about 75 km/s) that they either fragment or disperse entirely at high altitudes.

The more massive meteoroids (at least a ton in mass and a meter across) do make it to the surface, producing a crater such as the kilometer-wide Barringer Crater shown in Figure 8.17. From the size of this crater, we can estimate that the meteoroid responsible must have had a mass of about 200,000 tons. Only 25 tons of iron meteorite fragments have been found at the crash site. The remaining mass must have been scattered by the explosion at impact, broken down by subsequent erosion, or buried in the ground.

Currently, Earth is scarred with nearly 100 craters larger than 0.1 km in diameter. Most of these are so heavily eroded by weather and distorted by crustal activity that they can be identified only in satellite photography, as shown in Figure 14.17. Fortunately, such major collisions between Earth and large meteoroids are thought to be rare events now. Researchers believe that, on average, they occur only once every few hundred thousand years (see *Interlude 14-1*).

The orbits of large meteorites that survive their plunge through Earth's atmosphere can be reconstructed in a manner similar to that used to determine the orbits of meteor showers. In most cases, their computed orbits do indeed intersect the asteroid belt, providing the strongest evidence we have that they were once part of the belt before being redirected, probably by a collision with another asteroid, into

Figure 14.17 This photograph, taken from orbit by the U.S. *Skylab* space station, clearly shows the ancient impact basin that forms Quebec's Manicouagan Reservoir. A large meteorite landed there about 200 million years ago. The central floor of the crater rebounded after the impact, forming an elevated central peak. The lake, 70 km in diameter, now fills the resulting ring-shaped depression.

the Earth-crossing orbit that led to the impact with our planet. (See also *Interlude 14-3*.)

Not all meteoroid encounters with Earth result in an impact. One of the most recent meteoritic events occurred in central Siberia on June 30, 1908 (Figure 14.18). The presence of only a shallow depression as well as a complete lack of fragments implies that this Siberian intruder exploded several kilometers above the ground, leaving a blasted depression at ground level but no well-formed crater. Recent calculations suggest that the object in question was a rocky meteoroid about 30 m across. The explosion, estimated to have been equal in energy to a 10-megaton nuclear detonation, was heard hundreds of kilometers away and produced measurable increases in atmospheric dust levels all across the Northern Hemisphere.

METEORITE PROPERTIES

7 One feature that distinguishes the small micrometeoroids, which burn up in Earth's atmosphere, from larger meteoroids, which reach the ground, is composition. The average density of meteoritic fireballs too small to reach the ground (but which can be captured by high-flying aircraft)

is about 500–1000 kg/m^3. Such a low density is typical of comets, which are made of loosely packed ice and dust. In contrast, the meteorites that reach Earth's surface are often much denser—up to 5000 kg/m^3—suggesting a composition more like that of the asteroids. Meteorites like the ones shown in Figure 14.19 have received close scrutiny from planetary scientists—prior to the Space Age, they were the only type of extraterrestrial matter we could touch and examine in terrestrial laboratories.

Most meteorites are rocky in composition (Figure 14.20a), although a few percent are composed mainly of iron and nickel (Figure 14.20b). Their basic composition is much like that of the inner planets and the Moon, except that some of their lighter elements—such as hydrogen and oxygen—appear to have boiled away long ago when the bodies from which the meteorites originated were molten. Some meteorites show clear evidence of strong heating at some time in their past, most likely indicating that they originated on a larger body that either experienced some geological activity or was partially melted during the collision that liberated the fragments that eventually became the meteorites. Others show no such evidence and probably date from the formation of the solar system.

Most primitive of all are the *carbonaceous* meteorites, so called because of their relatively high carbon content. They are black or dark gray, and they may well be related to the carbon-rich C-type asteroids that populate the outer asteroid belt. (Similarly, the silicate-rich stony meteorites are probably associated with the inner S-type asteroids.) Many carbonaceous

Figure 14.18 The Tunguska event of 1908 leveled trees over a vast area. Although the impact of the blast was tremendous and its sound audible for hundreds of kilometers, the Siberian site was so remote that little was known about the event until scientific expeditions arrived to study it many years later.

Figure 14.19 (a) A large meteorite on display at the Hayden Planetarium in New York serves as a jungle gym for curious children in this photograph from the 1930s. (b) The Wabar meteorite, discovered in the Arabian desert. Although small fragments of the original meteor had been collected more than a century before, the 2000-kg main body was not found until 1965.

meteorites contain significant amounts of ice and other volatile substances, and they are usually rich in organic molecules.

Finally, almost all meteorites are *old*. Direct radioactive dating shows most of them to be between 4.4 and 4.6 billion years old—roughly the age of the oldest lunar rocks. Meteorites, along with some lunar rocks, comets, and perhaps the planet Pluto, provide essential clues to the original state of matter in the solar neighborhood.

Figure 14.20 (a) A stony meteorite often has a dark fusion crust, created when its surface is melted by the tremendous heat generated during passage through the atmosphere. (b) Iron meteorites, much rarer than stony ones, usually contain some nickel as well. Most such meteorites show characteristic crystalline patterns when their surfaces are cut, polished, and etched with weak acid.

INTERLUDE 14-3 Lunar and Martian Meteorites on Earth

Although the vast majority of large meteorites probably come from the asteroid belt, some are thought to have a different origin. The figure below shows a 0.03-kg meteorite, named ALH81005, which was discovered in the Antarctic ice fields. (The peculiar name results from this being the fifth rock found in 1981 in the Allan Hills region of Antarctica. For scale, the small cube at left is 1 cm wide.) Chemical analysis reveals that the composition of this meteorite differs significantly both from conventional (that is, asteroid-belt) meteorites and from rocks originating on Earth. However, it has many similarities to rocks returned from the lunar highlands by *Apollo* astronauts.

These *lunar meteorites* are thought to have been blasted off the Moon's surface by violent impacts long ago: the dozen or so known lunar meteorites seem to have originated in at least six separate events, some perhaps on the Moon's far side. The fragments had enough energy to escape the Moon's gravity and go into orbit around Earth, eventually falling to Earth's surface much like any other meteorite. ALH81005's interior is a jumble of fused rock fragments, just as would be expected if it had been blasted off the lunar surface by some violent event. The small white inclusions evident in the figure match the composition of rocks retrieved from the Moon's highlands.

Even more intriguing, scientists now think that a handful of rocks found on Earth may well have come from Mars! (See also *Interlude 10-2.*) Huge impacts millions of years ago,

possibly those responsible for some of the craters still visible on the Martian surface, have apparently provided us with about 12 free samples of that planet. The second figure shows an 8-kg rock, designated EETA79001, a meteorite found in the Elephantine Moraine region of Antarctica in 1979. (Again, the cube at lower left is 1 cm on a side.)

Although we cannot be certain that this meteorite hails from Mars, the evidence is reasonably strong. First, the relative abundances of some elements, such as nitrogen and argon trapped within the rock, are identical with those in the Martian atmosphere. Second, this meteorite is much younger than most other meteorites—no more than 1.3 billion years old, based on the time since the rock last solidified. (Some of the 12 Mars meteorites are as young as 170 million years.) Finally, like several of the Mars meteorites, EETA79001 has texture suggesting that it formed in lava flows, perhaps like those shown in Figure 10.3(a). The complexity of its rock types—both fine- and coarse-grain lava—suggest that it probably formed in two separate eruptions from the same volcanic event on Mars. Also embedded in the rock are dark veins of shocked glass, likely formed in the impact that blasted it off Mars some 15 million years ago.

We are left wondering how many other lunar and Martian rocks now sit on museum shelves and in university labs awaiting proper identification. We can also only speculate if any of the rocks in our own backyards have been blown off alien worlds far, far away.

Chapter Review

SUMMARY

More than 7000 **asteroids** (p. 314) have been cataloged. Most orbit in a broad band called the **asteroid belt** (p. 314) between the orbits of Mars and Jupiter. They are probably primal rocks that never clumped together to form a planet.

The masses of the asteroids are small, totaling less than 1/10 the mass of Earth's Moon. The largest asteroids are a few hundred kilometers across. Most are much smaller. Asteroids are classified according to the properties of their reflected light. Brighter S-type (silicate) asteroids dominate the inner asteroid belt, whereas darker C-type (carbonaceous) asteroids are more plentiful in the outer regions. They are believed to have changed little since the solar system formed. Smaller asteroids tend to be irregular in shape, and may have experienced violent collisions in the past. The few asteroid densities that have been reliably measured are generally consistent with the preceding classification. The best views of asteroids have been obtained by unmanned space probes during the 1990s.

A few **Earth-crossing asteroids** (p. 316) have orbits that intersect Earth's orbit and will probably collide with our planet one day. The **Trojan asteroids** (p. 317) share Jupiter's orbit, remaining 60° ahead of or behind that planet as it moves around the Sun. The **Kirkwood gaps** (p. 318) in the main asteroid belt have been cleared by Jupiter's gravity.

Comets (p. 319) are fragments of icy material that normally orbit far from the Sun. Unlike the orbits of most other bodies in the solar system, comet orbits are often highly elongated and not confined to the ecliptic plane. Most comets are thought to reside in the **Oort cloud** (p. 321), a vast "reservoir" of cometary material, tens of thousands of astronomical units across, completely surrounding the Sun. A very small fraction of comets happen to have highly elliptical orbits that bring them into the inner solar system. Comets with orbital periods less than about 200 years are thought to originate not in the Oort cloud but in the **Kuiper belt** (p. 321), a broad band lying roughly in the ecliptic plane, beyond the orbit of Neptune.

As a comet approaches the Sun its surface ice begins to vaporize. We see the comet by the sunlight reflected from the dust and vapor released. The **nucleus** (p. 319), or core, of a comet may be only a few kilometers in diameter. It is surrounded by a **coma** (p. 319) of dust and gas. Surrounding this is an extensive invisible **hydrogen envelope** (p. 319). Stretching behind the comet is a long **tail** (p. 319), formed by the interaction between the cometary material and the solar wind. The **ion tail** (p. 320) consists of ionized gas particles and always points directly away from the Sun. The **dust tail** (p. 320) is less affected by the solar wind and has a somewhat curved shape.

Spacecraft visited Halley's comet in 1986 and studied its nucleus. All other comet studies have been indirect, usually based on spectroscopic measurements. Comets are icy, dusty bodies, sometimes called "dirty snowballs," that are believed to be left-over material unchanged since the formation of the solar system. Their masses are comparable to the masses of small asteroids.

Comets and stray asteroids are responsible for most of the cratering on the various worlds in the solar system. Earth is still subject to these sorts of collisions. The most recent large impact occurred in 1908, when an asteroid apparently exploded several miles above Siberia. Comet Shoemaker–Levy 9 struck Jupiter in 1994, causing violent explosions in that planet's atmosphere.

Meteors (p. 326), or "shooting stars," are bright streaks of light that flash across the sky as a **meteoroid** (p. 326), a piece of interplanetary debris, enters Earth's atmosphere. If any of the meteoroid reaches the ground, it is called a **meteorite** (p. 326). The major difference between meteoroids and asteroids is their size. The dividing line between them is conventionally taken to be 100 m.

Each time a comet rounds the Sun, some cometary material becomes dislodged, forming a **meteoroid swarm** (p. 326)—a group of small **micrometeoroids** (p. 327) following the comet's original orbit. If Earth happens to pass through the comet's orbit, a **meteor shower** (p. 327) occurs. Larger meteoroids are probably pieces of material chipped off asteroids following collisions in the asteroid belt. Meteorite composition is thought to mirror the composition of the asteroids, and the few orbits that have been determined are consistent with an origin in the asteroid belt. Some meteorites show evidence of heating, but the oldest ones do not. Most meteorites are between 4.4 and 4.6 billion years old.

SELF-TEST: TRUE OR FALSE?

_____ **1.** Most asteroids move on almost circular orbits.

_____ **2.** The Apollo asteroids have perihelion distances of less than 1 A.U.

_____ **3.** The C-type asteroids are so named because of their heavily cratered surfaces.

_____ **4.** S-type asteroids are more common in the inner part of the asteroid belt.

_____ **5.** The least reflective asteroids are C-type.

_____ **6.** The Kirkwood gaps are two broad zones within the asteroid belt, lying around 2.5 and 3.0 A.U. from the Sun and each about 0.1 A.U. wide, in which no asteroids are found.

_____ **7.** Some comets travel up to 50,000 A.U. from the Sun.

_____ **8.** Cometary orbits always lie close to the ecliptic plane.

_____ **9.** Halley's comet spends most of its time beyond the orbit of Saturn.

_____ **10.** The Oort cloud is the large cloud of gas surrounding a comet while it is near the Sun.

_____ **11.** Tails of comets always lie along the path of the orbit.

_____ **12.** The light known as a meteor is due to the literal burning up of a meteoroid in Earth's atmosphere.

_____ **13.** Some meteorites found on Earth originally came from the Moon or Mars.

_____ **14.** Comets are the sources of meteor showers.

_____ **15.** Astronomers have succeeded in tracing the orbits of some meteorites back into the asteroid belt.

SELF-TEST: FILL IN THE BLANK

1. Asteroids are generally _____ in composition.

2. The asteroid belt, where most asteroids are found, lies between the orbits of _____ and _____.

3. The largest asteroids are _____ of kilometers in diameter; the smallest (by definition) are only _____ meters across.

4. The Trojan asteroids share an orbit with _____.

5. The Kirkwood gaps are caused by an orbital _____ with Jupiter.

6. Comets are generally _____ in composition.

7. The process of sublimation is one in which a solid turns into a _____.

8. Comets orbits are highly _____.

9. The nucleus of a comet is typically _____ kilometers across.

10. Passage of a comet near the Sun may leave a _____ moving in the comet's orbit.

11. The Kuiper belt is the source of the _____ comets.

12. Meteoroids are mostly fragments of _____ and _____.

13. When a meteoroid enters Earth's atmosphere, we see a _____.

14. Meteoroids that impact a planet or moon are called _____.

15. The oldest meteorites are about _____ years old.

REVIEW AND DISCUSSION

1. Describe the differences among the Trojan, Apollo, and Amor asteroids.

2. How have the best photographs of asteroids been obtained?

3. What are the Kirkwood gaps? How did they form?

4. Compare and contrast the C-type and S-type asteroids.

5. What are comets like when they are far from the Sun? What happens when they enter the inner solar system?

6. Where are most comets found?

7. Describe the various parts of a comet while it is near the Sun.

8. What are the typical ingredients of a comet nucleus?

9. What are some possible fates of comets?

10. Explain the difference between a meteor, a meteoroid, and a meteorite.

11. What causes a meteor shower?

12. What do meteorites reveal about the age of the solar system?

13. Why can comets approach the Sun from any direction, whereas asteroids generally orbit close to the ecliptic plane?

14. Why do meteorites contain information about the early solar system, yet Earth does not?

15. What might be the consequences if a 10-km-diameter meteorite struck Earth today?

PROBLEMS

1. The largest asteroid, Ceres, has a radius 0.073 that of Earth and a mass 0.00015 Earth's mass. How much would a 100-kg astronaut weigh there?

2. You are standing on the surface of a spherical asteroid 10 km in diameter, of density 3000 kg/m³. Can you throw a small rock into circular orbit around it? Give the speed required in km/s and mph.

3. Can you find a simple orbital resonance with Jupiter that can account for the small Kirkwood gap evident in Figure 14.6(a) at semi-major axis around 2.7 A.U.?

4. The asteroid Icarus (Figure 14.4) has perihelion of 0.2 A.U. and orbital eccentricity of 0.69. Calculate its orbital semi-major axis and aphelion distance from the Sun. Do these figures, by themselves, *necessarily* imply that Icarus will one day collide with Earth?

5. (a) Using the data given in the text, calculate Dactyl's orbital period as it moves around Ida. (b) *NEAR*'s orbit around Icarus will have periapsis (closest approach) of about 45 km from the asteroid's center. If the mass of Eros is 2×10^{16} kg and *NEAR*'s orbit has an eccentricity of 0.25, calculate the spacecraft's orbital period.

6. Using Kepler's laws of planetary motion (see Section 2.4), calculate the orbital period of a comet with a perihelion distance of 0.5 A.U. and aphelion in the Oort cloud, at a distance of 50,000 A.U. from the Sun. (b) As comet Hale–Bopp passed by the Sun, nongravitational forces changed its orbital period from 4200 years to 2400 years. By what factor did the comet's semi-major axis change? Given that the perihelion remained unchanged at 0.914 A.U., calculate the old and new orbital eccentricities. (c) A short-period comet has a perihelion distance of 1 A.U. and an orbital period of 125 years. What is its maximum distance from the Sun?

7. Astronomers estimate that comet Hale–Bopp lost mass at an average rate of about 350,000 kg/s during the time it spent close to the Sun—a total of about 100 days. Estimate the total amount of mass lost and compare it with the comet's estimated mass of 5×10^{15} kg.

8. It has been hypothesized that Earth is under continuous bombardment by house-sized "minicomets" with typical diameters of 10 m, at the rate of some 30,000 per day. Assuming spherical shapes and average densities of 100 kg/m³, calculate the total mass of material reaching Earth each year. Compare the total mass received in the past 1 billion years (assuming all rates were the same in the past) with Earth's mass.

9. A particular comet has a total mass of 10^{13} kg, 95 percent of which is ice and dust. The remaining 5 percent is in the form

of rocky fragments with an average mass of 100 g. How many meteoroids would you expect to find in the swarm formed by the breakup of this comet?

10. It is observed that the number of asteroids or meteoroids of a given diameter is roughly inversely proportional to the square of the diameter. Approximating the actual distribution of asteroids in this way—a single 1000-km body (Ceres), one hundred 100-km bodies, ten thousand 10-km asteroids, and so on—and assuming constant densities of 3000 kg/m^3, calculate the total mass found in the form of 1000-km bodies, 100-km bodies, 10-km bodies, and 1-km bodies.

PROJECTS

1. The only way to tell an asteroid from a star is to watch it over several nights; you can detect its movement in front of the star background. The astronomy magazines *Sky and Telescope* and *Astronomy* often publish charts for especially prominent asteroids. Look for the asteroids Ceres, Pallas, or Vesta. They are the brightest asteroids. Use the chart to locate the appropriate star field. Aim binoculars at that location in the sky; you may be able to pick out the asteroid from its location in the chart. If you can't, make a rough drawing of the entire field. Come back a night or two later, and look again. The "star" that has moved is the asteroid.

2. Although a spectacular naked-eye comet comes along only about once a decade, fainter comets can be seen with binoculars and telescopes in the course of every year. *Sky and Telescope* often runs a "Comet Digest" column announcing the whereabouts of comets. A comprehensive list of periodic comets expected to return in a given year can be found in Guy Ottewell's Astronomical Calendar. This calendar contains a wealth of other sky information as well, including monthly star charts. At this writing, it costs $15 a year and can be purchased from:

   Astronomical Workshop
   Furman University
   Greenville, South Carolina 29613
   (803) 294-2208

3. There are a number of major meteor showers every year, but if you plan to watch one, be sure to notice the phase of the Moon. Bright moonlight or city lights can obliterate a meteor shower. A common misconception about meteor watching is that most meteors are seen in the direction of the shower's radiant point. It's true that if you trace the paths of the meteors backward in the sky, they all can be seen to come from the radiant. But most meteors don't become visible until they are 20 or 30° from the radiant. Meteors can appear in all parts of the sky! Just relax and let your eyes rove among the stars. You will generally see many more meteors in the hours before dawn than in the hours after sunset. Why do you suppose meteors have different brightnesses? Can you detect their variety of colors? Watch for meteors that appear to "explode" as they fall. Watch for vapor trails that linger after the meteor itself has disappeared.

15 THE FORMATION OF THE SOLAR SYSTEM

The Birth of Our World ✵www

LEARNING GOALS

Studying this chapter will enable you to:

1 Summarize the major features that a theory of solar system origins has to explain.

2 Outline the process by which planets form as natural by-products of star formation.

3 Explain the role played by dust in the currently accepted model of solar system formation.

4 Account for the differences between the terrestrial and the jovian planets.

5 Discuss the role of collisions in determining specific characteristics of the solar system.

(Opposite page, background) An imaginary triple-star system, viewed from the vicinity of a Jupiter-like planet. The planet's parent star resembles our Sun, hence its yellowish appearance. The other two stars are much hotter, and appear bluish. The four inserts show, from top to bottom, a possible evolutionary sequence that gave rise to the triple-star system.

(Inset A) Neighboring, dark galactic clouds contracting, their central regions heating enough to emit a reddish glow.

(Inset B) Formation of huge, rotating disks of glowing matter around young stars ("protostars") foster further heating at their cores.

(Inset C) Two of the protostars become gravitationally bound to each other, interacting somewhat violently; the third (foreground) protostar drifts farther away, allowing it to evolve more calmly.

(Inset D) The third protostar continues to evolve slowly, with planet-sized clumps ("protoplanets") beginning to emerge within the disk.

Completing these chapters on the planets, you may be struck by the vast range of physical and chemical properties found in the solar system. The planets present a long list of interesting features and bizarre peculiarities, and the list grows even longer when we consider the characteristics of their moons. Every object has its idiosyncrasies, some of them due to particular circumstances, many others the result of planetary evolution. Each time a new discovery is made, we learn a little more about the properties and history of our planetary system. Our astronomical neighborhood may seem more like a great junkyard than a smoothly running planetary system. Can we really make any sense of the entire collection of solar system matter? Is there some underlying principle that unifies the knowledge we have gained? The answer, as we will see, is yes.

15.1 Modeling the Origin of the Solar System

The origin of the planets and their moons is a complex and as yet incompletely solved puzzle, although the basic outlines of the process are quite well understood. Most of our knowledge of the solar system's formative stages has emerged from studies of interstellar gas clouds, fallen meteorites, and Earth's Moon, as well as from the various planets observed with ground-based telescopes and planetary space probes. Ironically, studies of Earth itself do not help much because information about our planet's early stages eroded away long ago. Meteorites and comets provide perhaps the most useful information, for nearly all have preserved within them traces of solid and gaseous matter from the earliest times.

Despite the recent widely publicized discoveries of planets orbiting other stars (see *Interlude 15-1*), astronomers still have little detailed information on their properties, and so far there is no firm evidence for planets like Earth anywhere beyond our solar system. For that reason, our theories of planet formation still concentrate on the planetary system in which we live. Bear in mind, however, that no part of the scenario we will describe in the paragraphs that follow is in any way unique to our own system. The same basic processes could have occurred—and, many astronomers believe, probably *did* occur—during the formative stages of most of the stars in our galaxy. It is part of the job of the planetary scientist to distinguish between those properties of the solar system that are inherent—that is, that were imposed at formation—and those that must have evolved since the solar system formed. In this chapter we draw together all the planetary data we have amassed and show how the regularities—and the irregularities—of the solar system can be explained by a single comprehensive theory.

MODEL REQUIREMENTS

1 Any theory of the origin and architecture of our planetary system must adhere to the known facts. We know of nine outstanding properties of our solar system as a whole. They may be summarized as follows.

1. *Each planet is relatively isolated in space.* The planets exist as independent bodies at progressively larger distances from the central Sun; they are not bunched together. In

very rough terms, each planet tends to be twice as far from the Sun as its next inward neighbor.

2. *The orbits of the planets are nearly circular.* In fact, with the exceptions of Mercury and Pluto, which we will argue are special cases, each planetary orbit closely describes a perfect circle.

3. *The orbits of the planets all lie in nearly the same plane.* The planes swept out by the planets' orbits are accurately aligned to within a few degrees. Again, Mercury and Pluto are slight exceptions.

4. *The direction in which the planets orbit the Sun (counterclockwise as viewed from above Earth's North Pole) is the same as the direction in which the Sun's rotates on its axis.* Virtually all the large-scale motions in the solar system (other than comet orbits) are in the same plane and in the same sense. The plane is that of the Sun's equator, and the sense is that of the Sun's rotation.

5. *The direction in which most planets rotate on their axis is roughly the same as the direction in which the Sun rotates on its axis.* This property is less general than the one just described for revolution, as three planets—Venus, Uranus, and Pluto—do not share it.

6. *Most of the known moons revolve about their parent planets in the same direction that the planets rotate on their axes.*

7. *Our planetary system is highly differentiated.* The inner terrestrial planets are characterized by high densities, moderate atmospheres, slow rotation rates, and few or no moons. By contrast, the jovian planets, farther from the Sun, have low densities, thick atmospheres, rapid rotation rates, and many moons.

8. *The asteroids are very old and exhibit a range of properties not characteristic of either the inner or the outer planets or their moons.* The asteroid belt shares, in rough terms, the bulk orbital properties of the planets. However, it appears to be made of primitive, unevolved material, and the meteorites that strike Earth are the oldest rocks known.

9. *The comets are primitive, icy fragments that do not orbit in the ecliptic plane and reside primarily at large distances from the Sun.*

All these observed facts, taken together, strongly suggest a high degree of order within our solar system. The whole system is not a random assortment of objects spinning or orbiting this way or that. Consequently, it hardly seems possible that our solar system could have formed by the slow accumulation of already-made interstellar "planets" casually captured by our Sun over the course of billions of years. The overall architecture of our solar system is too neat, and the ages of its members too uniform, to be the result of random chaotic events. The overall organization points toward a single formation, an ancient but one-time event, 4.6 billion years ago. A convincing theory that explains all the nine features just listed has been a goal of astronomers for many centuries.

ADDITIONAL CONSIDERATIONS

It is equally important to recognize what our theory of the solar system does *not* have to explain. There is plenty of scope for planets to evolve after their formation, so circumstances that have developed since the initial state of the solar system was established need not be included in our list. Examples are Mercury's 3:2 spin–orbit coupling, Venus's runaway greenhouse effect, the Moon's synchronous rotation, the emergence of life on Earth and its absence on Mars, the Kirkwood gaps in the asteroid belt, and the rings and atmospheric appearance of the jovian planets. There are many more. Indeed, all the properties of the planets for which we have already provided an *evolutionary* explanation need not be included as items that our theory must account for at the outset.

In addition to its many regularities, our solar system also has many notable *irregularities*, some of which we have already mentioned. Far from threatening our theory, however, these irregularities are important facts for us to consider in shaping our explanations. For example, it is necessary that the explanation for the solar system not insist that *all* planets rotate in the same sense or have *only* prograde moons, because that is not what we observe. Instead, the theory of the solar system should provide strong reasons for the observed planetary characteristics yet be flexible enough to allow for and explain the deviations, too. And, of course, the existence of the asteroids and comets that tell us so much about our past must be an integral part of the picture. That's quite a tall order, yet many researchers now believe that we are close to that level of understanding.

15.2 The Condensation Theory

NEBULAR CONTRACTION

2 One of the earliest heliocentric models of solar system formation is termed the **nebular theory**, which may be traced back to the seventeenth-century French philosopher René Descartes. In this model, a large cloud of interstellar gas began to collapse under the influence of its own gravity. As it contracted, it became denser and hotter, eventually forming a star—the Sun—at its center. While all this was going on the outer, cooler, parts of the cloud formed a giant swirling region of matter, creating the planets and their moons essentially as by-products of the star-formation process. This swirling mass destined to become our solar system is usually referred to as the **solar nebula**.

In 1796 the French mathematician-astronomer Pierre Simon de Laplace tried to develop the nebular model in a quantitative way. He was able to show mathematically that the conservation of angular momentum (see *More Precisely 15-1*) demands that an interstellar cloud like the hypothetical solar nebula must spin faster as it contracts. A decrease in the size of a rotating mass must be balanced by an increase in its rotational speed.

The increase in rotation speed, in turn, must have caused the nebula's *shape* to change as it collapsed. In Chapter 11 we saw how a spinning body tends to develop a bulge around its middle. ∞ (Sec. 11.1) The rapidly spinning nebula behaved in exactly this way. As shown in Figure 15.1, the fragment eventually flattened into a pancake-shaped primitive solar system. If we now suppose that planets formed out of this

(a)

(b)

Figure 15.1 (a) Conservation of angular momentum demands that a contracting, rotating cloud (a) must spin faster as its size decreases. (b) Eventually, the primitive solar system came to resemble a giant pancake. The large blob at the center would ultimately become the Sun.

Size of Pluto's Orbit

(a)

R I V U X G

(b)

Figure 15.2 (a) A computer-enhanced view of a disk of warm matter surrounding the star Beta Pictoris. The top image is the actual data taken at visible wavelengths; the bottom image is a false-color rendition that accentuates the details and makes the structure clearer. In both images the overwhelmingly bright central star has been removed in order to let us see the much fainter disk surrounding it. The disk is nearly edge-on to our line of sight. It is made mostly of microscopic dust grains of ices and silicate particles (see Section 15.3) and is illuminated by the reflected light of the central star. For scale, the dimension of Pluto's orbit (78 A.U.) has been drawn adjacent to the images. (b) An artist's conception of the disk of clumped matter, showing the warm disk with a young star at the center and several comet-sized or larger bodies already forming at large radii. The colors are thought to be accurate—generally speaking, if you know the temperature and the density of the nebula, then you can derive its color. At the outer edges of the disk the temperature is low, and the color is a dull red. Progressing inward, the colors brighten and shift to a more yellowish tint as the temperature increases. Mottled dust is seen throughout—such protoplanetary regions are probably very dirty.

Video

spinning material, we can already begin to understand the origin of some of the large-scale architecture observed in our planetary system today, such as the circularity of the planets' orbits and the fact that they move in nearly the same plane.

Astronomers are fairly confident that the solar nebula formed such a disk because similar disks have been observed (or inferred) around other stars. Figure 15.2(a) shows a visible-light image of the region around a star called Beta Pictoris, lying about 50 light years from the Sun. When the light from Beta Pictoris itself is suppressed and the resulting image enhanced by a computer, a faint disk of warm matter (viewed almost edge-on here) can be seen. This particular disk is roughly 500 A.U. across—about 10 times the diameter of Pluto's orbit. Astronomers believe that Beta Pictoris is a very young star, perhaps only 100 million years old, and that we are witnessing it pass through an evolutionary stage similar to the one our own Sun experienced some 4.6 billion years ago.

The nebular theory is an example of an **evolutionary theory**, which describes the development of the solar system as a series of gradual and natural steps, understandable in terms of well-established physical principles. Evolutionary theories may be contrasted with **catastrophic theories**—theories that invoke accidental or unlikely celestial events in order to interpret observations.* Scientists usually do not like to invoke catastrophes to explain the universe. However, as we will see, there are instances where pure chance has played a critical role in determining the present state of the solar system.

*A good example of such a theory is the collision hypothesis, which imagines that the planets were torn from the Sun by a close encounter with a passing star. This hypothesis enjoyed some measure of popularity during the nineteenth century, in large part due to the inability of other theories to account for the observed properties of the solar system, but no scientist takes it seriously today. Aside from its extreme improbability, it is completely unable to explain the orbits, the rotations, or the composition of the planets and their moons.

THE ROLE OF DUST

3 Laplace imagined that as the spinning solar nebula contracted, it left behind a series of concentric rings, each of which would eventually become a planet orbiting a central **protosun**—a hot ball of gas well on its way to becoming the Sun. Each ring then clumped into a **protoplanet**—a forerunner of a genuine planet. The description of the collapse and flattening of the solar nebula is essentially correct, but when modern astronomers used computers to study the more subtle aspects of the problem, some fatal flaws were found in Laplace's nebular picture. Calculations show that a ring of the sort envisaged in his theory would probably not form and, even if it did, would not condense to form a planet in any case. In fact, computer calculations predict just the opposite: the rings would tend to disperse. The protoplanetary matter would be too warm, and no one ring would have enough mass to bind its own matter into a ball.

The model currently favored by most astronomers is a more sophisticated version of the nebular theory. Known as the **condensation theory**, it combines the good features of the old nebular theory with new information about interstellar chemistry to avoid most of the old theory's problems. The key new ingredient in the modern picture is the presence of **interstellar dust** in the solar nebula. Astronomers now recognize that the space between the stars is strewn with microscopic dust grains, an accumulation of the ejected matter of many long-dead stars (see Chapter 22). These dust particles probably formed in the cool atmospheres of old stars, then grew by accumulating more atoms and molecules from the interstellar gas within the Milky Way Galaxy. The end result is that our entire galaxy is littered with miniature chunks of icy and rocky matter having typical sizes of about 10^{-5} m. Figure 15.3 shows one of many such dusty regions found in the vicinity of the Sun.

Dust grains play an important role in the evolution of any gas. Dust helps to cool warm matter by efficiently radiating its heat away in the form of infrared radiation, reducing the pressure (which is just proportional to the gas temperature) and allowing the gas to collapse more easily under the influence of gravity. Furthermore, the dust grains greatly speed up the process of collecting enough atoms to form a planet. They act as **condensation nuclei**—microscopic platforms to which other atoms can attach, forming larger and larger balls of matter. This is similar to the way that raindrops form in Earth's atmosphere; dust and soot in the air act as condensation nuclei around which water molecules cluster.

PLANET FORMATION

Modern models trace the formative stages of our solar system along the following broad lines. Imagine a dusty interstellar cloud fragment measuring about a light year across. Intermingled with the preponderance of hydrogen and helium atoms in the cloud is some heavy-element gas and dust. Some external influence, such as the passage of another interstellar cloud or perhaps the explosion of a nearby star,

Figure 15.3   Interstellar gas and dark dust lanes mark this region of star formation. The dark cloud known as Barnard 86 (left) flanks a cluster of young blue stars called NGC 6520 (right). Barnard 86 may be part of a larger interstellar cloud that gave rise to these stars.

starts the fragment contracting, down to a size of about 100 A.U. As the cloud collapses, it rotates faster and begins to flatten (just as described in the old nebular theory). By the time it has shrunk to 100 A.U., the solar nebula has already formed an extended, rotating disk (Figure 15.4a, b; see also Figures 15.1b and 15.2b).

According to the condensation theory, the planets formed in three stages (Figure 15.4c–e). Early on, dust grains in the solar nebula formed condensation nuclei around which matter began to accumulate. This vital step greatly hastened the critical process of forming the first small clumps of matter. Once these clumps formed, they grew rapidly by sticking to other clumps. (Imagine a snowball thrown through a fierce snowstorm, growing bigger as it encounters more snowflakes.) As the clumps grew larger, their surface areas increased and consequently the rate at which they swept up new material accelerated. They gradually grew into objects of pebble size, baseball size, basketball size, and larger.

Eventually, this process of **accretion**—the gradual growth of small objects by collision and sticking—created objects a few hundred kilometers across. By that time, their gravity was strong enough to sweep up material that would otherwise not have collided with them, and their rate of growth became faster still. At the end of this first stage, the solar system was made up of hydrogen and helium gas and millions of **planetesimals**—objects the size of small moons, having gravitational fields just strong enough to affect their neighbors.

In the second phase of the accretion process, gravitational forces between the planetesimals caused them to collide and merge, forming larger and larger objects. Because larger objects have stronger gravity, the rich became richer in the early solar system, and eventually almost all the planetesimal material was swept up into a few large protoplanets—the accumulations of matter that would eventually evolve into the planets we know today. Figure 15.5 shows a computer simulation of accretion in the inner solar system. Notice how,

(a)

(b)

(c)

(d)

(e)

(f)

as the number of bodies decreases, the orbits of the remainder become more widely spaced and more nearly circular.

As the protoplanets grew, another process became important. The strong gravitational fields produced many high-speed collisions between planetesimals and protoplanets. These collisions led to **fragmentation**, as small objects broke into still smaller chunks, which were then swept up by the protoplanets. Not only did the rich get richer, but the poor were mostly driven to destruction! Some of these fragments produced the intense meteoritic bombardment we know occurred during the early evolution of the planets and moons, as we have seen repeatedly in the last few chapters. A relatively small number of 10–100-km fragments escaped capture by a planet or a moon to become the asteroids and comets.

Mathematical modeling, like the calculation shown in Figure 15.5, indicates that after about 100 million years, the primitive solar system had evolved into nine protoplanets, dozens of protomoons, and the big protosolar mass at the center. Computer simulations generally reproduce the increasing spacing between the planets ("Bode's law"), although the reasons for the regularity seen in the actual planetary spacing remain unclear. ∞ (*Interlude 6-1*) Roughly a billion years more were required to sweep the system reasonably clear of interplanetary trash. This was the billion-year period that saw the heaviest meteoritic bombardment, tapering off as the number of planetesimals decreased. ∞ (Sec. 8.4)

The four largest protoplanets became large enough to enter a third phase of planetary development: sweeping up large amounts of gas from the solar nebula to form what would ultimately become the jovian planets. The smaller, inner protoplanets never reached that point, and as a result their masses remained relatively low. Alternatively, the jovian planets might have formed through instabilities in the cool outer regions of the solar nebula, mimicking on small scales the collapse of the initial interstellar cloud. In this scenario, the jovian protoplanets formed directly, skipping the initial accretion stage. In either case, these first protoplanets had gravitational fields strong enough to scoop up more of the remaining gas and dust in the solar nebula, allowing them to grow into the gas giants we see today. Their large size reflects their head start in the accretion process.

Many of the moons of the planets (but not our own—see Chapter 8) presumably also formed through accretion but on a smaller scale, in the gravitational field of their

Figure 15.4 The condensation theory of planet formation (not drawn to scale; Pluto not shown). (a), (b) The solar nebula contracts and flattens into a spinning disk. The large blob in the center will become the Sun. Smaller blobs in the outer regions may become jovian planets. (c) Dust grains act as condensation nuclei, forming clumps of matter that collide, stick together, and grow into moon-sized planetesimals. (d) Strong winds from the still-forming Sun expel the nebular gas. (e) Planetesimals continue to collide and grow. (f) Over the course of a hundred million years or so, planetesimals form a few large planets that travel in roughly circular orbits.

Figure 15.5   Accretion in the inner solar system: initially, many moon-sized planetesimals orbited the Sun. Over the course of a hundred million years or so, they gradually collided and coalesced, forming a few large planets in roughly circular orbits.

parent planets. ∞ (Sec. 8.8) Once the nebular gas began to accrete onto the large jovian protoplanets, conditions probably resembled a miniature solar nebula, with condensation and accretion continuing to occur. The large moons of the outer planets almost certainly formed in this way. Some of the smaller moons may have been "chipped off" their parent planets during collisions with asteroids; others may be captured asteroids themselves.

15.3 The Differentiation of the Solar System

4 We can understand the basic differences in content and structure between the terrestrial and the jovian planets using the condensation theory of the solar system's origin. Indeed, it is in this context that the adjective *condensation* derives its true meaning. To see why the planets' composition depends on location in the solar system, it is necessary to consider the temperature structure of the solar nebula.

THE ROLE OF HEAT

As the primitive solar system contracted under the influence of gravity it heated up as it flattened into a disk. The density and temperature were greatest near the central protosun and much lower in the outlying regions. Detailed calculations indicate that the gas temperature near the core of the contracting system was several thousand kelvins. At a

distance of 10 A.U., out where Saturn now resides, the temperature was only about 100 K.

In the warmer regions of the cloud, dust grains broke apart into molecules, and they in turn split into excited atoms. Because the extent to which the dust was destroyed depended on the temperature, it also depended on location in the solar nebula. Most of the original dust in the inner solar system disappeared at this stage, but the grains in the outermost parts probably remained largely intact.

The destruction of the dust in the hot inner portion of the solar nebula introduced an important new ingredient into the theoretical mix, one that we omitted from our earlier account of the accretion process. With the passage of time, the gas radiated away its heat and the temperature decreased at all locations, except in the very core, where the Sun was forming. Everywhere beyond the protosun, new dust grains began to condense (or crystallize) from their hotter gas phase to their cooler solid phase, much as raindrops, snowflakes, and hailstones condense from moist, cooling air here on Earth. It may seem strange that although there was plenty of interstellar dust early on, it was mostly destroyed, only to form again later. However, a critical change had occurred. Initially, the nebular gas was uniformly peppered with dust grains. When the dust re-formed later, the distribution of grains was very different.

Figure 15.6 plots the temperature in various parts of the primitive solar system just prior to the onset of the accretion stage. At any given location, the only materials to condense out were those able to survive the temperature there. As marked on the figure, in the innermost regions, around Mercury's present orbit, only metallic grains could

INTERLUDE 15-1 The Discovery of Planets Beyond the Solar System

Many claims of evidence of extrasolar planets have been reported (and even published) over the past few decades, but virtually none have been confirmed and most have been discredited. Only within the past few years have we seen genuine advances in this fascinating astronomical specialty. These advances have come not through any dramatic scientific or technical breakthrough but through steady improvements in both telescope and detector technology and computerized data analysis.

It is not yet possible to image any of these newly discovered planets. The techniques used to find them are indirect, based on analysis of light from the parent star, not light from the unseen planet. As a planet orbits a star, gravitationally pulling first one way and then the other, the star wobbles slightly. The more massive the planet or the less massive the star, the greater the star's movement. If the wobble happens to occur along our line of sight to the star, then we see small fluctuations in the star's velocity, which can be measured using the Doppler effect. ∞ (Sec. 3.5) Alternatively, if the wobble is predominantly perpendicular to our line of sight, then the star's position in the sky changes slightly from night to night. The star's motion is very small, which is why unambiguous measurements have been so hard to obtain. However, *both* types of wobble have now been seen and confirmed.

The first figure shows two sets of data that betray the presence of planets. The top graph shows the line-of-sight velocity of the star 51 Pegasi, a near-twin to our Sun lying some 40 light years away. These data were acquired in 1994 by Swiss astronomers using the 1.9-m telescope at Haute-Provence Observatory in France. The regular 50-m/s fluctuations in the star's velocity have been confirmed by several groups of astronomers and imply that a planet at least half the mass of Jupiter orbits 51 Pegasi with a period of just 4.2 days. The bottom graph in the figure plots the declination (see *More Precisely 1-2*) of the star Lalande 21185, corrected for the motion of Lalande through the Milky Way Galaxy and the motion of

Earth around the Sun over the last half century. A 30-year-period wobble at the level of about 0.01″ can be seen—about the amount of wobble that would be produced in our own Sun by the planets of the solar system. These observations have triggered a minirevolution in astronomy and have doubtless paved the way for a flood of new planet searches (and reinterpretation of old data) in the next few years.

Currently, these techniques can be used only to detect planets that are at least as large as Jupiter. Thus far, none of the reports of nearby extrasolar planets has claimed an object of Earth dimensions. In fact, many of the findings to date indicate "hot Jupiters"—gigantic planets surprisingly close (within 1 A.U.) to their parent star. Furthermore, only one such unseen planet has been found (and confirmed) in

form. It was simply too hot for anything else to exist. A little farther out, at about 1 A.U., it was possible for rocky, silicate grains to form, too. Beyond about 3 or 4 A.U., water ice could exist, and so on, with the condensation of more and more material possible at greater and greater distances from the Sun. The composition of the material that could condense out at any given radius would ultimately determine the types of planets that formed there.

Figure 15.6 Theoretically computed variation of temperature across the primitive solar nebula. In the hot central regions only metals could condense out of the gaseous state to form grains. At greater distances from the central protosun the temperature was lower, so rocky and icy grains could also form. The labels indicate the minimum radii at which grains of various types could condense out of the nebula.

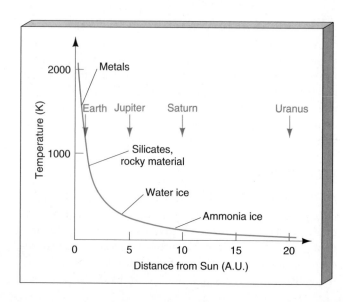

each system, although tentative analysis of the Lalande data suggests that there may be as many as three Jupiter-sized planets, each with a period of several years. So do not think that the newly discovered planetary systems closely resemble our own solar system. At least so far, they don't—and that has theorists puzzled.

The second figure represents the latest challenge to theory. This *HST* image shows what may be our first-ever view of an extrasolar planet—the tiny dot at bottom left (indicated by the arrow) is thought by some researchers to be a Jupiter-sized object escaping from the double-star system (top center) where it apparently formed. Theorists speculate that the planet might have been ejected following a close encounter with one of its two parent stars; the long filament

of gas may mark the planet's path. The problem is that the stars are estimated to be just 300,000 years old, which is far too little time for a jovian world to form by accretion. However, if the planet formed directly, through instabilities in the prestellar nebula (Sec. 15.2), then both its unstable orbit and its rapid formation may be explained.

The chart below summarizes all the confirmed extrasolar planets found as of mid-1998. Most appear to fall into the hot-Jupiter category, although some may well turn out to be *brown dwarfs*—"failed stars" having insufficient mass to become true stars (see *Interlude 19–1*). The dividing line between genuine Jupiter-like planets and brown dwarfs is unclear, but it is thought to be around five Jupiter masses.

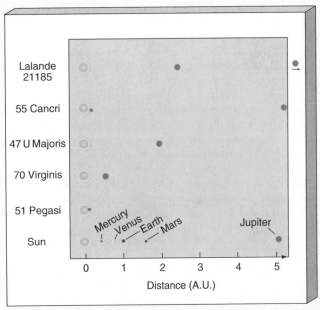

THE JOVIAN PLANETS

In the middle and outer regions of the primitive planetary system, beyond about 5 A.U. from the center, the temperature was low enough for the condensation of several abundant gases into solid form. After hydrogen and helium, the most common materials in the solar nebula (as they are today in the universe as a whole) were the elements carbon, nitrogen, and oxygen. The most common chemical compounds were those containing those elements—specifically, water vapor, ammonia, and methane. As we have seen, these compounds are still the primary constituents of jovian atmospheres.

At temperatures of a few hundred kelvins or less, these gases condensed out of the nebula. Consequently, the ancestral fragments destined to become the cores of the jovian planets were formed under cold conditions out of low-density, icy material. The planetesimals that formed at

these distances were predominantly composed of ice. Because more material could condense out of the solar nebula at these radii than in the inner regions near the protosun, accretion began sooner, with more resources to draw on. The outer planets grew rapidly to the point where they could accrete nebular gas, not just grains, and eventually formed the hydrogen-rich jovian worlds we see today.

With the formation of the four giant jovian planets, the remaining planetesimals were subject to those planets' strong gravitational fields. Over a period of hundreds of millions of years and after repeated "gravity assists" from the giant planets, especially Uranus and Neptune, many of the interplanetary fragments in the outer solar system were flung into orbits taking them far from the Sun. Astronomers believe that those fragments now make up the Oort cloud, whose members occasionally visit the inner solar system as comets. ∞ (Sec. 14.2) During this period many icy planetesimals were also

MORE PRECISELY 15-1 The Concept of Angular Momentum

Most celestial objects rotate. Planets, moons, stars, and galaxies all have some *angular momentum*, which we can define as the tendency of a body to keep spinning or moving in a circle. Angular momentum is as important a property of an object as its mass or its energy.

Consider, first, a simpler motion—*linear momentum*, which is defined as the product of an object's mass and its speed:

$$\text{linear momentum} = \text{mass} \times \text{speed}.$$

This is the tendency of an object to keep moving in a straight line in the absence of external forces. Consider a truck and a bicycle rolling equally fast down a street. Each has some linear momentum, but you would obviously find it easier to stop the less massive bicycle. Although the two vehicles have the same speed, the truck has more momentum. We see that the linear momentum of an object depends on the mass of that object. It also depends on the speed. If two bicycles were rolling down the street at different speeds, the slower one could be stopped more easily.

Angular momentum is an analogous property of objects that are rotating or revolving. However, in addition to mass and speed, the angular momentum also depends on the way in which the object's mass is distributed. The farther the mass is from the object's axis of rotation, the greater the angular momentum. For example, if you whirl a ball with constant speed at the end of a string, its angular momentum will depend directly on the length of the string—the longer the string, the greater the ball's angular momentum. We can therefore say, loosely, that

$$\text{angular momentum} = \text{mass} \times \text{speed} \times \text{"size,"}$$

where "size" is a quantity that depends not only on the object's dimensions but also on the distribution of its mass. For the simple case of the ball on a string, it is just the length of the string (the radius of the ball's "orbit"). Note that the "speed" for linear momentum refers to motion in a straight line. For angular momentum, "speed" refers to spinning or orbital motion.

According to Newton's laws of motion, both types of momentum—linear and angular—must be conserved at all times. In other words, both linear and angular momentum must remain constant before, during, and after a physical change in any object (so long as no external forces act). For example, if a spherical object having some spin begins to contract, the relationship above demands that it spin faster, so that the product mass × speed × size remains constant. The sphere's mass does not change during the contraction, yet the size of the object clearly decreases. Its rotation speed must therefore increase in order to keep the total angular momentum unchanged.

Figure skaters use the principle of angular-momentum conservation. They spin faster by drawing in their arms (as shown in the figure below) and slow down by extending them. Here the mass of the human body remains the same, but its lateral size changes, causing the body's circular speed to change to conserve angular momentum.

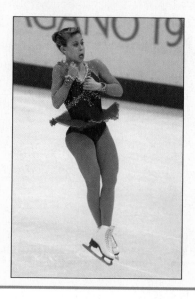

---

deflected into the inner solar system, where they played an important role in the evolution of the inner planets.

A key prediction of this model is that some of the original planetesimals should have remained behind, forming the broad band known as the Kuiper belt, lying beyond the orbit of Neptune. ∞ (Sec. 14.2) In 1993 several such asteroid-sized objects were discovered, lying between 30 and 35 A.U. from the Sun and lending strong support to the condensation theory. Over 60 Kuiper belt objects, having diameters ranging from 100 to 400 km, are now known. (See also *Interlude 15-2*.)

THE TERRESTRIAL PLANETS

In the inner regions of the primitive solar system, condensation from gas to solid began when the average temperature was about 1000 K. The environment there was too hot for ices to survive. Many of the abundant heavier elements, such as silicon, iron, magnesium, and aluminum, combined with oxygen to produce a variety of rocky materials. Planetesimals in the inner solar system were therefore rocky in nature, as were the protoplanets and planets they ultimately formed.

These heavier materials condensed into grains in the outer solar system, too, of course. However, they would have been vastly outnumbered by the far more abundant light elements there. The outer solar system is not deficient in heavy elements. The inner solar system is *underrepresented in light material*. Here we have another reason why the jovian planets grew so much bigger than the terrestrial worlds. The inner regions of the nebula had to wait for the temperature to drop so that a few rocky grains could appear and begin the accretion process, but the outer regions might not have had to wait at all. Accretion in the outer solar system began almost with the formation of the disk itself.

Very abundant light elements such as hydrogen and helium, as well as any other gases that failed to condense into solids, would have escaped from the terrestrial protoplanets or, more likely, were simply never accreted from the solar nebula. The inner planets' surface temperature was too high, and their gravity too low, to capture and retain those gases. Where then did Earth's volatile gases, particularly water, and the recently discovered ice on Mercury and the Moon, come from? The answer seems to be that icy fragments—comets—from the outer solar system, deflected into eccentric orbits by the jovian planets' gravity, participated in the meteoritic bombardment of the newly born inner planets, supplying them with water *after* their formation.

The myriad rocks of the asteroid belt between Mars and Jupiter failed to accumulate into a planet. Probably, nearby Jupiter's huge gravitational field caused them to collide too destructively to coalesce. Strong Jupiter tides on the planetesimals in the belt would also have hindered the development of a protoplanet. The result is a band of planetesimals, still colliding and occasionally fragmenting, but never coalescing into a larger body—surviving witnesses to the birth of the planets.

CLEANING UP THE DEBRIS

Most of the planetesimals left over after the major planets formed eventually collided with a planet or were ejected into the Oort cloud. Little solid material remained. But what of the gas that made up most of the original cloud? Why don't we see it today throughout the planetary system? In the outer solar system, some (but not all) of that gas was swept up into planets. But that did not occur in the inner regions, where the terrestrial protoplanets never became massive enough to accrete such light material. Instead, the newly formed Sun took a hand.

All young stars apparently experience a highly active evolutionary stage known as the *T Tauri* phase (Figure 15.7; see also Chapter 19), during which their radiation and stellar winds are very intense. Any gas remaining between the planets was blown away into interstellar space by the solar wind and the Sun's radiation pressure when the Sun entered this phase, just before nuclear burning started at its center. Afterward, all that remained were protoplanets and planetesimal fragments, ready to continue their long evolution into the solar system we know today.

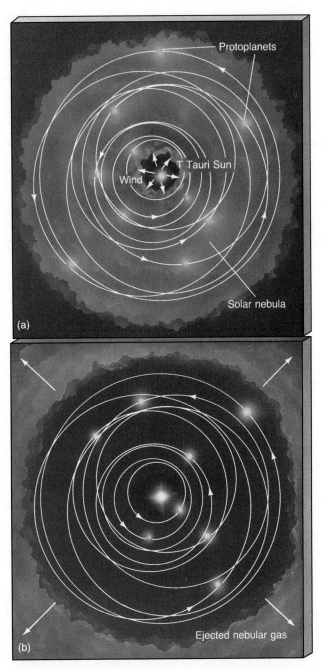

Figure 15.7 (a) Strong stellar winds from newly born stars are responsible for sweeping away any dust and gas left over from the star formation process, (b) leaving only planets and planetesimals behind.

15.4 The Role of Catastrophes

5 The condensation theory accounts for the nine "characteristic" points listed at the start of this chapter. Specifically, the planets' orbits are circular (2), in the same plane (3), and in the same direction as the Sun's rotation on its axis (4) as a direct consequence of the nebula's shape and rotation. The rotation of the planets (5) and the orbits of the

moon systems (6) are due to the tendency of the smaller-scale eddies to inherit the nebula's overall sense of rotation. The growth of planetesimals throughout the nebula, with each protoplanet ultimately sweeping up the material near it, accounts for point (1), which is the fact that the planets are widely spaced (even if the theory does not explain the regularity of the spacing). The heating of the nebula and the Sun's ignition resulted in the observed differentiation (7), and the debris from the accretion–fragmentation stage naturally accounts for the asteroids (8) and comets (9).

INTERLUDE 15-2 The Kuiper Belt and the Search for a Tenth Planet

No one has ever observed any comets in the faraway Oort cloud—they are just too small and dim for us to see from Earth. But in the 1990s such faint objects began to be inventoried in the relatively nearby Kuiper belt, just beyond Neptune's orbit. ∞ (Sec. 14.2) Ground-based telescopes have led the way in the painstaking work to capture the meager amounts of sunlight reflected from dozens of such dark objects orbiting in the outer Solar System. The largest of these objects is only a few hundred kilometers across, yet there must be many much smaller icy fragments in cold storage. In many ways, the Kuiper belt must resemble the asteroid belt, the collection of rocky debris found mainly between Mars and Jupiter. ∞ (Sec. 14.1)

As of mid-1998, the current count of trans-Neptunian, Kuiper belt objects was 62. They range in diameter from 100 to 400 km, which is considerably smaller than either Pluto (2300 km) or its moon Charon (1100 km). ∞ (Sec. 13.9) Estimates of the total number of such objects larger than 100 km run into the tens of thousands, so the combined mass of all the debris in the Kuiper belt could be hundreds of times larger than the inner asteroid belt (though still less

than the mass of Earth). Even with large ground-based telescopes equipped with sensitive CCD detectors, it is unlikely that objects much smaller than 100 km (which almost certainly are out there, and probably number in the billions) will be found in the near future. (Astronomers using the *Hubble Space Telescope* have reported finding numerous such smaller objects, but none has been confirmed after repeated tries.)

Much of this observational work on the Kuiper belt began as a search for a tenth planet. In the process, astronomers have begun to wonder if the Kuiper reservoir may be the source of more than just short-period comets. For example, Pluto, its moon Charon, and the Neptunian moon Triton all reside in or near the Kuiper belt. Furthermore, as we saw in Chapter 13, these three oddballs have very similar physical properties and differ greatly from their neighboring planets and moons. Could this peculiar trio be the last survivors of a much larger set of big Kuiper belt objects? If so, then Pluto doesn't deserve the status of a genuine planet. Ironically, work that began as a search to find a tenth planet may yet succeed in reducing the number of true solar system planets to eight.

We stressed earlier that an important aspect of any solar system theory is its ability to allow for the possibility of imperfections—deviations from the otherwise well ordered scheme of things. In the condensation theory that capacity is provided by the randomness inherent in the encounters that ultimately combined the planetesimals into protoplanets. As the numbers of large bodies decreased and their masses increased, individual collisions acquired greater and greater importance. The effects of these collisions can still be seen today in many parts of the solar system—for example, the large craters on many of the moons we have studied thus far.

Having started with nine regular points to explain, we end with eight irregular solar system features that still fall within the theory's scope. It is impossible to test any of these assertions directly, but it is reasonable to suppose that some (or even all) of the following "odd" aspects of the solar system can be explained in terms of collisions late in the formative stages of the protoplanetary system. Not all astronomers believe all these explanations; however, most would accept at least some.

1. Mercury's exceptionally large nickel–iron core may be the result of a collision between two partially differentiated protoplanets. The cores may have merged, and much of the mantle material may have been lost. ∞ (Sec. 8.7)

2. Two large bodies could have merged to form Venus, giving it its abnormally low rotation rate. ∞ (Sec. 9.2)

3. The Earth–Moon system may have formed from a collision between the proto-Earth and a Mars-sized object. ∞ (Sec. 8.8)

4. A late collision with a large planetesimal may have caused Mars's curious north–south asymmetry and ejected much of the planet's atmosphere. ∞ (Sec. 10.4)

5. The tilted rotation axis of Uranus may have been caused by a grazing collision with a sufficiently large planetesimal, or by a merger of two smaller planets. ∞ (Sec. 13.3)

6. Uranus's moon Miranda may have been almost destroyed by a planetesimal collision, accounting for its bizarre surface terrain. ∞ (Sec. 13.6)

7. Interactions between the jovian protoplanets and one or more planetesimals may account for the irregular moons of those planets and, in particular, Triton's retrograde motion. ∞ (Sec. 13.6)

8. Pluto may simply be a large representative of the Kuiper belt, and the Pluto–Charon system may be the result of a collision or near-miss between two icy planetesimals before most were ejected by interactions with the jovian planets. ∞ (Sec. 13.9)

15.5 The Angular Momentum Problem

A possible weak link in the condensation theory is sometimes known as the **angular momentum problem**. Although our Sun contains about 1000 times more mass than all the planets combined, it possesses a mere 0.3 percent of the total angular momentum of the solar system. Jupiter, for example, has a lot more angular momentum than does our Sun—in fact, about 60 percent of the solar system's angular momentum. All told, the four jovian planets account for well over 99 percent of the total angular momentum of the solar system. By comparison, the lighter (and closer) terrestrial planets have negligible angular momentum.

The problem here is that all mathematical models predict that the Sun should have been spinning very rapidly during the earliest epochs of the solar system and should command most of the solar system's angular momentum, basically because it contains most of the mass. However, as we have just seen, the reverse is true. Indeed, if all the planets' orbital angular momentum were transferred to the Sun, it would spin on its axis about 100 times as fast as it does at present.

Many researchers speculate that the solar wind, moving away from the Sun into interplanetary space, carried away much of the Sun's initial angular momentum. The early Sun probably produced more of a dense solar gale than the relatively gentle "breezes" now measured by our spacecraft. High-velocity particles leaving the Sun followed the solar magnetic field lines. As the rotating magnetic field of the Sun tried to drag those particles around with it they acted as a brake on the Sun's spin. Although each particle boiled off the Sun carries only a tiny amount of the Sun's angular momentum with it, over the course of nearly 5 billion years

the vast numbers of escaping particles could have robbed the Sun of most of its initial spin. Even today, our Sun's rotation rate continues to slow.

Other researchers prefer to solve the Sun's momentum problem by assuming that the primitive solar system was much more massive than the present-day system. They argue that the accretion process was not entirely successful during the system's formative stages. Matter not captured by the Sun or the planets might well have transported much angular momentum back into interstellar space as it escaped. This proposal is difficult to test, because the escaped matter would be well beyond the range of our current space probes. Perhaps the remote Oort cloud of innumerable comets is the "escaped" matter.

Despite some minor controversy as to how this angular momentum quandary can best be resolved, nearly all astronomers agree that some version of the condensation theory is correct. The details have yet to be fully worked out, but the broad outlines of the processes involved are quite firmly established. Our planet is a by-product of the formation of the Sun. We might reasonably now ask what preceded the Sun, and what circumstances led to the collapse of the solar nebula in the first place. To answer these questions, we must widen the scope of our studies. We will find it necessary to understand the workings not only of the stars and the gas between them but also of the Galaxy in which they reside.

Chapter Review

SUMMARY

Our solar system is an orderly place, making it unlikely that the planets were simply captured by the Sun. The overall organization points toward formation as the product of an ancient, one-time event, 4.6 billion years ago. An ideal theory of the solar system should provide strong reasons for the observed characteristics of the planets yet be flexible enough to allow for deviations.

In the **nebular theory** (p. 339) of the formation of the solar system, a large cloud of dust and gas—the **solar nebula** (p. 339)—began to collapse under its own gravity. As it did so, it began to spin faster, to conserve angular momentum, eventually forming a disk. **Protoplanets** (p. 341) formed in the disk and became planets, and the central **protosun** (p. 341) eventually evolved into the Sun.

The nebular theory is an example of an **evolutionary theory** (p. 340), in which the properties of the solar system evolved smoothly into their present state. In a **catastrophic theory** (p. 340), changes occur abruptly, as the result of accident or chance.

The **condensation theory** (p. 341) builds on the nebular theory by incorporating the effects of particles of **interstellar dust** (p. 341), which helped cool the nebula and acted as **condensation nuclei** (p. 341), allowing the planet-building process to begin.

Small clumps of matter grew by **accretion** (p. 341), gradually sticking together and growing into moon-sized **planetesimals** (p. 341), whose gravitational fields were strong enough to accelerate the accretion process. Competing with accretion in the solar nebula was **fragmentation** (p. 342), the breaking up of small bodies following collisions with larger ones. Eventually, only a few planet-sized objects remained. The planets in the outer solar system became so large that they could capture the hydrogen and helium gas in the solar nebula, forming the jovian worlds.

The condensation theory can explain the basic differences between the jovian and terrestrial planets because the temperature of the solar nebula would be expected to decrease with increasing distance from the Sun. At any given location, the temperature would determine which materials could condense out of the nebula and so control the composition of any planets forming there. The terrestrial planets are rocky because they formed in the hot inner regions of the solar nebula, near the Sun, where only rocky and metallic materials condensed out. Farther out, the nebula was cooler, and ices of water and ammonia could also form, leading to the observed differences in composition between the inner and outer solar system.

When the Sun became a star, its strong winds blew away any remaining gas in the solar nebula. Many leftover planetesimals were ejected into the Oort cloud by the gravitational fields of the outer planets. They now occasionally revisit our part of the solar system as comets. In the inner solar system, light elements such as

hydrogen and helium would have escaped into space. Much, if not all, of Earth's water was carried to our world by comets deflected from the outer solar system.

The asteroid belt is a collection of planetesimals that never managed to form a planet, probably because of Jupiter's gravitational influence. Many "odd" aspects of the solar system may conceivably be explained in terms of collisions late in the formation stages of the protoplanetary system.

The **angular momentum problem** (p. 348) is that although the Sun contains virtually all the solar system's mass, it accounts for almost none of the angular momentum. It is believed that the solar wind or ejected planetesimals carried off the Sun's initially high angular momentum, allowing its spin to slow to the rate observed today.

SELF-TEST: TRUE OR FALSE?

The following nine questions present properties of the solar system that any model of solar system formation must explain. Which are correctly stated and which are not?

_____ **1.** Each planet is relatively isolated in space.

_____ **2.** The orbits of the planets are not circular but significantly elliptical.

_____ **3.** The orbits of the planets all lie near the ecliptic plane.

_____ **4.** The direction of planetary revolution is in the same direction as the Sun's rotation.

_____ **5.** Planetary rotation is always in the same direction as the Sun's rotation.

_____ **6.** Moons do not usually revolve in the same direction as their parent planet rotates.

_____ **7.** The planetary system is highly differentiated.

_____ **8.** Asteroids were recently formed from the collision and breakup of an object orbiting within the asteroid belt.

_____ **9.** Most comets have short periods and orbit close to the ecliptic plane.

True or False?

_____ **10.** Water could not have condensed out any closer than 3 or 4 A.U. from the Sun.

_____ **11.** The accretion process occurred faster in the inner part of the solar system than it did in the outer regions.

_____ **12.** The condensation theory does not offer an explanation of the highly tilted rotation axis of Uranus.

_____ **13.** The condensation theory does not offer an explanation of the formation of the Moon.

_____ **14.** Random collisions, inherently a part of the condensation theory, can explain many of the odd properties found among some solar system objects.

_____ **15.** Astronomers know of no planetary systems other than our own.

SELF-TEST: FILL IN THE BLANK

1. The condensation theory, which currently is used to explain the formation of the solar system, is actually just a refined version of the older _____ theory.

2. In the condensation theory, astronomers realized the critical role played by _____ in starting the formation of small clumps of matter.

3. Initially, the accretion of matter into larger bodies occurred through _____ between particles in the solar nebula.

4. By the time planetesimals had formed, the accretion process was accelerated by the effect of _____.

5. In the final stage of accretion, the largest protoplanets were able to attract large quantities of _____ from the solar nebula.

6. The temperature of the inner part (out to 1 or 2 A.U.) of the solar nebula was well above the boiling point of _____.

7. Unlike the terrestrial planets, the planetesimals that formed the jovian planets were made up of _____ material.

8. High-speed collisions between planetesimals often led to _____ rather than accretion.

9. The large number of leftover planetesimals formed beyond about 5 A.U. were destined to become _____.

10. The water now found on Earth was probably brought here by _____.

11. The reason the planetesimals of the asteroid belt did not form a larger object was probably the gravitational influence of _____.

12. The _____ phase of the early Sun cleared out excess gas not used in planet formation.

13. Angular momentum depends on the mass, speed, and _____ of an object.

14. The Sun's angular momentum is now much _____ than it was when the Sun was a protostar and is now much _____ than the total orbital angular momentum of all the planets.

15. _____ played an important role in determining many of the irregularities in the solar system.

REVIEW AND DISCUSSION

1. List six properties of the solar system that any model of its formation must be able to explain.

2. Explain the difference between evolutionary theories and catastrophic theories of the solar system's origin.

3. Describe the basic features of the nebular theory of solar system formation.

4. Give three examples of how the nebular theory explains some observed features of the present-day solar system.

5. Explain the difference between angular momentum and linear momentum.

6. What is the key ingredient in the modern condensation theory of the solar system's origin that was missing or unknown in the nebular theory?

7. Why are the jovian planets so much larger than the terrestrial planets?

8. What solar system objects, still observable today, resulted from the process of fragmentation?

9. What influence did Earth's location in the solar nebula have on its final composition?

10. How did the temperature structure of the solar nebula determine planetary composition?

11. Why could Earth not have formed out of material containing water? How might Earth's water have gotten here?

12. What happened in the early solar system when the Sun became a T Tauri star?

13. What is the modern explanation for the formation of the Kuiper belt and the Oort cloud?

14. How do modern astronomers attempt to explain the angular momentum problem in light of modern theories of solar system formation?

15. Describe a possible history of a single comet now visible from Earth, starting with its birth in the solar nebula somewhere near the planet Jupiter.

PROBLEMS

1. The orbital angular momentum of a planet in a circular orbit is simply the product of its mass, its orbital speed, and its distance from the Sun. (a) Compare the orbital angular momenta of Jupiter, Saturn, and Earth. (b) Calculate the orbital angular momentum of an Oort cloud comet, of mass 10^{13} kg, moving in a circular orbit 50,000 A.U. from the Sun.

2. The rotational angular momentum of a spinning body is proportional to the product of its mass, its *angular* speed (in revolutions per day, say), and the *square* of its radius. An interstellar cloud fragment of 0.1 light year diameter is rotating slowly, at a rate of 1 revolution per million years, as it begins to collapse. Assuming that the mass remains constant, estimate what the rotation period will be when the cloud has shrunk to the size of the solar nebula, 100 A.U. across.

3. By what factor would Earth's rotational angular momentum (see problem #2) change if the planet's spin rate were to double? By what factor would Earth's orbital angular momentum change if the planet's distance from the Sun were to double (assuming that the orbit remained circular)?

4. We can make a rough model of accretion in the inner solar nebula by imagining a 1-km-diameter body moving at a relative speed of 500 m/s through a collection of similar bodies having a roughly uniform spatial density of 10^{-10} bodies per cubic kilometer. Neglecting any gravitational forces (and hence assuming that the body moves in a straight line until it collides with something), estimate how long, on average, will be needed for it to experience a collision.

5. Consider a planet growing by accretion of material from the solar nebula. As it grows, its density remains roughly constant. Does the force of gravity at its surface increase, decrease, or stay the same? Specifically, what would happen to the surface gravity and escape speed as the radius of the planet doubled? Give reasons for your answer.

6. How many 100-km-diameter rocky (3500 kg/m³) planetesimals would have been needed to form Earth?

7. Two asteroids, each of mass 10^{18} kg, orbit near the center of the asteroid belt in the plane of the ecliptic on circular paths of radii 2.8 and 2.9 A.U., respectively. Calculate the gravitational force between them at closest approach, and compare it with the tidal force that would be exerted by Jupiter if that planet happened also to be at closest approach at that time.

8. According to Figure 15.6, the temperature in the early solar nebula at a distance of 1 A.U. from the Sun was about 1100 K. Based on the discussion of surface temperature in Section 7.2, estimate the factor by which the Sun's current energy output would have to increase in order for Earth's present temperature to have this value.

9. A typical comet contains some 10^{13} kg of water ice. How many comets would have to strike Earth in order to account for the roughly 2×10^{21} kg of water presently found on our planet? If this amount of water accumulated over a period of 0.5 billion years, how frequently must Earth have been hit by comets during that time?

10. How many comet-sized planetesimals would have been needed to form Pluto?

16 THE SUN

Our Parent Star *www*

LEARNING GOALS

Studying this chapter will enable you to:

1 Summarize the overall properties of the Sun.

2 Explain how energy travels from the solar core, through the interior, and out into space.

3 Name the Sun's outer layers and describe what those layers tell us about the Sun's surface composition and temperature.

4 Discuss the nature of the Sun's magnetic field and its relationship to the various types of solar activity.

5 Outline the process by which energy is produced in the Sun's interior.

6 Explain how observations of the Sun's core challenge our present understanding of solar physics.

(Opposite page) This spectacular image of the Sun was made by capturing X rays emitted by our star's most active regions. It was taken by a camera on a rocket lofted shortly before the total solar eclipse of July 1991. (Note the shadow of the Moon approaching from the west, at top.) The brightest regions have temperatures of about 3 million kelvins.

Living in the solar system, we have the chance to study, at close range, perhaps the most common type of cosmic object—a star. Our Sun is a star, and a fairly average star at that, but with one unique feature: it is very close to us—some 300,000 times closer than our next nearest neighbor, Alpha Centauri. Whereas Alpha Centauri is 4.3 light years distant, the Sun is only 8 light minutes away from us. Consequently, astronomers know far more about the properties of the Sun than about any of the other distant points of light in the universe. A good fraction of all our astronomical knowledge is based on modern studies of the Sun. Just as we studied our parent planet, Earth, to set the stage for our exploration of the solar system, we now study our parent star, the Sun, as the next step in our exploration of the universe.

16.1 The Sun in Bulk

The Sun is the sole source of light and heat for the maintenance of life on Earth. It is a **star**, a glowing ball of gas held together by its own gravity and powered by nuclear fusion at its center. In its physical and chemical properties the Sun is very similar to most other stars, regardless of when and where they formed. Indeed, our Sun appears to be a rather "typical" star, lying right in the middle of the observed ranges of stellar mass, radius, brightness, and composition. Far from detracting from the interest in the Sun, this very mediocrity is one of the main reasons that astronomers study it—they can apply knowledge of solar phenomena to so many other stars in the universe.

PHYSICAL PROPERTIES

The Sun Data box at right lists some basic orbital and physical solar data. The Sun's radius, roughly 700,000 km, is most directly determined by measuring its angular size (0.5°) and then employing elementary geometry. ∞ (Sec. 1.5). The Sun's mass, 2.0×10^{30} kg, follows from application of Newton's laws of motion and gravity to the observed orbits of the planets. ∞ (*More Precisely 2-3*) The average solar density derived from these quantities, approximately 1400 kg/m³, is quite similar to that of the jovian planets and about one-quarter the average density of Earth.

The solar rotation period is found by timing sunspots ∞ (Sec. 2.5) and other surface features as they traverse the solar disk. These observations indicate that the Sun rotates in about a month, but it does not do so as a solid body. Instead, it spins *differentially*—faster at the equator and slower at the poles, like Jupiter and Saturn.

The Sun's surface temperature is measured by applying the radiation laws to the observed solar spectrum. ∞ (Sec. 3.4) The distribution of solar radiation has the approximate shape of a blackbody curve for an object at about 5800 K. The average solar temperature obtained in this way is known as the Sun's *effective temperature*.

Having a radius of more than 100 Earth radii, a mass of more than 300,000 Earth masses, and a surface temperature well above the melting point of any known material, the Sun is clearly a body very different from any other we have encountered so far.

SUN DATA

Greatest angular diameter, as seen from Earth	32.5′
Mass	1.99×10^{30} kg 332,000 (Earth = 1)
Equatorial radius	696,000 km 109 (Earth = 1)
Mean density	1410 kg/m³ 0.255 (Earth = 1)
Surface gravity	274 m/s² 28.0 (Earth = 1)
Escape speed	618 km/s
Sidereal rotation period	25.4 solar days (25.1 at equator, 34.4 at poles)
Axial tilt	7.25° (relative to ecliptic)
Surface temperature	5780 K (effective temperature)
Luminosity	3.85×10^{26} W

Figure 16.1 This photograph of the Sun shows a sharp solar limb, although our star, like all stars, is made of a gradually thinning gas. The edge appears sharp because the solar photosphere is so thin.

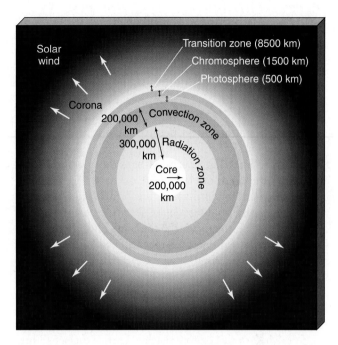

Figure 16.2 The main regions of the Sun, not drawn to scale, with some physical dimensions labeled.

OVERALL STRUCTURE

The Sun has a surface of sorts—not a solid surface (the Sun contains no solid material) but rather that part of the brilliant gas ball we perceive with our eyes or view through a heavily filtered telescope. This "surface"—the part of the Sun that emits the radiation we see—is called the **photosphere**. Its radius (listed in our table as the radius of the Sun) is about 700,000 km. However, the thickness of the photosphere is probably no more than 500 km, less than 0.1 percent of the radius, which is why we perceive the Sun as having a well-defined, sharp edge (Figure 16.1).

The main regions of the Sun are illustrated in Figure 16.2 and summarized in Table 16.1. Just above the photosphere is the Sun's lower atmosphere, called the **chromo-**

sphere, about 1500 km thick. From 1500 km to 10,000 km above the top of the photosphere lies a region called the **transition zone**, where the temperature rises dramatically. Above 10,000 km, and stretching far beyond, is a tenuous (thin), hot upper atmosphere, the solar **corona**. At still greater distances, the corona turns into the *solar wind*, which flows away from the Sun and permeates the entire solar system. ⚭ (Sec. 6.5)

Extending down some 200,000 km below the photosphere is the **convection zone**, a region where the material of the Sun is in constant convective motion. Below the convection zone lies the **radiation zone**, where solar energy is transported toward the surface by radiation rather than by

REGION	INNER RADIUS (km)	TEMPERATURE (K)	DENSITY (kg/m³)	DEFINING PROPERTIES
Core	0	15,000,000	150,000	Energy generated by nuclear fusion
Radiation zone	200,000	7,000,000	15,000	Energy transported by electromagnetic radiation
Convection zone	500,000	2,000,000	150	Energy carried by convection
Photosphere	696,000*	5800	2×10^{-4}	Electromagnetic radiation can escape—the part of the Sun we see
Chromosphere	696,500*	4500	5×10^{-6}	Cool lower atmosphere
Transition zone	698,000*	8000	2×10^{-10}	Rapid temperature increase
Corona	706,000	1,000,000	10^{-12}	Hot, low-density upper atmosphere
Solar wind	10,000,000	2,000,000	10^{-23}	Solar material escapes into space and flows outward through the solar system

TABLE 16.1 The Standard Solar Model

These radii are based on the accurately determined radius of the photosphere. The other radii quoted are approximate, round numbers.

convection. The term *solar interior* is often used to mean both the radiation and convection zones. The central **core**, roughly 200,000 km in radius, is the site of powerful nuclear reactions that generate the Sun's enormous energy output.

LUMINOSITY

The properties of size, mass, density, rotation rate, and temperature are familiar from our study of the planets. But the Sun has an additional property, perhaps the most important of all from the point of view of life on Earth—it *radiates* a great deal of energy into space, uniformly (we assume) in all directions. By holding a light-sensitive device—a solar cell, perhaps—perpendicular to the Sun's rays, we can measure how much solar energy is received per square meter of surface area every second. Imagine our detector as having a surface area of 1 m² and as being placed at the top of Earth's atmosphere. The amount of solar energy reaching this surface each second is a quantity known as the **solar constant.** Its value is approximately 1400 watts per square meter (W/m²). Most of this energy reaches Earth's surface. Thus, for example, a sunbather's body having a total surface area of about 0.5 m² receives solar energy at a rate of nearly 700 watts, roughly equivalent to the output of a typical electric room heater or about ten 75-W light bulbs.

Let us now ask about the *total* amount of energy radiated in all directions from the Sun, not just the small fraction intercepted by our detector or by Earth. Imagine a three-dimensional sphere that is centered on the Sun and is just large enough that its surface intersects Earth's center (Figure 16.3). The sphere's radius is 1 A.U., and its surface area is therefore $4\pi \times (1 \text{ A.U.})^2$, or approximately 2.8×10^{23} m². Multiplying the rate at which solar energy falls on each square meter of the sphere (that is, the solar constant) by the total surface area of our imaginary sphere, we can determine the total rate at which energy leaves the Sun's surface. This quantity is known as the **luminosity** of the Sun. It turns out to be just under 4×10^{26} W.

The Sun is an enormously powerful source of energy. *Every second*, it produces an amount of energy equivalent to the detonation of about 100 billion 1-megaton nuclear bombs. Put another way, the solar luminosity is equivalent to 4 trillion trillion 100-W light bulbs shining simultaneously—about 10^{19} dollars' worth of energy radiated every second (at current U.S. rates)!

16.2 The Solar Interior

MODELING THE STRUCTURE OF THE SUN

Lacking any direct measurements of the solar interior, astronomers must use more indirect means to probe the inner workings of our parent star. To accomplish this, they construct mathematical models of the Sun, combining all available observations with theoretical insight into solar physics. The result is the **Standard Solar Model**, which has gained widespread acceptance among astronomers.

To test and refine the Standard Solar Model, astronomers are eager to obtain information about the solar interior. However, we have little or no direct information about conditions below the solar photosphere, so we must rely on more indirect techniques. In the 1960s it was discovered that the surface of the Sun vibrates like a complex set of bells. These vibrations, illustrated in Figure 16.4, are the result of internal pressure ("sound") waves that reflect off the photosphere and repeatedly cross the solar interior. Because these waves can penetrate deep inside the Sun, analysis of their surface patterns allows scientists to study conditions far below the Sun's surface. This process is similar to the way in which seismologists study the interior of Earth by observing the P- and S-waves produced by earthquakes. ∞ (Sec. 7.3) For this reason, study of solar surface patterns is usually called **helioseismology**, even though solar pressure waves have nothing whatever to do with solar seismic activity—there is no such thing.

The most extensive study of solar oscillations is the ongoing GONG (Global Oscillations Network Group) project. By making continuous observations of the Sun from many clear sites around Earth and, most recently, from space (see *Interlude 16-1*), solar astronomers can obtain uninterrupted high-quality solar data spanning many days and even

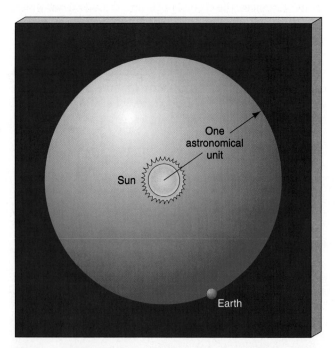

Figure 16.3 We can draw an imaginary sphere around the Sun so that the sphere's edge passes through Earth's center. The radius of this imaginary sphere equals one astronomical unit. By multiplying the sphere's surface area by the solar constant, we can measure the Sun's luminosity, the amount of energy it emits each second.

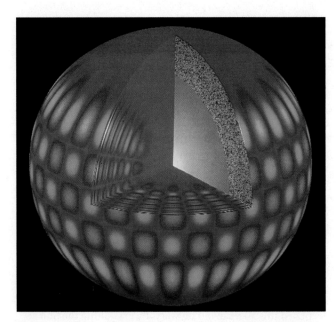

Figure 16.4 The Sun has been found to vibrate in a very complex way. By observing the motion of the solar surface, scientists can determine the wavelength and the frequencies of the individual waves and deduce information about the solar interior not obtainable by other means. The alternating patches represent gas moving down (red) and up (blue). (See also *Interlude 16-1*.)

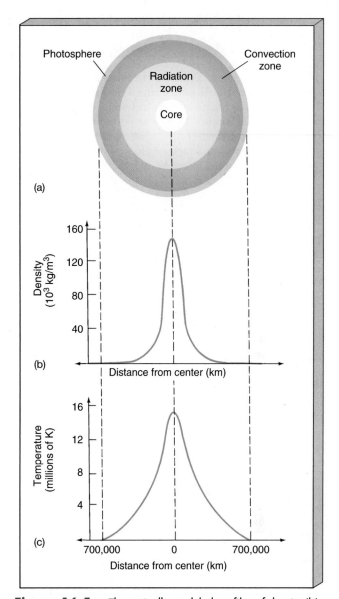

Figure 16.5 Theoretically modeled profiles of density (b) and temperature (c) for the solar interior, presented for perspective in (a). All three parts describe a cross-sectional cut through the center of the Sun.

weeks—almost as though Earth were not rotating and the Sun never set. Analysis of these data provides important additional information about the temperature, density, rotation, and the convective state of the solar interior, allowing detailed comparisons between theory and reality to be made.

Figure 16.5 shows the solar density and temperature according to the Standard Solar Model, plotted as functions of distance from Sun's center. Notice how the density drops rather sharply at first, then decreases more slowly near the solar photosphere, some 700,000 km from the center. The variation in density is large, ranging from a core value of about 150,000 kg/m^3, 20 times the density of iron, to an intermediate value (at 350,000 km) of about 1000 kg/m^3, the density of water, to an extremely small photospheric value of 2×10^{-4} kg/m^3, about 10,000 times less dense than air at the surface of Earth. Because the density is so high in the core, roughly 90 percent of the Sun's mass is contained within the inner half of its radius. The solar density continues to decrease out beyond the photosphere, reaching values as low as 10^{-23} kg/m^3 in the far corona—about as thin as the best vacuum physicists can create in laboratories on Earth.

The solar temperature also decreases with increasing radius, but not so rapidly as the density. Computer models indicate a temperature of about 15,000,000 K at the core, consistent with the minimum 10,000,000 K needed to initiate the nuclear reactions known to power most stars, decreasing to the observed value of about 5800 K at the photosphere.

ENERGY TRANSPORT

2 The very hot solar interior ensures violent and frequent collisions among gas particles. Particles move in all directions at high speeds, bumping into one another unceasingly. In and near the core, the extremely high temperatures guarantee that the gas is completely ionized. Recall from Chapter 4 that under less extreme conditions, atoms absorb photons that can boost their electrons to more excited states. ∞ (Sec. 4.2) With no electrons left on atoms to capture the photons, however, the deep solar interior is quite transparent to radiation. Only occasionally does a photon encounter and scatter off a free electron or proton. As a result, the energy produced by nuclear reactions in the core trav-

INTERLUDE 16-1 SOHO: Eavesdropping on the Sun

Throughout the few decades of the Space Age, various nations, led by the United States, have sent spacecraft to all but two of the major bodies in the solar system. One of these as-yet-unexplored bodies is Pluto, the most distant planet from the Sun, which has never been visited by a robot orbiter or even a flyby craft—and such a mission is not likely anytime soon. The other unexplored body is the Sun.

Currently, the next best thing to a dedicated reconnoitering spacecraft is the *Solar and Heliospheric Observatory* (*SOHO*), which since its launch in late 1995 has been radioing back to Earth volumes of new data—and a few new puzzles—about our parent star.

SOHO is a billion-dollar mission operated primarily by the European Space Agency. The 2-ton robot is now on-station about 1.5 million km sunward of Earth—about 1 percent of the distance from Earth to the Sun. This is the so-called L_1 Lagrange point, where the gravitational pull of the Sun and Earth are precisely equal—a good place to park a monitoring platform. ∞ (Sec. 14.1) There *SOHO* looks unblinkingly at our star 24 hours a day, eavesdropping on the Sun's surface, atmosphere, and interior. This automated vehicle carries a dozen instruments, capable of measuring almost everything from the Sun's corona and magnetic field to its solar wind and internal vibrations.

SOHO is positioned just beyond the region of space that is influenced by Earth's magnetic field, so its instruments can study cleanly the charged particles of the solar wind—the matter that escapes the Sun at high speed, flowing out into the corona and beyond. Coordinating these in situ measurements with *SOHO* images of the Sun itself, astronomers now think they can follow solar magnetic field loops expanding and breaking as the Sun prepares itself for mass ejections several days before they actually occur. Given that such coronal storms can play havoc with communications, power grids, satellite electronics, and other human activity, the prospect of having accurate forecasts of disruptive solar events is a welcome development.

Section 16.2 discussed how astronomers can "take the pulse" of the Sun by measuring its complex rhythmic motions. *SOHO* also has the ability to study the weak sound waves that echo and resonate inside the Sun and can map these vibrations with much higher resolution than was previously possible. It does so not by sensing sound itself but by watching the Sun's surface move up and down ever so slightly. The Sun's "loudest" vibrations are extremely low pitched (0.003 Hz, or one oscillation every 5 minutes), more like rolling rumbles, just as we might expect from such a huge and massive object.

The accompanying figure is not an image but a radial-velocity map of the Sun's surface (at much higher resolution than the diagram in Figure 16.4), obtained by measuring the Doppler shift of different portions of the solar photosphere. Bright areas are moving toward us, dark areas away. Observations of this sort enable researchers to refine models of the solar interior. For example, they have pinpointed the exact boundary between the convection zone and the radiation zone at 71.3 percent of the radius of the photosphere. The hope is to extend this work to discover how sharply defined the core is and just where its outer boundary lies.

Perhaps the most puzzling solar problem regards the temperature of the corona. How can it be so hot compared to the Sun's surface—as though the air above a candle's flame were much hotter than the flame itself? It doesn't make sense that the corona is a million degrees or so, when the solar photosphere, which is much closer than the corona to the vastly hotter core, is a mere 5800 K. If anything, *SOHO* has deepened this 50-year-old mystery. By examining the Sun at ultraviolet wavelengths, it has found that the very hottest matter in the upper solar atmosphere is even hotter than was previously thought. Ultraviolet spectral lines emitted by some ions are very sensitive to their surroundings, and *SOHO's* spectroscopic observations of highly excited oxygen ions have demonstrated that at least some parts of the corona have temperatures closer to 100 million kelvins, nearly a hundred times higher than earlier estimates.

els outward toward the surface in the form of radiation with relative ease.

As we move outward from the core, the temperature falls, atoms collide less frequently and less violently, and more and more electrons manage to remain bound to their parent nuclei. With more and more atoms retaining electrons that can absorb the outgoing radiation, the gas in the interior changes from being relatively transparent to being almost totally opaque. By the outer edge of the radiation zone, 200,000 km below the photosphere, *all* the photons produced in the Sun's

Figure 16.6 Physical transport of energy in the Sun's convection zone. We can visualize the upper interior as a boiling, seething sea of gas. Each convective loop is about 1000 km across. The convective cells are arranged in tiers with cells of progressively smaller size as the surface is neared. (This is a highly simplified diagram; there are many different cell sizes, and they are not so neatly arranged.)

core have been absorbed. Not one of them reaches the surface. But what happens to the energy they carry?

The photons' energy must travel beyond the Sun's interior. That we see sunlight—visible energy—proves that energy escapes. That energy reaches the surface by *convection*—the same basic physical process we saw in our study of Earth's atmosphere, although it operates in a very different environment in the Sun. ∞ (Sec. 7.2) Hot solar gas physically moves outward while cooler gas above it sinks, creating a characteristic pattern of convection cells. All through the convection zone, energy is transported to the surface by physical motion of the solar gas. (Remember that there is no physical movement of material when radiation is the energy-transport mechanism; convection and radiation are *fundamentally different* ways in which energy can be transported from one place to another.)

In reality, the zone of convection is much more complex than we have just described. As illustrated in Figure 16.6, there is a hierarchy of convection cells, organized in tiers at different depths. The deepest tier, about 200,000 km below the photosphere, is thought to contain large cells some tens of thousands of kilometers in diameter. Heat is then successively carried upward through a series of progressively smaller sized cells, stacked one on another until, at a depth of about 1000 km, the individual cells are about 1000 km across. The top of this uppermost tier of convection is the visible surface, where astronomers can directly observe the cell sizes. Information about convection below that level is inferred mostly from computer models of the solar interior.

At some distance from the core the solar gas becomes too thin to sustain further upwelling by convection. Theory suggests that this distance roughly coincides with the photospheric surface we see. Convection does not proceed into the solar atmosphere. There is simply not enough gas there. The density is so low that the gas becomes transparent once again, and radiation once again becomes the mechanism of energy transport. Photons reaching the photosphere escape more or less freely into space, and the photosphere emits thermal radiation, like any other hot object. The photosphere is narrow, and the "edge" of the Sun sharp, because this transition from opacity to complete transparency is very rapid. Just below the bottom of the photosphere the gas is

still convective, and radiation does not reach us directly. A few hundred kilometers higher, the gas is too thin to emit or absorb any significant amount of radiation.

GRANULATION

Figure 16.7 is a high-resolution photograph of the solar surface taken with instruments aboard NASA's *Skylab* space station as it orbited above much of Earth's atmosphere in the mid-1970s. The visible surface is highly mottled, or **granulated**, with regions of bright and dark gas known as *granules*. Each bright granule measures about 1000 km across—comparable in size to a continent on Earth—and has a lifetime of between 5 and 10 minutes. Together, several million granules constitute the top layer of the convection zone, immediately below the photosphere.

Each granule forms the topmost part of a solar convection cell. Spectroscopic observation within and around the bright regions shows direct evidence for the upward motion of gas as it "boils" up from within. This evidence proves that convection really does occur just below the

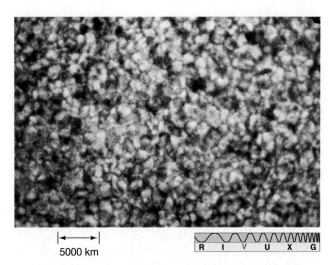

|←——→|
5000 km

R I V U X G

Figure 16.7 *Skylab* photograph of the granulated solar photosphere. Typical solar granules are comparable in size to Earth's continents. The bright portions of the image are regions where hot material is upwelling from below. The dark regions correspond to cooler gas that is sinking back down into the interior.

photosphere. Spectral lines detected from the bright granules appear slightly bluer than normal, indicating Doppler-shifted matter approaching us at about 1 km/s. ∞ (Sec. 3.5) Conversely, spectroscopes focused on the darker portions of the granulated photosphere show the same spectral lines to be redshifted, indicating matter moving away from us.

The brightness variations of the granules result strictly from differences in temperature. The upwelling gas is hotter and therefore emits more radiation than the cooler downwelling gas. The adjacent bright and dark gases appear to contrast considerably, but in reality their temperature difference is less than about 500 K. Careful measurements also reveal a much larger scale flow on the solar surface. **Supergranulation** is a flow pattern quite similar to granulation except that supergranulation cells measure some 30,000 km across. As with granulation, material upwells at the center of the cells, flows across the surface, then sinks down again at the edges. Scientists believe that supergranules are the imprint on the photosphere of a deeper tier of large convective cells, like those depicted in Figure 16.6.

16.3 The Solar Atmosphere

COMPOSITION

3 Astronomers can glean an enormous amount of information about the Sun from an analysis of the absorption lines that arise in the photosphere and lower atmosphere. ∞ (Sec. 4.4) Figure 16.8 (see also Figure 4.4) is a detailed spectrum of the Sun, obtained for a small portion of the range of visual wavelengths, from 360 to 690 nm. Notice the intricate dark Fraunhofer absorption lines superposed on the background continuous spectrum.

As discussed in Chapter 4, spectral lines arise when electrons in atoms or ions make transitions between states of well-defined energies, emitting or absorbing photons of specific energies (that is, wavelengths or colors) in the process. ∞ (Sec. 4.2) However, to explain the spectrum of the Sun (and, indeed, the spectra of all stars), we must modify slightly our earlier description of the formation of absorption lines. We explained these lines in terms of cool foreground gas intercepting light from a hot background source. In actuality, both the bright background and the dark absorption lines in Figure 16.8 form in roughly the same location in the Sun—the solar photosphere and lower chromosphere. To understand how line formation occurs, let's reconsider the solar energy-emission process in a little more detail.

Below the photosphere, the solar gas is sufficiently dense, and interactions among photons, electrons, and ions sufficiently common, that radiation cannot escape directly into space. In the solar atmosphere, however, the probability that a photon will escape without further interaction with matter depends on its energy. If that energy happens to correspond to some electronic transition in one of the atoms or ions present in the gas, then the photon may be absorbed again be-

Figure 16.8 A detailed spectrum of our Sun in a portion of the visible domain shows thousands of Fraunhofer spectral lines, which indicate the presence of some 67 different elements in various stages of excitation and ionization in the lower solar atmosphere.

fore it can travel very far—the more elements present of the type suitable for absorption, the lower the escape probability. Conversely, if the photon's energy does not coincide with any such transition, then the photon cannot interact further with the gas, and it leaves the Sun headed for interstellar space, or perhaps the detector of an astronomer on Earth.

Thus, as illustrated in Figure 16.9, when we look at the Sun, we are actually peering down into the solar atmosphere to a depth that depends on the wavelength of the light under consideration. Photons with wavelengths well away from any absorption feature tend to come from deep in the photosphere, whereas those at the centers of absorption lines, being much more likely to interact with matter as they travel through the solar gas, mainly escape from higher, cooler levels. The lines are darker than their surroundings because the temperature at the level of the atmosphere where they form is lower than the 5800 K temperature at the base of the photosphere, where most of the continuous emission originates. (Recall that by Stefan's law, the brightness of a radiating object depends on its temperature—the cooler the gas, the less energy it radiates.) ∞ (Sec. 3.4) The existence of the Fraunhofer lines is direct evidence that the temperature in the Sun's atmosphere decreases with height above the photosphere.

Tens of thousands of spectral lines have been observed and cataloged in the solar spectrum. In all, some 67 elements have been identified in the Sun, in various states of ionization and excitation. ∞ (Sec. 4.2) More elements probably exist there, but they are present in such small quantities that our instruments are simply not sensitive enough to detect them. Table 16.2 lists the 10 most common ele-

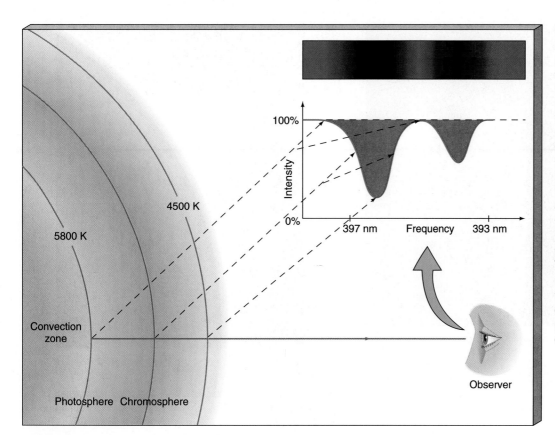

Figure 16.9
Formation of solar absorption lines. Photons with energies well away from any atomic transition can escape from relatively deep in the photosphere, but those with energies close to a transition are more likely to be reabsorbed before escaping, so the ones we see on Earth tend to come from higher, cooler levels in the solar atmosphere. The inset shows a close-up tracing of two of the thousands of solar absorption lines, those produced by calcium at about 395 nm.

TABLE 16.2 The Composition of the Sun

ELEMENT	ABUNDANCE (percentage of total number of atoms)	ABUNDANCE (percentage of total mass)
Hydrogen	91.2	71.0
Helium	8.7	27.1
Oxygen	0.078	0.97
Carbon	0.043	0.40
Nitrogen	0.0088	0.096
Silicon	0.0045	0.099
Magnesium	0.0038	0.076
Neon	0.0035	0.058
Iron	0.0030	0.14
Sulfur	0.0015	0.040

ments in the Sun. Notice that hydrogen is by far the most abundant element, followed by helium. This distribution is just what we saw on the jovian planets, and it is what we will find for the universe as a whole.

Strictly speaking, spectral analysis allows us to draw conclusions only about the part of the Sun where the lines form—the photosphere and chromosphere. However, most astronomers believe that, with the exception of the solar core (where nuclear reactions are steadily changing the com-

position—see Section 16.5), the data in Table 16.2 are representative of the entire Sun.

THE CHROMOSPHERE

Above the photosphere lies the cooler chromosphere, the inner part of the solar atmosphere. This region emits very little light of its own and cannot be observed visually under normal conditions. The photosphere is just too bright, dominating the chromosphere's radiation. The relative dimness of the chromosphere results from its low density—large numbers of photons simply cannot be emitted by a tenuous gas containing very few atoms per unit volume. Still, although it is not normally seen, astronomers have long been aware of the chromosphere's existence. Figure 16.10 shows the Sun during an eclipse in which the photosphere—but not the chromosphere—is obscured by the Moon. The chromosphere's characteristic reddish hue is plainly visible. This coloration is due to the red Hα (hydrogen alpha) emission line of hydrogen, which dominates the chromospheric spectrum. ∞ (Sec. 4.2)

The chromosphere is far from tranquil. Every few minutes, small solar storms erupt, expelling jets of hot matter known as *spicules* into the Sun's upper atmosphere (Figure 16.11). These long, thin spikes of matter leave the Sun's surface at typical speeds of about 100 km/s, reaching several thousand kilometers above the photosphere. Spicules are not spread evenly across the solar surface. Instead, they cover only about 1 percent of the total area, tending to accumulate around the edges of supergranules. The Sun's magnetic field is also known to be

Figure 16.10 This photograph of a total solar eclipse shows the solar chromosphere, a few thousand kilometers above the Sun's surface.

3000 km

Figure 16.11 Solar spicules, short-lived narrow jets of gas that typically last mere minutes, can be seen sprouting up from the solar chromosphere in this Hα image of the Sun. The spicules are the thin, dark, spikelike regions. They appear dark against the face of the Sun because they are cooler than the solar photosphere.

somewhat stronger than average in those regions. Scientists speculate that the downwelling material there tends to strengthen the solar magnetic field and that spicules are the result of magnetic disturbances in the Sun's churning outer layers.

THE TRANSITION ZONE AND CORONA

During the brief moments of an eclipse, if the Moon's angular size is large enough that both the photosphere and the chromosphere are blocked, the ghostly solar corona can be seen (Figure 16.12). With the photospheric light removed, the pattern of spectral lines changes dramatically. The intensities of the usual lines alter, suggesting changes in composition or temperature, or both, the spectrum shifts from absorption to emission, and an entirely new set of spectral lines suddenly appears. These new coronal (and in some cases chromospheric) lines were first observed during eclipses in the 1920s. For years afterward, some researchers (for want of any better explanation) attributed them to a new nonterrestrial element, which they dubbed "coronium."

We now recognize that these new spectral lines do not indicate any new kind of atom. Coronium does not exist. Rather, the new lines arise because atoms in the corona have lost several more electrons than atoms in the photosphere—that is, the coronal atoms are much more highly ionized. Therefore, their internal electronic structures, and hence their spectra, are quite different from the structure and spectra of atoms and ions in the photosphere. For example, astronomers have identified coronal lines corresponding to iron ions with as many as 13 of their normal 26 electrons missing. In the photosphere, most iron atoms have lost only 1 or 2 of their electrons.

Video

Figure 16.12 When both the photosphere and the chromosphere are obscured by the Moon during a solar eclipse, the faint corona becomes visible. This photograph shows clearly the emission of radiation from the solar corona.

The cause of this extensive electron stripping is the high coronal temperature. The degree of ionization inferred from spectra observed during solar eclipses tells us that the gas temperature of the upper chromosphere exceeds that of the photosphere. Furthermore, the temperature of the solar corona, where even more ionization is seen, is higher still.

Figure 16.13 shows how the temperature of the Sun's atmosphere varies with altitude. The temperature decreases to a minimum of about 4500 K some 500 km above the photosphere, after which it rises steadily. About 1500 km above the photosphere, the gas temperature begins to rise rapidly, reaching more than 1,000,000 K at an altitude of 10,000 km. Thereafter, it remains roughly constant.

The cause of this rapid temperature rise is not fully understood. The temperature profile runs contrary to intuition—moving away from a heat source, we would normally expect the heat to diminish, but this is not the case for the lower atmosphere of the Sun. The corona must have another energy source. Astronomers now believe that magnetic disturbances in the solar photosphere—a little like spicules, but on a much larger scale—are ultimately responsible for heating the corona (Section 16.4).

THE SOLAR WIND

Electromagnetic radiation and fast-moving particles—mostly protons and electrons—escape from the Sun all the time. The radiation moves away from the photosphere at the speed of light, taking 8 minutes to reach Earth. The particles travel more slowly, although at the still considerable speed of about 500 km/s, reaching Earth in a few days. This constant stream of escaping solar particles is the **solar wind**.

The solar wind results from the high temperature of the corona. About 10,000,000 km above the photosphere, the coronal gas is hot enough to escape the Sun's gravity, and it begins to flow outward into space. At the same time, the solar atmosphere is continuously replenished from below. If that were not the case, the corona would disappear in about a day. The Sun is, in effect, "evaporating"—constantly shedding mass through the solar wind. The wind is an extremely thin medium, however. Although it carries

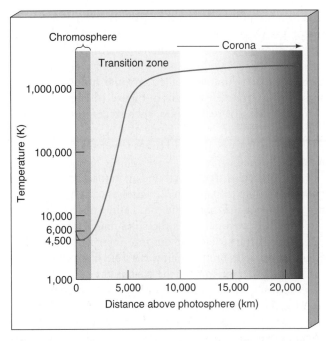

Figure 16.13 The change of gas temperature in the lower solar atmosphere is dramatic. The minimum temperature marks the outer edge of the chromosphere. Beyond that, the temperature rises sharply in the transition zone, finally leveling off at over 1,000,000 K in the corona.

away about a million tons of solar matter each second, less than 0.1 percent of the Sun has been lost since the solar system formed 4.6 billion years ago.

THE SUN IN X RAYS

What sort of radiation is emitted by a gas of 1,000,000 K? Unlike the 5800 K photosphere, which emits most strongly in the visible part of the electromagnetic spectrum, the hotter coronal gas radiates at much higher frequencies—primarily in X rays. ∞ (Sec. 3.4) For this reason, X-ray telescopes have become important tools in the study of the solar corona. Figure 16.14 shows X-ray images of the Sun. The full corona extends well beyond the regions shown, but the density of coronal particles emitting the radiation

Figure 16.14 Images of X-ray emission from the Sun observed by the *Skylab* space station. These frames were taken at 1-day intervals. Note the dark, boot-shaped coronal hole traveling from left to right, where the X-ray observations outline in dramatic detail the abnormally thin regions through which the high-speed solar wind streams forth.

diminishes rapidly with distance from the Sun. The intensity of X-ray radiation farther out is too dim to be seen here.

In the mid-1970s, instruments aboard NASA's *Skylab* space station revealed that the solar wind escapes mostly through solar "windows" called **coronal holes**. The dark area moving from left to right in Figure 16.14 represents a coronal hole. Not really holes, such structures are simply deficient in matter—vast regions of the Sun's atmosphere where the density is about 10 times lower than the already tenuous, normal corona.

Coronal holes are underabundant in matter because the gas there is able to stream freely into space at high speeds, driven by disturbances in the Sun's atmosphere and magnetic field. In coronal holes the solar magnetic field lines extend from the surface far out into interplanetary space. Charged particles tend to follow the field lines, so they can escape. In other regions of the corona the solar magnetic field lines stay close to the Sun, keeping charged particles near the surface and inhibiting the outward flow of the solar wind (just as Earth's magnetic field tends to prevent the incoming solar wind from striking Earth), so the density remains relatively high.

The largest coronal holes can be hundreds of thousands of kilometers across. Structures of this size are seen only a few times each decade. Smaller holes—perhaps only a few tens of thousand kilometers in size—are much more common, appearing every few hours.

16.4 The Active Sun

4 Most of the Sun's luminosity results from continuous emission from the photosphere. This radiation arises from what we call the **quiet Sun**—the underlying predictable star that blazes forth day after day. This steady behavior contrasts with the sporadic, unpredictable radiation of the **active Sun**, a much more irregular component of our star's energy output, characterized by explosive, unpredictable behavior. This aspect of solar radiation contributes little to the Sun's total luminosity and has little effect on the evolution of the Sun as a star, but it does affect us directly here on Earth (see *Interlude 16-2*). The size and duration of coronal holes are strongly influenced by the level of solar activity, as is the strength of the solar wind.

SUNSPOTS

Figure 16.15 is an optical photograph of the entire Sun, showing numerous dark blemishes on the surface. First studied in detail by Galileo, these "spots" provided one of the first clues that the Sun was not a perfect unvarying creation but a place of constant change. ∞ (Sec. 2.5) These dark areas are called **sunspots**. They typically measure about 10,000 km across, about the size of Earth. As shown in the figure, they often occur in groups. At any given time, the Sun may have hundreds of sunspots, or it may have none at all.

Studies of sunspots show an *umbra*, or dark center, surrounded by a grayish *penumbra*. The close-up views in Figure 16.16 shows each of these dark areas and the brighter undisturbed photosphere nearby. This gradation in darkness is really a gradual change in photospheric temperature—sunspots are simply *cooler* regions of the photospheric gas. The temperature of the umbra is about 4500 K, compared with the penumbra's 5500 K. The spots, then, are certainly composed of hot gases. They seem dark only because they appear against an even brighter background (the 5800 K photosphere). If we could magically remove a sunspot from the Sun (or just block out the rest of the Sun' emission), the spot would glow brightly, just like any other hot object having a temperature of roughly 5000 K.

Sunspots are not steady. Most change their size and shape, and all come and go. Figure 16.17 shows a time sequence in which several spots varied—sometimes growing, sometimes dissipating—over a period of several days. Individual spots may last anywhere from 1 to 100 days. A large group of spots typically lasts 50 days.

SOLAR MAGNETISM

What causes a sunspot? Why is it cooler than the surrounding photosphere? The answers to these questions involve the Sun's magnetism. As we saw in Chapter 4, analysis of spectral lines can yield information about the magnetic field where they originate. ∞ (Sec. 4.4) The magnetic field in a typical sunspot is about 1000 times greater than the field in neighboring, undisturbed photospheric regions (which is itself several times stronger than Earth's magnetic field). Scientists believe that sunspots are cooler than their surroundings because these abnormally strong fields tend to block (or redirect) the convective flow of hot gas, which is normally toward the surface of the Sun.

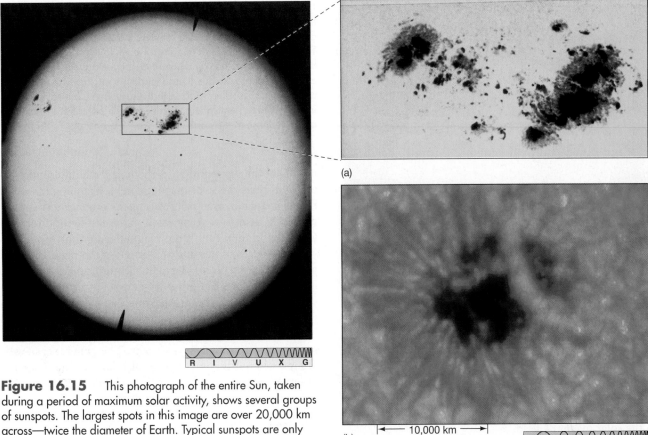

(a)

(b) |← 10,000 km →|

Figure 16.15 This photograph of the entire Sun, taken during a period of maximum solar activity, shows several groups of sunspots. The largest spots in this image are over 20,000 km across—twice the diameter of Earth. Typical sunspots are only about half this size.

Figure 16.16 (a) An enlarged photograph of the largest pair of sunspots in Figure 16.15. Each spot consists of a cool, dark inner region called the umbra, surrounded by a warmer, brighter region called the penumbra. The spots appear dark because they are slightly cooler than the surrounding photosphere. (b) A high-resolution, true-color image of a single sunspot shows details of its structure as well as much surface granularity surrounding it. The spot is about the size of Earth.

Another indicator of the magnetic nature of sunspots is their grouping. They almost always come in pairs, and the magnetic fields observed in the two members of any pair are always opposite to one another—the members of the pair are said to have opposite *magnetic polarities*. As illustrated in Figure 16.18, magnetic field lines emerge from the interior through one member of a sunspot pair, loop through the solar atmosphere, then reenter the solar surface through the other spot. What's more, *all* the sunspot pairs in the same solar hemisphere (north or south) at any instant have the same magnetic configuration—if the magnetic field lines are directed into the Sun in one leading spot (measured in the direction of the Sun's rotation), they are inwardly directed in all leading spots in that hemisphere. In

Figure 16.17 The evolution of some sunspots and lower chromospheric activity over a period of 12 days. The sequence runs from left to right. An Hα filter was used to make these photographs, taken from the *Skylab* space station. An arrow follows one set of sunspots over the course of a week as they are carried around the Sun by its rotation.

Figure 16.18 Sunspot pairs are linked by magnetic field lines. The Sun's magnetic field emerges from the surface through one member of the pair and reenters the Sun through the other. The leading members of all sunspot pairs in the solar northern hemisphere have the same polarity—if the magnetic field lines are directed into the Sun in one leading spot, they are inwardly directed in all leading spots in that hemisphere. The same is true in the southern hemisphere, except that the polarities are always opposite those in the north.

rapidly at the equator than at the poles, the differential rotation distorts the solar magnetic field, wrapping it around the solar equator, eventually causing the original north–south magnetic field to reorient itself in an east–west direction. Convection then causes the magnetized gas to upwell toward the surface, twisting and tangling the magnetic field pattern. In some places, the field becomes kinked like a knot in a garden hose, causing it to increase in strength. Occasionally, the field strength becomes so great that it overwhelms the Sun's gravitational field and a "tube" of field lines bursts out of the surface and loops through the lower atmosphere, forming a sunspot pair. The general east–west organization of the underlying solar field accounts for the observed polarities of the pairs in each hemisphere.

THE SOLAR CYCLE

Not only do sunspots come and go with time, but their numbers and distribution across the face of the Sun also change in a fairly regular fashion. Centuries of observations

the other hemisphere at the same time all sunspot pairs have the opposite polarity. Despite the irregularity of the sunspots themselves, these correlations suggest a high degree of order in the solar magnetic field.

The Sun is gaseous and rotates differentially, and these facts radically affect the character of solar magnetism. As illustrated in Figure 16.19, because the Sun rotates more

Figure 16.19 This diagram illustrates how the Sun's differential rotation wraps and distorts the solar magnetic field. Occasionally, the field lines burst out of the surface and loop through the lower atmosphere, thereby creating a sunspot pair. The underlying pattern of the solar field lines explains the observed pattern of sunspot polarities. (If the loop happens to occur on the limb of the Sun and is seen against the blackness of space, we see a phenomenon called a prominence, described in Section 16.4.)

INTERLUDE 16-2 Solar–Terrestrial Relations

Our Sun has often been worshipped as a god with power over human destinies. Obviously, the steady stream of solar energy arriving at our planet every day is essential to our lives. But over the past century there have also been repeated claims of a correlation between the Sun's activity and Earth's weather. Only recently, however, has the subject become scientifically respectable—that is, more natural than supernatural.

In fact, there do seem to be some correlations between the 22-year solar cycle (two sunspot cycles, with oppositely directed magnetic fields) and periods of climatic dryness here on Earth. For example, near the start of the past eight cycles, there have been droughts in North America—at least within the middle and western plains from South Dakota to New Mexico. The most recent of these droughts, which typically last 3 to 6 years, came in the late 1950s. The one expected in the 1980s, however, did not occur as clearly as anticipated.

Other possible Sun–Earth connections include a link between solar activity and increased atmospheric circulation on our planet. As circulation increases, terrestrial storm systems deepen, extend over wider ranges of latitude, and carry more moisture. The relationship is complex, and the subject controversial, because no one has yet shown any physical mechanism (other than the Sun's heat, which does not vary much during the solar cycle) that would allow solar activity to stir our terrestrial atmosphere. Without a better understanding of the physical mechanism involved, none of these effects can be incorporated into our weather forecasting models.

Solar activity may also influence long-term climate on Earth. For example, the Maunder minimum seems to correspond fairly well to the coldest years of the so-called Little Ice Age that chilled northern Europe during the late 1600s. ∞ (Sec. 16.4) The accompany-

ing "winter" scene actually captured the summer season in 17th-century Holland. How the active Sun, and its abundance of sunspots, may affect Earth's climate is a frontier problem in terrestrial climatology.

One correlation that is definitely established, and also better understood, is that between solar flares and geomagnetic activity at Earth. The extra radiation and particles thrown off by flares impinge on Earth's environment, overloading the Van Allen belts, causing brilliant auroras in our atmosphere and degrading our communication networks. These disturbances have been known for many years, but only recently have we associated them with the solar wind. We are only beginning to understand how the radiation and particles emitted by solar flares also interfere with terrestrial radars, power networks, and other technological equipment. Some power outages on Earth are actually caused not by increased customer demand or malfunctioning equipment but by flares on the Sun!

We cannot now predict when and where solar flares will occur. However, it would certainly be to our advantage to be able to do so, as this aspect of the active Sun affects our lives. This is a very fertile area of astronomical research and one for which there are clear terrestrial applications.

have established a clear **sunspot cycle**. Figure 16.20(a) shows the number of sunspots observed each year during the twentieth century. The average number of spots reaches a maximum every 11 or so years, then falls off almost to zero before the cycle begins afresh.

The latitudes at which sunspots appear vary as the sunspot cycle progresses. Individual sunspots do not move up or down in latitude, but new spots appear closer to the equator as older ones at higher latitudes fade away. Figure 16.20(b) is a plot of observed sunspot latitude as a function of time. At the start of each cycle, at the time of **solar minimum**, only a few spots are seen. They are generally confined to two narrow zones about 25° to 30° north and south of the solar equator. Approximately 4 years into the cycle, around the time of **solar maximum**, the number of spots has increased markedly. They are found

within about 15° to 20° of the equator. Finally, by the end of the cycle, at solar minimum again, the total number has fallen again, and most sunspots lie within about 10° of the solar equator. The beginning of each new cycle appears to overlap the end of the last.

Complicating this picture further, the 11-year sunspot cycle is actually only half of a longer 22-year **solar cycle**. During the first 11 years of the solar cycle, the leading spots of all the pairs in the northern hemisphere have the same polarity, while spots in the southern hemisphere have the opposite polarity (Figure 16.18). These polarities then reverse their signs for the next 11 years, so the full solar cycle takes 22 years to repeat.

Astronomers believe that the Sun's magnetic field is both generated and amplified by the constant stretching,

Figure 16.20 (a) Annual number of sunspots throughout the twentieth century, showing the 5-year average of the annual data to make long-term trends more evident. The (roughly) 11-year solar cycle is clearly visible. At the time of solar minimum, hardly any sunspots are seen. About 4 years later, at solar maximum, as many as 100–200 spots are observed per year. The most recent solar maximum occurred in 1990. (b) Sunspots cluster at high latitudes when solar activity is at a minimum. They appear at lower latitudes as the number of sunspots peaks. Finally, they are prominent near the Sun's equator as solar minimum is again approached.

(a)

twisting, and folding of magnetic field lines that results from the combined effects of differential rotation and convection. The theory is essentially the same "dynamo" theory that accounts for the magnetic fields of Earth and the jovian planets, except that the solar dynamo operates much faster, and on a much larger scale. ∞ (Sec. 7.4) One prediction of this theory is that the Sun's magnetic field should rise to a maximum, then fall to zero and reverse itself in a more or less periodic way, just as observed. Solar surface activity, such as the sunspot cycle, simply follows the variations in the magnetic field. The changing numbers of sunspots and their migration to lower latitudes are both consequences of the strengthening and eventual decay of the field lines as they become more and more tightly wrapped around the solar equator.

Figure 16.21 plots all sunspot data recorded since the invention of the telescope. As can be seen, the 11-year "periodicity" of the solar sunspot cycle is far from regular. Not only does the period range from 7 to 15 years, but the sunspot cycle has disappeared entirely in the relatively recent past. The lengthy period of solar inactivity that extended from 1645 to 1715 is called the *Maunder minimum*, after the British astronomer who drew attention to these historical records. The corona was apparently also less prominent during total solar eclipses around that time, and Earth aurorae were sparse throughout the late seventeenth century. Lacking a complete understanding of the solar cycle, we cannot easily explain how it could shut down entirely. Most astronomers suspect

changes in the Sun's convection zone and/or its rotation pattern, but the specific cause of the Sun's century-long variations remains a mystery.

ACTIVE REGIONS

Sunspots are relatively quiescent aspects of solar activity. However, the photosphere surrounding them occasionally erupts violently, spewing forth large quantities of energetic particles into the surrounding corona. The sites of these ex-

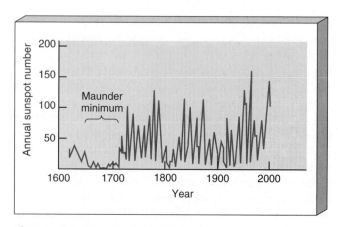

Figure 16.21 Number of sunspots occurring each year over the past four centuries. Note the absence of spots during the late seventeenth century.

R I V U X G

(b)

R I V U X G

Figure 16.22 (a) The looplike structure of this prominence clearly reveals the magnetic field lines connecting the two members of a sunspot pair. (b) This particularly large solar prominence was observed by ultraviolet detectors aboard the *Skylab* space station in 1979. (See also Figures 16.18 and 16.19.)

Video

plosive events are known simply as **active regions**. Most pairs or groups of sunspots have active regions associated with them. Like all other aspects of solar activity, these phenomena tend to follow the solar cycle and are most frequent and violent around the time of solar maximum.

Figure 16.22 shows two solar **prominences**—loops or sheets of glowing gas ejected from an active region on the solar surface, moving through the inner parts of the corona under the influence of the Sun's magnetic field. Magnetic instabilities in the strong fields found in and near sunspot groups may cause the prominences, although the details are still not completely understood. Observations like Figure 16.22(a) clearly show streams of hot ionized gas soaring high into the solar atmosphere, following the arching magnetic field lines between members of a sunspot pair (see also Figure 16.18).

Quiescent prominences persist for days or even weeks, hovering high above the photosphere, suspended by the Sun's magnetic field. *Active prominences* come and go much more erratically, changing their appearance in a matter of hours or surging up from the solar photosphere, then immediately falling back on themselves. A typical solar prominence measures some 100,000 km in extent, nearly 10 times the diameter of planet Earth. Prominences as large as that shown in Figure 16.22(b) (which traversed almost half a million kilometers of the solar surface) are less common and usually appear only at times of greatest solar activity. The largest prominences can release up to 10^{25} joules of energy, counting both particles and radiation—

not much compared with the total solar luminosity of 4×10^{26} W but still enormous by terrestrial standards. (All the power plants on Earth would take a billion years to produce this much energy.)

Flares are another type of solar activity observed low in the Sun's atmosphere near active regions. Also the result of magnetic instabilities, flares, like that shown in Figure 16.23, are even more violent (and even less well understood) than prominences. They often flash across a region of the Sun in minutes, releasing enormous amounts of energy as they go. Observations made by the *Solar Maximum* satellite demonstrated that X-ray and ultraviolet emissions are especially intense in the extremely compact hearts of flares, where temperatures can reach 100,000,000 K. So energetic are these cataclysmic explosions that some researchers have likened flares to bombs exploding in the lower regions of the Sun's atmosphere.

A major flare can release as much energy as the largest prominences, but in a matter of minutes or hours rather than days or weeks. Unlike the gas that makes up the characteristic loop of a prominence, the particles produced by a flare are so energetic that the Sun's magnetic field is unable to hold them and shepherd them back to the surface. Instead, the particles are simply blasted into space by the violence of the explosion.

THE CHANGING SOLAR CORONA

The solar corona also varies in step with the sunspot cycle. The photograph of the corona in Figure 16.12 shows the quiet Sun, at sunspot minimum. The corona is fairly reg-

(a)

⊢————⊣ 10,000 km

R I V U X G

(b)

R I V U X G

Figure 16.23 (a) Much more violent than a prominence, a solar flare is an explosion on the Sun's surface that sweeps across an active region in a matter of minutes, accelerating solar material to high speeds and blasting it into space. Visible here as the white, snakelike feature at the center, this flare extends across some 30,000 km. (b) A similar flare is seen from the side in this dramatic photograph, taken through a red Hα filter.

ular in appearance and appears to surround the Sun more or less uniformly. Compare this image with Figure 16.24, which was taken in 1991, close to the most recent peak of the sunspot cycle. The active corona is much more irregular in appearance and extends farther from the solar surface. The "streamers" of coronal material pointing away from the Sun are characteristic of this phase.

Astronomers think that the corona is heated primarily by solar surface activity, particularly prominences and flares, which can inject large amounts of energy into the upper

R I V U X G

Figure 16.24 Photograph of the solar corona during the July, 1991 eclipse, at the peak of the sunspot cycle. At these times, the corona is much less regular and much more extended than at sunspot minimum (compare Figure 16.12). Astronomers believe that coronal heating is caused by surface activity on the Sun. The changing shape and size of the corona are the direct result of variations in prominence and flare activity over the course of the solar cycle.

solar atmosphere, greatly distorting its shape. Extensive disturbances often move through the corona above an active site in the photosphere, distributing the energy throughout the coronal gas. Given this connection, it is hardly surprising that both the appearance of the corona and the strength of the solar wind are closely correlated with the solar cycle.

16.5 The Heart of the Sun

5 Compared with familiar terrestrial events—storms, tidal waves, tornadoes, even violent volcanic eruptions and earthquakes—the spots and flares of the active Sun are enormously energetic. The much greater *steady* emission from the inactive quiet Sun simply staggers the imagination. The Sun is somehow able to produce huge amounts of energy, and according to Earth's fossil record, it has been doing so for the last several billion years. What powers the Sun? What forces are at work in the Sun's core to produce such energy? By what process does the Sun shine, day after day, year after year, eon after eon? Answers to these questions are central to all astronomy. Without them, we can understand neither the physical existence of stars and galaxies in the universe nor the biological existence of life on Earth.

SOLAR ENERGY PRODUCTION

In round numbers, the Sun's luminosity is 4×10^{26} W and its mass is 2×10^{30} kg. We can quantify how efficiently the Sun generates energy by dividing the solar luminosity by the solar mass:

$$\frac{\text{solar luminosity}}{\text{solar mass}} = 2 \times 10^{-4} \text{ W/kg}.$$

This simply means that, on average, every kilogram of solar material yields about 0.2 milliwatts of energy—0.0002 joules (J) of energy every second. This is not much energy—a piece of burning wood generates about a million times more energy per unit mass per unit time than does our Sun. But there is an important difference: the wood will not burn for billions of years.

To appreciate the magnitude of the energy generated by our Sun, we must consider not the ratio of the solar luminosity to the solar mass but instead the total amount of energy generated by each gram of solar matter *over the entire lifetime of the Sun as a star*. This is easy to do. We simply multiply the rate at which the Sun generates energy by the age of the Sun, about 5 billion years. We obtain a value of 3×10^{13} J/kg. This is the average amount of energy radiated by every gram of solar material since the Sun formed. It represents a *minimum* value for the total energy radiated by the Sun, for more energy will be needed for every additional day the Sun shines. Should the Sun endure for another 5 billion years (as is predicted by theory), we would have to double this value.

This energy-to-mass ratio value is very large. At least 60 trillion joules (on average) of energy must arise from every kilogram of solar matter to power the Sun throughout its lifetime. But the generation of energy is not explosive, releasing large amounts of energy in a short period of time. Instead, it is slow and steady, providing a uniform and long-lived rate of energy production. Only one known energy-generation mechanism can conceivably power the Sun in this way. That process is nuclear **fusion**—the combining of light nuclei into heavier ones.

NUCLEAR FUSION

We can represent a typical fusion reaction symbolically as

$$\text{nucleus 1} + \text{nucleus 2} \rightarrow \text{nucleus 3} + \text{energy}.$$

For powering the Sun, the most important piece of this equation is the energy produced. Where does it come from?

The key point is that during a fusion reaction, the total mass *decreases*—the mass of nucleus 3 is less than the combined masses of nuclei 1 and 2. To understand the consequences of this, we use a very important law of modern physics—the **law of conservation of mass and energy**. Albert Einstein showed at the beginning of the twentieth century that matter and energy are interchangeable. One can be converted into the other, in accordance with Einstein's famous equation, $E = mc^2$. To determine the amount of energy corresponding to a given mass, simply multiply it by the square of the speed of light (*c* in the equation). For example, the energy equivalent of 1 kg of matter is $1 \times (3 \times 10^8)^2$, or 9×10^{16} J. The speed of light is so large that even small amounts of mass translate into enormous amounts of energy.

The law of conservation of mass and energy states that the *sum* of mass and energy (properly converted to the same units) must always remain constant in any physical process. There are no known exceptions. According to this law, an object can literally disappear, provided that some energy appears in its place. If magicians really made rabbits disappear, the result would be a flash of energy equaling the product of the rabbit's mass and the square of the speed of light—enough to destroy the magician, everyone in the audience, and probably all of the surrounding state as well! In the case of fusion reactions in the solar core, the energy is produced primarily in the form of electromagnetic radiation. The light we see coming from the Sun means that the Sun's mass must be slowly but steadily decreasing with time.

THE PROTON–PROTON CHAIN

All atomic nuclei are positively charged, so they repel one another. Furthermore, by the inverse-square law, the closer two nuclei come to one another, the greater is the repulsive force between them. ∞ (Sec. 3.2) How then do nuclei—two protons, say—ever manage to fuse into anything heavier? The answer is that if they collide at high enough speeds, one proton can momentarily plow deep into the other, eventually coming within the exceedingly short range of the **strong nuclear force**, which binds nuclei together (see *More Precisely 16-1*). At distances of less than about 10^{-15} m, the attraction of the nuclear force overwhelms the electromagnetic repulsion, and fusion occurs. Speeds in excess of a few hundred kilometers per second, corresponding to a gas temperature of 10^7 K or more, are needed to slam protons together fast enough to initiate fusion. Such conditions are found in the core of the Sun and at the centers of all stars.

The fusion of two protons is illustrated schematically in Figure 16.25. In effect, one of the protons turns into a neutron, creating new particles in the process, and combines with the other proton to form a **deuteron**, the nucleus of a special form of hydrogen called *deuterium*. Deuterium (also referred to as "heavy hydrogen") differs from ordinary

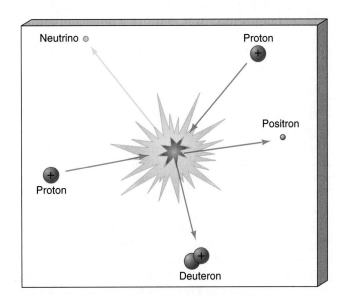

Figure 16.25 Two protons collide violently, initiating the chain of nuclear fusion that powers the Sun.

MORE PRECISELY 16-1 Fundamental Forces

Our studies of nuclear reactions have uncovered new ways in which matter can interact with matter at a subatomic level. Let's pause to consider in a slightly more systematic fashion the relationships among the various forces of nature.

As best we can tell, the behavior of all matter in the universe—from elementary particles to clusters of galaxies—is ruled by just four (or fewer) basic forces, which are *fundamental* to everything in the universe. In a sense, the search to understand the nature of the universe is the quest to understand the nature of these forces.

The *gravitational force* is probably the best known. Gravity binds galaxies, stars, and planets together and holds humans on the surface of Earth. As we saw in Chapter 2, its magnitude decreases with distance, according to an inverse-square law. ∞ (Sec. 2.7) Its strength is also proportional to the masses of each of the two objects involved. Thus, the gravitational field of an atom is extremely weak, but that of a galaxy, consisting of huge numbers of atoms, is very powerful. Gravity is by far the weakest of the forces of nature, but its effect accumulates as we move to larger and larger volumes of space, and nothing can cancel its attractive pull. As a result, gravity is the dominant force in the universe on all scales larger than that of Earth.

The *electromagnetic force* is another of nature's basic agents. Any particle having a net electric charge, such as an electron or a proton in an atom, exerts an electromagnetic force on any other charged particle. The everyday things we see around us are held together by this force. Like gravity, its strength also decreases with distance according to an inverse-square law. ∞ (Sec. 3.2) However, for subatomic particles, electromagnetism is much stronger than gravity. For example, the electromagnetic force between two protons exceeds their gravitational attraction by a factor of about 10^{36}. Unlike gravity, electromagnetic forces can repel (between like charges) as well as attract (between opposite charges). Positive and negative charges tend to neutralize each other, greatly diminishing their net electromagnetic

influence. Above the microscopic level most objects are in fact very close to being electrically neutral. Thus, except in unusual circumstances, the electromagnetic force is relatively unimportant on macroscopic scales.

A third fundamental force of nature is simply termed the *weak nuclear force*. It is much weaker than electromagnetism, and its influence is somewhat more subtle. The weak force governs the emission of radiation from some radioactive atoms; the emission of a neutrino during the first stage of the proton–proton reaction is also the result of a weak interaction. It is now known that the weak force is not really a separate force at all but just a form of the electromagnetic force. Thus, physicists often speak of the "electroweak force." However, when acting in its "weak" mode, the electroweak force does not obey the inverse-square law. Its effective range is less than the size of an atomic nucleus, about 10^{-15} m.

Strongest of all the forces is the *strong nuclear force*. It binds atomic nuclei together and governs the generation of energy in the Sun and all other stars. Like the weak force, and unlike the forces of gravity and electromagnetism, the strong force operates only at very close range. It is unimportant outside a distance of a hundredth of a millionth of a millionth (10^{-14}) of a meter. However, within this range (for example, in atomic nuclei), it binds particles with enormous strength. In fact, it is the range of the strong force that determines the typical sizes of atomic nuclei. Only when two protons are brought within about 10^{-15} m of one another can the attractive strong force overcome their electromagnetic repulsion.

Not all particles are subject to all types of force. All particles interact through gravity because all have mass. However, only charged particles interact electromagnetically. Protons and neutrons are affected by the strong force, but electrons are not. Under the right circumstances, the weak force can affect any type of subatomic particle, regardless of its charge.

---

hydrogen by virtue of an extra neutron in its nucleus. We can represent this reaction as follows:

proton + proton → deuteron + positron + neutrino.

The **positron** particle in this reaction is a positively charged electron. Its properties are identical to those of a normal negatively charged electron, except for its positive charge. Scientists call the electron and the positron a "matter–antimatter pair"—the positron is said to be the *antiparticle* of the electron. These newly created positrons find themselves in the midst of a sea of electrons, with which they interact immediately and violently. The particles and antiparticles annihilate one another, producing pure energy in the form of gamma-ray photons.

The final product of the reaction is a particle known as a **neutrino**, a word derived from the Italian for "little neutral one." Neutrinos carry no electrical charge and are of very low mass—at most 1/100,000 the mass of an electron, which itself has only 1/2000 the mass of a proton. (The exact mass of the

neutrino is uncertain, although recent evidence suggests that it is not zero.) They move at (or nearly at) the speed of light and interact with hardly anything. They can penetrate, without stopping, several light years of lead. Their interactions with matter are governed by the **weak nuclear force**, also described in *More Precisely 16-1*. Despite their elusiveness, however, neutrinos can be detected with carefully constructed instruments (see Section 16.6).

Nuclei such as normal hydrogen and deuterium, containing the same number of protons but different numbers of neutrons, represent different forms of the same element—they are known as **isotopes** of that element. Usually, there are about as many neutrons in a nucleus as protons, but the exact number of neutrons can vary, and most elements can exist in a number of isotopic forms. To avoid confusion when talking about isotopes of the same element, nuclear physicists attach a number to the symbol representing the element. This number indicates the total number of particles (protons plus neutrons) in its nucleus. Ordinary hydrogen is denoted by ^{1}H, deuterium by ^{2}H. Normal helium (2 protons plus 2 neutrons) is ^{4}He (also referred to as helium-4), and so on. We will adopt this convention for the rest of this book. We can now rewrite the preceding proton–proton reaction as

$$^1H + {^1H} \rightarrow {^2H} + \text{positron} + \text{neutrino}. \qquad \text{(I)}$$

This equation is labeled (I) because the production of a deuteron by the fusion of two protons is the first step in the fusion process powering most stars. It is the start of the **proton–proton chain**. Gargantuan quantities of protons are fused in this way within the core of the Sun each second. The positrons produced by this reaction quickly encounter electrons and are annihilated, releasing energy in the form of gamma rays.

The next step in solar fusion is the formation of an isotope of helium. A proton interacts with the deuteron particle produced in step (I):

$$^2H + {^1H} \rightarrow {^3He} + \text{energy}. \qquad \text{(II)}$$

This step begins as soon as deuterons appear. The reaction product is an isotope of helium—helium-3 (^{3}He)—lacking one

of the neutrons contained in the normal helium-4 nucleus. Energy is also emitted, again in the form of gamma-ray photons.

The third and final step in the proton–proton chain involves the production of a nucleus of helium-4. Helium-4 (^{4}He) comes about most often through the fusion of two of the helium-3 nuclei created in step (II):

$$^3He + {^3He} \rightarrow {^4He} + {^1H} + {^1H} + \text{energy}. \qquad \text{(III)}$$

The result is a helium-4 nucleus plus two more protons.

The *net* effect of steps (I) through (III) is this: four hydrogen nuclei (protons; six consumed, two returned) combine to create one helium-4 nucleus, plus some gamma-ray radiation and two neutrinos. The whole process is illustrated in Figure 16.26. Symbolically, we have

$$4(^1H) \rightarrow {^4He} + \text{energy} + 2 \text{ neutrinos}.$$

The gamma rays are slowly degraded in energy as they pass through the solar interior. Photons are repeatedly absorbed by electrons and ions, then reemitted at a wavelength that reflects the temperature of the surrounding gas, in accordance with Wien's law. ∞ (Sec. 3.4) Thus, as the radiation makes its way to the surface through ever-cooler

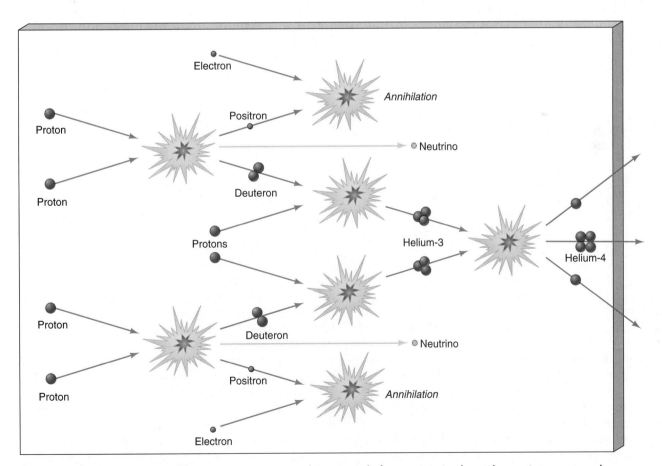

Figure 16.26   Diagram of the entire proton–proton chain. A total of six protons (and two electrons) are converted into two protons, one helium-4 nucleus, and two neutrinos. The two leftover protons are available as fuel for new proton–proton reactions, so the net effect is that four protons are fused to form one helium-4 nucleus. Energy, in the form of gamma rays, is produced in each reaction.

layers, its wavelength decreases. Eventually, it leaves the surface in the form of visible light. The neutrinos escape unhindered into space. The helium stays put in the core.

This basic process is responsible for the light emitted by almost all the stars we see. Alternative reactions leading to the same final result exist (see *More Precisely 20-1*), but they are relatively rare in stars like the Sun.

ENERGY GENERATED BY THE PROTON–PROTON CHAIN

Now let's calculate the energy produced in the fusion process and compare it with the 6×10^{13} J/kg needed to account for the Sun's luminosity. Careful laboratory experiments have determined the masses of all the particles involved in the conversion of four protons into a helium-4 nucleus: the total mass of the protons is 6.6943×10^{-27} kg, the mass of the helium-4 nucleus is 6.6466×10^{-27} kg, and the neutrinos are virtually massless. The difference between the total mass of the protons and the helium nucleus is 0.048×10^{-27} kg—not much mass, but easily measurable.

Multiplying the vanished mass by the square of the speed of light yields 4.3×10^{-12} J—this is the energy produced by the fusion of 6.7×10^{-27} kg (the rounded-off mass of the four protons) of hydrogen into helium. It follows that fusion of 1 kg of hydrogen would generate 6.4×10^{14} J—more than enough to power the Sun, even for another 5 billion years. To fuel the Sun's present energy output, hydrogen must be fused into helium in the core at a rate of 600 million tons per second.

16.6 Observations of Solar Neutrinos

6 Theorists are quite sure that the proton–proton chain operates in the Sun. However, because the gamma-ray energy released by the nuclear reactions just described is transformed into visible and infrared radiation by the time it emerges from the Sun, astronomers have no *direct* electromagnetic evidence of conditions in the solar core. However, the neutrinos that arise as by-products of the proton–proton chain *do* escape from the Sun. Interacting with virtually nothing, they leave at or near the speed of light, escaping into space a few seconds after being created in the core. They offer, at least in principle, the possibility of probing directly the conditions at the heart of the Sun.

Of course, the fact that they can pass through the entire Sun without interacting also makes neutrinos very difficult to detect on Earth. Nevertheless, they do interact a little more strongly with some elements—chlorine and gallium, for example—than with others, and this knowledge can be used in the construction of Earth-based neutrino-detection devices. Occasionally, a neutrino from the Sun will encounter a chlorine-37 nucleus, converting it into a nucleus of argon-37, or one will interact with a nucleus of gallium-71, turning it into germanium-71.

Figure 16.27 This swimming-pool-sized detector was a "neutrino telescope" of sorts, buried underground in a South Dakota gold mine.

Figure 16.27 shows a photograph of the first experiment designed to detect solar neutrinos. Built in the late 1960s by a team of researchers from Brookhaven National Laboratory, it consisted of a large tank filled with 400,000 liters (about 100,000 gallons) of a chlorine-containing chemical—the common cleaning fluid used by dry cleaners. It was sited near the bottom of the Homestake gold mine in South Dakota. At 1.5 km below ground level, the experimenters could be reasonably sure of avoiding interference from other sources, as most subatomic particles are unable to penetrate Earth's crust to such a depth. They left their tank in the mine for months at a time, periodically checking to see if any of the chlorine had been converted into argon, which would signal the absorption of a neutrino.

Given the size of the detector and the physical conditions in the Sun's core implied by the Standard Solar Model, about one solar neutrino of the roughly 10^{16} that streamed through the tank each day should have been detected. The experiment did succeed in detecting solar neutrinos—in itself a remarkable achievement—but the numbers were not as great as predicted. Over the course of the entire experiment, neutrinos were detected about twice per week, on average, not once per day. The neutrino deficit persisted over two decades of almost continuous monitoring (until the experiment was terminated in 1993), and though scores of technicians examined every facet of the Homestake instrument for instrumental flaws, they found none. This disagreement between theory and observation is known as the **solar neutrino problem**.

Other, more recent experiments, having quite different detector designs, have also found significantly fewer neutrinos than predicted by theory. Researchers using a detector located in Kamioka, Japan, reported only about 45 percent of the expected number of neutrinos, and two more sensitive experiments—the *Soviet–American Gallium Experiment* (SAGE, for short) and the U.S.–European GALLEX collaboration—each using the element gallium to capture solar neutrinos, find a roughly 50 percent shortfall. The Homestake and Kamioka experiments were actually sensitive not to neutrinos produced in reaction (I) but to those created by a much less probable sequence of events, occurring only about 0.25 percent of the time, leaving some room for theorists to maneuver in their attempts to explain the observations. However, SAGE and GALLEX can detect neutrinos produced by reaction (I)—the initial step in the proton–proton chain—so they provide a much more direct probe of energy generation in the solar core.

The four neutrino-detection experiments just described disagree somewhat in their measurements of the precise extent of the deficit, but each of them sees significantly fewer than the expected number of solar neutrinos. The inescapable conclusion is that although solar neutrinos are observed (and in fact their measured energies do lie in the range predicted by the Standard Solar Model), there is a real discrepancy between the Sun's theoretical neutrino output and the neutrinos we detect on Earth. How can we explain this contradiction? If, as we think, the detectors are working correctly, there are really only two possibilities. Either neutrinos are not produced as frequently as we think, or not all of them make it to Earth. Let us now consider these alternatives in turn.

If the temperature in the solar core were lower, then the number of neutrinos predicted by theory would also be lower. If the center of the Sun were about 10 percent cooler than in the Standard Solar Model—about 13,500,000 K—the neutrino emission would be reduced to a level consistent with observations. But lowering the temperature would also lower the Sun's luminosity, and most theorists agree that the numerical models could not be in error by as much as 1,500,000 K while remaining consistent with all other solar observations. In addition, observations by the GONG group (discussed earlier) seem to rule out a central temperature

below 15,000,000 K. Most astronomers regard it as quite unlikely that the resolution of the solar neutrino problem will be found in the nuclear physics of the Sun's interior.

Instead, the properties of the neutrinos themselves may provide the answer. If neutrinos do have a minute amount of mass, it may be possible for them to change their properties, even to transform into other particles, during their 8-minute flight from the solar core to Earth, through a process generally known as **neutrino oscillations**. In this picture, neutrinos are produced in the Sun at the rate required by the Standard Solar Model, but some of them turn into something else (they are said to "oscillate" into other particles) on their way to Earth and so go undetected. In June 1998, the Kamioka group reported what many experts regard as compelling experimental evidence of neutrino oscillations (and hence of nonzero neutrino masses). However, the observed oscillations do *not* involve neutrinos of the type produced in the Sun, so they do not by themselves explain the solar neutrino deficit. Several groups have attempted to detect oscillations in the "solar-type" neutrinos produced by nuclear reactors here on Earth, so far without success.

Where are the missing neutrinos? Is the proton–proton chain operating as we think? Do we really know what processes are at work deep in the hearts of stars? For now, the mystery of the solar neutrinos remains unsolved, although most physicists favor the neutrino-oscillation explanation. Virtually all researchers concur—or at least hope—that the correct interpretation of the solar neutrino problem will not tear apart the theoretical fabric of the proton–proton chain. Most believe that the description of solar fusion we have presented is this chapter is basically right; our understanding of neutrino physics just needs to be fine-tuned. But should drastic measures be needed to solve the solar puzzle, we may yet have to return to the drawing board to answer one of the most fundamental scientific questions of all: How does a star shine?

Chapter Review WWW

SUMMARY

A **star** (p. 354) is a glowing ball of gas held together by its own gravity and powered by nuclear fusion at its center. The main interior regions of the Sun are the **core** (p. 356), where nuclear reactions generate energy, the **radiation zone** (p. 355), where the energy travels outward in the form of electromagnetic radiation, and the **convection zone** (p. 355), where the Sun's matter is in constant convective motion.

The sharp solar disk visible from Earth marks the solar **photosphere** (p. 355)—the thin surface layer from which the Sun's light is

emitted. Above the photosphere lies the **chromosphere** (p. 355), which is separated from the solar **corona** (p. 355) by a thin **transition zone** (p. 355) in which the temperature increases from a few thousand to around a million kelvins.

The Sun's **luminosity** (p. 356) is the total amount of energy radiated from the solar surface per second. It is determined by measuring the **solar constant** (p. 356)—the amount of solar radiation reaching each square meter at Earth's distance from the

Sun—and multiplying that amount by the area of an imaginary sphere of radius 1 A.U.

The mathematical model that best fits the observed properties of the Sun is the **Standard Solar Model** (p. 356). Studies of **helioseismology** (p. 356)—oscillations of the solar surface caused by sound waves in the interior—provide further insight into the Sun's structure.

The effect of the solar convection zone can be seen on the surface in the form of **granulation** (p. 359) of the photosphere. As hotter (and therefore brighter) gas rises and cooler (dimmer) gas sinks, a characteristic "mottled" appearance results. Lower levels in the convection zone also leave their mark in the form of larger patterns called **supergranulation** (p. 360).

Most of the absorption lines seen in the solar spectrum are produced in the upper photosphere and the chromosphere. Studies of these allow scientists to determine the Sun's composition and the temperature structure of the solar atmosphere.

At about 10–15 solar radii, the gas in the corona is hot enough to escape the Sun's gravity, and the corona begins to flow outward as the **solar wind** (p. 363). Most of the solar wind flows from low-density regions of the corona called **coronal holes** (p. 364).

The steady component of the Sun's energy production is known as the **quiet Sun** (p. 364). Superimposed on that is the much more erratic emission of the **active Sun** (p. 364). Solar activity is generally associated with disturbances in the Sun's magnetic field. **Sunspots** (p. 364) are Earth-sized regions on the solar surface that are a little cooler than the surrounding photosphere. They are regions of intense magnetism.

Both the numbers and locations of sunspots vary in an 11-year **sunspot cycle** (p. 367). At **solar minimum** (p. 367), only a few spots are typically seen, and they lie far from the solar equator. At **solar maximum** (p. 367), the number of spots is much greater, and they generally lie much closer to the equator. The sunspot cycle is quite irregular. Its length varies from 7 to 15 years. There have been times in the past when no sunspots were seen for long periods. The overall direction of the solar magnetic field reverses from one sunspot cycle to the next. The 22-year cycle that results when the direction of the field is taken into account is called the **solar cycle** (p. 367).

Solar activity tends to be concentrated in **active regions** (p. 369) associated with sunspot groups. **Prominences** (p. 369) are loop- or sheetlike structures produced when hot gas ejected by activity on the solar surface interacts with the Sun's magnetic field. The more intense **flares** (p. 369) are violent surface explosions that blast particles and radiation into interplanetary space.

The Sun generates energy by "burning" hydrogen into helium in its core by the process of **nuclear fusion** (p. 371). Nuclei are held together by the **strong nuclear force** (p. 371). When four protons overcome their electromagnetic repulsion and are converted into a helium nucleus in the **proton–proton chain** (p. 373), some mass is lost. The **law of conservation of mass and energy** (p. 371) requires that this mass appear as energy, eventually resulting in the light we see. Very high temperatures are needed for fusion to occur.

Some particles produced during the solar fusion process are the **positron** (p. 372), or anti-electron, which quickly annihilates with electrons in the Sun's core to generate gamma rays, the **deuteron** (p. 371), an **isotope** (p. 372) of hydrogen consisting of a proton and a neutron, and the **neutrino** (p. 372), a near-massless particle that escapes from the Sun without any further interactions once it is created in the core. Neutrinos interact via the **weak nuclear force** (p. 372).

Despite their elusiveness, it is possible to detect a small fraction of the neutrinos streaming from the Sun. The observations lead to the **solar neutrino problem** (p. 374)—substantially fewer neutrinos are observed than are predicted by theory. The resolution to this problem is unclear. A leading explanation is that **neutrino oscillations** (p. 375) convert some neutrinos into other (undetected) particles en route from the Sun to Earth.

SELF-TEST: TRUE OR FALSE?

_____ **1.** The Sun is a rather normal star.

_____ **2.** The Sun's average density is greater than that of Earth.

_____ **3.** The Sun's diameter is about 10 times that of Earth.

_____ **4.** Observations of sunspots indicate that the Sun rotates differentially.

_____ **5.** In the solar convection zone, the gas is partly ionized.

_____ **6.** Convection involves cool gas rising toward the solar surface, and hot gas sinking into the interior.

_____ **7.** Most solar absorption lines spectrum are produced in the corona.

_____ **8.** There are as many absorption lines in the solar spectrum as there are elements present in the Sun.

_____ **9.** The faintness of the chromosphere is a direct result of its low temperature.

_____ **10.** The temperature of the solar corona decreases with increasing radius.

_____ **11.** Sunspots are regions of intense magnetic fields.

_____ **12.** Prominences are large flames erupting from the burning surface of the Sun.

_____ **13.** Positrons are the antiparticles of electrons.

_____ **14.** Nuclei are held together by the strong force.

_____ **15.** Neutrinos are hypothetical particles that are believed to exist but have never been detected experimentally.

SELF-TEST: FILL IN THE BLANK

1. The part of the Sun we actually see is called the _____.

2. Traveling outward from the surface, the two main regions of the solar atmosphere are the _____ and the _____.

3. Below the solar surface, in order of increasing depth, lie the _____ zone, the _____ zone, and the _____.

4. _____ seen on the surface of the Sun is evidence of solar convection.

5. The Sun appears to have a well-defined edge because the thickness of the _____ is only 0.1 percent of the solar radius.

6. The most abundant element in the Sun is _____.

7. The second most abundant element in the Sun is _____.

8. The two most abundant elements in the Sun make up about _____ percent of its composition (by number of atoms.).

9. The gas in the corona is highly _____.

10. Sunspots appear dark because they are _____ than the surrounding gas of the photosphere.

11. The sunspot cycle is _____ years long; the solar cycle is _____ years long.

12. The entire solar luminosity is produced in the _____ (give the region) of the Sun.

13. The net result of the proton–proton chain is that _____ protons are fused into a nucleus of _____, two _____ are emitted, and energy is released in the form of _____.

14. Energy is released in the proton–proton chain because mass is _____ in the process.

15. The solar neutrino problem is the fact that astronomers observe too _____ neutrinos coming from the Sun.

REVIEW AND DISCUSSION

1. Name and briefly describe the main regions of the Sun.

2. How massive is the Sun, compared with Earth?

3. How hot is the solar surface? The solar core?

4. How do scientists construct models of the Sun?

5. What is helioseismology, and what does it tell us about the Sun?

6. Describe how energy generated within the Sun reaches Earth.

7. Why does the Sun appear to have a sharp edge?

8. Give the history of "coronium," and tell how it increased our understanding of the Sun.

9. What is the solar wind?

10. What is the cause of sunspots, flares, and prominences?

11. What fuels the Sun's enormous energy output?

12. What are the ingredients and the end result of the proton–proton chain in the Sun? Why is energy released in the process?

13. Why are scientists trying so hard to detect solar neutrinos?

14. Describe some possible solutions to the solar neutrino problem.

15. What would we observe on Earth if the Sun's internal energy source suddenly shut off? How long do you think it might take—minutes, days, years, millions of years—for the Sun's light to begin to fade? Repeat the question for solar neutrinos.

PROBLEMS

1. Use the reasoning presented in Section 16.1 to calculate the value of the "solar constant" on Jupiter.

2. Use Wien's law (Section 3.4) to determine the wavelength corresponding to the peak of the blackbody curve (a) in the core of the Sun, where the temperature is 10^7 K, (b) in the solar convection zone (10^5 K), and (c) just below the solar photosphere (10^4 K). What form (visible, infrared, X ray, etc.) does the radiation take in each case?

3. The largest-amplitude solar sound waves have periods of about 5 minutes. This is the time taken for the waves to cross from one side of the Sun to the other and back. Calculate the average speed of the wave. Compare the wave period with the orbital period of an object moving just above the solar photosphere.

4. If convected solar material moves at 1 km/s, how long does it take to flow across the 1000-km expanse of a typical granule? Compare this with the roughly 10-minute lifetimes observed for most solar granules.

5. Use Stefan's law (flux $\propto T^4$, where T is the temperature in kelvins; see Section 3.4) to calculate how much less energy (as a fraction) is emitted per unit area of a 4500-K sunspot than from the surrounding 5800-K photosphere.

6. The Sun's differential rotation is responsible for wrapping the solar magnetic field around the Sun (Figure 16.19). Using the data presented in the Sun Data table on p. 354, calculate how long it takes for material at the solar equator to "lap" the material near the poles—that is, to complete one extra trip around the Sun's rotation axis.

7. How long does it take for the Sun to convert one Earth mass of hydrogen into helium?

8. Assuming constant luminosity, calculate how long it would take the Sun to radiate its own mass into space.

9. The solar wind carries mass away from the Sun at a rate of about 900,000 kg/s. Compare this rate with the rate at which the Sun loses mass in the form of radiation.

10. The entire reaction sequence shown in Figure 16.26 generates 4.3×10^{-12} joules of electromagnetic energy and releases two neutrinos. Assume that neutrino oscillations transform half of these neutrinos into other particles by the time they travel 1 A.U. from the Sun. Estimate the total number of solar neutrinos passing through Earth each second.

PROJECTS

The projects given here all require a special solar filter. Such filters are easily purchased from various sources.

NEVER LOOK DIRECTLY AT THE SUN WITHOUT A FILTER!

1. An appropriately filtered telescope will easily show you sunspots. Count the number of sunspots you see on the Sun's surface. Notice that sunspots often come in pairs or groups. Come back and look again a few days later and you'll see that the Sun's rotation has caused spots to move, and the spots themselves have changed. If a sufficiently large sunspot (or, more likely, sunspot group) is seen, continue to watch it as the Sun rotates. It will be out of view for about 2 weeks. Can you determine the rotation of the Sun from these observations?

2. Solar granulation is not too hard to see. The atmosphere of Earth is most stable in the morning hours. Observe the Sun on a cool morning, 1 or 2 hours after it has risen. Use high magnification and look initially at the middle of the Sun's disk. Can you see changes in the granulation pattern? They are there, but are not always obvious or easy to see.

3. View some solar prominences and flares. Hydrogen-alpha filters are commercially available for small telescopes are quite expensive, but many science departments will have one. You can often see prominences and flares even during times of sunspot minimum. You are actually viewing the chromosphere rather than the photosphere, so the Sun looks quite different from its normal appearance.

17 MEASURING THE STARS

Giants, Dwarfs, and the Main Sequence

WWW

LEARNING GOALS

Studying this chapter will enable you to:

1 Explain how stellar distances are determined.

2 Discuss the motions of the stars through space and how these motions are measured from Earth.

3 Explain how physical laws are used to estimate stellar sizes.

4 Distinguish between luminosity and brightness and explain how stellar luminosity is determined.

5 Explain how stars are often classified according to their colors, surface temperatures, and spectral characteristics, and tell why such a classification is useful.

6 State how an H–R diagram is constructed, and summarize the properties of the different types of stars that such a diagram helps us to identify.

7 Explain how the masses of stars are measured and how mass is related to other stellar properties.

8 Distinguish between open and globular star clusters, and explain why the study of clusters is important to astronomers.

(Opposite page) Even in our own local region of the cosmos, the number of stars is virtually beyond our ability to count. Relatively few of them have actually been studied in detail. Yet we have learned an enormous amount about the stars in general and the range of properties they exhibit—their masses, their temperatures, their luminosities, even their ages and (as we shall see in subsequent chapters) their destinies.

We have now studied Earth, the Moon, the solar system, and the Sun. To continue our inventory of the contents of the universe, we must move away from our local environment into the depths of space. In this chapter we take a great leap in distance and consider stars in general. Our primary goal is to comprehend the nature of the stars that make up the constellations, as well as the myriad more distant stars we cannot perceive with our unaided eyes. Rather than studying their individual peculiarities, however, we will concentrate on determining the physical and chemical properties they share. There is order in the legions of stars scattered across the sky. Like comparative planetology in the solar system, comparing and cataloging the stars play vital roles in furthering our understanding of the Galaxy and the universe we inhabit.

17.1 The Distances to the Stars

STELLAR PARALLAX

1 Recall from Chapter 1 how we can use *parallax* to measure distances to terrestrial and solar system objects. Parallax is an object's apparent shift relative to some more distant background as the observer's point of view changes. ∞ (Sec. 1.5) To measure parallax, we must observe the object from either end of some baseline and measure the angle through which the line of sight to the object shifts. In astronomical contexts, we determine the parallax by comparing photographs made from the two ends of the baseline.

As the distance to the object increases or the baseline shrinks, the parallax becomes smaller and therefore harder to measure. Accordingly, a large baseline is essential for measuring the distance to a very remote object. The stars are so far away from us that even Earth's diameter is too short to use as a baseline in determining their distance. Their apparent shift, as seen from different points on Earth, is too small to measure. However, by comparing observations made of a star at different times of the year, as shown in Figure 17.1, we effectively extend the baseline to the diameter of Earth's orbit around the Sun, 2 A.U. Only with this enormously longer baseline do some stellar parallaxes become measurable. As indicated in the figure, a star's parallactic angle—or, more commonly, just its "parallax"—is conventionally defined to be *half* its apparent shift relative to the background as we move from one side of Earth's orbit to the other.

The parallaxes of even the closest stars are very small, so astronomers generally find it convenient to measure parallax in arc seconds rather than in degrees. If we ask at what distance a star must lie in order for its observed parallax to be exactly 1″, we get an answer of 206,265 A.U., or 3.1×10^{16} m. Astronomers call this distance 1 **parsec** (1 pc), from "*par*allax in arc *sec*onds." Because parallax decreases as distance increases, we can relate a star's parallax to its distance by the following simple formula:

$$\text{distance (in parsecs)} = \frac{1}{\text{parallax (in arc seconds)}}.$$

Thus, a star with a measured parallax of 1″ lies at a distance of 1 pc from the Sun. The parsec is defined so as to make the conversion between distance and parallactic angle easy. An object with a parallax of 0.5″ lies at a distance of 2 pc; an

Figure 17.1 (a) The geometry of stellar parallax. For observations made 6 months apart, the baseline is twice the Earth–Sun distance, or 2 A.U. (b) The parallactic angle is usually measured photographically. (The shift is greatly exaggerated in this drawing.)

object with a parallax of 0.1″ lies at 10 pc, and so on. One parsec is approximately equal to 3.3 light years.

OUR NEAREST NEIGHBORS

The closest star to Earth (excluding the Sun) is called Proxima Centauri. This star is a member of a triple-star system (three separate stars orbiting one another, bound together by gravity) known as the Alpha Centauri complex. Proxima Centauri displays the largest known stellar parallax, 0.76″, which means that it is about 1.3 pc away—about 270,000 A.U., or 4.3 light years. That's the *nearest* star to Earth—at almost 300,000 times the distance from Earth to the Sun! This is a fairly typical interstellar distance in the Milky Way Galaxy.

Vast distances can sometimes be grasped by means of analogies. Imagine Earth as a grain of sand that orbits a golfball-sized Sun at a distance of about 1 m. The nearest star,

also a golfball-sized object, is then more than 270 *kilo*meters away. Except for the other planets in our solar system, themselves ranging in size from grains of sand to small marbles and all lying within 50 m of the "Sun," nothing else of consequence exists in the 270 km separating the two stars. Such is the void of interstellar space.

The next nearest neighbor to the Sun beyond the Alpha Centauri system is called Barnard's Star. Its parallax is 0.55″, so it lies at a distance of 1.8 pc, or 6.0 light years—370 km in our model. Figure 17.2 is a map of our nearest galactic neighbors—the 30 or so stars lying within 4 pc of Earth.

As mentioned in Chapter 5, ground-based images of stars are generally smeared out into a disk of radius 1″ or so by turbulence in Earth's atmosphere. ∞ (Sec. 5.3) However, astronomers have special equipment that can routinely measure stellar parallaxes of 0.03″ or less, corresponding to stars within about 30 pc (100 light years) of Earth. Several

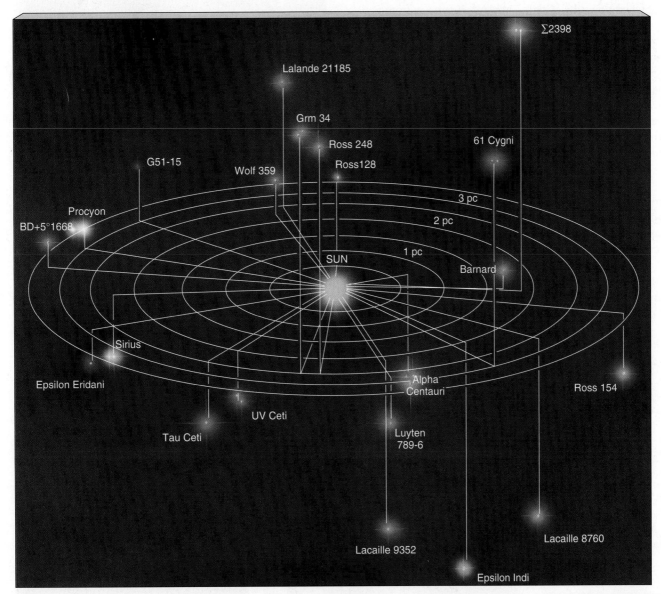

Figure 17.2 A plot of the 30 closest stars to the Sun, projected so as to reveal their three-dimensional relationships. Notice that many are members of multiple-star systems. All lie within 4 pc (about 13 light years) of Earth.

thousand stars lie within this range, most of them much dimmer than the Sun and invisible to the naked eye. As discussed in more detail in *Interlude 17-1*, recently released data from the European *Hipparcos* satellite have extended the range of accurately measured parallaxes to well over 100 pc, encompassing nearly a million stars. Even so, the vast majority of stars in our galaxy are far more distant than this.

17.2 Stellar Motion

2 In addition to the apparent motion caused by parallax, stars have real motion, too. In other words, stars travel through space. This stellar motion has two components. ∞ (Sec. 3.5) The *transverse component* measures a star's motion perpendicular to our line of sight—in other words, its motion across the sky. The *radial component* measures a star's movement along our line of sight—toward us or away from us.

The annual movement of a star across the sky, as seen from Earth (and corrected for parallax), is called **proper motion**. It describes the transverse component of a star's velocity relative to the Sun. Like parallax, proper motion is measured in terms of angular displacement. Since the angles involved are typically very small, proper motion is usually expressed in arc seconds per year. Stars' velocities through space can be quite large—tens or even hundreds of kilometers per second. However, because of their great distances, it usually takes many years for us to be able to discern their movement.

Figure 17.3 compares two photographs of the sky around Barnard's Star. They were made on the same day of the year, but 22 years apart. As the photographs show, Barnard's Star moved during this interval. If the two images were superimposed, the images of the other stars in the field of view would coincide, but those of Barnard's Star would not—Barnard's Star moved during this time interval. Because Earth was at the same point in its orbit when these photographs were taken, the displacement cannot be due to parallax caused by Earth's motion around the Sun. We conclude that the observed displacement indicates real space motion of Barnard's Star relative to the Sun.

Careful measurements show that Barnard's Star moved 227″ during the 22-year interval. The proper motion of Barnard's Star is therefore 227″/22 years, or 10.3″/yr. This

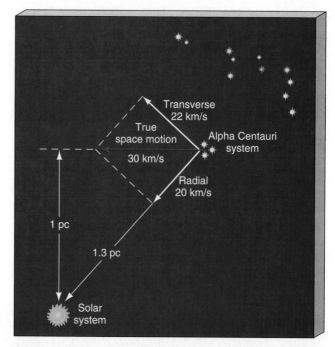

Figure 17.4 The motion of the Alpha Centauri star system drawn relative to our solar system. The transverse component of the velocity has been determined by observing the system's proper motion. The radial component is measured using the Doppler shift of lines in Alpha Centauri's spectrum. The true space velocity, indicated by the red arrow, results from the combination of the two.

is the largest known proper motion of any star. Only a few hundred stars have proper motions greater than 1″/yr.

A star's transverse velocity is easily calculated once its proper motion and its distance are known. At the distance of Barnard's Star (1.8 pc), an angle of 10.3″ corresponds to a physical displacement of 0.00009 pc, or about 2.8 billion km. Barnard's Star takes a year (3.2×10^7 s) to travel this distance, so its transverse velocity is 2.8 billion km/3.2×10^7 s, or 88 km/s.

As another example, consider the three-dimensional motion of our nearest neighbor, the Alpha Centauri system, sketched in Figure 17.4 in relation to our own solar system. Alpha Centauri's proper motion has been measured, relative to more distant background stars, at about 3.5″/yr. At Alpha Centauri's distance of 1.3 pc, this implies a transverse veloci-

Figure 17.3
Comparison of two photographic plates taken 22 years apart shows evidence of real space motion for Barnard's Star (denoted by an arrow).

ty of 22 km/s. We can determine the other component of motion—the radial velocity—using the Doppler effect. Spectral lines from Alpha Centauri are slightly blueshifted, allowing astronomers to measure the star system's radial velocity (relative to the Sun) as 20 km/s toward us. ∞ (Sec. 3.5)

What is the true space motion of Alpha Centauri? Will this alien system collide with our own some time in the future? The answer is no—Alpha Centauri's transverse velocity will steer it well clear of the Sun. We can combine the transverse and radial velocities according to the Pythagorean theorem. The total velocity is $\sqrt{(22^2 + 20^2)}$, or about 30 km/s, in the direction shown by the horizontal red arrow in Figure 17.4. As that figure indicates, Alpha Centauri will get no closer to us than about 1 pc, and that won't happen until 280 centuries from now.

17.3 Stellar Sizes

DIRECT AND INDIRECT MEASUREMENTS

3 Most stars are unresolved points of light in the sky, even when viewed through the largest telescopes. However, a few are big, bright enough, and close enough to allow us to measure their sizes *directly*. In an optical technique known as *speckle interferometry*, many short-exposure images of a star, each too brief for Earth's turbulent atmosphere to smear it out into a seeing disk, are combined to make a high-resolution map of the star's surface. In some cases, the results are detailed enough to allow a few surface features to be distinguished (Figure 17.5). As adaptive optics techniques continue to improve (see Chapter 5), it is becoming possible to combine the individual images in real time, again allowing very-high-resolution stellar images to be made. ∞ (Sec. 5.3)

By measuring a star's angular size and knowing its distance from the Sun, astronomers can determine its radius by simple geometry. Optical astronomers have directly measured the sizes of a few dozen stars in this way. However, most stars are too distant or too small for this to be possible. Their sizes must be inferred by other means.

Recall from Chapter 3 that the radiation emitted by any hot body is governed by Stefan's law, which states that the energy emitted per unit area per unit time by a hot body—the body's *energy flux*—increases as the fourth power of the temperature. ∞ (Sec. 3.4) To determine the star's *luminosity*, we must multiply by the star's surface area, which is proportional to the square of the stellar radius. ∞ (Sec. 16.1) Combining these proportionalities, we have

$$\text{luminosity} \propto \text{radius}^2 \times \text{temperature}^4,$$

where $\propto$ is the symbol representing proportionality. This **radius–luminosity–temperature relationship** is important because it demonstrates that knowledge of a star's luminosity and temperature can yield an estimate of its radius—an *indirect* determination of stellar size.

GIANTS AND DWARFS

Let's consider some examples to clarify these ideas. The star known as Mira (Omicron Ceti) has a surface temperature of about 3000 K and a luminosity of 1.6×10^{29} W. Thus, its surface temperature is half, and its luminosity about 400 times, the corresponding quantities for our Sun. The

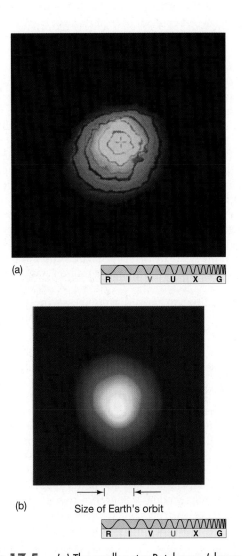

(a)

R I V U X G

(b)

Size of Earth's orbit

R I V U X G

Figure 17.5   (a) The swollen star Betelgeuse (shown here in false color) is close enough for us to directly resolve its size, along with some surface features thought to be storms similar to those that occur on the Sun. Betelgeuse is such a huge star (300 times the size of the Sun) that its photosphere spans roughly the size of Mars's orbit. Most of the surface features discernible here are larger than the entire Sun. (The dark lines are drawn contours, not part of the star itself.) (b) An ultraviolet view of Betelgeuse, in false color, as seen by a European camera onboard the *Hubble Space Telescope*. The bright spot at the bottom right of the star is more than 10 times bigger than Earth and at least 2000 K hotter than the surrounding 3000 K photosphere. Most astronomers regard it as a storm of some sort, perhaps similar to those that occur during the active phase of our Sun.

INTERLUDE 17-1 The Hipparcos Mission

In 1989, shortly before the large, complex, and expensive *Hubble Space Telescope* was launched by NASA into Earth orbit, a small, simple, and cheap satellite was also deployed by the European Space Agency. Called *Hipparcos*—a tortured attempt to honor the ancient Greek astronomer Hipparchus of Nicea, who made the first known star map, by creating an acronym out of *High Precision Parallax Collecting Satellite*—this spacecraft was initially thought to be lost in space. *Hipparcos* went into the wrong orbit, its onboard engine failed to fire, and its solar panels were heavily pelted with particles in Earth's Van Allen belts. However, before it fell silent in 1993, the engineers managed to control the errant spacecraft, and ESA scientists revamped its mission. Nearly a decade after its launch, its massive database was cataloged and released.

The heart of *Hipparcos* (shown here before launch) was a small 29-cm mirror designed to determine the positions of stars very accurately. By measuring those positions from two vantage points in orbit (above Earth's turbulent atmosphere), the satellite was able to determine stellar parallaxes with about 10 times greater accuracy than is possible using telescopes on the ground. Since volume scales as the cube of the size, *Hipparcos*'s prime mission was to measure accurate distances for roughly a million stars out to some 200 pc.

In addition to measuring distances to nearby stars with unprecedented accuracy, some other highlights of the *Hipparcos* mission included the following:

■ In about a million years, the star named Gliese 710 may pose a hazard to planet Earth. Currently this red dwarf star lies 20 pc away in the direction of the Ophiuchus constellation, but its rapid line-of-sight motion will eventually take it within 0.3 pc (less than a light year) of the Sun. As it passes perhaps right through the Oort cloud of comets surrounding our solar system, this dim dwarf star will shine as brilliantly as the red giant Betelgeuse does now. It may also perturb the Oort cloud significantly, sending a shower of comets hurtling in our direction. Astronomers estimate that no star has come this close to the Sun in the past 10 million years or so.

■ Parallax calibrates other "yardsticks" used by astronomers to measure distances in increasingly larger realms and eventually to estimate the size and scale of the entire observable universe. As a result, *Hipparcos* has affected distances far beyond those it was able to measure directly in our stellar neighborhood. In particular, recalculation of distances to a certain class of variable stars (called Cepheids) has put astronomy's entire distance scale on a much firmer foundation.

■ Because of these recalibrations, some distances have changed a little, others need substantial revision. For example, the Large Magellanic Cloud is now thought to be about 10 percent farther away than had previously been estimated. Likewise, the distance to the Andromeda Galaxy has been increased by about a third; its distance is now pegged at 0.9 Mpc (or 2.9 million light years). All distances throughout this text reflect the new values as determined by *Hipparcos*.

■ The new data have also increased the size, and hence the age, of the universe. The best number for that age, based on *Hipparcos* measurements, now seems to be about 12–13 billion years. This is the age used in this book.

■ *Hipparcos* also measured the colors (i.e., temperatures) and apparent brightnesses (and hence luminosities, since the distances are known) of the stars it surveyed. Based on these new data, a major revision of the Hertzsprung–Russell diagram (a key plot of stellar properties used throughout our text) implies that the oldest stars are a little younger than was previously thought. This new analysis of stellar evolutionary tracks seems to reconcile the age of the universe with that of the oldest stars. Prior to the *Hipparcos* mission, astronomers had faced the embarrassing problem that some stars seemed to be older than the universe itself. No more—this major breakthrough in a heretofore log-jammed subject may well be *Hipparcos*'s greatest contribution to astronomy.

The *Hipparcos* mission was not glamorous. This little satellite sent back no pictures for European citizens (who paid for it) and others to marvel at, and it grabbed few headlines in the world's press. Rather, its catalogs, comprising an astonishing 1000 gigabytes of data, resemble huge phonebooks—unexciting to look at, but of enormous value for accessing the vital statistics of stars and the entire universe beyond.

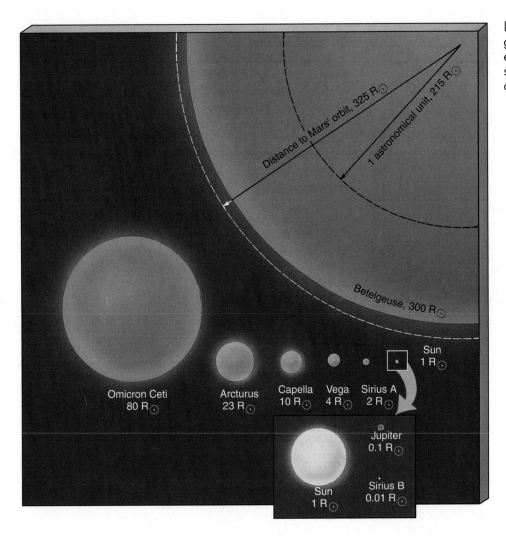

Figure 17.6 Star sizes vary greatly. Shown here are the estimated sizes of several well-known stars, including a few of those discussed in this chapter.

Distance to Mars' orbit, 325 R$_\odot$

1 astronomical unit, 215 R$_\odot$

Betelgeuse, 300 R$_\odot$

Sun 1 R$_\odot$

Omicron Ceti 80 R$_\odot$

Arcturus 23 R$_\odot$

Capella 10 R$_\odot$

Vega 4 R$_\odot$

Sirius A 2 R$_\odot$

Sun 1 R$_\odot$

Jupiter 0.1 R$_\odot$

Sirius B 0.01 R$_\odot$

radius–luminosity–temperature relationship, rearranged into the more convenient form

$$\text{radius} \propto \frac{\sqrt{\text{luminosity}}}{\text{temperature}^2},$$

then implies that the star's radius is $\sqrt{400}/0.5^2 = 80$ times that of our Sun. If our Sun were this large, its photosphere would extend as far as the orbit of Mercury. A star as large as Mira is known as a *giant*. More precisely, **giants** are stars having radii between 10 and 100 times that of the Sun. Even larger stars, ranging up to 1000 solar radii in size, are known as **supergiants**. Because the color of any 3000 K object is red, Mira is a **red giant**.

Now consider Sirius B—a faint companion to Sirius A, the brightest star in the night sky. Sirius B's surface temperature is roughly 24,000 K, four times that of the Sun. Its total luminosity is 10^{25} W, about 0.04 the solar value. Substituting these quantities into our equation, we obtain a radius of $\sqrt{0.04}/4^2 = 0.01$ solar radii—roughly the size of Earth. Sirius B is much hotter but smaller and far less luminous than our Sun. Such a star is known as a *dwarf*. In astronomical parlance, the term **dwarf** refers to any star of radius comparable to or smaller than the Sun (including the

Sun itself). Because any 24,000 K object glows white, Sirius B is an example of a **white dwarf**.

The radii of the vast majority of stars (mostly measured using the radius–luminosity–temperature relationship) range from less than 0.01 to over 100 times the radius of the Sun. Figure 17.6 illustrates the estimated sizes of a few well-known stars.

17.4 Luminosity and Apparent Brightness

4 Luminosity is an *intrinsic* property of a star—it does not depend in any way on the location or motion of the observer. It is sometimes referred to as the star's **absolute brightness**. However, when we look at a star, we see not its luminosity but rather its **apparent brightness**—the amount of energy striking a unit area of some light-sensitive surface or device (such as a CCD chip or a human eye) per unit time. Apparent brightness is a measure not of a star's luminosity but of the *energy flux* produced by the star, as seen from Earth; it depends on our *distance* from the star. In this section, we discuss in more detail how these important quantities are related to one another.

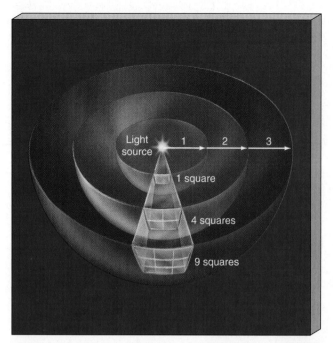

Figure 17.7 As it moves away from a source such as a star, radiation is steadily diluted while spreading over progressively larger surface areas (depicted here as sections of spherical shells). Thus, the amount of radiation received by a detector (the source's apparent brightness) varies inversely as the square of its distance from the source.

Animation

ANOTHER INVERSE-SQUARE LAW

Figure 17.7 shows light leaving a star and traveling through space. Moving outward, the radiation passes through imaginary spheres of increasing radius surrounding the source. The amount of radiation leaving the star per unit time—the star's luminosity—is constant, so the farther the light travels from the source, the less energy passes through each unit of area. Think of the energy as being spread out over an ever-larger area and therefore spread more thinly, or "diluted," as it expands into space. Because the area of a sphere grows as the square of the radius, the energy per unit area—the star's apparent brightness—is inversely proportional to the square of the distance from the star. Doubling the distance from a star makes it appear 2^2, or 4, times dimmer. Tripling the distance reduces the apparent brightness by a factor of 3^2, or 9, and so on.

Of course, the star's luminosity also affects its apparent brightness. Doubling the luminosity doubles the energy crossing any spherical shell surrounding the star and hence doubles the apparent brightness. We can therefore say that the apparent brightness of a star is directly proportional to the star's luminosity and inversely proportional to the square of its distance:

$$\text{apparent brightness (energy flux)} \propto \frac{\text{luminosity}}{\text{distance}^2}.$$

Thus, two identical stars can have the same apparent brightness if (and only if) they lie at the same distance from Earth.

However, as illustrated in Figure 17.8, two nonidentical stars can also have the same apparent brightness if the more luminous one lies farther away. A bright star (that is, one having large apparent brightness) is a powerful emitter of radiation (high luminosity), is near Earth, or both. A faint star (small apparent brightness) is a weak emitter (low luminosity), is far from Earth, or both.

Determining a star's luminosity is a twofold task. First, the astronomer must determine the star's apparent brightness by measuring the amount of energy detected through a telescope in a given amount of time. Second, the star's distance must be measured—by parallax for nearby stars and by other means (to be discussed later) for more distant stars. The luminosity can then be found using the inverse-square law. This is basically the same reasoning we used earlier in our discussion of how astronomers measure the solar luminosity. ∞ (Sec. 16.1)

THE MAGNITUDE SCALE

Instead of measuring apparent brightness in SI units (for example, watts per square meter, the unit used for the solar constant in Chapter 16), optical astronomers find it more convenient to work in terms of a construct called the **magnitude scale**. The scale dates from the second century B.C., when the Greek astronomer Hipparchus ranked the naked-eye stars into six groups. The brightest stars were categorized as first magnitude. The next brightest stars were labeled second magnitude, and so on, down to the faintest stars visible to the naked eye, which were classified as sixth magnitude. The range 1 (brightest) through 6 (faintest)

Figure 17.8 Two stars A and B of different luminosity can appear equally bright to an observer on Earth if the brighter star B is more distant than the fainter star A.

spanned all the stars known to the ancients. Notice that a *large* magnitude means a *faint* star.

When astronomers began using telescopes with sophisticated detectors to measure the light received from stars, they quickly discovered two important facts about the magnitude scale. First, the 1–6 magnitude range defined by Hipparchus spans about a factor of 100 in apparent brightness—a first-magnitude star is approximately 100 times brighter than a sixth-magnitude star. Second, the physiological characteristics of the human eye are such that each magnitude change of 1 corresponds to a factor of about 2.5 in apparent brightness. In other words, to the human eye a first-magnitude star is roughly 2.5 times brighter than a second-magnitude star, which is roughly 2.5 times brighter than a third-magnitude star, and so on. (By combining factors of 2.5, you can confirm that a first-magnitude star is indeed $(2.5)^5 \approx 100$ times brighter than a sixth-magnitude star.)

Modern astronomers have modified and extended the magnitude scale in a number of ways. First, we now *define* a change of 5 in the magnitude of an object (going from magnitude 1 to magnitude 6, say, or from magnitude 7 to magnitude 2) to correspond to *exactly* a factor of 100 in apparent brightness. Second, because we are really talking about apparent (rather than absolute) brightnesses, the numbers in Hipparchus's ranking system are called **apparent magnitudes**. Third, the scale is no longer limited to whole numbers—a star of apparent magnitude 4.5 is intermediate in apparent brightness between a star of apparent magnitude 4 and one of apparent magnitude 5. Finally, magnitudes outside the range 1–6 are allowed—very bright objects can have apparent magnitudes much less than 1, and very faint objects can have apparent magnitudes far greater than 6.

Figure 17.9 illustrates the apparent magnitudes of some astronomical objects, ranging from the Sun, at −26.8, to the faintest object detectable by the *Hubble* or *Keck* telescopes, an object having an apparent magnitude of 30—about as faint as a firefly seen from a distance equal to Earth's diameter.

Apparent magnitude measures a star's apparent brightness when seen at the star's actual distance from the Sun. To compare intrinsic, or absolute, properties of stars, however, astronomers imagine looking at all stars from a standard distance of 10 pc. There is no particular reason to use 10 pc—it is simply convenient. A star's **absolute magnitude** is its apparent magnitude when it is placed at a distance of 10 pc from the observer. Because the distance is fixed in this definition, absolute magnitude is a measure of a star's absolute brightness, or luminosity.

When a star farther than 10 pc away from us is moved to a point 10 pc away, its apparent brightness increases and hence its apparent magnitude decreases. Stars more than 10 pc from Earth therefore have apparent magnitudes that are greater than their absolute magnitudes. For example, if a star at a distance of 100 pc were moved to the standard 10-pc distance, its distance would decrease by a factor of 10, so (by the inverse-square law) its apparent brightness would in-

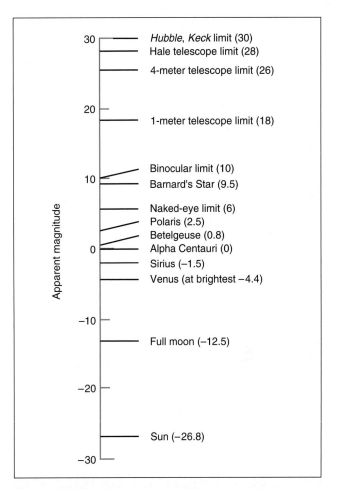

Figure 17.9 Apparent magnitudes of some astronomical objects. The original magnitude scale was defined so that the brightest stars in the night sky had magnitude 1, and the faintest stars visible to the naked eye had magnitude 6. It has since been extended to cover much brighter and much fainter objects. An increase of 1 in apparent magnitude corresponds to a decrease in apparent brightness of a factor of approximately 2.5.

crease by a factor of 100. Its apparent magnitude (by definition) would therefore decrease by 5. In other words the star's absolute magnitude exceeds its apparent magnitude by 5.

For stars closer than 10 pc, the reverse is true. An extreme example is our Sun. Because of its proximity to Earth, it appears very bright and thus has a large negative apparent magnitude (Figure 17.9). However, the Sun's absolute magnitude is 4.8. If the Sun were moved to a distance of 10 pc from Earth, it would be only slightly brighter than the faintest stars visible in the night sky.

The numerical difference between a star's absolute and apparent magnitudes is a measure of the distance to the star (see *More Precisely 17-1*). Knowledge of a star's apparent magnitude and of its distance allows us to compute its absolute magnitude. Alternatively, knowledge of a star's apparent and absolute magnitudes allows us to determine its distance.

17.5 Temperature and Color

Looking at the night sky, you can tell at a glance which stars are hot and which are cool. In Figure 17.10, which shows the constellation Orion as it appears through a small telescope, the colors of the cool red star Betelgeuse and the hot blue star Rigel are clearly evident. (Note, by the way, that these colors are intrinsic properties of the stars, and have *nothing* to do with Doppler redshifts or blueshifts.) However, to obtain these stars' temperatures (3000 K for Betelgeuse and 15,000 K for Rigel), more detailed observations are required.

Astronomers can determine a star's surface ("effective") temperature by measuring its apparent brightness (radiation intensity) at several frequencies, then matching the observations to the appropriate blackbody curve. ∞ (Sec. 3.4) In the case of the Sun, the theoretical curve that best fits the emission describes a 5800-K emitter. ∞ (Sec. 16.1) The same technique works for any other star, regardless of its distance from Earth.

Because the basic shape of the blackbody curve is so well understood, astronomers can estimate a star's temperature using as few as *two* measurements at selected wavelengths (which is fortunate, as detailed spectra of faint stars are often difficult and time consuming to obtain). This is accomplished through the use of telescope filters that block out all radiation except that within specific wavelength ranges. For example, a B (blue) filter rejects all radiation except for a certain range of violet to blue light. Defined by international agreement to extend from 380 to 480 nm, this range corresponds to wavelengths to which photographic film happens to be most sensitive. Similarly, a V (visual) filter passes only radiation within the 490 to 590 nm range (green to yellow), corresponding to that part of the spectrum to which human eyes are particularly sensitive. Many other filters are also in routine use—a U (ultraviolet) filter covers the near ultraviolet, and infrared filters span longer-wavelength parts of the spectrum.

Figure 17.11 shows how the B and V filters admit different amounts of light for objects of different temperatures. In curve (a), corresponding to a very hot 30,000-K emitter, considerably more radiation (about twice as much) is received through the B filter than through the V filter. In curve (b), the temperature is 10,000 K, and the B and V intensities are about the same. In the cool 3000-K curve (c), about five times more energy is received in the V range than in the B range. In each case, it is possible to reconstruct the entire blackbody curve on the basis of only those two measurements because no other blackbody curve can be drawn through both measured points. To the extent that a star's spectrum is well approximated as a blackbody, measurements of the B and V intensities are enough to specify the star's blackbody curve and thus yield its surface tempera-

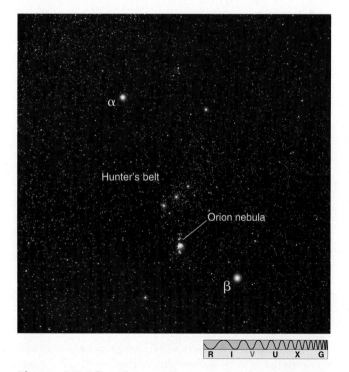

Figure 17.10 The constellation Orion, as it would appear through a small telescope or binoculars. The different colors of the member stars are easily distinguished. The bright red star at the upper left is Betelgeuse (α); the bright blue-white star at the lower right is Rigel (β). (Compare with Figure 1.6.) The scale of the photograph is about 3° across.

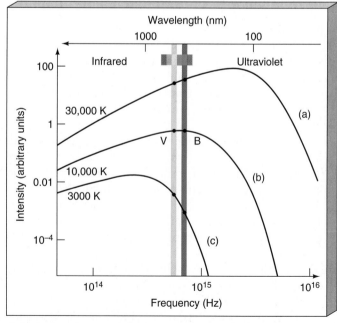

Figure 17.11 Blackbody curves for three different temperatures, along with the locations of the B (blue) and V (visual) filters. Star (a) is very hot—30,000 K—so its B intensity is considerably greater than its V intensity. Star (b) has about the same B and V readings and appears white. Its temperature is about 10,000 K. Star (c) is red. Its V intensity greatly exceeds the B value, and its temperature is 3000 K.

MORE PRECISELY 17-1   More on The Magnitude Scale

All the discussion of luminosity, brightness, and energy flux (Section 17.4) can equally well be written in terms of magnitudes. Here we present a little more detail on the "magnitude versions" of two important topics—stellar luminosity and the inverse-square law.

Recall that a star's *absolute magnitude* is the apparent magnitude we would measure if the star were located at a standard distance of 10 pc from us. Absolute magnitude is equivalent to luminosity—an intrinsic property of a star. Given that the Sun's absolute magnitude is 4.8, we can construct a "conversion chart" (shown at right) relating these two quantities. Since an *increase* in brightness by a factor of 100 corresponds, by definition, to a *reduction* of a star's magnitude by 5 units, it follows that a star with luminosity 100 times that of the Sun has absolute magnitude $4.8 - 5 = -0.2$, and an 0.01 solar luminosity star has absolute magnitude $4.8 + 5 = 9.8$. We can fill in the gaps by realizing that one magnitude corresponds

to a factor of $100^{1/5} \approx 2.512$, two magnitudes to $100^{2/5} \approx 6.310$, and so on. A factor of 10 in brightness corresponds to 2.5 magnitudes. You can use this chart to convert between solar luminosities and absolute magnitudes in many of the figures in this and later chapters.

The *inverse-square law* can also be recast in these terms. Increasing the distance to a star by a factor of 10 decreases its energy flux by a factor of 100 (by the inverse square law) and hence increases its apparent magnitude by 5 units. Increasing the distance by a factor of 100 increases the apparent magnitude by 10. Since absolute magnitude is a measure of the energy flux received from a star at a distance of 10 pc, we can write:

$$\text{apparent magnitude} - \text{absolute magnitude} =$$
$$5 \log_{10}\left(\frac{\text{distance}}{10 \text{ pc}}\right).$$

(The *logarithm* function is defined by the property that if $a = \log_{10}(b)$, then $b = 10^a$.) The difference between the apparent and absolute magnitudes of an object is usually referred to as its *distance modulus*. In terms of distance modulus (DM), we can rewrite the preceding equation as:

$$\text{distance} = 10 \text{ pc} \times 10^{\text{DM}/5}.$$

This doesn't look much like the inverse-square law presented in the text! Nevertheless, it contains exactly the same information, and it is in widespread use in this form throughout astronomy.

---

ture. Operationally, astronomers go about determining a star's temperature by measuring its **color index** (often referred to simply as the "color"), defined as the ratio of its B to V intensities (or, equivalently, the *difference* between the star's apparent magnitudes measured through the B and V filters). Table 17.1 lists a more detailed description of the

color index, the surface temperature derived from it, the dominant color perceived in the absence of filters, and some stellar examples. This type of non-spectral-line analysis, in which a star's intensity is measured through each of a set of standard filters, is known as **photometry** (literally meaning "light measurement").

TABLE 17.1   Stellar Colors and Temperatures

COLOR INDEX		SURFACE TEMPERATURE (K)	COLOR	FAMILIAR EXAMPLES
B intensity/V intensity	B magnitude – V magnitude			
1.3	−0.31	30,000	electric blue	
1.2	−0.24	20,000	blue	Rigel
1.00	0.00	10,000	white	Vega, Sirius
0.72	0.35	7,000	yellow-white	Canopus
0.55	0.65	6,000	yellow	Sun, Alpha Centauri
0.33	1.2	4,000	orange	Arcturus, Aldebaran
0.21	1.7	3,000	red	Betelgeuse, Barnard's Star

17.6 The Classification of Stars

5 Astronomers can use color and temperature to classify stars reasonably well, but they often use a more detailed classification scheme. This scheme incorporates additional knowledge of stellar physics obtained through spectroscopy, the study of spectral-line radiation.

DETAILED SPECTRA

Figure 17.12 compares the spectra of several different stars, arranged in order of decreasing surface temperature (as determined from measurements of their colors). All the spectra extend from 400 to 700 nm, and each shows a series of dark absorption lines superimposed on a background of continuous color, like those of the Sun. ∞ (Sec. 16.3) However, the precise patterns of lines show many differences. Some stars display strong lines in the long-wavelength part of the spectrum (to the left in Figure 17.10). Other stars have their strongest lines at short wavelengths (to the right). Still others show strong absorption lines spread across the whole visible spectrum. What do these differences tell us?

Although spectral lines of many elements are present with widely varying strengths, the differences among the spectra in Figure 17.12 are not due to differences in composition. Detailed spectral analysis indicates that the seven stars shown have very similar elemental abundances—all are more or less solar in makeup. Instead, the differences are due almost entirely to the stars' *temperatures*. The spectrum at the top of Figure 17.12 is exactly what we would expect from a star with solar composition and a surface temperature of about 30,000 K, the second from a 20,000-K star, and so on, down to the 3000-K star at the bottom.

The spectra of stars with surface temperatures exceeding 25,000 K usually show *strong* absorption lines of singly ionized helium and multiply ionized heavier elements, such as oxygen, nitrogen, and silicon (these latter lines are not shown in the spectra of Figure 17.12). These strong lines are not seen in the spectra of cooler stars because only very hot stars can excite and ionize these tightly bound atoms. In contrast, the hydrogen absorption lines in the spectra of very hot stars are relatively *weak*. The reason is not a lack of hydrogen—it is by far the most abundant element in all stars. At these high temperatures, however, much of the hydrogen is ionized, so there are few intact hydrogen atoms to produce strong spectral lines. What's more, the few hydrogen atoms that have managed to retain their single electron are mostly in such highly excited states that their spectral lines are invisible. ∞ (Sec. 4.4; *More Precisely 4-2*)

In cooler stars, with surface temperatures of around 10,000 K, hydrogen is responsible for the strongest absorption lines. This temperature is just right for electrons to move frequently between hydrogen's second and third orbitals, producing the characteristic red Hα (hydrogen alpha) line at 656.3 nm (see Section 4.2, particularly Figure 4.10). Lines of tightly bound atoms—

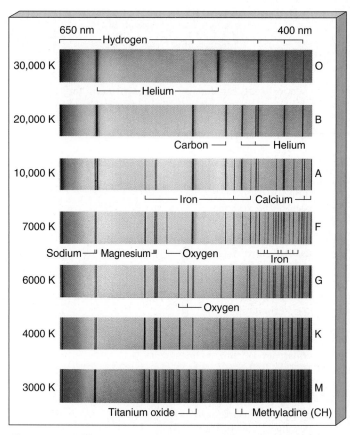

Figure 17.12 Comparison of spectra observed for seven different stars having a range of surface temperatures. The hottest stars, at the top, show lines of helium and multiply ionized heavy elements. In the coolest stars, at the bottom, helium lines are not seen, but lines of neutral atoms and molecules are plentiful. At intermediate temperatures, hydrogen lines are strongest. The actual compositions of all seven stars are about the same.

elements such as helium, oxygen, and nitrogen, which need lots of energy for excitation or ionization—are rarely observed in the spectra of 10,000-K stars, whereas lines from more loosely bound atoms—such as calcium and titanium—are fairly common.

Even cooler stars, with surface temperatures of only a few thousand kelvins, show extremely weak hydrogen lines. As with the hottest stars, cooler stars do not have much electron traffic between the second and higher orbitals in hydrogen atoms. But unlike the hottest stars, in which most hydrogen atoms are highly excited or ionized, the hydrogen lines in cool stars are weak because most of the electrons are in the ground state. The most intense lines are due to weakly excited heavy atoms; no lines from ionized elements are seen. In fact, the average energy of the photons leaving the surface of the coolest stars is less than that needed to destroy some molecules, and many of the observed absorption lines are produced by molecules rather than by atoms (note the lines of the molecules titanium oxide and methyladine at the bottom of Figure 17.12).

TABLE 17.2 Stellar Spectral Classes

SPECTRAL CLASS	SURFACE TEMPERATURE (K)	PROMINENT ABSORPTION LINES	FAMILIAR EXAMPLES
O	30,000	Ionized helium strong; multiply ionized heavy elements; hydrogen faint	
B	20,000	Neutral helium moderate; singly ionized heavy elements; hydrogen moderate	Rigel (B8)
A	10,000	Neutral helium very faint; singly ionized heavy elements; hydrogen strong	Vega (A0), Sirius (A1)
F	7,000	Singly ionized heavy elements; neutral metals; hydrogen moderate	Canopus (F0)
G	6,000	Singly ionized heavy elements; neutral metals; hydrogen relatively faint	Sun (G2), Alpha Centauri (G2)
K	4,000	Singly ionized heavy elements; neutral metals strong; hydrogen faint	Arcturus (K2), Aldebaran (K5)
M	3,000	Neutral atoms strong; molecules moderate; hydrogen very faint	Betelgeuse (M2), Barnard's Star (M5)

SPECTRAL CLASSIFICATION

Stellar spectra like those shown in Figure 17.12 were obtained for many stars well before the start of the twentieth century as observatories around the world amassed spectra from stars in both hemispheres of the sky. Between 1880 and 1920, researchers correctly identified some of the observed spectral lines on the basis of comparisons between stellar lines and those obtained in the laboratory. Those workers, though, had no firm understanding of how the lines were produced. Modern atomic theory had not yet been developed, so the correct interpretation of the line strengths, as just described, was impossible at the time.

Lacking full understanding of how atoms produce spectra, early workers classified stars primarily according to their hydrogen-line intensities. They adopted an A, B, C, D,… scheme in which A stars, with the strongest hydrogen lines, were thought to have more hydrogen than did B stars, and so on. The classification extended as far as the letter P.

In the 1920s, scientists began to understand the intricacies of atomic structure and the causes of spectral lines. Astronomers quickly realized that stars could be more meaningfully classified according to their surface temperature. Instead of adopting an entirely new scheme, however, they chose to shuffle the existing alphabetical categories—those based on the strength of hydrogen lines—into a new sequence based on temperature. In the modern scheme, the hottest stars are designated O, because they have very weak absorption lines of hydrogen and were classified toward the end of the original scheme. In order of decreasing temperature, the original letters now run O, B, A, F, G, K, M. (The other letter classes have been dropped.) These stellar designations are called **spectral classes** (or *spectral types*). Use the time-honored mnemonic "**O**h, **B**e **A** **F**ine **G**uy, **K**iss **M**e" to remember them in the correct order.

Astronomers further subdivide each lettered spectral classification into 10 subdivisions, denoted by the numbers 0–9. By convention, the lower the number, the hotter the star. Thus, for example, our Sun is classified as a G2 star (a little cooler than G1 and a little hotter than G3), Vega is a type A0, Barnard's Star is M5, Betelgeuse is M2, and so on. Table 17.2 lists the main properties of each stellar spectral class for the stars presented in Table 17.1.

We should not underestimate the importance of the early work in classifying stellar spectra. Even though the original classification was based on erroneous assumptions, the painstaking accumulation of large quantities of accurate data paved the way for rapid improvements in understanding once a theory came along that explained the observations.

17.7 The Hertzsprung–Russell Diagram

6 Astronomers use luminosity and surface temperature to classify stars in much the same way that height and weight serve to classify the bulk properties of human beings. We know that people's height and weight are well correlated—tall people tend to weigh more than short ones. We might naturally wonder if the two basic stellar properties are also related in some way.

Figure 17.13 plots luminosity versus temperature for a few well-known stars. Figures of this sort are called *Hertzsprung–Russell* diagrams, or **H–R diagrams**, after Danish astronomer Ejnar Hertzsprung and U.S. astronomer Henry Norris Russell, who independently pioneered the use of such plots in the second decade of the twentieth century. The vertical scale, expressed in units of the solar luminosity

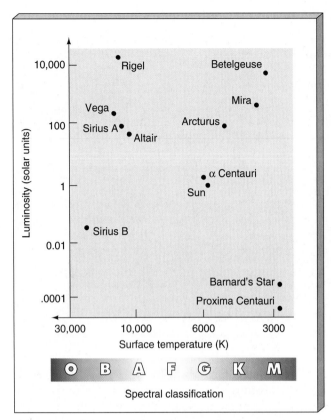

Figure 17.13 A plot of luminosity against surface temperature (or spectral classification), known as an H–R diagram, is a useful way to compare stars. Plotted here are the data for some stars mentioned earlier in the text. The Sun, of course, has a luminosity of 1 solar unit. Its temperature, read off the bottom scale, is 5800 K—a G-type star. Similarly, the B-type star Rigel, at top left, has a temperature of about 15,000 K and a luminosity more than 10,000 times that of the Sun. The M-type star Proxima Centauri, at bottom right, has a temperature of 3000 K and a luminosity less than 1/10,000 that of the Sun.

$(3.9 \times 10^{26}$ W), extends over a large range, from 10^{-4} to 10^4; the Sun appears right in the middle of the luminosity range, at a luminosity of 1. Surface temperature is plotted on the horizontal axis, although in the unconventional sense temperature increasing to the *left* (so that the spectral sequence O, B, A,... reads left to right). To change the horizontal scale so that temperature would increase conventionally to the right would play havoc with historical precedent.

As we have just seen, astronomers often use a star's *color* to measure its temperature. Indeed, the spectral classes plotted along the horizontal axis of spectral class in Figure 17.13 are equivalent to the B/V color index. Also, because astronomers commonly express a star's luminosity as an absolute magnitude, stellar magnitude instead of stellar luminosity could be plotted vertically (see *More Precisely 17-1*). For these reasons, many astronomers refer to diagrams such as Figure 17.13 as **color–magnitude diagrams**. In this book, however, we will cast our discussion strictly in terms of temperature and luminosity measurements.

THE MAIN SEQUENCE

The few stars plotted in Figure 17.13 give little indication of any particular connection between stellar properties. However, as Hertzsprung and Russell plotted more and more stellar temperatures and luminosities they found that a relationship does in fact exist. Stars are *not* uniformly scattered across the H–R diagram. Instead, most are confined to a fairly well defined band stretching diagonally from top left (high-temperature, high-luminosity) to bottom right (low-temperature, low-luminosity). In other words, cool stars tend to be faint (less luminous) and hot stars tend to be bright (more luminous). This band of stars spanning the H–R diagram is known as the **main sequence**.

Figure 17.14 shows a more systematic study of stellar properties, covering the 80 or so stars that lie within 5 pc of the Sun. As more points are included in the diagram the main sequence "fills up," and the pattern becomes more evident. The vast majority of stars in the immediate vicinity of the Sun lie on the main sequence.

The surface temperatures of main sequence stars range from about 3000 K (spectral class M) to over 30,000 K (spectral class O). This relatively small temperature range—only a factor of 10—is determined mainly by the rates at which nuclear reactions occur in stellar cores. In contrast, the observed range in luminosities is very large, covering some eight orders of magnitude (that is, a factor of 100 million) and ranging from 10^{-4} to 10^4 times the luminosity of the Sun.

Astronomers can use the radius–luminosity–temperature relationship ($L \propto R^2 T^4$; Section 17.3) to estimate the radii (R) of main-sequence stars from their temperatures (T) and luminosities (L). They find that in order to account for the observed range in luminosities, stellar radii must also vary along the main sequence. The faint, red M-type stars in the bottom right of the H–R diagram are only about 1/10 the size of the Sun, whereas the bright, blue O-type stars in the upper left are about 10 times larger. The oblique dashed lines in Figure 17.14 represent stars having the same radii (so $L \propto T^4$ along these lines). By including such lines, we can indicate stellar temperatures, luminosities, and radii on a single plot.

We see a very clear trend as we traverse the main sequence from top to bottom. At one end, the stars are large, hot, and bright. Because of their size and color, they are referred to as **blue giants**. The very largest are called **blue supergiants**. At the other end, stars are small, cool, and faint. They are known as **red dwarfs**. Our Sun lies right in the middle.

Figure 17.15 shows an H–R diagram for a different group of stars—the 100 stars of known distance having the greatest apparent brightness, as seen from Earth. Notice the much larger number of very luminous stars at the upper end of the main sequence than at the lower end in Figure 17.15. The reason for this excess of blue giants is simple—we can see very luminous stars a long way off. The stars shown in Figure 17.15 are scattered through a much greater volume of space than those in Figure 17.14, but the sample is heavily biased toward the brightest objects. In fact, of the 20 brightest stars in the sky,

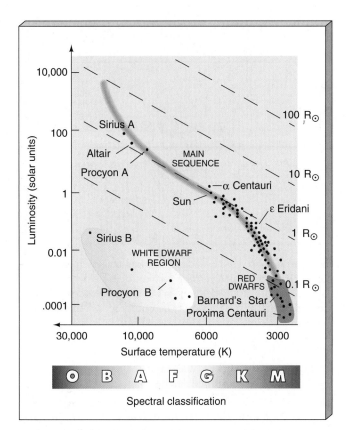

Figure 17.14 Most stars have properties within the shaded region known as the main sequence. The points plotted here are for stars lying within about 5 pc of the Sun. The diagonal lines correspond to constant stellar radius, so that stellar size can be represented on the same diagram as luminosity and temperature. (Recall that ⊙ stands for the Sun.)

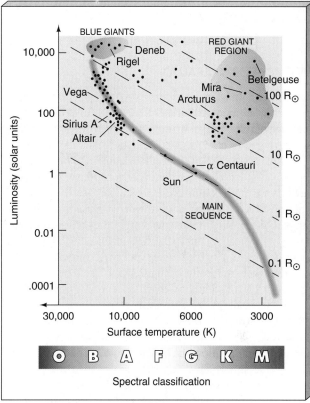

Figure 17.15 An H–R diagram for the 100 brightest stars in the sky. Such a plot is biased in favor of the most luminous stars—which appear toward the upper left—because we can see them more easily than we can the faintest stars. (Compare this with Figure 17.14, which shows only the closest stars.)

only 6 lie within 10 pc of us; the rest are visible, despite their great distances, because of their high luminosities.

If very luminous blue giants are overrepresented in Figure 17.15, then low-luminosity red dwarfs are surely underrepresented. In fact, no dwarfs appear on that diagram. This absence is not surprising, because such low-luminosity stars are very difficult to observe from Earth. They are just too faint, and they radiate much of their energy in the invisible, infrared part of the electromagnetic spectrum. However, in the 1970s, astronomers began to realize that they had greatly underestimated the number of red dwarfs in the Galaxy. As hinted at by the H–R diagram in Figure 17.14, which shows an unbiased sample of stars in the solar neighborhood, red dwarfs are actually the most common type of star in the sky. They probably account for upward of 80 percent of all stars in the universe.

WHITE DWARFS AND RED GIANTS

Most stars lie on the main sequence. However, some of the points plotted in Figures 17.13–17.15 clearly do not. One such point in Figure 17.13 represents Sirius B, a white dwarf whose surface temperature (24,000 K) is about four times that of the Sun and whose luminosity is about 0.04 times

the solar value. A few more such faint A-type stars can be seen in Figure 17.14 in the bottom left-hand corner of the H–R diagram. This region, known as the **white dwarf region**, is marked on Figure 17.14.

Also shown in Figure 17.13 is Mira (Omicron Ceti), whose surface temperature (3000 K), is about half that of the Sun and whose luminosity is some 400 times greater than the Sun's. Another point represents Betelgeuse (Alpha Orionis), the ninth brightest star in the sky, a little cooler than Mira but more than 30 times brighter. The upper right-hand corner of the H–R diagram, where these stars plot (marked on Figure 17.15), is called the **red giant region**. No red giants are found within 5 pc of the Sun (Figure 17.14), but many of the brightest stars seen in the sky are in fact red giants (Figure 17.15). Red giants are relatively rare, but they are so bright that they are visible to very great distances. They form a third distinct class of stars in the H–R diagram, very different in their properties from both main-sequence stars and white dwarfs.

About 90 percent of all stars in our solar neighborhood, and probably a similar percentage elsewhere in the universe, are main-sequence stars. About 9 percent of stars are white dwarfs, and 1 percent are red giants.

17.8 Extending the Cosmic Distance Scale

SPECTROSCOPIC PARALLAX

In Chapter 2 we introduced the first "rung" on a ladder of distance-measurement techniques that will ultimately carry us to the edge of the observable universe. That rung is radar ranging on the inner planets. ∞ (Sec. 2.6) It establishes the scale of the solar system to great accuracy and, in doing so, defines the astronomical unit. Earlier in this chapter we discussed a second rung in the cosmic distance ladder—stellar parallax—which is based on the first. Now, having used the first two rungs to determine the distances and other physical properties of many nearby stars, we can use that knowledge to construct a third rung in the ladder: **spectroscopic parallax**.* As illustrated schematically in Figure 17.16, this new rung expands our cosmic field of view still deeper into space.

We have already discussed the connections between absolute brightness (luminosity), apparent brightness (energy flux), and distance. Knowledge of a star's apparent brightness and distance allows us to determine its luminosity using the inverse-square law. But we can also turn the problem around. If we somehow knew a star's luminosity and then measured its apparent brightness, the inverse-square law would give us its distance from Earth.

Consider an analogy. Most of us have a rough idea of the approximate brightness and size of a red traffic signal. Suppose we are driving down an unfamiliar street and see a red traffic light in the distance. Our knowledge of the intrinsic luminosity of the light often enables us immediately to make a mental estimate of its distance. A normal traffic light that is relatively dim must be quite distant (assuming it's not just dirty); a bright one must be relatively close. *A measurement of the apparent brightness of a light source, combined with some knowledge of its intrinsic properties, can yield an estimate of the source's distance.* For a star, the trick is to find an independent measure of the luminosity without knowing the distance. The H–R diagram can provide just that.

The main sequence represents a fairly close correlation between temperature and luminosity for most stars, with the exception of a few giants and dwarfs. Thus, the main sequence tells us about the *average* properties of most stars. Let's imagine for a moment that the main sequence is a *line*, rather than a somewhat fuzzy band, in the H–R diagram and that *all* stars lie on the main sequence. From a star's spectrum, we can determine its surface temperature or spectral type. If the star lies on the main sequence, then there is only one possible luminosity corresponding to that temperature. We can read the star's luminosity directly off a graph such as Figure 17.14, and then determine its distance

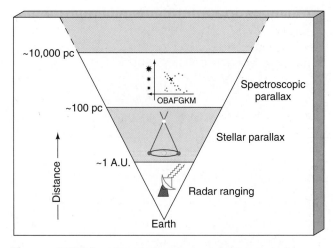

Figure 17.16 Knowledge of a star's luminosity and apparent brightness can yield an estimate of its distance. Astronomers use this third "rung" in our distance ladder, called spectroscopic parallax, to measure distances as far out as individual stars can be clearly discerned—several thousand parsecs.

by measuring the energy flux at Earth and using the inverse-square law. The existence of the main sequence allows us to make a connection between an easily measured quantity (temperature) and the star's luminosity, which would otherwise be unknown. The term *spectroscopic parallax* refers to this whole process of using stellar spectra to infer distances.

Spectroscopic parallax can be used to determine stellar distances out to several thousand parsecs. Beyond that, spectra and colors of individual stars are difficult to obtain. The "standard" main sequence is obtained from H–R diagrams of stars whose distances can be measured by (geometric) parallax, so the method of spectroscopic parallax is calibrated using nearby stars. Note that, in using this method, we are applying the "principle of mediocrity" that we discussed in Chapter 2. ∞ (Interlude 2-2) Specifically, we are assuming (without proof) that distant stars are basically similar to nearby stars, and that *they fall on the same main sequence as nearby stars.* Only by making this assumption can we expand the boundaries of our distance-measurement techniques.

Of course, the main sequence is not really a line in the H–R diagram: it has some thickness. For example, the luminosity of a main-sequence G2-type star (such as the Sun) can range from about 0.5 to 1.5 times the luminosity of the Sun. The main reason for this range is the variation in stellar composition and age from place to place in the Galaxy. As a result, there is an uncertainty in the luminosity obtained by this method and so some uncertainty in the distance. Distances obtained by spectroscopic parallax are probably accurate to no better than 25 percent. Although this may not seem very accurate—a cross-country traveler in the United States would hardly be impressed to be told that the best estimate of the distance between Los Angeles and New York is somewhere between 3000 and 5000 km— it illustrates the point that in astronomy even something as simple as the distance to another star can be very difficult to

---

*This unfortunate name is very misleading, as the method has nothing in common with stellar (geometric) parallax other than its use as a means of determining stellar distances.

measure. Still, an estimate with an uncertainty of ±25 percent is far better than no estimate at all.

LUMINOSITY CLASS

If a star happens to be a red giant or a white dwarf, its distance determined by spectroscopic parallax will be incorrect. We could simply argue that since roughly 90 percent of all stars are on the main sequence, the assumption that a star is a main-sequence star is valid 9 out of 10 times, but, in fact, astronomers can do much better. Recall from Chapter 4 that the *width* of a spectral line can provide information on the *density* of the gas where the line formed. ∞ (Sec. 4.4) The atmosphere of a red giant is much less dense than that of a main-sequence star, and this in turn is much less dense than the atmosphere of a white dwarf. By studying the width of a star's spectral lines, astronomers can usually tell with a high degree of confidence whether or not it is on the main sequence.

Over the years, astronomers have developed a system for classifying stars according to the width of their spectral lines. Because the line width is particularly sensitive to density in the stellar photosphere, and the atmospheric density in turn is well correlated with luminosity, the class in which a star is categorized has come to be known as its **luminosity class**. This classification provides a means for astronomers to distinguish supergiants from giants, giants from main-sequence stars, and main-sequence stars from white dwarfs by studying a single spectral property—the line broadening—of the radiation received.

The standard stellar luminosity classes are given in Table 17.3. Their locations on the H–R diagram are indicated in Figure 17.17. Now we have a way of specifying a star's location in the diagram in terms of properties that are measurable by purely spectroscopic means; spectral type and luminosity class define a star just as surely as do temperature and luminosity. The full specification of a star's spectral properties includes its luminosity class. For example, the Sun, on the main sequence, is of class G2V, Vega is A0V, the red dwarf Barnard's Star is M5V, the red supergiant Betelgeuse is M2Ia, and so on.

Consider, for example, a K7-type star (Table 17.4) with a surface temperature of approximately 4000 K. If the widths of the star's spectral lines tell us that it lies on the main sequence (that is, it is a K7V star), then its luminosity is about

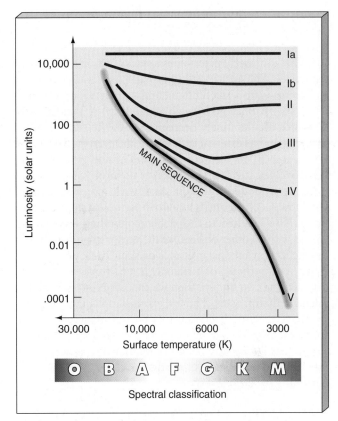

Figure 17.17 Stellar luminosity classes in the H–R diagram. Note that a star's location can be specified by its spectral type and luminosity class instead of by its temperature and luminosity.

0.1 times the solar value. If its spectral lines are observed to be narrower than lines normally found in main-sequence stars, the star may be recognized as a giant, with a luminosity 20 times that of the Sun (Figure 17.17). If the lines are very narrow, the star may instead be classified as a K7Ib supergiant, brighter by a further factor of 150, at 3000 solar luminosities. In this way the observed width of the star's spectral lines translates directly into a measure of the star's physical state. Knowledge of luminosity classes allows us to use spectroscopic parallax with some confidence that we are not accidentally counting a red giant or a white dwarf as a main-sequence star and making a huge error in our distance estimate as a result.

TABLE 17.3 Stellar Luminosity Classes

CLASS	DESCRIPTION
Ia	Bright supergiants
Ib	Supergiants
II	Bright giants
III	Giants
IV	Subgiants
V	Main-sequence stars/dwarfs

TABLE 17.4 Variation in Stellar Properties within a Spectral Class

SURFACE TEMPERATURE (K)	LUMINOSITY (solar luminosities)	RADIUS (solar radii)	OBJECT
4000	0.1	0.7	K7V (main-sequence star)
4000	20	10	K7III (giant)
4000	3000	100	K7Ib (supergiant)

17.9 Stellar Mass

7 What ultimately determines a star's position on the main sequence? The answer is its *mass* and its *composition*. Mass and composition are fundamental properties of any star. They are set once and for all at the time of a star's birth. Together they uniquely determine both the star's internal structure and its external appearance and even (as we will see in Chapter 20) its future evolution. The ability to measure these two key stellar properties is of the utmost importance if we are to understand how stars work. We have already seen how spectroscopy is used to determine a star's composition. Now let's turn to the problem of finding a star's mass.

As with all other objects, we measure a star's mass by observing its gravitational influence on some nearby body—another star, perhaps, or a planet. If we know the distance between the two bodies, then we can use Newton's laws to calculate their masses. The few extrasolar planetary systems that have recently been detected are still not well enough studied to provide independent stellar mass measurements, and we are a long way from placing our own spacecraft in orbit around other stars. Nevertheless, there are ways of determining stellar masses.

BINARY STARS

Most stars are members of *multiple-star systems*—groups of two or more stars in orbit around one another. The majority of stars are found in **binary-star systems**, which consist of two stars in orbit about their common center of mass, held together by their mutual gravitational attraction. Other stars are members of triple, quadruple, or even more complex systems. Most complex of all are the *star clusters*, to be discussed in Section 17.10. The Sun is not part of a multiple-star system; if it has anything at all uncommon about it, it may be its lack of stellar companions.

Astronomers classify binary-star systems (or simply *binaries*) according to their appearance from Earth and the ease with which they can be observed. **Visual binaries** have

Figure 17.18   The periods and separations of binary stars can be observed directly if each star is clearly seen. This binary star is known as *Kruger 60*. At left is a schematic diagram of the orbit, at right actual photographs taken in the years indicated.

widely separated members that are bright enough to be observed and monitored separately, as shown in Figure 17.18. The more common **spectroscopic binaries** are too distant to be resolved into separate stars, but they can be indirectly perceived by monitoring the back-and-forth Doppler shifts of their spectral lines as the stars orbit each other. Recall that motion toward an observer blueshifts the lines, and motion away from the observer redshifts them. ∞ (Sec. 3.5)

In a *double-line* spectroscopic binary, two distinct sets of spectral lines—one for each component star—shift back and forth as the stars move. Because we see particular lines alternately approaching and receding, we know that the objects emitting the lines are in orbit. In the more common *single-line* systems, such as that shown in Figure 17.19, one star is too faint for its spectrum to be distinguished, so only one set of

Figure 17.19   Binary properties can be determined indirectly by measuring the periodic Doppler shift of one star relative to the other as they move in their orbits. The diagram shows a so-called single-line system, in which only one spectrum (from the brighter component) is visible. (The observer is situated to the left of the diagram. Reference lab spectra are shown at top and bottom.)

Figure 17.20 If two stars in a binary system happen to eclipse one another, information on their radii and masses can be obtained by observing the periodic decrease in starlight as one passes in front of the other.

lines is observed to shift back and forth. This shifting means that the detected star must be in orbit around another star, even though the companion cannot be observed directly.

In the much rarer **eclipsing binaries**, the orbital plane of the pair of stars is almost edge-on to our line of sight. In this situation, depicted in Figure 17.20, we observe a periodic decrease of starlight as one component passes in front of the other. By studying the variation of the light from the binary system—the binary's **light curve**—astronomers can derive detailed information not only about the stars' orbits and masses but also about their radii.

These categories of binary-star systems are not mutually exclusive. For example, a single-line spectroscopic binary may also happen to be an eclipsing system. In that case astronomers can use the eclipses to gain extra information about the fainter member of the pair. Occasionally, two unrelated stars just happen to lie close together in the sky, even though they are actually widely separated. These *optical doubles* are just chance superpositions and carry no useful information about stellar properties.

MASS DETERMINATION

By observing the actual orbits of the stars, or the back-and-forth motion of the spectral lines, or the dips in the light curve—whatever information is available—astronomers can measure the binary's orbital period. Observed periods span a broad range—from hours to centuries. How much additional information can be extracted depends on the type of binary involved.

If the distance to a *visual* binary is known, the semimajor axis of its orbit can be determined directly, by simple geometry. Knowledge of the binary period and orbit semimajor axis is all we need to determine the combined mass of the component stars, using the modified form of Kepler's third law. ∞ (Sec. 2.7) Since the orbits of both components can be separately tracked, it is also possible to determine the individual component masses. Recall from Section 2.7 that in any system of orbiting objects, each object orbits the common center of mass. Measuring the distance from each star

to the center of mass of a visual binary yields the ratio of the stellar masses. Knowing both the sum of the masses and the ratio of the masses, we can then find the mass of each star.

For *spectroscopic* binaries, it is not possible to determine the semi-major axis directly. Doppler-shift measurements give us information on the orbital velocities of the member stars, but only on their radial components. As a result, we cannot determine the *inclination* of the orbit to our line of sight, and this imposes a limitation on how much information we can obtain—simply put, we cannot distinguish between a slow-moving binary seen edge-on and a fast-moving binary seen almost face-on (so that only a small component of the orbital motion is along the line of sight).

For a double-line spectroscopic system, individual radial velocities, and hence the ratio of the component masses, can be determined, but the uncertainty in the orbital inclination means that only lower limits on the individual masses can be obtained. For single-line systems, even less information is available, and only a fairly complicated relation between the component masses (known as the *mass function*) can be derived. However, if, as is often the case, the mass of the brighter component can be determined by other means (for example, if it is recognized as a main sequence star of a certain spectral class—see Figure 17.21), a lower limit can then be placed on the mass of the fainter, unseen star.

Finally, if a spectroscopic binary happens also to be an *eclipsing* system, then the uncertainty in the inclination is removed, as the binary is known to be edge-on, or very nearly so. In that case, both masses can be determined for a double-line binary. For a single-line system, the mass function is simplified to the point where the mass of the unseen component is known if the brighter component can be identified.

Despite all these qualifications, individual component masses have been obtained for many nearby binary systems. Virtually all we know about the masses of stars is based on such observations. As a simple example, consider the nearby visual binary system made up of the bright star Sirius A and its faint companion Sirius B. Their orbital period is 50 years and their orbital semi-major axis is 20 A.U.—7.5″ at a distance of 2.7 pc—implying that the sum of their masses is 3.2 times the mass of the Sun. Further study of the orbit shows that Sirius A has roughly twice the mass of its companion. It follows that the masses of Sirius A and Sirius B are 2.1 and 1.1 solar masses, respectively.

DEPENDENCE OF STELLAR PROPERTIES ON MASS

Figure 17.21 is a schematic H–R diagram showing how stellar mass varies along the main sequence. There is a clear progression from low-mass red dwarfs to high-mass blue giants. With few exceptions, main-sequence stars range in mass from about 0.1 to 20 times the mass of the Sun. The hot O- and B-type stars are generally about 10 to 20 times more massive than our Sun. The coolest K- and M-type stars contain only a few tenths of a solar mass. Because all other stellar properties are set once a star's mass is known, we can say that *the mass of a star at the time of formation determines its location on the main sequence.*

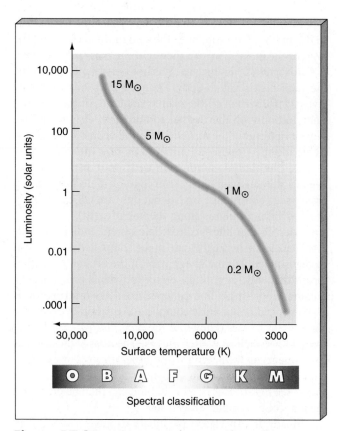

Figure 17.21 Mass, more than any other stellar property, determines a star's position on the main sequence. Stars that form with low mass will be cool and faint; they lie at the bottom of the main sequence. Very massive stars are hot and bright; they lie at the top of the main sequence.

Figure 17.22 illustrates how a main-sequence star's radius and luminosity depend on its mass. The two plots, called the *mass–radius* and *mass–luminosity* relations, are based on observations of binary-star systems. Along the main sequence, both radius and luminosity increase with mass. As a (very rough) rule of thumb, radius increases in direct proportion to mass, whereas luminosity increases much faster—more like the *cube* of the mass. For example, a 2–solar mass main-sequence star has a radius roughly twice that of the Sun and a luminosity of 8 (2^3) solar luminosities; a 0.2–solar mass main-sequence star has a radius of about 0.2 solar radii and a luminosity of 0.008 (0.2^3) solar luminosities.

STELLAR LIFETIMES

The rapid rate of nuclear burning deep inside a star releases vast amounts of energy per unit time. How long can the fire continue to burn? We can estimate a star's lifetime simply by dividing the amount of fuel available (the mass of the star) by the rate at which the fuel is being consumed (the star's luminosity):

$$\text{stellar lifetime} \propto \frac{\text{stellar mass}}{\text{stellar luminosity}}.$$

Because the mass–luminosity relation tells us that a star's luminosity is roughly proportional to the cube of its mass, we can rewrite this expression to obtain, approximately,

$$\text{stellar lifetime} \propto \frac{1}{(\text{stellar mass})^2}.$$

For example, O and B stars have masses 10 to 20 times that of the Sun and luminosities thousands of times higher than the solar luminosity. Accordingly, these massive stars can survive only for short times. Their nuclear reactions proceed so rapidly that their fuel is quickly depleted despite their large masses. From the given proportionalities, we see that the lifetime of a 20–solar mass O star is roughly $20/20^3 = 1/400$ of the (10-billion-year) lifetime of the Sun, or about 25 million years. We can be sure that all the O and B stars we now observe are quite young—less than

(a)

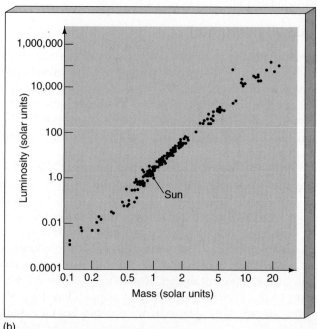

(b)

Figure 17.22 (a) Dependence of stellar radius on mass for main-sequence stars. The radius increases roughly in proportion to the mass over much of the range. (b) Dependence of luminosity on mass. The luminosity increases much faster than the mass.

TABLE 17.5 Key Properties of Some Well-Known Stars

STAR	SPECTRAL TYPE	MASS (solar masses)	CENTRAL TEMPERATURE (10^6 K)	LUMINOSITY (solar luminosities)	ESTIMATED LIFETIME (10^6 years)
Rigel	B8Ia	10	30	44,000	20
Sirius	A1V	2.3	20	23	1,000
Alpha Centauri	G2V	1.1	17	1.4	7,000
Sun	G2V	1.0	15	1.0	10,000
Proxima Centauri	M5V	0.1	5.0	0.00006	>1,000,000

a few tens of millions of years old. Massive stars older than that have already exhausted their fuel and no longer emit large amounts of energy. They have, in effect, died.

At the opposite end of the main sequence, the cooler K- and M-type stars have less mass than our Sun. With their low core densities and temperatures, their proton–proton reactions churn away rather sluggishly, much more slowly than those in the Sun's core. The small energy release per unit time leads to low luminosities for these stars, so they have very long lifetimes. Many of the K- and M-type stars now seen in the sky will shine on for at least another trillion years.

Table 17.5 compares some key properties of several well-known main-sequence stars, arranged in order of decreasing mass. Notice how little the central temperature differs from one star to another, and how large is the spread in stellar luminosities and lifetimes.

17.10 Star Clusters

8 When trying to obtain an H–R diagram for a distant region of the Galaxy, astronomers face a problem. In order to plot the diagram, we must know luminosities; and in order to know luminosities, we must know distances. Thus it would seem impossible to construct H–R diagrams for stars more distant than 100 pc or so, the maximum distance measurable by stellar parallax. (We can't use the method of spectroscopic parallax because that method *assumes* the properties of the main sequence.)

In some circumstances, however, it is possible to plot an H–R diagram for very distant stars even though their distances are not known. If we observe a group of stars that all lie at the same distance from us, then comparing *apparent* brightnesses is equivalent to comparing *absolute* brightnesses. Why? Because as the radiation travels toward Earth, the brightness of every star in the group is diminished by the same amount, according to the inverse-square law. By measuring and plotting apparent brightnesses, we can create a "relative" H–R diagram for the group that looks (apart from the numbers on the vertical axis) exactly the same as a "real" H–R diagram based on

actual luminosities. One such easily recognizable group of distant stars is called a **star cluster**.

Star clusters can include anywhere from a few dozen to a million stars in a region a few parsecs across. Astronomers believe that all the stars in a given cluster formed at the same time, out of the same cloud of interstellar gas, and under the same conditions. Thus, when we look at a star cluster, we are looking at a group of stars that all have the same age, are similar in composition, and lie in the same region of space, at essentially the same distance from Earth. Unlike the stars plotted in Figures 17.13–17.15, which differ in mass, age, and (to a lesser extent) chemical composition, the only factor distinguishing one cluster star from another is its mass.

Clusters are therefore almost ideal "laboratories" for stellar studies—not in the sense that astronomers can perform experiments on the stars in them, but because the properties of the stars are very tightly constrained. Hence theoretical models of star formation and evolution can be compared with reality without the major complications introduced by broad spreads in age, composition, and place of origin. Clusters are of central importance to astronomers who wish to understand how stars evolve in time.

OPEN CLUSTERS

Figure 17.23(a) shows a rather loose cluster—the Pleiades, or Seven Sisters—a well-known naked-eye object in the constellation Taurus. Individual stellar colors provide an estimate of the surface temperature of each star in that cluster. Luminosities follow directly from measurement of the apparent brightness and the cluster's distance (which in this case is known to be about 120 pc). Figure 17.23(b) shows the cluster H–R diagram obtained from these data. This type of cluster, found mainly in the strip across the sky known as the Milky Way, is called an **open cluster**. Open clusters typically contain from a few tens to a few hundred stars and are a few parsecs across.

The H–R diagram in Figure 17.23(b) shows stars throughout the main sequence—stars of all colors are represented. The blue stars must be relatively young, for, as we have seen, they burn their fuel rapidly. If all the stars in the cluster formed at the same time, then the red stars must be young too. Thus, even though we have no direct evidence

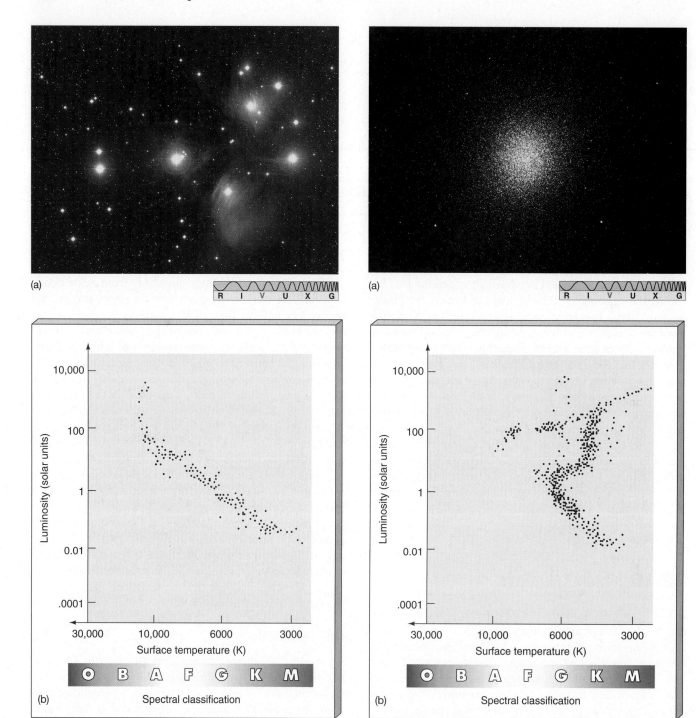

Figure 17.23 (a) The Pleiades cluster (also known as the Seven Sisters, or M45) lies about 120 pc from the Sun. The naked eye can see only its brightest stars. (b) The stars of this well-known open cluster yield an H–R diagram.

Figure 17.24 (a) The globular cluster Omega Centauri is approximately 5000 pc from Earth and spans some 40 pc in diameter. (b) An H–R diagram for many (but not all) of its stars.

of the cluster's birth, we can estimate its age as less than 25 million years, the lifetime of an O star. Other factors also hint at the cluster's youth. It contains a large amount of interstellar gas and dust not yet processed into stars or lost from the cluster. And it is abundant in heavy elements that (as we will see) could have been created only within the cores of many generations of ancient stars long since perished.

GLOBULAR CLUSTERS

A second type of stellar cluster, of which a representative is shown in Figure 17.24(a), is called a **globular cluster**. Globular clusters are much more tightly knit than the loose groups of stars that make up open clusters. All globular clusters are roughly spherical (which accounts for their name) and contain hundreds of thousands, and sometimes millions, of stars spread out over about 50 pc. As with open clusters, the entire assemblage is held together by gravity.

Figure 17.24(b) shows an H–R diagram for this cluster, which is called Omega Centauri. Notice its many differences from Figure 17.23(b)—globular clusters are a very different stellar environment from open clusters like the Pleiades. The distance to this cluster has been determined by a variation on the method of spectroscopic parallax but applied to the entire cluster rather than to individual stars. From calculations of the distance at which the apparent brightnesses of the cluster's stars taken as a whole best match theoretical models, the cluster is found to lie about 5000 pc from Earth.

The most outstanding feature of globular clusters is their lack of O- and B-type stars. Astronomers in the 1920s and 1930s, working with instruments incapable of detecting stars fainter than about 1 solar luminosity at such distances, and having no theory of stellar evolution to guide them, were very puzzled by the H–R diagrams they saw when they looked at globular clusters. Indeed, a comparison of just the top halves of the diagrams (so that the lower main sequences cannot be seen) reveals few similarities between Figures 17.23(b) and 17.24(b).

We now know that, although low-mass red stars and intermediate-mass yellow stars abound, globular clusters contain no main-sequence stars with masses greater than about 0.8 the mass of the Sun. (The A stars in this plot are stars at a much later stage in their evolution that happen to be passing through the location of the upper main sequence.) Apparently, globular clusters formed long ago; the more massive O through F stars have long since exhausted their nuclear fuel and disappeared from the main sequence (becoming the red giants and other luminous stars above the main sequence, as we will discuss in Chapter 20. Other factors confirm that globular clusters are old. For example, their spectra show few heavy elements, implying that these stars formed in the distant past when heavy elements were much less abundant than they are today.

On the basis of these and other observations, astronomers estimate that all globular clusters are at least 10 billion years old. They contain the oldest known stars in the Milky Way Galaxy. As such, globular clusters are considered to be remnants of the earliest stages of our Galaxy's existence.

We will never be able to watch a single star move through all its evolutionary phases. The lifetimes of humans—even of human civilizations—are far too short compared with the lifetimes of even the shortest-lived O and B stars. Instead, we must observe stars as they presently exist—through "snapshots" taken at specific moments in their life cycles. The H–R diagram is just such a snapshot. By studying stars of different ages or, even better, by studying stars in a cluster, in which the ages are known to be the same, we can patch together an understanding of a star's "life story" without having to follow a few individuals from birth to death. Such evolutionary studies will be the subject of the next few chapters.

Chapter Review

SUMMARY

The distances to the nearest stars can be measured by trigonometric parallax. A star with a parallax of one arc second (1) is 1 **parsec** (p. 380)—about 3.3 light years—away.

Stars have real motion through space as well as apparent motion as Earth orbits the Sun. A star's **proper motion** (p. 382), its true motion across the sky, is a measure of the star's velocity perpendicular to our line of sight. The star's radial velocity—along the line of sight—is measured by the Doppler shift of its spectral lines.

Only a few stars are large enough and close enough that their radii can be measured directly. The sizes of most stars are estimated indirectly through the **radius–luminosity–temperature relationship** (p. 383). Stars are categorized as **dwarfs** (p. 385) comparable in size to or smaller than the Sun, **giants** (p. 385) up to 100 times larger than the Sun, and **supergiants** (p. 385) more than 100 times larger than the Sun. In addition to "normal" stars such as the Sun, two other important classes of star are **red giants** (p. 385), which are large, cool, and luminous, and **white dwarfs** (p. 385), which are small, hot, and faint.

The **absolute brightness** (p. 385) of a star is equivalent to its luminosity. The **apparent brightness** (p. 385) of a star is the rate at which energy from the star reaches unit area of a detector. Apparent brightness falls off as the inverse square of the distance. Optical astronomers use the **magnitude scale** (p. 386) to express and compare stellar brightnesses. The greater the magnitude, the fainter the star; a difference of 5 magnitudes corresponds to a factor of 100 in brightness. **Apparent magnitude** (p. 387) is a measure of apparent brightness. The **absolute magnitude** (p. 387) of a star is the apparent magnitude it would have if placed at a standard distance of 10 pc from the viewer. It is a measure of the star's luminosity.

Astronomers often measure the temperatures of stars by measuring their brightnesses through two or more optical filters, then fitting a blackbody curve to the results. The **color index** (p. 389) of a star is the ratio of its brightnesses measured through two standard filters. The measurement of the amount of starlight received through each of a set of filters is called **photometry** (p. 389).

Astronomers classify stars according to the absorption lines in their spectra. The lines seen in the spectrum of a given star depend mainly on its temperature, and spectroscopic observations of stars provide an accurate means of determining both stellar temperatures and stellar composition. The standard stellar **spectral classes** (p. 391), in order of decreasing temperature, are O B A F G K M.

A plot of stellar luminosities versus stellar spectral classes (or temperatures) is called an **H–R diagram** (p. 391), or a **color-magnitude diagram** (p. 392). About 90 percent of all stars plotted on an H–R diagram lie on the **main sequence** (p. 392), which stretches from hot, bright **blue supergiants** (p. 392) and **blue giants** (p. 392), through intermediate stars such as the Sun, to cool, faint **red dwarfs** (p. 392). Most main-sequence stars are red dwarfs; blue giants are quite rare. About 9 percent of stars are in the **white dwarf region** (p. 393), and the remaining 1 percent are in the **red giant region** (p. 393).

By careful spectroscopic observations, astronomers can determine a star's **luminosity class** (p. 395), allowing them to distinguish main-sequence stars from red giants or white dwarfs of the same spectral type (or color). Once a star is known to be on the main sequence, measurement of its spectral type allows its luminosity to be estimated and its distance to be measured. This method of distance determination, which is valid for stars up to several thousand parsecs from Earth, is called **spectroscopic parallax** (p. 394).

Most stars are not isolated in space but instead orbit other stars in **binary-star systems** (p. 396). In a **visual binary** (p. 397), both stars can be seen and their orbit charted. In a **spectroscopic binary** (p. 396), the stars cannot be resolved, but their orbital motion can be detected spectroscopically. In an **eclipsing binary** (p. 397), the orbit is oriented in such a way that one star periodically passes in front of the other as seen from Earth and dims the light we receive. The binary's **light curve** (p. 397) is a plot of its apparent brightness as a function of time.

Studies of binary stars often allow stellar masses to be measured. The mass of a star determines its size, temperature, and brightness. Fairly well defined mass–radius and mass–luminosity relations exist for main-sequence stars. Hot blue giants are much more massive than the Sun; cool red dwarfs are much less massive. The lifetime of a star can be estimated by dividing its mass by its luminosity. High-mass stars burn their fuel rapidly and have much shorter lifetimes than the Sun. Low-mass stars consume their fuel slowly and may remain on the main sequence for trillions of years.

Many stars are found in compact groups known as **star clusters** (p. 399). **Open clusters** (p. 399), with a few hundred to a few thousand stars, are found mostly in the plane of the Milky Way. They typically contain many bright blue stars, indicating that they formed relatively recently. **Globular clusters** (p. 400) are found mainly away from the Milky Way plane and may contain millions of stars. They include no main-sequence stars much more massive than the Sun, indicating that they formed long ago. Globular clusters are believed to date from the formation of our Galaxy.

SELF-TEST: TRUE OR FALSE?

_____ **1.** One parsec is a little over 200,000 A.U.

_____ **2.** There are no stars within 1 pc of the Sun.

_____ **3.** Parallax can be used to measure stellar distances out to about 1000 pc.

_____ **4.** Most stars have radii between 0.1 and 10 times the radius of the Sun.

_____ **5.** Star A appears brighter than star B, as seen from Earth. Therefore, star A must be closer to Earth than star B.

_____ **6.** Star A and star B have the same absolute brightness (luminosity), but star B is twice as distant as star A. Therefore, star A appears four times brighter than star B.

_____ **7.** A magnitude 5 star looks brighter than a magnitude 2 star.

_____ **8.** Differences among stellar spectra are mainly due to differences in composition.

_____ **9.** Cool stars have very strong lines of hydrogen in their spectra.

_____ **10.** A G9 star is cooler than a G5 star.

_____ **11.** Red dwarfs lie in the lower left part of the H-R diagram.

_____ **12.** The brightest stars visible in the night sky are all found in the upper part of the H-R diagram.

_____ **13.** In a spectroscopic binary, the orbital motion of the component stars appears as variations in their radial velocities.

_____ **14.** Once a binary is recognized, it is always possible to determine the masses of both components.

_____ **15.** It is impossible to have a 1-billion-year-old O- or B-type main-sequence star.

SELF-TEST: FILL IN THE BLANK

1. Parallax measurements of the distances to the nearest stars use a baseline of _____.

2. The radial velocity of a star is determined by observing its _____ and using the _____ effect.

3. To determine the true space velocity of a star, its _____ , radial velocity, and _____ must all be measured.

4. The radius of a star can be indirectly determined if the _____ and _____ of the star are known.

5. The smallest stars normally plotted on the H-R diagram are _____.

6. The largest stars normally plotted on the H-R diagram are _____.

7. Observations of stars through B and V filters are used to determine stellar _____.

8. The hottest stars show little evidence of hydrogen in their spectra because hydrogen is mostly _____ at these temperatures.

9. The coolest stars show little evidence of hydrogen in their spectra because hydrogen is mostly _____ at these temperatures.

10. The Sun has a spectral type of _____.

11. The H-R diagram is a plot of _____ on the horizontal scale versus _____ on the vertical scale.

12. The band of stars extending from the top left of the H-R diagram to its bottom right is known as the _____.

13. _____ star systems are important for providing measurements of stellar masses.

14. Going from spectral type O to M along the main sequence, stellar masses _____.

15. The main-sequence lifetimes of high-mass stars are much _____ than the main-sequence lifetimes of low-mass stars.

REVIEW AND DISCUSSION

1. How is parallax used to measure the distances to stars?

2. What is a parsec? Compare it with the astronomical unit.

3. Explain two ways in which a star's real space motion translates into motion observable from Earth.

4. Describe some characteristics of red giant and white dwarf stars.

5. What is the difference between absolute and apparent brightness?

6. How do astronomers measure the temperatures of stars?

7. Briefly describe how stars are classified according to their spectral characteristics.

8. What information is needed to plot a star on the Hertzsprung–Russell diagram?

9. What is the main sequence? What basic property of a star determines where it lies on the main sequence?

10. How are distances determined using spectroscopic parallax?

11. Why does the H–R diagram constructed using the brightest stars differ so much from the diagram constructed using the nearest stars?

12. Which stars are most common in the Galaxy? Why don't we see many of them in H–R diagrams?

13. How can stellar masses be determined by observing binary star systems?

14. High-mass stars start off with much more fuel than low-mass stars. Why don't high-mass stars live longer?

15. Compare and contrast the properties of open star clusters and globular star clusters.

PROBLEMS

1. How far away is the star Spica, whose parallax is 0.013″? What would Spica's parallax be if it were measured from an observatory on Neptune's moon Triton as Neptune orbited the Sun?

2. A star lying 20 pc from the Sun has proper motion of 0.5″/yr. What is its transverse velocity? If the star's spectral lines are observed to be redshifted by 0.01 percent, calculate the magnitude of its three-dimensional velocity relative to the Sun.

3. (a) What is the luminosity of a star having three times the radius of the Sun and a surface temperature of 10,000 K? (b) A certain star has a temperature twice that of the Sun and a luminosity 64 times greater than the solar value. What is its radius, in solar units?

4. Two stars—A and B, of luminosities 0.5 and 4.5 times the luminosity of the Sun, respectively—are observed to have the same apparent brightness. Which one is more distant, and how much farther away is it than the other?

5. Calculate the solar energy flux (energy received per unit area per unit time) at a distance of 10 pc from the Sun.

6. Astronomical objects visible to the naked eye range in apparent brightness from faint sixth-magnitude stars to the Sun, with magnitude −27. What is the range in energy flux corresponding to this magnitude range?

7. A star has apparent magnitude 7.5 and absolute magnitude 2.5. How far away is it?

8. Two stars in an eclipsing spectroscopic binary are observed to have an orbital period of 10 days. Further observations reveal that the orbit is circular, with a separation of 0.5 A.U., and that one star is 1.5 times the mass of the other. What are the masses of the stars?

9. Given that the Sun's lifetime is about 10 billion years, estimate the life expectancy of (a) a 0.2–solar mass, 0.01–solar luminosity red dwarf (b) A 3–solar mass, 30–solar luminosity star, (c) A 10–solar mass, 1000–solar luminosity blue giant.

10. Assuming the mass–luminosity relation shown in Figure 17.22, estimate the mass of the faintest main-sequence star in Omega Centauri that could be observed by (a) a typical 1-m telescope and (b) the *Hubble Space Telescope*. (See also Figure 17.9.)

PROJECTS

1. Every winter, you can find an astronomy lesson in the evening sky. The Winter Circle is an asterism—or pattern of stars—made up six bright stars in five different constellations: Sirius, Rigel, Betelgeuse, Aldebaran, Capella, and Procyon. These stars span nearly the entire range of colors (and therefore temperatures) possible for normal stars. Rigel is a B star. Sirius is an A. Procyon is an F star. Capella is a G star. Aldebaran a K star. Betelgeuse is an M star. The color differences of these stars are easy to see. Why do you suppose there is no O star in the Winter Circle?

2. Summer is a good time to search with binoculars for open star clusters. Open clusters are generally found in the plane of the Galaxy. If you can see the hazy band of the Milky Way arcing across your night sky—in other words, if you are far from city lights and looking at an appropriate time of night and year—

you can simply sweep with your binoculars along the Milky Way. Numerous "clumps" of stars will pop into view. Many will turn out to be open star clusters.

3. Globular star clusters are harder to find. They are intrinsically larger, but they are also much farther away and therefore appear smaller in the sky. The most famous globular cluster visible from the Northern Hemisphere is M13 in the constellation Hercules, visible on spring and summer evenings. This cluster contains half a million or so of the Galaxy's most ancient stars. It may be glimpsed in binoculars as a little ball of light, located about one-third of the way from the star Eta to the star Zeta in the Keystone asterism of the constellation Hercules. Telescopes reveal this cluster as a magnificent, symmetrical grouping of stars.

18 THE INTERSTELLAR MEDIUM

Gas and Dust Among the Stars

LEARNING GOALS

Studying this chapter will enable you to:

1 Summarize the composition and physical properties of the interstellar medium.

2 Describe the characteristics of emission nebulae and explain their significance in the life cycle of stars.

3 Discuss the nature of dark interstellar clouds.

4 Specify the radio techniques used to probe the nature of interstellar matter.

5 Discuss the nature and significance of interstellar molecules.

(Opposite page) This is an infrared image of what may be the most luminous star known, about 7000 pc distant in the direction of the Galactic center. The "Pistol Star," named for the shape of the nebula surrounding it, has up to 10 million times the luminosity of the Sun and is big enough to fill the diameter of Earth's orbit. The nebula itself (shown mostly in red) extends about 1 pc across, and is thought to be energized by this one star. The star is estimated to be very young, about a million years old; it will not last long given its prodigous rate of emission, probably exploding as a supernova in about another million years. The image was taken with the newly installed infrared camera on the *Hubble Space Telescope*; it looks fuzzy because the camera is somewhat out of focus. It is possible that, with better resolution, this object may well be shown to be several smaller stars in a cluster.

Stars and planets are not the only inhabitants of our Galaxy. The space around us harbors invisible matter throughout the dark voids between the stars. The density of this interstellar matter is extremely low—approximately a trillion trillion times less dense than matter in either stars or planets, far more tenuous than the best vacuum attainable on Earth. Only because the volume of interstellar space is so vast does its mass amount to anything at all. So why bother to study this near-perfect vacuum? We do so for three important reasons. First, there is as much mass in the "voids" among the stars as there is in the stars themselves. Second, interstellar space is the region out of which new stars are born. Third, it is the region into which some old stars explode at death. It is one of the most significant crossroads through which matter passes in our universe.

18.1 Interstellar Matter

1 Figure 18.1 shows a large region of space, a much greater expanse of universal real estate than anything we have studied thus far. The bright regions are congregations of innumerable stars, some of whose properties we have just studied in Chapter 17. However, the dark areas are not simply "holes" in the stellar distribution. They are regions of space where *interstellar matter* obscures the light from stars beyond. Their very darkness means that they cannot be easily studied by the optical methods used for stellar matter. There is, quite simply, very little to see!

From Figure 18.1 it is evident that the dark interstellar matter is distributed very unevenly throughout space. In some directions the obscuring matter is largely absent, allowing astronomers to study objects literally billions of parsecs from the Sun. In other directions there are small amounts of interstellar matter, so the obscuration is moderate, preventing us from seeing objects more than a few thousand parsecs away, but still allowing us to study nearby stars. Still other regions are so heavily obscured that starlight from even relatively nearby stars is completely absorbed before reaching Earth. (See also *Interlude 18-1*.)

GAS AND DUST

The matter among the stars is called the **interstellar medium**. It is made up of two components—gas and dust—intermixed throughout all space.

Interstellar gas is made up mainly of individual atoms, of average size 10^{-10} m (0.1 nm) or so. The gas also contains some small molecules, no larger than about 10^{-9} m across. Regions containing such small particles are transparent to nearly all types of electromagnetic radiation, including ultraviolet, visible, infrared, and radio waves. Apart from numerous narrow atomic and molecular absorption lines, gas alone does not block radiation to any great extent.

Interstellar dust is more complex. It consists of clumps of atoms and molecules—not unlike chalk dust and the microscopic particles that make up smoke, soot, or fog. Light from distant stars cannot penetrate the densest accumulations of interstellar dust any more than a car's headlights can illuminate roadside objects in a thick Earth-bound fog. Comparisons of the dimming of starlight in interstellar space (over and above the effects of the inverse-square law) with the scattering of light in terrestrial fog indicate that the typical size of an interstellar dust particle—or **dust grain**—is about 10^{-7} m. The grains are thus comparable in size to the wavelength of visible light, and about 1000 times larger than interstellar gas particles.

Figure 18.1 A wide-angle photograph of a great swath of space, showing regions of brightness (vast fields of stars) as well as regions of darkness (obscuring interstellar matter). The field of view is roughly 30° across.

R I V U X G

The ability of a particle to scatter a beam of light depends on both the size of the particle and the wavelength of the radiation involved (see *More Precisely 7-1*). As a rule of thumb, only particles having diameters comparable to or larger than the wavelength can significantly influence the beam, and the amount of scattering produced by particles of a given size increases with decreasing wavelength. Consequently, dusty regions of interstellar space are transparent to long-wavelength radio and infrared radiation but opaque to shorter-wavelength optical, ultraviolet, and X-ray radiation. This dimming of starlight by interstellar matter is called **extinction**.

Because the interstellar medium is more opaque to short-wavelength radiation than to radiation of longer wavelengths, light from distant stars is preferentially robbed of its higher-frequency ("blue") components. Hence, in addition to being generally diminished in overall brightness, stars also tend to appear redder than they really are, an effect known as **reddening**.

As illustrated in Figure 18.2, extinction and reddening change a star's apparent brightness and color, but absorption lines in the original stellar spectrum are still recognizable in the radiation reaching Earth, so the star's spectral class can be determined. Astronomers can use this fact to study the interstellar medium. From a main-sequence star's spectral class astronomers learn its true luminosity and color. ∞ (Sec. 17.6) They then measure the degree to which the starlight has been affected by extinction and reddening en route to Earth, and this, in turn, allows them to estimate both the numbers and the sizes of interstellar dust particles along the line of sight to the star. By repeating these measurements for stars in many different directions and at many different distances from Earth, astronomers have built up a picture of the distribution and properties of the interstellar medium in the solar neighborhood.

TEMPERATURE AND DENSITY

The temperature of the interstellar gas and dust ranges from a few kelvins to a few hundred kelvins, depending on its proximity to a star or some other source of radiation. Generally, we can take 100 K as an average temperature of a typical dark region of interstellar space. Compare this with 273 K, at which water freezes, and 0 K, at which atomic and molecular motions all but cease. ∞ (*More Precisely 3-1*) Interstellar space is very cold.

Gas and dust are found everywhere in interstellar space—no part of our Galaxy is truly devoid of matter. However, the density of the interstellar medium is extremely low. The gas averages roughly 10^6 atoms per cubic meter—just 1 atom per cubic centimeter—although densities as great as 10^9 atoms/m^3 and as small as 10^4 atoms/m^3 have been found. Matter of such low density is far more tenuous than the best vacuum—about 10^{10} molecules/m^3—ever attained in laboratories on Earth. Interstellar dust is even rarer. On average, there is only 1 dust particle for every trillion or so atoms—just 10^{-6} dust particles per cubic meter, or 1000 per cubic *kilo*meter. Some parts of interstellar space are so thinly populated that harvesting all the gas and dust in a region the size of Earth would yield barely enough matter to make a pair of dice.

How can such fantastically sparse matter diminish light radiation so effectively? The key is size—interstellar space is

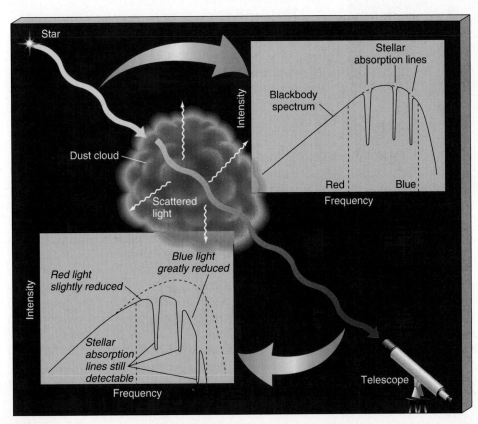

Figure 18.2 Starlight passing through a dusty region of space is both dimmed and reddened, but spectral lines are still recognizable in the light that reaches Earth.

INTERLUDE 18-1 Ultraviolet Astronomy and the "Local Bubble"

The ultraviolet is that part of the spectrum where we expect to witness events and see objects involving temperatures in the hundreds of thousands, even millions, of kelvins—hot regions like the seething atmospheres or eruptive flares of stars, and violent events, such as massive stars caught in the act of exploding, and active galaxies whose spinning hearts may harbor black holes. Yet ultraviolet astronomy has also contributed greatly to the study of the interstellar medium, by allowing a unique mapping of our local cosmic neighborhood. The line-of-sight technique for analyzing interstellar clouds, described in the text for optical observations, can also be applied to ultraviolet radiation. Astronomers search the ultraviolet spectra of distant, hot stars for narrow absorption lines caused by intervening interstellar clouds.

Short-wavelength ultraviolet radiation is effectively blocked by Earth's atmosphere. To study this region of the electromagnetic spectrum, instruments must be placed above our planet's atmospheric blanket. ∞ (Sec. 5.6) One of the most successful satellites yet launched—placed in orbit in 1978 and finally shut down in late 1996—is the *International Ultraviolet Explorer (IUE)*. More recently, the *Extreme Ultraviolet Explorer (EUVE)*, launched in 1992 and also still operational, has closed one of the few remaining gaps in the electromagnetic spectrum not well explored by other means. The photograph below shows the Science Operations Center at the University of California at Berkeley, where undergraduate students work alongside scientists and engineers to command the spacecraft and acquire its data.

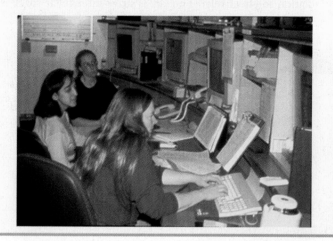

Only 25 years ago, astronomers had presumed that the gases filling the spaces among the stars would absorb virtually all short-wavelength UV radiation before it had a chance to reach Earth. But in 1975, during a historic linkup of the *Apollo* (U.S.) and *Soyuz* (USSR) space capsules, the onboard astronauts and cosmonauts performed a key experiment—they used a small telescope to detect extreme ultraviolet radiation from a few nearby, very hot stars. Shortly thereafter, theoretical ideas changed as astronomers began to realize that the interstellar gas is not all uniformly distributed in space. Rather, it is spread very unevenly, in cool dense clumps interspersed by regions of hot, low-density gas shaped like bubbles and tunnels.

Observations made by *IUE* of weak spectral lines from highly excited atoms showed that some regions of interstellar space are much thinner (5000 atoms/m³) and hotter (500,000 K) than previously expected. Part of the space among the dust clouds and the emission nebulae seems to contain extremely dilute yet seething plasma, probably the result of the concussion and expanding debris from stars that exploded long ago. Observations by *EUVE* and other space-based instruments have found that these superheated interstellar "bubbles," or *intercloud medium*, may extend far into interstellar space beyond our local neighborhood, and conceivably into the even vaster spaces among the galaxies.

The Sun seems to reside in one such low-density region—a huge cavity called the *Local Bubble*. Only because we live in this vast, low-density bubble can we detect so many stars in the extreme UV; the hot, tenuous interstellar gas is virtually transparent to this radiation.

The Local Bubble contains about 200,000 stars and extends for hundreds of trillions of kilometers, or nearly 100 pc. It was probably carved out by a supernova explosion that occurred in these parts roughly 100,000 years ago. ∞ (Sec. 21.3) Our hominid ancestors must have seen it—a stellar catastrophe as bright as the full Moon. But since it is no longer there for us to see, modern astronomers have given this supernova remnant the name Geminga, from an expression in Latin dialect meaning "it's not there." Very recently, however, pulsating X rays and gamma rays have been detected from a local pulsar—the dying core of a once-massive star that exploded nearby. ∞ (Sec. 22.2) Apparently, Geminga has at last been found.

---

vast. The typical distance between stars (1 pc or so in the vicinity of the Sun) is much, much greater than the typical size of the stars themselves (around 10^{-7} pc). Stellar and planetary sizes pale in comparison to the vastness of interstellar space. Thus, matter can accumulate, regardless of how thinly spread. For example, an imaginary cylinder 1 m² in cross section and extending from Earth to Alpha Centauri would contain more than 10 billion dust particles. Over huge distances, dust particles accumulate slowly but surely, to the point at which they can effectively block visible light and other short-wavelength radiation. Even though the density of matter there is very low,

interstellar space in the vicinity of the Sun contains about as much mass as exists in the form of stars.

Despite their rarity, dust particles make interstellar space a *relatively* dirty place. Earth's atmosphere, by comparison, is about a million times cleaner. Our air is tainted by only one dust particle for about every billion billion (10^{18}) atoms of atmospheric gas. If we could compress a typical parcel of interstellar space to equal the density of air on Earth, this parcel would contain enough dust to make a fog so thick that we would be unable to see our hand held at arm's length in front of us.

COMPOSITION

The composition of interstellar gas is reasonably well understood from spectroscopic studies of interstellar absorption lines formed when light from a distant star interacts with gas along the line of sight. ∞ (Sec. 4.2) The gas absorbs some of the stellar radiation in a manner that depends on its own temperature, density, and elemental abundance. The absorption lines thus produced contain information about dark interstellar matter, just as stellar absorption lines reveal the properties of stars. Because the interstellar absorption lines are produced by cold, low-density gas, astronomers can easily distinguish them from the much broader absorption lines formed in the star's hot lower atmosphere (see Section 18.3). ∞ (Sec. 4.4)

In most cases, the elemental abundances detected in interstellar gas mirror those found in other astronomical objects, such as the Sun, the stars, and the jovian planets. Most of the gas—about 90 percent—is atomic or molecular hydrogen; some 9 percent is helium, and the remaining 1 percent consists of heavier elements. The abundances of several of the heavy elements, such as carbon, oxygen, silicon, magnesium, and iron, are much lower in interstellar gas than in our solar system or in stars. The most likely explanation for this finding is that substantial quantities of these elements have been used to form the interstellar dust, taking them out of the gas and locking them up in a form that is much harder to observe.

In contrast to interstellar gas, the composition of interstellar dust is currently not very well known. We have some infrared evidence for silicates, graphite, and iron—the same elements that are underabundant in the gas—lending support to the theory that interstellar dust forms out of interstellar gas. The dust probably also contains some "dirty ice," a frozen mixture of ordinary water ice contaminated with trace amounts of ammonia, methane, and other chemical compounds. This composition is quite similar to that of cometary nuclei in our own solar system. ∞ (Sec. 14.2)

DUST SHAPE

Curiously, astronomers know the *shape* of interstellar dust particles better than their composition. Although the minute atoms in the interstellar gas are basically spherical, the dust particles are not. Individual dust grains are apparently elongated or rodlike, as shown in Figure 18.3. We can infer this because the light emitted by stars is dimmed and partially **polarized**, or aligned, by the dust.

Recall from Chapter 3 that light consists of electromagnetic waves composed of vibrating electric and magnetic fields. ∞ (Sec. 3.2) Normally, these waves are randomly oriented, and we say the radiation is *unpolarized*. Stars emit unpolarized radiation from their photospheres. Under some circumstances, however, the electric fields can become aligned—all vibrating in the same plane as the radiation moves through space. We then say the radiation is *polarized*. Polarization of starlight does not occur by chance.

|← 10⁻⁷ m →|

Figure 18.3   A diagram of a typical interstellar dust particle. The average size of such particles is only 1/10,000 of a millimeter, yet space contains enough of them to obscure our view in certain directions.

If the light detected by our telescope is polarized, it is because some interstellar matter lies between the emitting object and Earth. The polarization of starlight, then, provides another way to study the interstellar medium.

On Earth we can produce polarized light by passing unpolarized light through a Polaroid filter, which has specially aligned elongated molecules that allow the passage of only those waves having electric fields oriented in some specific direction (see Figure 18.4a). Other waves are absorbed and so do not pass through the filter. The alignment of the molecules determines which waves will be transmitted. In interstellar space, dust grains can act like the molecules in the Polaroid filter. If the starlight is polarized, astronomers can conclude that the interstellar dust particles must have an *elongated* shape (by analogy with the elongated molecules of the Polaroid filter) and that these molecules are aligned, as shown in Figure 18.4(b). Only then can the dust preferentially absorb certain waves, leaving the remainder (the ones we observe) polarized.

The alignment of the interstellar dust is the subject of intense research among astronomers. The current view, accepted by most, holds that the dust particles are affected by a weak interstellar magnetic field, perhaps a million times weaker than Earth's field. Each dust particle responds to the field in much the same way that small iron filings are aligned by an ordinary bar magnet. Measurements of the blockage and polarization of starlight thus yield information about the size and shape of interstellar dust particles, as well as about magnetic fields in interstellar space.

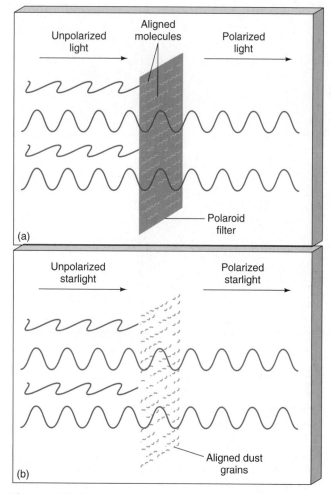

Figure 18.4 (a) Unpolarized light waves have randomly oriented electric fields. When the light passes through a Polaroid filter, only waves whose electric fields are oriented in a specific direction are transmitted, and the resulting light is polarized. (b) Aligned dust particles in interstellar space polarize radiation in a similar manner. Observations of the degree of polarization allow astronomers to infer the size, shape, and orientation of the particles.

18.2 Emission Nebulae

2 Figure 18.5 is a mosaic of photographs showing a region of the sky even larger than that shown in Figure 18.1. The bright areas are made up of myriad unresolved stars. The dark areas are vast pockets of dust, blocking from our view what would otherwise be a rather smooth distribution of bright starlight. The patchiness of the obscuration is plainly evident. From our vantage point on Earth this assemblage of stellar and interstellar matter stretches all the way across the sky. On a clear night it is visible to the naked eye as the Milky Way. In Chapter 23 we will come to recognize this band as the flattened disk, or *plane*, of our own Galaxy.

Figure 18.6 shows a 12°-wide swath of the galactic plane in the general direction of the constellation Sagittarius, as photographed from Earth. The field of view is mottled with stars and interstellar matter. In addition, several large fuzzy patches of light are clearly visible. These fuzzy objects, labeled M8, M16, M17, and M20, correspond to the eighth, sixteenth, seventeenth, and twentieth objects in a catalog compiled by Charles Messier, an eighteenth-century French astronomer.* Today they are known as **emission nebulae**—glowing clouds of hot interstellar matter.

OBSERVATIONS OF EMISSION NEBULAE

Historically, astronomers used the term **nebula** to refer to any "fuzzy" patch (bright or dark) on the sky—any region of space that was clearly distinguishable through a telescope but not sharply defined, unlike a star or a planet. We now know that many (though not all) nebulae are clouds of interstellar dust and gas. If they happen to obscure stars lying behind

---

*Messier was actually more concerned with making a list of celestial objects that might possibly be confused with comets, his main astronomical interest. However, the catalog of 109 "Messier objects" is now regarded as a much more important contribution to astronomy than any comets Messier discovered.

Figure 18.5 A mosaic of the plane of the Milky Way Galaxy. Photographed almost from horizon to horizon, and thus extending over nearly 180°, this band contains high concentrations of stars as well as interstellar gas and dust. The field of view is several times wider than that of Figure 18.1, whose outline is superimposed on this image.

them, we see a dark patch. If something within the cloud—a group of hot young stars, for example—causes it to glow, the nebula appears bright instead. The method of spectroscopic parallax applied to stars visible within the emission nebulae shown in Figure 18.6 indicates that their distances from Earth range from 900 pc (M20) to 1800 pc (M16). ∞ (Sec. 17.8) Thus, all four are near the limit of visibility for any object embedded in the dusty Galactic plane. M16, at the top left, is approximately 1000 pc from M20, near the bottom.

We can gain a better appreciation of these nebulae by examining progressively smaller fields of view. Figure 18.7 is an enlargement of the region near the bottom of Figure 18.6, showing M20 at the top and M8 at the bottom, only a few degrees away. Figure 18.8 is another enlargement of the top of Figure 18.7, presenting a close-up of M20 and its immediate environment. The total area of the close-up view displayed

Figure 18.6 A photograph of a small portion (about 12° across) of the Galactic plane shown in Figure 18.1, displaying higher-resolution evidence for stars, gas, and dust as well as several distinct fuzzy patches of light, known as emission nebulae. The plane of the Milky Way Galaxy is marked with a dashed line.

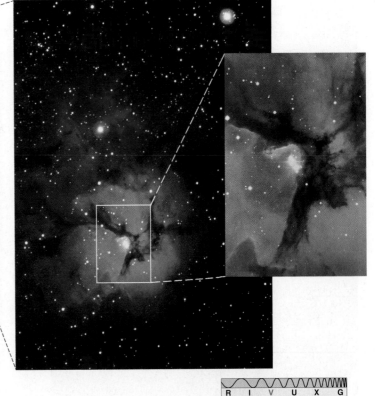

Figure 18.7 An enlargement of the bottom of Figure 18.6, showing M20 (top) and M8 (bottom) more clearly.

Figure 18.8 Further enlargements of the top of Figure 18.7, showing only M20 and its interstellar environment. The nebula itself (in red) is about 4 pc in diameter. It is often called the Trifid Nebula because of the dust lanes that trisect its midsection (insert). The blue region is unrelated to the red emission nebula and is caused by starlight reflected from intervening dust particles. It is called a *reflection nebula*.

measures some 10 pc across. Emission nebulae are among the most spectacular objects in the entire universe, yet they appear only as small, undistinguished patches of light when viewed in the larger context of the Milky Way, as in Figure 18.5. Perspective is crucial in astronomy.

These nebulae are regions of glowing, ionized gas. At or near the center of each is at least one newly formed hot O or B star producing copious amounts of ultraviolet light. As ultraviolet photons travel outward from the star they ionize the surrounding gas. As electrons recombine with nuclei they emit visible radiation, causing the gas to glow. Figure 18.9 shows enlargements of two of the other nebulae visible in Figure 18.6. Notice the predominant red coloration of the emitted radiation and the hot bright stars embedded within the glowing nebular gas.

The interaction between stars and gas is particularly striking in Figure 18.9(b). The three dark pillars visible in this spectacular *Hubble Space Telescope* image are part of the interstellar cloud from which the stars formed. The rest of the cloud in the vicinity of the new stars has already been dispersed by their radiation, in a process known as *photoevaporation*; the fuzz around the edges of the pillars, especially at top right and center, is the result of this process. As photoevaporation continues, it eats away the less dense material first, leaving behind delicate sculptures composed of the denser parts of the original cloud, just as wind and water create spectacular structures in Earth's deserts and shores by eroding away the softest rock. The process is a dynamic one—the pillars will eventually be destroyed, but probably not for another hundred thousand or so years.

(a)
(b)
(c)
(d)

R I V U X G

Video *Animation*

Figure 18.9 Enlargements of selected portions of Figure 18.8. (a) M16, the Eagle Nebula. (b) A *Hubble* image of huge columns of cold gas and dust inside M16, delicate sculptures created by the action of stellar ultraviolet radiation on the original cloud. In this image, green represents emission from hydrogen atoms, red, emission from singly ionized sulfur, and blue, emission from doubly ionized oxygen. (c) M8, the Lagoon Nebula. (d) A high-resolution view of the core of M8, a region known as the Hourglass. Notice in all cases the irregular shape of the emitting regions, the characteristic red color of the light, the bright stars within the gas, and the patches of obscuring dust.

The reddish hue of these nebulae—and in fact of all emission nebula—results when hydrogen atoms emit light in the red part of the visible spectrum. Specifically, it is caused by the emission of radiation at 656.3 nm—the Hα line discussed in Chapter 4. ∞ (Sec. 4.2) Other elements in the nebula also emit radiation as their electrons recombine, but because hydrogen is so plentiful, its emission usually dominates.

Woven through the glowing nebular gas, and plainly visible in Figures 18.7–18.9, are lanes of dark obscuring dust. Recent studies have demonstrated that these **dust lanes** are part of the nebulae and are not just unrelated dust clouds that happen to lie along our line of sight. This relationship is particularly evident in Figures 18.9(b) and (d), where regions of gas and dust are simultaneously silhouetted against background nebular emission and illuminated by foreground nebular stars.

Spectroscopists often refer to the *ionization state* of an atom by attaching a roman numeral to its chemical symbol—I for the neutral atom, II for a singly ionized atom (missing one electron), III for a doubly ionized atom (missing two electrons), and so on. Because emission nebulae are composed mainly of ionized hydrogen, they are often referred to as *HII regions*. Regions of space containing primarily neutral (atomic) hydrogen are known as *HI regions*.

NEBULAR SPECTRA

Most of the photons emitted by the recombination of electrons with atomic nuclei escape from the nebula. Unlike the ultraviolet photons originally emitted by the embedded stars, they do not carry enough energy to ionize the nebular gas, and they pass through the nebula relatively unhindered. Some eventually reach Earth, where we can detect them. Only through these lower-energy photons do we learn anything about emission nebulae.

Nebular spectra tell us a great deal about ionized interstellar gas. Because at least one hot star resides near the center of the nebula, we might think that the combined spectrum of the star and the nebula would be hopelessly confused. In fact, they are not. We can easily distinguish nebular spectra from stellar spectra because the physical conditions in stars and nebulae differ so greatly. In particular, emission nebulae are made of hot thin gas that, as we saw in Chapter 4, yields detectable *emission* lines. ∞ (Sec. 4.2) When our spectroscope is trained on a star, we see a familiar stellar spectrum, consisting of a blackbody-like continuous spectrum and absorption lines, together with superimposed emission lines from the nebular gas. When no star appears in the field of view, only the emission lines are seen.

Figure 18.10(b) is a typical nebular emission spectrum spanning part of the visible and near-ultraviolet wavelength interval. Numerous emission lines can be seen, and information on the nebula shown in Figure 18.10a can be extracted from all of them. The results of analyses of many nebular spectra show abundances close to those derived from observations of the Sun and other stars and elsewhere in the interstellar medium: hydrogen is about 90 percent abundant by number, followed by helium at about 9 percent; the heavier elements together make up the remaining 1 percent.

Unlike stars, nebulae are large enough for their actual sizes to be measurable by simple geometry. Coupling this size information with estimates of the amount of matter along our line of sight (as revealed by the nebula's total emission of light), we can find the nebula's density. Generally, emission nebulae have only a few hundred particles, mostly protons and electrons, in each cubic centimeter—a density some 10^{22} times lower than that of a typical planet. Spectral-line widths imply that the gas atoms and ions have temperatures around 8000 K. ∞ (Sec. 4.4) Table 18.1 lists some vital statistics for each of the nebulae shown in Figure 18.6.

(a)

(b)

Figure 18.10   (a) The Butterfly Nebula, a glowing patch of gas a few parsecs across. (b) Its emission spectrum, showing light intensity over the entire visible portion of the electromagnetic spectrum from red to deep violet.

TABLE 18.1 Some Nebular Properties

OBJECT	APPROX. DISTANCE (pc)	AVERAGE DIAMETER (pc)	DENSITY (10⁶ particles/m³)	MASS (solar masses)	TEMPERATURE (K)
M8	1200	14	80	2600	7500
M16	1800	8	90	600	8000
M17	1500	7	120	500	8700
M20	900	4	100	150	8200

"FORBIDDEN" LINES

When astronomers first studied the spectra of emission nebulae, they found many lines that did not correspond to anything observed in terrestrial laboratories. For example, in addition to the dominant red coloration just discussed, many nebulae also emit light with a characteristic green color (see Figure 18.11). The greenish tint of portions of this nebula puzzled astronomers in the early twentieth century and defied explanation in terms of the spectral lines known at the time, prompting speculation that the nebulae contained elements unknown on Earth. Some scientists even went so far as to invent the term "nebulium" for a new element,

much as the name helium came about when that element was first discovered in the Sun (recall also "coronium" from Chapter 16). ∞ (Sec. 16.3)

Later, with a fuller understanding of the workings of the atom, astronomers realized that these lines did in fact result from electron transitions within the atoms of familiar elements, but occurring under unfamiliar conditions not reproducible in laboratories. Astronomers now understand that the greenish tint in Figure 18.11(b) and (c) is caused by a particular electron transition in doubly ionized oxygen. However, the structure of oxygen is such that an ion in the higher-energy state for this transition tends to remain there for a very long time—many hours, in fact—before dropping back to the lower state and emitting a photon. Only if the ion is left undisturbed during this time, and not kicked into another energy state, will the transition actually occur and the photon be emitted.

In a terrestrial experiment no atom or ion is left undisturbed for long. Even in a "low-density" laboratory gas, there are many trillions of particles per cubic meter, and each particle experiences millions of collisions every second. The result is that an ion in the particular energy state that produces the peculiar green line in the nebular spectrum never has time to emit its photon in the lab—collisions kick it into some other state long before that occurs. For this reason the line is usually called **forbidden**, even

(a) (b) (c)

R I V U X G

Video

Figure 18.11 (a) The Orion Nebula (M42) lies some 450 pc from Earth. It is visible to the naked eye as the fuzzy middle "star" of Orion's sword. (b) Like all emission nebulae, it consists of hot, glowing gas powered by a group of bright stars in the center. In addition to the red Hα emission, parts of the nebula show a slight greenish tint, caused by a so-called forbidden transition in ionized oxygen. (c) A high-resolution, approximately true-color image shows rich detail in a region about 0.5 light year across. Structural details are visible down to a level of 0.1", or 6 light *hours*—a scale comparable to our solar system.

though it violates no laws of physics. It simply occurs on Earth with such low probability that it is never seen.

In an emission nebula the density is so low that collisions between particles are very rare. There is plenty of time for the excited ion to emit its photon, so the forbidden line is produced. Numerous forbidden lines are known in nebular spectra. They remind us once again that the environment in the interstellar medium is very different from conditions on Earth and warn us of the problems of extending our terrestrial experience to the study of interstellar space.

18.3 Dark Dust Clouds

3 Emission nebulae are only one small component of interstellar space. Most of space—in fact, more than 99 percent of it—is devoid of nebular regions and contains no stars. It is simply dark. Look again at Figure 18.5, or just ponder the evening sky. The dark regions are by far the most representative regions of interstellar space.

Within these dark voids among the nebulae and the stars lurks another type of astronomical object—the **dark dust cloud**. These clouds are cooler than their surroundings (with temperatures as low as a few tens of kelvins), and thousands or even millions of times denser. Along any given line of sight, cloud densities can range from 10^7 atoms/m^3 to more than 10^{12} atoms/m^3 (10^6 atoms/cm^3). These latter clouds are generally called *dense* interstellar

clouds by researchers, but even these densest interstellar regions are about as tenuous as the best laboratory vacuum. Still, it is because their density is much larger than the average value of 10^6 atoms/m^3 that we can distinguish clouds from the surrounding expanse of interstellar space.

These clouds bear little resemblance to terrestrial clouds. Most of them are bigger than our solar system, and some are many parsecs across. (Yet even so, they make up no more than a few percent of the entire volume of interstellar space.) Despite their name, these clouds are made up primarily of gas, just like the rest of the interstellar medium. However, their absorption of starlight is due almost entirely to the dust they contain.

OBSCURATION OF VISIBLE LIGHT

Figure 18.12(a) is an optical photograph of a typical interstellar dust cloud. Pockets of intense blackness mark regions where the dust and gas are especially concentrated and the light from background stars is completely obscured. This cloud takes its name from a nearby star, Rho Ophiuchi, and resides at the relatively nearby distance of about 300 pc. Measuring several parsecs across, this cloud is only a tiny part of the grand mosaic shown in Figure 18.5. Note especially the long "streamers" of (relatively) dense dust and gas. This cloud clearly is far from spherical. Indeed, most interstellar clouds are very irregularly shaped. The bright patches within the dark region are emission nebulae in the foreground. Some of them are part of the cloud itself, where newly formed stars near the surface have created a "hot spot"

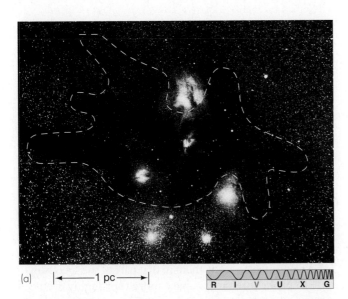

(a) |←——1 pc——→| R I V U X G

(b)   R I V U X G

Figure 18.12   (a) A typical dark dust cloud, known as Rho Ophiuchi, is "visible" only because it blocks the light coming from stars lying behind it. The approximate outline of the cloud is indicated by the dashed line. (b) An infrared map of the same region, to roughly the same scale. The infrared emission, and therefore the dust that produces it, displays a structure similar to the cloud's visual image. The very bright source of infrared radiation near the top of the cloud comes from a hot emission nebula, which can also be seen in the optical image. (The black diagonal streak at right is an instrumental effect.)

Figure 18.13 The Horsehead Nebula in Orion is a striking example of a dark dust cloud, silhouetted against the bright background of an emission nebula. The "neck" of the horse is about 0.25 pc across. The nebular region is roughly 1500 pc from Earth.

in the cold, dark gas. Others have no connection to the cloud and just happen to lie along our line of sight.

Like all dark dust clouds, the Rho Ophiuchi cloud is too cold to emit any visible light. However, it does radiate strongly at longer wavelengths. Figure 18.12(b) shows an infrared view of the same region, captured by sensitive detectors aboard the *Infrared Astronomy Satellite*. ∞ (Sec. 5.6)

These dark and dusty interstellar clouds are sprinkled throughout our Galaxy. We can study them at optical wavelengths only if they happen to block the light emitted by more distant stars or nebulae. The dark outline of Rho Ophiuchi and the dust lanes visible in Figures 18.8 and 18.9 are good examples of this obscuration. The dust is apparent only because it blocks the light coming from behind it. Figure 18.13 shows another striking example of a dark cloud—the Horsehead Nebula in Orion. This curiously shaped finger of gas and dust projects out from the much larger dark cloud in the bottom half of the image and stands out clearly against the red glow of a background emission nebula.

ABSORPTION SPECTRA

Astronomers first became aware of the true extent of dark interstellar clouds in the 1930s as they studied the optical spectra of distant stars. In addition to the wide absorption lines normally formed in stars' lower atmospheres, much narrower absorption lines were also detected. Figure 18.14(a) illustrates how light from a star may pass through several interstellar clouds on its way to Earth. These clouds need not be close to the star, and indeed they usually are not. Each absorbs some of the stellar radiation in a manner that depends on its own temperature, density, velocity, and elemental abundance. Figure 18.14(b) depicts part of a typical spectrum produced in this way.

The narrow absorption lines contain information about dark interstellar clouds, just as stellar absorption lines reveal the properties of stars, and nebular emission lines tell us about conditions in hot nebulae. By studying these lines, astronomers can probe the cold depths of interstellar space. In most cases the elemental abundances detected in interstellar clouds mirror those found in other astronomical objects—perhaps not surprising, since interstellar clouds are the regions that spawn emission nebulae and stars.

18.4 21-Centimeter Radiation

4 A basic problem with the optical technique just described is that we can examine interstellar clouds only along the line of sight to a distant star. To form an absorption line, there has to be a background source of radiation to absorb. The need to see stars through clouds also restricts this approach to relatively local regions, within a few thousand parsecs of Earth. Beyond that distance, stars are completely obscured, and optical observations are impossible. As we have seen, infrared observations provide a means of viewing the emission from some clouds, but they do not completely solve the problem, as only the denser, dustier clouds emit enough infrared radiation for astronomers to study them in that part of the spectrum.

To probe interstellar space more thoroughly, we need a more general, more versatile observational method—one that does not rely on conveniently located stars and nebulae. In short, we need a way to detect cold, neutral interstellar matter anywhere in space through its *own* radiation. This may sound impossible, but such an observational technique does in fact exist. The method relies on low-energy *radio* emissions produced by the interstellar gas itself.

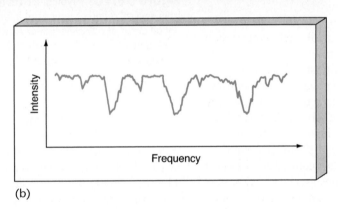

(b)

Figure 18.14 (a) A simplified diagram of some interstellar clouds between a hot star and Earth. Optical observations might show an absorption spectrum like that traced in (b). The wide, intense lines are formed in the star's hot atmosphere; narrower, weaker lines arise from the cold interstellar clouds. The smaller the cloud, the weaker the lines. The redshifts or blueshifts of the narrow absorption lines provide information on cloud velocities. The widths of all the spectral lines depicted here are greatly exaggerated for the sake of clarity.

Recall that a hydrogen atom has one electron orbiting a single-proton nucleus. Besides its orbital motion around the central proton, the electron also has some rotational motion—that is, *spin*—about its own axis. The proton also spins. This model parallels a planetary system, in which, in addition to the orbital motion of a planet about a central star, both the planet (electron) and the star (proton) rotate about their own axes.

The laws of physics dictate that there are exactly two possible spin configurations for a hydrogen atom in its ground state. The electron and proton can rotate in the same direction, with their spin axes parallel, or they can rotate with their axes antiparallel (that is, parallel, but oppositely oriented). Figure 18.15 shows these two configurations. The antiparallel configuration has slightly less energy than the parallel state.

All matter in the universe tends to achieve its lowest possible energy state, and interstellar gas is no exception. A slightly excited hydrogen atom with the electron and proton spinning in the same direction eventually drops down to the less energetic, opposite-spin state as the electron suddenly and spontaneously reverses its spin. As with any other such change, the transition from a high-energy state to a low-energy state releases a photon with energy equal to the energy difference between the two levels.

Because the energy difference between the two states is very small, the energy of the emitted photon is very low. Consequently, the wavelength of the radiation is rather long—in fact, about 21 cm, roughly the width of this book. That wavelength lies in the radio portion of the electromagnetic spectrum. Researchers refer to the spectral line that results from this hydrogen-spin-flip process as the

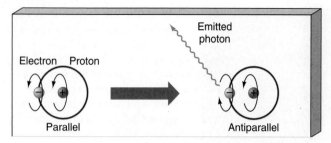

Figure 18.15 Diagram of a ground-level hydrogen atom changing from a higher-energy state (electron and proton spins are parallel) to a lower-energy state (spins are antiparallel). The emitted photon carries away an energy equal to the energy difference between the two spin states.

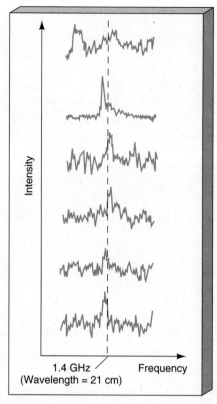

Figure 18.16 Typical 21-cm radio spectral lines observed from several different regions of interstellar space. The peaks do not all occur at a wavelength of exactly 21 cm, corresponding to a frequency of 1.4 GHz (gigahertz), because the gas in the Galaxy is moving with respect to Earth.

21-centimeter radiation. It provides a vital probe into any region of the universe containing atomic hydrogen gas. Figure 18.16 shows typical spectral profiles of 21-cm radio signals observed from several different regions of space. These tracings are the characteristic signatures of cold, atomic hydrogen in our Galaxy. Needing no visible starlight to help calibrate their signals, radio astronomers can observe *any* interstellar region that contains enough hydrogen gas to produce a detectable signal. Even the low-density regions between the dark clouds can be studied.

As can be seen in Figure 18.16, actual 21-cm lines are quite jagged and irregular, somewhat like nebular emission lines in appearance. These irregularities arise because there are usually numerous clumps of interstellar gas along any given line of sight. Each has its own density, temperature, radial velocity, and internal motion, so the intensity, width, and Doppler shift of the resultant 21-cm line vary from place to place. All these different lines are superimposed in the signal we eventually receive at Earth, and sophisticated computer analysis is generally required to disentangle them. The figures quoted earlier for the temperatures (100 K) and densities (10^6 atoms/m^3) of the regions among the dark dust clouds are based on 21-cm measurements; observations of the dark clouds themselves yield densities and temperatures in good agreement with those obtained by optical spectroscopy.

All interstellar atomic hydrogen emits 21-cm radiation. But if all atoms eventually fall into their lowest-energy configuration, why isn't all the hydrogen in the Galaxy in the lower-energy state by now? Why do we see 21-cm radiation today? The answer is that the energy difference between the two states is comparable to the energy of a typical atom at a temperature of 100 K or so. As a result, atomic collisions in the interstellar medium are energetic enough to boost the electron into the higher-energy configuration and so maintain comparable numbers of hydrogen atoms in either state. At any instant, any sample of interstellar hydrogen will contain many atoms in the upper level, and 21-cm radiation will always be emitted.

Of great importance, the wavelength of this characteristic radiation is much larger than the typical size of interstellar dust particles. Accordingly, this radio radiation reaches Earth completely unscattered by interstellar debris. The opportunity to observe interstellar space well beyond a few thousand parsecs, and in directions lacking background stars, makes 21-cm observations among the most important and useful in all astronomy.

18.5 Interstellar Molecules

4 5 In certain interstellar regions of cold (typically 20 K) neutral gas, densities can reach as high as 10^{12} particles/m^3. Until the late 1970s, astronomers regarded these regions simply as abnormally dense interstellar clouds, but it is now recognized that they belong to an entirely new class of interstellar matter. The gas particles in these regions are not in atomic form at all; they are molecules. Because of the predominance of molecules in these dense interstellar regions, they are known as **molecular clouds**. Only within recent years have astronomers begun to appreciate the vastness of these clouds. They literally dwarf even the largest emission nebulae, which were previously thought to be the most massive residents of interstellar space.

MOLECULAR SPECTRAL LINES

As noted in Chapter 4, molecules can become collisionally or radiatively excited, much like atoms. ∞ (Sec. 4.3) Furthermore, again like atoms, molecules eventually relax back to their ground states, emitting radiation in the process. The energy states of molecules are much more complex than those of atoms, however. Molecules, like atoms, can undergo internal electron transitions, but unlike atoms, they can also rotate and vibrate. They do so in specific ways, obeying the laws of quantum physics. Figure 18.17 depicts a simple molecule rotating rapidly—that is, in an excited rotational state. After a length of time that depends on its internal makeup, the molecule relaxes back to a slower rotational rate (a state of lower energy). This change causes a photon to be emitted, carrying an energy equal to the energy difference between the two rotational states involved. The energy differences between these states are generally very small, so the emitted radiation is usually in the radio range.

We are fortunate that molecules emit radio radiation, because they are invariably found in the densest and dustiest parts of interstellar space. These are regions where the absorption of shorter-wavelength radiation is enough to prohibit the use of ultraviolet, optical, and most infrared techniques that might ordinarily detect changes in the energy states of the molecules. Only low-frequency radio radiation can escape.

Why are molecules found only in the densest and darkest of the interstellar clouds? One possible reason is that the dust serves to protect the fragile molecules from the normally harsh interstellar environment—the same absorption that

prevents high-frequency radiation from getting out to our detectors also prevents it from getting in to destroy the molecules. Another possibility is that the dust acts as a catalyst that helps form the molecules. The grains provide both a place where atoms can stick and react and a means of dissipating any heat associated with the reaction, which might otherwise destroy the newly formed molecules. Probably the dust plays both roles. The close association between dust grains and molecules in dense interstellar clouds argues strongly in favor of this picture, but the details are still being debated.

MOLECULAR TRACERS

In mapping molecular clouds, radio astronomers are faced with a problem. Molecular hydrogen (H_2) is by far the most common constituent of these clouds, but unfortunately, despite its abundance, this molecule does not emit or absorb radio radiation. It emits only short-wavelength ultraviolet radiation, so it cannot easily be used as a probe of cloud structure. Nor are 21-cm observations helpful—they are sensitive only to *atomic* hydrogen, not to the *molecular* form of the gas. Theorists had expected H_2 to abound in these dense, cold pockets of interstellar space, but proof of its existence was hard to obtain. Only when spacecraft measured the ultraviolet spectra of a few stars located near the edges of some dense clouds was the presence of molecular hydrogen confirmed.

With hydrogen effectively ruled out as a probe of molecular clouds, astronomers must use observations of other molecules to study the dark interiors of these dusty regions. Molecules such as carbon dioxide (CO), hydrogen cyanide (HCN), ammonia (NH_3), water (H_2O), methyl alcohol (CH_3OH), formaldehyde (H_2CO), and about 60 others, some of them quite complex, are now known to exist in interstellar space.* These molecules are found only in very small quantities —they are generally 1 million to 1 billion times less abundant than H_2—but they are important as *tracers* of a cloud's structure and physical properties. They are thought to be produced by chemical reactions within molecular clouds. When we observe them, we know that the regions under study must also contain high densities of molecular hydrogen, dust, and other important constituents.

The rotational properties of different molecules often make them suitable as probes of regions with different physical properties. Formaldehyde may provide the most useful information on one region, carbon monoxide on another,

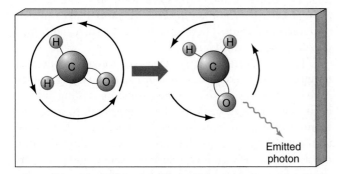

Figure 18.17 As a molecule changes from a rapid rotation (left) to a slower rotation (right), a photon is emitted that can be detected with a radio telescope. Depicted here is the formaldehyde molecule, H_2CO. The length of the curved arrows is proportional to the spin rate of the molecule.

*Some remarkably complex organic molecules have been found in the densest of the dark interstellar clouds, including formaldehyde (H_2CO), ethyl alcohol (CH_3CH_2OH), methylamine (CH_3NH_2), and formic acid (H_2CO_2). Their presence has fueled speculation about the origins of life, both on Earth and in the interstellar medium—especially since the recent report by radio astronomers of evidence that glycine (NH_2CH_2COOH), one of the key amino acids that form the large protein molecules in living cells, may also be present in interstellar space.

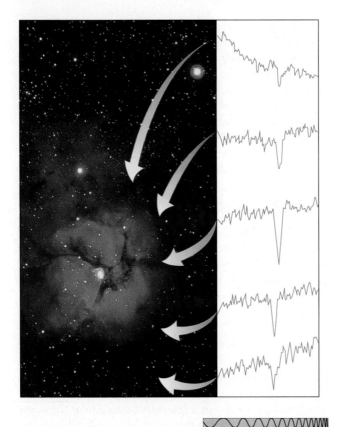

Figure 18.18 Spectra indicate that formaldehyde molecules exist around M20, as indicated by the arrows. The lines are most intense both in the dark dust lanes trisecting the nebula and in the dark regions beyond the nebula.

Figure 18.19 Contour map of the amount of formaldehyde near the M20 nebula, demonstrating how formaldehyde is especially abundant in the darkest interstellar regions. Other kinds of molecules have been found to be similarly distributed. The contour values increase from the outside to the inside, so the maximum density of formaldehyde lies just to the bottom right of the visible nebula. The green and red contours outline the intensity of the formaldehyde absorption lines at different rotational frequencies. The nebula itself is about 4 pc across.

and water on yet another, depending on the densities and temperatures of the regions involved. These data equip astronomers with a sophisticated spectroscopic "toolbox" for studying the interstellar medium.

For example, Figure 18.18 shows some of the sites where formaldehyde molecules have been detected near M20. At practically every dark area sampled between M16 and M8, this molecule is present in surprisingly large abundance (although it is still far less common than H_2). Analyses of spectral lines at many locations along the 12°-wide swath shown in Figure 18.6 indicate that the temperature and density are much the same in all the molecular clouds studied (50 K and 10^{11} molecules/m^3, on average). Figure 18.19 shows a contour map of the distribution of formaldehyde molecules in the immediate vicinity of the M20 nebula. It was made by observing radio spectral lines of formaldehyde at various locations and then drawing contours connecting regions of similar abundance. Notice that

the amount of formaldehyde (and, we assume, the amount of hydrogen) peaks in a dark region, well away from the visible nebula.

Radio maps of interstellar gas and infrared maps of interstellar dust reveal that molecular clouds do not exist as distinct and separate objects in space. Rather, they make up huge **molecular cloud complexes**, some spanning as much as 50 pc across and containing enough gas to make millions of stars like our Sun. There are about 1000 such giant complexes known in our Galaxy.

The very existence of molecules has forced astronomers to rethink and to reobserve interstellar space. In doing so, they have begun to realize that this active and interesting domain is far from the void suspected by theorists not so long ago. Regions of space recently thought to contain nothing more than galactic "garbage"—the cool, tenuous darkness among the stars—now play a critical role in our understanding of stars and the interstellar medium from which they are born.

Chapter Review

SUMMARY

The **interstellar medium** (p. 406) occupies the space among the stars. It is made up of cold (less than 100 K) gas, mostly atomic or molecular hydrogen and helium, and **dust grains** (p. 406). Interstellar dust is very effective at blocking our view of distant stars, even though the density of the interstellar medium is very low. The spatial distribution of interstellar matter is very patchy. The general diminution of starlight by dust is called **extinction** (p. 407). In addition, the dust preferentially absorbs short-wavelength radiation, leading to a distinct **reddening** (p. 407) of light passing interstellar clouds.

Interstellar dust is thought to be composed of silicates, graphite, iron, and "dirty ice." Interstellar dust particles are apparently elongated or rodlike. The **polarization** (p. 409) of starlight provides a means of studying them.

A **nebula** (p. 410) is a general term for any fuzzy bright or dark patch on the sky. **Emission nebulae** (p. 410) are extended clouds of hot, glowing interstellar gas. Associated with star formation, they result when hot O and B stars heat and ionize their surroundings. Studies of the emission lines produced by excited nebular atoms allow astronomers to measure the nebula's properties. Nebulae are often crossed by dark **dust lanes** (p. 413)—part of the larger cloud from which they formed.

Some excited atomic states take so long to emit a photon that the spectral lines associated with these transitions are never seen in terrestrial laboratories, where collisions always knock the atom into another energy state before it can emit any radiation. When these lines are seen in nebular spectra, they are called **forbidden lines** (p. 414).

Dark dust clouds (p. 415) are cold, irregularly shaped regions in the interstellar medium that diminish or completely obscure the light from background stars. Astronomers can learn about these clouds by studying the absorption lines they produce in starlight that passes through them.

Another way to observe cold, dark regions of interstellar space is through **21-centimeter radiation** (p. 418). Such radiation is produced whenever the electron in an atom of hydrogen reverses its spin, changing its energy very slightly in the process. This radio radiation is important because it is emitted by all cool atomic hydrogen gas, even if the gas is undetectable by other means. In addition, 21-cm radiation is not appreciably absorbed by the interstellar medium, so radio astronomers making observations at this wavelength can "see" to great distances.

The interstellar medium also contains many cold, dark **molecular clouds** (p. 419), which are observed mainly through the radio radiation emitted by the molecules they contain. Dust within these clouds probably both protects the molecules and acts as a catalyst to help them form. As with other interstellar clouds, hydrogen is by far the most common constituent, but molecular hydrogen happens to be very hard to observe. Astronomers usually study these clouds through observations of other "tracer" molecules that are less common but much easier to detect.

Astronomers believe that molecular clouds are likely sites of future star formation. Often, several molecular clouds are found close to one another, forming an enormous **molecular cloud complex** (p. 420) millions of times more massive than the Sun.

SELF-TEST: TRUE OR FALSE?

_____ **1.** Interstellar matter is quite evenly distributed throughout the Milky Way Galaxy.

_____ **2.** In the vicinity of the Sun there is about as much mass in the form of interstellar matter as in the form of stars.

_____ **3.** There is a lack of heavy elements in interstellar gas because they go into making interstellar dust.

_____ **4.** The fact that starlight becomes polarized as it passes through the interstellar medium tells us that interstellar dust particles are spherical in shape.

_____ **5.** A typical region of dark interstellar space has a temperature of about 500 K.

_____ **6.** An emission nebula is a cloud of dust reflecting the light of a nearby star cluster.

_____ **7.** Emission nebulae display spectra almost identical to those of the stars embedded in them.

_____ **8.** "Forbidden" emission lines can occur in emission nebulae because the density of interstellar gas there is extremely low.

_____ **9.** Because of the obscuration of visible light by interstellar dust, we can observe stars only within a few thousand parsecs of Earth in any direction.

_____ **10.** A typical dark dust cloud is many hundreds of parsecs across.

_____ **11.** Because of their low temperatures, dark dust clouds radiate mainly in the radio part of the electromagnetic spectrum.

_____ **12.** Most interstellar matter exists in the form of molecular clouds.

_____ **13.** 21-cm radiation provides astronomers with information on the density, temperature, and internal motions of interstellar gas.

_____ **14.** 21-cm radiation can pass unimpeded through the entire Milky Way Galaxy.

_____ **15.** Water, formaldehyde, carbon monoxide, and numerous organic molecules are all commonly found in molecular clouds.

SELF-TEST: FILL IN THE BLANK

1. The interstellar medium is made up of _____ and _____.
2. To scatter a beam of radiation, a particle must be _____ in size to the wavelength of the radiation.
3. Extinction is the _____ of starlight by interstellar _____.
4. The density of interstellar matter can be characterized as being very _____.
5. Interstellar gas is composed of 90 percent _____ and 9 percent _____.
6. The temperature of a typical emission nebula is about _____ K.
7. The process in which clouds of cool dense gas (such as in the Eagle Nebula) are eaten away by radiation from nearby hot young stars is called _____.

8. An HII region is another name for an _____.
9. Dark dust clouds can have temperatures as low as _____ K.
10. 21-cm radiation is emitted by _____ hydrogen.
11. 21-cm radiation results from a change in the _____ of the electron in a _____ atom.
12. Molecular clouds typically have temperatures of about _____ K.
13. Emissions from molecular clouds are in the _____ part of the electromagnetic spectrum.
14. The most common constituent of molecular clouds is molecular _____.
15. A molecular cloud complex may contain as much as _____ solar masses of gas.

REVIEW AND DISCUSSION

1. Give a brief description of the interstellar medium.
2. What is the composition of interstellar gas? What about interstellar dust?
3. How dense is interstellar matter between the stars, on average?
4. If space is a near-perfect vacuum, how can there be enough dust in it to block light?
5. How is interstellar matter distributed throughout space?
6. What are some methods that astronomers use to study interstellar dust?
7. What is an emission nebula?
8. Why do emission nebulae appear red in color photographs?
9. Give a brief description of a dark dust cloud.

10. What is 21-cm radiation? With what element is it associated?
11. Why can't 21-cm radiation be used to probe the interiors of molecular clouds?
12. How does a molecular cloud differ from other interstellar matter?
13. If our Sun were surrounded by a cloud of gas, would this cloud be an emission nebula? Why or why not?
14. Compare the reddening of stars by interstellar dust with the reddening of the setting Sun.
15. Explain what it means for a star's light to be polarized. How does the polarization of starlight provide a means of studying the interstellar medium?

PROBLEMS

1. The average density of interstellar gas within about 100 pc of the Sun (in the Local Bubble) is much lower than the value mentioned in the text—in fact, it is roughly 10^3 hydrogen atoms/m^3. Given that the mass of a hydrogen atom is 1.7×10^{-27} kg, calculate the total mass of interstellar matter contained within a volume equal to that of Earth.
2. Assuming the same average density as in the previous question, calculate the total mass of interstellar hydrogen contained within a cylinder of cross-sectional area 1 cm^2, extending from Earth to Alpha Centauri.
3. Assuming a density of 3000 kg/m^3, estimate the mass of the dust particle illustrated in Figure 18.3.
4. A beam of light shining through a dense molecular cloud is diminished in intensity by a factor of 2.5 for every 3 pc it travels. By what total factor is it reduced if the total thickness of the cloud is 60 pc?
5. Interstellar extinction is sometimes measured in magnitudes per kiloparsec (1 kpc = 1000 pc). Light from a star 250 pc away is observed to be diminished in intensity by a factor of 100 over and above the effect of the inverse-square law. What is the average interstellar extinction along the line of sight, in mag/kpc?
6. In order to carry enough energy to ionize a hydrogen atom, a photon must have a wavelength of less than 9.12×10^{-8} m

(91.2 nm). Using Wien's law (Section 3.4), calculate the temperature a star must have for the peak wavelength of its blackbody curve to equal this value.
7. Estimate the escape speeds near the edges of the four emission nebulae listed in Table 18.1, and compare them with the average speeds of hydrogen nuclei in those nebulae (see *More Precisely 8-1*). Do you think it is possible that the nebulae are held together by their own gravity?
8. A roughly spherical dark dust cloud has a radius of 10 pc and a density of 10^9 atoms/m^3. If there is one dust particle per trillion (10^{12}) atoms, the average mass of a dust particle is as in problem 3, and the gas contains 1 helium atom (of mass four times the mass of a hydrogen atom) for every 9 hydrogen atoms, calculate the total mass of gas and dust in the cloud.
9. Calculate the frequency of 21-cm radiation. If a group of interstellar clouds along the line of sight have radial velocities in the range 75 km/s (receding) to 50 km/s (approaching), calculate the range of frequencies and wavelengths over which the 21-cm line will be observed.
10. Calculate the radius of a spherical molecular cloud whose total mass equals the mass of the Sun. Assume a cloud density of 10^{12} hydrogen atoms per cubic meter.

PROJECTS

1. The constellation Orion the Hunter is prominent in the evening sky of winter. Its most noticeable feature is a short, straight row of three medium-bright stars: the famous belt of Orion. A line of stars extends from the eastmost star of the belt, toward the south. This line represents Orion's sword. Towards the bottom of the sword is the sky's most famous emission nebula, M42, the Orion Nebula.

   Observe the Orion Nebula with your eye, with binoculars, and with a telescope. What is its color? How can you account for this? With the telescope, try to find the Trapezium, a grouping of four stars in the center of M42. These are hot, young stars; their energy causes the Orion Nebula to glow.

2. Observe the Milky Way on a dark, very clear night. Is it a continuous band of light across the sky or is it mottled? The parts of the Milky Way that appear missing are actually dark dust clouds that are relatively near the Sun. Identify the constellations in which you see these clouds. Make a sketch and compare with a star atlas. Find other small clouds in the atlas and try to find them with your eye or with binoculars.

19 STAR FORMATION

A Traumatic Birth $\textit{www}$

LEARNING GOALS

Studying this chapter will enable you to:

1 Discuss the factors that compete against gravity in the process of star formation.

2 Summarize the sequence of events leading to the formation of a star like our Sun.

3 Explain how the process of star formation depends on stellar mass.

4 Describe some of the observational evidence supporting the modern theory of star formation.

5 Explain the nature of interstellar shock waves, and discuss their possible role in the formation of stars.

(Photo A) Disks and jets pervade the universe on many scales. Here, they seem to be a natural result of a rotating cloud of gas contracting to form a star. Matter falling onto the embryonic star creates a pair of high-speed jets of gas perpendicular to the star's flattened disk, carrying away heat and angular momentum that might otherwise prevent the birth of the star. This image shows a small region near the Orion Nebula known as HH1/HH2, whose twin jets have blasted outward for several trillion km (nearly half a light-year) before colliding with interstellar matter. (HH stands for Herbig–Haro, after the investigators who first cataloged such objects.) The next three photos show stellar jets ejected from three different very young stars. Reproduced here to scale, these images collectively depict the propagation of a jet through space.

(Photo B) This image of HH30, spanning approximately 250 billion km, or about 0.01pc, shows a thin jet (in red) emanating from a circumstellar disk (at left in grey) encircling a nascent star.

(Photo C) One of HH34's jets is longer, reaching some 600 billion km, yet remains narrow, with a beaded structure.

(Photo D) HH47 is more than a trillion km in length, or nearly 0.1pc. This photo shows one of its jets plowing through interstellar space, creating bow shocks in the process.

We now move from the interstellar medium—the gas and dust among the stars—back to the stars themselves. The next four chapters discuss the formation and evolution of stars. We have already seen that stars must evolve as they consume their fuel supply, and we have extensive observational evidence of stars at many different evolutionary stages. With the help of these observations, astronomers have developed a good understanding of stellar evolution—the complex changes experienced by stars as they form, mature, grow old, and die. We begin by studying the process of star formation, through which interstellar clouds of gas and dust are transformed into the myriad stars we see in the night sky.

19.1 Gravitational Competition

GRAVITY AND HEAT

1 How do stars form? What factors determine the masses, luminosities, and distribution of stars in our Galaxy? In short, what basic processes are responsible for the appearance of our nighttime sky? Simply stated, star formation begins when part of the interstellar medium—one of the cold, dark clouds discussed in Chapter 18—starts to collapse under its own weight. ∞ (Sec. 18.3) The cloud fragment heats up as it shrinks, and eventually its center becomes hot enough for nuclear burning to begin. At that point, the contraction stops, and a star is born. But what determines which interstellar clouds collapse? For that matter, since all clouds exert a gravitational pull, why didn't they all collapse long ago? To begin to answer these questions, let us consider a small portion of a large cloud of interstellar gas. We concentrate first on just a few atoms, as shown in Figure 19.1.

Even though the cloud's temperature is very low, each atom still has some random motion because of the cloud's heat. ∞ (More Precisely 3-1) Each atom is also influenced by the gravitational attraction of all its neighbors. The gravitational force is not large, however, because the mass of each atom is so small. When a few atoms accidentally cluster for an instant, as shown in Figure 19.1(b), their combined gravity is insufficient to bind them into a lasting, distinct clump of matter. This accidental cluster will disperse as quickly as it forms. The effect of heat—the random motion of the atoms—is much stronger than the effect of gravity.

Now let's concentrate on a larger group of atoms. Imagine, for example, 50, 100, 1000, even a million atoms, each gravitationally pulling on all the others. The force of gravity is now stronger than before. Will this many atoms exert a combined gravitational attraction strong enough to prevent the clump from dispersing again? The answer—at least under the conditions found in interstellar space—is still no.

The gravitational attraction of this mass of atoms is still far too weak to overcome the effect of heat.

We have already seen numerous instances of the competition between heat and gravity (see, for example, *More Precisely 8-1*). Since the temperature of a gas is simply a measure of the average speed of the atoms or molecules in it, the higher the temperature, the greater the average speed, and thus the higher the pressure of the gas. This is the main reason that the Sun and other stars don't collapse. The outward pressure of their heated gases exactly balances gravity's inward pull.

How many atoms must be accumulated in order for their collective pull of gravity to prevent them from dispersing back into interstellar space? The answer, even for a typical cool (100 K) cloud, is a truly huge number. Nearly 10^{57} atoms are required—much larger than the 10^{25} grains of sand on all the beaches of the world, even larger than the 10^{51} elementary particles that constitute all the atomic nuclei in our entire planet. There is simply nothing on Earth comparable to a star.

SOME COMPLICATING FACTORS

Heat is not the only factor that tends to oppose gravitational contraction. *Rotation*—that is, spin—can also compete with gravity's inward pull. As we saw in Chapter 15, a contracting cloud having even a small spin tends to develop a bulge around its midsection. ∞ (Sec. 15.2) As the cloud contracts, it must spin faster (to conserve its angular momentum), and the bulge grows—material on the edge tends to fly off into space. (Consider as an analogy mud flung from a rapidly rotating bicycle wheel.) Eventually, just as in Figure 15.1, the cloud forms a flattened, rotating disk.

For material to remain part of the cloud and not be spun off into space, a force must be applied—in this case, the force of gravity. The more rapid the rotation, the greater the tendency for the gas to escape, and the greater the gravitational force needed to retain it. It is in this sense that we can regard rotation as opposing the inward pull of gravity. Should

Figure 19.1 Motions of a few atoms within an interstellar cloud are influenced by gravity so slightly that their paths are hardly changed (a) before, (b) during, and (c) after an accidental, random encounter.

the rotation of a contracting gas cloud overpower gravity, the cloud would simply disperse. Thus, more mass is needed for a rapidly rotating interstellar cloud to contract to form a star than is needed for a cloud having no rotation at all.

Magnetism can also hinder a cloud's contraction. Just as Earth, the outer planets, and the Sun all have some magnetism, magnetic fields permeate most interstellar clouds. As a cloud contracts, it heats up, and atomic encounters become violent enough to (partly) ionize the gas. As we noted in Chapter 7 when discussing Earth's Van Allen belts and in Chapter 16 when discussing activity on the Sun, magnetic fields can exert electromagnetic control over charged particles. ∞ (Secs. 7.4, 16.4) In effect, the particles tend to become "tied" to the magnetic field—free to move *along* the field lines but inhibited from moving *perpendicular* to them.

As a result (Figure 19.2), interstellar clouds may contract in distorted ways. Because the charged particles and the magnetic field are linked, the field itself follows the contraction of a cloud. The charged particles literally pull the magnetic field toward the cloud's center in the direction perpendicular to the field lines. As the field lines are compressed, the magnetic field strength increases. In this way, the strength of magnetism in a cloud can become much larger than that normally permeating general interstellar space. The primitive solar nebula may have contained a strong magnetic field created in just this way.

Theory suggests that even small quantities of rotation or magnetism can compete quite effectively with gravity and can greatly alter the evolution of a typical gas cloud. Unfortunately, the interplay of these factors is not very well understood—both can lead to very complex behavior as a cloud contracts, and the combination of the two is extremely difficult to study theoretically. In this chapter we will gain an appreciation for the broad outlines of the star-formation process by neglecting these two complicating factors. Bear in mind, however, that both are probably important in determining the details.

19.2 The Formation of Stars Like the Sun

2 Star formation begins when gravity begins to dominate over heat, causing a cloud to lose its equilibrium and start to contract. Only after the cloud has undergone radical changes in its internal structure is equilibrium finally restored.

Table 19.1 lists seven evolutionary stages that an interstellar cloud goes through prior to becoming a main-sequence star like the Sun. These stages are characterized by varying central temperatures, surface temperatures, central densities, and radii of the prestellar object. They trace its progress from a quiescent interstellar cloud to a genuine star. The numbers given in Table 19.1 and the following discussion are valid *only* for stars of approximately the same mass as the Sun. In the next section we will relax this restriction and consider the formation of other stars.

STAGE 1: AN INTERSTELLAR CLOUD

The first stage in the star-formation process is a dense interstellar cloud—the core of a dark dust cloud or perhaps a molecular cloud. These clouds are truly vast, sometimes spanning tens of parsecs (10^{14}–10^{15} km) across. Typical temperatures are about 10 K throughout, with a density of perhaps 10^9 particles/m^3. Stage 1 clouds contain thousands of times the mass of the Sun, mainly in the form of cold atomic and molecular gas. (The dust they contain is important for cooling the cloud as it contracts and also plays a crucial role in planet formation, but it constitutes a negligible fraction of the total mass.) ∞ (Sec. 15.2)

If such a cloud is to be the birthplace of stars, it must become unstable and eventually break up into smaller pieces. The initial collapse occurs when a pocket of gas becomes gravitationally unstable. Perhaps it is squeezed

Figure 19.2 Magnetism can hinder the contraction of a gas cloud, especially in directions perpendicular to the magnetic field (solid lines). Frames (a), (b), and (c) trace the evolution of a slowly contracting interstellar cloud having some magnetism.

TABLE 19.1 Prestellar Evolution of a Solar-Type Star

STAGE	APPROXIMATE TIME TO NEXT STAGE (yr)	CENTRAL TEMPERATURE (K)	SURFACE TEMPERATURE (K)	CENTRAL DENSITY (particles/m³)	DIAMETER* (km)	OBJECT
1	2×10^6	10	10	10^9	10^{14}	Interstellar cloud
2	3×10^4	100	10	10^{12}	10^{12}	Cloud fragment
3	10^5	10,000	100	10^{18}	10^{10}	Cloud fragment/protostar
4	10^6	1,000,000	3000	10^{24}	10^8	Protostar
5	10^7	5,000,000	4000	10^{28}	10^7	Protostar
6	3×10^7	10,000,000	4500	10^{31}	2×10^6	Star
7	10^{10}	15,000,000	6000	10^{32}	1.5×10^6	Main-sequence star

*For comparison, recall that the diameter of the Sun is 1.4×10^6 km, while that of the solar system is roughly 1.5×10^{10} km.

by some external event, such as the pressure wave produced when a nearby O- or B-type star forms and ionizes its surroundings, or perhaps its supporting magnetic field leaks away as charged particles slowly drift across the confining field lines. Whatever the cause, theory suggests that once the collapse begins, fragmentation into smaller and smaller clumps of matter naturally follows, as gravitational instabilities continue to operate in the gas. As illustrated in Figure 19.3, a typical cloud can break up into tens, hundreds, even thousands, of fragments, each imitating the shrinking behavior of the parent cloud and contracting ever faster. The whole process, from a single quiescent cloud to many collapsing fragments, takes a few million years.

In this way, depending on the precise conditions under which fragmentation takes place, an interstellar cloud can produce either a few dozen stars, each much larger than our Sun, or a whole cluster of hundreds of stars, each comparable to or smaller than our Sun. There is little evidence of stars born in isolation, one star from one cloud. Most stars—perhaps even all stars—appear to originate as members of multiple systems or star clusters. The Sun, which is now found alone and isolated in space, probably escaped from the multiple-star system where it formed, after an encounter with another star or some much larger object (such as a molecular cloud).

STAGE 2: A COLLAPSING CLOUD FRAGMENT

The second stage in our evolutionary scenario represents the physical conditions in just one of the many fragments that develop in a typical interstellar cloud. A fragment destined to form a star like the Sun contains between one and two solar masses of material at this stage. Estimated to span a few hundredths of a parsec across, this fuzzy, gaseous blob is still about 100 times the size of our solar system. Its central density is now some 10^{12} particles/m³.

Even though it has shrunk substantially in size, the fragment's average temperature is not much different from that of the original cloud. The reason is that the gas constantly radiates large amounts of energy into space. The material of the fragment is so thin that photons produced within it easily escape without being reabsorbed by the cloud, so virtually all the energy released in the collapse is radiated away and does not cause any significant increase in temperature. Only at the center, where the radiation must traverse the greatest amount of material in order to escape, is there any appreciable temperature increase. The gas there may be as warm as 100 K by this stage. For the most part, however, the fragment stays cold as it shrinks.

The process of continued fragmentation is eventually stopped by the increasing density within the shrinking cloud.

Figure 19.3 As an interstellar cloud contracts, gravitational instabilities cause it to fragment into smaller pieces. The pieces themselves continue to collapse and fragment, eventually to form many tens or hundreds of separate stars.

As stage 2 fragments continue to contract, they eventually become so dense that radiation cannot get out easily. The trapped radiation causes the temperature to rise, the pressure to increase, and the fragmentation to cease.

STAGE 3: FRAGMENTATION CEASES

Several tens of thousands of years after it first began contracting, a typical stage 2 fragment has shrunk by the start of stage 3 to roughly the size of our solar system (still 10,000 times the size of our Sun). The inner regions have just become opaque to their own radiation and so have started to heat up considerably, as noted in Table 19.1. The central temperature has reached about 10,000 K—hotter than the hottest steel furnace on Earth. However, the temperature at the fragment's periphery has not increased much. It is still able to radiate its energy into space and so remains cool. The density increases much faster in the core of the fragment than at its periphery, so the outer portions of the cloud are both cooler and thinner than the interior. The central density by this time is approximately 10^{18} particles/m^3 (still only 10^{-9} kg/m^3 or so).

For the first time, our fragment is beginning to resemble a star. The dense, opaque region at the center is called a **protostar**—an embryonic object perched at the dawn of star birth. Its mass grows as more and more material rains down on it from outside, although its radius continues to shrink because its pressure is still unable to overcome the relentless pull of gravity. After stage 3, we can distinguish a "surface" on the protostar—its *photosphere*. Inside the photosphere, the protostellar material is opaque to the radiation it emits.* From here on, the surface temperatures listed in Table 19.1 refer to the photosphere and not to the "periphery" of the collapsing fragment, whose temperature remains low.

STAGE 4: A PROTOSTAR

As the protostar evolves, it shrinks, its density grows, and its temperature rises, both in the core and at the photosphere. Some 100,000 years after the fragment began to form, it reaches stage 4, where its center seethes at about 1,000,000 K. The electrons and protons ripped from atoms whiz around at hundreds of kilometers per second, yet the temperature is still short of the 10^7 K needed to ignite the proton–proton nuclear reactions that fuse hydrogen into helium. ∞ (Sec. 16.5) Still much larger than the Sun, our gassy heap is now about the size of Mercury's orbit. Heated by the material falling on it from above, its surface temperature has risen to a few thousand kelvins.

Knowing the protostar's radius and surface temperature, we can calculate its luminosity. Surprisingly, it turns out to be several thousand times the luminosity of the Sun. Even though the protostar has a surface temperature only about half that of the Sun, it is hundreds of times larger, making its total luminosity very large indeed—in fact, much greater than the luminosity of most main-sequence stars. Because nuclear reactions

have not yet begun, this luminosity is due entirely to the release of gravitational energy as the protostar continues to shrink, and material from the surrounding fragment (which we called the solar nebula back in Chapter 15) rains down on its surface.

By the time stage 4 is reached, our protostar's physical properties can be plotted on the Hertzsprung–Russell (H–R) diagram, as shown in Figure 19.4. Recall that an H–R diagram is a plot of two key stellar properties—surface temperature (increasing to the left) and luminosity (increasing upward). ∞ (Sec. 17.7) The luminosity scale in Figure 19.4 is expressed in terms of the solar luminosity (4×10^{26} W). Our G2-type Sun is plotted at a temperature of 6000 K and a luminosity of 1 unit. As before, the dashed diagonal lines in the H–R diagrams represent stellar radius, allowing us to follow the changes in a star's size as it evolves. At each phase of a star's evolution, its surface temperature and luminosity can be represented by a point on this diagram. The motion of that point as the star evolves is known as the star's **evolutionary track**. It is a graphical representation of a star's life.

The red track on Figure 19.4 depicts the approximate path followed by our interstellar cloud fragment since it became a protostar at stage 3 (which itself lies off the right-hand edge of the figure). This early evolutionary track is

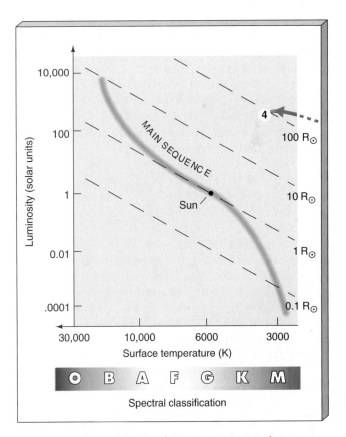

Figure 19.4 Diagram of the approximate evolutionary track followed by an interstellar cloud fragment prior to reaching the end of the Kelvin–Helmholtz contraction phase as a stage 4 protostar. (The circled numbers on this and subsequent plots refer to the prestellar evolutionary stages listed in Table 19.1 and described in the text.)

---

*Note that this is the same definition of "surface" that we used for the Sun in Chapter 16. ∞ (Sec. 16.1)

Figure 19.5 Artist's conception of the changes in an interstellar cloud during the early evolutionary stages outlined in Table 19.1. Shown are a stage 1 interstellar cloud; a stage 2 fragment; a smaller, hotter stage 3 fragment; and a stage 4/stage 5 protostar. (Not drawn to scale.) The duration of each stage, in years, is also indicated.

Animation

known as the *Kelvin–Helmholtz contraction phase*, after two European physicists (Lord Kelvin and Hermann von Helmholtz) who first studied the subject. Figure 19.5 is an artist's sketch of an interstellar gas cloud proceeding along the evolutionary path outlined so far.

Our protostar is still not in equilibrium. Even though its temperature is now so high that outward-directed pressure has become a powerful countervailing influence against gravity's continued inward pull, the balance is not yet perfect. The protostar's internal heat gradually diffuses out from the hot center to the cooler surface, where it is radiated away into space. As a result, the overall contraction slows, but it does not stop completely. From our perspective on Earth, this is quite fortunate: if the heated gas were somehow able to counteract gravity completely before the star reached the temperature and density needed to start nuclear burning in its core, the protostar would simply radiate away its heat and never become a true star. The night sky would be abundant in faint protostars, but completely lacking in the genuine article. Of course, there would be no Sun either, so it is unlikely that we, or any other intelligent life form, would exist to appreciate these astronomical subtleties.

After stage 4, the protostar on the H–R diagram moves down (toward lower luminosity) and slightly to the left (toward higher temperature), as shown in Figure 19.6. Its surface temperature remains almost constant, and it becomes less luminous as it shrinks. This portion of our protostar's evolutionary path running from point 4 to point 6 in Figure 19.6 is often called the *Hayashi track*, after C. Hayashi, a twentieth-century Japanese researcher.

Protostars on the Hayashi track often exhibit violent surface activity during this phase of their evolution, resulting in extremely strong protostellar winds, much denser than that of our Sun. As mentioned earlier in Section 15.3, this portion

of the evolutionary track is often called the **T Tauri star**, after T Tauri, the first "star" (actually protostar) to be observed in this stage of prestellar development.

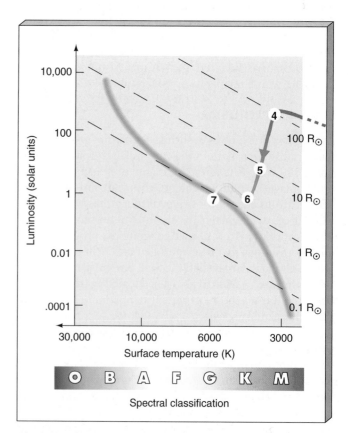

Figure 19.6 The changes in a protostar's observed properties are shown by the path of decreasing luminosity, from stage 4 to stage 6, often called the Hayashi track. At stage 7, the newborn star has arrived on the main sequence.

STAGE 5: PROTOSTELLAR EVOLUTION

By stage 5 on the Hayashi track, the protostar has shrunk to about 10 times the size of the Sun, its surface temperature is about 4000 K, and its luminosity has fallen to about 10 times the solar value. At this point, the central temperature has reached about 5,000,000 K. The gas is completely ionized by now, but the protons still do not have enough thermal energy to overwhelm their mutual electromagnetic repulsion and enter the realm of the nuclear binding force. The core is still too cool for nuclear burning to begin.

Events proceed more slowly as the protostar approaches the main sequence. The initial contraction and fragmentation of the interstellar cloud occurred quite rapidly, but by stage 5, as the protostar nears the status of a full-fledged star, its evolution slows. The cause of this slowdown is heat—even gravity must struggle to compress a hot object. The contraction is governed largely by the rate at which the protostar's internal energy can be radiated away into space. The greater this radiation of internal energy—that is, the more energy that moves through the star to escape from its surface, the faster the contraction occurs. As the luminosity decreases, so too does the contraction rate.

STAGE 6: A NEWBORN STAR

Some 10 million years after its first appearance, the protostar finally becomes a true star. By the bottom of the Hayashi track, at stage 6, when our roughly 1–solar mass object has shrunk to a radius of about 1,000,000 km, the contraction has raised the central temperature to 10,000,000 K, enough to ignite nuclear burning. Protons begin fusing into helium nuclei in the core, and a star is born. As shown in Figure 19.6, the star's surface temperature at this point is about 4500 K, still a little cooler than the Sun. Even though the newly formed star is slightly larger in radius than our Sun, its lower temperature means that its luminosity is somewhat less than (actually, about two-thirds of) the solar value.

STAGE 7: THE MAIN SEQUENCE AT LAST

Over the next 30 million years or so, the stage 6 star contracts a little more. In making this slight adjustment, the central density rises to about 10^{32} particles/m^3 (more conveniently expressed as 10^5 kg/m^3), the central temperature increases to 15,000,000 K, and the surface temperature reaches 6000 K. By stage 7, the star finally reaches the main sequence just about where our Sun now resides. Pressure and gravity are finally balanced, and the rate at which nuclear energy is generated in the core exactly matches the rate at which energy is radiated from the surface.

The evolutionary events just described occur over the course of some 40–50 million years. Although this is a long time by human standards, it is still less than 1 percent of the Sun's lifetime on the main sequence. Once an object begins fusing hydrogen and establishes a "gravity-in/pressure-out" equilibrium, it burns steadily for a very long time. The star's location on the H–R diagram will remain virtually unchanged for the next 10 billion years.

19.3 Stars of Different Masses

THE ZERO-AGE MAIN SEQUENCE

3 The numerical values and the evolutionary track just described are valid only for the case of a 1–solar mass star. The temperatures, densities, and radii of prestellar objects of other masses exhibit similar trends, but the numbers and the tracks differ, in some cases quite considerably. Perhaps not surprisingly, the most massive fragments within interstellar clouds tend to produce the most massive protostars and eventually the most massive stars. Similarly, low-mass fragments give rise to low-mass stars.

Figure 19.7 compares the theoretical pre–main-sequence track taken by the Sun with the corresponding evolutionary tracks of a 0.3–solar mass star and a 3–solar mass star. All three tracks traverse the H–R diagram in the same general manner, but cloud fragments that eventually form stars more massive than the Sun approach the main sequence along a higher track on the diagram, while those destined to form less massive stars take a lower track. The *time* required for an interstellar cloud to become a main-sequence star also depends strongly on its mass. The most massive cloud fragments heat up to the required 10 million K and become O stars in a mere million years, roughly 1/50 the time taken by the Sun. The opposite is the case for prestellar objects having masses less than that of our Sun. A typical M-type star, for example, requires nearly a billion years to form.

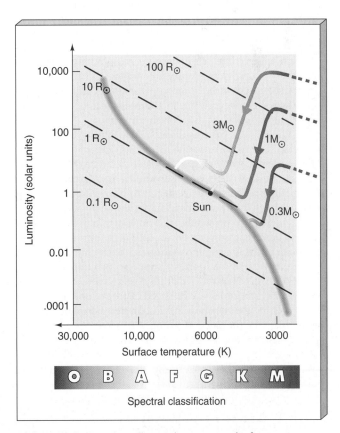

Figure 19.7 Prestellar evolutionary paths for stars more massive and less massive than our Sun.

Whatever the mass, the end point of the prestellar evolutionary track is the main sequence. A star is considered to have reached the main sequence when hydrogen burning begins in its core and the star's properties settle down to stable values. The main-sequence band predicted by theory is called the **zero-age main sequence** (ZAMS). It agrees quite well with main sequences observed for stars in the vicinity of the Sun and those observed for stars in more distant star clusters.

If all gas clouds contained precisely the same elements in exactly the same proportions, mass would be the sole determinant of a newborn star's location on the H–R diagram, and the zero-age main sequence would be a well-defined line rather than a broad band. However, the composition of a star affects its internal structure (mainly by changing the opacity of its outer layers), and this in turn affects both its temperature and its luminosity on the main sequence. Stars with more heavy elements tend to be cooler and slightly less luminous than stars that have the same mass but contain fewer heavy elements.

As a result, differences in composition between stars "blur" the zero-age main sequence into the broad band we observe.

It is important to realize that *the main sequence is itself not an evolutionary track—stars do not evolve along it*. Rather, it is just a "way station" on the H–R diagram where stars stop and spend most of their lives—low-mass stars at the bottom, high-mass stars at the top. Once on the main sequence, a star stays in essentially the same location in the H–R diagram during its whole time as a stage 7 object. (In other words, a star that arrives on the main sequence as, say, a G-type star can never "work its way up" to become a B- or an O-main-sequence blue supergiant, or move down to become an M-type red dwarf.) As we will see in Chapter 20, the next stage of stellar evolution occurs when a star moves away from the main sequence. A star leaving the main sequence and entering this next stage has pretty much the same surface temperature and luminosity it had when it arrived on the main sequence millions (or billions) of years earlier.

INTERLUDE 19-1 Observations of Brown Dwarfs

Cruelly put, brown dwarfs are "failed" stars—objects that formed through the contraction and fragmentation of an interstellar cloud, just like stars, but somehow never reached the critical mass of about 0.08 solar masses needed to initiate hydrogen fusion in their cores. We mentioned in the text that interstellar space could in principle contain large numbers of these small, dark, and very hard to detect objects. Recent observations of star-forming regions have reinforced that view by providing a plausible way in which the formation process could be stopped prior to the formation of a true star. Here we examine in a little more detail the observational evidence for these elusive objects.

Until late 1994 this discussion would have been very short indeed, as there was no firm observational evidence for the existence of any brown dwarfs at all! Since then, however, continuing advances in observational hardware and image-processing techniques have begun to bear fruit, and several likely brown dwarf candidates have now been identified.

Detecting a brown dwarf is no easy task. These objects are so faint that they are hard—although not impossible—to detect directly, and most progress to date has been made in finding them when they happen to orbit another star. (Recall from Chapter 17 that most stars are found in binary systems; the same may very well be true of brown dwarfs.) In searching for brown dwarfs in binaries, astronomers have used many of the same techniques employed in the search for extrasolar planets, as discussed more fully in Chapter 15. ∞ *(Interlude 15-1)* By measuring the tiny "wobbles" in a star's motion, either by carefully tracking its position in the sky or by measuring small variations in its radial velocity, researchers can infer not just the presence but also some of the physical properties of the star's unseen companion. The size of the wobbles is truly minuscule—only a few thousandths

of an arc second in position and a few tens of meters per second in radial velocity. Nevertheless, such precise measurements are now becoming possible.

The *Hubble* image below shows Gliese 623, a binary system containing a brown dwarf candidate (marked by an arrow) originally identified by radial velocity measurements. Astronomers hope that continued observations of this system will allow the low-mass companion's mass to be measured with sufficient accuracy to determine whether or not it really is a brown dwarf. (The "rings" in the image are instrumental artifacts.)

Actually, one might reasonably ask where the dividing line between brown dwarfs and Jupiter-sized planets really lies. The answer is not clear-cut and may depend on circumstances. Theoretical estimates indicate that objects less

FAILED STARS

Some cloud fragments are too small ever to become stars. The giant planet Jupiter is a good example. Jupiter contracted under the influence of gravity, and the resultant heat is still detectable, but the planet did not have enough mass for gravity to crush its matter to the point of nuclear ignition. ∞ (Sec. 11.3) It became stabilized by heat and rotation before the central temperature became hot enough to fuse hydrogen. Jupiter never evolved beyond the protostar stage. If Jupiter, or any of the other jovian planets, had continued to accumulate gas from the solar nebula, they might eventually have become stars (probably to the detriment of life on Earth), but virtually all the matter present during the formative stages of our solar system is gone now, swept away by the solar wind.

Low-mass interstellar gas fragments simply lack the mass needed to initiate nuclear burning. Rather than turning into stars, they will continue to cool, eventually becoming compact, dark "clinkers"—cold fragments of unburned matter—in interstellar space. On the basis of theoretical modeling, astronomers believe that the minimum mass of gas needed to generate core temperatures high enough to begin nuclear fusion is about 0.08 solar masses.

Vast numbers of Jupiter-like objects may well be scattered throughout the universe—fragments frozen in time somewhere along the Kelvin–Helmholtz contraction phase. Small, faint, and cool (and growing ever colder), they are known collectively as **brown dwarfs**. Our technology currently has great difficulty in detecting them, be they planets associated with stars or interstellar cloud fragments far from any star. We can telescopically detect stars and spectroscopically infer atoms and molecules, but although recent advances in observational hardware and image-processing techniques have now identified several likely brown dwarf candidates (see *Interlude 19-1*), astronomical objects of intermediate size outside our solar system remain very hard to see. Interstellar space could contain many cold, dark Jupiter-sized objects without our knowing it. Conceivably, they might even account for more mass than we observe in the form of stars and interstellar gas combined.

than about 10–20 times the mass of Jupiter cannot form by the usual contraction/fragmentation star-formation process, but the upper limit on the mass of a planet (which forms within a disk surrounding a protostar) is not well known. On the other hand, the theory of planet formation suggests that planets are likely to form in roughly circular orbits. ∞ (Sec. 15.2) Thus, the best we can say at present is that a low-mass (less than 10 Jupiter masses, say) object in a roughly circular orbit is most likely a planet, whereas a higher-mass object in an eccentric orbit is probably a brown dwarf. Whether the other two possible combinations (low mass/high eccentricity and high mass/low eccentricity) can occur, and how they will be classified if they are observed, remains to be seen.

Recent orbital measurements like those just described have so far produced two very strong brown dwarf candidates, both of them orbiting low-mass M-class stars, and at least three more possible brown dwarfs. (At present, the possibility that the latter three objects are planets cannot be completely ruled out.) In addition, in the past year, infrared and spectroscopic studies of stars have begun to find strong indications of isolated brown dwarfs. Infrared observations are a particularly effective way to search for brown dwarfs because these objects emit most of their radiation in the infrared part of the spectrum, while true stars tend to be brightest at optical or ultraviolet wavelengths.

The images at right show the binary star system Gliese 229, first identified as a possible brown dwarf by ground-based infrared observations (left), and recently imaged from space (right). The two objects are only 7″ apart; the fainter "star" has a luminosity only a few millionths that of the Sun and an estimated mass about 50 times that of Jupiter. (The bright diagonal streak in the latter image is caused by a hardware problem in the CCD chip used to record it.)

The presence of lithium, an element that is present in interstellar gas but that is rapidly depleted in a star once nuclear fusion begins, is another telltale indicator of brown dwarfs. Researchers using a spectrometer on the giant Keck telescope in Hawaii have so far identified two brown dwarf candidates by this means. Finally, several more candidates have been identified by combining multiple long-exposure images to reveal very faint, very red objects that had hitherto gone undetected.

The next few years may well see a flood of new observations of very-low-mass stars—perhaps even a revolution in our understanding of the entire star-formation process.

R I V U X G   R I V U X G

19.4 Observations of Cloud Fragments and Protostars

4 The evolutionary stages we have just described are derived from numerical experiments performed on high-speed computers. Table 19.1 and the evolutionary paths described in Figures 19.4, 19.6, and 19.7 are mathematical predictions of a multifaceted problem incorporating gravity, heat, rotation, magnetism, nuclear reaction rates, elemental abundances, and other physical conditions specifying the state of contracting interstellar clouds. Computer technology has enabled theorists to construct these models, but their accuracy is only partly known, in large part because it is difficult to test them observationally.

How then can we verify the theoretical predictions just outlined? The age of our entire civilization is much shorter than the time needed for a cloud to contract and form a star. We can never observe individual objects proceed through the full panorama of star birth. We can, however, observe many different objects—interstellar clouds, protostars, young stars approaching the main sequence—as they appear today at different stages of their evolutionary paths. Each observation is like part of a jigsaw puzzle. When properly oriented relative to all the others, the pieces can be used to build up a picture of the full life cycle of a star.

EVIDENCE OF CLOUD CONTRACTION

Prestellar objects at stages 1 and 2 are not yet hot enough to emit much infrared radiation, and certainly no optical radiation arises from their dark, cool interiors. The best way to study the early stages of cloud contraction and fragmentation is to observe the radio emission from interstellar molecules within these clouds. Consider again M20, the splendid emission nebula studied in Chapter 18. ∞ (Sec. 18.2) The brilliant region of glowing, ionized gas shown in Figure 18.8 is not our main interest here, however. Instead, the youthful O- and B-type stars that energize the nebula alert us to the general environment where stars are forming. Emission nebulae are signposts of star birth.

The region surrounding M20 contains galactic matter that seems to be contracting. The presence of (optically) invisible gas there was illustrated in Figure 18.20, which showed a contour map of the abundance of the formaldehyde (H_2CO) molecule. Formaldehyde and many other molecules are widespread in the vicinity of the nebula, especially throughout the dusty regions below and to the right of the emission nebula itself. Further analysis of the observations suggests that this region of greatest molecular abundance is also contracting and fragmenting, well on its way toward forming a star—or, more likely, a star cluster.

The interstellar clouds in and around M20 thus provide tentative evidence of three distinct phases of star formation, as shown in Figure 19.8. The huge, dark molecular cloud surrounding the visible nebula is the stage 1 cloud. Both its density and its temperature are low, about 10^8 particles/m³ and 20 K, respectively. Greater densities and temperatures typify

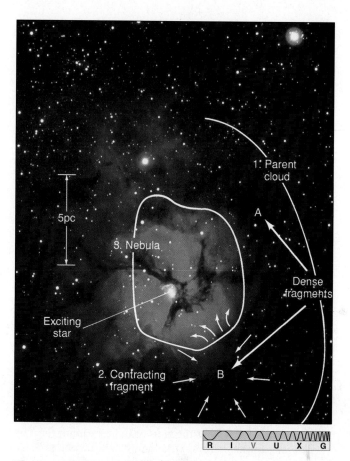

Figure 19.8 The M20 region shows observational evidence for three broad phases in the birth of a star: (1) the parent cloud (stage 1 of Table 19.1), (2) a contracting fragment (between stages 1 and 2), and (3) the emission nebula (M20 itself) resulting from the formation of one or more massive stars (stages 6 and 7).

smaller regions within this huge cloud. The totally obscured regions labeled A and B, where the molecular emission of radio energy is strongest, are such denser, warmer fragments. Here, the total gas density is observed to be at least 10^9 particles/m³, and the temperature is about 100 K. The Doppler shifts of the radio lines observed in the vicinity of region B indicate that this portion of M20, labeled "contracting fragment" in Figure 19.8, is contracting. Less than a light year across, this region has a total mass over 1000 times the mass of the Sun—considerably more than the mass of M20 itself. It lies somewhere between stages 1 and 2 of Table 19.1.

The third star-formation phase shown in Figure 19.8 is M20 itself. The glowing region of ionized gas results directly from a massive O-type star that formed there within the past million years or so. Because the central star is already fully formed, this final phase corresponds to stage 6 or 7 of our evolutionary scenario.

EVIDENCE OF CLOUD FRAGMENTS

Other parts of our Milky Way Galaxy provide sketchy evidence for prestellar objects in stages 3 through 5. The Orion complex, shown in Figure 19.9, is one such region. Lit from within by several O-type stars, the bright Orion Nebula is partly surrounded by a vast molecular cloud that extends well beyond the roughly 2 × 3 pc region bounded by the photograph in Figure 19.9(b).

The Orion molecular cloud harbors several smaller sites of intense radiation emitted by molecules deep within the core of the cloud fragment. Their extent, shown in Figures 19.9(c) and (e) measures about 10^{10} km, or 1/1000 of a light year, about the diameter of our solar system. The gas density of these smaller regions is about 10^{15} particles/m³, much denser than the surrounding cloud. Although their temperature cannot be estimated reliably, many researchers regard these regions as objects well on their way to stage 3. We cannot determine if these regions will eventually form stars like the Sun, but it does seem certain that these intensely emitting regions are on the threshold of becoming protostars.

EVIDENCE OF PROTOSTARS

In the hunt for and study of objects at more advanced stages of star formation, radio techniques become less useful because

stages 4, 5, and 6 have increasingly higher temperatures. By Wien's law, their emission shifts toward shorter wavelengths, and so these objects shine most strongly in the infrared. ∞ (Sec. 3.4) One particularly bright infrared emitter, known as the Becklin–Neugebauer object (see Figure 19.17c), was detected in the core of the Orion molecular cloud in the 1970s. Its luminosity is around 1000 times the luminosity of the Sun. Most astronomers agree that this warm, dense blob is a high-mass protostar, probably around stage 4.

Until the *Infrared Astronomy Satellite* was launched in the early 1980s, astronomers were aware only of giant stars forming in clouds far away. But *IRAS* showed that stars are forming much closer to home, and some of these protostars have masses comparable to that of our Sun. Figure 19.10 shows the premier example of a solar-mass protostar—

Figure 19.9   (a) The constellation Orion, with the region around its famous emission nebula marked by a rectangle. The Orion Nebula is the middle "star" of Orion's sword. The framed region is enlarged in part (b), suggesting how the nebula is partly surrounded by a vast molecular cloud. Various parts of this cloud are probably fragmenting and contracting, with even smaller sites forming protostars. The three frames at right show some of the evidence for those protostars. (c) A false-color radio image of some intensely emitting molecular sites. (d) A nearly real color visible image of embedded nebular "knots" thought to harbor protostars. (e) A high-resolution image of several young stars surrounded by disks of gas and dust where planets might ultimately form.

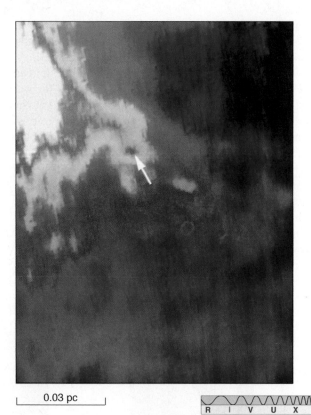

0.03 pc

R I V U X G

Figure 19.10 An infrared image of the nearby region containing the source Barnard 5 (indicated by the arrow). On the basis of its temperature and luminosity, Barnard 5 appears to be a protostar on the Hayashi track in the H–R diagram.

Barnard 5. Its infrared heat signature is that expected of an object on the Hayashi track, around stage 5.

The energy sources for some infrared objects seem to be luminous hot stars that are hidden from optical view by surrounding dark clouds. Apparently, some of the stars are already so hot that they emit large amounts of ultraviolet radiation, which is mostly absorbed by a "cocoon" of dust surrounding the central star. The absorbed energy is then reemitted by the dust as infrared radiation. These bright infrared sources are known as *cocoon nebulae*. Two considerations support the idea that the hot stars responsible for the clouds' heating have only recently ignited: (1) These dust cocoons are predicted to disperse quite rapidly once their central stars form, and (2) they are invariably found in the dense cores of molecular clouds. The central stars probably lie near stage 6.

PROTOSTELLAR WINDS

Protostars often exhibit strong winds. Radio and infrared observations of hydrogen and carbon monoxide molecules, again in the Orion cloud, have revealed gas expanding outward at velocities approaching 100 km/s. High-resolution interferometric observations have disclosed expanding knots of water emission within the same star-forming region and have linked the strong winds to the protostars themselves.

These winds may be related to the violent surface activity associated with many protostars.

Early in a protostar's life it may still be embedded in an extensive disk of nebular material in which planets are forming. ∞ (Sec. 15.2) When the protostellar wind begins to blow, it encounters less resistance in the directions perpendicular to the disk than in the plane of the disk. The result is known as a *bipolar flow*—two "jets" of matter are expelled in the directions of the poles of the protostar, as illustrated in Figure 19.11. As the protostellar wind gradually destroys the disk, blowing it away into space, the jets widen until, with the disk gone, the wind flows away from the star equally in all directions. Figure 19.12 shows a real bipolar flow, along with an artist's conception of the system producing it.

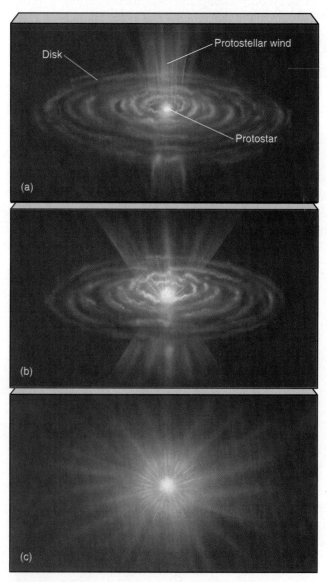

Figure 19.11 (a) When a protostellar wind encounters the disk of nebular gas surrounding the protostar, it tends to form a bipolar jet, preferentially leaving the system along the line of least resistance, which is perpendicular to the disk. (b) As the disk is blown away by the wind the jets fan out, eventually (c) merging into a spherical wind.

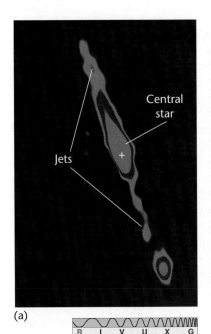

(a)

R I V U X G

Central star

Jets

+

(b)

Figure 19.12 (a) This false-color radio image shows two jets emanating from the young star system HH81-82 (whose position is marked with a cross at center). This is the largest stellar jet known, with a length of about 10,000 A.U. (The colors are coded in order of decreasing radio intensity: red, blue, green.) (b) An idealized artist's conception of a young star system, showing two jets flowing perpendicular to the disk of gas and dust rotating around the star. (See also the chapter-opening photos of more stellar jets on page 424.)

Animation

19.5 Shock Waves and Star Formation

5 The subject of star formation is in reality much more complicated than the preceding discussion suggests. Interstellar space is populated with many kinds of clouds, fragments, protostars, stars, and nebulae. They all interact in a complex fashion, and each type of object affects the behavior of all the others. For example, the presence of an emission nebula in or near a molecular cloud probably influences the evolution of the entire region. We can easily imagine expanding waves of matter driven outward by the high temperatures and pressures in the nebula. As the waves crash into the surrounding molecular cloud, interstellar gas tends to pile up and become compressed. Such a shell of gas, rushing rapidly through space, is known as a **shock wave**. It can push ordinarily thin matter into dense sheets, just as a plow pushes snow.

Many astronomers regard the passage of a shock wave through interstellar matter as the triggering mechanism needed to initiate star formation in a galaxy. Calculations show that when a shock wave encounters an interstellar cloud, it races around the thinner exterior of the cloud more rapidly than it can penetrate its thicker interior. Thus, shock waves do not blast a cloud from only one direction. They effectively squeeze it from many directions, as illustrated in Figure 19.13. Atomic bomb tests have experimentally demonstrated this squeezing—shock waves created in the blast tend to surround buildings, causing them to be blown together (imploded) rather than apart (exploded).

Once shock waves have begun the compression of an interstellar cloud, natural gravitational instabilities take over, dividing it into the fragments that eventually form stars. Figure 19.14 suggests how this mechanism might be at work

near M20. Note the correspondence between the shock-compressed region at lower right and the high-density molecular gas revealed by radio studies (Figure 18.19).

Emission nebulae are by no means the only generators of interstellar shock waves. At least two other driving forces are available—the deaths of old stars (planetary nebulae and supernovae, to be discussed in Chapters 20 and 21) and the spiral-arm waves that plow through the Milky Way (to be discussed in Chapter 23). Supernovae are by far the most

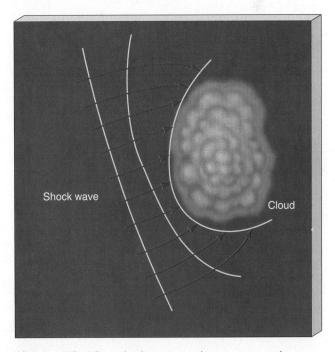

Shock wave

Cloud

Figure 19.13 Shock waves tend to wrap around interstellar clouds, compressing them to greater densities and thus possibly triggering star formation.

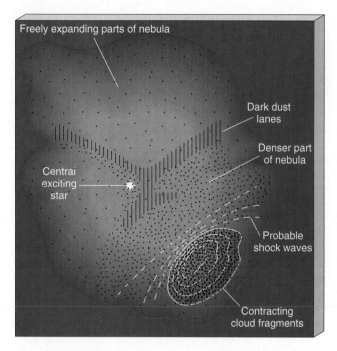

Figure 19.14 An artist's conception of a cloud fragment undergoing compression on the southerly edge of M20 as shock waves from the nebula penetrate the surrounding interstellar cloud.

energetic, and probably also the most efficient, way to pile up matter into dense clumps. However, they are relatively few and far between, so the other mechanisms may be more important overall in triggering star formation. Although the evidence is somewhat circumstantial, the presence of young (and thus quick-forming) O- and B-type stars in the vicinity of supernova remnants does suggest that the birth of stars is often initiated by the violent, explosive deaths of others.

This picture of shock-induced star formation is complicated by the fact that O- and B-type stars form quickly, live briefly, and die explosively. These massive stars, themselves perhaps born of a passing shock wave, may in turn create new shock waves, either through the expanding nebular gas produced by their births or by their explosive deaths. These new shock waves can produce "second-generation" stars, which in turn will explode and give rise to still more shock waves, and so on. As depicted in Figure 19.15, star formation resembles a chain reaction. Other, lighter, stars are also formed in the process, of course, but they are largely "along for the ride." It is the O- and B-type stars that drive the star-formation wave through the cloud. Observational evidence lends some support to this chain-reaction model. Groups of stars nearest molecular clouds do indeed appear to be the youngest, whereas those farther away seem to be older.

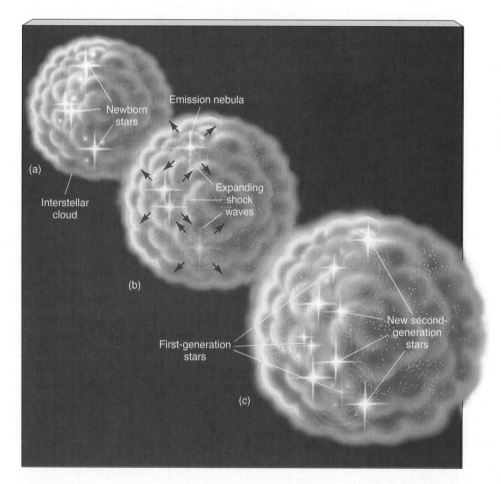

Figure 19.15 (a) Star birth and (b) shock waves lead to (c) more star births and more shock waves in a continuous cycle of star formation in many areas of our Galaxy. Like a chain reaction, old stars trigger the formation of new stars ever deeper into an interstellar cloud.

19.6 Emission Nebulae and Star Clusters

4 We have seen how a portion of an interstellar cloud can become unstable, collapse, and fragment into stars. Let us take a moment to ask what happens next—not to the newborn stars themselves (that will be the subject of the next three chapters), but to the Galactic environment in which they have just formed.

The end result of cloud collapse is a group of stars, all formed from the same parent cloud and lying in the same region of space—in other words, a *star cluster*. ∞ (Sec. 17.10) As a by-product of cluster formation, a certain amount of unused gas and dust remains. How many stars form, and of what type? How much gas is left over? What does the collapsed cloud look like once star formation has run its course?

At present, although the main stages in the formation of an individual star (stages 3–7) are becoming clearer, the answers to these more general questions (involving stages 1 and 2) are still very sketchy; they await a more thorough understanding of the star-formation process.

GAS DISPERSAL

In general, the more massive the collapsing region, the more stars are likely to form there. In addition, we know from observed H–R diagrams that low-mass stars are much more common than high-mass ones. ∞ (Sec. 17.7) The precise number of stars of any given mass or spectral type likely depends in a complex (and poorly understood) way on conditions within the parent cloud. The same is true of the *efficiency* of star formation—that is, the fraction of the total mass that actually finds its way into stars—which determines the amount of leftover material. However, if, as is usually the case, one or more O- or B-type stars form, their intense radiation and winds will cause the surrounding gas to disperse, leaving behind a young star cluster.

Figure 19.16(a) illustrates this process. They are contrast-enhanced views of part of the Eagle Nebula (Figure 18.9b), showing several narrow "fingers" of gas protruding from the end of the largest (rightmost) pillar in the earlier figure. The fingers remain behind as the rest of the

(a)

R I V U X G

Figure 19.16 (a) This enlargement of Figure 18.9(b) shows evidence of "evaporating gaseous globules" (known as EGGs)—star-forming fragments dense enough to survive the onslaught of ultraviolet radiation from newly formed hot, young stars in their vicinity. The EGGs lie at the ends of narrow fingers of gas projecting from the tip of the largest pillar. (b) The fingers survive because they lie in the EGGs' shadows and so are shielded from the radiation dispersing the rest of the cloud. From left to right, the sequence shows (1) a dense pocket of gas near the surface of the evaporating cloud, (2) the formation of a finger of gas in the EGG's shadow as the surrounding cloud evaporates, and (3) the finger beginning to detach from the cloud, and the central stellar or prestellar core of the EGG beginning to emerge as the EGG itself is slowly destroyed.

(b)

INTERLUDE 19-2 Eta Carinae

At the heart of the Carina emission nebula (shown at left in the figure) lies a remarkable object called Eta Carinae (right-hand images). With an estimated mass of around 100 times the mass of the Sun and a luminosity of 5 million times the solar value, Eta Carinae is the most massive star known. Formed probably only a few hundred thousand years ago, the short life of this star has been explosive indeed. In the mid-nineteenth century, Eta Carinae experienced an outburst that made it one of the brightest stars in the southern sky (even though it lies some 3000 pc away from Earth, a very long way compared to most of the bright stars visible in our night sky). The star released as much visible energy as a supernova explosion (see Chapter 21), yet it somehow survived the event. The object shown at bottom right below is the result of that outburst.

The bottom-right image, obtained by *HST* and carefully processed to reveal fine detail, is the highest-resolution view of the explosion obtained to date. Dust lanes, tiny

condensations in the outflowing material, and dark radial streaks of unknown origin all appear with exquisite clarity. The star itself is the white dot at the center of the image. The two ends of the "peanut" (at top right and bottom left) are blobs of ejected material racing away from the star at hundreds of kilometers per second—perhaps enough to expel the surrounding nebular gas and convert the Carina Nebula into the Carina Cluster. Perpendicular to the line joining these two blobs is a thin disk of gas, also moving outward at high speed.

The details of the events leading to this outburst are unclear; quite possibly such episodes of violent activity are the norm for supermassive stars. Smaller outbursts in the radio and X-ray parts of the spectrum have also been observed over the past few years. In addition to helping unravel the mysteries of Eta Carinae itself, astronomers hope that continued studies of this object will also provide insights into the less violent jets and other outflows associated with star formation.

cloud evaporates and disperses because their dense tips, which are thought to contain solar system–sized stage 3 fragments, shield the rest of the gas from ultraviolet radiation (Figure 19.16b). Eventually the dense fragments too are destroyed, and the stage 4 protostar emerges from its protective shell.

These observations clearly illustrate the important role played by environment in the star-formation process. The first massive stars to form tend to prevent the formation of additional high-mass stars by disrupting the environment in which stars are growing. This is one reason why low-mass stars are so much more common than high-mass stars; it also helps explain the existence of brown dwarfs, by providing a natural way in which star formation can stop before nuclear fusion begins in the growing stellar core. *Interlude 19-2* describes another system in which the gas-dispersal process is almost complete.

CLUSTER LIFETIMES

Until recently, the existence of star clusters within emission nebulae was largely conjecture—the stars cannot be seen optically because they are obscured by dust. However, infrared observations have now clearly demonstrated that stars really are found within star-forming regions! Figure 19.17 compares optical and infrared views of the central regions of the Orion Nebula. The optical image in Figure 19.17(a) shows

the Trapezium, the group of four bright stars responsible for ionizing the nebula; the false-color infrared image in Figure 19.17(c) reveals an extensive cluster of stars within and behind the visible nebula. The Becklin–Neugebauer object (Section 19.4) can be seen as the central yellow spot within this region; it is thought to be a dust-shrouded B star just beginning to form its own emission nebula. These remarkable images show many stages of star formation.

For every O or B giant, tens or even hundreds of G, K, and M dwarfs may form. Thus, even a modest emission nebula can give rise to a fairly extensive collection of stars. A typical open star cluster, like that shown in Figure 19.18, may measure 10 pc across and contain 1000 or more stars. Less massive, but more extended, clusters are usually known as **associations**. These typically contain no more than 100 bright stars but may span many tens of parsecs. Associations tend to be rich in very young stars. Those containing many pre–main-sequence T Tauri stars are known as *T associations*, whereas those with prominent O and B stars, such as the Trapezium in Orion, are called *OB associations*. It is quite likely that the main difference between associations and open clusters is simply the efficiency with which stars formed from the parent cloud.

Eventually, star clusters dissolve into individual stars. In some cases, the ejection of unused gas reduces a cluster's mass so much that it becomes gravitationally unbound and rapidly dissolves. In clusters that survive the early gas-loss phase,

Figure 19.17   Some views of the central regions of the Orion Nebula. (a) A short-exposure visible-light image (using a filter that is transparent only to certain emission lines of oxygen) shows the nebula itself and four bright O stars known as the Trapezium. (b) A magnified view of a smaller part of the nebula shows much irregular gas and dust but few obvious stars, which are hidden in the dust. (c) This short-exposure infrared image, acquired by the *Hubble* telescope in 1997, shows several faint red stars emerging from the nebular gas; the brightest star is known as the Becklin–Neugebauer object.

Figure 19.18 The Jewel Box cluster is a relatively young open cluster in the southern sky. Many bright stars appear in this image, but the cluster contains many more low-mass, less luminous stars. Because some red giants appear among the cluster's blue main-sequence stars, we can estimate the age of the cluster to be about 10 million years.

stellar encounters tend to eject the lightest stars from the cluster, just as the gravitational slingshot effect can propel spacecraft around the solar system. ∞ (*Interlude 6-2*) At the same time, the tidal gravitational field of the Milky Way Galaxy slowly strips outlying stars from the cluster. Occasional distant encounters with giant molecular clouds also tend to remove cluster stars; a near miss may even disrupt the cluster entirely.

Figure 19.19 M67, one of the oldest known open clusters, has survived for almost 5 billion years—an unusually long time for a star system near the plane of the Milky Way Galaxy.

As a result of all these influences, most open clusters break up in a few hundred million years, although the actual lifetime depends on the cluster's mass. Loosely bound associations may survive for only a few tens of millions of years, whereas some very massive open clusters such as M67, shown in Figure 19.19, are known from their H–R diagrams to be almost 5 billion years old. In a sense, only when a star's parent cluster has completely dissolved is the star-formation process really complete. The road from a gas cloud to a single, isolated star like the Sun is long and tortuous indeed!

Take another look at the nighttime sky. Ponder all that cosmic activity while gazing upward one clear, dark evening. After studying this chapter, you may find you have to modify your view of the night sky. Even the seemingly quiet nighttime darkness is dominated by continual change.

Chapter Review

SUMMARY

Stars form when an interstellar cloud collapses under its own gravity and breaks up into pieces comparable in mass to our Sun. Heat, rotation, and magnetism all compete with gravity to influence the cloud's evolution. The evolution of the contracting cloud—the changes in its temperature and luminosity—can be conveniently represented as an **evolutionary track** (p. 429) on the Hertzsprung–Russell diagram. A cold interstellar cloud containing a few thousand solar masses of gas can fragment into tens or hundreds of smaller clumps of matter, from which stars eventually form.

As a collapsing prestellar fragment heats up and becomes denser it eventually becomes a **protostar** (p. 429)—a warm, very luminous object that emits radiation mainly in the infrared portion of the electromagnetic spectrum. At this stage of its evolution, the protostar is also known as a **T Tauri star** (p. 430), after the first object of this type discovered.

Eventually, a protostar's central temperature becomes high enough for hydrogen fusion to begin, and the protostar becomes a star. For a star like the Sun, the whole formation process takes about 50 million years. More massive stars pass through similar stages, but much more rapidly. Stars less massive than the Sun take much longer to form. The **zero-age main sequence** (p. 432) is the region on the H–R diagram where stars lie when the formation process is over.

Mass is the key property for determining a star's characteristics and life span. The most massive stars have the shortest formation times and main-sequence lifetimes. At the other extreme, some low-mass fragments never reach the point of nuclear ignition.

The universe may be populated with a vast number of **brown dwarfs** (p. 433)—objects that are not massive enough to fuse hydrogen to helium in their interiors.

Many of the objects predicted by the theory of star formation have been observed in real astronomical objects. The dark interstellar regions near emission nebulae often provide evidence of cloud fragmentation and protostars. Radio telescopes are used for studying the early phases of cloud contraction and fragmentation; infrared observations allow us to see later stages of the process. Many well-known emission nebulae, lit by several O-type stars, are partially engulfed by molecular clouds, parts of which are probably fragmenting and contracting, with smaller sites forming protostars.

Protostellar winds encounter less resistance in the directions perpendicular to a protostar's disk. Thus they expel two jets of matter in the directions of the protostar's poles. As the protostellar wind gradually destroys the disk the jets widen until, with the disk gone, the wind flows away from the star equally in all directions.

Shock waves (p. 437) can compress other interstellar clouds and trigger star formation. Star birth and the production of shock waves are thought to produce a chain reaction of star formation in molecular cloud complexes.

A single collapsing and fragmenting cloud can give rise to hundreds or thousands of stars—a star cluster. Infrared observations have revealed young star clusters in several emission nebulae. Loosely bound groups of newborn stars are called stellar **associations** (p. 441). Eventually, star clusters break up into individual stars, although the process may take billions of years to complete.

SELF-TEST: TRUE OR FALSE?

_____ **1.** Given the typical temperatures found in interstellar space, a cloud containing as few as 1000 atoms has sufficient gravity for it to begin to collapse.

_____ **2.** Both rotation and magnetic fields act to accelerate the gravitational collapse of an interstellar cloud.

_____ **3.** The time a solar-type star spends forming is relatively short compared to the time it spends as a main-sequence star.

_____ **4.** Stars evolve along the main sequence.

_____ **5.** Most stars form as members of groups or clusters of stars.

_____ **6.** A stage 4 object has a luminosity about 1000 times that of the current Sun.

_____ **7.** As it evolves along the Hayashi track from stage 4 to stage 6, a prestellar object moves more or less horizontally across the H-R diagram.

_____ **8.** The rate of evolution of a stage 5 object is fast compared with the rates at previous stages.

_____ **9.** Brown dwarfs take a long time to form but will eventually arrive as stars on the lower main sequence.

_____ **10.** Stages 1 and 2 of star formation can be observed using optical telescopes.

_____ **11.** Shock waves produced from emission nebulae can initiate star formation in nearby molecular clouds.

_____ **12.** Shock waves for star formation can also be produced by large stars moving rapidly through the interstellar medium.

_____ **13.** In star formation, more G, K, and M stars form than O and B stars.

_____ **14.** The gas in an emission nebula eventually dissipates into space, leaving behind a star cluster.

_____ **15.** Star clusters eventually dissipate, leaving behind individual stars like the Sun.

SELF-TEST: FILL IN THE BLANK

1. Atoms in an interstellar cloud have random motions, with an average velocity determined by the cloud's _____.

2. A(n) _____ plots a star or protostar's changing location on the H-R diagram as the object evolves.

3. In stage 1 of prestellar evolution, a typical interstellar cloud has the following properties: temperature _____ K, size _____ pc, mass _____ solar masses.

4. In stage 2 of prestellar evolution, a contracting interstellar cloud _____ into smaller pieces.

5. During stage 3 of prestellar evolution, as each piece of the original interstellar cloud continues to contract, its central density and temperature _____.

6. At stage 4 of prestellar evolution, each piece of the interstellar cloud becomes a _____.

7. A stage 4 object is plotted in the _____ (upper/lower) _____ (right/left) part of the H-R diagram.

8. At stage 6 the central temperature of the object reaches _____ K.

9. At this temperature, a stage 6 object begins to _____.

10. At stage 7, the star has reached the _____.

11. The T Tauri phase of a star occurs during stage _____.

12. It takes a star like the Sun a total of about _____ million years to form.

13. Stars much more massive than the Sun take about _____ years to form; very low mass stars may take over _____ years.

14. Astronomers look for emissions at _____ wavelengths to identify interstellar clouds in stages 1 and 2.

15. At stages 4, 5, and 6, objects emit a great deal of radiation in the _____ part of the electromagnetic spectrum.

REVIEW AND DISCUSSION

1. Briefly describe the basic chain of events leading to the formation of a star like the Sun.

2. What is the role of heat in the process of stellar birth?

3. What is the role of rotation in the process of stellar birth?

4. What is the role of magnetism in the process of stellar birth?

5. What is an evolutionary track?

6. Why do stars tend to form in groups?

7. At what point does a star-forming cloud become a protostar? When does it become a full-fledged star?

8. What are brown dwarfs?

9. What are T Tauri stars?

10. Because stars live much longer than we do, how do astronomers test the accuracy of theories of star formation?

11. At what evolutionary stages must astronomers use radio and infrared radiation to study prestellar objects? Why can't they use visible light?

12. What is a shock wave? Of what significance are shock waves in star formation?

13. Explain the usefulness of the Hertzsprung–Russell diagram in studying the evolution of stars. Why can't evolutionary stages 1–3 be plotted on the diagram?

14. Compare the times necessary for the various stages in the formation of a star like the Sun. Why are some so short and others so long?

15. In the formation of a star cluster with a wide range of stellar masses, is it possible for some stars to die out before others have finished forming? Do you think this will have any effect on the cluster's formation?

PROBLEMS

1. In order for an interstellar gas cloud to contract, the average speed of its constituent atoms must be less than half the cloud's escape speed. Will a hydrogen cloud of mass 1000 solar masses, radius 10 pc, and temperature of 10 K begin to collapse? (See *More Precisely 8-1*.)

2. A certain interstellar cloud contains 10^{60} atoms. Hydrogen (mass per atom $= 1.7 \times 10^{-27}$ kg) accounts for 90 percent of the atoms, and the remainder are helium (each atom having four times the mass of a hydrogen atom). What is the cloud's mass, in solar masses (2×10^{30} kg)?

3. Use the radius–luminosity–temperature relation ($L \propto R^2 T^4$; see Section 17.3) to explain how a protostar's luminosity changes as it moves from stage 4 ($T = 3000$ K, $R = 2 \times 10^8$ km) to stage 6 ($T = 4500$ K, $R = 10^6$ km).

4. A protostar on the Hayashi track evolves from a temperature $T = 3500$ K and a luminosity of $L = 5000$ times that of the Sun to $T = 5000$ K and $L = 3$ solar units. What is its radius (a) at the start of the evolution? (b) At the end of the evolution? (c) When the luminosity is half the initial value?

5. Through how many magnitudes does a 3–solar mass star decrease in brightness as it evolves from stage 4 to stage 6? (See Figure 19.7.)

6. As a simple model of the final stage of star formation, imagine that between stages 6 and 7 a star's surface temperature increases with time at a constant rate while the luminosity remains constant at the stage 7 level. The stage 7 radius is equal to the solar value. Using the temperatures given in Table 19.1, calculate the star's radius at a time exactly halfway between these two stages.

7. What is the luminosity, in solar units, of a brown dwarf whose radius is 0.1 solar radii and whose surface temperature is 600 K (0.1 that of the Sun)?

8. A shock wave from a supernova explosion moves at a speed of about 5000 km/s. How long will such a disturbance take to cross a molecular cloud 20 pc in diameter?

9. A typical open cluster has a diameter of 10 pc; its component stars have an average speed of 1 km/s. Estimate the number of times a typical star orbits the center of the cluster in the 1 billion years it takes for the cluster to dissolve in the Galactic tidal field.

10. Approximating the gravitational field of our Galaxy as a mass of 10^{11} solar masses at a distance of 8000 pc (see Chapter 23), estimate the "tidal radius" of a 2000–solar mass star cluster—that is, the distance from the cluster center outside of which the Galactic tidal force overwhelms the cluster's gravity.

PROJECT

1. The Trifid Nebula, otherwise known as M20, is a place where new stars are forming. It has been called a "dark night revelation, even in modest apertures." An 8- to 10-inch telescope is needed to see the triple-lobed structure of the nebula. Ordinary binoculars reveal the Trifid as a hazy patch located in the constellation Sagittarius. This nebula is set against the richest part of the Milky Way, the edgewise projection of our own Galaxy around the sky. It is one of many wonders in this region of the heavens. What are the dark lanes in M20? Why are other parts of the nebula bright? There have been reports of large-scale changes occurring in this nebula in the last century and a half. The reports are based on old drawings, which show M20 looking slightly different from how it appears today. Do you think it possible for a cloud in space to undergo a change in appearance on a time scale of years, decades, or centuries?

20 STELLAR EVOLUTION

From Middle Age to Death

LEARNING GOALS

Studying this chapter will enable you to:

1 Explain why stars evolve off the main sequence.

2 Outline the events that occur after a Sun-like star exhausts the supply of hydrogen in its core.

3 Summarize the stages in the death of a typical low-mass star and describe the resulting remnant.

4 Contrast the evolutionary histories of high-mass and low-mass stars.

5 Discuss the observations that help verify the theory of stellar evolution.

6 Explain how the evolution of stars in binary systems may differ from that of isolated stars.

(Opposite page, background) These frames are the conceptions of noted space artist Dana Berry. The ten frames depict a sequence of the birth, evolution, and death of a binary-star system. The sequence starts with the large rendering of a 1–solar-mass star and a 4–solar-mass star in the process of formation and then proceeds clockwise from top left. Some highlights:
(Inset B) Nearing the end of its life, the 4–solar-mass star swells, spilling gas onto its companion and forming an accretion disk around it.
(Inset E) The red giant has reached the point where the gravity of the original two stars cannot contain the gas. The result is a gentle expulsion—forming a planetary nebula rich in oxygen gas (hence the green color).
(Inset G) Much later, an accretion bridge again joins the two stars as the lower mass star becomes a red giant.
(Inset I) The end point of the system is two white dwarfs of roughly equal mass circling each other forevermore.

After reaching the main sequence, a newborn star changes little in outward appearance for more than 90 percent of its lifetime. However, at the end of this period, as the star begins to run out of fuel and die, its properties once again change greatly. Aging stars travel along evolutionary tracks that take them far from the main sequence as they end their lives. In this and the next two chapters, we will study the evolution of stars during and after their main-sequence burning stages. We will find that the ultimate fate of a star depends primarily on its mass, although interactions with other stars can also play a decisive role, and that the final states of stars can be strange indeed. By continually comparing theoretical calculations with detailed observations of stars of all types, astronomers have refined the theory of stellar evolution into a precise and powerful tool for understanding the universe.

20.1 Leaving the Main Sequence

1 Most stars spend most of their lives on the main sequence. A star like the Sun, for example, after spending a few tens of millions of years in formation (stages 1–6 in Chapter 19), resides on or near the main sequence (stage 7) for 10 billion years before evolving into something else. ∞ (Sec. 19.2) That "something else" is the topic of this chapter.

Virtually all the low-mass stars that have ever formed still exist as stars. The coolest M stars—red dwarfs—burn so slowly that not one of them has yet left the main sequence. Some of them will burn steadily for a trillion years or more. ∞ (Sec. 17.9) Conversely, the most massive O and B stars evolve away from the main sequence after only a few tens of millions of years. Most of the high-mass stars that have ever existed perished long ago. Between these two extremes, many stars are observed in advanced stages of evolution, their properties quite different from when they formed. By combining these observations with theoretical models, astronomers have built up a comprehensive picture of how stars evolve.

On the main sequence, a star slowly fuses hydrogen into helium in its core. This process is called **core hydrogen burning**. In Chapter 16 we saw how the proton–proton fusion chain powers the Sun. ∞ (Sec. 16.5) *More Precisely 20-1* describes another sequence of nuclear reactions, of great importance in stars more massive than the Sun, that accomplishes the same basic result as the proton–proton chain, but in a very different way.

A star's equilibrium during the main-sequence phase is the result of a balance between gravity and pressure, in which pressure's outward push exactly counteracts gravity's inward pull (Figure 20.1). Eventually, however, as the hydrogen in the core is consumed, the balance starts to shift, and both the star's internal structure and its outward appearance begin to change: the star leaves the main sequence. You should keep Figure 20.1 in mind as you study the various stages of stellar evolution described next. Much of a star's complex behavior can be understood in these simple terms.

Once a star begins to evolve away from the main sequence, its days are numbered. The post–main-sequence stages of stellar evolution—the end of a star's life—depend critically on the star's mass. As a rule of thumb, we can say that low-mass stars die gently, whereas high-mass stars die catastrophically. The dividing line between these two very different outcomes lies around eight times the mass of the Sun,

Figure 20.1 In a steadily burning star on the main sequence, the outward pressure of hot gas balances the inward pull of gravity. This is true at every point within the star, guaranteeing its stability.

Pressure out
Gravity in

and in this chapter we will refer to stars of more than 8 solar masses as "high-mass" stars. Realize, though, that within both the "high-mass" and the "low-mass" (that is, less than 8 solar masses) categories there are substantial variations.

Rather than dwelling on the many details, we will concentrate on a few representative evolutionary sequences. We begin by considering the evolution of a fairly low mass star like the Sun. The stages described in the next few sections pertain to the Sun as it nears the end of its fusion cycle 5 billion years from now. In fact, most of the qualitative features of the discussion apply to any low-mass star, although the exact numbers vary considerably. Later, we will broaden our discussion to include all stars, large and small.

20.2 The Evolution of a Sun-like Star

The surface of a main-sequence star like the Sun occasionally erupts in flares and spots, but for the most part the star does not experience sudden, large-scale changes in its properties.

Its average temperature remains fairly constant while its luminosity increases very slowly with time. The Sun has roughly the same surface temperature as it had when it formed nearly 5 billion years ago, and is some 30 percent brighter than it was at that time.

THE HELIUM CORE

This stable state of affairs cannot continue indefinitely. Eventually, drastic changes occur in the star's interior structure. After approximately 10 billion years of steady core hydrogen burning, a Sun-like star begins to run out of fuel. The situation is a little like that of an automobile cruising effortlessly along a highway at a constant speed of 55 mph for many hours, only to have the engine suddenly cough and sputter as the gas gauge reaches empty. Unlike automobiles, though, stars are not easy to refuel.

As nuclear burning proceeds, the composition of the star's interior changes. Figure 20.2 illustrates the increase in helium abundance and the corresponding decrease in hydrogen abundance that take place in the stellar core as the star ages. Three cases are shown: (a) the chemical composition of the original core, (b) the composition after 5 billion years, and (c) the composition after 10 billion years. Case (b) represents approximately the present state of our Sun.

The star's helium content increases fastest at the center, where temperatures are highest and the burning is fastest. The helium content also increases near the edge of the core, but more slowly because the burning rate is less rapid there. The inner, helium-rich region becomes larger and more hydrogen deficient as the star continues to shine. Eventually, hydrogen becomes completely depleted at the center, the nuclear fires there cease, and the location of principal burning moves to higher layers in the core. An inner core of nonburning pure helium starts to grow.

Without nuclear burning to maintain it, the outward-pushing gas pressure weakens in the helium inner core; however, the inward pull of gravity does not. Once the outward push against gravity is relaxed—even a little—structural changes in the star become inevitable. As soon as hydrogen becomes substantially depleted, about 10 billion years after the star arrived on the main sequence (Figure 20.2c), the helium core begins to contract.

If more heat could be generated, then the core might possibly return to equilibrium. For example, if helium in the core were to begin fusing into some heavier element, then all would be well once again. Energy would be created as a by-product of helium burning, and the necessary outward-pushing gas pressure would be reestablished. But the helium at the center cannot burn—not yet, anyway. Despite its high temperature, the core is far too cold to fuse helium into anything heavier.

Recall from Chapter 16 that a minimum temperature of about 10^7 K is needed to fuse hydrogen into helium. Only above that temperature do colliding hydrogen nuclei (that is, protons) have enough speed to overwhelm the repulsive electromagnetic force between them. ∞ (Sec. 16.5) Because

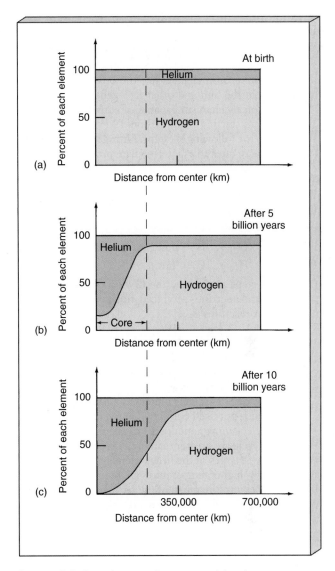

Figure 20.2 Theoretical estimates of the changes in a Sun-like star's composition. Hydrogen (yellow) and helium (orange) abundances are shown (a) at birth, on the zero-age main sequence; (b) after 5 billion years; and (c) after 10 billion years. At stage (b) only about 5 percent of the star's total mass has been converted from hydrogen into helium. This change speeds up as the nuclear burning rate increases with time.

helium nuclei, with two protons each, carry a greater positive charge, their electromagnetic repulsion is larger, and even higher temperatures are needed to cause them to fuse—at least 10^8 K. A core composed of helium at 10^7 K thus cannot generate energy through fusion.

The shrinkage of the helium core releases gravitational energy, driving up the central temperature and heating the overlying layers. The higher temperatures—now well over 10^7 K (but still less than 10^8 K)—cause hydrogen nuclei to fuse even more rapidly than before. Figure 20.3 depicts this situation, in which hydrogen is burning at a furious rate in a shell surrounding the nonburning inner core of helium "ash" in the center. This phase is usually known as the **hydrogen-shell-burning stage**. The hydrogen shell generates energy

MORE PRECISELY 20-1 The CNO Cycle

The proton–proton chain is not the only nuclear process operating in the Sun and other late-generation stars. Another fusion mechanism capable of converting hydrogen into helium, starting from carbon-12 (^{12}C), proceeds according to the following six steps. Nitrogen (N) and oxygen (O) nuclei are created as intermediate products:

$$^{12}\text{C} + {^1}\text{H} \rightarrow {^{13}}\text{N} + \text{energy}$$
$$^{13}\text{N} \rightarrow {^{13}}\text{C} + \text{positron} + \text{neutrino}$$
$$^{13}\text{C} + {^1}\text{H} \rightarrow {^{14}}\text{N} + \text{energy}$$
$$^{14}\text{N} + {^1}\text{H} \rightarrow {^{15}}\text{O} + \text{energy}$$
$$^{15}\text{O} \rightarrow {^{15}}\text{N} + \text{positron} + \text{neutrino}$$
$$^{15}\text{N} + {^1}\text{H} \rightarrow {^{12}}\text{C} + {^4}\text{He}$$

These six steps are termed the *CNO cycle*. Aside from the radiation and neutrinos produced, notice that the sum total of these six reactions is

$$^{12}\text{C} + 4({^1}\text{H}) \rightarrow {^{12}}\text{C} + {^4}\text{He}.$$

In other words, the net result is the fusion of four protons into a single helium-4 nucleus, just as in the proton–proton chain. The carbon-12 acts merely as a *catalyst*, an agent of change that is not itself consumed in the reaction.

The electromagnetic forces of repulsion operating in the CNO cycle are greater than in the proton–proton chain because the charges of the heavy-element nuclei are larger. Accordingly, higher temperatures are required to propel the heavy nuclei into the realm of the strong nuclear force and to ignite fusion. The accompanying figure presents a numerical estimate of the energy released in the Sun by the proton–proton chain and the CNO cycle, each as a func-

tion of gas temperature. The proton–proton chain dominates at lower temperatures, up to about 16 million K. Above this temperature, the CNO cycle is the more important fusion process in stars of solar composition.

According to our theoretical models of the Sun, the temperature of the solar core is 15 million K, so these curves indicate that the proton–proton cycle is the dominant source of solar energy (notice that each step on the vertical scale corresponds to a *factor* of 100 in energy generation). The CNO cycle contributes no more than 10 percent of the observed solar radiation. However, stars more massive than our Sun often have core temperatures much higher than 20 million K, making the CNO cycle the dominant energy-production mechanism.

faster than did the original main-sequence star's hydrogen-burning core, and the shell's energy production continues to increase as the helium core continues to shrink. Strange as it may seem, the star's response to the disappearance of the fire at its center is to get brighter!

THE RED-GIANT BRANCH

Conditions in the aging star have clearly changed from the equilibrium that characterized it as a main-sequence object. The helium core is unbalanced and shrinking. The rest of the core is also unbalanced, fusing hydrogen into helium at an ever-increasing rate. The gas pressure exerted by this enhanced hydrogen burning increases, forcing the star's nonburning outer layers to increase in radius. Not even gravity can stop

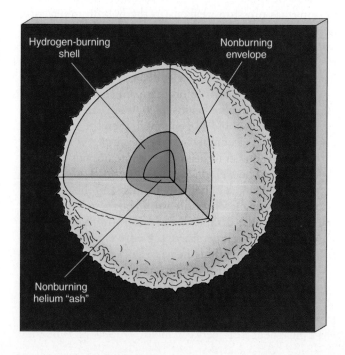

Figure 20.3 As a star's core loses more and more of its hydrogen the hydrogen in the shell surrounding the nonburning helium ash burns ever more violently.

them. While the core is shrinking and heating up, the overlying layers are expanding and cooling. The star is on its way to becoming a red giant. The change from normal main-sequence star to elderly red giant takes about 100 million years.

We can trace these large-scale changes on the H–R diagram. Figure 20.4 shows the path away from the main sequence, labeled as stage 7. The star first evolves to the right on the diagram, its surface temperature dropping while its luminosity increases only slightly. The star's roughly horizontal path from its main-sequence location (stage 7) to stage 8 on the figure is called the **subgiant branch**. By stage 8, the star's radius has increased to about three times the radius of the Sun.

The surface temperature at stage 8 has fallen to the point at which much of the interior is opaque to the radiation from within. Beyond this point, convection carries the core's enormous energy output to the surface. One consequence is that the star's surface temperature remains nearly constant between stages 8 and 9. The nearly vertical path followed by the star between stages 8 and 9 is known as the **red-giant branch** of the H–R diagram. By stage 9, the giant's luminosity is many hundreds of times the solar value, and its radius is around 100 solar radii.

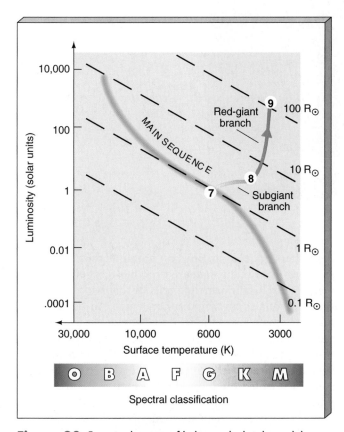

Figure 20.4   As the core of helium ash shrinks and the intermediate stellar layers expand, the star leaves the main sequence (stage 7). At stage 8 the star is on its way to becoming a red giant. The star continues to brighten and grow as it ascends the red-giant branch to stage 9, the top of the red-giant branch. As in Chapter 19, the diagonal lines correspond to stars of constant radius, allowing us to gauge the changes in the size of our star.

Figure 20.5 compares the relative sizes of a G star like our Sun and a stage 9 red giant. It also indicates the stages through which the star will evolve. The red giant is huge—about the size of Mercury's orbit. In contrast, its helium core is surprisingly small—only about 1/1000 the size of the entire star, making the core just a few times larger than Earth.

The density at the center of a red giant is enormous. Continued shrinkage of the red giant's core has compacted its helium gas to approximately 10^8 kg/m^3. Contrast this value with the 10^{-3} kg/m^3 in the giant's outermost layers, with the 5000 kg/m^3 average density of Earth, and with the 150,000 kg/m^3 in the present core of the Sun. About 25 percent of the mass of the entire star is packed into its planet-sized core. Perhaps the most famous red giant is the naked-eye star Betelgeuse in the constellation Orion (shown in Figure 17.10). Despite its great distance from Earth (about 150 pc), its enormous luminosity, 10^4 times that of the Sun, makes it one of the brightest stars in the night sky.

HELIUM FUSION

2 Should the unbalanced state of a red-giant star continue, the core would eventually collapse, and the rest of the star would slowly drift into space. The forces and pressures at work inside a red giant would literally pull it apart. However, this simultaneous shrinking and expanding cannot continue indefinitely. A few hundred million years after a solar-mass star leaves the main sequence, something else happens—helium begins to burn in the core. By the time the central density has risen to about 10^8 kg/m^3 (at stage 9), the temperature has reached the 10^8 K needed for helium to fuse into carbon, and the central fires reignite.

The reaction that transforms helium into carbon occurs in two steps. First, two helium nuclei come together to form a nucleus of beryllium-8 (^{8}Be). Beryllium-8 is a very unstable isotope that would normally break up into two helium nuclei in about 10^{-12} s. However, at the densities in the core of a red giant, it is very likely that the beryllium-8 nucleus will encounter another helium nucleus before this occurs, fusing with it to form carbon-12 (^{12}C). This is the second step of the helium-burning reaction. In part, it is because of the electrostatic repulsion between beryllium-8 (containing four protons) and helium-4 (containing two) that the temperature must rise to 10^8 K before this reaction can take place.

Symbolically, we can represent this next stage of stellar fusion as follows:

$$^4\text{He} + {}^4\text{He} \rightarrow {}^8\text{Be} + \text{energy},$$

$$^8\text{Be} + {}^4\text{He} \rightarrow {}^{12}\text{C} + \text{energy}.$$

Helium-4 nuclei are traditionally known as *alpha particles*. The term dates from the early days of nuclear physics, when the true nature of these particles was unknown. Because three alpha particles are required to get from helium-4 to carbon-12, the foregoing reaction is usually called the **triple-alpha process**.

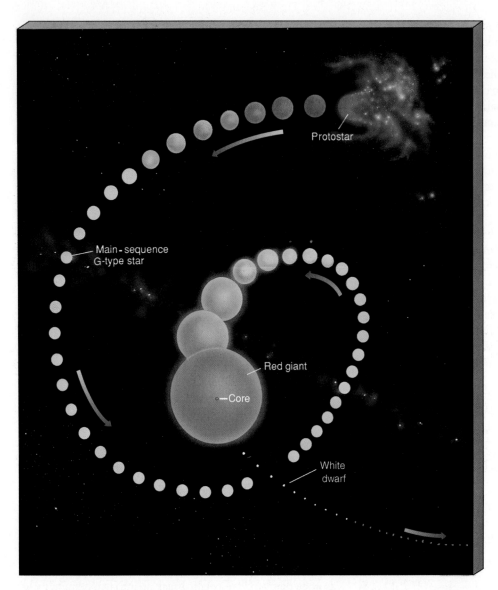

Figure 20.5 Diagram of the relative sizes and colors of a normal G-type star (such as our Sun) in its formative stages, on the main sequence, and while passing through the red-giant and white-dwarf stages. At maximum swelling, the red giant is approximately 70 times the size of its main-sequence parent; the core of the giant is about 1/15 the main-sequence size and would be barely discernible if this figure were drawn to scale. The length of time spent in the various stages—protostar, main-sequence star, red giant, and white dwarf—is roughly proportional to the length of this imaginary trek through space. The star's brief stay on the horizontal branch is not shown here.

THE HELIUM FLASH

For stars comparable in mass to the Sun, there is a major complication when helium fusion begins. At the high densities found in the core, the gas has entered a new state of matter whose properties are governed by the laws of quantum mechanics rather than by those of classical physics. Up to now we have been concerned primarily with the nuclei—protons, alpha particles, and so on—that make up virtually all the star's mass and participate in the reactions that generate its energy. However, the star contains another important constituent—a vast sea of electrons stripped from their parent nuclei by the ferocious heat in the stellar interior. At this stage in our story, these electrons play a critical role in determining the star's evolution.

Under the conditions found in the stage 9 red giant core, a rule of quantum mechanics known as the *Pauli exclusion principle* (after Wolfgang Pauli, one of the found-ing fathers of quantum physics) prohibits the electrons in the core from being squeezed too close together. In effect, the exclusion principle tells us that we can think of the electrons as tiny rigid spheres that can be squeezed relatively easily up to the point of contact but become virtually incompressible thereafter. This condition is known (for historical reasons) as *electron degeneracy*, and the pressure associated with the contact of the tiny electron spheres is called **electron degeneracy pressure**. It has nothing to do with the thermal pressure (due to the star's heat) that we have been studying up to now. In our red-giant core, the pressure resisting the force of gravity is supplied almost entirely by degenerate electrons. Hardly any of the core's support results from "normal" thermal pressure, and this has dramatic consequences once the helium begins to burn.

Under normal ("nondegenerate") circumstances, the core could react to and accommodate the onset of helium burning, but in its degenerate state the burning becomes unstable, with explosive consequences. In a star supported by thermal pressure, the increase in temperature produced by the onset of helium fusion would lead to an increase in pressure. The gas would then expand and cool, reducing the burning rate and reestablishing equilibrium. In the electron-supported core of a solar-mass red giant, however, the pressure is largely *independent* of the temperature. When burning starts and the temperature increases, there is no corresponding rise in pressure, no expansion of the gas, no drop in the temperature, and no stabilization of the core. Instead, the core is unable to respond to the rapidly changing conditions within it. The pressure remains more or less unchanged as the nuclear reaction rates increase, and the temperature rises rapidly in a runaway explosion called the **helium flash**.

For a few hours, the helium burns ferociously, like an uncontrolled bomb. Eventually, the flood of energy released by this period of runaway fusion heats the core to the point at which normal thermal pressure once again dominates. Finally able to react to the energy dumped into it by helium burning, the core expands, its density drops, and equilibrium is restored as the inward pull of gravity and the outward push of gas pressure come back into balance. The core, now stable, begins to burn helium into carbon at temperatures well above 10^8 K.

The helium flash terminates the giant star's ascent on the red-giant branch of the H–R diagram. Yet despite the explosive detonation of helium in the core, the flash does *not* increase the star's luminosity. On the contrary, the helium flash produces a rearrangement of the core that ultimately results in a *reduction* in the energy output. On the H–R diagram, the star jumps from stage 9 to stage 10, a stable state with steady helium burning in the core. As indicated in Figure 20.6, the surface temperature is now higher than it was on the red-giant branch, but the luminosity is considerably less than at the helium flash. This adjustment in the star's properties occurs quite quickly—in about 100,000 years.

At stage 10 our star is now stably burning helium in its core and fusing hydrogen in a shell surrounding it. It resides in a well-defined region of the H–R diagram known as the **horizontal branch**, where core-helium-burning stars remain for a time before resuming their journey around the H–R diagram. The star's specific position within this region is determined mostly by its mass—not its original mass, but whatever mass remains after its ascent of the red-giant branch. The two masses differ because during the red-giant stage strong stellar winds eject large amounts of matter from a star's surface (see *Interlude 20-1*). As much as 20–30 percent of the original stellar mass may escape during this period. It so happens that more massive stars have lower surface temperatures at this stage, but all stars have roughly the same luminosity after the helium flash. As a result, stage 10 stars tend to lie along a horizontal line on the H–R diagram, with more massive stars to the right, less massive ones to the left.

Figure 20.6 After its large increase in luminosity while ascending the red-giant branch is terminated by the helium flash, our star settles down into another equilibrium state at stage 10, on the horizontal branch.

THE CARBON CORE

The nuclear reactions in our star's helium core burn on, but not for long. Whatever helium exists in the core is rapidly consumed. The triple-alpha helium-to-carbon fusion reaction—like the proton–proton and CNO-cycle hydrogen-to-helium reactions before it—proceeds at a rate that increases very rapidly with temperature. At the extremely high temperatures found in the horizontal-branch core, the helium fuel doesn't last long—no more than a few tens of millions of years after the initial flash.

As helium fuses to carbon a new inner core of carbon ash forms, and phenomena similar to the earlier buildup of helium ash begin to occur. Now helium becomes depleted at the very center of the star, and eventually fusion ceases there. In response, the nonburning carbon core shrinks and heats as gravity pulls it inward, causing the hydrogen- and helium-burning rates in the overlying layers of the core to increase. The star now contains a shrinking carbon core surrounded by a helium-burning shell, which is in turn surrounded by a hydrogen-burning shell. The outer envelope of the star—the nonburning layers surrounding the core—expands, much as it did earlier during the first red-giant stage. By the time it reaches stage 11, the star has become

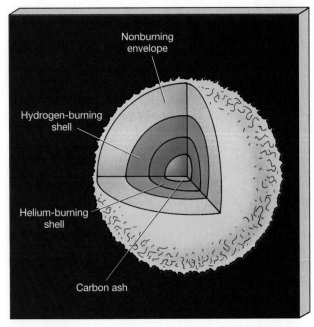

Figure 20.7 Within a few million years after the onset of helium burning, carbon ash accumulates in the inner core of a star, above which hydrogen and helium are still burning in concentric shells.

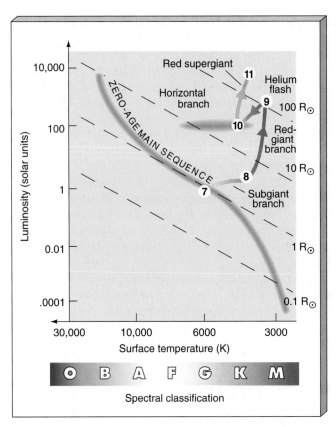

Figure 20.8 A carbon-core star reascends the giant branch of the H–R diagram—this time on a track called the asymptotic giant branch—for the same reason it evolved there the first time around: lack of nuclear burning at the core causes contraction of the core and expansion of the overlying layers.

a swollen red giant for a second time. Figure 20.7 depicts the star's interior structure during this time.

The star's second ascent of the giant branch is shown in Figure 20.8. To distinguish this second track from the first red-giant stage, this phase is sometimes known as the **asymptotic giant branch**. The burning rates at the center are much fiercer this time around, and the star's radius and luminosity increase to values even greater than those reached at the helium flash on the first ascent. Our star is now a **red supergiant**. The carbon core continues to shrink, driving the hydrogen-burning and helium-burning shells to higher and higher temperatures and luminosities.

COMPARING THEORY WITH REALITY

Table 20.1 summarizes the key stages through which a solar-mass star evolves. It is a continuation of Table 19.1, except that the density units have been changed from particles per cubic meter to the more convenient kilograms per cubic meter, and we now express sizes as radii rather than diameters. The numbers in the "Stage" column refer to the evolutionary stages noted in the figures and discussed in the text.

All the H–R diagrams and evolutionary tracks presented so far are theoretical constructs based largely on computer models of the interior workings of stars. Before continuing our study of stellar evolution, let's take a moment to compare our models with actual observations. Figure 20.9 shows a real H–R diagram, drawn using the stars of the old globular cluster M3. The similarity between theory and observation is striking—stars in each of the evolutionary stages 7–11 can be seen, in numbers consistent with the theoretical models. (The points

in Figure 20.9 are "shifted" a little to the left relative to Figure 20.8 because of composition differences between stars such as the Sun and stars in globular clusters—globular cluster stars tend to be slightly hotter than solar-type stars of the same mass.) Astronomers place great confidence in the theory of stellar evolution precisely because its predictions are so often found to be in excellent agreement with plots of real stars.

20.3 The Death of a Low-Mass Star

THE FIRES GO OUT

3 As our red supergiant ascends the asymptotic giant branch, its envelope swells while its core, too cool for further nuclear burning, continues to contract. If the central temperature could become high enough for carbon fusion to occur, still heavier products could be synthesized, and the newly generated energy might again support the star, restoring for a time the equilibrium between gravity and heat. For solar-mass stars, however, this does not occur. The temperature never reaches the 600 million K needed for a new round of nuclear

TABLE 20.1 Evolution of a Sun-like Star

STAGE	APPROXIMATE TIME TO NEXT STAGE (yr)	CENTRAL TEMPERATURE (10^6 K)	SURFACE TEMPERATURE (K)	CENTRAL DENSITY (kg/m^3)	RADIUS (km)	RADIUS (solar radii)	OBJECT
7	10^{10}	15	6000	10^5	7×10^5	1	Main-sequence star
8	10^8	50	4000	10^7	2×10^6	3	Subgiant branch
9	10^5	100	4000	10^8	7×10^7	100	Helium flash
10	5×10^7	200	5000	10^7	7×10^6	10	Horizontal branch
11	10^4	250	4000	10^8	4×10^8	500	Asymptotic giant branch
12	10^5	300	100,000	10^{10}	10^4	0.01	Carbon core
	—		3000	10^{-17}	7×10^8	1000	Planetary nebula*
13	—	100	50,000	10^{10}	10^4	0.01	White dwarf
14	—	Close to 0	Close to 0	10^{10}	10^4	0.01	Black dwarf

Values refer to the envelope.

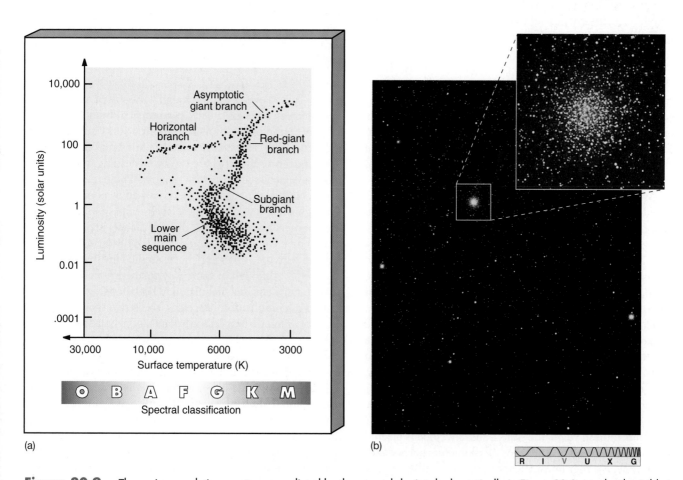

(a)

(b)

Figure 20.9 The various evolutionary stages predicted by theory and depicted schematically in Figure 20.8 are clearly visible in this H–R diagram (a) of an old star cluster—the globular cluster M3. The faintest main-sequence stars are not shown in this diagram, which was constructed using ground-based observations, because observational limitations make it difficult to determine the apparent brightness of low-luminosity stars in the cluster. These difficulties also manifest themselves in the thickness of the main sequence, which is due almost entirely to observational error. (b) Wide-angle photograph showing M3 as it appears in the night sky. The inset is a more detailed view of the cluster itself; its field of view is a few parsecs across.

reactions to occur. The red supergiant is now very close to the end of its nuclear-burning lifetime.

Before the carbon core can attain the incredibly high temperatures needed for carbon ignition, its density reaches a point beyond which it cannot be compressed further. At about 10^{10} kg/m³, the electrons in the core once again become degenerate, the contraction of the core ceases, and its temperature stops rising. This stage (stage 12 in Table 20.1) represents the maximum compression that the star can achieve—there is simply not enough matter in the overlying layers to bear down any harder.

The core density at this stage is extraordinarily high. A single cubic centimeter of core matter would weigh 1000 kg on Earth—a ton of matter compressed into a volume about the size of a grape. Yet despite the extreme compression of the core, the central temperature is "only" about 300 million K. Some oxygen is formed via reactions between carbon and helium at the inner edge of the helium-burning shell:

$$^{12}C + {}^{4}He \rightarrow {}^{16}O + energy,$$

but collisions among nuclei are neither frequent nor violent enough to create any heavier elements. For all practical purposes, the central fires go out once carbon has formed.

PLANETARY NEBULAE

Our aged stage 12 star is now in quite a predicament. Its inner carbon core no longer generates energy. The outer-core shells continue to burn hydrogen and helium, and as more and more of the inner core reaches its final, high-density state, the zone of nuclear burning increases in intensity. Meanwhile, the envelope continues to expand and cool.

Around this time, the burning becomes very unstable. The helium-burning shell is subject to a series of explosive *helium-shell flashes*. These flashes are caused by the enormous pressure there and the extreme sensitivity of the triple-alpha burning rate to small changes in temperature. The flashes produce large fluctuations in the intensity of the radiation reaching the star's outermost layers, causing them to pulsate more and more violently.

Compounding the star's problems, its surface layers are also becoming unstable. As the temperature drops to the point at which electrons can recombine with nuclei to form atoms, each recombination produces additional photons, which tend to push the outer envelope to greater and greater distances from the core. As shown in Figure 20.10, the radius of the star oscillates more and more violently. In less than a few million years, the star's outer envelope is ejected into space at a speed of a few tens of kilometers per second.

In time, a rather unusual looking object results. We say unusual because the "star" now has two distinct parts, both of which constitute stage 12 of Table 20.1. At the center is a small, well-defined core of mostly carbon ash. Hot and dense, only the outermost layers of this core still fuse helium into carbon and oxygen. Well beyond the core lies a spherical shell of cooler and thinner matter—the ejected envelope of the

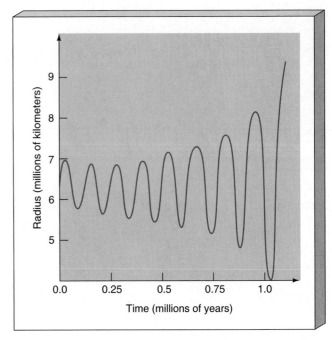

Figure 20.10 Buffeted by helium-shell flashes from within and subject to the destabilizing influence of recombination, the outer layers of a red giant become unstable and enter into a series of growing pulsations. Eventually, the envelope is ejected and forms a planetary nebula.

giant—spread over a volume roughly the size of our solar system. Such an object is called a **planetary nebula**. Some well-known examples are shown in Figures 20.11 and 20.12. In all, some 1000 planetary nebulae are known in our Galaxy.

The term *planetary* here is very misleading, for these objects have no association with planets. The name originated in the eighteenth century when, viewed at poor resolution through small telescopes, these shells of gas looked to some astronomers like the circular disks of planets in our solar system. The term *nebula* is also a little confusing, as it suggests kinship with the emission nebulae studied in Chapter 18. ∞ (Sec. 18.2) Although in some ways planetary nebulae do resemble some emission nebulae, and both undergo similar ionization–recombination processes, these two types of objects are very different. Not only are planetary nebulae much smaller than emission nebulae, they are also associated with much older stars. Emission nebulae are the signposts of recent stellar birth. Planetary nebulae indicate impending stellar death.

The "ring" of a planetary nebula is in reality a three-dimensional shell of warm, glowing gas completely surrounding the core. Its halo-shaped appearance is only an illusion. The shell is a complete envelope that has been expelled from around the core, but we can see it only at the edges, where emitting matter accumulates along our line of sight. As illustrated in Figure 20.11(b), the shell is virtually invisible in the direction of the core. Few planetary nebulae are quite as regular as this simple picture might suggest, however. Figure 20.12 shows two systems in which the details of the gas-ejection process have evidently played an important role in determining the planetary nebula's shape and appearance.

Figure 20.11 A planetary nebula is an object with a small dense core (central blue-white star) surrounded by an extended shell (or shells) of glowing matter. (a) The Ring Nebula in the constellation Lyra, a classic example of a planetary nebula, is about 1500 pc from us. It is about 0.2 pc in diameter—much larger than our solar system—but because of its great distance, its apparent size is only about 1/100 that of the full Moon, and it is too dim to see well with the naked eye. (See also the images on p. 54.) (b) The appearance of the planetary nebula can be explained once we realize that the shell of glowing gas around the central core is actually quite thin. There is very little gas along the line of sight between the observer and the central star (path A), so that part of the shell is invisible. Near the edge of the shell, however, there is more gas along the line of sight (paths B and C), so the observer sees a glowing ring. (c) The Helix Nebula appears to the eye as a small star with a halo around it. About 140 pc from Earth and 0.6 pc across, its apparent size in the sky is roughly half that of the full Moon. (All the other stars visible in the photo are foreground or background objects, unrelated to the planetary nebula.)

Figure 20.12 (a) The NGC6826 Nebula more clearly shows the shell-like structure of the expanding gases that make up a planetary nebula. It resides some 700 pc away in the constellation Cygnus. (b) The Cat's Eye Nebula is an example of a much more complex planetary nebula. Intricate structures, including concentric gas shells, jets of high-speed gas, and shock-induced knots of gas are all visible. As usual, red indicates the presence of excited hydrogen. The nebula is about 1000 pc away, in the constellation Draco. It may have been produced by a pair of binary stars (unresolved at the center) that have both shed planetary nebulae.

INTERLUDE 20-1 Mass Loss from Giant Stars

Astronomers now know that stars of all spectral types are active and have stellar winds. Consider the highly luminous, hot, blue O- and B-type stars, which have by far the strongest winds. Satellite and rocket observations have shown that their wind speeds may reach 3000 km/s. The result is a yearly mass loss sometimes exceeding 10^{-6} solar mass per year. Over the relatively short span of 1 million years, these stars blow a tenth of their total mass—more than an entire solar mass of material—into space. These powerful stellar winds, driven directly by the pressure of the intense ultraviolet radiation emitted by the stars themselves, hollow out vast cavities in the interstellar gas.

The accompanying figure shows a *Hubble Space Telescope* image of the supergiant star AG Carinae—50 times more massive than the Sun and a million times brighter—shedding its outer atmosphere. It is shown here puffing out vast clouds of gas and dust. (The star, at center, is intentionally obscured to show more clearly the surrounding nebulae; the bright vertical line is also an artifact.)

Observations made with radio, infrared, and optical telescopes have shown that luminous cool stars (for example, K- and M-type red giants) also lose mass at rates comparable to those of the luminous hot stars. Red-giant wind velocities, however, are much lower, averaging merely 30 km/s. They carry roughly as much mass into space as do O-star winds because their densities are generally much greater. Because luminous red stars are inherently cool objects (with surface temperatures of only about 3000 K), they emit virtually no ultraviolet radiation, so the mechanism driving the winds must differ from that in luminous hot stars. We can only surmise that gas turbulence or magnetic fields or both in the atmospheres of the red giants are somehow responsible. The surface conditions in red giants are in some ways

similar to those in T Tauri protostars, which are also known to exhibit strong winds. Possibly the same basic mechanism—violent surface activity—is responsible for both.

Unlike winds from hot stars, winds from these cool stars are rich in dust particles and molecules. Nearly all stars eventually evolve into red giants, so these winds provide a major source of new gas and dust to interstellar space. These stellar winds provide a vital link in the cycle of star formation and the evolution of the interstellar medium.

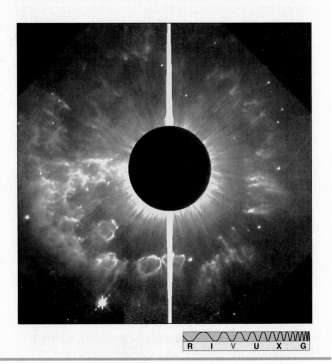

R I V U X G

The planetary nebula continues to spread out with time, becoming more diffuse and cooler, gradually dispersing into interstellar space. In doing so, it enriches the interstellar medium with atoms of helium, carbon, and oxygen dredged up by convection from the depths of the core into the envelope during the star's final years.

WHITE DWARFS

The carbon core, the stellar remnant at the center of the planetary nebula, continues to evolve. Formerly concealed by the atmosphere of the red-giant star, the core becomes visible as the envelope recedes. Several tens of thousands of years are needed for the core to appear from behind the veil of expanding gas. The core is very small. By the time the envelope is ejected as a planetary nebula, it has shrunk to about the size of Earth (in some cases it may be even smaller than our planet). Its mass is about half the mass of the Sun. Shining only by stored heat, not by nuclear reactions, this small "star" has a white-hot surface when it first becomes

visible, although it appears dim because of its small size. The core's temperature and size give rise to its new name—*white dwarf*. This is stage 13 of Table 20.1. The approximate path followed by the star on the H–R diagram as it evolves from stage 11 red supergiant to stage 13 white dwarf is shown in Figure 20.13.

Not all white-dwarf stars are found as the cores of planetary nebulae. Several hundred have been discovered "naked" in our Galaxy, their envelopes expelled to invisibility (or perhaps stripped away by a binary companion—as discussed shortly) long ago. Figure 20.14 shows an example of a white dwarf, Sirius B, that happens to lie particularly close to Earth; it is the faint binary companion of the much brighter and better-known Sirius A. ∞ (Sec. 17.3) Some properties of Sirius B are listed in Table 20.2. With more than the mass of the Sun packed into a volume smaller than Earth, Sirius B's density is about a million times greater than anything familiar to us in the solar system. (In fact, Sirius B has an unusually high mass for a white dwarf—it is believed to be the evolutionary

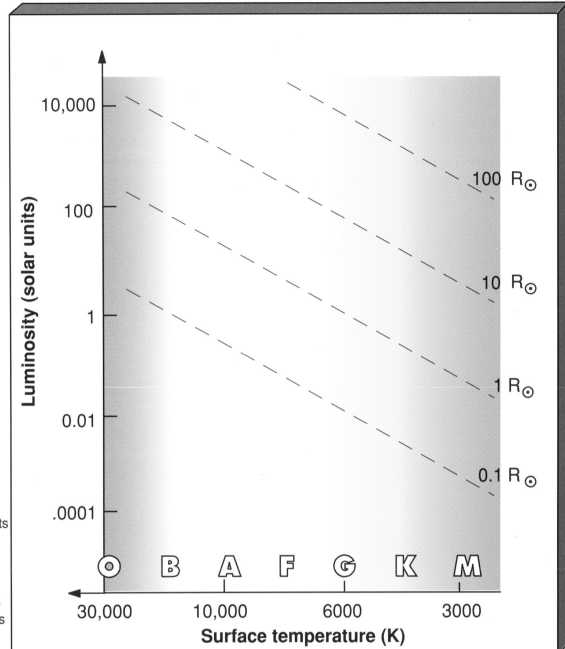

The H-R diagram plots stars by luminosity (vertical axis) and temperature, or spectral class (horizontal axis). The dashed diagonal lines are lines of constant radius.

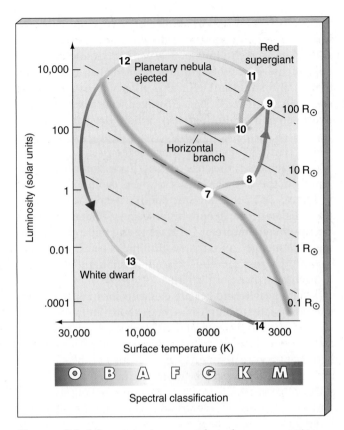

Figure 20.13 A star's passage from the horizontal branch (stage 10) to the white-dwarf stage (stage 13) by way of the asymptotic giant branch creates an evolutionary path that cuts across the entire H–R diagram.

Animation

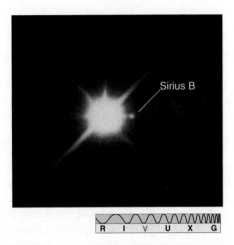

Figure 20.14 Sirius B (the speck of light at right) is a white-dwarf star, a companion to the much larger and brighter star Sirius A. (The "spikes" on the image of Sirius A are not real; they are artifacts caused by the support struts of the telescope.)

densities in the star (from the white-dwarf stage on), the resistance of electrons to being squeezed together—the same electron degeneracy that prevailed in the red-giant core around the time of the helium flash—holds the star up, even as its temperature drops almost to absolute zero. As the dwarf cools, it remains about the size of Earth.

20.4 Evolution of Stars More Massive than the Sun

4 High-mass stars evolve much faster than their low-mass counterparts. The more massive a star, the more ravenous is its fuel consumption and the shorter its main-sequence lifetime. The Sun will spend a total of some 10 billion years on the main sequence, but a 5–solar mass B-type star will remain there for only a hundred million years. A 10–solar mass O-type star will depart in just 20 million years or so. This trend toward much faster evolution for more massive stars continues even after the main sequence. All evolutionary changes happen much more rapidly for high-mass stars because their larger mass and stronger gravity generate more heat, speeding up all phases of stellar evolution.

Stars leave the main sequence for one basic reason—they run out of hydrogen in their cores. As a result, the early stages of stellar evolution beyond the main sequence are qualitatively the same in all cases: main-sequence hydrogen burning in the core (stage 7) eventually gives way to the formation of a nonburning, collapsing helium core surrounded by a hydrogen-burning shell (stages 8 and 9). A high-mass star leaves the main sequence on its journey toward the red-giant region with an internal structure quite similar to that of its low-mass cousin. Thereafter, their evolutionary tracks diverge.

product of a star roughly four times the mass of the Sun. *Interlude 20-2* discusses another possible peculiarity of Sirius B's evolution.)

Once an isolated star becomes a white dwarf, its evolution is over. (As we will see in Chapter 21, white dwarfs in binary systems may have further activity in store.) It continues to cool and dim with time, following the white-yellow-red track near the bottom of the H–R diagram of Figure 20.13, eventually becoming a **black dwarf**—a cold, dense, burned-out ember in space. This is stage 14 of Table 20.1, the graveyard of stars.

The cooling dwarf does not shrink much as it fades away, however. Even though its heat is leaking away into space, gravity does not compress it further. At the enormously high

TABLE 20.2 Sirius B—A Nearby White Dwarf

Mass	1.1 solar masses
Radius	0.008 solar radii (5500 km)
Luminosity (total)	0.04 solar luminosities (1.6×10^{25} W)
Surface temperature	24,000 K
Average density	3×10^9 kg/m^3

Figure 20.15 compares the post–main-sequence evolution of three stars having masses 1, 4, and 15 times the mass of the Sun. Note that, whereas stars like the Sun ascend the red giant branch almost vertically, higher-mass stars move nearly horizontally across the H–R diagram after leaving the upper main sequence. Their luminosities stay roughly constant as their radii increase and their surface temperatures drop.

In stars having more than about 2.5 times the mass of the Sun, helium burning begins smoothly and stably, *not* explosively—there is no helium flash. Calculations indicate that the more massive a star, the lower its core density when the temperature reaches the 10^8 K necessary for helium ignition and the smaller the contribution to the pressure from degenerate electrons. As a result, above 2.5 solar masses, the unstable core conditions described earlier do not occur. The 4–solar mass red giant in Figure 20.15 remains a red giant as helium starts to fuse into carbon. There is no sudden jump to the horizontal branch and no subsequent reascent of the giant branch. Instead, the star loops smoothly back and forth near the top of the H–R diagram.

A much more important divergence occurs at approximately 8 solar masses—the dividing line between high and low mass mentioned in Section 20.1. A low-mass star never achieves the 600 million K needed to fuse carbon nuclei, and so it ends its life as a carbon–oxygen white dwarf. A high-mass star, however, can fuse not only hydrogen and helium but also carbon, oxygen, and even heavier elements as its inner core continues to contract and its central temperature continues to rise. The burning rate accelerates as the core evolves. Evolution proceeds so rapidly in the 15–solar mass star whose evolution is shown in Figure 20.15 that the star doesn't even reach the red-giant region before helium fusion begins. The star achieves a central temperature of 10^8 K while still quite close to the main sequence, and its evolutionary track continues smoothly across the H–R diagram, seemingly unaffected by each new phase of burning.

With heavier and heavier elements forming at an ever-increasing rate, the high-mass stars shown in Figure 20.15 are very close to the ends of their lives. We will discuss their fate in more detail in the next chapter, but suffice it to say here that they are destined to die in a violent explosion soon after carbon and oxygen begin to fuse in their cores. These stars evolve so rapidly that, for most practical observational purposes, high-mass stars explode and die soon after they leave the main sequence.

20.5 Observing Stellar Evolution in Star Clusters

5 Star clusters provide excellent test sites for the theory of stellar evolution. Every star in a given cluster formed at the same time, from the same interstellar cloud, with virtually the same composition. Only the mass varies from one star to another. This allows us to check the accuracy of our theoretical models in a very straightforward way. Having studied in some detail the evolutionary tracks of individual stars, let us now consider how their collective appearance changes in time.

In Chapter 17 we saw how astronomers estimate the ages of star clusters by determining which of their stars have already left the main sequence. ∞ (Sec. 17.10) In fact, the main-sequence lifetimes that go into those age measurements represent only a tiny fraction of the data obtained from theoretical models of stellar evolution. Starting from the zero-age main sequence, astronomers can predict exactly how a newborn cluster should look at any subsequent time. Although we cannot see into the interiors of stars to test our models, we can compare stars' outward appearances with theoretical predictions. The agreement—in detail—between theory and observation is remarkably good.

We begin our study shortly after the cluster's formation, with the upper main sequence already fully formed

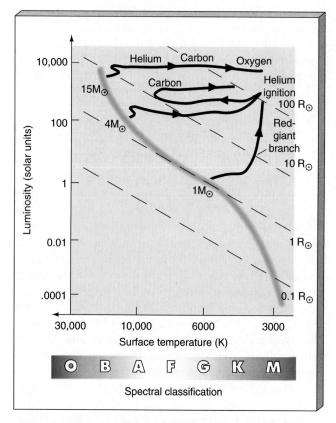

Figure 20.15 Evolutionary tracks for stars of 1, 4, and 15 solar masses (shown only up to the point of the helium flash in the low-mass cases). Low-mass stars ascend the giant branch almost vertically, whereas high-mass stars move roughly horizontally across the H–R diagram from the main sequence into the red-giant region. The most massive stars experience smooth transitions into each new burning stage. No helium flash occurs for stars more massive than about 2.5 solar masses. Some points are labeled with the element that has just started to fuse in the inner core.

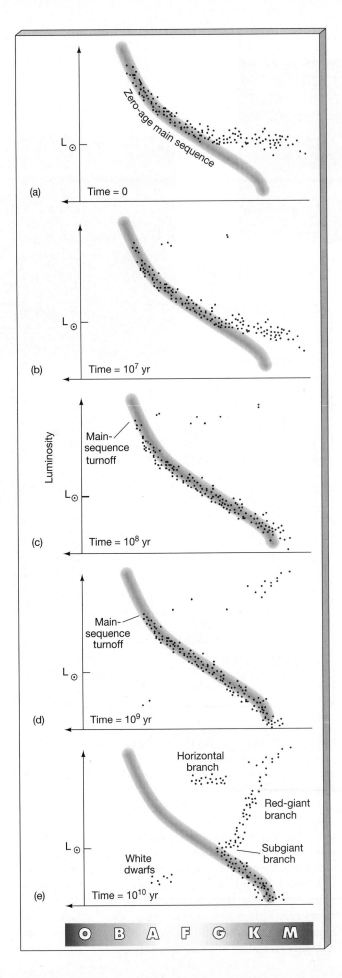

Figure 20.16 The changing H–R diagram of a hypothetical star cluster. (a) Initially, stars on the upper main sequence are already burning steadily while the lower main sequence is still forming. (b) At 10^7 years, O-type stars have already left the main sequence, and a few red giants are visible. (c) By 10^8 years, stars of spectral type B have evolved off the main sequence. More red giants are visible, and the lower main sequence is almost fully formed. (d) At 10^9 years, the main sequence is cut off at about spectral type A. The subgiant and red-giant branches are just becoming evident, and the formation of the lower main sequence is complete. A few white dwarfs may be present. (e) At 10^{10} years, only stars less massive than the Sun still remain on the main sequence. The cluster's subgiant, red-giant, horizontal, and asymptotic giant branches are all discernible. Many white dwarfs have now formed.

and burning steadily, and lower-mass stars just beginning to arrive on the main sequence, as shown in Figure 20.16(a). The appearance of the cluster at this early stage is dominated by its most massive stars—the bright blue supergiants. Now let's follow the cluster forward in time and see how its H–R diagram evolves.

Figure 20.16(b) shows the appearance of our cluster's H–R diagram after 10 million years. The most massive O-type stars have evolved off the main sequence. Most have already exploded and vanished, as just discussed, but one or two may still be visible as red giants. The remaining cluster stars are largely unchanged in appearance—their evolution is slow enough that little happens to them in such a relatively short period of time. The cluster's H–R diagram shows the main sequence slightly cut off, along with a rather poorly defined red-giant region. Figure 20.17 shows the twin open clusters h and χ (the Greek letter chi) Persei, along with their combined H–R diagram. Comparing Figure 20.17(b) with such diagrams as those in Figure 20.16, astronomers estimate the age of this pair of clusters to be about 10 million years.

After 100 million years (Figure 20.16c) stars brighter than type B5 or so (about 4–5 solar masses) have left the main sequence, and a few more red supergiants are visible. By this time most of the cluster's low-mass stars have finally arrived on the main sequence, although the dimmest M stars may still be in their contraction phase. The appearance of the cluster is now dominated by bright B stars and brighter red giants.

At any time during the evolution the cluster's original main sequence is intact up to some well-defined stellar mass, corresponding to the stars that are just leaving the main sequence at that instant. We can imagine the main sequence being "peeled away" from the top down, with fainter and fainter stars turning off and heading for the giant branch as time goes on. Astronomers refer to the high-luminosity end of the observed main sequence as the **main-sequence turnoff**. The mass of the star that is just evolving off the main sequence at any moment is known as the *turnoff mass*.

At 1 billion years, the main-sequence turnoff mass is around 2 solar masses, corresponding roughly to spectral

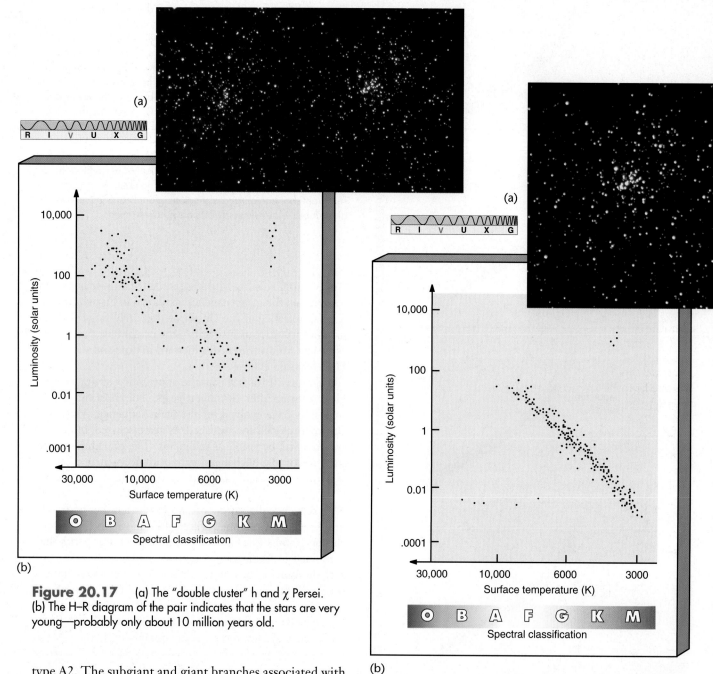

Figure 20.17 (a) The "double cluster" h and χ Persei. (b) The H–R diagram of the pair indicates that the stars are very young—probably only about 10 million years old.

Figure 20.18 (a) The Hyades cluster, a relatively young group of stars visible to the naked eye. (b) The H–R diagram for this cluster is cut off at about spectral type A, implying an age of about 500 million years.

type A2. The subgiant and giant branches associated with the evolution of low-mass stars are just becoming visible, as indicated in Figure 20.16(d). The formation of the lower main sequence is now complete. In addition, the first white dwarfs have just appeared, although they are often too faint to be observed at the distances of most clusters. Figure 20.18 shows the Hyades open cluster and its H–R diagram. The H–R diagram appears to lie between Figures 20.16(c) and 20.16(d), suggesting that the cluster's age is about 500 million years.

At 10 billion years, the turnoff point has reached solar-mass stars, of spectral type G2. The subgiant and giant branches are now clearly discernible (see Figure 20.16e), and the horizontal and asymptotic giant branches appear as distinct regions in the H–R diagram. Many white dwarfs

are also present in the cluster. Although stars in all these evolutionary stages are also present in the 1-billion-year-old cluster shown in Figure 20.16(d), they are few in number—typically only a few percent of the total number of stars in the cluster. Also, because they evolve so rapidly, they spend very little time in these regions. Low-mass stars are much more numerous and evolve more slowly, so their evolutionary tracks are more easily detected.

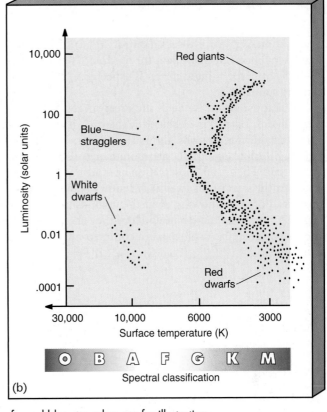

Figure 20.19 (a) The southern globular cluster 47 Tucanae. (b) Fitting its main-sequence turnoff and its giant and horizontal branches to theoretical models gives 47 Tucanae an age of about 11 billion years, making it one of the oldest known objects in the Milky Way Galaxy. The inset is a high-resolution ultraviolet image of 47 Tucanae's core region, taken with the *Hubble Space Telescope* and showing many "blue stragglers"— massive stars lying on the main sequence above the turnoff point, resulting perhaps from the merging of binary star systems (see also Figure 20.9). The points representing white dwarfs, red dwarfs, and blue stragglers are for illustration only—they have been added to the original data set, based on *Hubble* observations of this and other clusters.

Figure 20.19 shows the globular cluster 47 Tucanae. By carefully adjusting their theoretical models until the cluster's main sequence, subgiant, red-giant, and horizontal branches are all well matched, astronomers have determined its age to be roughly 11 billion years, a little older than our hypothetical cluster in Figure 20.16(e). In fact, globular cluster ages determined in this way show a remarkably small spread. All the globular clusters in our Galaxy appear to have formed between about 10 and 12 billion years ago.

Recent *Hubble* observations of nearby globular clusters have also revealed for the first time the white dwarf sequences long predicted by theory but previously too faint to detect at such large distances. Figure 20.20(b) shows an *HST* image of white dwarfs in a cluster called M4, lying 2100 pc from Earth. To illustrate our point, we have combined in Figure 20.19(b)

Figure 20.20
(a) The globular cluster M4, as seen through a large ground-based telescope on Kitt Peak Mountain. This is the closest globular cluster to us, at 2100 pc away; it spans some 16 pc. (b) A peek at M4's suburbs by the *Hubble* telescope shows nearly a hundred white dwarfs within a small 0.4-pc area, some of the brightest ones are circled in blue.

both the newer white-dwarf data for M4 and the older main-sequence/red-giant data for 47 Tuc. The complete H–R diagram for a single cluster is expected to look qualitatively similar.

Stellar evolution is one of the great success stories of astrophysics. Like all good scientific theories, it makes definite testable predictions about the universe while remaining flexible enough to incorporate new discoveries as they occur. Theory and observation have advanced hand in hand. At the start of the twentieth century many scientists despaired of ever knowing even the compositions of the stars, let alone why they shine and how they change. Today, the theory of stellar evolution is a cornerstone of modern astronomy.

20.6 The Evolution of Binary-Star Systems

6 We have noted that most stars in our Galaxy are not isolated objects but are actually members of binary-star systems. However, our discussion of stellar evolution has so far focused exclusively on isolated stars. This prompts us to ask how membership in a binary-star system changes the evolutionary tracks we have just described. Indeed, because

nuclear burning occurs deep in the core, does the presence of a stellar companion have any significant effect at all? Perhaps not surprisingly, the answer depends on the distance between the two stars in question.

For a binary system whose component stars are very widely separated—that is, the distance between the stars is greater than perhaps a thousand stellar radii—the two stars evolve more or less independently of one another, each following the track appropriate to an isolated star of its particular mass. However, if the two stars are closer, then the gravitational pull of one may strongly influence the envelope of the other. In that case the physical properties of both may deviate greatly from those calculated for isolated single stars.

As an example, consider the star Algol (Beta Persei, the second brightest star in the constellation Perseus). By studying its spectrum and the variation in its light intensity, astronomers have determined that Algol is actually a binary (in fact, an eclipsing double-lined spectroscopic binary, as described in Chapter 17), and they have measured its properties very accurately. ∞ (Sec. 17.9) Algol consists of a 3.7–solar mass main-sequence star of spectral type B8 (a blue giant) with a 0.8–solar mass red subgiant companion moving in a circular orbit around it. The stars are 4 million km apart and have an orbital period of about 3 days.

A moment's thought reveals that there is something odd about these findings. On the basis of our earlier discussion, the more massive main-sequence star should have evolved

INTERLUDE 20-2 Learning Astronomy from History

Sirius A, the brighter of the two objects shown in Figure 20.14, appears twice as luminous as any other visible star, excluding the Sun. Its absolute brightness is not very great, but because its distance from us is small (less than 3 pc), its apparent brightness is very large. Sirius has been prominent in the nighttime sky since the beginning of recorded history. Cuneiform texts of the ancient Babylonians refer to the star as far back as 1000 B.C., and historians know that the star strongly influenced the agriculture and religion of the Egyptians of 3000 B.C.

Even though a star's evolution takes such a long time, we might have a chance to detect a slight change in Sirius because the recorded observations of this star go back several thousand years. The chances for success are improved in this case because Sirius A is so bright that even the naked-eye observations of the ancients should be reasonably accurate. Interestingly, recorded history does suggest that Sirius A has changed in appearance, but the observations are confusing. Every piece of information about Sirius recorded between the years 100 B.C. and A.D. 200 claims that this star was *red*. (No earlier records of its color are known.) In contrast, modern observations now show it to be white or bluish white—definitely *not* red.

If these reports are accurate, then Sirius has apparently changed from red to blue-white in the intervening years.

But according to the theory of stellar evolution, no star should be able to change its color in this way in such a short time. A color change such as this should take at least several tens of thousands of years, and perhaps a lot longer. It should also leave some evidence of its occurrence.

Astronomers have offered several explanations for the rather sudden change in Sirius A. These include the suggestions that (1) some ancient observers were wrong and other scribes copied them; (2) a Galactic dust cloud passed between Sirius A and Earth some 2000 years ago, reddening the star much as Earth's dusty atmosphere often reddens our Sun at dusk; and (3) the companion to Sirius A, Sirius B, was a red giant and the dominant star of this double-star system 2000 years ago but has since expelled its planetary nebular shell to reveal the white-dwarf star that we now observe.

Each of these explanations presents problems. How could the color of the sky's brightest star be incorrectly recorded for hundreds of years? Where is the intervening Galactic cloud now? Where is the shell of the former red giant? We are left with the uneasy feeling that the sky's brightest star doesn't fit particularly well into the currently accepted scenario of stellar evolution.

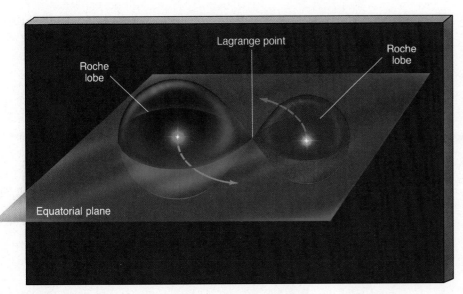

Figure 20.21 Each star in a binary system can be pictured as being surrounded by a "zone of influence," or Roche lobe, inside of which matter may be thought of as being "part" of that star. The two teardrop-shaped Roche lobes meet at the Lagrange point between the two stars. Outside the Roche lobes, matter may flow onto either star with relative ease.

faster than the less massive component. If the two stars formed at the same time (as is assumed to be the case), there should be no way that the 0.8–solar mass star could be approaching the giant stage first. Either our theory of stellar evolution is seriously in error, or something has modified the evolution of the Algol system. Fortunately for theorists, the latter is the case.

As sketched in Figure 20.21, each star in a binary system is surrounded by its own teardrop-shaped "zone of influence," inside of which its gravitational pull dominates the effects of both the other star and the overall rotation of the binary. Any matter within that region "belongs" to the star. It cannot easily flow onto the other component or out of the system. Outside the two regions, it is possible for gas to flow toward either star relatively easily. The two teardrop-shaped regions are called **Roche lobes**, after Edouard Roche, the French mathematician who first studied the binary-system problem in the nineteenth century and whose work we have already encountered in the context of planetary rings. ∞ (Sec. 12.4)

The Roche lobes of the two stars meet at a point on the line joining them—the inner Lagrange point (L_1), which we saw in Chapter 14 when discussing asteroid motions in the solar system. ∞ (Sec. 14.1) This Lagrange point is a place where the gravitational pulls of the two stars exactly balance the rotation of the binary system. The greater the mass of one component, the larger is its Roche lobe and the farther from its center (and the closer to the other star) is the Lagrange point.

Normally, both stars lie well within their respective Roche lobes, and such a binary system is said to be *detached*, as in Figure 20.22(a). However, as a star evolves off the main sequence and moves toward the giant branch, it is possible for its radius to become so large that it overflows its Roche lobe. Its gas begins to flow onto the companion through the

Lagrange point. The binary in this case is said to be *semide-tached* (Figure 20.22b). Because matter is flowing from one star onto the other, semidetached binaries are also known as **mass-transfer binaries**. If, for some reason, the other star also overflows its Roche lobe (either because of stellar evolution or because so much extra material is dumped onto it), the surfaces of the two stars merge. The binary system then consists of two nuclear-burning stellar cores surrounded by a single continuous common envelope—a **contact binary**, shown in Figure 20.22(c).

In a binary system in which the two stars are very close together, neither star has to evolve far off the main sequence before it overflows its Roche lobe and mass transfer begins. In a wide binary, however, both stars may evolve all the way up the giant branch without either surface ever reaching the Lagrange point, and they evolve just as though they were isolated. Depending on the stars involved and their orbital separations, there are many different possibilities for the eventual outcome of the evolution. Binary evolution is a very complex subject. Let's make these ideas more definite by returning to the question of how the binary star Algol may have reached its present state.

Astronomers believe that Algol started off as a detached binary. For reference, let us label the component that is now the 0.8–solar mass subgiant as star 1 and the 3.7–solar mass main-sequence star as star 2. Initially, star 1 was the more massive of the two, having perhaps three times the mass of the Sun. It thus evolved off the main sequence first. Star 2 was originally a less massive star, perhaps comparable in mass to the Sun. As star 1 ascended the giant branch, it overflowed its Roche lobe and gas began to flow onto star 2. This had the effect of reducing the mass of star 1 and increasing that of star 2, which in turn caused the Roche lobe of star 1 to shrink as its gravity decreased. As a result, the rate at which star 1 overflowed its Roche lobe

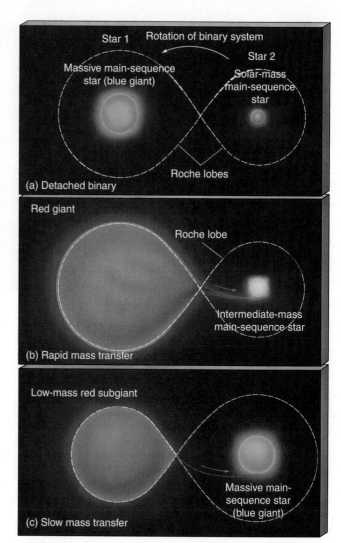

Figure 20.22 (a) In a detached binary, each star lies within its respective Roche lobe. (b) In a semidetached binary, one of the stars fills its Roche lobe and transfers matter onto the other, which still lies within its own Roche lobe. (c) In a contact or common-envelope binary, both stars have overflowed their Roche lobes, and a single star with two distinct nuclear-burning cores results.

Figure 20.23 The evolution of the binary star Algol. (a) Initially, Algol was probably a detached binary made up of two main-sequence stars—a relatively massive blue giant and a less massive companion similar to the Sun. (b) As the more massive component (star 1) evolved off the main sequence it expanded to fill and eventually overflow its Roche lobe, transferring large amounts of matter onto its smaller companion (star 2). (c) Today, star 2 is the more massive of the two, but it is on the main sequence. Star 1 is still in the subgiant phase and fills its Roche lobe, causing a steady stream of matter to pour onto its companion.

increased, and a period of unstable *rapid mass transfer* ensued, transporting most of star 1's envelope onto star 2. Eventually, the mass of star 1 became less than that of star 2. Detailed calculations show that the rate of mass transfer dropped sharply at that point, and the stars entered the relatively stable state we see today. These changes in Algol's components are illustrated in Figure 20.23.

Being part of a binary system has radically altered the evolution of both stars in the Algol system. The original high-mass star 1 is now a low-mass subgiant, while the roughly solar mass star 2 is now a massive blue main-se-

quence star. The removal of mass from the envelope of star 1 may prevent it from ever reaching the helium flash. Instead, its naked core may eventually be left behind as a *helium white dwarf*. In a few tens of millions of years, star 2 will itself begin to ascend the giant branch and fill its own Roche lobe. If star 1 is still a subgiant or a giant at that time, a contact binary system will result. If, instead, star 1 has by then become a white dwarf, a new mass-transferring period—with matter streaming from star 2 back onto star 1—will begin. In that case (as we will see), Algol may have a very active and violent future in store.

Just as molecules exhibit few of the physical or chemical properties of their constituent atoms, binaries can display types of behavior that are quite different from either of their component stars. The Algol system is a fairly simple example of binary evolution, yet it gives us an idea of the sorts of complications that can arise when two stars evolve interdependently. A significant fraction of all the binary stars in the Galaxy will pass through some sort of mass-transfer or common-envelope phase. In this chapter we have seen one possible result of mass transfer involving main-sequence stars. We will return to this subject in the next two chapters, when we continue our discussion of stellar evolution and the strange states of matter that may result.

Chapter Review

SUMMARY

Stars spend most of their lives on the main sequence, in the **core-hydrogen-burning** (p. 448) phase of stellar evolution, stably fusing hydrogen into helium at their centers. Stars leave the main sequence when the hydrogen in their cores is exhausted. For the Sun, which is about halfway through its main-sequence lifetime, this stage will occur about 5 billion years from now. Low-mass stars evolve much more slowly than the Sun, and high-mass stars evolve much faster.

When the central nuclear fires in the interior of a solar-mass star cease, the helium in the star's core is still too cool to fuse into anything heavier. With no internal energy source, the helium core is unable to support itself against its own gravity and begins to shrink. The star at this stage is in the **hydrogen-shell-burning** (p. 449) phase, in which the nonburning helium at the center is surrounded by a layer of burning hydrogen. The energy released by the contracting helium core heats the hydrogen-burning shell, greatly increasing the nuclear reaction rates there. As a result, the star becomes much brighter while the envelope expands and cools. A low-mass star like the Sun moves off the main sequence on the H–R diagram first along the **subgiant branch** (p. 451), then almost vertically up the **red-giant branch** (p. 451).

As the helium core contracts, it heats up. Eventually, it reaches the point at which helium begins to fuse into carbon. The net effect of the fusion reactions is that three helium nuclei (or alpha particles) combine to form a nucleus of carbon in the **triple-alpha process** (p. 451).

In a star like the Sun, conditions at the onset of helium burning are such that the electrons in the core have become degenerate—they can be thought of as tiny, hard spheres that once brought into contact, present stiff resistance to being compressed any further. This **electron degeneracy pressure** (p. 452) makes the core unable to "react" to the new energy source, and helium burning begins explosively, in the **helium flash** (p. 453). The flash expands the core and reduces the star's luminosity, sending it onto the **horizontal branch** (p. 453) of the H–R diagram. The star now has a core of burning helium surrounded by a shell of burning hydrogen.

As helium burns in the core it forms an inner core of nonburning carbon. The carbon core shrinks and heats the overlying burning layers, and the star once again becomes a red giant. It reenters the red-giant region of the H–R diagram along the **asymptotic giant branch** (p. 454), becoming an extremely luminous **red supergiant** (p. 454) star.

The core of a low-mass star never becomes hot enough to fuse carbon. Such a star continues to ascend the asymptotic giant branch until its envelope is ejected into space as a **planetary nebula** (p. 456). At that point the core becomes visible as a hot, faint, and extremely dense white dwarf. The planetary nebula diffuses into space, carrying helium and carbon into the interstellar medium. The white dwarf cools and fades, eventually becoming a cold **black dwarf** (p. 459).

Evolutionary changes happen more rapidly for high-mass stars than for low-mass stars because larger mass results in higher central temperatures. High-mass stars do not experience a helium flash and do attain central temperatures high enough to fuse carbon. They form heavier and heavier elements in their cores, at a more and more rapid pace, and eventually die explosively.

The theory of stellar evolution can be tested by observing star clusters, all of whose stars formed at the same time. As time goes by, the most massive stars evolve off the main sequence first, then the intermediate-mass stars, and so on. At any instant, no stars with masses above the cluster's **main-sequence turnoff** (p. 461) mass remain on the main sequence. Stars below this mass have not yet evolved into giants and so still lie on the main sequence. By comparing a particular cluster's main-sequence turnoff mass with theoretical predictions, astronomers can measure the age of the cluster.

Stars in binary systems can evolve quite differently from isolated stars because of interactions with their companions. Each star is surrounded by a teardrop-shaped **Roche lobe** (p. 465), which defines the region of space within which matter "belongs" to the star. As a star in a binary evolves into the giant phase it may overflow its Roche lobe, forming a **mass-transfer binary** (p. 465) as gas flows from the giant onto its companion. If both stars overflow their Roche lobes, a **contact binary** (p. 465) results. Stellar evolution in binaries can produce states not achievable in single stars. In a sufficiently wide binary, both stars evolve as though they were isolated.

SELF-TEST: TRUE OR FALSE?

_____ **1.** "Low-mass" stars are conventionally taken to have masses less than about 8 solar masses.

_____ **2.** All the red dwarf stars that ever formed are still on the main sequence today.

_____ **3.** Once on the main sequence, gravity is no longer important in determining a star's internal structure.

_____ **4.** The Sun will get brighter as it begins to run out of fuel in its core.

_____ **5.** As a star evolves away from the main sequence it gets larger.

_____ **6.** As a star evolves away from the main sequence it gets hotter.

_____ **7.** As a red giant the Sun will have a core that is smaller than it was when the Sun was on the main sequence.

_____ **8.** When helium starts to fuse inside a solar-mass red giant, it does so slowly at first; the rate of fusion increases gradually over many years.

_____ **9.** With the onset of helium fusion a red giant gets brighter.

_____ **10.** A planetary nebula is the disk of matter around a star that will eventually form a planetary system.

_____ **11.** For a high-mass star there is no helium flash.

_____ **12.** High-mass stars can fuse carbon and oxygen in their cores.

_____ **13.** A star cluster with an age of 100 million years will still contain many O-type stars.

_____ **14.** In a binary star system it is never possible for the lower-mass star to be more evolved than the higher-mass companion.

_____ **15.** In a mass-transfer binary one of the stars has filled its Roche lobe.

SELF-TEST: FILL IN THE BLANK

1. A main-sequence star doesn't collapse because of the outward _____ produced by hot gases in the stellar interior.

2. The Sun will leave the main sequence in about _____ years.

3. While a star is on the main sequence _____ is slowly depleted and _____ builds up in the core.

4. A temperature of at least _____ is needed to fuse helium.

5. At the end of its main-sequence lifetime, a star's core starts to _____.

6. When helium fuses, it produces _____ and releases _____.

7. Just before helium fusion begins in the Sun, the core's outward pressure will be provided mainly by electron _____.

8. As a star ascends the asymptotic giant branch its _____ core is shrinking.

9. As a red supergiant the Sun will eventually become about _____ times its present size.

10. The various stages of stellar evolution predicted by theory can be tested using observations of stars in _____.

11. By the time the envelope of a red supergiant is ejected, the core has shrunk down to a diameter of about _____.

12. A typical white dwarf has the following properties: about half a solar mass, fairly _____ surface temperature, small size, and _____ luminosity.

13. As time goes by, the temperature and the luminosity of a white dwarf both _____.

14. As a star cluster ages, the luminosity of the main-sequence turnoff _____.

15. Whether being a member of a binary star system will affect the evolution of a star depends largely on the _____ of the two stars in the binary.

REVIEW AND DISCUSSION

1. Why don't stars live forever? Which types of stars live the longest?

2. What is main-sequence equilibrium?

3. How long can a star like the Sun keep burning hydrogen in its core?

4. Why is the depletion of hydrogen in the core of a star such an important event?

5. What makes an ordinary star become a red giant?

6. How big (in A.U.) will the Sun become when it enters the red-giant phase?

7. How long does it take for a star like the Sun to evolve from the main sequence to the top of the red-giant branch?

8. What is the helium flash?

9. Describe an important way in which winds from red giant stars are linked to the interstellar medium.

10. How do the late evolutionary stages of high-mass stars differ from those of low-mass stars?

11. What is a planetary nebula? Why do many planetary nebulae appear as rings?

12. What are white dwarfs? What is their ultimate fate?

13. Do many black dwarfs exist in the Galaxy?

14. What are the Roche lobes of a binary system?

15. Why is it odd that the binary system Algol consists of a low-mass red giant orbiting a high-mass main-sequence star? How did Algol come to be in this configuration?

PROBLEMS

1. The Sun will evolve off the main sequence when roughly 10 percent of its hydrogen has been fused into helium. Using the data given in Section 16.5, calculate the total amount of mass destroyed (that is, converted into energy) and the total energy released.

2. Use the radius–luminosity–temperature relation ($L \propto R^2 T^4$; see Section 17.3) to calculate the radius of a red supergiant with temperature 3000 K (half the solar value) and luminosity 10,000 solar luminosities. How many planets of our solar system would this star engulf? (b) Repeat your calculation for a 12,000 K (twice the temperature of the Sun), 0.0004 solar luminosity white dwarf.

3. A main sequence star at a distance of 20 pc is barely visible through a certain telescope. The star subsequently ascends the giant branch, during which time its temperature drops by a factor of 3 and its radius increases 100-fold. What is the new maximum distance at which the star will still be visible using the same telescope?

4. The Sun will reside on the main sequence for 10^{10} years. If the luminosity of a main-sequence star is proportional to the cube of the star's mass, what mass star is just now leaving the main sequence in a cluster that formed 400 million years ago?

5. A Sun-like star goes through its most rapid luminosity change between stages 8 and 9, when the luminosity increases by about a factor of 100 in 10^5 years. On average, how rapidly does the star's absolute magnitude change, in magnitudes per year? Do you think this change would be noticeable in a distant star within a human lifetime?

6. Calculate the average density of a red-giant core of mass 0.25 solar mass and radius 15,000 km. Compare this with the average density of the giant's envelope, if its mass is 0.5 solar mass and its radius is 0.5 A.U. Compare each with the average core density of the Sun.

7. How many years are spent by a Sun-like star during its post–main-sequence evolution? (Use Table 20.1 and consider stages 8 through 12.) What is the probability that observers will happen to catch a star during one of its rapid periods of evolution—say, between stages 8 and 9—if they examine a large number of randomly chosen post–main-sequence stars? How would the odds change if all stars (including those on the main sequence) were included in the search?

8. How long will it take the Sun's planetary nebula, expanding at a speed of 50 km/s, to reach the orbit of Neptune? How long to reach the nearest star?

9. What are the escape speed and surface gravity of Sirius B? (See Table 20.2.)

10. From the discussions presented in *More Precisely 2-3* and *More Precisely 15-1*, it may be shown that the angular momentum of a circular binary, of separation R and component masses m_1 and m_2, is proportional to $m_1 \times m_2 \times \sqrt{R}$ (if the total mass is held fixed). A circular binary has component masses one and two times the mass of the Sun, respectively, and an orbital period of 2 years. (a) What is its orbital separation (semi-major axis)? (b) Mass transfer moves 0.2 solar mass of material from the more massive to the less massive star, keeping the total mass of the system fixed and conserving angular momentum. If the binary remains circular, calculate its new separation and orbital period.

PROJECTS

1. You can tour the Galaxy without ever leaving Earth, just by looking up. In the winter sky, you'll find the red supergiant Betelgeuse in the constellation Orion. It's easy to see because it's one of the brightest stars visible in our night sky. Betelgeuse is a variable star, with a period of about 6.5 years. Its brightness changes as the star expands and contracts. At maximum size, Betelgeuse fills a volume of space that would extend from the Sun beyond the orbit of Jupiter. Betelgeuse is thought to be about 10 to 15 times more massive than our Sun. It is probably between 4 and 10 million years old—and in the final stages of its evolution.

A similar star can be found shining prominently in midsummer. This is the red supergiant Antares in the constellation Scorpius. Depending on the time of year, can you find one of these stars? Why are these stars red?

2. Find a library that has the *Astrophysical Journal*. Find an article from the late 1950s and 1960s that gives the photometry of a star cluster like the Pleiades or Hyades. Plot a color–magnitude diagram (V vs. B-V; see Section 17.6). Determine the V magnitude of the main-sequence turnoff, and hence estimate the age of the cluster. Compare your age with that given in the article.

21 STELLAR EXPLOSIONS

Novae, Supernovae, and the Formation of the Heavy Elements WWW

LEARNING GOALS

Studying this chapter will enable you to:

1   Explain how white dwarfs in binary systems can become explosively active.

2   Summarize the sequence of events leading to the violent death of a massive star.

3   Describe the two types of supernovae, and explain how each is produced.

4   Describe the observational evidence that supernovae have occurred in our galaxy.

5   Explain the origin of elements heavier than helium, and discuss the significance of these elements for the study of stellar evolution.

(Opposite page, background) This is one of the sharpest images of a nova, shown here ejecting hot gas at high speeds from a binary-star system. Known as T Pyxidis, this recurrent nova periodically erupts about every 20 years. It resides nearly 2000 pc away in the dim southern constellation Pyxis. However, instead of a smooth shell of gaseous debris spread across a region of space roughly a light year across, the image reveals blobs of glowing matter, each roughly the size of our solar system, possibly the result of the newly erupted matter plowing into the debris of older, "fossilized" matter from previous eruptions.

(Inset A) An artist's conception helps us to visualize one possible explanation of T Pyxidis's periodic outbursts. Here, a white-dwarf star (upper left) orbits about a cool red-giant star; an elliptical orbit regularly brings the dwarf close to the giant.

(Inset B) The red giant spills some matter onto its companion, forming an accretion disk (lower right), drawn here as a swirling, flattened, blue vortex of hot gas.

(Inset C) As matter continues transferring, causing the process to become unstable, the disk erupts—at first gently as plasma escapes in bipolar flows (or geysers) perpendicular to the disk.

(Inset D) As the companion becomes overloaded with in-falling matter, spontaneous nuclear fusion occurs on or near its surface, resulting in a full-scale nova. The process repeats periodically, causing a "recurrent nova."

What fate awaits a star when it runs out of fuel? For a low-mass star, the white-dwarf stage is not necessarily the end of the road. The potential exists for further violent activity if a binary companion can provide additional fuel. High-mass stars—whether or not they are members of binaries—are also destined to die explosively, releasing vast amounts of energy, creating many heavy elements, and scattering the debris throughout interstellar space. These cataclysmic explosions may trigger the formation of new stars, continuing the cycle of stellar birth and death. In this chapter we will study in more detail the processes responsible for these cataclysmic explosions and the mechanisms that create heavy elements from which we ourselves are made.

21.1 Life after Death for White Dwarfs

1 Although most stars shine steadily day after day, year after year, some change dramatically in brightness over very short periods of time. One type of star, called a **nova** (plural: *novae*), may increase enormously in brightness—by as much as a factor of 10,000 or more—in a matter of days.

The word *nova* means "new" in Latin, and to early observers, these stars did indeed seem new, as they appeared suddenly in the night sky. Astronomers now recognize that a nova is not a new star at all. It is instead a white dwarf—a normally very faint star—undergoing an explosion on its surface that results in a rapid, temporary increase in luminosity. Figure 21.1 illustrates the brightening of a typical nova. Novae eventually fade back to normal, usually after a few weeks or months. On average, two or three novae are observed each year. Astronomers also know of many *recurrent novae*—stars that have been observed to "go nova" several times over the course of a few decades.

What could cause such an explosion on a faint, dead star? The energy involved is far too great to be explained by flares or other surface activity, and as we saw in the previous chapter, there is no nuclear activity in the dwarf's interior. ∞ (Sec. 20.3) To understand what happens, we must reconsider the fate of a low-mass star after it enters the white-dwarf phase.

We noted in Chapter 20 that the white-dwarf stage represents the end point of a star's evolution. Subsequently, the star simply cools, eventually becoming a black dwarf—a burned-out ember in interstellar space. This scenario is quite correct for an *isolated* star, such as our Sun. However, should the star be part of a *binary* system, an important new possibility exists. If the distance between the two stars is small enough, then the dwarf's tidal gravitational field can pull matter—primarily hydrogen and helium—away from the surface of its main-sequence or giant companion, as illustrated in Figure 21.2. The system becomes a mass-transferring binary, similar to those discussed in Chapter 20. A stream of gas leaves the companion through the Lagrange point and flows onto the dwarf. ∞ (Sec. 20.6)

As it builds up on the white dwarf's surface, the stolen gas becomes hotter and denser. Eventually its temperature exceeds 10^7 K, and the hydrogen ignites, fusing into helium at a furious rate (Figures 21.3a–d). This surface burning stage is as brief as it is violent. The star suddenly flares up in luminosity then fades away as some of the fuel is exhausted and the remainder is blown off into space. If the event happens to be visible from Earth, we see a nova. Figure 21.3(e) is a photograph of a nova apparently caught in the act of expelling mass from its surface.

The initial flare-up of luminosity from a nova declines in time, and eventually the star returns to its normal, pre-explosion appearance. The luminosity and temperature of a nova are not usually plotted on an H–R diagram, however. Instead, the change in luminosity is plotted in the form

(a)

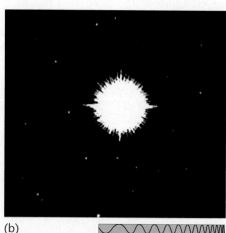
(b)

R I V U X G

Figure 21.1 A nova is a star that suddenly increases enormously in brightness, then slowly fades back to its original luminosity. Novae are the result of explosions on the surfaces of faint white-dwarf stars, caused by matter falling onto their surfaces from the atmosphere of a larger binary companion. Shown is Nova Herculis in (a) March 1935 and (b) May 1935, after brightening by a factor of 60,000.

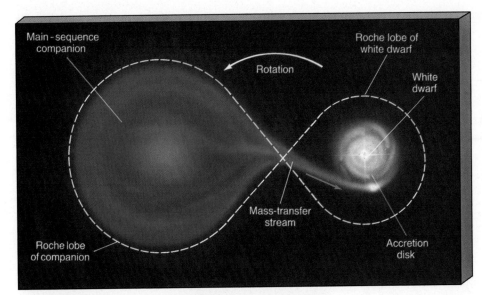

Figure 21.2 A white dwarf in a semidetached binary system may be close enough to its companion that its gravitational field can tear material from the companion's surface. Compare with Figure 20.21. Notice that unlike in the earlier figure, the matter does not fall directly onto the white dwarf's surface. Instead, as discussed a little later in the text, it forms an "accretion disk" of gas spiraling down onto the dwarf.

of a *light curve*, like that shown in Figure 21.4. Such curves show the dramatic rise in luminosity over a few days, followed by the much slower decay over the course of several months. The decline in brightness results from the expansion and cooling of the dwarf's surface layers as they are blown into space. Studies of the details of these curves provide astronomers with a wealth of information about both the dwarf and its binary companion.

Because of the binary's rotation, material leaving the companion does not fall directly onto the dwarf. Instead, it "misses" the compact star, loops around behind it, and goes into orbit around it, forming a swirling, flattened disk of matter called an **accretion disk** (shown in Figure 21.2). Due to the effects of viscosity (that is, friction) within the gas, the orbiting matter in the disk drifts gradually inward, its temperature increasing steadily as it spirals down onto the dwarf's surface. The inner part of the accretion disk becomes so hot that it radiates strongly in the visible, the ultraviolet, and even the X-ray portions of the electromagnetic spectrum. In many systems the disk outshines the white dwarf itself and is the main source of the light emitted between nova outbursts. X rays from the hot disk are routinely

(a) (b) (c) (d)

(e)

Figure 21.3 In this artist's conception, material accumulates on a white dwarf's surface after being accreted from a companion star (a) and then ignites in hydrogen fusion as a nova outburst (b and c). Part of the surface gas is ejected into space in the form of "bubbles" of hot plasma; the rest relaxes back down onto the accretion disk (d). The real photo in (e) corresponds roughly to the events depicted by the artist in frame (c). This nova is called Nova Persei (1901). (See also the chapter opening photo and art sequence A–D on page 470.)

R   I   V   U   X   G

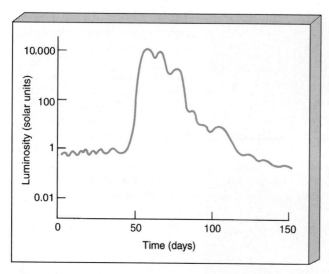

Figure 21.4 The light curve of a typical nova. The rapid rise and slow decline in the light received from the star, as well as the maximum brightness attained, are in good agreement with the explanation of the nova as a nuclear flash on a white dwarf's surface.

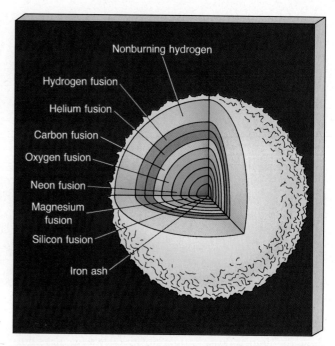

Figure 21.5 Cutaway diagram of the interior of a highly evolved star of mass greater than 8 solar masses. The interior resembles the layers of an onion, with shells of progressively heavier elements burning at smaller and smaller radii and at higher and higher temperatures.

observed in many Galactic novae. The point at which the infalling stream of matter strikes the accretion disk often forms a turbulent "hot spot," causing detectable fluctuations in the light emitted by the binary system.

A nova represents one way in which a star in a binary system can extend its "active lifetime" well into the white-dwarf stage. Recurrent novae can, in principle, repeat their violent outbursts many dozens, if not hundreds, of times. But even more extreme possibilities exist at the end of stellar evolution. Vastly more energetic events may be in store, given the right circumstances.

21.2 The End of a High-Mass Star

2 A low-mass star—one with a mass of less than about eight solar masses—never becomes hot enough to burn carbon in its core. It ends its life as a carbon white dwarf. A high-mass star, however, can fuse not just hydrogen and helium but also carbon, oxygen, and even heavier elements as its inner core continues to contract and its central temperature continues to rise. ∞ (Sec. 20.4) The burning rate accelerates as the core evolves. Can anything stop this runaway process? Is there a stable "white-dwarf–like" state at the end of the evolution of a high-mass star? What is its ultimate fate? To answer these questions, we must look more carefully at fusion in massive stars.

FUSION OF HEAVY ELEMENTS

Figure 21.5 is a cutaway diagram of the interior of a highly evolved star of large mass. Note the numerous layers where various nuclei burn. As the temperature increases with

depth, the ash of each burning stage becomes the fuel for the next stage. At the relatively cool periphery of the core, hydrogen fuses into helium. In the intermediate layers, shells of helium, carbon, and oxygen burn to form heavier nuclei. Deeper down reside neon, magnesium, silicon, and other heavy nuclei, all produced by nuclear fusion in the layers overlying the core. The core itself is composed of iron. We will study the key reactions in this burning chain in more detail later in this chapter.

As each element is burned to depletion at the center, the core contracts, heats up, and starts to fuse the ash of the previous burning stage. A new inner core forms, contracts again, heats again, and so on. Through each period of stability and instability, the star's central temperature increases, the nuclear reactions speed up, and the newly released energy supports the star for ever-shorter periods of time. For example, in round numbers, a star 20 times more massive than the Sun burns hydrogen for 10 million years, helium for 1 million years, carbon for 1000 years, oxygen for 1 year, and silicon for a week. Its iron core grows for less than a day.

COLLAPSE OF THE IRON CORE

Once the inner core begins to change into iron, our high-mass star is in trouble. Nuclear fusion involving iron does not produce energy—iron nuclei are so compact that energy cannot be extracted by combining them into heavier elements. In effect, iron plays the role of a fire extinguisher, damping the inferno in the stellar core. With the appearance of substantial quantities of iron, the central fires cease for the last time, and the star's internal support begins to dwindle. The star's foundation is destroyed, and its equilibrium is gone forever. Even though the temperature in the iron core has reached several billion kelvins by this stage, the enormous inward gravitational pull of matter ensures catastrophe in the very near future. Gravity overwhelms the pressure of the hot gas, and the star implodes, falling in on itself.

The core temperature rises to nearly 10 billion K. At these temperatures, individual photons, according to Wien's law, have tremendously high energies, enough to split iron into lighter nuclei and then to break those lighter nuclei apart until only protons and neutrons remain. ∞ (Sec. 3.4) This process is known as *photodisintegration* of the heavy elements in the core. In less than a second, the collapsing core undoes all the effects of nuclear fusion that occurred during the previous 10 million years! But to split iron and lighter nuclei into smaller pieces requires a lot of energy. After all, this splitting is just the opposite of the fusion reactions that generated the star's energy during earlier times. Photodisintegration *absorbs* some of the core's thermal energy—in other words, it cools the core and so reduces the pressure. As nuclei are destroyed, the core of the star becomes even less able to support itself against its own gravity. The collapse accelerates.

Now the core consists entirely of simple elementary particles—electrons, protons, neutrons, and photons—at enormously high densities, and it is still shrinking. As the core density continues to rise, the protons and electrons are crushed together, forming neutrons and neutrinos:

$$p + e \rightarrow n + \text{neutrino}.$$

This process is sometimes called the *neutronization* of the core. Recall from our discussion in Chapter 16 that the neutrino is an extremely elusive particle that interacts hardly at all with matter. ∞ (Sec. 16.5) Even though the central density by this time may have reached 10^{12} kg/m³ or more, most of the neutrinos produced by neutronization pass through the core as if it weren't there. They escape into space, carrying away energy as they go.

The disappearance of the electrons and the escape of the neutrinos make matters even worse for the core's stability. There is now nothing to prevent it from collapsing all the way to the point at which the neutrons come into contact with one another, at the incredible density of about 10^{15} kg/m³. At this point the neutrons in the shrinking core play a role similar in many ways to that of the electrons in a white dwarf. When far apart they offer little resistance to compression, but when brought into contact, they produce enormous pressures that strongly oppose further gravitational collapse. This *neutron degeneracy pressure*, akin to the electron degeneracy pressure that operates in red giants and white dwarfs, finally begins to slow the collapse. ∞ (Sec. 20.2) By the time the collapse is actually halted, however, the core has overshot its point of equilibrium, and may reach densities as high as 10^{17} or 10^{18} kg/m³ before turning around and beginning to reexpand. Like a fast-moving ball hitting a brick wall, the core becomes compressed, stops, then rebounds—with a vengeance!

The events just described do not take long. Only about a second elapses from the start of the collapse to the "bounce" at nuclear densities. At that point the core rebounds. An enormously energetic shock wave sweeps through the star at high speed, blasting all the overlying layers—including the heavy elements outside the iron inner core—into space. Although the details of how the shock reaches the surface and destroys the star are still uncertain, the end result is not. The star explodes, in one of the most energetic events known in the universe (see Figure 21.6). For a period of a few days the exploding star may rival in brightness the entire galaxy in which it resides. This spectacular death rattle of a high-mass star is known as a **core-collapse supernova**.

Figure 21.6 A supernova called SN1987A (arrow) was exploding near this nebula (30 Doradus) at the moment the photograph on the right was taken. The photograph on the left is the normal appearance of the star field. (See *Interlude 21-1*.)

21.3 Supernova Explosions

NOVAE AND SUPERNOVAE

3 Let's compare a supernova with a nova. Like a nova, a **supernova** is a star that suddenly increases dramatically in brightness, then slowly dims again, eventually fading from view. The exploding star is commonly called the supernova's *progenitor*. In some cases, supernovae light curves can appear quite similar to those of novae, so a distant supernova can look a lot like a nearby nova—so much so, in fact, that the difference between the two was not fully appreciated until the 1920s. But novae and supernovae are now known to be very different phenomena. Supernovae are much more energetic events, driven by very different underlying physical processes.*

Well before they understood the causes of either novae or supernovae, astronomers knew of clear observational differences between them. The most important of these is that a supernova is more than a million times brighter than a nova. A supernova produces a burst of light billions of times brighter than the Sun, reaching that brightness within just a few hours after the start of the outburst. The total amount of electromagnetic energy radiated by a supernova during the few months it takes to brighten and fade away is roughly 10^{43} J—nearly as much energy as the Sun will radiate during its *entire* 10^{10}-year lifetime! (Enormous as this energy is, however, it pales in comparison with the energy emitted in the form of *neutrinos*, which may be 100 times greater.)

A second important difference is that the same star may become a nova many times, but a star can become a supernova only once. This fact was unexplained before astronomers knew the precise nature of novae and supernovae, but it is easily understood now that we understand how and why these explosions occur. The nova accretion–explosion cycle described earlier can take place over and over again, but a supernova destroys the star involved, with no possibility of a repeat performance.

In addition to the distinction between novae and supernovae, there are also important observational differences *among* supernovae. Some supernovae contain very little hydrogen, according to their spectra, whereas others contain a lot. Also, the light curves of the hydrogen-poor supernovae are qualitatively different from those of the hydrogen-rich ones, as illustrated in Figure 21.7. Based on these observations, astronomers divide supernovae into two classes, known simply as Type I and Type II. **Type I supernovae**, the hydrogen-poor kind, have a light curve somewhat similar in shape to that of typical novae. **Type II supernovae**, whose spectra show lots of hydrogen, usually have a characteristic "plateau" in the light curve a few

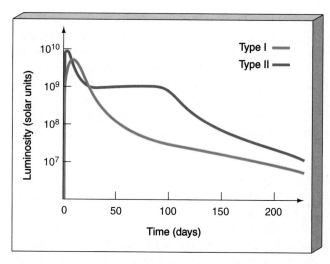

Figure 21.7 The light curves of typical Type I and Type II supernovae. In both cases the maximum brightness or intensity can reach nearly 10 billion Suns, but there are characteristic differences in the falloff of the luminosity after the initial peak. Type I light curves somewhat resemble those of novae (Figure 21.4). Type II curves have a characteristic bump in the declining phase.

months after the maximum. Observed supernovae are divided roughly equally between these two categories.

CARBON-DETONATION SUPERNOVAE

What is responsible for these differences among supernovae? Is there more than one way in which a supernova explosion can occur? The answer is yes. To understand the alternative supernova mechanism, we must return to the processes that cause novae and consider the long-term consequences of their accretion–explosion cycle.

Nova explosions eject matter from a white dwarf's surface, but they do not necessarily expel or burn all the material that has accumulated since the last outburst. In other words, there is a tendency for the dwarf's mass to increase slowly with each new nova cycle. As its mass grows and the internal pressure required to support its weight rises, the white dwarf can enter into a new period of instability—with disastrous consequences.

Recall that a white dwarf is held up not by thermal pressure (heat) but by the degeneracy pressure of electrons that have been squeezed so close together that they have effectively come into contact with one another. ∞ (Sec. 20.3) However, there is a limit to the pressure that these electrons can exert. Consequently, there is a limit to the mass of a white dwarf, above which electrons cannot provide the pressure needed to support the star. Detailed calculations show that the maximum mass of a white dwarf is about 1.4 solar masses, a mass often called the *Chandrasekhar mass*, after the Indian astronomer Subramanyan Chandrasekhar, whose work in theoretical astrophysics earned him a Nobel Prize in physics in 1983.

Note that, when discussing novae and supernovae, astronomers tend to blur the distinction between the observed event (the sudden appearance and brightening of an object in the sky) and the process responsible for it (a violent explosion in or on a star). The two terms can have either meaning, depending on context.

If an accreting white dwarf exceeds the Chandrasekhar mass, the pressure of degenerate electrons in its interior becomes unable to withstand the pull of gravity, and the star immediately starts to collapse. Its internal temperature rapidly rises to the point at which carbon can fuse into heavier elements. Carbon fusion begins everywhere throughout the white dwarf almost simultaneously, and the entire star explodes in another type of supernova—a so-called **carbon-detonation supernova**—comparable in violence to the "implosion" supernova associated with the death of a high-mass star, but very different in cause. In an alternative and (some astronomers think) possibly more common scenario, two white dwarfs in a binary system may collide and merge to form a massive, unstable star. The end result is the same—a carbon-detonation supernova.

We can now understand the differences between Type I and Type II supernovae. The explosion resulting from the detonation of a carbon white dwarf, the descendant of a low-mass star, is a supernova of Type I. Because this conflagration stems from a system containing virtually no hydrogen, we can readily see why the spectrum of a Type I supernova shows little evidence of that element. The appearance of the light curve (as we will soon see) results almost entirely from the radioactive decay of unstable heavy elements produced in the explosion itself.

The implosion–explosion of the core of a massive star, described earlier, produces a Type II supernova. Detailed computer models indicate that the characteristic shape of the Type II light curve is just what would be expected from the expansion and cooling of the star's outer envelope as it is blown into space by the shock wave sweeping up from below. The expanding material consists mainly of unburned gas—hydrogen and helium—so it is not surprising that those elements are strongly represented in the supernova's observed spectrum. (See *Interlude 21-1* for an account of a recent Type II supernova that confirmed many basic theoretical predictions while also forcing astronomers to revise the details of their models.)

Figure 21.8 summarizes the processes responsible for the two different types of supernovae. We emphasize that, despite the similarity in the total amounts of energy involved, Type I and Type II supernovae are unrelated to one another. They occur in stars of very different types, under very different circumstances. All high-mass stars become Type II (core-collapse) supernovae, but only a tiny fraction of low-mass stars evolve into white dwarfs that ultimately explode as Type I (carbon-detonation) supernovae. However, there are far more low-mass stars than high-mass stars, so by a remarkable coincidence the two types of supernova occur at roughly the same rate.

(a) Type I Supernova

Binary star system

White dwarf

Planetary nebula

Accretion disk

Red giant

Growing white dwarf

Detonation

Time

(b) Type II Supernova

Helium, carbon

Hydrogen

Normal star fusion

Heavy elements

Hydrogen

Iron core

Massive star imploding

Hydrogen

Core rebound

Remnant core

Shock wavw

Explosion

Figure 21.8 Type I and Type II supernovae have different causes. These sequences depict the evolutionary history of each type. (a) A Type I supernova usually results when a carbon-rich white dwarf pulls matter onto itself from a nearby red-giant companion. (b) A Type II supernova occurs when the core of a more massive star collapses, then rebounds in a catastrophic explosion.

R I V U X G   R I V U X G   R I V U X G

Figure 21.9 This remnant of an ancient supernova is called the Crab Nebula (or M1 in the Messier catalog). It resides about 1800 pc from Earth and has an angular diameter about one-fifth that of the full Moon. Because its debris is scattered over a region of "only" 2 pc, the Crab is considered to be a young supernova remnant. In A.D. 1054 Chinese astronomers observed the supernova explosion itself. The center frame shows the Crab in visible light. The left and right frames, to the same scale, show the Crab Nebula in the radio and ultraviolet, respectively.

SUPERNOVA REMNANTS

4 We have plenty of evidence that supernovae have occurred in our Galaxy. Occasionally, the explosions themselves are visible from Earth (see *Interlude 21-2*). In many other cases we can detect their glowing remains, or **supernova remnants**. One of the best-studied supernova remnants is known as the Crab Nebula, shown in Figure 21.9. Its brightness has greatly dimmed now, but the original explosion in the year A.D. 1054 was so brilliant that manuscripts of ancient Chinese and Middle Eastern astronomers claim that its brightness greatly exceeded that of Venus and—according to some (possibly exaggerated) accounts—even rivaled that of the Moon. For nearly a month, this exploded star reportedly could be seen in broad daylight. Native Americans also left engravings of the event in the rocks of what is now the southwestern United States.

The Crab Nebula certainly has the appearance of exploded debris. Even today, the knots and filaments give a strong indication of past violence (and continuing activity—see *Interlude 21-3*). In fact, astronomers have proved that this matter was ejected from some central explosion. Doppler-shifted spectral lines indicate that the nebula—the envelope of the high-mass star that exploded to create this Type II supernova—is expanding into space at several thousand kilometers per second. A vivid illustration of this fact is provided by Figure 21.10, which was made by superimposing a positive image of the Crab Nebula taken in 1960 and a negative image taken in 1974. If the gas were not in motion, the positive and negative images would overlap perfectly, but they

do not. The gas moved outward in the intervening 14 years. Running the motion backward in time, astronomers have found that the explosion must have occurred about nine centuries ago, consistent with the Chinese observations.

The nighttime sky harbors many relics of stars that blew up long ago. Figure 21.11 is another example. It shows the Vela supernova remnant, whose expansion velocities imply that its central star exploded around 9000 B.C. It lies only 500 pc away from Earth. Given its proximity, it may have

R I V U X G

Figure 21.10 Positive and negative photographs of the Crab Nebula taken 14 years apart do not superimpose exactly, indicating that the gaseous filaments are still moving away from the site of the explosion.

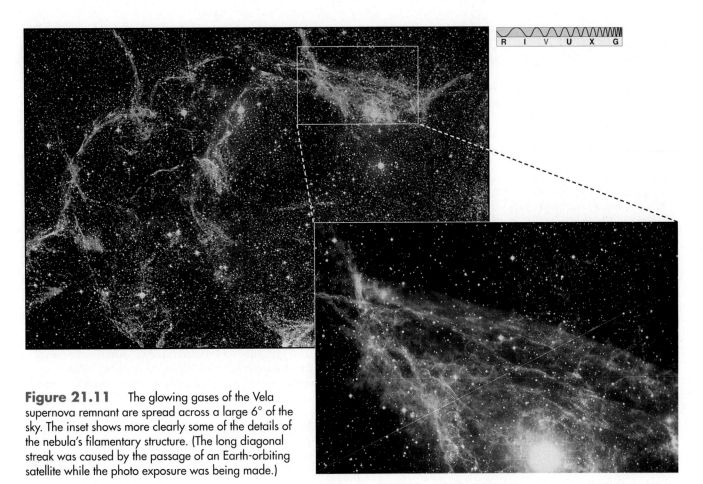

R I V U X G

Figure 21.11   The glowing gases of the Vela supernova remnant are spread across a large 6° of the sky. The inset shows more clearly some of the details of the nebula's filamentary structure. (The long diagonal streak was caused by the passage of an Earth-orbiting satellite while the photo exposure was being made.)

been as bright as the Moon for several months. We can only speculate what impact such a bright supernova might have had on the myths, religions, and cultures of Stone Age humans when it first appeared in the sky.

Although hundreds of supernovae have been observed in other galaxies during the twentieth century, no one has ever observed with modern equipment a supernova in our own Galaxy. A viewable Milky Way star has not exploded since Galileo first turned his telescope to the heavens almost four centuries ago (see *Interlude 21-2*). Now known as Tycho's supernova, this last supernova observed in our Galaxy caused a worldwide sensation in Renaissance times. The sudden appearance and subsequent fading of this very bright object in the year 1572 helped shatter the Aristotelian idea of an unchanging universe.

Based on stellar evolutionary theory, astronomers calculate that an observable supernova ought to occur in our Galaxy every 100 years or so. Because the brilliance of a nearby supernova might rival that of a full Moon, it seems unlikely that astronomers could have missed any since the last one nearly four centuries ago. Our local neighborhood of the Milky Way seems long overdue for a supernova. Unless massive stars explode much less frequently than predicted by the theory of stellar evolution, we should be treated to a (relatively) nearby version of nature's most spectacular cosmic event any day now.

21.4 The Formation of the Elements

5 Up to now we have studied nuclear reactions mainly for their role in stellar energy generation. Now let's consider them again, but this time as the processes responsible for creating much of the world in which we live. The evolution of the elements, combining nuclear physics with astronomy, is a very complex subject and a very important problem in modern astronomy.

TYPES OF MATTER

We currently know of 112 different elements, ranging from the simplest—hydrogen, containing 1 proton—to the most complex, discovered in 1996, with 112 protons in its nucleus. (See Appendix Table 2.) All elements exist in several different *isotopic* forms, each isotope having the same number of protons but a different number of neutrons. We often think of the most common or stable isotope as being the "normal" form of an element. Some elements, and many isotopes, are radioactively unstable, meaning that they eventually decay into other, more stable, nuclei.

The 81 stable elements found on Earth make up the overwhelming bulk of matter in the universe. In addition,

INTERLUDE 21-1 Supernova 1987A *Videos*

In 1987, astronomers were treated to a spectacular supernova in the Large Magellanic Cloud (LMC), a small satellite galaxy orbiting our own. ∞ (Sec. 24.2) Observers in Chile first saw the explosion on February 24, and within a few hours, nearly all Southern Hemisphere telescopes and every available orbiting spacecraft were focused on the object. It was officially named SN1987A. (The SN stands for "supernova," 1987 gives the year, and A identifies it as the first supernova seen that year.) This was one of the most dramatic changes observed in the universe in nearly 400 years. A 15-solar-mass B supergiant star with the catalog name SK-69°202 exploded, outshining for a few weeks all the other stars in the LMC combined, as shown in the "before" and "after" images of Figure 21.6.

Because the LMC is relatively close to Earth and because the explosion was detected so soon after it occurred, SN1987A has provided astronomers with a wealth of detailed information on supernovae, allowing them to make key comparisons between theoretical models and observational reality. By and large, the theory of stellar evolution described in the text has held up very well. Still, SN1987A did hold some surprises.

According to its hydrogen-rich spectrum, the supernova was of Type II—the core-collapse type—as expected for a high-mass parent star such as SK-69°202. But according to Figure 20.16 (which was computed for stars in our own galaxy), the parent star should have been a red supergiant at the time of the explosion—not a blue supergiant, as was actually observed. This unexpected finding caused theorists to scramble in search of an explanation. It now seems that, relative to young stars in the Milky Way, the parent star's envelope was deficient in heavy elements. This deficiency had little effect on the evolution of the core and on the supernova explosion, but it did change the star's evolutionary track on the H–R diagram. Unlike a Milky Way star of the same mass, once helium ignited in the core of SK-69°202, the star shrank and looped back toward the main sequence. The star had just begun to return to the right on the H–R diagram following the ignition of carbon, with a surface temperature of around 20,000 K, when the rapid chain of events leading to the supernova occurred.

The light curve of SN1987A, shown here, also differed somewhat from the "standard" Type II shape (see Figure 21.7). The peak brightness was less than 1/10 the expected value.

For a few days after its initial detection, the supernova faded as it expanded and cooled rapidly. After about a week the surface temperature had dropped to about 5000 K, at which point electrons and protons near the expanding surface recombined into atomic hydrogen, making the surface layers less opaque and allowing more radiation from the interior to leak out. As a result, the supernova brightened rapidly as it grew. The temperature of the expanding layers reached a peak in late May, by which point the radius of the expanding photosphere was about 2×10^{10} km—a little larger than our solar system. Subsequently, the photosphere cooled as it expanded, and the luminosity dropped as the internal supply of heat from the explosion leaked away into space.

Much of the preceding description would apply equally well to a Type II supernova in our own Galaxy. The differences between the SN1987A light curve shown here and the Type II light curve in Figure 21.7 are mainly the result of the (relatively) small size of SN1987A's parent star. The peak luminosity of SN1987A was less than that of a "normal" Type II supernova because SK-69°202 was small and quite tightly bound by gravity. A lot of the energy emitted in the form of visible radiation (and evident in Figure 21.7)

---

10 radioactive elements—including radon and uranium—also occur naturally on our planet. Even though their half-lives (the time required for half the nuclei to decay into something else) of these elements are very long (millions or even billions of years, typically), their steady decay means that they are scarce on Earth, in meteorites, and in lunar samples. ∞ (*More Precisely 7-2*) They are not observed in stars—there is just too little of them to produce detectable spectral lines.

Besides these 10 naturally occurring radioactive elements, 19 more radioactive elements have been artificially produced under special conditions in nuclear laboratories on Earth. The debris collected after nuclear weapons tests also contains traces of some of these elements. Unlike the naturally occurring radioactive elements, these artificial ones decay into other elements quite quickly (in much less than a million years). Consequently, they too are extremely rare in nature. Two other elements round out our list. Promethium is a stable element that is found on our planet only as a by-product of nuclear laboratory experiments. Technetium is an unstable element that is found in stars but does not occur naturally on Earth.

ABUNDANCE OF MATTER

How and where did all these elements form? Were they always present in the universe, or were they created after the universe formed? Since the 1950s, astronomers have was used up in expanding SN1987A's stellar envelope, so far less was left over to be radiated into space. Thus, SN1987A's luminosity during the first few months was lower than expected, and the early peak evident in Figure 21.7 did not occur. The peak in the SN1987A light curve at about 80–100 days actually corresponds to the "plateau" in the Type II light curve in Figure 21.7.

About 20 hours before the supernova was detected optically, a brief (13-second) burst of neutrinos was simultaneously recorded by underground detectors in Japan and the United States. As discussed in the text, the neutrinos are predicted to arise when electrons and protons in the star's collapsing core merge to form neutrons. The neutrinos preceded the light because they escaped during the collapse, whereas the first light of the explosion was emitted only after the supernova shock had plowed through the body of the star to the surface. In fact, theoretical models consistent with these observations suggest that vastly more energy was emitted in the form of neutrinos than in any other form. The supernova's neutrino luminosity was many tens of thousands of times greater than its optical energy output.

Despite some unresolved details in SN1987A's behavior, detection of this neutrino pulse is considered to be a brilliant confirmation of theory. This singular event—the detection of neutrinos—may well herald a new age of astronomy. For the first time, astronomers have received information from beyond the solar system by radiation outside the electromagnetic spectrum.

Theory predicts that the expanding remnant of SN1987A will be large enough to be resolvable by optical telescopes in a few years. The accompanying photograph was taken by the *Hubble Space Telescope* in late 1996. It shows the unresolved remnant (at center) surrounded by a much larger shell of glowing gas (in yellow). Scientists reason that the progenitor star expelled this shell during its red giant phase, some 40,000 years before the explosion. The image we see results from the initial flash of ultraviolet light from the supernova hitting the ring and causing it to glow brightly. In about 10 years the debris from the explosion itself will strike the ring, making it a temporary but intense

source of X rays. In 1998, the fastest-moving ejecta already began doing so.

This overexposed photo also shows the core debris moving outward toward the ring. The four insets, taken over a 24-month period, directly resolve material expanding at nearly 3000 km/s. These images also revealed, to everyone's surprise, two additional faint rings (in red) that might be radiation sweeping across the hourglass-shaped bubble of gas. Why the gas should exhibit this odd structure is unclear.

Buoyed by the success of stellar-evolution theory and armed with firm theoretical predictions of what should happen next, astronomers eagerly await future developments in the story of this remarkable object.

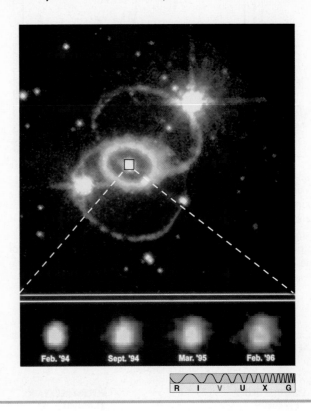

come to realize that the hydrogen and most of the helium in the universe are *primordial*—that is, these elements date from the very earliest times. All other elements in our universe result from **stellar nucleosynthesis**—that is, they were formed by nuclear fusion in the hearts of stars.

To test this idea we must consider not just the list of different *kinds* of elements and isotopes but also their observed *abundances*, shown in Figure 21.12. This curve is derived largely from spectroscopic studies of stars, including the Sun. The essence of the figure is summarized in Table 21.1, which combines all the known elements into eight distinct groups based on the number of nuclear particles (protons and neutrons) that they contain. (All isotopes of all elements are included in both Table 21.1 and Figure

21.12, although only a few elements are marked by dots and labeled in the figure.) Any theory proposed for the creation of the elements must reproduce these observed abundances. The most obvious feature is that the heavy elements are much less abundant than most light elements. However, the many peaks and troughs evident in Figure 21.12 also represent important constraints.

HYDROGEN AND HELIUM BURNING

Let's begin by reviewing the reactions leading to heavy-element production at various stages of stellar evolution. Stellar nucleosynthesis begins with the proton–proton chain studied in Chapter 16. ∞ (Sec. 16.5) Provided that

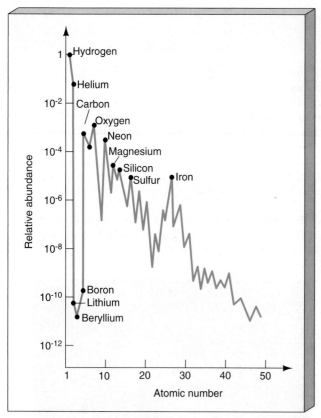

Figure 21.12 A summary of the cosmic abundances of the elements and their isotopes, expressed relative to the abundance of hydrogen. The horizontal axis shows atomic number—the number of protons in the nucleus. Notice how many common terrestrial elements are found on "peaks" of the distribution, surrounded by elements that are tens or hundreds of times less abundant. Notice especially the large peak around the element iron.

the temperature is high enough—at least 10^7 K—a series of nuclear reactions occur, ultimately forming a nucleus of ordinary helium (^{4}He) from four protons (^{1}H):

$$4(^1\text{H}) \rightarrow {}^4\text{He} + 2 \text{ positrons} + 2 \text{ neutrinos} + \text{energy.}$$

Recall that the positrons immediately interact with nearby free electrons, producing high-energy gamma rays through matter–antimatter annihilation. The neutrinos rapidly escape, carrying energy with them but playing no direct role in nucleosynthesis. The validity of these reactions has been directly confirmed in nuclear experiments conducted in laboratories around the world during recent decades. In massive stars, the CNO cycle (*Interlude 20-1*) may greatly accelerate the hydrogen-burning process, but the basic 4-protons-to-1-helium-nucleus reaction, illustrated in Figure 21.13, is unchanged.

As helium builds up in the core of a star, the burning ceases, and the core contracts and heats up. When the temperature exceeds about 10^8 K, helium nuclei can overcome their mutual electrical repulsion, leading to the *triple-alpha reaction*, which we discussed in Chapter 20: ∞ (Sec. 20.2)

$$3(^4\text{He}) \rightarrow {}^{12}\text{C} + \text{energy.}$$

The net result of this reaction is that three helium-4 nuclei are combined into one carbon-12 nucleus (Figure 21.14), releasing energy in the process.

CARBON BURNING AND HELIUM CAPTURE

At higher and higher temperatures, heavier and heavier nuclei can gain enough energy to overcome the electrical

TABLE 21.1	Cosmic Abundances of the Elements	
ELEMENTAL GROUP OF PARTICLES	**PERCENT ABUNDANCE BY NUMBER***
Hydrogen (1 nuclear particle)	90
Helium (4 nuclear particles)	9
Lithium group (7–11 nuclear particles)	0.000001
Carbon group (12–20 nuclear particles)	0.2
Silicon group (23–48 nuclear particles)	0.01
Iron group (50–62 nuclear particles)	0.01
Middle-weight group (63–100 nuclear particles)	0.00000001
Heaviest-weight group (over 100 nuclear particles)	0.000000001

*The total does not equal 100 percent because of uncertainties in the helium abundance. All isotopes of all elements are included.

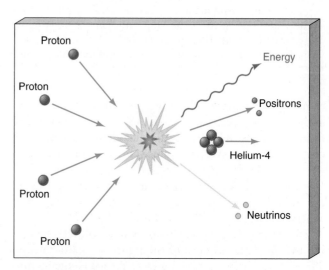

Figure 21.13 Diagram of the basic proton–proton hydrogen-burning reaction. Four protons combine to form a nucleus of helium-4, releasing energy in the process.

INTERLUDE 21-2 Nearby Supernovae

Only six Galactic supernovae have been recorded in the past 1000 years. The accompanying figure shows their positions in the Milky Way. They are labeled by the year in which they first appeared. The supernova Cassiopeia A (Cas A) apparently went unnoticed optically, although modern radio studies suggest that the first light from the explosion should have reached Earth midway through the seventeenth century. The combined radio–optical–X-ray image is dramatic proof that this supernova remnant, although invisible optically, makes quite an impact at other wavelengths. (Blue is radio, red is optical, and green is X ray.)

Most astronomers assume that many more stars than these six have blown up in our galaxy. Why haven't we seen them? Possibly because they were too distant to be detected by the naked eye, or most likely because dark dust clouds in the Galactic plane kept them from our view. Each of the Milky Way supernovae mapped in the first figure is in our "neighborhood"—that is, in our quadrant of the Galaxy—

and each is at least 100 pc above or below the Galactic plane. Studies of the rate at which supernovae occur suggest that we can expect one within 100 pc of our Sun only every 500,000 years. Thus, a truly "close" supernova would be a rare event indeed. Humanity may be destined to see all supernovae from a distance.

Despite their rarity, nearby supernovae may conceivably play an important role in determining the development of life on Earth. A supernova at a distance of a few parsecs would bombard our planet with high-energy radiation for a period of several months, possibly causing substantial long-term atmospheric changes, particularly in the ozone layer. Some scientists have even gone so far as to suggest that such an event might have been responsible for episodes of mass extinction that are known to have occurred hundreds of millions of years ago and in which, according to the fossil record, over 95 percent of all life on our planet vanished in a very short period of time. (Recall, though, that this is not necessarily the only "astronomical" explanation for mass extinctions on our planet—see *Interlude 14-1*.)

R I V U X G

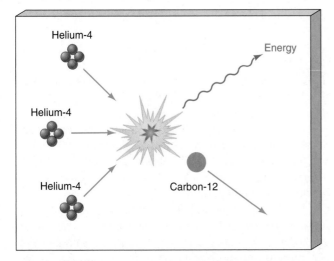

Figure 21.14 Diagram of the basic triple-alpha helium-burning reaction occurring in post–main-sequence stars. Three helium-4 nuclei combine to form carbon-12.

repulsion between them. At about 6×10^8 K (reached only in the cores of stars much more massive than the Sun), carbon nuclei can fuse to form magnesium, as depicted in Figure 21.15(a):

$$^{12}\text{C} + {}^{12}\text{C} \rightarrow {}^{24}\text{Mg} + \text{energy}.$$

However, because of the rapidly mounting nuclear charges—that is, the increasing number of protons in the nuclei—fusion reactions between any nuclei larger than carbon require such high temperatures that they are actually quite uncommon in stars. The formation of most heavier elements occurs by way of an easier path. For example, the repulsive force between two carbon nuclei is three times greater than the force between a nucleus of carbon and one of helium. Thus, carbon–helium fusion occurs at a lower temperature than that at which carbon–carbon fusion occurs. As we saw in Section 20.3, at temperatures

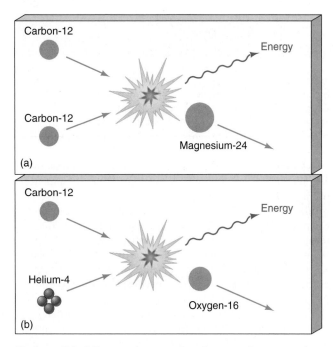

Figure 21.15 Carbon can form heavier elements (a) by fusion with other carbon nuclei or, more commonly, (b) by fusion with a helium nucleus.

above 2×10^8 K, a carbon-12 nucleus colliding with a helium-4 nucleus can produce oxygen-16:

$$^{12}\text{C} + {}^4\text{He} \rightarrow {}^{16}\text{O} + \text{energy}.$$

If any helium-4 is present, this reaction, shown in Figure 21.15(b), is much more likely to occur than the carbon–carbon reaction.

Similarly, the oxygen-16 thus produced may fuse with other oxygen-16 nuclei at a temperature of about 10^9 K to form sulfur-32,

$$^{16}\text{O} + {}^{16}\text{O} \rightarrow {}^{32}\text{S} + \text{energy},$$

but it is much more probable that an oxygen-16 nucleus will capture a helium-4 nucleus (if one is available) to form neon-20:

$$^{16}\text{O} + {}^4\text{He} \rightarrow {}^{20}\text{Ne} + \text{energy}.$$

The second reaction is more likely because it requires a lower temperature than that necessary for oxygen–oxygen fusion.

Thus, as the star evolves, heavier elements tend to form through **helium capture** rather than by fusion of like nuclei. Because these helium-capture reactions are so much more common, elements with nuclear masses of 4 units (that is, helium itself), 12 units (carbon), 16 units (oxygen), 20 units (neon), 24 units (magnesium), and 28 units (silicon) stand out as prominent peaks in Figure 21.12, our chart of cosmic abundances. Each element is built by combining the preceding element and a helium-4 nucleus as the star evolves.

IRON FORMATION

Helium capture is by no means the only type of nuclear reaction occurring in evolved stars. As nuclei of many different kinds accumulate, a great variety of reactions become possible. In some, protons and neutrons are freed from their parent nuclei and are absorbed by others, resulting in new nuclei with masses intermediate between those formed by helium capture. Laboratory studies confirm that common nuclei, such as fluorine-19, sodium-23, phosphorus-31, and many others, are created in this way. Their abundances, however, are not so great as those produced directly by helium capture, simply because the helium-capture reactions are much more common in stars. For this reason many of these elements (those with masses not divisible by 4, the mass of a helium nucleus) are found in the troughs of Figure 21.12.

Around the time silicon-28 appears in the core of a star, a competitive struggle begins between the continued capture of helium to produce even heavier nuclei and the tendency of more complex nuclei to break down into simpler ones. The cause of this breakdown is heat. By now the star's core temperature has reached the unimaginably large value of 3 billion K, and the gamma rays associated with that temperature have enough energy to break a nucleus apart, as illustrated in Figure 21.16(a). This is the same process of photodisintegration that will ultimately accelerate the star's iron core in its final collapse toward a Type II supernova.

Under the intense heat, some silicon-28 nuclei break apart into seven helium-4 nuclei. Other nearby nuclei that have not yet photodisintegrated may capture some or all of these helium-4 nuclei, leading to the formation of still heavier elements (Figure 21.1b). The process of photodisintegration provides raw material that allows the helium-capture process to proceed to greater masses. The process continues, with some heavy nuclei being destroyed and others increasing in mass. In succession the star forms sulfur-32, argon-36, calcium-40, titanium-44, chromium-48, iron-52, and nickel-56. The chain of reactions building from silicon-28 up to nickel-56 is

$$^{28}\text{Si} + 7({}^4\text{He}) \rightarrow {}^{56}\text{Ni} + \text{energy}.$$

This two-step process—photodisintegration followed by the direct capture of some or all of the resulting helium-4 nuclei (or alpha particles)—is often called the *alpha process*.

Nickel-56 is unstable. It decays rapidly, first into cobalt-56, then into a stable iron-56 nucleus. Any unstable nucleus will continue to decay until stability is achieved, and iron-56 is the stablest of all nuclei. Thus, the alpha process leads inevitably to the buildup of iron in the stellar core.

Iron's 26 protons and 30 neutrons are bound together more strongly than the particles in any other nucleus. Iron is said to have the greatest *nuclear binding energy* of any element. Any nucleus with more or fewer protons or neutrons has less nuclear binding energy and is not quite so stable as the iron-56 nucleus. This enhanced stability of iron explains why some of the heavier nuclei in the iron group are more abundant than many lighter nuclei (see Table 21.1 and Figure 21.12)—nuclei tend to "accumulate" near iron as stars evolve.

MAKING ELEMENTS BEYOND IRON

If the alpha process stops at iron, how did heavier elements, such as copper, zinc, and gold, form? To form them, some nuclear process other than helium capture must have been involved. That other process is **neutron capture**—the formation of heavier nuclei by the absorption of neutrons.

Deep in the interiors of highly evolved stars, conditions are ripe for neutron capture to occur. Neutrons are produced as "by-products" of many nuclear reactions, so there are many of them present to interact with iron and other nuclei. Neutrons have no charge, so there is no repulsive barrier for them to overcome in combining with positively charged nuclei. As more and more neutrons join an iron nucleus its mass continues to grow.

Figure 21.16 (a) At high temperatures heavy nuclei (such as silicon, shown here) can be broken apart into helium nuclei by high-energy photons. (b) Other nuclei can capture the helium nuclei—or alpha particles—thus produced, forming heavier elements by the so-called alpha process. This process continues all the way to the formation of iron.

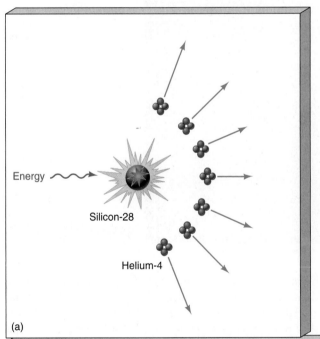

(a)

Adding neutrons to a nucleus—iron, for example—does not change the element. Rather, a more massive isotope is produced. Eventually, however, so many neutrons are added to the nucleus that it becomes unstable and then decays radioactively to form a stable nucleus of some other element. The neutron-capture process then continues. For example, an iron-56 nucleus can capture a single neutron (n) to form a relatively stable isotope, iron-57:

$$^{56}\text{Fe} + \text{n} \rightarrow {}^{57}\text{Fe}.$$

This reaction may be followed by another neutron capture:

$$^{57}\text{Fe} + \text{n} \rightarrow {}^{58}\text{Fe},$$

producing another relatively stable isotope, iron-58. Iron-58 can capture yet another neutron to produce an even heavier isotope of iron:

$$^{58}\text{Fe} + \text{n} \rightarrow {}^{59}\text{Fe}.$$

Iron-59 is known from laboratory experiments to be radioactively unstable. It decays in about a month into cobalt-59, which is stable. The neutron-capture process then resumes: cobalt-59 captures a neutron to form the unstable cobalt-60, which in turn decays to nickel-60, and so on.

Each successive capture of a neutron by a nucleus typically takes about a year, so most unstable nuclei have plenty of time to decay before the next neutron comes along. Researchers usually refer to this "slow" neutron-capture mechanism as the s-process. It is the origin of the copper and silver in the coins in our pockets, the lead in our car batteries, the gold (and the zirconium) in the rings on our fingers.

MAKING THE HEAVIEST ELEMENTS

The s-process explains the synthesis of stable nuclei up to and including bismuth-209, the heaviest known nonradioactive nucleus, but it cannot account for the heaviest nuclei, such as thorium-232, uranium-238, or plutonium-242. Any attempt to form elements heavier than bismuth-209 by slow neutron capture fails because the new nuclei decay back to bismuth as fast as they form. There must be

(b)

INTERLUDE 21-3 The Crab Nebula in Motion

The Crab Nebula is one of the most fascinating—and instructive—cosmic objects anywhere. Within this remarkable region of space we find a wide variety of physical processes at work, all the result of a supernova explosion that occurred almost a thousand years ago. A famous physics professor at MIT once taught a course exclusively on this object, claiming that if you understand all that transpires within the Crab, then you have mastered much of modern physics. Within this remarkable gas cloud, extending for a couple of parsecs and residing about 1800 pc away, we find applications of particle physics, nuclear physics, electromagnetism, thermodynamics, condensed-matter physics, plasma physics, and gravitational physics, to name but a few.

Now we are treated to new observations suggesting that the Crab Nebula is even more dynamic than had previously been thought. Apparently, things are changing rapidly among the debris, right before our eyes, giving astronomers a rare chance to study evolutionary changes that actually occur during a single human lifetime.

The first figure consists of two frames. On the left is a true-color, visible-light photo taken with the 5-m Hale telescope on Mount Palomar. ∞ (Sec. 5.1) This image is similar to the center frame of Figure 21.9. The visible light arises in two distinct ways. The outermost filaments seen in red, yellow, and green are literally glowing from the heat and violence of the explosion long ago. However, the faint bluish light toward the center of the nebula arises from a nonthermal process that releases energy as rapidly moving electrons spiral around lines of magnetic field. (This process is not studied in the text until Chapter 25, but you can skip

yet another nuclear mechanism that produces the very heaviest nuclei. This process is called the *r-process* (where *r* stands for "rapid," in contrast to the "slow" s-process we just described). The r-process operates very quickly, occurring (we think) literally during the supernova explosion that signals the death of a massive star.

During the first 15 minutes of the supernova blast, the number of free neutrons increases dramatically as heavy nuclei are broken apart by the violence of the explosion. Unlike the s-process, which stops when it runs out of stable nuclei, the neutron-capture rate during the supernova is so great that even unstable nuclei can capture many neutrons before they have time to decay. Jamming neutrons into light- and middle-weight nuclei, the r-process is responsible for the creation of the heaviest known elements. The

heaviest of the heavy elements, then, are actually born *after* their parent stars have died. However, because the time available for synthesizing these heaviest nuclei is so brief, they never become very abundant. Elements heavier than iron (see Table 21.1) are a billion times less abundant than hydrogen and helium.

OBSERVATIONAL EVIDENCE FOR STELLAR NUCLEOSYNTHESIS

The modern picture of element formation involves many different types of nuclear reactions occurring at many different stages of stellar evolution, from main-sequence stars all the way to supernovae. Light elements—from hydrogen to iron—are built first by fusion, then by alpha capture, with ahead and look at Figure 25.15 to get the gist of it.) The filamentary structure of the wispy debris leaves little doubt that this is the remnant of an exploded star.

The right frame is a recent image acquired by the *Hubble* telescope, covering the portion of the first image contained within the white box. The image was taken at a wavelength of around 550 nm; the red color was added artificially during computer enhancement.

The central "engine" at the heart of the Crab Nebula is a pulsar—a rapidly spinning, compact starlike object that managed to survive the supernova explosion that created the nebula itself. (We will discuss these strange objects in more detail in Chapter 22.) ∞ (Sec. 22.2) The pulsar can be seen in this image as the left member of the pair of stars near the center of the frame. The pulsar powers the Crab today by accelerat-

ing elementary particles into the nebular debris, causing knots and wisps of energetic matter to stream away from the core.

In the second set of images here we can almost see in action the tiny pulsar powering the Crab Nebula (again, the left star of the pair at upper center). These three images were taken sequentially over the course of a few months. The field of view is even smaller field than in the right frame on p. 486—we are now virtually inside the Crab, witnessing the minute changes among the nebular gases near the site of the ancient explosion. Astronomers can now watch the changes in the shock-driven features that brighten and fade, over and over, like ripples in a pond. The light-year-long ripples in the Crab, seen throughout the bottom half of these frames, are moving outward from the pulsar at about half the speed of light.

proton and neutron capture filling in the gaps. Elements beyond iron form by neutron capture and radioactive decay. We are reassured of the basic soundness of our theories by three convincing pieces of evidence.

First, the rate at which various nuclei are captured and the rate at which they decay are known from laboratory experiments. When these rates are incorporated into detailed computer models of the nuclear processes occurring in stars and supernovae, the resulting elemental abundances agree extremely well, point by point, with the observational data presented in Figure 21.12 and Table 21.1. The match is remarkably good for elements up through iron and is still fairly close for heavier nuclei. Although the reasoning is indirect, the agreement between theory and observation is so striking that most astronomers regard it as very strong

evidence in support of the entire theory of stellar evolution and nucleosynthesis.

Second, the presence of one particular nucleus—technetium-99—provides direct evidence that heavy-element formation really does occur in the cores of stars. Laboratory measurements show that the technetium nucleus has a radioactive half-life of about 200,000 years. This is a very short time astronomically speaking. No one has ever found even traces of naturally occurring technetium on Earth because it all decayed long ago. The observed presence of technetium in the spectra of many red-giant stars implies that it must have been synthesized through neutron capture—the only known way that technetium can form—within the past few hundred thousand years. Otherwise, we would not observe it. Many astronomers consider the spec-

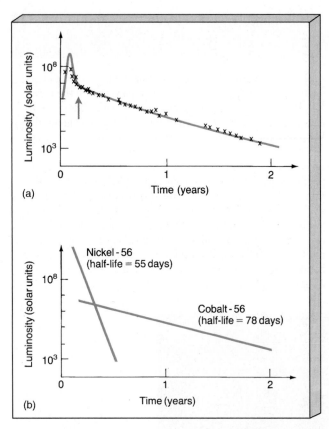

(a)

(b)

Figure 21.17 (a) The light curve of a Type I supernova, showing not only the dramatic increase and slow decrease in luminosity but also the characteristic change in the rate of decay about 2 months after the explosion (after the time indicated by the arrow). This particular supernova occurred in the faraway galaxy IC4182 in 1938. The crosses are the actual observations of the supernova's light. (b) Theoretical calculations of the light emitted by the radioactive decay of nickel-56 and cobalt-56 produce a light curve very similar to those actually observed in real supernova explosions, lending strong support to the theory of stellar nucleosynthesis.

troscopic evidence for technetium as proof that the s-process really does operate in evolved stars.

Third, the study of typical light curves from Type I supernovae indicates that radioactive nuclei form as a result of the explosion. Figure 21.17(a) (see also Figure 21.7) displays the dramatic rise in luminosity at the moment of explosion and the characteristic slower decrease in brightness. Depending on the initial mass of the exploded star, the luminosity takes from several months to many years to decrease to its original value, but the *shape* of the decay curve is nearly the same for all exploded stars. These curves have two distinct features. After the initial peak, the luminosity first declines rapidly, then decreases at a slower rate. This change in the luminosity decay invariably occurs about 2 months after the explosion, regardless of the intensity of the outburst.

We can explain the two-stage decline of the luminosity curve in Figure 21.17(a) in terms of the radioactive decay of

unstable nuclei, notably nickel-56 and cobalt-56, produced in abundance during the early moments of the supernova explosion. From theoretical models of the explosion we can calculate the amounts of these elements expected to form, and we know their half-lives from laboratory experiments. Because each radioactive decay produces a known amount of visible light, we can then determine how the light emitted by these unstable elements should vary in time. The result is in very good agreement with the observed light curve in Figure 21.17(b)—the luminosity of a Type I supernova is entirely consistent with the decay of about 0.6 solar masses of nickel-56. More direct evidence for the presence of these unstable nuclei was first obtained in the 1970s, when a gamma-ray spectral feature of decaying cobalt-56 was identified in a supernova observed in a distant galaxy.

21.5 The Cycle of Stellar Evolution

The theory of stellar nucleosynthesis can naturally account for the observed differences in heavy-element abundance between the old globular cluster stars and stars now forming in our galaxy. Even though an evolved star continuously creates new heavy elements in its interior, the composition changes are largely confined to the core, and the star's spectrum gives little indication of events within. Convection may carry some reaction products (such as the technetium observed in many red giants) from the core into the envelope but, the outer layers largely retain the composition with which the star originally formed. Only at the end of the star's life are its newly created elements released and scattered into space.

Thus, the spectra of the *youngest* stars show the *most* heavy elements, because each new generation of stars increases the concentration of these elements in the interstellar clouds from which the next generation forms. The photosphere of a recently formed star contains a much greater abundance of heavy elements than that of a star that formed long ago. Knowledge of stellar evolution allows astronomers to estimate the ages of stars from spectroscopic studies, even when the stars are isolated and not members of any cluster. ∞ (Sec. 20.5)

In the past three chapters we have seen all the ingredients that make up the complete cycle of star formation and evolution in our Galaxy. Let us now briefly summarize that process, which is illustrated in Figure 21.18.

1. Stars form when part of an interstellar cloud is compressed beyond the point at which it can support itself against its own gravity. The cloud collapses and fragments, forming a cluster of stars. The hottest stars heat and ionize the surrounding gas, sending shock waves through the surrounding cloud, possibly triggering new rounds of star formation.

2. Within the cluster, stars evolve. The most massive stars evolve fastest, creating heavy elements in their cores and spewing them forth into the interstellar medium in supernova explosions. The lighter stars take longer to evolve, but they too can create heavy elements and may contribute to the "seeding" of interstellar space when they shed their envelopes as planetary nebulae.

3. The creation and dispersal of new heavy elements are accompanied by further shock waves. Their passage simultaneously enriches the interstellar medium and compresses it into further star formation.

In this way, although some material is used up in each cycle—turned into energy or locked up in low-mass stars—the Galaxy continuously recycles its matter. Each new round of formation creates stars with more heavy elements than the preceding generation had. From the old, metal-poor globular clusters to the young, metal-rich open clusters, we observe this enrichment process in action. Our Sun is the product of many such cycles. We ourselves are another. Without the heavy elements synthesized in the hearts of stars, life on Earth would not exist.

Figure 21.18   The cycle of star formation and evolution continuously replenishes the Galaxy with new heavy elements and provides the driving force for the creation of new generations of stars.

Chapter Review

SUMMARY

A **nova** (p. 472) is a star that suddenly increases greatly in brightness, then slowly fades back to its normal appearance over a period of months. It results when a white dwarf in a binary system draws hydrogen-rich material from its companion. The gas builds up on the white dwarf's surface, eventually becoming hot and dense enough for the hydrogen to burn explosively, temporarily causing a large increase in the dwarf's luminosity.

The matter flowing from the companion star does not fall directly onto the surface of the dwarf. Instead it goes into orbit around it, forming an **accretion disk** (p. 473). Friction within the disk causes the gas to spiral slowly inward, heating up and glowing brightly as it nears the dwarf's surface.

Stars more massive than about 8 solar masses are able to attain high enough central temperatures to burn carbon and heavier nuclei. As they burn, their cores form a layered structure consisting of burning shells of successively heavier elements. A non-burning core of iron builds up at the center.

Iron is special in that its nuclei can neither be fused together nor split apart to produce energy. As a result, stellar nuclear burning stops at iron. As a star's iron core grows in mass it eventually becomes unable to support itself against gravity and begins to collapse. At the enormous densities and temperatures produced during the collapse, iron nuclei are broken down into their constituent particles—protons and neutrons. The protons combine with electrons to form more neutrons. Eventually, when the core has become so dense that the neutrons are effectively brought into physical contact with one another, their resistance to further squeezing stops the collapse and the core rebounds, sending a violent shock wave out through the rest of the star. The star is blown to pieces in a **core-collapse supernova** (p. 475).

Astronomers classify **supernovae** (p. 476) into two broad categories: Type I and Type II. These classes differ by their light curves and by their composition. **Type I supernovae** (p. 476) are hydrogen poor and have a light curve similar in shape to that of a nova. **Type II supernovae** (p. 476) are hydrogen rich and have a characteristic bump in the light curve a few months after maximum. A Type II supernova is a core-collapse supernova. A Type I supernova occurs when a carbon–oxygen white dwarf in a binary system exceeds about

1.4 solar masses (the Chandrasekhar mass)—the maximum mass that can be supported against gravity by electron degeneracy pressure. The star collapses and explodes as its carbon ignites. This type of supernova is called a **carbon-detonation supernova** (p. 477).

Theory predicts that a supernova visible from Earth should occur within our Galaxy about once a century, although none has been observed in the last 400 years. We can see evidence of a past supernova in the form of a **supernova remnant** (p. 478), a shell of exploded debris surrounding the site of the explosion and expanding into space at thousands of kilometers per second.

All elements heavier than helium formed by **stellar nucleosynthesis** (p. 481)—the production of new elements by nuclear reactions in the cores of evolved stars. Elements beyond carbon tend to form by **helium capture** (p. 484) rather than by the fusion of two heavy nuclei. Therefore, nuclei whose masses are a multiple of the mass of a helium nucleus tend to be more common than others.

At high enough core temperatures, photodisintegration breaks apart some heavy nuclei, providing helium-4 nuclei for the synthesis of even more massive elements, leading to a buildup of iron-56 in the core. Elements beyond iron form by **neutron capture** (p. 485) in the cores of evolved stars. With no repulsive electromagnetic barrier to overcome, neutrons can easily combine with nuclei. During a supernova explosion, rapid neutron capture occurs, producing the heaviest nuclei of all.

Comparisons between theoretical predictions of element production and observations of element abundances in stars and supernovae provide strong support for the theory of stellar nucleosynthesis.

The processes of star formation, evolution, and explosion form a cycle that constantly enriches the interstellar medium with heavy elements and sows the seeds of new generations of stars. Without the elements produced in supernovae, life on Earth would be impossible.

SELF-TEST: TRUE OR FALSE?

_____ **1.** A nova is a sudden outburst of light coming from an old main-sequence star.

_____ **2.** Some, but not all, novae occur in binary star systems.

_____ **3.** It takes less and less time to fuse heavier and heavier elements inside a high-mass star.

_____ **4.** In a core-collapse supernova, the outer part of the core rebounds from the inner, high-density core, destroying the entire outer part of the star.

_____ **5.** Most of the energy released during a supernova is emitted in the form of neutrinos.

_____ **6.** A Type I supernova is the same as a nova, but it appears much brighter because it occurs closer to us.

_____ **7.** The spectrum of a Type II supernova shows the presence of lots of hydrogen.

_____ **8.** It is possible that a recurrent nova will eventually result in a core-collapse supernova.

_____ **9.** Once the process gets underway, the core of a massive star collapses in about 1 second.

_____ **10.** Different isotopic forms of an element differ only in the number of protons in the nucleus.

_____ **11.** Carbon can fuse with helium more easily than it can fuse with another carbon atom because there is more helium than carbon in stars.

_____ **12.** The s-process is one in which heavy elements are formed from silicon.

_____ **13.** The r-process occurs only during the first few minutes of a supernova explosion.

_____ **14.** Stellar nucleosynthesis can account for the existence of all elements except hydrogen and helium.

_____ **15.** Stellar nucleosynthesis causes the abundance of heavy elements in interstellar gas and newborn stars to increase with time.

SELF-TEST: FILL IN THE BLANK

1. In a semidetached binary consisting of a white dwarf and a main-sequence or giant companion, matter leaving the companion forms an _____ disk around the dwarf.

2. A nova explosion is due to _____ fusion on the _____ of a white dwarf.

3. In the collapsing core of a massive star, photodisintegration is caused by the very high _____ in the core.

4. When a proton and an electron are forced together, they combine to form a _____ and a _____ .

5. Core collapse in a massive star is eventually stopped by _____ degeneracy pressure.

6. A Type _____ supernova occurs when the core of a massive star implodes and rebounds.

7. A Type _____ supernova occurs when a white dwarf exceeds the Chandrasekhar mass.

8. The maximum mass for a white dwarf is roughly _____ solar masses.

9. The two types of supernovae can be distinguished observationally by their spectra and by their _____.

10. The first evidence of supernova 1987A was a burst of _____.

11. Supernova 1987A surprised astronomers because its progenitor star was a _____ instead of a _____ at the moment of the explosion.

12. Elements differing by 4 mass units, such as carbon, oxygen, neon, and silicon, are produced in stars by _____.

13. By the time silicon appears in the core of a massive star, the temperature has reached _____ , sufficient to break apart heavy nuclei.

14. Neutron capture is responsible for the formation of elements heavier than _____.

15. Because of stellar nucleosynthesis, the spectra of old stars show _____ heavy elements than those of young stars.

REVIEW AND DISCUSSION

1. Under what circumstances will a binary star produce a nova?

2. What is a light curve? How can it be used to identify a nova or a supernova?

3. What occurs in a massive star to cause it to explode?

4. What are the observational differences between Type I and Type II supernovae?

5. How do the mechanisms that cause Type I and Type II supernovae explain their observed differences?

6. Roughly how often would we expect a supernova to occur in our own Galaxy? How often would we expect to see a Galactic supernova?

7. What evidence is there that many supernovae have occurred in our Galaxy?

8. How can astronomers estimate the age of an isolated star?

9. What proof do astronomers have that heavy elements are formed in stars?

10. As a star evolves, why do heavier elements tend to form by helium capture rather than by fusion of like nuclei?

11. Why do the cores of massive stars evolve into iron, not heavier elements?

12. How are nuclei heavier than iron formed?

13. What is the r-process? When and where does it occur?

14. Why was supernova 1987A so important?

15. Why are neutrino detectors important to the study of supernovae?

PROBLEMS

1. Estimate how close an 0.5–solar mass white dwarf must come to the center of a 2–solar mass subgiant of radius 10 times that of the Sun in order for the white dwarf's tidal field to strip matter from the companion's surface.

2. A certain telescope could just detect the Sun at a distance of 10,000 pc. What is the maximum distance at which it could detect a nova having a peak luminosity of 10^5 solar luminosities?

3. Repeat the previous calculation for a supernova having a peak luminosity 10^{10} times that of the Sun.

4. At what distance would a supernova of absolute magnitude -20 look as bright as the Sun? Would you expect a supernova to occur that close to us?

5. A (hypothetical) supernova at a distance of 20 pc has an absolute magnitude of -20. Compare its apparent magnitude with that of (a) the full Moon. (b) Venus at its brightest. (See Figure 17.9.)

6. A supernova's energy is often compared to the total energy output of the Sun over its lifetime. Using the Sun's current energy output, calculate its total energy output, assuming it has a 10^{10} year main-sequence lifetime. How does this compare with the energy released by a supernova?

7. The Crab Nebula is now about 1 pc in radius. If it was observed to explode in A.D. 1054, roughly how fast is it expanding? (Assume a constant expansion rate. Is that a reasonable assumption?)

8. As we will see in Chapter 23, the star-forming portion of our Galaxy consists of a highly flattened circular disk about 30 kpc (30,000 pc) in diameter. Interstellar extinction limits our view to within a radius of about 3 kpc of the Sun. If supernovae occur in the Galaxy roughly once every 30 years, on average, and are uniformly spread throughout the disk, calculate how often we should expect to see a supernova.

9. Assuming the data in the previous question and taking all supernovae, for simplicity, to have absolute magnitude -20, calculate how often we should expect to observe a supernova brighter than the full Moon (apparent magnitude -12.5).

10. Based on the data in Table 21.1, estimate the total masses of "iron-group" and "heaviest-weight group" elements in the Sun. Compare these with Earth's mass.

PROJECTS

1. In 1758 the French comet hunter Charles Messier discovered the sky's most legendary supernova remnant, now called M1, or the Crab Nebula. An 8-inch telescope reveals the Crab's oval shape, but it will appear faint. It is located northwest of Zeta Tauri, the star that marks the southern tip of the horns of Taurus the Bull. A 10-inch or larger telescope reveals some of its famous filamentary structure.

2. In the *Handbook of Chemistry and Physics*, available in any library reference section, look up the table of isotopes. Pick one or more isotopes and follow their decay into a final stable isotope. For example, choose cobalt-59, formed in the s-process. Note how the isotope decays, what is emitted, and the half-life of the decays. Try this exercise for uranium-235, uranium-238, and plutonium-239.

22 NEUTRON STARS AND BLACK HOLES

Strange States of Matter ✦www✦

LEARNING GOALS

Studying this chapter will enable you to:

1 Describe the properties of neutron stars and explain how these strange objects are formed.

2 Explain the nature and origin of pulsars and account for their characteristic radiation.

3 List and explain some of the observable properties of neutron-star binary systems.

4 Describe how black holes are formed and discuss their effects on matter and radiation in their vicinity.

5 Relate the phenomena that occur near black holes due to the warping of space around them.

6 Discuss the difficulties in observing black holes and explain some of the ways in which the presence of a black hole might be detected.

(Opposite page, background) Black holes can neither be seen nor drawn easily. In this artist's conception of the vicinity of a black hole, the disk-shaped region is a rapidly whirling collection of hot gas and dust about to fall into the black hole. The hole itself is represented merely by a black dot at the center of the accretion disk.

(Inset A) An artist's view of a nearly flat, spinning torus of heated gas, called an accretion disk. For a typical stellar black hole, the disk has solar-system dimensions.

(Inset B) A closeup view of the core of the accretion disk, showing a truncated cusp of very hot gas about to make its way down into the central black hole.

(Inset C) The anatomy of a hypothetical black hole would show its accretion disk to the left and right, with the hole itself still only a black point at the very center. The white vertical spike is a geyser (or jet) of matter shot away from *near* the hole perpendicular to the innermost part of the disk.

(Inset D) A closeup view of the previous cutaway drawing of the accretion disk, the jet, and the black hole at its heart. At this scale, the stellar black hole itself (still shown as a black dot in the magnified inset) is about 30 kilometers across—and on this scale, astronomers are very uncertain of what black holes really look like.

*O*ur study of stellar evolution has led us to some very unusual and unexpected objects. Red giants, white dwarfs, and supernova explosions surely represent extreme states of matter completely unfamiliar to us here on Earth. Yet stellar evolution can have even more bizarre consequences. The strangest states of all result from the catastrophic implosion–explosion of stars much more massive than our Sun. The almost unimaginable violence of a supernova explosion may bring into being objects so extreme in their behavior that they require us to reconsider some of our most hallowed laws of physics. They open up a science fiction writer's dream of fantastic phenomena. They may even one day force scientists to construct a whole new theory of the universe.

22.1 Neutron Stars

1 What remains after a supernova explosion? Is the entire progenitor star blown to bits and dispersed throughout interstellar space, or does some portion of it survive? For a Type I (carbon-detonation) supernova, most astronomers regard it as quite unlikely that any central remnant is left after the explosion. The entire star is shattered by the blast. However, for a Type II supernova, involving the implosion and subsequent rebound of a massive star's iron core, theoretical calculations indicate that part of the star may survive. ∞ (Sec. 21.2) The explosion destroys the parent star, but it may leave a tiny ultracompressed remnant at its center. Even by the high-density standards of a white dwarf, however, the matter within this severely compressed core is in a very strange state, unlike anything we are ever likely to find (or create) on Earth.

Recall from Chapter 21 that during the moment of implosion of a massive star—just prior to the supernova explosion itself—the electrons in the core violently smash into the protons there, forming neutrons and neutrinos. ∞ (Sec. 21.2) The neutrinos leave the scene at (or nearly at) the speed of light, accelerating the collapse of the neutron core, which continues to contract until its particles come into contact. At that point, neutron degeneracy pressure causes the central portion of the core to rebound, creating a powerful shock wave that races outward through the star, violently expelling matter into space.

The key point here is that the shock wave does not start at the very center of the collapsing core. The innermost part of the core—the region that bounces—remains intact as the shock wave it causes destroys the rest of the star. After the violence of the supernova has subsided, this ball of neutrons is all that is left. Researchers colloquially call this core remnant* a **neutron star**, although it is not a star in any true sense of the word—all its nuclear reactions have ceased forever.

Neutron stars are extremely small and very massive. Composed purely of neutrons packed together in a tight ball about 20 km across, a typical neutron star is not much bigger than a small asteroid or a terrestrial city (see Figure 22.1), yet its mass is greater than that of the Sun. With so much mass squeezed into such a small volume, neutron stars are incredibly dense. Their average density can reach 10^{17} or even 10^{18} kg/m^3, nearly a billion times denser than a white dwarf. A single thimbleful of neutron-star material would weigh 100 million tons—about as much as a good-sized terrestrial mountain. For comparison, the density of a normal atomic nucleus is about 3×10^{17} kg/m^3. In a sense, we can think of a neutron star as a single enormous nucleus, with an atomic mass of around 10^{57}.

Neutron stars are solid objects. Provided that a sufficiently cool one could be found, you might even imagine standing on it. However, this would not be easy, as a neutron star's gravity is extremely powerful. A 70-kg (150-pound) human would weigh the Earth-equivalent of about 1 billion kg (1 million tons). The severe pull of a neutron star's gravity would flatten you much thinner than this piece of paper!

Astronomers commonly use the term remnant *to mean whatever remains of a star's inner core after evolution has ended. Such objects are small and compact—no larger than Earth. They should not be confused with supernova remnants, which are the aftermath of supernova explosions: glowing clouds of debris scattered across many parsecs of interstellar space.* ∞ *(Sec. 21.3)*

Figure 22.1 Neutron stars are not much larger than many of Earth's major cities. In this fanciful comparison, a typical neutron star sits alongside Manhattan Island.

In addition to large mass and small size, newly formed neutron stars have two other very important properties. First, they *rotate* extremely rapidly, with periods measured in fractions of a second. This is a direct result of the law of conservation of angular momentum (Chapter 15), which tells us that any rotating body must spin faster as it shrinks. ∞ (*More Precisely 15-1*) Even if the core of the progenitor star were initially rotating quite slowly (once every couple of weeks, say, as is observed in many upper main-sequence stars), it would be spinning a few times per second by the time it had reached a diameter of 20 km. Second, newborn neutron stars have very strong *magnetic fields*. The original field of the progenitor star is amplified by the collapse of the core because the contracting material squeezes the magnetic field lines closer together, creating a magnetic field trillions of times stronger than Earth's field.

In time, theory indicates, our neutron star will spin more and more slowly as it radiates its energy into space, and its magnetic field will diminish. However, for a few million years after its birth, these two properties combine to provide the primary means by which this strange object can be detected and studied.

22.2 Pulsars

2 Can we be sure that objects as strange as neutron stars really exist? The answer is a confident yes. The first observation of a neutron star occurred in 1967, when Jocelyn Bell, a graduate student at Cambridge University, made a surprising discovery. She observed an astronomical object emitting radio radiation in the form of rapid *pulses*. Each pulse consisted of a 0.01-second (s) burst of radiation, after which there was nothing. Then, 1.34 s later, another pulse would arrive. The time interval between pulses was astonishingly uniform—so accurate, in fact, that the repeated emissions could be used as a very precise clock. Figure 22.2 is a recording of part of the radio radiation from the pulsating object Bell discovered.

Many hundreds of these pulsating objects are now known in our Milky Way Galaxy. They are called **pulsars**. Each has its own characteristic pulse period and duration.

The pulse periods of some pulsars are so stable that they are by far the most accurate natural clocks known in the universe—more accurate even than the best atomic clocks on Earth. In some cases the period is predicted to change by only a few seconds in a million years.

Most pulsars emit their pulses in the form of radio radiation, but some have been observed to pulse in the visible, X-ray, and gamma-ray parts of the spectrum as well. Whatever types of radiation are produced, these electromagnetic flashes at different frequencies are all synchronized—that is, they occur at regular, repeated time intervals—as we would expect if they arose from the same object. The period of most pulsars is usually short—ranging from about 0.03 to 0.3 s, corresponding to a flashing rate of between 3 and 30 times per second. The human eye is insensitive to such rapid flashes, making it impossible to observe the flickering of a pulsar with the naked eye or even using a large telescope. Fortunately, instruments can record pulsations of light that the human eye cannot detect.

A few pulsars are clearly associated with supernova remnants, although not all such remnants have a detectable pulsar within them. Figure 22.3(a) shows a pair of optical photographs of the Crab pulsar, at the center of the Crab supernova remnant. ∞ (Sec. 21.3) In the left frame, the pulsar is off; in the right frame, it is on. Figure 22.3(b) shows that the Crab also pulses in X rays. By observing the speed and direction of the Crab's ejected matter, astronomers can work backward to pinpoint the location in space at which the explosion must have occurred and where the supernova core remnant should be located. That is precisely the region of the Crab Nebula from which the pulsating signals arise. The Crab pulsar is evidently all that remains of the once-massive star whose supernova was observed in 1054.

When Bell made her discovery in 1967, she did not know what she was looking at. Indeed, no one at the time knew what a pulsar was. The explanation of pulsars as spinning neutron stars won Bell's thesis advisor, Anthony Hewish, the 1974 Nobel Prize in physics. Hewish reasoned that the only physical mechanism consistent with such precisely timed pulsations is a small, rotating source of radiation. Only rotation can cause the high degree of regularity of the observed pulses, and only a small object can account for the sharpness of each pulse. Radiation emitted from different regions of an object larger than a few tens of kilo-

Figure 22.2 Pulsars emit periodic bursts of radiation. This recording shows the regular change in the intensity of the radio radiation emitted by the first such object known. It was discovered in 1967. Some of the pulses are marked by arrows.

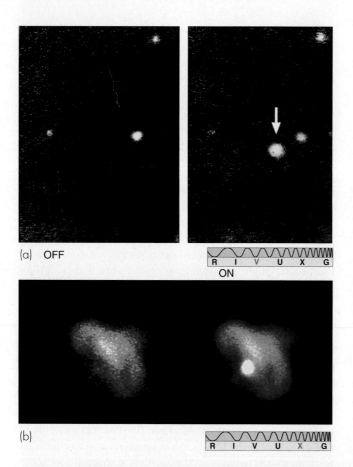

(a) OFF

ON

(b)

Figure 22.3 The pulsar in the core of the Crab Nebula blinks on and off about 30 times each second. (a) In this pair of closely spaced optical images, the pulsing can be seen clearly. (b) The same phenomenon is also detected in X rays.

meters across would arrive at Earth at slightly different times, blurring the pulse profile. The best current model describes a pulsar as a compact, spinning neutron star that periodically flashes radiation toward Earth.

Figure 22.4 outlines the important features of this pulsar model. Two "hot spots" on the surface of a neutron star, or in the magnetosphere just above the surface, continuously emit radiation in a narrow "searchlight" pattern. These spots are most likely localized regions near the neutron star's magnetic poles, where charged particles, accelerated to extremely high energies by the star's rotating magnetic field, emit radiation along the star's magnetic axis. The hot spots radiate more or less steadily, and the resulting beams sweep through space, like a revolving lighthouse beacon, as the neutron star rotates. Indeed, this pulsar model is often known as the **lighthouse model**. If the neutron star happens to be oriented such that the beam of its pulses sweeps across Earth, we see the pulses. The beams are observed as a series of rapid pulses—each time one of the beams sweeps past Earth, a pulse is seen. The period of the pulses is the star's rotation period.

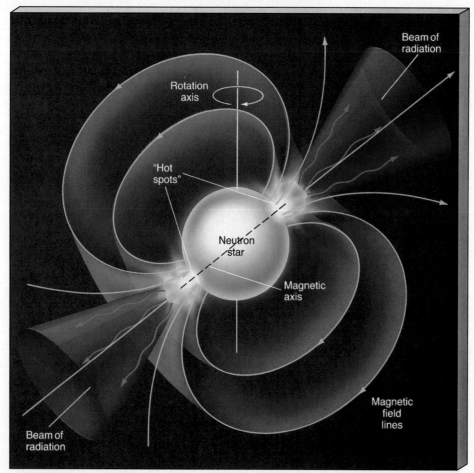

Figure 22.4 This diagram of the "lighthouse model" of neutron-star emission accounts for many of the observed properties of pulsars. Charged particles, accelerated by the magnetism of the neutron star, flow along the magnetic field lines, producing radiation that beams outward.

All pulsars are neutron stars, but not all neutron stars are pulsars, for two reasons. First, the two ingredients that make the neutron star pulse—rapid rotation and strong magnetic field—both diminish with time, so the pulses gradually weaken and become less frequent. Theory indicates that within a few tens of millions of years, the pulsations all but stop. Second, even a young, bright pulsar is not necessarily visible from Earth. The pulsar beam depicted in Figure 22.4 is relatively narrow—perhaps only a few degrees across. Only if the neutron star happens to be oriented in just the right way do we see pulses. When we can see the pulses from Earth, we call the body a pulsar.

Given our current knowledge of star formation, stellar evolution, and neutron stars, pulsar observations are consistent with the idea that every high-mass star dies in a Type II supernova, leaving a neutron star behind, and that *all* neutron stars emit beams of radiation, just like the pulsars we actually see.

22.3 Neutron-Star Binaries

3 We noted in Chapter 17 that most stars are not single but instead are members of binary systems. ∞ (Sec. 17.9) Although many pulsars are known to be isolated (that is, not part of any binary), there is strong evidence that at least some do have binary companions, and the same is true of neutron stars in general (that is, even the ones not seen as pulsars).

X-RAY SOURCES

The late 1970s saw several important discoveries about neutron stars in binary-star systems. Numerous X-ray sources were discovered near the central regions of our galaxy and also near the centers of a few rich star clusters. Some of these sources, known as **X-ray bursters**, emit much of their energy in violent eruptions, each thousands of times more luminous than our Sun, but lasting only a few seconds. A typical burst is shown in Figure 22.5.

This X-ray emission is thought to arise on or near neutron stars that are members of binary systems. Matter torn from the surface of the (main-sequence or giant) companion by the neutron star's strong gravitational pull accumulates on the neutron star's surface. As in the case of white-dwarf accretion (see Chapter 21), the material does not fall directly onto the surface. Instead, as illustrated in Figure 22.6(a), it forms an accretion disk. (Compare with Figure 21.2, which depicts the white-dwarf equivalent.) The gas goes into a tight orbit around the neutron star, then slowly spirals inward. The inner portions of the accretion disk become extremely hot, releasing a steady stream of X rays.

As gas builds up on the neutron star's surface its temperature rises due to the pressure of overlying material. Eventually, it becomes hot enough to fuse hydrogen. The result is a sudden period of rapid nuclear burning that releases a huge amount of energy in a brief but intense flash of X rays—an X-ray burst. After several hours of renewed accumulation, a fresh layer of matter produces the next burst. Thus, an X-ray burst is much like a nova explosion on a white dwarf, but occurring on a far more violent scale because of the neutron star's much stronger gravity. ∞ (Sec. 21.1)

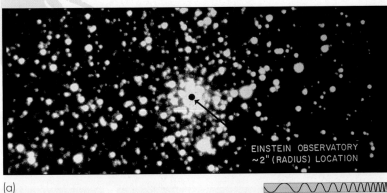

(a)

EINSTEIN OBSERVATORY
~2" (RADIUS) LOCATION

R I V U X G

Figure 22.5 An X-ray burster produces a sudden, intense flash of X rays, followed by a period of relative inactivity lasting as long as several hours. Then another burst occurs. The bursts are thought to be caused by explosive nuclear burning on the surface of an accreting neutron star, similar to the explosions on a white dwarf that give rise to novae. (a) An optical photograph of the star cluster Terzan 2, showing a 2" dot at the center where the X-ray bursts originate. (b) X-ray images taken before and during the outburst. The most intense X rays correspond to the position of the dot shown in frame (a).

BEFORE

DURING

(b)

R I V U X G

Not all the infalling gas makes it onto the neutron star surface, however. In at least one case—an object known as SS433*—we have direct observational evidence that some material is instead shot completely out of the system at enormously high speeds. SS433 expels more than one Earth mass of material every year in the form of two oppositely directed narrow jets moving roughly perpendicular to the disk. Observations of the Doppler shifts of optical emission lines produced within the jets themselves imply speeds of almost 80,000 km/s—over 25 percent of the speed of light! As the jets interact with the interstellar medium they emit radio radiation, as shown in Figure 22.6(b).

Jets of this sort are apparently quite common in astronomical systems in which an accretion disk surrounds a

---

The name simply identifies it as the 433rd entry in a particular catalog of stars with strong optical emission lines.

compact object (such as a neutron star or a black hole). They are believed to be produced by the intense radiation and magnetic fields near the inner edge of the disk, although the details of their formation are still uncertain. Incidentally, they are *not* the "lighthouse" beams of radiation from the neutron star itself (Figure 22.4) that can result in a pulsar. Although SS433 is the only stellar object currently known to produce jets, we will see examples of similar phenomena on much larger scales in later chapters. One of the most important aspects of SS433 is that we can actually study both the disk and the jets instead of simply having to assume their existence, as in more distant cosmic objects.

GAMMA-RAY BURSTS

Discovered serendipitously in the late 1960s by military satellites looking for violators of the Nuclear Test Ban Treaty, and first made public in the 1970s, **gamma-ray bursts** have developed into one of the deepest mysteries in astronomy today. Observationally, the bursts consist of bright, irregular flashes of gamma rays typically lasting only a few seconds, with substantial variations from one observed burst to another (Figure 22.7a). Until recently, it was thought that gamma-ray bursts were basically "scaled-up" versions of X-ray bursters, in which matter accreted from the binary companion experienced even more violent nuclear burning, accompanied by the release of the more energetic gamma rays. However, it now appears that this is not the case.

Civilian satellites such as the *Compton Gamma-Ray Observatory* (*GRO*) detect gamma-ray bursts at the rate of about one a day. ∞ (Sec. 5.6) Figure 22.7(b) shows an all-sky plot of the positions of about a thousand bursts detected during the first three years of *Compton's* operation. Unlike similar maps at all wavelengths (such as Figure 5.34e), which

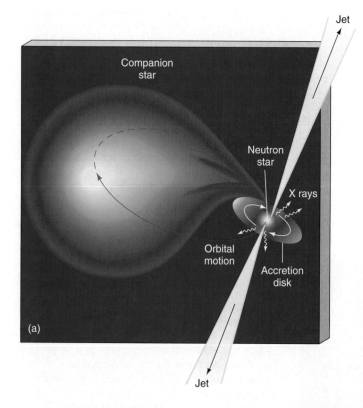

(a)

Figure 22.6 (a) Matter flows from a normal star toward a compact neutron-star companion and falls toward the surface in an accretion disk. As the gas spirals inward under the neutron star's intense gravity it heats up, becoming so hot that it emits X rays. In at least one instance—the peculiar object SS433—some material may be ejected in the form of two high-speed jets of gas. (b) False-color radiographs of SS433, made at monthly intervals (left to right), show the jets moving outward and the central source rotating under the gravitational influence of the companion star.

(b)

R I V U X G

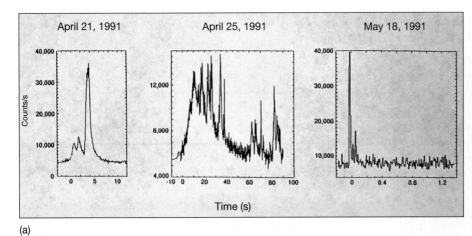

April 21, 1991   April 25, 1991   May 18, 1991

Time (s)

(a)

Figure 22.7   (a) Plots of intensity versus time (in seconds) for some gamma-ray bursts. Note the substantial differences between them. Some bursts are irregular and spiky, whereas others are much more smoothly varying. Whether this wide variation in burst appearance means that more than one physical process is at work is presently unknown. (b) Positions on the sky of the first 1000 gamma-ray bursts detected by *GRO*; BATSE stands for Burst and Transient Source Experiment. The bursts appear to be distributed isotropically (uniformly) across the entire sky.

1000 BATSE gamma-ray bursts

(b)

generally show a distinct correlation with the plane of the Milky Way, the gamma-ray bursts plotted here are distributed randomly across the sky (their distribution is said to be "isotropic"). The bursts seemingly never repeat at the same location, show no obvious clustering, and appear unaligned with any known large-scale structure, near or far.

The isotropy of the *GRO* data has convinced most astronomers that the bursts must originate at so-called cosmological distances—that is, far beyond our Milky Way galaxy. In addition, *Compton* finds faint bursts to be less frequent than would be expected from an infinite, unbounded population of such sources. Apparently there is an outer limit of some sort in their distribution in space. However, it is unclear whether the bursts are associated with the isotropic spread of distant galaxies—or whether they are something else yet again, outside of galaxies and somewhere in intergalactic space.

If the bursts really do originate very far away in space, that must mean that their intrinsic power is enormous; otherwise they wouldn't be detectable by our equipment, even on-station above Earth's atmosphere. At cosmological distances, each burst must generate more energy than a typical supernova explosion—in only a few seconds!

Until 1997 there was one major problem with this line of reasoning: the actual distances to the gamma-ray bursts were completely unknown. As a result, the inference that the bursts were very distant, and hence very energetic, was indirect, based mainly on their isotropic distribution. The

distances to the bursts were difficult to determine because astronomers were unable to find any optical counterparts or any kind of association with known cosmic objects. Part of the problem is that gamma rays are far too penetrating to be focused by conventional optics; consequently, *GRO*'s burst positions are uncertain by several arc degrees.

All this changed on May 8, 1997, thanks to a remarkable combination of gamma-ray, X-ray, and optical observations of the gamma-ray burst GRB970508. The Italian–Dutch *BeppoSAX* satellite, launched in 1996, recorded the burst in both the gamma- and X-ray regions of the spectrum. The importance of the X-ray observations is that they allowed a much more accurate determination of the burst's location on the sky—in fact, to within a few arc minutes. That was enough for ground-based optical astronomers to look for and find GRB970508's optical afterglow. For the first time, astronomers had observed an optical counterpart to a gamma-ray burst.

The burst's optical spectrum, obtained using the Keck telescope, revealed a very important piece of information. Several absorption lines of iron and magnesium were identified, but they were redshifted by almost a factor of 2 in wavelength. Such redshifts, as we will see in Chapter 24, are the result of the expansion of the universe, and they are clear proof—a "smoking gun," if you will—that at least this gamma-ray burst, and presumably all others, really did occur at cosmological distances. The events responsible for GRB970508 apparently occurred more than 2000 *mega-parsecs* from Earth. Figure 22.8 is an optical photograph of another gamma-ray burst, called GRB970228—one of the first such images that associates a gamma-ray burst with a potential host galaxy. In May 1998, astronomers announced a distance measurement for a second gamma-ray burst (GRB971214, which occurred on December 14, 1997), and again identified a host galaxy. According to the redshift of lines observed in that galaxy's spectrum, this burst was almost twice as far away as GRB970508, implying an enormous luminosity—perhaps as much as 100 times greater than that of a typical supernova. These observations lend strong support to the idea that gamma-ray bursts originate in faraway galaxies throughout the universe.

Figure 22.8 Optical photograph, in false color, of gamma-ray burst GRB970228. The optical counterpart of the gamma-ray burst (marked by the arrow) is a white blob to the upper left of image center; to the lower right is an extended object thought to be the burst's host galaxy.

Prior to the measurement of GRB970508's distance, some theorists had attempted to explain gamma-ray bursts in terms of nearby—and hence much less energetic—events, perhaps in the halo of our own galaxy (see Chapter 23) or even in the Oort cloud surrounding the Sun. ∞ (Sec. 14.2) However, it now seems clear that gamma-ray bursts are both very distant and extremely energetic. What's more, the millisecond flickering in the bursting gamma rays detected by *Compton* imply that, whatever their origin, all their energy must come from an extremely small volume—in fact, no larger than a *few hundred kilometers* across.

The reasoning is as follows. If the emitting region were, say, 300,000 km—1 light second—across, even an instantaneous change in intensity at the source would be smeared out over a time interval of 1 s as seen from Earth, because light from the far side of the object would take 1 s longer to reach us than light from the near side. For the gamma-ray variation not to be blurred by the light-travel time, the source cannot be more than 1 light millisecond, or 300 km, in diameter.

One leading mechanism for explaining gamma-ray bursts is the true end point of a binary-star system. Suppose that both members of the binary evolve to become neutron stars. As the system continues to evolve, gravitational radiation (see *Interlude 22-1*) is released and the two ultradense stars will spiral in toward each other. Once they are within a few kilometers of each another, coalescence is inevitable. Such a merger will likely produce a violent explosion, comparable in energy to a supernova, and this could conceivably explain the flashes of gamma rays we observe. However, the huge luminosity of the December 1997 burst may be too much even for this model to explain. Evidently, the final word on these enigmatic objects has yet to be written.

MILLISECOND PULSARS

In the mid-1980s an important new category of pulsars was found—a class of very rapidly rotating objects called **millisecond pulsars**. Several dozen millisecond pulsars are currently known in the Milky Way galaxy. These objects spin hundreds of times per second (that is, their pulse period is a few milliseconds, 0.001 s). This speed is about as fast as a typical neutron star can spin without flying apart. In some cases, the star's equator is moving at more than 20 percent of the speed of light. This speed suggests a phenomenon bordering on the incredible—a cosmic object of kilometer dimensions, more massive than our Sun, spinning almost at breakup speed, making nearly 1000 complete revolutions *every second*. Yet the observations and their interpretation leave little room for doubt.

The story of these remarkable objects is further complicated because many of them are found in globular clusters. This is odd because globular clusters are known to be very old—10 billion years, at least. Yet Type II supernovae (the kind that create neutron stars) are associated with massive stars that explode within a few tens of *millions* of years after their formation, and no stars have formed in any globular cluster since the cluster itself came into being. Thus, no new neutron star has been produced in a globular cluster in a very long time. But, as mentioned earlier, the pulsar produced in a supernova explosion is expected to slow down in only a few million years. After 10 billion years its rotation should have all but ceased. The rapid rotation of the pulsars found in globular clusters cannot be a relic of their birth. These objects must have been spun up—that is, had their rotation rates increased—by some other, much more recent, mechanism.

The most likely explanation for the high rotation rate of these objects is that the neutron star has been spun up by drawing in matter from a companion star. As matter spirals down onto the star's surface in an accretion disk it provides the "push" needed to make the neutron star spin faster (see Figure 22.9).

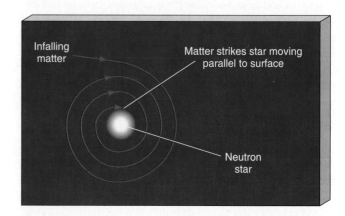

Figure 22.9 Gas from a companion star spirals down onto the surface of a neutron star. As the infalling matter strikes the star it moves almost parallel to the surface, so it tends to make the star spin faster. Eventually, this process can result in a millisecond pulsar—a neutron star spinning at the incredible rate of hundreds of revolutions per second.

INTERLUDE 22-1 Gravity Waves

Electromagnetic waves are common, everyday phenomena. Whether they are radio, infrared, visible, ultraviolet, X-ray, or gamma-ray radiation, all electromagnetic waves involve periodic changes in the strengths of electric and magnetic fields. ∞ (Sec. 3.2) They move through space and transport energy. Any accelerating charged particle, such as an electron in a broadcasting antenna or on the surface of a star, generates electromagnetic waves.

The modern theory of gravity—Einstein's theory of relativity (see *More Precisely 22-1*)—also predicts waves that move through space. A *gravity wave* is the gravitational counterpart of an electromagnetic wave. *Gravitational radiation* results from changes in the strength of a gravitational field. In principle, any time an object of any mass accelerates, a gravity wave should be emitted at the speed of light. The passage of a gravity wave should produce small distortions in the space through which it passes. Gravity is an exceedingly weak force compared with electromagnetism, so these distortions are expected to be very small—in fact, much smaller than the diameter of an atomic nucleus for the waves that may be produced by any sources located in the Milky Way galaxy—yet many researchers believe that these distortions should be measurable. So far, though, no one has succeeded in detecting gravity waves. However, their detection would provide very strong support for the theory of relativity, so scientists are eager to search for them.

Theorists are still debating which kinds of astronomical objects should produce gravity waves detectable on Earth. Leading candidates include (1) the merger of a binary-star system, (2) the collapse of a star into a black hole, and (3) the collision of two black holes or of two neutron stars. Because each of these possibilities involves the acceleration of huge masses, the strength of the gravitational fields should change drastically and rapidly in each case. Other astronomical objects are also expected to emit gravity waves, but only changes involving large masses will produce waves intense enough to be observed.

Of the three candidates, the first one probably presents the best chance for detecting gravity waves, at least for the present. Binary-star systems should emit gravitational radiation as the component stars orbit each other. As energy escapes in the form of gravity waves the two stars slowly spiral toward each other, orbiting more rapidly and emitting even more gravitational radiation. This runaway situation can lead to the decay and eventual merger of close binary systems in a relatively short time (which, in this case, means tens or hundreds of millions of years). As we saw in the text (Section 22.3), these mergers may be the cause of gamma-ray bursts, so gravitational radiation may provide an alternative means of studying these violent and mysterious phenomena.

Such a slow but steady decay in the orbit of a binary system has in fact been detected. In 1974 radio astronomer Joseph Taylor and his student Russell Hulse at the University of Massachusetts discovered a very unusual binary system. Both components are neutron stars, and one is observable from Earth as a pulsar. This system has become known as the *binary pulsar*. Measurements of the periodic Doppler shift of the pulsar's radiation prove that its orbit is slowly shrinking. Furthermore, the rate at which the orbit is shrinking is exactly what would be predicted by relativity theory if the energy were being carried off by gravity waves. Even though the gravity waves themselves have not yet been detected, the binary pulsar is regarded by most astronomers as a very strong piece of evidence in favor of general relativity. Taylor and Hulse received the 1993 Nobel Prize in physics for their discovery.

Gravity waves should contain a great deal of information about the physical events in some of the most exotic regions of space. In 1992, funding was approved for an ambitious gravity-wave observatory called LIGO—short for Laser Interferometric Gravity-wave Observatory. Twin detectors, one in the state of Washington, the other in Louisiana, will use laser beams to measure the extremely small distortions of space produced by gravitational radiation. The accompanying figure shows the complex arrangement of test masses, hanging from wires and fitted with mirrors, that lies at the heart of the LIGO detector. The laser beams will be used to measure the tiny motions of these masses that will result should a gravity wave pass by. They should be capable of detecting gravity waves from many Galactic and extragalactic sources. Work began on building the detectors in 1995; they should be operational by 2000. If successful, the discovery of gravity waves could herald a new age in astronomy, in much the same way that invisible electromagnetic waves, unknown a century ago, revolutionized classical astronomy and led to the field of modern astrophysics.

Theoretical calculations indicate that this process can spin the star up to breakup speed in about a hundred million years. Subsequently, an encounter with another star may eject the neutron star from the binary, or the pulsar's radiation may destroy its companion, so an isolated millisecond pulsar results. This general picture is supported by the finding that of the 40 or so millisecond pulsars seen in globular clusters, 10 are known currently to be members of binary systems. These numbers are quite consistent with the rate at which binaries can be broken up by encounters with other cluster members.

Thus, although a pulsar like the Crab is the direct result of a supernova explosion, millisecond pulsars are the product of a two-stage process. The neutron star was formed in an ancient supernova, billions of years ago. Only relatively recently, through interaction with a binary companion, has it achieved the rapid spin that we observe today. Once again, we see how members of a binary system can evolve in quite different ways from single stars. Notice that the scenario of accretion onto a neutron star from a binary companion is the same scenario that we just used to explain the existence of X-ray bursters. In fact, the two phenomena are very closely linked. Many X-ray bursters may be on their way to becoming millisecond pulsars.

The way in which a neutron star can become a member of a binary system is the subject of active research, because the violence of a supernova explosion would be expected to blow the binary apart in many cases. Only if the supernova progenitor lost a lot of mass before the explosion would the binary be likely to survive. Alternatively, by interacting with an existing binary and displacing one of its components, a neutron star may become part of a binary system *after* it is formed, as depicted in Figure 22.10. Astronomers are eagerly searching the skies for more millisecond pulsars to test their ideas.

PULSAR PLANETS

Radio astronomers can capitalize on the precision with which pulsar signals repeat themselves to make extremely accurate measurements of pulsar motion. In January 1992, radio astronomers at the Arecibo Observatory found that the pulse period of a recently discovered millisecond pulsar lying some 500 pc from Earth varies in an unexpected but quite regular way. Careful analysis of the data has revealed that the period fluctuates on two distinct time scales—one of 67 days, the other of 98 days. The changes in the pulse period are small—less than one part in 10^7—but repeated observations have confirmed their reality.

Figure 22.10 A neutron star can encounter a binary made up of two low-mass stars, ejecting one of them and taking its place. This mechanism provides a means of forming a binary system with a neutron-star component (which may later evolve into a millisecond pulsar) without having to explain how the binary survived the supernova explosion that formed the neutron star.

The leading explanation for these fluctuations holds that they are caused by the Doppler effect as the pulsar wobbles back and forth in space. But what causes the wobble? The Arecibo group believes it is the result of the combined gravitational pulls of not one but *two* planets, each about three times the mass of Earth! One orbits the pulsar at a distance of 0.4 A.U. and the other at a distance of 0.5 A.U. Their orbital periods are 67 and 98 days, respectively, matching the timing of the fluctuations. In April 1994 the group announced further observations that not only confirmed their earlier findings but also revealed the presence of a *third* body, with mass comparable to Earth's Moon, orbiting only 0.2 A.U. from the pulsar.

These remarkable results constitute the first definite evidence of planet-sized bodies outside our solar system. A few other millisecond pulsars have since been found with similar behavior. However, it is unlikely that any of these planets formed in the same way as our own. Any planetary system orbiting the pulsar's progenitor star was almost certainly destroyed in the supernova explosion that created the pulsar. As a result, scientists are still uncertain about how these planets came into being. One possibility involves the binary companion that provided the matter necessary to spin the pulsar up to millisecond speeds. Possibly the pulsar's intense radiation and strong gravity destroyed the companion, then spread its matter out into a disk (a little like the solar nebula) in whose cool outer regions the planets might have condensed.

Astronomers have been searching for decades for planets orbiting main-sequence stars like our Sun, on the assumption that planets are a natural by-product of star formation. ∞ (Sec. 15.2) Only very recently have these searches begun to yield positive results, although some of these findings remain controversial. ∞ (*Interlude 15-1*) It is ironic that the first (and only) Earth-sized planets to be found outside the solar system orbit a dead star and have little or nothing in common with our own world.

22.4 Black Holes

4 Neutron stars are peculiar objects. Nevertheless, theory predicts that they are in equilibrium, as are most other stars. For neutron stars, however, equilibrium does not mean a balance between the inward pull of gravity and the outward pressure of hot gas. Instead, as we have seen, the outward force is provided by the pressure of tightly packed neutrons. Squeezed together, the neutrons form a hard ball of matter that not even gravity can compress further. Or do they? Is it possible that given enough matter packed into a small enough volume, the collective pull of gravity can eventually crush any opposing pressure? Can gravity continue to compress a massive star into an object the size of a planet, a city, a pinhead—even smaller? The answer, apparently, is yes.

THE FINAL STAGE OF STELLAR EVOLUTION

We have seen that the eventual evolution of a star depends critically on its mass. Low-mass stars leave behind a compact remnant known as a white dwarf. ∞ (Sec. 20.3) High-mass stars can also produce a compact remnant, in the form of a neutron star. The laws of physics make specific predictions about the masses of these core remnants. A white dwarf must be less than about 1.4 solar masses—the so-called Chandrasekhar mass, beyond which the electrons cannot support the core against its own gravity. ∞ (Sec. 21.3) Similarly, neutron stars resulting from supernovae have masses between about 1.4 and 3 solar masses.* The lower limit of 1.4 solar masses comes from stellar evolution: the iron core of an evolved star must exceed the Chandrasekhar mass for core collapse to begin and a supernova to occur. The (quite uncertain) upper limit of 3 solar masses is the neutron-star equivalent of the Chandrasekhar mass—beyond 3 solar masses not even tightly packed neutrons can withstand the star's gravitational pull.

In fact, we know of *no* force that can counteract gravity beyond the point at which neutron degeneracy pressure is overwhelmed. If enough material is left behind after a supernova, as may happen in the case of an extremely massive progenitor star, gravity finally wins, and the central core collapses forever. As the core shrinks, the gravitational pull in its vicinity eventually becomes so great that even light itself is unable to escape. The resultant object therefore emits no light, no radiation, no information whatsoever. Astronomers call this bizarre end point of stellar evolution, in which a massive core remnant collapses in on itself and vanishes forever, a **black hole**.

Newtonian mechanics—up to now our reliable and indispensable tool in understanding the universe—cannot adequately describe conditions in or near black holes. ∞ (*More Precisely 2-2*) To comprehend these collapsed objects we must turn instead to the modern theory of gravity, Einstein's **general theory of relativity** (see *More Precisely 22-1*). The Einsteinian description of the universe is equivalent to Newton's for situations encountered in everyday life, but the two theories diverge radically in circumstances where speeds approach the speed of light and in regions of intense gravitational fields.

Despite the fact that general relativity is necessary for a proper description of black-hole properties, we can still usefully discuss some aspects of these strange bodies in more or less Newtonian terms. Let's reconsider the familiar Newtonian concept of escape speed—the speed needed for one object to escape from the gravitational pull of another—supplemented by two key facts from relativity: (1) nothing can travel faster than the speed of light, and (2) all things, *including light*, are attracted by gravity.

ESCAPE SPEED

To explore how gravity attracts even light, let's reconsider the concept of escape speed—the speed needed for one object—a molecule, a baseball, a rocket, whatever—to escape the gravitational pull of another. In Chapter 2 we noted that escape speed is proportional to the square root of a body's mass divided by the square root of its radius. ∞ (Sec. 2.7) Earth's radius is 6400 km, and the escape speed from Earth's surface is just over 11 km/s.

Now consider a hypothetical experiment in which Earth is squeezed on all sides by a gigantic vise. As our planet shrinks under the pressure its mass remains the same. The escape speed increases because the radius is decreasing. Suppose Earth were compressed to one-fourth its present size. The proportionality just mentioned for escape speed then predicts that the escape speed would double (because $1/\sqrt{1/4} = 2$). Any object escaping from this hypothetically compressed Earth would need a speed of at least 22 km/s.

Imagine compressing Earth some more. Squeeze it, for example, by an additional factor of 1000, making its radius hardly more than a kilometer. Now a speed of about 630 km/s would be needed to escape. Compress Earth still further, and the escape speed continues to rise. If our hypothetical vise were to squeeze Earth hard enough to crush its radius to about a centimeter, then the speed needed to escape its surface would reach 300,000 km/s. But this is no ordinary speed—it is the speed of light, the fastest speed allowed by the laws of physics as we currently know them.

Thus, if by some fantastic means the entire planet Earth could be compressed to less than the size of a grape, the escape speed would exceed the speed of light. And because nothing can exceed that speed, the compelling conclusion is that nothing—absolutely nothing—could escape from the surface of such a compressed body. Even radiation—radio waves, visible light, X rays, photons of all wavelengths—would be unable to escape the intense gravity of our reshaped Earth. With no photons leaving, our planet would be invisible and uncommunicative—no signal of any sort could be sent to the universe beyond. The origin of the term *black hole* becomes clear. For all practical purposes, such a supercompact Earth could be said to have disappeared from the universe! Only its gravitational field would remain behind, betraying the presence of its mass, now shrunk to a point.[†]

THE EVENT HORIZON

Astronomers have a special name for the critical radius at which the escape speed from an object would equal the speed of light and within which the object could no longer

---

The dividing lines at 1.4 and 3 solar masses are somewhat uncertain because they ignore the effects of magnetism and rotation, both of which are surely present in the cores of evolved stars. Because these effects can compete with gravity, they influence the evolution of stars. ∞ (Sec. 19.1) In addition, we do not know for certain how the basic laws of physics might change in regions of very dense matter that is both rapidly spinning and strongly magnetized. However, we expect that these dividing lines will shift generally upward when magnetism and rotation are included, because even larger amounts of mass will then be needed for gravity to compress stellar cores into neutron stars or black holes.

[†] *In fact, we now know that regardless of the composition or condition of the object that formed the hole, only three physical properties can be measured from the outside—the hole's mass, charge, and angular momentum. All other information is lost once the infalling matter crosses the event horizon. Thus, only three numbers are required to describe completely a black hole's outward appearance. In this chapter we will consider only holes that formed from nonrotating, electrically neutral matter. Such objects are completely specified once their masses are known.*

MORE PRECISELY 22-1 Einstein's Theories of Relativity

Albert Einstein won a Nobel Prize in 1921 for his explanation of the photoelectric effect, as described in Chapter 4. ∞ (*More Precisely 4-1*) However, he is probably best known for his two *theories of relativity*, the successors to Newtonian mechanics that form the foundation of twentieth-century physics.

The *special theory of relativity* (or just *special relativity*), proposed by Einstein in 1905, deals with the preferred status of the speed of light. We have noted that the speed of light *c* is the maximum speed attainable in the universe. But there is more to it than that. In 1887 a fundamental experiment carried out by two American physicists, A. A. Michelson and E. W. Morley, demonstrated a further important and unique aspect of light—that the measured speed of a beam of light is *independent* of the motion of the observer or the source. No matter what our speed may be relative to the source of the light, we always measure precisely the same value for *c*—299,792.458 km/s.

A moment's thought leads us to the realization that this is a decidedly nonintuitive statement. For example, if we were traveling in a car moving at 100 km/h and we fired a bullet forward with a speed of 1000 km/h relative to the car, an observer standing at the side of the road would see the bullet pass by at 100 + 1000 = 1100 km/h, as illustrated in the accompanying figure. However, if we were traveling in a rocket ship at 1/10 the speed of light, 0.1*c*, and we shone a searchlight beam ahead of us, the Michelson–Morley experiment tells us that an outside observer would measure the speed of the beam not as 1.1*c*, as the preceding example would suggest, but as *c*. The rules that apply to particles moving at or near the speed of light are different from those we are used to in everyday life.

Special relativity is the mathematical framework that allows us to extend the familiar laws of physics from low speeds (that is, speeds much less than *c*, which are often referred to as *nonrelativistic*)

be seen. It is the **Schwarzschild radius**, after Karl Schwarzschild, the German scientist who first studied its properties. The Schwarzschild radius of any object is simply proportional to its mass. For Earth, it is 1 cm; for Jupiter, at about 300 Earth masses, it is about 3 m; for the Sun, at 300,000 Earth masses, it is 3 km. For a 3–solar mass stellar core remnant, the Schwarzschild radius is about 9 km. As a convenient rule of thumb, the Schwarzschild radius of an object is simply 3 km multiplied by the object's mass, measured in solar masses. *Every* object has a Schwarzschild radius. It is the radius to which the object would have to be compressed for it to become a black hole. Put another way, a black hole is an object that happens to lie within its own Schwarzschild radius.

The surface of an imaginary sphere with radius equal to the Schwarzschild radius and centered on a collapsing star is called the **event horizon**. It defines the region within which no event can ever be seen, heard, or known by anyone outside. Even though there is no matter of any sort associated with it, we can think of the event horizon as the "surface" of a black hole.

A 1.4–solar mass neutron star has a radius of about 10 km and a Schwarzschild radius of 4.2 km. If we were to keep increasing the star's mass, the star's Schwarzschild radius would grow, although its actual physical radius would not. In fact the radius of a neutron star *decreases* slightly with increasing mass. By the time our neutron star's mass exceeded about 3 solar masses, it would lie just within its own event horizon, and it would collapse of its own accord. It would not stop shrinking at the Schwarzschild radius—the event horizon is not a physical boundary of any kind, just a communications barrier. The remnant would shrink right past it to ever-diminishing size on its way to being crushed to a point.

Thus, provided that at least 3 solar masses of material remain behind after a supernova explosion, the remnant core will collapse catastrophically, diving below the event horizon in less than a second. The core simply "winks out," to very high (or *relativistic*) speeds, comparable to *c*. Relativity is equivalent to Newtonian mechanics when objects move much more slowly than the speed of light, but it differs greatly in its predictions at relativistic velocities. For example, special relativity predicts that a rapidly moving spacecraft will appear to contract in the direction of its motion, its clocks will appear to run slow, and its mass will appear to increase. Despite their somewhat nonintuitive nature, all the theory's predictions have been repeatedly verified to very high accuracy. Today special relativity is at the heart of all physical science. No scientist seriously doubts its validity.

General relativity is what results when gravity is included in the framework of special relativity. In 1915 Einstein made the connection between special relativity and gravity with the following famous "thought experiment." Imagine that you are enclosed in an elevator with no windows, so that you cannot directly observe the outside world, and the elevator is floating in space. You are weightless. Now suppose that you begin to feel the floor press up against your feet. Weight has apparently returned. There are two possible explanations for this, shown in the accompanying diagram. A large mass could have come nearby, and you are feeling its downward gravitational attraction, *or* the elevator has begun to accelerate upward and the force you feel is that exerted by the elevator as it accelerates you at the same rate. The crux of Einstein's argument is this: there is *no* experiment that you can perform within the elevator, without looking outside, that will let you distinguish between these two possibilities.

Thus, Einstein reasoned, there is no way to tell the difference between a gravitational field and an accelerated frame of reference (such as the rising elevator in the thought experiment). Gravity can therefore be incorporated into special relativity as a general acceleration of all particles. However, another major modification to the theory of special relativity must be made. Central to relativity is the

notion that space and time are not separate quantities but instead must be treated as a single entity—*spacetime*. To incorporate the effects of gravity, the mathematics forces us to the conclusion that spacetime has to be *curved*.

In general relativity, then, gravity is a manifestation of curved spacetime. There is no such thing as a "gravitational field," in the Newtonian sense. Instead, objects move as they do because they follow the curvature of spacetime, and this curvature of spacetime is determined by the amount of matter present. We will explore some of the consequences of this view of gravity in more detail in the text.

disappearing and becoming a small dark region from which nothing can escape—a literal black hole in space. According to theory, this is the fate of any star whose main-sequence mass exceeds about 20–30 times the mass of the Sun.

PHOTON ORBITS

An alternative way of seeing the significance of the Schwarzschild radius is to consider what happens to rays of light emitted at different distances from the event horizon of a black hole. Imagine moving a light source closer and closer to the hole, as shown in Figure 22.11. Let's suppose that the source emits radiation uniformly in all directions.

At large distances from the black hole (Figures 22.11a,b) essentially all the radiation eventually escapes into space. Only that portion of the beam that is aimed directly at the hole is captured. As the light source moves closer, however, the effect of the hole's strong gravity becomes evident. Some photons that would have missed the hole if light trav-

eled in straight lines are instead deflected onto paths that cross the event horizon, and these photons are trapped by the black hole.

At 1.5 Schwarzschild radii (Figure 22.11c), exactly half the radiation emitted by our light source escapes into space. Photons emitted perpendicular to the line joining the source to the center of the hole move in circular orbits at this radius, never escaping the hole's gravity but never crossing the event horizon. The surface on which these photons move as they travel on their circular paths is called the *photon sphere*.

Closer still (Figure 22.11d), the amount of deflection continues to increase, and the fraction of the radiation that escapes into space steadily decreases until, close to the event horizon, only a thin sliver of our beam is able to escape. At the event horizon itself, the sliver vanishes completely. All light rays from that surface, whatever their initial direction, are destined to enter the hole. The black hole's immense gravitational field dominates the trajectories of all particles, even photons, in its vicinity.

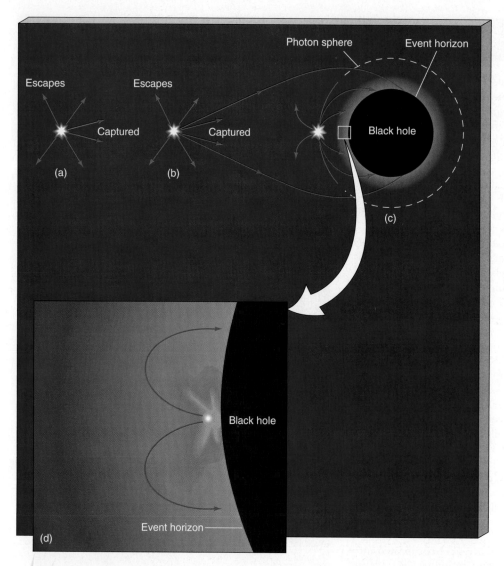

Figure 22.11 This light source, moving closer and closer to a black hole, emits radiation in all directions. (a) At large distances from the hole, most of the light (marked in blue) escapes into space. (b) As the distance decreases, more and more of the radiation is deflected by the black hole's gravity onto paths that intersect the event horizon, and this radiation is trapped (shown in red). (c) At a distance of 1.5 Schwarzschild radii, photons can orbit the hole on circular trajectories, which mark the photon sphere. (d) Eventually, when the source reaches the horizon, all the light it emits is destined to enter the hole.

22.5 Black Hole Properties

5 Modern notions about black holes rest squarely on the theory of relativity. Although white dwarfs and (to a lesser extent) neutron stars can be adequately described by the classical Newtonian theory of gravity, only the modern Einsteinian theory of relativity can properly account for the bizarre physical properties of black holes.

A central concept of general relativity (see *More Precisely 22-1*) is this: matter—all matter—tends to "warp" or curve space in its vicinity. Objects such as planets and stars react to this warping by changing their paths. In the Newtonian view of gravity, particles move on curved trajectories because they feel a gravitational force. In Einsteinian relativity, those same particles move on curved trajectories because they are following the curvature of space produced by some nearby massive object. The more the mass, the greater the warping. Close to a black hole, the gravitational field becomes overwhelming and the curvature of space extreme. At the event horizon itself, the curvature is so great that space "folds over" on itself, causing objects within to become trapped and disappear.

RUBBER SHEETS AND CURVED SPACE

Some props may help you visualize the curvature of space near a black hole. Bear in mind, however, that these props are not real but only tools to help you grasp some exceedingly strange concepts.

First, imagine a pool table with the tabletop made of a thin rubber sheet rather than the usual hard felt. As Figure 22.12 suggests, such a rubber sheet becomes distorted when a heavy weight, such as a rock, is placed on it. The otherwise flat sheet sags (becomes distorted), especially near the rock. The heavier the rock, the larger the distortion. Trying to play pool, you would quickly find that balls passing near the rock were deflected by the curvature of the tabletop. In much the same way, both matter *and radiation* are deflected by the curvature of space near a star. For example, Earth's orbital path is governed by the relatively gentle curvature of space created by our Sun. The more massive the object, the more the space surrounding it is curved.

Figure 22.12 A pool table made of a thin rubber sheet sags when a weight is placed on it. Likewise, space is bent, or warped, in the vicinity of any massive object.

Let's consider another analogy. Imagine a large extended family of people living on a huge rubber sheet—a sort of gigantic trampoline. Deciding to hold a reunion, they converge on a given place at a given time. As shown in Figure 22.13, one person remains behind, not wishing to attend. He keeps in touch with his relatives by means of "message balls" rolled out to him (and back from him) along the surface of the sheet. These message balls are the analog of radiation carrying information through space.

As the people converge, the rubber sheet sags more and more. Their accumulating mass creates an increasing amount of space curvature. The message balls can still reach the lone person far away in nearly flat space, but they arrive less frequently as the sheet becomes more and more warped and stretched—as shown in Figures 22.11(b) and (c)—and the balls have to climb out of a deeper and deeper well. Finally, when enough people have arrived at the appointed spot, the mass becomes too great for the rubber to support them. As illustrated in Figure 22.13(d), the sheet pinches off into a "bubble," compressing the people into oblivion and severing their communications with the lone survivor outside. This final stage represents the formation of an event horizon around the party.

Right up to the end—the pinching off of the bubble—two-way communication is possible. Message balls can reach the outside from within (but at a slower and slower rate as the rubber stretches), and messages from outside can get in without difficulty. However, once the event horizon (the bubble) forms, balls from the outside can still fall in, but they can no longer be sent back out to the person left behind, no matter how fast they are rolled. They cannot make it past the "lip" of the bubble in Figure 22.13(d). This analogy (very) roughly depicts how a black hole warps space completely around on itself, isolating its interior from the rest of the universe. The essential ideas—the slowing down and eventual cessation of outward-going signals and the one-way nature of the event horizon once it forms—all have clear parallels in the case of stellar black holes.

COSMIC CLEANERS? NO

Black holes are *not* cosmic vacuum cleaners. They don't cruise around interstellar space, sucking up everything in sight. The orbit of an object near a black hole is essentially the same as its orbit near a star of the same mass. Only if the object happens to pass within a few Schwarzschild radii (perhaps 50 or 100 km for a typical black hole formed in a supernova explosion) of the event horizon is there any significant difference between its actual orbit and the one predicted by Newtonian gravity and described by Kepler's laws. From a distance, the main observational difference is that an object orbiting a black hole appears to orbit a dark, empty region of space.

Black holes, then, do not go out of their way to drag in matter. However, if some matter does happen to fall

Figure 22.13 Any mass causes the rubber sheet (space) to be curved. As people assemble at the appointed spot on the sheet the curvature grows progressively larger, as shown in frames (a), (b), and (c). The blue arrows represent some directions in which information can be transmitted from place to place. The people are finally sealed inside the bubble (d), forever trapped and cut off from the outside world.

MORE PRECISELY 22-2 Tests of General Relativity

Special relativity is the most thoroughly tested and most accurately verified theory in the history of science. General relativity, however, is on somewhat less firm experimental ground.

The problem with verifying general relativity is that its effects on Earth and in the solar system—the places where we can most easily perform tests—are very small. Just as special relativity produces major departures from Newtonian mechanics only when velocities approach the speed of light, general relativity predicts large departures from Newtonian gravity only when extremely strong gravitational fields are involved—in effect, when orbit speeds and escape velocities become relativistic.

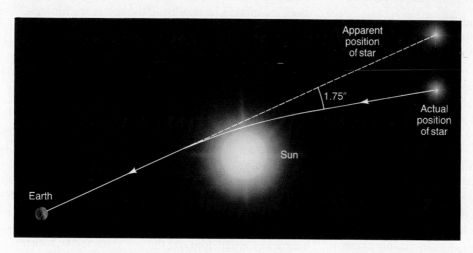

We will encounter other experimental and observational tests of general relativity elsewhere in this chapter. Here we consider just two "classical" tests of the theory. These tests are solar system observations that helped ensure acceptance of Einstein's theory. Later, more accurate measurements confirmed and strengthened these results. Bear in mind, however, that there are no known tests of general relativity in the "strong-field" regime—that part of the theory that predicts black holes, for example—so the full theory has never been experimentally tested.

At the heart of general relativity is the premise that everything, including light, is affected by gravity because of the curvature of spacetime. Shortly after he published his theory in 1915, Einstein noted that light from a star should be deflected by a measurable amount as it passes the Sun. The closer to the Sun the light comes, the more it is deflected. Thus, the maximum deflection should occur for a ray that just grazes the solar surface. Einstein calculated that the deflection angle should be 1.75"—a small, but detectable, amount. Of course, it is normally impossible to see stars close to the Sun. During a solar eclipse, however, when the Moon blocks the Sun's light, the observation becomes possible, as illustrated in the first figure.

In 1919 a team of observers, led by the British astronomer Sir Arthur Eddington, succeeded in measuring the deflection of starlight during an eclipse. The results were

in excellent agreement with the prediction of general relativity. Virtually overnight Einstein became world famous. His previous major accomplishments notwithstanding, this single prediction assured him a permanent position as the best-known scientist on Earth! Recently, the high-precision *Hipparcos* satellite (*Interlude 17-1*) has observed shifts in the apparent positions of many stars, even those whose line of sight is far from the Sun. The shifts are exactly as predicted by Einstein's theory.

Another prediction of general relativity is that planetary orbits should deviate slightly from the perfect ellipses of Kepler's laws. Again, the effect is greatest where gravity is strongest—that is, closest to the Sun. Thus, the largest relativistic effects are found in the orbit of Mercury. Relativity predicts that Mercury's orbit is not a closed ellipse. Instead, its orbit should rotate slowly, as shown in the (highly exaggerated) diagram at right. The amount of rotation is very small—only 43" per century—but Mercury's orbit is so well charted that even this tiny effect is measurable.

In fact, the observed rotation rate is 574" per century, much greater than that predicted by relativity. However, when other (nonrelativistic) gravitational influences, primarily the perturbations due to the other planets, are taken into account, the rotation is in complete agreement with the foregoing prediction.

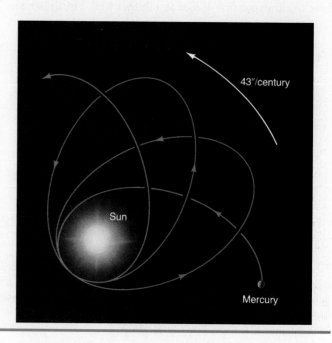

into one—if its orbit happens to take it too close to the event horizon—it will be unable to get out. Black holes are like turnstiles, permitting matter to flow in only one direction—inward. A black hole's mass never decreases (but see *More Precisely 22-3*). Because a black hole will accrete at least a little material from its surroundings, its mass tends to increase over time. The black hole's size is proportional to its mass, so the radius of the event horizon grows with time.

COSMIC HEATERS? YES

Matter flowing into a black hole is subject to great tidal stress. An unfortunate person falling feet first into a solar-mass black hole would find herself stretched enormously in height and squeezed unmercifully laterally. She would be torn apart even before she reached the event horizon, for the pull of gravity would be much stronger at her feet (which are closer to the hole) than at her head. The tidal forces at work in and near a black hole are the same phenomenon responsible for ocean tides on Earth and the spectacular volcanoes on Io. The only difference is that the tidal forces near a black hole are far stronger than any force we know in the solar system.

As shown in Figure 22.14, a similar fate awaits any kind of matter falling into a black hole. Whatever falls in—gas, people, space probes—is vertically stretched and horizontally squeezed, in the process being accelerated to high speeds. The net result of all this stretching and squeezing is

Figure 22.14   Any matter falling into the clutches of a black hole will become severely distorted and heated. This sketch shows an imaginary planet being pulled apart by the gravitational tides of the black hole.

numerous and violent collisions among the torn-up debris, causing a great deal of frictional heating of the infalling matter. Material is simultaneously torn apart and heated to high temperatures as it plunges into the hole.

The rapid heating of matter by tides and collisions is so efficient that, prior to submersion below the hole's event horizon, the matter emits radiation on its own accord. For a black hole of solar mass, the energy is expected to be emitted in the form of X rays. In effect, the gravitational energy of matter outside the black hole is converted into heat while that matter falls toward the hole. Once the hot matter falls below the event horizon, its radiation is no longer detectable—it never leaves the hole. Contrary to what we might expect from an object whose defining property is that nothing can escape from it, the region surrounding a black hole is expected to be a *source* of energy. *More Precisely 22-3* describes another, quite different, way in which a black hole can produce energy.

22.6 Space Travel Near Black Holes

One safe way to study a black hole would be to go into orbit around it, well beyond the disruptive influence of the hole's strong tidal forces. After all, Earth and the other planets of our solar system all orbit the Sun without falling into it and without being torn apart. The gravity field around a black hole is basically no different.

However, even from a stable circular orbit, a close investigation of the hole would be unsafe for humans. Human endurance tests conducted on astronauts of the United States and the former Soviet Union indicate that the human body cannot withstand stress greater than about 10 times the pull of gravity on Earth's surface. This breaking point would occur about 3000 km from a 10–solar mass black hole (which, recall, would have a 30-km event horizon). Closer than that, the tidal effect of the hole would tear a human body apart.

APPROACHING THE EVENT HORIZON

Let's instead send an imaginary indestructible astronaut—a mechanical robot, say—in a probe toward the center of the hole, as illustrated in Figure 22.15. Watching from a safe distance in our orbiting spacecraft, we can then examine the nature of space and time near the hole. Our robot will be a useful explorer of theoretical ideas, at least down to the event horizon. After that boundary is crossed, there is no way for the robot to return any information about its findings.

Suppose, for example, our robot has an accurate clock and a light source of known frequency mounted on it. From our safe vantage point far outside the event horizon, we could use telescopes to read the clock and measure the frequency of the light we receive. What might we discover?

Figure 22.15 Hypothetical robot astronauts can travel toward a black hole while performing experiments that humans, farther away, can monitor in order to learn something about the nature of space near the event horizon.

We would find that the light from the robot would become more and more redshifted as the robot neared the event horizon. Even if the robot used rocket engines to remain motionless, the redshift would still be detected. The redshift is not caused by motion of the light source. It is not the result of the Doppler effect arising as the robot falls into the hole. Rather, it is a redshift induced by the black hole's gravitational field, clearly predicted by Einstein's general theory of relativity and known as **gravitational redshift**.

We can explain the gravitational redshift as follows. According to general relativity, photons are attracted by gravity. As a result, in order to escape from a source of gravity, photons must expend some energy: they have to work to get out of the gravitational field. They don't slow down at all—photons always move at the speed of light—they just lose energy. Because a photon's energy is proportional to the frequency of its radiation, light that loses energy must have its frequency reduced (or, conversely, its wavelength lengthened). As illustrated in Figure 22.16, radiation coming from the vicinity of a gravitating object will be redshifted by an amount depending on the strength of the gravitational field.

As photons traveled from the robot's light source to the orbiting spacecraft they would become gravitationally redshifted. From our standpoint on the orbiting spacecraft, a green light, say, would become yellow and then red as the robot astronaut neared the black hole. From the robot's perspective, the light would remain green. As the robot got closer to the event horizon, the radiation from its light source would become undetectable with optical telescopes. The radiation reaching us in the

orbiting spacecraft would by then be lengthened so much that infrared and then radio telescopes would be needed to detect it. Closer still to the event horizon, the radiation emitted as visible light from the robot probe would be shifted to wavelengths even longer than conventional radio waves by the time it reached us.

Light emitted *from the event horizon itself* would be gravitationally redshifted to infinitely long wavelengths. In other words, each photon would use all its energy trying to escape from the edge of the hole. What was once light (on the robot) would have no energy left on arrival at the safely orbiting spacecraft. Theoretically, this radiation would reach us—still moving at the speed of light—but with zero energy. The light radiation originally emitted would be redshifted beyond our perception.

What about the robot's clock? Assuming that we could read it, what time would it tell? Would there be any observable change in the rate at which the clock ticked as it moved deeper into the hole's gravitational field? From the safely orbiting spacecraft, we would find that any clock close to the hole would appear to tick more *slowly* than an equivalent clock on board the spacecraft. The closer the clock came to the hole, the slower it would appear to run. The clock closest to the hole would operate slowest of all. On reaching the event horizon, the clock would seem to stop altogether. It would be as if the robot astronaut had found immortal-

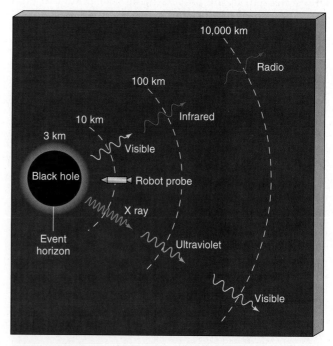

Figure 22.16 As a photon escapes from the strong gravitational field close to a black hole it must expend energy to overcome the hole's gravity. This energy does not come from a change in the speed at which it travels (this is always 300,000 km/s, even under these extreme conditions). Rather, the photon "gives up" energy by increasing its wavelength. Thus, the photon's color changes. This figure shows the effect on two beams of radiation, one of visible light and one of X rays, emitted from a space probe as it nears a 1-solar mass black hole.

MORE PRECISELY 22-3   Black-Hole Evaporation

Some attempts to understand gravity on a microscopic scale suggest that black holes may not be entirely black after all. Applying what they know of subatomic physics, scientists now believe it possible that some matter and radiation can escape from a black hole. Here's how. The laws of quantum physics allow a process known as *pair creation* to occur anywhere in space: a particle and its antiparticle—an electron and a positron, say—can come into being spontaneously, literally formed out of nothing. This, of course, violates one of the most cherished laws of physics, the law of conservation of mass and energy (recall from Section 16.5 that mass and energy are equivalent, related to each other by Einstein's famous equation $E = mc^2$), but this violation is permitted if the "books are balanced" by the disappearance of the particles (by mutual annihilation) within a short enough period of time. In effect, the rules can be broken, so long as they are repaired before anyone notices.

Most of the time, pairs of particles appear and disappear so rapidly that energy is conserved on all macroscopic scales. However, should pair creation happen near a black hole, as illustrated in the diagram, it is possible for one of the two particles to cross the event horizon *before* it meets and annihilates its partner. The other particle would then be free to leave the scene, making the black hole appear to the outside world as a source of matter or radiation. The energy required to create the new particle ultimately comes from the black hole. Because energy and mass are equivalent, this means that the hole must decrease in mass as it radiates. Thus, black holes do not last forever—they slowly "evaporate." This possibility was first realized by Cambridge University mathematician Stephen Hawking. The radiation from a black hole is known as *Hawking radiation*.

A remarkable result, also discovered by Hawking's group, is that the spectrum of Hawking radiation is described by exactly the same curve that characterizes emission from any hot body. Black holes emit blackbody radiation! The temperature of the radiation turns out to be inversely related to the mass of the hole. Big black holes are very cold, whereas small black holes are hot. A hole the mass of the Sun would emit radiation at a temperature of 10^{-6} K; one the mass of a mountain—about 10^{12} kg—would have a temperature of some 10^{12} K. Knowing a black hole's temperature T and surface area (of the event horizon) A, we can calculate its luminosity L in exactly the same way as for stars: $L \propto AT^4$. ∞ (Sec. 17.3)

A black hole radiates energy (and hence mass) into space. As the hole radiates, its mass drops and its temperature increases. That is, the black hole's temperature is inversely proportional to its mass, and its area is proportional to the square of its mass. The black hole, then, increases its luminosity as it evaporates. The increased luminosity, in turn, leads to a faster loss of mass. This runaway situation eventually ends violently, and the black hole explodes in a burst of gamma rays.

The lifetime of a hole depends on its mass. For a 1–solar mass black hole, the explosion is predicted to occur after about 10^{70} years! Astronomers today hardly expect to observe such an event. Thus, the issue of evaporation is moot for the black holes described in this chapter; astronomers do not expect ever to observe either their slow decay (which begins the moment they form) or their eventual explosion.

However, very small black holes, with masses of about 10^{12} kg, should have lifetimes roughly equal to the current age of the universe. Although we know of no way in which such objects could be created in the universe today, it is conceivable that conditions in the very earliest epochs of the universe might have been just right to compress pockets of matter into miniature black holes. Such black holes would have Schwarzschild radii of about 10^{-15} m, comparable to the size of a subatomic particle. If they exist, they should be exploding right now. Attempts have been made to observe the resultant gamma rays, so far without success.

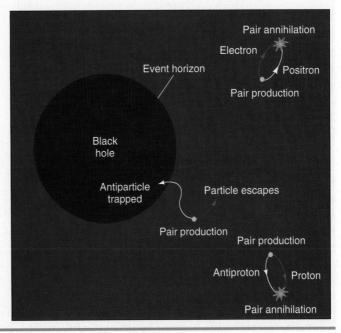

---

ity! All action would become virtually frozen in time. Consequently, an external observer would never actually witness an infalling astronaut sink below the event horizon. Such a process would appear to take forever.

This apparent slowing down of the robot's clock is known as **time dilation**. It is another clear prediction of general relativity, and in fact it is closely related to the gravitational redshift. To see this connection, imagine that

we use our light source as a clock, with the passage of a wave crest (say) constituting a "tick." The clock thus ticks at the frequency of the radiation. As the wave is redshifted, the frequency drops, and fewer wave crests pass the distant observer each second—the clock appears to slow down. This thought experiment demonstrates that the redshift of the radiation and the slowing of the clock are one and the same.

From the point of view of the indestructible robot, however, relativity theory predicts no strange effects at all. To the infalling robot, the light source hasn't reddened, and the clock keeps perfect time. In the robot's frame of reference, everything is normal. Nothing prohibits it from approaching within the Schwarzschild radius of the hole. No law of physics constrains an object from passing through an event horizon. There is no barrier at the event horizon and no sudden lurch as it is crossed; it is only an imaginary boundary in space. Travelers passing through the event horizon of a sufficiently massive hole (such as might lurk in the heart of our own galaxy, as we will see) might not even know it—at least until they tried to get out!

Most astronomical objects' gravitational fields are far too weak to produce any significant gravitational redshift, although in many cases the effect can still be measured. Exceedingly delicate laboratory experiments on Earth have succeeded in detecting the tiny gravitational redshift produced by even our own planet's weak gravity. Sunlight is redshifted by only about a hundredth of an angstrom. A few white-dwarf stars do show some significant reddening of their emitted light, however. Their radii are much smaller than that of our Sun, so their surface gravity is very much stronger than the Sun's. Neutron stars should show a substantial shift in their radiation, but it is currently difficult to disentangle the effects of gravity and magnetism on the observed signals.

DEEP DOWN INSIDE

No doubt you are wondering what lies within the event horizon of a black hole. The answer is simple: no one really knows. However, the question raises some very fundamental issues that lie at the forefront of modern physics.

Can an entire star simply shrink to a point and vanish? General relativity predicts that without some agent to compete with gravity, the core remnant of a high-mass star will collapse all the way to a point at which both its density and its gravitational field become infinite—a so-called **singularity**. However, we should not take this prediction of infinite density too literally. Singularities always signal the breakdown of the theory producing them. In other words, the present laws of physics are simply inadequate to describe the final moments of a star's collapse.

As it stands today, the theory of gravity is incomplete because it does not incorporate a proper (that is, a quantum-mechanical) description of matter on very small scales. As our collapsing stellar core shrinks to smaller and smaller radii we eventually lose our ability even to describe, let alone predict, its behavior. Perhaps matter trapped in a black hole never actually reaches a singularity. Perhaps it just approaches this bizarre state, in a manner that we will someday understand as the subject of *quantum gravity*—the merger of general relativity with quantum mechanics—develops.

Having said that, we can at least estimate how small the core can get *before* current theory fails. It turns out that by the time that stage is reached, the core is already much smaller than any elementary particle. Thus, although a complete description of the end-point of stellar collapse may well require

a major overhaul of the laws of physics, for all practical purposes the prediction of collapse to a point is valid. Even if a new theory somehow succeeds in doing away with the central singularity, it is very unlikely that the external appearance of the hole, or the existence of its event horizon, will change. Any modifications to general relativity are expected to occur only on submicroscopic scales, not on the macroscopic (kilometer-sized) scale of the Schwarzschild radius.

Singularities are places where the rules break down, and some very strange things may occur near them. Many possibilities have been envisaged—gateways into other universes, time travel, the creation of new states of matter—but none of them has been proved, and certainly none of them has been observed. Because these regions are places where science fails, their presence causes serious problems for many of our cherished laws of physics, from causality (the idea that cause should precede effect, which runs into immediate problems if time travel is possible), to energy conservation (which is violated if material can hop from one universe to another through a black hole).

Disturbed by the possibility of such chaos in science, some relativists have even proposed a "principle of cosmic censorship": Nature always hides *any* singularity, such as that found at the center of a black hole, inside an event horizon. In that case, even though physics fails, that breakdown cannot affect us outside, so we are safely insulated from any effects the singularity may have. What would happen if we one day found a so-called *naked singularity* somewhere, a singularity uncloaked by an event horizon? Would relativity theory still hold there? For now, we just don't know.

What sense are we to make of black holes? Do black holes, and all the strange phenomena that go on in and around them, really exist? The basis for understanding these weird phenomena is the relativistic concept that mass warps space—which has already been found to be a good representation of reality, at least for the weak gravitational fields produced by stars and planets (see *More Precisely 22-2*). The larger the mass concentration, the greater the warping, and thus, apparently, the stranger the observational consequences. Although general relativity is not proven, there is presently no reason to disbelieve it, and black holes are one of its most striking predictions. So long as general relativity stands as the correct theory of gravity in the universe, black holes are real.

22.7 Observational Evidence for Black Holes

6 Theoretical ideas aside, is there any observational evidence for black holes? Can we prove that these strange invisible objects really do exist?

STELLAR TRANSITS?

One way in which we might think of detecting a black hole would be to observe it transit (pass in front of) a star. Unfortunately, such an event would be extremely hard to

Figure 22.17 The gravitational bending of light around the edges of a small, massive black hole makes it impossible to observe the hole as a black dot superimposed against the bright background of its stellar companion.

see; the 12,000-km planet Venus is barely noticeable when transiting the Sun, so a 10-km-wide object moving across the image of a faraway star would be completely invisible with either current equipment or any equipment available in the foreseeable future.

Actually, this observation is not even as clear-cut as just suggested. Even if we were close enough to the star to resolve the disk of the transiting black hole, the observable effect would not be a black dot superimposed on a bright background. The background starlight would be deflected as it passed the black hole on its way to Earth, as indicated in Figure 22.17. The effect is the same as the bending of distant starlight around the edge of the Sun, a phenomenon that has been repeatedly measured during solar eclipses throughout the last several decades (see *More Precisely 22-2*). With a black hole, much larger deflections would occur. As a result, the image of a black hole in front of a bright companion star would show not a neat, well-defined black dot but rather a fuzzy image virtually impossible to observe, even from nearby.

BLACK HOLES IN BINARY SYSTEMS

A much better way to find black holes is to look for their effects on other objects. Our galaxy harbors many binary-star systems in which only one object can be seen. Recall from our study of binary-star systems in Section 17.9 that we need to observe the motion of only one star to infer the existence of an unseen companion and measure some of its properties. In the majority of cases, the invisible companion is simply small and dim, nothing more than an M star hidden in the glare of an O or B partner, or perhaps shrouded by dust or other debris, making it invisible to even the best available equipment. In either case, the invisible object is not a black hole.

A few close binary systems, however, have peculiarities suggesting that one of their members may be a black hole. Some of the most interesting observations, made during the 1970s and 1980s by Earth-orbiting satellites, revealed binary systems in which the invisible member emits large amounts of X rays. The mass of the emitting object is measured as several solar masses, so we know it is not simply a small, dim star, and radiation pressure from the binary members makes circumstellar debris an unlikely explanation for its invisibility.

One particular binary system drawing much attention lies in the constellation Cygnus. Figure 22.18(a) shows the area of the sky in Cygnus where astronomers have reasonably good evidence for a black hole. The rectangle outlines the celestial

(a)   R I V U X G   (b)   R I V U X G

Figure 22.18 (a) The brightest star in this photograph is a member of a binary system whose unseen companion, called Cygnus X-1, is thought to be a good candidate for a black hole. (b) X rays emitted by the Cygnus X-1 source were analyzed by changing them into electronic signals that were then viewed on a video screen, from which this picture was taken. (The field of view here is outlined by the rectangle in the previous figure.)

system of interest, some 2000 pc from Earth. The black-hole candidate is an X-ray source called Cygnus X-1, discovered by the *Uhuru* satellite in the early 1970s. The main observational features of this binary system are as follows:

1. The visible companion of the X-ray source—a blue B-type supergiant with the catalog name HDE226868—was identified a few years after Cyg X-1 was discovered. Assuming that the companion lies on the main sequence, we know that its mass must be around 25 times the mass of the Sun.

2. The binary system has an orbital period of 5.6 days. Combining this information with further spectroscopic measurements of the visible component's orbital speed, astronomers estimate the total mass of the binary system to be around 35 solar masses, implying that Cygnus X-1 has a mass of about 10 times the mass of the Sun. ∞ (Sec. 17.9)

3. Other spectroscopic studies suggest that hot gas is flowing from the bright star toward an unseen companion.

4. X-ray radiation emitted from the immediate neighborhood of Cygnus X-1 implies the presence of very high temperature gas, perhaps as hot as several million kelvins (see Figure 22.18b).

5. Rapid time variations of this X-ray radiation imply that the size of the X-ray-emitting region of Cygnus X-1 must be less than a few hundred kilometers across. The reasoning is the same as in the discussion of gamma-ray bursts in Section 22.3. X rays from Cygnus X-1 have been observed to vary in intensity on time scales as short as a millisecond. For this variation not to be blurred by the travel time of light across the source, Cygnus X-1 cannot be more than 1 light millisecond, or 300 km, in diameter.

These general properties suggest that the invisible X-ray-emitting companion could be a black hole. The X-ray-emitting region is likely an accretion disk formed as matter drawn from the visible star spirals down onto the unseen component. The rapid variability of the X-ray emission indicates that the unseen component must be compact—a neutron star or a black hole. The mass limit of the dark component argues for the latter, as astronomers believe that neutron stars' masses cannot exceed about 3 solar masses. Figure 22.19 is an artist's conception of this intriguing object. As shown, most of the gas drawn from the visible star ends up in a donut-shaped accretion disk of matter. As the gas flows toward the black hole it becomes superheated and emits the X rays we observe, just before they are trapped forever below the event horizon.

HAVE BLACK HOLES BEEN DETECTED?

A few other black-hole candidates are known. For example, the third X-ray source ever discovered in the Large Magellanic Cloud—called LMC X-3—is an invisible object that, like Cygnus X-1, orbits a bright companion star. LMC X-3's visible companion seems to be distorted into the shape of an egg by the unseen object's intense gravitational pull. Reasoning similar to that applied to Cygnus X-1 leads to the conclusion that the compact object LMC X-3 has a mass nearly 10 times that of the Sun, making it too massive to be anything but a black hole. The X-ray binary system A0620-00 has been found to contain an invisible compact object of mass 3.8 times the mass of the Sun.

In total, there are perhaps half a dozen known objects that may turn out to be black holes, although Cygnus X-1, LMC X-3, and A0620-00 have the strongest claims. But how sure are we that these objects are black holes? Cygnus X-1 and the other suspected black holes in binary systems all have masses relatively close to the dividing line separating neutron stars from black holes. Given the present uncertainties in both observations and theory, might they conceivably be merely dim, dense neutron stars and not black holes at all?

Most astronomers do not regard this as a likely possibility, but it highlights a problem. There is presently no

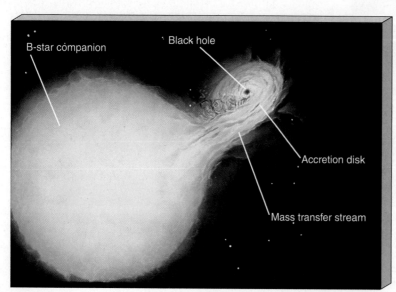

B-star companion

Black hole

Accretion disk

Mass transfer stream

Figure 22.19 Artist's conception of a binary system containing a large, bright, visible star and an invisible, X-ray-emitting black hole. This drawing is based on data obtained from detailed study of Cygnus X-1.

Animation

clear observational test that can unambiguously distinguish a 10–solar mass black hole from, say, a 10–solar mass neutron star (if one could somehow exist). Both objects would affect a companion star's orbit in the same way; both would tear mass from its surface, and both would form an accretion disk around themselves that would emit intense X rays (although some researchers believe that the accretion disks may differ sufficiently in their detailed properties that the nature of the central object may be identifiable from observations). The radiation from the surface of the neutron star itself, which would distinguish it from a black hole, might in some cases be so weak that it would be impossible to detect against the emission from the disk. We rule out the neutron star on purely theoretical grounds (on the basis of our understanding of neutron star masses), but we have no hard observational evidence to support our theory.

The argument really proceeds by elimination. Loosely stated, it goes: "Object X is compact and very massive. We don't know of anything else that can be that small and that massive. Therefore, object X is a black hole." We will see other instances later in this book in which astronomers use similar reasoning to infer the existence of very massive black holes in the hearts of galaxies. Still, some astronomers are troubled that the black-hole category has in some ways become a catch-all for things that have no other reasonable explanation.

So have stellar black holes really been discovered? The answer is probably yes. Skepticism is healthy in science, but only the most stubborn astronomers (and some do exist!) would take serious issue with the reasoning that supports the case for black holes. Can we guarantee that future modifications to the theory of compact objects will not invalidate our arguments? No, but similar statements could be made in many areas of astronomy—indeed, about any theory in any area of science. We conclude that, strange as they are, black holes have been detected in our galaxy. Perhaps some day, future generations of space travelers will visit Cygnus X-1 or LMC X-3 and (carefully!) test these conclusions firsthand. Until then, we will have to continue to rely on improving theoretical models and observational techniques to guide our discussions of these mysterious objects.

Chapter Review

SUMMARY

A core-collapse supernova may leave behind a remnant, an ultra-compressed ball of material called a **neutron star** (p. 494). The processes that form neutron stars ensure that these stars are rapidly rotating and strongly magnetized at birth.

Pulsars (p. 495) are objects that appear to emit regular bursts of electromagnetic energy. The accepted explanation for pulsars is the **lighthouse model** (p. 496), in which a rotating neutron star sends a beam of energy into space. If the beam sweeps past Earth, we see a pulsar. The pulse period is the rotation period of the neutron star.

A neutron star in a close binary system can draw matter from its companion, forming an accretion disk. The material in the disk heats up even before it reaches the neutron star, and the disk is usually a strong source of X rays. As gas builds up on the star's surface it eventually becomes hot enough to fuse hydrogen. As with a nova explosion on a white dwarf, when hydrogen burning starts on a neutron star, it does so explosively. An **X-ray burster** (p. 497) results. Even more energetic are **gamma-ray bursts** (p. 498), which may result from the violent merger of neutron stars in distant binary systems.

The rapid rotation of the inner part of the accretion disk causes the neutron star to spin faster as new gas arrives on its surface. The eventual result is a very rapidly rotating neutron star—a **millisecond pulsar** (p. 500). Many millisecond pulsars are found in the hearts of old globular clusters. They cannot have formed recently, so they must have been spun up by interactions with other stars.

Careful analysis of the radiation received has shown that some pulsars are orbited by planet-sized objects. The origin of these "pulsar planets" is still uncertain.

The upper limit on the mass of a neutron star is about 3 solar masses. Beyond that mass the star can no longer support itself against its own gravity, and it must collapse. No known force can prevent the material from collapsing all the way to a pointlike **singularity** (p. 512), a region of extremely high density where the known laws of physics break down. Surrounding the singularity, at a distance of a few kilometers for a solar-mass object, is a region of space from which even light cannot escape—a **black hole** (p. 503). Astronomers believe that the most massive stars form black holes, rather than neutron stars, after they explode in a supernova.

Conditions in and near black holes cannot be described by Newtonian mechanics. A proper description involves the **theories of relativity** (p. 503) developed by Albert Einstein early in the twentieth century. Even relativity theory fails right at the singularity, however.

The "surface" of a black hole is the **event horizon** (p. 504); its distance from a singularity is called the **Schwarzschild radius** (p. 504). At the event horizon the escape speed equals the speed of light. Within this distance, nothing can escape. Photons passing too close to a black hole are deflected onto paths that cross the event horizon and become trapped.

Relativity theory describes gravity in terms of a warping, or bending, of space by the presence of mass. The more mass, the greater the warping. All particles—including photons—respond to that warping by moving along curved paths. A black hole is a region where the warping is so great that space folds back on itself, cutting off the interior of the hole from the rest of the universe.

To a distant observer the clock on a spaceship falling into a black hole would show **time dilation** (p. 511)—it would appear to slow down as the ship approached the event horizon. The observer would never see the ship reach the surface of the hole. At the same time, light leaving the ship would be subject to **gravitational redshift** (p. 510) as it climbed out of the hole's intense gravitational field. Light emitted just at the event horizon would be redshifted to infinite wavelength. Both phenomena are predictions of the theory of relativity. The gravitational redshifts due to both Earth and the Sun are very small but have been detected experimentally.

Once matter falls into a black hole, it can no longer communicate with the outside. However, on its way in, it can form an accretion disk and emit X rays just as in the neutron-star case. The best candidates for black holes are binary systems in which one component is a compact X-ray source. Cygnus X-1, a well-studied X-ray source in the constellation Cygnus, is a long-standing black-hole candidate. Studies of orbital motions imply that the compact objects are too massive to be neutron stars, leaving black holes as the only alternative.

SELF-TEST: TRUE OR FALSE?

_____ **1.** The density of a neutron star is comparable to the density of an atomic nucleus.

_____ **2.** As a result of their high masses and small sizes, neutron stars have only weak gravitational pulls at their surfaces.

_____ **3.** Newly formed neutron stars have extremely strong magnetic fields.

_____ **4.** A millisecond pulsar is actually a very old neutron star that has been recently spun up by interaction with a neighbor.

_____ **5.** Millisecond pulsars are found only in globular clusters.

_____ **6.** Planet-sized bodies will never be found around a pulsar, because the supernova that formed the pulsar would have destroyed any planets in the system.

_____ **7.** Nothing can travel faster than the speed of light.

_____ **8.** All things, except light, are attracted by gravity.

_____ **9.** A black hole is an object whose escape speed equals or exceeds the speed of light.

_____ **10.** Although visible light cannot escape from a black hole, high-energy radiation, like gamma rays, can escape.

_____ **11.** It is possible for a beam of light to go into orbit around a black hole.

_____ **12.** If you could touch it, the surface of a black hole, the event horizon, would be very hard.

_____ **13.** The laws of physics break down near the center of a black hole.

_____ **14.** X rays are emitted by matter accreting onto a stellar-mass black hole.

_____ **15.** Thousands of black holes have now been identified.

SELF-TEST: FILL IN THE BLANK

1. No remnant remains after the explosion of a _____ supernova.

2. A typical neutron star is _____ km in diameter.

3. Neutron stars may be characterized as having _____ rates of rotation and _____ magnetic fields.

4. Pulsars were discovered through observations in the _____ part of the electromagnetic spectrum.

5. Typical pulsar periods range from _____ to _____. (Give numbers and units.)

6. The pulse period of pulsar radiation tells us the _____ of the neutron star emitting the radiation.

7. All millisecond pulsars are now, or once were, members of _____ star systems.

8. X-ray bursters result from accretion of material from a binary companion onto a _____ star.

9. According to general relativity, space is warped, or curved, by _____.

10. If the Sun were magically to turn into a black hole of the same mass, Earth's orbit would _____.

11. The radius of the event horizon of a black hole is called the _____.

12. The region of extremely high density at the center of a black hole is called a _____.

13. Photons _____ energy as they escape from a gravitational field.

14. Black holes of stellar origin are believed to have been discovered in several _____ systems.

15. Scientists infer that the energy-emitting region in Cygnus X-1 must be very small because of the rapid _____ of the radiation received.

REVIEW AND DISCUSSION

1. How does the way in which a neutron star forms determine some of its most basic properties?

2. What would happen to a person standing on the surface of a neutron star?

3. Why aren't all neutron stars seen as pulsars?

4. What are X-ray bursters?

5. Describe the line of reasoning that led astronomers to conclude that gamma-ray bursts are very distant and very energetic.

6. What is the favored explanation for the rapid spin rates of millisecond pulsars?

7. Why did astronomers not expect to find a pulsar with a planetary system?

8. What does it mean to say that the measured speed of a light beam is independent of the motion of the observer?

9. Use your knowledge of escape speed to explain why black holes are said to be "black."

10. Why is it so difficult to test the predictions of general relativity? Describe two tests of the theory.

11. What would happen to someone falling into a black hole?

12. What is the principle of cosmic censorship? Do you think it is a sound scientific principle?

13. What makes Cygnus X-1 a good black-hole candidate?

14. Imagine that you had the ability to travel at will through the Galaxy. Explain why you would discover many more neutron stars than those known to observers on Earth. Where would you be most likely to find these objects?

15. Do you think that planet-sized objects discovered in orbit around a pulsar should be called planets? Why or why not?

PROBLEMS

1. The angular momentum of a solid body (see *More Precisely 15-1*) is proportional to its angular velocity times the square of its radius. Using the law of conservation of angular momentum, estimate how fast a collapsed stellar core would spin if its initial spin rate was 1 revolution per day and its radius decreased from 10,000 km to 10 km.

2. What would your mass be if you were composed entirely of neutron-star material, of density 3×10^{17} kg/m^3? (Assume that your average density is 1000 kg/m^3.) Compare this with the mass of (a) the Moon. (b) a typical 1-km asteroid.

3. Calculate the surface gravity and escape speed of 1.4–solar mass neutron star with a radius of 10 km.

4. Use the radius–luminosity–temperature relation to calculate the luminosity of a 10-km-radius neutron star for temperatures of 10^5 K, 10^7 K, and 10^9 K. What do you conclude about the visibility of neutron stars? Could the brightest of them be plotted on our H–R diagram?

5. A gamma-ray detector of area 0.5 m^2 observing a gamma-ray burst records photons having total energy 10^{-8} joules. If the burst occurred 1000 Mpc away, calculate the total amount of energy it released (assuming that the energy was emitted isotropically). How would this figure change if the burst occurred 10,000 pc away instead, in the halo of our Galaxy? What if it occurred within the Oort cloud of our own solar system, at a distance of 50,000 A.U.?

6. A 10-km-radius neutron star is spinning 1000 times per second. Calculate the speed of a point on its equator, and com-

pare it with the speed of light. (Consider the equator as the circumference of a circle, and recall that circumference = $2\pi r$.) Also calculate the orbital speed of a particle in a circular orbit just above the neutron star's surface.

7. Supermassive black holes are believed to exist in the centers of some galaxies. What would be the Schwarzschild radii of black holes of 1 million and 1 billion solar masses, respectively? How does the first black hole compare in size with the Sun? How does the second compare in size with the solar system?

8. Calculate the tidal force on a 2-m-tall human falling feet first into a 1–solar mass black hole—that is, compute the difference in the forces acting on his head and his feet just before his feet cross the event horizon. Repeat the calculation for a 1–million solar mass black hole and for a 1 billion–solar mass black hole. Compare these forces with the person's weight on Earth of 600 N. (See *More Precisely 2-2*.)

9. What are the mass and Schwarzschild radius of a black hole having a blackbody temperature equal to that of the Sun—6000 K? (See *More Precisely 22-3*.) What is the black hole's luminosity?

10. Using the data given in the text (assume the upper limit on the stated range for the black-hole mass), calculate the orbital separation of Cyg X-1. If the companion star's radius is 20 million km, verify (approximately) that the black hole's tidal field is sufficient to draw matter from the companion's surface.

PROJECTS

1. Many amateur astronomers enjoy turning their telescopes on the ninth-magnitude companion to Cygnus X-1, the sky's most famous black-hole candidate. Because none of us can see in X rays, no sign of anything unusual can be seen. Still, it's fun to gaze toward this region of the heavens and contemplate Cygnus X-1's powerful energy emission and strange properties.

Even without a telescope, it is easy to locate the region of the heavens where Cygnus X-1 resides. The constellation Cygnus contains a recognizable star pattern, or asterism, in the shape of a large cross. This asterism is called the Northern Cross. The star in the center of the crossbar is called Sadr. The star at the bottom of the cross is called Albireo. Approximately midway along an imaginary line between Sadr and Albireo lies the star Eta Cygni. Cygnus X-1 is located slightly less than 0.5° from this star. With or without a telescope, sketch what you see.

2. Set up a demonstration of the densities of various astronomical objects—an interstellar cloud, a star, a terrestrial planet, a white dwarf, and a neutron star. Select a common object that is easily held in your hand, something that would be familiar to anyone—an apple, for example. For the lowest densities, calculate how large a volume would contain the object's equivalent mass. For high densities, calculate how many of the objects would have to be fit into a standard volume, such as 1 cm^3. This volume is better for this project than 1 m^3 because most people do not appreciate how large a volume 1 m^3 is. Present your demonstration to your class or to some other group of students. Tell them about each astronomical object and how it comes by its density.

23 THE MILKY WAY GALAXY

A Grand Design *www*

LEARNING GOALS

Studying this chapter will enable you to:

1 Describe the overall structure of the Milky Way Galaxy and specify how the various regions differ from one another.

2 Explain the importance of variable stars in determining the size and shape of our Galaxy.

3 Describe the orbital paths of stars in different regions of the Galaxy and explain how these motions are accounted for by our understanding of how the Galaxy formed.

4 Discuss some possible explanations for the existence of the spiral arms observed in our own and many other galaxies.

5 Explain what studies of galactic rotation reveal about the size and mass of our Galaxy and discuss the possible nature of dark matter.

6 Describe some of the phenomena observed at the center of our Galaxy.

(Opposite page, background) The varying interrelationships among the many components of matter in our Milky Way Galaxy comprise a sort of "galactic ecosystem." Its evolutionary balance may be as complex as that of life in a tidal pool or a tropical rainforest. Here, stars abound throughout the Lagoon Nebula, a rich stellar nursery about 1200 pc from Earth.

(Inset A) An emission nebula, the North American Nebula, glowing amidst a field of stars, its red color produced by the emission of light from vast clouds of hydrogen atoms.

(Inset B) An open cluster, the Jewel Box, containing many young, blue stars.

(Inset C) A typical globular cluster, 47 Tucanae. This true-color image reveals its dominant member stars to be elderly red giants.

(Inset D) A true-color *Hubble* image of part of the Cygnus Loop—a supernova remnant, the remains of a colossal stellar explosion that occurred about 15,000 years ago.

Looking up on a dark, clear night, we are struck by two aspects of the night sky. The first is a fuzzy band of light—the Milky Way—that stretches across the heavens. From the Northern Hemisphere this band is most easily visible in the summertime, arcing high above the horizon. Its full extent forms a great circle that encompasses the entire celestial sphere. Away from that glowing band, however, our second impression is that the nighttime sky seems more or less the same in all directions. Bunches of stars cluster here and there, but overall, apart from the band of the Milky Way, the evening sky looks pretty uniform. Yet this is only a local impression. Ours is a rather provincial view. When we consider much larger volumes of space, on scales far, far greater than the distances between neighboring stars, a new level of organization becomes apparent as the large-scale structure of the Milky Way Galaxy is revealed.

23.1 Our Parent Galaxy

1 A **galaxy** is a gargantuan collection of stellar and interstellar matter—stars, gas, dust, neutron stars, black holes—isolated in space and held together by its own gravity. Astronomers are aware of literally millions of galaxies beyond our own. The particular galaxy we happen to inhabit is known as the **Milky Way Galaxy**, or just *the Galaxy*, with a capital *G*.

Our Sun lies in a part of the Galaxy known as the **Galactic disk**—an immense, circular, flattened region containing most of our Galaxy's luminous stars and interstellar matter (and virtually everything we have studied so far in this book). We do not need sophisticated astronomical equipment to verify this statement. Our own unaided eyes will suffice. Figure 23.1 illustrates how, viewed from within, the Galactic disk appears as a band of light stretching across our night sky, a band known as the *Milky Way*. As indicated in the figure, if we look in a direction away from the Galactic disk (red arrows), relatively few stars lie in our field of view. However, if our line of sight happens to lie within the disk (white and blue arrows), we see so many stars that their light merges into a continuous blur.

Paradoxically, although we can study individual stars and interstellar clouds near the Sun in great detail, our location within the Galactic disk makes deciphering our Galaxy's large-scale structure from Earth a very difficult task—a little like trying to unravel the layout of paths, bushes, and trees in a city park without being able to leave one particular park bench. In some directions the interpretation of what we see is ambiguous and inconclusive. In others, foreground objects completely obscure our view of what lies beyond, but we cannot move around them to get a better look. As a result, astronomers who study the Milky Way Galaxy are often guided in their efforts by comparisons with more distant, but much more easily observable, systems.

Figures 23.2 and 23.3 show three galaxies thought to resemble our own in overall structure. Figure 23.2 is the Andromeda Galaxy, the nearest major galaxy to the Milky Way Galaxy, lying about 900 kpc (nearly 3 million light years) away. Andromeda's apparent elongated shape is a consequence of the angle at which we happen to view it. In fact, this galaxy, like our own, consists of a circular galactic disk of matter that fattens to a **galactic bulge** at the center. The disk and bulge are embedded in a roughly spherical ball of faint old stars known as the **galactic halo**. These three basic

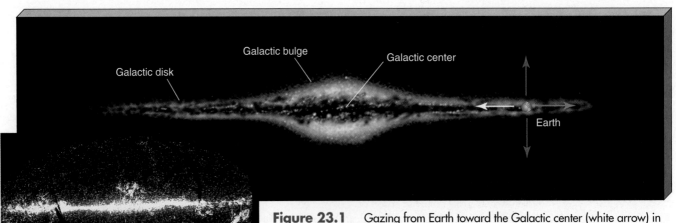

Figure 23.1 Gazing from Earth toward the Galactic center (white arrow) in this artist's conception, we see myriad stars stacked up within the thin band of light known as the Milky Way. Looking in the opposite direction (blue arrow), we still see the Milky Way band, but now it is fainter because our position is far from the Galactic center, with the result that we see more stars when looking toward the center than when looking in the opposite direction. Looking perpendicular to the disk (red arrows), we see far fewer stars. The inset is an enhanced infared satellite view of the sky all around us; the white band is the disk of our Milky Way Galaxy, which can be seen with the naked eye from very dark locations on Earth.

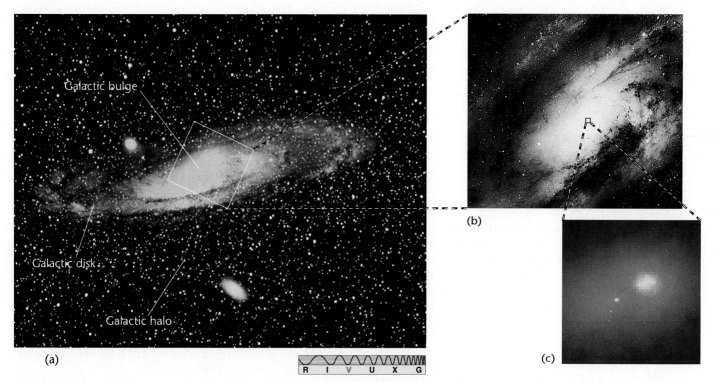

(a)

(b)

(c)

Figure 23.2 (a) The Andromeda Galaxy probably resembles fairly closely the overall layout of our own Milky Way Galaxy. The disk and bulge are clearly visible in this image, which is about 30,000 pc across. The faint halo stars cannot be seen. The white stars sprinkled all across this image are not part of Andromeda's halo; they are foreground stars in our own Galaxy, lying in the same part of the sky as Andromeda, but about a thousand times closer. (b) More detail within the inner parts of the galaxy. (c) The galaxy's peculiar—and still unexplained—double core; this inset covers a region only 15 pc across.

(a)

(b)

Figure 23.3 (a) This galaxy, catalogued as NGC 6744 and seen nearly face-on, is somewhat similar in its overall structure to our own Milky Way Galaxy and Andromeda. (b) The galaxy NGC 891 happens to be oriented in such a way that we see it edge-on, allowing us to see clearly its disk and central bulge.

galactic regions are indicated on the figure (the halo stars are so faint that they cannot be discerned here; see Figure 23.10). Figure 23.3(a) and (b) shows views of two other galaxies—one seen face-on, the other edge-on—that illustrate these points more clearly.

23.2 Measuring the Milky Way

Before the twentieth century, astronomers' conception of the cosmos differed markedly from the modern view. The fact that we live in just one of many enormous "islands" of matter separated by even larger tracts of apparently empty space was completely unknown, and the clear distinction between "our Galaxy" and "the universe" did not exist. The twin ideas that (1) the Sun is not at the center of the Galaxy and (2) the Galaxy is not at the center of the universe required both time and hard observational evidence before they gained widespread acceptance.

STAR COUNTS

In the late eighteenth century, long before the distances to any stars were known, the English astronomer William Herschel tried to estimate the size and shape of our Galaxy simply by counting how many stars he could see in different directions in the sky. Assuming that all stars were of about equal brightness, he concluded that the Galaxy was a somewhat flattened, roughly disk-shaped collection of stars lying in the plane of the Milky Way, with the Sun at its *center* (Figure 23.4). Subsequent refinements to this approach led to essentially the same picture; early in the twentieth century some workers went so far as to estimate the dimensions of this "Galaxy" as about 10 kpc in diameter by 2 kpc thick.

Today the Milky Way Galaxy is known to be several tens of kiloparsecs across, and the Sun lies far from the center. How could the older picture have been so flawed? The answer is that the earlier observations were made in the visible part of the electromagnetic spectrum, and astronomers failed to take into account the (then unknown) absorption of visible light by interstellar gas and dust. ∞ (Sec. 18.1) Only in the 1930s did astronomers begin to realize the true extent and importance of the interstellar medium.

Any objects in the Galactic disk that are more than a few kiloparsecs away from us are hidden from our view by the effects of interstellar absorption. The apparent falloff in the density of stars with distance in the plane of the Milky Way is thus not a real thinning of their numbers in space but simply a consequence of the murky environment in the Galactic disk. The long "fingers" in Herschel's map are directions in which the obscuration happens to be a little less severe than in others. However, because some obscuration occurs in all directions in the disk, the falloff is roughly similar no matter which way we look, and so the Sun appears to be more or less at the center. The horizontal extent of Figure 23.4 corresponds approximately to the span of the blue and white arrows in Figure 23.1.

In contrast, radiation coming to us from above or below the plane of the Galaxy, where there is less gas and dust along the line of sight, arrives on Earth relatively unscathed. There is still some patchy obscuration, but the Sun happens to lie in a location where the view out of the disk is largely unimpeded by nearby interstellar clouds.

SPIRAL NEBULAE AND GLOBULAR CLUSTERS

We have just seen how astronomers' attempts to probe the Galactic disk by optical means are frustrated by the effects of the interstellar medium, whereas looking in other direc-

Figure 23.4   Eighteenth-century English astronomer William Herschel constructed this "map" of the Galaxy by counting the numbers of stars he saw in different directions in the sky. He assumed that all stars were of roughly equal luminosity and, within the confines of the Galaxy, were uniformly distributed in space. Our Sun (marked by the large yellow dot) appears to lie near the center of the distribution, and the long axis of the diagram lies in the plane of the Galactic disk.

tions, out of the Milky Way plane, we can see to much greater distances. During the first quarter of the twentieth century, studies of the large-scale structure of our Galaxy focused on two particularly important classes of objects, both found mainly away from the Milky Way. The first are *globular clusters*, those tightly bound swarms of old, reddish stars we first met in Chapter 17. ∞ (Sec. 17.10) About 140 are now known. The second were known at the time as *spiral nebulae*. Examples are shown in Figures 23.2(a) and 23.3(a). We know them today as **spiral galaxies**, comparable in size to our own.

Early twentieth-century astronomers had no means of determining the distances to any of these objects. They are far too distant to have any observable parallax and, with the technology of the day, main-sequence stars (after the discovery of the main sequence in 1911) could not be clearly identified and measured. For these reasons neither of the techniques discussed in Chapter 17 was applicable. ∞ (Secs. 17.1, 17.8) As a result, even the most basic properties—size, mass, and stellar and interstellar content—of globular clusters and spiral nebulae were unknown. It was assumed that the globular clusters lay within our own Galaxy, which was thought at the time to be relatively small (using the size estimates just mentioned). The locations of the spiral nebulae were much less clear.

Knowing the distance to an object is vitally important to understanding its true nature. As an example, consider again the Andromeda "nebula" (Figure 23.2). In the late nineteenth century, when improved telescopes and photographic techniques allowed astronomers to obtain images showing detail comparable to Figure 23.2(a), the newly released photographs caused great excitement among astronomers, who thought they were seeing the formation of a star from a swirling gaseous disk! Comparing Figure 23.2(a) with the figures in Chapter 15 (see especially Figure 15.1b), we can perhaps understand how such a mistake could be made—*if* we believed we were looking at a relatively close, star-sized object. Far from demonstrating that Andromeda was distant and large, the new observations seemed to confirm that it was just a small part of our own Galaxy.

Further observations soon made it clear that Andromeda is not a star-forming region. Andromeda's parallax is too small to measure, indicating that it must be at least several hundred parsecs from Earth, and, even at 100 pc—which we now know is vastly less than Andromeda's true distance—an object the size of the solar nebula would be impossible to resolve and simply would not look like Figure 23.2(a). (For another, much more recent example of how distance measurements affect our theoretical understanding of observational data, see Section 22.3.)

During the first three decades of the twentieth century, both the size of our Galaxy and the distances to the spiral nebulae were hotly debated in astronomical circles. One school of thought maintained that the spiral nebulae were relatively small systems contained within our Galaxy. Other astronomers held that the spirals were much larger objects,

lying far outside the Milky Way Galaxy and comparable to it in size. However, with no firm distance information, both arguments were quite inconclusive. Only with the discovery of a new distance-measurement technique—which we discuss next—was the issue finally settled in favor of the latter view. However, in the process, astronomers' conception of our own Galaxy changed radically and forever.

A NEW YARDSTICK

2 An important by-product of the laborious effort to catalog stars around the turn of the twentieth century was the systematic study of **variable stars**. These are stars whose luminosity changes with time—some quite erratically, others more regularly. Only a small fraction of stars fall into this category, but those that do are of great astronomical significance.

We encountered several examples of variable stars in earlier chapters. Often, the variability is the result of membership in a binary system. Eclipsing binaries, novae, and Type I supernovae are examples. Novae and supernovae collectively are called *cataclysmic variables* because of their sudden, large changes in brightness. In other instances, however, the variability is a basic trait of a star and is not dependent on its being a part of a binary system. We call such a star an *intrinsic variable*.

A particularly important class of intrinsic variables is the class known as **pulsating variable stars**,* which vary cyclically in luminosity in very characteristic ways. Two types of pulsating variable stars that have played central roles in revealing both the true extent of our Galaxy and the distances to our galactic neighbors are the **RR Lyrae** and **Cepheid** variables. Following long-standing astronomical practice, the names come from the first star of each class to be discovered—in this case the variable star labeled RR in the constellation Lyra and the variable star Delta Cephei, the fourth brightest star in the constellation Cepheus, respectively.

RR Lyrae and Cepheid variable stars are recognizable by the characteristic shapes of their light curves. RR Lyrae stars all pulsate in essentially similar ways (Figure 23.5a), with only small differences in period between one RR Lyrae variable and another. Observed periods range from about 0.5 to 1 day. Cepheid variables also pulsate in distinctive ways (the regular "sawtooth" pattern in Figure 23.5b), but different Cepheids can have very different pulsation periods, ranging from about 1 to 100 days. The period of any given RR Lyrae or Cepheid variable is, to high accuracy, the same from one cycle to the next. The key point is that pulsating variable stars can be recognized and identified *just by observing the variations in the light they emit*.

Why do Cepheids and RR Lyrae variables pulsate? The basic mechanism was first suggested by the British astrophysicist Sir Arthur Eddington in 1941. The structure of any star is determined in large part by how easily radiation

---

Note that these stars have nothing whatsoever to do with pulsars.

can travel from the core to the photosphere—that is, by the *opacity* of the interior, the degree to which the gas hinders the passage of light through it. If the opacity rises, the radiation becomes trapped, the internal pressure increases, and the star "puffs up." If the opacity falls, radiation can escape more easily, and the star shrinks. According to theory, under certain circumstances a star can become unbalanced and enter a state in which the flow of radiation causes the opacity first to rise—making the star expand, cool, and diminish in luminosity—and then to fall, leading to the pulsations we observe.

Thus, pulsating variable stars are normal stars experiencing a brief period of instability as a natural part of stellar evolution. The conditions necessary to cause pulsations are not found in main-sequence stars; they occur in evolved post–main-sequence stars as they pass through a region of the Hertzsprung–Russell diagram known as the *instability strip* (Figure 23.6). When a star's temperature and luminosity place it in this strip, the star becomes internally unstable, and both its temperature and its radius vary in a regular way, causing the pulsations we observe: as the star brightens, its surface becomes hotter and its radius shrinks; as its luminosity decreases, the star expands and cools. As we learned in Chapter 20, high-mass stars evolve across the upper part of the H–R diagram; when their evolutionary tracks take them into the instability strip, they become Cepheid variables. ∞ (Sec. 20.4) RR Lyrae variables are low-mass horizontal-branch stars that lie within the lower portion of the instability strip. ∞ (Sec. 20.2)

The importance of these stars to Galactic astronomy lies in the fact that once we recognize a star as being of the

(a)

(b)

(c)

Figure 23.5 (a) Light curve of the pulsating variable star RR Lyrae. All RR Lyrae-type variables have essentially similar light curves, with periods of less than a day. (b) The light curve of a Cepheid variable star called WW Cygni, having a period of several days. (c) This Cepheid is shown here (boxed) on successive nights, near its maximum and minimum brightness; two photos, one from each night were superimposed and then slightly displaced.

Figure 23.6 Pulsating variable stars are found in the instability strip of the H–R diagram. As a high-mass star evolves through the strip it becomes a Cepheid variable. Low-mass horizontal-branch stars in the instability strip are RR Lyrae variables.

RR Lyrae or Cepheid type, we can infer its luminosity, and that in turn allows us to measure its distance. The distance calculation is precisely the same as presented in Chapter 17 during our discussion of spectroscopic parallax. ∞ (Sec. 17.8) Comparing the star's (known) luminosity with its (observed) apparent brightness yields an estimate of its distance, by the inverse-square law: ∞ (Sec. 17.4)

$$\text{apparent brightness} \propto \frac{\text{luminosity}}{\text{distance}^2}$$

In this way astronomers can use pulsating variables as a means of determining distances, both within our own Galaxy and far beyond.

How do we infer a variable star's luminosity? For RR Lyrae stars, this is simple. All such stars have basically the same luminosity (averaged over a complete pulsation cycle)—about 100 times that of the Sun—so once a variable star is recognized as being of the RR Lyrae type, its luminosity is immediately known. For Cepheids, we make use of a close correlation between average luminosity and pulsation period, discovered in 1908 by Henrietta Leavitt of Harvard University (see *Interlude 23-1*) and known simply as the **period–luminosity relationship**. Cepheids that vary slowly—that is, have long periods—have large luminosities; conversely, short-period Cepheids have low luminosities.

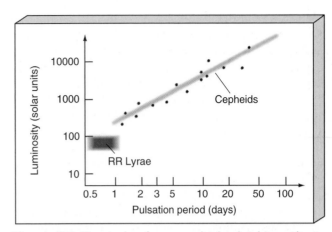

Figure 23.7 A plot of average absolute brightness (that is, luminosity) versus pulsation period for a group of Cepheid variable stars. The two properties are quite tightly correlated. The pulsation periods of some RR Lyrae variables are also shown.

Figure 23.7 illustrates the period–luminosity relationship for Cepheids found within a thousand parsecs or so of Earth. Astronomers can plot such a diagram for relatively nearby stars because they can measure the distances using stellar or spectroscopic parallax. Once distances are known, the luminosities of those stars can be calculated. We know

A large portion of the early research in observational astronomy focused on monitoring stellar luminosities and analyzing stellar spectra. Much of this pioneering work was done using photographic methods. What is not so well known is that most of the labor was accomplished by women. Around the turn of the twentieth century, a few dozen dedicated women—assistants at the Harvard College Observatory—created an enormous data base by observing, sorting, measuring, and cataloging photographic information that helped form the foundation of modern astronomy. Some of them went far beyond their duties in the lab to make several of the basic astronomical discoveries often taken for granted today.

This 1910 photograph shows several of those women carefully examining star images and measuring variations in luminosity or wavelengths of spectral lines. In the cramped quarters of the Harvard Observatory these women inspected image after image to collect a vast body of data on hundreds of thousands of stars. Note the plot of stellar luminosity changes pasted on the wall at the left. The cyclical pattern is so regular that it likely belongs to a Cepheid variable. Known as "computers" (there were no electronic devices then), these women were paid 25 cents an hour.

Beginning in 1880, these workers started a survey of the skies that would be carried on for a century. Their first major accomplishment was a catalog of the brightnesses and spectra of tens of thousands of stars, published in 1890 under the direction of Williamina Fleming. On the basis of this compilation, several of these women made fundamental con-

tributions to astronomy. In 1897 Antonia Maury undertook the most detailed study of stellar spectra to that time, enabling Hertzsprung and Russell independently to develop what is now called the H–R diagram. In 1898 Annie Cannon proposed the spectral classification system (described in Chapter 17) that is now the international standard for categorizing stars. ∞ (Sec. 17.6) In 1908 Henrietta Leavitt discovered the period–luminosity relationship for Cepheid variable stars, which later allowed astronomers to recognize our Sun's true position in our Galaxy, as well as our Galaxy's true place in the universe.

of no exceptions to the period–luminosity relationship, and it is consistent with theoretical calculations of pulsations in evolved stars. Consequently, we assume that it holds for all Cepheids, near and far. Thus, a simple measurement of a Cepheid variable's pulsation period immediately tells us its luminosity—we just read it off the plot in Figure 23.7. (The roughly constant luminosities of the RR Lyrae variables are also indicated in the figure.)

This distance-measurement technique works well provided the variable star can be clearly identified and its pulsation period measured. With Cepheids, this method allows astronomers to estimate distances out to about 15 million parsecs, enough to take us all the way to the nearest galaxies. The less luminous RR Lyrae stars are not so easily seen as Cepheids, so their useful range is not as great. However, they are much more common so, within their limited range, they are actually more useful than Cepheids.

Figure 23.8 extends our cosmic distance ladder, begun in Chapter 2 with radar ranging in the solar system and expanded in Chapter 17 to include stellar and spectroscopic parallax, by adding variable stars as a fourth method of determining distance. Note that, because the period–luminosity relationship is calibrated using nearby stars, this latest rung inherits any and all uncertainties and errors present in the lower levels. Uncertainties also arise from the "scatter" shown in Figure 23.7. Although the overall connection between period and luminosity is unmistakable, the individual data points do not quite lie on a straight line. Instead, a range of possible luminosities corresponds to any measured period.

THE SIZE AND SHAPE OF OUR GALAXY

Many RR Lyrae variables are found in globular clusters. Early in the twentieth century the American astronomer Harlow Shapley used observations of RR Lyrae stars to make two very important discoveries about the Galactic globular cluster system. First, he showed that most globular clusters reside at great distances—many thousands of parsecs—from the Sun. Second, by measuring the direction and distance of each cluster, he was able to determine their three-dimensional distribution in space. In this way Shapley demonstrated that the globular clusters map out a truly gigantic, and roughly *spherical*, volume of space, about 30 kpc across.* However, the center of the distribution lies nowhere near our Sun. It is located nearly 8 kpc away from us, in the direction of the constellation Sagittarius.

In a brilliant intellectual leap, Shapley realized that the distribution of globular clusters maps out the true extent of stars in the Milky Way Galaxy—the region that we now call the Galactic halo. (Since Shapley's time astronomers have identified many individual halo stars that do not belong to any globular cluster.) The hub of this vast collection of mat-

The Galactic globular cluster system and the Galactic halo of which it is a part are somewhat flattened in the direction perpendicular to the disk, but the degree of flattening is quite uncertain. The halo is certainly much less flattened than the disk, however.

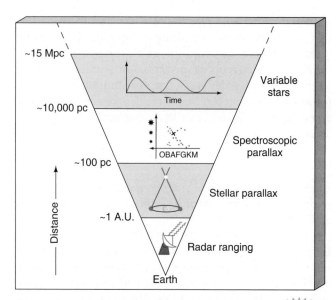

Figure 23.8 Application of the period–luminosity relationship for Cepheid variable stars allows us to determine distances out to about 15 Mpc with reasonable accuracy.

Animation

ter, 8 kpc from the Sun, is the **Galactic center**. As illustrated in Figure 23.9, we live in the suburbs of this huge ensemble, in the Galactic disk—the thin sheet of young stars, gas, and dust that cuts through the center of the halo.

Shapley's bold interpretation of the globular clusters as defining the overall distribution of stars in our Galaxy was an

Figure 23.9 Our Sun does not coincide with the center of the very large collection of globular clusters. Instead, more globular clusters are found in one direction than in any other. The Sun resides closer to the edge of the collection, which measures roughly 30 kpc across. We now know that the globular clusters outline the true distribution of stars in the Galactic halo.

enormous step forward in human understanding of our place in the universe. Five hundred years ago, Earth was considered the center of all things. Copernicus argued otherwise, demoting our planet to an undistinguished place removed from the center of the solar system. In Shapley's time, as we have just seen, the prevailing view was that our Sun was the center not only of the Galaxy but also of the universe. Shapley showed otherwise. With his observations of globular clusters he simultaneously increased the size of our Galaxy by almost a factor of 10 over earlier estimates and banished our parent Sun to its periphery, virtually overnight!

Curiously, Shapley's dramatic revision of the size of the Milky Way Galaxy and our place in it only strengthened his erroneous opinion that the spiral nebulae were part of our Galaxy and that our Galaxy was essentially the entire universe. He regarded as beyond belief the idea that there could be other structures as large as our Galaxy. Only in the late 1920s was the Copernican principle extended to the Galaxy itself, when American astronomer Edwin Hubble observed Cepheids in the Andromeda Galaxy and finally succeeded in measuring its distance.

23.3 The Large-Scale Structure of Our Galaxy

THE SPATIAL DISTRIBUTION OF STARS

Figure 23.10 illustrates the very different spatial distributions of the disk and halo components of the Milky Way Galaxy. As just mentioned, the Sun lies about 8 kpc from the Galactic center. Based on optical, infrared, and radio observations of stars, gas, and dust found within a thousand or so parsecs of the Sun, astronomers estimate that the disk in the vicinity of the Sun is relatively thin—"only" 300 pc thick, or about 1/100 of the Galactic diameter. Don't be fooled, though. Even if you could travel at the speed of light, it would take you 1000 years to traverse the thickness of the Galactic disk. The disk may be thin compared with the Galactic diameter, but it is huge by human standards.

Actually, the thickness of the Galactic disk depends on the kinds of objects measured. Young stars and interstellar gas are more tightly confined to the plane than are stars like the Sun, and solar-type stars in turn are more tightly confined than are older K- and M-type dwarfs. The reason for this is that stars form in interstellar clouds close to the disk plane but then tend to drift out of the disk over time, mainly due to their interactions with other stars and molecular clouds. Thus, as stars age, their abundance above and below the disk plane slowly increases. Note that these considerations do not apply to the Galactic halo, whose ancient stars and globular clusters extend far above and below the Galactic plane. As we will see in a moment, the halo is a remnant of an early stage of our Galaxy's evolution and predates the formation of the disk.

Recently, improved observational techniques have revealed an intermediate category of Galactic stars, midway between the old halo and the younger disk, both in age and in spatial distribution. Consisting of stars with estimated ages in the range of 7–10 billion years, this **thick disk** component of the Milky Way Galaxy measures some 2–3 kpc from top to bottom. Its thickness is also too great to be explained by the slow drift just described. Like the halo, it appears to be a vestige of our Galaxy's distant past.

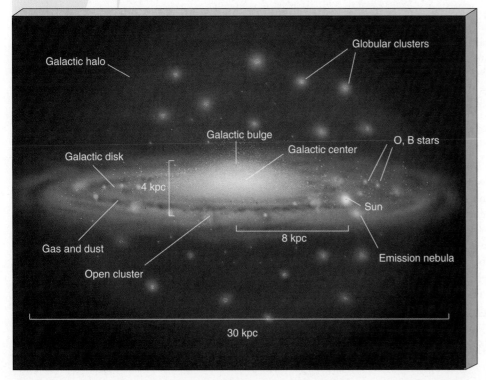

Figure 23.10 Artist's conception of a (nearly) edge-on view of the Milky Way Galaxy, showing the distributions of young blue stars, open clusters, old red stars, and globular clusters.

R I V U X G

Figure 23.11   A wide-angle far-infrared image of the disk and bulge of the Milky Way Galaxy, as observed by the *Cosmic Background Explorer (COBE)* satellite.

Also shown in Figure 23.10 is our Galaxy's central bulge, measuring roughly 6 kpc across in the plane of the Galactic disk by 4 kpc perpendicular to the disk plane. Obscuration by interstellar dust makes it difficult to study the detailed structure of the Galactic bulge in optical images of the Milky Way (see, for example, Figure 18.5, which would clearly show a large portion of the bulge were it not for interstellar absorption). However, at longer wavelengths, which are less affected by interstellar matter, a much clearer picture emerges (Figure 23.11; compare with Figure 23.3b). Detailed measurements of the motion of gas and stars in and near the bulge imply that it is football shaped, about half as wide as it is long, with the long axis of the football lying in the Galactic plane.

STELLAR POPULATIONS

Aside from their shapes, the three components of the Galaxy—disk, bulge, and halo—have several other properties that distinguish them from one another. First, the halo contains essentially *no* gas or dust—just the opposite of the disk and bulge, in which interstellar matter is common. Second, there are clear differences in both *appearance* and *composition* between disk, bulge, and halo stars. Stars in the Galactic bulge and halo are found to be distinctly *redder* than stars found in the disk. Observations of other spiral galaxies also show this trend—the blue-white tint of the disk and the yellowish coloration of the bulge are evident in Figures 23.2(a) and 23.3(a).

All the bright, blue stars visible in our sky are part of the Galactic disk, as are the young open star clusters and star-forming regions. In contrast, the cooler, redder stars—including those found in the old globular clusters—are more uniformly distributed throughout the disk, bulge, and halo. Galactic disks appear bluish because main-sequence O and B blue supergiants are very much brighter than G, K, and M dwarfs, even though the dwarfs are present in far greater numbers.

The explanation for the marked difference in stellar content between disk and halo is this: whereas the gas-rich Galactic disk is the site of ongoing star formation and so contains stars of all ages, all the stars in the Galactic halo are *old*. The absence of dust and gas in the halo means that no new stars are forming there, and star formation apparently ceased long ago—at least 10 billion years in the past, judging from the types of halo stars we now observe. (Recall from Chapter 20 that most globular clusters are thought to be between 10 and 12 billion years old.) ∞ (Sec. 20.5) The gas density is very high in the inner part of the Galactic bulge, making this region the site of vigorous ongoing star formation, and both very old and very young stars mingle there. The bulge's gas-poor outer regions have properties more similar to those of the halo.

Support for this picture comes from studies of the spectra of halo stars, which indicate that these stars are far less abundant in heavy elements (that is, elements heavier than helium) than are nearby stars in the disk. In Chapter 21 we saw how each successive cycle of star formation and evolution enriches the interstellar medium with the products of stellar nucleosynthesis, leading to a steady increase in heavy elements with time. ∞ (Sec. 21.5) Thus, the scarcity of these elements in halo stars is consistent with the view that the halo formed long ago.

Astronomers often refer to young disk stars as *Population I* stars and to old halo stars as *Population II* stars. The idea of two stellar "populations" dates from the 1930s, when the differences between disk and halo stars first became clear. The names are something of an oversimplification, as there is actually a continuous variation in stellar ages throughout the Milky Way Galaxy, not a simple division of stars into two distinct "young" and "old" categories. Nevertheless, the terminology is still widely used.

ORBITAL MOTION

3 Now let's turn our attention to the *dynamics* of the Milky Way Galaxy—that is, to the motion of the stars, dust, and gas it contains. Are the internal motions of our Galaxy's members chaotic and random, or are they part of some gigantic "traffic pattern"? The answer depends on our perspective. The motion of stars and clouds we see on small scales (within a few tens of parsecs of the Sun) seems random, but on larger scales (hundreds or thousands of parsecs) the motion is much more orderly.

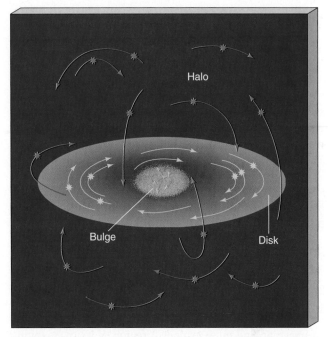

Figure 23.12 Stars and interstellar clouds in the neighborhood of the Sun show systematic Doppler motions. This information tells us that the disk of the Galaxy is spinning in a well-ordered way. These four Galactic quadrants are drawn (as dashed lines) to intersect at the Sun, not at the Galactic center, because it is from the viewpoint of our own planetary system that the observations are made. The longer the arrow, the greater the angular speed of the disk material. Because the Sun orbits faster than stars and gas at larger radii, we are drawing away from material at top left and gaining on that at top right, resulting in the Doppler shifts indicated. Similarly, stars and gas in the bottom left quadrant are gaining on us, while material at bottom right is pulling away.

Figure 23.13 Stars in the Galactic disk move in orderly, circular orbits about the Galactic center. In contrast, halo stars have orbits with largely random orientations and eccentricities. The orbit of a typical halo star takes it high above the Galactic disk, then down through the disk plane, then out the other side and far below the disk. The orbital properties of bulge stars are intermediate between those of disk stars and those of halo stars.

As we look around the Galactic disk in different directions, a clear pattern of motion emerges (Figure 23.12). Radiation received from stars and interstellar gas clouds in the upper right and the lower left quadrants of Figure 23.12 is generally *blueshifted*. At the same time, the interstellar regions sampled in the upper left quadrant and the lower right quadrant are *redshifted*. In other words, some regions of the Galaxy (in the blueshifted directions) are approaching the Sun, while others (the redshifted ones) are receding from us.

Careful study of the data leads to the conclusion that the entire Galactic disk is *rotating* about the Galactic center. In the vicinity of the Sun, the orbital speed is about 220 km/s. At the Sun's distance of 8 kpc from the Galactic center, material takes about 225 million years—an interval of time sometimes known as 1 *Galactic year*—to complete one circuit. At other distances from the center the rotation period is different—shorter closer to the center, longer at greater distances—that is, the Galactic disk rotates not as a solid object but *differentially*. Similar differential rotation is observed in Andromeda (and, in fact, in all other spiral galaxies).

This picture of orderly circular orbital motion about the Galactic center applies only to the Galactic disk. Stars in the Galactic halo and bulge are not so well behaved. The old globular clusters in the halo and the faint, reddish individual stars in both the halo and the bulge do *not* share the disk's well-defined rotation. Instead, their orbits are largely random.* Although they do orbit the Galactic center, they move in all directions, their paths filling an entire three-dimensional volume rather than a nearly two-dimensional disk. At any given distance from the Galactic center, bulge or halo stars move at speeds comparable to the disk's rotation speed at that radius but in *all* directions, not just one. Their orbits carry these stars repeatedly through the disk plane and out the other side. Figure 23.13 contrasts the motion of bulge and halo stars with the much more regular orbits of stars in the Galactic disk. Some well-known stars in the vicinity of the Sun—the bright giant Arcturus, for example—are actually halo stars that are "just passing through" the disk on orbits that take them far above and below the Galactic plane.

Halo stars do, in fact, have some net rotation about the Galactic center, but the rotational component of their motion is overwhelmed by the larger random component. The motion of bulge stars also has a rotational component, larger than that of the halo but still smaller than the random component of stellar motion in the bulge.

23.4 The Formation of the Milky Way Galaxy

3 Table 23.1 compares some key properties of the three basic components of the Galaxy. Is there some evolutionary scenario that can naturally account for the Galactic structure we see today? The answer is that there is, and it takes us all the way back to the birth of our Galaxy, 10–15 billion years ago. Not all the details are agreed upon by all astronomers, but the overall picture is now fairly widely accepted. For simplicity we confine our discussion here to the Galactic disk and halo. In many ways the bulge is intermediate in its properties between these two extremes.

Figure 23.14 illustrates the current view of our Galaxy's evolution, starting (not unlike the star-formation scenario outlined in Chapter 19) from a contracting cloud of pre-galactic gas. ∞ (Sec. 19.1) When the first Galactic stars and globular clusters formed, the gas in our Galaxy had not yet accumulated into a thin disk. Instead, it was spread out over an irregular, and quite extended, region of space, spanning many tens of kiloparsecs in all directions. When the first stars formed, they were distributed throughout this volume. Their distribution today (the Galactic halo) reflects that fact—it is an imprint of their birth. Many astronomers believe that the very first stars formed even earlier, in smaller systems that later merged to create our Galaxy (Figure 23.14a); the present-day halo would look the same in either case.

During the past 10–15 billion years, rotation has flattened the gas in our Galaxy into a relatively thin disk. Physically, this process is similar to the flattening of the solar nebula during the formation of the solar system, as described in Chapter 15, except on a vastly larger scale. ∞ (Sec. 15.2) Star formation in the halo ceased billions of years ago when the raw materials fell to the Galactic plane. Ongoing star formation in the disk gives it its bluish tint, but the halo's short-lived blue stars have long since burned out, leaving only the long-lived red stars that give it its characteristic pinkish glow. The Galactic halo is ancient, whereas the disk is full of youthful activity. The thick disk, with its intermediate-age stars, may represent an intermediate stage of star formation that occurred while the gas was still flattening into the plane.

The chaotic orbits of the halo stars are also explained by this theory. When the halo developed, the irregularly shaped Galaxy was rotating only very slowly, so there was no strongly preferred direction in which matter tended to move. As a result, halo stars were free to travel along nearly any path once they formed (or when their parent systems merged). As the Galactic disk formed, however, conservation of angular momentum caused it to spin more rapidly. Stars forming from the gas and dust of the disk inherit its rotational motion and so move on well-defined, circular orbits. Again, the thick disk's properties suggest that it formed while gas was still sinking to the Galaxy's midplane and had not yet reached its final (present-day) rotation rate.

In principle, the structure of our Galaxy bears witness to the conditions that created it. In practice, however, the interpretation of the observations is made difficult by the sheer complexity of the system we inhabit and by the many competing physical processes that have modified its appearance since it formed. As a result, the early stages of the Milky Way are still very poorly understood. We will return to the subject of galaxy formation in Chapter 24.

23.5 Spiral Structure

RADIO MAPS OF THE MILKY WAY

If we want to look beyond our immediate neighborhood and study the full extent of the Galactic disk, we cannot rely on optical observations, as interstellar absorption severely

---

TABLE 23.1 Overall Properties of the Galactic Disk, Halo, and Bulge

GALACTIC DISK	GALACTIC HALO	GALACTIC BULGE
Highly flattened	Roughly spherical—mildly flattened	Somewhat flattened and elongated in the plane of the disk ("football shaped")
Contains both young and old stars	Contains old stars only	Contains both young and old stars; more old stars at greater distances from the center
Contains gas and dust	Contains no gas and dust	Contains gas and dust, especially in the inner regions
Site of ongoing star formation	No star formation during the last 10 billion years	Ongoing star formation in the inner regions
Gas and stars move in circular orbits in the Galactic plane	Stars have random orbits in three dimensions	Stars have largely random orbits but with some net rotation about the Galactic center
Spiral arms	No obvious substructure	Ring of gas and dust near center; Galactic nucleus
Overall white coloration, with blue spiral arms	Reddish in color	Yellow-white

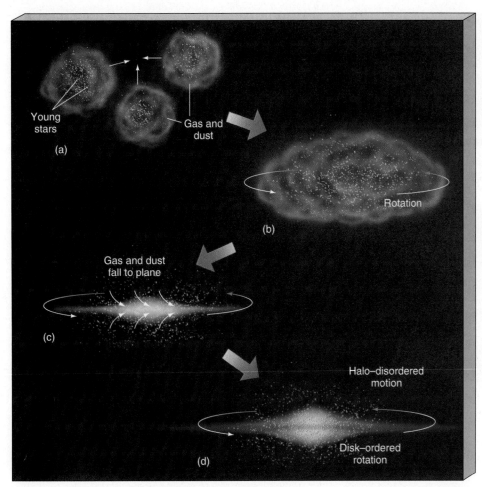

Figure 23.14 (a) The Milky Way Galaxy possibly formed via the merger of several smaller systems. (b) Astronomers reason that, early on, our Galaxy was irregularly shaped, with gas distributed throughout its volume. When stars formed during this stage, there was no preferred direction in which they moved and no preferred location in which they were found. (c) In time, rotation caused the gas and dust to fall to the Galactic plane and form a spinning disk. The stars that had already formed were left behind, forming the halo. (d) New stars forming in the disk inherit its overall rotation and so orbit the Galactic center on ordered, circular orbits.

limits our vision. In the 1950s, astronomers developed a very important tool to explore the distribution of gas in our Galaxy—spectroscopic radio astronomy.

The keys to observing Galactic interstellar gas are the 21-cm radio emission line produced by atomic hydrogen and the many radio molecular lines formed in molecular cloud complexes. ∞ (Sec. 18.4) Long-wavelength radio waves are largely unaffected by interstellar dust, so they travel more or less unimpeded through the Galactic disk, allowing us to "see" to great distances. Because hydrogen is by far the most abundant element in interstellar space, the 21-cm signals are strong enough that a large portion of the disk can be observed in this way. As noted in Chapter 18, observations of spectral lines from "tracer" molecules, such as carbon monoxide, allow us to study the distribution of the densest interstellar clouds. ∞ (Sec. 18.5)

Interstellar gas in the Galactic disk exhibits an organized pattern on a grand scale. According to radio studies, the center of the gas distribution coincides roughly with the center of the globular-cluster system, about 8 kpc from the Sun. (In fact, this figure is derived most accurately from radio observations of the distribution of Galactic gas, the center of which is normally taken to define the center of our Galaxy.) Near the center the gas in the disk fattens markedly in the Galactic bulge. Radio-emitting gas has been observed out to at least 50 kpc from the Galactic center.

Over much of the inner 20 kpc or so of the disk the gas is confined within about 100 pc of the Galactic plane. Beyond that distance the gas distribution spreads out somewhat, to a thickness of several kiloparsecs, and shows definite signs of being "warped," possibly because of the gravitational influence of a pair of nearby galaxies (to be discussed in Chapter 24; see also Figure 23.15).

Radio studies provide perhaps the best direct evidence that we live in a spiral galaxy. Figure 23.15 is an artist's conception (based on observational data) of the appearance of our Galaxy as seen from far above the disk, clearly showing our Galaxy's **spiral arms**, pinwheel-like structures originating close to the Galactic bulge and extending outward throughout much of the Galactic disk. One of these arms, as best we can tell, wraps around a large part of the disk and contains our Sun. Notice, incidentally, the scale markers on Figures 23.9, 23.10 and 23.15: the Galactic globular-cluster distribution (Figure 23.9), the luminous stellar component of the disk (Figure 23.10), and the known spiral structure (Figure 23.15) all have roughly the *same* diameter—about 30 kpc.

PERSISTENCE OF THE SPIRAL ARMS

4 The spiral arms in our Galaxy are made up of much more than just interstellar gas and dust. Studies of the Galactic disk within a kiloparsec or so of the Sun indicate that

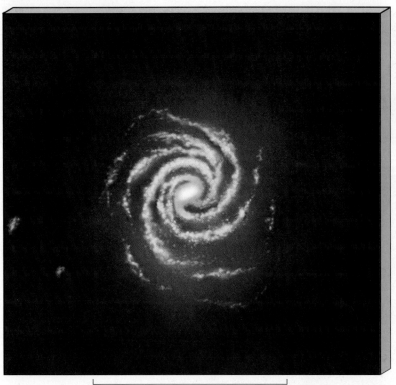

Figure 23.15 An artist's conception of our Milky Way Galaxy seen face-on. This illustration is based on data accumulated by legions of astronomers during the past few decades, including radio maps of gas density in the Galactic disk. Painted from the perspective of an observer 100 kpc above the Galactic plane, the spiral arms are at their best-determined positions. All the features are drawn to scale (except for the oversized yellow dot near the top, which represents our Sun). The two small blotches to the left are dwarf galaxies, called the Magellanic Clouds. We will study them in Chapter 24.

30 kpc

young stellar and prestellar objects—emission nebulae, O and B stars, and recently formed open clusters—are also distributed in a spiral pattern that closely follows the distribution of interstellar clouds. The obvious conclusion is that the spiral arms are the part of the Galactic disk where star formation takes place. The brightness of the young stellar objects just listed is the main reason that the spiral arms of other galaxies are easily seen from afar (Figure 23.3a).

A central problem facing astronomers trying to understand spiral structure is how that structure persists over long periods of time. The basic issue is simple: we know that the inner parts of the Galactic disk rotate more rapidly than do the outer regions. Thus stars in the Galactic disk do not move smoothly together, but ceaselessly change their positions relative to one another as they orbit the Galactic center. This differential rotation makes it impossible for any large-scale structure "tied" to the disk material to survive. Figure 23.16 shows how a spiral pattern consisting always of the same group of stars and gas clouds would necessarily "wind up" and disappear within a few hundred million years. Yet spiral arms clearly do exist in our own galaxy, and their prevalence in other disk galaxies suggests that they last for considerably longer than this. Thus, whatever the spiral arms are, they *cannot* simply be dense star-forming regions orbiting along with the rest of the Galactic disk.

How then do the Galaxy's spiral arms retain their structure over long periods of time in spite of differential rotation? A leading explanation for the existence of spiral arms holds that they are **spiral density waves**—coiled waves of gas compression that move through the Galactic disk,

squeezing clouds of interstellar gas and triggering the process of star formation as they go. ∞ (Sec. 19.5) The spiral arms we observe are defined by the denser-than-normal clouds of gas the density waves create and by the new stars formed as a result of the spiral waves' passage.

This explanation of spiral structure avoids the problem of differential rotation because the wave pattern is not tied to any particular piece of the Galactic disk. The spirals we see are merely patterns moving through the disk, not great masses of matter being transported from place to place. The density wave moves through the collection of stars and gas making up the disk just as a sound wave moves through air or an ocean wave passes through water, compressing different parts of the disk at different times. Even though the rotation rate of the disk material varies with distance from the Galactic center, the wave itself remains intact, defining the Galaxy's spiral arms.

In fact, over much of the visible portion of the Galactic disk (within about 15 kpc of the center), the spiral wave pattern is predicted to rotate *more slowly* than the stars and gas. Thus, Galactic material catches up with the wave, is temporarily slowed down and compressed as it passes through, then continues on its way. (For a more down-to-earth example of an analogous process, see *Interlude 23-2*.)

As shown in Figure 23.17, the slowly moving spiral density wave is outrun by the faster-moving disk material. As gas enters the arm from behind, the gas is compressed and forms stars. Dust lanes mark the regions of highest-density gas. The most prominent stars—the bright O and B blue giants—live for only a short time, so young stellar

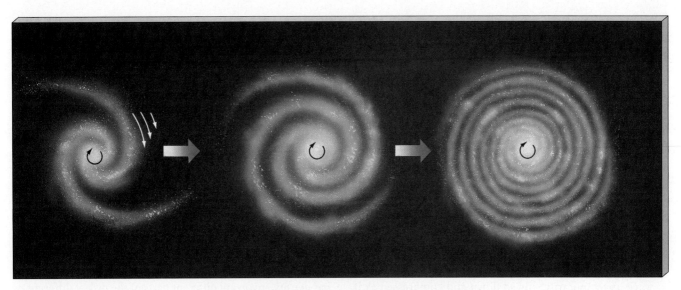

Figure 23.16   The disk of our Galaxy rotates differentially—stars close to the center take less time to orbit the Galactic center than those at greater distances. If spiral arms were somehow tied to the material of the Galactic disk, this differential rotation would cause the spiral pattern to wind up and disappear in a few hundred million years. Spiral arms would be too short-lived to be consistent with the numbers of spiral galaxies we observe today.

associations, emission nebulae, and open clusters with long main sequences are found only within the arms, near their birth sites, just ahead of the dust lanes. The brightness of these young systems emphasizes the spiral structure. Farther downstream, ahead of the spiral arms, we see mostly older stars and star clusters. These have had enough time

since their formation to outdistance the wave and pull away from it. Over millions of years their random individual motions, superimposed on the overall rotation around the Galactic center, distort and eventually destroy their original spiral configuration, and they become part of the general disk population.

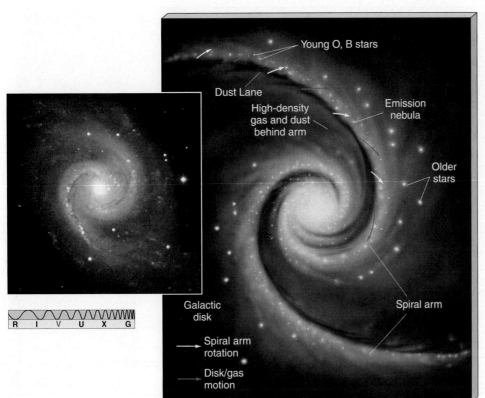

Figure 23.17   Density-wave theory holds that the spiral arms seen in our own and many other galaxies are waves of gas compression and star formation moving through the material of the galactic disk. In the painting at left, gas motion is indicated by red arrows, and arm motion is indicated by white arrows. Gas enters an arm from behind, is compressed, and forms stars. The spiral pattern is delineated by dust lanes, regions of high gas density, and newly formed O and B stars. The inset shows the spiral galaxy NGC 1566, which displays many of the features described in the text. Note, incidentally, that although both spirals here have two arms each, astronomers are not completely certain how many arms make up the spiral structure in our own Galaxy (see Figure 23.15). The theory makes no strong predictions on this point.

INTERLUDE 23-2 Density Waves

In the late 1960s American astrophysicists C. C. Lin and Frank Shu proposed a way in which spiral arms in the Galaxy could persist for many Galactic rotations. They argued that the arms themselves contain no "permanent" matter. They should not be viewed as assemblages of stars, gas, and dust moving intact through the disk—those would quickly be destroyed by differential rotation. Instead, a spiral arm should be envisaged as a *density wave*—a wave of compression and expansion sweeping through the Galaxy.

A wave in water builds up material temporarily in some places (crests) and lets it down in others (troughs). Similarly, as the spiral density wave encounters Galactic matter, the gas is compressed to form a region of slightly higher than normal density. Galactic material enters the wave, is temporarily slowed down and compressed as it passes through, then continues on its way. This compression triggers the formation of new stars and nebulae. In this way the spiral arms are formed and re-formed repeatedly, without wrapping up. Lin and Shu showed that the process can in fact maintain a spiral pattern for very long periods of time.

The accompanying figure illustrates the formation of a density wave in a much more familiar context—a traffic jam on a highway, triggered by the presence of a repair crew moving slowly down the road. As cars approach the crew they slow down temporarily, then speed up again as they pass the worksite and continue on their way. The result, as might be reported by a high-flying traffic heli-copter, is a region of high traffic density, concentrated around the location of the work crew and moving with it. An observer on the side of the road, however, sees that the jam never contains the same cars for very long. Cars constantly catch up to the bottleneck, move slowly through it, then speed up again, only to be replaced by more cars arriving from behind.

The traffic jam is analogous to the region of high stellar density in a Galactic spiral arm. Just as the traffic density wave is not tied to any particular group of cars, the spiral arms are not attached to any particular piece of disk material. Stars and gas enter a spiral arm, slow down for a while, then continue on their orbits around the Galactic center. The result is a moving region of high stellar and gas density, involving different parts of the disk at different times. Notice also that, just as in our Galaxy, the wave moves more slowly than, and independently of, the overall traffic flow.

We can extend our traffic analogy a little further. Most drivers are well aware that the effects of a such a tie-up can persist long after the road crew responsible for it has stopped work and gone home for the night. Similarly, spiral density waves can continue to move through the disk even after the disturbance that originally produced them has long since subsided. According to spiral density wave theory, this is precisely what has happened in the Milky Way. Some disturbance in the past produced the wave, which has been moving through the Galactic disk ever since.

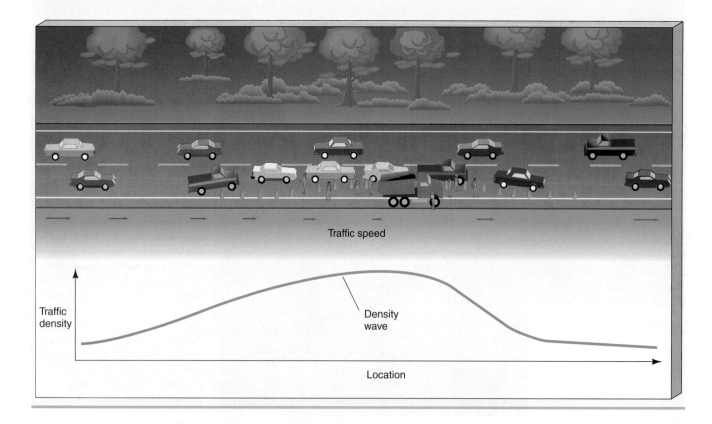

Figure 23.18 (a) Self-propagating star formation. In this theory of the formation of spiral arms, the shock waves produced by the formation and later evolution of a group of stars provide the trigger for new rounds of star formation. We have used supernova explosions to illustrate the point here, but the formation of emission nebulae and planetary nebulae are also important. (b) This process may well be responsible for the partial spiral arms seen in some galaxies, such as NGC 300, shown here in true color. The distinct blue appearance derives from the vast numbers of young stars that pepper its ill-defined spiral arms.

(a)

(b)

An alternative possibility is that the formation of stars drives the waves, instead of the other way around. Imagine a row of newly formed massive stars somewhere in the disk. As these stars form, the emission nebulae that appear around them send shock waves through the surrounding gas, possibly triggering new star formation. Similarly, when the stars explode in supernovae, more shocks are formed. ∞ (Sec. 21.2) As illustrated in Figure 23.18(a), the formation of one group of stars thus provides the mechanism for the creation of more stars. Computer simulations suggest that it is possible for the "wave" of star formation thus created to take on the form of a partial spiral and for this pattern to persist for some time. However, this process, sometimes known as *self-propagating star formation*, can produce only pieces of spirals, as are seen in some galaxies (Figure 23.18b). It apparently cannot produce the galaxywide spiral arms seen in other galaxies and present in our own. It may well be that there is more than one process at work in the spectacular spirals we see.

An important question (but one that unfortunately is not answered by either of the two theories just described) is: Where do these spirals come from? What was responsible for generating the density wave in the first place or for creating the line of newborn stars whose evolution drives the advancing spiral arm? Scientists speculate that (1) the gravitational effects of our satellite galaxies (the Magellanic Clouds, to be discussed in Chapter 24), (2) instabilities in the gas near the Galactic bulge, or (3) the possible barlike asymmetry within the bulge itself may have had a big enough influence on the disk to get the process going.

The first possibility is supported by growing evidence that many other spiral galaxies seem to have experienced gravitational interactions with neighboring systems in the relatively recent past (see Chapter 24). However, many astronomers still regard the other two possibilities as equally likely. For example, they point to *isolated* spirals, whose structure clearly cannot be the result of an external interaction. The fact is that we still don't know for sure how galaxies— including our own—acquire such beautiful spiral arms.

23.6 The Mass of the Milky Way Galaxy

5 We can measure our Galaxy's mass by studying the motions of gas clouds and stars in the Galactic disk. Recall from Chapter 2 that Kepler's third law (as modified by Newton) connects the orbital period, orbit size, and masses of any two objects in orbit around each other: ∞ (Sec. 2.7)

$$\text{total mass (solar masses)} = \frac{\text{orbit size (A.U.)}^3}{\text{orbit period (years)}^2}.$$

As we saw earlier, the distance from the Sun to the Galactic center is about 8 kpc, and the Sun's orbital period is 225 million years. Substituting these numbers into the preceding equation, we obtain a mass of almost 10^{11} solar masses—100 *billion* times the mass of our Sun. The Milky Way Galaxy is truly enormous, in mass as well as in size.

But what mass have we just measured? When we performed the analogous calculation in the case of a planet orbiting the Sun (Section 1.7), there was no ambiguity: neglecting the planet's mass, the result of our calculation was the mass of the Sun. However, the Galaxy's matter is not concentrated at the Galactic center (as the Sun's mass is concentrated at the center of the solar system); instead, Galactic matter is distributed over a large volume of space. Some of it lies inside the Sun's orbit (that is, within 8 kpc of the Galactic center), and some lies outside, at large distances from both the Sun and the center of the Galaxy. What portion of the Galaxy's mass controls the Sun's orbit? Isaac Newton answered this question three centuries ago: the Sun's orbital period is determined by the portion of the Galaxy that lies *within the orbit of the Sun*. This is the mass computed from the foregoing equation.

DARK MATTER

To determine the mass of the Galaxy on larger scales—that is, to find how much matter is contained within spheres of progressively larger radii—we must measure the orbital motion of stars and gas farther from the Galactic center than is the Sun. Astronomers have found that the most effective way to do this is to make radio observations of gas in the Galactic disk, because radio waves are relatively unaffected by interstellar absorption and allow us to probe to great distances, far beyond the Sun's orbit. On the basis of these studies, radio astronomers have determined our Galaxy's rotation rate at various distances from the Galactic center. The resultant plot of rotation speed versus distance from the center (Figure 23.19) is called the Galactic **rotation curve.**

Knowing the Galactic rotation curve, we can now repeat our earlier calculation to compute the total mass that lies within any given distance from the Galactic center. We find, for example, that the mass within about 15 kpc from the center—the volume defined by the globular clusters and the known spiral structure—is roughly 2×10^{11} solar masses, about twice the mass contained within the Sun's orbit. Does the distribution of matter in the Galaxy "cut off" at this point, where the luminosity drops off sharply? Surprisingly, it does not.

If most of the matter in the Galaxy were contained within the edge of the visible structure, Newton's laws of motion predict that the orbital speed of stars and gas beyond 15 kpc would decrease with increasing distance from the Galactic center, just as the orbital speeds of the planets diminish as we move outward from the Sun. The dashed line in Figure 23.19 indicates what the rotation curve would look like in that case. However, the true rotation curve is quite different. Far from falling off at larger distances, it *rises* slightly out to the limits of our measurement capabilities. This implies that the amount of mass contained within successively larger radii continues to grow beyond the orbit of the Sun, apparently out to a distance of at least 40 or 50 kpc. According to the preceding equation, the amount of mass within 40 kpc is approximately 6×10^{11} solar masses. Since 2×10^{11} solar masses lie within 15 kpc of the Galactic center, we have to conclude that at least twice as much mass lies *outside* the luminous part of our

Figure 23.19 The rotation curve for the Milky Way Galaxy plots rotation speed versus distance from the Galactic center. We can use this curve to compute the mass of the Galaxy that lies within any given radius. The dashed curve is the rotation curve expected if the Galaxy "ended" abruptly at a radius of 15 kpc, the limit of most of the known spiral structure and the globular cluster distribution. The fact that the red curve does not follow this dashed line, but instead stays well above it, indicates that there must be additional matter beyond that radius.

R I V U X G

Figure 23.20 Very sensitive visible observations with the *Hubble Space Telescope* have apparently ruled out faint red-dwarf stars as candidates for dark matter. The object shown here, the globular cluster NGC 6397, is one of many regions searched in the Milky Way. The inset, 0.4 pc on a side, shows a high-resolution *Hubble* view. The scores of diamonds have been overlaid at positions where red dwarfs might (statistically) have been expected if they did indeed make up the dark matter, but they were not found.

galaxy—the part made up of stars, star clusters, and spiral arms—as lies inside!

Based on these observations of the Galactic rotation curve, astronomers now believe that the luminous portion of the Milky Way Galaxy—the region outlined by the globular clusters and by the spiral arms—is merely the "tip of the Galactic iceberg." Our Galaxy is in reality very much larger. The luminous region is surrounded by an extensive, invisible **dark halo**, which dwarfs the inner halo of stars and globular clusters and extends well beyond the 15-kpc radius once thought to represent the limit of our Galaxy. But what is the composition of this dark halo? We do not detect enough stars or interstellar matter to account for the mass that our computations tell us must be there. We are inescapably drawn to the conclusion that most of the mass in our Galaxy exists in the form of invisible **dark matter**, which we presently simply do not understand.

The term *dark* here does not refer just to matter undetectable in visible light. The material has (so far) escaped detection at *all* wavelengths, from radio to gamma rays. Only by its gravitational pull do we know of its existence. Dark matter is not hydrogen gas (atomic or molecular), nor is it made up of ordinary stars. Given the amount of matter that must be accounted for, we would have been able to detect it with present-day equipment if it were in either of those forms. Its nature and its consequences for the evolution of galaxies and the universe are among the most important questions in astronomy today.

Many candidates have been suggested for this dark matter, although none is proven. Black holes may supply some of the unseen mass, although their very existence is still debated, and very few candidates exist. ∞ (Sec. 22.7) However, given that black holes are the evolutionary products of (relatively rare) massive stars, it is unlikely that

there could be enough of them to hide large amounts of Galactic matter. Currently among the strongest "stellar" contenders are brown dwarfs—low-mass prestellar objects that never reached the point of core nuclear burning—white dwarfs, and faint, low-mass red dwarfs. ∞ (Secs. 19.3, 20.3) These objects could in principle exist in great numbers throughout the Galaxy yet would be exceedingly hard to see.

Recent *Hubble Space Telescope* observations of globular clusters seem to argue against at least the last of these possibilities. Figure 23.20 shows a *Hubble* image of a relatively nearby globular cluster—one close enough that very faint red dwarfs could have been detected if any existed. The *Hubble* data suggest that there is a cutoff at about 0.2 solar masses, below which stars form much less frequently than had previously been supposed. As a result, very-low-mass stars seem to be unexpectedly rare in our Galaxy.

A radically different alternative is that the dark matter is made up of exotic *subatomic particles* that pervade the entire universe. Many theoretical astrophysicists believe that these particles could have been produced in abundance during the very earliest moments of our universe. If the particles have survived to the present day, there might be enough of them to account for all the dark matter we believe must be out there. This idea is hard to test, however, because any particles of this nature that might exist would be very hard to detect. Several detection experiments have been attempted, so far without success.

THE SEARCH FOR STELLAR DARK MATTER

Recently, researchers have obtained insight into the distribution of stellar dark matter by using a key element of Albert Einstein's theory of general relativity (see Chapter 22)—the

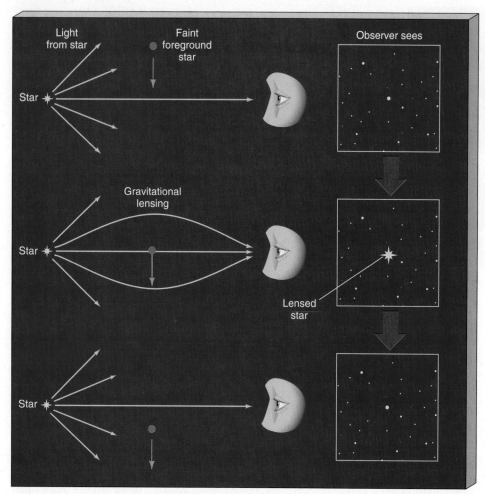

Figure 23.21 Gravitational lensing by a faint foreground object (such as a brown dwarf) can temporarily cause a background star to brighten significantly, providing a means of detecting otherwise invisible stellar dark matter.

prediction that a beam of light can be deflected by a gravitational field, which has already been verified in the case of starlight that passes close to the Sun. ∞ *(More Precisely 22-2)* Although this effect is small, it has the potential for making otherwise invisible stellar objects observable from Earth. Here's how.

Imagine looking at a distant star as a faint foreground object (such as a brown dwarf or a white dwarf) happens to cross your line of sight. As illustrated in Figure 23.21, the intervening object deflects a little more starlight than usual toward you, resulting in a temporary, but quite substantial, *brightening* of the distant star. In some ways, the effect is like the focusing of light by a lens, and the process is known as **gravitational lensing**. The foreground object is referred to as a *gravitational lens*. The amount of brightening and the duration of the effect depend on the mass, distance, and speed of the lensing object. Typically, the apparent brightness of the background star increases by a factor of 2 to 5 for a period of several weeks. Thus, even though the foreground object cannot be seen directly, its effect on the light of the background star makes it detectable. (In Chapter 25 we will encounter other instances of gravitational lensing in the universe, but on very much larger scales.)

Of course, the probability that one star will pass almost directly in front of another, as seen from Earth, is extremely small. But by observing millions of stars every few days over a period of years (using automated telescopes and high-speed computers to reduce the burden of coping with so much data), astronomers have been able to see enough of these events to let them estimate the amount of stellar dark matter in the Galactic halo. The technique represents an exciting new means of probing the structure of our Galaxy. The first lensing events were reported in late 1993. Subsequent observations are consistent with lensing by low-mass white dwarfs and suggest that such stars could account for at least half of the dark matter inferred from dynamical studies.

Bear in mind, though, that the identity of the dark matter is not necessarily an all-or-nothing proposition. It is perfectly conceivable that more than one type of dark matter exists. For example, it is quite possible that most of the dark matter in the inner (visible) parts of galaxies is in the form of brown dwarfs and very-low-mass stars, while the dark matter farther out may be primarily in the form of exotic particles. We will return to this perplexing problem in later chapters.

23.7 The Galactic Center

6 Theory predicts that the Galactic bulge, and especially the region close to the Galactic center, should be densely populated with billions of stars. However, we are unable to see this region of our Galaxy—the interstellar medium in the Galactic disk shrouds what otherwise would be a stunning view. Figure 23.22 shows the (optical) view we do have of the region of the Milky Way toward the Galactic center, in the general direction of the constellation Sagittarius.

With the help of infrared and radio techniques we can peer more deeply into the central regions of our Galaxy than we can by optical means. Infrared observations (Figure 23.23a) indicate that the heart of our Galaxy harbors roughly 50,000 stars per cubic parsec. That's a stellar density about a million times greater than in our solar neighborhood, high enough that stars must experience frequent close encounters and even collisions. Infrared radiation has also been detected from what appear to be huge clouds rich in dust. In addition,

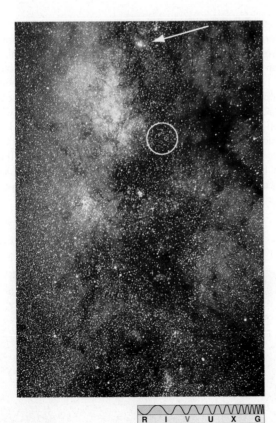

R I V U X G

Figure 23.22   A photograph of stellar and interstellar matter in the direction of the Galactic center. Because of heavy obscuration, even the largest optical telescopes can see no farther than 1/10 the distance to the center. The M8 nebula (arrow) can be seen at extreme top center. The field is roughly 20°, top to bottom, and is a continuation of the bottom part of Figure 18.6. The circle indicates the location of the center of our Galaxy.

radio observations indicate a ring of molecular gas nearly 400 pc across, containing some 30,000 solar masses of material and rotating at about 100 km/s. The origin of this ring is unclear, although researchers suspect that the gravitational influence of our Galaxy's elongated, rotating bulge may well be involved. The ring surrounds a bright radio source that marks the Galactic center.

High-resolution radio observations show more structure on small scales. Figure 23.23(b) shows the bright radio source Sagittarius A, which lies at the center of the circle in Figure 23.22 and within the boxed region in Figure 23.23(a) and, we think, at the center of our Galaxy. On a scale of about 100 pc, extended filaments can be seen. Their presence suggests to many astronomers that strong magnetic fields operate in the vicinity of the center, creating structures similar in appearance to (but much larger than) those observed on the active Sun. On even smaller scales (Figure 23.23c) the observations indicate a rotating ring or disk of matter only a few parsecs across.

What could cause all this activity? An important clue comes from the Doppler broadening of infrared spectral lines emitted from the central swirling whirlpool of gas. The extent of the broadening indicates that the gas is moving very rapidly. In order to keep this gas in orbit, whatever is at the center must be extremely massive—a million solar masses or more. Given the twin requirements of large mass and small size, a leading contender is a black hole. The hole itself is not the source of the energy, of course. Instead, the vast accretion disk of matter being drawn toward the hole by the enormous gravity emits the energy as it falls in, just as we saw (on a much smaller scale) in Chapter 22 when we discussed X-ray emission from neutron stars and stellar-mass black holes ∞ (Secs. 22.3, 22.7) The strong magnetic fields are thought to be generated within the accretion disk as matter spirals inward, and may act as "particle accelerators," creating the extremely high-energy particles detected on Earth as *cosmic rays* (see *Interlude 23-3*). Astronomers have reason to suspect that similar events are occurring at the centers of many other galaxies.

Figure 23.24 places these findings into a simplified perspective. Each frame is centered on the Galaxy's core, and each increases in resolution by a factor of 10. Frame (a) renders the Galaxy's overall shape, as painted in Figure 23.15. The scale of this frame measures about 100 kpc from top to bottom. Frame (b) spans a distance of 10 kpc from top to bottom and is nearly filled by the great circular sweep of the innermost spiral arm. Moving in to a 1-kpc span, frame (c) depicts the 400-pc ring of matter mentioned earlier. The dark blobs represent giant molecular clouds, the pink patches emission nebulae associated with star formation within those clouds.

In frame (d), at 100 pc, a pinkish region of ionized gas surrounds the reddish heart of the Galaxy. The source of energy producing this vast ionized cloud is assumed to be related to the activity in the Galactic center. Frame (e), spanning 10 pc, depicts the tilted, spinning whirlpool of hot

(a)

(b)

(c)

Figure 23.23   (a) An infrared image of the region around the center of our Galaxy shows many bright stars packed into a relatively small volume. The average density of matter in this region is estimated to be about a million times that in the solar neighborhood. (b) The central portion of our Galaxy, as seen in the radio part of the spectrum. This image shows a region about 200 pc across surrounding the Galactic center (which lies within the bright blob at bottom right). The long-wavelength radio emission cuts through the Galaxy's dust, providing an image of matter in the immediate vicinity of the Galaxy's center. (c) The spiral pattern of radio emission arising from Sagittarius A, the very center of the Galaxy. The data suggest a rotating ring of matter only 5 pc across.

(a)

(b)

(c)

(10^4 K) gas that marks the center of our Galaxy. The innermost part of this gigantic whirlpool is painted in frame (f), in which a swiftly spinning, white-hot disk of gas with temperatures in the millions of kelvins nearly engulfs a massive black hole too small in size to be pictured (even as a minute dot) on this scale.

At the very center of our Galaxy is a remarkable object with the odd-sounding name Sgr A* (pronounced "saj-ay-star"). By the standards of the active galaxies to be studied in Chapter 25, this compact galactic nucleus is not particularly energetic. Still, radio observations made during the past two decades, along with more recent X- and gamma-ray observations, suggest that it is nevertheless a pretty violent place. Its total energy output (at all wavelengths) is estimated to be 10^{33} W, which is more than a million times that of the Sun.

VLBI observations using radio telescopes arrayed from Hawaii to Massachusetts imply that Sgr A* cannot be much larger than 10 A.U., and it is probably a good deal smaller than that. This size is consistent with the view that the energy source is a massive black hole, although the case is not airtight. The main alternatives are that Sgr A* could be a multiple supernova of some sort (but we detect no expansion of debris), a rapidly moving neutron star (but Sgr A* seems to be more or less fixed relative to the surrounding galaxy), a large star cluster (but the energy-emitting region seems far too small), or even an ordinary neutron star embedded within a very unusual radio source (but that seems too ad hoc).

Some recent findings support the black hole picture. Infrared spectroscopic observations of two dozen stars within a few arc seconds of Sgr A* indicate that their orbital speeds are greater the closer they are to the center—just as one would expect for matter orbiting near a large black hole. In addition, an adaptive-optics infrared imaging camera has spotted a bright source very close to Sgr A* that seems to vary with a 10-minute period; this could be a hot spot on the accretion disk that circles the purported hole. The accumulated data imply that Sgr A* contains about 1–2 million solar masses. Even with this large mass, if Sgr A* is a genuine black hole, the size of its event horizon is still only 0.02 A.U. Such a small region, 8 kpc away, is currently unresolvable with any telescope now in existence.

If our knowledge of the Galaxy's center seems sketchy, that's because it *is* sketchy. Astronomers are still deciphering the clues hidden within its invisible radiation. We are only beginning to appreciate the full magnitude of this strange new realm deep in the heart of the Milky Way.

Figure 23.24 Six artist's conceptions, each centered on the Galactic center and each increasing in resolution by a factor of 10. Frame (a) shows the same scene as Figure 23.15. Frame (f) is a rendition of a vast whirlpool within the innermost parsec of our Galaxy. The data imaged in Figure 23.23 do not closely match these artistic renderings because the Figure 23.23 view is parallel to the Galactic disk, whereas these six paintings portray an idealized version perpendicular to that disk.

(d)   (e)   (f)

INTERLUDE 23-3 Cosmic Rays

In addition to stars, gas, dust, and dark matter, one other type of matter populates our Galaxy, namely, cosmic-ray particles, or cosmic rays for short. Cosmic rays continuously collide with Earth, and they make up our only samples of matter from outside the solar system.

In recent decades high-altitude balloons, rockets, and satellites have enabled astronomers to discover the chemical composition and the energies of cosmic-ray particles. Cosmic rays are actually not "rays" at all. They are subatomic particles. Nearly 90 percent are protons, the nuclei of hydrogen atoms. The nuclei of heavier atoms—helium and a long list of others—make up 9 percent. Electrons amount to 1 percent. The abundances of elements that make up the cosmic rays are close to those we see throughout the Galaxy.

Cosmic-ray particles are very energetic. By some unknown mechanism(s), cosmic rays are accelerated to extremely high velocities, much higher than any velocities we can achieve in terrestrial laboratories. Virtually all cosmic rays—even the heaviest—travel very close to the speed of light. When a galactic cosmic-ray particle hits our atmosphere, it creates a series of cascades or "showers" of many lower-energy particles. These lower-energy "secondary" particles in turn collide with objects, including humans, on Earth's surface virtually all the time. Your body is being peppered with them right now.

The accompanying photograph shows an actual track of a cosmic-ray particle—in this case, the nucleus of a sulfur atom (shown in red in this false-color image). Entering from the left, the particle collides with the nucleus of an atom in the photographic emulsion, producing a fluorine nucleus (green), other nuclear fragments (blue), and a multitude of subatomic particles (yellow).

On the basis of the numbers of cosmic rays hitting Earth, astronomers estimate the number density of cosmic rays in interstellar space to be on the order of 10^{-3} particle/m^3. That makes them rare, although still more numerous than interstellar dust particles.

What is the source of our Galaxy's cosmic rays? The answer is currently uncertain. The circuitous and complex paths taken by the cosmic-ray particles while traversing the Galaxy, with its tangled magnetic field, prevent us from pinpointing their source by observing their direction of arrival at Earth. Indeed, careful searches have failed to show any preferential direction for cosmic rays. They seem to arrive from all directions with equal likelihood. Candidates for the origin of cosmic rays include (1) violent events at the Galactic center, where strong electromagnetic forces accelerate these particles to high energies; (2) supernova explosions, described in Chapter 21; and (3) the vicinity of neutron stars and black holes, as discussed in Chapter 22.

Most researchers currently favor the second of these explanations, although each of these candidate events probably contributes some particles to the observed hodgepodge of cosmic rays in our Galaxy. Some of the highest-energy cosmic rays may even originate at great distances, in the hearts of galaxies far beyond our own.

Chapter Review

SUMMARY

A **galaxy** (p. 520) is a huge collection of stellar and interstellar matter isolated in space and bound together by its own gravity. Because we live within it, the **Galactic disk** (p. 520) of our own **Milky Way Galaxy** (p. 520) appears as a broad band of light across the sky, a band called the Milky Way. Near the center, the Galactic disk thickens into the **Galactic bulge** (p. 520). The disk is surrounded by a roughly spherical **Galactic halo** (p. 520) of old stars and star clusters. Our Galaxy, like many others visible in the sky, is a **spiral galaxy** (p. 523).

The halo can be studied using **variable stars** (p. 523), whose luminosity changes with time. **Pulsating variable stars** (p. 523) vary in brightness in a repetitive and predictable way. Two types of pulsating variable stars of great importance to astronomers are **RR Lyrae** variables (p. 523) and **Cepheid** variables (p. 523), whose characteristic light curves make them easily recognizable. All RR Lyrae stars have roughly the same luminosity. Astronomers can determine the luminosity of Cepheids by measuring the pulsation period and using the **period–luminosity relationship** (p. 525), a simple correlation between period and absolute brightness. The brightest Cepheids can be seen at distances of millions of parsecs, extending the cosmic distance ladder well beyond our own Galaxy. RR Lyrae stars are fainter but much more numerous, making them very useful within the Milky Way.

In the early twentieth century, Harlow Shapley used RR Lyrae stars to determine the distances to many of the Galaxy's globular clusters. He found that the clusters have a roughly spherical distribution in space, but the center of the sphere lies far from the Sun. The globular clusters are now known to map out the true extent of the luminous portion of the Milky Way Galaxy. The center of their distribution is close to the **Galactic center** (p. 526), which lies about 8 kpc from the Sun.

Disk and halo stars differ in their spatial distributions, ages, colors, and orbital motion. The luminous portion of our Galaxy has a diameter of about 30 kpc. The halo lacks gas and dust, so no stars are forming there. All halo stars are old. The gas-rich disk is the site of current star formation and contains many young stars. Stars and gas within the Galactic disk move on roughly circular orbits around the Galactic center. Stars in the halo and bulge move on largely random three-dimensional orbits that pass repeatedly through the disk plane but have no preferred orientation. Halo stars appeared early on, before the Galactic disk took shape, when there was still no preferred orientation for their orbits. As the gas and dust formed a rotating disk, stars that formed in the disk inherited its overall spin and so moved on circular orbits in the Galactic plane, as they do today.

In the vicinity of the Sun the Galactic disk is about 300 pc thick. Young stars, gas, and dust are more narrowly confined; older stars have a broader distribution. Intermediate between the young disk and the old halo, in both age and spatial distribution, are the stars of the **thick disk** (p. 527), which is about 2–3 kpc thick.

Attempts to map out the Galactic disk by optical observations are defeated by interstellar absorption. Astronomers use radio observations to explore the Galactic disk because radio waves are largely unaffected by interstellar dust. Regions where most of the hydrogen is in atomic form may be studied using 21-cm radiation. Regions where the gas is mostly molecular are usually studied by observing radio emission lines from "tracer" molecules, such as carbon monoxide. The gas distribution fattens near the center into the Galactic bulge. Radio-emitting gas has been detected in the disk at up to 50 kpc from the Galactic center.

Radio observations clearly reveal the extent of our Galaxy's **spiral arms** (p. 531). The spiral arms in spiral galaxies are regions of the densest interstellar gas and are the places where star formation is taking place. The spirals cannot be "tied" to the disk material, as the disk's differential rotation would have wound them up long ago. Instead, they may be **spiral density waves** (p. 532) that move through the disk, triggering star formation as they pass by. Alternatively, the spirals may arise from self-propagating star formation, when shock waves produced by the formation and evolution of one generation of stars triggers the formation of the next.

The Galactic **rotation curve** (p. 536) plots the orbital speed of matter in the disk versus distance from the Galactic center. By applying Newton's laws of motion, astronomers can determine the mass of the Galaxy. They find that the Galactic mass continues to increase beyond the radius defined by the globular clusters and the spiral structure we observe. The rotation curves of our own and other galaxies show that many, if not all, galaxies have an invisible **dark halo** (p. 537) containing far more mass than the visible portion of the galaxies. The **dark matter** (p. 537) making up these dark halos is of unknown composition. Leading candidates include low-mass stars and exotic subatomic particles. Recent attempts to detect stellar dark matter have used the fact that a faint foreground object can occasionally pass in front of a more distant star, deflecting the star's light and causing its apparent brightness to increase temporarily. This deflection is called **gravitational lensing** (p. 538).

Astronomers working at infrared and radio wavelengths have uncovered evidence of energetic activity within a few parsecs of the Galactic center. The leading explanation is that a black hole 1–2 million times more massive than the Sun resides at the heart of our Galaxy.

SELF-TEST: TRUE OR FALSE?

____ **1.** Cepheids can be used to determine the distances to the nearest galaxies.

____ **2.** RR Lyrae stars are a type of cataclysmic variable.

____ **3.** The Galactic halo contains about as much gas and dust as the Galactic disk.

____ **4.** The Galactic disk contains only old stars.

____ **5.** Population I objects are found only in the Galactic halo.

____ **6.** Up until the 1930s, the main error made in determining the size of the Galaxy was due to an incorrectly calibrated method of determining stellar distances.

____ **7.** Astronomers use 21-cm radiation to study Galactic molecular clouds.

____ **8.** Radio techniques are capable of mapping the entire Galaxy.

____ **9.** In the neighborhood of the Sun, the Galaxy's spiral density wave rotates more slowly than the overall Galactic rotation.

____ **10.** The mass of the Galaxy is determined by counting stars.

____ **11.** Dark matter is now known to be due to large numbers of black holes.

____ **12.** A million–solar mass black hole could account for the unusual properties of the Galactic center.

____ **13.** Cosmic rays are very energetic photons.

____ **14.** Most of the mass of our Galaxy exists in the form of dark matter.

____ **15.** The Galactic center has been extensively studied at visible and ultraviolet wavelengths.

SELF-TEST: FILL IN THE BLANK

1. One difficulty in studying our own galaxy in its entirety is that we live _____.

2. Herschel's attempt to map the Milky Way by counting stars led to an inaccurate estimate of the Galaxy's size because he was unaware of _____.

3. The highly flattened, circular part of the Galaxy is called the Galactic _____.

4. The roughly spherical region of faint old stars and globular clusters in which the rest of the Galaxy is embedded is the Galactic _____.

5. Cepheids and RR Lyrae stars are observed to vary in _____ with periods of days to months.

6. Cepheid pulsational periods range from _____ to_____.

7. Cepheids and RR Lyrae variables lie in a region of the H–R diagram called the _____.

8. According to the period–luminosity relation, the longer the pulsation period of a Cepheid, the _____ its luminosity.

9. Harlow Shapley determined the distances to the globular clusters using observations of _____.

10. The Sun lies roughly _____ pc from the Galactic center.

11. The orbital speed of the Sun around the Galactic center is _____.

12. The orbits of halo objects are _____ in direction.

13. The original cloud of gas from which the Galaxy formed probably had a size and shape similar to the present Galactic _____.

14. Rotational velocities in the outer part of the Galaxy are _____ than would be expected on the basis of observed stars and gas, indicating the presence of _____.

15. Observations of the _____ of infrared spectral lines indicate that gas near the Galactic center is orbiting at extremely high speeds.

REVIEW AND DISCUSSION

1. What are spiral nebulae? How did they get that name?

2. How are Cepheid variables used in determining distances?

3. Roughly how far out into space can we use Cepheids to measure distance?

4. What important discoveries were made early in this century using RR Lyrae variables?

5. Why are the central regions of our Galaxy best studied using radio telescopes?

6. Of what use is radio astronomy in the study of Galactic structure?

7. Contrast the motions of disk and halo stars.

8. Explain why Galactic spiral arms are believed to be regions of recent and ongoing star formation.

9. Describe what happens to interstellar gas as it passes through a spiral density wave.

10. What is self-propagating star formation?

11. What do the red stars in the Galactic halo tell us about the history of the Milky Way?

12. What does the rotation curve of our Galaxy tell us about its total mass?

13. What evidence is there for that dark matter in the Galaxy?

14. What are some possible explanations for dark matter?

15. Why do astronomers believe that a supermassive black hole lies at the center of the Milky Way?

PROBLEMS

1. Calculate the angular diameter of a prestellar nebula, of radius 100 A.U., lying 100 pc from Earth. Compare this with the roughly 6° diameter of the Andromeda galaxy (Figure 23.2a).

2. What is the greatest distance at which an RR Lyrae star of absolute magnitude 0 could be seen by a telescope capable of detecting objects as faint as 20th magnitude?

3. A typical Cepheid variable is 100 times brighter than a typical RR Lyrae star. How much farther away than RR Lyrae stars can Cepheids be used as distance-measuring tools?

4. The *Hubble Space Telescope* can see a star like the Sun at a distance of 100,000 pc. The brightest Cepheids have luminosities 30,000 times greater than that of the Sun. How far away can *HST* see these Cepheids?

5. Calculate the proper motion (in arc seconds/year) of a globular cluster with a transverse velocity (relative to the Sun) of 200 km/s and a distance of 3 kpc. Do you think that this motion is measurable?

6. Calculate the total mass of the Galaxy lying within 20 kpc of the Galactic center if the rotation speed at that radius is 240 km/s.

7. Using the data presented in Figure 23.19, calculate how long it takes the Sun to "lap" stars orbiting 15 kpc from the Galactic center. How long does it take matter at 5 kpc to lap us?

8. A density wave made up of two spiral arms is moving through the Galactic disk. At the 8-kpc radius of the Sun's orbit around the Galactic center, the wave's speed is 120 km/s, and the Galactic rotation speed is 220 km/s. Calculate how many times the Sun has passed through a spiral arm since the Sun formed 4.6 billion years ago.

9. Given the data in the previous question and the fact that O stars live at most 10 million years before exploding as supernovae, calculate the maximum distance at which an O star (orbiting at the Sun's distance from the Galactic center) can be found from the density wave in which it formed.

10. Material at an angular distance of 0.1″ from the Galactic center is observed to have an orbital speed of 1100 km/s. If the Sun's distance to the Galactic center is 8 kpc, and the material's orbit is circular and is seen edge-on, calculate the mass of the object around which the material is orbiting.

PROJECT

1. If you are far from city lights, look for a hazy band of light arching across the sky. This is our edgewise view of the Milky Way Galaxy. The Galactic center is located in the direction of the constellation Sagittarius, highest in the sky during the summer, but visible from spring through fall. Look at the band making up the Milky Way and notice dark regions; these are relatively nearby dust clouds. Sketch what you see. Look for faint fuzzy spots in the Milky Way and note their positions in your sketch. Draw in the major constellations for reference. Compare your sketch with a map of the Milky Way in a star atlas. Did you discover most of the dust clouds? Can you identify the faint fuzzy spots?

24 NORMAL GALAXIES

The Large-Scale Structure of the Universe

LEARNING GOALS

Studying this chapter will enable you to:

1 Describe the basic properties of the main types of normal galaxies.

2 Discuss the distance-measurement techniques that enable astronomers to map the universe beyond the Milky Way.

3 Summarize what is known about the large-scale distribution of galaxies in the universe.

4 Describe some of the methods used to determine the masses of distant galaxies.

5 Explain why astronomers think that most of the matter in the universe is invisible.

6 Discuss some theories of how galaxies form and evolve.

7 State Hubble's law and explain how it is used to derive distances to the most remote objects in the observable universe.

(Opposite page) This stunning image takes us far out into space, and far back into time. Known as the Hubble Deep Field, it shows thousands of galaxies of many shapes and colors. By "deep," we mean dim and distant—for this is the faintest photograph ever made; it reaches 30th magnitude, or about four billion times fainter than can be seen by the human eye. To create it, the *Hubble Space Telescope* exposed its electronic detectors for about a hundred hours, over the course of ten days. Astronomers have likened this photo to a geologist's "core sample" of the Earth's crust; it covers an area only about one one-hundredth of the full Moon. The big ground-based Keck telescope is finding that the small blue shards are among the most distant objects ever seen; perhaps galaxies caught in the act of formation. In all, the image implies that there are about 40 billion galaxies in the observable universe—not far from what astronomers had estimated before the image was taken.

Much of our knowledge of the workings of our own Galaxy is based on observations of other galaxies. We know of literally millions of galaxies beyond our own—many smaller than it, some comparable in size, a few much larger. All are vast, gravitationally bound assemblages of stars, gas, dust, dark matter, and radiation, separated from us by almost incomprehensibly large distances. Even a modest-sized galaxy harbors more stars than the number of people who have ever lived on Earth. The light we receive tonight from the most distant galaxies was emitted long before Earth existed. By comparing and classifying the properties of galaxies, astronomers have begun to understand their complex dynamics. By mapping out their distribution in space, astronomers trace out the immense realms of the universe. The galaxies remind us that our position in the universe is no more special than that of a boat adrift at sea.

24.1 Hubble's Galaxy Classification

1 Figure 24.1 shows a vast expanse of space lying about 100 million pc from Earth. Almost every patch or point of light in this figure is a separate galaxy—several hundred can be seen in just this one photograph. Over the years, astronomers have accumulated similar images of many millions of galaxies. We begin our study of these enormous accumulations of matter simply by considering their appearance on the sky.

Seen through even a small telescope, images of galaxies look distinctly nonstellar. They have fuzzy edges, and many are quite elongated—not at all like the sharp, point-like images normally associated with stars. Although it is difficult to tell from the photograph, some of the blobs of light in Figure 24.1 are actually spiral galaxies like the Milky Way Galaxy and Andromeda. Others, however, are definitely not spirals—no disks or spiral arms can be seen. Even when we take into account their different orientations in space, galaxies do *not* all look the same.

The American astronomer Edwin Hubble was the first to categorize galaxies in a comprehensive way. Working with the then recently completed 2.5-m optical telescope on Mount Wilson in California in 1924, he classified the galaxies he saw into four basic types—*spirals, barred spirals, ellipticals,* and *irregulars*—solely on the basis of appearance. Many modifications and refinements have been incorporated over the years, but the basic **Hubble classification scheme** is still widely used today.

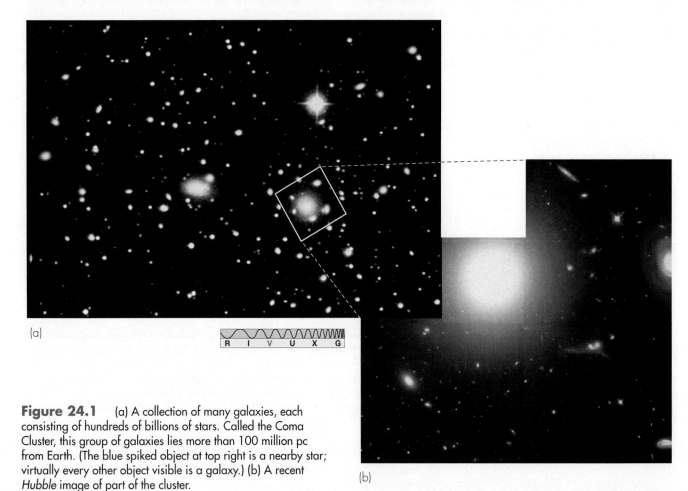

(a)

R I V U X G

(b)

Figure 24.1 (a) A collection of many galaxies, each consisting of hundreds of billions of stars. Called the Coma Cluster, this group of galaxies lies more than 100 million pc from Earth. (The blue spiked object at top right is a nearby star; virtually every other object visible is a galaxy.) (b) A recent *Hubble* image of part of the cluster.

SPIRALS

We saw several examples of **spiral galaxies** in Chapter 23—for example, our own Milky Way Galaxy and our neighbor Andromeda. All galaxies of this type contain a flattened galactic disk in which spiral arms are found, a central galactic bulge, and an extended halo of faint, old stars. ∞ (Sec. 23.3) The stellar density (that is, the number of stars per unit volume) is greatest in the galactic nucleus, at the center of the bulge. However, within this general description, spiral galaxies exhibit a wide variety of shapes, as illustrated in Figure 24.2.

In Hubble's scheme, a spiral galaxy is denoted by the letter S and classified as type "a," "b," or "c" according to the size of its central bulge—Type Sa galaxies having the largest bulges, Type Sc the smallest. The size of the bulge is quite well correlated with the tightness of the spiral pattern (although the correspondence is not perfect). Type Sa spiral galaxies, with large central bulges, tend to have tightly wrapped, almost circular, spiral arms. Type Sb galaxies, with smaller bulges, typically have more open spiral arms. Type Sc spirals, with the smallest bulges, often have loose, poorly defined spiral structure. The arms also tend to become more "knotty," or clumped, in appearance as the spiral pattern becomes more open.

Much of our description of the large-scale structure of the Milky Way Galaxy in Chapter 23 applies to spiral galaxies in general. The bulges and halos of spiral galaxies contain large numbers of reddish old stars and globular clusters, similar to those observed in our own Galaxy and in Andromeda. Most of the light from spirals, however, comes from A- through G-type stars in the disk, giving these galaxies an overall whitish glow. The spiral arms appear bluish because of the presence of bright blue O and B stars there. We assume that thick disks exist, too, but their faintness makes this assumption hard to confirm—the thick disk in the Milky Way contributes only a percent or so of our Galaxy's total light. ∞ (Sec. 23.3)

Like the disk of the Milky Way, the flat galactic disks of typical spiral galaxies are rich in gas and dust. The 21-cm radio radiation emitted by spirals betrays the presence of interstellar gas, and detailed photographs clearly show obscuring dust in many systems. Stars are forming within the spiral arms, where the interstellar medium is densest, and the arms contain numerous emission nebulae and newly formed O and B stars, as discussed in Chapter 23. ∞ (Sec. 23.5) Type Sc galaxies contain the most interstellar gas and dust, Sa galaxies the least. The Sc galaxy NGC 2997 shown in Figure 24.2(c) clearly shows the preponderance of interstellar gas, dust, and young blue stars tracing the spiral pattern. Spirals are not necessarily young galaxies, however. Like our own Galaxy, they are simply rich enough in interstellar gas to provide for continued stellar birth.

Most spirals are not seen face-on as in Figure 24.2. Many are tilted with respect to our line of sight, making their spiral structure hard to discern, However, we do not need to see spiral arms to classify a galaxy as a spiral. The presence of the disk, with its gas, dust, and newborn stars, is sufficient. For example, the galaxy shown in Figure 24.3 is classified as a spiral because of the clear line of obscuring dust seen along its midplane.

A variation of the spiral category in Hubble's classification scheme is the **barred-spiral galaxy**. The barred spirals differ from ordinary spirals mainly by the presence of an elongated "bar" of stellar and interstellar matter passing through the center and extending beyond the bulge, into the disk. The spiral arms project from near the ends of the bar rather than from the bulge (as they do in normal spirals). Barred spirals are designated by the letters SB and are subdivided, like the ordinary spirals, into categories

Figure 24.2
Variation in shape among spiral galaxies. As we progress from type Sa to Sb to Sc, the bulges become smaller while the spiral arms tend to become less tightly wound.

M81   Type Sa   M51   Type Sb   NGC 2997   Type Sc

R I V U X G

Figure 24.3 The Sombrero Galaxy, a spiral system seen edge-on. Officially cataloged as M104, this galaxy has a dark band composed of interstellar gas and dust. The large size of this galaxy's central bulge marks it as type Sa, even though its spiral arms cannot be seen.

SBa, SBb, and SBc, depending on the size of the bulge. Again like ordinary spirals, the tightness of the spiral pattern is correlated with bulge size. Figure 24.4 shows the variation among barred-spiral galaxies. In the case of the SBc category, it is often hard to tell where the bar ends and the spiral arms begin.

Astronomers often cannot distinguish between spirals and barred spirals, especially when a galaxy happens to be oriented with its galactic plane nearly edge-on toward Earth, as in Figure 24.3. Because of the physical and chemical similarities of spiral and barred-spiral galaxies, some researchers do not even bother to distinguish between them. Other researchers, however, regard the differences in their structures as very important, arguing that these differences suggest basic dissimilarities in the conditions that led to the formation of the galaxies eons ago. The recent findings that the bulge of our own Galaxy is elongated suggest that the Milky Way is a barred spiral, of type SBb or SBc. ∞ (Sec. 23.3)

ELLIPTICALS

The next major category in the Hubble scheme contains the **elliptical galaxies**. Unlike the spirals, ellipticals have no spiral arms and, in most cases, no flattened galactic disk—in fact, they often exhibit little internal structure of any kind. As with spirals, the stellar density increases sharply in the central nucleus. Ellipticals range in shape from highly elongated to nearly circular in appearance. Denoted by the letter E, these systems are subdivided according to how elliptical they are. The most circular are designated E0, slightly flattened systems are labeled E1, and so on, all the way to the most elongated ellipticals, of type E7 (Figure 24.5).

There is a large range in both the size and the number of stars contained in elliptical galaxies. The largest elliptical galaxies are much larger than our own Milky Way Galaxy. These *giant ellipticals* can range up to a few megaparsecs across and contain trillions of stars. At the other extreme, *dwarf ellipticals* may be as small as 1 kpc in diameter and contain only a few million stars. The substantial observational differences between giant and dwarf ellipticals have led many astronomers to conclude that these galaxies are members of separate classes, with quite different formation histories and stellar content. The dwarfs are by far the most common type of ellipticals, outnumbering their brighter counterparts by about 10 to 1. However, most of the *mass* that exists in the form of elliptical galaxies is contained in the larger systems.

Lack of spiral arms is not the only difference between spirals and ellipticals. Most ellipticals also contain little or no gas and dust. The 21-cm radio emission from neutral hydrogen gas is, with few exceptions, completely absent, and no obscuring dust lanes are seen. In most cases there is no evidence of young stars or of any ongoing star formation. Like the halo of our own Galaxy, ellipticals are made up mostly of old, reddish, low-mass stars. Indeed, having no disk, gas, or

NGC 3992   Type SBa   NGC1433   Type SBb   NGC1300   Type SBc

Figure 24.4 Variation in shape among barred-spiral galaxies. The variation from SBa to SBc is similar to that for the spirals in Figure 24.2, except that now the spiral arms begin at either end of a bar through the galactic center.

(a) M49   Type E1   (b) M84   Type E3

Figure 24.5 Variation in shape among elliptical galaxies. (a) The E1 galaxy M49 is nearly circular in appearance. (b) M84 is a slightly more elongated elliptical galaxy. It is classified as E3. Both these galaxies lack spiral structure, and neither shows evidence of interstellar matter.

dust, elliptical galaxies are, in a sense, "all halo." Again as in the halo of our Galaxy, the orbits of stars in ellipticals are disordered, exhibiting little or no overall rotation; objects move in all directions, not in regular, circular paths as in our Galaxy's disk. Apparently all, or nearly all, of the interstellar gas within elliptical galaxies was swept up into stars (or out of the galaxy) long ago, before a disk had a chance to form, leaving stars in random orbits, with no loose gas and dust for the creation of future generations of stars.

Some giant ellipticals are exceptions to many of the foregoing general statements about elliptical galaxies, as they have been found to contain disks of gas and dust in which stars are forming. Astronomers speculate that these galaxies may really be otherwise "normal" ellipticals that have collided and merged with a companion spiral system (although this explanation is not yet universally accepted). We will return later in this chapter to the important role played by galactic collisions in determining the appearance of the systems we observe today (*Interlude 24-1*).

Intermediate between the E7 ellipticals and the Sa spirals in the Hubble classification is a class of galaxies that show evidence of a thin disk and a flattened bulge but that contain no gas and no spiral arms. Two such objects are shown in Figure 24.6. These galaxies are known as **S0 galaxies** if no bar is evident and **SB0 galaxies** if a bar is present. They look a little like spirals whose dust and gas have been stripped away, leaving behind a stellar disk. Observations in recent years have shown that many normal elliptical galaxies have faint disks within them, like the S0 galaxies. As with the S0s, the origin of these disks is uncertain, but some researchers suspect that the S0s and ellipticals actually form a continuous sequence, along which the bulge-to-disk ratio varies smoothly.

IRREGULARS

The final galaxy class identified by Hubble is a catch-all category—**irregular galaxies**—so named because their visual appearance does not allow us to place them into any of the other categories just discussed. Irregulars tend to be rich in interstellar matter and young, blue stars, but they lack any regular structure, such as well-defined spiral arms or central bulges. They are divided into two subclasses—Irr I galaxies and Irr II galaxies. The Irr I galaxies often look like misshapen spirals. The much rarer Irr II galaxies, in addition to their irregular shape, have other peculiarities, often exhibiting a distinctly explosive or filamentary appearance. Figure 24.7 shows some examples of these strangely shaped galaxies. Their appearance once led astronomers to suspect that vio-

(a) NGC 1201   Type S0   (b) NGC 2859   Type SB0

Figure 24.6 (a) S0 galaxies contain a disk and a bulge but no interstellar gas and no spiral arms. They are in many respects intermediate between E7 ellipticals and Sa spirals in their properties. (b) SB0 galaxies are similar to S0 galaxies, except for a bar of stellar material extending beyond the central bulge.

(a)   (b)   (c)

R I V U X G

Figure 24.7 Some irregular (Irr II) galaxies. (a) The oddly shaped galaxies NGC 4485 and NGC 4490 may be close to each other and interacting gravitationally. (b) The galaxy M82 seems to show an explosive appearance, although interpretations remain uncertain. (c) Many irregular galaxies are small and dim, but this one, NGC 4449, is comparable in both size and luminosity to the Milky Way Galaxy.

lent events had occurred within them. However, it now seems more likely that in some (but probably not all) cases we are seeing the result of a close encounter, or even a collision, between two previously "normal" systems.

Irregular galaxies tend to be smaller than spirals but somewhat larger than dwarf ellipticals. They typically contain between 10^8 and 10^{10} stars. The smallest are called *dwarf irregulars*. As with elliptical galaxies, the dwarf type is the most common irregular. Dwarf ellipticals and dwarf irregulars occur in approximately equal numbers and together make up the vast majority of galaxies in the universe. They are often found close to a larger "parent" galaxy.

Figure 24.8 shows the **Magellanic Clouds**, a famous pair of Irr I galaxies that orbit the Milky Way Galaxy. They are shown to proper scale in Figure 23.15. Studies of Cepheid variables within the Clouds show them to be approximately 50 kpc from the center of our Galaxy. The Large Cloud contains about 6 billion solar masses of material and is a few kiloparsecs across. Both Magellanic Clouds contain lots of gas, dust, and blue stars (and the recent, well-documented supernova discussed in *Interlude 21-2*), indicating youthful activity and ongoing star formation. Both also contain many old stars and several old globular clusters, so we know that star formation has been going on there for a very long time.

Figure 24.8 The Magellanic Clouds are prominent members of the sky in the Southern Hemisphere. Named for the sixteenth-century Portuguese explorer Ferdinand Magellan, whose round-the-world expedition first brought word of these fuzzy patches of light to Europe, they are dwarf irregular (Irr I) galaxies, gravitationally bound to our own Milky Way Galaxy. They orbit our Galaxy and accompany it on its trek through the cosmos. (a) The Clouds' relationship to each other in the southern sky. Both the Small (b) and the Large Magellanic Cloud (c) have distorted, irregular shapes, although some observers claim they can discern a single spiral arm in the Large Cloud.

R I V U X G

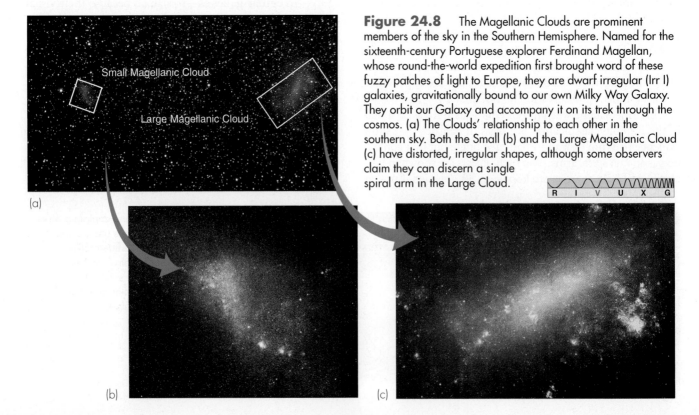

(a)

Small Magellanic Cloud

Large Magellanic Cloud

(b)   (c)

TABLE 24.1 Basic Galaxy Properties by Type

	SPIRAL/BARRED SPIRAL (S, SB)	ELLIPTICAL (E)	IRREGULAR (Irr)
Shape and structural properties	Highly flattened disk of stars and gas, containing spiral arms and thickening to a central bulge; Sa and Sba galaxies have largest bulges, the least obvious spiral structure, and roughly spherical stellar halos SB galaxies have an elongated central "bar" of stars and gas	No disk Stars smoothly distributed through an ellipsoidal volume ranging from nearly spherical (E0) to very flattened (E7) in shape No obvious substructure other than a dense central nucleus	No obvious structure; Irr II galaxies often have "explosive" appearance
Stellar content	Disks contain both young and old stars; halos consist of old stars only	Contain old stars only	Contain both young and old stars
Gas and dust	Disks contain substantial amounts of gas and dust; halos contain little of either	Contain little or no gas and dust	Very abundant in gas and dust
Star formation	Ongoing star formation in spiral arms	No significant star formation during the last 10 billion years	Vigorous ongoing star formation
Stellar motion	Gas and stars in disk move in circular orbits around the galactic center; halo stars have random orbital orientations in three dimensions	Stars have random orbital orientations in three dimensions.	Stars and gas have very irregular orbits.

Over the years, radio studies have hinted at a possible bridge of hydrogen gas connecting the Milky Way to the Magellanic Clouds, but more observational research is needed to establish this link beyond doubt. It is possible that the tidal force of the Milky Way tore this stream of gas from the Clouds the last time their orbits brought them close to our Galaxy. Of course, gravity works both ways, and many researchers believe that the forces exerted by the Clouds may in turn be responsible for distorting our Galaxy, warping and thickening the outer parts of the Galactic disk. ∞ (Sec. 23.5)

AN H–R DIAGRAM FOR GALAXIES?

Table 24.1 summarizes the basic characteristics of the various galaxy types. When he first developed his classification scheme, Hubble arranged the galaxy types into the "tuning fork" diagram shown in Figure 24.9. His aim in doing this was simply to indicate similarities in appearance among galaxies, not to suggest that any connection might exist, but some astronomers have since speculated that perhaps the diagram has deeper significance. Figure 24.9 certainly suggests an evolutionary connection of some sort among galaxy

Figure 24.9 Hubble's "tuning fork" diagram, showing his basic galaxy classification scheme. The placement of the four basic galaxy types—ellipticals, spirals, barred spirals, and irregulars—in the diagram is suggestive, but no "evolutionary track" along the sequence (in either direction) is proven.

types, but does one really exist? The answer, as best we can tell, is no.

Despite many attempts, no one has ever succeeded in explaining the observed properties of normal galaxies in evolutionary terms. Isolated normal galaxies do *not* evolve from one type to another. Spirals are not ellipticals with arms, nor are ellipticals spirals that have somehow lost their star-forming disks. In short, astronomers know of no parent-child relationship among normal galaxies. However, the subject of galaxy evolution is still poorly understood.

24.2 The Distribution of Galaxies in Space

3 Now that we have seen some of their basic properties, let us ask how galaxies are spread through the expanse of the universe beyond the Milky Way Galaxy. Are they scattered everywhere throughout intergalactic space all the way out to the very limits of the observable universe, or is there a boundary beyond which galaxies no longer exist? To begin to answer this question, we must first know the *distances* to the galaxies.

EXTENDING THE DISTANCE SCALE

2 Astronomers estimate that some 40 billion galaxies exist in the observable universe. Some reside close enough for the Cepheid variable technique to work—astronomers have detected and measured the periods of Cepheids in galaxies as far away as 15 Mpc (see Figure 24.10). ∞ (Sec. 23.2) However, most known galaxies lie much farther away. Cepheid variable stars in very distant galaxies simply cannot be observed well enough, even through the world's most sensitive telescopes, to allow us to measure their luminosity and period. To extend our distance-measurement ladder, therefore, we must find some new object to study. What individual objects are bright enough for us to observe at great distances?

One way in which researchers have tackled this problem is through observations of **standard candles**—easily recognizable astronomical objects whose luminosities are confidently known. The basic idea is very simple. Once an object is identified as a standard candle—by its appearance or by the shape of its light curve, say—its luminosity can be estimated. Comparison of this luminosity with the object's apparent brightness then yields the object's distance and hence the distance to the galaxy in which it resides. Note that the Cepheid variable technique relies on identical reasoning; however, the term *standard candle* tends to be applied only to very bright objects. To be most useful, a standard candle must (1) have a narrowly defined luminosity, so that the uncertainty in estimating its brightness is small, and (2) be bright enough to be seen at large distances.

Over the years astronomers have explored the use of many types of objects as standard candles—novae, emission nebulae, planetary nebulae, globular clusters, Type I (carbon-detonation) supernovae, even entire galaxies have been employed. Not all have been equally useful, however; some

Figure 24.10 This sequence of six snapshots chronicles the periodic changes in a Cepheid variable star in the spiral galaxy M100, a member of the Virgo Cluster of galaxies. The Cepheid appears at the center of each inset, taken at the different times indicated in 1994. The star looks like a square because of the high magnification by the digital CCD camera—we are seeing individual pixels of the image. The 24th-magnitude star varies by about a factor of 2 in brightness every 7 weeks. (Compare with the image of M100 in *Interlude 5-1*.)

Figure 24.11 A galaxy's rotation causes some of the radiation it emits to be blueshifted and some to be redshifted (relative to what the emission would be from a nonrotating source). From a distance, when the radiation from the galaxy is combined into a single beam and analyzed spectroscopically, the redshifted and blueshifted components combine to produce a broadening of the galaxy's spectral lines. The amount of broadening is a direct measure of the rotation speed of the galaxy.

have larger intrinsic spreads in their luminosities than others, making them less useful for distance-measuring purposes. Planetary nebulae and Type I supernovae have proved particularly reliable. The latter have remarkably consistent peak luminosities and are bright enough to be seen and identified out to distances of hundreds of megaparsecs. The small luminosity spread of Type I supernovae is a direct consequence of the circumstances in which these violent events occur. As discussed in Chapter 21, an accreting white dwarf explodes when it reaches a well-defined critical mass at which carbon fusion begins. ∞ (Sec. 21.3) The magnitude of the explosion is relatively insensitive to the details of how the white dwarf formed, or how it subsequently reached critical mass, with the result that all such supernovae have quite similar observational properties.*

An important alternative to standard candles was discovered in the 1970s when astronomers found a close correlation between the rotational speeds and the luminosities of spiral galaxies within a few tens of megaparsecs of the Milky Way Galaxy. Rotation speed is a measure of a spiral galaxy's total mass, so it is perhaps not surprising that this property is related to luminosity. ∞ (Sec. 23.6) What *is* surprising is how tight the correlation is. The **Tully–Fisher relation**, as it is now known (after its discoverers), allows us

to obtain a remarkably accurate estimate of a spiral galaxy's luminosity simply by observing how fast it rotates. As usual, comparing the galaxy's (known) luminosity with its (observed) apparent brightness yields its distance.

To see how the method is used, imagine we are looking edge-on at a distant spiral galaxy and observing one particular emission line, as illustrated in Figure 24.11. Radiation from the side of the galaxy where matter is generally approaching us is blueshifted by the Doppler effect. Radiation from the other side, which is receding from us, is redshifted by a similar amount. The overall effect is that line radiation from the galaxy is "smeared out," or broadened, by the galaxy's rotation. The faster the rotation, the greater the amount of broadening. By measuring the amount of broadening, we can determine the galaxy's rotation speed. Once we know that, the Tully–Fisher relation tells us the galaxy's luminosity.

The particular line normally used in these studies actually lies in the radio part of the spectrum. It is the 21-cm line of cold, neutral hydrogen in the galactic disk. ∞ (Sec. 18.4) It is used in preference to optical lines because (1) optical radiation is strongly absorbed by dust in the disk under study and (2) the 21-cm line is normally very narrow, making the broadening easier to observe. In addition, astronomers often use *infrared*, rather than optical, luminosities to avoid absorption problems caused by dust, both in our own Galaxy and in others.

The Tully–Fisher relation can be used to measure distances to spiral galaxies out to about 200 Mpc, beyond which the line broadening becomes increasingly difficult to measure accurately. A somewhat similar connection exists for

---

*Recall from Chapter 21 that a Type II supernova also occurs when a growing stellar core—this time at the center of a massive star—reaches a critical mass. ∞ (Sec. 21.2) However, the outward appearance of the explosion can be significantly modified by the amount of stellar material through which the blast wave must travel before it reaches the star's surface, resulting in a greater spread in observed luminosities.

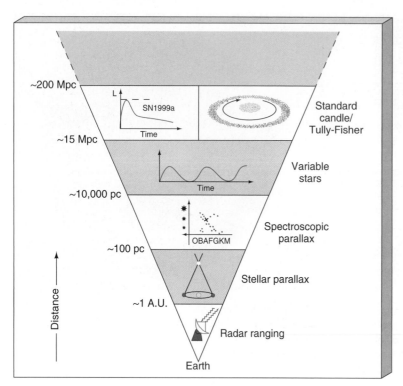

Figure 24.12   An inverted pyramid summarizes the distance techniques used to study different realms of the universe. The techniques shown in the bottom four layers—radar ranging, stellar parallax, spectroscopic parallax, and variable stars— take us as far as the nearest galaxies. To go farther, we must use new techniques, the method of standard candles and the Tully–Fisher relation, each based on distances determined by the four lowest techniques.

elliptical galaxies, linking the broadening of a galaxy's spectral lines (which, in the case of an elliptical, measures the average random velocity of the stars in the galaxy) and the galaxy's size. By measuring the broadening, astronomers can determine the galaxy's true size, which is then compared with the apparent size to give the distance. ∞ (Sec. 1.5) These methods bypass many of the standard candles just described and so provide independent means of determining distances to faraway objects.

Standard candles and the Tully–Fisher relation share the fifth rung of our cosmic distance ladder (Figure 24.12). Just as with the lower rungs, we calibrate the properties of these new techniques using distances measured by more local means. In this way, the distance-measurement process "bootstraps" itself to greater and greater distances. However, at the same time, the errors and uncertainties in each step accumulate, so the distances to the farthest objects are the least well known.

GALAXY CLUSTERS

Figure 24.13 sketches the locations of all the known major astronomical objects within about 1 Mpc of the Milky Way Galaxy. Our Galaxy appears with its three known satellite galaxies—the two Magellanic Clouds and another recently discovered companion (still unnamed, but designated "Sagittarius" in the figure) lying almost within our own Galactic plane. The Andromeda Galaxy, lying 900 kpc from us, is also shown. Two galactic neighbors of Andromeda are shown in Figure 24.14; M33 is a spiral, and M32 is a dwarf elliptical, easily seen in Figure 23.2(a) to the bottom right of Andromeda's central bulge.

All told, some 20 galaxies populate our Galaxy's neighborhood. Three of them (the Milky Way, Andromeda, and M33) are spirals; the remainder are dwarf irregular and ellipticals. Together, these galaxies form the **Local Group**—a new level of structure in the universe above the scale of our Galaxy. As indicated in Figure 24.13, the Local Group's diameter is roughly 1 Mpc. The Milky Way Galaxy and Andromeda are by far its largest members. The combined gravity of the galaxies in the Local Group binds them together, like stars in a star cluster, but on a millionfold larger scale. More generally,

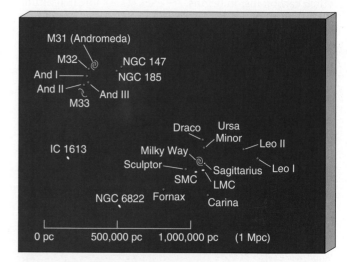

Figure 24.13   Diagram of the Local Group, made up of some 20 galaxies within approximately 1 Mpc of our Milky Way Galaxy. Only a few are spirals; most of the rest are dwarf elliptical or irregular galaxies. Spirals are shown in blue, ellipticals in pink, and irregulars in white.

Figure 24.14   Two well-known neighbors of the Andromeda Galaxy (M31): (a) the spiral galaxy M33 and (b) the dwarf elliptical galaxy M32 (also visible in Figure 23.2a, a larger-scale view of the Andromeda system).

R I V U X G

a group of galaxies held together by their mutual gravitational attraction is called a **galaxy cluster**.

Moving beyond the Local Group, we come to the next large concentration of galaxies, the Virgo Cluster (Figure 24.15). It lies about 18 Mpc from the Milky Way Galaxy. Like the Local Group, the Virgo Cluster is held together by the mutual gravitational attraction of its member systems. Unlike the Local Group, however, the Virgo Cluster does not contain a mere 20 galaxies. Instead, it houses approximately 2500 galaxies, each containing 100 billion or so individual stars. Those galaxies are bound together in a tightly knit group about 3 Mpc across. Figure 24.16 illustrates the approximate locations of Virgo and several other well-defined clusters in our cosmic neighborhood, lying within about 30 Mpc of the Milky Way Galaxy.

Many thousands of galaxy clusters have now been identified and cataloged, and they come in many shapes and sizes. Large, "rich" clusters, like Virgo, contain many thousands of individual galaxies distributed fairly smoothly in space. Small clusters, such as the Local Group, contain only a few galaxies and are quite irregular in shape. A small fraction of galaxies are not members of any cluster. They are apparently isolated systems, moving alone through intercluster space.

CLUSTERS OF CLUSTERS

Does the universe have even greater groupings of matter, or do galaxy clusters top the cosmic hierarchy? Most astronomers believe that the galaxy clusters themselves are

R I V U X G

Figure 24.15   The central region of the Virgo Cluster of galaxies, about 18 Mpc from Earth. Several large spiral and elliptical galaxies can be seen. The galaxy near the center is a giant elliptical known as M86.

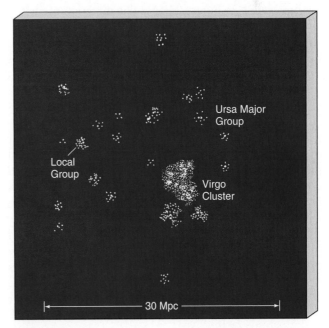

Local Group

Ursa Major Group

Virgo Cluster

◄———— 30 Mpc ————►

Figure 24.16   Schematic diagram of the location of several galaxy clusters in our part of the universe. Our Milky Way Galaxy is just one of these dots, and our Local Group just one of the clusters.

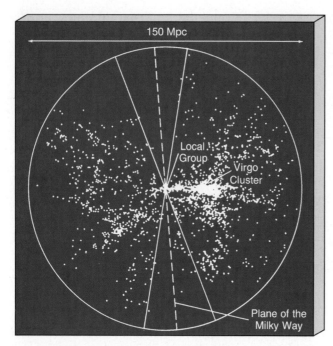

Figure 24.17 The Local Supercluster. Each of the 2200 points shown represents a galaxy, and the Milky Way Galaxy is at the center of the diagram. The Virgo Cluster and the plane of our own Galaxy are marked. Our Galaxy is seen edge-on. Its dust obscures our views to the top and the bottom, resulting in two empty V-shaped regions on the map.

clustered, forming titanic agglomerations of matter known as **superclusters**.

Figure 24.17 shows the *Local Supercluster*, containing the Local Group, the Virgo Cluster, and most of the other clusters shown in Figure 24.16. (The direction of view is roughly perpendicular to that shown in the previous figure, so many of the clusters noted there are difficult to discern here.) Each point represents a galaxy, and the diagram is centered on the Milky Way Galaxy. The perspective is such that the disk of our Galaxy is seen edge-on and runs vertically up the page. The two nearly empty V-shaped regions at the top and bottom of the figure are not devoid of galaxies—they are simply obscured from view by the dust in our own Galaxy's plane. The total mass of the entire Local Supercluster is about 10^{15} solar masses. By now it should come as no surprise that the Local Group is not found at its heart, which lies within the Virgo Cluster. Instead, we live far off in the periphery, some 18 Mpc from the center.

The Local Supercluster contains a huge number of individual galaxies—several tens of thousands—yet the great majority of known galaxy clusters and superclusters lie far beyond its edge. Figure 24.18 is a long-exposure photograph of one such remote cluster. This rich cluster is far outside the limit of the circle shown in Figure 24.17. It is only one of many large and distant groups of galaxies scattered throughout the observable universe. The Coma Cluster, shown in Figure 24.1, is another. On and on, the picture is much the same. The farther we peer

R I V U X G

Figure 24.18 The galaxy cluster CL1358+62 contains huge numbers of galaxies and resides roughly a billion parsecs from Earth. Virtually every patch of light in this photograph is a separate galaxy. Thanks to the high resolution of the *Hubble Space Telescope*, we can now discern, even at this great distance, spiral structure in some of the galaxies. We also see many galaxies in collision—some tearing matter from one another, others merging into single systems.

into deep space, the more galaxies, clusters of galaxies, and superclusters we see.

Is there structure on even larger scales? As we will see in a moment, the answer is still yes. However, before we make our final leap in distance to the limits of the visible universe, let's pause for a moment to take stock of some important properties of galaxies and galaxy clusters and consider a few current ideas on how galaxies form and evolve.

24.3 Galaxy Masses

4 How can we measure the masses of systems as large as galaxies and galaxy clusters? Surely, we can neither count all their stars nor estimate their interstellar content very well. Galaxies are just too complex to take direct inventory of their material makeup. Instead, we must rely on indirect techniques. As usual, we turn to Newton's law of gravity.

MASS MEASUREMENT

Astronomers can calculate the masses of some spiral galaxies by determining their *rotation curves*, which plot rotation speed, obtained by measuring the Doppler shift of various spectral lines, versus distance from the galactic center. ∞ (Sec. 23.6) The mass within any given radius then follows directly from Newton's laws. Some rotation curves for nearby spirals are shown in Figure 24.19. They imply masses ranging from about 10^{11} to 5×10^{11} solar masses within about 25 kpc of the center—quite comparable to the results obtained for our own Galaxy using the same technique. Distant galaxies are generally too far away for such detailed curves to be drawn. Nevertheless, by observing the broadening of spectral lines—as

Figure 24.19 Rotation curves for some nearby spiral galaxies indicate masses of a few hundred billion times the mass of the Sun. The corresponding curve for our own Galaxy (Figure 23.19) is marked in red for comparison.

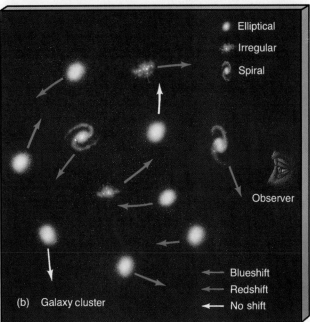

Figure 24.20 (a) In a binary galaxy system, galaxy masses may be estimated by observing the orbit of one galaxy about the other. (b) In a galaxy cluster, cluster masses are measured by observing the motion of many galaxies in the cluster and then estimating how much mass is needed to prevent the cluster from flying apart.

discussed earlier in the context of the Tully–Fisher relation—we can still measure the overall rotation speed. Estimating the galaxy's size then leads to an estimate of its mass.

This approach is useful only for measuring the mass lying within about 50 kpc of a galaxy's center—the extent of the electromagnetic emission from stellar and interstellar material. To probe farther from the center, galactic astronomers turn to binary and multiple systems of galaxies, much as stellar astronomers study binary stars to determine stellar masses. If we knew the orbit size and period of a galaxy–galaxy binary system, the combined mass of the pair could be inferred from Kepler's third law, just as described in Chapter 17. ∞ (Sec. 17.9) However, there is a problem here. Unlike the stellar case, we cannot watch the galaxies travel even a small fraction of their entire orbit. Instead, the orbital period is usually simply estimated on the basis of two observations: (1) the velocities of the galaxies measured along our line of sight (Figure 24.20a) and (2) the present distance between the two galaxies. Masses obtained in this way, then, are fairly uncertain. However, we can combine many such measurements to obtain quite reliable *statistical* information about galaxy masses.

From investigations using these methods, we find that most normal spirals (the Milky Way Galaxy included) and large ellipticals contain between 10^{11} and 10^{12} solar masses of material. Irregular galaxies often contain less mass, about 10^8 to 10^{10} times that of the Sun. Dwarf ellipticals and irregulars can contain as little as 10^6 to 10^7 solar masses of material.

We can use another statistical technique to derive the combined mass of all the galaxies within a galaxy cluster. As depicted in Figure 24.20(b), each galaxy within a cluster moves relative to all other cluster members. As with binary galaxies, we cannot watch galaxies move around within a cluster—typical orbit periods are billions of years. But we can estimate the cluster's mass simply by determining, using

Newtonian mechanics, how massive it must be in order to bind its galaxies gravitationally. Typical cluster masses obtained in this way lie in the range of 10^{13}–10^{14} solar masses. Notice that this calculation gives us *no* information whatsoever about the masses of individual galaxies. It tells us only about the *total* mass of the entire cluster.

DARK MATTER IN THE UNIVERSE

3 5 Radio observations indicate that the rotation curves of many spiral galaxies, such as those shown in Figure 24.19, remain flat (that is, do not decline and may even rise slightly)

far beyond the visible image of the galaxy. We conclude that these spiral galaxies—and perhaps all other spiral galaxies as well—must have invisible dark halos similar to that surrounding the Milky Way Galaxy. ∞ (Sec. 23.6) Depending on how far out these halos extend (and observations appear to indicate that they reach at least as far as 50 kpc from the center of a Milky Way–sized galaxy), spiral galaxies may contain 3 to 10 times more mass than can be accounted for in the form of visible matter. Some studies of elliptical galaxies suggest similarly large dark halos surrounding them too, but these observations are still far from conclusive.

Astronomers find an even greater discrepancy when they study galaxy clusters. Calculated cluster masses range from 10 to nearly 100 times the mass suggested by the light emitted by individual cluster galaxies. Stated another way, a lot more mass than we can see is needed to bind galaxy clusters. The problem of dark matter exists, then, not just in our own Milky Way Galaxy but also in other galaxies and, to an even greater degree, in galaxy clusters as well. It most likely applies to the entire universe. In that case, we must accept the fact that *upward of 90 percent of the universe is composed of dark matter*. As noted in Chapter 23, this matter is not just dark in the visible portion of the spectrum—it is invisible at *all* electromagnetic wavelengths.

As discussed in Chapter 23, many possible explanations for the dark matter have been suggested, ranging from stellar remnants of various sorts to exotic subatomic particles. ∞ (Sec. 23.6) Whatever its nature, however, the dark matter in clusters apparently cannot be simply the accumulation of smaller amounts of dark matter within individual galaxies. Even including the galaxies' dark halos, we still cannot account for all the dark matter in galaxy clusters. As

we look on larger and larger scales, we find that a larger and larger fraction of the matter in the universe is dark.

INTRACLUSTER GAS

Could the additional dark matter be diffuse intergalactic matter existing among the galaxies within the clusters—that is, intracluster gas? Until the late 1970s, astronomers had no observational evidence for intergalactic matter, either inside or outside galaxy clusters. Then satellites orbiting above Earth's atmosphere detected substantial amounts of X-ray radiation in the direction of many galaxy clusters. Figure 24.21 shows false-color X-ray images of two such clusters. The X-ray-emitting region is centered on, and comparable in size to, the visible image of the cluster. These X-ray observations demonstrated for the first time the existence of large amounts of invisible hot gas—at more than 10 million K—within galaxy clusters.

How much matter have the X-ray satellites found? The observations suggest that at least as much matter—and in a few cases substantially more—exists within clusters in the form of hot gas as is visible in the form of stars. This is a lot of material, but it still doesn't solve the dark-matter problem. To account for the total masses of galaxy clusters implied by dynamical studies, we would have to find from 10 to 100 times more mass in gas than in stars. In the mid-1990s, *ROSAT* observations of gas within the Virgo cluster of galaxies actually compounded the problem (Figure 24.22). The gas was so hot that the amount of dark matter needed to bind it to the cluster and prevent it from dispersing into intercluster space was far greater than had previously been suspected to exist!

Figure 24.21 (a) X-ray image of Abell 85, an old, distant cluster of galaxies, taken by the *Einstein* X-ray satellite observatory. The cluster's X-ray emission is shown in orange. The green graphs display a smooth, peaked intensity profile centered on the cluster but not associated with individual galaxies. (b) The contour map of X rays is superimposed on an optical photo, showing its X rays peaked on Abell 85's central supergiant galaxy. Images like these demonstrated for the first time that the space among the galaxies within galaxy clusters is filled with superheated gas. (c) A *ROSAT* X-ray image of hot gas within another cluster of galaxies (called Abell 2256). The cluster is nearly a billion parsecs from Earth and measures about 3 Mpc in diameter.

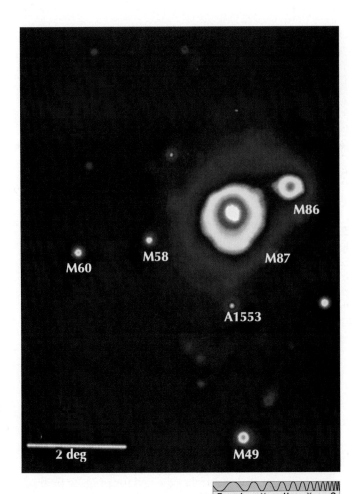

Figure 24.22 This combined optical and X-ray (*ROSAT*) image shows hot (10^7 K) X-ray-emitting gas (false-colored blue, increasing to yellow and red) within the rich Virgo Cluster of galaxies (see also Figure 24.15). The amount of mass required to bind this gas (especially that surrounding M87, shown in blue) to the cluster greatly exceeds the total mass of the galaxies themselves.

Astronomers have no evidence for gas beyond the well-defined galaxy clusters. "Extracluster" space is apparently empty of luminous material. Evidently, when the clusters formed eons ago, they did so very efficiently, sweeping up all the matter within any given region of the universe.

24.4 Galaxy Formation and Evolution

⑥ As mentioned earlier, astronomers know of no evolutionary connection among the various types of galaxies in the Hubble classification scheme. How then did those different galaxies come into being? To address this question, we must understand how galaxies formed. Unfortunately, the theory of galaxy formation is still very much in its infancy, and no definitive answers yet exist. Do we understand galaxy formation as well as, say, star formation? The answer is a resounding *no*! The theory is presently unable to answer completely even a basic question such as why spirals and ellipticals exist at all; it certainly cannot predict the different evolutionary tracks that galaxies may take.

There are several good reasons for our lack of understanding of galaxy formation: galaxies are much more complex than stars, they are harder to observe, and the observations are harder to interpret. We have only a partial understanding, and no observations, of the conditions in the universe immediately preceding galaxy formation, quite unlike the corresponding situation for stars. ∞ (Sec. 18.5) Finally, whereas stars almost never collide with one another, with the result that single stars and binaries evolve in isolation, galaxies may suffer many collisions and mergers during their lives (see *Interlude 24-1*), making it much harder to decipher their pasts. Yet despite all these difficulties, some general ideas have begun to gain widespread acceptance, and we can offer a few insights into the processes responsible for the galaxies we see.

MERGERS AND ACQUISITIONS

The seeds of galaxy formation were sown in the very early universe, when small density fluctuations in the primordial matter began to grow. For now, let us begin our discussion with these "pregalactic" blobs of gas already formed. The masses of these fragments were quite small—perhaps only a few million solar masses, comparable to the masses of the smallest present-day dwarf galaxies, which may in fact be remnants of this early time. So where did the larger galaxies we find today come from? The key point in our current understanding is the realization that galaxies grow by repeated *merging* of smaller objects, as illustrated in Figure 24.23. Contrast this process with that of star formation, in which a large cloud fragments into smaller pieces that eventually become stars. ∞ (Sec. 19.2)

Theoretical evidence for this scenario is provided by computer simulations of the early universe, which clearly show merging taking place. Further strong support comes from recent observations that indicate that galaxies at large distances from us (well over 2000 Mpc away, meaning that the light we see was emitted more than 6 billion years ago) appear distinctly smaller and more irregular than those found nearby. Figures 24.23(b) and 24.24 show some of these images, as does the chapter opening image. The vague bluish patches are separate small galaxies, each containing only a few percent of the mass of the Milky Way Galaxy. Their irregular shape is thought to be the result of galaxy mergers; the bluish coloration comes from young stars that formed during the merger process.

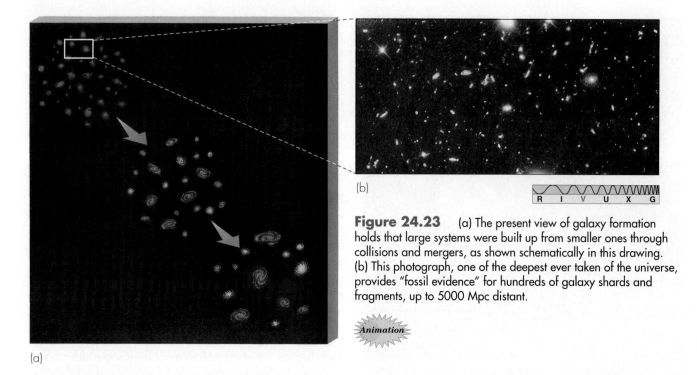

(a)

(b)

Figure 24.23 (a) The present view of galaxy formation holds that large systems were built up from smaller ones through collisions and mergers, as shown schematically in this drawing. (b) This photograph, one of the deepest ever taken of the universe, provides "fossil evidence" for hundreds of galaxy shards and fragments, up to 5000 Mpc distant.

Animation

Analysis of Figure 24.23(b) reveals a group of 18 gigantic "star clusters," all lying in the same region of space, some 600 kpc across, about 3000 Mpc from Earth (see Section 24.5 for more on how such large distances are measured). The 18 rich star clusters—far larger than the conventional open and globular clusters seen in nearby galaxies, yet smaller than typical galaxies like our own—are the small bluish blobs toward the upper left quadrant of the

figure. They are shown in more detail in Figure 24.25; each frame is a small part of the larger image, spanning about 1/10 the size of the Milky Way Galaxy. Each blob seems to contain several billion stars spread throughout a distorted spheroid a few hundred parsecs in diameter—just about the size and scale expected for pregalactic fragments. Their decidedly bluish tint suggests that active star formation is already under way. We see them as they were

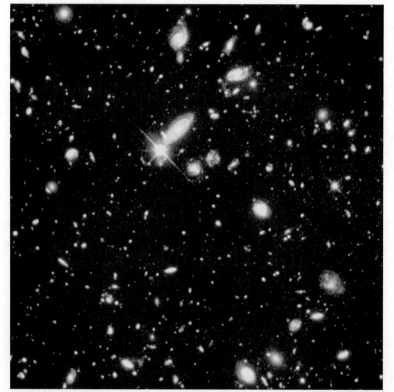

Figure 24.24 Numerous small, irregularly shaped young galaxies can be seen in this very deep optical image. Known as the Hubble Deep Field, this image, made with a total exposure of approximately 100 hours, captured objects as faint as 30th magnitude. Redshift measurements (see Section 24.5) indicate that some of these galaxies lie thousands of megaparsecs from Earth. Their size, color, and irregular appearance support the theory that galaxies grew by mergers and were smaller and less regular in the past. The field of view is about 1' across. (See also the chapter-opening image on page 546.)

Figure 24.25 Enlargements of selected portions of Figure 24.23(b) reveal rich (billion-star) "star clusters," all lying within a relatively small volume of space (some 600 kpc across). Their proximity to one another suggests that we may be seeing a group of pregalactic fragments about to merge to form a galaxy. The events pictured took place about 10 billion years ago.

nearly 10 billion years ago, apparently poised to merge into a galaxy-sized object. If this interpretation is correct, then these images lend strong support to the "bottom-up" view of galaxy formation.

If galaxies form by repeated mergers, how might we account for the differences between spirals and ellipticals? The answer is still unclear. One apparently very important factor is just when and where stars first appeared—in the original blobs, during a merger, or later—and how much gas was used up or ejected from the young galaxy in the process. If many stars formed early on and little gas was left over, an elliptical galaxy would be a likely outcome, with many old stars on random orbits and no gas to form a central disk. Alternatively, if a lot of gas remained, it would tend to sink to a central plane and form a rotating disk—in other words, a spiral galaxy would result. However, it is not known what determines the time, the place, or the rate of star formation. For this reason, it is still an open question whether spirals and ellipticals form in basically the same environment or if they tend to form in different places.

We do have some important clues to guide us, though. For example, spiral galaxies are relatively rare in regions of high galaxy density, such as the central regions of rich galaxy clusters. Is this because they simply tended not to form there, or is it because their disks are so fragile that they are easily destroyed by collisions and mergers, which are more common in dense galactic environments? Computer simulations suggest that collisions between spiral galaxies can indeed destroy the spirals' disks, ejecting much of the gas into intergalactic space (creating the hot intracluster gas noted in Section 24.3) and leaving behind objects that look very much like ellipticals. Recent observations of interacting galaxies appear to support this scenario. Figure 24.26 shows an example of this phenomenon. Additional evidence comes from the observed fact that spirals seem to be more common at larger distances (that is, in the past), implying that their numbers are decreasing with time, presumably as the result of collisions.

However, nothing in this area of astronomy is clear cut. We know of numerous isolated elliptical galaxies in low-density regions of the universe, which are hard to explain as the result of mergers. Apparently some, but *not* all, ellipticals formed in this way.

GALAXY INTERACTIONS

A related, and very important, question is: When did galaxy formation *stop*? Astronomers are divided on this issue. Some maintain that there was a fairly well defined time in the past—given by the age of the globular clusters in our own Galaxy, for example—by which most formation was over. Others point out that many galaxies show evidence of mergers and the accumulation of smaller satellite galaxies over an extended period of time—even up to the present day. These astronomers suggest that the many galaxy interactions we observe today (see *Interlude 24-1*) are just part of the

Figure 24.26 The peculiar (Irr II) galaxy NGC 1275 contains a system of long filaments that seem to be exploding outward into space. Its blue blobs, as revealed by the *Hubble Space Telescope*, are probably young globular clusters formed by the collision of two galaxies.

Contemplating the congested confines of a rich galaxy cluster (such as Virgo or Coma, with thousands of member galaxies orbiting within a region a few megaparsecs across), we might expect that collisions among galaxies would be common. Gas particles collide in our atmosphere, and hockey players collide in the rink—do galaxies in clusters collide too? The answer is yes.

This stunning image of three galaxies probably shows the aftermath of a bull's-eye collision of one of the two small galaxies at right (which one is still unknown) with the larger one at left. The result is the Cartwheel Galaxy, about 150 Mpc from Earth, its halo of young stars resembling a vast ripple in a pond. The ripple is most likely a density wave created by the passage of the smaller galaxy through the disk of the larger one (see Section 23.5). The resulting disturbance is now spreading outward from the impact region, creating new stars as it goes.

Whether some galaxies are genuinely colliding or only experiencing a close encounter cannot easily be determined. No human will ever witness an entire collision, for it would last many millions of years. However, computer simulations of these systems display formations remarkably similar to the real thing. The particular simulation shown below began with two spiral galaxies, but the details of the original structure have been largely obliterated by the collision. Notice the similarity to the real image of NGC 4038/4039 (the black and white image at right), the so-called Antennae galaxies, which show extended tails as well as double galactic centers only a few hundred parsecs across. Star formation induced by the collision is clearly traced by the blue light from thousands of young, hot stars, evident in the high-resolution *Hubble* image at right. (The *Hubble* field of view is outlined on the black and white image.) The simulations suggest that ultimately the two galaxies will merge into one.

Direct observational evidence now indicates that galaxies in clusters apparently collide quite often. In the smaller groups, the galaxies' speeds are low enough that interacting galaxies tend to "stick together," and mergers, as shown in the computer simulation, are a very common outcome. In

R I V U X G

same process begun when the first fragments merged. In this view, galaxy formation is still occurring today. In either case, astronomers have ample evidence that galaxies evolve in response to external factors, even long after formation.

We now know that spiral galaxies have huge, invisible dark halos surrounding them, and we strongly suspect that all galaxies have similar halos. Consider two galaxies orbiting each other—a binary galaxy system. As they orbit, the galaxies interact with each other's halos, one galaxy stripping halo material from the other by tidal forces. The freed matter is either redistributed within a common envelope or is entirely lost from the binary system. This interaction between the halos changes the orbits of the galaxies, which tend to spiral toward each other, eventually merging. If one galaxy of the pair happens to have a much lower mass than the other, the process is colloquially termed *galactic cannibalism*. Such cannibalism may explain why supermassive galaxies are often found at the cores of rich galaxy clusters. Having dined on their companions, they now lie

R I V U X G

Curiously, although a collision may wreak havoc on the large-scale structure of the galaxies involved, it has essentially no effect on the individual stars they contain. The stars within each galaxy just glide past one another. Although we have plenty of photographic evidence for galaxy collisions, no one has ever witnessed or photographed a collision between two stars. Stars do collide in other circumstances—in the dense central cores of galactic nuclei and globular clusters, or as a result of stellar evolution in binary systems—but stellar collisions are a very rare consequence of galaxy interactions.

To understand why individual stars do not collide when galaxies collide, recall that the galaxies within a typical cluster are bunched together fairly tightly. The distance between adjacent galaxies in a given cluster averages a few hundred thousand parsecs, which is only about 10 times greater than the size of a typical galaxy. Galaxies simply do not have that much room to roam around without bumping into one another. By contrast, stars within a galaxy are spread out much more thinly. The average distance between stars within a galaxy is a few parsecs, which is millions of times greater than the size of a typical star. When two galaxies collide, the star population merely doubles for a time, and the stars continue to have so much space that they do not run into one another. The stellar and interstellar contents of each galaxy are certainly rearranged, and the resultant burst of star formation may indeed be spectacular from afar, but from the point of view of the stars, it's clear sailing.

larger groups, galaxies move faster and tend to pass through one another without sticking. Since the early 1980s it has become increasingly clear, on the basis of both observations and numerical simulations, that collisions can have very large effects on the galaxies involved. The stellar and interstellar contents of each galaxy are rearranged, and the merged interstellar matter very likely experiences shock waves that trigger widespread bursts of star formation (see also Figures 24.26 and 24.29). Some researchers go so far as to suggest that most galaxies have been strongly influenced by collisions, in many cases in the relatively recent past.

Figure 24.27   This computer-enhanced, false-color composite optical photograph of the galaxy cluster Abell 2199 is thought to show an example of galactic cannibalism. The large central galaxy of the cluster (itself 120 kpc along its long axis) is displayed with a superimposed "window." (This results from a shorter time exposure, which shows only the brightest objects that fall within the frame.) Within the core of the large galaxy are several smaller galaxies (the three bright yellow images at center) apparently already "eaten" and now being "digested" (that is, being torn apart and becoming part of the larger system). Other small galaxies swarm on the outskirts of the swelling galaxy, almost certainly to be eaten, too.

at the center of the cluster, waiting for more food to arrive. Figure 24.27 is a remarkable combination of images that has apparently captured this process at work.

Now consider two interacting disk galaxies, one a little smaller than the other but each having a mass comparable to that of the Milky Way Galaxy. As shown in the computer-generated frames of Figure 24.28, the smaller galaxy can substantially distort the larger one, causing spiral arms to appear where none existed before. The entire event requires several hundred million years—a span of evolution that supercomputers can model in minutes.

The final frame of Figure 24.28 looks remarkably similar to the double galaxy shown in Figure 24.2(b). Shown there are two galaxies with sizes, shapes, and velocities corresponding very closely to those in the computer simulation. The magnificent spiral galaxy is M51, popularly known as the Whirlpool Galaxy, about 10 Mpc from Earth. Its smaller companion is an irregular galaxy that may have drifted past M51 millions of years ago. Did this smaller galaxy cause the spiral structure we see in M51? Does the model mirror reality? Perhaps. We need more evidence from other galaxies to confirm the accuracy of these and similar simulations. Still, the computer simulation does demonstrate a plausible way in which two galaxies may have

Time ▶

Figure 24.28 Galaxies can change their shapes long after their formation. In this computer-generated sequence, two galaxies closely interact over several hundred million years. The smaller galaxy, in red, has gravitationally distorted the larger galaxy, in blue, changing it into a spiral galaxy. Compare the result of this supercomputer simulation with Figure 24.2(b), a photograph of M51 and its small companion.

interacted millions of years ago and how spiral arms may be created or enhanced as a result.

It now seems that many of the most spectacular changes that occur in galaxies result from interactions with other galaxies. Astronomers have cataloged numerous **starburst galaxies**, such as the one shown in Figure 24.29, where violent events, possibly a near-collision with a neighbor, appear to have rearranged the galaxy's internal structure and triggered a sudden, intense burst of star formation in the recent past. Such close encounters are random events and do not seem to represent any genuine evolutionary sequence linking all spirals to all ellipticals and irregulars. However, it *is* clear that many galaxies and galaxy clusters have evolved greatly since they first formed long ago.

R I V U X G

Figure 24.29 This interacting galaxy pair (IC 694, at left, and NGC 3690) shows starbursts now under way in both galaxies—hence the bluish tint. Such intense, short-lived bursts probably last for no more than a few tens of millions of years—a small fraction of a typical galaxy's lifetime.

24.5 Hubble's Law

27 Now let's turn our attention to the large-scale *motions* of galaxies and galaxy clusters. Within a galaxy cluster, individual galaxies move in essentially random ways. You might expect that, on the largest possible scales, the clusters themselves would also have random, disordered motion—some clusters moving this way, some that. This is not the case at all, however. On the largest scales, galaxies that are not members of a cluster, and galaxy clusters alike, display very ordered motion.

UNIVERSAL RECESSION

In 1912 the American astronomer Vesto M. Slipher, working under the direction of Percival Lowell, discovered that virtually every spiral galaxy he observed had a redshifted spectrum—it was apparently *receding* from our Galaxy. In fact, with the exception of a few nearby systems, *every* known galaxy is part of a general motion away from us in all directions. Individual galaxies that are not part of galaxy clusters are steadily receding. Galaxy clusters too have an overall recessional motion, although their individual member galaxies move randomly with respect to one another. (Consider a jar full of fireflies that has been thrown into the air. The fireflies within the jar, like the galaxies within the cluster, have random motions due to their individual whims, but the jar as a whole, like the galaxy cluster, has some directed motion as well.)

Figure 24.30 shows the optical spectra of several galaxies, arranged in order of increasing distance from the Milky Way Galaxy. These spectra are redshifted, indicating that the galaxies are steadily receding. Furthermore, the extent of the redshift increases progressively from top to bottom in the figure. Because the galaxies' distances from us also increase from top to bottom, we conclude that there is a connection between Doppler shift and distance: the greater the distance, the greater the redshift. This trend holds for nearly all galaxies in the universe. (Two galaxies within our Local Group, including Andromeda, and a few galaxies in the Virgo

Figure 24.30 Optical spectra (on the left) of several galaxies (on the right). Both the extent of the redshift (denoted by the horizontal yellow arrows) and the distance from the Milky Way Galaxy to each galaxy (numbers in center column) increase from top to bottom. The vertical yellow arrow in the top spectrum highlights a particular spectral feature (a pair of dark absorption lines). The horizontal yellow arrows indicate how this feature shifts to longer wavelengths in spectra of more distant galaxies. The lines at top and bottom of each spectrum are laboratory references.

Cluster display blueshifts and so have some motion toward us, but this results from their random motions within their parent clusters. Recall the fireflies in the jar.)

Figure 24.31(a) shows recessional velocity plotted versus distance for the galaxies of Figure 24.30. Figure 24.31(b) is a similar plot for some more galaxies within about 1 billion pc of Earth. Plots like these were first made by Edwin Hubble in the 1920s and now bear his name—*Hubble diagrams*. The data points generally fall close to a straight line, indicating that a simple relationship connects recessional velocity and distance: the rate at which a galaxy recedes is *directly proportional* to its distance from us. This rule is called **Hubble's law**. We could construct such a diagram for any group of galaxies, provided we could determine their distances and velocities. The universal recession described by the Hubble diagram is sometimes called the *Hubble flow*.

To distinguish recessional redshift from redshifts caused by motion *within* an object—for example, galaxy orbits within

a cluster or explosive events in a galactic nucleus—the redshift resulting from the Hubble flow is called the **cosmological redshift**. Objects that lie so far away that they exhibit a large cosmological redshift are said to be at *cosmological distances*—distances comparable to the scale of the universe.

Hubble's law is an *empirical discovery*—that is, a discovery based strictly on observational results. Its central relationship—a statistical correlation between recessional velocity and distance—is well documented as far as galaxy distances can be reliably determined, but no law of nature demands that all galaxies recede, and no law of physics requires that a link exist between velocity and distance. In that sense, Hubble's "law" is not really a law at all. It is simply a convenient way of noting the *observational* fact that any galaxy's recessional velocity is directly proportional to its distance from us.

The recessional motions of the galaxies prove that the cosmos is not steady and unchanging on the largest scales. Its contents are in constant relative motion, and the motion is

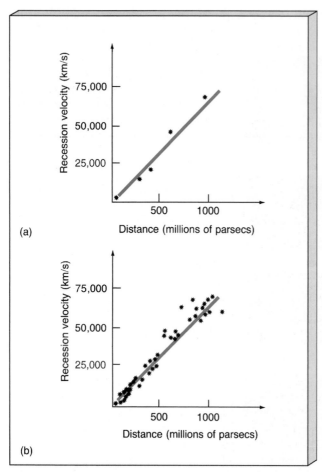

(a)

(b)

Figure 24.31 Plots of recessional velocity versus distance (a) for the galaxies shown in Figure 24.30 and (b) for numerous other galaxies within about 1 billion pc of Earth.

not random. The universe is expanding—and expanding in an ordered fashion. In short, it is evolving. Before going any further, however, let's be clear on *what* is expanding and what is not. Hubble's law does *not* mean that humans, Earth, the solar system, or even individual galaxies are physically increasing in size. These groups of atoms, rocks, planets, and stars are held together by their own internal forces and are not themselves getting bigger. Only the largest framework of the universe—the ever-increasing distances separating the galaxies and the galaxy clusters—is expanding.

Hubble's law has some fairly obvious and dramatic implications. If nearly all galaxies show recessional velocity according to Hubble's law, then doesn't that mean that they all started their journey from a single point? If we could run time backward, wouldn't all the galaxies fly back to this one point, perhaps the site of some explosion in the remote past? The answer is yes. In Chapters 26 and 27 we explore the ramifications of the Hubble flow for the past and future evolution of our universe. For the remainder of this chapter, however, we set aside its cosmic implications and use Hubble's law simply as a convenient distance-measuring tool.

HUBBLE'S CONSTANT

The constant of proportionality between recessional velocity and distance in Hubble's law is known as **Hubble's constant**. It is usually denoted by the symbol H_0. The data shown in Figure 24.31 then obey the equation

$$\text{recessional velocity} = H_0 \times \text{distance}.$$

The value of Hubble's constant is the slope of the straight line—recessional velocity divided by distance—in Figure 24.31(b). Reading the numbers off the graph, this value comes to roughly 65,000 km/s divided by 1000 Mpc, or 65 km/s/Mpc (kilometers per second per megaparsec, the most commonly used unit for H_0). Astronomers continually strive to refine the accuracy of the Hubble diagram and the resulting estimate of H_0 because Hubble's constant is one of the most fundamental quantities of nature—it specifies the rate of expansion of the entire cosmos.

The precise value of Hubble's constant is still the subject of considerable debate. In the 1970s, astronomers obtained a value of around 50 km/s/Mpc, using a chain of standard candles to extend their observations to large distances. However, in the early 1980s, after the Tully–Fisher technique had become fairly well established, other researchers used it to obtain a measurement of H_0 that was largely independent of methods relying on standard candles. From observations of galaxies within about 150 Mpc, the latter group deduced a value of $H_0 \approx 90$ km/s/Mpc, a result inconsistent with the earlier measurements (even allowing for the estimated uncertainties involved). For some reason, the distances obtained using the Tully–Fisher method were only about half those determined by using standard candles, so the measured value of Hubble's constant nearly doubled using this approach.

Subsequent determinations of H_0 by other researchers, using different galaxies and a variety of distance-measurement techniques, have yielded results mostly within the range 45–90 km/s/Mpc, and most astronomers would be quite surprised if the true value of H_0 turned out to lie outside this range. The recent revision to the cosmic distance scale by the *Hipparcos* mission, by increasing our distance estimates, has correspondingly decreased the measured value of H_0. ∞ (*Interlude 17-1*) We will adopt $H_0 = 65$ km/s/Mpc as the best current estimate of Hubble's constant for the remainder of the text, but you should bear in mind that there is still considerable uncertainty—and debate—among astronomers as to the true value of this very important number. (The distances in Figures 24.30 and 24.31 are based on older observations, but they have been adjusted to reflect the new *Hipparcos* findings.)

THE COSMIC DISTANCE SCALE

One very important application of Hubble's law is its use as a means of determining distances. Using Hubble's law, we can derive the distance to a remote object simply by measuring the object's recessional velocity. An astronomer measures the redshift of the object's spectral lines. The extent of the shift is then converted to velocity, using the Doppler relationship discussed in Chapter 3. ∞ (Sec. 3.5) Knowing the object's velocity, the astronomer then finds its distance by using the plot of Figure 24.31(b). Notice, however, that the uncertainty in Hubble's constant translates directly into a similar uncertainty in the distance determined by this method.

Using Hubble's law in this way tops our inverted pyramid of distance-measurement techniques (Figure 24.32). This sixth method simply assumes that Hubble's law holds. If this assumption is correct, Hubble's law enables us to measure great distances in the universe—so long as we can obtain an object's spectrum, we can determine how far away the object is.

Many redshifted objects have recessional motions that are a substantial fraction of the speed of light. The most distant object thus far observed in the universe has the catalog name 0140 + 326RD1. Its extremely high redshift implies a recessional velocity 95 percent that of light. At that speed, ultraviolet spectral lines are Doppler shifted all the way into the far infrared! Hubble's law implies that, solely on the basis of its observed redshift, 0140 + 326RD1 lies some 5500 Mpc away from us (see *More Precisely 25-1*). This object resides as close to the limits of the observable universe as astronomers have yet been able to probe.

The speed of light is finite. It takes time for light or any kind of radiation to travel from one point in space to another. The radiation that we now see from distant objects originated long ago. Incredibly, the radiation that astronomers now detect from 0140 + 326RD1 left that object almost 9.5 bil-

lion years ago, well before our planet, our Sun, and perhaps even our Galaxy came into being. (See again *More Precisely 25-1* for reconciliation of this time with the present distance of 18 billion light years implied by the previous paragraph.)

LARGE-SCALE STRUCTURE

3 Using Hubble's law, we can complete our census of the large-scale distribution of galaxies. One of the most extensive surveys of the universe yet undertaken has been carried out by astronomers at Harvard University. Using Hubble's law as their distance indicator, these researchers are compiling a catalog of the positions and redshifts of all galaxies within about 250 Mpc of our Galaxy. This is an extremely painstaking task—even with a large telescope, it takes a long time to obtain a detailed spectrum of a distant galaxy. Rather than trying to cover the entire sky at once, the team elected to map the universe in a series of wedge-shaped "slices," each 6° thick, starting in the northern sky. The first slice, shown in Figure 24.33, covered a region of the sky containing the Coma Cluster (Figure 24.1), which happens to lie in a direction almost perpendicular to our Galaxy's plane.

The most striking feature of maps such as this is that the distribution of galaxies on very large scales is decidedly nonrandom. The galaxies appear to be arranged in a network of strings, or filaments, surrounding large, relatively empty regions of space known as **voids**. The biggest voids measure some 100 Mpc across. For a time they were the largest objects in the universe known to astronomers. The most likely explanation for the voids and filamentary structure in Figure 24.33 is that the galaxies and galaxy clusters are

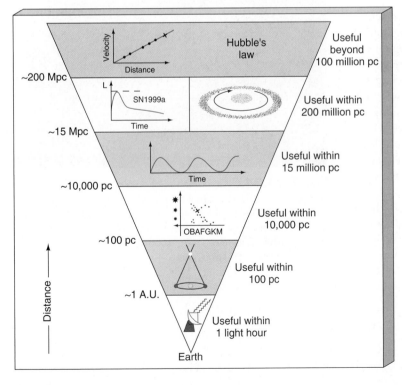

Figure 24.32 Hubble's law tops the hierarchy of distance-measurement techniques. It is used to find the distances of astronomical objects all the way out to the limits of the observable universe.

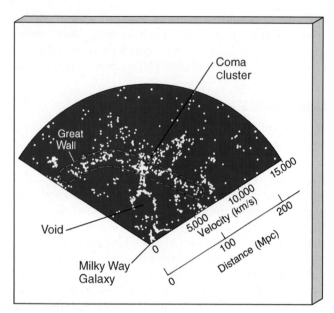

Figure 24.33   The first slice of a survey of the universe, covering 1057 galaxies out to an approximate distance of 200 Mpc, clearly shows that galaxies and clusters are not randomly distributed on large scales. Instead, they appear to have a filamentary structure, surrounding vast, nearly empty voids. The distances shown assume $H_0 = 65$ km/s/Mpc.

spread across the surfaces of vast "bubbles" in space. The voids are the interiors of these gigantic bubbles. The galaxies seem to be distributed like beads on strings only because of the way our slice of the universe cuts through the bubbles. Like suds on soapy water, these bubbles fill the entire universe. The densest clusters and superclusters lie in regions where several bubbles meet.

The notion that the filaments are just the intersection of the survey slice with much larger structures (the bubble surfaces) was confirmed when the next three slices of the survey, lying above and below the first, were completed. The region of Figure 24.33 indicated by the red outline was found to continue through both the other slices, so we know that it covers *at least* 36° on the sky perpendicular to the outline in the figure. This extended sheet of galaxies, which has come to be known as the *Great Wall*, measures at least 70 Mpc (out of the plane of the page) by 200 Mpc (across the page). For now at least, it is one of the largest known structures in the universe. Figure 24.34 combines all the available data with results of other surveys of the southern sky. (The large empty regions are either obscured by our Galaxy or not yet mapped.) The Great Wall can be seen arcing around the left side of the figure.

For a time, theorists tried to explain voids as gigantic cavities hollowed out by the explosions of extremely massive stars early in the history of the universe, but this explanation does not work. The energy requirements would have been just too great, and the explosions would have had other observable consequences (mostly related to the radiation they would have released) that simply are not seen. Instead, most theorists now believe that all structure on scales larger than a few megaparsecs traces its origin directly to the conditions found in the very earliest stages of the universe (Chapter 27). As such, these studies of large-scale structure may be vital to our efforts to understand the origin and nature of the cosmos itself.

Are there still larger structures in the universe? Only with more extensive surveys will we know for sure. All large-scale surveys performed to date have detected structures comparable in size to the region surveyed—in other words, they have found objects as large as the largest object they could have hoped to find. Observers and theorists alike are eager to find out how much farther this trend will continue. With several new larger-scale surveys nearing completion, some answers may be in hand early in the 21st century.

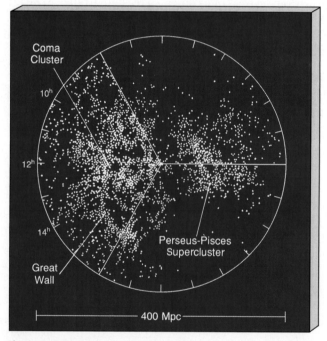

Figure 24.34   Combination of data from several redshift surveys of the universe reveal the extent of large-scale structure within 200–300 Mpc of the Sun. The arc on the left is the Great Wall. The empty regions are mostly areas obscured by our Galaxy. Positions for more than 4500 galaxies are plotted here. We assume $H_0 = 65$ km/s/Mpc.

Chapter Review

SUMMARY

The **Hubble classification scheme** (p. 548) divides galaxies into several classes, depending on their appearance. **Spiral galaxies** (p. 549) have flattened disks, central bulges, and spiral arms. They are further subdivided on the basis of the size of the bulge and the tightness of the spiral structure. The halos of these galaxies consist of old stars, whereas the gas-rich disks are the sites of ongoing star formation. **Barred-spiral galaxies** (p. 549) contain an extended "bar" of material projecting beyond the central bulge.

Elliptical galaxies (p. 550) have no disk and contain no gas or dust. In most cases, they consist entirely of old stars. They range in size from dwarf ellipticals, which are much less massive than the Milky Way Galaxy, to giant ellipticals, which may contain trillions of stars. **S0 and SB0 galaxies** (p. 551) have properties intermediate between those of ellipticals and spirals. They have extended halos and stellar disks and bulges (and bars, in the SB0 case) but little or no gas and dust.

Irregular galaxies (p. 551) are galaxies that do not fit into either of the other categories. Some may be the result of galactic collisions or close encounters. Many irregulars are rich in gas and dust and are the sites of vigorous star formation. The **Magellanic Clouds** (p. 552), two small systems that orbit the Milky Way Galaxy, are examples of this type of galaxy.

Astronomers often use **standard candles** (p. 554) as distance-measuring tools. These are objects that are easily recognizable (by their light curves, spectra, or some other directly observable characteristic) and whose luminosities are known to lie in some reasonably well-defined range. Comparing luminosity and apparent brightness, astronomers determine the distance using the inverse-square law. Type I supernovae are particularly useful for measuring distances to faraway galaxies. They are bright and hence easily seen, and have a relatively narrow spread in luminosity. An alternative is the **Tully–Fisher relation** (p. 555), an empirical correlation between rotational velocity and luminosity in spiral galaxies. By measuring the rotation speed of a spiral and using this relationship, astronomers can determine the galaxy's luminosity and hence its distance.

The Milky Way, Andromeda, and several other smaller galaxies form the **Local Group** (p. 556), a small **galaxy cluster** (p. 557). Galaxy clusters consist of a collection of galaxies orbiting one another, bound together by their own gravity. The nearest large galaxy cluster to the Local Group is known as the Virgo Cluster. Galaxy clusters themselves tend to clump together into **super-clusters** (p. 558). The Virgo Cluster, the Local Group, and several other nearby clusters form the Local Supercluster.

The masses of nearby spiral galaxies can be determined by studying their rotation curves. For more distant spirals, masses can be inferred from observations of the broadening of their spectral lines. On larger scales, astronomers use studies of binary galaxies and galaxy clusters to obtain statistical mass estimates of the galaxies involved. As in the Milky Way Galaxy, measurements of the masses of other galaxies and of galaxy clusters reveal the presence of large amounts of dark matter that is presently undetectable at any electromagnetic wavelength. The fraction of dark matter apparently grows as the scale under consideration increases. Large amounts of hot X-ray-emitting gas have been detected among the galaxies in many clusters, but not enough to account for the dark matter inferred from dynamical studies.

Researchers know of no evolutionary sequence that links spiral, elliptical, and irregular galaxies, and the process of galaxy formation is still only poorly understood. There is growing evidence that large galaxies formed by the merger of smaller ones in a process that may be continuing today. Collisions and mergers of galaxies play very important roles in galactic evolution. Interactions between galaxies appear to be very common. A **starburst galaxy** (p. 566) may result when a galaxy experiences a close encounter with a neighbor. The strong tidal distortions caused by the encounter compress galactic gas, resulting in a widespread burst of star formation.

Distant galaxies are observed to be receding from the Milky Way at rates that increase proportional to their distances from us. This relationship between recessional speed and distance is called **Hubble's law** (p. 567). The constant of proportionality in the law is **Hubble's constant** (p. 568). Its value is believed to lie between 45 and 90 km/s/Mpc, with most astronomers now favoring a value closer to the lower end of this range. Astronomers use Hubble's law to determine distances to the most remote objects in the universe. The redshift associated with the Hubble expansion is called the **cosmological redshift** (p. 567).

On very large scales, galaxies and galaxy clusters are not spread randomly throughout space. Instead, they are arranged on the surfaces of enormous "bubbles" of matter surrounding vast low-density regions called **voids** (p. 569). The origin of this structure is thought to be closely related to conditions in the very earliest epochs of the universe.

SELF-TEST: TRUE OR FALSE?

____ **1.** Barred-spiral galaxies have the same properties as normal spirals, except for the "bar" feature.

____ **2.** Elliptical galaxies do not contain a flattened disk.

____ **3.** There is no interstellar dust in elliptical galaxies, but there are substantial amounts of interstellar gas.

____ **4.** Most ellipticals contain only old stars.

____ **5.** Most galaxies are spirals.

____ **6.** Irregular galaxies, although small, have lots of star formation taking place in them.

____ **7.** Spiral galaxies evolve into ellipticals.

____ **8.** Type I supernovae can be used to determine distances to galaxies.

____ **9.** Every galaxy is a member of some galaxy cluster.

____ **10.** Galaxy collisions can occur, but they are extremely rare.

_____ **11.** Galaxy collisions have little or no effect on the stars and interstellar gas in the galaxies involved.

_____ **12.** Distant galaxies appear to be much larger than those nearby.

_____ **13.** A typical galaxy cluster has a mass of about 10^{11} solar masses.

_____ **14.** Most galaxies appear to be receding from the Milky Way Galaxy.

_____ **15.** Hubble's law can be used to determine distances to the farthest objects in the universe.

SELF-TEST: FILL IN THE BLANK

1. Galaxies are categorized by their _____ classification.

2. Spiral galaxies with tightly wrapped spiral arms tend to have _____ central bulges.

3. Spiral galaxies of type _____ have the least amount of gas; type _____ have the most.

4. Elliptical galaxies generally contain far _____ gas and dust than do irregulars.

5. The Milky Way Galaxy, the Andromeda Galaxy, and 18 other galaxies form a small cluster known as the _____.

6. In the Tully–Fisher relation, a galaxy's luminosity is found to be related to the _____ of its 21-cm line.

7. The diameter of the Local Supercluster is about _____ .

8. When galaxies collide, the star formation rate often _____ .

9. Galaxy mass determinations from rotation curves, line broadening, and binary galaxies all make use of _____ laws.

10. Dark matter may make up as much as _____ percent of the entire universe.

11. Intergalactic gas in galaxy clusters emits large amounts of energy in the form of _____.

12. By which process do galaxies form: fragmentation (large objects breaking up into small) or mergers (small objects accumulating into large)? _____

13. Hubble's law is a correlation between the redshifts and the _____ of galaxies.

14. Hubble's constant is believed to lie in the range _____ to _____ km/s/Mpc.

15. The largest known structures in the universe, such as voids and the Great Wall, have sizes on the order of _____.

REVIEW AND DISCUSSION

1. In what sense are elliptical galaxies "all halo"?

2. Describe the four rungs in the distance-measurement ladder used to determine the distance to a galaxy lying 5 Mpc away.

3. Describe the contents of the Local Group. How much space does it occupy compared to the volume of the Milky Way?

4. How is the Tully–Fisher relation used to measure distances to galaxies?

5. What is the Virgo Cluster?

6. Describe two techniques for measuring the mass of a galaxy.

7. Why do astronomers believe that galaxy clusters contain more mass than we can see?

8. Why are galaxies at great distances from us generally smaller and bluer than nearby galaxies?

9. What evidence do we have that galaxies collide with one another?

10. Describe the role of collisions in the formation and evolution of galaxies.

11. Do you think that collisions between galaxies constitute "evolution" in the same sense as the evolution of stars?

12. What is Hubble's law?

13. How is Hubble's law used by astronomers to measure distances to galaxies?

14. What is the most likely range of values for Hubble's constant? Why is the exact value uncertain?

15. What are voids? What is the distribution of galactic matter on very large (more than 100 Mpc) scales?

PROBLEMS

1. A supernova of luminosity 1 billion times the luminosity of the Sun is used as a standard candle to measure the distance to a faraway galaxy. From Earth the supernova appears as bright as the Sun would appear from a distance of 10 kpc. What is the distance to the galaxy?

2. A Cepheid variable star in the Virgo cluster has an absolute magnitude of −5 and is observed to have an apparent magnitude of 26.3. Use these figures to calculate the distance to the Virgo cluster.

3. The Andromeda Galaxy is approaching our Galaxy with a radial velocity of 266 km/s. Given the galaxies' present separation of 930 kpc, and neglecting both the transverse component of the velocity and the effect of gravity in accelerating the motion, estimate when the two galaxies will collide.

4. Calculate the average speed of hydrogen nuclei (protons) in a gas of temperature 50 million K (see *More Precisely 8-1*). Compare this with the average orbital speed of 1000 km/s of galaxies in a typical galaxy cluster.

5. Based on the data in Figure 24.19, estimate the amount of line broadening (maximum minus minimum wavelength) of the 656.3 nm Hα line observed in the galaxy NGC 4984.

6. Two galaxies are orbiting each other at a distance of 500 kpc. Their orbital period is estimated to be 30 billion years. Use Kepler's law (as stated in Section 23.6) to find the total mass of the pair.

7. Use Kepler's third law (Section 23.6) to estimate the mass required to keep a galaxy moving at 1000 km/s in a circular orbit or radius 3 Mpc around the center of a galaxy cluster. Given the approximations involved in determining this figure, do you think this is a good estimate of the cluster's true mass?

8. In a galaxy collision, two similar-sized galaxies pass through each other with a combined relative velocity of 1500 km/s. If each galaxy is 100 kpc across, how long will the event last?

9. According to Hubble's law, with $H_0 = 65$ km/s/Mpc, what is the recessional velocity of a galaxy at a distance of 200 Mpc? How far away is a galaxy whose recessional velocity is 4000 km/s? How do these answers change if $H_0 = 50$ km/s/Mpc? If $H_0 = 80$ km/s/Mpc?

10. According to Hubble's law, with $H_0 = 65$ km/s/Mpc, how long will it take for the distance from the Milky Way Galaxy to the Virgo Cluster to double?

PROJECTS

1. Look for a copy of the *Atlas of Peculiar Galaxies* by Halton Arp. It is available in book form or on laser disk. Search for examples of interacting galaxies of various types: (1) tidal interactions, (2) starburst galaxies, (3) collisions between two spirals, and (4) collisions between a spiral and an elliptical. For (1) look for galactic material pulled away from a galaxy by a neighboring galaxy. Is the latter galaxy also tidally distorted? In (2) the surest sign of starburst activity are bright knots of star formation. In what type(s) of galaxies do you find starburst activity? For (3) and (4) how do collisions differ depending on the types of galaxies involved. What typically happens to a spiral galaxy after a near-miss or collision? Do ellipticals suffer the same fate?

2. Look for the Virgo Cluster of galaxies. An 8-inch telescope is a perfect size for this project, although a smaller telescope will also work. The constellation Virgo is visible from the United States during much of fall, winter, and spring. To locate the center of the cluster, first find the constellation Leo. The eastern part of Leo is composed of a distinct triangle of stars, Denebola (β), Chort (θ), and Zosma (δ). Go from Chort to Denebola in a straight line east and continue on the same distance as between the two stars and you will be approximately at the center of the Virgo Cluster. Look for the following Messier objects that make up some of the brightest galaxies in the cluster: M49, 58, 59, 60, 84, 86, 87 (giant elliptical thought to have a massive black hole at its center), 89, and 90. Examine each galaxy for unusual features; some have very bright nuclei.

25 ACTIVE GALAXIES AND QUASARS

Limits of the Observable Universe

LEARNING GOALS

Studying this chapter will enable you to:

1 Specify the basic differences between active and normal galaxies.

2 Describe the important features of Seyfert and radio galaxies.

3 Explain what drives the central engine thought to power all active galaxies.

4 Describe the observed properties of quasars and discuss the special properties of the radiation they emit.

5 Discuss the place of active galaxies in current theories of galactic evolution.

(Opposite page, background) This artist's conception depicts one possible scenario for the "central engine" of an active galaxy. Two jets of matter are shown moving outward perpendicular to a flattened accretion disk surrounding a supermassive black hole.

(Inset A) Numerical simulations performed on supercomputers can help us visualize key events in the universe. Here a uniform distribution of some thousand stars is spread over a dimension of several hundred parsecs.

(Inset B) As the simulation proceeds, the hypothetical stars congregate preferentially toward the core of the star cluster, forming a "luminosity cusp" of light.

(Inset C) If we could see inside the cusp, an accretion disk would probably be evident, swirling around a giant black hole.

(Inset D) As the accretion disk rotates, tilts, and acquires more matter, jets of matter emerge to cool the environment surrounding the hole.

*O*ur journey from the Milky Way to the Great Wall in the past two chapters has widened our cosmic field of view by a factor of 10,000, yet the galaxies that make up the structures we see show remarkable consistency in their properties. The overwhelming majority of galaxies fit neatly into the Hubble Classification Scheme, showing few, if any, unusual characteristics. However, sprinkled through the mix of normal galaxies, even relatively close to the Milky Way Galaxy, are some that are decidedly abnormal in their properties. Although their optical appearance is often quite ordinary, these abnormal galaxies emit huge amounts of energy—far more than a normal galaxy— mostly in the invisible part of the electromagnetic spectrum. Observing such objects at great distances, we may be seeing some of the formative stages of our own galactic home.

25.1 Beyond the Local Realm

1 Astronomers have recognized and cataloged spiral, elliptical, and irregular galaxies as far away as several hundred megaparsecs from Earth. Beyond this distance, galaxies appear so faint that it is difficult to discern their shapes. Consequently, their types are largely unknown. Nevertheless, according to their observed redshifts (and Hubble's law), we know that many galaxies lie well beyond this distance, in the farthest reaches of the observable universe. But what kinds of objects are they? Are they normal galaxies— close relatives of the galaxies that populate our local neighborhood—or are they somehow different? The answer is that although most distant galaxies appear basically normal in their properties, some are distinctly different from galaxies found in our cosmic backyard: they are more luminous— more *active*—than the typical galaxy found close to home. These energetic objects, some of them hundreds, even thousands, of times more luminous than the Milky Way Galaxy, are known collectively as **active galaxies**.

In addition to their greater overall luminosities, active galaxies differ fundamentally from normal galaxies in the *character* of the radiation they emit. Most of a normal galaxy's radiated energy is emitted in or near the visible portion of the electromagnetic spectrum, much like the radiation from ordinary stars. Indeed, to a large extent, the light we see from a normal galaxy is just the accumulated light of its many component stars. For example, our entire Milky Way has a luminosity of about 10^{37} W at optical frequencies—20 billion Suns' worth of radiation—but only 10^{31} W at radio frequencies—a million times less. By contrast, as illustrated in Figure 25.1, the radiation observed emitted by active galaxies does *not* peak at optical frequencies—far more energy is emitted at longer wavelengths than in the visible range. The radiation from active galaxies is *inconsistent* with what we would expect if it were the combined radiation of myriad stars. The radiation is thus said to be *nonstellar* in nature.

The two most important categories of active galaxies are *Seyfert galaxies* and *radio galaxies*, although other classes exist (see, for example, *Interlude 25-1*). Even more extreme in their properties are the *quasars*. Astronomers conventionally distinguish between active galaxies and quasars based on their spectra, appearance, and distance from us. At visible wavelengths active galaxies often *look* like fairly normal galaxies. Indeed, for many purposes, we can think of active galaxies as

being otherwise "normal" systems (that is, emitting visible radiation) that happen also to be extremely intense sources of radio and/or infrared radiation. Quasars, on the other hand, are mostly so far away that little internal structure can be discerned. However, this distinction between active galaxies and quasars is largely historical, dating from the days when the connection between them was not understood. As we will see, most astronomers now believe that quasars are simply an early stage of galaxy formation and that there is really no sharp dividing line between quasars and active galaxies. Many researchers include quasars in the "active galaxy" category.

Not all active galaxies are distant, and only a small fraction of all distant galaxies are active. Some active galaxies are found locally, scattered among the normal galaxies that make up most of our cosmic neighborhood, and many faraway "normal luminosity" galaxies are known. As a general rule, though, active galaxies are more common at greater distances, and the most active objects lie farthest from Earth.

We might wonder whether this predominance of energetic objects with large redshifts is just an observational effect, resulting from our inability to detect relatively faint normal galaxies at great distances. After all, the apparent brightness

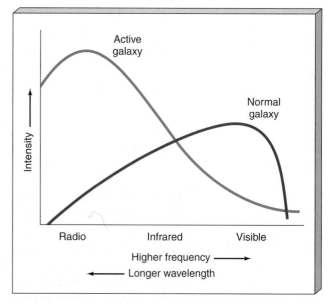

Figure 25.1   The nature of the energy emitted from a normal galaxy differs from that of an active galaxy. (This plot illustrates the general run of intensity for all galaxies of a particular type and does not represent any one individual galaxy.)

of any astronomical object decreases as the square of its distance from us, so even with the very best telescopes, we would expect to observe only the more energetic and powerful galaxies in remote regions of space. Although this observational bias does play a role, it turns out that it can only partly explain the apparent predominance of energetic objects among those having large redshifts. Bright active galaxies really are more common at great distances.

Physical conditions were undoubtedly different at earlier times from what they are now. Perhaps, then, we should not be surprised that remote astronomical objects, which long ago emitted the radiation we observe today, differ from nearby objects, which emitted their radiation much more recently. What is surprising—in fact, astounding—is the *amount* of energy radiated by some of the most luminous objects. Their tremendous power, nonstellar radiation, and abundance at great distances suggest to many astronomers that the universe was once a much more violent place than it is today.

25.2 Seyfert Galaxies

2 In 1943 Carl Seyfert, an American optical astronomer studying spiral galaxies from Mount Wilson Observatory, discovered the type of active galaxy that now bears his name. **Seyfert galaxies** are a class of astronomical objects whose properties lie between those of normal galaxies like the Milky Way and those of the most violent active galaxies known. This fact suggests to some astronomers that Seyferts represent an evolutionary link between these two extremes. The spectral lines of Seyfert galaxies are usually substan-

tially redshifted, telling us that most Seyferts reside at large distances (hundreds of megaparsecs) from us. ∞ (Sec. 24.5) However, a few lie just 20 or 30 Mpc away.

Figure 25.2 shows two optical images of a typical Seyfert galaxy. A casual glance at a long-exposure photograph of a Seyfert (Figure 25.2a) reveals nothing strange. Superficially, Seyferts resemble normal spiral galaxies. However, closer study of Seyferts reveals some peculiarities not found in normal spirals.

First, maps of Seyfert energy emission show that nearly all the radiation stems from a small central region known as the **galactic nucleus**. This region lies at the center of the overexposed white patch in Figure 25.2(a); it is shown in more detail in Figure 25.2(b). Astronomers suspect that a Seyfert nucleus may be quite similar to the center of a normal galaxy, such as the Milky Way or the Andromeda Galaxy, but with one very important difference: the nucleus of a Seyfert is 10,000 times brighter than the center of our Galaxy. Indeed, the brightest Seyfert nuclei are 10 times more energetic than the *entire* Milky Way.

Second, Seyfert galaxies emit their radiation in two broad frequency ranges. The stars in the Seyfert's galactic disk and spiral arms produce about the same amount of *visible* radiation as those of a normal spiral galaxy. However, most of the energy from the Seyfert's nucleus is emitted in the form of *invisible* radio and infrared radiation, which cannot be explained as coming from stars—it must be nonstellar in origin.

Third, Seyfert spectral lines bear little or no resemblance to those produced by ordinary stars, although they do have many similarities to the spectral lines observed toward the center of our own Galaxy. ∞ (Sec. 25.7) Seyfert spectra contain strong emission lines of highly ionized heavy

(a)   (b)

Figure 25.2   Photographs of a Seyfert galaxy (NGC 5728) (a) from the ground and (b) from Earth orbit. The enlarged view in (b) of two cone-shaped beams of light shows a group of glowing blobs near the galaxy's core, perhaps illuminated by radiation arising from the accretion disk of a black hole. This object is about 40 Mpc away—one of the closest active galaxies known.

elements, especially iron. The lines are very broad, indicating either that the galaxy's gases are tremendously hot (more than 10^8 K) or that they are rotating very rapidly (at about 1000 km/s) around some central object. ∞ (Sec. 4.5) The first possibility can be ruled out, since such a high temperature would cause all the gas to be ionized, in which case no spectral lines would be produced. Thus, the broadening indicates rapid internal motion in the nucleus.

Finally, extensive monitoring of Seyfert radiation over long periods of time has shown that the energy emission often varies over time. Figure 25.3 shows an example of luminosity variations for a typical Seyfert. Such radiative changes are unlike anything found in the Milky Way or in any other normal galaxy. A Seyfert's luminosity can double or halve within a fraction of a year.

These rather rapid fluctuations lead us to conclude that the source of energy emissions must be quite compact. As mentioned in Chapter 22 (in the context of luminosity variations in neutron stars and black holes), for astronomers to be able to detect a coherent variation in brightness within a certain time interval, the source of radiation must be smaller in size than the distance traveled by light during that interval. ∞ (Sec. 22.3) Otherwise, the intensity variations would be blurred, not sharp, as observed. Simply put, an object cannot "flicker" in less time than radiation takes to cross it. Because the rise and fall of a Seyfert's radiation usually occurs within one year, we conclude that the emitting region must be less than 1 light year across—an extraordinarily small region, considering the amount of energy emanating from it.

High-resolution interferometric radio maps of Seyfert cores generally confirm this reasoning. ∞ (Sec. 5.5)

The rapid time variability and large radio and infrared luminosities observed in Seyferts together imply violent nonstellar activity in their nuclei. This activity may well be similar in *nature* to processes occurring at the center of our own Galaxy, but its *magnitude* is thousands of times greater than the comparatively mild events within our own Galaxy's heart. ∞ (Sec. 23.7)

25.3 Radio Galaxies

2 As the name suggests, **radio galaxies** are active galaxies that emit most of their energy in the radio portion of the electromagnetic spectrum. They differ from Seyferts not only in the wavelengths at which they radiate most strongly (Seyferts tend to radiate at shorter radio wavelengths and infrared wavelengths; radio galaxies radiate at longer radio wavelengths) but also in both the appearance and the extent of their emitting regions. Whereas the nonstellar emission from Seyfert galaxies always comes from a small central nucleus, the nonstellar emitting regions of radio galaxies can be much larger—hundreds of kiloparsecs across in some cases.

CORE–HALO RADIO GALAXIES

One common type of radio galaxy is often called a *core–halo* radio galaxy. As illustrated in Figure 25.4, the energy from such an object comes mostly from an extremely small central nucleus (which radio astronomers refer to as the *core*) less than 1 pc across, with weaker emission coming from an

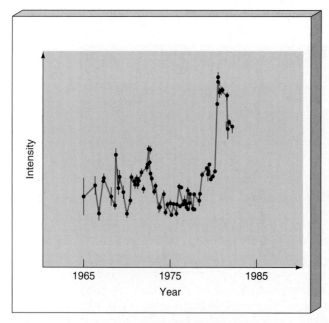

Figure 25.3 The irregular variations of a particular Seyfert galaxy's luminosity over a period of two decades. Because this Seyfert, called 3C 84, emits most strongly in the radio part of the electromagnetic spectrum, these observations were made with large radio telescopes. The optical and X-ray luminosities vary as well.

Figure 25.4 Radio contour map of a typical core–halo radio galaxy. The radio emission from such a galaxy comes from a bright central nucleus, or core, surrounded by an extended, less intense halo. The radio map is superimposed on an optical image of the galaxy and some of its neighbors, shown previously in Figure 24.15.

Figure 25.5 The giant elliptical galaxy M87 (also called Virgo A) is displayed here at several different wavelengths. (a) A long optical exposure of its halo and embedded central region. (b) A short optical exposure of its core and an intriguing jet of matter, on the same scale as (a). (c) A radio image of its jet, on a somewhat expanded scale compared with (b). The red dot at left marks the bright nucleus of the galaxy; the red and yellow blob near the center of the image corresponds to the bright "knot" visible in the jet in (b). (d) A near-infrared image of the jet, at roughly the same scale as (c).

extended region called the *halo* surrounding the nucleus. The halo typically measures about 50 kpc across, similar in size to the surrounding visible galaxy, which is usually elliptical and often quite faint.* The radio luminosity from the nucleus can be as great as 10^{37} W—about the same as the emission from a Seyfert nucleus and comparable to the output from our entire Milky Way Galaxy at all wavelengths.

Figure 25.5 presents several images of a nearby core—halo radio galaxy. This object is a giant elliptical galaxy

known as M87—the 87th object in Messier's catalog. (We can be sure that this eighteenth-century Frenchman had no idea what he was really looking at. Nor perhaps do we!) M87 is roughly 18 Mpc distant, a prominent member of the Virgo Cluster, and one of the closest active galaxies. Its nearness and interesting activity have made it one of the most intensely studied of all astronomical objects.

A long time-exposure (Figure 25.5a) shows a large fuzzy ball of light—a fairly normal looking type E1 elliptical galaxy whose full extent is about 100 kpc across (considerably larger than the view shown in the figure). A shorter time-exposure of M87 (Figure 25.5b), capturing only the galaxy's bright inner regions, reveals a long thin *jet* of matter ejected from

*The term visible galaxy *is commonly used to refer to the components of an active galaxy that emit visible "stellar" radiation, as opposed to the nonstellar "active" component of the galaxy's emission.*

M87's center. The jet is about 2 kpc long and is traveling outward at very high speed, possibly as much as half the speed of light. Computer enhancement shows that the jet is made up of a series of distinct "blobs" more or less evenly spaced along its length. This high-speed jet, which emits energy at the rate of almost 10^{35} W, has been imaged in the radio and infrared regions of the spectrum (Figures 25.5c and d) as well as in the visible region. Jets such as this are a very common feature of active galaxies. As we will see, they play a vital role in our understanding of these energetic objects.

LOBE RADIO GALAXIES

Many radio galaxies are not of the core–halo type. Very little of their radio emission arises from the compact central nucleus. Instead, most of this emission comes from huge extended regions called **radio lobes**—roundish clouds of gas up to 1 Mpc across, lying well beyond the visible part of the galaxy. For this reason, these objects are known as *lobe radio galaxies*.

The radio lobes of all lobe radio galaxies are truly enormous. From end to end, the radio lobes are more than 10 times the size of the Milky Way Galaxy, comparable in size to the entire Local Group. The lobes emit no visible light, but their radio luminosity can range from 10^{36} to 10^{38} W— between 1/10 and 10 times the total energy emitted by our Galaxy. Several of these strange objects are located relatively nearby, so we can study them at close range. One such system, known as Centaurus A, is shown in Figure 25.6. It lies only 4 Mpc from Earth.

Figure 25.7 superposes another representation of Figure 25.6 on an optical image of Centaurus A, to show the relationship between the visible and radio emission. In visible light, Centaurus A is a rather peculiar-looking object, apparently an E2 galaxy bisected by an irregular band of dust. Numerical simulations suggest that this system is probably

Figure 25.6   Lobe radio galaxies, such as Centaurus A shown here, have giant radio-emitting regions extending a million parsecs or more beyond the central galaxy. The lobes cannot be imaged in visible light and are observable only with radio telescopes. (The lobes are filled with false color to indicate decreasing intensity from red to yellow to green to blue.)

the result of a merger between an elliptical galaxy and a smaller spiral galaxy about 500 million years ago. The radio lobes are roughly symmetrically placed, jutting out from the center of the visible galaxy roughly perpendicular to the dust lane. The elliptical galaxy itself is very large—some 500 kpc in diameter.

The lobes of radio galaxies vary in size and shape from galaxy to galaxy, but they are aligned with the center of the visible galaxy in nearly all cases. This alignment suggests that the lobes consist of material that was somehow ejected in oppo-

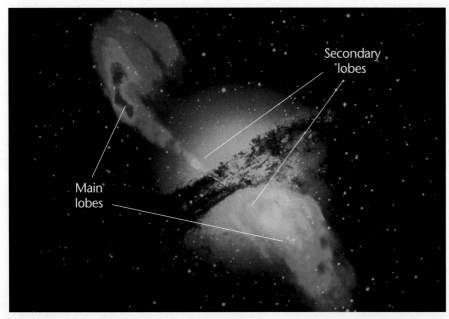

Secondary lobes

Main lobes

Figure 25.7   An optical photograph of Centaurus A, one of the most massive and peculiar galaxies known and believed to be the result of a galaxy collision some 500 million years ago. The pastel false colors mark the radio emission shown in Figure 25.6, in this case more recently acquired and with higher resolution. Note that the radio lobes emit no visible light.

(a)

R I V U X G

(b)

R I V U X G

Figure 25.8 (a) Cygnus A also appears to be two galaxies in collision, although it is not completely clear that that is what is really happening. (b) On a much larger scale, this cosmic object displays radio-emitting lobes on each side of the optical image. To put these images into proper perspective, the optical galaxy in (a) is about the size of the small dot at the center of the "radiograph" in (b). Notice the thin line of radio-emitting material joining the right lobe to the central galaxy. The distance from one lobe to the other is approximately a million light years.

site directions by violent events in the galactic nucleus. In the case of Centaurus A, this argument is strengthened by the presence of an additional pair of secondary lobes, smaller than the main lobes (about 50 kpc in length, marked in Figure 25.7) and closer to the visible galaxy. Both pairs of lobes share the same high degree of linear alignment. Astronomers believe that the inner lobes were expelled from the nucleus by the same basic process as the outer ones, but more recently, so they have not had time to travel as far. Still higher resolution studies reveal the presence of a roughly 1-kpc-long jet in the center of Centaurus A, again aligned with the larger lobes.

If material is ejected from the nucleus at close to the speed of light (which seems likely), it follows that Centaurus A's outer lobes were created a few hundred million years ago, quite possibly around the time of the collision thought to be responsible for the galaxy's peculiar optical appearance. Apparently, some violent process at the center of Centaurus A—most probably triggered by the merger—started up around that time and has been intermittently firing jets of matter out into intergalactic space ever since.

Further evidence in favor of the interpretation of radio lobes as material ejected from the center of a galaxy is provided by another object, Cygnus A, shown as an optical image in Figure 25.8(a) and as high-resolution radio map in Figure 25.8(b). Note the filamentary structure evident in the radio lobes and the thin, radio-emitting line joining the right lobe to the center of the visible galaxy (the dot at the center of the radio image). These features strongly suggest that we are seeing two oppositely directed, narrow jets of material running into the intracluster gas filling the galaxy cluster of which Cygnus A is a member, then slowing down and spreading out to form the radio lobes. ∞ (Sec. 24.3)

In some systems, known as *head–tail* radio galaxies, the lobes seem to form a "tail" behind the main galaxy. For example, the lobes of radio galaxy NGC 1265, shown in Figure 25.9, appear to be "swept back" by some onrushing wind and, indeed, this is the most likely explanation for

Figure 25.9 (a) Radiograph, in false color, of the active "head–tail" galaxy NGC 1265. (b) The same radio data, in contour form, superposed on the optical image of the galaxy. Astronomers reason that this object is moving rapidly through space, trailing a "tail" behind as it goes—a little like a comet, but on a vastly larger scale.

(a)

R I V U X G

(b)

R I V U X G

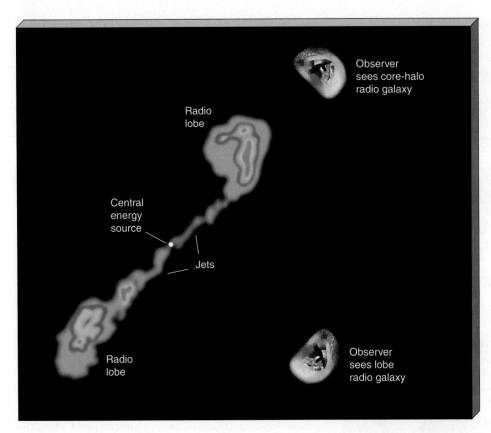

Figure 25.10 A central energy source produces high-speed jets of matter that interact with intergalactic gas to form radio lobes. The system may appear to us as either a lobe or a core–halo radio galaxy, depending on our location with respect to the jets and lobes.

the galaxy's appearance. If NGC 1265 were at rest, it would be just another double-lobe source, perhaps looking quite similar to Centaurus A. However, the galaxy is traveling through the intergalactic medium of its parent galaxy cluster (known as the Perseus Cluster), and the outflowing matter forming the lobes tends to be left behind as the galaxy moves.

Radio galaxies share many characteristics with Seyfert galaxies. Both types of active galaxies emit comparably large amounts of energy, and there is good evidence that the energy source in both is a compact region at the center of an otherwise relatively normal looking galaxy. In lobe radio galaxies, that energy is fired out from the nucleus in the form of narrow, high-speed jets of matter that travel into the intergalactic medium and become extended lobes far from the center of the galaxy. As a result, the energy from a lobe radio galaxy is ultimately emitted (in the form of radio radiation) from a region well outside the visible galaxy. It may be that the differences between core–halo and lobe radio galaxies is largely a matter of perspective (Figure 25.10). If we view the jets and lobes from the side, we see a lobe radio galaxy, but if we view the jet almost head-on—in other words, looking *through* the lobe—we see a core–halo system. If that is so, then in *all* cases studied so far the central compact nucleus is the place where the energy is actually produced. Let us now consider the current view of the "engine" that powers all this activity.

25.4 The Central Engine of an Active Galaxy

3 The behavior of active galaxies is contrary to that expected from vast collections of stars. The lobe radio galaxies in particular, with their huge energy emission from far beyond the optical galaxy, are among the most powerful objects in the universe. Can we explain this enormous nonstellar energy output in terms of known physics? Remarkably, the answer is yes. The present consensus among astronomers is that, despite the great differences in appearance, Seyferts and radio galaxies may share a common energy-generation mechanism. The energy can be reprocessed into many different forms before it is finally emitted into intergalactic space, but the basic engine is probably the same in both cases.

As a class, active galaxies (and quasars too, as we will see) show some of or all the following properties.

1. They have *high luminosities*, generally greater than the 10^{37} W characteristic of a fairly bright normal galaxy like the Milky Way.
2. Their energy emission is mostly *nonstellar*—it cannot be explained as the combined radiation of even trillions of stars.
3. Their energy output can be highly *variable*, implying that it is emitted from a small central nucleus much less than a parsec across.

INTERLUDE 25-1 BL Lac Objects

In 1929 an object thought to be a variable star was discovered in the constellation Lacerta. Astronomers gave it the two-letter code BL, so it became known as BL Lacertae, or BL Lac for short. Not until the 1970s, when it became clear that BL Lac was a strong radio source, did anyone question its classification as a star. As more objects like BL Lac were found, astronomers began to realize that they had stumbled across a whole new class of extragalactic sources. *BL Lac objects* seem starlike in a telescope and vary greatly in luminosity over time. They are also powerful compact radio sources, sometimes colloquially referred to as "blazars."

Astronomers originally placed BL Lac objects into a special class because they displayed no spectral lines. Their redshifts could not be determined, and their distances were unknown. However, recent discoveries of extremely faint spectral lines strongly suggest that BL Lac objects are extremely distant, making them nearly as luminous as quasars (discussed later in the text). Further careful observations have shown that blazars reside at the centers of relatively normal elliptical galaxies. We consider these galaxies active because of the tremendous strength of the blazar radio emission.

The weak spectral lines may be the key needed to unravel the nature of the BL Lac objects. Evidently, the thermal emission from any stars they contain, which would provide spectral lines, is swamped by their nonthermal synchrotron radiation (see Section 25.4), which does not contain spectral features. Other types of active galaxies show a mixture of thermal and nonthermal radiation, so it is more difficult to study the nonthermal radiation alone. BL Lac objects, then, may offer astronomers a chance to study the "bare machine" (which generates the nonthermal radiation) powering these active galaxies.

In some ways the BL Lac objects seem to represent a "link" or transition phase between radio galaxies and quasars (see Section 25.5). The luminosities of the compact radio sources inside BL Lac objects span the whole range from the relatively weak cores observed in radio galaxies to the stronger ones found in quasars. Perhaps there exists an evolutionary sequence of some sort, in which the ancient quasars use up their fuel, turning into the weaker yet more erratic BL Lac objects, which subsequently become the rather inactive cores of radio galaxies. An alternative possibility is that BL Lacs—and perhaps also the weaker core–halo radio sources—*are* just lobe radio galaxies, but radio galaxies in which the jet happens to be pointing nearly straight at us. We would then be seeing the galaxy *through* the nearer radio lobe. Only further observations and better statistics will tell us for sure which (if either) of these two possibilities is correct.

4. They often exhibit *jets* and other signs of explosive activity.
5. Their optical spectra may show broad emission lines, indicative of *rapid internal motion* within the energy-producing region.

The principal questions then are: How can such vast quantities of energy arise from these relatively small regions of space? Why is so much of the energy radiated at low frequencies, especially in the radio and infrared? And what is the origin of the extended radio-emitting lobes and jets? Let us first consider how the energy is produced.

ENERGY PRODUCTION

To develop a feeling for the enormous emissions of active galaxies, consider for a moment an object with a luminosity of 10^{38} W. In and of itself, this energy output is not inconceivably large. The brightest giant ellipticals are comparably powerful. Thus, some 10^{12} stars—a few normal galaxies' worth of material—could *equivalently* power a typical active galaxy. The difficulty arises when we consider that in an active galaxy this energy production is packed into an object much less than a parsec in diameter!

It is difficult to imagine how several Milky Way Galaxies could be compressed into a space no larger than a parsec. Even if we could somehow squeeze that much mass into such a volume, it would immediately collapse to form a huge black hole, and none of the light it produced could escape. Thus, even neglecting its nonstellar spectrum, the total energy output of an active galaxy simply cannot be explained as the combined energy of many stars. We must think of something else.

The twin requirements of large energy generation and small physical size bring to mind our discussion of X- and gamma-ray sources in Chapter 22. ∞ (Secs. 22.3, 22.7) The presence of the jet in M87 and the ejection of matter to form radio lobes in Centaurus A and Cygnus A strengthen the connection. Recall that the best current explanation for those "small-scale" phenomena involves the accretion of material onto a compact object—a neutron star or a black hole. Large amounts of energy are produced as matter spirals down onto the central object, and high-speed jets may well be a common by-product of the process (see Figure 22.6). In Chapter 23, we suggested that a similar mechanism, involving a *supermassive black hole*—one with a mass of around a million suns—may also be responsible for the energetic radio and infrared emission observed at the center of our own Galaxy. ∞ (Sec. 23.7)

As illustrated in Figure 25.11, the leading model for the central engine of active galaxies is essentially a scaled-up version of the same accretion process, only now the black holes involved have masses between a few million and a billion times the mass of the Sun. As with its smaller-scale counterparts, infalling gas forms an accretion disk and spirals down toward the hole. It is heated to high temperatures by friction within the disk and emits large amounts of radiation as a

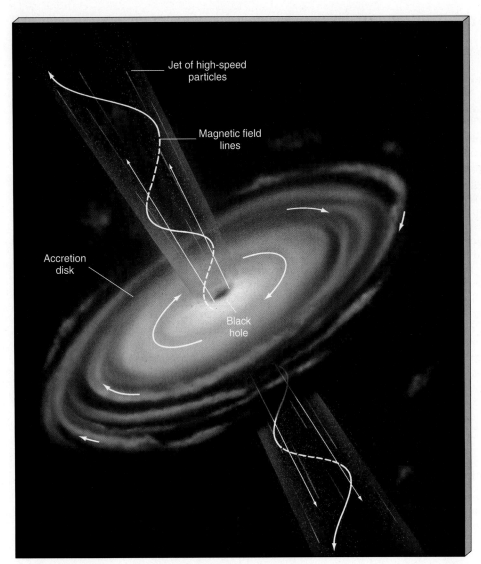

Figure 25.11 The leading theory for the energy source in active galactic nuclei (and quasars) holds that these objects are powered by material accreting onto a supermassive black hole. As matter spirals toward the hole it heats up, producing large amounts of energy. At the same time high-speed beams of gas may be ejected perpendicular to the accretion disk, forming the jets and lobes seen in many active objects. Magnetic fields generated in the disk are carried by the jets out to the radio lobes, where they play a crucial role in producing the observed radiation.

result. In this case, however, the origin of the accreted gas is not a binary companion, as in stellar X-ray sources, but entire stars and clouds of interstellar gas that come too close to the hole and are torn apart by its strong gravity.

Accretion is extremely efficient at converting infalling mass (in the form of gas) into energy (in the form of electromagnetic radiation). Detailed calculations indicate that as much as 10 or 20 percent of the total mass–energy of the infalling matter can be radiated away before it crosses the hole's event horizon and is lost forever. Since the total mass–energy of a star like the Sun—the mass times the speed of light squared—is about 2×10^{47} joules, if follows that the 10^{38}-W luminosity of a bright active galaxy can be accounted for by the consumption of only 1 solar mass of gas per decade by a billion–solar mass black hole. ∞ (22.4) Less luminous active galaxies would require correspondingly less fuel—for example, a 10^{36}-W Seyfert galaxy would devour only one star's worth of material every thousand years.

In this picture the small size of the emitting region is a direct consequence of the compact nature of the central black hole. Even a billion–solar mass hole has a radius of only 3×10^9 km, or 10^{-4} pc—about 20 A.U.—and theory suggests that the part of the accretion disk responsible for most of the emission is much less than a parsec across. Instabilities in the accretion disk can cause fluctuations in the energy released, leading to the variability observed in many objects. The broadening of the spectral lines observed in the nuclei of Seyferts (and in quasars) results from the rapid orbital motion of the gas in the hole's intense gravity.

Recent observations of galaxies in the Virgo Cluster by the *Hubble Space Telescope* lend strong support to this general picture. Figure 25.12 shows an image of a disk of gas and dust apparently feeding a possible black hole at the core of a giant elliptical galaxy. As expected from the theory just described, the disk is perpendicular to the huge jets emanating from the center of the active galaxy.

Hubble has also allowed astronomers to probe the fine details of the Virgo Cluster's most prominent object—the huge M87 galaxy (Figure 25.5)—and what they have found is again in excellent agreement with the idea that the energy is produced by accretion onto a large black hole. At M87's distance, *Hubble's* resolution of 0.05″ corresponds to a distance

Figure 25.12 (a) A combined optical/radio image of the giant elliptical galaxy NGC 4261, in the Virgo Cluster, shows a white visible galaxy at center, from which red-orange (false color) radio lobes extend for about 60 kpc. (b) A close-up photograph of the galaxy's core reveals a 100-pc-diameter disk surrounding a bright hub that presumably harbors a black hole.

Animation

of about 5 pc, so we are still far from seeing the (solar system–sized) central black hole itself, but the improved "circumstantial" evidence has convinced many doubters of the correctness of the theory. Figure 25.13 shows imaging and spectroscopic data that suggest the existence of a rapidly rotating disk of matter orbiting the galaxy's center, perpendicular to the jet. Measurements of the gas velocity on opposite sides of the disk indicate that the mass within a few parsecs of the center is approximately 3×10^9 solar masses—we assume that this is the mass of the central black hole.

Even more compelling evidence for a supermassive black hole comes from radio studies. Using the Very Long Baseline Array—a continent-wide network of 10 radio telescopes—a U.S.–Japanese team was able to achieve angular resolution hundreds of times better than that attainable with the *Hubble Space Telescope*. VLBA observations of NGC 4258, a spiral galaxy about 6 Mpc away, have uncovered a group of molecular clouds swirling in an organized fashion about the galaxy's core. The astronomers have detected the red- and blueshifts of water-vapor spectral lines in those faraway

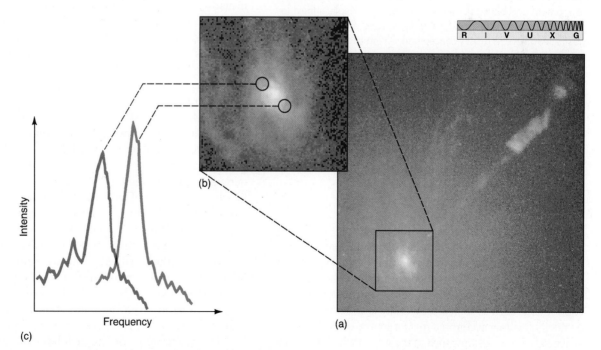

Figure 25.13 Imaging and spectroscopic observations of M87 support the idea of a rapidly whirling accretion disk at its core. (a) An image of the central region of M87, similar to that shown in Figure 25.5(d), shows its bright core and jet. The scale is comparable to the scale of Figure 25.5(c). (b) A magnified view of the core suggests a spiral swarm of stars, gas, and dust. (c) Spectral-line features observed on opposite sides of the core show opposite Doppler shifts, implying that the material there is coming toward us on one side and moving away from us on the other. The strong implication is that an accretion disk spins perpendicular to the jet, and that at its center is a black hole having some 3 billion times the mass of the Sun.

Animation

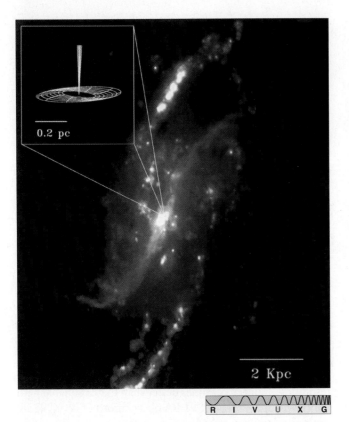

Figure 25.14 A network of radio telescopes has probed the core of the spiral galaxy NGC 4258, shown here in the light of mostly hydrogen emission. Within the innermost 0.2 pc (inset), observations of Doppler-shifted molecular clouds (designated by red, green, and blue dots) show that they obey Kepler's third law perfectly, and have revealed a slightly warped disk of rotating gas (shown here in artist's conception). At the center of the disk presumably lurks a huge black hole.

clouds. As depicted in Figure 25.14, the pattern revealed by the clouds is that of a slightly warped and spinning disk centered precisely on the galaxy's heart. The rotation velocities imply the presence of more than 40 million solar masses all packed within a region less than 0.2 pc across.

ENERGY EMISSION

The model just described is now widely accepted as the correct picture of how active galaxies and quasars generate their enormous power. It provides a natural explanation of the observed facts and has the added advantage of being essentially similar to the processes thought to power smaller-scale phenomena, such as stellar X-ray sources and normal galactic nuclei. Having accounted for the source of the energy in active galaxies, let us now turn to the way in which it is eventually emitted into intergalactic space.

In general, the theoretically predicted radiation spectra for matter accreting onto a black hole do not match well with the spectra observed in active galaxies. In order to account for the details of the spectra of Seyfert galaxies, it is necessary to modify the mechanism just described by adding the assump-

tion that the energy emitted from the accretion disk is "reprocessed"—that is, absorbed and reemitted at infrared and longer wavelengths—by gas and dust surrounding the nucleus.

A different reprocessing mechanism is responsible for the spectra of the jets and lobes seen in many active galaxies, especially radio galaxies. The jets consist of material (mainly protons and electrons) blasted out into space—and out of the visible portion of the galaxy entirely—from the inner regions of the disk. The details of how jets form remain uncertain, but there is a growing consensus among theorists that jets are a common feature of accretion flows, large and small. The jets also contain strong *magnetic fields* (shown in Figure 25.11), possibly generated by the swirling gas motion within the disk; the fields accompany the gas as it leaves the galaxy.

As sketched in Figure 25.15(a), whenever a charged particle (here an electron) encounters a magnetic field, it tends to spiral around the field lines. We have encountered this idea several times previously, in a variety of different contexts (see, for example, the discussion of planetary magnetospheres in Chapters 7 and 11, or solar activity in Chapter 16. As the particles whirl around, they emit electromagnetic radiation, as discussed in Chapter 3. ∞ (Sec. 3.2) The faster the particles move, or the stronger the magnetic field, the greater the amount of energy radiated. In most cases, the fastest-moving particles are the low-mass electrons, so they are responsible for most of the radiation we observe.

The electromagnetic radiation produced in this way—called **synchrotron radiation**, after the type of particle accelerator in which it was first observed—is *nonthermal* in nature: there is no link between the emission and the temperature of the radiating object, so the radiation is not described by a blackbody curve. Instead, its intensity decreases with increasing frequency, as shown in Figure 25.15(b). This is just what is needed to explain the overall spectrum of radiation recorded from active galaxies (compare Figure 25.15b with Figure 25.1). Observations of the radiation received from the jets and radio lobes in active galaxies are completely consistent with this process.

Eventually, the jet is slowed and stopped by the intergalactic medium, the flow becomes turbulent, and the magnetic field grows tangled. The result is a giant radio lobe, like those pictured in Figures 25.6–25.8, emitting virtually all its energy in the form of synchrotron radiation. Even though the radio emission comes from an enormously extended volume of space that dwarfs the visible galaxy, the *source* of the energy is still the (relatively) tiny accretion disk—a billion billion times smaller in volume than the radio lobe—lying at the galactic center. The jets serve merely as a conduit to transport energy from the nucleus, where it is generated, into the lobes, where it is finally radiated into space.

The existence of the inner lobes of Centaurus A and the blobs in M87's jet imply that jet formation may be an intermittent process. There is also evidence to suggest that much of, if not all, the activity observed in nearby active galaxies could have been sparked by recent interaction with a neighbor. Many nearby active galaxies (Centaurus A, for example) appear to have been "caught in the act" of inter-

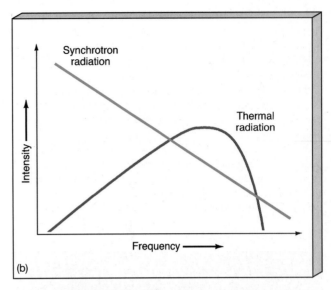

Figure 25.15 (a) Charged particles, especially fast electrons (red), emit synchrotron radiation (blue) while spiraling along magnetic field lines (black lines). This process is not confined to active galaxies. It occurs, on smaller scales, when charged particles interact with magnetism in Earth's Van Allen belts (see Chapter 7), when charged matter arches above sunspots on the Sun (see Chapter 16), in the vicinity of neutron stars (see Chapter 22), and at the center of our own Galaxy (see Chapter 23). (b) Variation in the intensity of thermal and synchrotron (nonthermal) radiation with frequency. Thermal radiation, described by a blackbody curve, peaks at some frequency that depends on the temperature of the source. Nonthermal synchrotron radiation, by contrast, is most intense at low frequencies. It is independent of the temperature of the emitting object. Compare this figure with Figure 25.1.

acting with another galaxy, suggesting that the fuel supply can sometimes be turned on by a companion. Just as tidal forces can trigger star formation in starburst galaxies (Chapter 24), they may also divert gas and stars into the galactic nucleus, triggering an outburst that may last for millions or even billions of years. ∞ (Sec. 24.4)

25.5 Quasi-stellar Objects

THE DISCOVERY OF QUASARS

4 In the early days of radio astronomy, many radio sources were detected for which no corresponding visible object was known. By 1960 several hundred such sources were listed in the *Third Cambridge Catalog*, and astronomers were scanning the skies in search of optical counterparts. Their job was made difficult both by the low resolution of the radio observations (which meant that the observers did not know exactly where to look) and by the faintness of these objects at visible wavelengths.

In 1960, astronomers detected what appeared to be a faint blue star at the location of the radio source 3C 48 (the 48th object on the Cambridge list) and obtained its spectrum. Containing many unknown broad emission lines, the unusual spectrum defied interpretation. 3C 48 remained a unique curiosity until 1962, when another similar-looking, and similarly mysterious, faint blue object with "odd" spectral lines was discovered and identified with the radio source 3C 273. Several of these peculiar objects are shown in Figure 25.16.

The following year saw a breakthrough, when astronomers realized that the strongest unknown lines in 3C 273's spectrum were simply familiar spectral lines of hydrogen redshifted by a very unfamiliar amount—about 16 percent! This large redshift indicated a recessional velocity of about 44,000 km/s (see *More Precisely 25-1*). Figure 25.17 shows the spectrum of 3C 273. Some prominent emission lines, and the extent of their redshift, are marked. Once the nature of the strange spectral lines was known, astronomers quickly found a similar explanation for the spectrum of 3C 48. Its 37 percent redshift implied that it is receding from Earth at almost one-third the speed of light.

These huge speeds mean that neither of the two objects can possibly be members of our Galaxy. Applying the Hubble law (with our adopted value of the Hubble constant $H_0 = 65$ km/s/Mpc), we obtain distances of 660 Mpc for 3C 273 and 1340 Mpc for 3C 48. (*More Precisely 25-1* discusses in more detail how these distances are determined and what they mean). Clearly not stars (with such enormous redshifts), these objects became known as *quasi-stellar radio sources* ("quasi-stellar" simply means "starlike"). The term has been shortened to **quasars**. Because we now know that not all such highly redshifted, starlike objects are strong radio sources, the term **quasi-stellar object** (or QSO) is more common today. However, the name quasar persists, and we will continue to use it here.

OBSERVED PROPERTIES OF QUASARS

The most striking characteristic of the several hundred quasars now known is that their spectra all show large redshifts, ranging from 0.06 (that is, a 6 percent increase in

Figure 25.16 (a) Optical photograph of 3C 275, one of the first quasars discovered. Its starlike appearance shows no obvious structure and gives little outward indication of this object's enormous luminosity. However, 3C 275 has a much larger redshift than any of the other stars or galaxies in this image; it is about 2 billion pc away. (b) Optical image of a field of quasars (marked), including QSO 1229+204, one of the most powerful quasars yet discovered, shown enlarged in (c). Its starlike appearance shows only a hint of structure and gives little outward indication of the object's enormous luminosity. Like 3C 275, its distance from Earth is about 2000 Mpc.

Animation

(a)

R I V U X G

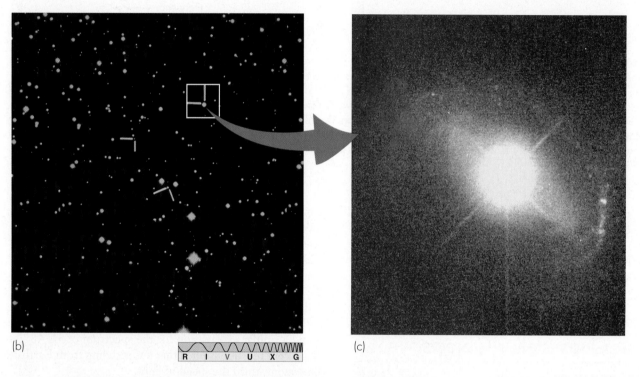

(b)

R I V U X G

(c)

Figure 25.17 Optical spectrum of the distant quasar 3C 273. Notice both the redward shifts and the widths of the three hydrogen spectral lines marked as Hβ, Hγ, and Hδ. The redshift indicates the quasar's enormous distance. The width of the lines implies rapid internal motion within the quasar itself. (Note that in this figure, red is to the right and blue is to the left.)

Figure 25.18 Although quasars are the most luminous objects in the universe, they are often rather unimpressive in appearance. In this optical image, a distant quasar (marked by an arrow) is seen close (on the sky) to a nearby spiral galaxy. The quasar's much greater distance makes it appear much fainter than the galaxy.

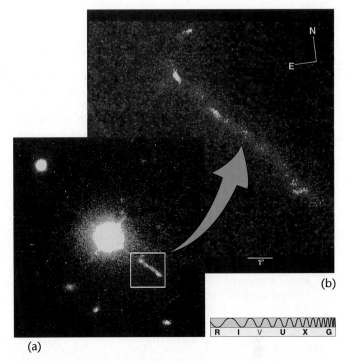

Figure 25.19 (a) The bright quasar 3C 273 displays a luminous jet of matter, but the main body of the quasar is starlike in appearance. (b) The jet extends for about 30 kpc and can be seen better in this high-resolution image.

wavelength) up to the current maximum of about 5. (See again *More Precisely 25-1* for an explanation of the meaning of redshifts greater than 1.) Thus, *all* quasars lie at large distances from us—the closest is 240 Mpc away, the farthest nearly 5500 Mpc (Table 25.1). The majority of quasars lie more than 1000 Mpc from Earth. We therefore see most quasars as they existed long ago—they represent the universe as it was in the distant past.

Thus, despite their unimpressive optical appearance—see for example, Figure 25.18, which compares a quasar and a spiral galaxy that happens to lie close to it on the sky—the large distances implied by quasar redshifts mean that these faint "stars" are in fact the brightest known objects in the universe! 3C 273, for example, has a luminosity of about 10^{40} W. More generally, quasars range in luminosity from around 10^{38} W—about the same as the brightest radio galaxies—up to nearly 10^{42} W. A value of 10^{40} W, comparable to 20 trillion Suns or 1000 Milky Way Galaxies, is fairly typical. Thus quasars outshine the brightest normal and active galaxies by about a factor of 1000.

Quasars display many of the same general properties as active galaxies. Their radiation is nonthermal, and some show evidence of jets and extended emission features (although few quasars are more than a ball of luminous fuzz in visible-light images). Figure 25.19 is an optical photograph of 3C 273. Notice the jet of luminous matter, reminiscent of the jet in M87, extending nearly 3 kpc from the quasar itself. Often, as shown in Figure 25.20, quasar radio radiation arises from regions lying beyond the bright central core, much like the

emission from core–halo and lobe radio galaxies. In other cases the radio emission is confined to the central optical image. Quasars have been observed in the radio, infrared, optical, ultraviolet, and X-ray parts of the electromagnetic spectrum,

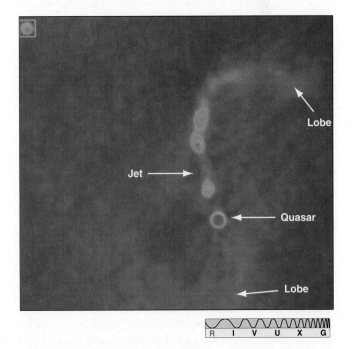

Figure 25.20 Radio image of the quasar 2300-189 showing radio jets feeding faint radio lobes. The bright (red) central object is the quasar, some 400 Mpc away.

MORE PRECISELY 25-1 Relativistic Redshifts and Look-back Time

When discussing very distant objects such as quasars, astronomers usually talk about their redshifts rather than their distances. Indeed, it is very common for researchers to speak of an event occurring "at" a certain redshift—meaning that the light received today from that event is redshifted by the specified amount. Of course, because of Hubble's law, redshift and distance are equivalent to each other. However, redshift is the "preferred" quantity, because it is a directly observable property of an object, whereas distance is derived from redshift using Hubble's constant, whose value is not accurately known. (In the next chapter we will see another reason why astronomers favor the use of redshift in studies of the cosmos.)

The redshift of a beam of light is, by definition, the *fractional* increase in its wavelength resulting from the recessional motion of the source. ∞ (Sec. 3.5) Thus, a redshift of 1 corresponds to a *doubling* of the wavelength. From the formula for the Doppler shift presented in Chapter 3, the redshift of radiation received from a source moving away from us with speed v is given by

$$\text{redshift} = \frac{\text{observed wavelength} - \text{true wavelength}}{\text{true wavelength}}$$

$$= \frac{\text{recessional velocity } v}{\text{speed of light } c}.$$

Let's illustrate this with two examples, rounding the speed of light c to 300,000 km/s. A galaxy at a distance of 100 Mpc has a recessional velocity (by Hubble's law) of 65 km/s/Mpc × 100 Mpc = 6500 km/s. Its redshift therefore is 6500 km/s ÷ 300,000 km/s = 0.022. Conversely, an object with a redshift of 0.05 has a recessional velocity of 0.05 × 300,000 km/s = 15,000 km/s and hence a distance of 15,000 km/s divided by 65 km/s/Mpc = 230 Mpc.

Unfortunately, while the foregoing equation is quite correct for low velocities, it does not take into account the

effects of relativity. As we saw in Chapter 22, the rules of everyday physics have to be modified when velocities begin to approach the speed of light, and the formula for the Doppler shift is no exception. ∞ (*More Precisely 22-1*) In particular, although our formula is valid for velocities much less than the speed of light, when $v = c$, the redshift is not 1, as the equation suggests, but is in fact *infinite*. In other words, radiation received from an object moving away from us at nearly the speed of light would be redshifted to almost infinite wavelength.

Thus do not be alarmed to find that many quasars have redshifts greater than 1. This does not mean that they are receding faster than light! It simply means that their recessional velocities are *relativistic*—comparable to the speed of light—and the preceding simple formula is not applicable. Table 25.1 presents a conversion chart relating redshift, recession speed, and present distance. The column headed "v/c" gives recession velocities based on the Doppler effect, taking relativity properly into account (but see Section 26.2 for a more correct interpretation of the redshift). All values are based on reasonable assumptions and are usable even for $v \approx c$. We take Hubble's constant to be 65 km/s/Mpc. The conversions in the table are used consistently throughout this text. Notice that the entries in the table agree very well with our earlier simple formulas for small redshifts (less than a few percent, corresponding to objects lying within 100 or 200 Mpc of Earth), they differ greatly for larger distances.

Because the universe is expanding, the "distance" to a galaxy is not very well defined—do we mean the distance when the galaxy emitted the light we see today, or the present distance (as listed in the table), even though we do not see the galaxy as it is today, or is some other measure more appropriate? Largely because of this ambiguity, astronomers prefer to work in terms of a quantity known as the *look-back time* (shown in the last column of Table 25.1), which is simply how long ago an object emitted the radiation we see

---

and some have even been found to emit gamma rays. However, most quasars emit most of their energy in the infrared.

In addition to their own strongly redshifted spectra, many quasars also show additional absorption features that are redshifted by substantially less than the lines from the quasar itself. For example, the quasar known as PHL 938 has an emission-line redshift of 1.955, placing it at a distance of some 3900 Mpc, but it also shows three sets of absorption lines, with redshifts of 1.949, 1.945, and 0.613, respectively. The first two sets may well come from high-speed gas within the quasar itself (the differences in recession velocity are only a few hundred kilometers per second), but the third is interpreted as arising from intervening gas that is much closer to us (only about 2000 Mpc away), which explains why it has a smaller redshift than the quasar itself. The most likely possibility is that this gas is part of an otherwise invisible galaxy lying along the line of sight. Quasar spectra, then, afford astronomers a means of probing previously undetected parts of the universe.

Many quasars have been observed to vary irregularly, and by large factors, in brightness over periods of months, weeks, days, or (in some cases) even hours, in many parts of the electromagnetic spectrum. The same reasoning as we used earlier for active galaxies leads to the conclusion that the region generating the energy must be very *small*—not too much larger than our solar system in some cases. *More Precisely 25-2* discusses another curious aspect of quasar variability that may have much to tell us about the details of the energy source.

QUASAR ENERGY GENERATION AND LIFETIMES

5 Quasars have all the properties described earlier for active galaxies—large luminosities, nonthermal emission, jets, lobes, and rapid variability (implying small size)—so it should come as no surprise that the best current explanation of the quasar engine is basically a scaled-up version of the today. Researchers talk frequently about redshifts and sometimes about look-back times, but hardly ever of the distances to high-redshift objects.

For nearby sources the look-back time is numerically equal to the distance in light years—the light we receive tonight from a galaxy at a distance of 100 million light years was emitted 100 million years ago. However, for more distant objects, the look-back time and the present distance in light years differ because of the expansion of the universe, and the divergence increases dramatically with increasing redshift. For example, a galaxy now located 15 billion light years from Earth was much closer to us when it emitted the light we now see. Consequently, its light has taken considerably less than 15 billion years—in fact, only about 8.8 billion years—to reach us.

As a simple analogy, imagine an ant crawling across the surface of an expanding balloon at a constant speed of 1 cm/s relative to the balloon's surface. After 10 seconds the ant may think it has traveled a distance of 10 cm, but an outside observer with a tape measure will find that it is actually more than 10 cm from its starting point (measured along the surface of the balloon), simply because of the balloon's expansion. Furthermore, the difference between the actual distance and 10 cm depends on the details of the balloon's expansion during the journey—the more rapid the expansion, the greater the disparity. In exactly the same way the present distance to a galaxy with a given redshift depends in detail on how the universe expanded in the past. We will see in Chapter 26 that the details of the past expansion are not well known, so the distance is not well determined. The distances given in the accompanying table are used consistently throughout this book, but realize that they are subject to considerable uncertainty.

Finally, notice that, according to the table, the recessional velocity equals the speed of light—and the redshift becomes infinite—for objects that emitted their radiation

about 10 billion years ago. We will examine the reasons for and implications of this fact in the next chapter.

TABLE 25.1   Redshift and Recessional Velocity

REDSHIFT	v/c	PRESENT DISTANCE (Mpc)	PRESENT DISTANCE (10^6 ly)	LOOK-BACK TIME (millions of years)
0.000	0.000	0	0	0
0.010	0.010	46	149	149
0.025	0.025	113	369	365
0.050	0.049	222	725	708
0.100	0.095	429	1401	1336
0.200	0.180	804	2622	2400
0.250	0.220	974	3177	2853
0.500	0.385	1693	5523	4570
0.750	0.508	2251	7346	5697
1.0	0.600	2702	8815	6483
1.5	0.724	3390	11062	7492
2.0	0.800	3899	12720	8099
3.0	0.882	4612	15048	8775
4.0	0.923	5099	16637	9132
5.0	0.946	5459	17809	9347
10.0	0.984	6443	21022	9754
50.0	0.999	7933	25882	10001
100.0	1.000	8307	27101	10019
∞	1.000	9224	30096	10029

mechanism powering lower-luminosity active galaxies—accretion onto a supermassive black hole.

A 10^8– or 10^9–solar-mass black hole can emit enough energy to power even the brightest (10^{38}-W) radio galaxy by swallowing stars and gas at the relatively modest rate of 1 star every 10 years. To power a 10^{40}-W quasar, which is 100 times brighter, the hole simply consumes 100 times more fuel—10 stars per year. The reprocessing mechanisms that convert the quasar's power into the radiation we actually detect—namely, the ejection of matter in jets and lobes and the reemission of radiation by surrounding gas and dust—probably operate in much the same manner as the mechanisms described earlier for Seyferts and for radio galaxies. The most likely explanation for the large luminosities of quasars is simply that there was more fuel available at very early times, perhaps left over from the formation of the galaxies in which the quasars reside. At the distances of most quasars, the galaxies themselves cannot easily be seen. Often, only their intensely bright nuclei are visible from Earth.

In this picture the brightest known quasars devour about 1000 solar masses of material every year. A simple calculation indicates that if they kept up this rate of energy production for the roughly 10-billion-year age of the universe, a total of 10^{13} stars would have had to be destroyed. Unless the galaxies housing quasars are much larger than any other galaxy we know of, most of the quasar's parent galaxy would be completely consumed, and the universe should contain many 10^{13}–solar mass black holes—"burned-out" quasars. We have no evidence of the existence of any such objects.

One way around this problem is to suppose that a quasar spends only a fairly short period of time in this highly luminous phase—perhaps a few tens of millions of years. There is theoretical evidence to suggest that black holes tend to eat out "cavities" at the centers of their host galaxies, effectively cutting off their fuel supply through their own greed. Alternatively, as with

MORE PRECISELY 25-2 Faster-Than-Light Velocities?

The compact sources of emission within radio galaxies and quasars vary not only in intensity but also in structure. High-resolution maps of some quasars, made using very-long-base-line interferometry, display dramatic changes in structure, often on time scales as short as months. For example, the accompanying illustrations show three radio images of the core of the well-studied "nearby" quasar 3C 273, made several years apart. The interior of this quasar is dominated by two large blobs of gas, which move over the course of time. Knowing the distance to 3C 273 (about 660 Mpc) and measuring the angle through which the blobs moved in the course of 3 years (about 0.002″), astronomers have calculated the blobs' velocities. Astonishingly, the result is nearly 10 times the speed of light!

The notion that the speed of light is the highest attainable velocity is central to modern physics. Scores of predictions made assuming this fact to be true have been verified to high accuracy since Einstein first published his theory of relativity early in the twentieth century (*More Precisely 22-1*). Astronomers almost universally agree that some reinterpretation of these apparent *superluminal* (that is, faster than light) quasar motions is needed.

One alternative assumes that quasars are not at cosmological distances from us. If quasars were relatively local, the angular motion of the blobs would not imply high physical velocities, and the problem of explaining how the individual blobs move faster than light would not exist. But if the quasars are local and not distant, the observed redshifts of their spectral lines cannot be a distance indicator, and we are forced to find some other explanation for the large quasar redshifts. As discussed in more detail in *Interlude 25-2*, few astronomers are prepared to make that assumption.

Several alternative solutions have been proposed to account for the apparent superluminal motion of the blobs without requiring the quasars to be local or the speeds to be truly faster than light. The most straightforward model suggests that the observed changes in quasar structure are not caused by actual motions of the interior blobs at all but rather by variations of the radio intensity of *stationary* blobs. In other words, the interiors of quasars may resemble the blinking lights that sometimes appear on movie theater marquees. These marquee bulbs blink on and off in a programmed sequence that gives the illusion of motion, but the individual bulbs are of course stationary. Likewise, blinking radio sources within the quasar may suggest motion where none really exists. A variation on this model explains the motions as a different kind of illusion—a projection effect, produced by blobs moving almost precisely along our line of sight at slightly less than the speed of light. Calculations of how these blobs would look from Earth indicate that they could in fact appear to be moving faster than light.

None of the alternative models is simple. They all require peculiar geometries, and no one model is agreed upon by all researchers. We still lack a complete explanation of the puzzling phenomenon of superluminal velocities. However, the existing models demonstrate that although the interiors of quasars are exceedingly complex and not terribly well understood, there is no compelling need to discard the laws of physics in order to explain them.

nearby active galaxies, the high luminosities may be the result of interactions between galaxies in the early universe. The fact that quasars have been observed in some distant galaxy clusters argues in favor of this latter view. For a radically different view of quasars, however, see *Interlude 25-2*.

QUASAR "MIRAGES"

In 1979, astronomers were surprised to discover what appeared to be a binary quasar—two quasars with exactly the same redshift and very similar spectra, separated by only a few arc seconds on the sky. Remarkable as the discovery of such a binary would have been, the truth about this pair turned out to be even more amazing. Closer study of the quasars' radio emission revealed that they were not in fact two distinct objects. Instead, they were two separate images of the *same* quasar! Optical views of such a *twin quasar* are shown in Figure 25.21.

What could produce such a "doubling" of a quasar image? The answer is *gravitational lensing*—the deflection and focusing of light from a background object by the gravity of some foreground body (Figure 25.22). In Chapter 23 we saw how lensing by brown dwarfs or other compact

(a)

(b)

R I V U X G

Figure 25.21 (a) This "twin" quasar (designated AC114 and located about 2 billion pc away) is not two separate objects at all. Instead, the two large blobs (at upper left and lower right) are images of the same object, created by a gravitational lens. These two blobs of light have striking symmetry. The lensing galaxy itself is probably not visible in this image—the two objects near the center of the frame are thought to be unrelated galaxies in a foreground cluster. (b) A larger perspective of AC114 in the night sky.

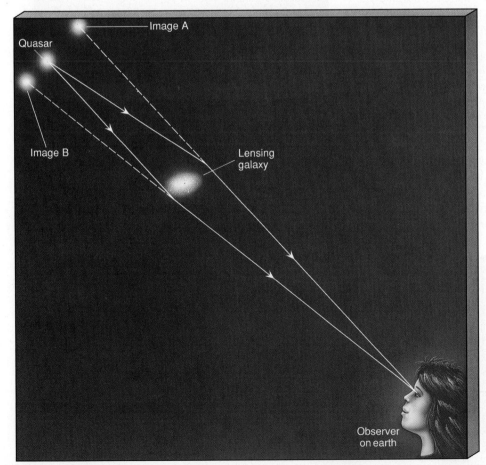

Figure 25.22 When light from a distant object passes close to a galaxy or cluster of galaxies along the line of sight, the image of the background object (here, the quasar) can sometimes be split into two or more separate images (here, A and B). The foreground object is a gravitational lens.

INTERLUDE 25-2 Could Quasars Be Local?

When quasars were first discovered, their large luminosities and small sizes troubled many astronomers. In the 1960s and 1970s no known mechanism could account for the generation of 1000 Milky Ways' worth of energy within a region comparable in size to the solar system. The idea that supermassive black holes may be quite common in the cores of galaxies was unknown.

In response to these problems some astronomers, notably the respected observer Halton Arp and the equally reputable theorist Geoffrey Burbidge, sought an alternative explanation for quasars. Instead of believing that these objects were at cosmological distances and so very luminous, these researchers argued that perhaps there was an alternative, *noncosmological*, explanation for their great redshifts. Quasars could then be relatively nearby and hence much less bright. For example, if 3C 273 were 100 times closer—only 6.6 Mpc, not 660 Mpc, away—the inverse-square law says that the luminosity required to account for its observed apparent brightness would be reduced by a factor of 10,000. ∞ (Sec. 17.4) In that case, 3C 273 would radiate "only" 10^{36} W (1/10 the energy output of our Galaxy). This is still a lot of energy, but it is perhaps more easily explainable in terms of "familiar" stellar events: the formation of high-mass stars, supernovae, and so on.

Arp has reported many examples of instances where galaxies and quasars are found close together on the sky but have very different, conflicting redshifts. Figure 25.18 is an example of such an alignment. He argues that there are simply too many of these "coincidences," where a foreground galaxy lies in nearly the same direction as a supposed background quasar, for the distant-quasar hypothesis to be correct. Instead, he claims, the quasars must be *physically* close to the galaxies, and the redshifts have some other, noncosmological, explanation.

Arp and coworkers have gone further, citing instances where neighboring *galaxies* have conflicting redshifts, such

as the system in the first image here, where one of the five galaxies (Stephen's Quintet) has a redshift very different from those of the others. Similarly, in the second image (which is shown as a photographic negative to bring out the faint structure), a small galaxy appears to be connected to a larger galaxy (NGC 7603) having a very different redshift by a faint "bridge" of gas, reinforcing the view that the two really are close together in space. Such findings call into question both the cosmological interpretation of *all* galactic redshifts and Hubble's law itself.

Most astronomers would argue that the claims of conflicting redshifts are not statistically significant. In other words, given the numbers of known galaxies and quasars, accidental superpositions on the sky should be quite com-

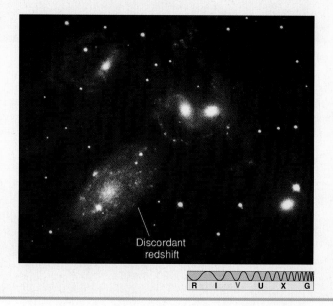

Discordant redshift

R I V U X G

objects in the halo of the Milky Way may temporarily cause the light from a distant star to be amplified, allowing astronomers to detect otherwise invisible stellar dark matter in our Galaxy. ∞ (Sec. 23.6) In the case of quasars the idea is the same, except that the foreground lensing object

is an entire galaxy or galaxy cluster, and the deflection of the light is so great (an arc second or so) that several separate images of the quasar may be formed, as illustrated in Figure 25.23. About two dozen such gravitational lenses are presently known. As telescopes probe the universe with

(a)

(b)

R I V U X G

Figure 25.23 (a) The "Einstein Cross," a multiple-imaged quasar. In this *Hubble* view, spanning only a couple of arc seconds, four separate images of the same quasar have been produced by the galaxy at the center. (b) An artist's conception of what might be occurring here, with Earth at right and the distant quasar at left.

monplace, and the observed quasar–galaxy and galaxy–galaxy alignments are quite consistent with pure chance. The apparent bridges may simply be photographic or image-processing defects. Usually, a lot of computer enhancement is needed to bring out the images, and there is ample opportunity for "features" to appear where none really exist. Finally, Hubble's law is well established for galaxies within a few hundred megaparsecs, and some quasars have been found in galaxy clusters, *sharing* the redshift of their neighbors. Thus, at least some quasar redshifts are known to be cosmological, so the violations of Hubble's law that Arp

claims exist appear only quite selectively—if they really do exist at all.

If quasars are actually local, no convincing explanation has ever been advanced for their redshifts. If they are simply moving at high speeds, then we are led to the new "energy problem" of explaining how they were accelerated to such speeds. If quasars were fired out of galaxies—one alternative explanation for their great speed—we would expect some to be moving toward us, and hence blueshifted. So why are they *all* redshifted? If quasars are all "local" and come from our own Galaxy, then the Milky Way is special—a major violation of the principle of mediocrity and current scientific dogma—or the quasars must be so close (and thus quite faint) that we should have observed the motion of some of them across the sky. Arp's own supposed correlations between other galaxies and quasars would also argue against a Milky Way origin. In short, no matter where we try to put the quasars, their redshifts pose problems. An alternative attempt to explain the redshift as a gravitational redshift suffered by light as it climbs out from the vicinity of a black hole (see Chapter 22) also fails because it cannot account for the observed widths of quasar spectral lines. ∞ (Sec. 22.6)

There is no clear observational evidence for conflicting quasar redshifts, and no satisfactory mechanism for producing the redshifts by noncosmological means. Furthermore, as we have seen, there really is no "quasar luminosity problem" anymore. Quasar luminosities can be explained by the same mechanism that powers active galaxies, with only a modest increase in fuel consumption and certainly without posing a serious challenge to the laws of physics. Consequently, the overwhelming majority of astronomers hold that quasar redshifts are cosmological in origin and that quasars really are the most distant objects known in the universe. Of course, there are many instances in the history of astronomy where the majority has later been proved totally wrong!

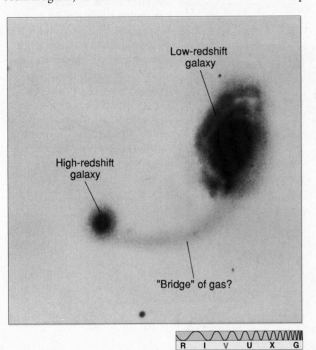

Low-redshift galaxy

High-redshift galaxy

"Bridge" of gas?

R I V U X G

greater and greater sensitivity, astronomers are beginning to realize that gravitational lenses are relatively common, if odd, features of the cosmos.

The existence of these multiple images provides astronomers with several useful observational tools. First, the lensing tends to amplify the light of the quasar, making it easier to observe. Second, because the light rays forming the images usually follow paths of different lengths, there is often a *time delay*, ranging from several days to several years, between them. This delay provides advance notice of explosive events, such as sudden flare-ups in the quasar's brightness—if one image flares up, astronomers know that in time the others will too, so they have a second chance to study the event. The time delay also permits astronomers to determine the *distance* to the lensing galaxy by carefully timing the measurements. If enough lenses can be found, this method may provide a reliable alternative means of measuring Hubble's constant that is independent of any of the techniques discussed in Chapter 24.

Third, *microlensing*—lensing by individual stars in the foreground galaxy—can cause large fluctuations in a quasar's brightness. ∞ (Sec. 23.6) Microlensing allows astronomers to study the stellar content of the lensing galaxy. Finally, by studying the lensing of background quasars and galaxies by foreground galaxy clusters, astronomers can obtain a better understanding of the distribution of dark matter in those clusters, an issue that has great bearing on the large-scale structure of the cosmos. Figures 25.24(a) and (b) show how the images of faint background galaxies are bent into arcs by the gravity of nearby galaxy clusters. ∞ (Sec. 24.2) The degree of bending allows the total mass of the cluster (including the dark matter) to be measured.

The (mostly blue) loop- and arc-shaped features visible in Figure 25.24(b) are thought to be multiple images of a single, distant (unseen) object, lensed by the foreground galaxy cluster (the yellow-reddish blobs in the image). Careful analysis demonstrates the nearly identical shape of each of the blue features, strongly suggesting that they come from a single

(a)

(b)

R I V U X G

Figure 25.24 (a) This spectacular example of gravitational lensing shows more than a hundred faint arcs from very distant galaxies. The wispy pattern spread across the intervening galaxy cluster (A2218, several billion parsecs distant) resembles a spider's web, but it is really an illusion caused by the gravitational field of the lensing cluster, which deflects the light from background galaxies and distorts their appearance. By measuring the extent of this distortion, astronomers can estimate the mass of the intervening cluster. (b) An approximately true-color image of the galaxy cluster known only by its catalog name, 0024+1654, residing some 1.5 billion pc away. The reddish-yellow blobs are mostly elliptical normal galaxy members of the cluster, concentrated toward the center of the image. The bluish looplike features are thought to be images of a single background galaxy.

object, thought to lie nearly twice as far away from us as the intervening cluster. Given the looplike shape and blue color, the distant object may be a normal spiral or ring-shaped galaxy. Astronomers are now trying to delicately "reassemble" an undistorted image of the distant object by tracing its elongated arcs back to a common point.

25.6 Active Galaxy Evolution

5 In Chapters 23 and 24 we addressed the issue of evolutionary change in normal galaxies. ∞ (Secs. 23.4, 24.4) Let us now briefly consider the possibility of evolutionary links among active galaxies and between normal and active galaxies. We emphasize that this section is really mostly speculation. Although the consensus is that galaxies began to form some 9–10 billion years ago (corresponding to a redshift of 5, the upper limit on the measured redshifts of known quasars—see Table 25.1) and that quasars were an early stage of galaxy evolution, the details of the connections among different types of active and normal galaxies remain uncertain.

Most quasars are very distant, indicating that they were more common in the past than they are today. At the same time, normal-looking galaxies seem to be less common in the distant past. These two pieces of evidence suggest to many astronomers that when galaxies first formed, they probably were quasars. This opinion is strengthened by the

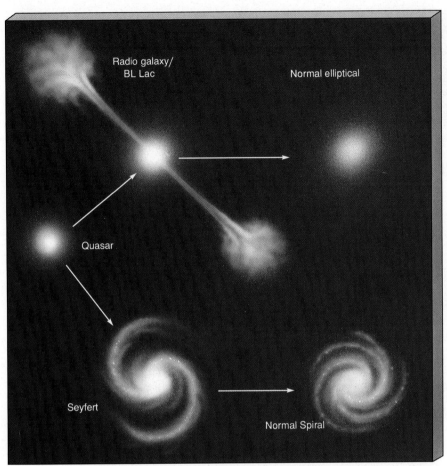

Figure 25.25 A possible evolutionary sequence for galaxies, beginning with the highly luminous quasars, decreasing in violence through the radio and Seyfert galaxies, and ending with normal spirals and ellipticals. The central black holes that powered the early activity are still there at later times; they simply run out of fuel as time goes on.

fact that the same black-hole energy-generation mechanism can account for the luminosity of quasars, active galaxies, and the central regions of normal galaxies like our own. Large black holes do not simply vanish, at least in the 10–20 billion years that the universe has existed (see *More Precisely 22-3*). Thus, the presence of supermassive black holes in the centers of many, if not all, normal galaxies is consistent with the idea that they started off as quasars, then "wound down" to become the relatively quiescent objects we see today.

In this picture, the gradual reduction in violence from a quasar to a Seyfert galaxy to a normal spiral, for example, occurs primarily because the fuel supply is reduced as the galaxy evolves. A similar sequence might connect quasars to BL Lac objects (*Interlude 25-1*) to radio galaxies to normal ellipticals. These possible evolutionary connections among the active and normal galaxies are illustrated in Figure 25.25. Adjacent objects along this sequence are nearly indistinguishable from one another. For example, weak quasars share some characteristics with some very active galaxies, and the feeblest active galaxies often resemble the most explosive normal galaxies.

If we accept this appealing (but still unproven) view, we can construct the following possible scenario for the evolution of galaxies in the universe. Galaxies began to form about 10 billion years ago. The early round of massive star formation that may have expelled galactic gas and possibly helped determine a galaxy's Hubble type—spiral or elliptical—could also have given rise to many large, stellar-mass black holes, which sank to the center of the still-forming galaxy and merged into a supermassive black hole there. Alternatively, the supermassive hole may have formed directly by gravitational collapse of the dense central regions of the protogalaxy. Whatever the cause, large black holes appeared at the centers of many galaxies at a time when there was still plenty of fuel available to power them, resulting in many highly luminous quasars. The brightest quasars—the ones we see from Earth—were those with the greatest fuel supply.

Young galaxies at this early stage were much fainter than their bright quasar cores. As a result, until very recently, astronomers were hard-pressed to discern any galactic structure in quasar images. Although computer-enhanced, ground-based observations hinted at wispy "fuzz" in some quasar images, evidence for host galaxies surrounding quasars was inconclusive. In 1996 several groups of astronomers used the *Hubble Space Telescope* to take up the challenge. Early results were ambiguous, but after removing the bright quasar cores from the *Hubble* images and carefully reanalyzing the remnant light, researchers have reported that in every case studied—a total of about a dozen quasars so far—there appears to be a normal host galaxy

Figure 25.26 This long-exposure *Hubble* image of a distant quasar clearly shows the young host galaxy in which the quasar resides, lending strong support to the view that quasars represent an early, very luminous phase of galaxy evolution. The quasar has the catalog name PG0052+251 and resides roughly 500 Mpc from Earth.

enveloping the quasar. Figure 25.26 shows one of the longest exposures ever taken of a quasar. Even without sophisticated computer processing, the host can clearly be seen. Indeed, there is even some resemblance to a normal spiral galaxy, with thin arms extending away from the core.

As the galaxy developed and the black hole used up its fuel the luminosity of the central nucleus diminished.

Although still active, the quasar no longer completely overwhelmed the emission from the surrounding stars. The result was an active galaxy—a radio galaxy or a Seyfert—still emitting a lot of energy but now with a definite "stellar" component in its spectrum.

The central activity continued to decline. Eventually, only the surrounding galaxy remained visible—a normal galaxy, like the majority of those we now see around us. Today the black holes that generated so much youthful energy lie dormant in galactic cores, producing only a relative trickle of radiation. Occasionally, two nearby normal galaxies may interact with each other, causing a flood of new fuel to be directed toward the central black hole of one or both. The engine starts up for a while, giving rise to the nearby active galaxies we observe.

Should this picture be correct—and the confirmation of host galaxies surrounding many quasars lends strong support to this view—then many normal galaxies, including perhaps our own Milky Way Galaxy, were once brilliant quasars. Perhaps some alien astronomer, thousands of megaparsecs away, is at this very moment observing our Galaxy—seeing it as it was billions of years ago—and is commenting on its enormous luminosity, nonstellar spectrum and high-speed jets, and wondering what exotic physical process can possibly account for its violent activity!

When they were first discovered, active galaxies and quasars seemed to present astronomers with insurmountable problems. For a time their dual properties of enormous energy output and small size appeared incompatible with the known laws of physics and threatened to overturn our modern view of the universe. Yet the problems were eventually solved, and the laws of physics remain intact. Far from jeopardizing our knowledge of the cosmos, these violent phenomena have become part of the thread of understanding that binds our own Galaxy to the earliest epochs of the universe in which we live.

Chapter Review ✹www

SUMMARY

Active galaxies (p. 576) are much more luminous than normal galaxies and have spectra that are nonstellar in nature, indicating that the energy they emit is not simply the accumulated light of many stars. Most of the energy from active galaxies is emitted in the radio and infrared parts of the electromagnetic spectrum. The fraction of observed galaxies that display activity increases with increasing distance from the Milky Way, indicating that galaxies were generally more active in the past than they are today.

A **Seyfert galaxy** (p. 577) looks like a normal spiral except that it has an extremely bright central **galactic nucleus** (p. 577), whose luminosity can in many cases exceed that of the rest of the galaxy. Spectral lines from Seyfert nuclei are very broad, indicating rapid internal motion. In addition, Seyfert luminosities can vary by large amounts in fractions of a year, implying that the region emitting most of the radiation is much less than 1 light year across.

Radio galaxies (p. 578) are active galaxies that emit most of their energy in the radio part of the spectrum. They are generally comparable to the Seyferts in total energy output. Unlike Seyferts, they are usually associated with elliptical galaxies. In a core–halo radio galaxy, the energy is emitted from a small central nucleus, as in a Seyfert. In a lobe radio galaxy, the energy comes from enormous **radio lobes** (p. 580) that dwarf the visible galaxy and lie far outside it. The lobes are usually symmetrically placed with respect to the center of the visible galaxy.

Many active galaxies have high-speed jets of matter shooting out from their central nuclei. In lobe radio galaxies, astronomers believe that the jets transport energy from the nucleus, where it is generated, to the lobes, where it is radiated into space. The jets often appear to be made up of distinct "blobs" of gas, suggesting that the process generating the energy is intermittent.

The generally accepted explanation for the observed properties of active galaxies is that the energy is generated by accretion of galactic gas onto a supermassive (billion–solar mass) black hole lying at the center of the nucleus. As material spirals down toward the hole it heats up and releases enormous amounts of energy. The small size of the accretion disk explains the compact extent of the emitting region, and the high-speed orbits of gas in the hole's intense gravity account for the rapid motion observed. Typical active-galaxy luminosities require the consumption of about one solar mass of material every few years.

Some of the infalling matter is blasted out into space, producing magnetized jets that create and feed the extended radio lobes. Charged particles spiraling around the magnetic field lines produce **synchrotron radiation** (p. 586) whose spectrum is consistent with the nonstellar radiation observed in radio galaxies and Seyferts.

Quasars (p. 587), or **quasi-stellar objects** (p. 587), were first discovered as starlike radio sources with unknown broad spectral lines. In the early 1960s, astronomers realized that the unfamiliar lines are actually those of familiar elements, but redshifted to wavelengths much longer than normal. Even the closest quasars lie at great distances from us. They are the most luminous objects known.

Quasars exhibit the same basic features as active galaxies, and astronomers believe that their power source is also basically the same. This source—a black hole—must be scaled-up in a quasar to a consumption rate of many stars per year. If that is the case, the brightest quasars consume so much fuel that their energy-emitting lifetimes must be relatively short. Quasars probably represent a brief phase of violent activity early in the life of a galaxy.

Some quasars have been observed to have double or multiple images. These result from gravitational lensing, in which the gravitational field of a foreground galaxy or galaxy cluster bends and focuses the light from the more distant quasar. Analysis of this bending provides a means of determining the masses of galaxy clusters—including the dark matter—far beyond the optical images of the galaxies themselves.

Quasars, active galaxies, and normal galaxies may represent an evolutionary sequence. When galaxies began to form, conditions might have been suitable for the formation of large black holes at their centers. If a lot of gas was available at those early times, a highly luminous quasar could have been the result. As the fuel supply diminished, the quasar dimmed, and the galaxy in which it was embedded became visible as an active galaxy. At even later times the fuel supply declined to the point at which the nucleus became virtually inactive, and a normal galaxy was all that remained.

SELF-TEST: TRUE OR FALSE?

_____ **1.** Active galaxies can emit thousands of times more energy than our own Galaxy.

_____ **2.** The "extra" radiation emitted by active galaxies is due to the tremendous number of stars they contain.

_____ **3.** Active galaxies emit most radiation at optical wavelengths.

_____ **4.** The spectrum of an active galaxy is not well described by a blackbody curve.

_____ **5.** Most core–halo radio galaxies are spirals.

_____ **6.** The size of a billion–solar mass black hole is about 20 A.U.

_____ **7.** Nearby active galaxies are most likely the result of interactions between galaxies.

_____ **8.** A redshift greater than 1 implies a recessional velocity greater than the speed of light.

_____ **9.** All quasars are far away.

_____ **10.** Quasar luminosities are so large that astronomers have no idea what process can account for them.

_____ **11.** Other than a small amount of visible light, quasars emit all their radiation at radio wavelengths.

_____ **12.** Many nearby normal galaxies may once have been quasars.

____ **13.** Quasars emit much more energy than normal galaxies.

____ **14.** The quasar stage of a galaxy ends because the central black hole uses up all the matter around it.

____ **15.** Astronomers still have no evidence that quasars are found in the hearts of young galaxies.

SELF-TEST: FILL IN THE BLANK

1. Active galaxies are more common at _____ distances.

2. Active galaxies generally emit more radiation at _____ wavelengths.

3. Seyfert galaxies appear like normal spirals, but with a very bright galactic _____.

4. In a core–halo radio galaxy, most of the radio radiation is emitted from the _____.

5. Lobe radio galaxies emit radio radiation from regions that are typically much _____ in size than the optical galaxy.

6. Radio lobes are always found aligned with the _____ of the optical galaxy.

7. For all types of active galaxies the original source of the tremendous energy emitted is the galactic _____.

8. The energy source of an active galaxy is unusual in that there is a large amount of energy emitted from a region less than _____ in diameter. (Give size and unit.)

9. The mass of the black hole responsible for energy production in the active galaxy M87 is thought to be approximately _____ solar masses.

10. The amount of mass that must be consumed by a supermassive black hole to provide the energy for an active galaxy is about _____ per _____.

11. In visible light, quasars have a _____ appearance.

12. The distance to a quasar in light years is not simply equal to the time in years since the quasar emitted the light we see, because of the _____ of the universe.

13. Quasar spectra were understood when it was discovered that their radiation is _____ by an unexpectedly large amount.

14. The fact that a typical quasar would consume an entire galaxy's worth of mass in 10 billion years suggests that quasar lifetimes are relatively _____.

15. The image of a distant quasar can be split into several images because of gravitational lensing by a foreground _____ along the line of sight.

REVIEW AND DISCUSSION

1. Name two basic differences between normal galaxies and active galaxies.

2. Describe some of the basic properties of Seyfert galaxies.

3. What distinguishes a core–halo radio galaxy from a lobe radio galaxy?

4. What is the evidence that the radio lobes of some active galaxies consist of material ejected from the galaxy's center?

5. What conditions result in a head–tail radio galaxy?

6. Briefly describe the leading model for the central engine of an active galaxy.

7. How is the process of synchrotron emission related to observations of active galaxies?

8. What was it about the spectra of quasars that was so unexpected and surprising?

9. Why do astronomers prefer to speak in terms of redshifts rather than distances to faraway objects?

10. How do we know that quasars are extremely luminous?

11. How are the spectra of distant quasars used to probe the space between us and them?

12. How are BL Lac objects related to other active galaxies?

13. What evidence do we have that quasars represent an early stage of galaxy evolution?

14. What happened to the energy source at the center of a quasar?

15. What are the arguments for and against the idea that quasars may in fact be nearby and not at cosmological distances?

PROBLEMS

1. A Seyfert galaxy is observed to have broadened emission lines indicating an orbital speed of 1000 km/s at a distance of 1 pc from its center. Assuming circular orbits, use Kepler's laws (see Section 23.6) to estimate the mass within this 1-pc radius.

2. Calculate the energy flux—that is, the energy received per unit area per unit time—that would be observed at Earth from a 10^{37}-W Seyfert nucleus located at the Galactic center, neglecting the effects of interstellar extinction. Using the data presented in Appendix 3, Table 4, compare this energy flux with that received from Sirius A, the brightest star in the night sky. From what you know about active galaxy energy emission, is it reasonable to ignore interstellar extinction?

3. Centaurus A—from one radio lobe to the other—spans about 1 Mpc. It lies at a distance of 4 Mpc from Earth. What is the angular size of Centaurus A? Compare this value with the angular diameter of the Moon.

4. Assuming a jet speed of 0.75 c, calculate the time taken for material in Cygnus A's jet to cover the 500 kpc between the galaxy's nucleus and its radio-emitting lobes.

5. Assuming the same efficiency as indicated in the text, calculate the amount of energy an active galaxy would generate if it consumed one Earth mass of material every day. Compare this value with the luminosity of the Sun.

6. Based on the data presented in the text, calculate the orbital speed of material orbiting at a distance of 0.5 pc from the center of M87.

7. A certain quasar has a redshift of 0.17 and the same apparent brightness as the Sun would have if it were placed at a distance of 500 pc. Assuming a Hubble constant of 65 km/s/Mpc (and neglecting the complications described in *More Precisely 25-1*), calculate the quasar's luminosity.

8. Assuming an energy-generation efficiency (that is, the ratio of energy released to total mass–energy available) of 10 percent, calculate how long a 10^{41}-W quasar can shine if a total of 10^{10} solar masses of fuel is available.

9. Light from a distant quasar passes 10 Mpc from the center of an intervening galaxy cluster before being deflected to a detector on Earth. If Earth, the cluster, and the quasar are all aligned, the quasar is 750 Mpc away, and the cluster lies midway between Earth and the quasar, calculate the angle through which the galaxy cluster bends the quasar's light.

10. Light from a distant star is deflected by 1.75″ as it grazes the Sun. ∞ (*More Precisely 22-2*) Given that the deflection angle is proportional to the mass of the gravitating body and is inversely proportional to the minimum distance between the light ray and the body, calculate the mass of the galaxy cluster in the previous question.

PROJECTS

Here are three observing projects that are increasingly challenging.

1. In the previous chapter you were given directions for finding the Virgo Cluster of galaxies. M87, in the central part of this cluster, is the nearest core–halo radio galaxy. M87 has coordinates RA = 12^h 30.8^m, dec = +12° 24′. At magnitude 8.6 it should not be difficult to find in an 8-inch telescope. Its distance is roughly 20 Mpc. Describe its nucleus; compare what you see with other nearby ellipticals in the Virgo Cluster.

2. NGC 4151 is the brightest Seyfert galaxy. Its coordinates are RA = 12^h 10.5^m, dec = +39° 24′, and it can be found below the Big Dipper in Canes Venatici. At magnitude 10–12 (it is variable), it should be visible in an 8-inch telescope, but it will be challenging to find. Its distance is 13.5 Mpc. As in the case of M87, describe its nucleus and compare with what you have seen for other galaxies.

3. 3C 273 is the nearest and brightest quasar. However, that does not mean it will be easy to find and see! Its coordinates are RA = 12^h 29.2^m, dec = +2° 03′. It is located in the southern part of the Virgo Cluster but is not associated with it. At magnitude 12–13 (again, it is variable), it may require a 10- or 12-inch telescope to see, but try it first with an 8-inch. It should appear as a very faint star. The significance of seeing this object is that it is 640 Mpc distant. The light you are seeing left this object over 2 billion years ago! 3C 273 is the most distant object observable with a small telescope.

If you can find the three objects listed here, you have started to become an accomplished observer!

26 COSMOLOGY

The Big Bang and the Fate of the Universe *www*

LEARNING GOALS

Studying this chapter will enable you to:

1 State the cosmological principle and explain its significance.

2 Explain how the approximate age of the universe is determined, and discuss the uncertainties involved.

3 Summarize the leading evolutionary models of the universe.

4 Discuss the factors that determine whether the universe will expand forever.

5 Explain the relationship between the future of the universe and the overall geometry of space.

6 Describe the cosmic microwave background radiation, and explain its importance to our understanding of cosmology.

(Opposite page, background) By virtually all accounts, the universe began in a fiery explosion some 10–20 billion years ago. Out of this maelstrom emerged all the energy that would later form galaxies, stars, and planets (depicted here in an artist's rendering). The story of the origin and fate of all these systems—and especially of the entire universe itself—comprises the subject of cosmology.

(Inset A) These three images of spiral galaxies lying at different distances from Earth capture representative views of what the universe was like at much earlier times. At top, about 10 billion years ago, the galaxy's spiral shape is hardly recognizable. At middle, about 8 billion years ago, the spiral is still vague, though starburst activity is evident in the outer regions. At bottom, some 5 billion years ago, the galaxy's spiral features are more prominent, with much star formation in its arms.

(Inset B) Elliptical galaxies also suggest evolution with time—indeed, evolution more rapid than that of spiral galaxies. These images of different galaxies (from top to bottom, 10, 8, and 5 billion years ago, respectively) suggest that many ellipticals had already taken on their eventual shapes early in the universe. In contrast to spirals, ellipticals seem to have made all their stars long, long ago.

Our field of view now extends for billions of parsecs into space and billions of years back in time. We have asked and answered many questions about the structure and evolution of planets, stars, and galaxies. At last we are in a position to address the central issues of the biggest puzzle of all: How big is the universe? How long has it been around, and how long will it last? What was its origin, and what will be its fate? Is the universe a one-time event, or does it recur and renew itself, in a grand cycle of birth, death, and rebirth? How and when did matter, atoms, our Galaxy form? These are basic questions, but they are hard questions. Many cultures have asked them, in one form or another, and have developed their own cosmologies—theories about the nature, origin, and destiny of the universe—to answer them. In this and the next chapter we will see how modern scientific cosmology addresses these important issues and what it has to tell us about the universe we inhabit. After more than 10,000 years of civilization, science may be ready to provide some insight regarding the ultimate origin of all things.

26.1 The Universe on the Largest Scales

THE END OF STRUCTURE

The universe shows structure on every scale we have examined so far. Subatomic particles form nuclei and atoms. Atoms form planets and stars. Stars form star clusters and galaxies. Galaxies form galaxy clusters, superclusters, and even larger structures—voids, filaments and sheets that stretch across the sky. From the quarks in a proton to the galaxies in the Great Wall, we can trace a hierarchy of "clustering" of matter from the very smallest to the very largest scales. It is natural to ask: Does the clustering ever end? Is there some scale on which the universe can be regarded as more or less smooth and featureless? Perhaps surprisingly, given the trend we have just described, most astronomers think the answer is yes.

We saw in Chapter 24 how surveys of the universe have revealed the existence of structures as large as 200 Mpc across. ∞ (Sec. 24.5) Yet, although they cover wide areas of the sky and enormous volumes of space, these studies are still relatively "local," in the sense that they span only a few percent of the distance to the farthest quasars (which lie over 5000 Mpc from Earth). A major obstacle to extending these wide-angle surveys to much greater distances is the sheer observational effort of measuring the redshifts of all the galaxies within larger and larger volumes of space.

An alternative approach is to narrow the field of view to only a few small patches of the sky but to study very distant galaxies within those patches. The volume surveyed then becomes a long, thin "pencil beam" extending deep into space rather than a wide swath through the local universe. Figure 26.1 presents the results of one such survey, showing the distribution of galaxies in two directions perpendicular to the plane of our Galaxy, out to a distance of about 2000 Mpc.

Although the numbers of galaxies fall off at large distances—mainly because very distant galaxies are very hard to see—a distinctive "on–off" pattern of galaxies, looking a little like a picket fence, can be seen. It is thought that the gaps between the "pickets" are voids like those seen closer

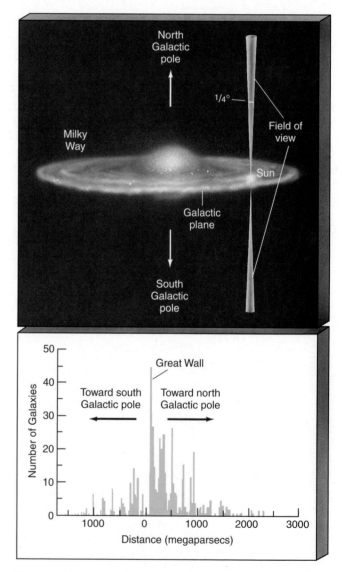

Figure 26.1 The results of a deep "pencil-beam" survey of two small portions of the sky in opposite directions from Earth perpendicular to the Galactic plane. The graph shows the number of galaxies at different distances from us, out to a distance of about 2000 Mpc. The distinctive "picket fence" appearance seems to show voids and sheets of galaxies on scales of 100 or 200 Mpc but gives no indication of any larger structure.

to home and that the pickets themselves are places where the line of sight intersects sheets like the Great Wall. The data seem to indicate that the largest structures in the local universe are only 100–200 Mpc across—no voids or clumps of galaxies much larger than that are seen. In short, there is presently *no* evidence for structure in the universe on scales greater than about 200 Mpc. *Interlude 26-1* discusses a recent very deep survey that has generated a great deal of excitement among astronomers.

On the basis of these rather sketchy data, theoretical insight, and not a little philosophical preference, *cosmologists*—astronomers who study the large-scale structure and dynamics of the universe—assume that the universe is roughly **homogeneous** on scales greater than a few hundred megaparsecs. In other words, if we imagine taking a huge cube—300 Mpc on a side, say—and placing it anywhere in the universe, as illustrated in Figure 26.2, the number of galaxies it enclosed would be pretty much the same—around 100,000, excluding the faint dwarf ellipticals and irregulars—no matter *where* in the universe we put it. Some of the galaxies would be clustered and clumped into fairly large structures, and some would not, but the total number would not vary much as the cube was moved from place to place. In short, the universe looks *smooth* on the largest scales.

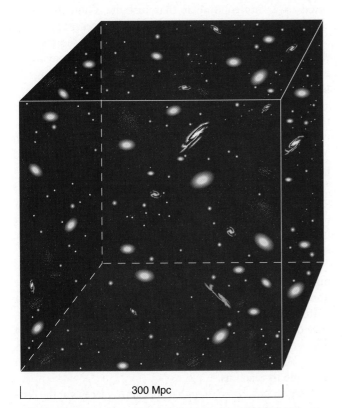

Figure 26.2 Diagram of galaxies contained within an enormous cube, 300 Mpc on a side. Cosmologists believe that the contents of such a cube would look pretty much the same no matter where we placed the cube in the universe. (This artist's illustration is not drawn to scale; for accurate scaling see the deep-space photos in Figures 24.18, 24.23, and 24.24.)

300 Mpc

THE COSMOLOGICAL PRINCIPLE

1 Cosmic homogeneity is the first of two major assumptions that cosmologists make when studying the large-scale structure of the universe. Observations suggest that it is true, but it is by no means proven. The second assumption, also supported by observational evidence and theoretical reasoning, is that the universe is **isotropic**—that is, it looks the same in any direction. Isotropy is on much firmer observational ground than homogeneity. Apart from regions of the sky that are obscured by our Galaxy, the universe *does* look much the same in all directions, at any wavelength, provided we look far enough. In other words, any deep pencil-beam survey of the sky should count about the same number of galaxies as the study mentioned previously, regardless of which patch of the sky is chosen.

The assumptions of homogeneity and isotropy form the foundation of modern **cosmology**—the study of the structure and evolution of the entire universe. Together, these twin pillars of cosmology are known as the **cosmological principle**. No one really knows if this principle is absolutely correct. All that we can say is that, so far, it seems consistent with observations. From this point on we will simply assume that it holds—at least on large enough scales. Should it turn out to be incorrect—for example, were a structure a few thousand megaparsecs across to be discovered tomorrow—then some of the discussion that follows would be on very shaky ground indeed!

The cosmological principle has very far reaching implications. For example, it implies that there can be no edge to the universe, because that would violate the assumption of homogeneity. Furthermore, it implies that there is no *center*, because that would mean that the universe would not be the same in all directions from any noncentral point, a violation of the assumption of isotropy. Thus, this single principle strongly limits what the overall geometry of the universe can be. The cosmological principle is the ultimate expression of the principle of mediocrity. It states not only that are we not central to the universe but that *no one* can be central, because *the universe has no center!*

26.2 The Expanding Universe

OLBERS'S PARADOX

You may not have thought of it before in these terms, but every time you go outside at night and notice that the sky is dark, you are making a profound cosmological observation. Here's why.

According to the cosmological principle, the universe is homogeneous and isotropic. Let's assume that it is also infinite in spatial extent and unchanging in time—precisely the view of the universe that prevailed until the early part of the twentieth century. On average, then, the universe is uniformly populated with galaxies filled with stars. In that case, when you look up at the night sky, your line of sight must *eventually*

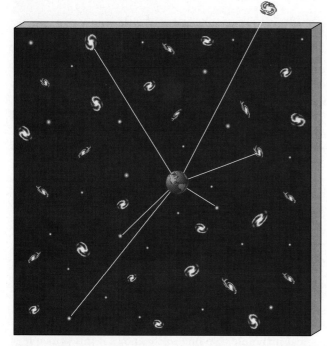

Figure 26.3 If the universe were homogeneous, isotropic, infinite in extent, and unchanging, any line of sight from Earth should eventually run into a star, and the entire night sky should be bright. This obvious contradiction of the facts is known as Olbers's paradox.

encounter a star, as illustrated in Figure 26.3. The star may lie at an enormous distance, in some remote galaxy, but the laws of probability dictate that, sooner or later, any line drawn outward from Earth will run into a bright stellar surface.

This has a dramatic implication. No matter where you look, the sky should be as bright as the surface of a star—the entire night sky should be as brilliant as the surface of the Sun! The obvious difference between this prediction and the actual appearance of the night sky is known as **Olbers's paradox**, after the nineteenth-century German astronomer, Heinrich Olbers, who popularized the idea.

What is the resolution to this paradox? Why is it dark at night? We have accepted the cosmological principle, so we believe that the universe is homogeneous and isotropic. We must conclude, then, that one of the other two assumptions (or both)—that the universe is infinite in extent and unchanging in time—is false. Either the universe is finite in extent, or it evolves in time. In fact, the answer involves a little of each and is intimately tied to the behavior of the universe on the largest scales.

THE BIRTH OF THE UNIVERSE

2 We saw in Chapter 24 that all the galaxies in the universe are rushing away from us in a manner described by Hubble's law:

$$\text{recession velocity} = H_0 \times \text{distance},$$

where $H_0 \approx 65$ km/s/Mpc is Hubble's constant. ∞ (Sec. 24.5) We used this relation as a convenient means of deter-

mining the distances to galaxies and quasars, but it is much more than that.

Assuming that all velocities have remained constant in time, we can ask a simple question: How long has it taken for any given galaxy to reach its present distance from us? The answer follows from Hubble's law. The time taken is simply the distance traveled divided by the velocity, so

$$\text{time} = \frac{\text{distance}}{\text{velocity}}$$

$$= \frac{\text{distance}}{H_0 \times \text{distance}} \quad \begin{array}{l}\text{(using Hubble's law}\\\text{for the velocity)}\end{array}$$

$$= 1/H_0.$$

For $H_0 = 65$ km/s/Mpc, this value turns out to be about 15 billion years. Notice that the time is *independent* of the distance—galaxies twice as far away are moving twice as fast, so the time they took to cross the intervening distance is the same.

Hubble's law therefore implies that at some time in the past—15 billion years ago, according to the foregoing simple calculation—*all* the galaxies in the universe lay right on top of one another. In fact, astronomers believe that *everything* in the universe—matter and radiation alike—was confined to a single point at that instant. Then the point exploded, flying apart at high speeds. The present locations and velocities of the galaxies are a direct consequence of that primordial blast. This gargantuan explosion, involving everything in the universe, is known as the **Big Bang**. As best we can tell, it marked the beginning of the universe.

Thus, by measuring Hubble's constant, we can estimate the age of the universe to be $1/H_0 \approx 15$ billion years. The range of possible error in this age is considerable, both because Hubble's constant is not known precisely and because the assumption that galaxies moved at constant speed in the past is not a very good one—in fact, galaxies moved faster in the past; their motion has been slowed by the effects of gravity. We will refine our estimate in a moment, but regardless of the details, the critical fact here is that the age of the universe is *finite*.

This then is the explanation of why the sky is dark at night. Olbers's paradox is resolved by the evolution of the universe itself. Whether the universe is actually finite or infinite in extent is irrelevant, at least as far as the appearance of the night sky is concerned. We see only a *finite* part of it—the region lying within roughly 15 billion light years of us. What lies beyond is unknown—its light has not yet had time to reach us.

Realize that, even though it appears to place us at the center of the expansion, Hubble's law does *not* violate the cosmological principle in any way. This can be understood from Figure 26.4, which shows how observers on five separate galaxies might perceive the motion of their neighbors. For simplicity, the galaxies are taken to be equally spaced, 100 Mpc apart, and they are separating in accordance with Hubble's law with

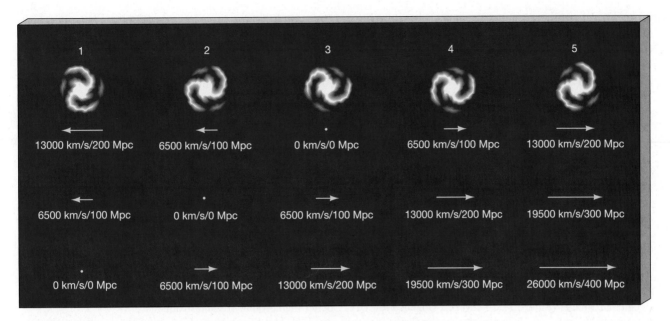

Figure 26.4   Hubble's law is the same no matter who makes the measurements. The top set of numbers are the distances and recessional velocities as seen by an observer on the middle of five galaxies, galaxy 3. The bottom two sets are from the points of view of observers on galaxies 2 and 1, respectively. In all cases, the same Hubble law holds.

H_0 = 65 km/s/Mpc. The first pair of numbers beneath each galaxy represents its distance and recessional velocity as measured by the observer on the middle galaxy, number 3.

Now consider how the expansion looks from the point of view of the observer on galaxy 2. Galaxy 4, for example, is moving with velocity 6500 km/s to the right relative to galaxy 3, and galaxy 3 in turn is moving at 6500 km/s to the right as seen by observer 2. Therefore, galaxy 4 is moving at a velocity of 13,000 km/s to the right as seen by the observer on galaxy 2. The distances and velocities that would be measured by observer 2 are noted in the second row. Similarly, the measurements made by an observer on galaxy 1 are noted in the third row. *Each* observer sees an overall expansion described by Hubble's law, and the constant of proportionality—Hubble's constant—is the same in all cases. Far from singling out any one observer as central, Hubble's law is in fact the *only* expansion law consistent with the cosmological principle.

WHERE WAS THE BIG BANG?

Let's pause for a moment to take stock of what we have just learned and to shift in a very important way our view of the expansion of the universe as described by Hubble's law. We know *when* the Big Bang occurred. Is there any way of telling *where?* The cosmological principle indicates that the universe is the same everywhere, yet we have just seen that the observed recession of the galaxies described by Hubble's law implies that all the galaxies exploded from a point some time in the past. Wasn't that point, then, different from the rest of the universe, violating the assumption of homogeneity expressed in the cosmological principle? The answer is a definite *no!*

To understand why there is no "center" to the expansion, we must make a great leap in our perception of the universe. If we were to imagine the Big Bang as simply an enormous explosion that spewed matter out into space, ultimately to form the galaxies we see, then the foregoing reasoning would be quite correct—there would be a center and an edge, and the cosmological principle would not apply. But the Big Bang was *not* an explosion in an otherwise featureless, empty universe. The only way that we can have Hubble's law hold *and* retain the cosmological principle is to realize that the Big Bang involved the entire universe—not just the matter and radiation within it, but the universe *itself*. In other words, the galaxies are not flying apart into the rest of the universe. The universe itself is expanding. Like raisins in a loaf of raisin bread that move apart as the bread expands in an oven, the galaxies are just along for the ride.

Let's reconsider some of our earlier statements in light of this new perspective. We now recognize that Hubble's law describes the expansion of the universe itself. Although galaxies have some small-scale, individual random motions, on average they are *not* moving with respect to the fabric of space—any such overall motion would pick out a "special" direction in space and violate the assumption of isotropy. On the contrary, the portion of the galaxies' motion that makes up the Hubble flow is really an expansion of space itself. The expanding universe remains homogeneous at all times. There is no "empty space" beyond the galaxies into which they rush. At the time of the Big Bang, the galaxies did not reside at a point located at some well-defined place within the universe. The *entire universe* was a point. That point was in no way different from the rest of the universe; that point *was* the universe. Therefore, there was no one point where the Big Bang "happened"—because the Big Bang involved the entire universe, it happened *everywhere* at once.

To illustrate these ideas, imagine an ordinary balloon with coins taped to its surface, as shown in Figure 26.5.

(Better yet, do the experiment yourself.) The coins represent galaxies, and the two-dimensional surface of the balloon represents the "fabric" of our three-dimensional universe. The cosmological principle applies to the balloon because every point on the balloon looks pretty much

INTERLUDE 26-1 A Stunning View of Deep Space

As we probe deep into space we are looking far back in time. Telescopes are time machines, and astronomers are historians. Now, thanks to the *Hubble Space Telescope*, we have the deepest, most detailed optical view of extragalactic space yet obtained. Called the Hubble Deep Field, it is shown here in approximately true color. It was assembled from nearly 300 exposures taken during more than 100 orbits over the course of 10 consecutive days, for a total exposure time of about 100 hours. All the data were combined to make this one spectacular image. (Figure 24.24 shows a smaller view of the same field.) The field is in a region of the sky near the Big Dipper that happens to lie in a direction perpendicular to the congested plane of the Milky Way.

Virtually every blob of light in this image is a galaxy. (A few clear exceptions are local stars in our own Milky Way, such as the bright star with the artificial spikes at bottom left of the image.) In all, approximately 1600 galaxies of all types can be counted in this one image. Hundreds of these galaxies are so

faint that they have never been seen before. The image extends to 30th magnitude, making the Hubble Deep Field the most sensitive photograph ever taken. ∞ (Sec. 17.4) The image is "deep" in the sense that it allows us to see very faint objects. The word *deep* should not be construed to mean that this image goes beyond the realm of ground-based telescopes. *Hubble* does not see significantly farther into the universe than any other large telescope, but it does see more sensitively; hence it is able to pick up light from very dim objects and often to resolve them better than ground-based instruments.

The Hubble Deep Field covers only a tiny part of the entire sky. The field of view shown here is just 2' across. To map the entire sky to this depth would take *Hubble* nearly a million years! One way to put this image into a larger perspective is to think of it as a core sample, much like the sample a geologist might take on Earth. By examining the dirt and rocks in their samples, geologists try to reconstruct the history of events that occurred as Earth evolved. Likewise, by the same as every other. Imagine yourself as a resident of one of the three coin "galaxies" in the leftmost frame, and note your position relative to your neighbors. As the balloon inflates (that is, as the universe expands), the other galaxies recede from you; more distant galaxies recede more rapidly. (Notice, incidentally, that the coins themselves do *not* expand along with the balloon, any more than people, planets, stars, or galaxies—all of which are held together by their own internal forces—expand along with the universe.)

studying the galaxies in the Hubble Deep Field, astronomers seek to understand better the history of the universe.

The image shows many galaxies of many ages. Disentangling young, distant galaxies from old, nearby ones is tricky, but as a general rule of thumb (and perhaps a little counterintuitively, given that we have gone to some lengths to stress that distant objects are characterized by large redshifts) the blue objects are probably the farthest away (since we see their stars in their blazing youth), whereas red objects tend to be closer (we see them by the light of their old, reddened stars).

Of particular interest are the small, faint smudges of blue light in the image; they may well be the faraway protogalaxies that originally gave rise to the galaxies we see today. If this is so, then some of the faintest galaxies in this image existed when the universe was a mere billion years old, eons before the birth of our solar system. Only spectroscopic observations can tell if the dim smudges are truly distant. The big ground-based telescopes in Hawaii, Chile, and on Kitt Peak are better suited

than *Hubble* to determine redshifts, and this work is now under way. The second image shows the same field of view at left but with some redshifts marked—note the enormous spread in distances (see Table 25.1) evident in even the small sample of redshifts measured so far.

Although the Hubble Deep Field covers but a speck of the sky, astronomers consider it to be representative of the typical spread of galaxies in space. This is another way of stating the cosmological principle—statistically, the universe on the largest scales should look pretty much the same in all directions. Extrapolating the contents of this new image over the entire sky, astronomers then estimate that the total number of galaxies in the observable universe is approximately 40 billion. Contrary to sensationalized news accounts, this is not an astonishingly larger number of galaxies than previously thought. Rather, it is midway between the 10 billion and 100 billion galaxies that astronomers for decades have reasoned should exist in the observable universe.

R I V U X G

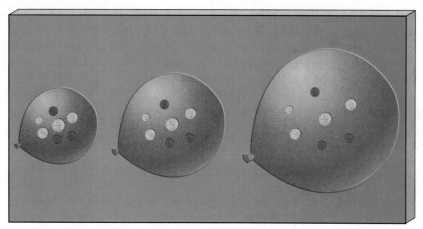

Figure 26.5 The coins taped to the surface of a spherical balloon recede from one another as the balloon inflates (left to right). Similarly, galaxies recede from one another as the universe expands. As the coins recede, the distance between any two of them increases, and the rate of increase of this distance is proportional to the distance between them. Thus, the balloon expands according to Hubble's law.

Regardless of which galaxy you chose to consider, you would see all the other galaxies receding from you. Nothing is special or peculiar about the fact that all the galaxies are receding from you. Such is the cosmological principle: no observer anywhere in the universe has a privileged position. There is no center to the expansion and no position that can be identified as the location from which the universal expansion began. Everyone sees an overall expansion described by Hubble's law, with the same value of Hubble's constant in all cases.

Now imagine letting the balloon deflate. This corresponds to running the universe backward from the present time to the Big Bang. *All* the galaxies (coins) would arrive at the same place at the same time—at the instant the balloon reached zero size. But there is no one point on the balloon that could be said to be *the* place where that occurred. The entire balloon expanded from a point, just as the Big Bang encompassed the entire universe and expanded from a point.

This analogy has its shortcomings. The main difficulty here is that we see the balloon, which in our illustration we imagined as two dimensional, expanding into the third dimension of space. This might suggest that the three-dimensional universe is expanding "into" some fourth spatial dimension. It is not, so far as we know. At the very least, if higher spatial dimensions are involved, they are not relevant to our theory of the universe.

THE COSMOLOGICAL REDSHIFT

This view of the expanding universe requires us to reinterpret the cosmological redshift. ∞ (Sec. 24.5) Up to now we have explained the redshift of galaxies as a Doppler shift, a consequence of their motion relative to us. However, we have just argued that the galaxies are *not* in fact moving with respect to the universe, in which case the Doppler interpretation is incorrect. The true explanation is that as a photon moves through space its wavelength is influenced by the expansion of the universe. In a sense we can think of the photon as being attached to the expanding fabric of space, so its wavelength expands along with the universe, as illustrated in Figure 26.6. Although it is common practice in astronomy to refer to the cosmological redshift in terms of recessional velocity, bear in mind that, strictly speaking, this is not the right thing to do. The cosmological redshift is a consequence of the changing size of the universe—it is *not* related to velocity at all.

The redshift of a photon measures the amount by which the universe has expanded since that photon was emitted. For example, when we measure the light from a quasar to have been redshifted by a factor of 5, that means that the light was emitted at a time when the universe was just one-fifth its present size (and that we are observing the quasar as it was at that time). In general, the larger a photon's redshift, the smaller the universe was at the time the photon was emitted, and so the longer ago that emission occurred. Because the

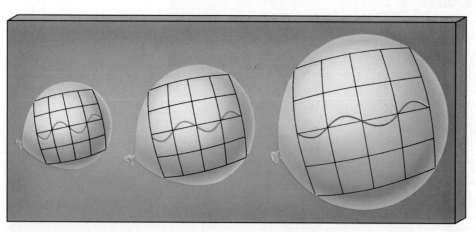

Figure 26.6 As the universe expands, photons of radiation are stretched in wavelength, giving rise to the cosmological redshift.

universe expands with time and redshift is related to that expansion, cosmologists routinely use redshift as a convenient means of expressing time. ∞ (*More Precisely 25-1*)

RELATIVITY AND THE UNIVERSE

These concepts are difficult to grasp. The notion of the entire universe expanding from a point—with *nothing*, not even space and time, outside—takes some getting used to. Nevertheless, this picture of the universe lies at the heart of modern cosmology. The description of the universe itself (not just its contents) as a dynamic, evolving object is far beyond the capabilities of Newtonian mechanics, which we have used almost everywhere throughout this book. ∞ (Sec. 2.7) Instead, the more powerful techniques of Einstein's *general relativity*, with its built-in notions of warped space and dynamic spacetime, are needed.

We encountered general relativity in Chapter 22 (see especially *More Precisely 22-1*) when we discussed the strange properties of black holes. We can loosely summarize its description of the universe by saying that the presence of matter or energy causes a curvature of space (more correctly, space*time*) and that the curved trajectories of freely falling particles within warped space are what Newton thought of as orbits under the influence of gravity. The amount of curvature depends on the amount of matter present, and the orbits of particles in turn depend on the curvature. Put less formally, "space tells matter how to move, and matter tells space how to curve." In a homogeneous universe, the curvature of space must be uniform.

Must we keep this difficult notion of a warped, expanding, homogeneous, universe constantly in the forefront of our thoughts if we are to comprehend the evolution of the cosmos? Perhaps surprisingly, given that relativity is the only theory that properly describes the large-scale behavior of the universe, the answer is no. Although general relativity predicts some curious consequences for the overall *geometry* of space on large scales, as we will see in Section 26.5, the *dynamics* of the universe can be understood using simple concepts that would have been thoroughly familiar to Newton.

Consider two neighboring points A and B in the expanding universe, as pictured in Figure 26.7. From the perspective of point A, every other point, including B, is rushing away from it, in accordance with Hubble's law. How does the distance between A and B change in time? Does it just keep growing, or does its rate of increase eventually slow down and stop? From a Newtonian viewpoint, we might expect that the overall gravitational pull of the universe would tend to slow the expansion, just as Earth's gravity tends to slow the upward motion of an object projected from the surface. We have just argued that Newtonian physics is inadequate to describe the large-scale dynamics of the universe. However, a surprising prediction of general relativity is that the Newtonian picture gives the right result! The motion of point B relative to point A is *exactly* the same as if A were at the center of a planet and B were a projectile fired vertically upward from its surface. Thus, we *can* discuss

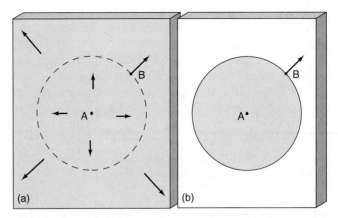

Figure 26.7 The relative motion of any two points in the expanding universe (a) can be addressed as a problem in Newtonian mechanics, even though Einstein's theory of general relativity is needed to explain why it is correct to do so. If the rest of the universe is ignored—as in part (b)—then the Newtonian calculation of B's motion relative to A gives the same answer as a calculation using general relativity.

the expansion of the universe in simple Newtonian terms, but we need general relativity to justify our doing so.

BEFORE THE BIG BANG

The Big Bang was a *singularity* in space and time—an instant when the present laws of physics say the universe had zero size and infinite temperature and density. As we saw in Chapter 22, where we discussed the singularities at the center of black holes, these predictions should not be taken too literally. ∞ (Sec. 22.6) The presence of these singularities signals that, under extreme conditions, the theory—in this case, general relativity—making the predictions has broken down.

At present no theory exists to let us penetrate the singularity at the start of the universe. We have no means of describing these earliest of times, so we have no way of answering the question, What came *before* the Big Bang? Indeed, given the laws of physics as we currently know them, the question itself may be meaningless. The Big Bang represented the beginning of the entire universe—mass, energy, space, *and* time came into being at that instant. Without time, the notion of "before" does not exist. Consequently, some cosmologists maintain that asking what happened before the Big Bang is like asking what lies north of the North Pole! Others disagree, arguing that one day the proper theory will explain the singularity and allow us to answer the question of what came before.

Although we cannot extend our theory of the universe all the way back to the Big Bang itself, we can come very close. Theorists estimate that the "known" physics of today is adequate to describe the universe since about 10^{-43} s after the Big Bang. For the purposes of this chapter, at least, 10^{-43} is close enough to zero that we need not worry about this very early, and completely unknown, epoch.

26.3 The Fate of the Universe

3 At present the universe is expanding. Will that expansion continue forever? And if not, what will happen next, and when? These are absolutely fundamental questions concerning the fate of the entire universe. Yet, remarkably, now that we know we can confidently apply our familiar Newtonian concepts of motion under gravity to the behavior of the universe, we can address these issues by considering a simpler and much more familiar problem.

CRITICAL DENSITY

4 Consider a rocket ship launched from the surface of a planet. What is the likely outcome of that motion? There are basically two possibilities, depending on the speed of the ship. If the launch speed is high enough, it will exceed the planet's escape speed, and the ship will never return to the surface. ∞ (Sec. 2.7) The speed will diminish because of the planet's gravitational pull, but it will never reach zero. The spacecraft leaves the planet on an *unbound* trajectory, as illustrated in Figure 26.8(a). Alternatively, if the ship's launch speed is lower than the escape speed, it will reach a maximum distance from the planet, then fall back to the surface. Its *bound* trajectory is shown in Figure 26.8(b).

Similar reasoning applies to the expansion of the universe. Reconsider Figure 26.7, but now imagine that A and B are galaxies at some known distance from each other, with their present relative velocity given by Hubble's law. The same two basic possibilities exist for these galaxies as for our spacecraft—the distance between them can increase forever, or it can increase for a while and then start to decrease. What's more, the cosmological principle says that, whatever the outcome, it must be the same for *any* two galaxies—in other words, the same statement applies to the universe *as a whole*. Thus, as illustrated in Figure 26.9, the universe has only two options: it can continue to expand forever—an *unbound* universe—or the present expansion will someday stop and turn around into a contraction—a *bound* universe.

The middle curve on Figure 26.9 marks the dividing line between these two possibilities. It shows a *marginally bound* universe that expands forever, but at an ever-decreasing rate, analogous to our rocket ship's leaving the planet with precisely the escape speed. The three curves are drawn so that they all pass through the same point at the present time. All are possible descriptions of the universe given its present size and expansion rate.

What determines which of these possibilities will actually occur? The answer is the *density* of the universe. In all cases gravity decelerates the expansion over time. The more matter there is—the denser the universe—the more "pull" there is against the expansion, just as the more mass a planet has, the less likely it is that a rocket ship can escape. In a high-density universe, there is enough mass to stop the expansion and cause a recollapse—the universe is bound. A low-density universe, conversely, is unbound and will expand forever. The dividing line between these two outcomes, the density corresponding to a marginally bound universe, is called the **critical density**. Its value depends on Hubble's constant—more matter is required to bind a more rapidly

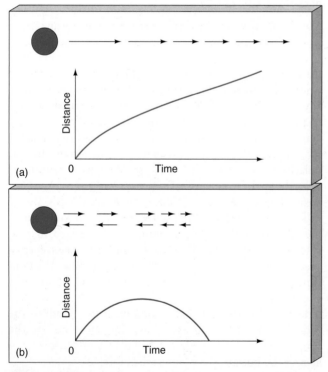

Figure 26.8 (a) A spacecraft (arrow) leaving a planet (blue ball) with a speed greater than the escape velocity leaves on an unbound trajectory (top). The graph (bottom) shows the distance between the ship and the planet as a function of time. (b) If the launch speed is less than the escape velocity, the ship eventually drops back to the planet. Its distance from the planet first rises, then falls.

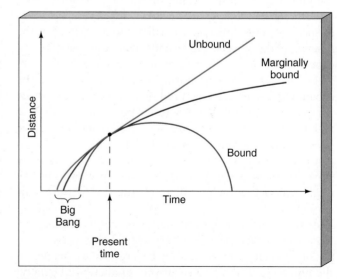

Figure 26.9 Distance between two galaxies as a function of time in each of the three possible universes discussed in the text: unbound, bound, and marginally bound. The point where the three curves touch represents the present time.

expanding universe. For $H_0 = 65$ km/s/Mpc, the present critical density turns out to be 8×10^{-27} kg/m^3 (which we will round off to 10^{-26} kg/m^3). That's an extraordinarily low density—about five hydrogen atoms per cubic *meter*, a volume the size of a typical household closet. In more "cosmological" terms, it corresponds to about one Milky Way Galaxy (excluding the dark matter) per cubic megaparsec.

TWO FUTURES

If the universe emerged from the Big Bang with a density above the critical value, then it contains enough matter to halt its own expansion, and the recession of the galaxies will eventually stop. At some time in the future, astronomers everywhere—on any planet within any galaxy—will announce that the radiation received from nearby galaxies is no longer redshifted. (The light from *distant* galaxies will still be redshifted, however, because we will see them as they were in the past, at a time when the universe was still expanding.) The bulk motion of the universe, and of the galaxies within, will be stilled—at least momentarily.

The expansion may stop, but the pull of gravity will not. The universe will begin to contract. Nearby galaxies will begin to show blueshifts, and both the density and the temperature of the universe will start to rise as matter collapses back onto itself. As illustrated in Figure 26.10(a), the universe will recollapse to a point, requiring just as much time to fall back as it took to rise. First galaxies, then stars will collide with increasing frequency and violence as the available space diminishes and the entire universe shrinks toward a superdense, superhot singularity much like the one from which it originated. The cosmos will ultimately experience a "heat death," in which all matter and life are destined to be incinerated. Some astronomers call the final collapse of this high-density universe the "Big Crunch."

Cosmologists do not know what will happen to the universe if it ever reaches the point of collapse. We cannot penetrate forward in time beyond the singularity at the Big Crunch any more than we can probe backward past the Big Bang—the laws of physics as we presently understand them are simply inadequate to describe these extreme conditions. However, some theorists speculate that with both density and temperature increasing as the contraction nears completion, the pressure may somehow be sufficient to overcome gravity, pushing the universe back out into another cycle of expansion. As depicted in Figure 26.10(b), the universe may not simply end—instead it may "bounce." A hypothetical universe having many—perhaps infinitely many—cycles of expansion and contraction could be the result. Bear in mind, though, that any discussion of the universe outside of the current cycle is pure speculation.

A quite different fate awaits the universe if its density is below the critical value. In that case its density always has been, and always will be, too small for gravity to cause it to recontract. As illustrated in Figure 26.10(c), such a low-density universe will expand forever. In this scenario the galaxies will continue to recede forever, their radiation weakening with increasing distance. In time an observer on Earth will see no galaxies in the sky beyond the Local Group (which is not itself expanding), even with the most powerful telescope. The rest of the observable universe will appear dark, the distant galaxies too faint to be seen. Eventually, the Milky Way and the Local Group too will peter out as their fuel supply is consumed. This universe will ultimately experience a "cold death." All radiation, matter, and life are eventually destined to freeze.

In the intermediate, critical-density case the universe contains just enough matter eventually to halt the expansion—but only after an infinitely long time. This universe will also expand forever, as indicated in Figure 26.10(c).

(a)

(b)

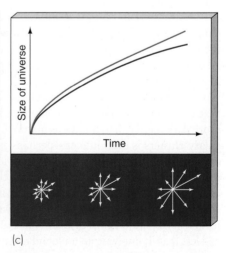
(c)

Figure 26.10 (a) A high-density universe has a beginning, an end, and a finite lifetime. The lower frames illustrate its evolution, from explosion to maximum size to recollapse. (b) An oscillating universe has neither a beginning nor an end. Each expansion–contraction phase ends in a "bounce" that becomes the "Big Bang" of the next expansion. There is currently no information on whether this can actually occur. (c) A low-density universe expands forever from its explosive beginning. The upper curve represents a universe with density less than the critical value. The lower curve represents a universe with density exactly equal to the critical value.

26.4 Will the Universe Expand Forever?

Is there any way for us to determine which of these models actually describes our universe (that is, apart from just waiting to find out)? Will the universe end as a small, dense point much like that from which it began? Or will it expand forever? Fortunately, we live at a time when astronomers are subjecting these questions to observational tests. Unfortunately, the results of these observations are still inconclusive.

THE DENSITY OF THE UNIVERSE

The most straightforward approach is to try to estimate the universe's average density, because density is the basic quantity that distinguishes one model from the other. As just noted, for $H_0 = 65$ km/s/Mpc, the critical density that separates the two possible futures is about 10^{-26} kg/m^3. Cosmologists conventionally call the ratio of the actual density to the critical value the *cosmic density parameter* and denote it by the symbol Ω_0 ("omega nought"). In terms of this quantity, then, a critical universe has $\Omega_0 = 1$. A universe with Ω_0 less than 1 will expand forever; one with Ω_0 greater than 1 will recollapse.

How can we determine the average density of the universe? On the face of it, it would seem simple—just measure the average mass of the galaxies residing within some large parcel of space, calculate the volume of that space, and compute the total mass density. When astronomers do this, they usually find a little less than 10^{-28} kg/m^3 in the form of luminous matter. Largely independent of whether the chosen region contains many scattered galaxies or only a few rich galaxy clusters, the resulting density is about the same, within a factor of 2 or 3. Galaxy counts thus yield Ω_0 to be about 0.01. If that measure were correct, then the universe would expand forever.

But there is an important additional consideration. We have noted (Chapters 23 and 24) that most of the matter in the universe is *dark*—it exists in the form of invisible material that has been detected only through its gravitational effect in galaxies and galaxy clusters. ∞ (Secs. 23.6, 24.3) We currently do not know what the dark matter is, but we *do* know that it is there. Galaxies may contain as much as 10 times more dark matter than luminous material, and the figure for galaxy clusters is even higher—perhaps as much as 95 percent of the total mass in clusters is invisible. Even though we cannot see it, dark matter contributes to the average density of the universe and plays its part in opposing the expansion. Including all the dark matter that is known to exist in galaxies and galaxy clusters increases the value of Ω_0 to 0.2 or 0.3—still less than 1, implying an unbound universe, but a lot closer to 1 than the 0.01 value calculated using visible matter only.

Unfortunately, the distribution of dark matter on larger scales is not very well known. We can infer its presence in galaxies and galaxy clusters, but we are largely ignorant of its extent in superclusters, voids, or other larger structures.

However, there are indications that it accounts for an even greater fraction of the mass on large scales than it does in galaxy clusters. For example, observations of gravitational lensing by galaxy clusters suggest that dark matter may be even more extensive than is indicated by the motion of galaxies within the clusters. ∞ (Sec. 25.5) Furthermore, optical and infrared observations of the overall motion of galaxies in the local supercluster suggest the presence of a nearby huge accumulation of mass known as the *Great Attractor*, with a total mass of about 10^{17} solar masses and a size of 100–150 Mpc. If the current best estimates of the size and mass of this gargantuan object are correct, its average density may be quite close to the critical value.

Thus it is quite conceivable that invisible matter may account for as much as 99 percent of the total mass in the universe. In that case, the vast "voids" are not empty at all—they are huge seas of invisible matter, and the visible galaxies are merely insignificant "islands" of brightness within them. If the total amount of dark matter lurking in the darkness beyond the galaxy clusters exceeds the luminous mass by a factor of 100 or more, Ω_0 may even exceed 1, and the universe will recollapse. This is why it is so important to search for reservoirs of invisible matter beyond the galaxies. The measured value of Ω_0 has steadily increased over the past 20 years as larger and larger regions of the universe have been surveyed.

COSMIC DECELERATION

Determining the mass density of the universe is an example of a local measurement that provides an estimate of Ω_0. But the result we obtain depends on just how local our measurement is—the larger the scale, the larger the result we obtain. In an attempt to get around this problem, astronomers have devised alternative methods that rely instead on *global* measurements, which cover a much larger portion of the observable universe. The idea is that global tests should indicate the universe's *overall* density, not just its value in our vicinity.

One such global method attempts to measure the deceleration of the universe directly by observing faraway galaxies. The universe expanded faster in the past, and the greater the average density of the universe, the greater the deceleration has been. The issue facing astronomers is to determine the *rate* at which the expansion is slowing, because this translates directly into an estimate of the density.

Astronomers cannot hope to measure the cosmic slowdown by watching the motion of any one galaxy. But because the expansion is slowing, objects at great distances—that is, objects that emitted their radiation long ago—should appear to be receding *faster* than Hubble's law predicts. Figure 26.11 illustrates this concept. If the universal expansion were constant in time, recessional velocity and distance would be related by a straight line—the solid line in the figure. However, in a decelerating universe, the recessional velocity is greater at large distances (that is, at times in the distant past). This velocity difference is greater for a denser universe, in which gravity has been more effective at slowing the initial expansion.

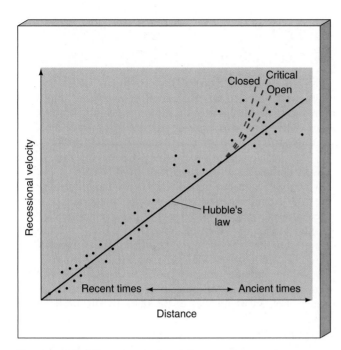

Figure 26.11 An idealized diagram showing how we might detect evidence for a deceleration of the universe by observing a departure from Hubble's law (solid line). The dashed curves show the expected departure for different evolving models of the universe (exaggerated for clarity.)

As indicated in the figure, many distant galaxies do seem to have recessional velocities substantially greater than Hubble's law would predict. Unfortunately, these galaxies are generally very faint and accurate measurements of their properties are difficult. The uncertainties in their measured distances (which are much greater than the uncertainties in the velocities) mean that we currently cannot distinguish among the various models based on these observations.

In 1998, astronomers announced the first results from a long-term, systematic survey of distant Type-I (carbon detonation) supernovae. ∞ (Sec. 21.3) Recall from Section 24.2 that Type-I supernovae have a remarkably narrow spread in luminosities, making them particularly useful as distance-measurement tools. Far from clarifying the picture of cosmic deceleration, however, these findings seem to indicate that the expansion of the universe is not slowing, but actually *accelerating!* According to the supernova data, galaxies at large distances are receding *less* rapidly than Hubble's law would predict. These observations, if confirmed, are inconsistent with the standard Big Bang model presented in this chapter, and may necessitate a major revision of our view of the cosmos. *Interlude 27-1* discusses one possible interpretation of these results.

So, what is the ultimate fate of the universe? The answer is still not known with absolute assurance. However, although there is a large uncertainty in the value of Ω_0, most astronomers would probably agree that it lies between 0.1 and 1. As best we can tell, given the current data, the universe is destined to expand forever.

THE AGE OF THE UNIVERSE

In Section 26.2, when we estimated the age of the universe from the accepted value of Hubble's constant, we made the assumption that the expansion speeds of the galaxies were constant in the past. However, as we have now found, this is not the case. At least within the standard Big Bang model, the effects of gravity have slowed the universe's expansion over time, so, regardless of which evolutionary model turns out to be correct, the universe must have expanded *faster* in the past than it does today. The assumption of a constant expansion rate therefore leads to an overestimate of the universe's age—the universe is actually younger than the 15 billion years we calculated earlier. How much younger depends on how much deceleration has occurred.

Figure 26.12 illustrates this point. It is similar to Figure 26.9, except that here we have added a line corresponding to constant expansion at the present rate—a completely *empty* universe that is 15 billion years old. Because all the models we have discussed lie below this line, we can see graphically that the true age of the universe is indeed less than the age obtained assuming a constant expansion speed. In the special case of critical density, the age of the universe happens to be particularly easy to calculate—it is *two-thirds* of the foregoing value, or 10 billion years. (This was the assumption underlying Table 25.1 in *More Precisely 25-1*.) A low-density, unbound universe is older than 10 billion years (but still less than 15 billion years old); a high-density, bound universe is younger.

Impressive though it is that we can pin down the age of the universe to within less than a factor of 2, these numbers highlight a continuing problem in astronomy. Unless the universe is of quite low density, and hence closer to 15 billion

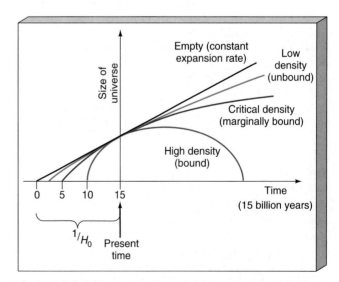

Figure 26.12 As the density of the universe increases, its deceleration increases, too. The universe contains some matter, so whatever the model, its trajectory on this graph will lie below the line for the constant-expansion-rate empty universe. Thus, the age of the universe is always *less* than $1/H_0$. The true age decreases for larger values of the present density.

INTERLUDE 26-2 The Age Controversy Revisited

In Section 26.4 we note a potentially serious discrepancy in the estimates of the age of the universe yielded by two independent branches of astronomy. For $H_0 = 65$ km/s/Mpc, cosmologists calculate that the likely range of cosmic ages is 10–15 billion years, with the upper limit ($1/H_0$) corresponding to a uniformly expanding universe devoid of matter, and the lower limit (two-thirds of this value) to a marginally bound universe. In part because observations seem to indicate a cosmic density of at least a few tenths the critical value, and for theoretical reasons to be discussed in Section 27.4, a value closer to the lower limit is favored by many cosmologists. However, this estimate may be at odds with the ages of the oldest star clusters in our galaxy, as determined by analyses of the turn-off point for main-sequence stars on the H–R diagram. ∞ (Sec. 20.5) For example, the age of the globular cluster 47 Tucanae, according to this method, is *at least* 12 billion years. Hence the paradox at hand: some stars seem to be older than the universe itself.

This problem is not really a new one. Despite its notoriety in the media lately, the essence of the problem has been around for well over a hundred years. In the 1860s Lord Kelvin sought to assess the age of Earth by making two assumptions—that Earth probably formed at the same time as the Sun, and that the Sun shone probably because it burned some chemical, such as wood or gasoline. The first assumption was a good one (indeed, we make it today), but the second one was most definitely not—the Sun, most assuredly, is not made of wood! In any case, based on the amount of sunlight radiated by our star, Kelvin came up with an estimate for the age of the Sun, and hence Earth, of at most 100 million years. But this value conflicted with the amount of time needed by Charles Darwin to explain the fossil record in terms of evolution by natural selection; long-dead life forms seemed to be at least several hundred million years old. How could life on Earth be older than Earth itself?

This early age discrepancy eventually went away. Once radioactivity was discovered around the turn of the twentieth century, geologists learned to measure the ages of rocks directly. They found that Earth was several billion years old—old enough to explain Darwin's fossils. Much later, physicists unraveled the truth about solar energy generation and realized that it could sustain the Sun's output for many billions of years, and the paradox was fully resolved. ∞ (Sec. 16.5)

The problem resurfaced in the 1930s. At issue then was Edwin Hubble's first measurements of H_0. ∞ (Sec. 24.5) Owing to observational uncertainties in the brightnesses of the galaxies, and also to calibration errors in the analyses of the acquired data, he found a very high value for H_0 of around 500 km/s/Mpc, meaning that $1/H_0 \approx 2$ billion years—less than the age of Earth. Again, the discrepancy gradually went away. This time what helped most was painstaking work by many astronomers: improved observations of more galaxies, careful measurements of their brightnesses, and better analyses of the data. By the 1950s the value of H_0 had diminished fivefold, and therefore $1/H_0$ increased fivefold, so the inferred age of the universe became around 10 billion years, safely older than Earth, and the problem dissipated—for a while.

The issue reemerged yet again in the 1980s. Some observers measured a value for H_0 of around 90 km/s/Mpc, implying $1/H_0 \approx 11$ billion years. At the same time, stellar evolution theorists maintained that the oldest globular clusters had to be at least 14 billion years old. Again, improved observations have partly alleviated this problem, although it has not yet been completely eliminated. Reanalysis of the abundance of helium in globular clusters, together with a general increase in estimates of their distances, suggest that the ages of these clusters may have been their ages overestimated by almost 20 percent, reducing the 14 billion year limit to the 12 billion quoted earlier. At the same time, a recalibration of the cosmic distance scale has resulted in an increase in galactic distance estimates, a corresponding reduction in H_0, and hence an increase in the "cosmological" age of the universe. ∞ (*Interlude 17-1*)

The value of H_0 is still being vigorously debated at observatories around the world. Some researchers continue to maintain that a high value, perhaps as much as 80 km/s/Mpc, is correct, in which case the age problem persists. Others, who have always favored a lower value—

years old, the age that we obtain from cosmology is uncomfortably close to the low end of the 10–12-billion-year age range implied by studies of globular clusters in our own Galaxy (see Chapters 17 and 20). ∞ (Secs. 17.10, 20.5) Because the Galaxy cannot be older than the universe, and because the density of the universe appears to be at least relatively close to the critical value (see also Section 27.4), we are forced to conclude that there may be a contradiction between these two major areas of astronomy.

If, as observations improve, Hubble's constant turns out to be closer to 50 or 60 km/s/Mpc, and if the upper limit on globular-cluster ages can be reduced, then the two age estimates may be reconciled. However, if Hubble's constant is

more like 70 or 80 km/s/Mpc, then the discrepancy may become a serious embarrassment to astronomy. *Interlude 26-2* explores this topic in a little more detail.

26.5 The Geometry of Space

5 We have reverted to the familiar notion of gravity, away from the more correct concept of warped spacetime, because speaking in terms of gravity makes our discussion of the evolution of the universe much easier to understand. However, general relativity makes some predictions that do not

50 km/s/Mpc or less—point out that in their universe, the latest "age problem" never existed in the first place! The choice of $H_0 = 65$ km/s/Mpc adopted in the text is a compromise value, favored by many moderates, between these extremes.

The accompanying image shows the latest battleground in the quest to determine H_0, the barred spiral galaxy NGC 1365 in the Fornax Cluster, another rich cluster of galaxies like the Virgo Cluster, lying about 15 Mpc away. The chevron-shaped insert at the right shows a true-color magnification of the outskirts of this galaxy, where about 50 Cepheid variable stars are being monitored by the *Hubble Space Telescope* in an attempt to measure accurately the galaxy's distance and hence the value of H_0.

This area of forefront research will continue, as will perhaps the controversy. Keep in mind, though, that both age estimates—of stars and of the universe—are very close to each other. Remarkably, we are trying to specify a couple of key cosmic numbers to well within a factor of 2, when many other cosmologically important numbers are known only to within a factor of 10. Don't lose sight of the larger significance of what's involved here: astronomers on Earth are today probing the universe and attempting to pin down one of the crucial parameters of the universe—its age. Our generation is likely to find the answer, and we are the ones who will then put this important number into the books for all our descendents to read. It will be one of the legacies that we leave to the future.

R I V U X G

have a simple description in Newtonian terms. Foremost among these is the fact that space is *curved*, with the degree of curvature determined by the amount of matter within it. The curvature must be the same everywhere (because of the cosmological principle), so there are only three distinct possibilities for the large-scale geometry of the universe, and they correspond to the possible futures we have just described. For more information on the different types of geometry involved, see *More Precisely 26-1*.

If the average density of the cosmos is above the critical value, space is curved so much that it bends back on itself and "closes off," making this bound universe *finite* in size. Such a universe is known as a **closed universe**. It is difficult to visu-

alize a three-dimensional volume uniformly arching back on itself in this way, but the two-dimensional version is well known: it is just the surface of a sphere, as with the balloon we discussed earlier. Figure 26.5, then, is the two-dimensional likeness of a three-dimensional closed universe. Like the surface of a sphere, a closed universe has no boundary, yet it is finite in extent.* One remarkable property of a closed universe

*Notice that for the sphere analogy to work, we must imagine ourselves as two-dimensional "flatlanders" who cannot visualize or experience in any way the third dimension perpendicular to the sphere's surface. Flatlanders are confined to the sphere's surface, just as we are confined to the three-dimensional volume of our universe.

Figure 26.13 In a closed universe a beam of light launched in one direction may return someday from the opposite direction after circling the universe, just as motion in a "straight line" on Earth's surface will eventually encircle the globe.

is illustrated in Figure 26.13. Just as a traveler on the surface of a sphere can keep moving forward in a straight line and eventually return to her starting point, a flashlight beam shone in some direction in space might eventually traverse the entire universe and return from the opposite direction!

The surface of a sphere curves, loosely speaking, "in the same direction" no matter which way we move from a given point. A sphere is said to have *positive curvature*. However, if the average density of the universe is below the critical value, the surface curves like a saddle. It has *negative curvature*. Most people have a good idea of what a saddle looks like—it curves "up" in one direction and "down" in another, but no one has ever seen a uniformly negatively curved surface, for the simple reason that it cannot be constructed in three-dimensional Euclidean space! It is just "too big" to fit. A low-density, unbound, saddle-curved universe is infinite in extent and is usually called an **open universe**.

The intermediate case, in which the density is precisely equal to the critical density, is the easiest to visualize. This universe, called a **critical universe**, has no curvature. It is said to be "flat," and it is infinite in extent. In this case, and *only* in this case, the geometry of space on large scales is precisely the familiar Euclidean geometry taught in high schools. Apart from its overall expansion, this is basically the universe that Newton knew.

Euclidean geometry—the geometry of flat space—is familiar to most of us because it is a good description of space in the vicinity of Earth. It is the geometry of everyday experience. Does this mean that the universe is flat, which would in turn mean that it has exactly the critical density?

The answer is no. Just as a flat street map is a good representation of a city, even though we know Earth is really a sphere, Euclidean geometry is a good description of space within the solar system, or even the Galaxy, because the curvature of the universe is negligible on scales smaller than about 1000 Mpc. Only on the very largest scales do the geometric effects we have just discussed become evident.

26.6 The Cosmic Microwave Background

6 Looking out into space is equivalent to looking back into time. ∞ *(More Precisely 25-1)* But how far back in time can we probe? Is there any way to study the universe beyond the most distant quasar? How close can we come to perceiving directly the edge of time, the very origin of the universe?

A partial answer to these questions was discovered by accident in 1964, during an experiment designed to improve the U.S. telephone system. As part of a project to identify and eliminate interference in satellite communications, Arno Penzias and Robert Wilson, two scientists at Bell Telephone Laboratories in New Jersey, were studying the Milky Way's emission at microwave (radio) wavelengths, using the horn-shaped antenna shown in Figure 26.14. In their data they noticed a bothersome background "hiss" that just would not go away—a little like the background static on an AM radio station. Regardless of where and when they pointed their antenna, the hiss persisted. Never diminishing or intensifying, the weak signal was detectable at any time of the day, any day of the year, apparently filling all space.

Figure 26.14 This "sugarscoop" antenna, originally built to communicate with Earth-orbiting satellites, was used in discovering the 3 K cosmic background radiation. Pictured are Robert Wilson (left) and Arno Penzias, who used the antenna to make the discovery.

What is the source of this radio noise? And why does it appear to come uniformly from all directions, unchanging in time? Unaware that they had detected a signal of great cosmological significance, Penzias and Wilson sought many different origins for the excess emission, including atmospheric storms, ground interference, equipment short circuits—even pigeon droppings inside the antenna! Eventually, after conversations with colleagues at Bell Labs and theorists at nearby Princeton University, the two experimentalists realized that the origin of the mysterious static was nothing less than the fiery creation of the universe itself. The radio hiss that Penzias and Wilson detected is now known as the **cosmic microwave background**. Their discovery won them the 1978 Nobel Prize in physics.

In fact, researchers had predicted the existence and general properties of the microwave background well before its discovery. As early as the 1940s, physicists had realized that in addition to being extremely dense, the early universe must also have been very *hot* and that shortly after the Big Bang the universe must have been filled with extremely high energy thermal radiation—gamma rays of very short wavelength. Researchers at Princeton had extended these ideas, reasoning that the frequency of this primordial radiation would have been redshifted from gamma ray, to X ray, to ultraviolet, eventually all the way into the radio range of the electromagnetic spectrum as the universe expanded and cooled (Figure 26.15). ∞ (Sec. 3.4) By the present time, they argued, this redshifted "fossil remnant" of the primeval fireball should have a temperature of no more than a few tens of kelvins—peaking in the microwave part of the spectrum. The Princeton group was in the process of con-

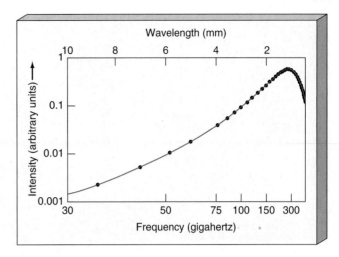

Figure 26.16   The intensity of the cosmic background radiation, as measured by the *COBE* satellite, agrees very well with that expected from theory. The curve is the best fit to the data, corresponding to a temperature of 2.735K. The experimental errors in this remarkably accurate observation are smaller than the dots representing the data points.

structing a microwave antenna to search for this radiation when Penzias and Wilson announced their discovery.

The Princeton researchers confirmed the existence of the microwave background and estimated its temperature at about 3 K. However, because of atmospheric absorption, this part of the electromagnetic spectrum happens to be difficult to observe from the ground, and it was 25 years before astronomers could demonstrate conclusively that the radiation was described by a blackbody curve. In 1989 the *Cosmic Background Explorer (COBE)* satellite measured the intensity of the microwave background at wavelengths straddling the peak of the curve, from half a millimeter up to about 10 cm. The results are shown in Figure 26.16. The solid line is the blackbody curve that best fits the *COBE* data. The near-perfect fit corresponds to a universal temperature of about 2.7 K.

A striking aspect of the cosmic microwave background is its high degree of *isotropy*. Its intensity is virtually constant (in fact, to about 1 part in 10^5) from one direction on the sky to another. This isotropy provides strong support for the assumption of the cosmological principle. It also provides us with a novel means of measuring Earth's "true" velocity through space, without reference to any neighboring galaxies or galaxy clusters. If we were precisely at rest with respect to the universal expansion (like the coin taped to the surface of the expanding balloon in Figure 26.5), then we would see the microwave background as almost perfectly isotropic, as illustrated in Figure 26.17(a). However, if we are moving with respect to that frame of reference, as in Figure 26.17(b), then the radiation from in front of us should be slightly blueshifted by our motion, while that from behind should be redshifted.

Thus, to a moving observer, the microwave background should appear a little hotter than average in front and

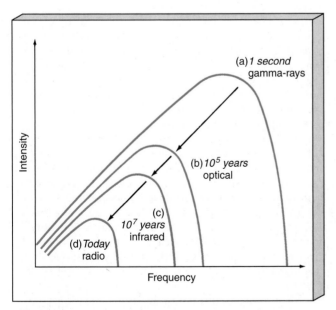

Figure 26.15   Theoretically derived blackbody curves for the universe (a) 1 second after the Big Bang, (b) 100,000 years after the Big Bang, (c) 10 million years after the Big Bang, and (d) at present, approximately 10 billion years after the Big Bang.

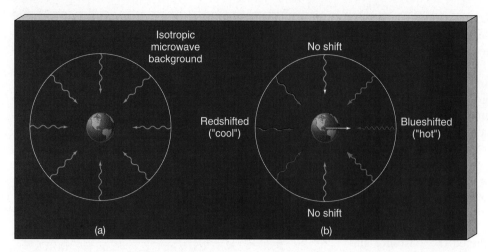

Figure 26.17 (a) To an observer at rest with respect to the expanding universe, the microwave background appears isotropic. (b) A moving observer measures "hot" blueshifted radiation in one direction (the direction of motion) and "cool" redshifted radiation in the opposite direction.

slightly cooler behind. Figure 26.18 shows a *COBE* map of the microwave background temperature over the entire sky. The blue regions are hotter than average, by about 0.0034 K; the red regions cooler by the same amount. The data indicate that Earth's velocity is about 400 km/s in the approximate direction of the constellation Leo. Even though the principle of relativity says that there is no preferred frame of reference, as the laws of physics look the same to

---

 MORE PRECISELY 26-1 Curved Space

Euclidean geometry is the geometry of flat space—the geometry taught in high schools everywhere. Set forth by one of the most famous of the ancient Greek mathematicians, Euclid, who lived around 300 B.C., it is the geometry of everyday experience. Houses are usually built with flat floors. Writing tablets and blackboards are also flat. We work easily with flat, straight objects, because the straight line is the shortest distance between any two points.

When we construct houses or any other straight-walled buildings on the surface of Earth, the other basic axioms of Euclid's geometry also apply: parallel lines never meet even when extended to infinity; the angles of any triangle always sum to 180°; the circumference of a circle equals π times its diameter. If these rules were not obeyed, walls and roof would never meet to form a house!

In reality, though, the geometry of Earth's surface is not really flat. It is curved. We live on the surface of a sphere, and on that *surface*, Euclidean geometry breaks down. Instead, the rules for the surface of a sphere are those of *Riemannian geometry*, named after the nineteenth-century German mathematician Georg Friedrich Riemann. For example, there are no parallel "straight" lines on a sphere . The analog of a straight line on a sphere's surface is a "great circle"—the arc where a plane passing through the center of the sphere intersects the surface. Any two such lines must eventually intersect. The sum of a triangle's angles, when drawn on the surface of a sphere, exceeds 180°—in the 90°–90°–90° triangle shown in the accompanying figure, the sum is actually 270°. And the circumference of a circle is less than π times its diameter.

We see that the curved surface of a sphere, governed by the spherical geometry of Riemann, differs greatly from the flat-space geometry of Euclid. These two geometries are approximately the same only if we confine ourselves to a small patch on the surface. If the patch is small enough compared with the sphere's radius, the surface looks "flat" nearby, and Euclidean geometry is approximately valid. This is why we can draw a usable map of our home, our city, even our state, on a flat sheet of paper, but an accurate map of the entire Earth must be drawn on a globe.

When we work with larger parts of Earth, we must abandon Euclidean geometry. World navigators are fully aware of this. Aircraft do not fly along what might appear on most maps as a straight-line path from one point to another. Instead, they follow a great circle on Earth's surface. On the curved surface of a sphere, such a path is always the shortest distance between two points. For example, a flight from Los Angeles to London does not proceed directly across the United States and the Atlantic Ocean, as you might expect from looking at a flat map. Instead, it goes far to the north, over Canada and Greenland, above the Arctic Circle, finally coming in over Scotland for a landing at London. This is the great circle route—the shortest path between the two cities, as you can easily see if you inspect a globe.

The "positively curved" space of Riemann is not the only possible departure from flat space. Another is the "negatively curved" space first studied by Nikolai Ivanovich Lobachevsky, a nineteenth-century Russian mathematician. In this geometry, there are an *infinite* number of lines through any given point parallel to another line, the sum of a triangle's angles is *less* than 180° (see the accompanying figure), and the circumference of a circle is *greater* than π times its diameter. This type of space is described by the surface of a curved saddle, rather than a flat plane or a curved sphere. It is a hard geometry to visualize!

Most of the local realm of the *three*-dimensional universe (including the solar system, the neighboring stars, and even our Milky Way Galaxy) is correctly described by Euclidean geometry.

Figure 26.18   A *COBE* map of the microwave sky reveals that the microwave background appears a little hotter in the direction of the constellation Leo and a little cooler in the opposite direction. The maximum temperature deviation from the average is about 0.0034 K, corresponding to a velocity of 400 km/s in the direction of Leo.

all observers, there *is* nevertheless a way to determine our absolute velocity with respect to the universe!

When we observe the microwave background, we are looking almost all the way to the very beginning of the uni-

verse. The photons that we receive as these radio waves today have not interacted with matter since the universe was a mere 100,000 years old, when, according to our models, it was less than 1/1000 of its present size. To probe further, back to the Big Bang itself, requires us to enter the world of nuclear and particle physics. The Big Bang was the biggest and the most powerful particle accelerator of all! In the next chapter we will see how studies of conditions in the primeval fireball may aid us in understanding the present-day structure and future evolution of the universe in which we live.

Chapter Review

SUMMARY

On scales larger than a few hundred megaparsecs, the universe appears roughly **homogeneous** (the same everywhere, p. 605) and **isotropic** (the same in all directions, p. 605). In **cosmology** (p. 605)—the study of the universe as a whole—researchers usually simply assume that the universe is homogeneous and isotropic. This assumption is known as the **cosmological principle** (p. 605). It implies that the universe cannot have a center or an edge.

If the universe were homogeneous, isotropic, infinite, and unchanging, then the night sky would be bright because any line of sight would eventually intercept a star. The fact that the night sky is dark is called **Olbers's paradox** (p. 606). Its resolution lies in the fact that, regardless of whether or not the universe is infinite, we see only a finite part of it from Earth—the region from which light has had time to reach us since the universe began.

Tracing the observed motions of galaxies back in time implies that some 15 billion years ago, the universe consisted of a single point that then began to expand rapidly, at the time of the **Big Bang** (p. 606). However, the galaxies are not simply flying apart into the rest of an otherwise empty universe. Space itself is expanding. The Big Bang did not happen at any particular location in space, because the entire universe was compressed to a point at that instant—the Big Bang happened everywhere at once. The cosmological redshift occurs as a photon's wavelength is "stretched" by cosmic expansion. The extent of the observed redshift is a direct measure of the expansion of the universe since the photon was emitted.

Although Einstein's theory of general relativity is needed for a full description of the cosmos, the dynamics of the universe can be understood using simple Newtonian concepts. The eventual fate of the universe depends on its density. If the density is greater than the **critical density** (p. 612), then there is enough cosmic matter to stop the expansion and cause the universe to recollapse. If the density is less than the critical value, the universe will expand forever. Luminous matter by itself contributes only about 1 percent of the critical density. When dark matter in galaxies and clusters is taken into account, the figure rises to 20 or 30 percent. The fraction of dark matter on larger scales is uncertain, but it may be even greater than is found in clusters. Most astronomers believe that the present density lies somewhere between 10 and 100 percent of the critical value.

For $H_0 = 65$ km/s/Mpc, the age of a critical-density universe is about 10 billion years. This age estimate may be in conflict with the ages of globular clusters derived from studies of stellar evolution. If the cluster ages are correct, then the density of the universe must be significantly less than critical, or Hubble's constant must be less than 65 km/s/Mpc.

General relativity provides a description of the geometry of the universe on the largest scales. The curvature of spacetime in a high-density universe is sufficiently large that the universe "bends back" on itself and is finite in extent, somewhat like the surface of a sphere. Such a universe is said to be a **closed universe** (p. 617). A low-density **open universe** (p. 618) is infinite in extent and has a "saddle-shaped" geometry. The intermediate-density **critical universe** (p. 618) will expand forever, but it is spatially flat.

The **cosmic microwave background** (p. 619) is isotropic blackbody radiation that fills the entire universe. Its present temperature is about 3 K. The existence of the microwave background is direct evidence that the universe expanded from a hot, dense state. As the universe has expanded, the initially high energy radiation has been redshifted to lower and lower temperatures.

SELF-TEST: TRUE OR FALSE?

____ **1.** Cosmic homogeneity can be tested observationally, but cosmic isotropy has no observational test.

____ **2.** Olbers's paradox asks, Why is the night sky dark?

____ **3.** Olbers's paradox is resolved by the cosmological redshift, which reduces the energy of photons received from very distant objects.

____ **4.** Hubble's law implies that the universe will expand forever.

____ **5.** The Big Bang is an expansion only of matter, not of space.

____ **6.** All points in space appear to be at the center of the expanding universe.

____ **7.** For $H_0 = 65$ km/s/Mpc, a critical-density universe is approximately 10 billion years old.

____ **8.** Current physics can describe the universe as far back in time as 1 second after the Big Bang, but no earlier.

____ **9.** The cosmological redshift is a direct measure of the expansion of the universe.

____ **10.** As it travels through space a photon's wavelength expands at the same rate as the universe is expanding.

____ **11.** According to the standard Big Bang model of the cosmos, the universe has only two possible futures: either it will continue to expand forever, or it will someday stop expanding and start to contract.

____ **12.** If the universe recollapses, it must rebound into a new expansion cycle similar to the present one.

____ **13.** A bound universe will ultimately end in a "cold death."

____ **14.** An unbound universe is geometrically flat.

____ **15.** The cosmic microwave background is the highly redshifted radiation of the early Big Bang.

SELF-TEST: FILL IN THE BLANK

1. Pencil-beam surveys suggest that the largest structures in space are no larger than about _____ Mpc in size.

2. Homogeneous means "the same _____."

3. Isotropic means "the same in all _____."

4. Together, the assumptions of cosmic homogeneity and isotropy are known as the _____.

5. If the universe had an edge, this would violate the assumption of _____.

6. If the universe had a center, this would violate the assumption of _____.

7. A Hubble constant of 65 km/s/Mpc gives a maximum age for the universe of _____ billion years.

8. _____ is slowing the expansion of the universe.

9. The _____ of the universe determines whether the universe will expand forever.

10. A _____ value for Hubble's constant yields an age for the universe incompatible with the ages of the oldest globular clusters.

11. Luminous matter makes up about _____ percent of the critical density.

12. By observing distant galaxies, astronomers have determined that the universe was expanding _____ long ago than it is today.

13. The surface of a sphere is a two-dimensional example of a _____ curved universe.

14. An open universe has a density _____ the critical value.

15. The temperature of the universe, as measured by the cosmic microwave background, is _____ K.

REVIEW AND DISCUSSION

1. What evidence do we have that there is no structure in the universe on very large scales? How large is "very large"?

2. What is the cosmological principle?

3. What is Olbers's paradox? How is it resolved?

4. Explain how an accurate measure of Hubble's constant can lead to an estimate of the age of the universe.

5. We appear to be at the center of the Hubble flow. Why doesn't this violate the cosmological principle?

6. Why isn't it correct to say that the expansion of the universe involves galaxies flying outward into empty space?

7. Where did the Big Bang occur?

8. How does the cosmological redshift relate to the expansion of the universe?

9. What measureable property of the universe determines whether or not it will expand forever?

10. Is there enough luminous matter to halt the current cosmic expansion?

11. Is there enough dark matter to halt the current cosmic expansion?

12. What is the significance of the cosmic microwave background?

13. Why does the temperature of the microwave background fall as the universe expands?

14. Many cultures throughout history have developed their own cosmologies. Do you think the modern scientific cosmology is more likely to endure than any other? Why or why not?

15. Even with the recent revision of the value of Hubble's constant, estimates of the age of the universe based on H_0 are still somewhat lower than estimates of the ages of many globular clusters in our own Galaxy (*Interlude 26-2*). How do you think astronomers should proceed in resolving this discrepancy?

PROBLEMS

1. What is the greatest distance at which a galaxy survey sensitive to objects as faint as 20th magnitude could detect a galaxy as bright as the Milky Way (absolute magnitude −20)?

2. If the entire universe were filled with Milky Way–like galaxies, with an average density of 1 per cubic megaparsec, calculate the total number of galaxies observable by the survey in the previous question, if it covered the entire sky.

3. Eight galaxies are located at the corners of a cube. The present distance from each galaxy to its nearest neighbor is 10 Mpc, and the entire cube is expanding according to Hubble's law, with $H_0 = 65$ km/s/Mpc. Calculate the recession velocity of one corner of the cube relative to the opposite corner.

4. According to the Big Bang theory described in this chapter, what is the maximum possible age of the universe if $H_0 = 50$ km/s/Mpc? 65 km/s/Mpc? 80 km/s/Mpc?

5. For a Hubble constant of 65 km/s/Mpc, the critical density is 8×10^{-27} kg/m^3. (a) How much mass does that correspond to within a volume of 1 cubic astronomical unit? (b) How large a cube would be required to enclose 1 Earth mass of material?

6. The Virgo Cluster is observed to have a recessional velocity of 1200 km/s. Assuming $H_0 = 65$ km/s/Mpc and a critical-den-

sity universe, calculate the total mass contained within a sphere centered on Virgo and just enclosing the Milky Way Galaxy. What is the escape speed from the surface of this sphere?

7. Assuming critical density, and using the distances presented in Table 25.1, estimate the total amount of matter in the observable universe. Express your answer (a) in kilograms, (b) in solar masses, and (c) in "galaxies," where 1 galaxy = 10^{11} solar masses.

8. The critical density is proportional to the square of Hubble's constant. If the critical density were equal to the known density of "normal" matter (not dark matter), about 10^{-28} kg/m^3, what would be the corresponding value of Hubble's constant? Is this within the currently accepted range of values?

9. What was the temperature of the cosmic microwave background at the epoch of quasar formation (at a redshift of 5)?

10. Using the information given in problem 6, calculate the distance from our location to the point that would one day become the center of the Virgo Cluster, at the time when the temperature of the microwave background was equal to the present surface temperature of the Sun.

PROJECTS

1. Make a model of a two-dimensional universe and examine Hubble's law on it. Find a balloon that will expand into a nice large sphere. Blow it up about halfway and mark dots all over its surface; the dots will represent galaxies. Choose one dot as your home galaxy. Using a measuring tape, measure the distances to various other galaxies, numbering the dots so you will not confuse them later. Now blow the balloon up to full size and measure the distances again, and find the new distances to each dot. Calculate the change in the distances for each galaxy; this is a measure of their velocity (change in position/change in time; the time is the same for all and is arbitrary). Plot their velocities versus their new distances as in

Figure 26.11. Do you get a straight-line correlation, i.e. a "Hubble" law? Does it matter which dot you choose as home? Demonstrate this to your class.

2. Write a paper on the philosophical differences between living in an open, closed, or flat universe. Are there aspects of any of these three possibilities that are hard to accept? It is quite possible that astronomers may determine within your lifetime which is correct. Do you have a preference?

3. Go to your library and read about the *steady-state universe*, which enjoyed some measure of popularity in the 1950s and 1960s. How does it differ from the standard Big Bang model? Why do you think the steady-state model is not widely accepted today?

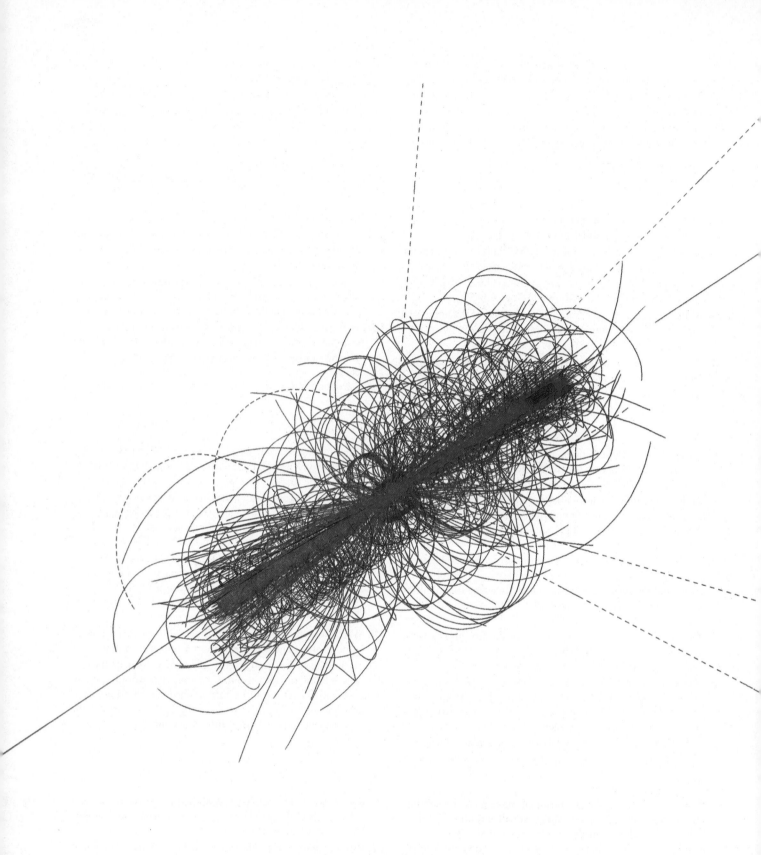

27 THE EARLY UNIVERSE

Toward the Beginning of Time ✺www✺

LEARNING GOALS

Studying this chapter will enable you to:

1 Describe the characteristics of the universe immediately after its birth.

2 Explain how matter emerged from the primeval fireball.

3 Enumerate the epochs in the evolutionary history of the universe and specify the major characteristics of each.

4 Explain how and when the simplest nuclei and atoms formed.

5 Summarize the horizon and flatness problems, and discuss the theory of cosmic inflation as a possible solution to these problems.

6 Explain the formation of large-scale structure in the cosmos, and discuss the observational evidence of our theories of structure formation.

(Opposite page) Astronomers are unable to observe the universe when it was very young, because truly far-away and long-ago events were engulfed in a sea of intense radiation. In the very early universe, no galaxies, stars, or planets had yet emerged. Only subatomic particles existed—not only the protons, neutrons and electrons we know today, but also, we think, various strange and exotic elementary particles predicted by current theory. Surprisingly, part of this melange of particles that characterized the early universe can now be studied here on Earth, in huge accelerators such as those of the Conseil Européen pour la Recherche Nucleaire (CERN) near Geneva, Switzerland. The odd image at left shows the violent collision of elementary particles, typical of events thought to have occurred about a trillionth of a second after the beginning of the universe. Two fast-moving protons (straight red lines) are shown colliding head on. Their encounter produces a multitude of charged particles, seen here curving in the magnetic field of the Large Hadron (proton) Collider at CERN. One rather massive particle thought to have played a unifying role in the early universe—a particle bizarrely named the Higgs boson—is depicted here amidst the debris, at the intersection of the straight green lines (which are yet other particles decaying in the aftermath of the violence). This image, however, is only a computer simulation; the Large Hadron Collider is still under construction, and the Higgs boson has not yet been seen by anyone. Whoever does eventually discover the Higgs (assuming it really exists) will doubtless win a free, all-expense-paid trip to Sweden—to pick up a Nobel Prize.

What were the conditions during the first few seconds of the universe, and how did those conditions change to give rise to the universe we see today? In studying the earliest moments of our universe, we enter a truly alien domain. As we move backward in time toward the Big Bang, our customary landmarks slip away one by one. Atoms vanish, then nuclei, then even the elementary particles themselves. In the beginning the universe consisted of pure energy, at unimaginably high temperatures. As it expanded and cooled, the ancient energy gave rise to the particles that make up everything we see around us today. Modern physics has now arrived at the point where it can reach back almost to the instant of the Big Bang itself, allowing scientists to unravel some of the mysteries of our own beginning.

27.1 Back to the Big Bang

MATTER AND RADIATION

1 On the very largest scales we can regard the universe as a roughly homogeneous mixture of matter and radiation. The overall density of matter is not known with certainty, but it is thought to be at least a few tenths of the critical density of about 10^{-26} kg/m^3, above which the universe will eventually recollapse, and below which it will expand forever. The universe is apparently open, but barely so.

The matter in the universe consists of the familiar building blocks of atoms—protons, neutrons, and electrons—as well as the mysterious dark matter, whose composition is still hotly debated by astronomers. Most of the radiation in the universe is in the form of the cosmic microwave background, the low-temperature (3 K) radiation field that fills all space. ∞ (Sec. 26.6) Surprisingly, although the microwave background radiation is very weak, it still contains more energy than has been emitted by all the stars and galaxies that have ever existed! The reason for this is that stars and galaxies, though very intense sources of radiation, occupy only a tiny fraction of space. When their energy is averaged out over the volume of the entire universe, it falls short of the energy of the microwave background by at least a factor of 10. For our current purposes, then, we can ignore much of the first 26 chapters of this book and regard the cosmic microwave background as the only significant form of radiation in the universe!

Is matter the dominant component of the universe, or does radiation also play an important role on large scales? In order to compare matter and radiation, we must first convert them to a "common currency"—either mass or energy. Let's choose to compare their masses. We can express the energy in the microwave background as an equivalent density by first calculating the number of photons in any cubic centimeter of space, then converting the total energy of these photons into a mass using the relation $E = mc^2$. ∞ (Sec. 16.5) When we do this, we arrive at an equivalent density for the microwave background of about 5×10^{-31} kg/m^3. Thus, *at the present moment* the density of matter (around 10^{-26} kg/m^3) in the universe far exceeds the density of radiation. In cosmological terminology, we say that we live in a **matter-dominated** universe.

Was the universe always matter dominated? To answer this question, we must ask how the densities of both matter

and radiation change as the universe expands. Both decrease, as the expansion dilutes the numbers of atoms and photons alike. But the radiation is also diminished in energy by the cosmological redshift, so its density falls *faster* than that of matter as the universe grows (see Figure 27.1). Conversely, as we look back in time, closer and closer to the Big Bang, the density of the radiation increases faster than that of matter. Accordingly, even though today the radiation density is much less than the matter density, there must have been a time in the past when they were equal. Before that time, radiation was the main constituent of the cosmos. The universe is said to have been **radiation dominated** then. Given our best estimates of the present densities, the crossover point—the time at which the densities of matter and radiation

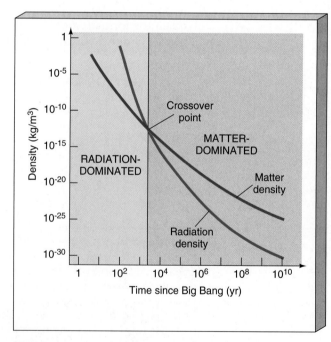

Figure 27.1 As the universe expanded, the number of both matter particles and photons per unit volume decreased. However, the photons were also reduced in energy by the cosmological redshift, reducing their equivalent mass, and hence their density, still further. As a result, the density of radiation fell faster than the density of matter as the universe grew. Tracing the curves back from the densities we observe today, we see that radiation must have dominated matter at early times—that is, at times before the crossover point.

were equal—occurred a few thousand years after the Big Bang, when the universe was about 20,000 times smaller than it is today. The temperature of the background radiation at that time was about 60,000 K, so it peaked well into the ultraviolet portion of the spectrum.

Throughout this book we have been concerned exclusively with the history of the universe long after it became matter dominated—the formation and evolution of galaxies, stars, and planets as the universe thinned and cooled toward the state we see today. In this chapter we consider some important events in the early, radiation-dominated universe, long before any star or galaxy existed, that played no less a role in determining the present condition of the cosmos.

PARTICLE PRODUCTION IN THE EARLY UNIVERSE

2 The very early universe was characterized by temperatures and densities far greater than anything we have encountered thus far, even in the hearts of supernovae. In order to understand the early universe, we must delve a little more deeply into the behavior of matter and radiation at very high temperatures.

The key to understanding events at very early times lies in a process called **pair production**, in which two photons give rise to a *particle–antiparticle* pair, as shown in Figure 27.2(a) for the particular case of electrons and positrons. In this way, matter can be created from radiation. The reverse process can also occur—a particle and its antiparticle can *annihilate* each other to produce energy in the form of electromagnetic radiation, as depicted in Figure 27.2(b). In other words, energy in the form of radiation can be freely converted into matter in the form of particles and antiparticles, and particles and antiparticles can be freely converted back into radiation, subject only to the law of conservation of mass and energy.

The higher the temperature of a radiation field, the greater the energy of the typical constituent photons, and the greater the masses of the particles than can be created by pair production. For any given particle, the critical temperature above which pair production is possible, and below which it is not, is called the particle's *threshold temperature*. The threshold temperature increases as the mass of the particle increases. For electrons it is about 6×10^9 K. For protons, which are nearly 2000 times more massive, it is just over 10^{13} K.

As an example of how pair production affected the composition of the early universe, consider the production of electrons and positrons as the universe expanded and cooled. At high temperatures—above about 10^{10} K—most photons had enough energy to form an electron or a positron, and pair production was commonplace. As a result, space seethed with electrons and positrons, constantly created from the radiation field and annihilating one another to form photons again. Particles and radiation are said to have been in *thermal equilibrium*—new particle–antiparticle pairs were created by pair production at the same rate as they annihilated one another. As the universe expanded and the temperature decreased, so did the average photon energy. By the time

Figure 27.2 (a) Two photons can produce a particle–antiparticle pair—in this case an electron and a positron—if their total energy exceeds the mass energy of the particles produced. (b) The reverse process is particle–antiparticle annihilation, in which an electron and positron destroy each other, vanishing in a flash of gamma rays. (c) Tracks in a particle detector allow us to visualize pair creation. Here a gamma ray, whose path is invisible because it is electrically neutral, arrives from the left; it dislodges an atomic electron and sends it flying (the longest path). At the same time it provides the energy to produce an electron–positron pair (the spiral paths, which curve in opposite directions in the detector's magnetic field because of their opposite electric charges).

the temperature had fallen below a billion or so kelvins, photons no longer had enough energy for pair production to occur, and only radiation remained. Figure 27.3 illustrates how this change took place.

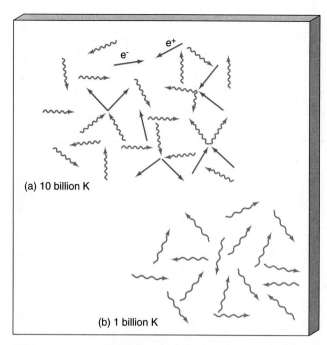

Figure 27.3 (a) At 10 billion K most photons have enough energy to create particle–antiparticle (electron–positron) pairs, so these particles exist in great numbers, in equilibrium with the radiation. (b) Below about 1 billion K, photons have too little energy for pair production to occur.

Pair production in the very early universe was directly responsible for all the matter that exists in the universe today. *Everything we see around us was created out of radiation as the cosmos expanded and cooled.* Because we are here to ponder the subject and we ourselves are made of matter, we know that some matter must have survived these early moments. For some reason there was a slight excess of matter over antimatter at early times. A small residue of particles that outnumbered their antiparticles was left behind as the temperature dropped below the threshold for creating them. With no antiparticles left to annihilate them, the number of particles has remained constant ever since. These survivors are said to have *frozen out* of the radiation field as the universe cooled.

The first hundred or so seconds of the universe's existence saw the creation of all of the basic "building blocks" of matter we know today—protons and neutrons froze out when the temperature dropped below 10^{13} K, when the universe was only 0.0001 s old. The lighter electrons froze out somewhat later, about a minute or so after the Big Bang, when the temperature fell below 10^9 K. This "matter-creation" phase of the universe's evolution ended when the electrons—the lightest known elementary particles—appeared out of the cooling primordial fireball. From that point on, matter has continued to evolve, clumping together into more and more complex structures, eventually forming the atoms, planets, stars, galaxies, and large-scale structure we see today, but no new matter has been created since that early time.

27.2 The Evolution of the Universe

3 For the first few thousand years after the Big Bang, the universe was small and dense and dominated by the effects of radiation. We will refer to this period as the *Radiation Era*. Some matter existed during this time, but it was a mere contaminant in the blinding radiation of the primeval Big Bang fireball. Afterwards matter came to dominate, in the *Matter Era*. Atoms, molecules, and galaxies formed as the universe cooled and thinned toward the state we see today.

Let's begin our study of the early universe by summarizing in broad terms the history of the cosmos, starting at the Big Bang. Table 27.1 and Figure 27.4 present the time, density, and temperature spans of eight major epochs in the development of the universe, along with a brief description of the main physical events that dominated the universe during each. The numbers in this table and figure result from pushing the known laws of physics as far back in time as we can. In the next few sections we will expand on some of these epochs in greater detail, but let's not lose sight of the big picture and the place of each epoch in it.

THE RADIATION ERA

The universe began with an explosion from an incredibly hot and dense state. Precisely what state, we cannot say. And why it exploded, we really don't know. To understand why the universe began expanding, or even more fundamentally, why the universe exists at all, is currently beyond science—there are simply no relevant data. Although ignorant of the moment of creation itself, theorists nevertheless believe that the physical conditions in the universe can be understood in terms of present-day physics back to an extraordinarily short time—a mere 10^{-43} s, in fact—after the Big Bang.

Why can't theorists push our knowledge back to the Big Bang itself? The answer is that we presently have no theory capable of describing the universe at these earliest of times. Theorists believe that under the extreme conditions of density and temperature within 10^{-43} s of the Big Bang, gravity and the other fundamental forces (electromagnetism, the strong force, and the weak force, as described in *More Precisely 27-1*) were indistinguishable from one another—a far cry from the radically different characters we see today. The four forces are said to have been *unified* at this early time—there was, in effect, only one force of nature.

The theory that combines quantum mechanics (the proper description of microscopic phenomena) with general relativity (which describes the universe on the largest scales) is generically known as quantum gravity. ∞ (Sec. 22.6) The period of time from the beginning to 10^{-43} s is often referred to as the *Planck epoch*, after Max Planck, one of the creators of quantum mechanics. Unfortunately, for now at least, there is no working theory of quantum gravity, so we simply cannot talk meaningfully about the universe during the Planck epoch.

TABLE 27.1 Major Epochs in the History of the Universe

ERA	EPOCH	TIME (after Big Bang)	DENSITY (kg/m³)	TEMPERATURE (K)	MAIN EVENTS
Radiation Era					
		0 s	∞	∞	
	Planck				Unknown physics; quantum gravity
		10^{-43} s	10^{95}	10^{32}	
	GUT				Strong, weak, and electromagnetic forces unified
		10^{-35} s	10^{75}	10^{27}	
	Hadron				Heavy and light particles all in thermal equilibrium
		10^{-4} s	10^{16}	10^{12}	
	Lepton				Only light particles still in thermal equilibrium; neutrinos decouple
		10^{2} s	10^{4}	10^{9}	
	Nuclear				Deuterium and helium formed by fusion of protons and neutrons during first 1000 s
		10^{3} yr ($\approx 3 \times 10^{10}$ s)	10^{-13}	6×10^{4}	
Matter Era					
		10^{3} yr ($\approx 3 \times 10^{10}$ s)	10^{-13}	6×10^{4}	
	Atomic				Matter begins to dominate; atoms form; electromagnetic radiation decouples
		10^{6} yr ($\approx 3 \times 10^{13}$ s)	10^{-19}	10^{3}	
	Galactic				Galaxies and larger-scale structure form
		10^{9} yr ($\approx 3 \times 10^{16}$ s)	3×10^{25}	10	
	Stellar				All galaxies have formed; stars continue to form
		$>10^{10}$ yr ($\approx 3 \times 10^{17}$ s)	10^{-26}	3	

By the end of the Planck epoch the temperature was around 10^{32} K and the universe was filled with radiation and a vast array of subatomic particles created by the mechanism of pair production. At around this time, gravity parted company with the other forces of nature—it became distinguishable from the "quantum" forces, and has remained so ever since. The strong, weak, and electromagnetic forces were still unified. The present-day theories that describe this unification are collectively known as **Grand Unified Theories**, or GUTs for short (see *More Precisely 27-1*). Accordingly, we will refer to this period of time as the *GUT epoch*.

Theory indicates that at temperatures below 10^{28} K, the strong nuclear force becomes distinguishable from the electroweak force (the unified weak and electromagnetic force). Once the universe had cooled to this temperature, about 10^{-35} s after the Big Bang, the GUT epoch ended. According to many GUTs, one important legacy of this

epoch may have been the appearance and subsequent freeze-out of a veritable "zoo" of very massive (and as yet unobserved) elementary particles which interact only very weakly with normal matter. These "exotic" particles are prime candidates for the dark matter of unknown composition known to exist in abundance both within galaxies and in the unseen depths of intergalactic space. (∞ Sec. 23.6, 24.3)

Our next major subdivision of the Radiation Era covers the period when all "heavy" elementary particles—that is, all the way down in mass to protons, neutrons, and their constituent quarks—were in thermal equilibrium with the radiation. We refer to this period as the *hadron epoch*, after the general term for particles that interact via the strong force.

The universe continued to expand and cool. At a temperature of about 10^{15} K, 10^{-10} s after the Big Bang, the weak and the electromagnetic components of the electroweak force began to display their separate characters. By about

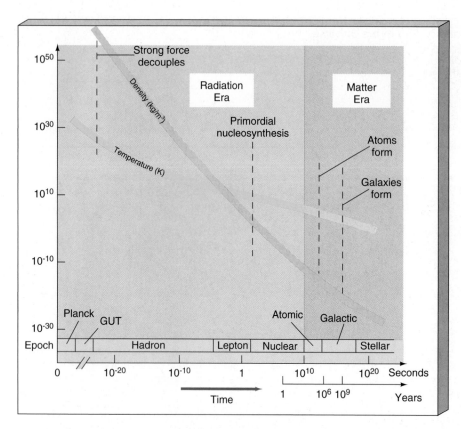

Figure 27.4 The average temperature and average density throughout the history of the universe. The epochs listed in Table 27.1 are indicated. (The numerical values for these plots are taken from Table 27.1.) Some key events in the history of the universe are also marked.

0.1 milliseconds (10^{-4} s) after the Big Bang, the temperature had dropped well below the 10^{13} K threshold for the creation of protons and neutrons (the lightest hadrons), and the hadron epoch ended. The main constituents of the universe were now lightweight particles—muons, (see *More Precisely 27-1*) electrons, neutrinos, and their antiparticles—all still in thermal equilibrium with the radiation. Compared with the numbers of these lighter particles, only very few protons and neutrons remained at this stage because most had been annihilated.

Electrons, muons, and neutrinos are collectively known as *leptons*, after the Greek word meaning "light" (that is, not heavy). Accordingly, this period in the history of the universe is known as the *lepton epoch*. Quite early on in this epoch, the rapidly thinning universe became transparent to neutrinos. These ghostly particles have been streaming freely through space ever since—most have not interacted with any other particle since the universe was a few seconds old. The lepton epoch ended when the universe was about 100 s old and the temperature fell to about 1 billion K—too low for electron–positron pair production to occur. The density of the universe by this time was about 10 times that of water.

The final significant event in the Radiation Era occurred when protons and neutrons began to fuse into heavier nuclei. At the start of this period, which we will call the *nuclear epoch*, the temperature was a few hundred million kelvins, and fusion occurred very rapidly, forming deuterium and helium in quick succession before conditions became too cool for further reactions to occur. By the time the universe was about 15 minutes old, much of the helium we observe today had been formed.

THE MATTER ERA

Time passed, the universe continued to expand and cool, and radiation gave way to matter as the dominant constituent of the universe. Our next major epoch extends in time from a few thousand years (the end of the Radiation Era) to about 1 million years after the Big Bang. As the primeval fireball diminished in intensity, a crucial change occurred—perhaps the most important change in the history of the universe. At the end of the nuclear epoch, radiation still overwhelmed matter. As fast as protons and electrons combined, radiation broke them apart again, preventing the formation of even simple atoms or molecules. However, as the universe expanded and cooled, the early dominance of radiation eventually ended. Once formed, atoms remained intact. We will call this period the *atomic epoch*.

The last two epochs together bring us to the current age of the universe. During these late stages, change happened at a much more sedate pace. By the time the universe was about a billion years old, galaxies and large-scale structure had formed. For the first time the visible universe departed from homogeneity on macroscopic scales. The largely uniform universe of the Radiation Era became a universe containing large agglomerations of matter. We call the period from 10^6 to 10^9 years after the Big Bang the *galactic epoch*. At its end, quasars were shining brightly and the first stars were burning and exploding, helping to determine the future shape of their parent galaxies. Since then, stars, planets, and life have appeared in the universe. This final *stellar epoch* has been the subject of the first 25 chapters of this book.

MORE PRECISELY 27-1 More on Fundamental Forces

In *More Precisely 16-1* we noted that the behavior of all matter in the universe is ruled by just *three fundamental forces*—gravity, the electroweak force (the unification of the electromagnetic and weak forces), and the strong (nuclear) force. In terrestrial laboratories these forces display very different properties from one another. Gravity and electromagnetism are long-range, inverse-square forces, whereas the strong and weak forces have very short ranges—10^{-15} and 10^{-17} m, respectively. Furthermore, the forces do not all affect the same particles. Gravity affects everything. The electromagnetic force affects only charged particles. The strong force operates between nuclear particles, such as protons and neutrons, but it does not affect electrons and neutrinos. The weak force shows up in certain nuclear reactions and radioactive decays. The strong force is 137 times stronger than the electromagnetic force, 100,000 times stronger than the weak force, and 10^{39} times stronger than gravity.

In fact, there is more structure below the level of the nucleus. Protons and neutrons are not truly "elementary" in nature but are actually made of subparticles called *quarks*. (The name derives from a meaningless word coined by novelist James Joyce in his book *Finnegans Wake*.) According to current theory, there are precisely six distinct types of quark in the universe, paired with six distinct types of lepton—the electron and two related "electron-like" particles (called muons and taus), and three types of neutrino. The most massive, and most elusive, of them—the so-called top quark—was discovered at the Fermi National Laboratory (Fermilab) in Illinois in April 1994. The strong nuclear force is itself just a manifestation of an even more basic force that binds quarks to one another.

On the face of it, one might not imagine that there could be any deep underlying connection between forces as dissimilar as those just described, yet there is growing evidence that they may really be just different aspects of a single basic phenomenon. In the 1960s, theoretical physicists succeeded in explaining the electromagnetic and weak forces in terms of the electroweak force. Shortly thereafter, the first attempts were made at combining the strong and electroweak forces into a single all-encompassing "superforce."

Theories that combine the strong and electroweak forces into one are generically known as *Grand Unified Theories*, or, less formally, GUTs. (Note that the term is plural—no one GUT has yet been proven to be "the" correct description of nature.) One general prediction of GUTs is that the three nongravitational forces are indistinguishable from one another only at enormously high energies, corresponding to temperatures in excess of 10^{28} K. Below that temperature the superforce splits into two, displaying its separate strong and electroweak aspects. In particle-physics parlance, we say that there is a "symmetry" between the strong and the electroweak forces that becomes broken at temperatures below 10^{28} K, allowing the separate characters of the two forces to become apparent. At "low" temperatures—less than about 10^{15} K, a range that includes almost everything we know on Earth and in the stars—

there is a second symmetry breaking, and the electroweak force splits to reveal its more familiar electromagnetic and weak nature.

The key predictions of the electroweak theory were experimentally verified in the 1970s, winning the theory's originators the 1979 Nobel Prize in physics. The GUTs have not yet been experimentally verified (or refuted), in large part because of the extremely high energies that must be reached in order to observe their predictions.

An important idea that has arisen from the realization that the strong and the electroweak forces can be unified is the notion of supersymmetry. It extends the idea of symmetry between fundamental forces to place all particles—those that feel forces (such as protons and electrons), and those that transmit those forces (such as photons and gluons)—on an equal footing. One particularly important prediction of this theory is that all particles should have so-called supersymmetric partners—extra particles that must exist in order for the theory to remain self-consistent. None of these new particles has ever been detected, yet many physicists are convinced of the theory's essential correctness. Experiments are planned that may soon provide evidence for supersymmetric partners to some of the known elementary particles.

Of particular interest to astronomers, these new particles, if they exist, would have been produced in abundance in the Big Bang and should still be around today. They are also expected to be very massive—at least a thousand times heavier than a proton. These so-called supersymmetric relics are among the current leading candidates for the dark matter in the universe (although it must be admitted that recent experimental failures to detect them has dampened some astronomers' early enthusiasm).

Efforts to include gravity within this picture have so far been unsuccessful. Gravitation has not yet been incorporated into a single "SuperGUT," in which all the fundamental forces are united. Some theoretical efforts to merge gravity with the other forces have tried to fit gravity into the quantum world by postulating extra particles—called *gravitons*—that transmit the gravitational force. However, this is a very different view of gravity from the geometric picture embodied in Einstein's general relativity, and combining the two into a consistent theory of quantum gravity has proved very difficult. Alternative approaches start from the geometric view and attempt to explain the basic forces of nature in terms of additional curved dimensions of spacetime. They, too, encounter serious problems. A theory currently under very active investigation seeks to interpret all particles and forces in terms of particular modes of vibration of submicroscopic objects known as *strings*. Some theorists feel that this approach, although very complex in its mathematical details, may offer the best hope of unifying the forces of nature.

However, at present, none of these theories has succeeded in making any definite statement about conditions in the very early universe. A complete theory of quantum gravity continues to elude researchers.

27.3 The Formation of Nuclei and Atoms

HELIUM FORMATION IN THE EARLY UNIVERSE

4 We now have all the ingredients needed to complete our story of the creation of the elements, begun in Chapter 21 but never quite finished. ∞ (Sec. 21.4) The theory of stellar nucleosynthesis accounts very well for the observed abundances of heavy elements in the universe, but there are discrepancies between theory and observations when it comes to the abundances of the light elements, especially helium. Simply put, there is far more helium in the universe—about 25 percent by mass—than can be explained by nuclear fusion in stars. The accepted explanation is that this base level of helium is *primordial*—that is, it was created during the early, hot epochs of the universe, before any stars had formed. The production of elements heavier than hydrogen by nuclear fusion shortly after the Big Bang is called **primordial nucleosynthesis**.

By about 100 s after the Big Bang, the temperature had fallen to about 1 billion K, and apart from "exotic" dark-matter particles, matter in the universe consisted of electrons, protons, and neutrons, with the protons outnumbering the neutrons by about 5 to 1. The stage was set for nuclear fusion to occur. Protons and neutrons combined to produce deuterium nuclei:

$$^1\text{H (proton)} + \text{neutron} \rightarrow {}^2\text{H (deuterium)} + \text{energy}.$$

Although this reaction must have occurred very frequently during the lepton epoch, the temperature then was still so high that the deuterium nuclei were broken apart by high-energy gamma rays as soon as they formed. The universe had to wait until it became cool enough for the deuterium to survive. This waiting period is sometimes called the *deuterium bottleneck*.

Only when the temperature of the universe fell below about 900 million K, roughly 2 minutes after the Big Bang, was deuterium at last able to form and endure. Once that occurred, the deuterium was quickly converted into heavier elements by numerous reactions:

$$^2\text{H} + {}^1\text{H} \rightarrow {}^3\text{He} + \text{energy},$$

$$^3\text{He} + \text{neutron} \rightarrow {}^4\text{He} + \text{energy},$$

$$^2\text{H} + {}^2\text{H} \rightarrow {}^4\text{He} + \text{energy},$$

along with many others. The result was that once the universe passed the deuterium bottleneck, fusion proceeded rapidly and large amounts of helium were formed. In just a few minutes most of the free neutrons were consumed, leaving a universe whose matter content was primarily hydrogen and helium. Figure 27.5 illustrates some of the reactions responsible for helium formation. Contrast it with Figure 16.28, which depicts how helium is formed today in the cores of main-sequence stars like the Sun.* ∞ (Sec. 16.5)

The proton–proton chain that powers the Sun played no significant role in primordial helium formation. The proton–proton reaction that starts the chain is very slow compared with the proton–neutron reaction discussed here and is important in the Sun only because the solar interior contains no free neutrons to make the latter reaction possible.

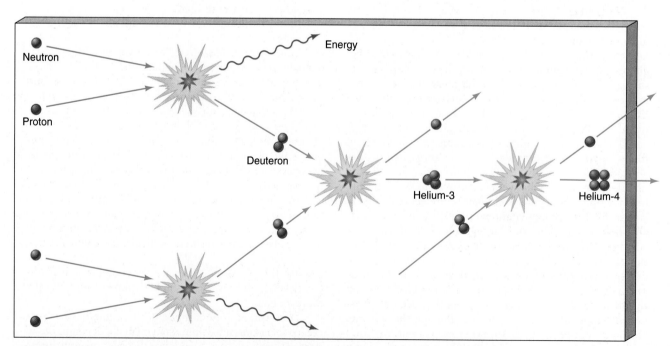

Figure 27.5 Some of the reaction sequences that led to the formation of helium in the early universe. Compare this figure with Figure 16.26, which depicts the proton–proton chain in the Sun.

We might imagine that fusion could have continued to create heavier and heavier elements, just as in the cores of stars, but this did not occur. In stars, the density and the temperature both *increase* slowly with time, allowing more and more massive nuclei to form, but in the early universe exactly the opposite was true. The temperature and density were both *decreasing* rapidly, making conditions less and less favorable for fusion as time went on. Even before the supply of neutrons was completely used up, the nuclear reactions had effectively ceased. Reactions between helium nuclei and protons may also have formed trace amounts of lithium (the next element beyond helium) by this time, but for all practical purposes, the expansion of the universe caused fusion to stop at helium. The brief epoch of primordial nucleosynthesis was over about 15 minutes after it began.

By the end of the nuclear epoch, some 1000 s after the Big Bang, the temperature of the universe was about 300 million K and the cosmic elemental abundance was set. Careful calculations indicate that about 1 helium nucleus had formed for every 12 protons remaining. Because a helium nucleus is four times more massive than a proton, helium accounted for about one quarter of the total mass of matter in the universe:

$$\frac{1 \text{ helium nucleus}}{12 \text{ protons} + 1 \text{ helium nucleus}} = \frac{4 \text{ mass units}}{12 \text{ mass units} + 4 \text{ mass units}}$$

$$= \frac{4}{16} = \frac{1}{4}.$$

The remaining 75 percent of the matter in the universe was hydrogen. It would be almost a billion years before nucleosynthesis in stars would change these numbers.

The foregoing calculation implies that all cosmic objects should contain *at least* 25 percent helium by mass. The figure for the Sun, for example, is about 28 percent. However, it is difficult to disentangle the contributions to the present-day helium abundance from primordial nucleosynthesis and later hydrogen burning in stars. Our best hope of determining the amount of primordial helium is to study the oldest stars known, since they formed early on, before stellar nucleosynthesis had had time to change significantly the helium content of the universe. Unfortunately, stars surviving from that early time are of low mass and hence quite cool, making the helium lines in their spectra very weak and hard to measure accurately. ∞ (Secs. 17.6, 17.9) Nevertheless, despite this uncertainty, the observations are generally consistent with the theory just described.

Bear in mind that while all this was going on matter was just an insignificant "contaminant" in the radiation-dominated universe. Radiation outmassed matter by about a factor of 5000 at the time helium formed. The existence of helium is very important in determining the structure and appearance of stars today, but its creation was completely *irrelevant* to the evolution of the universe at the time.

DEUTERIUM AND THE DENSITY OF THE COSMOS

During the nuclear epoch, although most deuterium was quickly fused into helium as soon as it formed, a small amount was left over when the primordial nuclear reactions ceased. Observations of deuterium—especially those made by orbiting satellites able to capture deuterium's strongest spectral feature, which happens to be emitted in the ultraviolet part of the spectrum—indicate a present-day abundance of about 2 deuterium nuclei for every 100,000 protons. However, unlike helium, deuterium is not produced to any significant degree in stars (in fact, deuterium tends to be destroyed in stars), so any deuterium we see today *must* be primordial.

This observation is of great importance to astronomers because it provides them with a sensitive method—and one that is completely *independent* of the techniques discussed in previous chapters—of probing the present-day density of matter in the universe. According to theory, as illustrated in Figure 27.6, the denser the universe is today, the more particles there were at early times to react with deuterium as it formed, and the less deuterium was left over when nucleosynthesis ended. Comparison of the observed deuterium abundance (marked on the figure) with the theoretical results implies a present-day density of *at most* 3×10^{-28} kg/m^3—only a few percent of the critical density. Thus, measurements of the cosmic deuterium abundance provide us with a remarkably firm estimate of $\Omega_0 \approx 0.03$.

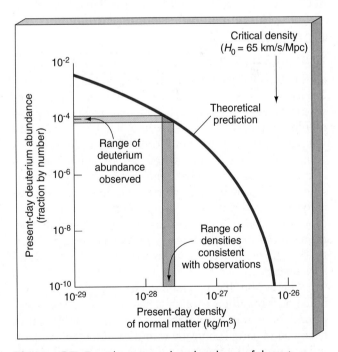

Figure 27.6 The present-day abundance of deuterium depends strongly on the amount of matter present at early times, and this, in turn, determines the present-day density of the universe. Thus, measuring the amount of deuterium in the universe gives us an estimate of the overall density of matter. The best deuterium measurements are marked; they imply that the density of matter in the universe is at most a few percent of the critical value.

But before we jump to the conclusion that the universe is open and will expand forever, we must make one very important qualification. Primordial nucleosynthesis as just described depends *only* on the presence of protons and neutrons in the early universe. Thus, measurements of the abundance of helium and deuterium tell us *only* about the density of "normal" matter—matter made up of protons and neutrons—in the cosmos. This finding has a momentous implication for the overall composition of the universe. As we saw earlier, astronomers have concluded, on the basis of studies of the motions of galaxies in clusters and superclusters, that Ω_0 is at least 0.2 or 0.3, and may possibly be much more. ∞ (Sec. 26.4) If this reasoning is correct, and if the density of normal matter is only a few percent of the critical value, then we are forced to admit that not only is most of the matter in the universe dark, but most of the dark matter is *not* composed of protons and neutrons.

Thus, the bulk of the matter in the universe apparently exists in the form of elusive subatomic particles (discussed as dark-matter candidates in Chapter 23), whose nature we do not fully understand and whose very existence has yet to be conclusively demonstrated in laboratory experiments. ∞ (Sec. 23.6) For the sake of brevity we will adopt from here on the convention that the term "dark matter" refers only to these unknown particles and not to "stellar" dark matter, such as black holes and white dwarfs (also discussed in Chapter 23), which are made of normal matter.

THE FIRST ATOMS

A few thousand years after the Big Bang, radiation ceased to be the dominant component of the universe. The Matter Era had begun. At the start of the atomic epoch, matter consisted of electrons, protons, helium nuclei (formed by primordial nucleosynthesis), and dark matter. The temperature was several tens of thousands of kelvins—far too hot for atoms of hydrogen to exist, although some helium ions may already have formed. During the next few hundred thousand years, a major change occurred. The universe expanded by another factor of 10, the temperature dropped to a few thousand kelvins, and electrons and nuclei combined to form neutral atoms. By the time the temperature had fallen to 4500 K, the universe consisted of atoms, photons, and dark matter.

The period during which nuclei and electrons combined to form atoms is often called the epoch of **decoupling**, for it was during this period that the radiation background parted company with normal matter. At early times, when matter was ionized, the universe was filled with large numbers of free electrons, which interacted frequently with electromagnetic radiation of all wavelengths. As a result, a photon could not travel far before encountering an electron and scattering off it. In effect, the universe was opaque to radiation (rather like the deep interior of a star like the Sun). Matter and radiation were strongly "tied," or *coupled*, to one another by these interactions.

When the electrons combined with nuclei to form atoms of hydrogen and helium, however, only certain wavelengths of radiation—the ones corresponding to the spectral lines of those atoms—could interact with matter. ∞ (Sec. 4.2) Radiation of other wavelengths could travel virtually forever without being absorbed. The universe became nearly *transparent*. From that time on, photons passed generally unhindered through space. As the universe expanded, the radiation simply cooled, eventually to become the microwave background we see today.

The microwave photons now detected on Earth have been traveling through the universe ever since they decoupled. Their last interaction with matter (at the epoch of decoupling) occurred when the universe was a few hundred thousand years old and roughly 1500 times smaller (and hotter) than it is today—that is, at a redshift of 1500. As illustrated in Figure 27.7, the epoch of atom formation created a kind of "photosphere" in the universe, completely surrounding Earth at a distance of approximately 9000 Mpc, the distance at which the photons last interacted before they decoupled. ∞ *(More Precisely 25-1)* On our side of the photosphere—that is, since decoupling—the universe is transparent. On the far side—before decoupling—it was opaque. Thus, by observing the microwave background, we are probing conditions in the universe almost all the way back in time to the Big Bang, in much the same way as studying sunlight tells us about the surface layers of the Sun.

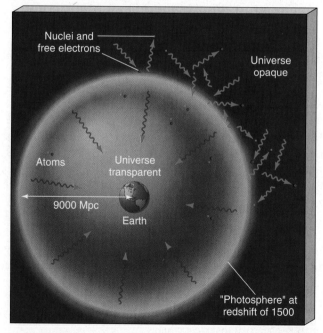

Figure 27.7 When atoms formed, the universe became virtually transparent to radiation. Thus, observations of the cosmic background radiation allow us to study conditions in the universe around a time at a redshift of 1500, when the temperature dropped below about 4500 K. For an explanation of how we can see a region of space 9000 Mpc (30 billion light years) away when the universe is just 10 billion years old, see *More Precisely 25-1*.

27.4 The Inflationary Universe

THE HORIZON AND FLATNESS PROBLEMS

5 In the late 1970s, cosmologists trying to piece together the evolution of the universe were confronted with two nagging problems that had no easy explanation within the standard Big Bang model. The resolution of these difficulties caused cosmologists to rethink their view of the very early universe.

The first problem is known as the **horizon problem**, and it concerns the remarkable *isotropy* of the cosmic microwave background. Recall that the temperature of this radiation is virtually constant, at about 2.7 K, in all directions. Imagine observing the microwave background in two opposite directions on the sky, as illustrated in Figure 27.8. As we have just seen, the radiation last interacted with matter in the universe around a redshift of 1500. Thus, in observing these two distant regions of the universe, marked A and B on the figure, we are studying regions that were separated by several million parsecs at the time that they emitted this radiation. The fact that the background radiation is isotropic to high accuracy means that regions A and B had very similar densities and temperatures at the time the radiation we see left them. The problem is, within the Big Bang theory as just described, there is no particular reason *why* these regions should be so similar to each other.

To take an everyday example, we all know that heat flows from regions of high temperature to regions of low temperature, but it takes time for this to occur. If we light a fire in one corner of a room, we have to wait a while for the other corners to warm up. Eventually, the room reaches a more or less uniform temperature, but only after the heat from the fire—or, more generally, the *information* that the fire is there—has had time to spread. Similar reasoning applies to regions A and B in Figure 27.8. They are separated by many megaparsecs, and there has not been time for information, which can travel no faster than the speed of light, to travel from one to the other. In cosmological parlance, the two regions are said to be outside each others' *horizon*. But if that is so, then how do they "know" that they are supposed to look the same? With no possibility of communication between them, the only alternative is that regions A and B simply started off looking alike—an assumption that cosmologists are very unwilling to make.

The second problem with the standard Big Bang model is called the **flatness problem**. Whatever the exact value of Ω_0, it appears to be quite close to 1—the density of the universe is fairly near the critical value needed for the expansion barely to continue forever. In terms of spacetime curvature, we can say that the universe is remarkably close to being flat. We say "remarkably" here because again there is no good reason *why* the universe should have formed with a density very close to critical. Why not a millionth of or a million times that value? Furthermore, as shown in Figure 27.9, a universe that starts off close to, but not exactly on, the critical curve soon deviates greatly from it, so if the universe is close to critical now, it must have been *extremely* close to critical in the past. For example, if $\Omega_0 = 0.1$ today, then the departure from the critical density at the time of nucleosynthesis would have been only 1 part in 10^{15} (a thousand trillion).

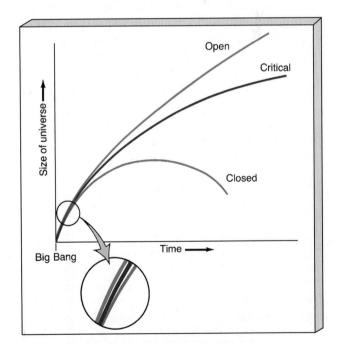

Figure 27.8 The horizon problem. The isotropy of the microwave background indicates that regions A and B in the universe were very similar to each other when the radiation we observe left them, but there has not been enough time since the Big Bang for them ever to have interacted physically with one another. Why then should they look the same?

Figure 27.9 The flatness problem. If the universe deviates even slightly from the critical case, that deviation will grow rapidly in time. For the universe to be as close to critical as it is today, it must have differed from the critical density in the past by only a tiny amount.

These observations constitute "problems" because cosmologists want to be able to *explain* the present condition of the universe, not just accept it "as is." They would prefer to resolve the horizon and flatness problems in terms of physical processes that could have taken a universe with no special properties and caused it to evolve into the cosmos we now see. The resolution of both problems takes us back in time even earlier than nucleosynthesis or the formation of any of the elementary particles we know today—back, in fact, almost to the instant of the Big Bang itself.

FREEZE-OUT

Grand Unified Teories predict that three of the four basic forces of nature—electromagnetism and the strong and weak nuclear forces—are in reality aspects of a single, all-encompasing "superforce" (see *More Precisely 27–1*). However, this unification is evident only at enormously high engergies, corresponding to temperatures in excess of 10^{28} K. At lower temperatures the superforce reveals its separate electromagnetic, strong, and weak characters.

An essential concept in quantum physics is the idea that forces between elementary particles are exerted, or *mediated*, by the exchange of another type of particle, generically called a *boson*. We might imagine the two particles as playing a rapid game of catch, using a boson as a ball, as illustrated in Figure 27.10. As the ball is thrown back and forth, the force is transmitted. In ordinary electromagnetism, the boson involved is the photon—a bundle of electromagnetic energy that always travels at the speed of light. The strong force is mediated by particles known as *gluons*. The electroweak theory includes a total of four bosons: the massless photon and three other massive particles, called (for historical reasons) W^+, W^-, and Z^0, all of which have been observed in laboratory experiments. Most of the particles we have encountered so far in this book—electrons, protons, neutrons, neutrinos—play "catch" with at least some of these "balls."

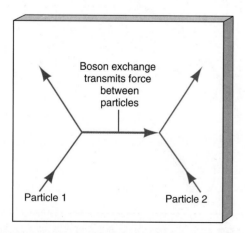

Figure 27.10 Forces between elementary particles are transmitted through the exchange of particles called bosons. As two particles interact, they exchange bosons, a little like playing catch with a submicroscopic ball.

Section 27.1 discussed how particles "froze out" of the universe as its temperature dropped below the threshold temperature for their creation by pair production. Now that we know that the basic forces of nature are also mediated by particles, we can understand—in general terms, at least—how the fundamental forces froze out too as the universe cooled. The W and Z particles responsible for the electroweak force have masses about 100 times the mass of a proton. The threshold temperature for their production—roughly 10^{15} K—is the temperature at which the weak and electromagnetic forces parted company near the end of the hadron epoch. According to the GUTs, the particle that unifies the strong and electroweak forces is extremely massive—at least 10^{15} times the mass of the proton, and possibly much more. It is because this particle is so massive that the unification of the strong and electroweak forces becomes evident only at extremely high temperatures.

The freezing out of the electroweak force at a temperature of 10^{15} K, some 10^{-10} s after the Big Bang, had little overall effect on the cosmic expansion. By contrast, the freeze-out of the strong force, a mere 10^{-35} s after the Big Bang, produced one of the strangest events in the history of the cosmos.

COSMIC INFLATION

As the temperature fell below 10^{28} K and the strong force appeared as for the first time a separate entity—a little like a gas liquefying or water freezing as the temperature drops—theory implies that the universe briefly entered a very odd, and unstable, high-energy state that physicists call the "false vacuum." In essence, the universe remained in the "unified" condition a little too long, like water that has been cooled below freezing but has not yet turned to ice. This had dramatic consequences. For a short while, empty space acquired an enormous *pressure*, which temporarily overcame the pull of gravity and accelerated the expansion of the universe at an enormous rate. The pressure remained constant as the cosmos expanded, and the acceleration grew more and more rapid with time—in fact, the size of the universe *doubled* every 10^{-34} s or so! This period of unchecked cosmic expansion, illustrated in Figure 27.11, is known as the epoch of **inflation**.

Eventually, the universe returned to the lower-energy "true vacuum" state. Regions of normal space began to appear within the false vacuum and rapidly spread to include the entire cosmos. With the return of the true vacuum, inflation stopped. The whole episode lasted a mere 10^{-32} s, but during that time the universe swelled in size by the incredible factor of about 10^{50}. When the inflationary phase ended, the grand unified force was gone forever. In its place were the more familiar electroweak and strong forces that operate around us in the low-temperature universe of today. With the normal vacuum restored, the universe once again resumed its (relatively) leisurely expansion, slowly decelerated by the effect of gravity. But a number of important changes had occurred that would have far-reaching ramifications for the evolution of the cosmos.

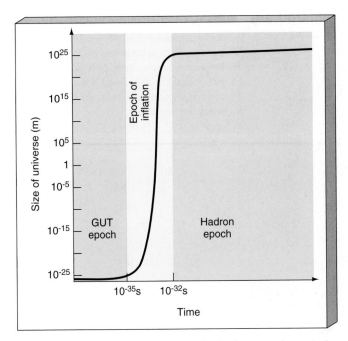

Figure 27.11 During the period of inflation at the end of the GUT epoch, the universe expanded enormously in a very short time. Afterward, it resumed its earlier "normal" expansion rate, except that the size of the cosmos was about 10^{50} times bigger than it was before inflation.

IMPLICATIONS FOR THE UNIVERSE

The inflationary epoch provides a natural solution to the horizon and flatness problems described earlier. The horizon problem is solved because inflation took regions of the universe that had already had time to communicate with one another—and so had established similar physical properties—and then dragged them far apart, well out of communications range of one another. Regions A and B in Figure 27.8 have been out of contact since 10^{-32} s after creation, but they were in contact before then. As illustrated in Figure 27.12, their properties are the same today because they were the same long ago, before inflation separated them.

Figure 27.12(a) shows a small piece of the universe just before the onset of inflation. The point that will one day become the site of the Milky Way Galaxy is at the center of the shaded region, which represents the portion of space "visible" to that point at that time—that is, there has been enough time since the Big Bang for light to have traveled from the edge of this region to its center. That entire region is more or less homogeneous because different parts of it have been able to interact with one another, so any initial differences between them have largely been smoothed out. The points A and B of Figure 27.8 are also marked. They lie within the homogeneous patch, so they have very similar properties. The actual size of the shaded region is about 10^{-26} m—only a trillionth the size of a proton.

Immediately after inflation, as shown in Figure 27.12(b), the homogeneous region has expanded by 50 orders of magnitude, to a diameter of about 10^{24} m, or 30 Mpc—larger than the largest supercluster. By contrast, the visible portion of the universe, indicated by the dashed line, has grown only by a factor of a thousand and is still microscopic in size. In effect, the universe expanded much faster than the speed of light during the inflationary epoch, so what was once well within the horizon now lies far beyond

Figure 27.12 Inflation solves the horizon problem by taking a small region of the very early universe, whose parts had already had time to interact with one another and which had thus already become homogeneous, and expanding it to enormous size. In (a), points A and B are well within the (shaded) homogeneous region of the universe centered on the eventual site of the Milky Way Galaxy. In (b), after inflation, A and B are far outside the horizon (indicated by the dashed line), so they are no longer visible from our location. Subsequently, the horizon expands faster than the universe, so that today (c) A and B are just reentering our field of view. They have similar properties now because they had similar properties before the inflationary epoch.

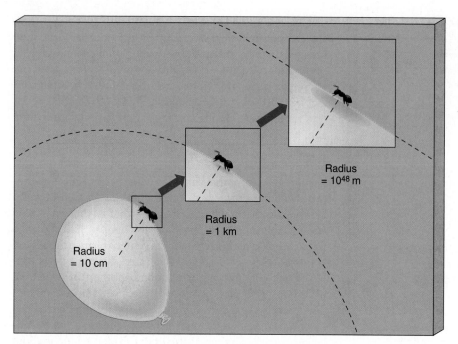

Figure 27.13 Inflation solves the flatness problem by taking a curved surface, here represented by the surface of the expanding balloon, and expanding it enormously in size. To an ant on the surface, the balloon looks virtually flat when the expansion is over.

it. In particular, points A and B are no longer visible, either to us or to each other, at this time. (Note that while the theory of relativity restricts matter and energy to speeds less than the speed of light, it imposes no such limit on the universe as a whole.)

Since then, the universe has expanded by a further factor of 10^{27}, so the size of the homogeneous region of space surrounding us is now about 10^{51} m (10^{28} Mpc)—10 trillion trillion times greater than the distance to the most distant quasar. As shown in Figure 27.12(c), the horizon has expanded faster than the universe, so points A and B are just now becoming visible again. As the portion of the universe now observable from Earth grows in time it remains homogeneous because our cosmic field of view is simply reexpanding into a region of the universe that was within our horizon long ago. We will have to wait a very long time—at least 10^{35} years—before the edge of the homogeneous patch surrounding us comes back into view.

To see how inflation solves the flatness problem, let's return to our earlier balloon analogy. Imagine that you are a 1-mm-long ant sitting on the surface of the balloon as it expands, as illustrated in Figure 27.13. When the balloon is just a few centimeters across, you can easily perceive the surface to be curved—its circumference is only a few times your own size. When the balloon expands to, say, a few meters in diameter, the curvature of the surface is less pronounced, but still perceptible. However, by the time the balloon has expanded to a few kilometers across, an "ant-sized" patch of the surface will look quite flat, just as the surface of Earth looks flat to us.

Now imagine that the balloon expands 100 trillion trillion trillion trillion times, as the universe did during the period of inflation. Your local patch of the surface is now completely indistinguishable from a perfectly flat plane, deviating from flatness by no more than 1 part in 10^{50}.

Exactly the same argument applies to the universe. Any curvature the universe may have had before inflation has been expanded so much that space is now virtually flat, at least on the scale of the observable universe (and, in fact, on much larger scales too).

Notice that this resolution to the flatness problem—the universe looks pretty flat because it *is* flat—has a very important consequence. Because the universe is flat, the density of matter must be exactly critical: $\Omega_0 = 1$. That means that there must be a lot of invisible matter in the universe beyond the clusters and the superclusters, filling the huge voids on the largest scales. And because we just saw that primordial nucleosynthesis implies that the density of normal matter is at most 3 percent of the critical value, it follows that the rest of the mass—at least 97 percent of all the matter in the universe—must be in the form of dark matter (whatever it may be).

When the idea of inflation was first suggested, most astronomers were very skeptical. Some still are. They point out that there is *no* observational evidence for a cosmic density as high as the critical value. However, as we saw earlier, there is ample evidence for dark matter in the universe with as much as 20 or 30 percent of the critical density, and estimates of Ω_0 do seem to increase as the scale under consideration increases. Furthermore, inflation is a clear prediction of a group of theories—the GUTs—that are becoming more and more firmly established as the standard description of matter at high energies. If the GUTs are correct, then inflation *must* have occurred. Finally, inflation provides a neat solution to two serious difficulties with the standard noninflationary Big Bang model. For all these reasons, inflation has become entrenched (albeit as an unwelcome guest in some cases) in most cosmologists' models of the universe. *Interlude 27-1* discusses a possibly related, and even more controversial, theory of cosmic expansion.

27.5 The Formation of Structure in the Universe

6 Just as stars form from inhomogeneities in interstellar clouds, galaxies, galaxy clusters, and larger structures are believed to have grown from small density fluctuations in the matter of the expanding universe. ∞ (Sec. 19.1) Given the conditions in the universe during the atomic and galactic epochs (Table 27.1), cosmologists calculate that regions of higher-than-average density that contained more than about a million times the mass of the Sun would have begun to contract. There was thus a natural tendency for million–solar mass "pregalactic" objects to form. In Chapter 24 we saw a little of how these pregalactic fragments might have interacted and merged to form galaxies. ∞ (Sec. 24.4) Here, we concern ourselves mostly with the formation of structure on much larger scales.

THE GROWTH OF INHOMOGENEITIES

By the early 1980s, cosmologists had come to realize that galaxies could not have formed from the contraction of inhomogeneities involving only *normal* matter. The following lines of reasoning led to this conclusion:

1. Calculations show that before decoupling (which occurred at a redshift of 1500), the intense background radiation would have prevented clumps of normal matter from contracting. Thus, any such clumps would have had to wait until *after* decoupling before their densities could start to increase.

2. Because radiation was "tied" to normal matter up until decoupling, any variations in the matter density at that time would have led to temperature variations in the cosmic background radiation—denser regions would have been a little hotter than less dense ones. The high degree of isotropy observed in the microwave background indicates that any density variations from one region of space to another during the epoch of decoupling must have been small—at most a few parts in 10^5. ∞ (Sec. 26.6)

3. Galaxies—or, at least, quasars—are known to have formed by a redshift of 5, and some theorists believe that in order to produce the densest galactic nuclei we see today, the formation process must have already been well established as long ago as a redshift of 10 or 20. ∞ (Sec. 25.6)

4. Because the contracting matter had to "fight" the general expansion of the universe, theory shows contracting pregalactic clumps could have increased in density by a factor of at most 50–100 in the time available (even with fairly optimistic assumptions). As a result, the small inhomogeneities permitted by observations of the microwave background could not have grown into galaxies in the time available—the universe would still have been almost perfectly homogeneous at the time when we know galaxies had already formed.

Put another way, if galaxies had grown from density fluctuations in the normal-matter component of the early universe, then the fluctuations would have had to be so large as to leave a clearly observable imprint on the cosmic microwave background. That imprint is not observed.

DARK MATTER

Fortunately for cosmology, the fact that most of the universe is made of *dark* matter, which has properties very different from those of normal matter, provides an alternative explanation for the large-scale structure we see today. Whatever its true nature, dark matter interacts only very weakly with normal matter and radiation. As a result, the density of initially high-density regions of dark matter has been increasing ever since matter first began to dominate the universe at a redshift of about 20,000. Thus, point 1 of our list does not apply because the dark matter started clumping well before decoupling (redshift 1500). Furthermore, because the dark matter is not directly tied to the radiation, dark-matter density inhomogeneities could have been quite large at the time of decoupling without having a correspondingly large effect on the microwave background, so the constraints of point 2 are also avoided. In short, dark matter could clump to form large-scale structure in the universe without running into the problems just described for normal matter.

In this picture (Figure 27.14), dark matter determines the overall distribution of mass in the universe and clumps to form the observed large-scale structure without violating any observational constraints on the microwave background. Then, at later times, normal matter is drawn by gravity into the regions of highest density, eventually forming galaxies and galaxy clusters. This picture explains why so much dark matter is found outside the visible galaxies. The luminous material is strongly concentrated near the density peaks and dominates the dark matter there, but the rest of the universe is essentially devoid of normal matter. Like foam on the crest of an ocean wave, the universe we see is only a tiny fraction of the total.

Given that the nature of the dark matter is still unknown, theorists have considerable freedom in choosing its properties when they attempt to simulate the formation of structure in the universe. Cosmologists distinguish between two basic types of dark matter on the basis of its temperature at the time when galaxies began to form. These types are known as **hot dark matter** and **cold dark matter**, respectively, and they lead to quite different kinds of structure in the present-day universe.

Hot dark matter consists of lightweight particles—much less massive than the electron. If neutrinos turn out to have a small mass, as now appears to be the case, they may be leading candidates for hot dark-matter particles. Simulations of a universe filled with hot dark matter indicate that large structures, such as superclusters and voids, form fairly naturally, but the computer models cannot account for the existence of structure on smaller scales. Small amounts of hot material tend to disperse, not clump together. Attempts to produce galaxies and clusters by other means after the formation of larger objects have been only partly successful, and most cosmologists

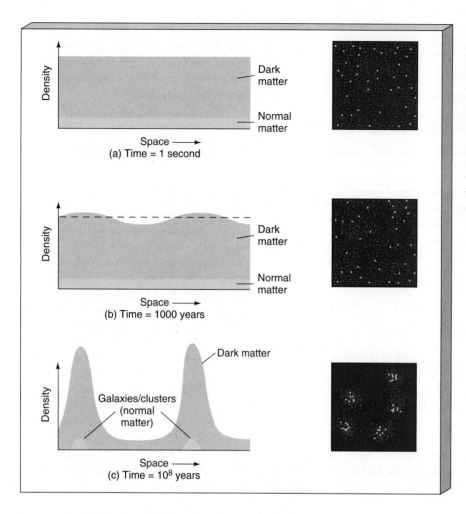

(a) Time = 1 second

(b) Time = 1000 years

(c) Time = 10⁸ years

Figure 27.14 The formation of structure in the cosmos. (a) The universe started out as a mixture of (mostly) dark and normal matter. (b) A few thousand years after the Big Bang, the dark matter began to clump. (c) Eventually the dark matter formed large structures (represented here by the two high-density peaks) into which normal matter flowed, ultimately to form the galaxies we see today. The three frames at right represent the densities of dark matter (red) and normal matter (yellow) graphed at left.

have concluded that models based purely on hot dark matter are unable to explain the observed structure of the universe.

Cold dark matter consists of very massive particles, possibly formed during the GUT era or even before. Computer simulations modeling the universe with these particles as the dark matter easily produce small-scale structure. With the understanding that galaxies form preferentially in the densest regions, and with some fine-tuning, these models can also be made to produce large-scale structure comparable to what is actually observed.

Perhaps the best results to date (that is, the results that agree most closely with observations) come from calculations in which a *mixture* of hot and cold dark matter is assumed. Figure 27.15 shows the results of a supercomputer simulation of a mixed-dark-matter universe. Compare these images with the real observations of nearby structure shown in Figures 24.33 and 24.34. Although calculations like this cannot *prove* that dark-matter models are the correct description of the universe, the similarities between the models and reality are certainly very striking.

THE MICROWAVE BACKGROUND

Because dark matter does not interact directly with photons, its density variations do not cause large (and easily observable) temperature variations in the microwave back-

ground. However, the radiation is influenced slightly by the *gravity* of the growing dark clumps, experiencing a slight gravitational redshift that varies from place to place depending on the dark-matter density. As a result, dark-matter models predict that there should be tiny "ripples" in the microwave background—temperature variations of only a few parts per million from place to place on the sky.

Until the late 1980s these ripples were too small to be accurately measured, although cosmologists were confident that they would be found. In 1992, after almost two years of careful observation, the *COBE* team (see Section 26.6) announced that the expected ripples had indeed been detected. The temperature variations are tiny—only 30–40 millionths of a kelvin from place to place in the sky—but they are there. The *COBE* results are displayed as a temperature map of the microwave sky in Figure 27.16(a). The temperature variation due to Earth's motion (see Figure 26.20) has been subtracted out, as has the radio emission from the Milky Way, and temperature deviations from the average are displayed.

Initially, it seemed that the inhomogeneities in the microwave background were not consistent with the "standard" dark-matter models that provided the best agreement with actual observations of structure in the present-day universe. The ripples seen by *COBE*, taken in conjunction with the models, appeared to imply too little structure on large scales—that is, the computer simulations predicted

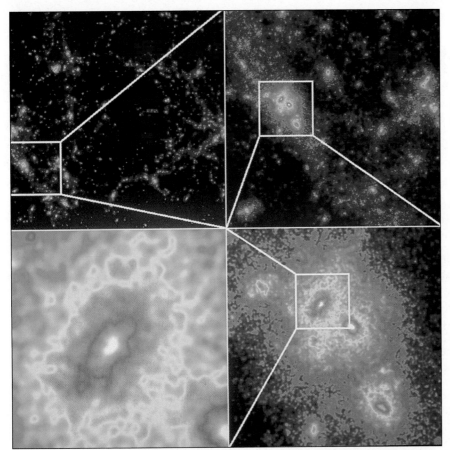

Figure 27.15 Successively magnified views of a $100 \times 100 \times 100$ Mpc cube in a simulated mixed-dark-matter universe with $\Omega_0 = 1$, showing the present-day structure that results from the growth of small density fluctuations in the very early universe. In this particular calculation, 20 percent of the total mass was assumed to be in the form of hot-dark-matter particles (actually, neutrinos); the rest was cold dark matter. Colors represent mass density, ranging from the cosmic mean (dark blue), through green, yellow, and red, to 100 times the mean (white). The enlargements zoom in on one particular galaxy in one particular small group of galaxies. The final frame (at bottom left) is roughly 1.5 Mpc across. Notice both the large-scale filamentary structure evident in the top two frames and the extensive dark-matter halos surrounding individual galaxies (the galaxies are roughly the white regions in the bottom two frames).

(a)

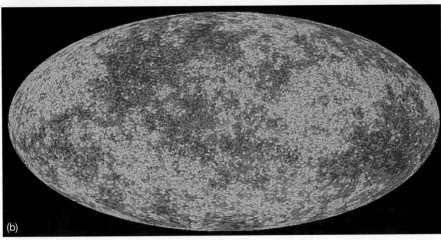
(b)

Figure 27.16 (a) *COBE* map of temperature fluctuations in the cosmic microwave background. Hotter-than-average regions are shown in red, cooler-than-average regions in blue. The total temperature range shown is ± 200 millionths of a kelvin. (b) Simulated map of microwave background temperature fluctuations corresponding to the simulation shown in Figure 27.15. Dark blue and red represent temperature variations of ± 200 millionths of a kelvin from the average, so this map can be compared more or less directly with the *COBE* map in part (a). However, the resolution here is about 0.5°—20 times sharper than the *COBE* map, and roughly the resolution expected in the next generation of satellite observations.

R   I   V   U   X   G

INTERLUDE 27-1   The Cosmological Constant

Even the greatest minds are fallible. The first scientist to apply general relativity to the universe was, not surprisingly, the theory's inventor, Albert Einstein. When he derived and solved the equations describing the behavior of the universe, Einstein discovered that they predicted a universe that evolved in time. But in 1917 neither he nor anyone else knew about the expansion of the universe, as described by Hubble's law, which would not be discovered for another 10 years. ∞ (Sec. 24.5) At the time, Einstein, like most scientists, believed that the universe was *static*—that is, unchanging and everlasting. The discovery that there was no static solution to his equations seemed to Einstein to be a near-fatal flaw in his new theory.

To bring his theory into line with his beliefs, Einstein tinkered with his equations, introducing a "fudge factor" now known as the *cosmological constant*. As illustrated in the accompanying figure, which shows the effect of introducing a cosmological constant into the equations describing the expansion of a critical-density universe, this factor allows many other solutions to Einstein's equations. One of these solutions describes a "coasting" universe, whose radius does in fact remain constant for an indefinite period of time, Einstein took this to be the static universe he expected.

Instead of predicting an evolving cosmos, which would have been one of relativity's greatest triumphs, Einstein yielded to a preconceived notion of the way the universe "should be," unsupported by observational evidence. Later, when the expansion of the universe was discovered and Einstein's equations—without the fudge factor—were found to describe it perfectly, he declared that the cosmological constant was the biggest mistake of his scientific career.

For many researchers—Einstein included—the main problem with the cosmological constant was (and still is) the fact that it had no clear physical interpretation. Einstein introduced it to fix what he thought was a problem with his equations, but he discarded it immediately once he realized that no problem actually existed. Scientists are very reluctant to introduce unknown quantities into their equations purely to make the results "come out right." As a result, the cosmological constant fell out of favor among astronomers for many years.

Since the early 1980s, however, the cosmological constant has made something of a comeback, and it now enjoys a measure of respectability in cosmological circles. One reason for this is that the leading theories of matter and radiation on very small scales (the GUTs described in *More Precisely 27-1*) now suggest that the universe really did go through an early phase when its evolution was determined by a "cosmological constant" of sorts. As discussed in Section 27.4, this idea is now firmly entrenched in many cosmologists' models of the universe. Unfortunately, although the GUTs make the idea of a cosmological constant more acceptable, they presently offer no useful means of estimating its value.

A second reason for the cosmological constant's resurgence in popularity is the fact that it offers a possible means of resolving a nagging problem in astronomy—the fact that the measured ages of the oldest stars in our Galaxy, some 12 billion years (Section 20.5), may be greater than the age of the universe, as determined from the Hubble expansion and our best estimates of the cosmic density (Section 26.4), especially if the universe is close to critical density (Section 27.4). *Interlude 26-2* discusses the conventional view of how this discrepancy may be resolved by improved observations. The cosmological constant provides a radical alternative. In addition, recent observations of distant Type-I supernovae, which seem to imply that the expansion of the universe is not

fewer superclusters, voids, Great Walls, and so on, than are actually seen. However, with some modifications to the details of the models, it now looks as though the disagreement is not so serious as it first seemed, and a growing number of cosmologists are coming to regard the *COBE* observations as confirmation of a central prediction of dark-matter theory.

The simulation shown in Figure 27.15 implies temperature fluctuations in the microwave background that agree very well with the present *COBE* observations, but theorists can (of course) go one stage further. Figure 27.16(b) presents a *prediction*, based on the same simulation, of the temperature fluctuations that should be observed by the next generation of (much higher resolution) satellite experiments. If our current understanding of the microwave background anisotropy is correct, then future maps should look qualitatively like this one.

If the *COBE* results hold up—as they are checked and rechecked by collaborators and competitors alike—they may

one day come to rank alongside the discovery of the microwave background itself in terms of their importance to the field of cosmology.

27.6 Toward Creation

As we mentioned earlier, our efforts to penetrate the Planck epoch are currently hampered by physicists' ignorance of how to incorporate the force of gravity into GUTs. No one has yet invented a "SuperGUT" that merges gravity and the grand unified force into a single, truly fundamental force at the energies characterizing the earliest part of the Planck epoch, although this is an area of very active research.

Many researchers feel that once we have in hand the proper description of quantum gravity, our understanding slowing down, but is actually accelerating (Section 26.4), are readily explained by cosmological models that include a cosmological constant. (See for example the light-green curve in the accompanying figure.)

As indicated in the figure, by carefully adjusting the value of the cosmological constant, theorists can construct mathematical models of the universe that did not expand so rapidly in the past and so are older than the measured value of H_0 suggests. With a suitably chosen cosmological constant, a Hubble constant of 65 km/s/Mpc in a critical-density universe can be reconciled with a cosmic age of 12 billion years, or even much more. In fact, with a sufficiently large cosmological constant, it is even possible to avoid a Big Bang altogether, although this possibility now seems to be quite firmly ruled out by observations of high-redshift quasars and galaxies. Notice, incidentally, that if the cosmological constant does indeed provide the explanation of the cosmic age paradox, it also implies that the universe will expand forever, regardless of the present value of the cosmic mass density.

But before we jump to any conclusions about the cosmo-

logical constant and overstate its potential role in solving the outstanding problems of astronomy, it is probably worth remembering the experience of its inventor and bearing in mind that—at least for now—its physical meaning remains completely unknown.

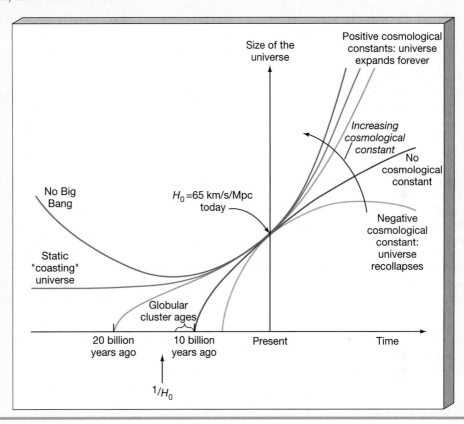

may automatically include a natural description of creation itself. It is even conceivable that the primal energy that formed the universe originally emerged from literally *nothing*. Even in a perfect vacuum—a region of space containing neither matter nor energy—virtual particle–antiparticle pairs are constantly appearing and disappearing within a time span too short to observe, causing natural *quantum fluctuations* to occur in empty space. We might be living in a sort of "self-creating universe" that erupted into existence spontaneously from just such a random quantum fluctuation! This sort of "statistical" creation of the primal cosmic energy from absolutely nothing has been dubbed "the ultimate free lunch."

Whether or not this scenario proves to be the correct description of the birth of the universe, it suggests a possible explanation for the original density fluctuations that eventually grew into today's large-scale structure. They might have been microscopic quantum fluctuations that

existed naturally in the universe during the GUT epoch, then grew to macroscopic size during the inflationary epoch. The fluctuations predicted by this idea agree quite well with the requirements of several cold-dark-matter models.

If this view is correct, then our Galaxy, the Sun, Earth, even life itself, are direct consequences of a series of random events that occurred during an unimaginably short period of time some 10–15 billion years ago. Of course, these ideas are very speculative. In the strict sense, they are not really science at all, as they violate one of the central tenets of the scientific method: they are practically impossible to test experimentally. Still, whether or not you find them philosophically acceptable or intellectually pleasing, they do illustrate just how far the scope of physics and astronomy has expanded in the twentieth century. Astronomy is a subject that links the very big and the very small. Nowhere is that more evident than in our study of the most important question of all—the origin of the universe.

Chapter Review *WWW*

SUMMARY

At present, the density of matter in the universe greatly exceeds the equivalent mass density of radiation. The universe is **matter dominated** (p. 626). The density of matter was much greater in the past, when the universe was smaller. However, because radiation is redshifted as the universe expands, the density of radiation was greater still. The early universe was **radiation dominated** (p. 626). During the first few minutes after the Big Bang, matter was formed out of the primordial fireball by the process of **pair production** (p. 627). In the early universe, matter and radiation were linked by this process. Particles "froze out" of the radiation background as the temperature fell below the threshold for creating them. The existence of matter today means that there must have been unequal amounts of matter and antimatter early on.

The physical state of the universe can be understood in terms of present-day physics back to about 10^{-43} s after the Big Bang. Before that, the four fundamental forces of nature—gravity, electromagnetism, the strong force, and the weak force—were all indistinguishable. There is presently no theory that can describe these extreme conditions. As the universe expanded and its temperature dropped, the forces became distinct from one another. First gravity, then the strong force, and then the weak and electromagnetic forces separated out.

Only a little of the helium observed in the universe today was formed in stars. Most of it was created by **primordial nucleosynthesis** (p. 632) in the early universe. Some deuterium was also formed at these early times, and it provides a sensitive indicator of the present density of the universe in the form of "normal" (as opposed to dark) matter. Studies of deuterium indicate that normal matter can account for at most 3 or 4 percent of the critical density. The remaining mass inferred from studies of clusters must then be made of dark matter, in the form of unknown particles formed at some very early epoch.

When the universe was about 1500 times smaller than it is today, the temperature became low enough for atoms to form. At that time, the (then-optical) radiation background **decoupled** (p. 634) from the matter. The universe became transparent. The photons that now make up the microwave background have been traveling freely through space ever since.

According to modern **Grand Unified Theories** (p. 629), when the three nongravitational forces of nature began to display their separate characters, about 10^{-35} s after the Big Bang, a brief period of rapid cosmic expansion called the epoch of **inflation** (p. 636) occurred, during which the size of the universe increased by a factor of about 10^{50}.

Cosmologists wonder why regions of the universe that have not had time to "communicate" with one another look so similar. This is called the **horizon problem** (p. 635). Inflation solves it by taking a small homogeneous patch of the early universe and expanding it enormously in size. The patch is still homogeneous, but it is now much larger than the portion of the universe we can see today.

Cosmologists also wonder why the density of the universe seems to be so near the critical value. This is called the **flatness problem** (p. 635). Inflation implies that the cosmic density is in fact exactly critical. If this is the case, then over 95 percent of the matter in the universe is dark.

The large-scale structure observed in the universe could not have formed out of density fluctuations in normal gaseous matter—there simply has not been enough time, given the twin constraints of the smoothness of the microwave background and the epoch at which the first galaxies and quasars are known to have formed. Instead, dark matter clumped and grew to form the "skeleton" of the structure now observed. Normal matter then flowed into the densest regions of space, eventually forming the galaxies we now see.

Cosmologists distinguish between **hot dark matter** (p. 639) and **cold dark matter** (p. 639), depending on its temperature at the end of the Radiation Era. In order to explain the observed large-scale structure in the universe, much of the dark matter must be cold. In 1992 the *COBE* satellite discovered the expected ripples in the cosmic microwave background associated with the clumping of dark matter at early times.

SELF-TEST: TRUE OR FALSE?

____ **1.** The light emitted from all the stars in the universe now far outshines the cosmic microwave background.

____ **2.** The time between the Big Bang and 10^{-43} s afterward cannot be described for lack of a theory of quantum gravity.

____ **3.** All elementary particle formation was completed by the time the universe was about 1 second old.

____ **4.** About 25 percent by mass of matter in the universe is primordial helium.

____ **5.** The present-day abundance of deuterium gives information on the density of normal matter.

____ **6.** Decoupling refers to interactions between matter and antimatter.

____ **7.** The present-day microwave background radiation originated just before the period of decoupling.

____ **8.** After decoupling, neutral atoms could finally exist.

____ **9.** The horizon problem relates to the isotropy of the microwave background radiation.

____ **10.** The flatness problem is the fact that the observed density of matter is unexpectedly different from the critical density.

____ **11.** Inflation was caused by the freeze-out of the strong force.

____ **12.** The universe grew in size by a factor of about 10,000 during the inflationary period.

____ **13.** If the theory of inflation is correct, then the density of the universe is exactly the critical density.

____ **14.** Physicists have detected cold-dark-matter particles in terrestrial laboratories.

____ **15.** Galaxies had already formed by the time the universe was about 1 million years old.

SELF-TEST: FILL IN THE BLANK

1. Comparing the mass density of radiation and matter, we find that, at the present time, _____ dominates.

2. Energetically speaking, radiation and matter were equally important approximately _____ years after the Big Bang.

3. In the process of pair production, two _____ interact to form a particle and an _____.

4. The temperature necessary to form electrons and positrons is about _____ K.

5. The temperature necessary to form protons and _____ is about _____ K.

6. When the universe was a few minutes old, nuclear fusion produced _____ and _____.

7. After about a million years, _____ had formed in the universe.

8. Elements heavier than helium were not formed primordially because, as time passed, the density and temperature of the universe _____.

9. The fact that the density of normal matter is so much less than the known density of matter in the universe implies that much of the dark matter must be _____.

10. The cosmic microwave background radiation last interacted with matter when the universe was about _____ times its present size.

11. If the theory of inflation is correct, then dark matter must make up about _____ percent of all matter in the universe.

12. Which had density fluctuations first, normal matter or dark matter? _____.

13. We know that density fluctuations in the normal-matter component of the early universe must have been very small because otherwise we would see their imprint on the _____.

14. Hot and cold dark matter differ in the masses of their particles. Cold dark matter consists of _____ particles.

15. Theory predicted tiny fluctuations in the _____ of the microwave background; the _____ satellite found them.

REVIEW AND DISCUSSION

1. For how long was the universe dominated by radiation? How hot was the universe when the dominance of radiation ended?

2. Why is our knowledge of the Planck epoch so limited?

3. When and how did the first atoms form?

4. Describe the universe at the end of the galactic epoch.

5. Why do all stars, regardless of their abundance of heavy elements, seem to contain at least one-quarter helium by mass?

6. Why didn't heavier and heavier elements form in the early universe, as they do in stars?

7. If large amounts of deuterium formed in the early universe, why do we see so little deuterium today?

8. How do measurements of the cosmic deuterium abundance provide a reliable estimate of Ω_0?

9. When did the universe become transparent to radiation?

10. What are GUTs?

11. How does cosmic inflation solve the horizon problem?

12. How does cosmic inflation solve the flatness problem?

13. What is the difference between hot and cold dark matter?

14. What is the connection between dark matter and the formation of large- and small-scale structures?

15. Why were the observations made by the *COBE* satellite so important to cosmology?

PROBLEMS

1. What was the equivalent mass density of the cosmic radiation field when the universe was one-thousandth its present size? *Hint:* don't forget the cosmological redshift!

2. Estimate the temperature needed for electron-positron pair production. The mass of an electron is 9.1×10^{-31} kg. Use $E = mc^2$ to find the energy (Section 16.5), $E = hf$ to find the frequency f of a photon having that energy (Section 4.2), and finally Wien's law to find the temperature for which a blackbody spectrum peaks at that fequency (Section 3.4). How does your answer compare with the threshold temperature given in the text?

3. Given that the threshold temperature for the production of electron–positron pairs is about 6×10^9 K and that a proton is 1800 times more massive than an electron, calculate the threshold temperature for proton–antiproton pairs.

4. By what factor did the volume of the universe increase during the epoch of primordial nucleosynthesis, from the time when deuterium could first survive until the time at which all nuclear reactions ceased? By what factor did the matter density of the universe decrease during this period?

5. From Table 25.1, the "photosphere" in the universe corresponding to the epoch of decoupling presently lies some 9000

Mpc from us (see Figure 27.7). How far away was the photosphere when the background radiation we see today was emitted?

6. Calculate the mass, in kilograms, of the particle that unifies the strong and electroweak forces, if it froze out at a temperature of 10^{28} K. Calculate the mass of a hypothetical particle that might unify gravity with the other forces, given that it froze out at the end of the Planck epoch (Table 27.1).

7. How many times did the universe double in size during the inflationary period if it expanded by a factor of 10^{50}?

8. What would be the radius of a million–solar mass patch of a homogeneous, critical-density universe (a) today? (b) At decoupling? (c) At the end of the nuclear epoch?

9. The "blobs" evident in Figure 27.17(a) are about 10° across. If those blobs represent clumps of matter around the time of decoupling, and assuming Euclidean geometry, estimate the size of the clumps (at the time of decoupling).

10. If dark-matter particles are remnants of the GUT era, they must have decoupled from the rest of the matter and radiation in the universe at the end of the GUT epoch. Repeat problem 5, but for dark-matter particles instead of electromagnetic photons, assuming a present photosphere distance of 9200 Mpc.

PROJECTS

1. Read the book *The First Three Minutes*, by Steven Weinberg. It is fairly nonmathematical in its presentation. What new results are presented in this chapter that were not known by Weinberg when he wrote the first edition of this book in 1977? How much progress has been made in understanding the very earliest epochs since that time?

2. Write a paper on the cosmological constant. It there presently any observational evidence favoring a nonzero value for this constant? What are the main theoretical reasons for including a cosmological constant in theories of the universe?

28 LIFE IN THE UNIVERSE

Are We Alone? ✦WWW✦

LEARNING GOALS

Studying this chapter will enable you to:

1 Summarize the process of cosmic evolution as it is currently understood.

2 Evaluate the chances of finding life in the solar system.

3 Summarize the various probabilities used to estimate the number of advanced civilizations that might exist in the Galaxy.

4 Discuss some of the techniques we might use to search for extraterrestrials and to communicate with them.

(Opposite page) The arrow of time, from the origin of the universe to the present and beyond, spans several major epochs of history. Cosmic evolution is the study of the many varied changes in the assembly and composition of energy, matter, and life in the thinning and cooling universe. Inserted along the arrow of time are seven "windows" outlining some of the key events in the history of the universe:

■ evolution of primal energy into elementary particles,
■ evolution of atoms into galaxies and stars,
■ evolution of stars into heavy elements,
■ evolution of elements into solid, rocky planets,
■ evolution of those elements into the molecular building blocks of life,
■ evolution of those molecules into life itself
■ evolution of advanced life forms into intelligence, culture, and technological civilization.

Are we unique? Is life on our planet the only example of life in the universe? These are difficult questions, for the subject of extraterrestrial life is one on which we have no data, but they are important questions, with profound implications for the human species. Earth is the only place in the universe where we know for certain that life exists. In this chapter we take a look at how humans evolved on Earth and then consider whether those evolutionary steps might have happened elsewhere. Having done that, we will assess the likelihood of our having Galactic neighbors and consider how we might learn about them if they exist.

28.1 Cosmic Evolution

1 Figure 28.1 (see also the chapter-opening art) identifies seven major phases in the history of the universe: *particulate, galactic, stellar, planetary, chemical, biological,* and *cultural* evolution. Together, these evolutionary stages make up the grand sweep of **cosmic evolution**—the continuous transformation of matter and energy that has led to the appearance of life and civilization on Earth. The first four phases represent, in reverse order, the contents of this book. We now expand our field of view beyond astronomy to include the other three.

From the Big Bang, to the formation of galaxies, to the birth of the solar system, to the emergence of life, to the evolution of intelligence and culture, the universe has evolved from simplicity to complexity. We are the result of an incredibly complex chain of events that spanned billions of years. Were those events random, making us

unique, or are they in some sense *natural*, so that technological civilization is inevitable? Put another way, are we alone in the universe, or are we just one among countless other intelligent life-forms in our Galaxy? In this chapter we consider the development of life on Earth and try to assess the likelihood of finding intelligent life elsewhere in the cosmos.

LIFE IN THE UNIVERSE

Before embarking on our study, we need a working definition of *life*. This seemingly simple task is not an easy one—the distinction between the living and the nonliving is not as obvious as we might at first think. Though most physicists would agree on the definitions of matter and energy, biologists have not arrived at a clear-cut definition of life. Generally speaking, scientists regard the following as characteristics of living organisms: (1) they can *react* to their environment and can often heal themselves when damaged; (2) they can *grow* by taking in nourishment from their surroundings and processing it into energy; (3) they can *reproduce*, passing along some of their own characteristics to their offspring; and (4) they have the capacity for genetic change and can therefore *evolve* from generation to generation so as to adapt to a changing environment.

These rules are not strict, and there is great leeway in interpreting them. Stars, for example, react to the gravity of their neighbors, grow by accretion, generate energy, and "reproduce" by triggering the formation of new stars, but no one would suggest that they are alive. A virus (see *Interlude 28-1*) is crystalline and inert when isolated from living organisms, but once inside a living system, it exhibits all the properties of life, seizing control of a living cell and using the cell's own genetic machinery to grow and reproduce. Most researchers now believe that the distinction between living and nonliving is more one of structure and complexity than a simple checklist of rules.

The general case in favor of extraterrestrial life is summed up in what are sometimes called the *assumptions of mediocrity*: (1) because life on Earth depends on just a few basic molecules, and (2) because the elements that make up these molecules are (to a greater or lesser extent) common to all stars, and (3) if the laws of science we know apply to the entire universe, as we have supposed throughout this book, then—given sufficient time—life must have originated elsewhere in the cosmos. The opposing view main-

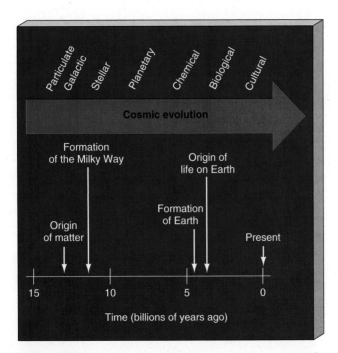

Figure 28.1 Some highlights of cosmic history are noted along this arrow of time, from the beginning of the universe to the present. Noted along the top of the arrow are the major phases of cosmic evolution.

INTERLUDE 28-1 The Virus

The central idea of chemical evolution is that life evolved from nonlife. But aside from insight based on biochemical knowledge and laboratory simulations of some key events on primordial Earth, do we have any direct evidence that life could have developed from nonliving molecules? The answer is yes. The smallest and simplest entity that sometimes appears to be alive is a virus. We say "sometimes" because viruses seem to have the attributes of both nonliving molecules and living cells. *Virus* is the Latin word for "poison," an appropriate name since viruses are often a cause of disease. Although they come in many sizes and shapes—a typical example is the polio virus, shown here magnified 300,000 times—all viruses are smaller than the size of a typical modern cell. Some are made of only a few thousand atoms. In terms of size, then, viruses seem to bridge the gap between cells that are living and molecules that are not.

Viruses contain some proteins and genetic information (in the form of DNA or the closely related molecule RNA) but not much else—none of the material by which living organisms normally grow and reproduce. How, then, can a virus be considered alive? When alone, it cannot; a virus is absolutely lifeless when isolated from living organisms. But when inside a living system, a virus has all the properties of life. Viruses come alive by transferring their genetic material into living cells. The genes of a virus seize control of a cell and establish themselves as the new master of chemical activity. Viruses grow and reproduce copies of themselves by using the genetic machinery of the invaded cell, often robbing the cell of its usual function. Some

viruses multiply rapidly and wildly, spreading the disease and—if unchecked—eventually killing the invaded organism. In a sense, then, viruses exist within the gray area between the living and the nonliving.

10μ

R I V U X G

---

tains that intelligent life on Earth is the product of a series of extremely fortunate accidents—astronomical, geological, chemical, and biological events unlikely to have occurred anywhere else in the universe. The purpose of this chapter is to examine some of the arguments for and against these viewpoints.

CHEMICAL EVOLUTION

What information do we have about the earliest stages of planet Earth? Unfortunately, not very much. Geological hints about the first billion years or so were largely erased by violent surface activity as volcanoes erupted and meteorites bombarded our planet; subsequent erosion by wind and water has seen to it that little evidence has survived to the present. Scientists believe that the early Earth was barren, with shallow, lifeless seas washing upon grassless, treeless continents. Outgassing from our planet's interior through volcanoes, fissures, and geysers produced an atmosphere rich in hydrogen, nitrogen, and carbon compounds and poor in free oxygen. As Earth cooled, ammonia, methane, carbon dioxide, and water formed. The stage was set for the appearance of life.

The surface of the young Earth was a very violent place. Natural radioactivity, lightning, volcanism, solar ultraviolet radiation, and meteoritic impacts all provided large amounts of energy that eventually shaped the ammonia, methane, carbon dioxide, and water into more complex molecules known as **amino acids** and **nucleotide bases**—organic (carbon-based) molecules that are the building blocks of life as we know it. Amino acids build *proteins*, and proteins control metabolism, the daily utilization of food and energy by means of which organisms stay alive and carry out their vital activities. Sequences of nucleotide bases form *genes*—parts of the DNA molecule—which direct the synthesis of proteins and thus determine the characteristics of the organism. These same genes also carry the organism's hereditary characteristics from one generation to the next in reproduction. In all living creatures on Earth—from bacteria to amoebas to humans—genes mastermind life, and proteins maintain it.

The idea that complex molecules could have evolved naturally from simpler ingredients found on the primitive Earth has been around since the 1920s. The first experimental verification was provided in 1953 when scientists Harold Urey and Stanley Miller, using laboratory equip-

Figure 28.2 This chemical apparatus is designed to synthesize complex biochemical molecules by energizing a mixture of simple chemicals. A mixture of gases (ammonia, methane, carbon dioxide, water vapor) is placed in the upper bulb to simulate the primordial Earth atmosphere and then energized by spark-discharge electrodes. After about a week, amino acids and other complex molecules are found in the trap at the bottom, which simulates the primordial oceans into which heavy molecules produced in the overlying atmosphere would have fallen.

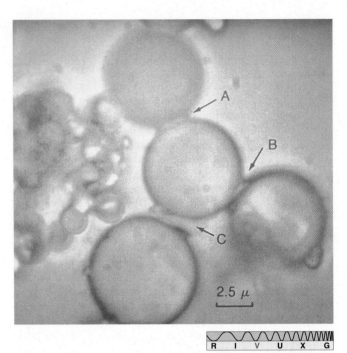

Figure 28.3 These carbon-rich, proteinlike droplets display the clustering of as many as a billion amino acid molecules in a liquid. Droplets can "grow," and parts of droplets can separate from the "parent" to become new individual droplets (as at A, B, C). The scale of 2.5 microns noted here is 1/4000 of a centimeter.

ment somewhat similar to that shown in Figure 28.2, took a mixture of the materials thought to be present on Earth long ago—a "primordial soup" of water, methane, carbon dioxide, and ammonia—and energized it by passing an electrical discharge ("lightning") through the gas. After a few days they analyzed their mixture and found that it contained many of the same amino acids found today in all living things on Earth. About a decade later, scientists succeeded in constructing nucleotide bases in a similar manner. These experiments have been repeated in many different forms, with more realistic mixtures of gases and a variety of energy sources, but always with the same basic outcomes.

Although none of these experiments has ever produced a living organism, or even a single strand of DNA, they do demonstrate conclusively that "biological" molecules can be synthesized by strictly *non*biological means, using raw materials available on the early Earth. More advanced experiments, in which amino acids are united under the influence of heat, have fashioned proteinlike blobs that behave to some extent like true biological cells. Such near-protein material resists dissolution in water (so it would remain intact when it fell from the primitive atmosphere into the ocean) and tends to cluster into small droplets called microspheres—a little like oil globules floating on the surface of water. Figure

28.3 shows some of these proteinlike microspheres. The walls of these laboratory-made droplets permit the inward passage of small molecules, which then combine within the droplet to construct more complex molecules too large to pass back out through the walls. As the droplets "grow," they tend to "reproduce," forming smaller droplets.

Can we consider these proteinlike microspheres to be alive? Almost certainly not. Most biochemists would say that the microspheres are not life itself, but they contain many of the basic ingredients needed to form life. The microspheres lack the hereditary molecule DNA. However, as illustrated in Figure 28.4, they do have similarities to ancient cells found in the fossil record. Thus, while no actual living cells have yet been created "from scratch" in any laboratory, many biochemists feel that the chain of events leading from simple nonbiological molecules almost to the point of life itself has been amply demonstrated.

Recently, however, a dissenting view has emerged. Some scientists have argued that Earth's primitive atmosphere might *not* in fact have been a particularly suitable environment for the production of complex molecules. Instead, they say, there may not have been sufficient energy available to power the chemical reactions, and the early atmosphere may not have contained enough raw material for the reactions to have become important in any case. These researchers suggest that much, if not all, of the organic material that combined to form the first living cells was produced in *interstellar*

R I V U X G

Figure 28.4 The photograph on the left, taken through a microscope, shows a fossilized organism found in sediments radioactively dated as 2 billion years old. This primitive system possesses concentric spheres or walls connected by smaller spheroids. The roundish fossils here measure about a thousandth of a centimeter. The photograph on the right, also taken through a microscope and on approximately the same scale, displays modern blue-green algae.

space and subsequently arrived on Earth in the form of interplanetary dust and meteors that did not burn up during their descent through the atmosphere.

Interstellar molecular clouds are known to contain very complex molecules, and large amounts of organic material were detected on comet Halley by space probes when Halley last visited the inner solar system (Section 14.2) Similarly complex molecules were observed on comet Hale–Bopp. ∞ (*Interlude 14-2*) Thus, the idea that organic matter is constantly raining down on Earth from space in the form of interplanetary debris is quite plausible. Whether or not this was the *primary* means by which complex molecules first appeared in Earth's oceans remains unclear. For now, the issue is unresolved.

DIVERSITY AND CULTURE

However the basic materials appeared on Earth, we know that life *did* appear. The fossil record chronicles how life on Earth became widespread and diversified over the course of time. The study of fossil remains shows the initial appearance about 3.5 billion years ago of simple one-celled organisms such as blue-green algae. These were followed about 2 billion years ago by more complex one-celled creatures, like the amoeba. Multicellular organisms such as sponges did not appear until about 1 billion years ago, after which there flourished a wide variety of increasingly complex organisms—insects, reptiles, mammals, and humans.

The fossil record leaves no doubt that biological organisms have changed over time—all scientists accept the reality of *biological evolution*. As conditions on Earth shifted and Earth's surface evolved, those organisms that could best take advantage of their new surroundings succeeded and thrived—often at the expense of those organisms that could not make the necessary adjustments and consequently became extinct. What led to these changes? Chance. An organism that happened to have a certain useful genetically determined trait—for example, the ability to run faster, climb higher, or even hide more easily—would find itself with the upper hand in a particular environment. This organism was therefore more likely to reproduce successfully, and its advantageous characteristic would then be more likely to be passed on to the next generation. The evolution of the rich variety of life on our planet, including human beings, occurred as chance mutations—changes in genetic structure—led to changes in organisms over millions of years.

What about the development of intelligence? Many anthropologists believe that, like any other highly advantageous trait, intelligence *is* strongly favored by natural selection. As humans learned about fire, tools, and agriculture, the brain became more and more elaborate. The social cooperation that went with coordinated hunting efforts was another important competitive advantage that developed as brain size increased. Perhaps most important of all was the development of language. Indeed, some anthropologists have gone so far as to suggest that human intelligence *is* human language. By communicating, individuals could signal one another while hunting food or seeking protection. Now our ancestors could share ideas as well as food and shelter. Experience, stored in the brain as memory, could be passed down from generation to generation. A new kind of evolution had begun, namely, *cultural evolution*, the changes in the ideas and behavior of society. Our more recent ancestors have created, within only the past 10,000 years or so, the entirety of human civilization.

To put all this into historical perspective, let's imagine the entire lifetime of Earth to be 46 years rather than 4.6 billion years. We have no reliable record of the first decade of our planet's existence. Life originated at least 35 years ago, when Earth was about 10 years old. Our planet's middle age is largely a mystery, although we can be sure that life continued to evolve and that generations of mountain chains and oceanic trenches came and went. Not until about 6 years ago did abundant life flourish throughout Earth's oceans. Life came ashore about 4 years ago, and plants and animals mastered the land only about 2 years ago. Dinosaurs reached their peak about 1 year ago, only to die suddenly about 4 months later. ∞ (*Interlude 14-1*) Humanlike apes changed into apelike humans only last week, and the latest ice ages occurred only a few days ago. *Homo sapiens*—our species—did not emerge until about 4 hours ago. Agriculture was invented within the last hour, and the Renaissance—along with all of modern science—is just 3 minutes old!

28.2 Life in the Solar System

2 Simple one-celled life-forms reigned supreme on Earth for most of our planet's history. It took time—a great deal of time—for life to emerge from the oceans, to evolve into simple plants, to continue to evolve into complex animals, and to develop intelligence, culture, and technology. Have those (or similar) events occurred elsewhere in the universe? Let's try to assess what little evidence we have on the subject.

LIFE AS WE KNOW IT

"Life as we know it" is generally taken to mean carbon-based life that originated in a liquid water environment—in other words, life on Earth. Is there any reason to suppose that such life might exist elsewhere in our solar system? The answer appears to be no. It seems that no environment in the solar system besides Earth is particularly well suited for sustaining Earthlike life.

The Moon and Mercury lack liquid water, protective atmospheres, and magnetic fields, and so these two bodies are subjected to fierce bombardment by solar ultraviolet radiation, the solar wind, meteoroids, and cosmic rays. Simple molecules could not possibly survive in such hostile environments. Venus has far too much protective atmosphere! Its dense, dry, scorchingly hot atmospheric blanket effectively rules it out as a possible abode for life.

The jovian planets have no solid surfaces (although some researchers maintain that life could have evolved in their atmospheres), and Pluto and most of the moons of the outer planets are too cold. However, the possibility of liquid water below Europa's icy surface has refueled speculation about the development of life there. ∞ (Sec. 11.5) Saturn's moon Titan, with its atmosphere of methane, ammonia, and nitrogen and possibly with some liquid on its surface, is conceivably a site where life might have arisen, although the results of the 1980 *Voyager 1* flyby suggest that Titan's frigid surface conditions are inhospitable to anything familiar to us. ∞ (Sec. 12.5)

What about the cometary and meteoritic debris that orbits within our solar system? Comets contain many of the basic ingredients for life—for instance, ammonia, methane, and water vapor—and although comets are frozen, their icy matter warms while nearing the Sun. Indeed, some heavy molecules have been observed in comet spectra. In addition, a small fraction of the meteorites that survive the plunge to Earth's surface do contain organic compounds. The Murchison meteorite (Figure 28.5), which fell near Murchison, Australia, in 1969, is a well-studied example. Located soon after crashing to the ground, this meteorite contains many of the well-known amino acids normally found in living cells. The moderately large molecules found in meteorites and in interstellar clouds are our only evidence that chemical evolution has occurred elsewhere in the universe. Most researchers regard this organic matter

R I V U X G

Figure 28.5 The Murchison meteorite contains relatively large amounts of amino acids and other organic material, indicating that chemical evolution of some sort has occurred beyond our own planet. In this magnified view of a meteorite fragment, the arrow points to a microscopic sphere of organic matter.

as prebiotic—that is, matter that could eventually lead to life but that has not yet done so.

The planet most likely to harbor life (or to have harbored it in the past) seems to be Mars. This planet seems harsh by Earth standards—liquid water is scarce, the atmosphere is thin, and the lack of magnetism and an ozone layer allows the solar high-energy particles and ultraviolet radiation to reach the surface unabated. But the Martian atmosphere was thicker, and the surface warmer and much wetter in the past. ∞ (Secs. 10.4, 10.5) In the hope that life might once have evolved on Mars as it did on Earth, the *Viking* lander carried a television camera to seek fossilized remnants of large plants or animals. No fossils of any kind were seen. *Mars Pathfinder* also surveyed part of the Martian surface, again without finding any evidence of present or past life on Mars. ∞ (Sec. 10.4) The *Viking* landers scooped up Martian soil (Figure 28.6) and tested for life by conducting chemical experiments designed to detect the waste gases and other products of metabolic activity, but no unambiguous evidence of Martian life has emerged. ∞ (Interlude 10-2)

The consensus among biologists and chemists today is that Mars does not house any life similar to that on Earth. However, some scientists think that a different type of biology may be operating on the Martian surface. They suggest that Martian microbes capable of eating and digesting oxygen-rich compounds in the Martian soil could also explain the *Viking* results. This speculation will be greatly strengthened if recent announcements of fossilized bacteria in meteorites originating on Mars are confirmed. ∞ (Interlude 10-2) In addition, microbial life may reside in more habitable regions on Mars, such as near the moist polar caps. It seems

R I V U X G

Figure 28.6 A trench dug by the "arm" of one of the *Viking* robots can be seen at the left of this photograph. Soil samples were scooped up and taken inside the robot, where instruments tested them for chemical composition and any signs of life.

Although the possibility of such alien life forms is a fascinating scientific problem, most biologists would argue that chemistry based on carbon and water is the one most likely to give rise to life. Carbon's flexible chemistry and water's wide liquid temperature range are just what are needed for life to develop and thrive. Silicon and ammonia seem unlikely to fare as well as the bases for advanced life-forms. Silicon's chemical bonds are weaker than those of carbon and may not be able to form complex molecules—an apparently essential aspect of carbon-based life. Also, the colder the environment, the less energy there is to drive biological processes. The low temperatures necessary for ammonia to be liquid might inhibit or even prevent completely the chemical reactions leading to the equivalent of amino acids and nucleotide bases.

Still, we must admit that we know next to nothing about noncarbon, nonwater biochemistries, for the very good reason that there are no examples of them to study experimentally. We can speculate about alien life-forms and try to make general statements about their properties, but we can say little of substance about them.

that a solid verdict regarding life on Mars will not be reached until we have thoroughly explored our intriguing neighbor.

ALTERNATIVE BIOCHEMISTRIES

It thus appears that no environment in the solar system besides Earth is particularly well suited for sustaining life—at least life similar to that on Earth. However, some scientists have suggested that different types of biology may be at work out there, ones that we cannot recognize and that we do not know how to test for. What other biologies might exist?

Some scientists have pointed out that the abundant element silicon has chemical properties somewhat similar to those of carbon and have suggested it as a possible alternative to carbon as the basis for living organisms. Ammonia (made of the common elements hydrogen and nitrogen) is sometimes put forward as a possible liquid medium in which life might develop, at least on a planet cold enough for ammonia to exist in the liquid state. Together or separately, these alternatives would surely give rise to organisms with radically different biochemistries from those we know on Earth. Conceivably, we might have difficulty even recognizing these organisms as alive.

28.3 Intelligent Life in the Galaxy

3 With humans apparently the only intelligent life in the solar system, we must broaden our search for extraterrestrial intelligence to other stars, perhaps even other galaxies. At such distances, though, we have little hope of detecting actual life with current equipment. Instead, we must ask: How likely is it that life in any form—carbon-based, silicon-based, water-based, ammonia-based, or something we cannot even dream of—exists? The word *likely* in the last sentence speaks of probabilities. Let's look at some numbers to develop statistical estimates of the probability of life elsewhere in the universe.

THE DRAKE EQUATION

An early approach to this statistical problem is usually known as the **Drake equation**, after the U.S. astronomer who pioneered this analysis:

| number of technological, intelligent civilizations now present in the Milky Way Galaxy | = | rate of star formation averaged over the lifetime of the Galaxy | × | fraction of those stars having planetary systems | × | average number of planets within those planetary systems that are suitable for life | × | fraction of those habitable planets on which life actually arises | × | fraction of those life-bearing planets on which intelligence evolves | × | fraction of those intelligent-life planets that develop technological society | × | average lifetime of a technologically competent civilization. |

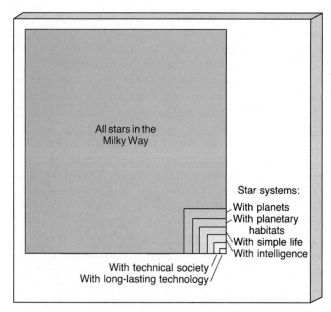

Figure 28.7 Of all the star systems in our Milky Way (represented by the largest box), progressively fewer and fewer have each of the qualities typical of a long-lasting technological society (represented by the smallest box at the lower right corner).

Several of the terms in this formula are largely a matter of opinion. We do not have nearly enough information to determine—even approximately—every term in the equation, so the Drake equation cannot give us a hard-and-fast answer. Its real value is that it subdivides a large and very difficult question into smaller pieces that we can attempt to answer separately. It provides the framework within which the problem can be addressed and parcels out the responsibility for the final solution among many different scientific disciplines. Figure 28.7 illustrates how, as our requirements become more and more stringent, only a small fraction of star systems in the Milky Way are likely to generate the advanced qualities specified by the combination of terms on the right-hand side of the equation.

Let's examine the terms in the equation one by one and make some educated guesses about their values. Bear in mind, though, that if you ask two scientists for their best estimates of any given term, you will likely get two very different answers!

RATE OF STAR FORMATION

We can estimate the average number of stars forming each year in the Galaxy simply by noting that at least 100 billion stars now shine in the Milky Way. Dividing this number by the 10-billion-year lifetime of the Galaxy, we obtain a formation rate of 10 stars per year. This may be an overestimate because we think that fewer stars are forming now than formed at earlier epochs of the Galaxy, when more interstellar gas was available. However, we do know that stars are forming today, and our estimate does not include stars that formed in the past and have since exploded, so our value

of 10 stars per year is probably reasonable when averaged over the lifetime of the Milky Way.

FRACTION OF STARS HAVING PLANETARY SYSTEMS

Many astronomers believe that planet formation is a natural result of the star-formation process. If the condensation theory (Chapter 15) is correct, and if there is nothing special about our Sun, as we have argued throughout this book, we would expect many stars to have at least one planet. ∞ (Sec. 15.2) Indeed, as we have seen, increasingly sophisticated observations indicate the presence of disks around young stars. Could these disks be protosolar systems? The condensation theory suggests they are.

No planets like our own have yet been *seen* orbiting any other star. The light reflected by an Earthlike planet circling even the closest star would be too faint to detect even with the very best equipment. The light would be lost in the glare of the parent star. Large orbiting telescopes may soon be able to detect Jupiter-sized planets orbiting the nearest stars, but even those huge planets will be barely visible. However, as described in *Interlude 15-1*, there is now strong indirect evidence that planets are orbiting other stars. Several groups of observers have reported detection of the back-and-forth or side-to-side motions produced in a parent star by an orbiting planet's gravitational pull. The planets found so far are Jupiter-sized rather than Earth-sized, but astronomers are confident that an Earthlike planet will one day be detected by these means.*

Accepting the condensation theory and its consequences, and without being either too conservative or naively optimistic, we assign a value near 1 to this term—that is, we believe that essentially all stars have planetary systems.

NUMBER OF HABITABLE PLANETS PER PLANETARY SYSTEM

Temperature, more than any other single quantity, determines the feasibility of life on a given planet. The surface temperature of a planet depends on two things: the planet's distance from its parent star and the thickness of its atmosphere. Planets with a nearby parent star (but not too close) and some atmosphere (though not too thick) should be reasonably warm, like Earth or Mars. Planets far from the star and with no atmosphere, like Pluto, will surely be cold by our standards. And planets too close to the star and with a thick atmosphere, like Venus, will be very hot indeed.

Figure 28.8 illustrates that a three-dimensional zone of "comfortable" temperatures—often called a *habitable zone*—surrounds every star. It represents the range of distances within which a planet of mass and composition similar to those of Earth would have a surface temperature between the freezing and boiling points of water. (Our Earth-based

---

Recall from Chapter 22 that Earth-mass planets have been observed orbiting some neutron stars. However, as we saw in that chapter, the formation of those planets was very different from that of Earth and the solar system. ∞ (Sec. 22.3) They are not thought to be likely candidates for the emergence of life.

Figure 28.8 The extent of the habitable zone is much larger around a hot star than around a cool one. For a star like the Sun (a G-type star), the zone extends from about 0.85 A.U. to 2.0 A.U. For an F-type star, the range is 1.2 to 2.8 A.U. For a faint M-type star only planets orbiting between about 0.02 and 0.06 A.U. would be habitable.

bias is plainly evident here!) The hotter the star, the larger this zone. For example, A- and F-type stars have a rather large habitable zone, but the size of the zone diminishes rapidly as we proceed through G-, K-, and M-type stars. O- and B-type stars are not considered here because they are not expected to last long enough for life to develop, even if they do have planets.

Three planets—Venus, Earth, and Mars—reside within the habitable zone surrounding our Sun. Venus is too hot because of its thick atmosphere and proximity to the Sun. Mars is a little too cold because its atmosphere is too thin and it is too far from the Sun. But if Venus had Mars's thin atmosphere, and if Mars had Venus's thick atmosphere, both of these nearby planets might conceivably have surface conditions resembling those on Earth.

To estimate the number of habitable planets per planetary system, we first take inventory of how many stars of each type shine in our Galaxy and calculate the sizes of their habitable zones. Then we eliminate binary-star systems because a planet's orbit within the habitable zone of a binary would likely be unstable, as illustrated in Figure 28.9. Given the known properties of binaries in our Galaxy, habitable planetary orbits would probably be unstable in most cases, so there would not be time for life to develop.

Taking all these factors into account, we assign a value of 1/10 to this term in our equation. In other words we believe that, on average, there is 1 potentially habitable planet for every 10 planetary systems that might exist in our Galaxy. Single F-, G-, and K-type stars are the best candidates.

FRACTION OF HABITABLE PLANETS ON WHICH LIFE ARISES

The number of possible combinations of atoms is incredibly large. If the chemical reactions that led to the complex molecules that make up living organisms occurred completely at random, then it is extremely unlikely that those molecules

Figure 28.9 In binary-star systems, planets are restricted to only a few kinds of orbits that are gravitationally stable. The orbit in case (a) is stable only if the planet lies very close to its "parent" star, so the gravity of the other star is negligible. Case (b) shows a planet circulating at a great distance about both stars in an elliptical orbit. This orbit is stable only if it lies far from both stars. Another possible, but unstable path, (c) interweaves between the two stars in a figure-eight pattern.

could have formed at all. In that case, life is extraordinarily rare, this term is close to zero, and we are probably alone in the Galaxy, perhaps even in the entire universe.

However, laboratory experiments (like the Urey–Miller experiment described earlier) seem to suggest that certain chemical combinations are strongly favored over others—that is, the reactions are not random. Of the billions upon billions of basic organic groupings that could possibly occur on Earth from the random combination of all sorts of simple atoms and molecules, only about 1500 actually do occur. Furthermore, these 1500 organic groups of terrestrial biology are made from only about 50 simple "building blocks" (including the amino acids and nucleotide bases mentioned earlier). This suggests that molecules critical to life may not be assembled by pure chance. Apparently, additional factors are at work at the microscopic level. If a relatively small number of chemical "evolutionary tracks" are likely to exist, then the formation of complex molecules—and hence, we assume, life—becomes much more likely, given sufficient time.

To assign a very low value to this term in the equation is to believe that life arises randomly and rarely. To assign a value close to 1 is to believe that life is inevitable, given the proper ingredients, a suitable environment, and a long enough period of time. No easy experiment can distinguish between these extreme alternatives, and there is little or no middle ground. To many researchers, the discovery of life on Mars, Europa, Titan, or some other object in our solar system would convert the appearance of life from an unlikely miracle to a virtual certainty throughout the Galaxy. We will take the optimistic view and adopt a value of 1.

FRACTION OF LIFE-BEARING PLANETS ON WHICH INTELLIGENCE ARISES

As with the evolution of life, the appearance of a well-developed brain is a very unlikely event if only random chance is involved. However, biological evolution through natural selection is a mechanism that generates apparently highly improbable results by singling out and refining useful characteristics. Organisms that profitably use adaptations can develop more complex behavior, and complex behavior provides organisms with the *variety* of choices needed for more advanced development.

One school of thought maintains that, given enough time, intelligence is inevitable. In this view, assuming that natural selection is a universal phenomenon, at least one organism on a planet will always rise to the level of "intelligent life." If this is correct, then the fifth term in the Drake equation equals or nearly equals 1. Others argue that there is only one known case of intelligence, and that case is life on Earth. For 2.5 billion years—from the start of life about 3.5 billion years ago to the first appearance of multicellular organisms about 1 billion years ago—life did not advance beyond the one-celled stage. Life remained simple and dumb, but it survived. If this latter view is correct, then the fifth term in our equation is very small, and we are faced with the depressing prospect that humans may be the smartest form of life anywhere in the Galaxy. As with the previous term, we will be optimistic and simply adopt a value of 1 here.

FRACTION OF PLANETS ON WHICH INTELLIGENT LIFE DEVELOPS AND USES TECHNOLOGY

To evaluate the sixth term of our equation, we need to estimate the probability that intelligent life eventually develops technological competence. Should the rise of technology be inevitable, this term is close to 1, given long enough periods of time. If it is not inevitable—if intelligent life can somehow "avoid" developing technology—then this term could be much less than 1. The latter possibility envisions a universe possibly teeming with intelligent civilizations but very few among them ever becoming technologically competent. Perhaps only one managed it—ours.

Again, it is difficult to decide conclusively between these two views. We don't know how many prehistoric Earth cultures failed to develop technology, or rejected its use. We do know that the roots of our present civilization arose independently at several different places on Earth, including Mesopotamia, India, China, Egypt, Mexico, and Peru. Because so many of these ancient cultures originated at about the same time, it is tempting to conclude that the chances are good that some sort of technological society will inevitably develop, given some basic intelligence and enough time.

If technology is inevitable, then why haven't other lifeforms on Earth also found it useful? Possibly the competitive edge given by intellectual and technological skills to humans, the first species to develop them, allowed us to dominate so rapidly that other species—gorillas and chimpanzees, for example—simply haven't had time to catch up. The fact that only one technological society exists on Earth does not imply that the sixth term in our Drake equation must be very much less than 1. On the contrary, it is precisely because *some* species will probably always fill the niche of technological intelligence that we will take this term to be close to 1.

AVERAGE LIFETIME OF A TECHNOLOGICAL CIVILIZATION

The reliability of the estimate of each term in the Drake equation declines markedly from left to right. For example, our knowledge of astronomy enables us to make a reasonably good stab at the first term, namely, the rate of star formation in our Galaxy, but it is much harder to evaluate some of the later terms, such as the fraction of life-bearing planets that eventually develop intelligence. The last term on the right-hand side of the equation, the longevity of technological civilizations, is totally unknown. There is only one known example of such a civilization—humans on planet Earth. Our own civilization has presently survived in its "technological" state for only about 100 years, and how long we will be around before a natural or human-made catastrophe (see *Interlude 14-1*) ends it all is impossible to tell.

One thing is certain: if the correct value for *any one term* in the equation is very small, then few technological civilizations now exist in the Galaxy. If the pessimistic view of the development of life or of intelligence is correct, then we are unique, and that is the end of our story. However, if both life and intelligence are inevitable consequences of chemical and biological evolution, as many scientists believe, and if intelligent life always becomes technological, then we can plug the higher, more optimistic values into the Drake equation. In that case, combining our estimates for the other six terms (and noting that $10 \times 1 \times 1/10 \times 1 \times 1 \times 1 = 1$), we can say

the number of technological, intelligent civilizations now present in the Milky Way Galaxy	=	the average lifetime of a techologically competent civilization, *in years*.

Thus, if civilizations typically survive for 1000 years, there should be 1000 of them currently in existence scattered throughout the Galaxy. If they live for a million years, on average, we would expect there to be a million advanced civilizations in the Milky Way, and so on.

28.4 The Search for Extraterrestrial Intelligence

4 Let us continue our optimistic assessment of the prospects for life and assume that civilizations enjoy a long stay on their parent planet once their initial technological "teething problems" are past. In that case, they are likely to be plentiful in the Galaxy. How might we become aware of their existence?

MEETING OUR NEIGHBORS

For concreteness, let us assume that the average lifetime of a technological civilization is 1 million years—only 1 percent of the reign of the dinosaurs, but 100 times longer than human civilization has survived thus far. Given the size and shape of our Galaxy, we can then estimate the average *distance* between these civilizations to be some 50 pc, or about 150 light years. Thus, any two-way communication with our neighbors—using signals traveling at or below the speed of light—will take at least 300 years (150 years for the message to reach the planet and another 150 years for the reply to travel back to us).

One obvious way to search for extraterrestrial life would be to develop the capability to travel far outside our solar system. However, this may never be a practical possibility. At a speed of 50 km/s, the speed of the current fastest unmanned space probes, the trip to our nearest neighbor, Alpha Centauri, would take about 25,000 years. The journey to the nearest civilization (assuming, as earlier, that a total of 1 million such cultures exist in the Galaxy) would take almost 1 million years. Interstellar travel at these speeds is clearly not feasible. Speeding up our ships to near the speed of light would reduce the travel time, but this is far beyond our present technology.

Actually, our civilization has already launched some interstellar probes, although they have no specific stellar destination. Figure 28.10 is a reproduction of a plaque mounted onboard the *Pioneer 10* spacecraft launched in the mid-1970s and now well beyond the orbit of Pluto, on its way out of the solar system. Similar information was also included aboard the *Voyager* probes launched in 1978. Although these spacecraft would be incapable of reporting back to Earth the news that they had encountered an alien culture, scientists hope that the civilization on the other end would be able to unravel most of its contents using the universal language of mathematics. The caption to Figure 28.10 notes how the aliens might discover from where and when the *Pioneer* and *Voyager* probes were originally launched.

Setting aside the many practical problems of establishing direct contact with extraterrestrials, some scientists have argued that it might not even be a particularly good idea. Our recent emergence as a technological civilization implies that we must be one of the least advanced technological intelligences in the entire Galaxy. Any other civilization that we discover, or that discovers us, will almost surely be more advanced than us. Consequently, a healthy degree of caution is warranted. If extraterrestrials behave even remotely like human civilizations on Earth, then the most advanced aliens may naturally try to dom-

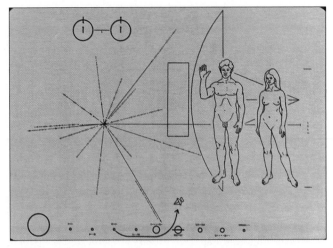

Figure 28.10 A replica of a plaque mounted on board the *Pioneer 10* spacecraft. The important features of the plaque include a scale drawing of the spacecraft, a man, and a woman; a diagram of the hydrogen atom undergoing a change in energy (top left); a starburst pattern representing various pulsars and the frequencies of their radio waves that can be used to estimate when the craft was launched (middle left); and a depiction of the solar system, showing that the spacecraft departed the third planet from the Sun and passed the fifth planet on its way into outer space (bottom). All the drawings have computer- (binary) coded markings from which actual sizes, distances, and times can be derived.

inate all others. The behavior of the "advanced" European cultures toward the "primitive" races they encountered on their voyages of discovery in the seventeenth, eighteenth, and nineteenth centuries should serve as a clear warning of the possible undesirable consequences of contact. Of course, the aggressiveness of Earthlings may not apply to extraterrestrials, but given the history of the one intelligent species we know, the cautious approach may be in order.

RADIO COMMUNICATION

A cheaper, and much more practical, alternative is to try to make contact with extraterrestrials using only electromagnetic radiation, the fastest known means of transferring information from one place to another. Because light and other high-frequency radiation are heavily scattered while moving through dusty interstellar space, long-wavelength radio radiation seems to be the natural choice. We would not attempt to broadcast to all nearby candidate stars, however—that would be far too expensive and inefficient. Instead, radio telescopes on Earth would listen *passively* for radio signals emitted by other civilizations. Some preliminary searches of selected nearby stars are now under way, thus far without success.

In what direction should we aim our radio telescopes? The answer to this question at least is fairly easy. On the basis of our earlier reasoning, we should target all F-, G-, and K-type stars in our vicinity. But are extraterrestrials broadcasting radio signals? If they are not, this search technique will obviously fail. If they are, how do we distinguish their artificially generated radio signals from signals naturally emitted by interstellar gas clouds? At what frequency should we tune our receivers? This depends on whether the signals are produced deliberately or are simply "waste radiation" escaping from a planet.

Consider how Earth would look at radio wavelengths to extraterrestrials. Figure 28.11 shows the pattern of radio signals we emit into space. From the viewpoint of a distant observer, the spinning Earth emits a bright flash of radio radiation every few hours. In fact, Earth is now a more intense radio emitter than the Sun. The flashes result from the periodic rising and setting of hundreds of FM radio stations and television transmitters. Each station broadcasts mostly parallel to Earth's surface, sending a great "sheet" of electromagnetic radiation into interstellar space, as illustrated in Figure 28.11(a). (The more common AM broadcasts are trapped below our ionosphere, so those signals never leave Earth.)

Because the great majority of these transmitters are clustered in the eastern United States and western Europe, a distant observer would detect blasts of radiation from Earth as our planet rotates each day (Figure 28.11b). This radiation races out into space, and has been doing so since the invention of these technologies nearly seven decades ago. Another civilization as advanced as ours might have constructed devices capable of detecting this radiation. If any sufficiently advanced (and sufficiently interested) civilization resides on a planet orbiting any of the thousand or so stars within about 65 light years (20 pc) of Earth, then we have already broadcast our presence to them.

THE WATER HOLE

Now let us suppose that a civilization has decided to assist searchers by actively broadcasting its presence to the rest of the Galaxy. At what frequency should we listen for such an extraterrestrial beacon? The electromagnetic spectrum is enormous; the radio domain alone is vast. To hope to detect a signal at some unknown radio frequency is like searching for a needle in a haystack. Are some frequencies more likely than others to carry alien transmissions?

Some basic arguments suggest that civilizations might well communicate at a wavelength near 20 cm. As we saw in Chapter 18, the basic building blocks of the universe, namely,

Figure 28.11 Radio radiation now leaks from Earth into space because of the daily activities of our technological civilization. (a) FM radio and television transmitters broadcast their energy parallel to Earth's surface, so they send a great "sheet" of electromagnetic radiation into interstellar space, producing the strongest signal in any given direction when they happen to lie on Earth's horizon, as seen from that direction. (The more common AM broadcasts are trapped below our ionosphere, so those signals never leave Earth.) (b) Because the great majority of transmitters are clustered in the eastern United States and western Europe, a distant observer would detect blasts of radiation from Earth as our planet rotates each day.

hydrogen atoms, naturally radiate at a wavelength of 21 cm. ∞ (Sec. 18.4) Also, one of the simplest molecules, hydroxyl (OH), radiates near 18 cm. Together, these two substances form water (H₂O). Arguing that water is likely to be the interaction medium for life anywhere and that radio radiation travels through the disk of our Galaxy with the least absorption by interstellar gas and dust, some researchers have proposed that the interval between 18 and 21 cm is the best wavelength range for civilizations to transmit or monitor. Called the **water hole**, this radio interval might serve as an "oasis" where all advanced Galactic civilizations would gather to conduct their electromagnetic business.

This water-hole frequency interval is only a guess, of course, but it is supported by other arguments as well. Figure 28.12 shows the water hole's location in the electromagnetic spectrum and plots the amount of natural emission from our Galaxy and from Earth's atmosphere. The 18- to 21-cm range lies within the quietest part of the spectrum, where the Galactic "static" from stars and interstellar clouds happens to be minimized. Furthermore, the atmospheres of typical planets are also expected to interfere least at these wavelengths. Thus the water hole seems like a good choice for the frequency of an interstellar beacon, although we cannot be sure of this reasoning until contact is actually achieved. A few radio searches are now in progress at frequencies in and around the water hole. So far, however, nothing resembling an extraterrestrial signal has been detected.

The space surrounding all of us could be, right now, flooded with radio signals from extraterrestrial civilizations. If only we knew the proper direction and frequency, we might be able to make one of the most startling discoveries

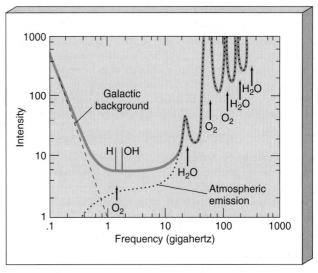

Figure 28.12   The "water hole" is bounded by the natural emission frequencies of the hydrogen (H) atom (21-cm wavelength) and the hydroxyl (OH) molecule (18-cm wavelength). The topmost solid (blue) curve sums the natural emissions of our Galaxy (dashed line on left side of diagram) and Earth's atmosphere (dotted line on right side of diagram denoted by various chemical symbols). This sum is minimized near the water-hole frequencies. Perhaps all intelligent civilizations conduct their interstellar communications within this quiet "electromagnetic oasis."

of all time. The result would likely provide whole new opportunities to study the cosmic evolution of energy, matter, and life throughout the universe.

Chapter Review

SUMMARY

The history of the universe can be divided into seven phases: particulate, galactic, stellar, planetary, chemical, biological, and cultural. **Cosmic evolution** (p. 648) is the continuous process that has led to the appearance of galaxies, stars, planets, and life on Earth.

Living organisms may be characterized by their ability to react to their environment, to grow by taking in nutrition from their surroundings, and to reproduce, passing along some of their own characteristics to their offspring.

Powered by natural energy sources, reactions between simple molecules in the oceans of the primitive Earth are believed to have led to the formation of **amino acids** (p. 649) and **nucleotide bases**

(p. 649), the basic molecules of life. Alternatively, some complex molecules may have been formed in interstellar space and then delivered to Earth by meteors or comets.

Organisms that can best take advantage of their new surroundings succeed at the expense of those organisms that cannot make the necessary adjustments. Intelligence is strongly favored by natural selection.

The best hope for life beyond Earth in the solar system is the planet Mars, although no evidence for living organisms has been found. Saturn's Titan may also be a possibility, but conditions there are harsh by terrestrial standards.

The **Drake equation** (p. 653) provides a means of estimating the probability of intelligent life in the Galaxy. The astronomical terms in the equation are the Galactic star-formation rate, the likelihood of planets, and the number of habitable planets. Chemical and biological terms are the probability that life appears and the probability that it subsequently develops intelligence. Cultural and political terms are the probability that intelligence leads to technology, and the lifetime of a civilization in the technological state.

Taking an optimistic view of the development of life and intelligence leads to the conclusion that the total number of technologically competent civilizations in the Galaxy is approximately equal to the lifetime of a typical civilization, expressed in years.

Even with optimistic assumptions, the distance to our nearest intelligent neighbor is likely to be many hundreds of parsecs.

Space travel is not presently a feasible means of searching for intelligent life. Current programs to discover extraterrestrial intelligence involve scanning the electromagnetic spectrum for signals. So far, no intelligent broadcasts have been received.

A technological civilization would probably "announce" itself to the universe by the radio and television signals it emits into space. Observed from afar, our planet would appear as a radio source with a 24-hour period, as different regions of the planet rise and set.

The **water hole** (p. 659) is a region in the radio range of the electromagnetic spectrum, near the 21-cm line of hydrogen and the 18-cm line of hydroxyl, where natural emissions from the Galaxy happen to be minimized. Many researchers regard this as the best part of the spectrum for communications purposes.

SELF-TEST: TRUE OR FALSE?

_____ **1.** The assumptions of mediocrity favor the existence of extraterrestrial life.

_____ **2.** The definition of life requires that to be considered "alive" you only must be able to reproduce.

_____ **3.** Miller and Urey conducted experiments in which they produced biological molecules from nonbiological molecules.

_____ **4.** Organic molecules important for life could have reached Earth's surface via comets.

_____ **5.** Organic molecules are known to exist only on Earth.

_____ **6.** Laboratory experiments have succeeded in creating living cells from nonbiological molecules.

_____ **7.** For most of the history of Earth, life consisted of only single-celled life-forms.

_____ **8.** The *Viking* landers on Mars discovered microscopic evidence of life but found no large fossil evidence.

_____ **9.** The rate of star formation in the Galaxy is reasonably well known.

_____ **10.** As yet there is no direct evidence that Earthlike planets orbit other stars.

_____ **11.** In estimating whether intelligence arises and develops technology, we have only life on Earth as an example.

_____ **12.** Dinosaurs existed on Earth for 10,000 times longer than human civilization has existed to date.

_____ **13.** Our civilization has launched probes into interstellar space.

_____ **14.** One disadvantage of interstellar radio communication is that we can do it only with another civilization that has a technology equal to or greater than our own.

_____ **15.** We have already broadcast our presence to our interstellar neighbors, if any.

SELF-TEST: FILL IN THE BLANK

1. Amino acids are the building blocks of _____

2. The naturally occurring molecules present on the young Earth included water, carbon dioxide, _____, and _____.

3. Two sources of energy for chemical reactions available on the young Earth were _____ and _____.

4. The fossil record clearly shows evidence of life dating back _____ years.

5. Multicellular organisms did not appear on Earth until about _____ years ago.

6. The Murchison meteorite was discovered to contain relatively large amounts of _____.

7. The Drake equation estimates the number of _____ in the Milky Way Galaxy.

8. Planets in binary star systems are not considered habitable because the planetary orbits are often _____.

9. The probabilities of the development of life and intelligence on Earth are both extremely _____ if random chance is the only evolutionary factor involved.

10. Direct contact between extraterrestrial life-forms may be impractical because of the large _____ between civilizations.

11. Radio communications over interstellar distances is practical because the signals travel at the speed of _____.

12. Radio waves can travel throughout the Galaxy because they are not blocked by interstellar _____.

13. Radio waves leaking away from Earth have now traveled out a distance of _____ light years.

14. Radio wavelengths between 18 and 21 cm are referred to as the _____.

15. A two-way communication with another civilization at a distance of 100 light years will require _____ years.

REVIEW AND DISCUSSION

1. Why is life difficult to define?

2. What is chemical evolution?

3. What is the Urey–Miller experiment? What important organic molecules were produced in this experiment?

4. What are the basic ingredients from which biological molecules formed on Earth?

5. How do we know anything at all about the early episodes of life on Earth?

6. What is the role of language in cultural evolution?

7. Where else, besides Earth, have organic molecules been found?

8. Where—besides the planet Mars—might we find signs of life in our solar system?

9. Do we know whether Mars ever had life at any time during its past? What argues in favor of the position that it may once have harbored life?

10. What is generally meant by "life as we know it"? What other forms of life might be possible?

11. How many of the terms in the Drake equation are known with any degree of certainty? Which factor is least well known?

12. What is the relationship between the average lifetime of Galactic civilizations and the possibility of our someday communicating with them?

13. How would Earth appear at radio wavelengths to extraterrestrial astronomers?

14. What are the advantages in using radio waves for communication over interstellar distances?

15. What is the water hole? What advantage does it have over other parts of the radio spectrum?

PROBLEMS

1. If Earth's 4.6-billion-year age were compressed to 46 years, as described in the text, what would be your age, in seconds? How long ago was the end of World War II? The Declaration of Independence? Columbus's discovery of the New World?

2. According to the inverse-square law, a planet receives energy from its parent star at a rate proportional to the star's luminosity and inversely proportional to the square of the planet's distance from the star (Section 17.4). According to Stefan's law, the rate at which the planet radiates energy into space is proportional to the fourth power of its surface temperature (Section 3.4). In equilibrium, the two rates are equal. Based on this information and the data presented in Figure 28.8, estimate the extent of the habitable zone surrounding a K star of luminosity 1/10 the luminosity of the Sun.

3. Based on the numbers presented in the text, and assuming an average lifetime of 5 billion years for suitable stars, estimate the total number of habitable planets in the Galaxy.

4. A planet orbits one component of a binary star system at a distance of 1 A.U. (see Figure 28.9a). If both stars have the same mass, and their orbit is circular, estimate the minimum distance between the stars for the tidal force due to the companion not to exceed a "safe" 0.01 percent of the gravitational force between the planet and its parent star.

5. Suppose that each of the "fraction" terms in the Drake equation turns out to have a value of 1/10, that stars form at an average rate of 20 per year, and that each star has exactly 1 habitable planet orbiting it. Estimate the present number of technological civilizations in the Milky Way Galaxy if the average lifetime of a civilization is (a) 100 years (b) 10,000 years (c) one million years.

6. Adopting the estimate from the text that the number of technological civilizations in the Milky Way Galaxy is equal to the average lifetime of a civilization, in years, it follows that the distance to our nearest neighbor decreases as the average lifetime increases. Assuming that civilizations are uniformly spread over a two-dimensional Galactic disk of radius 15 kpc, and all have the same lifetime, calculate the *minimum* lifetime for which a two-way radio communication with our nearest neighbor would be possible before our civilization ends. Repeat the calculation for a round-trip personal visit, using current-technology spacecraft that travel at 50 km/s.

7. How fast would a spacecraft have to travel in order to complete the trip from Earth to Alpha Centauri (a distance of 1.3 pc) and back in less than an average human lifetime (80 years, say)?

8. Assuming that there are 10,000 FM radio stations on Earth, each transmitting at a power level of 50 kW calculate the total radio luminosity of Earth in the FM band. Compare this value with the roughly 10^6 W radiated by the Sun in the same frequency range.

9. Convert the water hole's wavelengths to frequencies. For practical reasons, any search of the water hole must be broken up into channels, much like you find on a television, except these channels are very narrow in radio frequency, about 100 Hz wide. How many channels must astronomers search in the water hole?

10. There are 20,000 stars within 100 light years that are to be searched for radio communications. How long will the search take if 1 hour is spent looking at each star? What if 1 day is spent per star?

PROJECTS

1. Some people suggest that if extraterrestrial life is discovered, it will have a profound effect on people. Interview as many people as you can and ask the following two questions: (1) Do you believe that extraterrestrial life exists? (2) Why? From your results, try to decide whether there will be a profound effect on people if extraterrestrial life is discovered.

2. Conduct another poll, or do it at the same time you do the first one. Ask the following question: What one question would you like to ask an extraterrestrial life-form in a radio communication? How many responses do you receive that indicate

the person is very "Earth-centered" in thinking? How many responses suggest a lack of understanding of how alien an extraterrestrial life-form might be? Is your conclusion from the first project different or changed in any way?

3. The Drake equation should be able to "predict" at least one civilization in our Galaxy: us. Try changing the values of various factors so that you end up with at least one. What do these various combinations of factors imply about how life arises and develops? Are there some combinations that just don't make any sense?

APPENDIX 1

Scientific Notation

The objects studied by astronomers range in size from the smallest particles to the largest expanse of matter we know—the entire universe. Subatomic particles have sizes of about 0.000000000000001 meter, while galaxies typically measure some 1,000,000,000,000,000,000,000 meters across. The most distant known objects in the universe lie on the order of 100,000,000,000,000,000,000,000,000 meters from Earth.

Obviously, writing all those zeros is both cumbersome and inconvenient. More important, it is also very easy to make an error—write down one zero too many or too few and your calculations become hopelessly wrong! To avoid this, scientists always write large numbers using a shorthand notation in which the number of zeros following or preceding the decimal point is denoted by a superscript power, or *exponent*, of 10. The exponent is simply the number of places between the first significant (nonzero) digit in the number (reading from left to right) and the decimal point. Thus 1 is 10^0, 10 is 10^1, 100 is 10^2, 1000 is 10^3, and so on. For numbers less than 1, with zeros between the decimal point and the first significant digit, the exponent is negative: 0.1 is 10^{-1}, 0.01 is 10^{-2}, 0.001 is 10^{-3}, and so on. Using this notation we can shorten the number describing subatomic particles to 10^{-15} meter and write the number describing the size of a galaxy as 10^{21} meters.

More complicated numbers are expressed as a combination of a power of 10 and a multiplying factor. This factor is conventionally chosen to be a number between 1 and 10, starting with the first significant digit in the original number. For example, 150,000,000,000 meters (the distance from Earth to the Sun, in round numbers) can be more concisely written as 1.5×10^{11} meters, 0.000000025 meters as 2.5×10^{-8} meter, and so on. The exponent is simply the number of places the decimal point must be moved to the left to obtain the multiplying factor.

Some other examples of scientific notation are

■ the approximate distance to the Andromeda Galaxy
= 3,000,000 light years = 3×10^6 light years
■ the size of a hydrogen atom
= 0.00000000005 meter = 5×10^{-11} meter

■ the diameter of the Sun
= 1,392,000 kilometers = 1.392×10^6 kilometers
■ the U.S. national debt (as of June 1, 1998)
= \$5,545,414,000,000.00 = \$5.545414 trillion
= 5.545414×10^{12} dollars.

In addition to providing a simpler way of expressing very large or very small numbers, this notation also makes it easier to do basic arithmetic. The rule for multiplication of numbers expressed in this way is simple: just multiply the factors and add the exponents. Similarly for division: divide the factors and subtract the exponents. Thus, 3.5×10^{-2} multiplied by 2.0×10^3 is simply $(3.5 \times 2.0) \times 10^{-2+3}$ = 7.0×10^1—that is, 70. Again, 5×10^6 divided by 2×10^4 is just $(5/2) \times 10^{6-4}$, or 2.5×10^2 (= 250). Applying these rules to unit conversions, we have 200,000 nanometers is 200,000 $\times 10^{-9}$ meter (since 1 nanometer = 10^{-9} meter; see Appendix 2), or $2 \times 10^5 \times 10^{-9}$ meter, or $2 \times 10^{5-9} = 2 \times 10^{-4}$ meter. Verify these rules yourself with a few examples of your own. The advantages of this notation when considering astronomical objects will soon become obvious.

Scientists often use "rounded-off" versions of numbers, both for simplicity and for ease of calculation. For example, we will usually write the diameter of the Sun as 1.4×10^6 kilometers, instead of the more precise number given earlier. Similarly, Earth's diameter is 12,756 kilometers, or 1.2756×10^4 kilometers, but for "ballpark" estimates, we really don't need so many digits, and the more approximate number 1.3×10^4 kilometers will suffice. Very often, we perform rough calculations using only the first one or two significant digits in a number, and that may be all that is necessary to make a particular point. For example, to support the statement, "The Sun is much larger than Earth," we need say only that the ratio of the two diameters is roughly 1.4×10^6 divided by 1.3×10^4. Since 1.4/1.3 is close to 1, the ratio is approximately $10^6/10^4 = 10^2$, or 100. The essential fact here is that the ratio is much larger than 1; calculating it to greater accuracy (to get 109.13) would give us no additional *useful* information. This technique of stripping away the arithmetic details to get to the essence of a calculation is very common in astronomy, and we use it frequently throughout this text.

APPENDIX 2

Astronomical Measurement

Astronomers use many different kinds of units in their work, simply because no single system of units will do. Rather than the *Système Internationale* (SI), or the meter-kilogram-second (mks) metric system used in most high school and college science classes, many professional astronomers still prefer the older centimeter-gram-second (cgs) system. However, astronomers also commonly introduce new units when convenient. For example, when discussing stars, astronomers often use the mass and radius of the Sun as reference points. The solar mass, written as $M_\odot$, is equal to 2.0×10^{33} g, or 2.0×10^{30} kg (since 1 kg = 1000 g). The solar radius, $R_\odot$ is equal to 700,000 km, or 7.0×10^8 m (1 km = 1000 m). The subscript $\odot$ always stands for the Sun. Similarly, the subscript $\oplus$ always stands for Earth. In this book we use the units that astronomers commonly use in any given context, but we also give the "standard" SI equivalents where appropriate.

Of particular importance are the units of length astronomers use. On small scales, the *angstrom* (1 Å = 10^{-10} m = 10^{-8} cm), the *nanometer* (1 nm = 10^{-9} m = 10^{-7} cm), and the *micron* (1 μm = 10^{-6} m = 10^{-4} cm) are used. Distances within the solar system are usually expressed in terms of the *astronomical unit* (A.U.), the mean distance between Earth and the Sun. One A.U. is approximately equal to 150,000,000 km, or 1.5×10^{11} m. On larger scales, the *light year* (1 ly = 9.5×10^{15} m = 9.5×10^{12} km) and the *parsec* (1 pc = 3.1×10^{16} m = 3.1×10^{13} km = 3.3 ly) are commonly used. Still larger distances use the regular prefixes of the metric system: *kilo* for one thousand and *mega* for one million. Thus, 1 *kiloparsec* (kpc) = 10^3 pc = 3.1×10^{19} m, 10 megaparsecs (Mpc) = 10^7 pc = 3.1×10^{23} m, and so on.

Astronomers use units that make sense within a context, and as contexts change, so do the units. For example, we might measure densities in grams per cubic centimeter (g/cm^3), in atoms per cubic meter (atoms/m^3), or even in solar masses per cubic megaparsec ($M_\odot/Mpc^3$), depending on the circumstances. The important thing to know is that once you understand the units, you can convert freely from one set to another. For example, the radius of the Sun could equally well be written as $R_\odot$ = 6.96×10^8 m, or 6.96×10^{10} cm, or 109 $R_\oplus$, or 4.65×10^{-3} A.U., or even 7.36×10^{-8} ly—whichever happens to be most useful. Some of the more common units used in astronomy, and the contexts in which they are most likely to be encountered, appear below.

LENGTH:		
1 angstrom (Å)	= 10^{-10} m	atomic physics, spectroscopy
1 nanometer (nm)	= 10^{-9} m	
1 micron (μm)	= 10^{-6} m	interstellar dust and gas
1 centimeter (cm)	= 0.01 m	in widespread use throughout astronomy
1 meter (m)	= 100 cm	
1 kilometer (km)	= 1000 m = 10^5 cm	
Earth radius ($R_\oplus$)	= 6378 km	planetary astronomy
Solar radius ($R_\odot$)	= 6.96×10^8 m	solar system, stellar evolution
1 astronomical unit (A.U.)	= 1.496×10^{11} m	
1 light year (ly)	= 9.46×10^{15} m = 63,200 A.U.	galactic astronomy, stars and star clusters
1 parsec (pc)	= 3.09×10^{16} m = 3.26 ly	
1 kiloparsec (kpc)	= 1000 pc	galaxies, galaxy clusters, cosmology
1 megaparsec (Mpc)	= 1000 kpc	
MASS:		
1 gram (g)		in widespread use throughout astronomy
1 kilogram (kg)	= 1000 g	
Earth mass ($M_\oplus$)	= 5.98×10^{24} kg	planetary astronomy
Solar mass ($M_\odot$)	= 1.99×10^{30} kg	"standard" unit for all mass scales larger than Earth
TIME:		
1 second (s)		in widespread use throughout astronomy
1 hour (h)	= 3600 s	planetary and stellar scales
1 day (d)	= 86,400 s	
1 year (yr)	= 3.16×10^7 s	virtually all processes occurring on scales larger than a star

APPENDIX 3 Tables

TABLE 1 Some Useful Constants and Physical Measurements*

astronomical unit	$1\ \text{A.U.} = 1.496 \times 10^8\ \text{km}\ (1.5 \times 10^8\ \text{km})$
light year	$1\ \text{ly} = 9.46 \times 10^{12}\ \text{km}\ (10^{13}\ \text{km}; 6\ \text{trillion miles})$
parsec	$1\ \text{pc} = 3.09 \times 10^{13}\ \text{km} = 3.3\ \text{ly}$
speed of light	$c = 299{,}792.458\ \text{km/s}\ (3 \times 10^5\ \text{km/s})$
Stefan-Boltzmann constant	$\sigma\ \text{[Greek sigma]} = 5.67 \times 10^{-8}\ \text{W/m}^2 \cdot \text{K}^4$
Planck's constant	$h = 6.63 \times 10^{-34}\ \text{Js}$
gravitational constant	$G = 6.67 \times 10^{-11}\ \text{Nm}^2/\text{kg}^2$
mass of the Earth	$M_\oplus = 5.97 \times 10^{24}\ \text{kg}\ (6 \times 10^{24}\ \text{kg}; \text{about 6000 billion billion tons})$
radius of the Earth	$R_\oplus = 6378\ \text{km}\ (6500\ \text{km})$
mass of the Sun	$M_\odot = 1.99 \times 10^{30}\ \text{kg}\ (2 \times 10^{30}\ \text{kg})$
radius of the Sun	$R_\odot = 6.96 \times 10^5\ \text{km}\ (7 \times 10^5\ \text{km})$
luminosity of the Sun	$L_\odot = 3.90 \times 10^{26}\ \text{W}\ (4 \times 10^{26}\ \text{W})$
effective temperature of the Sun	$T_\odot = 5778\ \text{K}\ (5800\ \text{K})$
Hubble constant	$H_0 \approx 65\ \text{km/s/Mpc}$
mass of the electron	$m_e = 9.11 \times 10^{-31}\ \text{kg}$
mass of the proton	$m_p = 1.67 \times 10^{-27}\ \text{kg}$

The rounded-off values used in the text are shown in parentheses.

Conversions between English and Metric Units

1 inch	= 2.54 centimeters (cm)	1 mile	= 1.609 kilometers (km)
1 foot (ft)	= 0.3048 meters (m)	1 pound (lb)	= 453.6 grams (g) or .4536 kilograms (kg) (on Earth)

TABLE 2 Periodic Table of the Elements

Atomic number, Symbol of element, Atomic weight, Name of element (as shown in the example box: 2 / He / 4.003 / Helium)

1																	VIII A
1 H 1.0080 Hydrogen																	**2** He 4.003 Helium

	I A	II A											III A	IV A	V A	VI A	VII A	
2	3 Li 6.939 Lithium	4 Be 9.012 Beryllium											5 B 10.81 Boron	6 C 12.011 Carbon	7 N 14.007 Nitrogen	8 O 15.9994 Oxygen	9 F 18.998 Fluorine	10 Ne 20.183 Neon
3	11 Na 22.990 Sodium	12 Mg 24.31 Magnesium	III B	IV B	V B	VI B	VII B	VIII B		I B	II B		13 Al 26.98 Aluminum	14 Si 28.09 Silicon	15 P 30.974 Phosphorus	16 S 32.064 Sulfur	17 Cl 35.453 Chlorine	18 Ar 39.948 Argon
4	19 K 39.102 Potassium	20 Ca 40.08 Calcium	21 Sc 44.96 Scandium	22 Ti 47.90 Titanium	23 V 50.94 Vanadium	24 Cr 52.00 Chromium	25 Mn 53.94 Manganese	26 Fe 55.85 Iron / 27 Co 58.93 Cobalt / 28 Ni 58.71 Nickel		29 Cu 63.54 Copper	30 Zn 65.37 Zinc		31 Ga 69.72 Gallium	32 Ge 72.59 Germanium	33 As 74.92 Arsenic	34 Se 78.96 Selenium	35 Br 79.909 Bromine	36 Kr 83.80 Krypton
5	37 Rb 85.47 Rubidium	38 Sr 87.62 Strontium	39 Y 88.91 Yttrium	40 Zr 91.22 Zirconium	41 Nb 92.91 Niobium	42 Mo 95.94 Molybdenum	43 Tc (99) Technetium	44 Ru 101.1 Ruthenium / 45 Rh 102.90 Rhodium / 46 Pd 106.4 Palladium		47 Ag 107.87 Silver	48 Cd 112.40 Cadmium		49 In 114.82 Indium	50 Sn 118.69 Tin	51 Sb 121.75 Antimony	52 Te 127.60 Tellurium	53 I 126.90 Iodine	54 Xe 131.30 Xenon
6	55 Cs 132.91 Cesium	56 Ba 137.34 Barium	57 TO 71	72 Hf 178.49 Hafnium	73 Ta 180.95 Tantalum	74 W 183.85 Tungsten	75 Re 186.2 Rhenium	76 Os 190.2 Osmium / 77 Ir 192.2 Iridium / 78 Pt 195.09 Platinum		79 Au 197.0 Gold	80 Hg 200.59 Mercury		81 Tl 204.37 Thallium	82 Pb 207.19 Lead	83 Bi 208.98 Bismuth	84 Po (210) Polonium	85 At (210) Astantine	86 Rn (222) Radon
7	87 Fr (223) Francium	88 Ra 226.05 Radium	89 TO 103															

57 La 138.91 Lanthanum	58 Ce 140.12 Cerium	59 Pr 140.91 Praseodymium	60 Nd 144.24 Neodymium	61 Pm (147) Promethium	62 Sm 150.35 Samarium	63 Eu 151.96 Europium	64 Gd 157.25 Gadolinium	65 Tb 158.92 Terbium	66 Dy 162.50 Dysprosium	67 Ho 164.93 Holmium	68 Er 167.26 Erbium	69 Tm 168.93 Thullium	70 Yb 173.04 Ytterbium	71 Lu 174.97 Lutetium
89 Ac (227) Actinium	90 Th 232.04 Thorium	91 Pa (231) Protactinium	92 U 238.03 Uranium	93 Np (237) Neptunium	94 Pu (242) Plutonium	95 Am (243) Americium	96 Cm (247) Curium	97 Bk (249) Berkelium	98 Cf (251) Californium	99 Es (254) Einsteinium	100 Fm (253) Fermium	101 Md (256) Mendelevium	102 No (254) Nobelium	103 Lw (257) Lawrencium

A-3

TABLE 3A Planetary Orbital Data

PLANET	SEMI-MAJOR AXIS (A.U.)	(10⁶ km)	ECCENTRICITY e	PERIHELION (A.U.)	(10⁶ km)	APHELION (A.U.)	(10⁶ km)
Mercury	0.39	57.9	0.206	0.31	46.0	0.47	69.8
Venus	0.72	108.2	0.007	0.72	107.5	0.73	108.9
Earth	1.00	149.6	0.017	0.98	147.1	1.02	152.1
Mars	1.52	227.9	0.093	1.38	206.6	1.67	249.2
Jupiter	5.20	778.4	0.048	4.95	740.7	5.46	816
Saturn	9.54	1427	0.054	9.02	1349	10.1	1504
Uranus	19.19	2871	0.047	18.3	2736	20.1	3006
Neptune	30.07	4498	0.009	29.8	4460	30.3	4537
Pluto	39.48	5906	0.249	29.7	4437	49.3	7376

PLANET	MEAN ORBITAL SPEED (km/s)	SIDEREAL PERIOD (tropical years)	SYNODIC PERIOD (days)	INCLINATION TO THE ECLIPTIC (degrees)	GREATEST ANGULAR DIAMETER AS SEEN FROM EARTH (arc seconds)
Mercury	47.87	0.24	115.88	7.00	13
Venus	35.02	0.62	583.92	3.39	64
Earth	29.79	1.00	—	0.01	—
Mars	24.13	1.88	779.94	1.85	25
Jupiter	13.06	11.86	398.88	1.31	50
Saturn	9.65	29.42	378.09	2.49	21
Uranus	6.80	83.75	369.66	0.77	4.1
Neptune	5.43	163.7	367.49	1.77	2.4
Pluto	4.74	248.0	366.72	17.2	0.11

TABLE 3B Planetary Physical Data

PLANET	EQUATORIAL RADIUS (km)	(Earth = 1)	MASS (kg)	(Earth = 1)	MEAN DENSITY (kg/m³)	SURFACE GRAVITY (Earth = 1)	ESCAPE SPEED (km/s)
Mercury	2,440	0.38	3.30×10^{23}	0.055	5430	0.38	4.2
Venus	6,052	0.95	4.87×10^{24}	0.82	5240	0.91	10.4
Earth	6,378	1.00	5.97×10^{24}	1.00	5520	1.00	11.2
Mars	3,394	0.53	6.42×10^{23}	0.11	3930	0.38	5.0
Jupiter	71,492	11.21	1.90×10^{27}	317.8	1330	2.53	60
Saturn	60,268	9.45	5.68×10^{26}	95.16	690	1.07	36
Uranus	25,559	4.01	8.68×10^{25}	14.54	1270	0.91	21
Neptune	24,766	3.88	1.02×10^{26}	17.15	1640	1.14	24
Pluto	1,137	0.18	1.27×10^{22}	0.0021	2060	0.07	1.2

PLANET	SIDEREAL ROTATION PERIOD (solar days)*	AXIAL TILT (degrees)	SURFACE MAGNETIC FIELD (Earth = 1)	MAGNETIC AXIS TILT (degrees relative to rotation axis)	ALBEDO†	SURFACE TEMPERATURE‡ (K)	NUMBER OF MOONS
Mercury	58.6	0.0	0.011	<10	0.11	100 to 700	0
Venus	−243.0	177.4	<0.001		0.65	730	0
Earth	0.9973	23.45	1.0	11.5	0.37	290	1
Mars	1.026	23.98	0.001		0.15	180 to 270	2
Jupiter	0.41	3.08	13.89	9.6	0.52	124	16
Saturn	0.44	26.73	0.67	0.8	0.47	97	18
Uranus	−0.72	97.92	0.74	58.6	0.50	58	17
Neptune	0.67	29.6	0.43	46.0	0.5	59	8
Pluto	−6.39	118	?		0.6	40 to 60	1

*A negative sign indicates retrograde rotation.
† Fraction of sunlight reflected from surface.
‡ Temperature is effective temperature for Jovian planets.

TABLE 4   The Twenty Brightest Stars

NAME	STAR	SPECTRAL TYPE* A	SPECTRAL TYPE* B	PARALLAX (arc seconds)	DISTANCE (pc)	APPARENT VISUAL MAGNITUDE* A	APPARENT VISUAL MAGNITUDE* B
Sirius	α CMa	A1V	wd†	0.37	2.7	−1.46	+8.7
Canopus	α Car	F0Ib-II		0.033	30	−0.72	
Alpha Centauri	α Cen	G2V	K0V	0.77	1.3	−0.01	+1.3
Arcturus	α Boo	K2III		0.091	11	−0.06	
Vega	α Lyr	A0V		0.13	8.0	+0.04	
Capella	α Aur	GIII	M1V	0.071	14	+0.05	+10.2
Rigel	β Ori	B8Ia	B9	—	250	+0.14	+6.6
Procyon	α CMi	F5IV-V	wd†	0.29	3.5	+0.37	+10.7
Betelgeuse	α Ori	M2Iab		—	150	+0.41	
Achernar	α Eri	B5V		0.050	20	+0.51	
Hadar	β Cen	B1III	?	0.011	90	+0.63	+4
Altair	α Aql	A7IV-V		0.20	5.1	+0.77	
Acrux	α Cru	B1IV	B3	0.008	120	+1.39	+1.9
Aldebaran	α Tau	K5III	M2V	0.063	16	+0.86	+13
Spica	α Vir	B1V		0.013	80	+0.91	
Antares	α Sco	MIIb	B4V	0.008	120	+0.92	+5.1
Pollux	β Gem	K0III		0.083	12	+1.16	
Formalhaut	α PsA	A3V		0.14	7.0	+1.19	+6.5
Deneb	α Cyg	A2Ia		—	430	+1.26	
Mimosa	β Cru	B1IV		—	150	+1.28	

NAME	VISUAL LUMINOSITY* (Sun = 1) A	VISUAL LUMINOSITY* (Sun = 1) B	ABSOLUTE VISUAL MAGNITUDE* A	ABSOLUTE VISUAL MAGNITUDE* B	PROPER MOTION (arc seconds/yr)	TRANSVERSE VELOCITY (km/s)	RADIAL VELOCITY (km/s)
Sirius	23.5	0.04	+1.4	+11.6	1.33	17.0	−7.6‡
Canopus	1510		−3.1		0.02	2.8	+20.5
Alpha Centauri	1.56	0.46	+4.4	+5.7	3.68	22.7	−24.6
Arcturus	115		−0.3		2.28	119	−5.2
Vega	55.0		+0.5		0.34	12.9	−13.9
Capella	166	0.01	−0.7	+9.5	0.44	29	+30.2‡
Rigel	4.6×10^4	126	−6.8	−0.4	0.00	1.2	+20.7‡
Procyon	7.7	0.0006	+2.6	+13.0	1.25	20.7	−3.2‡
Betelgeuse	1.4×10^4		−5.5		0.03	21	+21.0‡
Achernar	219		−1.0		0.10	9.5	+19
Hadar	3800	182	−4.1	−0.8	0.04	17	−12‡
Altair	11.5		+2.2		0.66	16	−26.3
Acrux	3470	2190	−4.0	−3.5	0.04	24	−11.2
Aldebaran	105	0.0014	−0.2	+12	0.20	15	+54.1
Spica	2400		−3.6		0.05	19	+1.0‡
Antares	5500	115	−4.5	−0.3	0.03	17	−3.2
Pollux	41.7		+0.8		0.62	35	+3.3
Formalhaut	13.8	0.10	+2.0	+7.3	0.37	12	+6.5
Deneb	5.0×10^4		−6.9		0.003	6	−4.6‡
Mimosa	6030		−4.6		0.05	36	

*A and B columns identify individual components of binary systems.
† "wd" stands for "white dwarf."
‡ Average value of variable velocity.

TABLE 5 The Twenty Nearest Stars

NAME	SPECTRAL TYPE*		PARALLAX (arc seconds)	DISTANCE (pc)	APPARENT VISUAL MAGNITUDE*	
	A	B			A	B
Sun	G2V				−26.72	
Proxima Centauri	M5		0.772	1.30	+11.05	
Alpha Centauri	G2V	K0V	0.750	1.33	−0.01	+1.33
Barnard's Star	M5V		0.545	1.83	+9.54	
Wolf 359	M8V		0.421	2.38	+13.53	
Lalande 21185	M2V		0.397	2.52	+7.50	
UV Ceti	M6V	M6V	0.387	2.58	+12.52	+13.02
Sirius	A1V	wd†	0.377	2.65	−1.46	+8.3
Ross 154	M5V		0.345	2.90	+10.45	
Ross 248	M6V		0.314	3.18	+12.29	
ε Eridani	K2V		0.303	3.30	+3.73	
Ross 128	M5V		0.298	3.36	+11.10	
61 Cygni	K5V	K7V	0.294	3.40	+5.22	+6.03
ε Indi	K5V		0.291	3.44	+4.68	
Grm 34	M1V	M6V	0.290	3.45	+8.08	+11.06
Luyten 789-6	M6V		0.290	3.45	+12.18	
Procyon	F5IV-V	wd†	0.285	3.51	+0.37	+10.7
Σ 2398	M4V	M5V	0.285	3.55	+8.90	+9.69
Lacaille 9352	M2V		0.279	3.58	+7.35	
G51-15	MV		0.278	3.60	+14.81	

NAME	VISUAL LUMINOSITY* (Sun = 1)		ABSOLUTE VISUAL MAGNITUDE*		PROPER MOTION (arc seconds/yr)	TRANSVERSE VELOCITY (km/s)	RADIAL VELOCITY (km/s)
	A	B	A	B			
Sun	1.0		+4.85				
Proxima Centauri	0.00006		+15.5		3.86	23.8	−16
Alpha Centauri	1.6	0.45	+4.4	+5.7	3.68	23.2	−22
Barnard's Star	0.00045		+13.2		10.34	89.7	−108
Wolf 359	0.00002		+16.7		4.70	53.0	+13
Lalande 21185	0.0055		+10.5		4.78	57.1	−84
UV Ceti	0.00006	0.00004	+15.5	+16.0	3.36	41.1	+30
Sirius	23.5	0.04	+1.4	+11.2	1.33	16.7	−8
Ross 154	0.00048		+13.3		0.72	9.9	−4
Ross 248	0.00011		+14.8		1.58	23.8	−81
ε Eridani	0.30		+6.1		0.98	15.3	+16
Ross 128	0.00036		+13.5		1.37	21.8	−13
61 Cygni	0.082	0.039	+7.6	+8.4	5.22	84.1	−64
ε Indi	0.14		+7.0		4.69	76.5	−40
Grm 34	0.0061	0.00039	+10.4	+13.4	2.89	47.3	+17
Luyten 789-6	0.00014		+14.6		3.26	53.3	−60
Procyon	7.65	0.00055	+2.6	+13.0	1.25	2.8	−3
Σ 2398	0.0030	0.0015	+11.2	+11.9	2.28	38.4	+5
Lacaille 9352	0.013		+9.6		6.90	117	+10
G51-15	0.00001		+17.0		1.26	21.5	—

*A and B columns identify individual components of binary systems.
† "wd" stands for "white dwarf."

GLOSSARY

A

A ring One of three Saturnian rings visible from Earth. The A ring is farthest from the planet and is separated from the B ring by the Cassini division. (p. 272)

absolute brightness The apparent brightness a star would have if it were placed at a standard distance of 10 parsecs from Earth. (p. 385)

absolute magnitude The apparent magnitude a star would have if it were placed at a standard distance of 10 parsecs from Earth. (p. 387)

absorption line Dark line in an otherwise continuous bright spectrum, where light within one narrow frequency range has been removed. (p. 80)

acceleration The rate of change of velocity of a moving object. (p. 45)

accretion Gradual growth of bodies, such as planets, by the accumulation of other, smaller, bodies. (p. 341)

accretion disk Flat disk of matter spiraling down onto the surface of a neutron star or black hole. Often, the matter originated on the surface of a companion star in a binary system. (p. 473)

active galaxies The most energetic galaxies, which can emit hundreds or thousands of times more energy per second than the Milky Way, mostly in the form of long-wavelength nonthermal radiation. (p. 576)

active optics Collection of techniques used to increase the resolution of ground-based telescopes. Minute modifications are made to the overall configuration of an instrument as its temperature and orientation change, to maintain the best possible focus at all times. (p. 109)

active region Region of the photosphere of the Sun surrounding a sunspot group, which can erupt violently and unpredictably. During sunspot maximum, the number of active regions is also a maximum. (p. 369)

active Sun The unpredictable aspects of the Sun's behavior, such as sudden explosive outbursts of radiation in the form of prominences and flares. (p. 364)

adaptive optics Technique used to increase the resolution of a telescope by deforming the shape of the mirror's surface under computer control while a measurement is being taken, to undo the effects of atmospheric turbulence. (p. 110)

amino acids Organic molecules which form the basis for building the proteins that direct metabolism in living creatures. (p. 649)

amplitude The maximum deviation of a wave above or below zero point. (p. 58)

angular momentum problem The fact that the Sun, which contains nearly all of the mass of the solar system, accounts for just 0.3 percent of the total angular momentum of the solar system. This is an aspect of the solar system that any acceptable formation theory must address. (p. 348)

angular resolution The ability of a telescope to distinguish between adjacent objects in the sky. (p. 104)

annular eclipse Solar eclipse occurring at a time when the Moon is far enough away from Earth that it fails to cover the disk of the Sun completely, leaving a ring of sunlight visible around its edge. (p. 19)

aphelion The point on the elliptical path of an object in orbit about the Sun that is most distant from the Sun (p. 38)

Apollo asteroid *See* Earth-crossing asteroid.

apparent brightness The brightness that a star appears to have, as measured by an observer on Earth. (p. 385)

apparent magnitude The apparent brightness of a star, expressed using the magnitude scale. (p. 387)

arc degree Unit of angular measure. There are 360 arc degrees in one complete circle. (p. 9)

association Small grouping of (typically 100 or less) bright stars, spanning up to a few tens of parses across, usually rich in very young stars. (p. 441)

asteroid One of thousands of very small members of the solar system orbiting the Sun between the orbits of Mars and Jupiter. Often referred to as "minor planets." (p. 314)

asteroid belt Region of the solar system, between the orbits of Mars and Jupiter, in which most asteroids are found. (p. 314)

asthenosphere Layer of Earth's interior, just below the lithosphere, over which the surface plates slide. (p. 163)

astronomical unit (A.U.) The average distance of Earth from the Sun. Precise radar measurements yield a value for the A.U. of 149,603,500 km. (p. 39)

astronomy Branch of science dedicated to the study of everything in the universe that lies above Earth's atmosphere. (p. 4)

asymptotic giant branch Path on the Hertzsprung–Russell diagram corresponding to the changes that a star undergoes after helium burning ceases in the core. At this stage, the carbon core shrinks and drives the expansion of the envelope, and the star becomes a swollen red giant for a second time. (p. 454)

atmosphere Layer of gas confined close to a planet's surface by the force of gravity. (p. 148)

atom Building block of matter, composed of positively charged protons and neutral neutrons in the nucleus, surrounded by negatively charged electrons. (p. 83)

aurora Event which occurs when atmospheric molecules are excited by incoming charged particles from the solar wind, then emit energy as they fall back to their ground states. Aurorae generally occur at high latitudes, near the north and south magnetic poles. (p. 160)

autumnal equinox Date on which the Sun crosses the celestial equator moving southward, occurring on or near September 22. (p. 11)

B

B ring One of three Saturnian rings visible from Earth. The B ring is the brightest of the three, and lies just past the Cassini division, closer to the planet than the A ring. (p. 272)

barred-spiral galaxy Spiral galaxy in which a bar of material passes through the center of the galaxy, with the spiral arms beginning near the ends of the bar. (p. 549)

baseline The distance between two observing locations used for the purposes of triangulation measurements. The larger the baseline, the better the resolution attainable. (p. 21)

belt Dark, low-pressure region in the atmosphere of a jovian planet, where gas flows downward. (p. 246)

Big Bang Event that cosmologists consider the beginning of the universe, in which all matter and radiation in the entire universe came into being. (p. 606)

binary-star system A system which consists of two stars in orbit about their common center of mass, held together by their mutual gravitational attraction. Most stars are found in binary-star systems. (p. 396)

blackbody curve The characteristic way in which the intensity of radiation emitted by a hot object depends of frequency. The frequency at which the emitted intensity is highest is an indication of the temperature of the radiating object. Also referred to as the Planck curve. (p. 66)

black dwarf The end-point of the evolution of an isolated, low-mass star. After the white dwarf stage, the star cools to the point where it is a dark "clinker" in interstellar space. (p. 459)

black hole A region of space where the pull of gravity is so great that nothing—not even light—can escape. A possible outcome of the evolution of a very massive star. (p. 503)

blue giant Large, hot, bright star at the upper left end of the main sequence on the Hertzsprung–Russell diagram. Its name comes from its color and size. (p. 392)

blueshift Motion-induced changed in the observed wavelength from a source that is moving toward us. Relative approaching motion between the object and the observer causes the wavelength to appear shorter (and hence bluer) than if there were no motion at all. (p. 72)

blue supergiant The very largest of the large, hot, bright stars at the uppermost left end of the main sequence on the Hertzsprung–Russell diagram. (p. 392)

Bohr model First theory of the hydrogen atom to explain the observed spectral lines. This model rests on three ideas: that there is a state of lowest energy for the electron, that there is a maximum energy, beyond which the electron is no longer bound to the nucleus, and that within these two energies the electron can only exist in certain energy levels. (p. 83)

brown dwarf Remnants of fragments of collapsing gas and dust that did not contain enough mass to initiate core nuclear fusion. Such objects are then frozen somewhere along their pre-main-sequence contraction phase, continually cooling into compact dark objects. Because of their small size and low temperature they are extremely difficult to detect observationally. (p. 433)

brown oval Feature of Jupiter's atmosphere that appears only at latitudes near 20 degrees N, this structure is a long-lived hole in the clouds that allows us to look down into Jupiter's lower atmosphere. (p. 250)

C

C ring One of three Saturnian rings visible from Earth. The C ring lies closest to the planet and is relatively thin compared to the A and B rings. (p. 272)

carbon-detonation supernova *See* type-I supernova.

Cassegrain telescope A type of reflecting telescope in which incoming light hits the primary mirror and is then reflected upward toward the prime focus, where a secondary mirror reflects the light back down through a small hole in the main mirror, into a detector or eyepiece. (p. 102)

Cassini Division A relatively empty gap in Saturn's ring system, discovered in 1675 by Giovanni Cassini. It is now known to contain a number of thin ringlets. (p. 272)

catastrophic theory A theory that invokes statistically unlikely accidental events to account for observations. (p. 340)

celestial coordinates Pair of quantities—right ascension and declination—similar to longitude and latitude on Earth, used to pinpoint locations of objects on the celestial sphere. (p. 10)

celestial equator The projection of Earth's equator onto the celestial sphere. (p. 9)

celestial sphere Imaginary sphere surrounding Earth, to which all objects in the sky were once considered to be attached. (p. 8)

center of mass The "average" position in space of a collection of massive bodies, weighted by their masses. For an isolated system this point moves with constant velocity, according to Newtonian mechanics. (p. 49)

Cepheid variable Star whose luminosity varies in a characteristic way, with a rapid rise in brightness followed by a slower decline. The period of a Cepheid variable star is related to its luminosity, so a determination of this period can be used to obtain an estimate of the star's distance. (p. 523)

chaotic rotation Unpredictable tumbling motion that non-spherical bodies in eccentric orbits, such as Saturn's satellite Hyperion, can exhibit. No amount of observation of an object rotating chaotically will ever show a well-defined period. (p. 283)

charge-coupled device (CCD) An electronic devise used for data acquisition, composed of many tiny pixels, each of which records a buildup of charge to measure the amount of light striking it. (p. 108)

chromatic aberration The tendency for a lens to focus red and blue light differently, causing images to become blurred. (p. 101)

chromosphere The Sun's lower atmosphere, lying just above the visible atmosphere. (p. 355)

closed universe Geometry that the universe as a whole would have if the density of matter is above the critical value. A closed universe is finite in extent, and has no edge, like the surface of a sphere. It has enough mass to stop the present expansion, and will eventually collapse. (p. 617)

cold dark matter Class of dark-matter candidates made up of very heavy particles, possibly formed in the very early universe. (p. 639)

collecting area The total area of a telescope capable of capturing incoming radiation. The larger the telescope, the greater its collecting area, and the fainter the objects it can detect. (p. 103)

color index A convenient method of quantifying a star's color by comparing its apparent brightness as measured through different filters. If the star's radiation is well described by a black body spectrum, the ratio of its blue intensity (B) to its visual intensity (V) is a measure of the object's surface temperature. (p. 389)

color-magnitude diagram A way of plotting stellar properties, in which absolute magnitude is plotted against color index. (p. 392)

coma An effect occurring during the formation of an off-axis image in a telescope. Stars whose light enters the telescope at a large angle acquire comet-like tails on their images. The brightest part of a comet, often referred to as the "head." (p. 319)

comet A small body, composed mainly of ice and dust, in an elliptical orbit about the Sun. As it comes close to the Sun, some of its material is vaporized to form a gaseous head and extended tail. (p. 319)

comparative planetology The systematic study of the similarities and differences among the planets, with the goal of obtaining deeper insight into how the solar system formed and has evolved in time. (p. 131)

condensation nuclei Dust grains in the interstellar medium which act as seeds around which other material can cluster. The presence of dust was very important in causing matter to clump during the formation of the solar system. (p. 341)

condensation theory Currently favored model of solar system formation which combines features of the old nebular theory with new information about interstellar dust grains, which acted as condensation nuclei. (p. 341)

conservation of mass and energy A fundamental law of modern physics which states that the sum of mass and energy must always remain constant in any physical process. In fusion reactions, the lost mass is converted into energy, primarily in the form of electromagnetic radiation. (p. 371)

constellation A human grouping of stars in the night sky into a recognizable pattern. (p. 6)

contact binary A binary star system in which both stars have expanded to fill their Roche lobes and the surfaces of the two stars merge. The binary system now consists of two nuclear burning stellar cores surrounded by a continuous common envelope. (p. 465)

continuous spectrum Spectrum in which the radiation is distributed over all frequencies, not just a few specific frequency ranges. A prime example is the black body radiation emitted by a hot, dense body. (p. 78)

convection Churning motion resulting from the constant upwelling of warm fluid and the concurrent downward flow of cooler material to take its place. (p. 150)

convection zone Region of the Sun's interior, lying just below the surface, where the material of the Sun is in constant convection motion. This region extends into the solar interior to a depth of about 20,000 km. (p. 355)

Copernican revolution The realization toward the end of the sixteenth century that Earth is not at the center of the universe. (p. 35)

core The central region of Earth, surrounded by the mantle. (p. 148)

core The central region of any planet or star. (p. 356)

core-collapse supernova *See* type-II supernova.

core hydrogen burning The energy burning stage for main sequence stars, in which the helium is produced by hydrogen fusion in the central region of the star. A typical star spends up to 90% of its lifetime in hydrostatic equilibrium brought about by the balance between gravity and the energy generated by core hydrogen burning. (p. 448)

corona One of numerous large, roughly circular regions on the surface of Venus, thought to have been caused by upwelling mantle material causing the planet's crust to bulge outward. (p. 212)

corona The tenuous outer atmosphere of the Sun, which lies just above the chromosphere, and at great distances turns into the solar wind. (p. 355)

coronal hole Vast regions of the Sun's atmosphere where the density of matter is about 10 times lower than average. The gas there streams freely into space at high speeds, escaping the Sun completely. (p. 364)

cosmic distance scale Collection of indirect distance-measurement techniques that astronomers use to measure distances in the universe. (p. 21)

cosmic evolution The collection of the seven major phases of the history of the universe, namely galactic, stellar, planetary, chemical, biological, cultural, and future evolution. (p. 648)

cosmic microwave background The almost perfectly isotropic radio signal that is the electromagnetic remnant of the Big Bang. (p. 619)

cosmological principle Two assumptions which make up the basis of cosmology, namely that the universe is homogeneous and isotropic on sufficiently large scales. (p. 605)

cosmological redshift The component of the redshift of an object which is due only to the Hubble flow of the universe. (p. 567)

cosmology The study of the structure and evolution of the entire universe. (pp. 32, 605)

crater Bowl-shaped depression on the surface of a planet or moon, resulting from a collision with interplanetary debris. (p. 180)

critical density The cosmic density corresponding to the dividing line between a universe that recollapses and one that expands forever. (p. 612)

critical universe Universe in which the density of matter is exactly equal to the critical density. The universe is infinite in extent, and has zero curvature. The expansion will continue forever, but approach an expansion speed of zero. (p. 618)

crust Layer of Earth which contains the solid continents and the seafloor. (p. 148)

D

D ring Collection of very faint, thin rings, extending from the inner edge of the C ring down nearly to the cloud tops of Saturn. This region contains so few particles that it is completely invisible from Earth. (p. 275)

dark dust cloud A large cloud, often many parsecs across, which contains gas and dust in a ration of about 1012 gas atoms for every dust particle. Typical densities are a few tens or hundreds of millions of particles per cubic meter. (p. 415)

dark halo Region of a galaxy beyond the visible halo where dark matter is believed to reside. (p. 537)

dark matter Term used to describe the mass in galaxies and clusters whose existence we infer from rotation curves and other techniques, but which has not been confirmed by observations at any electromagnetic wavelength. (p. 537)

declination Celestial coordinate used to measure latitude above or below the celestial equator on the celestial sphere. (p. 10)

decoupling Event in the early universe when atoms first formed, and after which photons could propagate freely through space. (p. 634)

deferent A construct of the geocentric model of the solar system which was needed to explain observed planetary motions. A deferent is a large circle encircling Earth, on which an epicycle moves. (p. 33)

degree *See* arc degree.

density A measure of the compactness of the matter within an object, computed by dividing the mass of the object by its volume. Units are kilograms per cubic meter (kg/m^3), or grams per cubic centimeter (g/cm^3). (p. 132)

deuteron An isotope of hydrogen in which a neutron is bound to the proton in the nucleus. Often called "heavy hydrogen" because of the extra mass of the neutron. (p. 371)

differential rotation The tendency for a gaseous sphere, such as a jovian planet or the Sun, to rotate at a different rate at the equator than at the poles. More generally, a condition where the angular speed varies with location within an object. (p. 245)

differentiation Variation in the density and composition of a body, such as Earth, with low density material on the surface and higher density material in the core. (p. 156)

diffraction The ability of waves to bend around corners. The diffraction of light establishes its wave nature. (p. 58)

Doppler effect Any motion-induced change in the observed wavelength (or frequency) of a wave. (p. 71)

Drake equation Expression that gives an estimate of the probability that intelligence exists elsewhere in the galaxy, based on a number of supposedly necessary conditions for intelligent life to develop. (p. 653)

dust grain An interstellar dust particle, roughly 10-8 m in size,, comparable to the wavelength of visible light. (p. 406)

dust lane A lane of dark, obscuring interstellar dust in an emission nebula or galaxy. (p. 413)

dust tail The component of a comet's tail that is composed of dust particles. (p. 320)

dwarf Any star with radius comparable to, or smaller than, that of the Sun (including the Sun itself) (p. 385)

dynamo theory Theory that explains planetary and stellar magnetic fields in terms of rotating, conducting material flowing in an object's interior. (p. 160)

E

E ring A faint ring, well outside the main ring system of Saturn, which was discovered by Voyager, and is believed to be associated with volcanism on the moon Enceladus. (p. 275)

Earth-crossing asteroid An asteroid whose orbit crosses that of Earth. Earth-crossing asteroids are also called Apollo asteroids, after the first of the type was discovered. (p. 316)

earthquake A sudden dislocation of rocky material near Earth's surface. (p. 153)

eccentricity A measure of the flatness of an ellipse, equal to the distance between the two foci divided by the length of the major axis. (p. 37)

eclipse Event during which one body passes in front of another, so that the light from the occulted body is blocked. (p. 16)

eclipse season Times of the year when the Moon lies in the same plane as Earth and Sun, so that eclipses are possible. (p. 20)

eclipsing binary Rare binary-star system that is aligned in such a way that from Earth we observe one star pass in front of the other, eclipsing the other star. (p. 397)

ecliptic The apparent path of the Sun, relative to the stars on the celestial sphere, over the course of a year. (p. 11)

electric field A field extending outward in all directions from a charged particle, such as a proton or an electron. The electric field determines the electric force exerted by the particle on all other charged particles in the universe; the strength of the electric field decreases with increasing distance from the charge according to an inverse-square law. (p. 60)

electromagnetic radiation Another term for light, electromagnetic radiation transfers energy and information from one place to another. (p. 56)

electromagnetic spectrum The complete range of electromagnetic radiation, from radio waves to gamma rays, including the visible spectrum. All types of electromagnetic radiation are basically the same phenomenon, differing only by wavelength, and all move at the speed of light. (p. 63)

electromagnetism The union of electricity and magnetism, which do not exist as independent quantities, but are in reality two aspects of a single physical phenomenon. (p. 60)

electron An elementary particle with a negative electric charge, one of the components of the atom. (p. 59)

electron degeneracy pressure The pressure produced by the resistance of electrons to further compression once they are squeezed to the point of contact. (p. 452)

element Matter made up of one particular atom. The number of protons in the nucleus of the atom determines which element it represents. (p. 87)

ellipse Geometric figure resembling an elongated circle. An ellipse is characterized by its degree of flatness, or eccentricity, and the length of its long axis. In general, bound orbits of objects moving under gravity are elliptical. (p. 37)

elliptical galaxy Category of galaxy in which the stars are distributed in an elliptical shape on the sky, ranging from highly elongated to nearly circular in appearance. (p. 550)

emission line Bright line in a specific location of the spectrum of radiating material, corresponding to emission of light at a certain frequency. A heated gas in a glass container produces emission lines in its spectrum. (p. 79)

emission nebula A glowing cloud of hot interstellar gas. The gas glows as a result of one or more nearby young stars which ionize the gas. Since the gas is mostly hydrogen, the emitted radiation falls predominantly in the red region of the spectrum, because of the hydrogen-alpha emission line. (p. 410)

emission spectrum The pattern of spectral emission lines, produced by an element. Each element has its own unique emission spectrum. (p. 80)

Encke gap A small gap in Saturn's A ring. (p. 272)

epicycle A construct of the geocentric model of the solar system which was necessary to explain observed planetary motions. Each planet rides on a small epicycle whose center in turn rides on a larger circle (the deferent). (p. 33)

escape speed The speed necessary for one object to escape the gravitational pull of another. Anything that moves away from a gravitating body with more than the escape speed will never return. (p. 49)

event horizon Imaginary spherical surface surrounding a collapsing star, with radius equal to the Schwarzschild radius, within which no event can be seen heard, or known about by an outside observer. (p. 504)

evolutionary theory A theory which explains observations in a series of gradual steps, explainable in terms of well-established physical principles. (p. 340)

evolutionary track A graphical representation of a star's life as a path on the Hertzsprung–Russell diagram. (p. 429)

excited state State of an atom when one of its electrons is in a higher energy orbital than the ground state. Atoms can become excited by absorbing a photon of a specific energy, or by colliding with a nearby atom. (p. 83)

extinction The dimming of starlight as it passes through the interstellar medium. (p. 407)

F

F ring Faint narrow outer ring of Saturn, discovered by *Pioneer 11* in 1979. The F ring lies just inside the Roche limit of Saturn, and was found by *Voyager 1* to be made up of several ring strands apparently braided together. (p. 276)

flare Explosive event occurring in or near an active region on the Sun. (p. 369)

flatness problem One of two conceptual problems with the Standard Big Bang model, which is that there is no natural way to explain why the density of the universe is so close to the critical density. (p. 635)

fluidized ejecta The ejecta blankets around some Martian craters, which apparently indicate that the ejected material was liquid at the time the crater formed. (p. 227)

focus One of two special points within an ellipse, whose separation from each other indicates the eccentricity. In a bound orbit, planets orbit in ellipses with the Sun at one focus. (p. 37)

forbidden line A spectral line seen in emission nebulae, but not seen in laboratory experiments because, under laboratory conditions, collisions kick the electron in question into some other state before emission can occur. (p. 414)

force Action on an object that causes its momentum to change. The rate at which the momentum changes is numerically equal to the force. (p. 45)

fragmentation The breaking up of a large object into many smaller pieces (for example, as the result of high-speed collisions between planetesimals and protoplanets in the early solar system). (p. 342)

Fraunhofer lines The collection of over 600 absorption lines in the spectrum of the Sun, first categorized by Joseph Fraunhofer in 1812. (p. 81)

frequency The number of wave crests passing any given point in a unit time. (p. 58)

full Moon Phase of the Moon in which it appears as a complete circular disk in the sky. (p. 14)

fusion Mechanism of energy generation in the core of the Sun, in which light nuclei are combined, or fused, into heavier ones, releasing energy in the process. (p. 371)

G

galactic bulge Thick distribution of warm gas and stars around the galactic center. (p. 520)

galactic cannibalism A galaxy merger in which a larger galaxy consumes a smaller one. (p. 564)

galactic center The center of the Milky Way, or any other, galaxy. The point about which the disk of a spiral galaxy rotates. (p. 526)

galactic disk Flattened region of gas and dust that bisects the galactic halo in a spiral galaxy. This is the region of active star formation. (p. 520)

galactic halo Region of a galaxy extending far above and below the galactic disk, where globular clusters and other old stars reside. (p. 520)

galactic nucleus Small central high-density region of a galaxy. Almost all the radiation from active galaxies is generated within the nucleus. (p. 577)

galaxy Gravitationally bound collection of a large number of stars. The Sun is a star in the Milky Way Galaxy. (p. 520)

galaxy cluster A collection of galaxies held together by their mutual gravitational attraction. (p. 557)

Galilean satellites The four brightest and largest moons of Jupiter (Io, Europa, Ganymede, Callisto), named after Galileo Galilei, the 17th century astronomer who first observed them. (p. 244)

gamma ray Region of the electromagnetic spectrum, far beyond the visible spectrum, corresponding to radiation of very high frequency and very short wavelength. (p. 56)

gamma-ray burst Object that radiates tremendous amounts of energy in the form of gamma rays, possibly due to the collision and merger of two neutron stars initially in orbit around one another. (p. 498)

geocentric model A model of the solar system which holds that Earth is at the center of the universe and all other bodies are in orbit around it. The earliest theories of the solar system were geocentric. (p. 32)

giant A star with a radius between 10 and 100 times that of the Sun. (p. 385)

globular cluster Tightly bound, roughly spherical collection of hundreds of thousands, and sometimes millions, of stars spanning about 50 parsecs. Globular clusters are distributed in the halos around the Milky Way and other galaxies. (p. 400)

Grand Unified Theories Class of theories describing the behavior of the single force that results from unification of the strong, weak, and electromagnetic forces in the early universe. (p. 629)

granulation Mottled appearance of the solar surface, caused by rising (hot) and falling (cool) material in convective cells just below the photosphere. (p. 359)

gravitational field Field created by any object with mass, extending out in all directions, which determines the influence of that object on all others. The strength of the gravitational field decreases as the square of the distance. (p. 47)

gravitational lensing The effect induced on the image of a distant object by a massive foreground object. Light from the distant object is bent into two or more separate images. (p. 538)

gravitational red shift A prediction of Einstein's general theory of relativity. Photons lose energy as they escape the gravitational field of a massive object. Because a photon's energy is proportional to its frequency, a photon that loses energy suffers a decrease in frequency, which corresponds to an increase, or redshift, in wavelength. (p. 510)

gravity The attractive effect that any massive object has on all other massive objects. The greater the mass of the object, the stronger its gravitational pull. (p. 46)

Great Dark Spot Prominent storm system in the atmosphere of Neptune, located near the equator of the planet. The system is comparable in size to Earth. (p. 293)

Great Red Spot A large, high-pressure, long-lived storm system visible in the atmosphere of Jupiter. The Red Spot is roughly twice the size of Earth. (p. 246)

greenhouse effect The partial trapping of solar radiation by a planetary atmosphere, similar to the trapping of heat in a greenhouse. (p. 152)

ground state The lowest energy state that an electron can have within an atom. (p. 83)

H

heliocentric model A model of the solar system which is centered on the Sun, with Earth in motion about the Sun. (p. 34)

helioseismology The study of conditions far below the Sun's surface through the analysis of internal "sound" waves that repeatedly cross the solar interior. (p. 356)

helium capture The formation of heavy elements by the capture of a helium nucleus. For example, carbon can form heavier elements by fusion with other carbon nuclei, but it is much more likely to occur by helium capture, which requires less energy. (p. 484)

helium flash An explosive event in the post-main-sequence evolution of a low-mass star. When helium fusion begins in a dense stellar core, the burning is explosive in nature. It continues until the energy released is enough to expand the core, at which point the star achieves stable equilibrium again. (p. 453)

helium precipitation Mechanism responsible for the low abundance of helium of Saturn's atmosphere. Helium condenses in the upper layers to form a mist, which rains down toward Saturn's interior, just as water vapor forms into rain in the atmosphere of Earth. (p. 271)

Hertzsprung–Russell diagram A plot of luminosity against temperature (or spectral class) for a group of stars. (p. 391)

high-energy telescope Telescope designed to detect X and gamma radiation (p. 120)

highlands Relatively light-colored regions on the surface of the Moon which are elevated several kilometers above the maria. Also called terrae. (p. 180)

homogeneity Assumed property of the universe such that the number of galaxies in an imaginary large cube of the universe is the same no matter where in the universe the cube is placed. (p. 605)

horizon problem One of two conceptual problems with the standard Big Bang model, which is that some regions of the universe which have very similar properties are too far apart to have exchanged information in the age of the universe. (p. 635)

horizontal branch Region of the Hertzsprung–Russell diagram where post-main sequence stars again reach hydrostatic equilibrium. At this point, the star is burning helium in its core, and hydrogen in a shell surrounding the core. (p. 453)

hot dark matter A class of candidates for the dark matter in the universe, composed of lightweight particles, such as neutrinos, much less massive than the electron. (p. 639)

Hubble Classification scheme Method of classifying galaxies according to their appearance, developed by Edwin Hubble. (p. 548)

Hubble's constant The constant of proportionality which gives the relation between recessional velocity and distance in Hubble's law. (p. 568)

Hubble's law Law that relates the observed velocity of recession of a galaxy to its distance from us. The velocity of recession of a galaxy is directly proportional to its distance away. (p. 567)

hydrogen envelope An invisible sheath of gas engulfing the coma of a comet, usually distorted by the solar wind, and extending across millions of kilometers of space. (p. 319)

hydrogen shell burning Fusion of hydrogen in a shell that is driven by contraction and heating of the helium core. Once hydrogen is depleted in the core of a star, hydrogen burning stops and the core contracts due to gravity, causing the temperature to rise, heating the surrounding layers of hydrogen in the star, and increasing the burning rate there. (p. 449)

hydrosphere Layer of Earth which contains the liquid oceans and accounts for roughly 70 percent of Earth's total surface area. (p. 148)

I

image The optical representation of an object produced when light from the object is reflected or refracted by a mirror or lens. (p. 98)

inertia The tendency of an object to continue moving at the same speed and in the same direction, unless acted upon by a force. (p. 45)

inflation Short period of unchecked cosmic expansion early in the history of the universe. During inflation, the universe swelled in size by a factor of about 10^{50}. (p. 636)

infrared Region of the electromagnetic spectrum just outside the visible range, corresponding to light of a slightly longer wavelength than red light. (p. 56)

infrared telescope Telescope designed to detect infrared radiation. Many such telescopes are designed to be lightweight so that they can be carried above (most of) Earth's atmosphere by balloons, airplanes, or satellites. (p. 117)

inner core The central part of Earth's core, believed to be solid, and composed mainly of nickel and iron. (p. 155)

intensity A basic property of electromagnetic radiation that specifies the amount or strength of the radiation. (p. 65)

intercrater plains Regions on the surface of Mercury that do not show extensive cratering, but are relatively smooth. (p. 193)

interference The ability of two or more waves to interact in such a way that they either reinforce or cancel each other. (p. 59)

interferometer Collection of two or more telescopes working together as a team, observing the same object at the same time and at the same wavelength. The effective diameter of an interferometer is equal to the distance between its outermost telescopes. (p. 116)

interferometry Technique in widespread use to dramatically improve the resolution of radio and infrared maps. Several telescopes observe the object simultaneously, and a computer analyzes how the signals interfere with each other. (p. 116)

interstellar dust Microscopic dust grains that populate the space between stars, having their origins in the ejected matter of long-dead stars. (p. 341)

interstellar medium The matter between stars, composed of two components, gas and dust, intermixed throughout all of space. (p. 406)

inverse-square law The law that a field follows if its strength decreases with the square of the distance. Fields that follow the inverse square law decrease rapidly in strength as the distance increases, but never quite reach zero. (p. 47)

Io plasma torus Doughnut-shaped region of energetic ionized particles, emitted by the volcanoes on Jupiter's moon Io, and swept up by Jupiter's magnetic field. (p. 257)

ion Atom that has had at least one of its electrons removed. (p. 83)

ion tail Thin stream of ionized gas that is pushed away from the head of a comet by the solar wind. It extends directly away from the Sun. Often referred to as a plasma tail. (p. 320)

ionosphere Layer in Earth's atmosphere above about 100 km where the atmosphere is significantly ionized and conducts electricity. (p. 148)

irregular galaxy A galaxy which does not fit into any of the other major categories in the Hubble classification scheme. (p. 551)

isotopes Nuclei containing the same number of protons but different numbers of neutrons. Most elements can exist in several isotopic forms. A common example of an isotope is deuterium, which differs from normal hydrogen by the presence of an extra neutron in the nucleus. (p. 372)

isotropy Assumed property of the universe such that the universe looks the same in every direction. (p. 605)

J

jovian planet One of the four giant outer planets of the solar system, resembling Jupiter in physical and chemical composition. (p. 135)

K

Kepler's Laws of Planetary Motion Three laws, based on precise observations of the motions of the planets by Tycho Brahe, which summarize the motions of the planets about the Sun. (p. 37)

Kirchhoff's laws Three rules governing the formation of different types of spectra. (p. 81)

Kirkwood gaps Gaps in the spacings of orbital semi-major axes of asteroids in the asteroid belt, produced by dynamical resonances with nearby planets, especially Jupiter. (p. 318)

Kuiper belt A region in the plane of the solar system outside the orbit of Neptune where most short-period comets are thought to originate. (p. 321)

L

Lagrange point One of five special points in the plane of two massive bodies orbiting one another, where a third body of negligible mass can remain in equilibrium. (p. 283)

lava dome Volcanic formation formed when lava oozes out of fissures in a planet's surface, creating the dome, and then withdraws, causing the crust to crack and subside. (p. 212)

light *See* electromagnetic radiation.

light curve The variation in brightness of a star with time. (p. 397)

lighthouse model The leading explanation for pulsars. A small region of the neutron star, near one of the magnetic poles, emits a steady stream of radiation which sweeps past Earth each time the star rotates. The period of the pulses is the star's rotation period. (p. 496)

light year The distance that light, moving at a constant speed of 300,000 km/s, travels in one year. One light year is about 10 trillion kilometers. (p. 4)

line of nodes The line of intersection of the Moon's orbit with the ecliptic plane. (p. 20)

lithosphere Earth's crust and a small portion of the upper mantle that make up Earth's plates. This layer of Earth undergoes tectonic activity. (p. 163)

Local Group The small galaxy cluster that includes the Milky Way Galaxy. (p. 556)

luminosity One of the basic properties used to characterize stars, luminosity is defined as the total energy radiated by star each second, at all wavelengths. (p. 356)

luminosity class A classification scheme which groups stars according to the width of their spectral lines. For a group of stars with the same temperature, luminosity class differentiates between supergiants, giants, main-sequence stars and subdwarfs. (p. 395)

lunar eclipse Celestial event during which the moon passes through the shadow of Earth, temporarily darkening its surface. (p. 16)

lunar phase The appearance of the Moon at different points along its orbit. (p. 14)

M

Magellanic Clouds Two small irregular galaxies that are gravitationally bound to the Milky Way Galaxy. (p. 552)

magnetic field Field which accompanies any changing electric field, and governs the influence of magnetized objects on one another. (p. 60)

magnetosphere A zone of charged particles trapped by a planet's magnetic field, lying above the atmosphere. (p. 148)

magnitude scale A system of ranking stars by apparent brightness, developed by the Greek astronomer Hipparchus. Originally, the brightest stars in the sky were categorized as being of first magnitude, while the faintest stars visible to the naked eye were classified as sixth magnitude. The scheme has since been extended to cover stars and galaxies too faint to be seen by the unaided eye. Increasing magnitude means fainter stars, and a difference of 5 magnitudes corresponds to a factor of 100 in apparent brightness. (p. 386)

main sequence Well-defined band on the Hertzsprung–Russell diagram, on which most stars are found, running from the top left of the diagram to the bottom right. (p. 392)

main-sequence turnoff Special point on the Hertzsprung–Russell diagram for a cluster, indicative of the cluster's age. If all the stars in the cluster are plotted, the lower mass stars will trace out the main sequence up to the point where stars begin to evolve off the main sequence toward the red giant branch. The point where stars are just beginning to evolve off is the main-sequence turnoff. (p. 461)

mantle Layer of Earth just interior to the crust. (p. 148)

mare Relatively dark-colored and smooth region on the surface of the Moon. (plural: maria) (p. 180)

mass A measure of the total amount of matter contained within an object. (p. 45)

mass-luminosity relation The dependence of the luminosity of a main-sequence star on its mass. The luminosity increases roughly as the mass raised to the third power. (p. 398)

mass-radius relation The dependence of the radius of a main sequence star on its mass. The radius rises roughly in proportion to the mass. (p. 398)

mass-transfer binary *See* semi-detached binary.

matter-dominated universe A universe in which the density of matter exceeds the density of radiation. The present-day universe is matter-dominated. (p. 626)

mesosphere Region of Earth's atmosphere lying between the stratosphere and the ionosphere, 50–80 km above Earth's surface. (p. 148)

meteor Bright streak in the sky, often referred to as a "shooting star," resulting from a small piece of interplanetary debris entering Earth's atmosphere and heating air molecules, which emit light as they return to their ground states. (p. 326)

meteor shower Event during which many meteors can be seen each hour, caused by the yearly passage of Earth through the debris spread along the orbit of a comet. (p. 327)

meteorite Any part of a meteoroid that survives passage through the atmosphere and lands on the surface of Earth. (p. 326)

meteoroid Chunk of interplanetary debris prior to encountering Earth's atmosphere. (p. 326)

meteoroid swarm Pebble-sized cometary fragments dislodged from the main body, moving in nearly the same orbit as the parent comet. (p. 326)

micrometeoroid Relatively small chunks of interplanetary debris ranging from dust particle size to pebble-sized fragments. (p. 327)

Milky Way Galaxy The spiral galaxy in which the Sun resides. The disk of our Galaxy is visible in the night sky as the faint band of light known as the Milky Way. (p. 520)

millisecond pulsar A pulsar whose period indicates that the neutron star is rotating nearly 1000 times each second. The most likely explanation for these rapid rotators is that the neutron star has been spun up by drawing in matter from a companion star. (p. 500)

molecular cloud A cold, dense interstellar cloud which contains a high fraction of molecules. It is widely believed that the relatively high density of dust particles in these clouds plays an important role in the formation and preservation of the molecules. (p. 419)

molecular cloud complex Collection of molecular clouds that spans as much as 50 parsecs and may contain enough material to make millions of Sun-sized stars. (p. 420)

molecule A tightly bound collection of atoms held together by the atoms' electromagnetic fields. Molecules, like atoms, emit and absorb photons at specific wavelengths. (p. 88)

moon A small body in orbit about a planet. (p. 130)

N

nebula General term used for any "fuzzy" patch on the sky, either light or dark. (p. 410)

nebular theory One of the earliest models of solar system formation, dating back to Descartes, in which a large cloud of gas began to collapse under its own gravity to form the Sun and planets. (p. 339)

neutrino Virtually massless and chargeless particle that is one of the products of fusion reactions in the Sun. Neutrinos move at close to the speed of light, and interact with matter hardly at all. (p. 372)

neutrino oscillations Possible solution to the solar neutrino problem, in which the neutrino has a very tiny mass. In this case, the correct number of neutrinos can be produced in the solar core, but on their way to Earth, some can "oscillate," or become transformed into other particles, and thus go undetected. (p. 375)

neutron An elementary particle with roughly the same mass as a proton, but which is electrically neutral. Along with protons, neutrons from the nuclei of atoms. (p. 87)

neutron capture The primary mechanism by which very massive nuclei are formed in the violent aftermath of a supernova. Instead of fusion of like nuclei, heavy elements are created by the addition of more and more neutrons to existing nuclei. (p. 485)

neutron star A dense ball of neutrons that remains at the core of a star after a supernova explosion has destroyed the rest of the star. Typical neutron stars are about 20 km across, and contain more mass than the Sun. (p. 494)

new Moon Phase of the moon during which none of the lunar disk is visible. (p. 14)

Newtonian mechanics The basic laws of motion, postulated by Newton, which are sufficient to explain and quantify virtually all of the complex dynamical behavior found on Earth and elsewhere in the universe. (p. 45)

Newtonian telescope A reflecting telescope in which incoming light is intercepted before it reaches the prime focus and is deflected into an eyepiece at the side of the instrument. (p. 102)

north celestial pole Point on the celestial sphere directly above Earth's north pole. (p. 9)

nova A star that suddenly increases in brightness, often by a factor of as much as 10,000, then slowly fades back to its original luminosity. A nova is the result of an explosion on the surface of a white dwarf star, caused by matter falling onto its surface from the atmosphere of a binary companion. (p. 472)

nuclear fusion *See* fusion.

nucleotide base An organic molecule, the building block of genes that pass on hereditary characteristics from one generation of living creatures to the next. (p. 649)

nucleus Dense, central region of an atom, containing both protons and neutrons, and orbited by one or more electrons. (p. 83)

nucleus The solid region of ice and dust that composes the central region of the head of a comet. (p. 319)

O

Olber's paradox A thought experiment suggesting that if the universe were homogeneous, infinite, and unchanging, the entire night sky would be as bright as the surface of the Sun. (p. 606)

Oort cloud Spherical halo of material surrounding the solar system, out to a distance of about 50,000 A.U. where most comets reside. (p. 321)

opacity A quantity that measures a material's ability to block electromagnetic radiation. Opacity is the opposite of transparency. (p. 65)

open cluster Loosely bound collection of tens to hundreds of stars, a few parsecs across, generally found in the plane of the Milky Way. (p. 399)

open universe Geometry that the universe would have if the density of matter were less than the critical value. In an open universe there is not enough matter to halt the expansion of the universe. An open universe is infinite in extent. (p. 618)

outer core The outermost part of Earth's core, believed to be liquid and composed mainly of nickel and iron. (p. 155)

outflow channel Surface feature on Mars, evidence that liquid water once existed there in great quantity, believed to be the relics of catastrophic flooding about 3 billion years ago. Found only in the equatorial regions of the planet. (p. 230)

ozone layer Layer of Earth's atmosphere at an altitude of 20 to 50 km where incoming ultraviolet solar radiation is absorbed by oxygen, ozone, and nitrogen in the atmosphere. (p. 150)

P

pair production Process in which two photons of electromagnetic radiation give rise to a particle-antiparticle pair. (p. 627)

parallax The apparent motion of a relatively close object with respect to a more distant background as the location of the observer changes. (p. 22)

parsec The distance at which a star must lie in order for its measured parallax to be exactly 1 arc second; 1 parsec equals 206,000 A.U. (p. 380)

partial eclipse Celestial event during which only a part of the occulted body is blocked from view. (p. 16)

penumbra Portion of the shadow cast by an eclipsing object in which the eclipse is seen as partial. (p. 17)

penumbra The outer region of a sunspot, surrounding the umbra, which is not as dark, and not as cool as the central region. (p. 364)

perihelion The closest approach to the Sun of any object in orbit about it. (p. 38)

period The time needed for an orbiting body to complete one revolution about another body. (p. 39)

period-luminosity relation A relation between the pulsation period of a Cepheid variable and its absolute brightness. Measurement of the pulsation period allows the distance of the star to be determined. (p. 525)

permafrost Layer of permanently frozen water ice believed to lie just under the surface of Mars. (p. 229)

photoelectric effect Experiment concerning the detection of electrons from a metal surface, whose speed off the surface was dependent on the frequency of light striking the surface. The theoretical explanation rests on viewing light as made up of photons, or individual "bullets" of energy. (p. 84)

photometer A device that measures the total amount of light received in all or part of the image. (p. 103)

photometry Branch of observational astronomy in which the brightness of a source is measured through each of a set of standard filters. (p. 389)

photon Individual packet of electromagnetic energy that makes up electromagnetic radiation. (p. 85)

photosphere The visible surface of the Sun, lying just above the uppermost layer of the Sun's interior, and just below the chromosphere. (p. 355)

pixel One of many tiny picture elements, organized into an array, making up a digital image. (p. 108)

Planck curve *See* blackbody curve.

planet One of nine major bodies that orbit the Sun, visible to us by reflected sunlight. (p. 130)

planetary nebula The ejected envelope of a red giant star, spread over a volume roughly the size of our solar system. (p. 456)

planetary ring system Material organized into thin, flat rings encircling a giant planet, such as Saturn. (p. 272)

planetesimal Term given to objects in the early solar system that had reached the size of small moons, at which point their gravitational fields were strong enough to begin to influence their neighbors. (p. 341)

plate tectonics The motions of regions of Earth's lithosphere, which drift with respect to one another. Also known as continental drift. (p. 162)

polarization The alignment of the electric fields of emitted photons, which are generally emitted with random orientations. (p. 409)

positron Atomic particle with properties identical to those a negatively charged electron, except for its positive charge. This positron is the antiparticle of the electron. Positrons and electrons annihilate one another when they meet, producing pure energy in the form of gamma rays. (p. 372)

precession The slow change in the direction of the rotation axis of a spinning object, caused by some external gravitational influence. (p. 14)

primary atmosphere The chemical components that would have formed Earth's atmosphere. (p. 153)

prime focus The point in a reflecting telescope where the mirror focuses incoming light to a point. (p. 98)

primordial nucleosynthesis The production of elements heavier than hydrogen by nuclear fusion in the high temperatures and densities which existed in the early universe. (p. 632)

Principle of Cosmic Censorship A proposition to separate the unexplained physics near a singularity from the rest of the well-behaved universe. The principle states that nature always hides any singularity, such as a black hole, inside an event horizon, which insulates the rest of the universe from seeing it. (p. 512)

prominence Loop or sheet of glowing gas ejected from an active region on the solar surface, which then moves through the inner parts of the corona under the influence of the Sun's magnetic field. (p. 369)

proper motion The angular movement of a star across the sky, as seen from Earth, measured in seconds of arc per year. This movement is a result of the star's actual motion through space. (p. 382)

proton An elementary particle, carrying a positive electric charge, a component of all atomic nuclei. The number of protons in the nucleus of an atom dictates what type of atom it is. (p. 59)

proton–proton chain The chain of fusion reactions, leading from hydrogen to helium, that powers main-sequence stars. (p. 373)

protoplanet Clump of material, formed in the early stages of solar system formation, that was the forerunner of the planets we see today. (p. 341)

protostar Stage in star formation when the interior of a collapsing fragment of gas is sufficiently hot and dense that it becomes opaque to its own radiation. The protostar is the dense region at the center of the fragment. (p. 429)

protosun The central accumulation of material in the early stages of solar system formations, the forerunner of the present-day Sun. (p. 341)

Ptolemaic model Geocentric solar system model, developed by the second century astronomer Claudius Ptolemy. It predicted with great accuracy the positions of the then known planets. (p. 33)

pulsar Object that emits radiation in the form of rapid pulses with a characteristic pulse period and duration. Charged particles, accelerated by the magnetic field of a rapidly rotating neutron star, flow along the magnetic field lines, producing radiation that beams outward as the star spins on its axis. (p. 495)

pulsating variable star A star whose luminosity varies in a predictable, periodic way. (p. 523)

Q

quantization The fact that light and matter on small scales behave in a discontinuous manner, and manifest themselves in the form of tiny "packets" of energy, called quanta. (p. 83)

quarter Moon Lunar phase in which the Moon appears as a half disk. (p. 14)

quasar Star-like radio source with an observed redshift that indicates an extremely large distance from Earth. (p. 587)

quasi-stellar object (QSO) *See* quasar.

quiet Sun The underlying predictable elements of the Sun's behavior, such as its average photospheric temperature, which do not change in time. (p. 364)

R

radar Acronym for RAdio Detection And Ranging. Radio waves are bounced off an object, and the time taken for the echo to return indicates its distance. (p. 44)

radial motion Motion along a particular line of sight, which induces apparent changes in the wavelength (or frequency) of radiation received. (p. 71)

radiation A way in which energy is transferred from place to place in the form of a wave. Light is a form of electromagnetic radiation. (p. 56)

radiation darkening The effect of chemical reactions that result when high-energy particles strike the icy surfaces of objects in the outer solar system. The reactions lead to a build-up of a dark layer of material. (p. 296)

radiation-dominated universe Early epoch in the universe, when the equivalent density of radiation in the cosmos exceeded the density of matter. (p. 626)

radiation zone Region of the Sun's interior where extremely high temperatures guarantee that the gas is completely ionized. Photons only occasionally interact with electrons, and travel through this region with relative ease. (p. 355)

radio Region of the electromagnetic spectrum corresponding to radiation of the longest wavelengths. (p. 56)

radio galaxy Type of active galaxy that emits most of its energy in the form of long-wavelength radiation. (p. 578)

radio lobe Roundish extended region of radio-emitting gas, lying well beyond the center of a radio galaxy. (p. 580)

radio telescope Large instrument designed to detect radiation from space at radio wavelengths. (p. 110)

radioactivity The release of energy by rare, heavy elements when their nuclei decay into lighter nuclei. (p. 157)

radius-luminosity-temperature relationship A mathematical proportionality, arising from Stefan's Law, which allows astronomers to indirectly determine the radius of a star once its luminosity and temperature are known. (p. 383)

red dwarf Small, cool faint star at the lower-right end of the main sequence on the Hertzsprung–Russell diagram. (p. 392)

red giant A giant star whose surface temperature is relatively low, so that it glows red. (p. 385)

red-giant branch The section of the evolutionary track of a star corresponding to intense hydrogen shell burning, which drives a steady expansion and cooling of the outer envelope of the star. As the star gets larger in radius and its surface temperature cools, it becomes a red giant. (p. 451)

red giant region The upper right hand corner of the Hertzsprung–Russell diagram, where red-giant stars are found. (p. 393)

redshift Motion-induced change in the wavelength of light emitted from a source moving away from us. The relative recessional motion causes the wave to have an observed wavelength longer (and hence redder) than it would if it were not moving. (p. 72)

red supergiant An extremely luminous red star. Often found on the asymptotic giant branch of the Hertzsprung–Russell diagram. (p. 454)

reddening Dimming of starlight by interstellar matter, which tends to scatter higher-frequency (blue) components of the radiation more efficiently than the lower-frequency (red) components. (p. 407)

reflecting telescope A telescope which uses a mirror to gather and focus light from a distant object. (p. 98)

refracting telescope A telescope which uses a lens to gather and focus light from a distant object. (p. 98)

refraction The tendency of a wave to bend as it passes from one transparent medium to another. (p. 98)

residual cap Portion of Martian polar ice caps that remains permanently frozen, undergoing no seasonal variations. (p. 232)

retrograde motion Backward, westward loop traced out by a planet with respect to the fixed stars. (p. 32)

revolution Orbital motion of one body about another, such as Earth about the Sun. (p. 11)

right ascension Celestial coordinate used to measure longitude on the celestial sphere. The zero point is the position of the Sun at the vernal equinox. (p. 10)

rille A ditch on the surface of the Moon where molten lava flowed in the past. (p. 192)

ringlet Narrow region in Saturn's planetary ring system where the density of ring particles is high. Voyager discovered that the rings visible from Earth are actually composed of tens of thousands of ringlets. (p. 274)

Roche limit Often called the tidal stability limit, the Roche limit gives the distance from a planet at which the tidal force, due to the planet, between adjacent objects exceeds their mutual attraction. Objects within this limit are unlikely to accumulate into larger objects. The rings of Saturn occupy the region within Saturn's Roche limit. (p. 274)

Roche lobe An imaginary surface around a star. Each star in a binary system can be pictured as being surrounded by a tear-shaped zone of gravitational influence, the Roche lobe. Any material within the Roche lobe of a star can be considered to be part of that star. During evolution, one member of the binary star can expand so that it overflows its own Roche lobe, and begins to transfer matter onto the other star. (p. 465)

rotation Spinning motion of a body about an axis. (p. 8)

rotation curve Plot of the orbital speed of disk material in a galaxy against its distance from the galactic center. Analysis of rotation curves of spiral galaxies indicates the existence of dark matter. (p. 536)

RR Lyrae star Variable star whose luminosity changes in a characteristic way. All RR Lyrae stars have more or less the same average luminosity. (p. 523)

runaway greenhouse effect A process in which the heating of a planet leads to an increase in its atmosphere's ability to retain heat and thus to further heating, causing extreme changes in the temperature of the surface and the composition of the atmosphere. (p. 218)

runoff channel River-like surface feature on Mars, evidence that liquid water once existed there in great quantities. They are found in the southern highlands, and are thought to have been formed by water that flowed nearly 4 billion years ago. (p. 230)

S

S0 galaxy Galaxy which shows evidence of a thin disk and a bulge, but which has no spiral arms and contains little or no gas. (p. 551)

SB0 galaxy An S0-type galaxy whose disk shows evidence of a bar. (p. 551)

scarp Surface feature on Mercury believed to be the result of cooling and shrinking of the crust forming a wrinkle on the face of the planet. (p. 193)

Schwarzschild radius The distance from the center of an object such that, if all the mass compressed within that region, the escape speed would equal the speed of light. Once a stellar remnant collapses within this radius, light cannot escape and the object is no longer visible. (p. 504)

scientific method The set of rules used to guide science, based on the idea that scientific "laws" be continually tested, and modified or replaced if found inadequate. (p. 42)

seasonal cap Portion of Martian polar ice caps that is subject to seasonal variations, growing and shrinking once each Martian year. (p. 232)

seasons Changes in average temperature and length of day that result from the tilt of Earth's (or any planet's) axis with respect to the plane of its orbit. (p. 11)

secondary atmosphere The chemicals that composed Earth's atmosphere after the planet's formation, once volcanic activity outgassed chemicals from the interior. (p. 153)

seeing A term used to describe the ease with which good telescopic observations can be made from Earth's surface, given the blurring effects of atmospheric turbulence. (p. 107)

seeing disk Roughly circular region on a detector over which a star's pointlike images is spread, due to atmospheric turbulence. (p. 107)

seismic wave A wave that travels outward from the site of an earthquake through Earth. (p. 153)

semi-major axis One half of the major axis of an ellipse. The semi-major axis is the way in which the size of an ellipse is usually quantified. (p. 37)

Seyfert galaxy Type of active galaxy whose emission comes from a very small region within the nucleus of an otherwise normal-looking spiral system. (p. 577)

shepherd satellite Satellite whose gravitational effect on a ring helps preserve the ring's shape. Examples are two satellites of Saturn, Prometheus and Pandora, whose orbits lie on either side of the F ring. (p. 276)

shield volcano A volcano produced by repeated nonexplosive eruptions of lava, creating a gradually sloping, shield-shaped low dome. Often contains a caldera at its summit. (p. 212)

shock wave Wave of matter, which may be generated by a new-born star or supernova, which pushes material outward into the surrounding molecular cloud. The material tends to pile up, forming a rapidly moving shell of dense gas. (p. 437)

sidereal day The time needed between successive risings of a given star. (p. 11)

sidereal month Time required for the Moon to complete one trip around the celestial sphere. (p. 14)

sidereal year The time required for the constellations to complete once cycle around the sky and return to their starting points, as seen from a given point on Earth. Earth's orbital period around the Sun is one sidereal year (p. 12)

singularity A point in the universe where the density of matter and the gravitational field are infinite, such as at the center of a black hole. (p. 512)

solar constant The amount of solar energy reaching Earth per unit area per unit time, approximately 1400 W/m^2. (p. 356)

solar core The region at the center of the Sun, with a radius of nearly 200,000 km, where powerful nuclear reactions generate the Sun's energy output. (p. 356)

solar cycle The 22-year period that is needed for both the average number of spots and the Sun's magnetic polarity to repeat themselves. The Sun's polarity reverses on each new 11-year sunspot cycle. (p. 367)

solar day The period of time between the instant when the Sun is directly overhead (i.e. at noon) to the next time it is directly overhead. (p. 9)

solar eclipse Celestial event during which the new Moon passes directly between Earth and the Sun, temporarily blocking the Sun's light. (p. 16)

solar interior The region of the Sun between the solar core and the photosphere. (p. 356)

solar maximum Point of the sunspot cycle during which many spots are seen. They are generally confined to regions in each hemisphere, between about 15–20 degrees latitude. (p. 367)

solar minimum Point of the sunspot cycle, during which only a few spots are seen. They are generally confined to narrow regions in each hemisphere at about 25–30 degrees latitude. (p. 367)

solar nebula The swirling gas surrounding the early Sun during the epoch of solar system formation, also referred to as the primitive solar system. (p. 339)

solar neutrino problem The discrepancy between the theoretically predicted flux of neutrinos streaming from the Sun as a result of fusion reactions in the core and the flux which is actually observed. The observed number of neutrinos is only about half the predicted number. (p. 374)

solar system The Sun and all the bodies that orbit it—Mercury, Venus, Earth, Mars, Jupiter, Saturn, Uranus, Neptune, Pluto, their moons, the asteroids, and the comets. (p. 131)

solar wind An outward flow of fast-moving charged particles from the Sun. (p. 363)

south celestial pole Point on the celestial sphere directly above Earth's south pole. (p. 9)

spectral class Classification scheme, based on the strength of stellar spectral lines, which is an indication of the temperature of a star. (p. 391)

spectrograph Instrument used to produce detailed spectra of stars. Usually, a spectrograph records a spectrum on a photographic plate, or more recently, in electronic form on a computer. (p. 78)

spectrometer Instrument used to produce detailed spectra of stars. Usually, a spectrograph records a spectrum on a photographic plate, or more recently, in electronic form on a computer. (p. 103)

spectroscope Instrument used to view a light source so that it is split into its component colors. (p. 78)

spectroscopic binary A binary-star system which from Earth appears as a single star, but whose spectral lines show back-and-forth Doppler shifts as two stars orbit one another. (p. 396)

spectroscopic parallax Method of determining the distance to a star by measuring its temperature and then determining its absolute brightness by comparing with a standard Hertzsprung–Russell diagram. The absolute and apparent brightness of the star give the star's distance from Earth. (p. 394)

spectroscopy The study of the way in which atoms absorb and emit electromagnetic radiation. Spectroscopy allows astronomers to determine the chemical composition of stars. (p. 81)

speed of light The fastest possible speed, according to the currently known laws of physics. Electromagnetic radiation exists in the form of waves or photons moving at the speed of light. (p. 62)

spin-orbit resonance State that a body is said to be in if its rotation period and its orbital period are related in some simple way. (p. 186)

spiral arm Distribution of material in a galaxy in a pinwheel-shaped design apparently emanating from near the galactic center. (p. 531)

spiral density wave A wave of matter formed in the plane of planetary rings, similar to ripples on the surface of a pond, which wrap around the rings forming spiral patterns similar to grooves in a record disk. Spiral density waves can lead to the appearance of ringlets. (p. 275)

spiral density wave A proposed explanation for the existence of galactic spiral arms, in which coiled waves of gas compression move through the galactic disk, triggering star formation. (p. 532)

spiral galaxy Galaxy composed of a flattened, star-forming disk component which may have spiral arms and a large central galactic bulge. (pp. 523, 549)

standard candle Any object with an easily recognizable appearance and known luminosity, which can be used in estimating distances. Supernovae, which all have the same peak luminosity (depending on type) are good examples of standard candles and are used to determine distances to other galaxies. (p. 554)

Standard Solar Model A self-consistent picture of the Sun, developed by incorporating the important physical processes that are believed to be important in determining the Sun's internal structure, into a computer program. The results of the program are then compared with observations of the Sun, and modifications are made to the model. The Standard Solar Model, which enjoys widespread acceptance, is the result of this process. (p. 356)

star A glowing ball of gas held together by its own gravity and powered by nuclear fusion in its core. (p. 354)

star cluster A grouping of anywhere from a dozen to a million stars which formed at the same time from the same cloud of interstellar gas. Stars in clusters are useful to aid our understanding of stellar evolution because, within a given cluster, stars are all roughly the same age and chemical composition, and lie at roughly the same distance from Earth. (p. 399)

starburst galaxy Galaxy in which a violent event, such as near-collision, has caused a sudden, intense burst of star formation in the recent past. (p. 566)

Stefan's law Relation that gives the total energy emitted per square centimeter of its surface per second by an object of a given temperature. Stefan's law shows that the energy emitted increases rapidly with an increase in temperature, proportional to the temperature raised to the fourth power. (p. 67)

stellar nucleosynthesis The formation of heavy elements by the fusion of lighter nuclei in the hearts of stars. Except for hydrogen and helium, all other elements in our universe resulted from stellar nucleosynthesis. (p. 481)

stellar occultation The dimming of starlight produced when a solar system object such as a planet, moon or ring, passes directly in front of a star. (p. 300)

stratosphere The portion of Earth's atmosphere lying above the troposphere, extending up to an altitude of 40 to 50 km. (p. 148)

strong nuclear force Short-range force responsible for binding atomic nuclei together. The strongest of the four fundamental forces of nature. (p. 371)

subgiant branch The section of the evolutionary track of a star that corresponds to changes that occur just after hydrogen is depleted in the core, and core hydrogen burning ceases. Shell hydrogen burning heats the outer layers of the star, which causes a general expansion of the stellar envelope. (p. 451)

summer solstice Point on the ecliptic where the Sun is at its northernmost point above the celestial equator, occurring on or near June 21. (p. 11)

sunspot An Earth-sized dark blemish found on the surface of the Sun. The dark color of the sunspot indicates that it is a region of lower temperature than its surroundings. (p. 364)

sunspot cycle The fairly regular pattern that the number and distribution of sunspots follows, in which the average number of spots reaches a maximum every 11 or so years then fall off to almost zero. (p. 367)

supercluster Grouping of several clusters of galaxies into a larger, but not necessarily gravitationally bound, unit. (p. 558)

supergiant A star with a radius between 100 and 1000 times that of the sun. (p. 385)

supergranulation Large-scale flow pattern on the surface of the Sun, consisting of cells measuring up to 30,000 km across, believed to be the imprint of large convective cells deep in the solar interior. (p. 360)

supernova Explosive death of a star, caused by the sudden onset of nuclear burning (type I), or an enormously energetic shock wave (type II). One of the most energetic events of the universe, a supernova may temporarily outshine the rest of the galaxy in which it resides. (p. 476)

supernova remnant The scattered glowing remains from a supernova that occurred in the past. The Crab Nebula is one of the best-studied supernova remnants. (p. 478)

synchrotron radiation Type of nonthermal radiation produced by high-speed charged particles, such as electrons, as they are accelerated in a strong magnetic field. (p. 586)

synchronous orbit State of an object when its period of rotations is exactly equal to its average orbital period. The Moon is in a synchronous orbit, and so presents the same face toward Earth at all times. (p. 183)

synodic month Time required for the Moon to complete a full cycle of phases. (p. 14)

T

T Tauri star Protostar in the late stages of formation, often exhibiting violent surface activity. T Tauri stars have been observed to brighten noticeably in a short period of time, consistent with the idea of rapid evolution during this final phase of stellar formation. (p. 430)

tail Component of a comet that consists of material streaming away from the main body, sometimes spanning hundreds of millions of kilometers. May be composed of dust or ionized gases. (p. 319)

telescope Instrument used to capture as many photons as possible from a given region of the sky and concentrate them into a focused beam for analysis. (p. 98)

temperature A measure of the amount of heat in an object, and an indication of the speed of the particles that comprise it. (p. 65)

terrae *See* highlands.

terrestrial planet One of the four innermost planets of the solar system, resembling Earth in general physical and chemical properties. (p. 134)

theories of relativity Einstein's theories, on which much of modern physics rests. Two essential facts of the theory are that nothing can travel faster than the speed of light, and that everything, including light, is affected by gravity. (p. 503)

thick disk Region of a spiral galaxy where an intermediate population of stars resides, younger than the halo stars, but older than stars in the disk. (p. 527)

tidal bulge Elongation of Earth caused by the difference between the gravitational force on the side nearest the Moon and the force on the side farthest from the Moon. The long axis of the tidal bulge points toward the Moon. (p. 168) More generally, the deformation of any body produced by the tidal effect of a nearby gravitating object.

tidal force The variation in one body's gravitational force from place to place across another body—for example, the variation of the Moon's gravity across Earth. (p. 168)

tides Rising and falling motion of terrestrial bodies of water, exhibiting daily, monthly and yearly cycles. Ocean tides on Earth are caused by the competing gravitational pull of the Moon and Sun on different parts of Earth. (p. 168)

time dilation A prediction of the theory of relativity, closely related to the gravitational reshift. To an outside observer, a clock lowered into a strong gravitational field will appear to run slow. (p. 511)

total eclipse Celestial event during which one body is completely blocked from view by another. (p. 16)

transition zone The region of rapid temperature increases that separates the Sun's chromosphere from the corona. (p. 355)

transverse motion Motion perpendicular to a particular line of sight, which does not result in Doppler shift in radiation received. (p. 71)

triangulation Method of determining distance based on the principles of geometry. A distant object is sighted from two well-separated locations. The distance between the two locations and the angle between the line joining them and the line to the distant object are all that are necessary to ascertain the object's distance. (p. 21)

triple-alpha process The creation of carbon-12 by the fusion of three helium-4 nuclei (alpha particles). Helium-burning stars occupy a region of the Hertzsprung–Russell diagram known as the horizontal branch. (p. 451)

Trojan asteroid One of two groups of asteroids which orbit at the same distance from the Sun as Jupiter, 60 degrees ahead and behind the planet. (p. 317)

tropical year The time interval between one vernal equinox and the next. (p. 11)

troposphere The portion of Earth's atmosphere from the surface to about 15 km. (p. 148)

Tully-Fisher relation A relation used to determine the absolute luminosity of a spiral galaxy. The rotational velocity, measured from the broadening of spectral lines, is related to the total mass, and hence the total luminosity. (p. 555)

21-centimeter radiation Radio radiation emitted when an electron in the ground state of a hydrogen atom flips its spin to become parallel to the spin of the proton in the nucleus. (p. 418)

type-I supernova One possible explosive death of a star. A white dwarf in a binary system can accrete enough mass that it cannot support its own weight. The star collapses and temperatures become high enough for carbon fusion to occur. Fusion begins throughout the white dwarf almost simultaneously and an explosion results. (p. 476)

type-II supernova One possible explosive death of a star, in which the highly evolved stellar core rapidly implodes and then explodes, destroying the surrounding star. (p. 476)

U

ultraviolet Region of the electromagnetic spectrum, just beyond the visible range, corresponding to wavelengths slightly shorter than blue light. (p. 56)

ultraviolet telescope A telescope that is designed to collect radiation in the ultraviolet part of the spectrum. Earth's atmosphere is partially opaque to these wavelengths, so ultraviolet telescopes are put on rockets, balloons or satellites to get high above most or all of the atmosphere. (p. 119)

umbra Central region of the shadow cast by an eclipsing body. (p. 17)

umbra The central region of a sunspot, which is its darkest and coolest part. (p. 364)

unbound An orbit which does not stay in a specific region of space, but where an object escapes the gravitational field of another. Typical unbound orbits are hyperbolic in shape. (p. 50)

universe The totality of all space, time, matter and energy. (p. 4)

V

Van Allen belts At least two doughnut-shaped regions of magnetically trapped charged particles high above Earth's atmosphere. (p. 158)

variable star A star whose luminosity changes with time. (p. 523)

vernal equinox Date on which the Sun crosses the celestial equator moving northward, occurring on or near March 21. (p. 11)

visible light The small range of the electromagnetic spectrum that human eyes perceive as light. The visible spectrum ranges from about 400 to 700 nm, corresponding to blue through red light. (p. 56)

visible spectrum The small range of the electromagnetic spectrum that human eyes perceive as light. The visible spectrum ranges from about 4000 to 7000 angstroms, corresponding to blue through red light. (p. 63)

visual binary A binary star system in which both members are resolvable from Earth. (p. 397)

void Large, relatively empty region of the universe around which superclusters and "walls" of galaxies are organized. (p. 569)

volcano Upwelling of hot lava from below Earth's crust to the planet's surface. (p. 156)

W

water hole The radio interval between 18 cm and 21 cm, the wavelengths at which hydroxyl (OH) and hydrogen (H) radiate, respectively, in which intelligent civilizations might conceivably send their communication signals. (p. 659)

wave A pattern that repeats itself cyclically in both time and space. Waves are characterized by the speed at which they move, their frequency, and their wavelength. (p. 57)

wave period The amount of time required for a wave to repeat itself at a specific point in space. (p. 57)

wavelength The distance from one wave crest to the next, at a given instant in time. (p. 57)

weak nuclear force Short-range force, weaker than both electromagnetism and the strong force, but much stronger than gravity, responsible for certain nuclear reations and radioactive decays. (p. 372)

weird terrain A region on the surface of Mercury of oddly rippled features. This feature is thought to be the result of a strong impact which occurred on the other side of the planet, and sent seismic waves traveling around the planet, converging in the weird region. (p. 193)

white dwarf A dwarf star with sufficiently high surface temperature that it glows white. (p. 385)

white dwarf region The bottom left-hand corner of the Hertzsprung–Russell diagram, where white dwarf stars are found. (p. 393)

white oval Light-colored region near the Great Red Spot in Jupiter's atmosphere. Like the red spot, such regions are apparently rotating storm systems. (p. 250)

Wien's law Relation between the wavelength at which a black body curve peaks and the temperature of the emitter. The peak wavelength is inversely proportional to the temperature, so the hotter the object, the bluer its radiation. (p. 67)

winter solstice Point on the ecliptic where the Sun is at its southernmost point below the celestial equator, occurring on or near December 21. (p. 11)

X

X ray Region of the electromagnetic spectrum corresponding to radiation of high frequency and short wavelengths, far beyond the visible spectrum. (p. 56)

X-ray burster X-ray source that radiates thousands of times more energy than our Sun, in short bursts that last only a few seconds. A neutron star in a binary system accretes matter onto its surface until temperatures reach the level needed for hydrogen fusion to occur. The result is a sudden period of rapid nuclear burning and release of energy. (p. 497)

Z

zero-age main sequence The region on the Hertzsprung–Russell diagram, as predicted by theoretical models, where stars are located at the onset of nuclear burning in their cores. (p. 432)

zodiac The twelve constellations on the celestial sphere through which the Sun appears to pass during the course of a year. (p. 12)

zonal flow Alternating regions of westward and eastward flow, roughly symmetrical about the equator of Jupiter, associated with the belts and zones in the planet's atmosphere. (p. 246)

zone Bright, high pressure region in the atmosphere of a jovian planet, where gas flows upward. (p. 246)

INDEX

PHOTO CREDITS

Chapter 1
CO1 J. Sanford/Astrostock-Sanford; AURA; NASA; AURA; Harvard Medical School   1.1 NASA   1.2 AURA   1.3 R. Dufour/Rice University   1.4 R. Williams/NASA   1.6 S. Westphal   1.8 AURA, Astronomical Society of the Pacific   1.14 UCO/Lick Observatory Publications Office   1.16 G. Schneider   1.17 G. Schneider   1.18 NOAA   1.19 G. Schneider

Chapter 2
CO2 Harvard College Observatory; Harvard College Observatory; Harvard College Observatory; Special Collections, Vassar College Libraries, Poughkeepsie, NY; Harvard College Observatory   2.1 English Heritage   2.2 G. Gerster/Comstock; J. Cornell   2.3 The Granger Collection   2.4 Museum of Science, Boston   2.7 E. Lessing/Art Resource, NY   2.9 E. Lessing/Art Resource, NY   2.10 The Royal Ontario Museum © ROM   2.13 Art Resource, NY   2.14 R. Beebe   2.15 AURA   2.17 The Granger Collection   Int. 2-2 G. Schneider

Chapter 3
CO3 AURA; Smithsonian Astrophysical Observatory   3.1 T. Hallas   3.4 Springer-Verlag GmbH & Co. KG   3.12 AURA; NASA; NASA; NRAO   3.15 Harvard-Smithsonian Center for Astrophysics; J. Moran; AURA; NASA

Chapter 4
CO4 Harvard College Observatory; Harvard College Observatory; Harvard College Observatory; Harvard College Observatory; Astronomical Society of the Pacific   4.3 Wabash Instrument Corporation   4.4 AURA   4.10 S. Heap/NASA   4.13 European Southern Observatory   4.15 Bausch & Lomb Inc./Personal Products Division

Chapter 5
CO5 D. Berry   5.6 Yerkes Observatory   5.8 Palomar Observatory/California Institute of Technology   5.9 European Southern Observatory   5.12 AURA   5.13 R. Wainscoat; R. Underwood/W. M. Keck Observatory   5.15 European Southern Observatory   5.16 AURA; R. Wainscoat/Peter Arnold, Inc.   5.17 NASA; NASA; NASA; AURA   5.18 B. Brandl (Max-Planck Institut)/Sky & Telescope Magazine/Sky Publishing Corporation; W. M. Keck Observatory   5.19 R. Ressmeyer/Starlight Collection, A Division of Corbis © 1994; MIT Lincoln Laboratory, Lexington, MA   5.20 NASA   5.21 D. Parker/Science Photo Library; T. Acevedo/NAIC; Cornell University Laboratory of Planetary Studies   5.22 MIT Haystack Observatory   5.23 AURA   5.24 NRAO   5.24 NRAO; AURA   5.26 Harvard College Observatory   5.27 Smithsonian Astrophysical Observatory; NASA   5.28 NASA; J. Sanford/Astrostock-Sanford   5.29 NASA   5.31 NASA   5.32 K. Dennerl & W. Voges/Max-Planck Institut fur Extraterrestrische Physik   5.33 NASA; C. Gro/NASA   5.34 NRAO; NASA; Lund Observatory; K. Dennerl & W. Voges; C. Gro/NASA   Int. 5-1 NASA; NASA; Hubble Space Telescope/NASA; D. Malin/Anglo-Australian Observatory

Chapter 6
CO6 NASA   6.1 Museo della Scienza, Florence, Italy/Scala/Art Resource, NY   6.2 Birr Scientific & Heritage Foundation   6.3 NASA   6.4 NASA   6.6 S. Westphal   6.8 NASA; J. Lodriguss   6.9 NASA   6.10 Sovfoto/Eastfoto   6.11 NASA   6.12 NASA   6.15 NASA   Int. 6-3 NASA

Chapter 7
CO7 The Planetarium Armagh–Northern Ireland/NASA; NASA; NASA; NASA; Edmund Scientific Co.   7.11 University Cooperation for Atmospheric Research, National Science Foundation; Anglo-Australian Observatory/NASA   7.13 P. Chesley/Tony Stone Images; H. Yamaguchi/Sygma   7.16 Earth Satellite Corporation/Science Photo Library/Photo Researchers, Inc.; L. Day/Black Star   7.17 P. Menzel/P. Menzel Photography   Int. 7-1 D. Rohr/Lamont Doherty Earth Observatory of Columbia University   M. P. 7-1 Joyce Photographics/Photo Researchers, Inc.   Earth Data box NASA

Chapter 8
CO8 NASA   8.1 J. Sanford/Science Photo Library/Photo Researchers, Inc.   8.3 R. Beebe   8.4 UCO/Lick Observatory Publications Office; UCO/Lick Observatory Publications Office; California Institute of Technology; California Institute of Technology   8.5 NASA   8.6 NASA   8.7 Palomar Observatory/California Institute of Technology   8.8 NASA   8.9 NASA   8.13 NASA   8.15 NASA   8.16 NASA   8.17 U.S. Geological Survey, U.S. Department of the Interior   8.18 NASA   8.19 NASA   8.20 NASA   8.21 NASA   8.22 NASA   8.26 W. Benz   8.27 U.S. Geological Survey, U.S. Department of the Interior   Int. 8-1 NASA   Int. 8-2 Washington Department of Defense   Moon Data box 8-1 NASA   Mercury Data box NASA

Chapter 9
CO9 NASA   9.1 J. Schad/Science Source/Photo Researchers, Inc.   9.2 UCO/Lick Observatory Publications Office   9.5 NOAO   9.6 Edmund Scientific Co.   9.7 NASA   9.8 NASA   9.9 Russian Space Agency; NASA   9.16 Russian Space Agency   9.18 NASA   Venus Data box NASA

Chapter 10
CO10 NASA   10.2 Pic-du-Midi Observatory (France, CNRS & Universite Paul Sabatier); D. Crisp/WFPC2 Team/NASA; NASA   10.3 NASA   10.4 NASA   10.5 NASA   10.6 NOAO   10.7, 10.8, 10.9, 10.10, 10.11, 10.12, 10.13, 10.14, 10.16, 10.17 NASA   Int. 10-1 Lowell Observatory; NASA; NASA   Int. 10-2 NASA   Mars Data box P. James, T. Clancy, S. Lee/NASA

Chapter 11
CO11 NASA   11.1 AURA; NASA   11.3, 11.7, 11.8, 11.9, 11.13, 11.14, 11.16, 11.17, 11.18, 11.19, 11.20, 11.21, 11.22, 11.23 NASA   Jupiter Data box NASA

Chapter 12
CO12 NASA   12.1, 12.2, 12.4, 12.8, 12.11, 12.12, 12.13, 12.14, 12.15, 12.17, 12.19, 12.20, 12.21 NASA   12.6 J. Westphal/NASA   Int. 12-1 NASA   Saturn Data box NASA

Chapter 13
CO13 NASA   13.1, 13.3 UCO/Lick Observatory Publications Office   13.2, 13.4, 13.7, 13.8 NASA   13.11 UCO/Lick Observatory Publications Office; NASA; NASA; NASA; NASA; NASA   13.12, 13.13, 13.14, 13.15, 13.16, 13.17, 13.18, 13.19 NASA   13.21 UCO/Lick Observatory Publications Office   13.22 U.S. Naval Observatory; NASA   Int. 13-1 NASA   Uranus Data box, Neptune Data box, Pluto Data box NASA

Chapter 14
CO14 J. Lodriguss   14.2 NASA   14.3 NASA   14.4 Palomar Observatory/California Institute of Technology   14.7 NOAO   14.8 U.S. Naval Observatory; NASA   14.11 California Institute of Technology, Mount Stromlo and Sliding Spring Observatories, Canberra, Australia   14.13 European Space Agency/Max Planck Institut (H.U. Keller)   14.14 NASA   14.15 P. Parviainen/Science Photo Library/Photo Researchers, Inc.   14.18 Sovfoto/Eastfoto   14.19 Corbis-Bettmann; J. Mandaville/Aramco World Magazine   14.20 Science Graphics   Int. 14-1 C. Butler/Astrostock-Sanford   Int. 14-2 NASA; H. Weaver/NASA   Int. 14-3 NASA

Chapter 15
CO15 D. Berry   15.2 NASA; D. Berry   15.3 D. Malin/Anglo-Australian Observatory   Int. 15-1 NASA   M. P. 15-1 D. Mills/AP Wide World Photos

Chapter 16
CO16 NASA   16.1 NOAO   16.4 National Solar Observatory   16.7 Palomar Observatory/California Institute of Technology   16.8 Palomar Observatory/California Institute of Technology   16.10 G. Schneider   16.11 NOAO   16.12 Sky Publishing Corporation   16.14 NASA   16.15 Palomar Observatory/California Institute of Technology   16.16 Palomar Observatory/California Institute of Technology; M. Penn/National Solar Observatory   16.17 NASA   16.22 National Solar Observatory; NASA   16.23 NASA; National Solar Observatory   16.24 National Solar Observatory   16.27 Brookhaven National Laboratory, Upton, NY   Int. 16-1 NASA   Int. 16-2 The Toledo Museum of Art, Toledo, OH   Sun Data box Kitt Peak National Observatory/NOAO

Chapter 17
CO17 J. Sanford/Astrostock-Sanford   17.3 Harvard College Observatory   17.5 NOAO; NASA/European Space Agency   17.10 J. Sanford/Astrostock-Sanford   17.18 Harvard College Observatory   17.23 NOAO   17.24 NOAO   Int. 17-1 European Space Agency

Chapter 18
CO18 D. Figer (UCLA)/NASA   18.1 Palomar Observatory/California Institute of Technology   18.5 Palomar Observatory/California Institute of Technology   18.6 Harvard College Observatory   18.7 Royal Observatory, Edinburgh/Science Photo Library/Photo Researchers, Inc.   18.8 NOAO; D. Malin/Anglo-Australian Observatory   18.9 NASA; AURA; AURA; NASA   18.10 Harvard College Observatory; AURA   18.11 NASA; UCO/Lick Observatory Publications Office; NASA   18.12 Harvard-Smithsonian Center for Astrophysics; NASA   18.13 Royal Observatory of Belgium; D. Malin/Anglo-Australian Observatory   18.18 NASA   18.19 NASA   Int. 18-1 NASA

Chapter 19
CO19 NASA   19.8 NOAO   19.9 J. Sanford/Astrostock-Sanford; AURA; Harvard-Smithsonian Center for Astrophysics; NASA; NASA   19.10 NASA   19.12 L. Rodriguez   19.16 J. Hester/NASA   19.17 UCO/Lick Observatory Publications Office; NASA; NASA   19.18 NOAO   19.19 AURA   Int. 19-1 NASA; S. Kulkarni   Int. 19-2 NOAO; NASA; J. Morse/University of Colorado/NASA

Chapter 20
CO20 D. Berry   20.9 B. Carney; D. Malin/Anglo-Australian Observatory   20.11 NOAO; D. Malin/Anglo-Australian Observatory   20.12 B. Balick (UW)/NASA; J. Harrington & K. Borkowski (University of Maryland)/NASA   20.14 Palomar Observatory/California Institute of Technology   20.17 NOAO   20.18 NASA   20.19 Space Telescope Science Institute; R. Saffer, D. Zurek/NASA   20.20 NASA/AURA   Int. 20-1 NASA

Chapter 21
CO21 M. Shara, R. Williams, R. Gilmozzi/NASA; D. Berry   21.1 UCO/Lick Observatory Publications Office   21.3 D. Berry; Palomar Observatory/California Institute of Technology   21.6 Space Telescope Science Institute   21.9 NOAO; NRAO; NASA   21.10 Harvard-Smithsonian Center for Astrophysics   21.11 D. Malin/Anglo-Australian Observatory   21.18 L. Chaisson   Int. 21-1 C. Burrows/NASA   Int. 21-2 C. Burrows, ESA/STSci/NASA   Int. 21-3 J. Hester/Arizona State University/NASA; NASA

Chapter 22
CO22 D. Berry   22.1 NASA   22.3 UCO/Lick Observatory Publications Office; Max-Planck Institut fur Extraterrestrische Physik   22.5 Harvard-Smithsonian Center for Astrophysics; HEAO-2/NASA   22.6 NRAO   22.7 NASA   22.8 NASA   22.18 Harvard-Smithsonian Center for Astrophysics; NASA   22.19 L. Chaisson   Int. 22-1 California Institute of Technology

Chapter 23
CO23 AURA; J. Sanford/Astrostock-Sanford; AURA; Space Telescope Science Institute; NASA   23.1 NASA   23.2 NASA; Palomar Observatory/California Institute of Technology; NASA   23.3 NOAO/AURA; Palomar Observatory/California Institute of Technology   23.5 Harvard College Observatory   23.11 NOAO   23.15 L. Chaisson   23.17 AURA   23.18 European Southern Observatory   23.20 NASA   23.22 Space Telescope Science Institute   23.23 NASA; NRAO; NRAO   23.24 L. Chaisson   Int. 23-1 Harvard College Observatory   Int. 23-3 C. Powell, P. Fowler, & D. Perkins/Science Photo Library/Photo Researchers, Inc.

Chapter 24
CO24 R. Williams/NASA   24.1 NASA   24.2 NOAO; D. Malin/Anglo-Australian Observatory; D. Malin/Anglo-Australian Observatory   24.3 AURA   24.4 Smithsonian Institution; Smithsonian Astrophysical Observatory; University of Hawaii Board of Regents   24.5 AURA   24.6 Palomar Observatory/California Institute of Technology   24.7 AURA; Palomar Observatory/California Institute of Technology; D. Malin   24.8 NOAO; Royal Observatory, Edinburgh/Science Photo Library/Photo Researchers, Inc.; Royal Observatory, Edinburgh/Science Photo Library/Photo Researchers, Inc.   24.10 NASA   24.13 Palomar Observatory   24.14 Palomar Observatory/NASA; Palomar Observatory/California Institute of Technology   24.15 Smithsonian Institution   24.18 NASA   24.21 AURA; Max-Planck-Institut fur Extraterrestrische Physik   24.22 NASA   24.23 R. Williams/NASA   24.24 NASA   24.25 NASA   24.26 NASA   24.27 Harvard College Observatory   24.28 J. Barnes & L. Hernquist   24.29 D. Malin   24.30 Palomar Observatory/California Institute of Technology   Int. 24-1 Smithsonian Astrophysical Observatory; AURA; B. Whitmore/NASA

Chapter 25
CO25 D. Berry   25.2 AURA; NASA   25.4 Harvard-Smithsonian Center for Astrophysics   25.5 AURA; AURA; NRAO; NASA   25.6 NRAO   25.7 J. Burns   25.8 NASA; NRAO   25.9 NRAO; Palomar Observatory/California Institute of Technology   25.12 Hubble Space Telescope/NASA   25.13 NASA   25.16 NOAO; Palomar Observatory/California Institute of Technology; NASA   25.16 Palomar Observatory/California Institute of Technology   25.17 Palomar Observatory/California Institute of Technology   25.18 Astronomical Society of the Pacific   25.19 AURA   25.20 NRAO   25.21 AURA   25.23 NASA; D. Berry   25.24 NASA; W. Colley/Space Telescope Science Institute   Int. 25-2 UCO/Lick Observatory Publications Office; Astronomical Society of the Pacific

Chapter 26
CO26 D. Berry; NASA   26.16 Bell Photographers, Inc.   26.18 NASA   Int. 26-1 R. Williams/NASA   Int. 26-2 NASA

Chapter 27
CO27 K. Sliwa, CERN/Atlas Collaboration   27.15 E. Bertschinger, MIT   27.17 NASA; E. Bertschinger, MIT

Chapter 28
CO28 D. Berry   28.3 Martin and Sally Fox Photography   28.5 Harvard-Smithsonian Center for Astrophysics   28.6 NASA   28.10 NASA   Int. 28-1 R. Williams

For video and animation credits, please see the CD-ROM.

ANSWERS TO THE SELF-TESTS

Chapter 1
True or False? 1. T 2. F 3. F 4. T 5. F 6. T 7. F
8. F 9. F 10. F 11. T 12. F 13. F 14. F 15. T
Fill in the Blank 1. galaxy 2. distance 3. axis
4. celestial sphere 5. stars 6. ecliptic 7. winter
solstice, lowest 8. celestial equator 9. 1/60
10. quarter 11. lunar 12. angular size 13. distance
from the Sun or radius of orbit 14. angular size
15. Earth

Chapter 2
True or False? 1. F 2. F 3. T 4. F 5. F 6. F 7. F
8. T 9. T 10. F 11. F 12. T 13. F 14. T 15. T
Fill in the Blank 1. calendar 2. Chinese 3. Islamic
4. retrograde 5. scientific method 6. Copernicus
7. Earth's 8. Tycho Brahe 9. ellipse, circle
10. square, cube 11. moons, phases, sunspots
12. radar 13. force 14. product, square 15. masses

Chapter 3
True or False? 1. T 2. F 3. T 4. F 5. F 6. T 7. F
8. F 9. F 10. T 11. F 12. T 13. F 14. T 15. T
Fill in the Blank 1. 300,000 2. wavelength
3. frequency 4. diffraction 5. electric, magnetic
6. 4000, 7000 Å or 400, 700 nm 7. red 8. radio,
infrared, visible 9. temperature 10. 0 11. 273
12. 6000 13. the 1200 K object 14. high 15. shorter

Chapter 4
True or False? 1. T 2. F 3. T 4. T 5. F 6. F 7. F
8. T 9. T 10. F 11. F 12. T 13. F 14. F 15. T
Fill in the Blank 1. prism 2. continuous 3. Sun
4. dense 5. cool 6. particle 7. photoelectric effect
8. positive, negative 9. ionized 10. absorbs
11. emits 12. difference 13. infrared 14. radio
15. broaden

Chapter 5
True or False? 1. F 2. F 3. F 4. T 5. F 6. F 7. T
8. T 9. F 10. T 11. T 12. F 13. F 14. F 15. F
Fill in the Blank 1. refracting 2. reflecting
3. reflecting 4. area or diameter 5. diameter,
wavelength 6. atmosphere 7. one 8. digital
9. resolution 10. reflecting 11. interferometer
12. Gamma ray, X-ray, ultraviolet 13. infrared
14. absorbed 15. images

Chapter 6
True or False? 1. F 2. T 3. T 4. F 5. F 6. F 7. T
8. T 9. F 10. F 11. F 12. T 13. F 14. F 15. T
Fill in the Blank 1. planets 2. asteroids 3. Mercury,
Pluto 4. terrestrial 5. jovian 6. Jupiter 7. rocky, icy
8. terrestrial 9. icy, outer 10. U.S., Mercury
11. radar 12. Mars 13. *Pioneer 10* 14. *Voyager 2*
15. gravitational

Chapter 7
True or False? 1. F 2. T 3. F 4. T 5. F 6. T 7. F
8. F 9. F 10. F 11. F 12. T 13. F 14. F 15. T
Fill in the Blank 1. 6400 (6378) 2. crust 3. nitrogen,
oxygen 4. convection 5. infrared 6. water, carbon
dioxide 7. volcanoes 8. life 9. solid, liquid
10. granite, basalt 11. molten 12. aurora 13. plate
tectonics 14. water 15. difference

Chapter 8
True or False? 1. F 2. T 3. F 4. F 5. F 6. T 7. F
8. T 9. T 10. F 11. T 12. F 13. T 14. F 15. T
Fill in the Blank 1. laser ranging 2. dawn, sunset
3. 1/4, 1/3 4. heavy elements 5. larger 6. closer to
the Sun 7. maria 8. impacts 9. radar 10. poles
11. 10 12. thicker 13. mantle 14. magnetic field
15. large, Mars-sized object

Chapter 9
True or False? 1. T 2. T 3. F 4. T 5. F 6. F 7. T
8. F 9. T 10. F 11. F 12. T 13. T 14. F 15. F
Fill in the Blank 1. spacecraft 2. density 3. slow or
retrograde 4. carbon dioxide 5. trace 6. greenhouse
effect 7. closer 8. radar 9. elevated continental-
sized formations 10. volcanoes 11. coronae
12. break up 13. young 14. volcanoes
15. temperature, pressure

Chapter 10
True or False? 1. T 2. F 3. T 4. F 5. F 6. T 7. F
8. F 9. F 10. F 11. T 12. T 13. F 14. F 15. F
Fill in the Blank 1. opposition 2. 1/2 3. 25
4. carbon dioxide 5. cratered 6. bulge 7. volcanoes
8. gravity 9. permafrost or frozen water 10. rain
water 11. flooding 12. billion 13. water
14. atmosphere 15. a meteorite

Chapter 11
True or False? 1. T 2. F 3. F 4. F 5. T 6. F 7. T
8. T 9. F 10. T 11. T 12. T 13. F 14. F 15. T
Fill in the Blank 1. density 2. hydrogen, helium
3. rotation 4. hurricanes 5. twice 6. zones, belts
7. twice 8. liquid 9. hydrogen 10. synchronously
11. 4, 16 12. Ganymede 13. water 14. volcanoes
15. Europa

Chapter 12
True or False? 1. T 2. F 3. F 4. F 5. T 6. F 7. F
8. F 9. T 10. T 11. T 12. F 13. T 14. T 15. F
Fill in the Blank 1. ring 2. ammonia 3. gravity
4. helium 5. helium precipitation 6. 3 7. A, B 8. B
9. Roche limit 10. shepherd 11. nitrogen
12. organic 13. temperature 14. E 15. Lagrange

Chapter 13
True or False? 1. F 2. F 3. T 4. F 5. T 6. T 7. F
8. T 9. T 10. T 11. T 12. T 13. T 14. F 15. F
Fill in the Blank 1. twice 2. perpendicular 3. methane
4. Uranus 5. Neptune 6. magnetic field 7. Miranda
8. 8 9. retrograde 10. eccentricity 11. atmosphere
12. Triton 13. Neptune 14. Triton 15. eclipses

Chapter 14
True or False? 1. F 2. T 3. F 4. T 5. T 6. F 7. T
8. F 9. T 10. F 11. F 12. F 13. T 14. T 15. T
Fill in the Blank 1. rocky 2. Mars, Jupiter
3. hundreds, 100 4. Jupiter 5. resonance 6. icy
7. gas 8. eccentric 9. a few 10. meteoroid swarm
11. short period comets 12. asteroids, comets
13. meteor 14. meteorites 15. 4.6 billion years

Chapter 15
True or False? 1. T 2. F 3. T 4. T 5. F 6. F 7. T
8. F 9. T 10. F 11. F 12. T 13. T 14. F 15. F
Fill in the Blank 1. nebular 2. interstellar dust
3. collisions 4. gravity 5. gas 6. water 7. icy
8. fragmentation 9. comets 10. comets 11. Jupiter
12. T Tauri 13. size 14. lower 15. collisions

Chapter 16
True or False? 1. T 2. F 3. F 4. T 5. T 6. F 7. F
8. F 9. F 10. F 11. T 12. F 13. T 14. T 15. F
Fill in the Blank 1. photosphere 2. chromosphere,
corona 3. convective, radiative, core 4. granules
5. photosphere 6. hydrogen 7. helium 8. 98 or 99
9. ionized 10. cooler 11. 11, 22 12. core 13. 4,
helium, neutrinos, gamma-rays 14. destroyed 15. few

Chapter 17
True or False? 1. T 2. T 3. F 4. F 5. F 6. T 7. F
8. F 9. F 10. F 11. F 12. T 13. T 14. F 15. F
Fill in the Blank 1. 2 A.U. 2. spectrum, Doppler
3. proper motion, distance 4. luminosity, temperature
5. white dwarfs 6. supergiants 7. temperature
8. ionized 9. in its lowest energy level 10. G2
11. spectral type, luminosity 12. main sequence
13. binary 14. decrease 15. shorter

Chapter 18
True or False? 1. F 2. T 3. F 4. F 5. F 6. F 7. F
8. T 9. T 10. F 11. F 12. F 13. T 14. T 15. T
Fill in the Blank 1. gas, dust 2. similar or larger
3. dimming, dust 4. low 5. hydrogen, helium
6. 8,000 7. photoevaporation 8. emission nebula
9. 10 or 20 10. atomic 11. spin, hydrogen 12. 20
13. radio 14. hydrogen 15. a million

Chapter 19
True or False? 1. F 2. F 3. T 4. F 5. T 6. F 7. F
8. F 9. F 10. F 11. T 12. F 13. T 14. T 15. T
Fill in the Blank 1. temperature 2. evolutionary track
3. 10, 10 or more, 1,000 or more 4. fragments
5. increase 6. protostar 7. upper, right 8. 10 million
9. fuse hydrogen 10. main sequence 11. 5 or 6 12. 50
13. one million, one billion 14. radio 15. infrared

Chapter 20
True or False? 1. T 2. T 3. F 4. T 5. T 6. F 7. T
8. F 9. F 10. F 11. T 12. T 13. F 14. F 15. T
Fill in the Blank 1. pressure 2. 5 billion
3. hydrogen, helium 4. 100 million K 5. contract
6. carbon, energy 7. degeneracy 8. carbon 9. 100
10. clusters 11. the Earth 12. high, low
13. decrease 14. decreases 15. separation

Chapter 21
True or False? 1. F 2. F 3. T 4. T 5. T 6. F 7. T
8. F 9. T 10. F 11. F 12. T 13. T 14. T 15. T
Fill in the Blank 1. accretion 2. hydrogen, surface
3. temperature 4. neutron, neutrino 5. neutron
6. II 7. I 8. 1.41 9. light curves 10. neutrinos
11. blue supergiant, red supergiant 12. helium
capture 13. 3 billion K 14. iron 15. fewer

Chapter 22
True or False? 1. T 2. F 3. T 4. F 5. F 6. F 7. T
8. F 9. T 10. F 11. T 12. F 13. T 14. T 15. F
Fill in the Blank 1. Type I 2. 20 3. high, strong
4. radio 5. 0.03 to 0.3 seconds 6. rotation 7. binary
8. neutron 9. matter or gravity 10. be unchanged
11. Schwarzschild radius 12. singularity 13. lose
14. binary star 15. fluctuations

Chapter 23
True or False? 1. T 2. F 3. F 4. F 5. F 6. F 7. F
8. T 9. T 10. F 11. F 12. F 13. F 14. F 15. F
Fill in the Blank 1. inside it 2. dust 3. disk 4. halo
5. luminosity 6. 1, 100 7. instability strip 8. brighter
9. RR Lyrae stars 10. 8,000 11. 220 km/s 12. random
13. halo 14. higher, dark matter 15. width

Chapter 24
True or False? 1. T 2. T 3. F 4. T 5. T 6. T 7. F
8. T 9. F 10. F 11. F 12. F 13. F 14. T 15. T
Fill in the Blank 1. Hubble 2. large 3. Sa, Sc 4. less
5. Local Group 6. broadening 7. 100 Mpc
8. increases 9. Kepler's Third 10. 90 11. X-rays
12. mergers 13. distance 14. 50 to 90 15. 100 to
200 Mpc

Chapter 25
True or False? 1. T 2. F 3. F 4. F 5. F 6. T 7. F
8. F 9. T 10. F 11. F 12. T 13. T 14. T 15. F
Fill in the Blank 1. large 2. long or infrared 3. nucleus
4. core 5. larger 6. nucleus or center 7. nucleus or
center 8. one parsec 9. 3 billion 10. one solar mass,
decade 11. starlike 12. expansion 13. redshifted
14. short 15. galaxy or cluster of galaxies

Chapter 26
True or False? 1. F 2. T 3. F 4. T 5. F 6. T 7. F
8. F 9. T 10. T 11. T 12. F 13. F 14. F 15. F
Fill in the Blank 1. 200 2. everywhere 3. directions
4. Cosmological Principle 5. homogeneity 6. isotropy
7. 15 8. gravity 9. density 10. high 11. one
12. faster 13. positively 14. lower than 15. 2.7

Chapter 27
True or False? 1. F 2. F 3. F 4. T 5. T 6. F 7. F
8. T 9. T 10. F 11. T 12. F 13. T 14. F 15. F
Fill in the Blank 1. matter 2. a few thousand
3. photons or gamma rays, antiparticle 4. 10^9
5. neutrons, 10^{13} 6. deuterium, helium 7. neutral
atoms 8. decreased 9. nonbaryonic (not protons or
neutrons) 10. 1/1500 11. 97 12. dark matter
13. cosmic microwave background 14. heavy
15. temperature, COBE

Chapter 28
True or False? 1. T 2. F 3. T 4. T 5. F 6. F 7. T
8. F 9. T 10. T 11. T 12. T 13. T 14. T 15. T
Fill in the Blank 1. proteins 2. ammonia, methane
3. any two of lightning, ultraviolet light, radioactivity,
volcanism, meteoritic impact 4. 3.5 billion
5. 1 billion 6. organic compounds 7. technical
civilizations 8. unstable 9. unlikely 10. distances
11. light 12. dust 13. 65 14. water hole 15. 200

These star maps show the brighter stars and the prominent constellations as they appear on the dates and at the times indicated. To use these maps, face the south and hold the book overhead with top of the map toward the north and the right-hand edge toward the west. The brightest stars are indicated by the star symbol (☆) and the names are indicated. (Star maps courtesy of Robert Dixon, *Dynamic Astronomy*, 6th ed., Prentice Hall, 1992.)

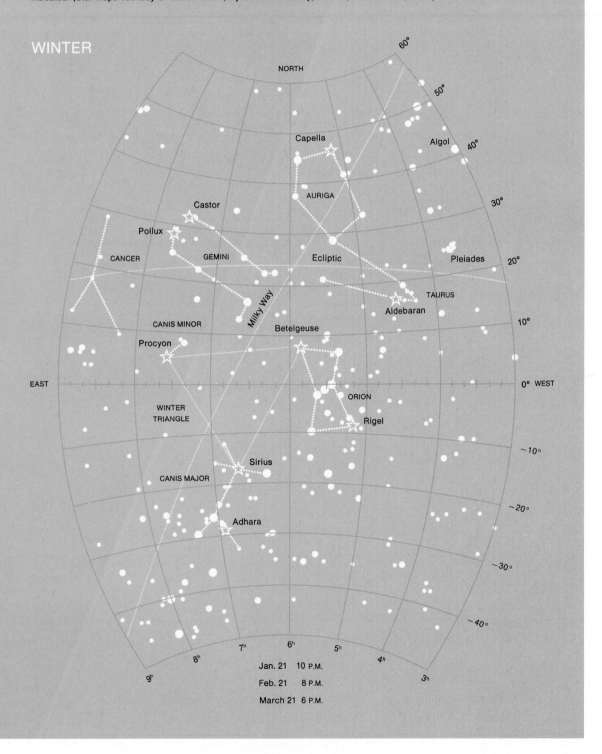

WINTER

Jan. 21   10 P.M.

Feb. 21   8 P.M.

March 21 6 P.M.

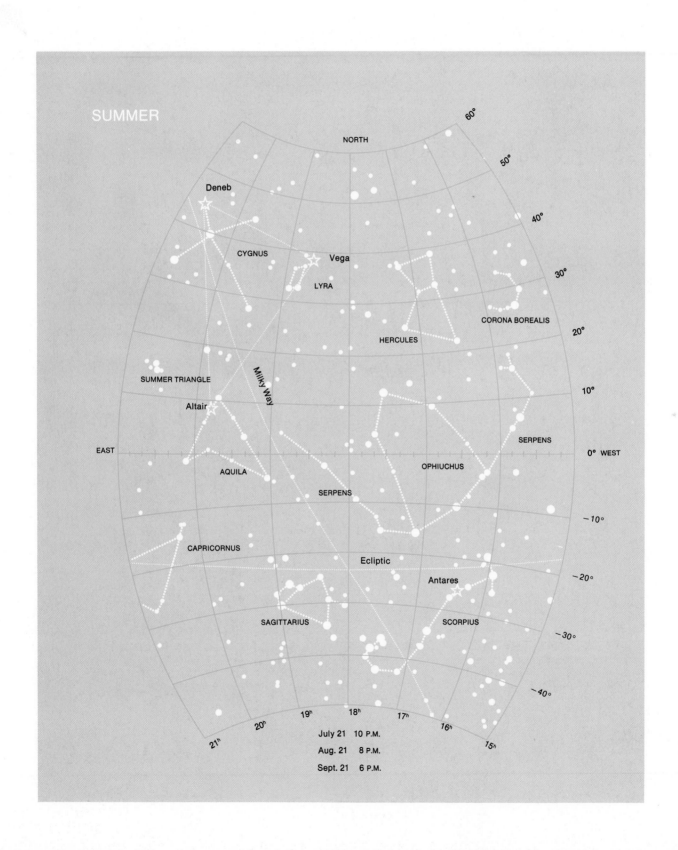

SUMMER

NORTH

Deneb

CYGNUS   Vega

LYRA

CORONA BOREALIS

HERCULES

SUMMER TRIANGLE

Altair

Milky Way

SERPENS

EAST   0° WEST

AQUILA   OPHIUCHUS

SERPENS

CAPRICORNUS

Ecliptic

Antares

SAGITTARIUS   SCORPIUS

60°
50°
40°
30°
20°
10°
−10°
−20°
−30°
−40°

21ʰ 20ʰ 19ʰ 18ʰ 17ʰ 16ʰ 15ʰ

July 21   10 P.M.

Aug. 21   8 P.M.

Sept. 21   6 P.M.

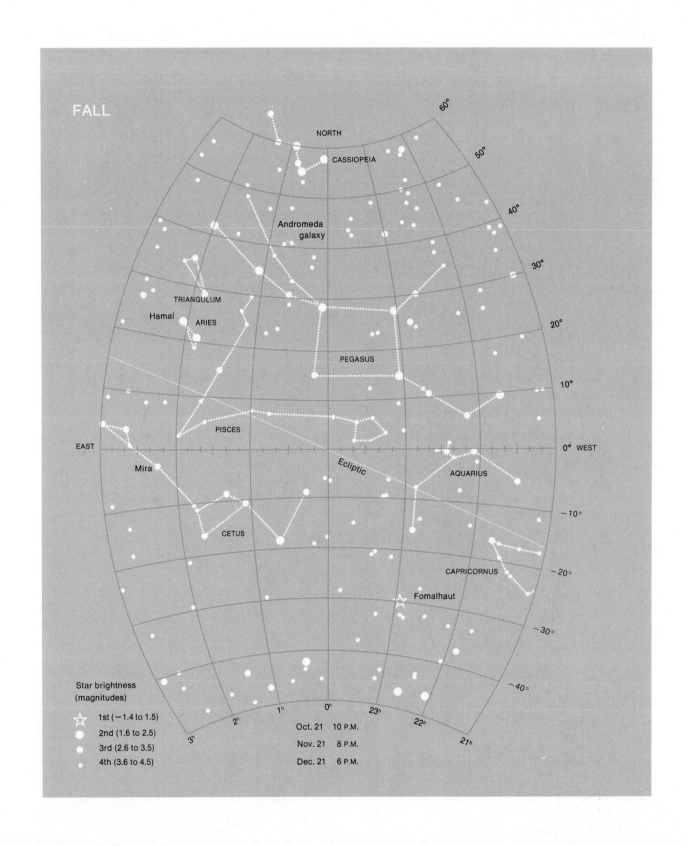

FALL

NORTH

60°

50°

CASSIOPEIA

40°

Andromeda
galaxy

30°

TRIANGULUM

Hamal

20°

ARIES

PEGASUS

10°

PISCES

EAST   0° WEST

Mira   Ecliptic   AQUARIUS

−10°

CETUS

−20°

CAPRICORNUS

Fomalhaut

−30°

−40°

Star brightness
(magnitudes)

1st (−1.4 to 1.5)

2nd (1.6 to 2.5)

3rd (2.6 to 3.5)

4th (3.6 to 4.5)

3ʰ   2ʰ   1ʰ   0ʰ   23ʰ   22ʰ   21ʰ

Oct. 21   10 P.M.

Nov. 21   8 P.M.

Dec. 21   6 P.M.

The Distance Pyramid

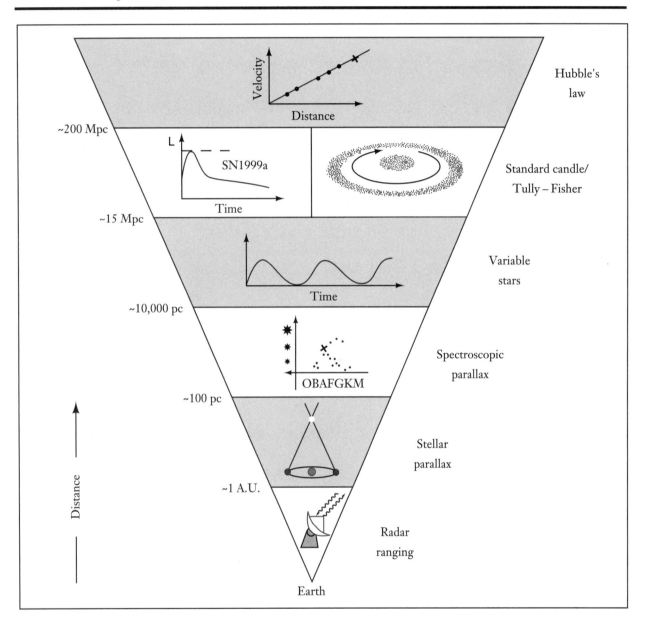

Main-Sequence Stellar Properties by Spectral Class

Spectral Class	Typical Surface Temperature	Color	Mass*	Luminosity*	Lifetime*	Familiar Examples
	(K)		($M_\odot$)	($L_\odot$)	(10^6 yr)	
O	>30,000	Electric Blue	>20	>100,000	<2	
B	20,000	Blue	8	3000	30	Spica (B1)
A	10,000	White	3	75	400	Vega (A0) Sirius (A1)
F	7,000	Yellow - white	1.5	4	4000	Procyon (F5)
G	6,000	Yellow	1.0	1.5	9000	Sun (G2) Alpha Centauri (G2)
K	4,000	Orange	0.5	0.1	60,000	Epsilon Eridani (K2)
M	3,000	Red	0.1	0.005	200,000	Proxima Centauri (M5) Barnard's Star (M5)

*Approximate values for stars of solar composition